Automotive Engineering

Note from the Publisher

This book has been compiled using extracts from the following books within the range of Automotive Engineering books in the Elsevier collection:

Blundell, M and Harty, D. (2004) *The Multibody Systems Approach to Vehicle Dynamics*, 9780750651127
Brown, J., Robertson, A.J. and Serpento, S. (2001) *Motor Vehicle Structures*, 9780750651349
Davies, G. (2003) *Materials for Automobile Bodies*, 9780750656924
Fenton, J. and Hodkinson, R. (2001) *Lightweight Electric/Hybrid Vehicle Design*, 9780750650922
Garrett, T.K., Newton, K. and Steels, W. (2000) *The Motor Vehicle 13e*, 9780750644495
Happian-Smith, J (2001) *Introduction to Modern Vehicle Design*, 9780750661294
Heisler, H. (1998) *Vehicle and Engine Technology*, 9780340691861
Martyr, A.J. and Plint, M.A. (2007) *Engine Testing 3e*, 9780750684392
Pacejka, H. (2005) *Tyre and Vehicle Dynamics*, 9780750669184
Reimpell, J., Stoll, H. and Betzler, J. (2001) *Automotive Chassis: Engineering Principles*, 9780750650540
Ribbens, W. (2003) *Understanding Automotive Electronics*, 9780750675994
Vlacic, L. and Parent, M. (2001) *Intelligent Vehicle Technologies*, 9780750650939

The extracts have been taken directly from the above source books, with some small editorial changes. These changes have entailed the re-numbering of Sections and Figures. In view of the breadth of content and style of the source books, there is some overlap and repetition of material between chapters and significant differences in style, but these features have been left in order to retain the flavour and readability of the individual chapters.

Units of measure

Units are provided in either SI or IP units. A conversion table for these units is provided at the front of the book.

Upgrade to an Electronic Version

An electronic version of *Automotive Engineering*, the *Automotive Engineering e-Mega Reference*, 9781856175784

- A fully searchable Mega Reference eBook, providing all the essential material needed by Automotive Engineers on a day-to-day basis.
- Fundamentals, key techniques, engineering best practice and rules-of-thumb at one quick click of a button
- Over 1,500 pages of reference material, including over 1,000 pages not included in the print edition

Go to http://www.elsevierdirect.com/9781856175777 and click on **Ebook Available**

Automotive Engineering

Powertrain, Chassis System and Vehicle Body

Edited by David A. Crolla

Amsterdam · Boston · Heidelberg · London · New York · Oxford
Paris · San Diego · San Francisco · Sydney · Tokyo
Butterworth-Heinemann is an imprint of Elsevier

Butterworth-Heinemann is an imprint of Elsevier
Linacre House, Jordan Hill, Oxford OX2 8DP, UK
30 Corporate Drive, Suite 400, Burlington, MA 01803, USA

First edition 2009

Copyright © 2009 Elsevier Inc. All rights reserved

No part of this publication may be reproduced, stored in a retrieval system or transmitted in any form or by any means electronic, mechanical, photocopying, recording or otherwise without the prior written permission of the publisher

Permissions may be sought directly from Elsevier's Science & Technology Rights Department in Oxford, UK: phone (+44) (0) 1865 843830; fax (+44) (0) 1865 853333; email: permissions@elsevier.com. Alternatively visit the Science and Technology website at www.elsevierdirect.com/rights for further information

Notice
No responsibility is assumed by the publisher for any injury and/or damage to persons or property as a matter of products liability, negligence or otherwise, or from any use or operation of any methods, products, instructions or ideas contained in the material herein. Because of rapid advances in the medical sciences, in particular, independent verification of diagnoses and drug dosages should be made

British Library Cataloguing in Publication Data
A catalogue record for this book is available from the British Library

Library of Congress Cataloguing-in-Publication Data
A catalog record for this book is available from the Library of Congress

ISBN: 978-1-85617-577-7

For information on all Butterworth-Heinemann publications
visit our web site at elsevierdirect.com

Printed and bound in the United States of America

09 10 11 11 10 9 8 7 6 5 4 3 2 1

Working together to grow
libraries in developing countries
www.elsevier.com | www.bookaid.org | www.sabre.org
ELSEVIER BOOK AID International Sabre Foundation

Contents

Section One

Introduction to engine design

Chapter 1.1

1.1

Piston-engine cycles of operation

Heinz Heisler

1.1.1 The internal-combustion engine

The piston engine is known as an internal-combustion heat-engine. The concept of the piston engine is that a supply of air-and-fuel mixture is fed to the inside of the cylinder where it is compressed and then burnt. This internal combustion releases heat energy which is then converted into useful mechanical work as the high gas pressures generated force the piston to move along its stroke in the cylinder. It can be said, therefore, that a heat-engine is merely an energy transformer.

To enable the piston movement to be harnessed, the driving thrust on the piston is transmitted by means of a connecting-rod to a crankshaft whose function is to convert the linear piston motion in the cylinder to a rotary crankshaft movement (Fig. 1.1-1). The piston can thus be made to repeat its movement to and fro, due to the constraints of the crankshaft crankpin's circular path and the guiding cylinder.

The backward-and-forward displacement of the piston is generally referred to as the *reciprocating* motion of the piston, so these power units are also known as reciprocating engines.

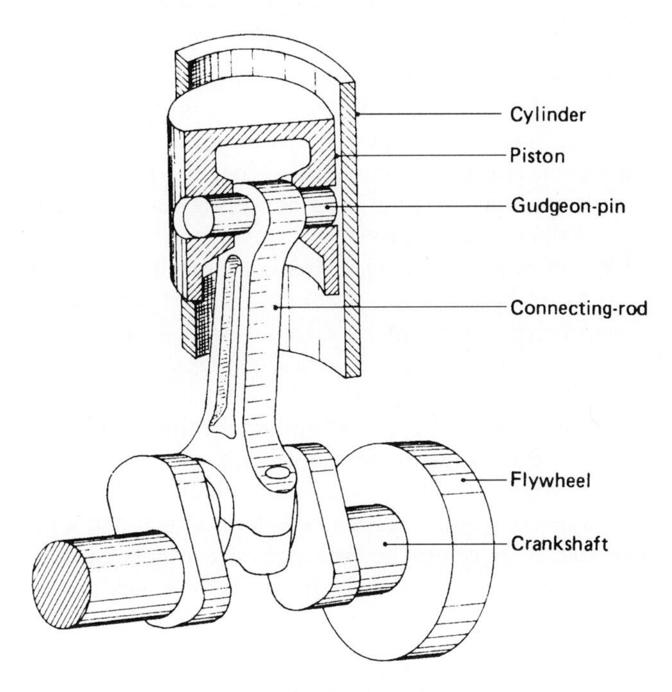

Fig. 1.1-1 Pictorial view of the basic engine.

1.1.1.1 Engine components and terms

The main problem in understanding the construction of the reciprocating piston engine is being able to identify and name the various parts making up the power unit. To this end, the following briefly describes the major components and the names given to them (Figs. 1.1-1 and 1.1-2).

Cylinder block This is a cast structure with cylindrical holes bored to guide and support the pistons and to harness the working gases. It also provides a jacket to contain a liquid coolant.

Cylinder head This casting encloses the combustion end of the cylinder block and houses both the inlet and exhaust poppet-valves and their ports to admit air–fuel mixture and to exhaust the combustion products.

Crankcase This is a cast rigid structure which supports and houses the crankshaft and bearings. It is usually cast as a mono-construction with the cylinder block.

Sump This is a pressed-steel or cast-aluminium-alloy container which encloses the bottom of the crankcase and provides a reservoir for the engine's lubricant.

Vehicle and Engine Technology, ISBN: 9780340691861
Copyright © 1998 Heinz Heisler. All rights of reproduction, in any form, reserved.

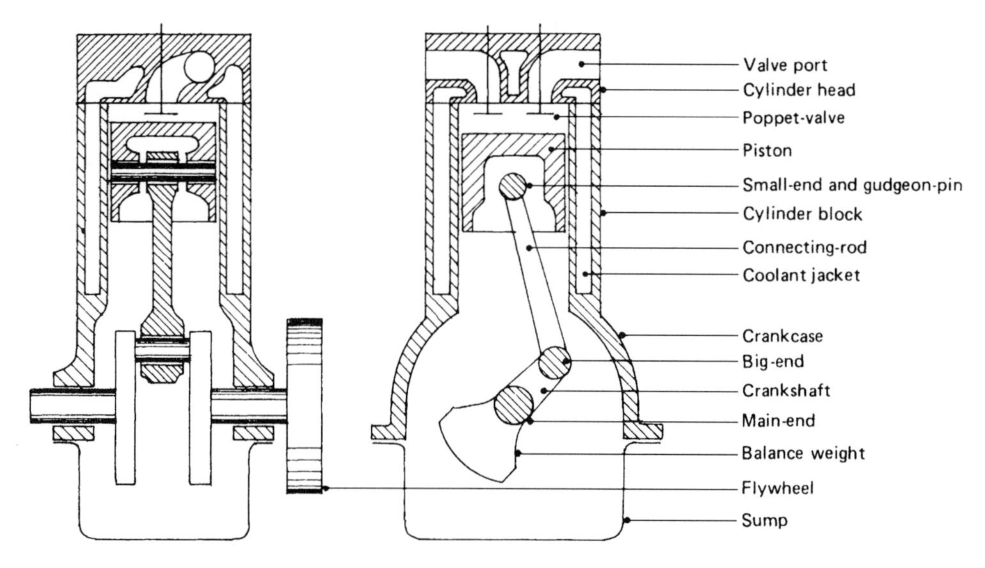

Fig. 1.1-2 Sectional view of the basic engine.

Piston This is a pressure-tight cylindrical plunger which is subjected to the expanding gas pressure. Its function is to convert the gas pressure from combustion into a concentrated driving thrust along the connecting-rod. It must therefore also act as a guide for the small-end of the connecting-rod.

Piston rings These are circular rings which seal the gaps made between the piston and the cylinder, their object being to prevent gas escaping and to control the amount of lubricant which is allowed to reach the top of the cylinder.

Gudgeon-pin This pin transfers the thrust from the piston to the connecting-rod small-end while permitting the rod to rock to and fro as the crankshaft rotates.

Connecting-rod This acts as both a strut and a tie link-rod. It transmits the linear pressure impulses acting on the piston to the crankshaft big-end journal, where they are converted into turning-effort.

Crankshaft A simple crankshaft consists of a circular-sectioned shaft which is bent or cranked to form two perpendicular crank-arms and an offset big-end journal. The unbent part of the shaft provides the main journals. The crankshaft is indirectly linked by the connecting-rod to the piston – this enables the straight-line motion of the piston to be transformed into a rotary motion at the crankshaft about the main-journal axis.

Crankshaft journals These are highly finished cylindrical pins machined parallel on both the centre axes and the offset axes of the crankshaft. When assembled, these journals rotate in plain bush-type bearings mounted in the crankcase (the main journals) and in one end of the connecting-rod (the big-end journal).

Small-end This refers to the hinged joint made by the gudgeon-pin between the piston and the connecting-rod so that the connecting-rod is free to oscillate relative to the cylinder axis as it moves to and fro in the cylinder.

Big-end This refers to the joint between the connecting-rod and the crankshaft big-end journal which provides the relative angular movement between the two components as the engine rotates.

Main-ends This refers to the rubbing pairs formed between the crankshaft main journals and their respective plain bearings mounted in the crankcase.

Line of stroke The centre path the piston is forced to follow due to the constraints of the cylinder is known as the line of stroke.

Inner and outer dead centres When the crankarm and the connecting-rod are aligned along the line of stroke, the piston will be in either one of its two extreme positions. If the piston is at its closest position to the cylinder head, the crank and piston are said to be at inner dead centre (IDC) or top dead centre (TDC). With the piston at its furthest position from the cylinder head, the crank and piston are said to be at outer dead centre (ODC) or bottom dead centre (BDC). These reference points are of considerable importance for valve-to-crankshaft timing and for either ignition or injection settings.

Clearance volume The space between the cylinder head and the piston crown at TDC is known as the clearance volume or the combustion-chamber space.

Crank-throw The distance from the centre of the crankshaft main journal to the centre of the big-end journal is known as the crank-throw. This radial length influences the leverage the gas pressure acting on the piston can apply in rotating the crankshaft.

Piston stroke The piston movement from IDC to ODC is known as the piston stroke and corresponds to the crankshaft rotating half a revolution or 180°. It is also equal to twice the crank-throw.

i.e. $L = 2R$
where L = piston stroke
and R = crank-throw

Thus a long or short stroke will enable a large or small turning-effort to be applied to the crankshaft respectively.

Cylinder bore The cylinder block is initially cast with sand cores occupying the cylinder spaces. After the sand cores have been removed, the rough holes are machined with a single-point cutting tool attached radially at the end of a rotating bar. The removal of the unwanted metal in the hole is commonly known as boring the cylinder to size. Thus the finished cylindrical hole is known as the cylinder bore, and its internal diameter simply as the bore or bore size.

1.1.1.2 The four-stroke-cycle spark-ignition (petrol) engine

The first internal-combustion engine to operate successfully on the four-stroke cycle used gas as a fuel and was built in 1876 by Nicolaus August Otto, a self-taught German engineer at the Gas-motoreufabrik Deutz factory near Cologne, for many years the largest manufacturer of internal-combustion engines in the world. It was one of Otto's associates – Gottlieb Daimler – who later developed an engine to run on petrol which was described in patent number 4315 of 1885. He also pioneered its application to the motor vehicle (Fig. 1.1-3).

Petrol engines take in a flammable mixture of air and petrol which is ignited by a timed spark when the charge is compressed. These engines are therefore sometimes called spark-ignition (S.I.) engines.

These engines require four piston strokes to complete one cycle: an air-and-fuel intake stroke moving outward from the cylinder head, an inward movement towards the cylinder head compressing the charge, an outward power stroke, and an inward exhaust stroke.

Induction stroke (Fig. 1.1-3(a)) The inlet valve is opened and the exhaust valve is closed. The piston descends, moving away from the cylinder head (Fig. 1.1-3(a)). The speed of the piston moving along the cylinder creates a pressure reduction or depression which reaches a maximum of about 0.3 bar below atmospheric pressure at one-third from the beginning of the stroke. The depression actually generated will depend on the speed and load experienced by the engine, but a typical average value might be 0.12 bar below atmospheric pressure. This depression induces (sucks in) a fresh charge of air and atomised petrol in proportions ranging from 10 to 17 parts of air to one part of petrol by weight.

An engine which induces fresh charge by means of a depression in the cylinder is said to be 'normally aspirated' or 'naturally aspirated'.

Compression stroke (Fig. 1.1-3(b)) Both the inlet and the exhaust valves are closed. The piston begins to ascend towards the cylinder head (Fig. 1.1-3(b)). The induced air-and-petrol charge is progressively compressed to something of the order of one-eighth to one-tenth of the cylinder's original volume at the piston's innermost position. This compression squeezes the air and atomised-petrol molecules closer together and not only increases the charge pressure in the cylinder but also raises the temperature. Typical maximum cylinder compression pressures will range between 8 and 14 bar with the throttle open and the engine running under load.

Power stroke (Fig. 1.1-3(c)) Both the inlet and the exhaust valves are closed and, just before the piston approaches the top of its stroke during compression, a spark-plug ignites the dense combustible charge (Fig. 1.1-3(c)). By the time the piston reaches the innermost point of its stroke, the charge mixture begins to burn, generates heat, and rapidly raises the pressure in the cylinder until the gas forces exceed the resisting load. The burning gases then expand and so change the piston's direction of motion and push it to its outermost position. The cylinder pressure then drops from a peak value of about 60 bar under full load down to maybe 4 bar near the outermost movement of the piston.

Exhaust stroke (Fig. 1.1-3(d)) At the end of the power stroke the inlet valve remains closed but the exhaust valve is opened. The piston changes its direction of motion and now moves from the outermost to the innermost position (Fig. 1.1-3(d)). Most of the burnt gases will be expelled by the existing pressure energy of the gas, but the returning piston will push the last of the spent gases out of the cylinder through the exhaust-valve port and to the atmosphere.

During the exhaust stroke, the gas pressure in the cylinder will fall from the exhaust-valve opening pressure (which may vary from 2 to 5 bar, depending on the engine speed and the throttle-opening position) to atmospheric pressure or even less as the piston nears the innermost position towards the cylinder head.

Cycle of events in a four-cylinder engine (Figs. 1.1-3(e)–(g)) Fig. 1.1-3(e) illustrates how the cycle of events – induction, compression, power, and exhaust – is phased in a four-cylinder engine. The relationship between cylinder pressure and piston stroke position over the four strokes is clearly shown in Figs. 1.1-3(f) and (g) and, by following the arrows, it can be seen that a figures of eight is repeatedly being traced.

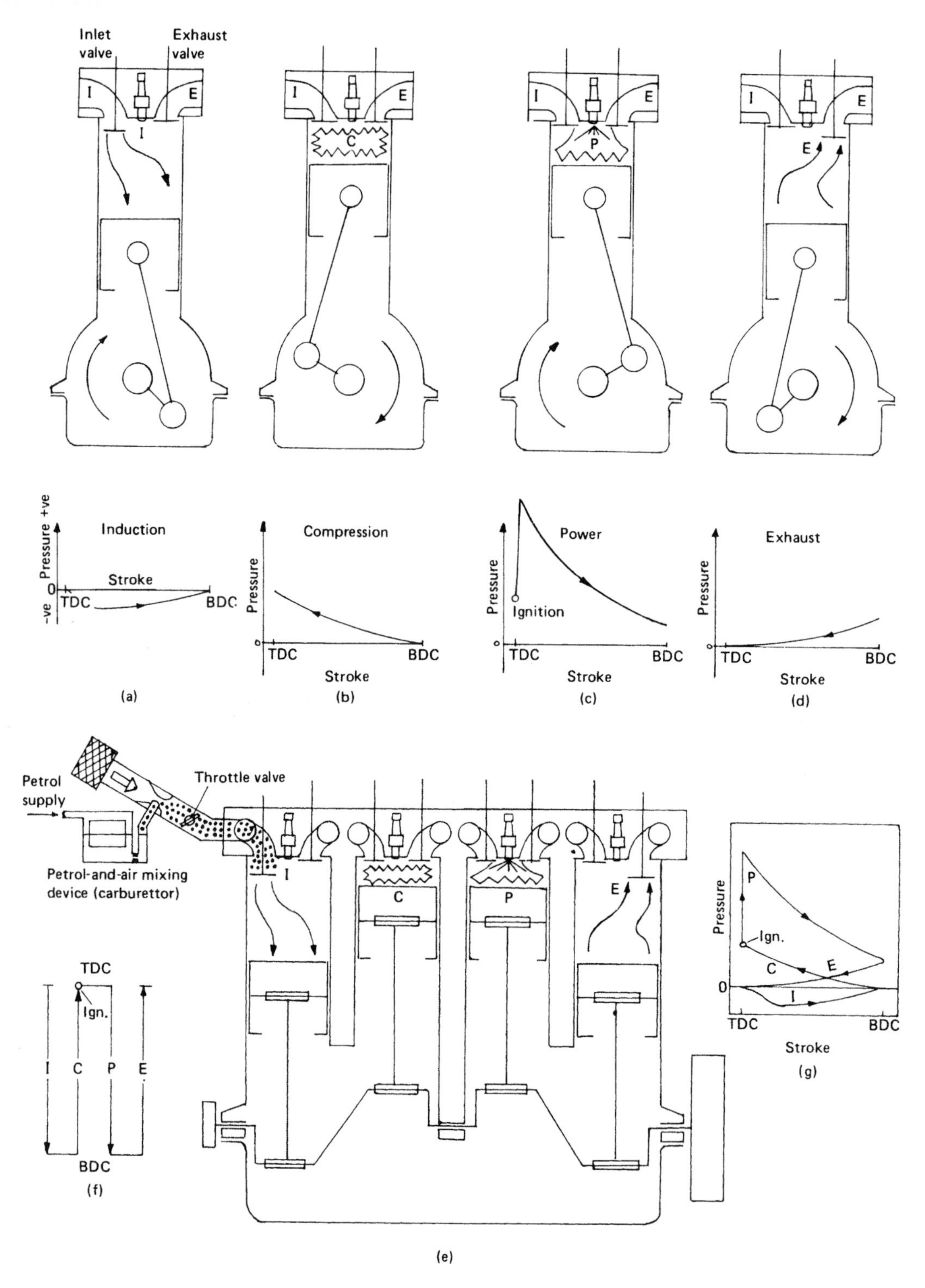

Fig. 1.1-3 Four-stroke-cycle petrol engine.

1.1.1.3 Valve timing diagrams

In practice, the events of the four-stroke cycle do not start and finish exactly at the two ends of the strokes – to improve the breathing and exhausting, the inlet valve is arranged to open before TDC and to close after BDC and the exhaust valve opens before BDC and closes after TDC. These early and late opening and closing events can be shown on a valve timing diagram such as Fig. 1.1-4.

Valve lead This is where a valve opens so many degrees of crankshaft rotation before either TDC or BDC.

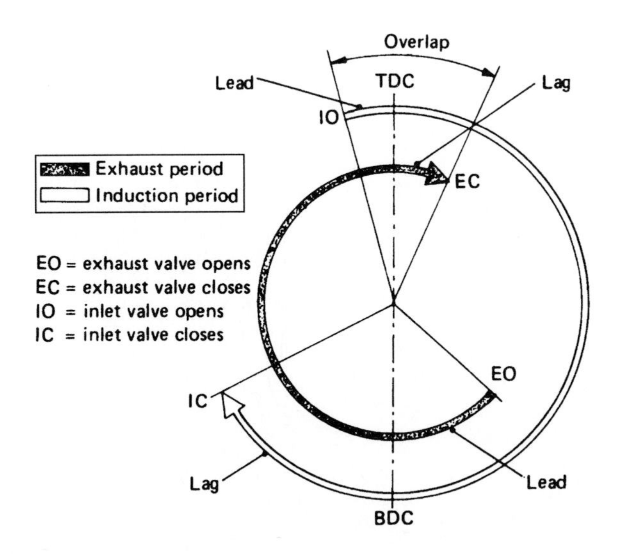

Fig. 1.1-4 Valve timing diagram.

Valve lag This is where a valve closes so many degrees of crankshaft rotation after TDC or BDC.

Valve overlap This is the condition when both the inlet and the exhaust valves are open at the same time during so many degrees of crankshaft rotation.

1.1.2 The two-stroke-cycle petrol engine

The first successful design of a three-port two-stroke engine was patented in 1889 by Joseph Day & Son of Bath. This employed the underside of the piston in conjunction with a sealed crank-case to form a scavenge pump ('scavenging' being the pushing-out of exhaust gas by the induction of fresh charge) (Fig. 1.1-5).

This engine completes the cycle of events – induction, compression, power, and exhaust – in one revolution of the crankshaft or two complete piston strokes.

Crankcase-to-cylinder mixture transfer (Fig. 1.1-5(a)) The piston moves down the cylinder and initially uncovers the exhaust port (E), releasing the burnt exhaust gases to the atmosphere. Simultaneously the downward movement of the underside of the piston compresses the previously filled mixture of air and atomised petrol in the crankcase (Fig. 1.1-5(a)). Further outward movement of the piston will uncover the transfer port (T), and the compressed mixture in the crankcase will then be transferred to the combustion-chamber side of the cylinder. The situation in the cylinder will then be such that the fresh charge entering the cylinder will push out any remaining burnt products of combustion – this process is generally referred to as cross-flow scavenging.

Cylinder compression and crankcase induction (Fig. 1.1-5(b)) The crankshaft rotates, moving the piston in the direction of the cylinder head. Initially the piston seals off the transfer port, and then a short time later the exhaust port will be completely closed. Further inward movement of the piston will compress the mixture of air and atomised petrol to about one-seventh to one-eighth of its original volume (Fig. 1.1-5(b)).

At the same time as the fresh charge is being compressed between the combustion chamber and the piston head, the inward movement of the piston increases the total volume in the crank-case so that a depression is created in this space. About half-way up the cylinder stroke, the lower part of the piston skirt will uncover the inlet port (I), and a fresh mixture of air and petrol prepared by the carburettor will be induced into the crankcase chamber (Fig. 1.1-5(b)).

Cylinder combustion and crankcase compression (Fig. 1.1-5(c)) Just before the piston reaches the top of its stroke, a spark-plug situated in the centre of the cylinder head will be timed to spark and ignite the dense mixture. The burning rate of the charge will rapidly raise the gas pressure to a maximum of about 50 bar under full load. The burning mixture then expands, forcing the piston back along its stroke with a corresponding reduction in cylinder pressure (Fig. 1.1-5(c)).

Considering the condition underneath the piston in the crankcase, with the piston initially at the top of its stroke, fresh mixture will have entered the crankcase through the inlet port. As the piston moves down its stroke, the piston skirt will cover the inlet port, and any further downward movement will compress the mixture in the crankcase in preparation for the next charge transfer into the cylinder and combustion-chamber space (Fig. 1.1-5(c)).

The combined cycle of events adapted to a three-cylinder engine is shown in Fig. 1.1-5(d). Figs. 1.1-5(e) and (f) show the complete cycle in terms of opening and closing events and cylinder volume and pressure changes respectively.

1.1.2.1 Reverse-flow (Schnuerle) scavenging

To improve scavenging efficiency, a loop-scavenging system which became known as the reverse-flow or (after its inventor, Dr E. Schnuerle) as the Schnuerle scavenging system was developed (Fig. 1.1-6). This layout has a transfer port on each side of the exhaust port, and these direct the scavenging charge mixture in a practically tangential direction towards the opposite cylinder wall. The two separate columns of the scavenging mixture meet and merge together at this wall to form one inward rising flow which turns under the cylinder head and then flows down on the entry side, thus forming a complete loop. With this form of porting, turbulence and intermixing of fresh fuel mixture with residual burnt gases will be minimal over a wide range of piston speeds.

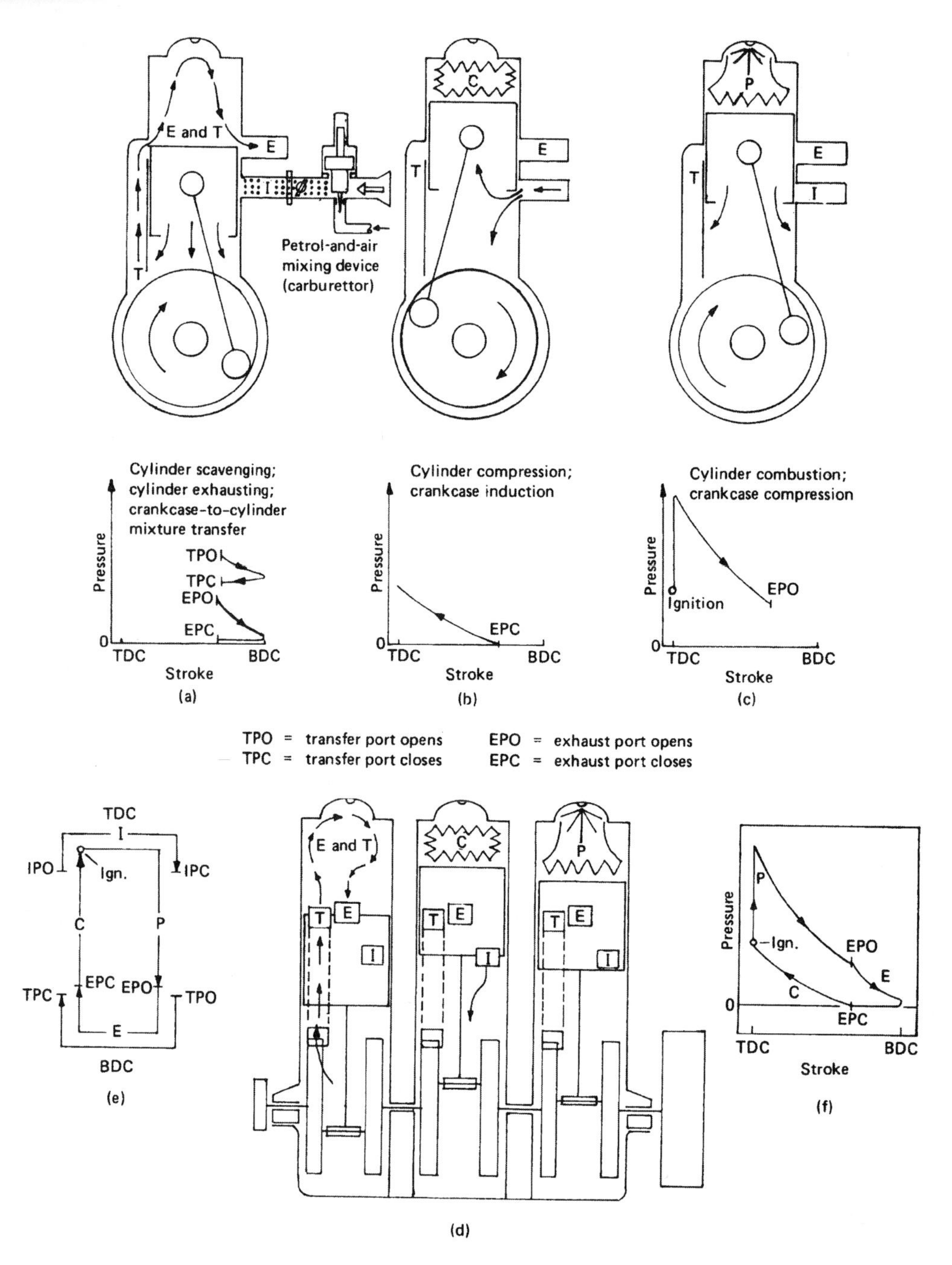

Fig. 1.1-5 Two-stroke-cycle petrol engine.

Note that in this particular design the charge mixture is transferred through ports formed in the piston skirt. Alternatively, extended transfer passages may be preferred so that the piston skirt plays no part in the timed transfer.

1.1.2.2 Crankcase disc-valve and reed-valve inlet charge control

An alternative to the piston-operated crankcase inlet port is to use a disc-valve attached to and driven by the crankshaft (Fig. 1.1-7(a)). This disc-valve is timed to open and close so that the fresh charge is induced to enter the crankcase as early as possible, and only at the point when the charge is about to be transferred into the cylinder is it closed. This method of controlling crankcase induction does not depend upon the piston displacement to uncover the port – it can therefore be so phased as to extend the filling period (Fig. 1.1-7).

A further method of improving crankcase filling is the use of reed-valves (Fig. 1.1-7(b)). These valves are not timed to open and close, but operate automatically when the pressure difference between the crankcase and the air intake is sufficient to deflect the reed-spring. In other

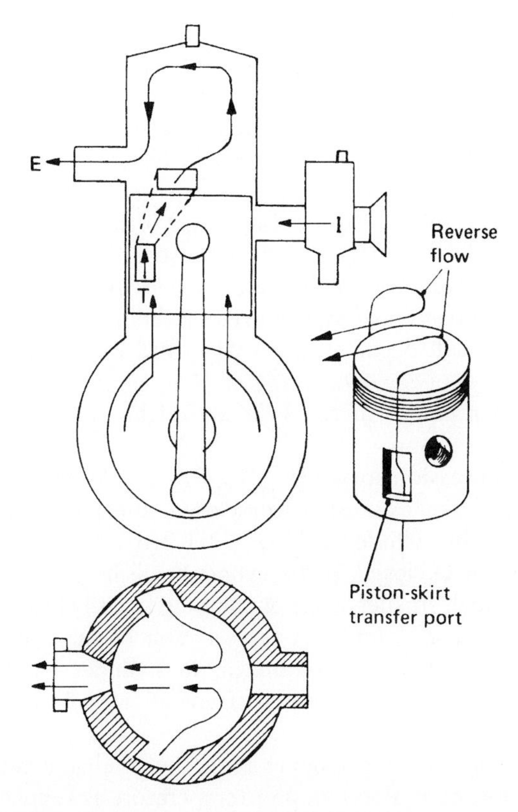

Fig. 1.1-6 Reverse flow or Schnuerle scavenging.

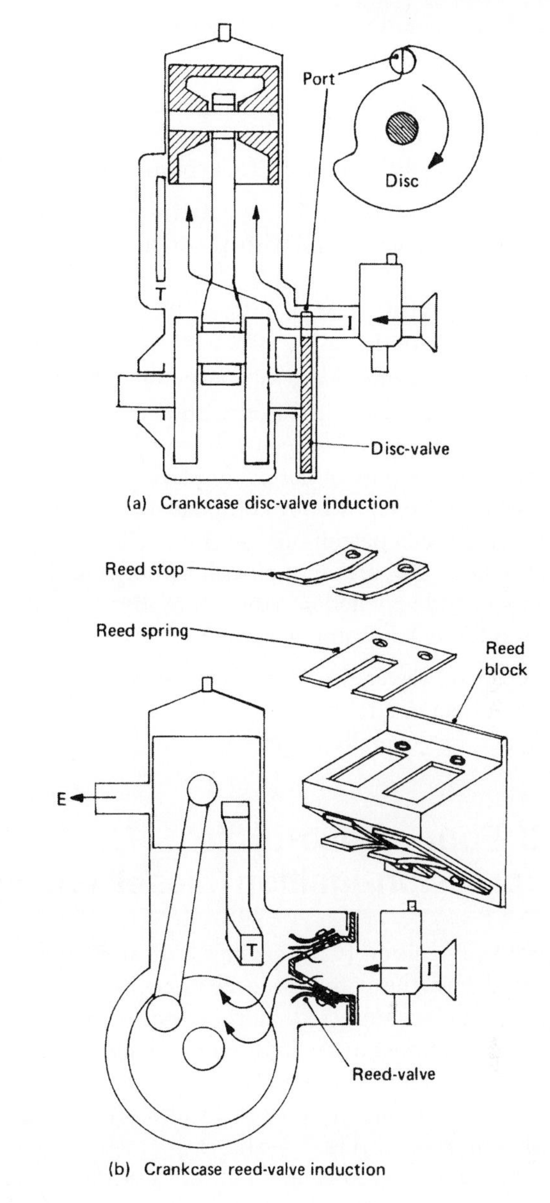

Fig. 1.1-7 Crankcase disc-valve and reed-valve induction.

words, these valves sense the requirements of the crankcase and so adjust their opening and closing frequencies to match the demands of the engine.

1.1.2.3 Comparison of two- and four-stroke-cycle petrol engines

The following remarks compare the main points regarding the effectiveness of both engine cycles.

a) The two-stroke engine completes one cycle of events for every revolution of the crankshaft, compared with the two revolutions required for the four-stroke engine cycle.

b) Theoretically, the two-stroke engine should develop twice the power compared to a four-stroke engine of the same cylinder capacity.

c) In practice, the two-stroke engine's expelling of the exhaust gases and filling of the cylinder with fresh mixture brought in through the crankcase is far less effective than having separate exhaust and induction strokes. Thus the mean effective cylinder pressures in two-stroke units are far lower than in equivalent four-stroke engines.

d) With a power stroke every revolution instead of every second revolution, the two-stroke engine will run smoother than the four-stroke power unit for the same size of flywheel.

e) Unlike the four-stroke engine, the two-stroke engine does not have the luxury of separate exhaust and induction strokes to cool both the cylinder and the piston between power strokes. There is therefore a tendency for the piston and small-end to overheat under heavy driving conditions.

f) Due to its inferior scavenging process, the two-stroke engine can suffer from the following:

i) inadequate transfer of fresh mixture into the cylinder,

ii) excessively large amounts of residual exhaust gas remaining in the cylinder,

iii) direct expulsion of fresh charge through the exhaust port.

These undesirable conditions may occur under different speed and load situations, which greatly influences both power and fuel consumption.

g) Far less maintenance is expected with the two-stroke engine compared with the four-stroke engine, but there can be a problem with the products of combustion carburising at the inlet, transfer, and exhaust ports.

h) Lubrication of the two-stroke engine is achieved by mixing small quantities of oil with petrol in proportions anywhere between 1:16 and 1:24 so that, when crankcase induction takes place, the various rotating and reciprocating components will be lubricated by a petroil-mixture mist. Clearly a continuous proportion of oil will be burnt in the cylinder and expelled into the atmosphere to add to unwanted exhaust emission.

i) There are fewer working parts in a two-stroke engine than in a four-stroke engine, so two-stroke engines are generally cheaper to manufacture.

1.1.3 Four-stroke-cycle compression-ignition (diesel) engine

Compression-ignition (C.I.) engines burn fuel oil which is injected into the combustion chamber when the air charge is fully compressed. Burning occurs when the compression temperature of the air is high enough to spontaneously ignite the finely atomised liquid fuel. In other words, burning is initiated by the self-generated heat of compression (Fig. 1.1-8).

Engines adopting this method of introducing and mixing the liquid fuel followed by self-ignition are also referred to as 'oil engines', due to the class of fuel burnt, or as 'diesel engines' after Rudolf Diesel, one of the many inventors and pioneers of the early C.I. engine. Note: in the United Kingdom fuel oil is known as 'DERV', which is the abbreviation of 'diesel-engine road vehicle'.

Just like the four-stroke-cycle petrol engine, the C.I. engine completes one cycle of events in two crankshaft revolutions or four piston strokes. The four phases of these strokes are (i) induction of fresh air, (ii) compression and heating of this air, (iii) injection of fuel and its burning and expansion, and (iv) expulsion of the products of combustion.

Induction stroke (Fig. 1.1-8(a)) With the inlet valve open and the exhaust valve closed, the piston moves away from the cylinder head (Fig. 1.1-8(a)).

The outward movement of the piston will establish a depression in the cylinder, its magnitude depending on the ratio of the cross-sectional areas of the cylinder and the inlet port and on the speed at which the piston is moving. The pressure difference established between the inside and outside of the cylinder will induce air at atmospheric pressure to enter and fill up the cylinder. Unlike the petrol engine, which requires a charge of air-and-petrol mixture to be drawn past a throttle valve, in the diesel-engine inlet system no restriction is necessary and only pure air is induced into the cylinder. A maximum depression of maybe 0.15 bar below atmospheric pressure will occur at about one-third of the distance along the piston's outward stroke, while the overall average pressure in the cylinder might be 0.1 bar or even less.

Compression stroke (Fig. 1.1-8(b)) With both the inlet and the exhaust valves closed, the piston moves towards the cylinder head (Fig. 1.1-8(b)).

The air enclosed in the cylinder will be compressed into a much smaller space of anything from 1/12 to 1/24 of its original volume. A typical ratio of maximum to minimum air-charge volume in the cylinder would be 16:1, but this largely depends on engine size and designed speed range.

During the compression stroke, the air charge initially at atmospheric pressure and temperature is reduced in volume until the cylinder pressure is raised to between 30 and 50 bar. This compression of the air generates heat which will increase the charge temperature to at least 600 °C under normal running conditions.

Power stroke (Fig. 1.1-8(c)) With both the inlet and the exhaust valves closed and the piston almost at the end of the compression stroke (Fig. 1.1-8(c)), diesel fuel oil is injected into the dense and heated air as a high-pressure spray of fine particles. Provided that they are properly atomised and distributed throughout the air charge, the heat of compression will then quickly vaporise and ignite the tiny droplets of liquid fuel. Within a very short time, the piston will have reached its innermost position and extensive burning then releases heat energy which is rapidly converted into pressure energy. Expansion then follows, pushing the piston away from the cylinder head, and the linear thrust acting on the piston end of the connecting-rod will then be changed to rotary movement of the crankshaft.

Exhaust stroke When the burning of the charge is near completion and the piston has reached the outermost position, the exhaust valve is opened. The piston then reverses its direction of motion and moves towards the cylinder head (Fig. 1.1-8(d)).

The sudden opening of the exhaust valve towards the end of the power stroke will release the still burning products of combustion to the atmosphere. The pressure energy of the gases at this point will accelerate their expulsion from the cylinder, and only towards the end of

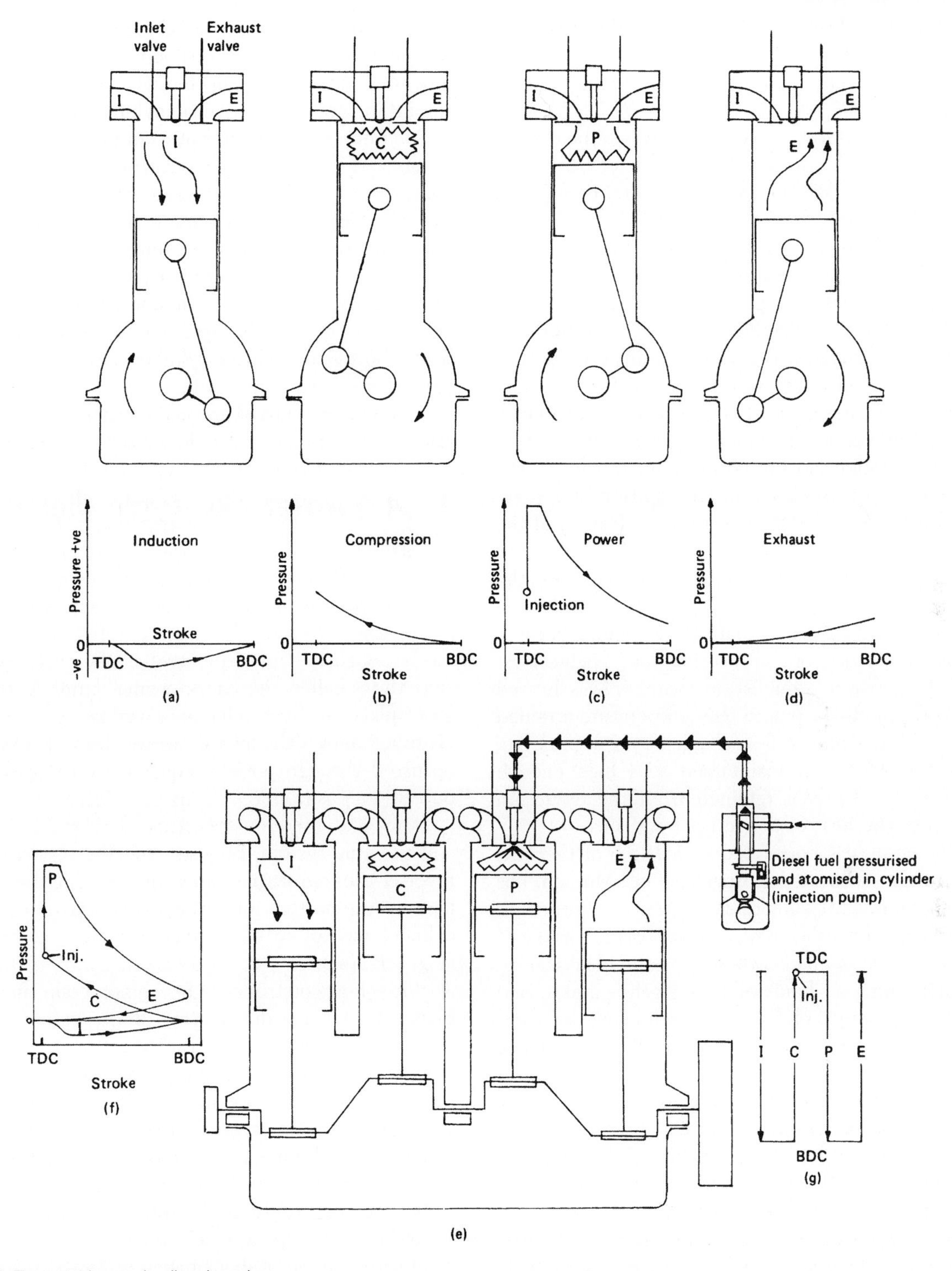

Fig. 1.1-8 Four-stroke-cycle diesel engine.

the piston's return stroke will the piston actually catch up with the tail-end of the outgoing gases.

Fig. 1.1-8(e) illustrates the sequence of the four operating strokes as applied to a four-cylinder engine, and the combined operating events expressed in terms of cylinder pressure and piston displacement are shown in Figs. 1.1-8(f) and (g).

1.1.3.1 Historical background to the C.I. engine

Credit for the origination of the C.I. engine is controversial, as eminent engineers cannot agree amongst themselves as to which of the patents by Herbert Akroyd-Stuart or Rudolf Diesel contributed most to the instigation and evolution of the high-speed C.I. engine burning heavy fuel oil. A brief summary of the background and achievements of these two pioneers is as follows.

Herbert Akroyd-Stuart, born 1864, was trained as an engineer in his father's works at Fenny Stratford, England. Between 1885 and 1890 he took out several patents for improvements to oil engines, and later, in conjunction with a Charles R. Binney of London, he took out patent number 7146 of 1890 describing the operation of his engine. Air alone was drawn into the cylinder and compressed into a separate combustion chamber (known as the vaporiser) through a contracted passage or bottleneck. A liquid fuel spray was then injected into the compressed air near the end of the compression stroke by means of a pump and a spraying nozzle. The combination of the hot chamber and the rise in temperature of the compressed air provided automatic ignition and rapid combustion at nearly constant volume – a feature of the C.I. engines of today.

These early engines were of low compression, the explosion taking place mainly due to the heat of the vaporiser chamber itself so that these engines became known as 'hot-bulb' or 'surface-ignition' engines. At starting, the separate combustion chamber was heated externally by an oil-lamp until the temperature attained was sufficient to ignite a few charges by compression. Then the chamber was maintained at a high enough temperature by the heat retained from the explosion together with the heat of the compressed air.

Rudolf Diesel was born in Paris in 1858, of German parents, and was educated at Augsburg and Munich. His works training was with Gebrü-der Sulzer in Winterthur. Dr Diesel's first English patent, number 7421, was dated 1892 and was for an engine working on the ideal Carnot cycle and burning all kinds of fuel – solid, liquid, and gas – but the practical difficulties of achieving this thermodynamic cycle proved to be far too much. A reliable diesel oil engine was built in 1897 after four years of experimental work in the Mashinen-fabrik Augsburg Nürnberg (MAN) workshops.

In this engine, air was drawn into the cylinder and was compressed to 35–40 bar. Towards the end of the compression stroke, an air blast was introduced into the combustion space at a much higher pressure, about 68–70 bar, thus causing turbulence in the combustion chamber. A three-stage compressor driven by the engine (and consuming about 10% of the engine's gross power) supplied compressed air which was stored in a reservoir. This compressed air served both for starting the engine and for air-injection into the compressed air already in the cylinder – that is, for blasting air to atomise the oil fuel by forcing it through perforated discs fitted around a fluted needle-valve injector. The resulting finely divided oil mist ignites at once when it contacts the hot compressed cylinder air, and the burning rate then tends to match the increasing cylinder volume as the piston moves outwards – expansion will therefore take place at something approaching constant pressure.

A summary of the combustion processes of Akroyd-Stuart and Diesel is that the former inventor used a low compression-ratio, employed airless liquid-fuel injection, and relied on the hot combustion chamber to vaporise and ignite the fuel; whereas Diesel employed a relatively high compression-ratio, adopted air-injection to atomise the fuel, and made the hot turbulent air initiate burning. It may be said that the modern high-speed C.I. engine embraces both approaches in producing sparkless automatic combustion – combustion taking place with a combined process of constant volume and constant pressure known as either the mixed or the dual cycle.

1.1.4 Two-stroke-cycle diesel engine

The pump scavenge two-stroke-cycle engine designed by Sir Dugald Clerk in 1879 was the first successful two-stroke engine; thus the two-stroke-cycle engine is sometimes called the Clerk engine. Uniflow scavenging took place – fresh charge entering the combustion chamber above the piston while the exhaust outflow occurred through ports uncovered by the piston at its outermost position.

Low- and medium-speed two-stroke marine diesels still use this system, but high-speed two-stroke diesels reverse the scavenging flow by blowing fresh charge through the bottom inlet ports, sweeping up through the cylinder and out of the exhaust ports in the cylinder head (Fig. 1.1-9(a)).

With the two-stroke-cycle engine, intake and exhaust phases take place during part of the compression and power stroke respectively, so that a cycle of operation is completed in one crankshaft revolution or two piston strokes. Since there are no separate intake and exhaust strokes, a blower is necessary to pump air into the cylinder for expelling the exhaust gases and to supply the cylinder with fresh air for combustion.

Scavenging (induction and exhaust) phase (Fig. 1.1-9(a)) The piston moves away from the cylinder head and, when it is about half-way down its stroke, the exhaust valves open. This allows the burnt gases to escape into the atmosphere. Near the end of the power stroke, a horizontal row of inlet air ports is uncovered by the piston lands (Fig. 1.1-9(a)). These ports admit pressurised air from the blower into the cylinder. The space above the piston is immediately filled with air, which now blows up the cylinder towards the exhaust valves in the cylinder head. The last remaining exhaust gases will thus be forced out of the cylinder into the exhaust system. This process

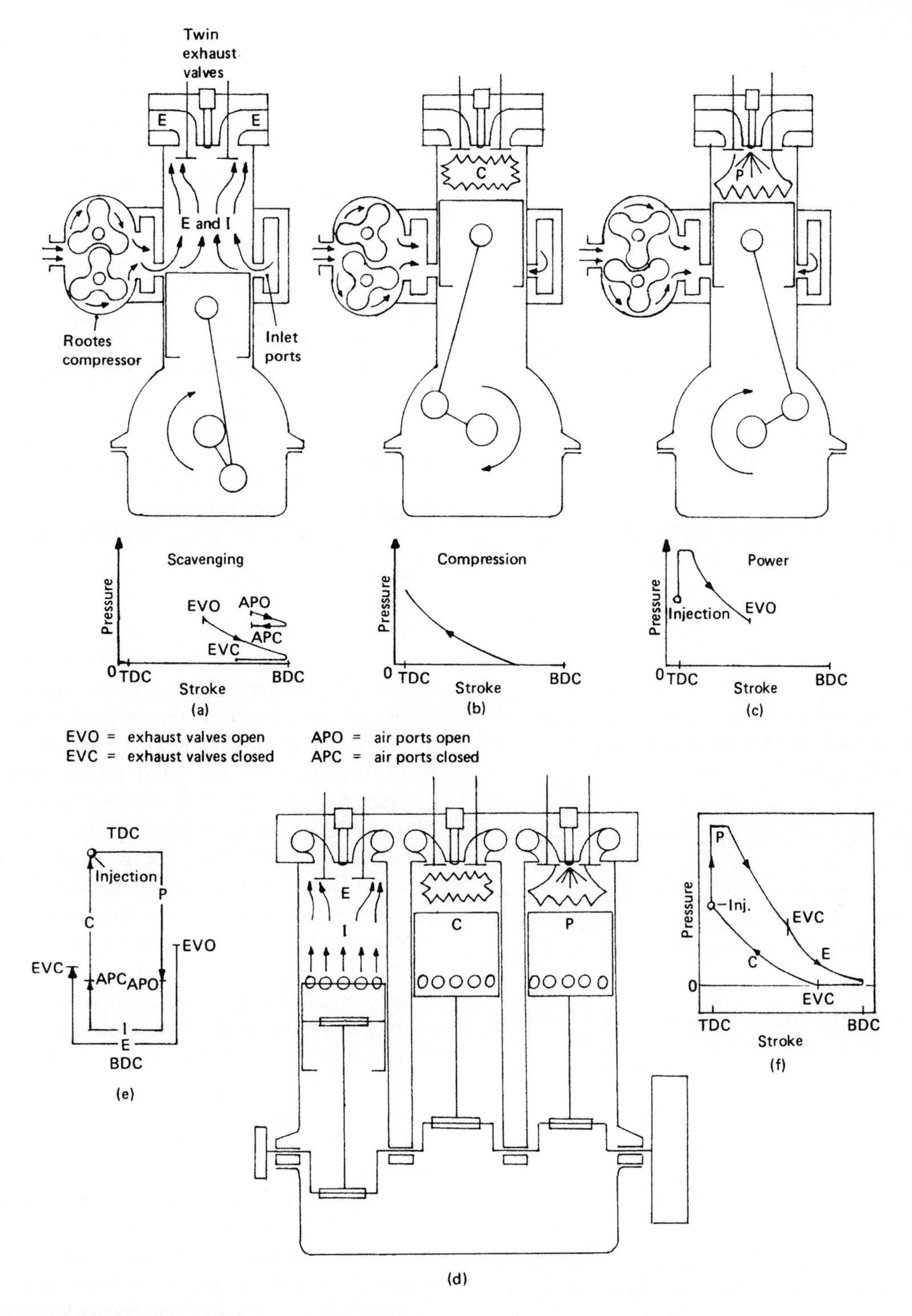

Fig. 1.1-9 Two-stroke-cycle diesel engine.

of fresh air coming into the cylinder and pushing out unwanted burnt gas is known as scavenging.

Compression phase (Fig. 1.1-9(b)) Towards the end of the power stroke, the inlet ports will be uncovered. The piston then reaches its outermost position and reverses its direction of motion. The piston now moves upwards so that the piston seals and closes the inlet air ports, and just a little later the exhaust valves close. Any further upward movement will now compress the trapped air (Fig. 1.1-9(b)). This air charge is now reduced to about 1/15 to 1/18 of its original volume as the piston reaches the innermost position. This change in volume corresponds to a maximum cylinder pressure of about 30–40 bar.

Power phase (Fig. 1.1-9(c)) Shortly before the piston reaches the innermost position to the cylinder head on its upward compression stroke, highly pressurised liquid fuel is sprayed into the dense intensely heated air charge (Fig. 1.1-9(c)). Within a very short period of time, the injected fuel droplets will vaporise and ignite, and rapid burning will be established by the time the piston is at the top of its stroke. The heat liberated from the charge will be converted mainly into gas-pressure energy which will expand the gas and so do useful work in driving the piston outwards.

An overall view of the various phases of operation in a two-stroke-cycle three-cylinder diesel engine is shown in Figs. 1.1-9(d), and Figs. 1.1-9(e) and (f) show the cycle of events in one crankshaft revolution expressed in terms of piston displacement and cylinder pressure.

1.1.4.1 Comparison of two- and four-stroke-cycle diesel engines

A brief but critical comparison of the merits and limitations of the two-stroke-cycle diesel engine compared with the four-stroke power unit is made below.

a) Theoretically, almost twice the power can be developed with a two-stroke engine compared with a four-stroke engine.

b) A comparison between a typical 12 litre four-stroke engine and a 7 litre two-stroke engine having the same speed range would show that they would develop similar torque and power ratings. The ratio of engine capacities for equivalent performance for these four-stroke and two-stroke engines would be 1.7:1.

c) In a four-stroke engine, the same parts generate power and empty and fill the cylinders. With the two-stroke engine, the emptying and filling can be carried out by light rotary components.

d) With a two-stroke engine, 40–50% more air consumption is necessary for the same power output; therefore the air-pumping work done will be proportionally greater.

e) About 10–20% of the upward stroke of a two-stroke engine must be sacrificed to emptying and filling the cylinder.

f) The time available for emptying and filling a cylinder is considerably less in a two-stroke-cycle engine – something like 33% of the completed cycle as compared to 50% in a four-stroke engine. Therefore more power will be needed to force a greater mass of air into the cylinder in a shorter time.

g) Compared with a two-stroke engine, more power is needed by the piston for emptying and filling the cylinder in a four-stroke engine, due to pumping and friction losses at low speeds. At higher engine speeds the situation is reversed, and the two-stroke's Rootes blower will consume proportionally more engine power – this could be up to 15% of the developed power at maximum speed.

h) With reduced engine load for a given speed, a two-stroke engine blower will consume proportionally more of the power developed by the engine.

i) A two-stroke engine runs smoother and relatively quietly, due to the absence of reversals of loading on bearings as compared with a four-stroke engine.

1.1.5 Comparison of S.I. and C.I. engines

The pros and cons of petrol and C.I. engines are now considered.

Fuel economy The chief comparison to be made between the two types of engine is how effectively each engine can convert the liquid fuel into work energy. Different engines are compared by their thermal efficiencies. Thermal efficiency is the ratio of the useful work produced to the total energy supplied. Petrol engines can have thermal efficiencies ranging between 20% and 30%. The corresponding diesel engines generally have improved efficiencies, between 30% and 40%. Both sets of efficiency values are considerably influenced by the chosen compression-ratio and design.

Power and torque The petrol engine is usually designed with a shorter stroke and operates over a much larger crankshaft-speed range than the diesel engine. This enables more power to be developed towards the upper speed range in the petrol engine, which is necessary for high road speeds; however, a long-stroke diesel engine has improved pulling torque over a relatively narrow speed range, this being essential for the haulage of heavy commercial vehicles.

At the time of writing, there was a trend to incorporate diesel engines into cars. This new generation of engines has different design parameters and therefore does not conform to the above observations.

Reliability Due to their particular process of combustion, diesel engines are built sturdier, tend to run cooler, and have only half the speed range of most petrol engines. These factors make the diesel engine more reliable and considerably extend engine life relative to the petrol engine.

Pollution Diesel engines tend to become noisy and to vibrate on their mountings as the operating load is reduced. The combustion process is quieter in the petrol engine and it runs smoother than the diesel engine. There is no noisy injection equipment used on the petrol engine, unlike that necessary on the diesel engine.

The products of combustion coming out of the exhaust system are more noticeable with diesel engines, particularly if any of the injection equipment components are out of tune. It is questionable which are the more harmful: the relatively invisible exhaust gases from the petrol engine, which include nitrogen dioxide, or the visible smoky diesel exhaust gases.

Safety Unlike petrol, diesel fuels are not flammable at normal operating temperature, so they are not a handling hazard and fire risks due to accidents are minimised.

Cost Due to their heavy construction and injection equipment, diesel engines are more expensive than petrol engines.

1.1.6 Engine-performance terminology

To enable intelligent comparisons to be made between different engines' ability to pull or operate at various speeds, we shall now consider engine design parameters and their relationship in influencing performance capability.

1.1.6.1 Piston displacement or swept volume

When the piston moves from one end of the cylinder to the other, it will sweep or displace air equal to the cylinder volume between TDC and BDC. Thus the full stroke movement of the piston is known as either the swept volume or the piston displacement.

The swept or displaced volume may be calculated as follows:

$$V = \frac{\pi d^2 L}{4000}$$

where V = piston displacement (cm^3)
$\pi = 3.142$
d = cylinder diameter (mm)
and L = cylinder stroke (mm)

1.1.6.2 Mean effective pressure

The cylinder pressure varies considerably while the gas expands during the power stroke. Peak pressure will occur just after TDC, but this will rapidly drop as the piston moves towards BDC. When quoting cylinder pressure, it is therefore more helpful to refer to the average or mean effective pressure throughout the whole power stroke. The units used for mean effective pressure may be either kilonewtons per square metre (kN/m^2) or bars (note: 1 bar = 100 kN/m^2).

1.1.6.3 Engine torque

This is the turning-effort about the crankshaft's axis of rotation and is equal to the product of the force acting along the connecting-rod and the perpendicular distance between this force and the centre of rotation of the crankshaft. It is expressed in newton metres (N m);

i.e. $T = Fr$
where T = engine torque (N m)
F = force applied to crank (N)
and r = effective crank-arm radius (m)

During the 180° crankshaft movement on the power stroke from TDC to BDC, the effective radius of the crank-arm will increase from zero at the top of its stroke to a maximum in the region of mid-stroke and then decrease to zero again at the end of its downward movement (Fig. 1.1-10). This implies that the torque on the power stroke is continually varying. Also, there will be no useful torque during the idling strokes. In fact some of the torque on the power stroke will be cancelled out in overcoming compression resistance and pumping losses, and the torque quoted by engine manufacturers is always the average value throughout the engine cycle.

The average torque developed will vary over the engine's speed range. It reaches a maximum at about mid-speed and decreases on either side (Fig. 1.1-11).

1.1.6.4 Engine power

Power is the rate of doing work. When applied to engines, power ratings may be calculated either on the basis of indicated power (i.p.), that is the power actually developed in the cylinder, or on the basis of brake power (b.p.), which is the output power measured at the crankshaft. The b.p. is always less than the i.p., due to frictional and pumping losses in the cylinders and the reciprocating mechanism of the engine.

Since the rate of doing work increases with piston speed, the engine's power will tend to rise with crankshaft speed of rotation, and only after about two-thirds of the engine's speed range will the rate of power rise drop off (Fig. 1.1-11).

The slowing down and even decline in power at the upper speed range is mainly due to the very short time available for exhausting and for inducing fresh charge into the cylinders at very high speeds, with a resulting reduction in the cylinders' mean effective pressures.

Different countries have adopted their own standardised test procedures for measuring engine performance, so slight differences in quoted output figures

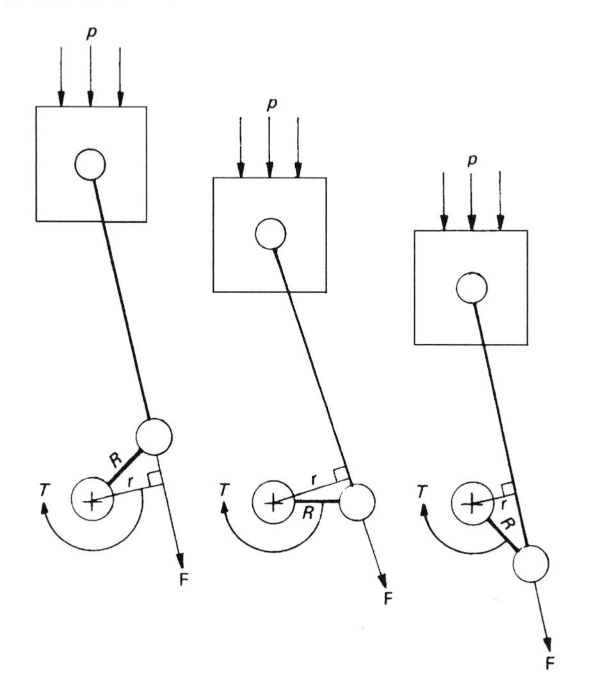

Fig. 1.1-10 Torque variation during crankshaft rotation (p = cylinder gas pressure; F = connecting-rod thrust; R = crank-throw; r = effective crank radius; T = turning-effort or torque).

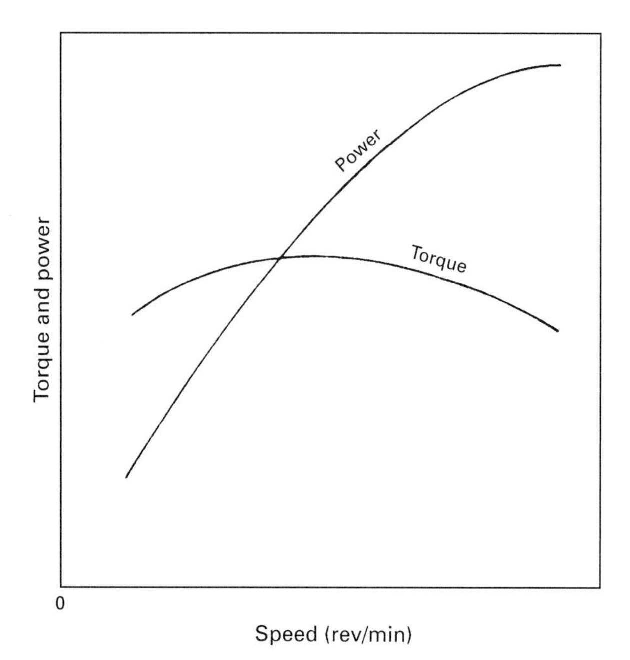

Fig. 1.1-11 Torque and power variation over engine speed range.

will exist. Quoted performance figures should therefore always state the standard used. The three most important standards are those of the American Society of Automotive Engineers (SAE), the German Deutsch Industrie Normale (DIN), and the Italian Commissione technica di Unificazione nell Automobile (CUNA).

The two methods of calculating power can be expressed as follows:

$$\text{i.p.} = \frac{pLANn}{60\,000}$$

where i.p. = indicated power (kW)
p = effective pressure (kN/m^2)
L = length of stroke (m)
A = cross-sectional area of piston (m^2)
N = crankshaft speed (rev/min)
and n = number of cylinders

$$\text{b.p.} = \frac{2\,\pi\,TN}{60\,000}$$

where b.p. = brake power (kW)
π = 3.142
T = engine torque (N m)
and N = crankshaft speed (rev/min)

The imperial power is quoted in horsepower (hp) and is defined in terms of foot pounds per minute. In imperial units one horsepower is equivalent to 33 000 ft lb per minute or 550 ft lb per second. A metric horsepower is defined in terms of Newton-metres per second and is equal to 0.986 imperial horsepower. In Germany the abbreviation for horsepower is PS derived from the translation of the words 'Pferd-Stärke' meaning horse strength.

The international unit for power is the watt, W, or more usually the kilowatt, kW, where 1 kW = 1000 W.

Conversion from watt to horsepower and vice versa is:

1 kW = 1.35 hp and 1 hp = 0.746 kW

1.1.6.5 Engine cylinder capacity

Engine sizes are compared on the basis of total cylinder swept volume, which is known as engine cylinder capacity. Thus the engine cylinder capacity is equal to the piston displacement of each cylinder times the number of cylinders,

$$\text{i.e. } V_E = \frac{Vn}{1000}$$

where V_E = engine cylinder capacity (litre)
V = piston displacement (cm^3)
and n = number of cylinders

Piston displacement is derived from the combination of both the cross-sectional area of the piston and its stroke. The relative importance of each of these dimensions can be demonstrated by considering how they affect performance individually.

The cross-sectional area of the piston crown influences the force acting on the connecting-rod, since the product of the piston area and the mean effective cylinder pressure is equal to the total piston thrust;

i.e. $F = pA$

where F = piston thrust (kN)
p = mean effective pressure (kN/m^2)
and A = cross-sectional area of piston (m^2)

The length of the piston stroke influences both the turning-effort and the angular speed of the crankshaft. This is because the crank-throw length determines the leverage on the crankshaft, and the piston speed divided by twice the stroke is equal to the crankshaft speed;

i.e. $N = \dfrac{v}{2L}$

where N = crankshaft speed (rev/min)
v = piston speed (m/min)
and L = piston stroke (m)

This means that making the stroke twice as long doubles the crankshaft turning-effort and halves the crankshaft angular speed for a given linear piston speed.

The above shows that the engine performance is decided by the ratio of bore to stroke chosen for a given cylinder capacity.

1.1.7 Compression-ratio

In an engine cylinder, the gas molecules are moving about at considerable speed in the space occupied by the gas, colliding with other molecules and the boundary surfaces of the cylinder head, the cylinder walls, and the piston crown. The rapid succession of impacts of many millions of molecules on the boundary walls produces a steady continuous force per unit surface which is known as pressure (Fig. 1.1-12).

When the gas is compressed into a much smaller space, the molecules are brought closer to one another. This raises the temperature and greatly increases the speed of the molecules and hence their kinetic energy, so more violent impulses will impinge on the piston crown. This increased activity of the molecules is experienced as increased opposition to movement of the piston towards the cylinder head.

The process of compressing a constant mass of gas into a much smaller space enables many more molecules to impinge per unit area on to the piston. When burning of the gas occurs, the chemical energy of combustion is rapidly transformed into heat energy which considerably increases the kinetic energy of the closely packed gas molecules. Therefore the extremely large number of molecules squeezed together will thus bombard the piston crown at much higher speeds. This then means that a very large number of repeated blows of considerable magnitude will strike the piston and so push it towards ODC.

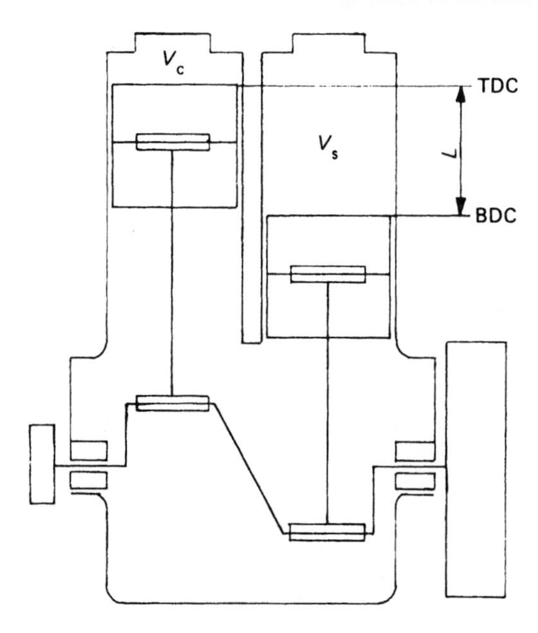

Fig. 1.1-12 Illustration of compression-ratio.

This description of compression, burning, and expansion of the gas charge shows the importance of utilising a high degree of compression before burning takes place, to improve the efficiency of combustion. The amount of compression employed in the cylinder is measured by the reduction in volume when the piston moves from BDC to TDC, the actual proportional change in volume being expressed as the compression-ratio.

The compression-ratio may be defined as the ratio of the maximum cylinder volume when the piston is at its outermost position (BDC) to the minimum cylinder volume (the clearance volume) with the piston at its innermost position (TDC) – that is, the sum of the swept and clearance volumes divided by the clearance volume,

i.e. $\mathrm{CR} = \dfrac{V_s + V_c}{V_c}$

where CR = compression ratio
V_s = swept volume (cm^3)
V_c = clearance volume (cm^3)

Petrol engines have compression-ratios of the order of 7:1 to 10:1; but, to produce self-ignition of the charge, diesel engines usually double these figures and may have values of between 14:1 and 24:1 for naturally aspirated (depression-induced filling) types, depending on the design.

Section Two

Engine testing

Chapter 2.1

2.1

Measurement of torque, power, speed and fuel consumption; acceptance and type tests, accuracy of the measurements

A.J. Martyr and M.A. Plint

2.1.1 Introduction

The torque produced by a prime mover under test is resisted and measured by the dynamometer to which it is connected. The accuracy with which a dynamometer measures both torque and speed is fundamental to all the other derived measurements made in the test cell.

In this chapter the principles of torque measurement are reviewed and then the types of dynamometer are reviewed in order to assist the purchaser in the selection of the most appropriate machine.

2.1.2 Measurement of torque: trunnion-mounted (cradle) machines

The essential feature of trunnion-mounted or cradled dynamometers is that the power absorbing element of the machine is mounted on bearings coaxial with the machine shaft and the torque is restrained and measured by some kind of transducer acting tangentially at a known radius from the machine axis.

Until the beginning of the present century, the great majority of new and existing dynamometers used this method of torque measurement. In traditional machines the torque measurement was achieved by physically balancing a combination of dead weights and a spring balance against the torque absorbed (Fig. 2.1-1). As the stiffness of the balance was limited, it was necessary to adjust its position depending on the torque, to ensure that the force measured was accurately tangential.

Modern trunnion-mounted machines, shown diagrammatically in Fig. 2.1-2, use a force transducer, almost invariably of the strain gauge type, together with an appropriate bridge circuit and amplifier. The strain gauge transducer or 'load cell' has the advantage of being extremely stiff, so that no positional adjustment is necessary, but the disadvantage of a finite fatigue life after a (very large) number of load applications. The backlash and 'stiction'-free mounting of the transducer between carcase and base is absolutely critical.

The trunnion bearings are either a combination of a ball bearing (for axial location) and a roller bearing or hydrostatic type. These bearings operate under

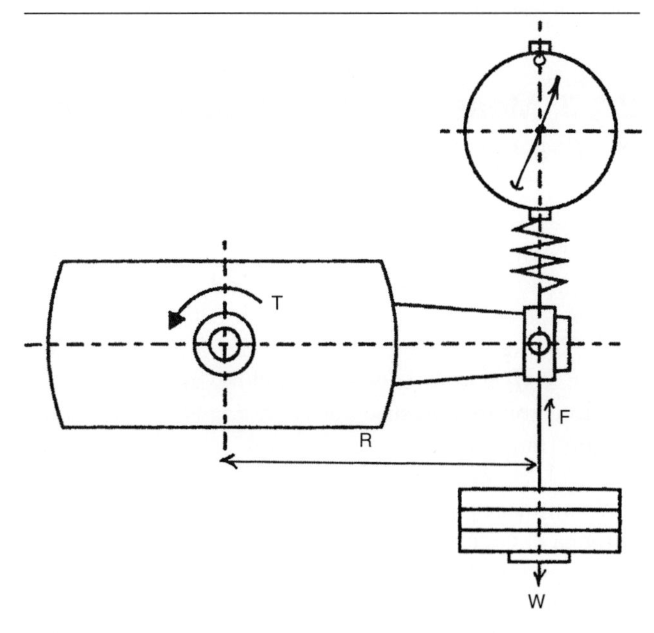

Fig. 2.1-1 Diagram of Froude type, trunnion-mounted, sluice-gate dynamometer measuring torque with dead weights and spring balance.

Engine Testing, 3rd edn; ISBN: 9780750684392
Copyright © 2007 Elsevier Ltd. All rights of reproduction, in any form, reserved.

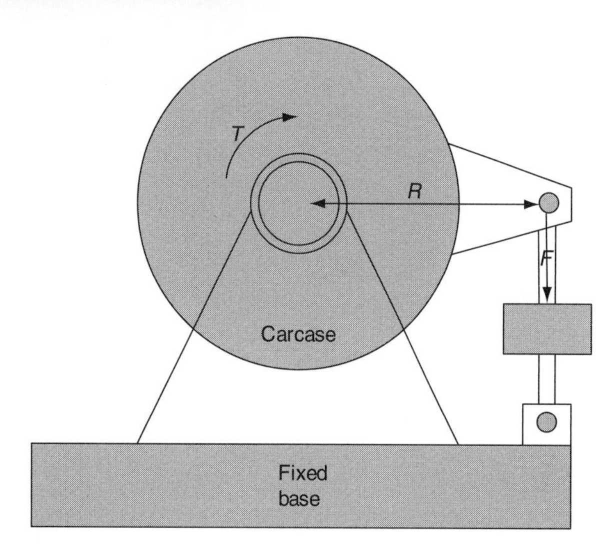

Fig. 2.1-2 Diagram of trunnion-mounted dynamometer measuring torque with a load cell.

unfavourable conditions, with no perceptible angular movement, and the rolling element type is consequently prone to brinelling, or local indentation of the races, and to fretting. This is aggravated by vibration that may be transmitted from the engine and periodical inspection and turning of the outer bearing race is recommended in order to avoid poor calibration. A Schenck dynamometer design (Fig. 2.1-3) replaces the trunnion bearings by two radial flexures, thus eliminating possible friction and wear, but at the expense of the introduction of torsional stiffness, of reduced capacity to withstand axial loads and of possible ambiguity regarding the true centre of rotation, particularly under side loading.

2.1.3 Measurement of torque using in-line shafts or torque flanges

A torque shaft dynamometer is mounted in the drive shaft between engine and brake device. It consists essentially of a flanged torque shaft fitted with strain gauges and designs are available both with slip rings and with RF signal transmission. Fig. 2.1-4 is a brushless torque shaft unit intended for rigid mounting.

More common in automotive testing is the 'disc' type torque transducer, commonly known as a torque flange (Fig. 2.1-5), which is a device that is bolted directly to the input flange of the brake and transmits data to a static antenna encircling it.

A perceived advantage of the in-line torque measurement arrangement is that it avoids the necessity, discussed below, of applying torque corrections under transient conditions of torque measurement. However, not only are such corrections, using known constants, trivial with modern computer control systems, but there are also important problems that may reduce the inherent accuracy of this arrangement.

For steady state testing, a well-designed and maintained trunnion machine will give more consistently auditable and accurate torque measurements than the inline systems; the justification for this statement can be listed as follows:

- The in-line torque sensor has to be oversized for the rating of its dynamometer and being oversized the resolution of the signal is lower. The transducer has to be overrated because it has to be capable of dealing with the instantaneous torque peaks of the engine which are not experienced by the load cell of a trunnion-bearing machine.
- The transducer forms part of the drive line and requires very careful installation to avoid the imposition of bending or axial stresses on the torsion sensing element from other components or its own clamping device.
- The in-line device is difficult to protect from temperature fluctuations within and around the drive line.
- Calibration checking of these devices is not as easy as for a trunnion-mounted machine; it requires a means of locking the dynamometer shaft in addition to the fixing of a calibration arm in a horizontal position without imposing bending stresses.
- Unlike the cradled machine and load cell, it is not possible to verify the measured torque of an in-line device during operation.

It should be noted that, in the case of modern alternating current (a.c.) dynamometer systems, the tasks of torque measurement and torque control may use different data acquisition paths. In some installations the control of the trunnion-mounted machine may use its own torque calculation and control system, while the test values are taken from an inline transducer such as a torque shaft.

2.1.4 Calibration and the assessment of errors in torque measurement

We have seen that in a conventional dynamometer, torque T is measured as a product of torque arm radius R and transducer force F.

Calibration is invariably performed by means of a *calibration arm*, supplied by the manufacturer, which is bolted to the dynamometer carcase and carries dead weights which apply a load at a certified radius. The manufacturer certifies the distance between the axis of the weight hanger bearing and an axis defined by a line

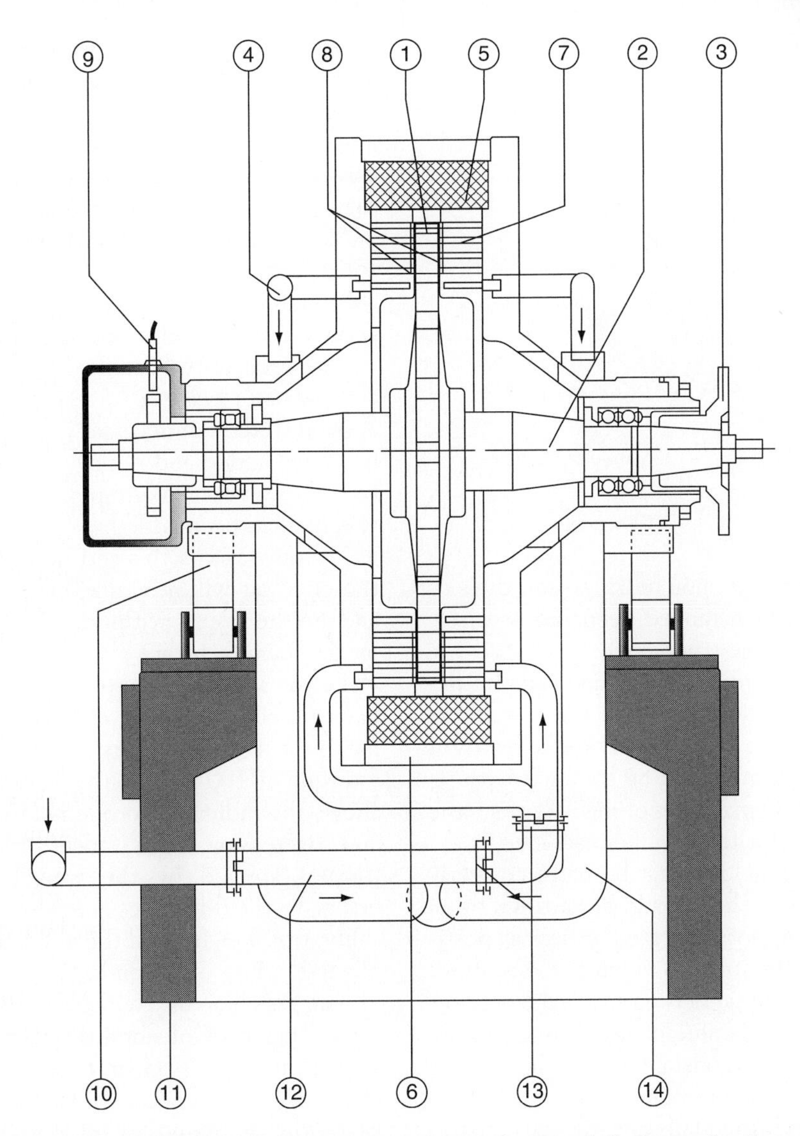

Fig. 2.1-3 Schenck, dry gap, disc type eddy-current dynamometer 1, rotor; 2, rotor shaft; 3, coupling flange; 4, water outlet with thermostat; 5, excitation coil; 6, dynamometer housing; 7, cooling chamber; 8, air gap; 9, speed pick-up; 10, flexure support; 11, base; 12, water inlet; 13, joint; 14, water outlet pipe.

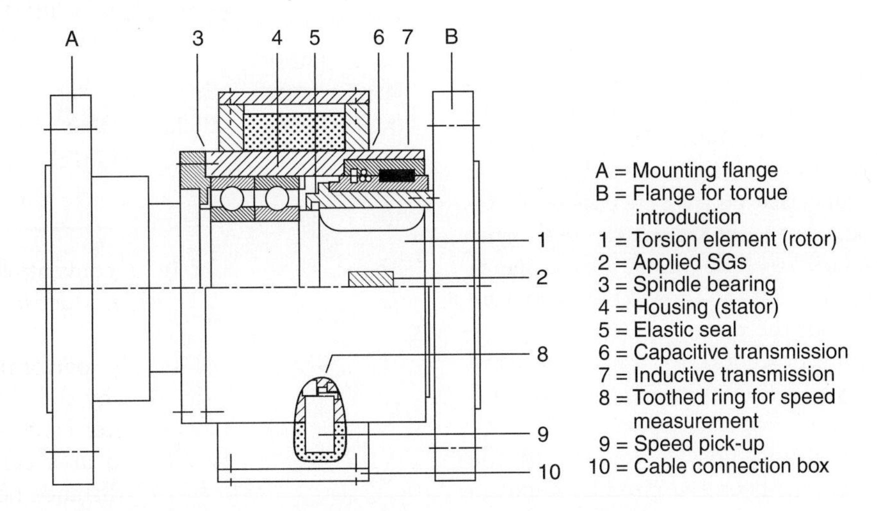

Fig. 2.1-4 Brushless torque-shaft for mounting in shaft-line between engine and 'brake'.

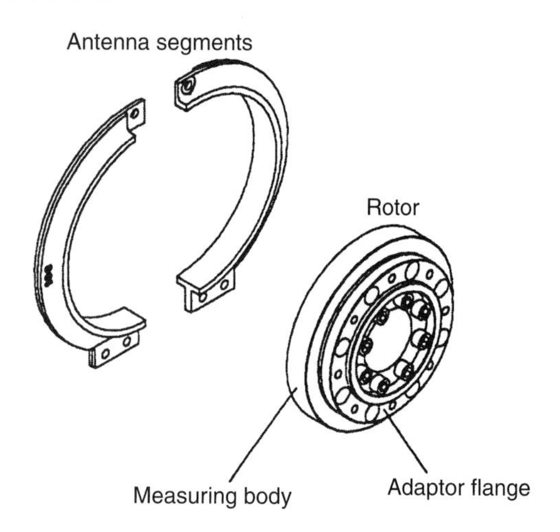

Fig. 2.1-5 Shaft-line components of a torque flange.

joining the centres of the trunnion bearings (not the axis of the dynamometer, which indeed need not precisely coincide with the axis of the trunnions).

There is no way, apart from building an elaborate fixture, in which the dynamometer user can check the accuracy of this dimension: he is entirely in the hands of the manufacturer. The arm should be stamped with its effective length. For R&D machines of high accuracy the arm should be stamped for the specific machine.

The 'dead weights' should in fact be more correctly termed 'standard masses'. They should be certified by an appropriate standards authority located as near as possible to the geographical location in which they are used. The force they exert on the calibration arm is the product of their mass and the local value of 'g'. This is usually assumed to be 9.81 m/s^2 and constant: in fact this value is only correct at sea level and a latitude of about 47° N. It increases towards the poles and falls towards the equator, with local variations. As an example, a machine calibrated in London, where $g = 9.81\ m/s^2$, will read 0.13 per cent high if recalibrated in Sydney, Australia and 0.09 per cent low if recalibrated in St Petersburg without correcting for the different local values of g.

These are not negligible variations if one is hoping for accuracies better than 1 per cent. The actual process of calibrating a dynamometer with dead weights, if treated rigorously, is not entirely straightforward. We are confronted with the facts that no transducer is perfectly linear in its response, and no linkage is perfectly frictionless. We are then faced with the problem of adjusting the system so as to ensure that the (inevitable) errors are at a minimum throughout the range.

A suitable calibration procedure for a machine using a typical strain-gauge load cell for torque measurement is as follows.

The dynamometer should not be coupled to the engine. After the system has been energized long enough to warm up the load cell output is zeroed with the machine in its normal no-load running condition (cooling water on, etc.) and the calibration arm weight balanced by equal and opposite force. Dead weights are then added to produce approximately the rated maximum torque of the machine. This torque is calculated and the digital indicator set to this value.

The weights are removed, the zero reading noted, and weights are added, preferably in 10 equal increments, the cell readings being noted. The weights are removed in reverse order and the readings again noted.

The procedure described above means that the load cell indicator was set to read zero before any load was applied (it did not necessarily read zero after the weights had been added and removed), while it was adjusted to read the correct maximum torque when the appropriate weights had been added.

We now ask: is this setting of the load cell indicator the one that will minimize errors throughout the range and are the results within the limits of accuracy claimed by the manufacturer?

Let us assume we apply this procedure to a machine having a nominal rating of 600 N m torque and that we have six equal weights, each calculated to impose a torque of 100 N m on the calibration arm. Table 2.1-1 shows the indicated torque readings for both increasing and decreasing loads, together with the calculated torques applied by the weights. The corresponding errors, or the differences between torque applied by the calibration weights and the indicated torque readings are plotted in Figs. 2.1-6 and 2.1-7.

The machine is claimed to be accurate to within ±0.25 per cent of nominal rating and these limits are shown. It will be clear that the machine meets the claimed limits of accuracy and may be regarded as satisfactorily calibrated.

Table 2.1-1 Dynamometer calibration (example taken from actual machine)

Mass (kg)	Applied torque (N m)	Reading (N m)	Error (N m)	Error (% reading)	Error (% full scale)
0	0	0.0	0.0	0.0	0.0
10	100	99.5	−0.5	−0.5	−0.083
30	300	299.0	−1.0	−0.33	−0.167
50	500	500.0	0.0	0.0	0.0
60	600	600.0	0.0	0.0	0.0
40	400	400.5	+0.5	+0.125	+0.083
20	200	200.0	0.0	0.0	0.0
0	0	0.0	0.0	0.0	0.0

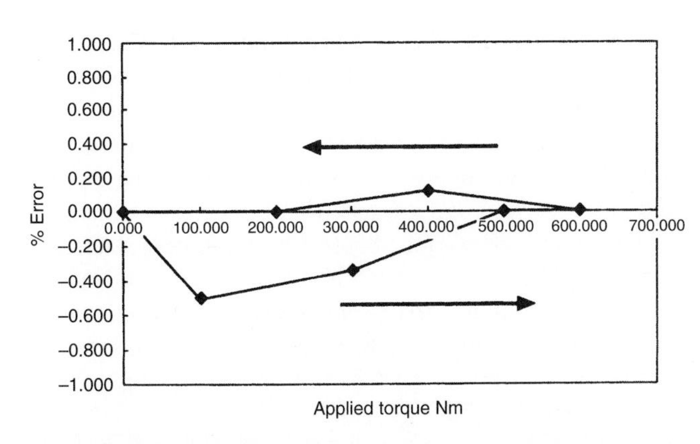

Fig. 2.1-6 Dynamometer calibration error as percentage of applied torque.

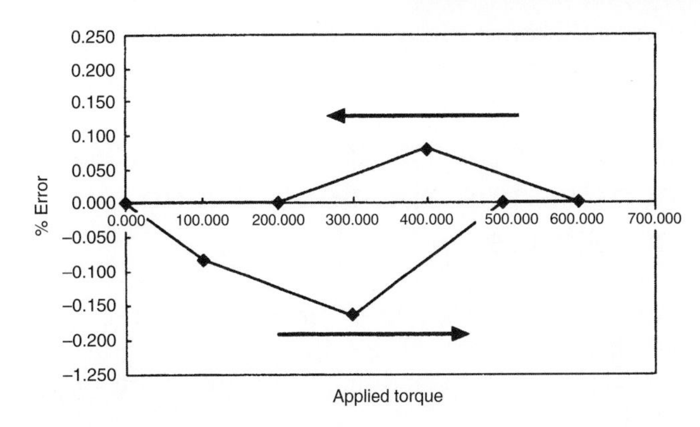

Fig. 2.1-7 Dynamometer calibration error as percentage of full scale.

It is usually assumed, though it is not necessarily the case, that hysteresis effects, manifested as differences between observed torque with rising load and with falling load, are eliminated when the machine is running, due to vibration, and it is a common practice when calibrating to knock the machine carcase lightly with a soft mallet after each load change to achieve the same result.

It is certainly not wise to assume that the ball joints invariably used in the calibration arm and torque transducer links are frictionless. These bearings are designed for working pressures on the projected area of the contact in the range 15 to 20 MN/m^2 and a 'stick slip' coefficient of friction at the ball surface of, at a minimum, 0.1 is to be expected. This clearly affects the effective arm length (in either direction) and must be relaxed by vibration.

Some large dynamometers are fitted with torque multiplication levers, reducing the size of the calibration masses. In increasingly litigious times and ever more stringent health and safety legislation, the frequent handling of multiple 20 or 25 kg weights may not be advisable. It is possible to carry out torque calibration by way of 'master' load cells or proving rings.* These devices have to be mounted in a jig attached to the dynamometer and give an auditable measurement of the force being applied on the target load cell by means of a hydraulic actuator. Such systems produce a more complex 'audit trail' in order to refer the calibration back to national standards.

It is important when calibrating an eddy-current machine that the water pressure in the casing should be at operational level, since pressure in the transfer pipes can give rise to a parasitic torque. Similarly, any disturbance to the run of electrical cables to the machine must be avoided once calibration is completed. Finally, it is possible, particularly with electrical dynamometers with forced cooling, to develop small parasitic torques due to air discharged non-radially from the casing. It is an easy matter to check this by running the machine uncoupled under its own power and noting any change in indicated torque.

Experience shows that a high grade dynamometer such as would be used for research work, after careful calibration, may be expected to give a torque indication that does not differ from the absolute value by more than about ±0.1 per cent of the full load torque rating of the machine.

Systematic errors such as inaccuracy of torque arm length or wrong assumptions regarding the value of *g* will certainly diminish as the torque is reduced, but other errors will be little affected: it is safer to assume a band of uncertainty of constant width. This implies, for example, that a machine rated at 400 Nm torque with an accuracy of ±0.25 per cent will have an error band of ± 1 N. At 10 per cent of rated torque, this implies that the true value may lie between 39 and 41 Nm. It is as well to match the size of the dynamometer as closely as possible with the rating of the engine.

All load cells used by reputable dynamometer manufacturers will compensate for changes in temperature, though their rate of response to a change may vary. They will not, however, be able to compensate for internal temperature gradients induced, for example, by air blasts from ventilation fans or radiant heat from exhaust pipes.

The subject of calibration and accuracy of dynamometer torque measurement has been dealt with in some detail, but this is probably the most critical measurement that the test engineer is called upon to make, and one for which a high standard of accuracy is expected but not easily achieved. Calibration and certification of the dynamometer and its associated system should be

* A proving ring is a hollow steel alloy ring whose distortion under a rated range of compressive loads is known and measured by means of an internal gauge.

carried out at the very least once a year, and following any system change or major component replacement.

2.1.5 Torque measurement under accelerating and decelerating conditions

With the increasing interest in transient testing it is essential to be aware of the effect of speed changes on the 'apparent' torque measured by a trunnion-mounted machine.

The basic principle is simple:

Inertia of dynamometer rotor I kg m^2
Rate of increase in speed ω rad/s^2
N rpm/s
Input torque to dynamometer T_1 N m
Torque registered by dynamometer T_2 N m

$$T_1 - T_2 = I\omega = \frac{2\pi NI}{60}\ \text{N m}$$
$$= 0.1047NI\ \text{N m}$$

To illustrate the significance of this correction, a typical eddy-current dynamometer capable of absorbing 150 kW with a maximum torque of 500 N m has a rotor inertia of 0.11 kg m^2. A direct current (d.c.) regenerative machine of equivalent rating has a rotational inertia of 0.60 kg m^2.

If these machines are coupled to an engine that is accelerating at the comparatively slow rate of 100 rpm/s the first machine will read the torque low *during the transient phase* by an amount:

$$T_1 - T_2 = 0.1047 \times 100 \times 0.11 = 1.15\ \text{N m}$$

while the second will read low by 6.3 Nm.

If the engine is decelerating, the machines will read high by the equivalent amount.

Much larger rates of speed change are demanded in some transient test sequences and this can represent a serious variation of torque indication, particularly when using high inertia dynamometers.

With modern computer processing of the data, corrections for these and other electrically induced transient effects can be made with software supplied by test plant manufacturers.

2.1.6 Measurement of rotational speed

Rotational speed of the dynamometer is measured either by a system using a toothed wheel and a pulse sensor within its associated electronics and display or, more recently, by use of an optical encoder system. While the pulse pick-up system is robust and, providing the wheel to transducer gap is correctly set and maintained, reliable, the optical encoders, which use the sensing of very fine lines etched on a small disk, need more care in mounting and operation. Since the commonly used optical encoders transmit over 1000 pulses per revolution, misalignment of its drive may show up as a sinusoidal speed change, therefore they are normally mounted as part of an accurately machined assembly forming part of the machine housing.

It should be remembered that with bidirectional dynamometers and modern electrical machines operating in four quadrants (Fig. 2.1-8), it is necessary to measure not only speed but also direction of rotation. Encoder systems can use separate tracks of their engraved disks to sense rotational direction. It is extremely important that the operator uses a common and clearly understood convention describing direction of rotation throughout the facility, particularly in laboratories operating reversible prime movers.

As with torque measurement, specialized instrumentation systems may use separate transducers for the measurement of speed or for the control of the dynamometer. In many cases, engine speed is monitored separately and in addition to dynamometer speed. The control system can use these two signals to shut down automatically in the case of a shaft failure.

Measurement of power, which is the product of torque and speed, raises the important question of sampling time. Engines never run totally steadily and the torque transducer and speed signals invariably fluctuate. An instantaneous reading of speed will not necessarily, or even probably, be identical with a longer-term average. Choice of sampling time and of the number of samples to be averaged is a matter of experimental design and compromise.

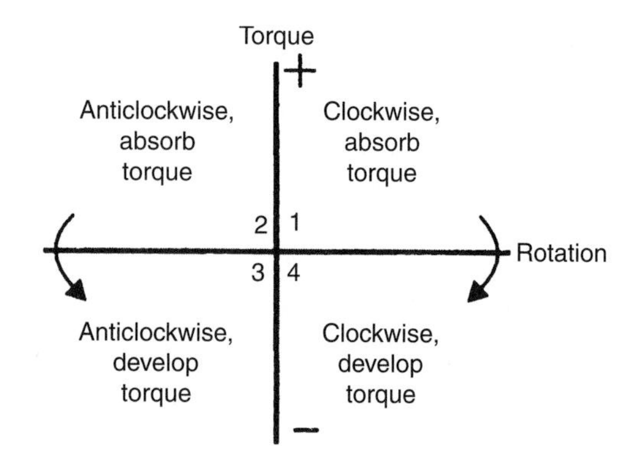

Fig. 2.1-8 Dynamometer operating quadrants.

2.1.7 Choice of dynamometer

Perhaps the most difficult question facing the engineer setting up a test facility is the choice of the most suitable dynamometer. In this part of the chapter the characteristics, advantages and disadvantages of the various types are discussed and a procedure for arriving at the correct choice is described.

The earliest form of dynamometer, the rope brake dates back to the early years of the last century. An extremely dangerous device, it was nevertheless capable of giving quite accurate measurements of power. Its successor, the Prony brake, also relied on mechanical friction and like the rope brake required cooling by water introduced into the hollow brake drum and removed by a scoop.

Both these devices are only of historical interest. Their successors may be classified according to the means adopted for absorbing the mechanical power of the prime mover driving the dynamometer.

2.1.8 Classification of dynamometers

1. ***Hydrokinetic or 'hydraulic' dynamometers (water brakes).*** With the exception of the disc dynamometer, all machines work on similar principles (Fig. 2.1-9).

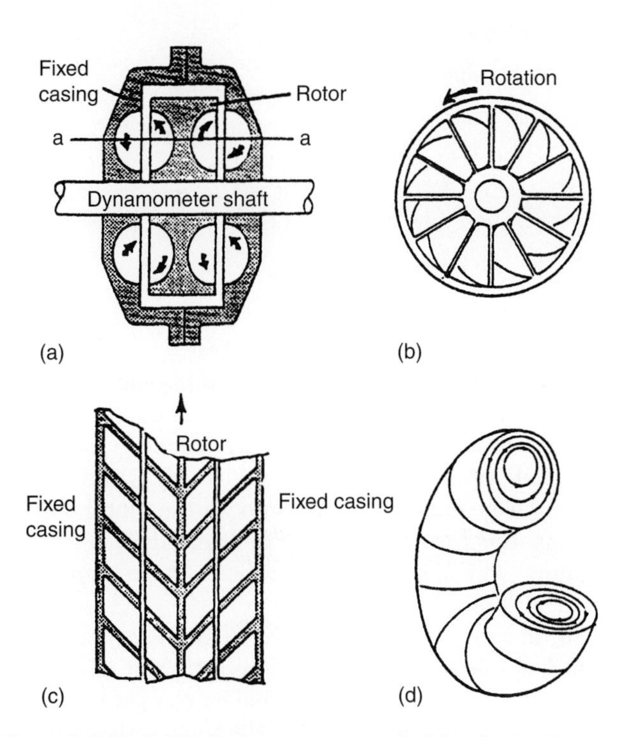

Fig. 2.1-9 Hydrokinetic dynamometer, principle of operation: (a) section through dynamometer; (b) end view of rotor; (c) development of section a–a of rotor and casing; (d) representation of toroidal vortex.

A shaft carries a cylindrical rotor which revolves in a watertight casing. Toroidal recesses formed half in the rotor and half in the casing or stator are divided into pockets by radial vanes set at an angle to the axis of the rotor. When the rotor is driven, centrifugal force sets up an intensive toroidal circulation as indicated by the arrows in Fig. 2.1-9a. The effect is to transfer momentum from rotor to stator and hence to develop a torque resistant to the rotation of the shaft, balanced by an equal and opposite torque reaction on the casing.

A forced vortex of toroidal form is generated as a consequence of this motion, leading to high rates of turbulent shear in the water and the dissipation of power in the form of heat to the water. The centre of the vortex is vented to atmosphere by way of passages in the rotor and the virtue of the design is that power is absorbed with minimal damage to the moving surfaces, either from erosion or from the effects of cavitation.

The machines are of two kinds, depending on the means by which the resisting torque is varied.

1(a) ***Constant fill machines:*** *the classical Froude or sluice plate design*, Fig. 2.1-10. In this machine, torque is varied by inserting or withdrawing pairs of thin sluice plates between rotor and stator, thus controlling the extent of the development of the toroidal vortices.

1(b) ***Variable fill machines,*** Fig. 2.1-11. In these machines, the torque absorbed is varied by adjusting the quantity (mass) of water in circulation within the casing. This is achieved by a valve, usually on the water outlet, associated with control systems of widely varying complexity. The particular advantage of the variable fill machine is that the torque may be varied much more rapidly than is the case with sluice plate control. Amongst this family of machines are the largest dynamometers ever made with rotors of around 5 m diameter. There are several designs of water control valve and valve actuating mechanisms depending on the range and magnitude of the loads absorbed and the speed of change of load required. For the fastest response, it is necessary to have adequate water available to fill the casing rapidly and it may be necessary to fit both inlet and outlet control valves with an integrated control system.

1(c) ***'Bolt-on' variable fill machines.*** These machines, available for many years in the USA, operate on the same principle as those described in 1(b) above, but are arranged to bolt directly on to the engine clutch housing or into the truck chassis. Machines are available for ratings up to about 1000 kW. In these machines, load is usually controlled by an inlet control valve associated with a throttled outlet. By nature of their simplified design and lower mass, these machines are not capable of the same level of speed holding or torque measurement as the more conventional 1(b) designs.

1(d) ***Disc dynamometers***. These machines, not very widely used, consist of one or more flat discs located

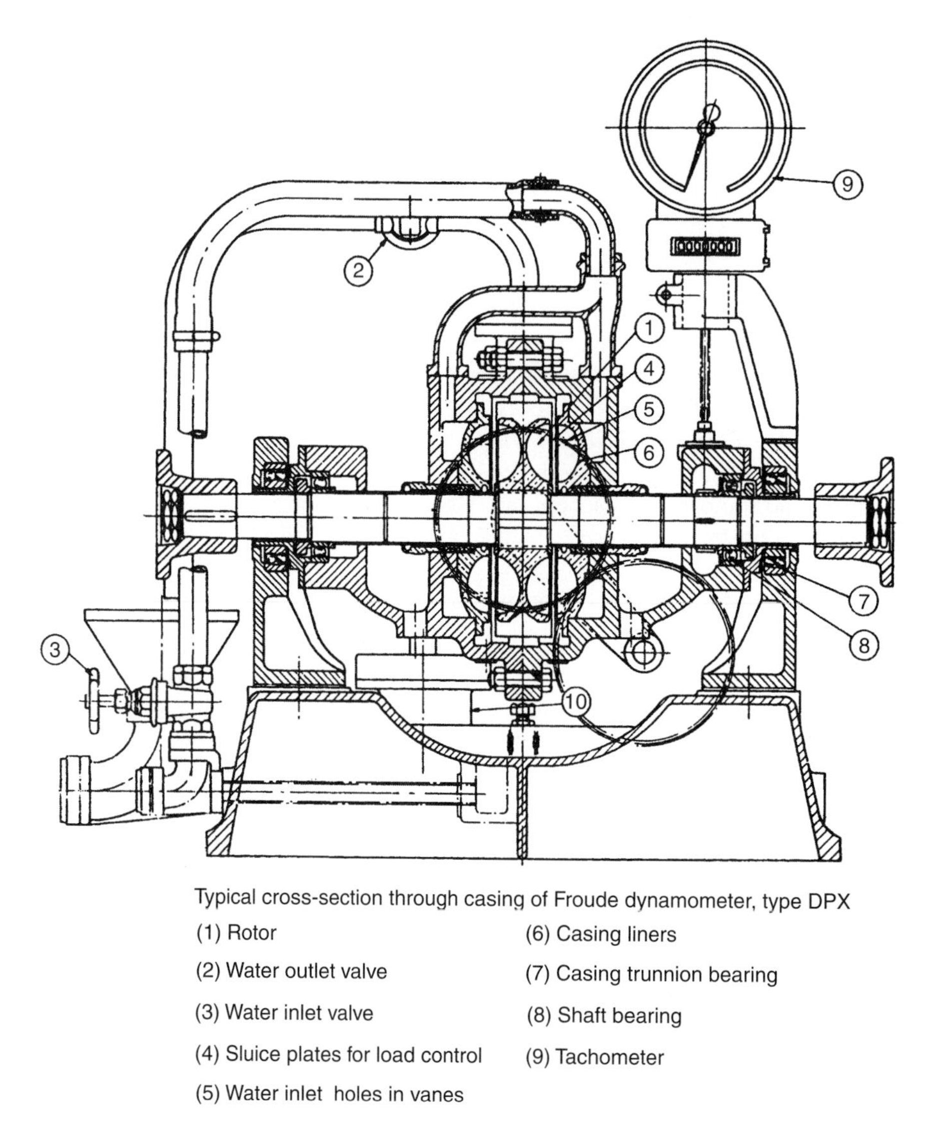

Fig. 2.1-10 Froude sluice-plate dynamometer.

between flat stator plates, with a fairly small clearance. Power is absorbed by intensive shearing of the water and torque is controlled as in variable fill machines. Disc dynamometers have comparatively poor low speed performance but may be built to run at very high speeds, making them suitable for loading gas turbines. A variation is the perforated disc machine, in which there are holes in the rotor and stators, giving greater power dissipation for a given size of machine.

2. ***Hydrostatic dynamometers.*** Not very widely used, these machines consist generally of a combination of a fixed stroke and a variable stroke positive displacement hydraulic pump/motor similar to that found in large off-road vehicle transmissions. The fixed stroke machine forms the dynamometer. An advantage of this arrangement is that, unlike most other, non-electrical machines, it is capable of developing full torque down to zero speed and is also capable of acting as a source of power to 'motor' the engine under test.

3. ***Electrical motor-based dynamometers.*** The common feature of all these machines is that the power absorbed is transformed into electrical energy, which is 'exported' from the machine via its associated 'drive' circuitry. The energy loss within both the motor and its drive in the form of heat is transferred to a cooling medium, which may be water or is more commonly forced air flow.

All motor-based dynamometers have associated with them large drive cabinets that produce heat and noise. The various sections of these cabinets contain high voltage/power devices and complex electronics; they have to be housed in suitable conditions which have a clean and non-condensing atmosphere with sufficient space for access and cooling. When planning a facility layout, the

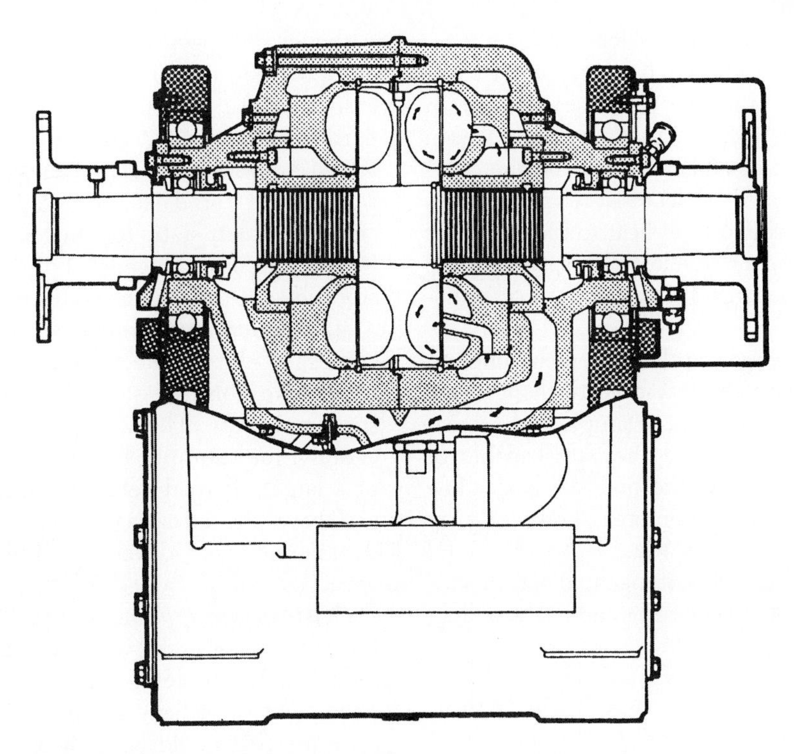

Fig. 2.1-11 Variable fill hydraulic dynamometer controlled by fast acting outlet valve at bottom of the stator.

designer should remember that these large and heavy cabinets have to be positioned after the building work has been completed. The position of the drives should normally be within 15 m of the dynamometer, but this should be minimized so far as is practical to reduce the high cost of the connecting power cables.

3(a) ***D.c. dynamometers.*** These machines consist essentially of a trunnion-mounted d.c. motor generator. Control is almost universally by means of a thyristor based a.c. → d.c. → a.c. converter.

These machines have a long pedigree in the USA, are robust, easily controlled, and capable of motoring and starting as well as of absorbing power. Disadvantages include limited maximum speed and high inertia, which can present problems of torsional vibration (see Chapter 2.1a) and limited rates of speed change. Because they contain a commutator, the maintenance of d.c. machines may be higher than those based on a.c. squirrel cage motors.

3(b) ***Asynchronous or a.c. dynamometers.*** These asynchronous machines consist essentially of an induction motor with squirrel cage rotor, the speed of which is controlled by varying the supply frequency. The modern power control stage of the control will invariably be based upon insulated gate bipolar transistor (IGBT) technology.

The squirrel cage rotor machines have a lower rotational inertia than d.c. machines of the same power and are therefore capable of better transient performance. Being based on an asynchronous motor they have proved very robust in service requiring low maintenance.

However, it is misleading to think that any motor's mechanical design may be used without adaptation as a dynamometer. During the first decade of their wide industrial use, it was discovered that several different dynamometer/motor designs suffered from bearing failures caused by an electrical arcing effect within the rolling elements; this was due to the fact that, in their dynamometer role, a potential difference developed between the rotor and the stator (ground). Ceramic bearing elements and other design features are now used to prevent such damage occurring.

3(c) ***Synchronous, permanent magnet dynamometers.*** The units represent the latest generation of dynamometer development and while using the same drive technology as the asynchronous dynamometers are capable of higher dynamic performance because of their inherently lower rotational inertia. It is this generation of machine that will provide the high dynamic test tools required by engine and vehicle system simulation in the test cell.

Acceleration rates of 160 000 rpm/s and air-gap torque rise times of less than 1 ms have been achieved, which makes it possible to use these machines as engine simulators where the full dynamic fluctuation speed and torque characteristic of the engine is required for drive line component testing.

3(d) ***Eddy-current dynamometers,*** Fig. 2.1-3. These machines make use of the principle of electromagnetic induction to develop torque and dissipate power. A toothed rotor of high-permeability steel rotates, with a fine clearance, between water-cooled steel loss plates. A magnetic field parallel to the machine axis is generated by two annular coils and motion of the rotor gives rise to changes in the distribution of magnetic flux in the loss plates. This in turn gives rise to circulating eddy currents and the dissipation of power in the form of electrical resistive losses. Energy is transferred in the form of heat to cooling water circulating through passages in the loss plates, while some cooling is achieved by the radial flow of air in the gaps between rotor and plates.

Power is controlled by varying the current supplied to the annular exciting coils and rapid load changes are possible. Eddy-current machines are simple and robust, the control system is simple and they are capable of developing substantial braking torque at quite low speeds. Unlike a.c. or d.c. dynamometers, however, they are unable to develop motoring torque.

There are two common forms of machine both having air circulating in the gap between rotor and loss (cooling) plates, hence 'dry gap':

1. Dry gap machines fitted with one or more tooth disc rotors. These machines have lower inertia than the drum machines and a very large installed user base, particularly in Europe. However, the inherent design features of their loss plates place certain operational restrictions on their use. It is absolutely critical to maintain the required water flow through the machines at all times; even a very short loss of cooling will cause the loss plates to distort leading to the rotor/plate gap closing with disastrous results. These machines must be fitted with flow detection devices interlocked with the cell control system; pressure switches should not be used since in a closed water system it is possible to have pressure without flow.
2. Dry gap machines fitted with a drum rotor. These machines usually have a higher inertia than the equivalent disc machine, but may be less sensitive to cooling water conditions.

Although no longer so widely used, an alternative form of eddy-current machine is also available. This employs a simple disc or drum design of rotor in which eddy currents are induced and the heat developed is transferred to water circulated through the gaps between rotor and stator. These 'wet gap' machines are liable to corrosion if left static for any length of time, have higher inertia, and have a high level of minimum torque, arising from drag of the cooling water in the gap.

4. *Friction dynamometers,* Fig. 2.1-12. These machines, in direct line of succession from the original rope brake, consist essentially of water-cooled, multidisc friction brakes. They are useful for low-speed applications, for example for measuring the power output of a large, off-road vehicle transmission at the wheels, and have the advantage, shared with the hydrostatic dynamometer, of developing full torque down to zero speed.

5. *Air brake dynamometers.* These devices, of which the Walker fan brake was the best-known example, are now largely obsolete. They consisted of a simple arrangement of radially adjustable paddles that imposed a torque that could be approximately estimated. They survive mainly for use in the field testing of helicopter engines, where high accuracy is not required and the noise is no disadvantage.

2.1.8.1 Hybrid and tandem dynamometers

For completeness, mention should be made of both a combined design that is occasionally adopted for cost reasons and the use of two dynamometers in line for special test configurations.

The d.c. or a.c. electrical dynamometer is capable of generating a motoring torque almost equal to its braking torque. However, the motoring torque required in engine testing seldom exceeds 30 per cent of the engine power

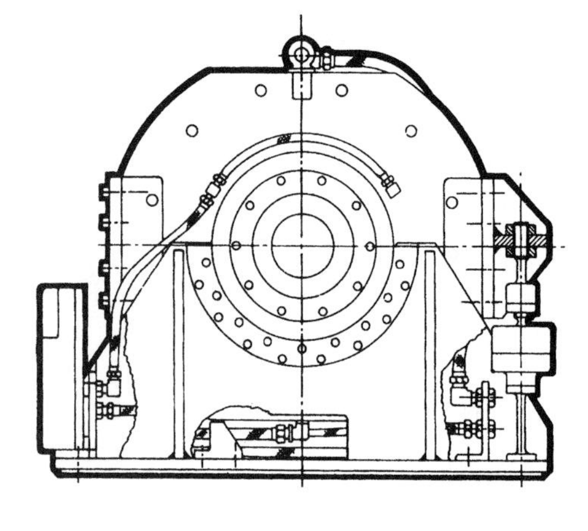

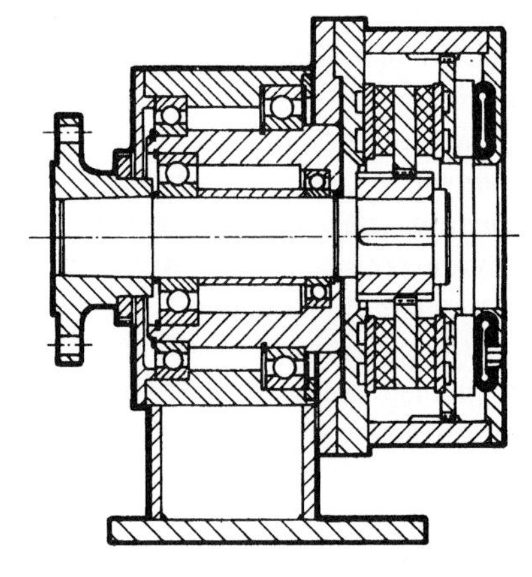

Fig. 2.1-12 Water-cooled friction brake used as a dynamometer.

output. Since, for equal power absorption, a.c. and d.c. machines are more expensive than other types, it is sometimes worth running an electrical dynamometer in tandem with, for example, a variable fill hydraulic machine. Control of these hybrid machines is a more complex matter and the need to provide duplicate services, both electrical power and cooling water, is a further disadvantage. The solution may, however, on occasion be cost-effective.

Tandem machines are used when the torque/speed envelope of the prime mover cannot be covered by a standard dynamometer, usually this is found in gas turbine testing when the rotational speed is too high for a machine fitted with a rotor capable of absorbing full rated torque. The first machine in line has to have a shaft system capable of transmitting the combine torques.

Tandem machines are also used when the prime mover is producing power through two contrarotating shafts as with some aero and military applications; in these cases the first machine in line is of a special design with a hollow rotor shaft to allow the housing of a quill shaft connecting the second machine.

2.1.8.2 One, two or four quadrant?

Fig. 2.1-8 illustrates diagrammatically the four 'quadrants' in which a dynamometer may be required to operate. Most engine testing takes place in the first quadrant, the engine running anticlockwise when viewed on the flywheel end. On occasions it is necessary for a test installation using a unidirectional water brake to accept engines running in either direction; one solution is to fit the dynamometer with couplings at both ends mounted on a turntable. Large and some 'medium speed' marine engines are usually reversible.

All types of dynamometer are naturally able to run in the first (or second) quadrant. Hydraulic dynamometers are usually designed for one direction of rotation, though they may be run in reverse at low fill state without damage. When designed specifically for bidirectional rotation they may be larger than a single-direction machine of equivalent power and torque control may not be as precise as that of the unidirectional designs. The torque measuring system must of course operate in both directions. Eddy-current machines are inherently reversible.

When it is required to operate in the third and fourth quadrants (i.e. for the dynamometer to produce power as well as to absorb it) the choice is effectively limited to d.c. or a.c. machines, or to the hydrostatic or hybrid machine. These machines are generally reversible and therefore operate in all four quadrants.

Table 2.1-2 Operating quadrants of dynamometer designs

Type of machine	Quadrant
Hydraulic sluice plate	1 or 2
Variable fill hydraulic	1 or 2
'Bolt on' variable fill hydraulic	1 or 2
Disc type hydraulic	1 and 2
Hydrostatic	1, 2, 3, 4
d.c. electrical	1, 2, 3, 4
a.c. electrical	1, 2, 3, 4
Eddy current	1 and 2
Friction brake	1 and 2
Air brake	1 and 2
Hybrid	1, 2, 3, 4

There is an increasing requirement for four-quadrant operation as a result of the growth in transient testing, with its call for very rapid load changes and even for torque reversals.

If mechanical losses in the engine are to be measured by 'motoring', a four-quadrant machine is obviously required.

A useful feature of such a machine is its ability also to start the engine. Table 2.1-2 summarizes the performance of machines in this respect.

2.1.9 Matching engine and dynamometer characteristics

The different types of dynamometer have significantly different torque-speed and power–speed curves, and this can affect the choice made for a given application. Fig. 2.1-13 shows the performance curves of a typical hydraulic dynamometer. The different elements of the performance envelope are as follows:

- Dynamometer full (or sluice plates wide open). Torque increases with square of speed, no torque at rest.
- Performance limited by maximum permitted shaft torque.
- Performance limited by maximum permitted power, which is a function of cooling water throughput and its maximum permitted temperature rise.
- Maximum permitted speed.

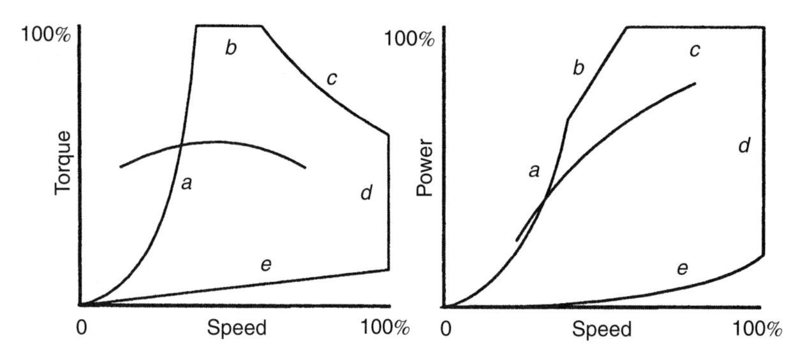

Fig. 2.1-13 Engine torque curves plotted on hydraulic dynamometer torque curves.

- Minimum torque corresponding to minimum permitted water flow.

Fig. 2.1-14 shows the considerably different performance envelope of an electrical machine, made up of the following elements:

- Constant torque corresponding to maximum current and excitation.
- Performance limited by maximum permitted power output of machine.
- Maximum permitted speed.

Since these are 'four-quadrant' machines, power absorbed can be reduced to zero and there is no minimum torque curve.

Fig. 2.1-15 shows the performance curves for an eddy-current machine, which lie between those of the previous two machines:

- Low speed torque corresponding to maximum permitted excitation.
- Performance limited by maximum permitted shaft torque.
- Performance limited by maximum permitted power, which is a function of cooling water throughput and its maximum permitted temperature rise.
- Maximum permitted speed.
- Minimum torque corresponding to residual magnetization, windage and friction.

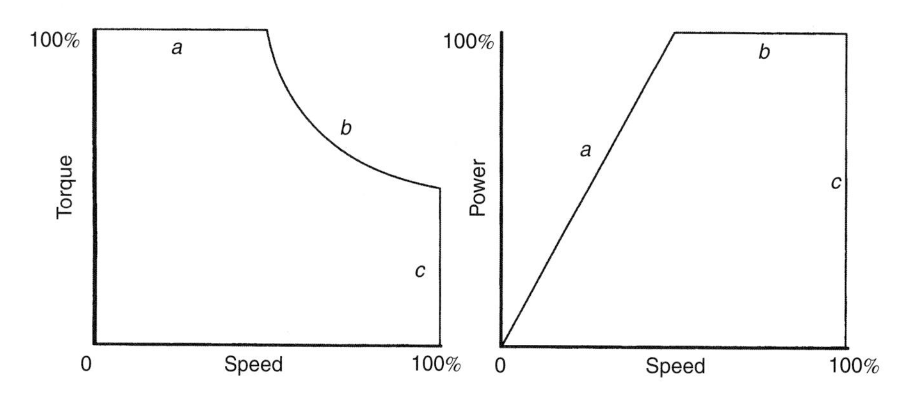

Fig. 2.1-14 Performance curve shapes of electrical motor-based dynamometers.

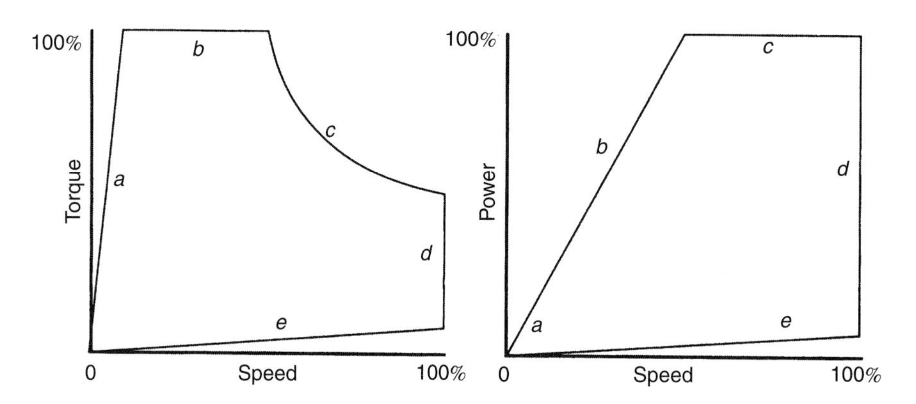

Fig. 2.1-15 Performance curve shapes of an eddy-current dynamometer.

In choosing a dynamometer for an engine or range of engines, it is essential to superimpose the maximum torque– and power–speed curves on to the dynamometer envelope. See the example in Fig. 2.1-13 which demonstrates a typical problem: the hydraulic machine is incapable of developing sufficient torque at the bottom end of the speed range.

For best accuracy, it is desirable to choose the smallest machine that will cope with the largest engine to be tested. Hydraulic dynamometers are generally able to deal with a moderate degree of overload and overspeed, but it is undesirable to run electrical machines beyond their rated limits: this can lead to damage to commutators, overheating and distortion of eddy-current loss plates.

Careful attention must also be given to the arrangements for coupling engine and dynamometer, see Chapter 2.1a.

2.1.10 Engine starting and cranking

Starting an engine when it is connected to a dynamometer may present the cell designer and operator with a number of problems, and is a factor to be borne in mind when selecting the dynamometer. If the engine is fitted with a starter motor, the cell system must provide the high current d.c. supply and associated switching; in the absence of an engine mounted starter a complete system to start and crank the engine must be available which compromises neither the torsional characteristics (see Chapter 2.1a) nor the torque measurement accuracy.

2.1.10.1 Engine cranking, no starter motor

The cell cranking system must be capable of accelerating the engine to its normal starting speed and, in most cases, of disengaging when the engine fires. A four-quadrant dynamometer, suitably controlled, will be capable of starting the engine directly. The power available from any four-quadrant machine will always be greater than that required, therefore excessive starting torque must be avoided by an alarm system otherwise an engine locked by seizure or fluid in a cylinder may cause damage to the drive line.

The preferred method of providing other types of dynamometer with a starting system is to mount an electric motor at the non-engine end of the dynamometer shaft, driving through an over-running or remotely engaged clutch, and generally through a speed-reducing belt drive. The clutch half containing the mechanism should be on the input side, otherwise it will be affected by the torsional vibrations usually experienced by dynamometer shafts. The motor may be mounted above, below or alongside the dynamometer to save cell length.

The sizing of the motor must take into account the maximum break-away torque expected, usually estimated as twice the average cranking torque, while the normal running speed of the motor should correspond to the desired cranking speed. The choice of motor and associated starter must take into account the maximum number of starts per hour that may be required, both in normal use and when dealing with a faulty engine. The running regime of the motor is demanding, involving repeated bursts at overload, with the intervening time at rest, and an independent cooling fan may be necessary.

Some modern diesel engines, when 'green',* require cranking at more than the normal starting speed, sometimes as high as 1200 rev/min, in order to prime the fuel system. In such cases a two-speed or fully variable speed starter motor may be necessary.

The system must be designed to impose the minimum parasitic torque when disengaged, since this torque will not be sensed by the dynamometer measuring system.

In some cases, to avoid this source of inaccuracy, the motor may be mounted directly on the dynamometer carcase and permanently coupled to the dynamometer shaft by a belt drive. This imposes an additional load on the trunnion bearings, which may lead to brinelling, and it also increases the effective moment of inertia of the dynamometer. However, it has the advantage that motoring and starting torque may be measured by the dynamometer system.

An alternative solution is to use a standard vehicle engine starter motor in conjunction with a gear ring carried by a 'dummy flywheel' carried on a shaft with separate bearings incorporated in the drive line, but this may have the disadvantage of complicating the torsional behaviour of the system.

2.1.10.2 Engine-mounted starter systems

If the engine is fitted with its own starter motor on arrival at the test stand, all that must be provided is the necessary 12 or 24 V supply. The traditional approach has been to locate a suitable battery as close as possible to the starter motor, with a suitable battery charger supply. This system is not ideal, as the battery needs to be in a suitably ventilated box, to avoid the risk of

* A green engine is one that has never been run. The rubbing surfaces may be dry, the fuel system may need priming, and there is always the possibility that it, or its control system, is faulty and incapable of starting.

accidental shorting, and will take up valuable space. Special transformer/rectifier units designed to replace batteries for this duty are on the market. They will include an 'electrical services box' to provide power in addition for ignition systems and diesel glow plugs. In large integrated systems there may be a bus bar system for the d.c. supplies.

The engine starter will be presented with a situation not encountered in normal service: it will be required to accelerate the whole dynamometer system in addition to the engine while a 'green' engine may exhibit a very high breakaway torque and require prolonged cranking at high speed to prime the fuel system before it fires.

2.1.10.3 Non-electrical starting systems

Diesel engines larger than the automotive range are usually started by means of compressed air, admitted to the cylinders by way of starting valves. In some cases it is necessary to move the crankshaft to the correct starting position, either by barring or using an engine-mounted inching motor. The test facility should include a compressor and a receiver of capacity at least as large as that recommended for the engine in service.

Compressed air or hydraulic motors are sometimes used instead of electric motors to provide cranking power but have no obvious advantages over a d.c. electric motor, apart from a marginally reduced fire risk in the case of compressed air, provided the supply is shut off automatically in the case of fire.

In Chapter 2.1a, attention is drawn to the possibility of overloading flexible couplings in the drive line during the starting process, and particularly when the engine first fires. This should not be overlooked.

2.1.11 Choice of dynamometer

Table 2.1-3 lists the various types of dynamometer and indicates their applicability for various classes of engine being tested in steady or mild transient states.

In most cases, several choices are available and it will be necessary to consider the special features of each type of dynamometer and to evaluate the relative

Table 2.1-3 Dynamometers: advantages and disadvantages of available types

Dynamometer type	Advantages	Disadvantages
Froude sluice plate	Obsolete, but many cheap and reconditioned models in use worldwide, robust	Slow response to change in load. Manual control not easy to automate
Variable fill water brakes	Capable of medium speed load change, automated control, robust and tolerant of overload. Available for largest prime-movers	'Open' water system required. Can suffer from cavitation or corrosion damage
'Bolt-on' variable fill water brakes	Cheap and simple installation. Up to 1000 kW	Lower accuracy of measurement and control than fixed machines
Disc type hydraulic	Suitable for high speeds	Poor low speed performance
Hydrostatic	For special applications, provides four quadrant performance	Mechanically complex, noisy and expensive. System contains large volumes of high pressure oil
d.c. electrical motor	Mature technology. Four quadrant performance	High inertia, commutator may be fire and maintenance risk
asychronous motor (a.c.)	Lower inertia than d.c. Four quadrant performance	Expensive. Large drive cabinet needs suitable housing
Permanent magnet motor	Lowest inertia, most dynamic four quadrant performance. Small size in cell	Expensive. Large drive cabinet needs suitable housing
Eddy current	Low inertia (disc type air gap). Well adapted to computer control. Mechanically simple	Vulnerable to poor cooling supply. Not suitable for sustained rapid changes in power (thermal cycling)
Friction brake	Special purpose applications for very high torques at low speed	Limited speed range
Air brake	Cheap. Very little support services needed	Noisy. Limited control accuracy
Hybrid	Possible cost advantage over sole electrical machine	Complexity of construction and control

importance of these in the particular case. These features are listed in Table 2.1-3 and other special factors are considered later.

2.1.12 Some additional considerations

The final choice of dynamometer for a given application may be influenced by some of the following factors:

1. The speed of response required by the test sequences being run: steady state, transient, dynamic or high dynamic. This will determine the technology and probably the number of quadrants of operation required.
2. Load factor. If the machine is to spend long periods out of use, the possibilities of corrosion must be considered, particularly in the case of hydraulic or wet gap eddy-current machines. Can the machine be drained readily? Should the use of corrosion inhibitors be considered?
3. Overloads. If it may be necessary to consider occasional overloading of the machine a hydraulic machine may be preferable, in view of its greater tolerance of such conditions. Check that the torque measuring system has adequate capacity.
4. Large and frequent changes in load. This can give rise to problems with eddy- current machines, due to expansion and contraction with possible distortion of the loss plates.
5. Wide range of engine sizes to be tested. It may be difficult to achieve good control and adequate accuracy when testing the smallest engines, while the minimum dynamometer torque may also be inconveniently high.
6. How are engines to be started? If a non-motoring dynamometer is favoured it may be necessary to fit a separate starter to the dynamometer shaft. This represents an additional maintenance commitment and may increase inertia.
7. Is there an adequate supply of cooling water of satisfactory quality? Hard water will result in blocked cooling passages and some water treatments can give rise to corrosion. This may be a good reason for choosing d.c. or a.c. dynamometers, despite extra cost.
8. Is the pressure of the water supply subject to sudden variations? Sudden pressure changes or regular pulsations will affect the stability of control of hydraulic dynamometers. Eddy-current and indirectly cooled machines are unaffected providing inlet flow does not fall below emergency trip levels.
9. Is the electrical supply voltage liable to vary as the result of other loads on the same circuit? With the exception of air brakes and manually controlled hydraulic machines, all dynamometers are affected by electrical interference and voltage changes.
10. Is it proposed to use a shaft docking system for coupling engine and dynamometer? Are there any special features or heavy overhung or axial loads associated with the coupling system? Such features should be discussed with the dynamometer manufacturer before making a decision. Some machines, notably the Schenck flexure plate mounting system, are not suited to taking axial loads.

The supplier of any new dynamometer should offer an acceptance test and basic training in operation, calibration and safety of the new machine. A careful check should be made on the level of technical support, including availability of calibration services, spares and local service facilities, offered by the manufacturer.

2.1a. Coupling the engine to the dynamometer

2.1a.1 Introduction

The selection of suitable couplings and shaft for the connection of the engine to the dynamometer is by no means a simple matter. Incorrect choice or faulty system design may give rise to a number of problems:

- torsional oscillations;
- vibration of engine or dynamometer;
- whirling of coupling shaft;
- damage to engine or dynamometer bearings;
- excessive wear of shaft line components;
- catastrophic failure of coupling shafts;
- engine starting problems.

This whole subject, the coupling of engine and dynamometer, can give rise to more trouble than any other routine aspect of engine testing, and a clear understanding of the many factors involved is desirable.

2.1a.2 The nature of the problem

The special feature of the problem is that it must be considered afresh each time an engine not previously encountered is installed. It must also be recognized that unsatisfactory torsional behaviour is associated with the whole system – engine, coupling shaft and dynamometer – rather than with the individual components, all of which may be quite satisfactory in themselves.

Problems arise partly because the dynamometer is seldom equivalent dynamically to the system driven by the engine in service. This is particularly the case with vehicle engines. In the case of a vehicle with rear axle drive, the driveline consists of a clutch, which may itself act as a torsional damper, followed by a gearbox, the characteristics of which are low inertia and some damping capacity. This is followed by a drive shaft and differential, itself having appreciable damping, two half shafts and two wheels, both with substantial damping capacity and running at much slower speed than the engine, thus reducing their effective inertia.

When coupled to a dynamometer this system, Fig. 2.1a-1, with its built-in damping and moderate inertia, is replaced by a single drive shaft connected to a single rotating mass, the dynamometer, running at the same speed as the engine. The clutch may or may not be retained.

Particular care is necessary where the moment of inertia of the dynamometer is more than about twice that of the engine. A further consideration that must be taken seriously concerns the effect of the difference between the engine mounting arrangements in the vehicle and on the testbed. This can lead to vibrations of the whole engine that can have a disastrous effect on the drive shaft.

2.1a.2.1 Overhung mass on engine and dynamometer bearings

Care must be taken when designing and assembling a shaft system that the loads imposed by the mass and unbalanced forces do not exceed the overhung weight limits of the engine bearing at one end and the dynamometer at the other. Steel adaptor plates required to adapt the bolt holes of the shaft to the dynamometer flange or engine flywheel can increase the load on bearings significantly and compromise the operation of the system. Dynamometer manufacturers produce tables showing the maximum permissible mass at a given distance from the coupling face of their machines; the equivalent details for most engines is more difficult to obtain, but the danger of overload should be kept in mind by all concerned.

2.1a.3 Background reading

The mathematics of the subject is complex and not readily accessible. Den Hartog[1] gives what is possibly the clearest exposition of fundamentals. Ker Wilson's classical treatment in five volumes[2] is probably still the best source of comprehensive information; his abbreviated version[3] is sufficient for most purposes. Mechanical Engineering Publications have published a useful practical handbook[4] while Lloyd's Register[5] gives rules for the design of marine drives that are also useful in the present context. A listing of the notation used is to be found at the end of this chapter.

2.1a.4 Torsional oscillations and critical speeds

In its simplest form, the engine–dynamometer system may be regarded as equivalent to two rotating masses connected by a flexible shaft, Fig. 2.1a-2. Such a system has an inherent tendency to develop torsional oscillations.

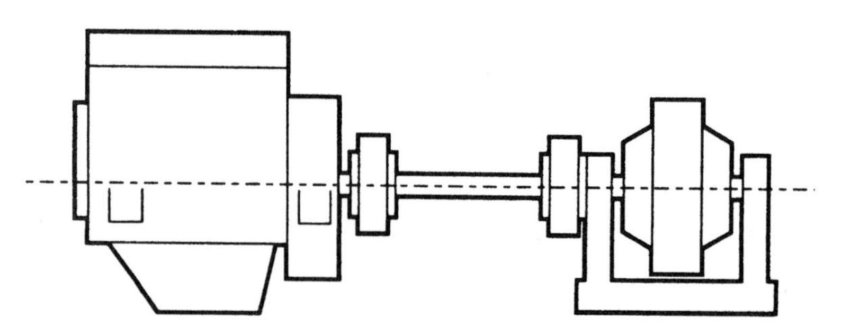

Fig. 2.1a-1 Simple form of dynamometer/engine drive line.

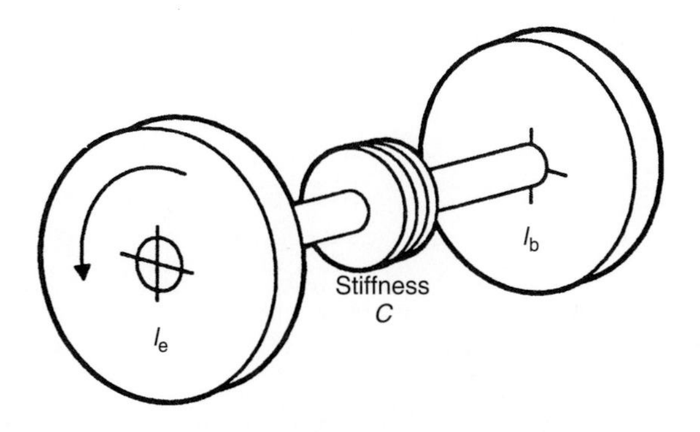

Fig. 2.1a-2 Two mass system (compare with Fig. 2.1a-1).

The two masses can vibrate 180° out of phase about a node located at some point along the shaft between them. The oscillatory movement is superimposed on any steady rotation of the shaft. The resonant or critical frequency of torsional oscillation of this system is given by:

$$n_c = \frac{60}{2\pi}\sqrt{\frac{C_c(I_e + I_b)}{I_e I_b}} \tag{1}$$

If an undamped system of this kind is subjected to an exciting torque of constant amplitude T_{ex} and frequency n, the relation between the amplitude of the induced oscillation θ and the ratio n/n_c is as shown in Fig. 2.1a-3.

At low frequencies, the combined amplitude of the two masses is equal to the static deflection of the shaft under the influence of the exciting torque, $\theta_0 = T_{ex}/C_s$. As the frequency increases, the amplitude rises and at $n = n_c$ it becomes theoretically infinite: the shaft may fracture or non-linearities and internal damping may prevent actual failure. With further increases in frequency the amplitude falls and at $n-\sqrt{2n_c}$ it is down to the level of the static deflection. Amplitude continues to fall with increasing frequency.

The shaft connecting engine and dynamometer must be designed with a suitable stiffness C_s to ensure that the critical frequency lies outside the normal operating range of the engine, and also with a suitable degree of damping to ensure that the unit may be run through the critical speed without the development of a dangerous level of torsional oscillation. Fig. 2.1a-3 also shows the behaviour of a damped system. The ratio θ/θ_0 is known as the dynamic magnifier M. Of particular importance is the value of the dynamic magnifier at the critical frequency, M_c. The curve of Fig. 2.1a-3 corresponds to a value $M_c = 5$.

Torsional oscillations are excited by the variations in engine torque associated with the pressure cycles in the individual cylinders (also, though usually of less importance, by the variations associated with the movement of the reciprocating components).

Fig. 2.1a-4 shows the variation in the case of a typical single cylinder four-stroke diesel engine. It is well known that any periodic curve of this kind may be synthesized from a series of *harmonic components*, each a pure sine wave of a different amplitude having a frequency corresponding to a multiple or submultiple of the engine speed and Fig. 2.1a-4 shows the first six components.

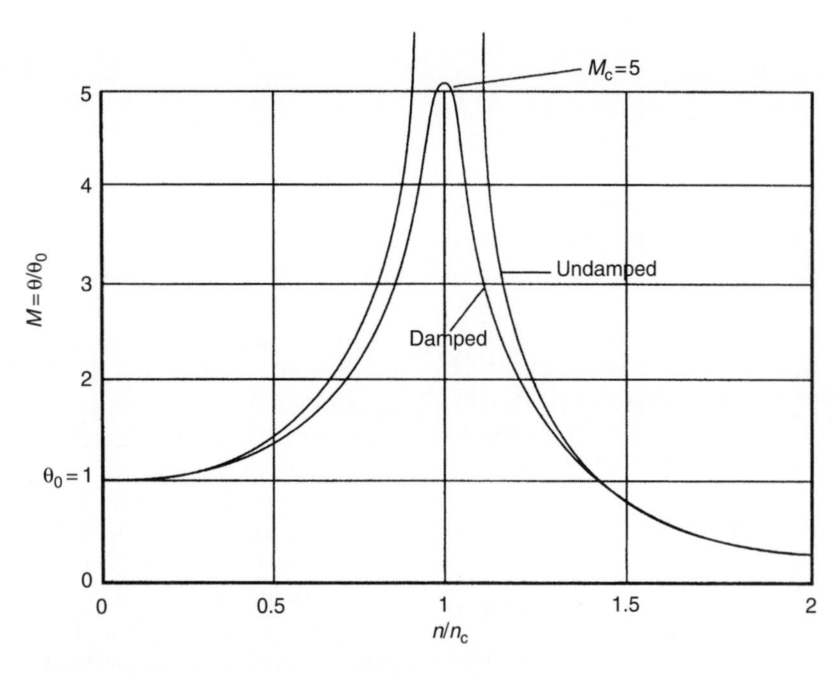

Fig. 2.1a-3 Relationship between frequency ratio, amplitude and dynamic amplifier M.

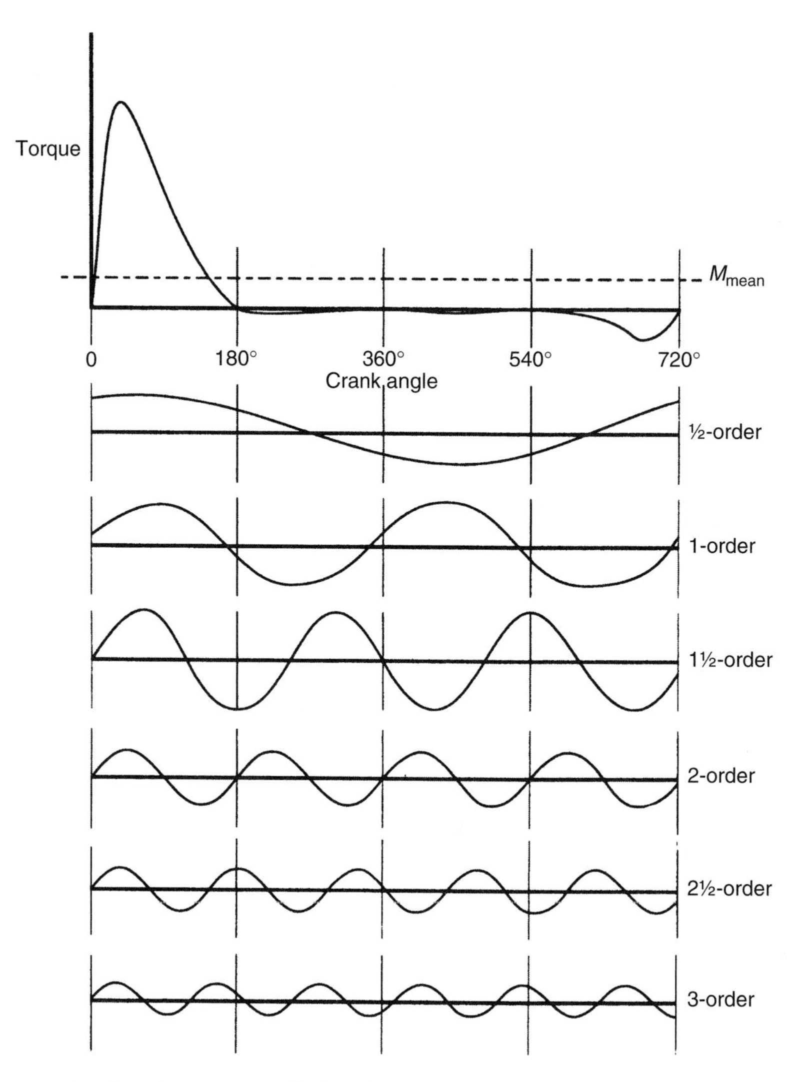

Fig. 2.1a-4 Harmonic components of turning moment, single cylinder four-stoke gasoline engine.

The *order* of the harmonic defines this multiple. Thus a component of order $N_0 = 1/2$ occupies two revolutions of the engine, $N_0 = 1$ one revolution and so on. In the case of a four cylinder four-stroke engine, there are two firing strokes per revolution of the crankshaft and the turning moment curve of Fig. 2.1a-4 is repeated at intervals of 180°. In a multicylinder engine, the harmonic components of a given order for the individual cylinders are combined by representing each component by a vector for the inertia forces. A complete treatment of this process is beyond the scope of this book, but the most significant results may be summarized as follows.

The first major critical speed for a multicylinder, in-line engine is of order:

$$N_0 = N_{CYL}/2 \text{ for a four-stroke engine} \quad (2a)$$

$$N_0 = N_{CYL} \text{ for a two-stroke engine} \quad (2b)$$

Thus, in the case of a four cylinder, four-stroke engine the major critical speeds are of order 2, 4, 6, etc. In the case of a six cylinder engine, they are of order 3, 6, 9, etc.

The distinction between a major and a minor critical speed is that in the case of an engine having an infinitely rigid crankshaft it is only at the major critical speeds that torsional oscillations can be induced. This, however, by no means implies that in large engines having a large number of cylinders, the minor critical speeds may be ignored.

At the major critical speeds the exciting torques T_{ex} of all the individual cylinders in one line act in phase and are thus additive (special rules apply governing the calculation of the combined excitation torques for Vee engines).

The first harmonic is generally of most significance in the excitation of torsional oscillations, and for engines of moderate size, such as passenger vehicle engines, it is generally sufficient to calculate the critical frequency

Table 2.1a-1 *p* factors

Order	$\frac{1}{2}$	1	$1\frac{1}{2}$	2	$2\frac{1}{2}$	3...	8
p factor	2.16	2.32	2.23	1.91	1.57	1.28	0.08

from eq. (1), then to calculate the corresponding engine speed from:

$$N_c = n_c/N_0 \tag{3}$$

The stiffness of the connecting shaft between engine and dynamometer should be chosen so that this speed does not lie within the range in which the engine is required to develop power.

In the case of large multicylinder engines, the 'wind-up' of the crankshaft as a result of torsional oscillations can be very significant and the two-mass approximation is inadequate; in particular, the critical speed may be overestimated by as much as 20 per cent and more elaborate analysis is necessary. The subject is dealt with in several different ways in the literature; perhaps the easiest to follow is that of Den Hartog.[1] The starting point is the value of the mean turning moment developed by the cylinder, M_{mean} (Fig. 2.1a-4). Values are given for a so-called '*p* factor', by which M_{mean} is multiplied to give the amplitude of the various harmonic excitation forces. Table 2.1a-1, reproduced from Den Hartog, shows typical figures for a four-stroke medium speed diesel engine.

Exciting torque:

$$T_{ex} = p.\,M_{mean} \tag{4}$$

The relation between M_{mean} and imep (indicated mean effective pressure) is given by:

$$\text{for a four-stroke engine} \quad M_{mean} = p_i.\frac{B^2S}{16}.10^{-4} \tag{5a}$$

$$\text{for a two-stroke engine} \quad M_{mean} = p_i.\frac{B^2S}{8}.10^{-4} \tag{5b}$$

Lloyd's Rulebook,[5] the main source of data on this subject, expresses the amplitude of the harmonic components rather differently, in terms of a 'component of tangential effort', T_m. This is a pressure that is assumed to act upon the piston at the crank radius $S/2$. Then exciting torque per cylinder:

$$T_{ex} = T_m\frac{\pi B^2}{4}\frac{S}{2} \times 10^{-4} \tag{6}$$

Lloyd's give curves of T_m in terms of the indicated mean effective pressure p_i and it may be shown that the values so obtained agree closely with those derived from Table 2.1a-1.

The amplitude of the vibratory torque T_v induced in the connecting shaft by the vector sum of the exciting torques for all the cylinders, $\sum T_{ex}$is given by:

$$T_v = \frac{\sum T_{ex}M_c}{(1 + I_e/I_b)} \tag{7}$$

The complete analysis of the torsional behaviour of a multicylinder engine is a substantial task, though computer programs are available which reduce the effort required. As a typical example, Fig. 2.1a-5 shows the 'normal elastic curves' for the first and second modes of torsional oscillation of a 16 cylinder Vee engine coupled to a hydraulic dynamometer. These curves show the amplitude of the torsional oscillations of the various components, relative to that at the dynamometer which is taken as unity. The natural frequencies are respectively $n_c = 4820$ c.p.m. and $n_c = 6320$ c.p.m. The curves form the basis for further calculations of the energy input giving rise to the oscillation. In the case of the engine under consideration, these showed a very severe fourth-order oscillation, $N_0 = 4$, in the first mode. (For an engine having eight cylinders in line the first major critical speed, from eq. (2a), is of order $N_0 = 4$.) The engine speed corresponding to the critical frequency of torsional oscillation is given by:

$$N_c = n_c/N_0 \tag{8}$$

giving, in the present case, $N_c = 1205$ rev/min, well within the operating speed range of the engine. Further calculations showed a large input of oscillatory energy at $N_0 = 4\frac{1}{2}$, a minor critical speed, in the second mode, corresponding to a critical engine speed of $6320/4\frac{1}{4} = 1404$ rev/min, again within the operating range. Several failures of the shaft connecting engine and dynamometer occurred before a safe solution was arrived at.

This example illustrates the need for caution and for full investigation in setting up large engines on the test bed.

It is not always possible to avoid running close to or at critical speeds and this situation is usually dealt with by the provision of torsional vibration dampers, in which the energy fed into the system by the exciting forces is absorbed by viscous shearing. Such dampers are commonly fitted at the non-flywheel end of engine crankshafts. In some cases it may also be necessary to consider their use as a component of engine test cell drive lines, when they are located either as close as possible to the engine flywheel, or at the dynamometer. The damper

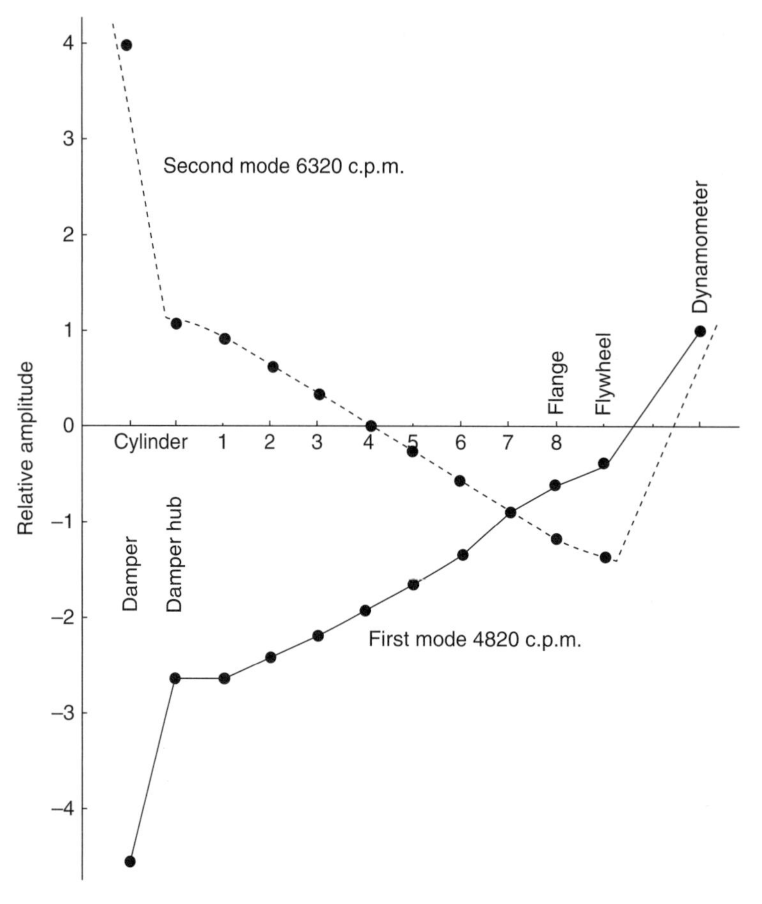

Fig. 2.1a-5 Normal elastic curves for a particular 16 cylinder V engine coupled to a hydraulic dynamometer (taken from an actual investigation).

must be 'tuned' to be most effective at the critical frequency and the selection of a suitable damper involves equating the energy fed into the system per cycle with the energy absorbed by viscous shear in the damper. This leads to an estimate of the magnitude of the oscillatory stresses at the critical speed. For a clear treatment of the theory, see the work by Den Hartog.[1]

Points to remember:

- As a general rule, it is good practice to avoid running the engine under power at speeds between 0.8 and 1.2 times critical speed. If it is necessary to take the engine through the critical speed, this should be done off load and as quickly as possible. With high inertia dynamometers the transient vibratory torque may well exceed the mechanical capacity of the drive line and the margin of safety of the drive line components may need to be increased.
- Problems frequently arise when the inertia of the dynamometer much exceeds that of the engine: a detailed torsional analysis is desirable when this factor exceeds 2. This situation usually occurs when it is found necessary to run an engine of much smaller output than the rated capacity of the dynamometer.
- the simple 'two mass' approximation of the engine-dynamometer system is inadequate for large engines and may lead to overestimation of the critical speed.

2.1a.5 Design of coupling shafts

The maximum shear stress induced in a shaft, diameter D, by a torque T N m is given by:

$$\tau = \frac{16T}{\pi D^3}\,\text{Pa} \tag{9a}$$

In the case of a tubular shaft, bore diameter d, this becomes:

$$\tau = \frac{16TD}{\pi(D^4 - d^4)}\,\text{Pa} \tag{9b}$$

For steels, the shear yield stress is usually taken as equal to 0.75 × yield stress in tension. A typical choice of material would be a nickel–chromium–molybdenum steel, to specification BS 817M40 (previously En 24) heat-treated to the 'T' condition.

The various stress levels for this steel are roughly as follows:

ultimate tensile strength not less than 850 MPa (55 t.s.i.)
0.1% proof stress in tension 550 MPa
ultimate shear strength 500 MPa
0.1% proof stress in shear 300 MPa
shear fatigue limit in reversed stress ±200 MPa

It is clear that the permissible level of stress in the shaft will be a small fraction of the ultimate tensile strength of the material.

The choice of designed stress level at the maximum rated steady torque is influenced by two principal factors.

2.1a.6 Stress concentrations, keyways and keyless hub connection

For a full treatment of the very important subject of stress concentration see Ref. 6. There are two particularly critical locations to be considered:

- At a shoulder fillet, such as is provided at the junction with an integral flange. For a ratio fillet radius/shaft diameter = 0.1 the stress concentration factor is about 1.4, falling to 1.2 for $r/d = 0.2$.
- At the semicircular end of a typical rectangular keyway, the stress concentration factor reaches a maximum of about 2.5× nominal shear stress at an angle of about 50° from the shaft axis. The authors have seen a number of failures at this location and angle.

Cyclic stresses associated with torsional oscillations is an important consideration and as, even in the most carefully designed installation involving an internal combustion engine, some torsional oscillation will be present, it is wise to select a conservative value for the nominal (steady state) shear stress in the shaft.

In view of the stress concentration inherent in shaft keyways and the backlash present that can develop in splined hubs, the use of keyless hub connection systems of the type produced by the Ringfeder Corporation are now widely used. These devices are supported by comprehensive design documentation; however, the actual installation process must be exactly followed for the design performance to be ensured.

Stress concentration factors apply to the cyclic stresses rather than to the steady state stresses. Fig. 2.1a-6 shows diagrammatically the Goodman diagram for a steel having the properties specified above. This diagram indicates approximately the relation between the steady shear stress and the permissible oscillatory stress. The example

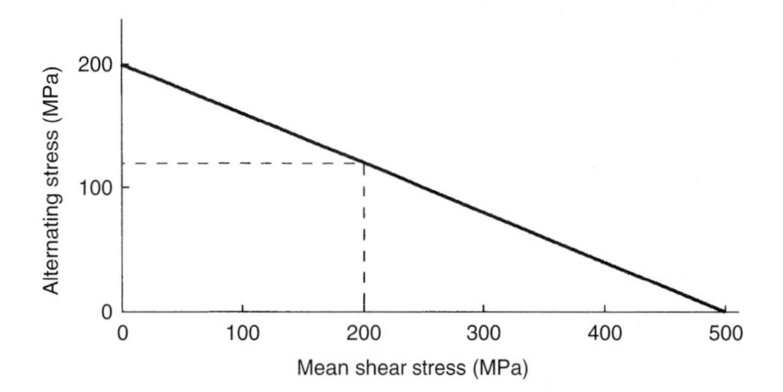

Fig. 2.1a-6 Goodman diagram, steel shaft in shear.

shown indicates that, for a steady torsional stress of 200 MPa, the accompanying oscillatory stress (calculated after taking into account any stress concentration factors) should not exceed ±120 MPa. In the absence of detailed design data, it is good practice to design shafts for use in engine test beds very conservatively, since the consequences of shaft failure can be so serious. A shear stress calculated in accordance with eq. (9) of about 100 MPa for a steel with the properties listed should be safe under all but the most unfavourable conditions. To put this in perspective, a shaft 100 mm diameter designed on this basis would imply a torque of 19 600 Nm, or a power of 3100 kW at 1500 rev/min.

The torsional stiffness of a solid shaft of diameter D and length L is given by:

$$C_s = \frac{\pi D^4 G}{32L} \tag{10a}$$

for a tubular shaft, bore d:

$$C_s = \frac{\pi (D^4 - d^4)}{32L} \tag{10b}$$

2.1a.7 Shaft whirl

The coupling shaft is usually supported at each end by a universal joint or flexible coupling. Such a shaft will 'whirl' at a rotational speed N_w (also at certain higher speeds in the ratio $2^2 N_w$, $3^2 N_w$, etc.).

The whirling speed of a solid shaft of length L is given by:

$$N_w = \frac{30\pi}{L^2}\sqrt{\frac{E\pi D^4}{64W_s}} \tag{11}$$

It is desirable to limit the maximum engine speed to about 0.8 N_w. *When using rubber flexible couplings it is essential to allow for the radial flexibility of these couplings, since this can drastically reduce the whirling speed.* It is sometimes the practice to fit self-aligning rigid steady bearings at the centre of flexible couplings in high-speed applications, but these are liable to give fretting problems and are not universally favoured.

As is well known, the whirling speed of a shaft is identical with its natural frequency of transverse oscillation. To allow for the effect of transverse coupling flexibility the simplest procedure is to calculate the transverse critical frequency of the shaft plus two half couplings from the equation:

$$N_t = \frac{30}{\pi}\sqrt{\frac{\kappa}{W}} \tag{12a}$$

where W=mass of shaft + half couplings and k = combined radial stiffness of the two couplings.

Then whirling speed N taking this effect into account will be given by:

$$\left(\frac{1}{N}\right)^2 = \left(\frac{1}{N_w}\right)^2 + \left(\frac{1}{N_t}\right)^2 \tag{12b}$$

2.1a.8 Couplings

The choice of the appropriate coupling for a given application is not easy: the majority of drive line problems probably have their origin in an incorrect choice of components for the drive line, and are often cured by changes in this region. A complete discussion would much exceed the scope of this book, but the reader concerned with drive line design should obtain a copy of Ref. 4, which gives a comprehensive treatment together with a valuable procedure for selecting the best type of coupling for a given application. A very brief summary of the main types of coupling follows.

2.1a.8.1 Quill shaft with integral flanges and rigid couplings

This type of connection is best suited to the situation where a driven machine is permanently coupled to the source of power, when it can prove to be a simple and reliable solution. It is not well suited to test bed use, since it is intolerant of relative vibration and misalignment.

2.1a.8.2 Quill shaft with toothed or gear type couplings

Gear couplings are very suitable for high powers and speeds, and can deal with relative vibration and some degree of misalignment, but this must be very carefully controlled to avoid problems of wear and lubrication. Lubrication is particularly important as once local tooth to tooth seizure takes place deterioration may be rapid and catastrophic. Such shafts are inherently stiff in torsion.

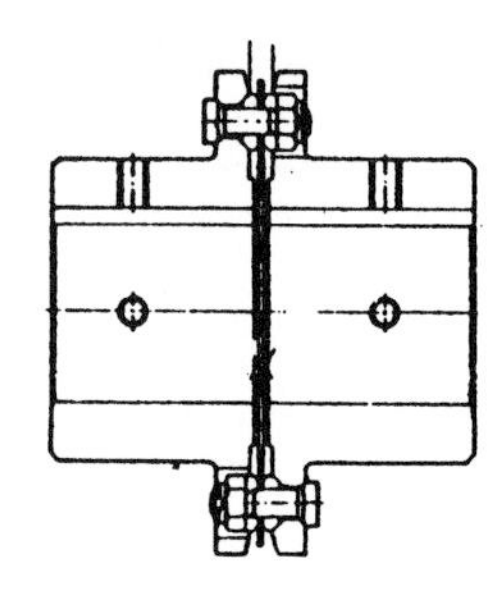

Fig. 2.1a-7 Multiple steel disc type flexible coupling.

2.1a.8.3 Conventional 'cardan shaft' with universal joints

These shafts are readily available from a number of suppliers, and are probably the preferred solution in the majority of cases. However, standard automotive type shafts can give trouble when run at speeds in excess of those encountered in vehicle applications. A correct 'built-in' degree of misalignment is necessary to avoid fretting of the needle rollers.

2.1a.8.4 Multiple membrane couplings

These couplings, Fig. 2.1a-7, are stiff in torsion but tolerant of a moderate degree of misalignment and relative axial displacement. They can be used for very high speeds.

2.1a.8.5 Elastomeric element couplings

There is a vast number of different designs on the market and selection is not easy. Ref. 8 may be helpful. The great advantage of these couplings is that their torsional stiffness may be varied widely by changing the elastic elements and problems associated with torsional vibrations and critical speeds dealt with (see the next section).

2.1a.9 Damping: the role of the flexible coupling

The earlier discussion leads to two main conclusions: the engine–dynamometer system is susceptible to torsional oscillations and the internal combustion engine is a powerful source of forces calculated to excite such oscillations. The magnitude of these undesirable disturbances in any given system is a function of the damping

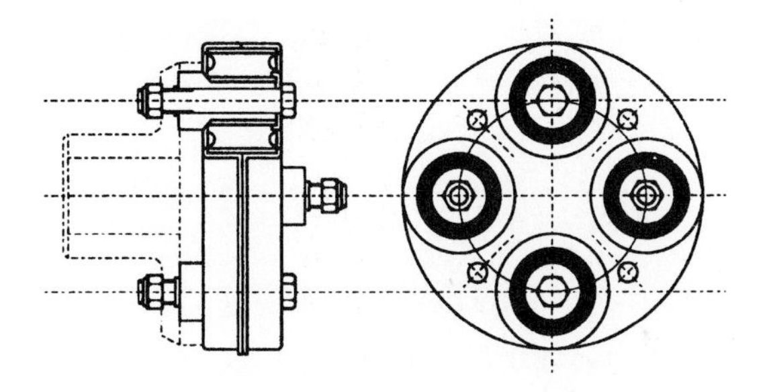

Fig. 2.1a-8 Rubber bush type torsionally resilient coupling.

capacities of the various elements: the shaft, the couplings, the dynamometer and the engine itself.

The couplings are the only element of the system, the damping capacity of which may readily be changed, and in many cases, for example with engines of automotive size, the damping capacity of the remainder of the system may be neglected, at least in an elementary treatment of the problem, such as will be given here.

The dynamic magnifier M (Fig. 2.1a-3) has already been mentioned as a measure of the susceptibility of the engine–dynamometer system to torsional oscillation. Now referring to Fig. 2.1a-1, let us assume that there are two identical flexible couplings, of stiffness C_c, one at each end of the shaft, and that these are the only sources of damping. Fig. 2.1a-8 shows a typical torsionally resilient coupling in which torque is transmitted by way of a number of shaped rubber blocks or bushes which provide torsional flexibility, damping and a capacity to take up misalignment. The torsional characteristics of such a coupling are shown in Fig. 2.1a-9.

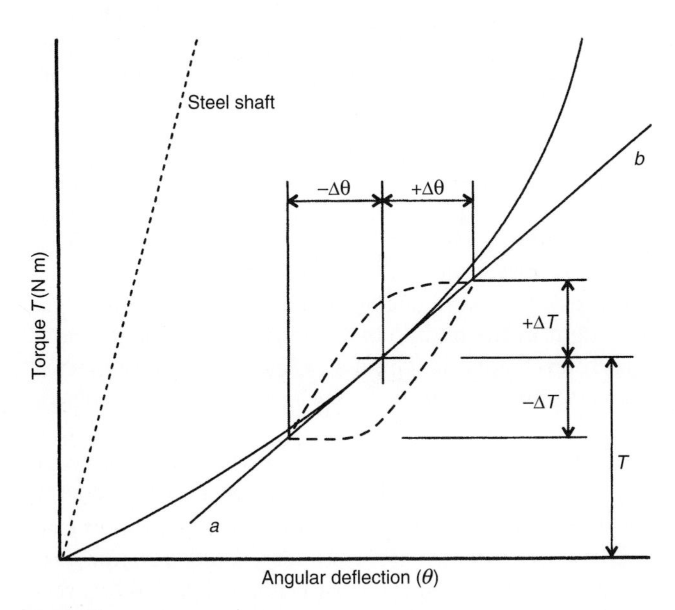

Fig. 2.1a-9 Dynamic torsional characteristic of multiple bush type coupling.

These differ in three important respects from those of, say, a steel shaft in torsion:

1. The coupling does not obey Hooke's law: the stiffness or coupling rate $C_c = \Delta T/\Delta\theta$ increases with torque. This is partly an inherent property of the rubber and partly a consequence of the way it is constrained.
2. The shape of the torque–deflection curve is not independent of frequency. Dynamic torsional characteristics are usually given for a cyclic frequency of 10 Hz. If the load is applied slowly the stiffness is found to be substantially less. The following values of the ratio dynamic stiffness (at 10 Hz) to static stiffness of natural rubber of varying hardness are taken from Ref. 4.

Shore (IHRD) hardness	40	50	60	70
$\dfrac{\text{Dynamic stiffness}}{\text{Static stiffness}}$	1.5	1.8	2.1	2.4

Since the value of C_c varies with the deflection, manufacturers usually quote a single figure which corresponds to the slope of the tangent *ab* to the torque–deflection curve at the rated torque, typically one third of the maximum permitted torque.

3. If a cyclic torque $\pm\Delta T$, such as that corresponding to a torsional vibration, is superimposed on a steady torque T, Fig. 2.1a-9, the deflection follows a path similar to that shown dotted. It is this feature, the hysteresis loop, which results in the dissipation of energy, by an amount ΔW proportional to the area of the loop that is responsible for the damping characteristics of the coupling.

Damping energy dissipated in this way appears as heat in the rubber and can, under adverse circumstances, lead to overheating and rapid destruction of the elements. The appearance of rubber dust inside coupling guards is a warning sign.

The damping capacity of a component such as a rubber coupling is described by the *damping energy ratio*:

$$\psi = \frac{\Delta W}{W}$$

This may be regarded as the ratio of the energy dissipated by hysteresis in a single cycle to the elastic energy corresponding to the wind-up of the coupling at mean deflection:

$$W = \frac{1}{2}T\theta = \frac{1}{2}T^2/C_c$$

The damping energy ratio is a property of the rubber. Some typical values are given in Table 2.1a-2.

Table 2.1a-2 Damping energy ratio ψ

Shore (IHRD) hardness	50/55	60/65	70/75	75/80
Natural rubber	0.45	0.52	0.70	0.90
Neoprene		0.79		
Styrene-butadiene (SBR)		0.90		

The dynamic magnifier is a function of the damping energy ratio: as would be expected a high damping energy ratio corresponds to a low dynamic magnifier. Some authorities give the relation:

$$M = 2\pi/\psi$$

However, it is pointed out in Ref. 2 that for damping energy ratios typical of rubber the exact relation:

$$\psi = (1 - e^{-2\pi/M})$$

is preferable. This leads to values of M shown in Table 2.1a-3, which correspond to the values of ψ given in Table 2.1a-2.

It should be noted that when several components, e.g. two identical rubber couplings, are used in series the dynamic magnifier of the combination is given by:

$$\left(\frac{1}{M}\right)^2 = \left(\frac{1}{M_1}\right)^2 + \left(\frac{1}{M_2}\right)^2 + \left(\frac{1}{M_3}\right)^2 + \cdots \quad (13)$$

(this is an empirical rule, recommended in Ref. 5).

2.1a.10 An example of drive shaft design

The application of these principles is best illustrated by a worked example. Fig. 2.1a-1 represents an engine coupled by way of twin multiple-bush type rubber couplings and an intermediate steel shaft to an eddy current dynamometer, with dynamometer starting.

Engine specification is as follows:

Four cylinder four-stroke gasoline engine
Swept volume 2.0 litre, bore 86 mm, stroke 86 mm

Table 2.1a-3 Dynamic magnifier M

Shore (IHRD) hardness	50/55	60/65	70/75	75/80
Natural rubber	10.5	8.6	5.2	2.7
Neoprene		4.0		
Styrene-butadiene (SBR)		2.7		

Table 2.1a-4 Service factors for dynamometer/engine combinations

Dynamometer type	Number of cylinders									
	Diesel					Gasolene				
	1/2	3/4/5	6	8	10+	1/2	3/4/5	6	8	10+
Hydraulic	4.5	4.0	3.7	3.3	3.0	3.7	3.3	3.0	2.7	2.4
Hyd. + dyno. start	6.0	5.0	4.3	3.7	3.0	5.2	4.3	3.6	3.1	2.4
Eddy current (EC)	5.0	4.5	4.0	3.5	3.0	4.2	3.8	3.3	2.9	2.4
EC + dyno. start	6.5	5.5	4.5	4.0	3.0	5.7	4.8	3.8	3.4	2.4
d.c. + dyno. start	8.0	6.5	5.0	4.0	3.0	7.2	5.8	4.3	3.4	2.4

Maximum torque	110 Nm at 4000 rev/min
Maximum speed	6000 rev/min
Maximum power output	65 kW
Maximum bmep	10.5 bar
Moments of inertia	$I_e = 0.25$ kg m^2
	$I_d = 0.30$ kg m^2

Table 2.1a-4 indicates a service factor of 4.8, giving a design torque of 110 × 4.8 = 530 Nm.

It is proposed to connect the two couplings by a steel shaft of the following dimensions:

Diameter $D = 40$ mm
Length $L = 500$ mm
Modulus of rigidity $G = 80 \times 10^9$ Pa

From eq. (9a), torsional stress $\tau = 42$ MPa, very conservative.

From eq. (10a)

$$C_s = \frac{\pi \times 0.04^4 \times 80 \times 10^9}{32 \times 0.5} = 40\ 200\ \text{N m/rad}$$

Consider first the case when rigid couplings are employed:

$$n_c = \frac{60}{2\pi}\sqrt{\frac{40\ 200 \times 0.55}{0.25 \times 0.30}} = 5185\ \text{c.p.m.}$$

For a four cylinder, four-stroke engine, we have seen that the first major critical occurs at order $N_0 = 2$, corresponding to an engine speed of 2592 rev/min. This falls right in the middle of the engine speed range and is clearly unacceptable. This is a typical result to be expected if an attempt is made to dispense with flexible couplings.

The resonant speed needs to be reduced and it is a common practice to arrange for this to lie between either the cranking and idling speeds or between the idling and minimum full load speeds. In the present case these latter speeds are 500 and 1000 rev/min, respectively. This suggests a critical speed N_c of 750 rev/min and a corresponding resonant frequency $n_c = 1500$ cycles/min.

This calls for a reduction in the torsional stiffness in the ratio:

$$\left(\frac{1500}{5185}\right)^2$$

i.e. to 3364 Nm/rad.

The combined torsional stiffness of several elements in series is given by:

$$\frac{1}{C} = \frac{1}{C_1} + \frac{1}{C_2} + \frac{1}{C_3} + \cdots \tag{14}$$

This equation indicates that the desired stiffness could be achieved by the use of two flexible couplings each of stiffness 7480 Nm/rad. A manufacturer's catalogue shows a multi-bush coupling having the following characteristics:

Maximum torque 814 N m (adequate)
Rated torque 170 N m
Maximum continuous vibratory torque ±136 Nm
Shore (IHRD) hardness 50/55
Dynamic torsional stiffness 8400 Nm/rad

Substituting this value in eq. (14) indicates a combined stiffness of 3800 Nm/rad. Substituting in eq. (1) gives $n_c = 1573$, corresponding to an engine speed of 786 rev/min, which is acceptable.

It remains to check on the probable amplitude of any torsional oscillation at the critical speed. Under no-load conditions, the imep of the engine is likely to be in the region of 2 bar, indicating, from eq. (5a), a mean turning moment $M_{mean} = 8$ Nm.

From Table 2.1a-1, p factor = 1.91, giving $T_{ex} = 15$ Nm per cylinder:

$$\sum T_{ex} = 4 \times 15 = 60\,\text{N m}$$

Table 2.1a-3 indicates a dynamic magnifier $M = 10.5$, the combined dynamic magnifier from eq. (13) = 7.4.

The corresponding value of the vibratory torque, from eq. (7), is then:

$$T_v = \frac{60 \times 7.4}{(1 + 0.25/0.30)} = \pm 242\,\text{N m}$$

This is in fact outside the coupling continuous rating of ±136 Nm, but multiple bush couplings are tolerant of brief periods of quite severe overload and this solution should be acceptable provided the engine is run fairly quickly through the critical speed. An alternative would be to choose a coupling of similar stiffness using SBR bushes of 60/65 hardness. Table 2.1a-3 shows that the dynamic magnifier is reduced from 10.5 to 2.7, with a corresponding reduction in T_v.

If in place of an eddy current dynamometer we were to employ a d.c. machine, the inertia I_b would be of the order of 1 kg m^2, four times greater.

This has two adverse effects:

1. Service factor, from Table 2.1a-4 increased from 4.8 to 5.8;
2. The denominator in eq. (7) is reduced from $(1 + 0.25/0.30) = 1.83$ to $(1 + 0.25/1.0) = 1.25$, corresponding to an increase in the vibratory torque for a given exciting torque of nearly 50 per cent.

This is a general rule: the greater the inertia of the dynamometer the more severe the torsional stresses generated by a given exciting torque.

An application of eq. (1) shows that for the same critical frequency the combined stiffness must be increased from 3364 Nm/rad to 5400 Nm/rad. We can meet this requirement by changing the bushes from Shore Hardness 50/55 to Shore Hardness 60/65, increasing the dynamic torsional stiffness of each coupling from 8400 Nm/rad to 14000 Nm/rad (in general, the usual range of hardness numbers, from 50/55 to 75/80, corresponds to a stiffness range of about 3:1, a useful degree of flexibility for the designer).

Eq. (1) shows that with this revised coupling stiffness n_c changes from 1573 cycles/min to 1614 cycles/min, and this should be acceptable. The oscillatory torque generated at the critical speed is increased by the two factors mentioned above, but reduced to some extent by the lower dynamic magnifier for the harder rubber, $M = 8.6$ against $M = 10.5$. As before, prolonged running at the critical speed should be avoided.

For completeness, we should check the whirling speed from eq. (11). The mass of the shaft per unit length is: $W_s = 9.80$ kg/m.

$$N_w = \frac{30\pi}{0.50^2}\sqrt{\frac{200 \times 10^9 \times \pi \times 0.04^4}{64 \times 9.80}} = 19\,100\ \text{r.p.m.}$$

The mass of the shaft + half couplings is found to be 12 kg and the combined radial stiffness 33.6 MN/m. From eq. (12a):

$$N_t = \frac{30}{\pi}\sqrt{\frac{33.6 \times 10^6}{12}} = 16\,000\ \text{r.p.m.}$$

then from eq. (12b), whirling speed = 12 300 rev/min, which is satisfactory.

Note, however, that, if shaft length were increased from 500 to 750 mm, whirling speed would be reduced to about 7300 rev/min, which is barely acceptable. This is a common problem, usually dealt with by the use of tubular shafts, which have much greater transverse stiffness for a given mass.

There is no safe alternative, when confronted with an engine of which the characteristics differ significantly from any run previously on a given test bed, to following through this design procedure.

2.1a.10.1 An alternative solution

The above worked example makes use of two multiple-bush type rubber couplings with a solid intermediate shaft. An alternative is to make use of a conventional propeller shaft with two universal joints, as used in road vehicles, with the addition of a coupling incorporating an annular rubber element in shear to give the necessary torsional flexibility. These couplings, Fig. 2.1a-10, are generally softer than the multiple bush type for a given torque capacity, but are less tolerant of operation near a critical speed. If it is decided to use a conventional universal joint shaft, the supplier should be informed of the maximum speed at which it is to run. This will probably be much higher than is usual in the vehicle and may call for tighter than usual limits on balance of the shaft.

2.1a.10.2 Shock loading of couplings due to cranking, irregular running and torque reversal

Systems for starting and cranking engines are described in Chapter 2.1, where it is emphasized that during engine starting severe transient torques can arise. These have been known to result in the failure of flexible couplings of apparently adequate torque capacity. The maximum torque that can be necessary to get a green engine over t.d.c. or that can be generated at first firing should be estimated and checked against maximum coupling capacity.

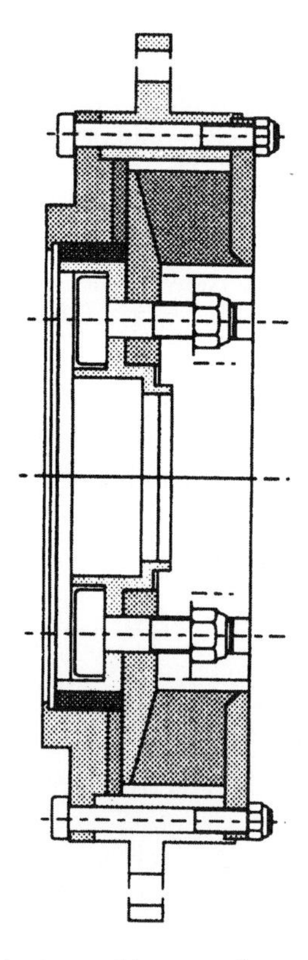

Fig. 2.1a-10 Annular type rubber coupling.

Irregular running or imbalance between the powers generated by individual cylinders can give rise to exciting torque harmonics of lower order than expected in a multicylinder engine and should be borne in mind as a possible source of rough running. Finally, there is the possibility of momentary torque reversal when the engine comes to rest on shutdown.

However, the most serious problems associated with the starting process arise when the engine first fires. Particularly when, as is common practice, the engine is motored to prime the fuel injection pump, the first firing impulses can give rise to severe shocks. Annular type rubber couplings, Fig. 2.1a-10, can fail by shearing under these conditions. In some cases, it is necessary to fit a torque limiter or slipping clutch to deal with this problem.

2.1a.10.3 Axial shock loading

Engine test systems that incorporate automatic shaft docking systems have to provide for the axial loads on both the engine and dynamometer imposed by such a system. In some high volume production facilities, an intermediate pedestal bearing isolates the dynamometer from both the axial loads of normal docking operation and for cases when the docking splines jam 'nose to nose'; in these cases the docking control system should be programmed to back off the engine, spin the dynamometer and retry the docking.

2.1a.10.4 Selection of coupling torque capacity

Initial selection is based on the maximum rated torque with consideration given to the type of engine and number of cylinders, dynamometer characteristics and inertia. Table 2.1a-4, reproduced by courtesy of Twiflex Ltd, shows recommended service factors for a range of engine and dynamometer combinations. The rated torque multiplied by the service factor must not exceed the permitted maximum torque of the coupling.

Other manufacturers may adopt different rating conventions, but Table 2.1a-4 gives valuable guidance as to the degree of severity of operation for different situations. Thus, for example, a single cylinder diesel engine coupled to a d.c. machine with dynamometer start calls for a margin of capacity three times as great as an eight

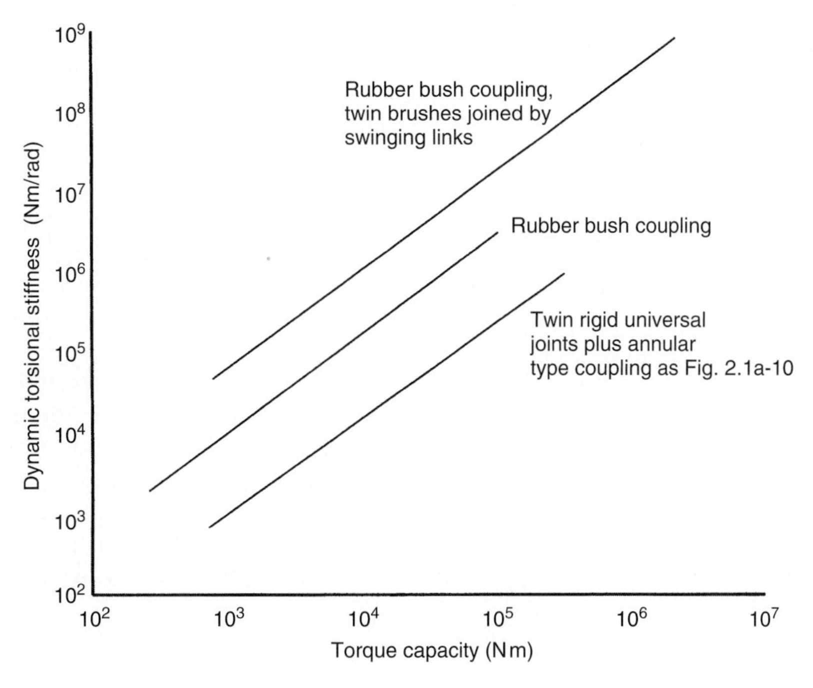

Fig. 2.1a-11 Ranges of torsional stiffness for different types of rubber coupling.

cylinder gasoline engine coupled to a hydraulic dynamometer.

Fig. 2.1a-11 shows the approximate range of torsional stiffness associated with three types of flexible coupling: the annular type as illustrated in Fig. 2.1a-10, the multiple bush design of Fig. 2.1a-8 and a development of the multiple bush design which permits a greater degree of misalignment and makes use of double-ended bushed links between the two halves of the coupling. The stiffness figures of Fig. 2.1a-11 refer to a single coupling.

2.1a.10.5 The role of the engine clutch

Vehicle engines are invariably fitted with a clutch and this may or may not be retained on the test bed. The advantage of retaining the clutch is that it acts as a torque limiter under shock or torsional vibration conditions. The disadvantages are that it may creep, particularly when torsional vibration is present, leading to ambiguities in power measurement, while it is usually necessary, when the clutch is retained, to provide an outboard bearing. Clutch disc springs may have limited life under test bed conditions.

2.1a.10.6 Balancing of drive line components

This is a matter which is often not taken sufficiently seriously and can lead to a range of troubles, including damage (which can be very puzzling) to bearings, unsatisfactory performance of such items as torque transducers, transmitted vibration to unexpected locations and serious drive line failures. Particular care should be taken in the choice of couplings for torque shaft dynamometers: couplings such as the multiple disc type, Fig. 2.1a-7, cannot be relied upon to centre these devices sufficiently accurately.

It has sometimes been found necessary to carry out in situ balancing of a composite engine drive line where the sum of the out of balance forces in a particular radial relationship causes unacceptable vibration; specialist companies with mobile plant exist to provide this service.

Conventional universal joint type cardan shafts are often required to run at higher speeds in test bed applications than is usual in vehicles; when ordering, the maximum speed should be specified and, possibly, a more precise level of balancing than the standard specified.

BS 5265, *Mechanical Balancing of Rotating Bodies*,[9] gives a valuable discussion of the subject and specifies 11 Balance Quality Grades. Drive line components should generally be balanced to Grade G 6.3, or, for high speeds, to grade G 2.5. The standard gives a logarithmic plot of the permissible displacement of the centre of gravity of the rotating assembly from the geometrical axis against speed. To give an idea of the magnitudes involved, G 6.3 corresponds to a displacement of 0.06 mm at 1000 rev/min, falling to 0.01 mm at 5000 rev/min.

2.1a.10.7 Alignment of engine and dynamometer

This is a fairly complex and quite important matter. For a full treatment and description of alignment techniques, see Ref. 4. Differential thermal growth and the movement of the engine on its flexible mountings when on load should be taken into account and if possible the mean position should be determined. The laser-based alignment systems now available greatly reduce the effort and skill required to achieve satisfactory levels of accuracy. In particular, they are able to bridge large gaps between flanges without any compensation for droop and deflection of arms carrying dial indicators, a considerable problem with conventional alignment methods.

There are essentially three types of shaft having different alignment requirements to be considered:

1. Rubber bush and flexible disc couplings should be aligned as accurately as possible as any misalignment encourages heating of the elements and fatigue damage.
2. Gear type couplings require a minimum misalignment of about 0.04° to encourage the maintenance of an adequate lubricant film between the teeth.
3. Most manufacturers of universal joint propeller shafts recommend a small degree of misalignment to prevent brinelling of the universal joint needle rollers. Note that it is essential, in order to avoid induced torsional oscillations, that the two yokes of the intermediate shaft joints should lie in the same plane.

Distance between end flanges can be critical, as incorrect positioning can lead to the imposition of axial loads on bearings of engine or dynamometer.

2.1a.10.8 Guarding of coupling shafts

Not only is the failure of a high speed coupling shaft potentially dangerous, as very large amounts of energy can be released, but it is also quite common. The ideal shaftguard will contain the debris resulting from a failure of any part of the drive line and prevent accidental or casual contact with rotating parts, while being sufficiently light and adaptable not to interfere with engine rigging and alignment. A guard system that is very inconvenient to use will eventually be misused or neglected.

A really substantial guard, preferably a steel tube not less than 6 mm thick, split and hinged in the horizontal plane for shaft access, is an essential precaution. The hinged 'lid' should be interlocked with the emergency stop circuit to prevent engine cranking or running while it is open. Many designs include shaft restraint devices loosely fitting around the tubular portion, made of wood or a non-metallic composite and intended to prevent a failing shaft from whirling; these should not be so close to the shaft as to be touched by it during normal starting or running otherwise they will be the cause of failure rather than a prevention of damage.

2.1a.11 Engine to dynamometer coupling: summary of design procedure

1. Establish speed range and torque characteristic of engine to be tested. Is it proposed to run the engine on load throughout this range?
2. Make a preliminary selection of a suitable drive shaft. Check that maximum permitted speed is adequate. Check drive shaft stresses and specify material. Look into possible stress raisers.
3. Check manufacturer's recommendations regarding load factor and other limitations.
4. Establish rotational inertias of engine and dynamometer and stiffness of proposed shaft and coupling assembly. Make a preliminary calculation of torsional critical speed from eq. (1). (In the case of large multicylinder engines consider making a complete analysis of torsional behaviour.)
5. Modify specification of shaft components as necessary to position torsional critical speeds suitably. If necessary, consider use of viscous torsional dampers.
6. Calculate vibratory torques at critical speeds and check against capacity of shaft components. If necessary specify 'no go' areas for speed and load.
7. Check whirling speeds.
8. Specify alignment requirements.
9. Design shaft guards.

2.1a.12 Flywheels

No treatment of the engine/dynamometer drive line would be complete without mention of flywheels which may form a discrete part of the shaft system. A flywheel is a device that stores kinetic energy. The energy stored may be precisely calculated and is instantly available. The storage capacity is directly related to the mass moment of inertia which is calculated by:

$$I = k \times M \times R^2$$

where:

I = moment of inertia (kg m^2)
k = inertia constant (dependent upon shape)

M = mass (kg)
R = radius of flywheel mass

In the case of a flywheel taking the form of a uniform disc, which is the common form found within dynamometer cells and chassis dynamometer designs:

$$I = \frac{1}{2}MR^2$$

The engine or vehicle test engineer would normally expect to deal with flywheels in two roles:

1. As part of the test object, as in the common case of an engine flywheel where it forms part of the engine/dynamometer shaft system and contributes to the system's inertial masses taken into account during a torsional analysis.
2. As part of the test equipment where one or more flywheels may be used to provide actual inertia that would, in 'real life', be that of the vehicle or some part of the engine driven system.

No mention of flywheels should be made without consideration of the safety of the application. The uncontrolled discharge of energy from any storage device is hazardous. The classic case of a flywheel failing by bursting is now exceptionally rare and invariably due to incompetent modification rather than the nineteenth century problems of poor materials, poor design or overspeeding.

In the case of engine flywheels, the potential danger in the test cell is the shaft system attached to it. This may be quite different in mass and fixing detail from its final application connection, and can cause overload leading to failure. Cases are on record where shock loading caused by connecting shafts touching the guard system due to excessive engine movement has created shock loads that have led to the cast engine flywheel fracturing, with severe consequential damage.

The most common hazard of test rig mounted flywheels is caused by bearing or clutch failure where consequential damage is exacerbated by the considerable energy available to fracture connected devices or because of the time that the flywheel and connected devices will rotate before the stored energy is dissipated and movement is stopped.

It is vital that flywheels are guarded in such a manner as to prevent absolutely accidental entrainment of clothing or cables, etc.

A common and easy to comprehend use of flywheels is as part of a vehicle brake testing rig. In these devices, flywheels supply the energy that has to be absorbed and dissipated by the brake system under test. The rig motor is only used to accelerate the chosen flywheel combinations up to the rotational speed required to simulate the vehicle axle speed at the chosen vehicle speed. Flywheel brake rigs have been made up to the size that can provide the same kinetic energy as fully loaded high speed trains. Flywheels are also used on rigs used to test automatic automotive gearboxes.

Test rig flywheel sets need to be rigidly and securely mounted and balanced to the highest practical standard. Multiples of flywheels forming a common system that can be engaged in different combinations and in any radial relationship require particular care in the design of both their base frame and individual bearing supports. Such systems can produce virtually infinite combinations of shaft balance and require each individual mass to be as well balanced and aligned on as rigid a base as possible.

2.1a.12.1 Simulation of inertia* versus iron inertia

Modern a.c. dynamometer systems and control software have significantly replaced the use of flywheels in chassis and engine dynamometer systems in the automotive industry. Any perceived shortcoming in the speed of response or accuracy of the simulation is usually considered to be of less concern than the mechanical simplicity of the electric dynamometer system and the reduction in required cell space.

Finally, it should be remembered that, unless engine rig flywheels are able to be engaged through a clutch, the engine starting/cranking system will have to be capable of accelerating engine, dynamometer and flywheel mass up to engine start speed.

2.1a.13 Notation

Frequency of torsional oscillation	n cycles/min
Critical frequency of torsional oscillation	n_c cycles/min
Stiffness of coupling shaft	C_s N m/rad
Rotational inertia of engine	I_e kg m^2
Rotational inertia of dynamometer	I_b kg m^2
Amplitude of exciting torque	T_x N m
Amplitude of torsional oscillation	θ rad
Static deflection of shaft	θ_0 rad
Dynamic magnifier	M
Dynamic magnifier at critical frequency	M_c
Order of harmonic component	N_o

* Some readers may object to the phrase 'simulation of inertia' since one is simulating the effects rather than the attribute, but the concept has wide industrial acceptance.

Number of cylinders	N_{cyl}	Transverse critical frequency	N_t cycles/min
Mean turning moment	M_{mean} Nm	Dynamic torsional stiffness of coupling	C_c Nm/rad
Indicated mean effective pressure	p_i bar	Damping energy ratio	ψ
Cylinder bore	B mm	Modulus of elasticity of shaft material	E Pa
Stroke	S mm	Modulus of rigidity of shaft material	G Pa
Component of tangential effort	T_m Nm	(for steel, $E = 200 \times 10^9$ Pa, $G = 80 \times 10^9$ Pa)	
Amplitude of vibratory torque	T_v Nm		
Engine speed corresponding to n_c	N_c rev/min		
Maximum shear stress in shaft	τ N/m^2		
Whirling speed of shaft	N_w rev/min		

References

1. Den Hartog, J.P. (1956) *Mechanical Vibrations*, McGraw-Hill, Maidenhead, UK.
2. Ker Wilson, W. (1963) *Practical Solution to Torsional Vibration Problems* (5 vols), Chapman and Hall, London.
3. Ker Wilson, W. (1959) *Vibration Engineering*, Griffin, London.
4. Neale, M.J., Needham, P. and Horrell, R. (1998) *Couplings and Shaft Alignment*, Mechanical Engineering Publications, London.
5. *Rulebook*, Chapter 8. Shaft vibration and alignment, Lloyd's Register of Shipping, London.
6. Pilkey, W.D. (1997) *Peterson's Stress Concentration Factors*, Wiley, New York.
7. Young, W.C. (1989) *Roark's Formulas for Stress and Strain*, McGraw-Hill, New York.
8. BS 6613 *Methods for Specifying Characteristics of Resilient Shaft Couplings*.
9. BS 5265 Parts 2 and 3 *Mechanical Balancing of Rotating Bodies*.

Further reading

BS 4675 Parts 1 and 2 *Mechanical Vibration in Rotating Machinery*.

BS 6861 Part 1 *Method for Determination of Permissible Residual Unbalance*.

BS 6716 *Guide to Properties and Types of Rubber*.

Nestorides, E.J. (1958) *A Handbook of Torsional Vibration*, Cambridge University Press, Cambridge.

Section Three

Engine emissions

Chapter 3.1

3.1

Emissions control

T.K. Garrett, K. Newton and W. Steels

Attention was first directed to atmospheric pollution in Los Angeles in 1947. Subsequently, in 1952, Dr Arie J. Haagen-Smit asserted on the basis of his research that, at least locally, it was due mainly to automotive exhaust emissions. It was subsequently said, however, that it would have cost the USA less to have moved Los Angeles than to have converted all their vehicles to reduce the obnoxious emissions to the levels now required by law! Japan was close behind the USA with emission control laws, and Europe has practically caught up.

Given complete combustion, each kg of hydrocarbon fuel when completely burnt produces mainly 3.1 kg of CO_2 and 1.3 kg of H_2O. Most of the undesirable exhaust emissions are produced in minute quantities (parts per million), and these are: oxides of nitrogen, generally termed NO_x, unburnt hydrocarbons (HCs), carbon monoxide (CO), carbon dioxide (CO_2), lead salts, polyaromatics, soots, aldehydes ketones and nitro-olefins. Of these, only the first three are of major significance in the quantities produced. However, concentrations in general could become heavier as increasing numbers of vehicles come onto our roads. By the end of the 1980s, CO_2 was beginning to cause concern, not because it is toxic but because it was suspected of facilitating the penetration of our atmosphere by ultraviolet rays emitted by the sun. Controversy has raged over lead salts, but no proof has been found that, in the quantities in which they are present in the atmosphere, they are harmful. For many years, manufacturers of catalytic converters pressed for unleaded petrol because lead deposits rapidly rendered their converters ineffective.

Carbon monoxide is toxic because it is absorbed by the red corpuscles of the blood, inhibiting absorption of the oxygen necessary for sustaining life. The toxicity of hydrocarbons and oxides of nitrogen, on the other hand, arises indirectly as a result of photochemical reactions between the two in sunlight, leading to the production of other chemicals.

There are two main oxides of nitrogen: nitric acid and nitrogen dioxide, NO and NO_2, of which the latter is of greatest significance as regards toxic photochemical effects. Under the influence of solar radiation, the NO_2 breaks down into NO + O, the highly reactive oxygen atom then combining with O_2 to make O_3, which of course is ozone. Normally, this would then rapidly recombine with the NO to form NO_2 again, but the presence of hydrocarbons inhibits this reaction and causes the concentration of ozone to rise. The ozone then goes on, in a complex manner, to combine with the other substances present to form chemicals which, in combination with moisture in an atmospheric haze, produce what has been described as the obnoxious smoky fog now known as smog.

Unburnt hydrocarbons can come from evaporation from the carburettor float chamber and fuel tank vent as well as from inefficient combustion due in different instances to faulty ignition, inadequate turbulence, poor carburation, an over-rich mixture, or long flame paths from the point of ignition. The relationships between emissions and the air:fuel ratio are illustrated in Fig. 3.1-1. Other factors are overcooling, large quench areas in the combustion chamber, the unavoidable presence of a quench layer of gas a few hundredths of a millimetre thick clinging to the walls of the combustion chamber, and quenching in crevices such as the clearance between the top land of the piston and the cylinder bore.

The Motor Vehicle, 13th edn; ISBN: 9780750644495
Copyright © 2000 Elsevier Ltd. All rights of reproduction, in any form, reserved.

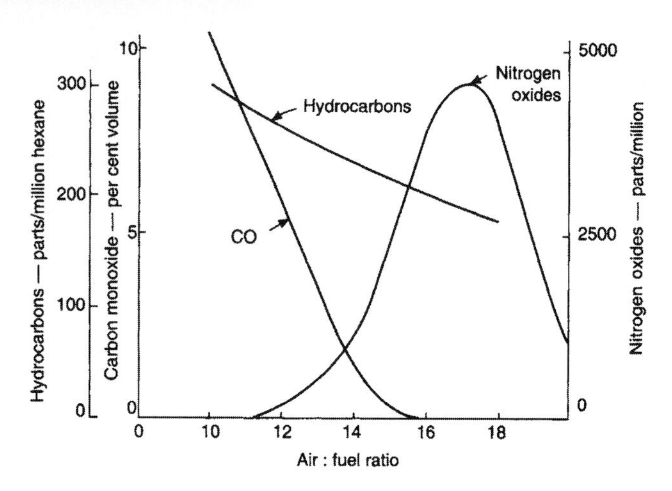

Fig. 3.1-1 Effect of air:fuel ratio on exhaust emissions.

3.1.1 Early measures for controlling emissions

A basic essential for spark ignition engine emission control is a carburettor or injection system capable of extreme accuracy in metering the fuel supply relative to the air entering the engine. Diesel engine emissions will be covered from Section 3.1.18 to thc cnd of thc chaptcr. All modem fuel injection systems have been developed from the outset specifically for accuracy of metering, for minimal emissions and best fuel economy. Irregular combustion must be avoided during idling and, on the overrun, the mixture must either be totally combustible or the fuel supply totally cut off. In the latter event a smooth return to normal combustion when the throttle is opened again is essential. Idling speeds are typically 750 rev/min with automatic and 550 rev/min with manual transmission.

A capsule sensitive to manifold depression could be used to retard the ignition in the slow-running condition, the manifold depression tapping being taken from a position immediately downstream from the edge of the throttle when it is closed. A centrifugal mechanism may retard the ignition from about 5° to 15°, while the depression capsule can further retard it by perhaps another 15°.

Sudden closing of the throttle, and the consequent rapid increase in the depression over the slow-running orifices, may draw off extra fuel that cannot be burned completely. To overcome this, a gulp valve has been developed for admitting extra air into the induction manifold in these circumstances.

Coolant thermostats have been set to open at higher temperatures for improved combustion in cold conditions. Also, thermostatically controlled air-intake valves deflect some air from over the exhaust manifold to mix with the remainder to maintain the overall temperature at about 40–45°C, thus assisting evaporation in cold weather.

In the 1970s much effort was devoted to the development of various stratified charge engines. By the 1980s, however, high-compression lean-burn systems had been the main practical outcome. With increasing pressures for fuel economy as a means of reducing CO_2 output, interest in stratified charge began to surface again in the early 1990s.

Positive crankcase ventilation (PCV), totally eliminated pollution originating from crankcase fumes, and at a modest cost (Section 3.1.3). By 1968, weakening the air:fuel ratio, retarding the spark timing, preheating the air passing into the engine intake, and, on some models, the installation of a pump to inject air for oxidising the HC and CO in the exhaust system reduced the total emissions about 39–41% by comparison with the 1960 cars. In absolute terms, emissions from General Motors (GM) cars, for example, had been reduced to 6.3 and 51 gm/mile, respectively, for HC and CO. However, there were still no controls on NO_x. New developments then being investigated included carbon canister systems for the temporary storage and subsequent combustion of evaporative emissions of fuel catalytic converters for controlling exhaust emissions.

3.1.2 Evolution of the US Federal test procedures

In 1970 the US Congress had adopted regulations requiring by 1975 a reduction of 90% on the then current emissions requirements. The Federal Environmental Protection Agency was formed and introduced a better method of sampling. Previously all the exhaust gas had been collected in one huge bag and then analysed. This had the disadvantage that it gave absolutely no indication of how the engine behaved under the different conditions of operation during the test; moreover, in some circumstances, some of the gases interacted in the bag, giving misleading results.

The new requirement was for collection into three bags, one for each main stage of the test (Fig. 3.1-2). The first, termed the *cold-transient stage*, comprised cycles 1 to 5 of the test, which represented the beginning of a journey starting from cold. Next came 13 cycles, representing the remainder of the journey with the engine warm and including some operation at high temperature. After this, the engine was shut down for 10 min, to represent a hot soak, before starting up again and repeating the first five cycles, for what is

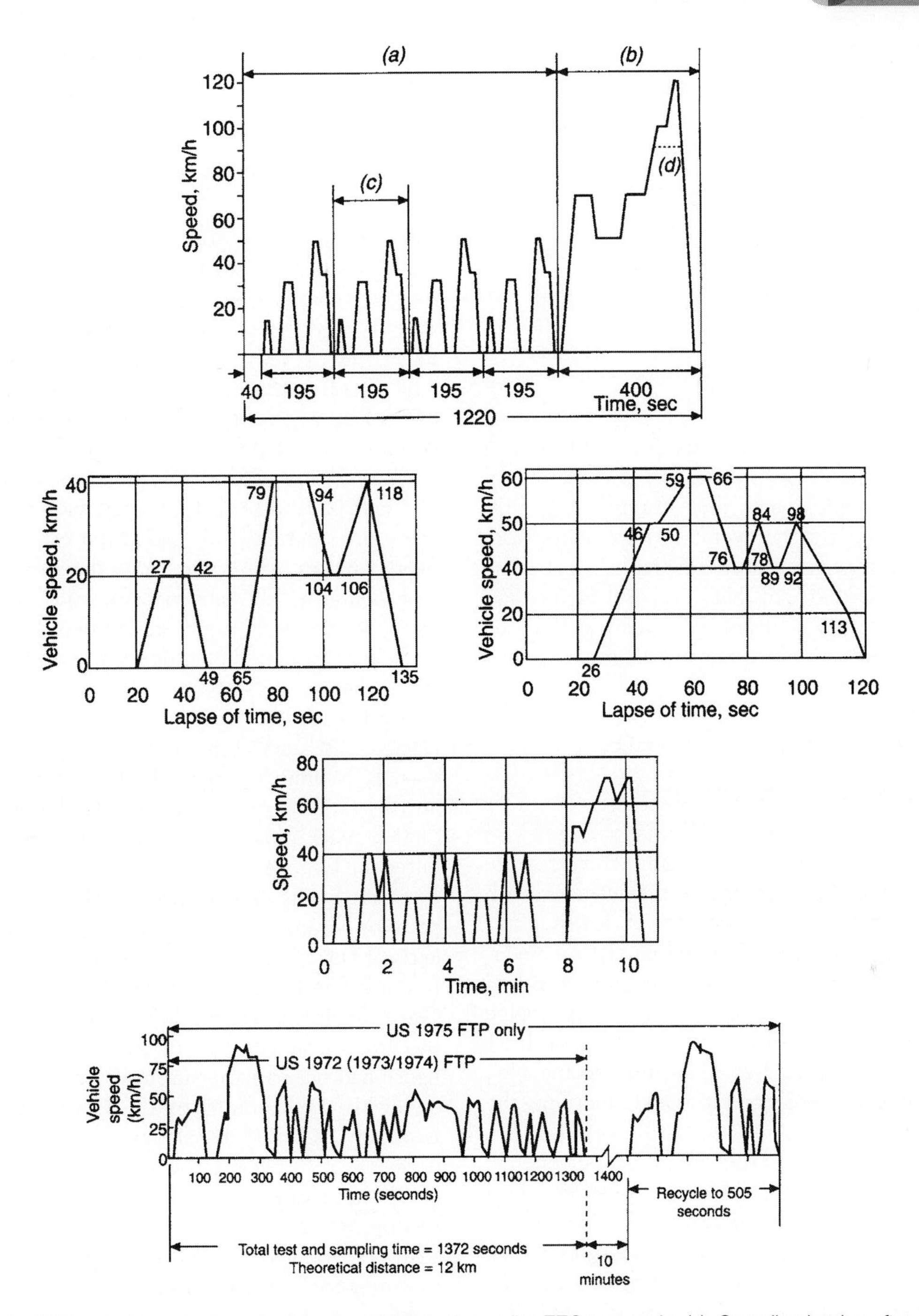

Fig. 3.1-2 *Top*, in 1992, a hot transient mode, (*b*) was added to the earlier EEC test cycle, (*a*). Sampling begins after 40 s. The lower limit, (*d*) is for vehicles incapable of attaining the higher maximum speed. Mid-left, the Japanese 10 mode cycle, which has to be run through 6 times. Mid-right, the Japanese 11 mode cycle to be run through 4 times. Below, Japanese hot transient test introduced in the early 1990s, comprising 24 s idle, the first three modes of its predecessor and a 15 mode high speed, or hot, test. *Bottom*, US Federal Test Program (FTP).

termed the *hot transient stage* of a journey started with a hot engine.

By applying weighting factors to alter the relative effects of the three bag analyses on the totals, the results of the test are easily adjusted to represent different characteristic types of operation. Obviously HC emissions are high for the period following starting from cold, while NO_x emissions are of little significance except under hot running conditions.

Conditions that encourage the generation of NO_x in the combustion chamber are principally temperatures above about 1350°C in the gas at high pressures and the length of dwell at those temperatures. Exhaust gas recirculation (EGR) was introduced to lower the

Table 3.1-1

Year	HC, CO and NO_x respectively gm/mile
1975	1.5, 15.0 and 3.1
1977–79	1.5, 15.0 and 2.0
1980	0.41, 7.0 and 2.0
1981 and beyond	0.41, 3.4 and 1.0

temperatures of combustion in cars for California in 1972, and extended nationwide in 1973, when legal limits, at 3.1 gm/mile, were first imposed on NO_x emissions. Subsequently, the overall requirements were progressively tightened as shown in Table 3.1-1. The 1981 regulations were so tight that, for diesel engines and innovative power units, a delay of 4 years had to be allowed for NO_x and up to 2 years on CO. Since then the regulations have been tightened periodically and clearly this process will continue.

3.1.3 Catalytic conversion

At this point while some other maunfacturers were promoting the lean-burn concept as the way forward, GM engineers, accepting the penalty of low Octane Number, opted for unleaded fuel and catalytic conversion for meeting regulations on both emissions and fuel economy, while avoiding adverse effects on engine durability. As a first step, all their car engines for 1971 were designed for fuel rated at 91 Motor Octane No., mainly by reducing compression ratios and modifying the valves and their seats.

They argued that unleaded fuel offers several benefits: first, the major source of particulate emissions, lead oxyhalide salts, is eliminated; secondly, there is a consequent reduction in combustion chamber deposits, which have the effect of thickening the boundary layers in the gas in the combustion chambers and this, by quenching them, encourages the formation of HC; thirdly, a further reduction in HC is obtained because of the additional oxidation that occurs in the exhaust system owing to the absence of lead additives and also because the lead salt deposits tend to cause deterioration of the NO_x control system by adversely affecting the flow characteristics of the EGR orifices; fourthly, maintenance of spark plugs, exhaust systems and the frequency of changing lubricating oil are all reduced by the elimination of the lead salts, as also of course is the generation of acids by the halide scavengers that have to be used with them; finally, because catalytic converters call for unleaded fuels controversy over the alleged toxic effects of lead salts in the environment was neatly side-stepped.

3.1.4 Two-way catalytic conversion

The emissions regulations for the 1975 model year required reductions of 87% in HC, 82% in CO and 24% in NO_x by comparison with 1960 levels. GM concluded that to meet these requirements while simultaneously improving not only economy but also driveability, both of which had deteriorated severely as a result of emission control by engine modifications, two-way catalytic converters were needed. The term two-way conversion implies oxidation of the two constituents in the exhaust, HC and CO, to form CO_2 and H_2O. Such a converter therefore contains only oxidation catalysts and, moreover, without oxygen in the exhaust cannot function. Consequently, the air–fuel mixture supplied to the engine must be at least stoichiometric or, better still, lean. Incidentally, the earlier practice of feeding air into the exhaust was intended primarily for burning the excess hydrocarbons during the first five cycles of the test after a cold start with engines equipped with carburettors. It is unnecessary with the accurate regulation of air:fuel ratio by computer-controlled injection.

If a spark plug were to fail, air–fuel mixture would enter the two-way catalytic converter and burn, seriously overheating the unit. Consequently, high-energy ignition systems became a necessary adjunct for the 1975 models, and copper-cored spark plugs were fitted to obviate cold fouling. The overall result of all these measures on the GM models was a reduction in fuel consumption of 28% by comparison with that of their 1974 cars. By 1977 this figure had been further improved by 48% and, by 1982, owing to the stimulus of the Corporate Average Fuel Economy (CAFE) legislation, by 103%. Incidentally, under the CAFE legislation the average fuel consumption of all cars marketed by each corporation in the USA had to improve in stages, from 18 mpg in 1978, by 1 mpg each year to 1980, then 2 mpg annually to 1983 and again by 1 mpg for 1984, and then to 0.5 mpg, to 27.5 mpg, for 1985.

3.1.5 The converter

Two-way catalytic converters comprise a container, usually of chromium stainless steel, and the catalysts and their supports, all enclosed in an aluminised steel heat shield (Fig. 3.1-3). Initially, the alumina pellet type of support for the catalyst was the most favoured because it had been developed to an advanced stage in other industries. The monolithic type (one-piece) did not go into regular production until 1977.

Fig. 3.1-3 The AC-Delco stainless steel housing for monolithic catalyst carriers is enclosed in an aluminised steel outer casing. Sandwiched between top halves of the outer and inner shells is heat insulation material. The perforations in the lower half of the outer shell, termed the grass shield, facilitate local cooling.

Either platinum (Pt) alone or platinum and palladium (P1) are used as catalysts. The cost of this noble metal content is of the order of 15 to 20 times that of the stainless steel shell that houses them, so other catalysts such as copper and chromium have been tried, with some success, but have not come into general use because they are prone to deterioration owing to attack by the sulphuric acid formed by combustion of impurities in the fuel. A typical two-way converter for an American car contains about 1.6 g of noble metals in the Pt:P1 ratio of 5:2.

3.1.6 Catalyst support

Considerable development effort has been devoted to monolithic catalyst supports in the form of one-piece extruded ceramic honeycomb structures having large surface areas on to which the noble metal catalysts are deposited. Gas flow paths through them are well defined and their mass is smaller than that of the pellet type, warming up more rapidly to their working temperature of about 550°C. In some applications, for ease of manufacture two such monoliths are installed in tandem in a single chamber (Fig. 3.1-4).

Pellet systems are, nevertheless, widely employed in the USA for trucks, where compactness is not an overriding requirement but durability under extremely adverse conditions is. The pellets are relatively insensitive to thermal stress because they can move to relieve it. Moreover, the hottest part of such a bed is near its centre, whereas that of a ceramic monolith is about 25 mm from its leading edge, accentuating thermal stress problems. Packaging for pellets, on the other hand, is more complex, so both assembly and servicing of the monolithic type are easier.

3.1.7 Metallic monoliths for catalytic converters

Another important aim is of course durability at both very high and rapidly changing temperatures. Ceramics do not satisfy all these requirements, so efforts were

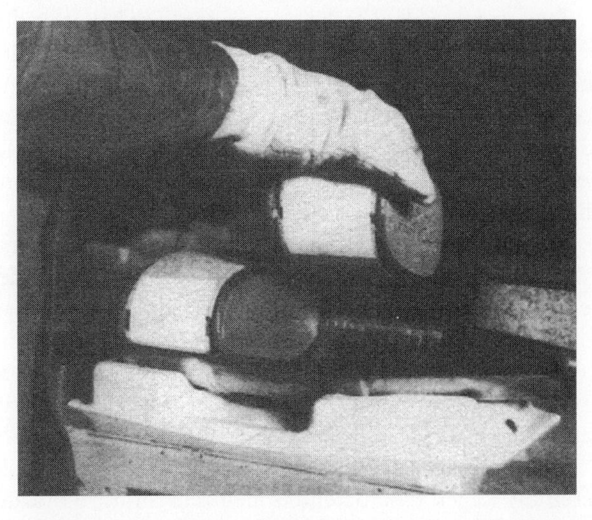

Fig. 3.1-4 Two monolithic catalyst carriers being assembled in series into their casing in the AC-Delco factory at Southampton.

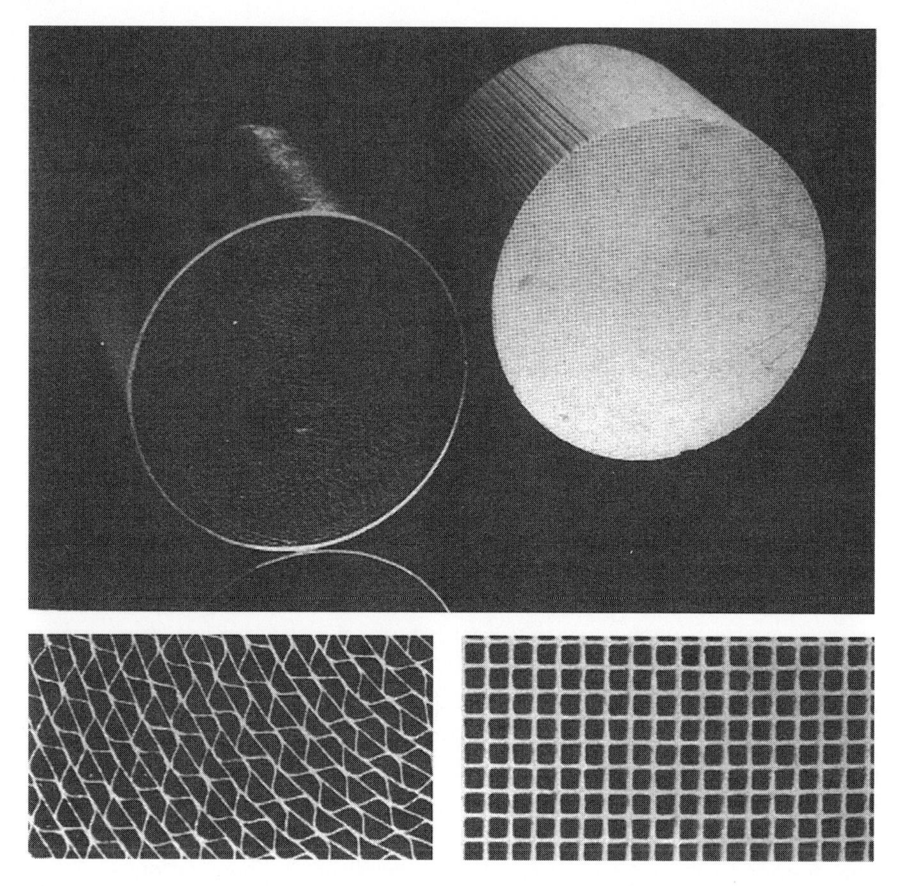

Fig. 3.1-5 Comparison between ceramic and metallic monoliths. The thermal stresses in monoliths comprising alternate strips of plain and corrugated metal foil in looped S-form are considerably lower than those that are simply spirally wound.

directed at producing acceptable metallic matrices. These offer advantages of compactness, minimum back-pressure in the exhaust system, rapid warm-up to the minimum effective operating temperature (widely termed the *light-off* temperature) which, for this type of monolith, is claimed to be about 250°C.

Two obstacles had to be overcome. First was the difficulty of obtaining adequate corrosion resistance with the very thin sections needed for both compactness and acceptably low back-pressure. Secondly, it was difficult to join the very thin sections while retaining the robustness necessary to withstand the severe thermal loading and fatigue.

By 1989, these problems had been solved by Emitec, a GKN-Unicardan company in Germany. They had developed a special stainless steel alloy, called Emicat, which they used in foil strips only 0.04 mm thick to construct the catalyst carriers in monolithic form. These are now made up into matrices comprising alternate plain and corrugated strips, wound in an S-form, as shown in Fig. 3.1-5. The matrices are inserted into steel casings and the whole assembly joined by a patented high-temperature brazing process. S-form matrices proved to be more durable than spirally wound cylindrical units.

Emicat is an Fe, 20% Cr, 5% Al, 0.05% Y alloy. Yttrium, chemical symbol Y, has a melting point of 1250°C. It is a metal but with a strong chemical resemblance to rare earths, with which it therefore is usually classified. Its oxide, Y_2O_3, forms on the surface of the foil and protects the substrate from further oxidation. At a content of 0.05% yttrium is very effective in enabling the alloy to withstand not only temperatures of up to 1100°C over long periods, but also the higher peaks that can be attained in catalytic converters in the event of a malfunction of the ignition system. Even better protection, however, can be had by increasing it up to 0.3%, though at higher cost.

The advantages obtained with the monoliths made of Emicat include: rapid warm-up; resistance to both thermal shock and rapid cyclical temperature changes up to well over 1300°C (both due to the good thermal conductivity of the material and low heat capacity of the assembly); minimal back-pressure, by virtue of the thin sections of the catalyst carrier foil (Fig. 3.1-5); compactness due to thinness of the sections and the absence of the mat needed around a ceramic monolith (to absorb its thermal expansion); large area of the catalyst exposed to the gas flow (owing to the high surface:volume ratio of

Table 3.1-2 Metal and ceramic monolith materials compared

Property	Metal	Ceramic
Wall thickness, mm	0.04	0.15–0.2
Cell density, cell/in^2	400	400
Clear cross section, %	91.6	67.1
Specific surface area, m^2/l	3.2	2.4
Thermal conductivity, W/m K	14–22	1–1.08
Heat capacity, kJ/kg K	0.5	1.05
Density, g/cm^3	7.4	2.2–2.7
Thermal expansion, $\Delta L/L$, 10^{-6} K	15	1

Note: Thicknesses and cross sections are of metals uncoated with catalyst.

the foil); and avoidance of local overheating, by virtue of both the compactness of the unit and the good thermal conductivity of the metal as compared with that of ceramic; and, finally, because the complete unit is directly welded into the exhaust system, the costs of assembling ceramic monoliths and their wire mesh or fibre mat elastic supports into their cans are avoided. The properties of the two types of converter are set out in Table 3.1-2.

3.1.8 Ford Exhaust Gas Ignition system for preheating catalysts

Generally, catalysts on ceramic monoliths do not become reasonably effective until they have attained a temperature of approximately 350°C, and are not fully so until a temperature of 450°C is attained. On average, two-thirds of all car journeys undertaken are less than 5 miles in length. Indeed, on a short journey, as much as 80% of the total emissions after starting with a cold engine are produced during the first 2 min, and the situation is even worse in cold climates and in urban conditions.

Ford have obtained catalyst light-off consistently in a few seconds by briefly burning a measured mixture of fuel and air in an afterburner just upstream of the catalyst. They have termed their system exhaust gas ignition (EGI).

Immediately after start-up from cold, three actions are initiated by the electronic control unit (ECU): the engine is run on a rich mixture; air is delivered to the afterburner by a pump which is electrically driven, so that it too can be controlled by the ECU; and sparks are continuously fired across the points of a plug situated in the afterburner. Because the mixture is rich, the exhaust gases contain not only unburned hydrocarbons and carbon monoxide, but also hydrogen. The hydrogen gas is highly inflammable, and therefore is utilised to light up the other constituents. Once alight, the mixture will continue to burn and generate the heat needed to bring the temperature of the catalyst up to its light-off value.

Hydrogen is one of the products of combustion of rich mixtures at high temperatures. The chemical processes, which are similar to those occurring during the production of water gas, are as follows:

$$CO + H_2O = CO_2 + H_2$$
$$CO_2 + H_2 = H_2O + CO$$

These two reactions alternate in the combustion chamber which is why, with the rich mixture, there is always some hydrogen in the exhaust, especially since some of the products of combustion are frozen by the cold walls of the chamber.

Alternative methods of expediting warm-up can be less satisfactory. For example, placing the catalytic converter close to the engine entails a risk of degradation of the catalyst due to overheating when the car is driven at high speed and load. The alternative of electric heating requires a current of about 500 A at 12 V, and therefore calls for a significant and costly uprating of the battery charging system.

3.1.9 Three-way conversion

By 1978 GM had developed a three-way converter, the term implying the conversion of a third component, namely the NO_x. Whereas two-way conversion is done in a single stage, three-way conversion calls for two stages. By 1980 it became necessary for meeting the stringent requirements for the control of NO_x in California and, by 1981, in the rest of the USA.

The additional catalytic bed contains rhodium (Rh) for reduction of the oxides of nitrogen. An outcome was an increase to about 3 g in the total noble metal content. In practice, with a 0.1% rich mixture, about 95% of the NO_x can be removed by such a catalytic converter. A reducing atmosphere is essential so the mixture must not be lean, and therefore the conversion of NO_x has to precede the oxidation of the HC and CO.

Oxygen released in the initial reduction process, in the Rh bed, immediately starts the second stage of the overall process while the exhaust gas is still in the first stage. The oxygen that remains unused then passes on into the Pt or Pt-P1 second stage of the converter, in a separate housing downstream of the first. Here extra air is supplied for completion of the oxidation. On the other hand, if what is termed a *dual-bed converter* is used, both stages are in a single housing, though in separate compartments between which is sandwiched a third chamber into which the extra air is pumped, to join the gas stream before it enters the Pt, or Pt-Pl, stage.

With three-way conversion, a closed-loop control system is essential, for regulating the supply of fuel accurately in relation to the mass air-flow into the engine. This entails installing an oxygen sensor in the exhaust, and an on-board microprocessor to exercise control, both to correct continuously for divergencies from the stoichiometric ratio and to ensure good driveability.

3.1.10 The electronic control system

In practice, the electronic control system has to be more complex than might be assumed from the preceding paragraph. When the engine is being cranked for starting it has to switch automatically from a closed- to an open-loop system, to provide a rich mixture. In this condition the air supply for the second converter bed is diverted to the exhaust manifold to oxidise the inevitable HC and CO content, thus avoiding a rapid rise in temperature in, and overloading of, the second stage of the converter. Owing to the low temperatures in the combustion chambers of the engine, NO_x production is minimal or even zero, so no conversion is required in the first stage.

During warm-up the mixture strength has to be modified for the transition from rich to stoichiometric mixture. However, to cater to heavy loading, such as acceleration uphill, it may again have to be enriched, perhaps with EGR in this condition to inhibit the formation of NO_x. The on-board microprocessor capabilities, therefore, must include control over idle speed, spark timing, EGR, purge of hydrocarbons from carbon canister vapour-traps, early evaporation of fuel by air intake heating, torque converter lock-up, and a fault-diagnosis system.

3.1.11 Warm-air intake systems

Apart from setting the coolant thermostat to open at higher temperatures to improve combustion in cold conditions, several manufacturers have introduced automatic control of the temperature of the air drawn into the carburettor. The GM system is built into a conventional air cleaner. There are two air valves, operated by vacuum-actuated diaphragm mechanisms, and controlled by a thermostat. One valve lets warm and the other cold air into the intake.

The thermostat, mounted on the air cleaner, senses the temperature inside, maintaining it at 40–45°C. This thermostat operates a two-way control valve directing either induction system depression or atmospheric pressure to the actuator, according to whether cool or warm air is required. The warm air is taken from a jacket around the exhaust manifold.

Ford have developed a similar system. In this case, however, the thermostat senses under-bonnet temperature, and provision is made, by means of a vacuum-actuated override, to enable maximum power output to be obtained during warm-up.

The Austin-Rover design is outstanding for its simplicity. It is a banjo-shaped pressed steel box assembly, the handle of which is represented by the air intake duct to the air cleaner. In both the upper and lower faces of the box is a large diameter port, and a flap valve is poised between them. This flap valve is mounted on a bimetal strip which, when hot, deflects to close one port and, when cold, to close the other. The latter port simply lets air at the ambient temperature into the intake, while the former is connected by a duct to a metal shroud over the exhaust manifold and therefore passes hot air into the intake. It follows that the temperature of the incoming mixture from both ports is regulated by its effect on the bimetal strip, which deflects the flap valve towards the hot or cold port, as necessary.

3.1.12 Evaporative emissions

The evaporative emissions are mostly hydrocarbons though, with some special fuels and those that have been modified to increase octane number, alcohols may also be present. In general, the vapour comes from four sources:

(1) Fuel tank venting system.

(2) Permeation through the walls of plastics tanks.

(3) The carburettor venting system.

(4) Through the crankcase breather.

Fumes from the fuel tank venting system are absorbed in carbon canisters which are periodically purged (Section 3.1.17). Permeation through the walls of plastics tanks is controlled by one of four methods. These are:

(1) Fluorine treatment.

(2) Sulphur trioxide treatment.

(3) Du Pont one-shot injection moulding (a laminar barrier treatment).

(4) Premier Fuel Systems method of lamination.

Plastics tanks are generally moulded by extrusion of what is termed a *parison* (a large-diameter tube), which is suspended in a female mould into which it is then blow-moulded radially outwards. The chemical treatments are applied internally, either in the parison or in the blow-moulding. With either procedure, problems arise owing to the toxicity of the barrier chemicals and in the disposal of the chemical waste.

Fig. 3.1-6 (*Left*) Sealed Housing for Evaporative Determination (SHED) installed alongside a chassis dynamometer (*right*) in a thermostatically controlled chamber at the Shell Thornton Research laboratories.

For the Du Pont laminar barrier technology, a modified one-shot extruder is used for producing the parison. It automatically injects into the high-density polyethylene (HDPE), which forms the walls of the tank, a barrier resin called Selar RB. This resin forms within the wall an impermeable layer of platelets, in the form of a layer about 4–5% of its total thickness. Full details have been published in *Automotive Fuels and Fuel Systems*, Vol. 1, by T.K. Garrett, Wiley.

Item (4) in the second list is a patented method of laminating a plastics tank, which first went into series production in 1994 for the Jaguar XJ 220. It is produced by vacuum moulding, and is designed to meet the Californian requirements which limit the evaporative emissions to 12 parts per million from the whole car in one hour, during the SHED test (Fig. 3.1-6). There are three laminations. The outer layer is a fabric impregnated with a high nitrile polyvinyl chloride (PVC), while the inner wall is of unreinforced high nitrile PVC. Sandwiched between them is the layer that forms the impermeable barrier. This is of fluorinated ethane propane (FTP, or Teflon), both faces of which are etched to facilitate bonding to the outer layers.

3.1.13 Crankcase emission control

About 55% of the hydrocarbon pollution is in the exhaust, crankcase emissions account for a further 25% and the fuel tank and carburettor evaporation makes up the other 20%. These figures, of course, vary slightly according to the ambient temperature. In general unburnt hydrocarbons from these two sources amount to no more than about 4–10% of the total pollutants.

Crankcase fumes are drawn into the induction manifold by a closed-circuit, positive ventilation system. One pipe is generally taken from the interior of the air filter to the rocker cover, and another from the crankcase to the induction manifold. Thus, air that has passed through a filter is drawn past the rocker gear into the crankcase and thence to the manifold, whence it is delivered into the cylinders, where any hydrocarbon fumes picked up from the crankcase are burnt.

There are three requirements for such a system: first, the flow must be restricted, to avoid upsetting the slow running condition; secondly, there must be some safeguard to prevent blow-back in the event of a backfire and, thirdly, the suction in the crankcase has to be limited. AC-Delco produce a valve for insertion in the suction line to meet these requirements. It comprises a spring-loaded disc valve in a cylindrical housing. When there is no suction – engine off, or backfire condition – the valve seats on a port at one end, completely closing it. With high depression in the manifold, slow running or overrun, the valve seats on a larger diameter port at the other end, and a limited flow passes through the holes which, because they are near its periphery, are covered when it seats on the smaller diameter port. Flow through the larger port is restricted by the valve stem projecting into it. In normal driving, the valve floats in equilibrium between the two seats, and air can pass through the clearance around its periphery as well as through the holes (Fig. 3.1-7).

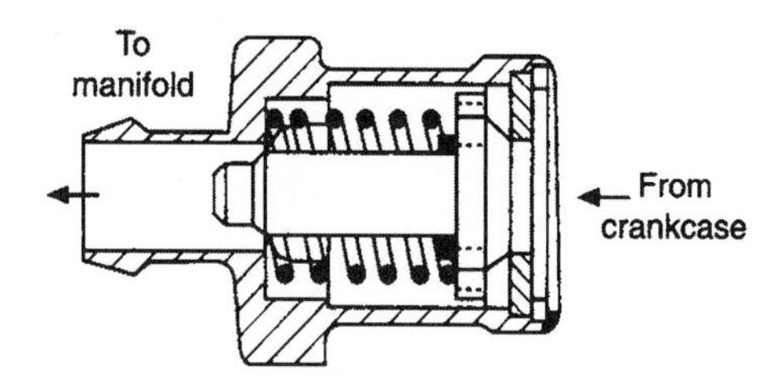

Fig. 3.1-7 AC-Delco crankcase ventilation control valve. With zero depression in the manifold, the valve seats on the right-hand orifice and with maximum depression on the left-hand one. In normal running it floats between the two.

3.1.14 Air injection and gulp valve

A version of the AC-Delco air injection system made in the UK comprises an engine-driven air pump delivering into an air manifold, and thence through a nozzle or nozzles into either the exhaust ports or, latterly, between the reducing and oxidising catalytic beds. At the junction between the delivery pipe and the manifold, there is a check valve, to prevent back-flow of exhaust gas into the pump and thence to the engine compartment: this could happen in the event of failure of the pump or its drive.

A pipe is also taken back from the check valve to a gulp valve. When the throttle is closed suddenly for rapid deceleration, this admits extra air to the induction tract to ensure that there is enough air passing into the cylinders to burn the consequent momentary surge of rich mixture and to prevent explosions in the exhaust system. The gulp valve comprises two chambers. Of these, the bottom one is divided into two by a spring-loaded flexible diaphragm, from the centre of which a stem projects upwards to actuate a valve in the upper chamber. Air from the pump enters at the top and, when the valve is open, passes out through a port in the side, to the induction manifold, (Fig. 3.1-8).

Manifold depression, introduced through a smaller port in the side of the housing, is passed into the chamber above the diaphragm. Therefore, a large transient increase in depression lifts the diaphragm, which subsequently returns under the influence of the spring. The

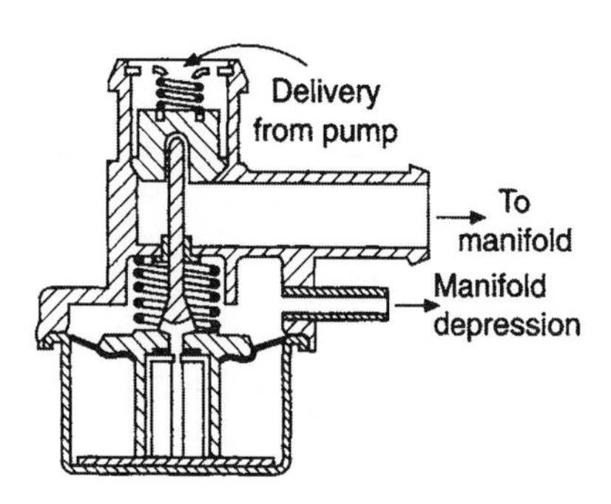

Fig. 3.1-8 AC-Delco gulp valve, for admitting extra air to the induction tract when the throttle is suddenly closed.

duration of the lift is determined by the size of a balance orifice in the centre of the diaphragm. A period of opening of about 1 and 4 seconds is generally adopted. Slower variation in depression, which occur in normal driving, are absorbed by flow through the balance orifice.

In the AC-Delco system, there is also a simple spring-loaded pressure relief valve, which can be equipped with a small air silencer. This limits the pump delivery pressure and prevents excess air being injected into the exhaust ports under high speed conditions of operation.

Another system is the Lucas-Smiths Man-Air-Ox. This also comprises an air pump, gulp valve, and check valve. As with the AC-Delco unit, there is an alternative to the gulp valve: a dump, or by-pass, valve can be used. This, instead of admitting a gulp of air to the induction tract, opens to atmosphere a delivery pipe from the pump. Thus, it effectively stops the supply through the nozzles, and prevents the possibility of explosions in the exhaust system. It, too, is suction controlled, but is a two-way valve. The sudden depression produced by the overrun condition lifts the valve, venting the air delivery to atmosphere, and at the same time seating the valve on the port through which the air was formerly being delivered to the manifold.

If air injection is employed for emission control it might also be utilised for cooling the exhaust valves. However, this is not favoured, because it entails the use of valves of special steel for avoiding oxidation. Afterburning in the exhaust gases may also be encouraged by the incorporation of ribs or other forms of hot spot in the exhaust manifold.

3.1.15 Air management valves

For engines equipped with catalytic converters something more than a simple gulp valve is needed. Consequently, comprehensive air management systems are employed to shut off completely the supply of air to the exhaust system, or to divert it from the catalytic converter into the exhaust manifold, during phases of operation in which a rich mixture is supplied to the engine. Otherwise, the excessive quantities of unburnt hydrocarbons passing through may either cause explosions in the exhaust system or overload, and thus overheat, the oxidising catalytic converters. Moreover, with carburetted engines the very high depression arising on sudden closure of the throttle tends to draw off excess fuel, which can similarly cause damage to the exhaust system, including the catalytic converter.

One of the simpler of the wide range of air management valves available is the Rochester products standardisation diverter valve (Fig. 3.1-9). During normal operation of the engine this valve is held open by a spring, so that the pump delivers air through it to the exhaust

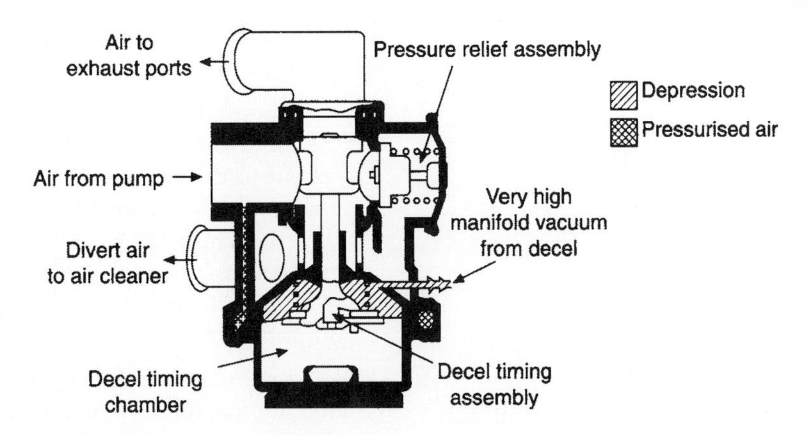

Fig. 3.1-9 Rochester standardised diverter valve, with high manifold depression applied to the diaphragm to divert air from its normal route, which is to the exhaust system, into the air cleaner.

manifold to burn off HC and CO. To counter the effects of over-fuelling due to the very high depression on sudden closure of the throttle, that depression is transmitted to the chamber above the diaphragm. This lifts the diaphragm against its return spring and thus closes a valve in the port through which the pump delivers air to the exhaust manifold, diverting the flow into the dirty side of the air cleaner, to silence the discharge. An orifice in the central plate of the diaphragm assembly allows the depression to bleed away from the diaphragm chamber, the duration of diversion of the air to the cleaner being therefore a function of the sizes of the bleed and depression signal orifices and the volumes of the upper and lower parts of the diaphragm chamber. At high rotational speeds of the engine the rate of supply of air exceeds requirements, so it is this time diverted by the opening of a pressure relief valve, to the air cleaner.

A variant of this system is the air intake control valve (Fig. 3.1-10). During sudden deceleration this valve, instead of closing the outlet to the exhaust manifold, simply opens a port to divert some of the air to the induction manifold instead of all of it into the air cleaner. This, by weakening the mixture, not only helps to reduce the HC and CO content of the exhaust gases but also reduces the rate of deceleration. At high engine speeds, again, air in excess of requirements is diverted by the pressure relief valve into the air cleaner.

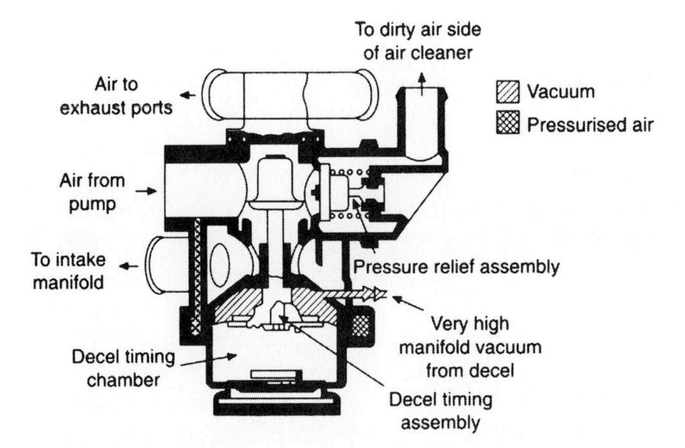

Fig. 3.1-10 Air intake control valve. Some of the air is also diverted into the induction manifold to reduce CO and HC output in the exhaust.

3.1.16 Some more complex valve arrangements

Where a computer-controlled catalytic system is installed, a solenoid-actuated air-switching valve is interposed between the output from the normal diverter valve and the delivery line to the exhaust manifold. When the system is operating with a cold engine, on the open-loop principle, the standardised diverter valve functions exactly as described in the previous section. However, when the engine is warm and the computer control switches to closed-loop operation, the solenoid-actuated valve diverts the output of air from the exhaust manifold to the second catalytic bed (Fig. 3.1-11), to enable the reducing catalyst in the first bed to function efficiently while, at the same time, to further the oxidation process in the second bed.

The functions of air management valves so far described have been:

(1) To direct air from the pump to the exhaust manifold during normal operation.

(2) To divert either all the air from the exhaust manifold to the air intake cleaner, or some of it to the induction manifold (the gulp-valve function) during deceleration.

(3) To divert the air excess to requirements, via a blow-off valve, back to the air cleaner.

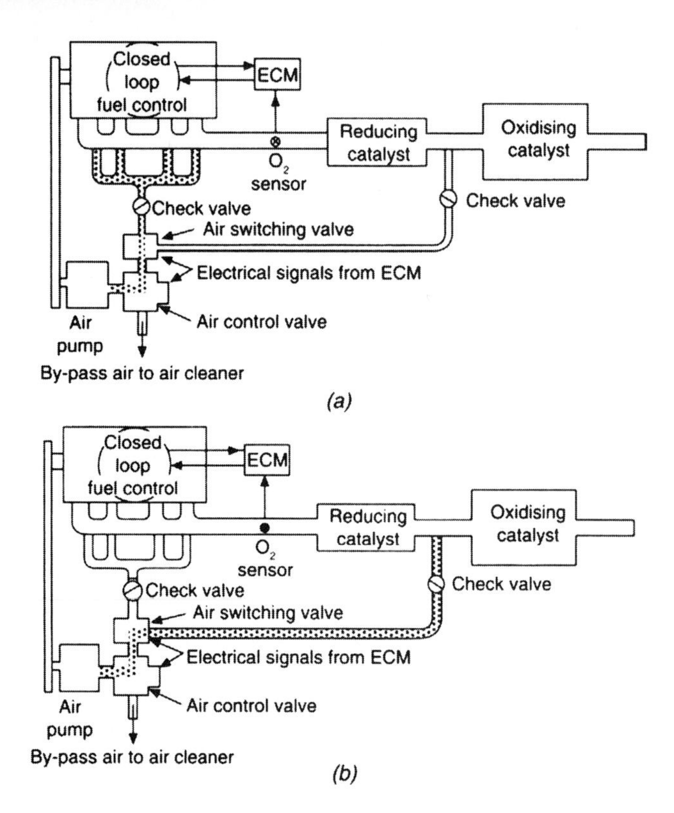

Fig. 3.1-11 System for switching air from the manifold (*a*) for open-loop operation to the oxidising catalytic converter, (*b*) for closed-loop operation.

(4) Where control is exercised by computer, the solenoid-actuated air switching valve diverts air from the exhaust to the second stage catalytic converter for closed-loop operation or back again to the exhaust manifold for open-loop operation.

There are also, however, even more complex air management valves. The additional functions they perform are:

(5) In response to a low or zero depression signal from the induction manifold, to divert the air pump output to the air cleaner when the engine is operating under heavy load from nearly to fully open throttle. This is to avoid the overheating of the catalytic converter that would occur if the excess hydrocarbon required to obtain maximum power output were to be oxidised in it.

(6) In response to a high depression signal from the induction manifold, to divert the air pump output to the air cleaner during normal road load conditions. This improves fuel consumption by decreasing exhaust back-pressure and, to a lesser degree, by reducing the power required to operate the pump. The air flow reverts to the exhaust manifold when the load on the engine increases and therefore the converter is needed to come back into operation to control the hydrocarbon emissions.

(7) By means of a solenoid-actuated valve, to enable air to be diverted by electronic control during any driving mode. With this arrangement, the diaphragm valve is actuated by high depression and a spring, in the normal way, except when the electronic control opens the solenoid valve to introduce air pump delivery pressure beneath the diaphragm to override its spring return mode.

There are other variants of these valves, but not enough space here to describe them all. Their modes of operation can be deduced from a study of Figs. 3.1-12 to 3.1-14. Which type of valve is selected depends

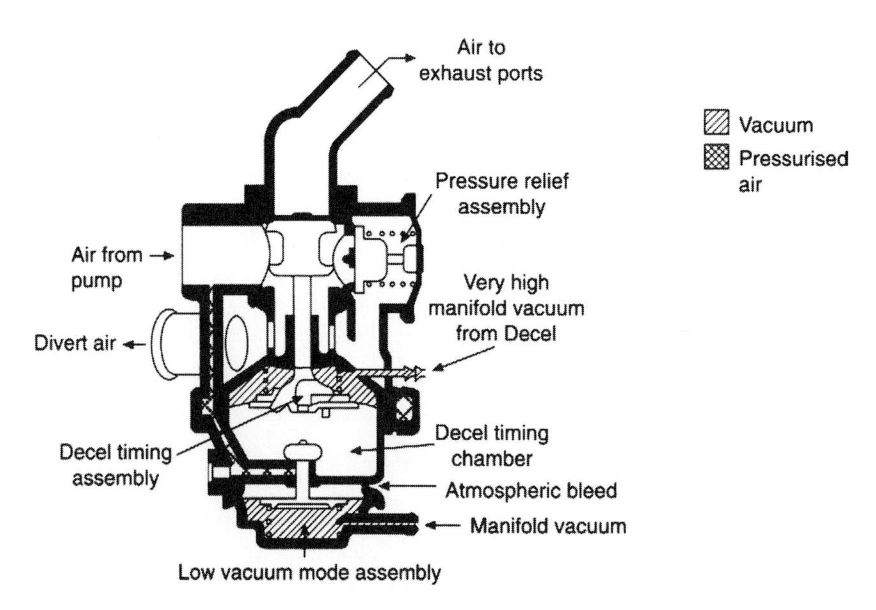

Fig. 3.1-12 The Rochester normal diverter valve diverts air from the oxidising converter to the air cleaner for operation at heavy load.

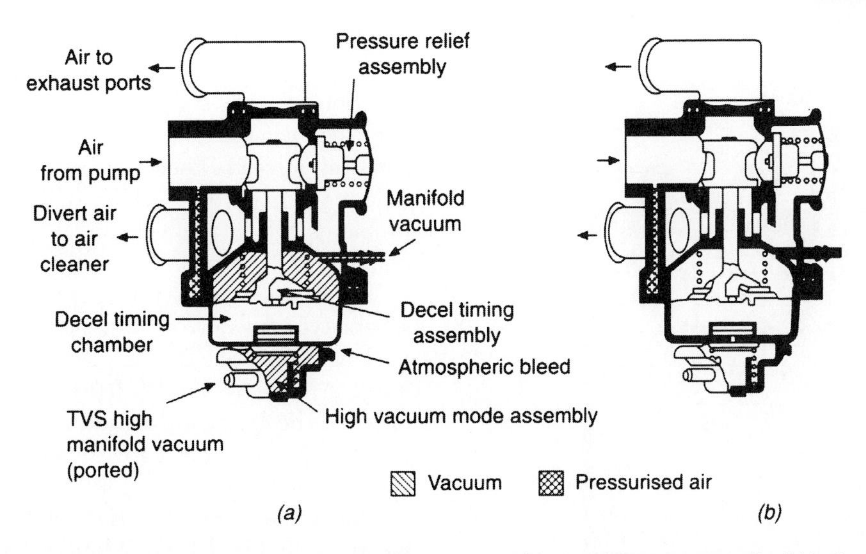

Fig. 3.1-13 This is what Rochester term their high vacuum air-control valve: (*a*) in the high vacuum mode; (*b*) in the low vacuum mode. On sudden closure of the throttle valve with the engine cold it operated in the same way as the standardised diverter valve in Fig. 3.1-8.

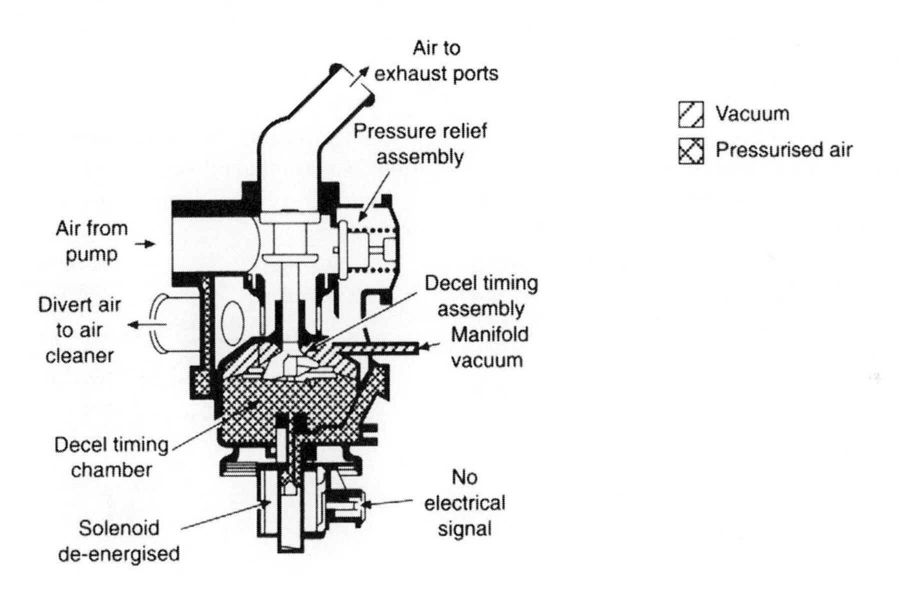

Fig. 3.1-14 The electrically actuated air-control valve, with solenoid energised.

mainly on the engine and emission-control system characteristics.

3.1.17 Vapour collection and canister purge systems

Carbon-filled canisters for storing vapours emitted from fuel tanks and float chambers are of either the open- or closed-bottom type. The filter in the base of the open-bottom type (Fig. 3.1-15) has to be changed regularly, generally every 24 months or 30 000 miles. On the other hand, the purge air for the closed-bottom type (Fig. 3.1-16) is drawn through the engine air intake before being conducted through a tube to the canister, so its filters do not need to be changed. A major advantage of the closed-bottom canister is that water or condensation cannot enter its base and, in cold weather, freeze in the filter, restricting the entry of purge air. Even so, canisters of the closed-bottom type, and also drawing air from atmosphere, are not unknown.

Either closed- or open-bottom canisters may have one inlet tube, through which both the fuel tank and carburettor float chamber fumes enter, and one purge tube. Alternatively, there may be separate inlet tubes for these two sources of fumes, making a total of three, as in Fig. 3.1-16. Generally, the carburettor float chamber vent tube, as in the illustration, is fitted with a diaphragm

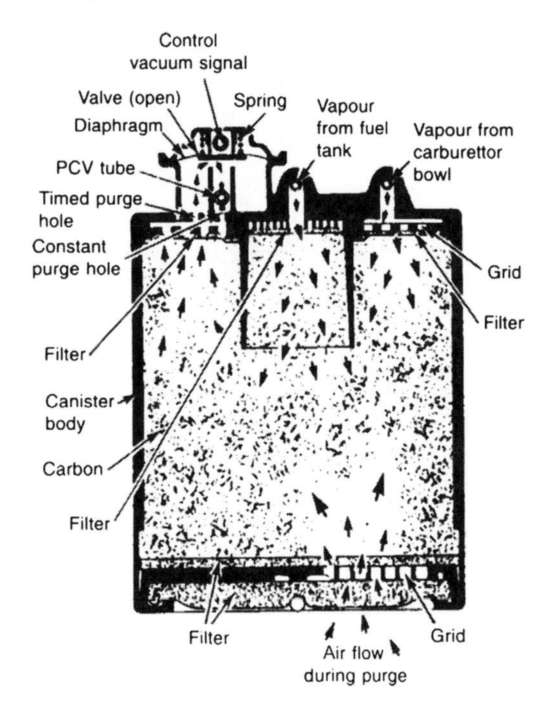

Fig. 3.1-15 A purge valve canister of the open-bottom type.

valve which is normally held open by a spring. However, when the engine is started, closure of the throttle uncovers a hole just downstream of it and transmits manifold depression to close the diaphragm valve, the venting of the float chamber then being effected directly to the carburettor air intake. Yet another option is to take the float chamber vent to a valve on the purge tube, for both the purging and diversion of the float chamber venting in the *timed* manner just described.

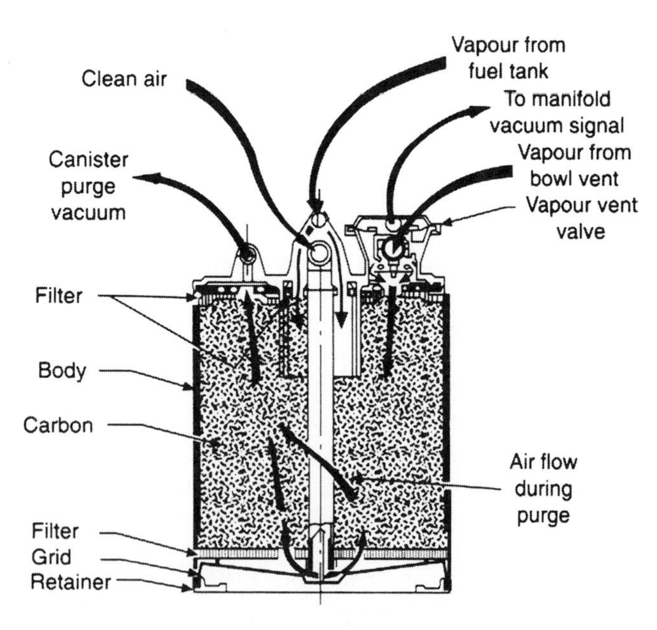

Fig. 3.1-16 Canisters of the closed-bottom type will not let water in at the base, where it could freeze and inhibit the purge action.

There may or may not be a central cylindrical or conical screen around the inlet from the tank vent. The function of such a screen is to ensure that the fumes entering from the tank are initially spread into the carbon particles, instead of being short circuited directly into the induction manifold, which could cause driveability problems when the vehicle is being operated slowly under very light load.

The canister purge tube may be fitted with either of several types of valve as alternatives to that in Fig. 3.1-15. Purging takes place through both the timed and constant purge orifices. When the depression over the diaphragm falls and the spring lowers the valve on its seat again, the flow continues, but at a greatly reduced rate, through the constant purge orifice.

One of the alternatives is the Rochester Products valve illustrated in Fig. 3.1-17. When the engine is not running, the spring holds the valve open, allowing the carburettor float chamber to vent into the canister. As soon as it is started, the *timed* manifold depression seats the valve so that the float chamber is then vented internally through ducts in the carburettor body to the air intake.

The valve in Fig. 3.1-18 performs two functions: it acts as both a carburettor vent valve, as described in the previous paragraph, and a purge valve, as in Fig. 3.1-15. When

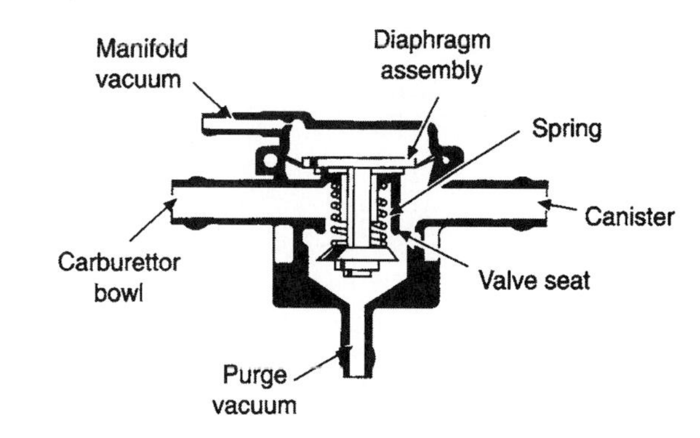

Fig. 3.1-17 The Rochester type 1 canister control valve functions as a simple vapour vent valve.

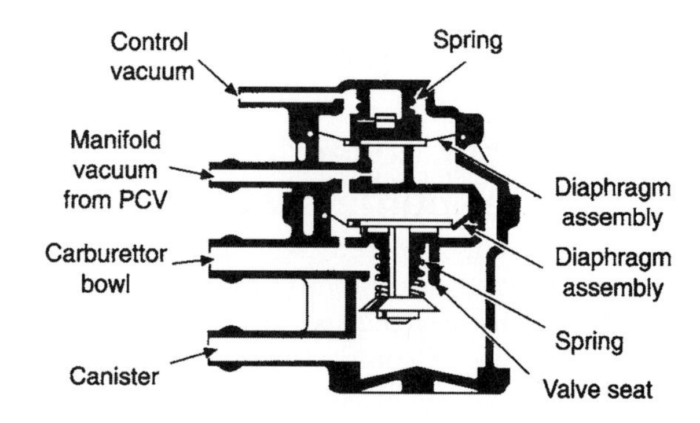

Fig. 3.1-18 Canister control valve type 2 performs the functions not only of the vapour vent valve but also of the purge valve as in Fig. 3.1-15.

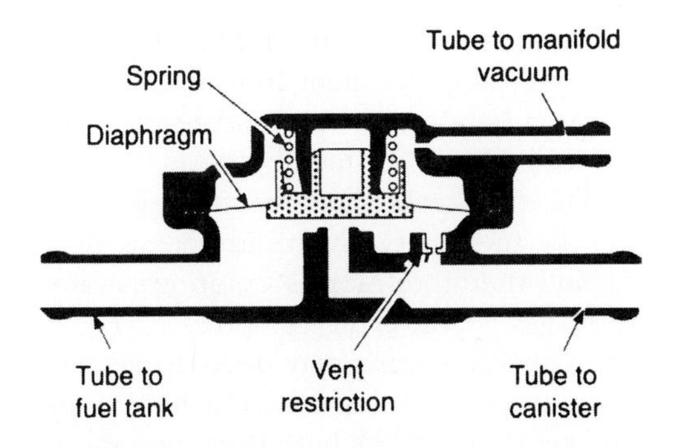

Fig. 3.1-19 Fuel tank pressure control valve. When the engine is not running, the valve is closed by the spring and the tank is then vented only through the restriction.

the engine is started, the lower diaphragm is lifted by the purge control system depression and this closes the valve serving the float chamber venting system. As the throttle is opened further, the depression signal, again *timed* by the passage of the throttle valve over a hole in the wall of the manifold, lifts the upper diaphragm valve to open it and purge the canister.

When a car is parked for long periods in hot sunshine there could be a tendency for large volumes of vapour to pass from the fuel tank into the canister. This would call for the use of an otherwise unnecessarily large capacity canister so, to avoid this, a fuel tank pressure control valve (Fig. 3.1-19) may be fitted. This is a diaphragm valve the port of which is normally closed by a spring. To allow for variations in volume of air and vapour in the tank with changes in ambient temperature, the tank is then vented through the restricted orifice alongside the port. Thus, when the engine is not running the major part of the fumes tends to be kept in the tank. When the engine is started the manifold depression opens the valve port, to vent the fuel tank to the canister, which of course will be cleared when the purge valve opens.

Yet another method of controlling the purge is available if an electronic control module is installed. A solenoid-controlled valve can be used and its opening and closing regulated by the computer.

3.1.18 Diesel engine emissions

Diesel and spark ignition engines produce the same emission. On the other hand, owing to the low volatility of diesel fuel relative to that of gasoline and the fact that carburettors are not employed, evaporative emissions are not so significant. Crankcase emissions, too, are of less importance, since only pure air is compressed in the cylinder and blow-by constitutes only a minute proportion of the total combustion gases produced during the expansion stroke.

Sulphur, which plays a major part in the production of particulates and smoke emissions is present in larger proportions in diesel fuel than in petrol. This is one of the reasons why the combustion of diesel fuel produces between 5 and 10 times more solid particles than that of petrol.

Because diesel power output is governed by regulating the supply of fuel without throttling the air supply, there is excess air and therefore virtually zero CO in the exhaust under normal cruising conditions. Reduction of NO_x on the other hand can be done only in an oxygen-free atmosphere, so a three-way catalytic converter is impracticable.

As a diesel engine is opened up towards maximum power and torque, NO_x output increases because of the higher combustion temperatures and pressures. At the same time HC, CO and sooty particulates also increase. However, because of the relatively low volatility of diesel fuel and the extremely short time available for evaporation, problems arise also when the engine is very cold. In the latter circumstances, under idling or light load conditions, the situation is aggravated by the fact that the minute volumes of fuel injected per stroke are not so well atomised as when larger volumes are delivered through the injector holes.

3.1.19 Reduction of emissions: conflicting requirements

Measures taken to reduce NO_x tend to increase the quantity of particulates and HC in the exhaust (Fig. 3.1-20). This is primarily because, while NO_x is reduced by lowering the combustion temperature, both soot and HC are burned off by increasing it. In consequence, some of the regulations introduced in Europe

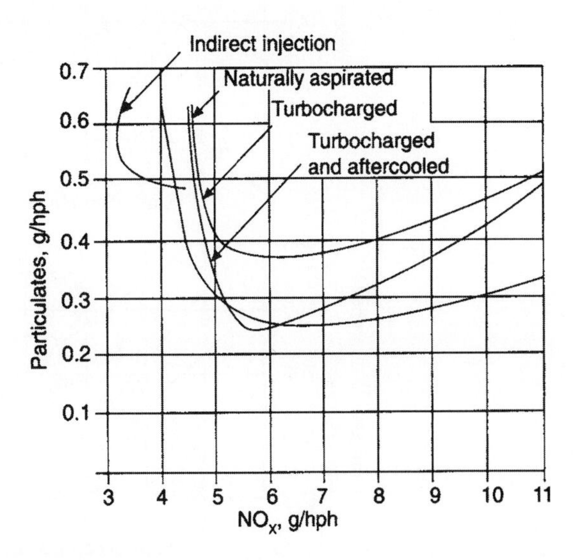

Fig. 3.1-20 Tests done in the late 1980s to demonstrate how NO_x increases when measures are taken to reduce particulates in different types of engine.

have placed limits on the total output of both NO_x and HC, instead of on each separately, leaving manufacturers free to obtain the best compromise between the two.

The problem of emission control, however, is not so severe as might be inferred from the last paragraph. Both NO_x output and heat to exhaust become significant only as maximum torque and power are approached. At lighter loads, the gases tend to be cooled because of both their excess air content and the large expansion ratio of the diesel engine. Since the proportion of excess air falls as the load increases, oxidising catalysts can be used without risk of overheating, even at maximum power output.

Fuel blending and quality have a profound effect on emissions. Since fuel properties and qualities are interrelated, it is generally unsatisfactory to vary one property unilaterally. Indeed, efforts to reduce one exhaust pollutant can increase others and adversely affect other properties.

3.1.20 Oxides of nitrogen, NO_x

To understand the effects of fuel properties on NO_x output certain basic facts must be borne in mind. First, it depends not only on the peak temperature of combustion but also on the rate of rise and fall to and from it. Secondly, the combustion temperatures depend on primarily the quantity and, to a lesser degree, the cetane number of the fuel injected.

Increasing the cetane number reduces the delay period, so the fuel starts to bum earlier, so higher temperatures and therefore more NO_x are generated while the burning gas is still being compressed (Fig. 3.1-21). However, a smaller quantity of fuel is injected before combustion begins, and this, by reducing the amount of fuel burning at around TDC, reduces the peak combustion temperature. The net result of the two effects is relatively little or even no change in NO_x output. An interesting feature in Fig. 3.1-22 is the enormous difference between the NO_x outputs from direct and indirect injection systems.

The popular concept that increasing fuel volatility reduces NO_x is an illusion: what happens in reality is that the weight of fuel injected is reduced, and the engine is therefore de-rated. Consequently, combustion temperatures are lowered. This is explained in more detail in Section 3.1.26, in connection with black smoke.

In the early 1990s, the overall output of NO_x from the diesel engine was, on average, between 5 and 10 times that of an equivalent gasoline power unit with a catalytic

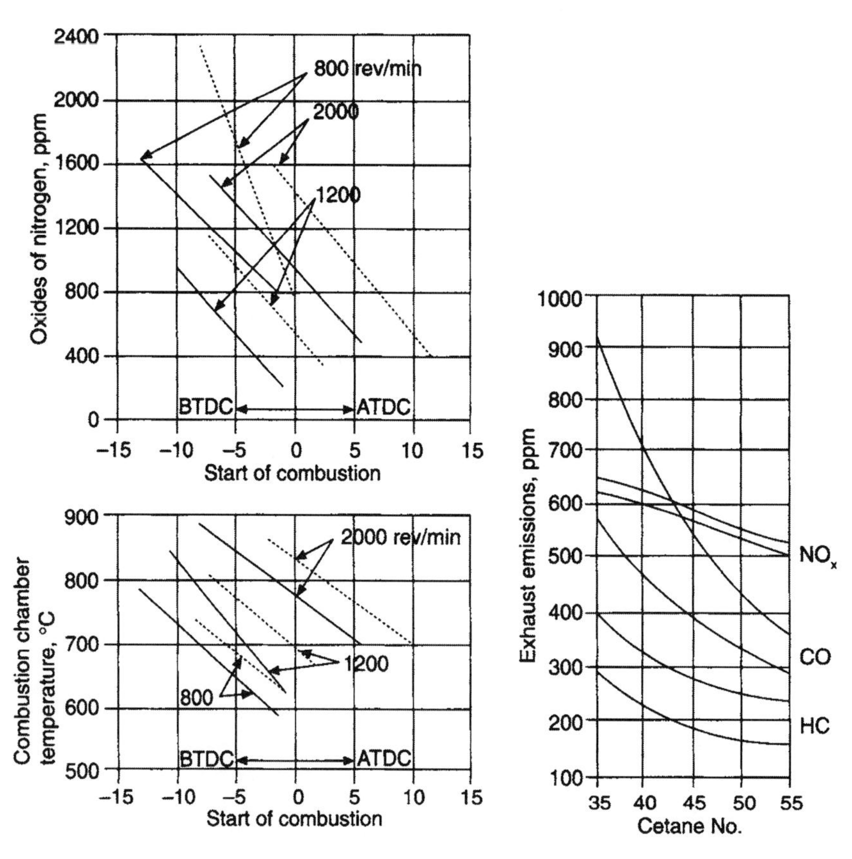

Fig. 3.1-21 (*Left*) Tests by BP showing how injection timing influences combustion and therefore NO_x output; (*right*) the influence of cetane number on the principal emissions.

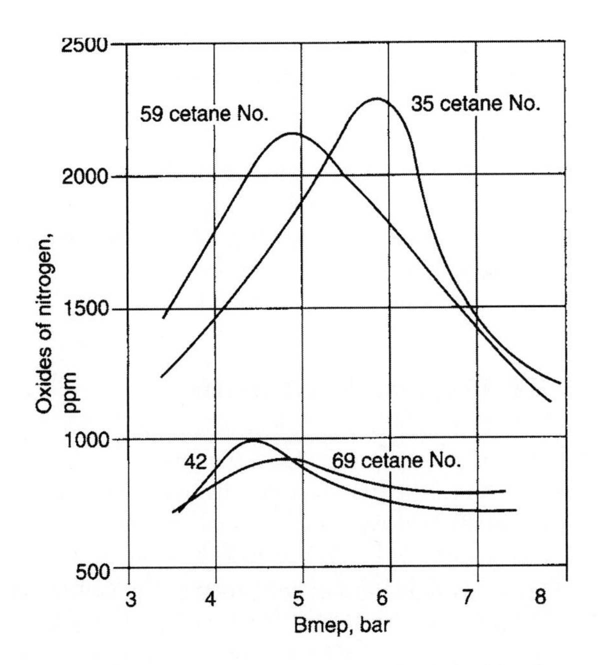

Fig. 3.1-22 NO_x emissions with direct and indirect injection.

converter, but this differential will be reduced as diesel combustion a control techniques improve. Efforts are being made to develop catalysts suitable for diesel application, but at the time of writing no satisfactory solution has been found.

Unfortunately, most of the current conventional methods of reducing NO_x also impair efficiency and therefore increase fuel consumption and therefore the output of CO_2. The relationship between NO_x output and fuel consumption is illustrated in Fig. 3.1-23. In general, NO_x tends to form most readily in fuel-lean zones around the injection spray.

EGR displaces oxygen that otherwise would be available for combustion and thus reduces the maximum temperature. However, it also heats the incoming charge, reduces power output, causes both corrosion and wear, and leads to smoke emission at high loads. For these reasons it has to be confined to operation at moderate loads. Electronic control of EGR is therefore desirable. Fortunately, heavy commercial vehicles are driven most of the time in the economical cruising range, maximum power and torque being needed mostly for brief periods.

Reduction of the rate of swirl is another way of reducing the output of NO_x. It increases the time required for the fuel to mix with the air, and therefore reduces the concentration of oxygen around the fuel droplets. Consequently, the temperature of combustion does not rise to such a high peak. Again, however, it also reduces thermal efficiency. Moreover, unless measures, such as increasing the number of holes in the injector nozzle and reducing their diameter, are taken to shorten the lengths of the sprays, more fuel tends to be deposited on the combustion chamber walls.

Delaying the start of injection has the effect of reducing peak temperatures, and therefore NO_x. This is because the combustion process builds up to its peak later in the cycle, when the piston is on its downward stroke and the gas is therefore being cooled by expansion. However, to get a full charge of fuel into the cylinder in the time remaining for it to be completely burnt, higher injection pressures are needed. Therefore, to avoid increasing the proportion of fuel sprayed on to the combustion chamber walls, the holes in the injector must again be smaller in diameter and larger in number.

Turbocharging increases the temperature of combustion by increasing both the temperature and quantity of air entering the cylinder. After-cooling, however, can help by removing the heat generated by both compression of the gas and conduction from the turbine. It also increases the density of the charge, and therefore thermal efficiency and power output. The net outcome of turbocharging with charge cooling, therefore, is generally an increase or, at worst, no reduction in thermal efficiency.

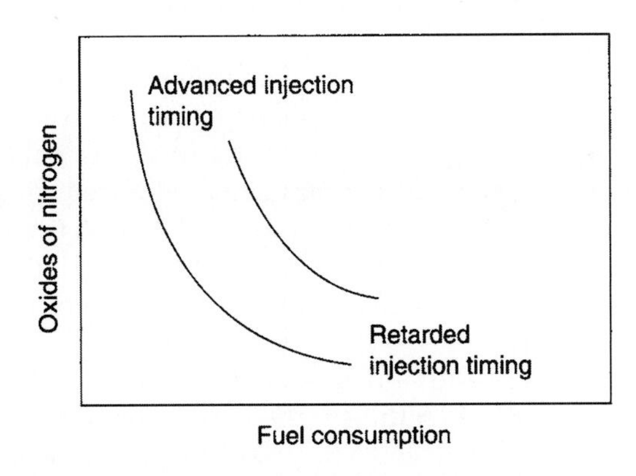

Fig. 3.1-23 Relationship between fuel consumption and NO_x emissions with (*left*) and without (*right*) charge cooling.

3.1.21 Unburnt hydrocarbons

Hydrocarbons in the exhaust are the principal cause of the unpleasant smell of a diesel engine, though the lubricating oil also makes a small contribution. There are three main reasons for this. First, at low temperatures and light loads, the mixture may be too lean for efficient burning so the precombustion processes during the ignition delay period are partially inhibited. This is why some of the mixture subsequently fails to burn.

Secondly, because of the low volatility of diesel fuel relative to petrol, and the short period of time available for it to evaporate before combustion begins, HCs are generated during starting and warming up from cold. In these circumstances, fuel droplets, together with water vapour produced by the burning of the hydrogen content of the remainder of the fuel, issue from the cold exhaust pipe in the form of what is generally termed *white smoke*, but which is in fact largely a mixture of fuel and water vapours. At about 10% load and rated speed, both HC and CO output are especially sensitive to fuel quality and, in particular, cetane number.

Thirdly, after cold starting and during warm-up, a higher than normal proportion of the injected fuel, failing to evaporate, is deposited on the combustion chamber walls. This further reduces the rate of evaporation of the fuel, so that it fails to be ignited before the contents of the chamber have been cooled, by expansion of the gases, to a level such that ignition can no longer occur. Similarly, the cooling effect of the expansion stroke when the engine is operating at or near full load can quench combustion in fuel-rich zones of the mixture. This is the fourth potential cause of HC emissions.

Unburnt HCs tend to become a problem also at maximum power output, owing to the difficulty under these conditions of providing enough oxygen to burn all the fuel. As fuel delivery is increased, a critical limit is reached above which first the CO and then the HC output rise steeply. Injection systems are normally set so that fuelling does not rise up to this limit, though the CO can be removed subsequently by a catalytic converter in the exhaust system.

Another potential cause of HCs is the fuel contained in the volume between the pintle needle seat and the spray hole or holes (the *sac volume*). After the injector needle has seated and combustion has ceased, some of the trapped fuel may evaporate into the cylinder. Finally, the crevice areas, for example between the piston and cylinder walls above the top ring, also contain unburnt or quenched fractions of semi-burnt mixture, Expanding under the influence of the high temperatures due to combustion and falling pressures during the expansion stroke, and forced out by the motions of the piston and rings, these vapours and gases find their way into the exhaust.

In general, therefore, the engine designer can reduce HC emissions in three ways. One is by increasing the compression ratio; secondly, the specific loading can be increased by installing a smaller, more highly rated, engine for a given type of operation; and, thirdly, by increasing the rate of swirl both to evaporate the fuel more rapidly and to bring more oxygen into intimate contact with it.

Reduction of lubricating oil consumption is another important aim as regards not only control of HCs but also, and more importantly, particulate emissions. Whereas oil consumption at a rate of 1% of that of fuel was, until the mid-1980s, been regarded as the norm, the aim now is generally nearer to 0.2%. Using a lubricant containing a low proportion of volatile constituents helps too.

Avoidance of cylinder-bore distortion can play a significant part in the reduction of oil consumption. The piston rings tend to ride clear over and therefore fail to sweep the oil out of the pools that collect in the hollows formed by distortion of the bores, thus reducing the effectiveness of oil control. Other means of reducing contamination by lubricating oil include improving the sealing around the inlet valve stems, the use of piston rings designed to exercise better control over the thickness of the oil film on the cylinder walls and, if the engine is turbocharged, reduction of leakage of oil from the turbocharger bearings into the incoming air.

3.1.22 Carbon monoxide

Even at maximum power output, there is as much as 38% of excess air in the combustion chamber. However, although carbon monoxide (CO) should not be formed, it may in fact be found in small quantities in the exhaust. The reason is partly that, in local areas of the combustion chamber, most of the oxygen has been consumed before injection ceases and, therefore, fuel injected into these areas cannot burn completely to CO_2.

3.1.23 Particulates

Regulations define particulates as anything that is retained, at an exhaust gas temperature of 52°C, by a filter having certain specified properties. They therefore include liquids as well as solids. Particle sizes range from 0.01 to 10 μm, the majority being well under 1.0 μm. While black smoke comprises mainly carbon, the heavier particulates comprise ash and other substances, some combined with carbon. The proportions, however, depend on types of engine, fuel and lubricant.

Measures appropriate for reducing the fuel and oil content of the particulates are the same as those already mentioned in connection with HC emissions (Section 3.1.21). The overall quantity of particulates can be reduced by increasing the injection pressure and reducing the size of the injector holes, to atomise the fuel better. This however, tends to increase the NO_x content. Increasing the combustion temperature helps to burn the loose soot deposited on the combustion chamber walls. Various measures have been taken to increase the temperature of these particulates, though mostly only experimentally. They include insulation by introducing an air gap, or some other form of thermal barrier, between the chamber and the remainder of the piston, and the incorporation of ceramic combustion chambers in the piston crowns.

Reduction of the sulphur content of the fuel also reduces particulates. Although the proportion of sulphate + water is shown in Table 3.1-3 as being only 2% of the total, if the insoluble sulphur compounds are added, this total becomes more like 25%. Because most measures

Table 3.1-3 Analyse, expressed in percentages, of particulates from different types of diesel engine

Engine type	Fuel-derived HC	Oil-derived HC	Insoluble ash	Sulphates + water
Ford 1.8 DI	15	13	70	2
Ford 1.8 IDI	48	20	30	2
Average DI HD turbocharged after-cooled engine	14	7	25	4

Note: Horrocks (Ford Motor Co.) differentiated between the carbon and other ash (at 41% and 13%, respectively), making the total 44%.

taken to reduce NO_x increase particulates, the most appropriate solution is to use fuels of high quality, primarily having low sulphur and aromatic contents and high cetane number. The relationship between fuel quality and particulate and NO_x output has been demonstrated by Volvo (Fig. 3.1-24).

A small proportion of the particulates is ash, most of which comes from burning the lubricating oil. Reduction of sulphur in the fuel reduces the need for including, in lubricating oils, additives that neutralise the acid products of combustion; these additives that are responsible for a significant proportion of the ash content.

Incidentally, sulphur compounds can also reduce the efficiency of catalytic converters for the oxidation of CO and HC. In so doing, they form hydrogen disulphide, which accounts for the unpleasant smell of the exhaust when fuels with high sulphur content are burnt in an engine having an exhaust system equipped with a catalytic converter.

An ingenious method of reducing visible particulates emitted from a turbocharged engine in a bus has been investigated by MAN. Compressed air from the vehicle braking system is injected in a controlled manner into the combustion chambers to bum off the carbon. This increases the exhaust gas energy content, and therefore compensates for turbocharger performance falling off under light load, including initially during acceleration and while gear changes are being made.

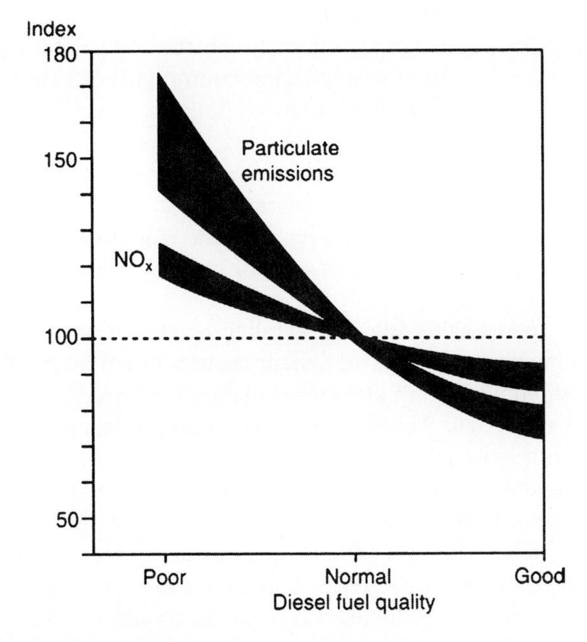

Fig. 3.1-24 Relationship between particulate emissions and fuel quality, as established by Volvo.

3.1.24 Particle traps

Basically, particle traps are filters, mostly catalytically coated to facilitate their regeneration (burning off of the particles that have collected). Many of the filters are extruded ceramic honeycomb type monoliths, though some are foamed ceramic tubes. The honeycomb ceramic monoliths generally differ from those used for catalytic conversion of the exhaust in gasoline engines, in that the passages through the honeycomb section are sealed alternately along their lengths with ceramic plugs (Fig. 3.1-25), so that the gas passes through their porous walls. These walls may be less than 0.5 mm thick.

Ceramic fibre wound on to perforated stainless steel tubes has also been used. These are sometimes termed candle or deep *bed*-type filters. Their pores are larger than those of the honeycomb type, and their wall thicknesses greater.

If catalyst assisted, regeneration is done mostly at temperatures around 600°C, and if not, at 900°C. Alternatively, special catalysts may be used to lower further the ignition temperature of the particles. These are mostly platinum and palladium, which are useful also for burning HCs and CO. Sulphate deposition adversely affects the platinum catalyst, but the formation of sulphates is largely inhibited by palladium. Copper has been used too, because it reduces the carbon particle ignition temperature to about 350°C, but it suffers the

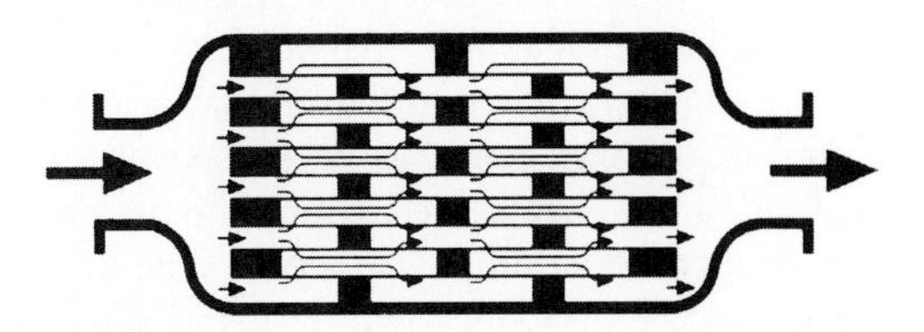

Fig. 3.1-25 Diagram showing how the gas flow is diverted in a particulate trap.

disadvantage of a limited life. Better results have been obtained with alloys of copper with silver, vanadium and titanium.

It is important to regenerate the filters reliably at the appropriate intervals, otherwise they will become overloaded with soot, which can then ignite and burn uncontrollably, developing excessive heat and destroying the filter. Even with normal burn off, however, local temperatures near the centre of the beds can be as high as 1200°C.

Some particulate filters rely solely on excess air in the exhaust for burning off the particulates. Others have extra air injected into them, generally at timed intervals, either from the brake system or by a compressor or blower. With air injection, burn-off can continue when the engine is operating at or near maximum power and therefore with little excess oxygen in the exhaust. At lighter engine loads and therefore with excess air present, once ignition has been initiated, combustion can continue without the burner.

With some systems the engine must be stopped while regeneration is in progress but with others, which are regulated by electronic control, either the paticulates are burnt off intermittently while the vehicle is running or there are two filters in parallel, with deflector valves directing the exhaust gases first through one filter, while the other is being regenerated, and then *vice versa*. It is also possible to program the electronic controller to bring both filters into operation simultaneously as maximum power and torque output is approached, to cater for the increased exhaust flow under these conditions. This enables smaller filters to be used.

Claimed efficiencies of particulate removal range from about 70–95%, but some of these claims are suspect. However, since the carbon content is easiest to burn off, up to 99% of it can be removed. Currently, the useful lives of the catalytically coated filters vary enormously from type to type.

Since we are in the early stages of development, the situation is in a continuous state of flux; some systems no doubt will soon fall by the wayside, while others will be developed further. Clearly, however, if particulates can be removed in the combustion chamber by using better quality fuel or taking measures to improve combustion, it will be to the advantage of all concerned, except of course the manufacturers of particulate filters!

All filter systems are bulky and present burn-off problems, including high thermal loading. None produced so far can be regarded as entirely satisfactory and, at the time of writing, none has been used on cars. Interestingly, VW have been investigating the possibility of using an iron-based additive to reduce the oxidation temperature of the deposits down to that of the exhaust gas. A Coming monolithic ceramic filter is used, and the additive, developed jointly with Pluto GmbH and Veba Oel, is carded under pressure in a special container. An electronic control releases additive automatically, as needed, into the fuel supply line, so that regeneration proceeds at temperatures as low as 200°C.

Of the systems currently available for commercial vehicles, those requiring the engine to be stopped during burn-off are unsuitable for any automotive applications other than city buses and large local delivery vehicles operated on regular schedules. The continuous and cyclic burn-off filters are generally even more bulky. On the other hand, many of these systems are claimed to serve also as a silencer. There is not enough space here to give details of proprietary filter systems, but most are described in *Automotive Fuels and Fuel Systems*, Vol. 2, by T.K. Garrett, Wiley.

3.1.25 Influence of fuel quality on diesel exhaust emissions

How individual emissions are influenced by different fuel properties have been summarised by the UK Petroleum Industry Association as follows:

NO_x	Increases slightly with cetane number. Decreases as aromatic content is lowered.
CO	No significant effects.
HC	Decreases slightly as cetane number increases. Decreases with density. Relationship with volatility inconsistent.
Black smoke	Increases with fuel density and decreases with aromatic content. Is not significantly affected by volatility. Increases with injection retard (e.g. for reducing NO_x).
Particulates	Reduced as volatility is lowered. Reduced as cetane number is lowered, though inconsistently. Unaffected by aromatic content. Reduced as sulphur content is lowered.

A good-quality fuel is generally regarded as one having a cetane number of 50 and a sulphur content of no greater than 0.05%.

In Sweden, Volvo have shown that by bringing the sulphur content down from 0.2 to 0.05% particulate emissions can be reduced by up to 20% and NO_x is also reduced. Furthermore, ignoring the effect of the tax on fuel prices, the cost of such a reduction is only about 2 pence per gallon whereas to obtain a commensurate improvement by reducing the aromatic content and increasing the cetane number would cost about 22 pence per gallon.

3.1.26 Black smoke

The effect of sulphur content on the formation of particulates has been covered in Section 3.1.23. Other factors include volatility and cetane number. As regards visibility, however, the carbon content is much more significant. Suggestions that volatility *per se* influences black smoke are without foundation. Smoke is reduced with increasing volatility for two reasons: the first is the correspondingly falling viscosity, and the second the associated rising API gravity of the fuel. A consequence of the first is increased leakage of fuel through the clearances around both the pumping elements and the injector needles and, of the second, the weight of the fuel injected falls. Therefore, for any given fuel pump delivery setting, the power output decreases with increasing volatility. In fact, the real influence of volatility depends on an extremely complex combination of circumstances, and varies with factors such as speed, load and type of engine.

The reason is that each engine is designed to operate at maximum efficiency over a given range of speeds and loads with a given grade of fuel. Therefore, at any given speed and load, a change of fuel might increase the combustion efficiency, yet at another speed and load the same change might reduce it. This is because a certain weight of fuel is required to produce a given engine power output so, if the API gravity is increased, a commensurately larger volume of fuel must be supplied, and this entails injection for a longer period which, for any given engine operating condition could have either a beneficial or detrimental effect on combustion efficiency. Similarly, the resultant change in droplet size and fuel penetration relative to the air swirl could have either a beneficial or detrimental effect.

The reason why the cetane number does not have a significant effect on the output of black smoke is simple. It is that smoke density is largely determined during the burning of the last few drops of fuel to be injected into the combustion chamber.

3.1.27 White smoke

White smoke is a mixture of partially vaporised droplets of water and fuel, the former being products of combustion and the latter arising because the temperature of the droplets fails to rise to that needed for ignition. It can be measured by passing the exhaust through a box, one side of which is transparent and the other painted matt black. A beam of light is directed through the transparent wall on to the matt black surface. If there is no white smoke, no light is reflected back to a sensor alongside the light source; the degree of reflection therefore is a function of the density of the white smoke. For testing fuels, the criterion is the time taken, after starting from a specified low temperature, for the smoke level to reduce to an acceptable level. After starting at 0°C, satisfactory smoke levels are generally obtainable with a Diesel Index of 57 and a cetane number of 53.5.

Section Four

Digital engine control

Chapter 4.1

4.1

Digital engine control systems

William Ribbens

4.1.1 Introduction

Traditionally, the term *powertrain* has been thought to include the engine, transmission, differential, and drive axle/wheel assemblies. With the advent of electronic controls, the powertrain also includes the electronic control system (in whatever configuration it has). In addition to engine control functions for emissions regulation, fuel economy, and performance, electronic controls are also used in the automatic transmission to select shifting as a function of operating conditions. Moreover, certain vehicles employ electronically controlled clutches in the differential (transaxle T/A) for traction control.

These electronic controls for these major powertrain components can either be separate (i.e., one for each component) or an integrated system regulating the powertrain as a unit.

This latter integrated control system has the benefit of obtaining optimal vehicle performance within the constraints of exhaust emission and fuel economy regulations. Each of the control systems is discussed separately beginning with electronic engine control. Then a brief discussion of integrated powertrain follows. This chapter concludes with a discussion of hybrid vehicle (HV) control systems in which propulsive power comes from an internal combustion engine (ICE) or an electric motor (EM) or a combination of both. The proper balance of power between these two sources is a very complex function of operating conditions and governmental regulations.

4.1.2 Digital engine control

This chapter explores some practical digital control systems. There is, of course, considerable variation in the configuration and control concept from one manufacturer to another. However, this chapter describes representative control systems that are not necessarily based on the system of any given manufacturer, thereby giving the reader an understanding of the configuration and operating principles of a generic representative system. As such, the systems in this discussion are a compilation of the features used by several manufacturers.

In fact, most modern engine control systems, such as discussed in this chapter, are digital. A typical engine control system incorporates a microprocessor and is essentially a special-purpose computer (or microcontroller).

Electronic engine control has evolved from a relatively rudimentary fuel control system employing discrete analog components to the highly precise fuel and ignition control through 32-bit (sometimes more) microprocessor-based integrated digital electronic powertrain control. The motivation for development of the more sophisticated digital control systems has been the increasingly stringent exhaust emission and fuel economy regulations. It has proven to be cost effective to implement the powertrain controller as a multimode computer-based system to satisfy these requirements.

A multimode controller operates in one of many possible modes, and, among other tasks, changes the various calibration parameters as operating conditions change in order to optimize performance. To implement multimode control in analog electronics it would be necessary to change hardware parameters (for example, via switching systems) to accommodate various operating conditions. In a computer-based controller, however, the control law and system parameters are changed via program (i.e., software) control. The hardware remains fixed but the software is reconfigured in accordance with operating conditions as determined by sensor measurements and switch inputs to the controller.

Understanding Automotive Electronics; ISBN: 9780750675994
Copyright © 2003 Elsevier Ltd; All rights of reproduction, in any form, reserved.

This chapter explains how the microcontroller under program control is responsible for generating the electrical signals that operate the fuel injectors and trigger the ignition pulses. This chapter also discusses secondary functions (including management of secondary air that must be provided to the catalytic converter exhaust gas recirculation (EGR) regulation and evaporative emission control).

4.1.3 Digital engine control features

The primary purpose of the electronic engine control system is to regulate the mixture (i.e., air–fuel), the ignition timing, and EGR. Virtually all major manufacturers of cars sold in the United States (both foreign and domestic) use the three-way catalyst for meeting exhaust emission constraints. For such cars, the air/fuel ratio is held as closely as possible to the stoichiometric value of about 14.7 for as much of the time as possible. Ignition timing and EGR are controlled separately to optimize performance and fuel economy.

Fig. 4.1-1 illustrates the primary components of an electronic engine control system. In this figure, the engine control system is a microcontroller, typically implemented with a specially designed microprocessor and operating under program control. Typically, the controller incorporates hardware multiply and ROM. The hardware multiply greatly speeds up the multiplication operation required at several stages of engine control relative to software multiplication routines, which are generally cumbersome and slow. The associated ROM contains the program for each mode as well as calibration parameters and lookup tables. The earliest such systems incorporated 8-bit microprocessors, although the trend is toward implementation with 32-bit microprocessors. The microcontroller under program control generates output electrical signals to operate the fuel injectors so as to maintain the desired mixture and ignition to optimize performance. The correct mixture is obtained by regulating the quantity of fuel delivered into each cylinder during the intake stroke in accordance with the air mass.

In determining the correct fuel flow, the controller obtains a measurement or estimate of the mass air flow (MAF) rate into the cylinder. The measurement is obtained using an MAF sensor. Alternatively, the MAF rate is estimated (calculated) using the speed–density method. This estimate can be found from measurement of the intake manifold absolute pressure (MAP), the revolutions per minute (RPM) and the inlet air temperature.

Using this measurement or estimate, the quantity of fuel to be delivered is determined by the controller in accordance with the instantaneous control mode. The quantity of fuel delivered by the fuel injector is determined by the operation of the fuel injector. A fuel injector is essentially a solenoid-operated valve. Fuel that is supplied to each injector from the fuel pump is supplied to each fuel injector at a regulated fuel pressure. When the injector valve is opened, fuel flows at a rate R_f (in gal/sec) that is determined by the (constant) regulated pressure and by the geometry of the fuel injector valve. The quantity of fuel F delivered to

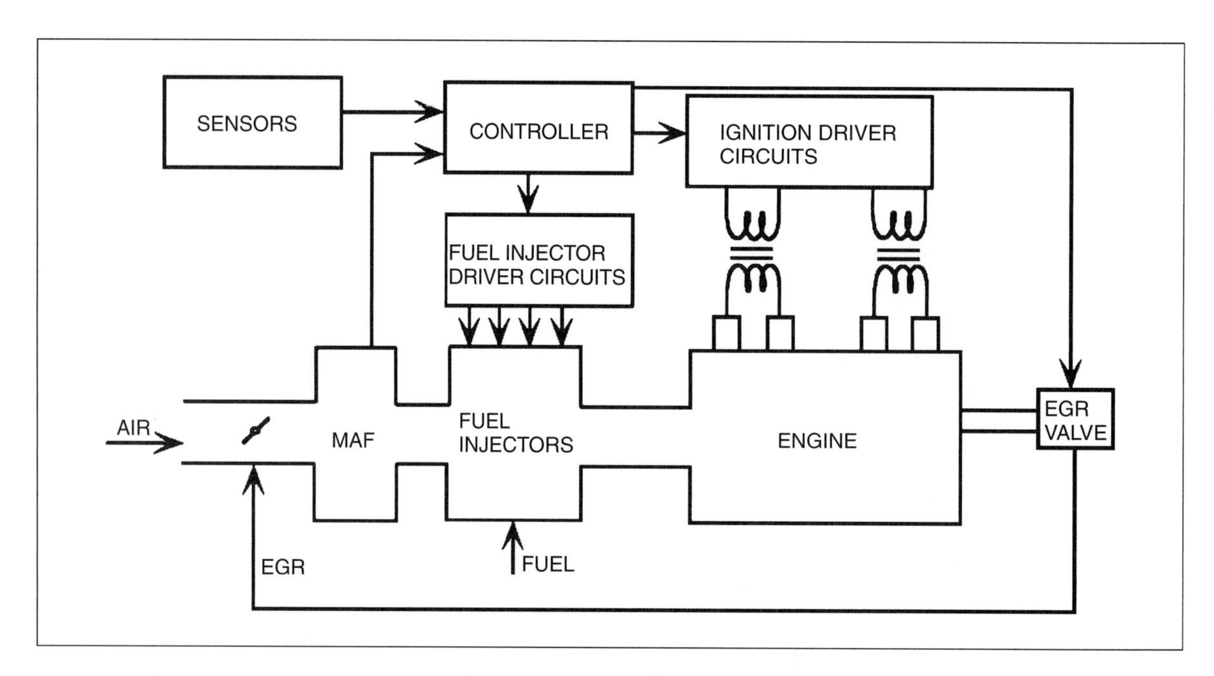

Fig. 4.1-1 Components of an electronically controlled engine.

any cylinder is proportional to the time T that this valve is opened:

$$F = R_f T$$

The engine control system, then, determines the correct quantity of fuel to be delivered to each cylinder (for a given operating condition) via measurement of MAF rate. The controller then generates an electrical signal that opens the fuel injector valve for the appropriate time interval T to deliver this desired fuel quantity to the cylinder such that a stoichiometric air/fuel ratio is maintained.

The controller also determines the correct time for fuel delivery to correspond to the intake stroke for the relevant cylinder. This timing is determined by measurements of crankshaft and camshaft position using sensors such as those described in Chapter 6.

4.1.4 Control modes for fuel control

The engine control system is responsible for controlling fuel and ignition for all possible engine operating conditions. However, there are a number of distinct categories of engine operation, each of which corresponds to a separate and distinct operating mode for the engine control system. The differences between these operating modes are sufficiently great that different software is used for each. The control system must determine the operating mode from the existing sensor data and call the particular corresponding software routine.

For a typical engine, there are seven different engine operating modes that affect fuel control: engine crank, engine warm-up, open-loop control, closed-loop control, hard acceleration, deceleration, and idle. The program for mode control logic determines the engine operating mode from sensor data and timers.

In the earliest versions of electronic fuel control systems, the fuel metering actuator typically consisted of one or two fuel injectors mounted near the throttle plate so as to deliver fuel into the throttle body. These throttle body fuel injectors (TBFIs) were in effect an electromechanical replacement for the carburetor. Requirements for the TBFIs were such that they only had to deliver fuel at the correct average flow rate for any given MAF. Mixing of the fuel and air, as well as distribution to the individual cylinders, took place in the intake manifold system.

The more stringent exhaust emissions regulations of the late 1980s and the 1990s have demanded more precise fuel delivery than can normally be achieved by TBFI. These regulations and the need for improved performance have led to timed sequential port fuel injection (TSPFI). In such a system there is a fuel injector for each cylinder that is mounted so as to spray fuel directly into the intake of the associated cylinder. Fuel delivery is timed to occur during the intake stroke for that cylinder.

The digital engine control system requires sensors for measuring the engine variables and parameters. Referring to Fig. 4.1-1, the set of sensors may include, for example, MAF, exhaust gas oxygen (EGO) concentration, and crankshaft angular position (CPS), as well as RPM, camshaft position (possibly a single reference point for each engine cycle), coolant temperature (CT), throttle plate angular position (TPS), intake air temperature, and exhaust pressure ratio (EPR) for EGR control.

In the example configuration of Fig. 4.1-1, fuel delivery is assumed to be TSPFI (i.e., via individual fuel injectors located so as to spray fuel directly into the intake port and timed to coincide with the intake stroke). Air flow measurement is via an MAF sensor. In addition to MAF, sensors are available for the measurement of EGO concentration, RPM, inlet air and CTs, throttle position, crankshaft (and possibly camshaft) position, and exhaust differential pressure (DP) (for EGR calculation). Some engine controllers involve vehicle speed sensors and various switches to identify brake on/off and the transmission gear, depending on the particular control strategy employed.

When the ignition key is switched on initially, the mode control logic automatically selects an engine start control scheme that provides the low air/fuel ratio required for starting the engine. Once the engine RPM rises above the cranking value, the controller identifies the "engine started" mode and passes control to the program for the engine warm-up mode. This operating mode keeps the air/fuel ratio low to prevent engine stall during cool weather until the engine CT rises above some minimum value. The instantaneous air/fuel is a function of CT. The particular value for the minimum CT is specific to any given engine and, in particular, to the fuel metering system. (Alternatively, the low air/fuel ratio may be maintained for a fixed time interval following start, depending on start-up engine temperature.)

When the CT rises sufficiently, the mode control logic directs the system to operate in the open-loop control mode until the EGO sensor warms up enough to provide accurate readings. This condition is detected by monitoring the EGO sensor's output for voltage readings above a certain minimum rich air/fuel mixture voltage set point. When the sensor has indicated rich at least once and after the engine has been in open loop for a specific time, the control mode selection logic selects the closed-loop mode for the system. (*Note*: other criteria may also be used.) The engine remains in the closed-loop mode until either the EGO sensor cools and fails to read a rich mixture for a certain length of time or a hard acceleration or deceleration occurs. If the sensor cools, the control mode logic selects the open-loop mode again.

During hard acceleration or heavy engine load, the control mode selection logic chooses a scheme that provides a rich air/fuel mixture for the duration of the acceleration or heavy load. This scheme provides maximum torque but relatively poor emissions control and poor fuel economy regulation as compared with a stoichiometric air/fuel ratio. After the need for enrichment has passed, control is returned to either open-loop or closed-loop mode, depending on the control mode logic selection conditions that exist at that time.

During periods of deceleration, the air/fuel ratio is increased to reduce emissions of HC and CO due to unburned excess fuel. When idle conditions are present, control mode logic passes system control to the idle speed control mode. In this mode, the engine speed is controlled to reduce engine roughness and stalling that might occur because the idle load has changed due to air conditioner compressor operation, alternator operation, or gearshift positioning from park/neutral to drive, although stoichiometric mixture is used if the engine is warm.

In modern engine control systems, the controller is a special-purpose digital computer built around a microprocessor. A block diagram of a typical modern digital engine control system is depicted in Fig. 4.1-2. The controller also includes ROM containing the main program (of several thousand lines of code) as well as rRAM for temporary storage of data during computation. The sensor signals are connected to the controller via an input/output (I/O) subsystem. Similarly, the I/O subsystem provides the output signals to drive the fuel injectors (shown as the fuel metering block of Fig. 4.1-2) as well as to trigger pulses to the ignition system (described later in this chapter). In addition, this solid-state control system includes hardware for sampling and analog-to-digital conversion such that all sensor measurements are in a format suitable for reading by the microprocessor.

The sensors that measure various engine variables for control are as follows:

MAF	Mass air flow sensor
CT	Engine temperature as represented by coolant temperature

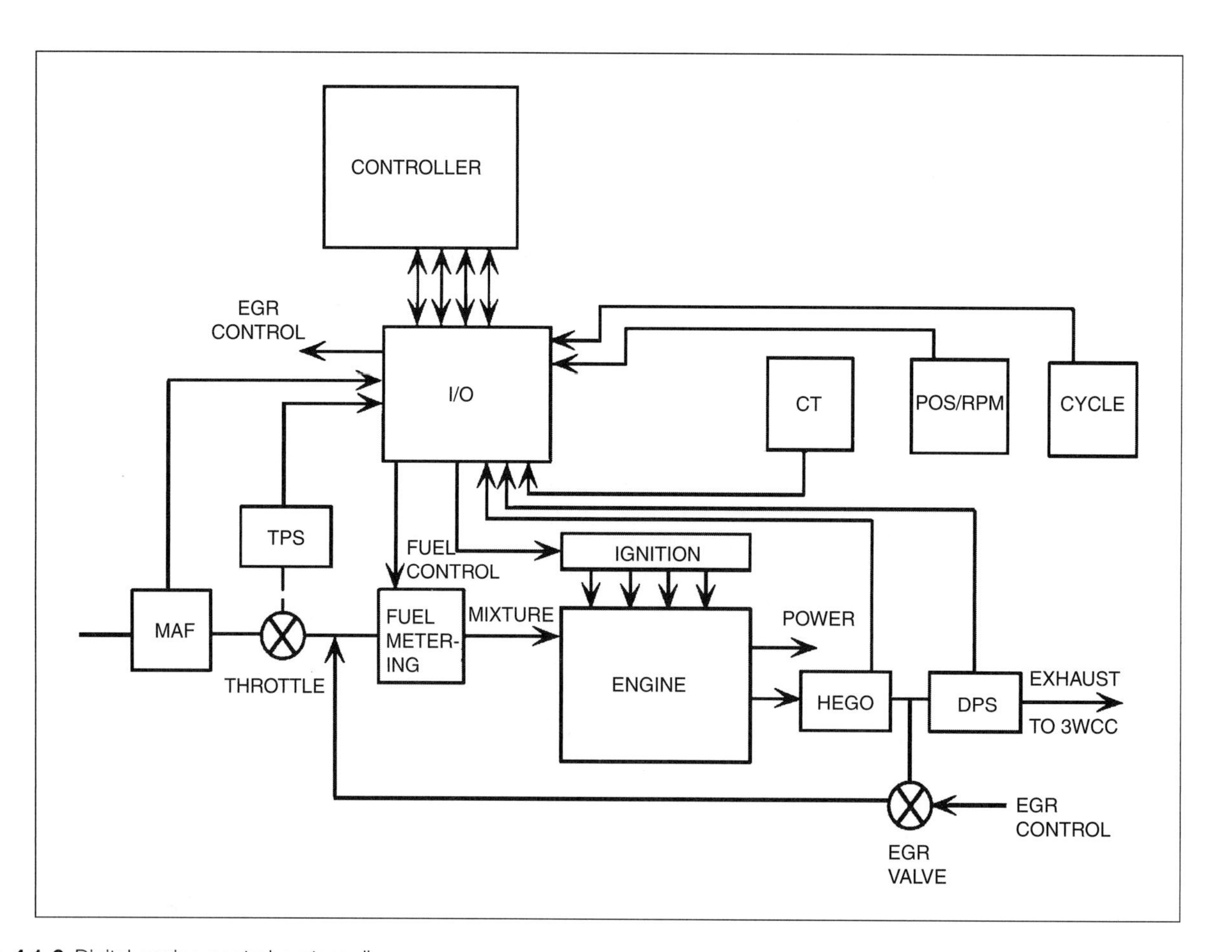

Fig. 4.1-2 Digital engine control system diagram.

HEGO	(One or two) heated EGO sensor(s)
POS/RPM	Crankshaft angular position and RPM sensor cycle camshaft position sensor for determining start of each engine cycle
TPS	Throttle position sensor
DPS	Differential pressure sensor (exhaust to intake) for EGR control

The control system selects an operating mode based on the instantaneous operating condition as determined from the sensor measurements. Within any given operating mode the desired air/fuel ratio $(A/F)_d$ is selected. The controller then determines the quantity of fuel to be injected into each cylinder during each engine cycle. This quantity of fuel depends on the particular engine operating condition as well as the controller mode of operation, as will presently be explained.

4.1.4.1 Engine crank

While the engine is being cranked, the fuel control system must provide an intake air/fuel ratio of anywhere from 2:1 to 12:1, depending on engine temperature. The correct $[A/F]_d$ is selected from an ROM lookup table as a function of CT. Low temperatures affect the ability of the fuel metering system to atomize or mix the incoming air and fuel. At low temperatures, the fuel tends to form into large droplets in the air, which do not burn as efficiently as tiny droplets. The larger fuel droplets tend to increase the apparent air/fuel ratio, because the amount of usable fuel (on the surface of the droplets) in the air is reduced; therefore, the fuel metering system must provide a decreased air/fuel ratio to provide the engine with a more combustible air/fuel mixture. During engine crank the primary issue is to achieve engine start as rapidly as possible. Once the engine is started the controller switches to an engine warm-up mode.

4.1.4.2 Engine warm-up

While the engine is warming up, an enriched air/fuel ratio is still needed to keep it running smoothly, but the required air/fuel ratio changes as the temperature increases. Therefore, the fuel control system stays in the open-loop mode, but the air/fuel ratio commands continue to be altered due to the temperature changes. The emphasis in this control mode is on rapid and smooth engine warm-up. Fuel economy and emission control are still a secondary concern.

A diagram illustrating the lookup table selection of desired air/fuel ratios is shown in Fig. 4.1-3. Essentially, the measured CT is converted to an address for the lookup table. This address is supplied to the ROM table via the system address bus (A/B). The data stored at this address in the ROM are the desired $(A/F)_d$ *for* that temperature. These data are sent to the controller via the system data bus (D/B).

There is always the possibility of a CT failure. Such a failure could result in excessively rich or lean mixtures, which can seriously degrade the performance of both the engine and the three-way catalytic converter (3wcc). One scheme that can circumvent a temperature sensor failure involves having a time function to limit the duration of the engine warm-up mode. The nominal time to warm the engine from cold soak at various temperatures is known. The controller is configured to switch from engine warm-up mode to an open-loop (warmed-up engine) mode after a sufficient time by means of an internal timer.

It is worthwhile at this point to explain how the quantity of fuel to be injected is determined. This method is implemented in essentially all operating modes and is described here as a generic method, even though each engine control scheme may vary somewhat from the following. The quantity of fuel to be injected during the intake stroke of any given cylinder (which we call F) is determined by the mass of air (A) drawn into that cylinder (i.e., the air charge) during that intake

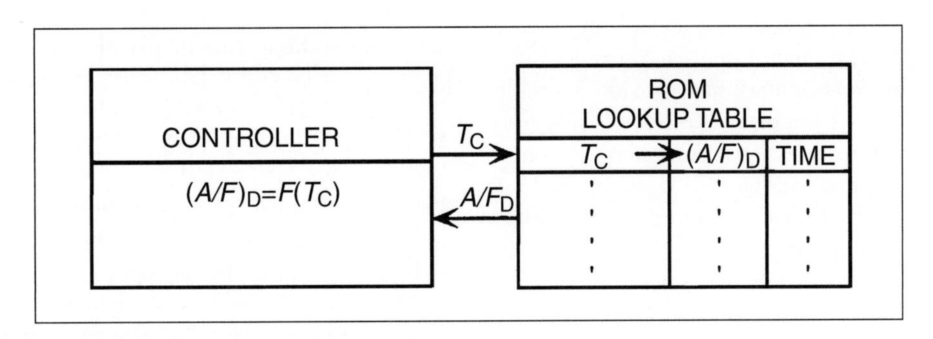

Fig. 4.1-3 Illustration of lookup table for desired air/fuel ratio.

stroke. That quantity of fuel is given by the air charge divided by the desired air/fuel ratio:

$$F = \frac{A}{(A/F)_d}$$

The quantity of air drawn into the cylinder, A, is computed from the MAF rate and the RPM. The MAF rate will be given in kg/sec. If the engine speed is RPM, then the number of revolutions/second (which we call r) is:

$$r = \frac{\text{RPM}}{60}$$

Then, the MAF is distributed approximately uniformly to half the cylinders during each revolution. If the number of cylinders is N then the air charge (mass) in each cylinder during one revolution is:

$$A = \frac{\text{MAF}}{r(N/2)}$$

In this case, the mass of fuel delivered to each cylinder is:

$$F = \frac{\text{MAF}}{r(N/2)(A/F)_d}$$

This computation is carried out by the controller continuously so that the fuel quantity can be varied quickly to accommodate rapid changes in engine operating condition. The fuel injector pulse duration T corresponding to this fuel quantity is computed using the known fuel injector delivery rate R_f:

$$T = \frac{F}{R_f}$$

This pulse width is known as the *base pulse width*. The actual pulse width used is modified from this according to the mode of operation at any time, as will presently be explained.

Corrections of the base pulse width occur whenever anything affects the accuracy of the fuel delivery. For example, low battery voltage might affect the pressure in the fuel rail that delivers fuel to the fuel injectors. Corrections to the base pulse width are then made using the actual battery voltage.

An alternate method of computing MAF rate is the speed–density method. Although this method has essentially been replaced by direct MAF measurements, there will continue to be a number of cars employing this method for years to come, so it is arguably worthwhile to include a brief discussion in this chapter. This method, which is illustrated in Fig. 4.1-4, is based on measurements of MAP, RPM, and intake air temperature T_i. The air density d_a is computed from MAP and T_i, and the volume flow rate R_v of combined air and EGR is computed from RPM and volumetric efficiency, the latter being a function of MAP and RPM. The volume rate for air is found by subtracting the EGR volume flow rate from the combined air and EGR. Finally, the MAF rate is computed as the product of the volume flow rate for air and the intake air density. Given the complexity of the speed–density method it is easy to see why automobile manufacturers would choose the direct MAF measurement once a cost-effective MAF sensor became available.

The speed–density method can be implemented either by computation in the engine control computer or via lookup tables. Fig. 4.1-5 is an illustration of the lookup table implementation. In this figure, three variables need to be determined: volumetric efficiency (n_v), intake density (d_a), and EGR volume flow rate (R_E). The volumetric efficiency is read from ROM with an address determined from RPM and MAP measurements. The intake air density is read from another section of ROM with an address determined from MAP and T_i measurements. The EGR volume flow rate is read from still another section of ROM with an address determined from DP and EGR valve position. These variables are combined to yield the MAF rate:

$$\text{MAF} = d_a\left[\left(\frac{\text{RPM}}{60}\right)\left(\frac{D}{2}\right)n_v - R_E\right]$$

where D is the engine displacement.

4.1.4.3 Open-loop control

For a warmed-up engine, the controller will operate in an open loop if the closed-loop mode is not available for any reason. For example, the engine may be warmed sufficiently but the EGO sensor may not provide a usable signal. In any event, it is important to have a stoichiometric mixture to minimize exhaust emissions as soon as possible. The base pulse width T_b is computed as described above, except that the desired air/fuel ratio $(A/F)_d$ is 14.7 (stoichiometry):

$$T_b = \frac{\text{MAF}}{r(N/2)(14.7)R_f} \quad \text{base pulse width}$$

4.1.4.4 Closed-loop control

Perhaps the most important adjustment to the fuel injector pulse duration comes when the control is in the closed-loop mode. In the open-loop mode the accuracy of the fuel delivery is dependent on the accuracy of the measurements of the important variables. However, any

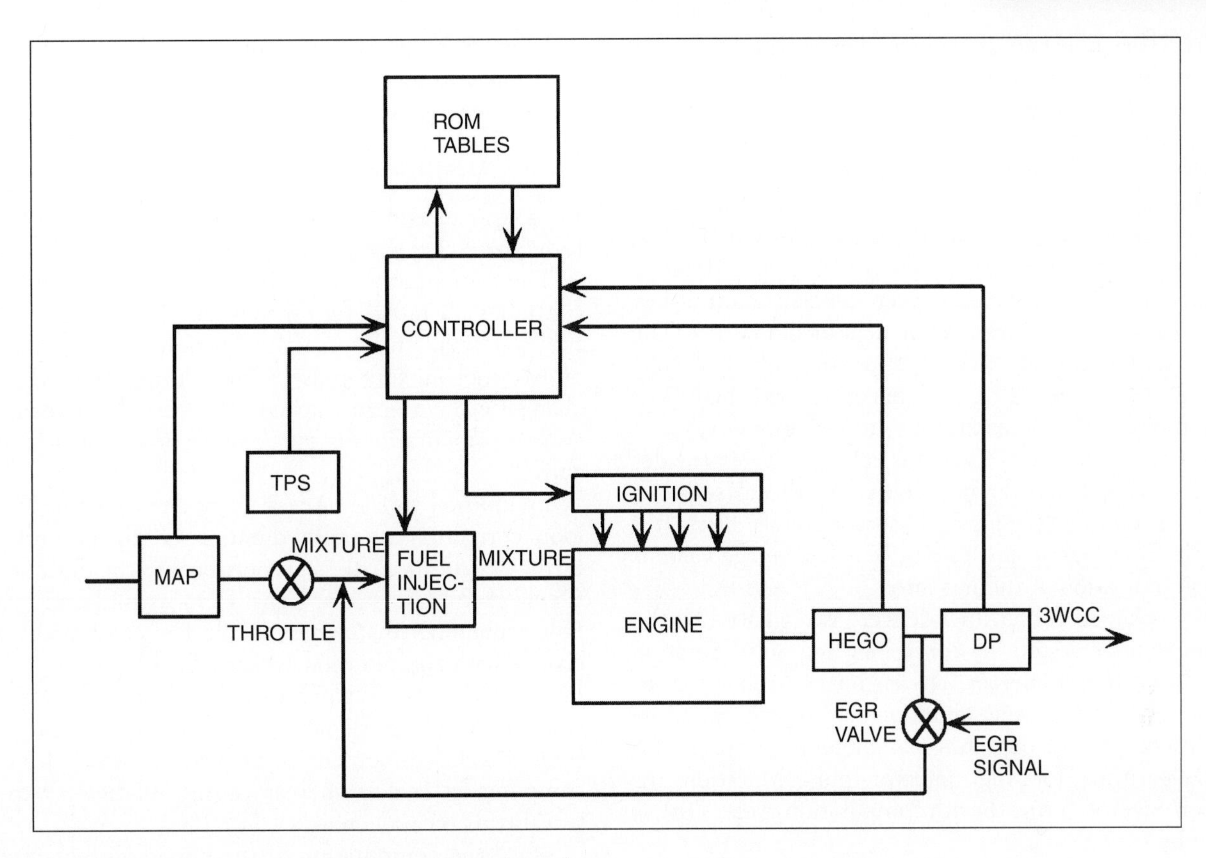

Fig. 4.1-4 Engine control system using the speed–density method.

physical system is susceptible to changes with either operating conditions (e.g., temperature) or with time (aging or wear of components).

In any closed-loop control system a measurement of the output variables is compared with the desired value for those variables. In the case of fuel control, the variables being regulated are exhaust gas concentrations of HC, CO, and NO_x. Although direct measurement of these exhaust gases is not feasible in production automobiles, it is sufficient for fuel control purposes to

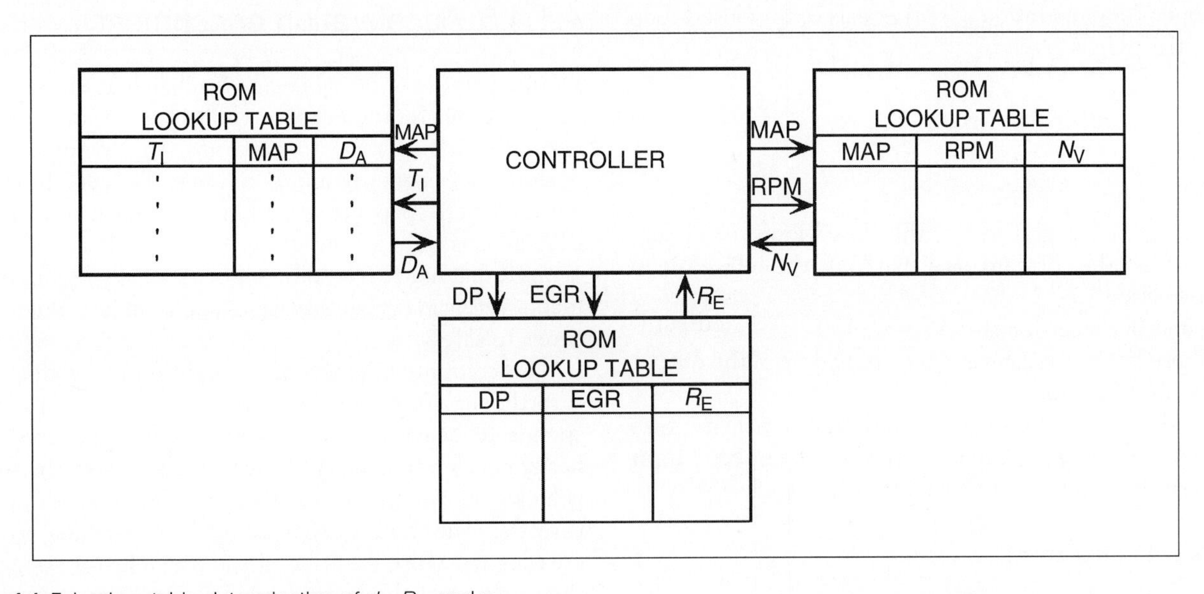

Fig. 4.1-5 Lookup table determination of d_a, R_E, and n_v.

measure the EGO concentration. These regulated gases can be optimally controlled with a stoichiometric mixture. The EGO sensor is, in essence, a switching sensor that changes output voltage abruptly as the input mixture crosses the stoichiometric mixture of 14.7.

The closed-loop mode can only be activated when the EGO (or HEGO) sensor is sufficiently warmed. That is, the output voltage of the sensor is high (approximately 1 volt) when the exhaust oxygen concentration is low (i.e., for a rich mixture relative to stoichiometry). The EGO sensor voltage is low (approximately 0.1 volt) whenever the exhaust oxygen concentration is high (i.e., for a mixture that is lean relative to stoichiometry).

The time-average EGO sensor output voltage provides the feedback signal for fuel control in the closed-loop mode. The instantaneous EGO sensor voltage fluctuates rapidly from high to low values, but the average value is a good indication of the mixture.

As explained earlier, fuel delivery is regulated by the engine control system by controlling the pulse duration (T) for each fuel injector. The engine controller continuously adjusts the pulse duration for varying operating conditions and for operating parameters. A representative algorithm for fuel injector pulse duration for a given injector during the nth computation cycle, $T(n)$, is given by:

$$T(n) = T_b(n) \times [1 + C_L(n)]$$

where

$T_b(n)$ is the base pulse width as determined from measurements of MAF rate and the desired air/fuel ratio

$C_L(n)$ is the closed-loop correction factor

For open-loop operation, $C_L(n)$ equals 0; for closed-loop operation, C_L is given by:

$$C_L(n) = \alpha I(n) + \beta P(n)$$

where

$I(n)$ is the integral part of the closed-loop correction

$P(n)$ is the proportional part of the closed-loop correction

α and β are constants

These latter variables are determined from the output of the EGO sensor.

Whenever the EGO sensor indicates a rich mixture (i.e., EGO sensor voltage is high), then the integral term is reduced by the controller for the next cycle,

$$I(n+1) = I(n) - 1$$

for a rich mixture.

Whenever the EGO sensor indicates a lean mixture (i.e., low output voltage), the controller increments $I(n)$ for the next cycle,

$$I(n+1) = I(n) + 1$$

for a lean mixture. The integral part of C_L continues to increase or decrease in a limit-cycle operation.

The computation of the closed-loop correction factor continues at a rate determined within the controller. This rate is normally high enough to permit rapid adjustment of the fuel injector pulse width during rapid throttle changes at high engine speed. The period between successive computations is the computation cycle described above.

In addition to the integral component of the closed-loop correction to pulse duration is the proportional term. This term, $P(n)$, is proportional to the deviation of the average EGO sensor signal from its mid-range value (corresponding to stoichiometry). The combined terms change with computation cycle as depicted in Fig. 4.1-6. In this figure the regions of lean and rich (relative to stoichiometry) are depicted. During relatively lean periods the closed-loop correction term increases for each computation cycle, whereas during relatively rich intervals this term decreases.

Once the computation of the closed-loop correction factor is completed, the value is stored in a specific memory location (RAM) in the controller. At the appropriate time for fuel injector activation (during the intake stroke), the instantaneous closed-loop correction factor is read from its location in RAM and an actual pulse of the corrected duration is generated by the engine control.

4.1.4.5 Acceleration enrichment

During periods of heavy engine load such as during hard acceleration, fuel control is adjusted to provide an enriched air/fuel ratio to maximize engine torque and neglect fuel economy and emissions. This condition of enrichment is permitted within the regulations of the EPA as it is only a temporary condition. It is well recognized that hard acceleration is occasionally required for maneuvering in certain situations and is, in fact, related at times to safety.

The computer detects this condition by reading the throttle angle sensor voltage. High throttle angle corresponds to heavy engine load and is an indication that heavy acceleration is called for by the driver. In some vehicles a switch is provided to detect wide open throttle. The fuel system controller responds by increasing the pulse duration of the fuel injector signal for the duration of the heavy load. This enrichment enables the engine to operate with a torque greater than that

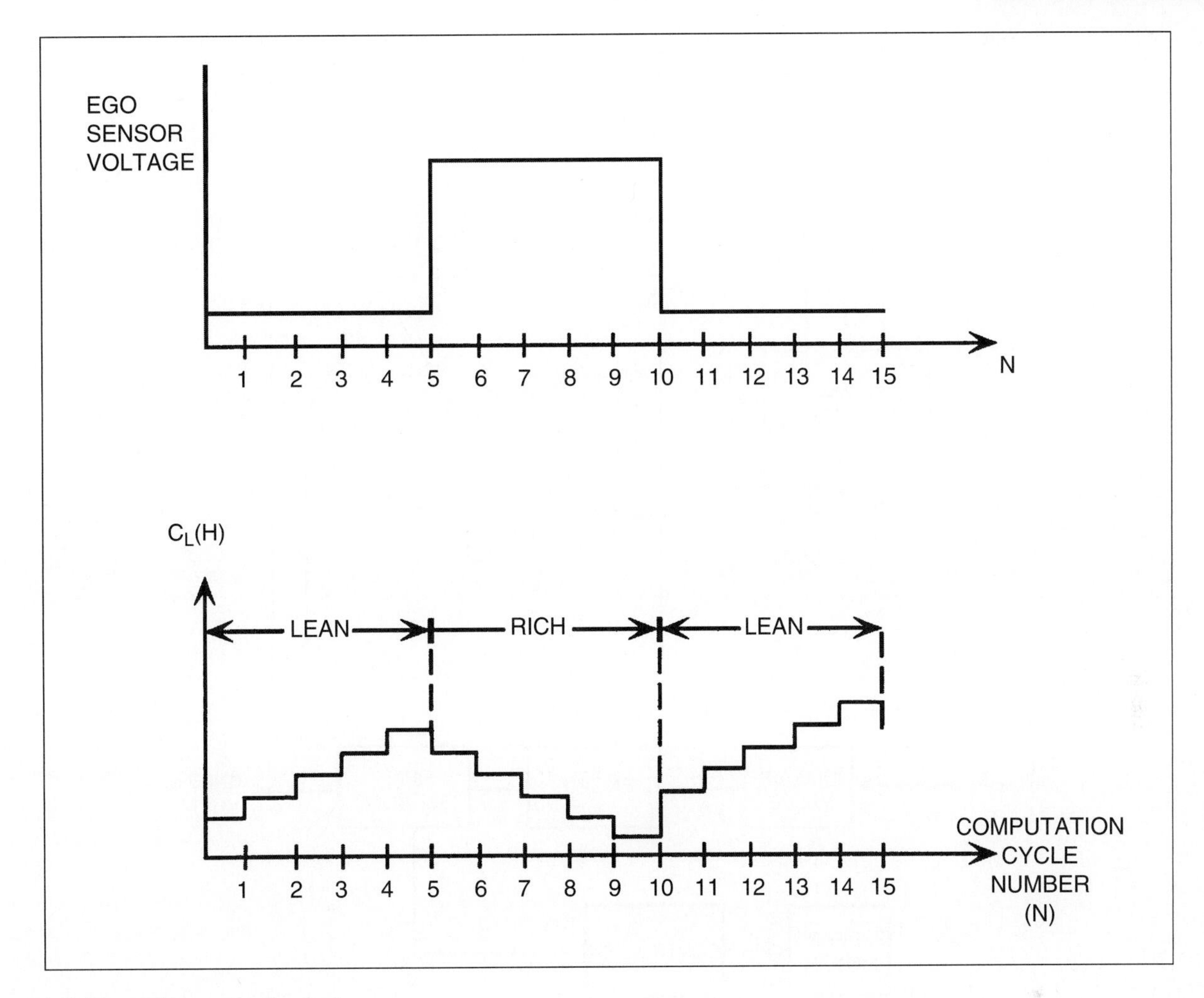

Fig. 4.1-6 Closed-loop correction factor.

allowed when emissions and fuel economy are controlled. Enrichment of the air/fuel ratio to about 12:1 is sometimes used.

4.1.4.6 Deceleration leaning

During periods of light engine load and high RPM such as during coasting or hard deceleration, the engine operates with a very lean air/fuel ratio to reduce excess emissions of HC and CO. Deceleration is indicated by a sudden decrease in throttle angle or by closure of a switch when the throttle is closed (depending on the particular vehicle configuration). When these conditions are detected by the control computer, it computes a decrease in the pulse duration of the fuel injector signal. The fuel may even be turned off completely for very heavy deceleration.

4.1.4.7 Idle speed control

Idle speed control is used by some manufacturers to prevent engine stall during idle. The goal is to allow the engine to idle at as low an RPM as possible, yet keep the engine from running rough and stalling when power-consuming accessories, such as air conditioning compressors and alternators, turn on.

The control mode selection logic switches to idle speed control when the throttle angle reaches its zero (completely closed) position and engine RPM falls below a minimum value, and when the vehicle is stationary. Idle speed is controlled by using an electronically controlled throttle bypass valve (Fig. 4.1-7a) that allows air to flow around the throttle plate and produces the same effect as if the throttle had been slightly opened.

There are various schemes for operating a valve to introduce bypass air for idle control. One relatively common method for controlling the idle speed bypass air uses a special type of motor called a *stepper motor*. A stepper motor moves in fixed angular increments when activated by pulses on its two sets of windings (i.e., open or close). Such a motor can be operated in either direction by supplying pulses in the proper phase to the windings. This is advantageous for idle speed control

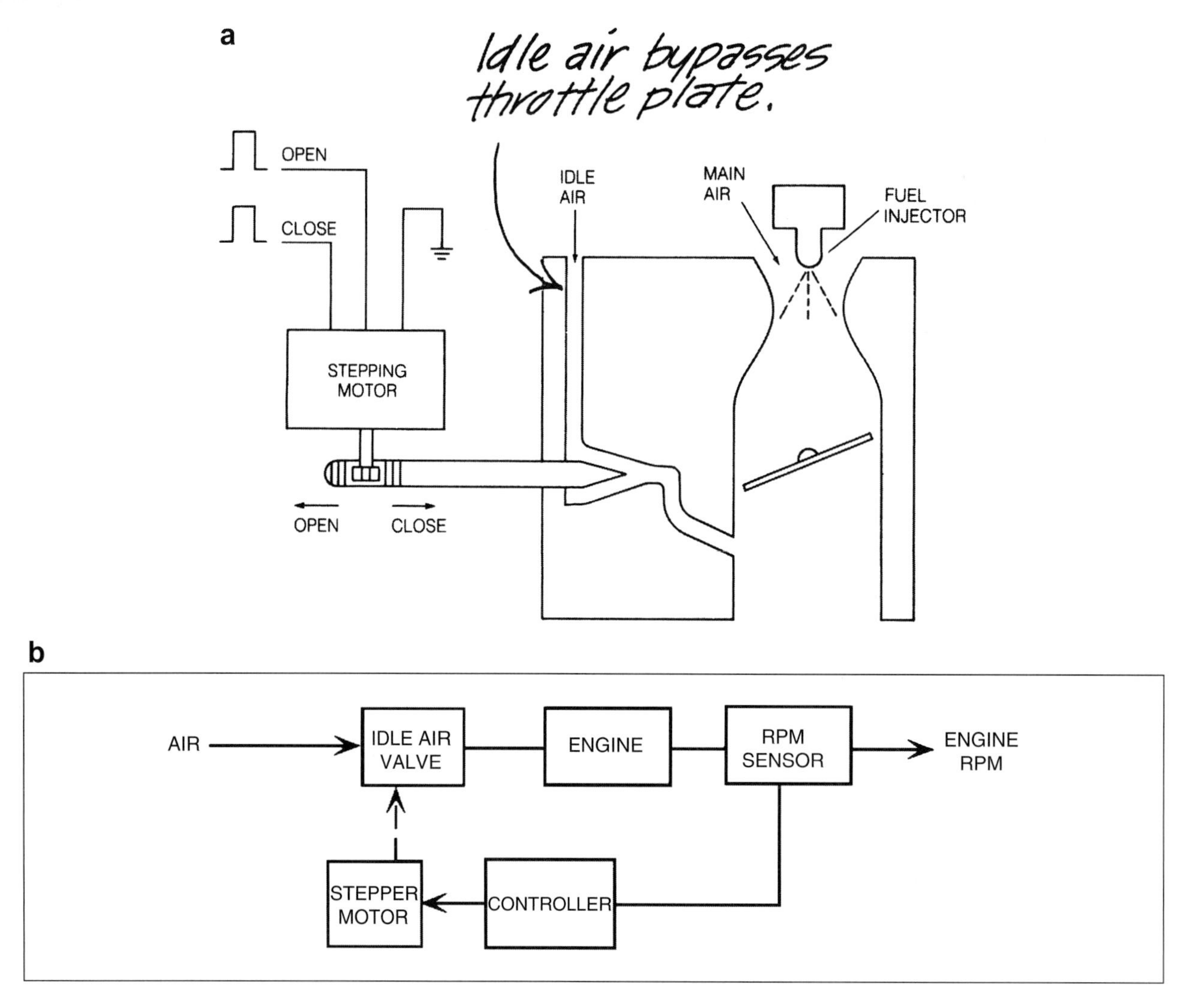

Fig. 4.1-7 Idle air control.

since the controller can very precisely position the idle bypass valve by sending the proper number of pulses of the correct phasing.

The engine control computer can precisely know the position of the valve in a number of ways. In one way the computer can send sufficient pulses to completely close the valve when the ignition is first switched on. Then it can send open pulses (phased to open the valve) to a specified (known) position.

A block diagram of a simplified idle speed control system is shown in Fig. 4.1-7b. Idle speed is detected by the RPM sensor, and the speed is adjusted to maintain a constant idle RPM. The computer receives digital on/off status inputs from several power-consuming devices attached to the engine, such as the air conditioner clutch switch, park-neutral switch, and the battery charge indicator. These inputs indicate the load that is applied to the engine during idle.

When the engine is not idling, the idle speed control valve may be completely closed so that the throttle plate has total control of intake air. During periods of deceleration leaning, the idle speed valve may be opened to provide extra air to increase the air/fuel ratio in order to reduce HC emissions.

4.1.5 EGR control

A second electronic engine control subsystem is the control of exhaust gas that is recirculated back to the intake manifold. Under normal operating conditions, engine cylinder temperatures can reach more than 3,000°F. The higher the temperature, the more is the chance that the exhaust will have NO_x emissions. A small amount of exhaust gas is introduced into the cylinder to replace normal intake air. This results in lower combustion temperatures, which reduces NO_x emissions.

The control mode selection logic determines when EGR is turned off or on. EGR is turned off during cranking, cold engine temperature (engine warm-up), idling, acceleration, or other conditions demanding high torque.

Since EGR was first introduced as a concept for reducing NO_x exhaust emissions, its implementation has gone through considerable change. There are in fact many schemes and configurations for EGR realization. We discuss here one method of EGR implementation that incorporates enough features to be representative of all schemes in use today and in the near future.

Fundamental to all EGR schemes is a passageway or port connecting the exhaust and intake manifolds. A valve is positioned along this passageway whose position regulates EGR from zero to some maximum value. Typically, the valve is operated by a diaphragm connected to a variable vacuum source. The controller operates a solenoid in a periodic variable-duty-cycle mode. The average level of vacuum on the diaphragm varies with the duty cycle. By varying this duty cycle, the control system has proportional control over the EGR valve opening and thereby over the amount of EGR.

In many EGR control systems the controller monitors the DP between the exhaust and intake manifold via a differential pressure sensor (DPS). With the signal from this sensor the controller can calculate the valve opening for the desired EGR level. The amount of EGR required is a predetermined function of the load on the engine (i.e., power produced).

A simplified block diagram for an EGR control system is depicted in Fig. 4.1-8a. In this figure the EGR valve is operated by a solenoid-regulated vacuum actuator (coming from the intake). The engine controller determines the required amount of EGR based on the engine operating condition and the signal from the DPS between intake and exhaust manifolds. The controller then commands the correct EGR valve position to achieve the desired amount of EGR.

4.1.6 Variable valve timing control

An earlier work introduced the concept and relative benefits of variable valve timing for improved volumetric efficiency. There it was explained that performance improvement and emissions reductions could be achieved if the opening and closing times (and ideally the valve lift) of both intake and exhaust valves could be controlled as a function of operating conditions. The mechanism for varying camshaft phasing, which is in production in certain vehicles, is used for varying exhaust camshaft phasing. This system improves volumetric efficiency by varying valve overlap from exhaust closing to intake opening. In addition to improving volumetric efficiency, this variable valve phasing can achieve desired EGR fraction.

The amount of valve overlap is directly related to the relative exhaust-intake camshaft phasing. Generally, minimal overlap is desired at idle. The desired optimal amount of overlap is determined during engine development as a function of RPM and load (e.g., by engine mapping).

The desired exhaust camshaft phasing is stored in memory (ROM) in the engine control system as a function of RPM and load. Then during engine operation the correct camshaft phasing can be found via table lookup and interpolation based on measurements of RPM and load. The RPM measurement is achieved using a noncontacting angular speed sensor. Load is measured either using an MAP sensor directly or it is computed from MAF as well as RPM.

Once the desired camshaft phasing has been determined, the engine control system sends an appropriate electrical control signal to a solenoid operated valve. The

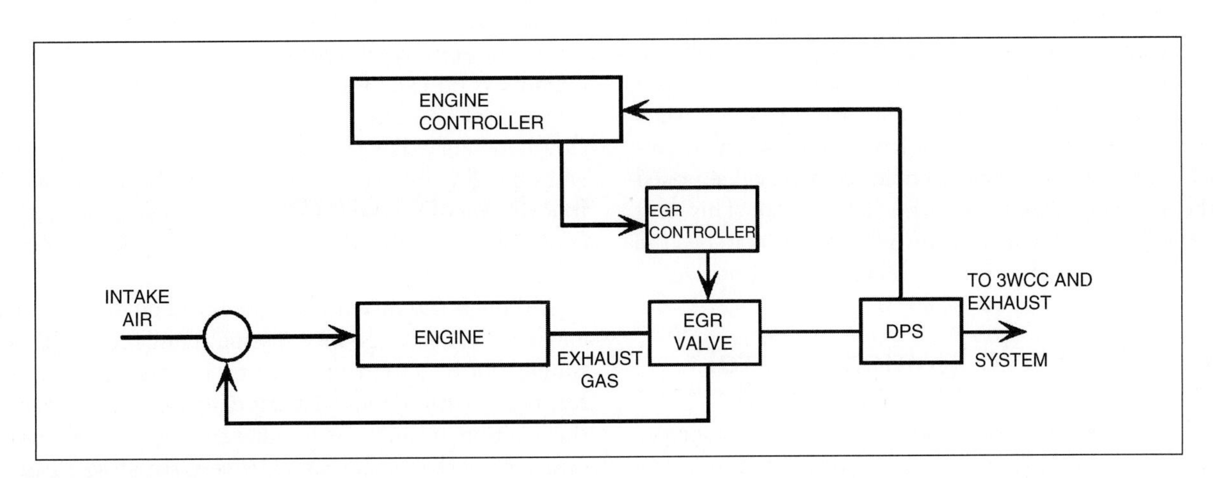

Fig. 4.1-8a EGR control.

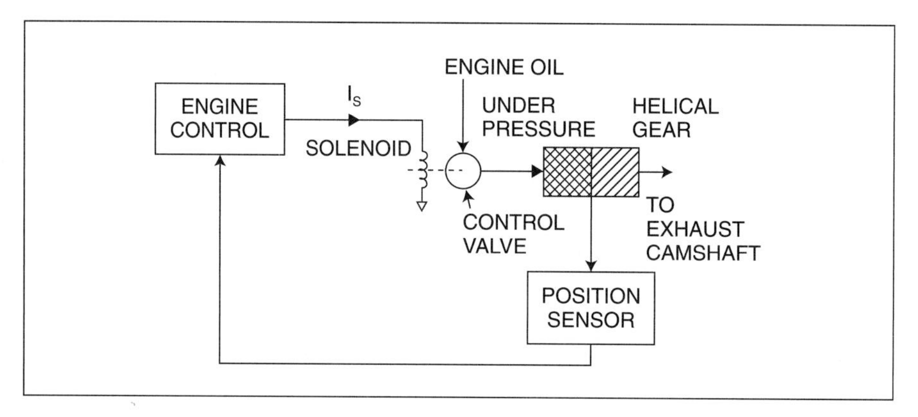

Fig. 4.1-8b Closed-loop control system.

camshaft phasing is regulated by the axial position of a helical spline gear. This axial position is determined by the pressure of (engine) oil action on one face of the helical spline gear acting against a spring. This oil pressure is regulated by the solenoid-operated valve.

Control of camshaft phasing can be either open loop or closed loop. In the open-loop case, the correct camshaft phasing depends on the relationship between axial position of the helical spline gear and the oil pressure/return spring relationship.

Since this cam closed phasing system is in fact a position control system, loop control of exhaust camshaft operation requires a measurement of camshaft phase relative to intake. This phase measurement can be accomplished by measuring axial displacement of the helical gear because there is a unique mechanical relationship between this axial displacement and exhaust camshaft phasing. Fig. 4.1-8b depicts a block diagram of a representative camshaft phasing control system.

The control system shown in Fig. 4.1-8b presumes a closed-loop control system. This VVT control system could also be an open-loop control system in which no position sensor would be required.

For the hypothetical control system, the engine control system calculates desired camshaft angular displacement (as a function of load and RPM) and compares that with actual angular displacement. The difference between these represents an error signal from which solenoid current I_s is determined. The solenoid regulates oil supplied to the chamber that moves the helical gear. This gear moves until actual exhaust camshaft phasing matches the desired value, thereby optimizing engine performance.

4.1.7 Electronic ignition control

An engine must be provided with fuel and air in correct proportions and the means to ignite this mixture in the form of an electric spark. Before the development of electronic ignition the traditional ignition system included spark plugs, a distributor, and a high-voltage ignition coil. The distributor would sequentially connect the coil output high voltage to the correct spark plug. In addition, it would cause the coil to generate the spark by interrupting the primary current (ignition points) in the desired coil, thereby generating the required spark. The time of occurrence of this spark (i.e., the ignition timing) in relation of the piston to top dead center (TDC) influences the torque generated.

In most present-day electronically controlled engines the distributor has been replaced by multiple coils. Each coil supplies the spark to either one or two cylinders. In such a system the controller selects the appropriate coil and delivers a trigger pulse to ignition control circuitry at the correct time for each cylinder. (*Note:* In some cases the coil is on the spark plug as an integral unit.)

Fig. 4.1-9a illustrates such a system as an example of a 4-cylinder engine. In this example a pair of coils provides the spark for firing two cylinders for each coil. Cylinder pairs are selected such that one cylinder is on its compression stroke while the other is on exhaust. The cylinder on compression is the cylinder to be fired (at a time somewhat before it reaches TDC). The other cylinder is on exhaust.

The coil fires the spark plugs for these two cylinders simultaneously. For the former cylinder, the mixture is ignited and combustion begins for the power stroke that follows. For the other cylinder (on exhaust stroke), the combustion has already taken place and the spark has no effect.

Although the mixture for modern emission-regulated engines is constrained by emissions regulations, the spark timing can be varied in order to achieve optimum performance within the mixture constraint. For example, the ignition timing can be chosen to produce the best possible engine torque for any given operating condition. This optimum ignition timing is known for any given

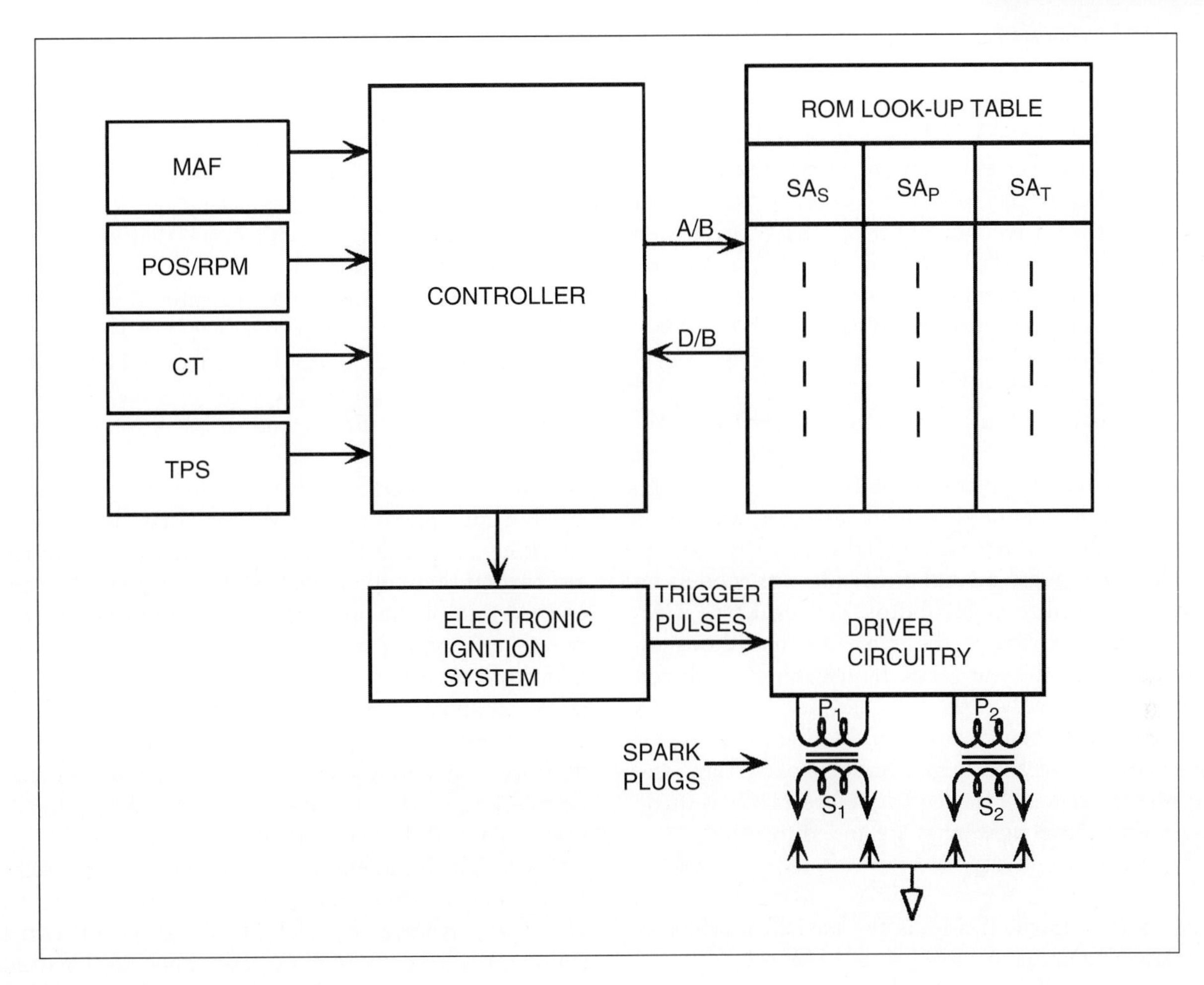

Fig. 4.1-9a Distributorless ignition system.

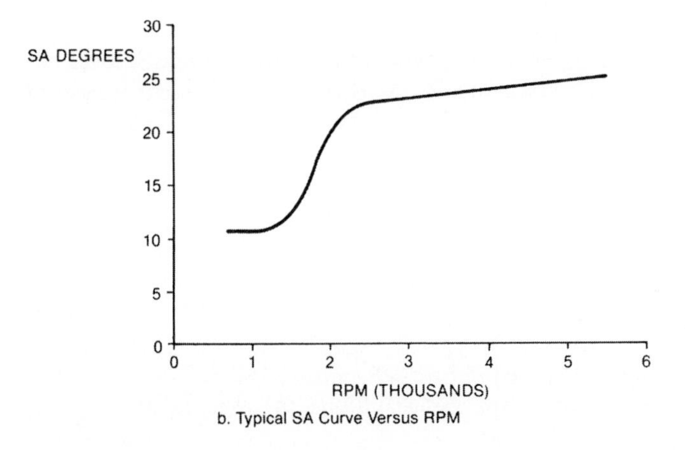

Fig. 4.1-9b Spark advance versus RPM.

engine configuration from studies of engine performance as measured on an engine dynamometer.

Fig. 4.1-9a is a schematic of a representative electronic ignition system. In this example configuration the spark advance (SA) value is computed in the main engine control (i.e., the controller that regulates fuel). This system receives data from the various sensors (as described above with respect to fuel control) and determines the correct SA for the instantaneous operating condition.

The variables that influence the optimum spark timing at any operating condition include RPM, manifold pressure (or MAF), barometric pressure, and CT. The correct ignition timing for each value of these variables is stored in an ROM lookup table. For example, the variation of SA with RPM for a representative engine is shown in Fig. 4.1-9b. The engine control system obtains readings from the various sensors and generates an address to the lookup table (ROM). After reading the data from the lookup tables, the control system computes the correct SA. An output signal is generated at the appropriate time to activate the spark.

In the configuration depicted in Fig. 4.1-9a, the electronic ignition is implemented in a stand-alone ignition module. This solid-state module receives the correct SA data and generates electrical signals that operate the coil driver circuitry. These signals are produced in response to timing inputs coming from crankshaft and camshaft signals (POS/RPM).

The coil driver circuits generate the primary current in windings P_1 and P_2 of the coil packs depicted in Fig. 4.1-9a. These primary currents build up during the so-called *dwell period* before the spark is to occur. At the correct time the driver circuits interrupt the primary currents via a solid-state switch. This interruption of the primary current causes the magnetic field in the coil pack to drop rapidly, inducing a very high voltage (20,000–40,000 volts) that causes a spark. In the example depicted in Fig. 4.1-9a, a pair of coil packs, each firing two spark plugs, is shown. Such a configuration would be appropriate for a 4-cylinder engine. Normally there would be one coil pack for each pair of cylinders.

The ignition system described above is known as a *distributorless ignition system* (DIS) since it uses no distributor. There are a number of older car models on the road that utilize a distributor. However, the electronic ignition system is the same as that shown in Fig. 4.1-9a, up to the coil packs. In distributor-equipped engines there is only one coil, and its secondary is connected to the rotary switch (or distributor).

In a typical electronic ignition control system, the total spark advance, *SA* (in degrees before TDC), is made up of several components that are added together:

$$SA = SA_S + SA_P + SA_T$$

The first component, SA_S, is the basic SA, which is a tabulated function of RPM and MAP. The control system reads RPM and MAP, and calculates the address in ROM of the SA_S that corresponds to these values. Typically, the advance of RPM from idle to about 1200 RPM is relatively slow. Then, from about 1200 to about 2300 RPM the increase in RPM is relatively quick. Beyond 2300 RPM, the increase in RPM is again relatively slow. Each engine configuration has its own SA characteristic, which is normally a compromise between a number of conflicting factors (the details of which are beyond the scope of this book).

The second component, SA_P is the contribution to SA due to manifold pressure. This value is obtained from ROM lookup tables. Generally speaking, the SA is reduced as pressure increases.

The final component, SA_T, is the contribution to SA due to temperature. Temperature effects on SA are relatively complex, including such effects as cold cranking, cold start, warm-up, and fully warmed-up conditions, and are beyond the scope of this book.

4.1.7.1 Closed-loop ignition timing

The ignition system described in the foregoing section is an open-loop system. The major disadvantage of open-loop control is that it cannot automatically compensate for mechanical changes in the system. Closed-loop control of ignition timing is desirable from the standpoint of improving engine performance and maintaining that performance in spite of system changes.

One scheme for closed-loop ignition timing is based on the improvement in performance that is achieved by advancing the ignition timing relative to TDC. For a given RPM and manifold pressure, the variation in torque with SA is as depicted in Fig. 4.1-10. One can see that advancing the spark relative to TDC increases the torque until a point is reached at which best torque is produced. This SA is known as *mean best torque*, or MBT.

When the spark is advanced too far, an abnormal combustion phenomenon occurs that is known as *knocking*. Although the details of what causes knocking are beyond the scope of this book, it is generally a result of a portion of the air–fuel mixture autoigniting, as opposed to being normally ignited by the advancing flame front that occurs in normal combustion following spark ignition. Roughly speaking, the amplitude of knock is proportional to the fraction of the total air and fuel mixture that autoignites. It is characterized by an abnormally rapid rise in cylinder pressure during combustion, followed by very rapid oscillations in cylinder pressure. The frequency of these oscillations is specific to a given engine configuration and is typically in the range of a few kilohertz. Fig. 4.1-11 is a graph of a typical cylinder pressure versus time under knocking conditions. A relatively low level of knock is arguably beneficial to performance, although excessive knock is unquestionably damaging to the engine and must be avoided.

One control strategy for SA under closed-loop control is to advance the spark timing until the knock level becomes unacceptable. At this point, the control system reduces the SA (retarded spark) until acceptable levels of knock are achieved. Of course, an SA control scheme based on limiting the levels of knocking requires a knock

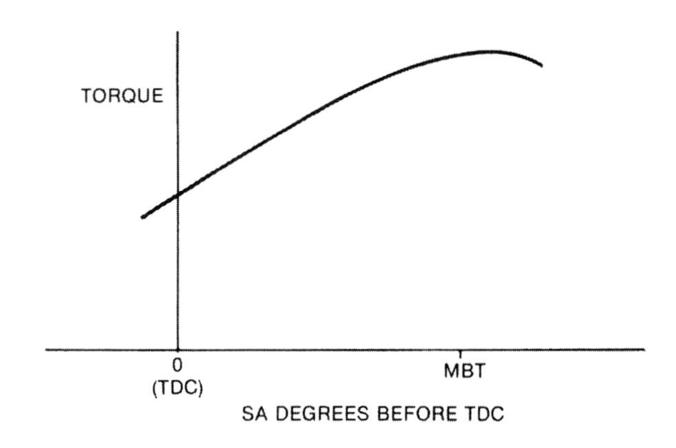

Fig. 4.1-10 Torque versus SA for typical engine.

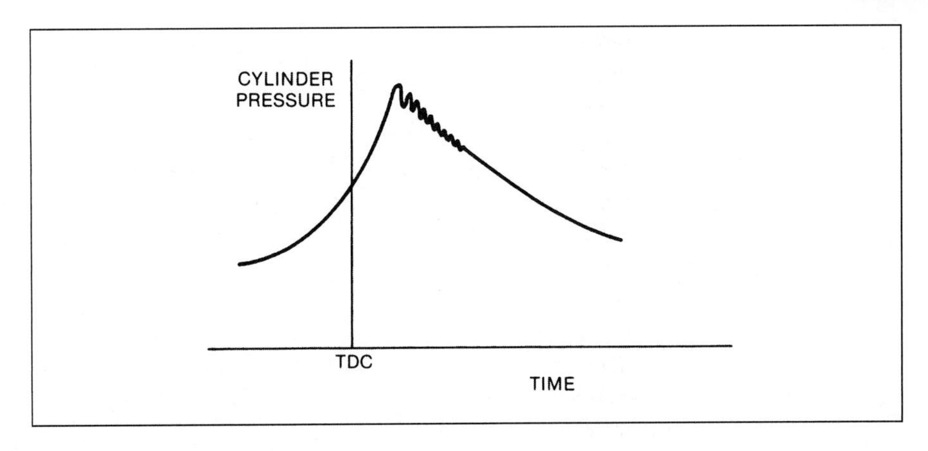

Fig. 4.1-11 Cylinder pressure (knocking condition).

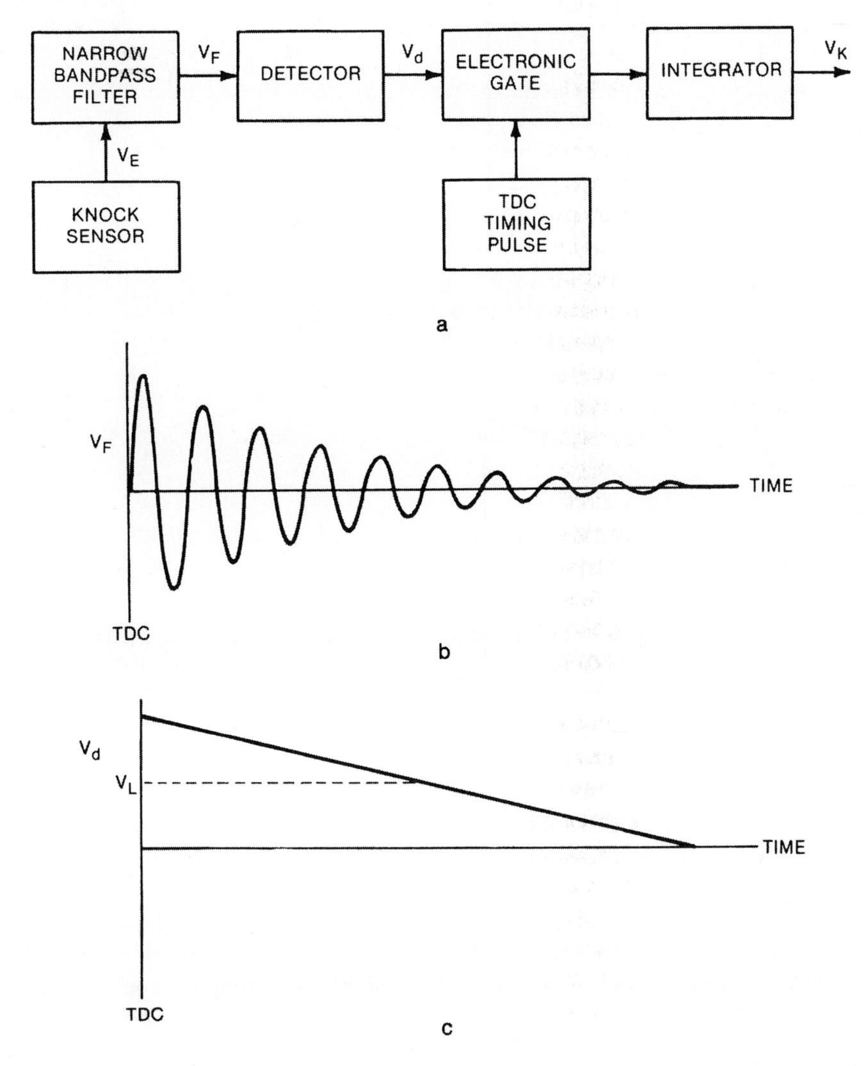

Fig. 4.1-12 Instrumentation and waveforms for closed-loop ignition control.

sensor. This sensor responds to the acoustical energy in the spectrum of the rapid cylinder pressure oscillations, as shown in Fig. 4.1-11.

Fig. 4.1-12 is a diagram of the instrumentation for measuring knock intensity. Output voltage V_E of the knock sensor is proportional to the acoustical energy in the engine block at the sensor mounting point. This voltage is sent to a narrow band-pass filter that is tuned to the knock frequency. The filter output voltage V_F is proportional to the amplitude of the knock oscillations,

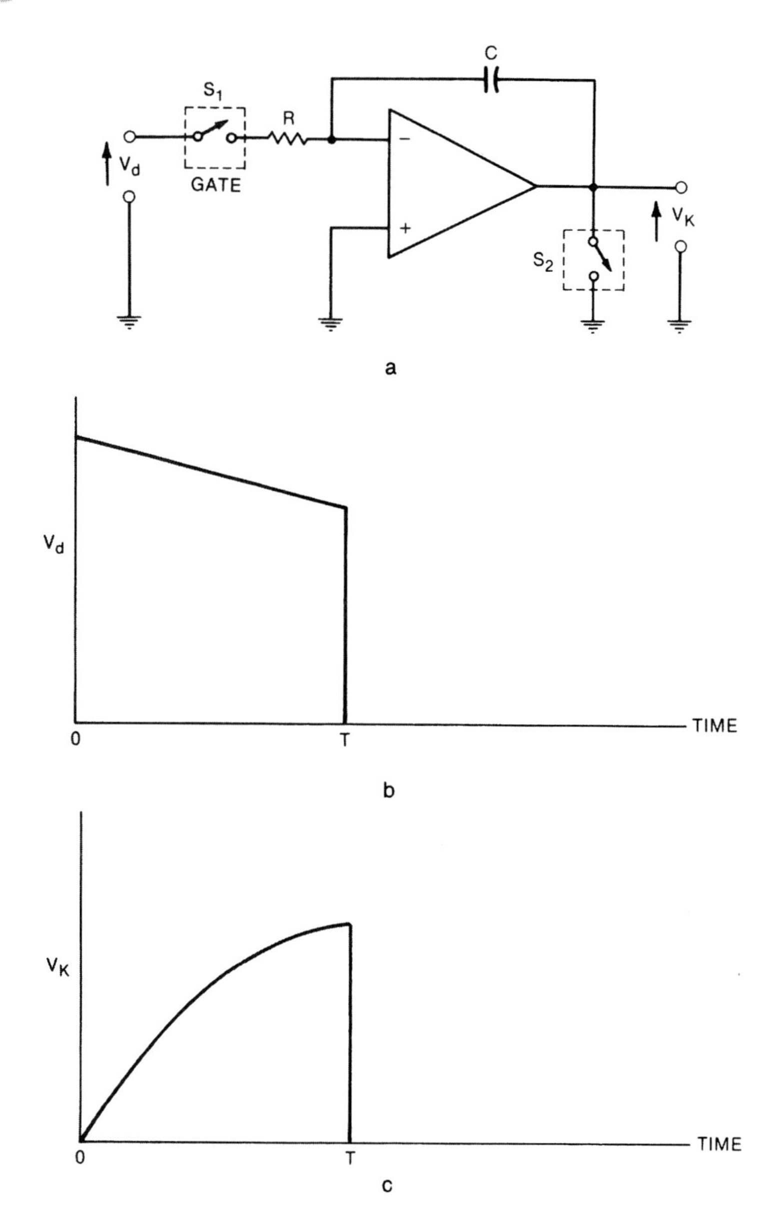

Fig. 4.1-13 Example integrator circuit diagram.

and is thus a "knock signal." The envelope voltage of these oscillations, V_d, is obtained with a detector circuit. This voltage is sent to the controller, where it is compared with a level corresponding to the knock intensity threshold. Whenever the knock level is less than the threshold, the spark is advanced; whenever it exceeds the threshold, the spark is retarded.

Following the detector in the circuit of Fig. 4.1-12 is an electronic gate that examines the knock sensor output at the time for which the knock amplitude is largest (i.e., shortly after TDC). The gate is, in essence, an electronic switch that is normally open, but is closed for a short interval (from 0 to T) following TDC. It is during this interval that the knock signal is largest in relationship to engine noise. The probability of successfully detecting the knock signal is greatest during this interval. Similarly, the possibility of mistaking engine noise for true knock signal is smallest during this interval.

The final stage in the knock-measuring instrumentation is integration with respect to time; this can be accomplished using an operational amplifier. For example, the circuit of Fig. 4.1-13 could be used to integrate the gate output. The electronic gate actually controls switches S_1 and S_2. The output voltage V_K at the end of the gate interval T is given by:

$$V_K = -(1/RC)\int_0^T V_d(t)dt$$

This voltage increases sharply (negative), reaching a maximum amplitude at the end of the gate interval, as

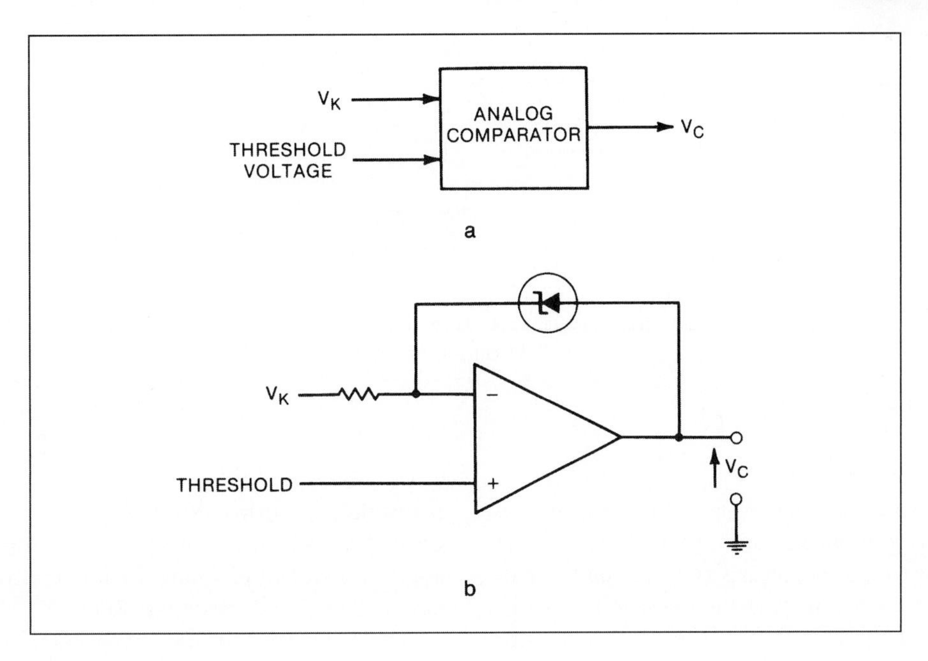

Fig. 4.1-14 Knock level detector circuit.

shown in Fig. 4.1-13, provided that knock occurs. However, if there is no knock, V_K remains near zero.

The level of knock intensity is indicated by voltage V_K (T) at the end of the gate interval. The spark control system compares this voltage with a threshold voltage (using an analog comparator) to determine whether knock has or has not occurred (Fig. 4.1-14). The comparator output voltage is binary valued, depending on the relative amplitude of V_K (T) and the threshold voltage. Whenever V_K (T) is less than the threshold voltage, the comparator output is low, indicating no knock. Whenever V_K (T) is greater than the threshold value, the comparator output is high, indicating knock.

Although this scheme for knock detection has shown a constant threshold, there are some production applications that have a variable threshold. The threshold in such cases increases with RPM because the competing noises in the engine increase with RPM.

4.1.7.2 SA correction scheme

Although the details of SA control vary from manufacturer to manufacturer, there are generally two classes of correction that are used: fast correction and slow correction. In the fast correction scheme, the SA is decreased for the next engine cycle by a fixed amount (typically from 5° to 10°) whenever knock is detected. Then the SA is advanced in one-degree increments every 5 to 20 crankshaft revolutions.

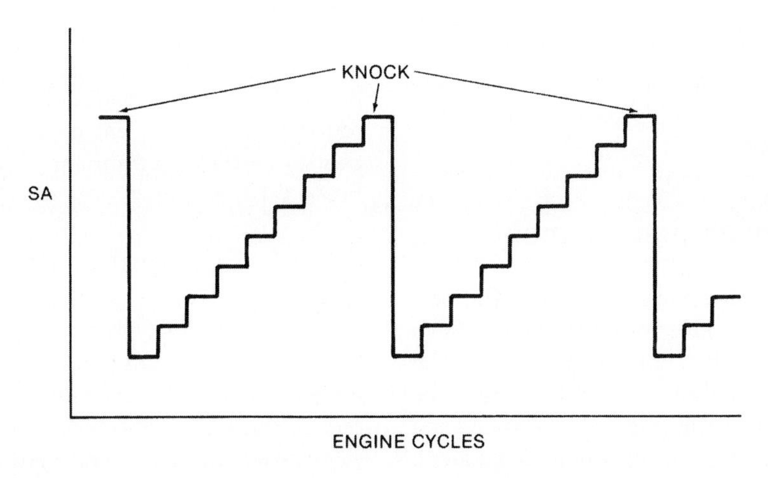

Fig. 4.1-15 Fast correction SA.

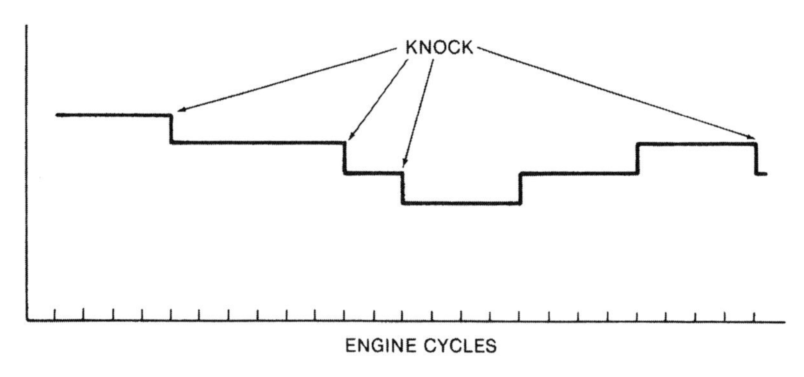

Fig. 4.1-16 Slow correction SA.

The fast correction ensures that minimum time is spent under heavy knocking conditions. Further, this scheme compensates for hysteresis (i.e., for one degree of SA to cause knocking, more than one degree must be removed to eliminate knocking). The fast correction scheme is depicted in Fig. 4.1-15.

In the slow correction scheme (Fig. 4.1-16), SA is decreased by one (or more) degree each time knock is detected, until no knocking is detected. The SA proceeds in one-degree increments after many engine cycles.

The slow correction scheme is more of an adaptive closed-loop control than is the fast correction scheme. It is primarily employed to compensate for relatively slow changes in engine condition or fuel quality (i.e., octane rating).

4.1.8 Integrated engine control system

Each control subsystem for fuel control, spark control, and EGR has been discussed separately. However, a fully integrated electronic engine control system can include these subsystems and provide additional functions. (Usually the flexibility of the digital control system allows such expansion quite easily because the computer program can be changed to accomplish the expanded functions.) Several of these additional functions are discussed in the following.

4.1.8.1 Secondary air management

Secondary air management is used to improve the performance of the catalytic converter by providing extra (oxygen-rich) air to either the converter itself or to the exhaust manifold. The catalyst temperature must be above about 200°C to efficiently oxidize HC and CO and reduce NO_X. During engine warm-up when the catalytic converter is cold, HC and CO are oxidized in the exhaust manifold by routing secondary air to the manifold. This creates extra heat to speed the warm-up of the converter and EGO sensor, enabling the fuel controller to go to the closed-loop mode more quickly.

The converter can be damaged if too much heat is applied to it. This can occur if large amounts of HC and CO are oxidized in the manifold during periods of heavy loads, which call for fuel enrichment, or during severe deceleration. In such cases, the secondary air is directed to the air cleaner, where it has no effect on exhaust temperatures.

After warm-up, the main use of secondary air is to provide an oxygen-rich atmosphere in the second chamber of the three-way catalyst, dual-chamber converter system. In a dual-chamber converter, the first chamber contains rhodium, palladium, and platinum to reduce NO_X and to oxidize HC and CO. The second chamber contains only platinum and palladium. The extra oxygen from the secondary air improves the converter's ability to oxidize HC and CO in the second converter chamber.

The computer program for the control mode selection logic can be modified to include the conditions for controlling secondary air. The computer controls secondary air by using two solenoid valves similar to the EGR valve. One valve switches air flow to the air cleaner or to the exhaust system. The other valve switches air flow to the exhaust manifold or to the converter. The air routing is based on engine CT and air/fuel ratio. The control system diagram for secondary air is shown in Fig. 4.1-17.

4.1.8.2 Evaporative emissions canister purge

During engine-off conditions, the fuel stored in the fuel system tends to evaporate into the atmosphere. To reduce these HC emissions, the fuel tank is sealed and evaporative gases are collected by a charcoal filter in a canister. The collected fuel is released into the intake

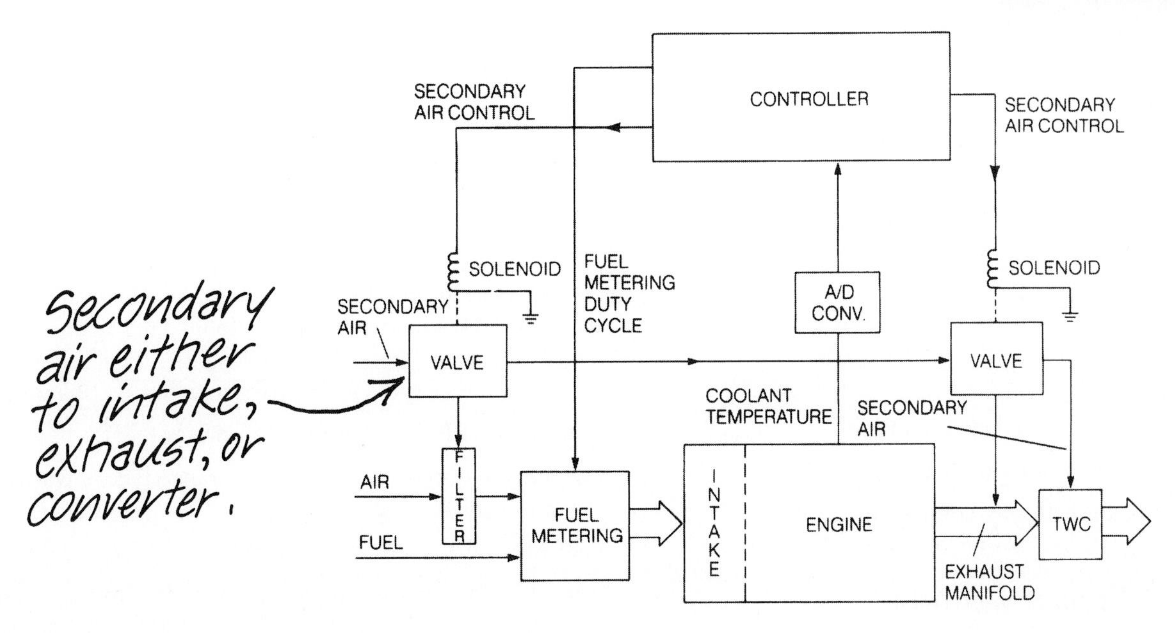

Fig. 4.1-17 Secondary air control system.

through a solenoid valve controlled by the computer. This is done during closed-loop operation to reduce fuel calculation complications in the open-loop mode.

4.1.8.3 Automatic system adjustment

Another important feature of microcomputer engine control systems is their ability to be programmed to adapt to parameter changes. Many control systems use this feature to enable the computer to modify lookup table values for computing open-loop air/fuel ratios. While the computer is in the closed-loop mode, the computer checks its open-loop calculated air/fuel ratios and compares them with the closed-loop average limit-cycle values. If they match closely, the open-loop lookup tables are unchanged. If the difference is large, the system controller corrects the lookup tables so that the open-loop values more closely match the closed-loop values. This updated open-loop lookup table is stored in separate memory (RAM), which is always powered directly by a car battery so that the new values are not lost while the ignition key is turned off. The next time the engine is started, the new lookup table values will be used in the open-loop mode and will provide more accurate control of the air/fuel ratio. This feature is very important because it allows the system controller to adjust to long-term changes in engine and fuel system conditions. This feature can be applied in individual subsystem control systems or in the fully integrated control system. If not available initially, it may be added to the system by modifying its control program.

4.1.8.4 System diagnosis

Another important feature of microcomputer engine control systems is their ability to diagnose failures in their control systems and alert the operator. Sensor and actuator failures or misadjustments can be easily detected by the computer. For instance, the computer will detect a malfunctioning MAF sensor if the sensor's output goes above or below certain specified limits, or fails to change for long periods of time. A prime example is the automatic adjustment system just discussed. If the open-loop calculations consistently come up wrong, the engine control computer may determine that one of the many sensors used in the open-loop calculations has failed.

If the computer detects the loss of a primary control sensor or actuator, it may choose to operate in a different mode until the problem is repaired. The operator is notified of a failure by an indicator on the instrument panel (e.g., check engine). Because of the flexibility of the microcomputer engine control system, additional diagnostic programs might be added to accommodate different engine models that contain more or fewer sensors. Keeping the system totally integrated gives the microcomputer controller access to more sensor inputs so they can be checked.

4.1.9 Summary of control modes

Now that a typical electronic engine control system has been discussed, let's summarize what happens in an integrated system operating in the various modes.

4.1.9.1 Engine crank (start)

The following list is a summary of the engine operations in the engine crank (starting) mode. Here, the primary control concern is reliable engine start.

1. Engine RPM at cranking speed.
2. Engine coolant at low temperature.
3. Air/fuel ratio low.
4. Spark retarded.
5. EGR off.
6. Secondary air to exhaust manifold.
7. Fuel economy not closely controlled.
8. Emissions not closely controlled.

4.1.9.2 Engine warm-up

While the engine is warming up, the engine temperature is rising to its normal operating value. Here, the primary control concern is rapid and smooth engine warm-up. A summary of the engine operations during this period follows:

1. Engine RPM above cranking speed at command of driver.
2. Engine CT rises to minimum threshold.
3. Air/fuel ratio low.
4. Spark timing set by controller.
5. EGR off.
6. Secondary air to exhaust manifold.
7. Fuel economy not closely controlled.
8. Emissions not closely controlled.

4.1.9.3 Open-loop control

The following list summarizes the engine operations when the engine is being controlled with an open-loop system. This is before the EGO sensor has reached the correct temperature for closed-loop operation. Fuel economy and emissions are closely controlled.

1. Engine RPM at command of driver.
2. Engine temperature above warm-up threshold.
3. Air/fuel ratio controlled by an open-loop system to 14.7.
4. EGO sensor temperature less than minimum threshold.
5. Spark timing set by controller.
6. EGR controlled.
7. Secondary air to catalytic converter.
8. Fuel economy controlled.
9. Emissions controlled.

4.1.9.4 Closed-loop control

For the closest control of emissions and fuel economy under various driving conditions, the electronic engine control system is in a closed loop. Fuel economy and emissions are controlled very tightly. The following is a summary of the engine operations during this period:

1. Engine RPM at command of driver.
2. Engine temperature in normal range (above warm-up threshold).
3. Average air/fuel ratio controlled to 14.7, ± 0.05.
4. EGO sensor's temperature above minimum threshold detected by a sensor output voltage indicating a rich mixture of air and fuel for a minimum amount of time.
5. System returns to open loop if EGO sensor cools below minimum threshold or fails to indicate rich mixture for given length of time.
6. EGR controlled.
7. Secondary air to catalytic converter.
8. Fuel economy tightly controlled.
9. Emissions tightly controlled.

4.1.9.5 Hard acceleration

When the engine must be accelerated quickly or if the engine is under heavy load, it is in a special mode. Now, the engine controller is primarily concerned with providing maximum performance. Here is a summary of the operations under these conditions:

1. Driver asking for sharp increase in RPM or in engine power, demanding maximum torque.
2. Engine temperature in normal range.
3. Air/fuel ratio rich mixture.
4. EGO not in loop.
5. EGR off.
6. Secondary air to intake.
7. Relatively poor fuel economy.
8. Relatively poor emissions control.

4.1.9.6 Deceleration and idle

Slowing down, stopping, and idling are combined in another special mode. The engine controller is primarily concerned with reducing excess emissions during deceleration, and keeping idle fuel consumption at a minimum. This engine operation is summarized in the following list.

1. RPM decreasing rapidly due to driver command or else held constant at idle.
2. Engine temperature in normal range.
3. Air/fuel ratio lean mixture.
4. Special mode in deceleration to reduce emissions.
5. Special mode in idle to keep RPM constant at idle as load varies due to air conditioner, automatic transmission engagement, etc.
6. EGR on.
7. Secondary air to intake.
8. Good fuel economy during deceleration.
9. Poor fuel economy during idle, but fuel consumption kept to minimum possible.

4.1.10 Improvements in electronic engine control

The digital engine control system in this chapter has been made possible by a rapid evolution of the state of technology. Some of this technology has been briefly mentioned in this chapter. It is worthwhile to review some of the technological improvements that have occurred in digital engine control in greater detail to fully appreciate the capabilities of modern digital engine control.

4.1.10.1 Integrated engine control system

One of the developments that has occurred since the introduction of digital engine control technology is the integration of the various functions into a single control unit. Whereas the earlier systems in many cases had separate control systems for fuel and ignition control, the trend is toward integrated control. This trend has been made possible, in part, by improvements in digital hardware and in computational algorithms and software. For example, one of the hardware improvements that has been achieved is the operation of the microprocessor unit (MPU) at higher clock frequencies. This higher frequency results in a reduction of the time for any given MPU computation, thereby permitting greater computational capability. This increased computational capability has made it possible, in turn, to have more precise control of fuel delivery during rapid transient engine operation.

Except for long steady cruise while driving on certain rural roads or freeways, the automobile engine is operated under changing load and RPM conditions. The limitations in the computational capability of early engine control systems restricted the ability of the controller to continuously maintain the air/fuel ratio at stoichiometry under such changing operating conditions. The newer, more capable digital engine control systems are more precise than the earlier versions at maintaining stoichiometry and therefore operate more of the time within the optimum window for the three-way catalytic converter.

Moreover, since the control of fuel and ignition requires, in some cases, data from the same sensor set, it is advantageous to have a single integrated system for fuel and ignition timing control. The newer engine controllers have the capability to maintain stoichiometry and simultaneously optimize ignition timing.

4.1.10.2 Oxygen sensor improvements

Improvements have also been made in the EGO sensor, which remains today as the primary sensor for

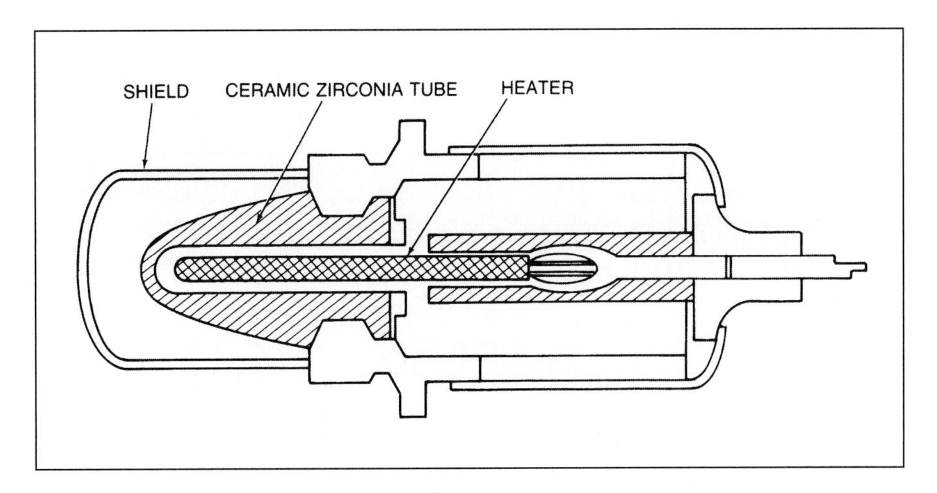

Fig. 4.1-18 Heated EGO sensor.

closed-loop operation in cars equipped with the three-way catalyst. As we have seen, the signal from the oxygen sensor is not useful for closed-loop control until the sensor has reached a temperature of about 300°C. Typically, the temperature of the sensor is too low during the starting and engine warm-up phase, but it can also be too low during relatively long periods of deceleration. It is desirable to return to closed-loop operation in as short a time as possible. Thus, the oxygen sensor must reach its minimum operating temperature in the shortest possible time.

An improved EGO sensor has been developed that incorporates an electric heating element inside the sensor, as shown in Fig. 4.1-18. This EGO sensor is known as the heated exhaust gas oxygen, or HEGO, sensor. The heat current is automatically switched on and off depending on the engine operating condition. The operating regions in which heating is applied are determined by the engine control system as derived from engine RPM and MAP sensors. The heating element is made from resistive material and derives heat from the power dissipated in the associated resistance. The HEGO sensor is packaged in such a way that this heat is largely maintained within the sensor housing, thereby leading to a relatively rapid temperature rise.

Normally, the heating element need only be turned on for cold-start operations. Shortly after engine start the exhaust gas has sufficient heat to maintain the EGO sensor at a suitable temperature.

4.1.10.3 Fuel injection timing

Earlier in this chapter, the fuel control methods and algorithms were explained for a sequential multipoint fuel injection system. In such a fuel control system, it was shown that a separate fuel injector is provided for each cylinder. The fuel injector for each cylinder is typically mounted in the intake manifold such that fuel is sprayed directly into the intake port of the corresponding cylinder during the intake stroke.

During the intake stroke, the intake valve is opened and the piston is moving down from TDC. Fig. 4.1-19 illustrates the timing for the fuel injectors for a 4-cylinder engine. It can be seen that in two complete engine revolutions (as indicated by the No. 1 cylinder position), all four injectors have been switched on for a time $T(n)$. This pulse duration results in delivery of the desired quantity of fuel for the nth engine cycle. This system provides for highly uniform fueling of all the cylinders and is superior in performance to either carburetors or TBFIs.

4.1.10.4 Automatic transmission control

The vast majority of cars and light trucks sold in the United States is equipped with automatic transmissions. The majority of these transmissions are controlled electronically. An automatic transmission consists of a torque converter and a sequence of planetary gear sets. Control of an automatic transmission consists of selecting the appropriate gear ratio from input shaft to output shaft as a function of operating condition. The operating condition in this case includes load, engine RPM, and vehicle speed (or equivalently RPM of the drive shaft). The gear ratio for the transmission is set by activating clutches on the components of the various planetary gear systems.

The relevant clutches are activated by the pressure of transmission fluid acting on piston-like mechanisms. The pressure is switched on at the appropriate clutch via solenoid-activated valves that are supplied with automatic transmission fluid under pressure.

The gear ratio for the planetary gear sets is uniquely determined by the combination of clutches that are activated. The electronic transmission controller determines the desired gear ratio from measurements of engine load

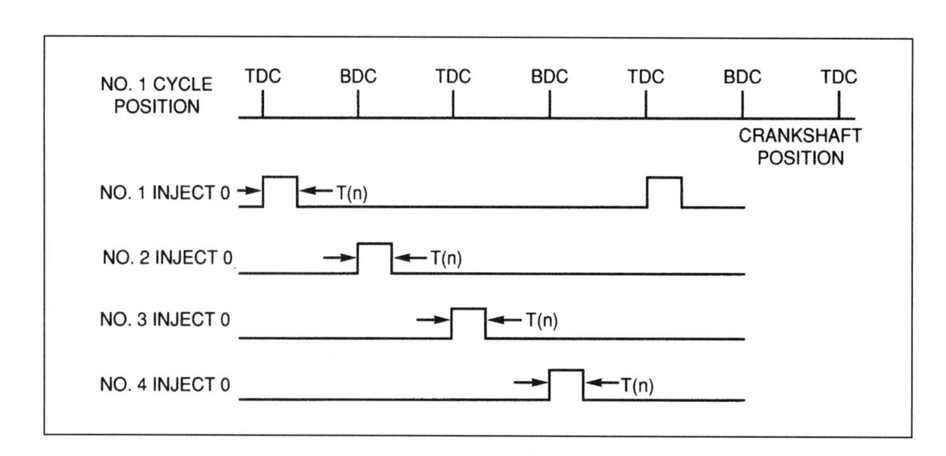

Fig. 4.1-19 Injector timing for 4-cylinder engine.

and RPM as well as transmission output shaft RPM. These RPM measurements are made using noncontacting angular speed sensors (usually magnetic in nature). Engine load can be measured directly from MAP or from MAF and from a somewhat complicated algorithm relating these measurements to desired gear ratio. Once this desired gear ratio is determined, the set of clutches to be activated is uniquely determined, and control signals are sent to the appropriate clutches.

Normally, the highest gear ratio (i.e., ratio of input shaft speed to output shaft speed) is desired when the vehicle is at low speed such as in accelerating from a stop. As vehicle speed increases from a stop, a switching level will be reached at which the next lowest gear ratio is selected. This switching (gear-changing) threshold is an increasing function of load (i.e., MAP).

At times (particularly under steady vehicle speed conditions), the driver demands increasing engine power (e.g., for heavy acceleration). In this case, the controller shifts to a higher gear ratio, resulting in higher acceleration than would be possible in the previous gear setting. The functional relationship between gear ratio and operating condition is often termed the "shift schedule."

4.1.10.5 Torque converter lock-up control

Automatic transmissions use a hydraulic or fluid coupling to transmit engine power to the wheels. Because of slip, the fluid coupling is less efficient than the nonslip coupling of a pressure-plate manual clutch used with a manual transmission. Thus, fuel economy is usually lower with an automatic transmission than with a standard transmission. This problem has been partially remedied by placing a clutch functionally similar to a standard pressure-plate clutch inside the torque converter of the automatic transmission and engaging it during periods of steady cruise. This enables the automatic transmission to provide fuel economy near that of a manual transmission and still retain the automatic shifting convenience.

Here is a good example of the ease of adding a function to the electronic engine control system. The torque converter locking clutch (TCC) is activated by a lock-up solenoid controlled by the engine control system computer. The computer determines when a period of steady cruise exists from throttle position and vehicle speed changes. It pulls in the locking clutch and keeps it engaged until it senses conditions that call for disengagement.

4.1.10.6 Traction control

It was earlier explained that the transmission output shaft is coupled to the drive axles via the differential. The differential is a necessary component of the drivetrain because the left and right drive wheels turn at different speeds whenever the car moves along a curve (e.g., turning a corner). Unfortunately, wherever there is a large difference between the tire/road friction from left to right, the differential will tend to spin the low friction wheel. An extreme example of this occurs whenever one drive wheel is on ice and the other is on dry road. In this case, the tire on the ice side will spin and the wheel on the dry side will not. Typically, the vehicle will not move in such circumstances.

Certain cars are equipped with so-called traction control devices that can overcome this disadvantage of the differential. Such cars have differentials that incorporate solenoid-activated clutches that can "lock" the differential, permitting power to be delivered to both drive wheels. It is only desirable to activate these clutches in certain conditions and to disable them during normal driving, permitting the differential to perform its intended task.

A traction control system incorporates sensors for measuring wheel speed and a controller that determines the wheel-slip condition based on these relative speeds. Wherever a wheel-spin condition is detected, the controller sends electrical signals to the solenoids, thereby activating the clutches to eliminate the wheel slip.

4.1.10.7 HV powertrain control

The propulsive power coming from an ICE and an EM, is the basic concept of an HV. The HV combines the low (ideally zero) emissions of an electric vehicle with the range and performance capabilities of ICE-powered cars. However, optimization of emissions performance and/or fuel economy is a complex control problem.

There are numerous issues and considerations involved in HV powertrain control, including the efficiencies of the ICE and EM as a function of operating condition; the size of the vehicle and the power capacity of the ICE and EM; the storage capacity and state of charge (SOC) of the battery pack; accessory load characteristics of the vehicle; and finally, the driving characteristics. With respect to this latter issue, it would be possible to optimize vehicle emissions and performance if the exact route, including vehicle speed, acceleration, deceleration, road inclination, and wind characteristics, could be programmed into the control memory before any trip were to begin. It is highly impractical to do such pre-programming. However, by monitoring instantaneous vehicle operation, it is possible to achieve good, though suboptimal, vehicle performance and emissions.

Depending on operating conditions, the controller can command pure electric vehicle operation, pure ICE

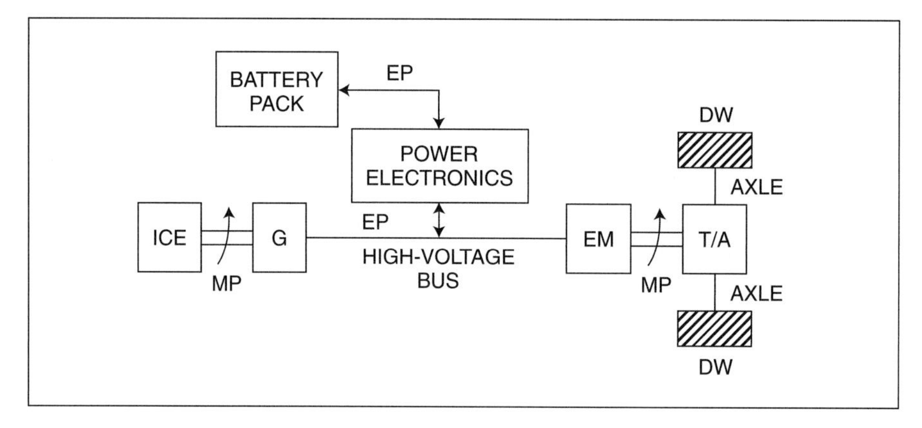

Fig. 4.1-20 SHV representation.

operation, or a combination. Whenever the ICE is operating, the controller should attempt to keep it at its peak efficiency.

Certain special operating conditions should be noted. For example, the ICE is stopped wherever the vehicle is stopped. Clearly, such stoppage benefits vehicle fuel economy and improves air quality when the vehicle is driven in dense traffic with long stoppages such as those that occur while driving in large urban areas.

There are two major types of HV vehicles depending on the mechanism for coupling the ICE and the EM. Fig. 4.1-20 is a schematic representation of one HV configuration known as a series hybrid vehicle (SHV). In this SHV, the ICE drives a generator and has no direct mechanical connection to the drive axles. The vehicle is propelled by the EM, which receives its input electrical power from a high-voltage bus (HVB). This bus, in turn, receives its power either from the engine-driven generator (for ICE propulsion) or from the battery pack (for EM propulsion), or from a combination of the two.

In this figure, mechanical power is denoted as MP and electrical power as EP. The mechanical connection from the EM to the T/A provides propulsive power to the drive wheels (DWs).

Fig. 4.1-21 is a schematic of an HV type known as a parallel hybrid. The parallel hybrid of Fig. 4.1-21a can operate with ICE alone by engaging both solenoid-operated clutches on either side of the EM but with no electrical power supplied to the EM. In this case, the MP supplied by the ICE directly drives the T/A, and the EM rotor spins essentially without any mechanical drag. This HV can also operate with the EM supplying propulsive power by switching off the ICE, disengaging clutch C_1, engaging clutch C_2, and providing electrical power to the EM from the HVB. Of course, if both ICE and EM are to produce propulsive power, then both clutches are engaged. Not shown in Fig. 4.1-21 is a separate controller for the brushless DC motor. Also not shown in this figure but discussed later in this section is the powertrain controller that optimizes performance and emissions for the overall vehicle and engages/disengages clutches as required.

The HV of Fig. 4.1-21b operates similarly to that of Fig. 4.1-21a except that mechanical power from ICE and EM are combined in a mechanism denoted coupler. For the system of Fig. 4.1-21b pure ICE propulsion involves engaging clutch C_1 disengaging clutch C_2, and providing no electrical power to the EM. Alternatively, pure EM propulsion involves disengaging clutch C_1, switching off the ICE, engaging clutch C_2, and providing electrical power to the EM via the HVB. Simultaneous ICE and EM propulsion involves running the ICE, providing electrical power to the EM, and engaging both clutches.

For either series or parallel HV, dynamic braking is possible during vehicle deceleration, with the EM acting as a generator. The EM/ generator supplies power to the HVB, which is converted to the low-voltage bus (LVB) voltage level by the power electronics subsystem. In this deceleration circumstance the energy that began as vehicle kinetic energy is recovered with the motor acting as a generator and is stored in the battery pack. This storage of energy occurs as an increase in the SOC of the battery pack. In addition to the lead acid battery in common use today, there are new energy storage means including nickel-metal-hydride (NiMH) and even special capacitors called ultra-caps. Each of these electrical energy storage technologies has advantages and disadvantages for HV application.

The battery pack has a maximum SOC that is fixed by its capacity. Dynamic braking is available as an energy recovery strategy as long as SOC is below its maximum value. Nevertheless, dynamic braking is an important part of HV fuel efficiency. It is the only way some of the energy supplied by the ICE and/or EM can be recovered instead of being dissipated in the vehicle brakes.

The storage of the energy recovered during dynamic braking requires that the corresponding electrical energy be direct current and at a voltage compatible with the

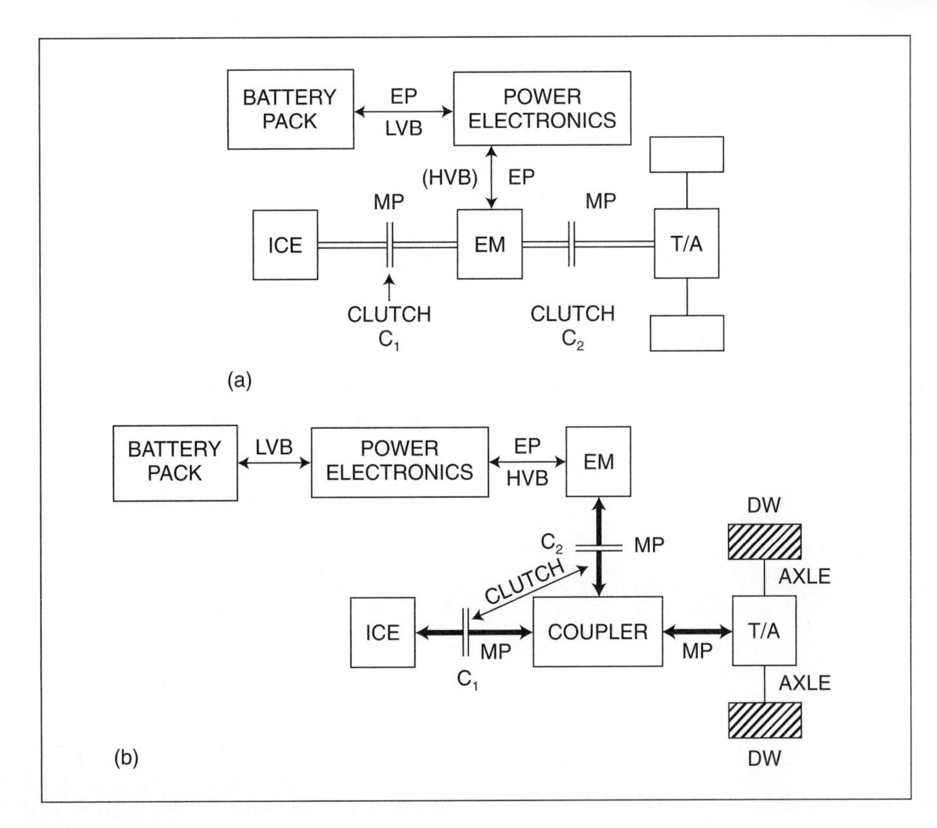

Fig. 4.1-21 Parallel hybrid schematic.

battery pack. The trend in the worldwide automotive industry is toward a 42-volt battery pack consisting, for example, of three 12-volt (rated) batteries connected in series. The 42-volt system receives this nominal rating since a fully charged (so-called 12-volt) storage battery as well as the LVB voltage is approximately 14 volts. Thus, the 42-volt terminology is a suitable way to represent this type of battery pack.

On the other hand, efficient EM operation is achieved for a much higher voltage level than the 42-volt LVB voltage. The desired HVB voltage for supplying the EM is something on the order of 250 volts.

Conversion of electrical power from one voltage level V_1 to a second V_2 is straightforward using a transformer as long as this power is alternating current. Fig. 4.1-22 schematically illustrates transformer structure and the conversion of voltages from one level to another.

A transformer consists of a core of magnetically permeable material (usually a steel alloy) around which a pair of closely wrapped coils are formed. One coil (termed the primary) consists of N_1 turns and the other (termed the secondary) consists of N_2 turns.

Assuming (arbitrarily) that AC electrical power comes from a source (e.g., an AC generator) at peak voltage V_1,

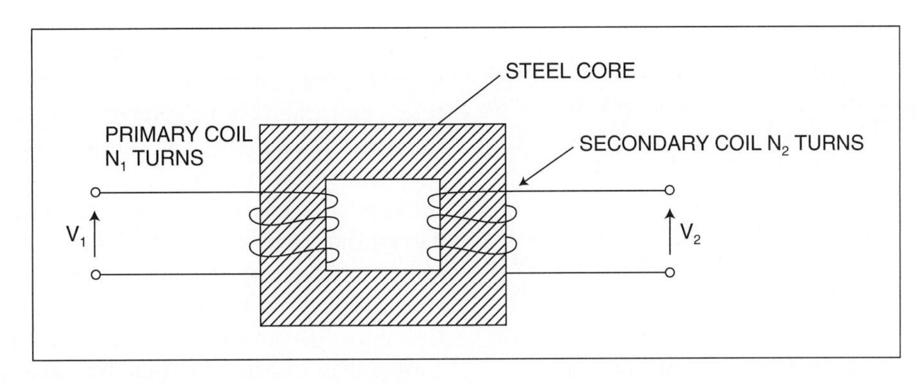

Fig. 4.1-22 Transformer structure and conversion of voltages.

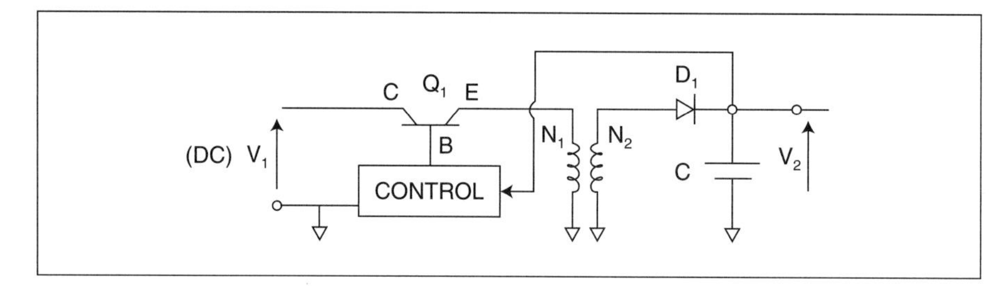

Fig. 4.1-23 DC to DC converter.

then the power flowing from the transformer secondary to a load will be at a peak voltage V_2 where

$$V_2 = (N_2/N_1)V_1$$

The validity of this simple model for a transformer depends on many factors, but for an introduction to transformer operating theory it is sufficient.

Conversion of DC electrical power from one voltage to another can be accomplished using a transformer only if the DC power is converted to AC power. Fig. 4.1-23 is a greatly simplified schematic of a DC to DC converter in which a transistor is used to convert an input DC signal to AC that is sent to a transformer for conversion to a different voltage.

The control electronics supplies a pulsating signal to the base B of transistor Q_1, alternately switching it on and off. When Q_1 is on (i.e., conducting), voltage V_1 is applied to the transformer primary (i.e., N_1). When Q_1 is off (i.e., nonconducting), transformer primary voltage is zero. In this case, the pulsating AC voltage that is alternately V_1 and 0 applied to the primary results in an AC voltage in the secondary that is essentially N_2/N_1 times the primary voltage. This secondary voltage is converted to DC by rectification using diode D_1 and filtering via capacitor C. The secondary voltage is fed back to the control electronics, which varies the relative ON and OFF times to maintain V_2 at the desired level.

A variation of the circuit of Fig. 4.1-23 appears in the power electronics module for conversion between the LVB and the HVB. Of course, the specific details of the relevant power electronics depend on the HV manufacturer.

Powertrain control for an HV is achieved using a multimode digital control system. It is somewhat more complicated than the digital engine control system discussed earlier in this chapter in that it must control an ICE as well as an EM. In addition, it must achieve the balance between ICE and EM power, and it must engage or disengage the solenoid-operated clutches (if present).

The inputs to this controller come from sensors that measure:

- Power demand from driver (accelerator pedal)
- SOC of battery pack
- Vehicle speed
- ICE RPM and load
- EM voltage and current

The system outputs include control signals to:

- ICE throttle position
- EM motor control inputs
- Clutch engage/disengage
- Switch ICE ignition on/off

In this vehicle, there is no direct mechanical link from the accelerator pedal to the throttle. Rather, the throttle position (as measured by a sensor) is set by the control system via an electrical signal sent to an actuator (motor) that moves the throttle in a system called drive-by-wire.

The control system itself is a digital controller using the inputs and outputs listed above and has the capability of controlling the hybrid powertrain in many different modes. These modes include starting from a standing stop, steady cruise, regenerative braking, recharging battery pack, and many others that are specific to a particular vehicle configuration.

In almost all circumstances, it is desirable for the ICE to be off at all vehicle stops. Clearly, it is a waste of fuel and an unnecessary contribution to exhaust emissions for an ICE to run in a stopped vehicle. Exceptions to this rule involve cold-weather operations in which it is desirable or even necessary to have some limited engine operations with a stopped vehicle. In addition, a low-battery SOC might call for ICE operation at certain vehicle stops.

When starting from a standing start, normally the EM propulsion is used to accelerate the car to desired speed, assuming the battery has sufficient charge. If charge is low, then the controller can engage the clutch to the ICE such that the EM can begin acceleration and at the same time crank the ICE to start it. Then, depending on the time that the vehicle is in motion, the ICE can provide propulsive power and/or battery charge power. Should the vehicle go to a steady cruise for engine operation near its optimum, then the control strategy normally is to switch off the electric power to the EM and power the vehicle solely with the ICE. In other cruise conditions, the controller can balance power between ICE and EM in a way that maximizes total fuel economy (subject to emission constraints).

For urban driving with frequent stops, the control strategy favors EM operation as long as SOC is sufficient. In this operating mode regenerative braking (in which energy is absorbed by vehicle deceleration), the recovered energy appears as increased SOC.

The various operating modes and control strategies for an HV depend on many factors, including vehicle weight; relative size and power capacity of ICE/EM; and exhaust emissions and fuel economy of the ICE (as installed in the particular vehicle). It is beyond the scope of this book to attempt to cover all possible operating modes for all HV configurations. However, the above discussion has provided background within which specific HV configurations' operating modes and control strategies can be understood.

Section Five

Transmissions

Chapter 5.1

5.1

Transmissions and driveline

Julian Happian-Smith

The aims of this chapter are to:

- demonstrate the need for transmission design and matching;
- give examples of common gearboxes and transmissions available for vehicle design;
- indicate the terminology and methods for transmission design;
- aid the designer to understand the elements of the analysis of transmission systems.

5.1.1 Introduction

This section introduces the transmission systems that can be found in today's passenger car. Of course, many car-derived components and systems can also be found in small commercial vehicles. Also, larger derivatives, which have much in common, can be found in heavy goods and public service vehicles. We have endeavoured to introduce the main transmission types and some areas of technology that can be found within the units. In this chapter, however, we can only hope to introduce the subject of transmissions to you. In order to make up for this brevity, we include references to other material so you are able to follow up any particular subject in greater detail.

It is probably worth stating that, in practice, the choice of transmission units for a particular vehicle is heavily influenced by what is in production and available. The cost of developing and, more importantly, installing the equipment to manufacture a new gearbox would be prohibitive for a small specialist vehicle manufacturer. Equally, producing a special transmission to support a specific model would also be difficult to justify even for a large vehicle manufacturer.

Current developments are extremely interesting as technology, particularly electronic control, is very much blurring the distinction between the conventional classes of transmission. For example, automatic transmissions (ATs) are often found now with a manual override function to allow the car to be driven using the gears selected by the driver. Conversely, manual gearboxes are having automation added to operate the clutch or shift the gears. These developments not only make the transmission interesting from an engineering perspective, but also create marketing features from an area of the vehicle often hidden from view and largely ignored by the buyer until it causes a problem.

5.1.1.1 Definitions

Transmission – This term can be used to describe one unit within the driveline of a vehicle, often the main gearbox, or as a general term for a number of units.

Driveline – This includes all of the assembly(s) between the output of the engine and the road wheel hubs.

Powertrain – Essentially the driveline and engine together, and may also be taken to include other related parts of the vehicle such as the exhaust or fuel system.

Automatic transmission – ATs come in various forms but have the common ability to change the ratio at which they are operating with no intervention from the driver.

Manual transmission – As the name suggests, drivers have to change the gear ratio setting rather than the transmission doing the job for them.

Introduction to Modern Vehicle Design; ISBN: 9780750650441
Copyright © 2001 Julian Happian-Smith. All rights of reproduction, in any form, reserved.

Continuously variable transmission (CVT) – CVTs are able to vary the ratio between input and output in a stepless manner rather than having a number of discrete ratios.

Infinitely variable transmission (IVT) – Essentially a CVT which has the additional ability to operate with zero output speed, hence negating the need for a separate starting device.

This chapter is going to look at the transmission systems used in cars. The rest of the driveline will not be considered in any detail; so there will be no detail on such things as axles or 4 × 4 transfer gearboxes.

5.1.2 What the vehicle requires from the transmission

According to some engine colleagues, the transmission is a large, expensive bracket to stop the engine from dragging on the road. However, we will, hopefully, demonstrate that transmissions are much more interesting than the other, less significant, part of the power train!

Essentially, the transmission or driveline takes the power from the engine to the wheels and, in doing so, actually makes the vehicle usable. The functions that enable this include:

- Allowing the vehicle to start from rest, with the engine running continuously.
- Leting the vehicle stop by disconnecting the drive when appropriate.
- Enabling the vehicle to start at varied rates, under a controlled manner.
- Varying the speed ratio between the engine and wheels.
- Allowing this ratio to change when required.
- Transmitting the drive torque to the required wheels.

The transmission needs to perform all of the above functions and others in a refined manner. The structural aspects of the transmission, predominantly the casing, often contribute significantly to the structure of the powertrain and the vehicle as a whole. This is important when it comes to engineering for the lowest noise and vibration. The stiffness of the powertrain assembly itself is important in determining the magnitude and the frequency of the vibrations at the source (the engine). This stiffness (and indeed the strength) can also be important to the integrity of the vehicle in a crash. Particularly with front-wheel-drive vehicles, the way in which the body collapses on impact has to be engineered very carefully, and the presence of a large rigid lump such as the powertrain has a critical influence on the way this occurs. The size, shape and orientation of the unit also affect the intrusion into the passenger space after an impact.

5.1.2.1 The layout of the vehicle

The position of the powertrain components within the vehicle has implications both for the engineering of the vehicle and the driveline components including the transmission itself. Effects include:

- the space available for the powertrain and how it is packaged within the vehicle including the location of ancillary components;
- the weight distribution, since the powertrain components are relatively heavy;
- the structure to support the powertrain and react against the driving torques;
- vehicle handling and ride both from weight distribution and the location of the driven wheel set;
- safety structure and passenger protection.

The choice of vehicle layout is determined principally by the target market sector and brand image that the vehicle is required to project. Possible alternatives include saloons, ranging from large luxury saloons to micro or town cars, sports coupés or convertibles, estate cars or off-highway vehicles. In many cases, the same vehicle platform will be used for several of these variants. The vehicle layout must also be sufficiently flexible to accommodate different engine and transmission options that are offered with many vehicles.

The main vehicle configurations in use are shown in Fig. 5.1-1. The most widely used currently is the 'standard' front-wheel-drive layout shown in Fig. 5.1-1(a). This has an engine mounted transversely to the vehicle axis with the transmission also transverse and in line with the engine. The differential can be incorporated into the transmission casing. Another possibility is shown in Fig. 5.1-1(b) with a longitudinal engine transmission assembly, again including a differential and the drive being taken to the front wheels. This configuration is used for larger front-wheel-drive vehicles where the size (i.e. length) of the engine gearbox assembly makes installation across the vehicle impossible. It also allows front-, rear- and four-wheel drive vehicles to be developed easily from the same vehicle platform as the engine installation and front structure of the vehicle can remain the same in each. The main alternative, however, is the classic front-engine rear-wheel-drive layout as in Fig. 5.1-1(c). The engine and transmission are still in line but mounted longitudinally with a connecting shaft to a separate rear mounted final drive and differential that are a part of the rear axle. A common variant amongst two-seater sports vehicles is shown in Fig. 5.1-1(d) with the engine and transmission transversely mounted to the rear of the vehicle and driving the rear wheels. If the engine is in front of the rear axle then this is usually referred to as a mid-engine layout. The final example shown in Fig. 5.1-1(e) is a four-wheel

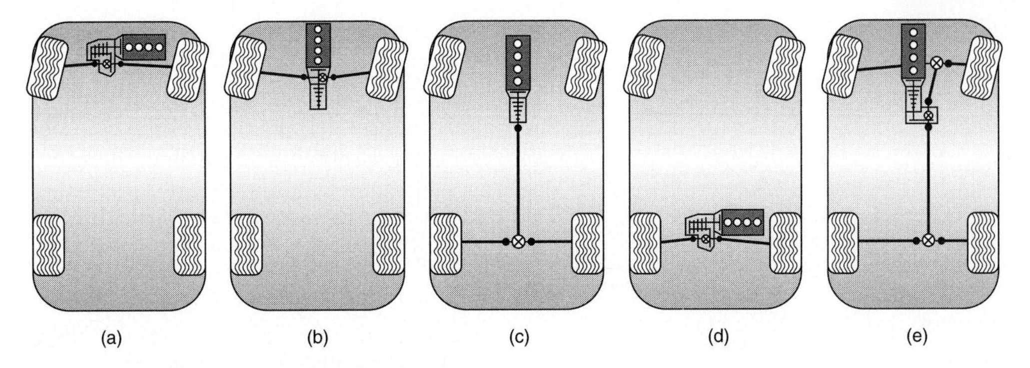

Fig. 5.1-1 Some typical vehicle/powertrain configurations.

(or all-wheel) drive power train frequently used in off-highway vehicles. The greater height of these vehicles allows the engine to be mounted above the front axle line with the front differential alongside. Variants of this also take the drive to the rear axle directly in line from the gearbox. These are normally differentiated by virtue of the transfer gearbox design. 'Double offset' being the one illustrated and 'single offset' where the drive to the rear axle is in line with the gearbox output shaft. It is also possible to derive four-wheel-drive configurations from the two-wheel-drive layouts. For example, adding a longitudinal propeller shaft from the front differential of the standard transverse layout (Fig. 5.1-1(a)) to an additional rear differential. There are many more front- and rear-wheel-drive variants than those included here, but these few account for the vast majority of vehicles on the road.

The vehicle layout adopted has consequences for the transmission itself and the necessary controls and interconnections. These include the opportunity for the differential to be included in the same casing as the transmission and eliminate the need for an additional housing. However, the transmission and differential gears must then share the same lubricating fluid. For manual gearboxes the routing of the gear-change linkage can be more complicated for the mid-engined (and other rear-engined) layouts. These also have greater complication in ancillary and cooling system layouts that are discussed in more detail in the environmental considerations in Section 5.1.6. There is also a particular fuel economy advantage for transverse layouts that do not have to turn the drive direction through a right angle. This eliminates a bevel gear set that is less efficient than parallel shaft transfer gears.

5.1.2.2 Starting from rest

As the internal combustion engine cannot provide torque at zero speed, a device is required in the transmission that will enable the vehicle to start from rest and, when propulsion is not required, to disengage the drive between the engine and road wheels. Several devices are used in automotive transmissions to achieve this:

- The single-plate dry friction clutch – used commonly with car manual gearboxes.
- The multi-plate, wet (oil immersed) clutch – frequently used in motorcycles, variable transmissions and some large, heavy-duty ATs.
- The fluid flywheel – rarely used today.
- The torque converter – used in the majority of ATs.
- Electromagnetic clutches – again used in some variable transmissions.

These devices are fitted between the engine output and transmission input. The design and application of the dry clutch and the torque converter are discussed in the sections on manual and ATs, respectively. It should be pointed out that a smaller multi-plate clutch is often used in ATs to disconnect or connect particular gears and hence allow the gear change required; these applications do not have the capacity of starting the vehicle from rest.

5.1.2.3 The vehicle requirement – what the powertrain has to deliver

If we consider the torque requirements (on the engine and driveline), there are a number of forces acting on the body of the vehicle that have to be overcome:

- The rolling resistance of the tyres.
- The aerodynamic drag of the vehicle body.
- Any resistance due to the climbing of an incline.
- Overcoming the inertia of the vehicle (as a whole) and the rotating parts, while the vehicle is accelerating.

This last point indicates that the engine also has to accelerate its own inertia; the effect of this is particularly significant in the lower gears.

Consider the first three of these that occur during steady-state conditions:

Total running resistance force $= F_{tot}$

$$F_{tot} = F_{Ro} + F_{Ae} = F_{Cl}$$

where F_{Ro} = rolling resistance $= fmg$

m = vehicle mass, kg

f = coefficient of rolling resistance – typically approx 0.013–0.015 for normal road – however, it does increase with speed.

$g = 9.81$ m/s^2 – gravitational acceleration

F_{Ae} = aerodynamic resistance $= 0.5\,\rho c_d A\,(\upsilon + \upsilon_h)^2$

ρ = air density – typically 1.2–1.3 kg/m^3 (the latter is at 'standard temperature and pressure')

c_d = drag coefficient, often around 0.3–0.4 for many cars.

A = frontal area of a vehicle in m^2

υ = vehicle speed, υ_h = headwind speed, in m/s

F_{Cl} = climbing resistance $= mg \sin \beta$

β = the gradient of the hill being climbed (degrees)

In addition to these, the engine also has to overcome any resistive forces from 'work' the vehicle may be doing, for example, towing a trailer. While operating off road, a vehicle will have to also overcome the resistance provided by the soft ground. This can vary greatly and depends on the type of soil, how wet it is and other factors such as how disturbed or compacted the ground is. These additional forces acting on a – vehicle can, in the extreme, be so large as to prevent the vehicle from moving, severely restrict the speed it is able to attain or exceed the available traction from the tyres.

Examples of how the rolling resistance and aerodynamic forces add up with increasing road speed, for a range of vehicles, are illustrated on Fig. 5.1-2. This assumes zero wind speed on level road. For example, if the vehicle were climbing an incline, the lines would move up by a constant amount. A few interesting things can be seen on this graph:

- Firstly, just compare the overall resistance of the different cars. It can be seen that both the overall magnitude and the difference between vehicles increase significantly with speed.
- Compare the difference between the older design of the Mini and the more recent Lupo; this becomes exaggerated at speed. The drag coefficients have a more pronounced effect as the speed increases.
- The very large load produced by the high weight combined with the large frontal area of the 4 × 4 vehicle.
- The difference between the medium and large can be seen to cross over as the speed increases. The heavier large car has the highest resistance load at low speeds, but then gains an advantage at higher speeds because of the better aerodynamics – this is almost certainly helped by the longer length of the body and the body style.

The load on the transmission

The total rolling resistance that has to be overcome is the load acting on the vehicle. This is seen as a torque requirement at the driving wheels(s), which can be calculated if the dynamic rolling radius of the tyre(s) is known:

$$\text{Torque at the wheel (N m)} = F_{tot} \times \text{rolling radius (m)}$$

Care should be taken as to how many wheels share the drive; hence, the torque seen by any one part of the

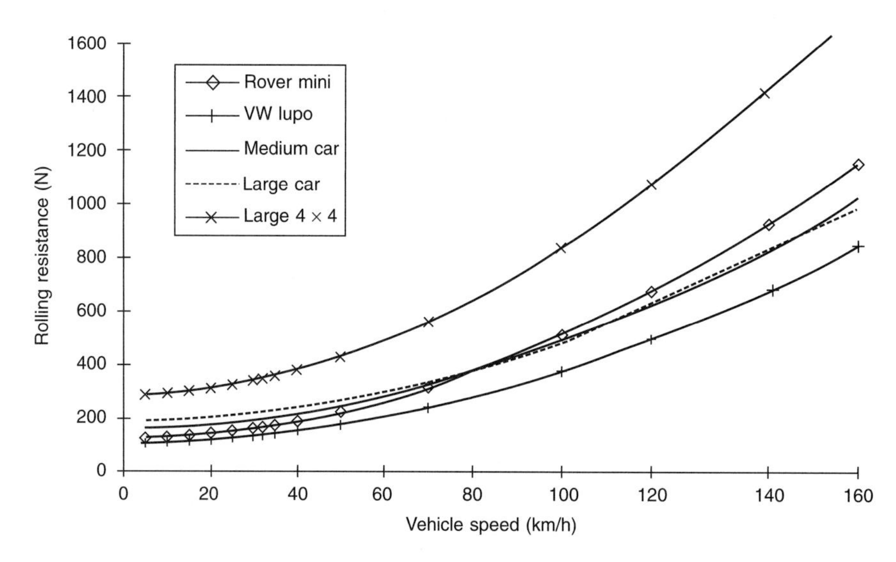

Fig. 5.1-2 Plot of total rolling resistance vs road speed (level road) – data shown for various vehicles.

driveline may not be the whole figure. The main gearbox, however, invariably sees the whole of the engine torque; so, the number of drive wheels can be ignored when considering this.

This torque value can then be calculated back up the driveline, taking account of the transmission ratio(s) and efficiency to give the torque required at the engine (at the clutch or end of the crank). By considering the rolling radius of the road wheel and the ratios in the transmission, the engine speed can also be calculated.

There are a variety of conditions at which the vehicle has to operate that determine the gear ratios to be chosen. These are likely to be modified by certain practical considerations within the transmission, but for this purpose, we can consider the initial requirements in order to determine the ratio set for the transmission.

5.1.2.4 Changing ratios – matching of the transmission to the vehicle

It is important to appreciate that the choice of gear ratios in a transmission is often dictated, in practice, by what is available or what is already in production. This situation occurs because of the large expense involved in engineering new gearsets, and installing or modifying the manufacturing plant to make the new parts. There are some cases that do necessitate a change, however. These may include a change in the engine, for example, from petrol to diesel, or a significant change to the weight of the vehicle in which the gearbox is to be installed. Obviously, the finances available within the vehicle manufacturer and the volumes involved will have a very large influence in this decision. Where changes can be accommodated, they may be limited to one or two gear ratios, leaving the intermediate ratios as is, hence not necessarily optimized. Finally, before looking at how the 'ideal' ratio may be chosen, the other limitation on ratio choice is the gear design itself. An example is a first gear pair where there could be a limit on how small the drive pinion might be in order to withstand the shock loading which can occur in the gearbox.

There are a number of decisions that need to be made when deciding what gear ratios should be fitted in a particular transmission unit. A similar process has to be done for manual transmissions, automatics and CVTs. There is more flexibility in an automatic or a CVT because of the effect of the torque converter and/or the shift map. These, in addition to the gear ratios, influence the effective, overall ratio at any point in the operating regime. The factors, which have to be taken into account, are:

- The performance requirements of the vehicle.
- The weight, rolling resistance and other parameters of the vehicle.
- The restrictions that exist on the design of the transmission.
- Packaging restrictions in the vehicle and on the engine ancillaries, if the casing has to be altered.
- Availability – as discussed above.

The performance of a vehicle is very rarely simply a matter of top speed and acceleration!

Selection of the lowest ratio – 1st gear

This governs the starting performance of the vehicle and will depend on:

- Gradient of hill required to be climbed – worst case.
- Gross (fully laden) weight of the vehicle.
- Weight of any trailer required.
- Characteristics of the engine at low engine revs – i.e. minimum engine speed for effective air inlet 'boost' on pressure-charged engines.

Selection of top gear ratio – typically 5th in passenger cars

- Engine characteristics.
- Economy requirements at cruise.
- In-gear performance – is the driver expected to change gear on overtaking?
- Top speed to be achieved in top or next gear (usually 4th) – is top gear an 'overdrive'?

The intermediate gears are usually spaced to provide an even, comfortable spread between these extremes. In theory, the ratios are often chosen to give constant speed or varying speed increments, between the gears (Fig. 5.1-3). By using constant speed increments, the engine would reduce by a consistent speed change each time the driver changed up. For example if a driver changed up while accelerating every time they reached, say 3000 rev/min, the engine speed would be the same after each gear shift. With variable speed increments, this would not be the case, usually meaning that the change in engine speed with each gearshift would get progressively smaller as higher gears were engaged. The following figures illustrate this. The 'upshift' points are shown as constant for illustration, although this is obviously not necessarily so in practice.

The particular vehicle requirements or limitations of the transmission selected can modify this spacing, for example, due to:

- Complexity requirements – existing ratio sets may limit choice on new vehicles, especially for lower-volume vehicles.
- In gear acceleration requirements – provision of particular characteristics at certain vehicle speeds, for example, achievement of 0–60 mph/100 kph without too many gear changes.
- Casing limitations on gear sizes.
- Emission and fuel economy requirements, i.e. engine conditions during the legislated drive cycle.
- Refinement issues at particular engine or driveline speeds.

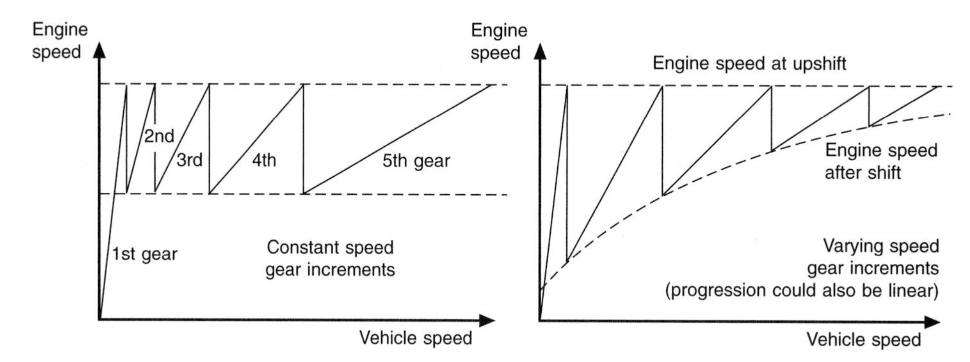

Fig. 5.1-3 Engine speed vs vehicle speed for differing gear ratio progressions.

All of these factors will influence the selection of the gear ratios in practice and possibly cause a compromise between the calculated, 'ideal' ratio set for a given car and what can be used on an existing vehicle.

Example of the considerations in matching a transmission to a vehicle

For this example, we will look at some of the factors which would need to be considered when designing a gearbox for a road car, in this case a large 4 × 4.

Consider the rolling resistance of the 4 × 4 vehicle in Fig. 5.1-2. Taking a rolling radius of 0.375 m for the tyre, the torque required at the wheel for any road speed within the range can be calculated. Consider Fig. 5.1-4 – this is a fuel consumption chart for a large petrol engine. (A line can be drawn through the top of the lines of constant fuel flow to indicate the max torque line.)

Taking some of the vehicle transmission details as:

Final drive ratio 4.2
Fifth gear 0.75
Fourth 1.0
Third 1.4

Plotting the engine conditions for 120 km/h (motorway) and assuming a loss of 5% in the transmission system give the engine conditions shown on the graph for the different gears. The required tractive force at this speed is 1100 N; this equates to a torque of 413 Nm (total) required at the wheels. In theory, this would be a nominal 103 Nm at each wheel in the case of our 4 × 4 example.

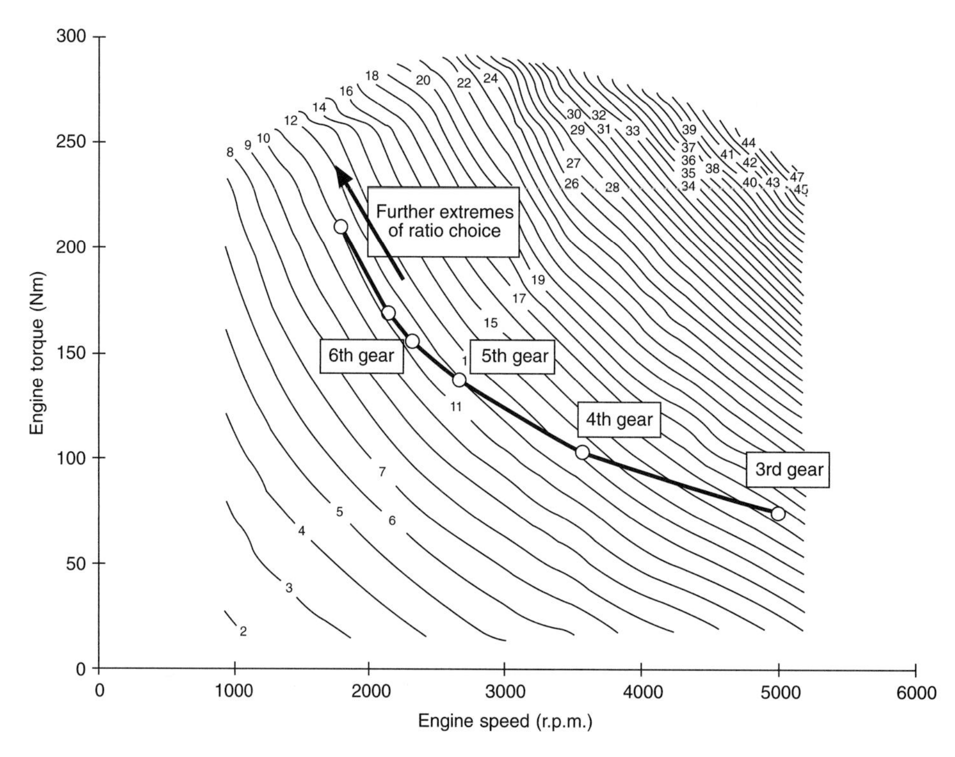

Fig. 5.1-4 Chart of fuel (mass) flow for a large petrol engine – also shows engine conditions for a range of gear ratios.

We can also calculate how fast the wheels, transmission and engine would be rotating at this speed. The rolling radius (above) means that the vehicle travels 1 km every 424 revolutions of the wheel. (Sometimes, it is easier to consider the speed of the wheel for a given road speed, in this case 7.1 rev/min per 1 km/h.) This means our wheels will be rotating at 852 rev/min at 120 km/h.

From these figures, the engine torque and speed at 120 km/h for the various gears quoted can be calculated. The operating points for the engine in the various gears show that as the vehicle changes up to 4th and 5th gear, the engine speed drops, the torque increases and the indicated fuel mass flow reduces. As we might expect, the vehicle uses less fuel in top than the lower gears.

What happens if we add an 'overdrive' sixth speed with a ratio of 0.6, or even 0.5? The line on the graph also indicates how the trend would continue if an overdrive ratio were to be added to the gearbox. The result indicates that, if taken too far, the fuel used would not necessarily continue to reduce. The engine conditions as the speed is reduced and the torque required increases are such that could find the engine to be unresponsive, requiring large throttle openings and even higher emissions due to the high engine load.

If we now consider how the tractive force ('effort') provided by the powertrain varies in each gear (by using the maximum torque values for the engine considered above). By taking account of the various gear ratios, the force provided at the road can be compared with the road load (rolling resistance). In Fig. 5.1-5, the original line from Fig. 5.1-2 has been added (again considering the 4 × 4 vehicle). An allowance has also been made for the force required to climb hills of various gradients; so additional rolling resistance lines have been added for the different gradients.

These graphs can be plotted easily for any vehicle/transmission/engine combination provided the basic information referred to above is known or can be estimated. The information provided is varied and useful; such as:

- The maximum speed attainable for different conditions and gears can be seen. In this example, we may expect the vehicle to go faster in 4th than 5th gear as the tractive effort line for level ground crosses the available force line for 5th at a lower speed than in 4th gear (and before we run out of available engine speed in the lower gear).
- The maximum gradient that the vehicle could be expected to climb in any one gear can be estimated. Here it could be assumed that 1 in 5 hills could just be climbed in 3rd gear – and at a maximum speed of about 80 km/h.
- Where the available force line is just above the required force, the close proximity of the two lines indicates that there is little, if any, available torque from the engine. So if the vehicle were on a 1 in 10 gradient at say 40 km/h in 4th gear, we might expect to be able to accelerate to nearer 120 km/h by looking at the graph. The two lines are quite close to each other, however, indicating that there is little additional torque available to accelerate the vehicle mass or accelerate the engine itself. At the very least we might expect the vehicle to be quite unresponsive.

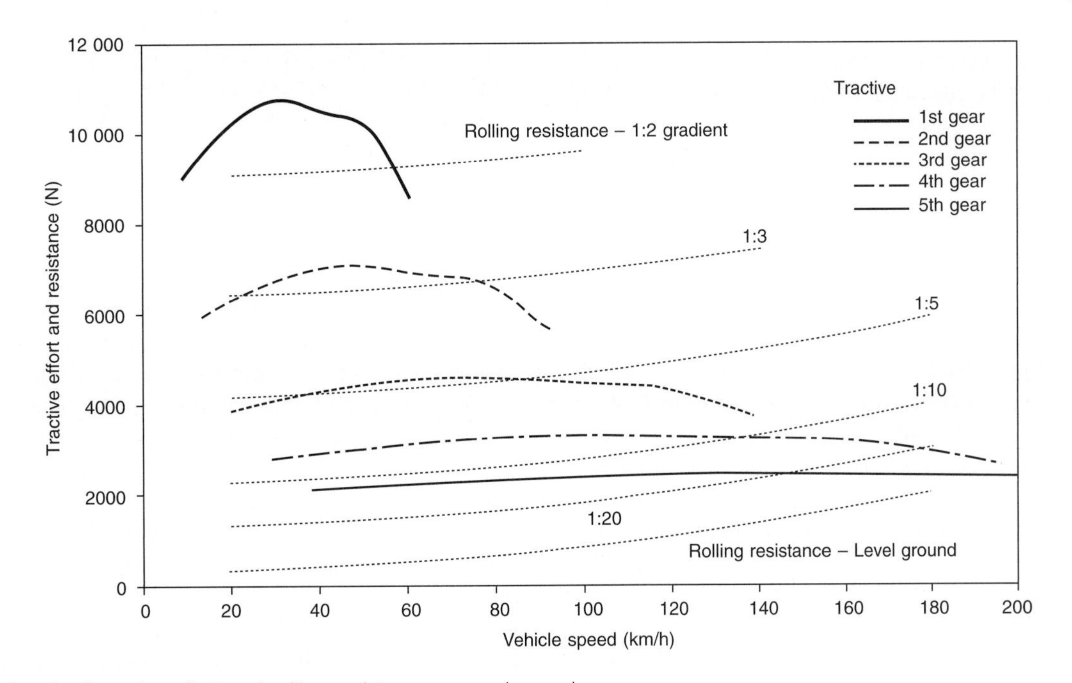

Fig. 5.1-5 Graph of tractive effort and rolling resistance vs road speed.

It should be noted that in the lower gears these graphs can indicate that very steep hills can be climbed. In practice, it may not be possible to actually start from rest on anything like these gradients because of the capacity of the clutch and the difficulty of achieving just the correct engine conditions. On two-wheel-drive vehicles, the available grip from the tyres can also be a limiting factor. Even on the 4 × 4 example we are considering, in practice, the low ratio in the transfer gearbox would be required at gradients much steeper than, say 1 in 3.

5.1.3 The manual gearbox

Most people who drive will be able to describe some aspects of one of these. As the name suggests, the driver has to change between one gear ratio and another, as the vehicle requires, when using this type of gearbox. The different gears have different ratios that allow different relative speeds between the engine and road wheels. There are several distinct types of these transmissions; including 'transverse' or 'transaxle' front-wheel-drive gearboxes and 'inline' gearboxes used in rear and four-wheel-drive vehicles. Four-wheel-drive vehicles will have an additional transmission unit on the rear of the gearbox to enable the drive of both front and rear axles.

Uses

Inline gearboxes are used in a wide range of vehicles from small passenger cars up to large trucks, while the vast majority of transverse gearboxes is used in passenger cars and small vans. It should be noted that manual gearboxes are nowhere near as common in the US and Japanese passenger car markets as they are in Europe. This is particularly the case with small to medium cars. In the past, the majority of larger passenger and commercial vehicles in Europe used manual gearboxes of one type or another. This particular area of the market is changing and becoming dominated by ATs.

Advantages

- Usually have high mechanical efficiency.
- Arguably the most fuel-efficient type of transmission, although this depends on the driver selecting the most appropriate gear.
- Relatively cheap to produce – possibly only half of the equivalent automatic.
- Light weight – typically 50–70% of the equivalent automatic weight.
- Smaller and hence usually easier to package in the vehicle.

Disadvantages

- Some driver skill required – ask anyone who only drives autos!
- Emissions and fuel consumption can be heavily influenced by the driver's gear selection.
- Clutch operation and changing gears can be tiring, especially when in heavy traffic.
- Not suitable for all drivers; controls on larger vehicles can be heavy and most require some dexterity during operation.

5.1.3.1 The front-wheel-drive passenger car gearbox

Fig. 5.1-6 is a cross section of a 'typical' front-wheel-drive transmission. The features it contains are typical of those found in many, if not all, gearboxes. The essential elements are the three shafts that take the drive from the engine (on the centre line of the crankshaft) to the output of the gearbox. From here, the driveshafts connect the gearbox to the (driven) wheel hubs. As mentioned in Section 5.1.2.1, this configuration of gearbox can also be used in rear-wheel-drive, 'mid-engined' vehicles, the installations being very similar. Two of the obvious features of this gearbox are the integral differential and the final drive gear pair. The function of the differential is described in Section 5.1.6. Because of the final drive, the overall ratio of the front-wheel-drive gearbox has typical reduction ratios of around 12:1 in 1st gear and around 3:1 in top gear. In comparison, the typical rear-wheel-drive gearbox will have a reduction of 3 or 4:1 in 1st and an overdrive ratio of around 0.8 in top gear. Fig. 5.1-7 is a schematic of the front-wheel-drive gearbox to show the drive path in the gearbox.

Drive passes from the engine, via the clutch to the input shaft. The various gear pairs then transmit torque to the intermediate shaft. The final drive pinion is on or part of this shaft and in turn drives the final drive wheel. The final drive wheel is part of the differential assembly. Each of the driveshafts is connected by a spline to the side gears in the differential. This allows the two driveshafts (and hence the wheels) to rotate at differing speeds to each other although the average speed will always be the same as the final drive wheel. It should be noted that the three shafts rarely lie in a single plane, and, viewed from the end the shafts, would lie in a '*V*' shape. In this way, the centre distance and relative position between the input and output shafts can be designed to suit the installation.

Drive is only engaged between the input and intermediate shafts using one set of gears at any one time. For example, when 1st gear is engaged, only that particular gear pair carries the drive between the two shafts. This is achieved using the synchromesh assemblies, which can be seen in the section illustration above. In this example, the synchromesh assemblies are on the intermediate shaft, although they can be positioned on the input shaft.

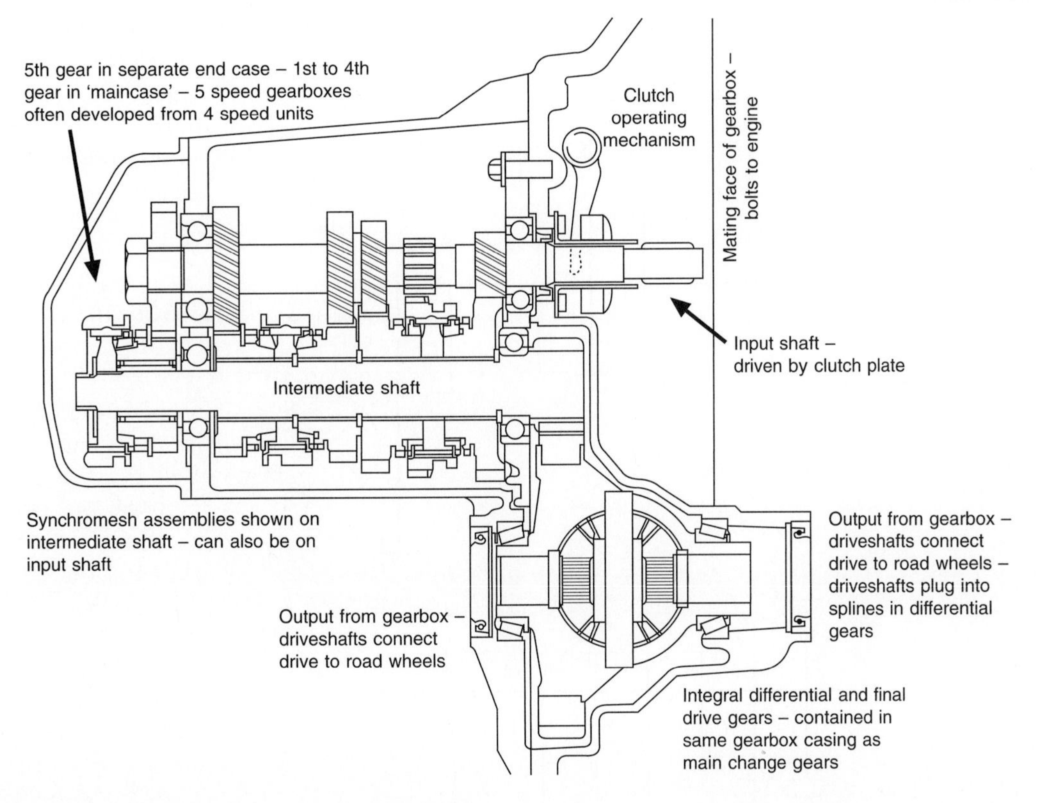

Fig. 5.1-6 Cross section of a front-wheel-drive manual gearbox.

The gearbox illustrated in this example is the simplest form of front-wheel-drive gearbox. Some gearboxes have two intermediate shafts, which, although more complex (and hence tend to be more expensive), do shorten the design which can sometimes allow easier packaging in the vehicle.

5.1.3.2 The rear wheel drive car and commercial gearbox

Two key features distinguish the rear-wheel-drive gearbox from the front-wheel-drive gearbox discussed above; the lack of a differential, and the absence of the final drive reduction gear pair. Both of these will be found within a different unit in the vehicle and this can be seen in the diagrams in the section on vehicle/power train layout. As stated earlier, the ratios are typically from 4:1 to about 0.8:1. As discussed below, 4th gear is often a direct drive between the input and output and so gives a ratio of 1:1. In some gearboxes, however, this direct drive is provided in 5th gear to give a more efficient drive in that gear (there are no friction losses in the gear mesh points). This means that the final drive needs to be adjusted accordingly to provide the same overall gearing and the ratios in the lower gears will be higher to give the same ratio spread – a ratio of around 5:1 rather than 4:1 in first gear perhaps. Fig. 5.1-8 is a cross section of a typical rear-wheel-drive gearbox, while Fig. 5.1-9 shows a schematic of the gear layout.

It can be seen from the diagrams that the gearboxes have several separate sets of gears with various ratios. With the gearbox in a particular gear, the power follows one of the possible paths through the gearbox. When the driver changes gear, the power will flow along an alternative path. The schematic diagram below shows a simple four-speed gearbox to illustrate the concept.

In these gearboxes, the input shaft is driven, via the clutch, by the engine. The constant gears are continually in mesh and the input shaft always drives the layshaft. With no gear selected, i.e. neutral, all the gears on the mainshaft are free to rotate on the shaft and no drive can pass to the mainshaft. The synchromesh mechanism (discussed later) allows either the gears on the mainshaft to be connected to the shaft or the input and mainshaft to be connected together – shown below in Fig. 5.1-10. This connection by the synchromesh allows the various gears to provide the different ratios of speed between the input and output shafts.

5.1.3.3 Gearchanging and the synchromesh

The mechanism usually used in modern manual gearboxes to allow the gears to be changed (when the clutch

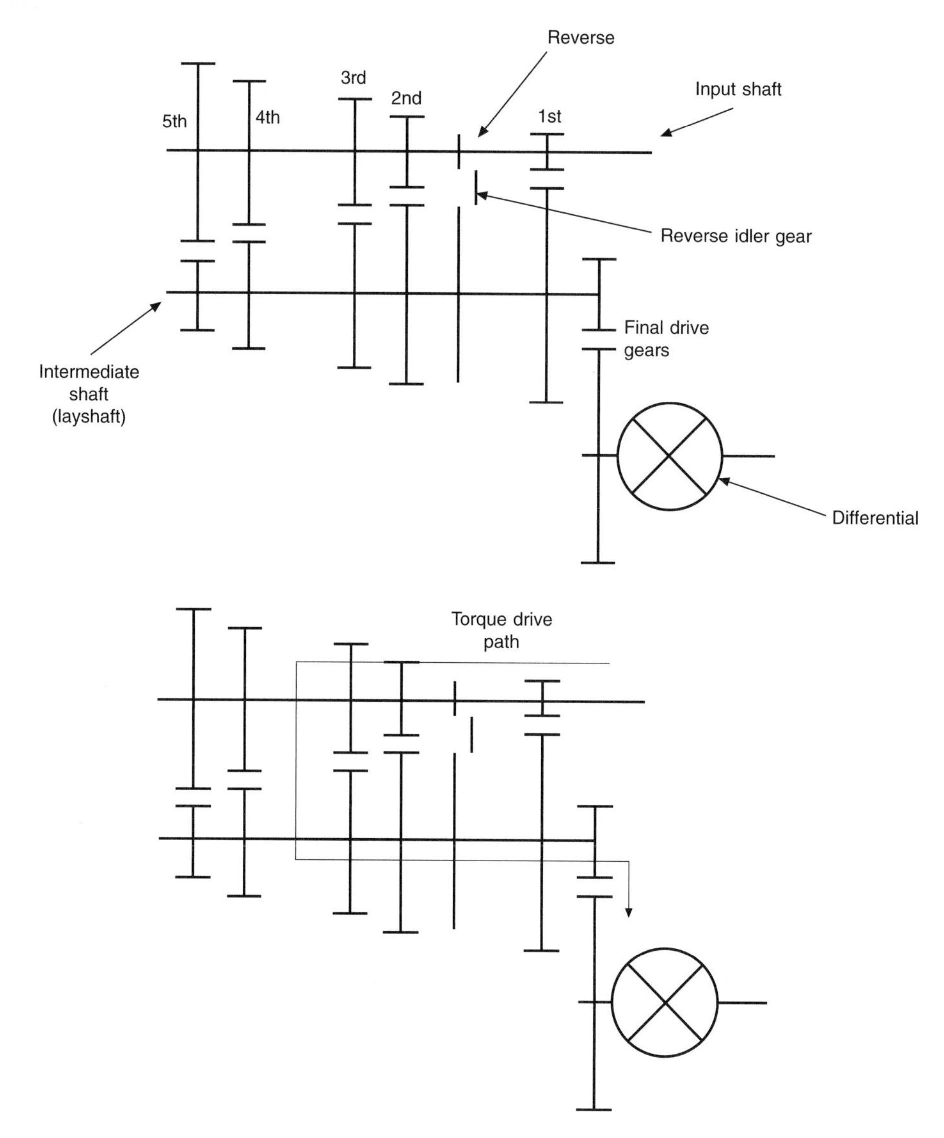

Fig. 5.1-7 Schematic of a front-wheel-drive gearbox gear layout.

disengages the drive from the engine) is the synchromesh. This essentially allows the speed of the components to be matched before they are connected together as the new gear is engaged. If you consider the earlier diagram of a transverse gearbox (Fig. 5.1-6), it can be seen that the synchromesh assemblies are on the intermediate shaft. This is not a requirement, and they could easily be on the input shaft or split between the two. Similarly, the synchromesh mechanisms can be on the layshaft or mainshaft in a three-shaft, rear-wheel-drive gearbox.

A good starting point for synchromesh design is the paper written by Socin and Walters (1968). This not only explains the function of the synchromesh used in many gearboxes, but also follows the main calculations that may be required during the design of single-cone synchromesh.

In considering the function of a synchromesh, we need to refer back to the section on manual gearboxes above.

You may remember that, in this schematic, the drive goes via the layshaft, then the mainshaft gears and onto the mainshaft. Remember, the mainshaft gears are not permanently attached to the mainshaft – you can see this in the diagram of a manual gearbox above. Looking at the 1st and 2nd mainshaft gears for a moment, you can work out that when you are in 1st gear, the 1st mainshaft gear is coupled to and driving the mainshaft. When in 2nd gear, the 2nd mainshaft gear is doing the same job. If you think about it, you can work out that the mainshaft, because it is connected to the wheels, carries on rotating at more or less the same speed both before and after any gearshift. Therefore, if the driver changes gear, say from 1st to 2nd, the layshaft, input shaft and clutch have to

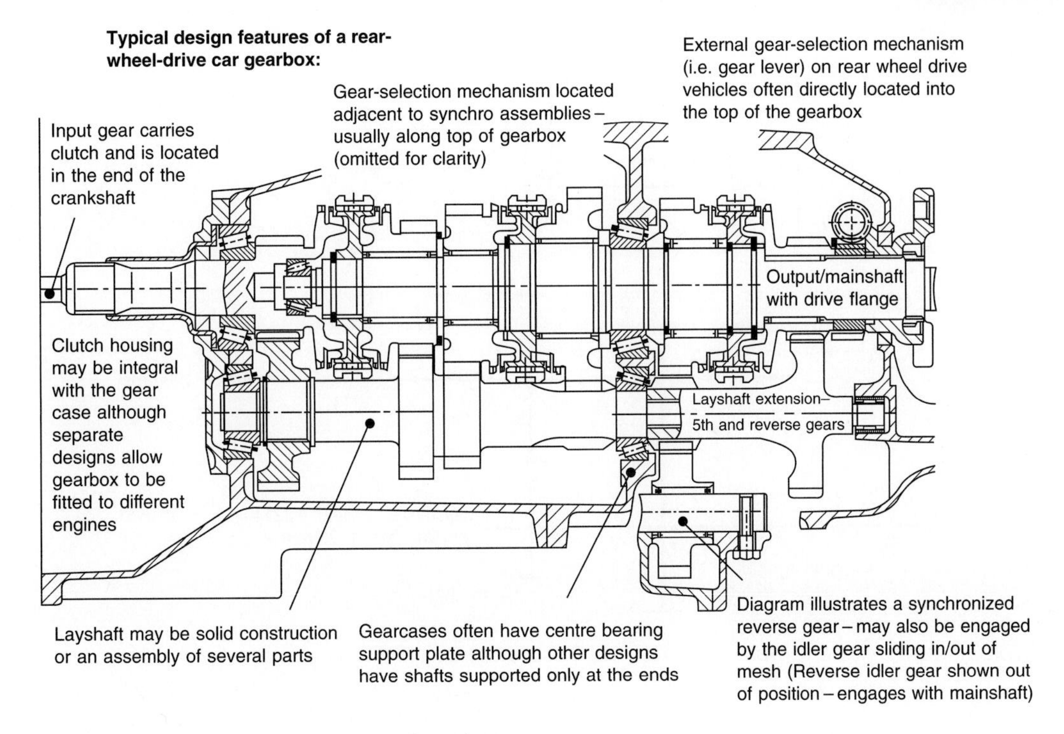

Fig. 5.1-8 Cross section of a rear-wheel-drive manual gearbox.

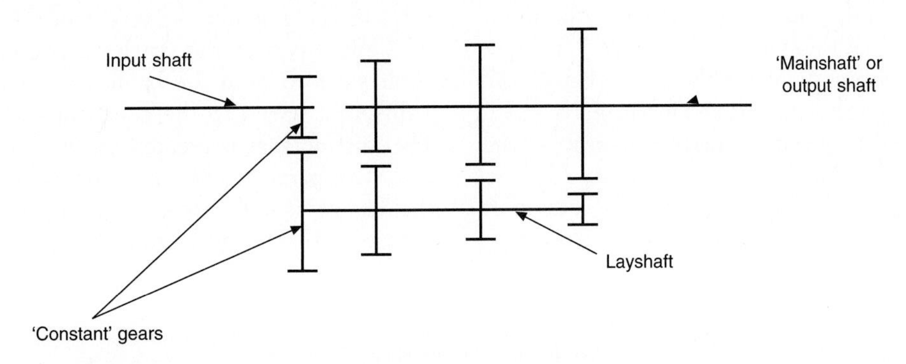

Fig. 5.1-9 Schematic of a gear train layout in a rear-wheel-drive gearbox.

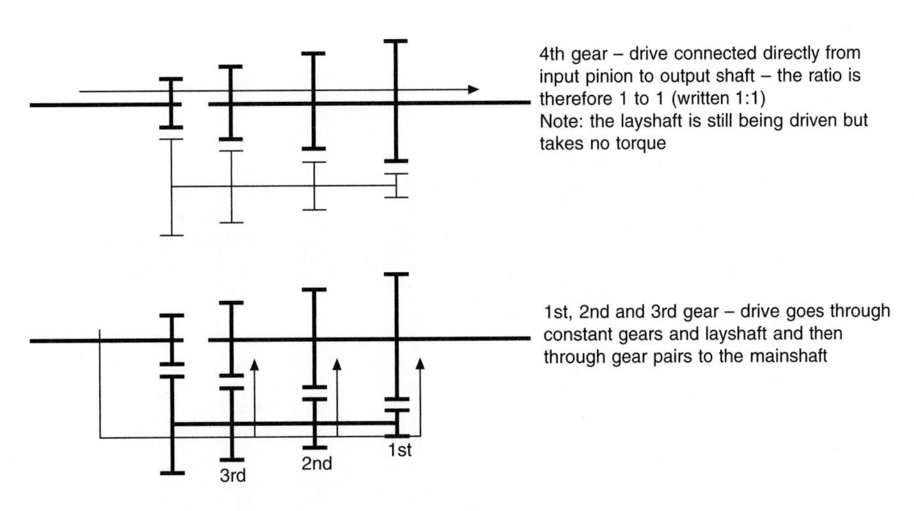

Fig. 5.1-10 Diagrams to illustrate different gears selected in a simplified gearbox.

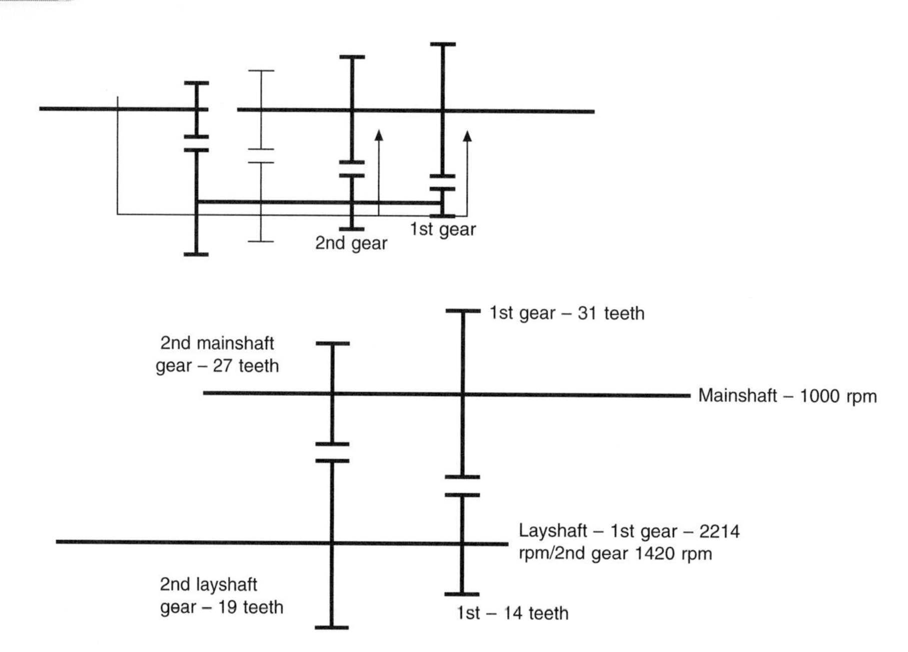

Fig. 5.1-11 Diagram to show the gears involved and ratio implications of a gearchanging between 1st and 2nd gear.

slow down so that the 2nd mainshaft gear can go at the same speed as the mainshaft. (See Fig. 5.1-11.)

It is this speeding up and slowing down of the input side of the gearbox that the synchromesh does. The work done by the synchromesh assembly is to change the speed of the inertia on the layshaft and input shaft, which includes the clutch driven plate. The large majority of the inertia is found in the clutch plate.

The diagram below (Fig. 5.1-12) is a cross section of single- and triple-cone synchromesh. Synchromesh is effectively a cone brake device, and the action of the driver in applying force to the sleeve allows the two sides of the assembly to tend towards the same speed. In the first part of the synchronizing process, the sleeve applies a load onto the baulk ring, which is rotating at a different speed to the cone onto which it is pushed. The friction that is created by this causes the speed of the two parts to become the same. At this point, the sleeve is able to move past the baulk ring and engage with the gear, and the gearshift is complete.

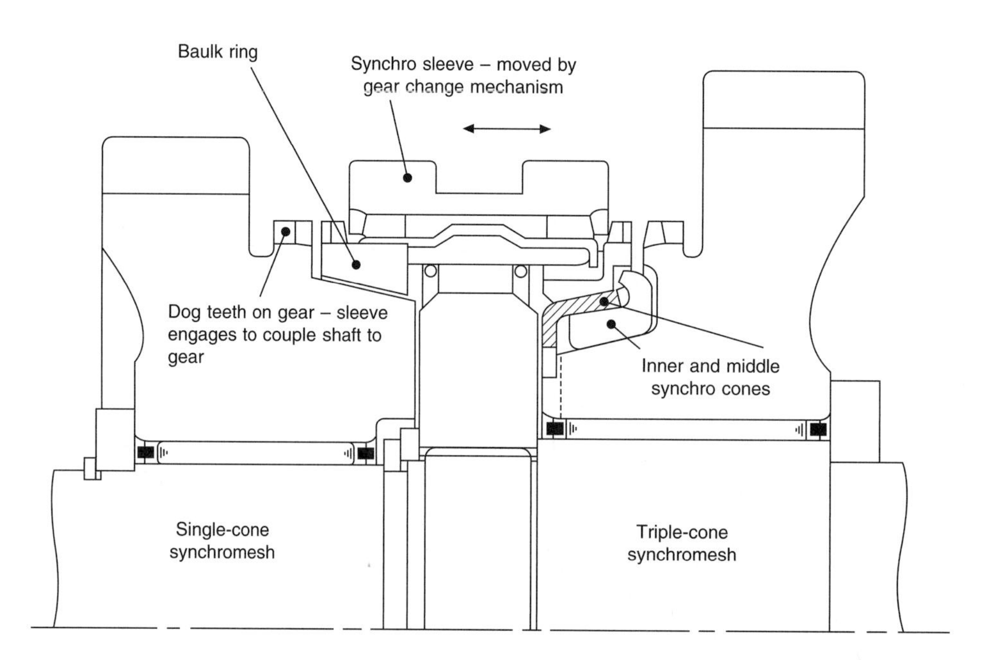

Fig. 5.1-12 Cross section of a synchromesh.

The detail of this process is covered in the paper referred to above.

5.1.3.4 Gear ratios – how they are achieved

Section 5.1.2 discussed why different ratios were needed and how they were selected. Given the overall design package of a transmission, the next task is to design the gear pairs within the casing to achieve the required ratios between the input and output shaft.

As an illustration of how the ratios in a gearbox are achieved, included below are the gear tooth numbers used in a version of the Land Rover LT77 manual gearbox. This was used in the Land Rover, Range Rover Classic and a number of other vehicles in both two- and four-wheel-drive versions.

The constant gears:

- 22 teeth on the input shaft/33 on the layshaft 'constant' gear (driven gear)
- Ratio – 0.666, i.e. layshaft rotates at 0.666 of the speed of the input shaft (slower)

3rd gear:

- 29 teeth on layshaft/27 teeth on the mainshaft gear
- Ratio – 1.074, i.e. mainshaft rotates at 1.074 of the speed of the layshaft (quicker)
- Combining the ratios gives 0.666 × 1.074 = 0.715, i.e. mainshaft/output shaft rotates at 0.715 of the input speed (slower)
- The inverse of this is normally quoted to cause a bit of confusion! So you would see 3rd gear quoted as being 1/0.715, which is 1.397

2nd gear:

- 19 teeth on layshaft/27 teeth on the mainshaft gear
- Ratio – 0.704, (slower)
- Combining the ratios gives 0.666 × 0.704 = 0.469, i.e. mainshaft/output shaft rotates at 0.469 of the input speed (slower)
- As with 3rd, by convention this is quoted as 2.132 – (1/0.469)

1st gear:

- 14 teeth on layshaft/31 teeth on the mainshaft gear
- Ratio – 0.452, i.e. mainshaft rotates at 0.452 of the speed of the layshaft (slower)
- Combining the ratios gives 0.666 × 0.452 = 0.301, i.e. mainshaft/output shaft rotates at 0.301 of the input speed (slower)
- As with 2nd and 3rd by convention this is quoted as 3.322 – (1/0.301)

5th gear:

Fifth gear on the LT77 is an 'overdrive' gear. This means that the output shaft of the gearbox rotates faster than the input. The numbers work out as:

- 37 teeth on layshaft/19 teeth on the mainshaft gear
- Ratio – 1.947, i.e. mainshaft rotates at 1.947 times quicker than the layshaft
- Combining the ratios gives 0.666 × 1.947 = 1.297, i.e. mainshaft/output shaft rotates at 1.297 times the input speed
- The inverse of this is again normally quoted, so you would see 5th gear quoted as being 1/1.297, which is 0.771

5.1.3.5 The clutch

In vehicles there is a requirement for a device to provide a coupling from the engine crankshaft to the transmission. This allows the engine to be started and run without the vehicle moving and the vehicle to be started from rest under control at various rates of acceleration. In manual gearboxes, the drive from the engine also has to be disconnected during gear changes.

Clutches are associated with manual gearboxes and are normally operated by the driver. Recent developments in both the passenger car and commercial truck market have meant that an increasing number of vehicles are automating the function of the clutch. The vast majority, however, remains controlled mechanically by the driver.

The spring pressure clamps the pressure plate onto the driven plate and the flywheel; with the assembly like this, the drive is passed from the engine to transmission (see Fig. 5.1-13). When the driver depresses the clutch pedal, the movement is passed to the release bearing by either hydraulics or cable and the release bearing then pushes or pulls the diaphragm spring (depending on whether the clutch is a 'push' or 'pull' design), see Fig. 5.1-14. The outer part of the release bearing is held (by the release lever) so it does not rotate, and the inner

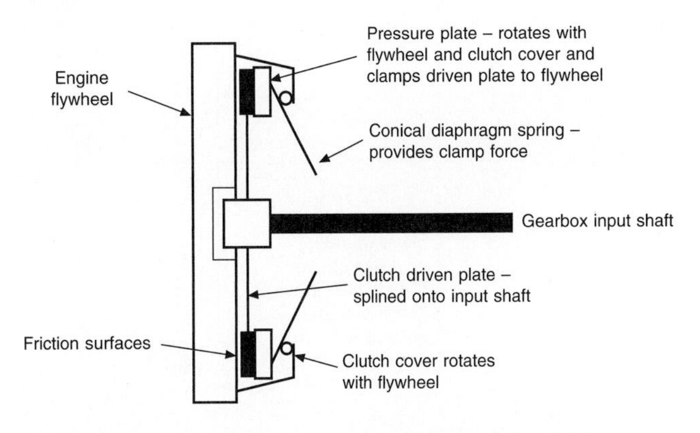

Fig. 5.1-13 Diagram of a conventional single-plate clutch.

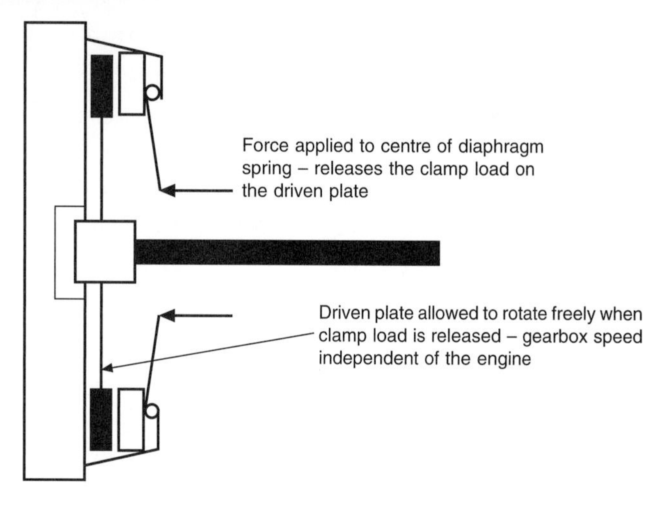

Fig. 5.1-14 Diagram of clutch when released.

race of the bearing rotates with the diaphragm spring and clutch cover. The force applied and travel of the bearing are determined by the lever ratio built into the clutch pedal, hydraulics (or cable) and release lever. There is often significant development in this area as the loads and movement of the pedal have to be as low as possible for the driver while ensuring the clutch will operate correctly throughout the life of the clutch plate. As the clutch plate wears, the position of the various components and the load applied by the spring vary. Obviously, the load has to always be sufficient to clamp the driven plate and not allow any slip, and yet fully clear to allow the plate to rotate freely when the clutch pedal is depressed.

Fig. 5.1-15 shows the whole clutch system. By depressing the diaphragm spring, the pressure on the driven plate is released. This actuation of the spring is achieved by hydraulics, cable or, on older vehicles, may have been mechanical linkage. When the load on the cover plate is released, the driven plate is allowed to rotate freely inside the assembly and the drive to the transmission is disconnected.

5.1.3.6 Automated manual transmission

With the introduction of a number of vehicles recently, automation of synchromesh, 'manual' transmissions is becoming more popular. The reason for the development of these transmissions is twofold; firstly, they can show an economy benefit over both manual and ATs. This is because they are more efficient than automatics and can be programmed to change gear more effectively than most drivers would. Secondly, automated manual transmissions are gaining in popularity in the performance car market, probably because of the links to Formula 1 racing and as a result of clever marketing! Examples include: BMW M3, MMC Smart, VW Lupo and Alfa 156.

These developments started some time ago with the introduction of automated clutches on several vehicles including the Renault Twingo, Saab 900 Sensonic and Ferrari. These cars retained the normal gear lever but

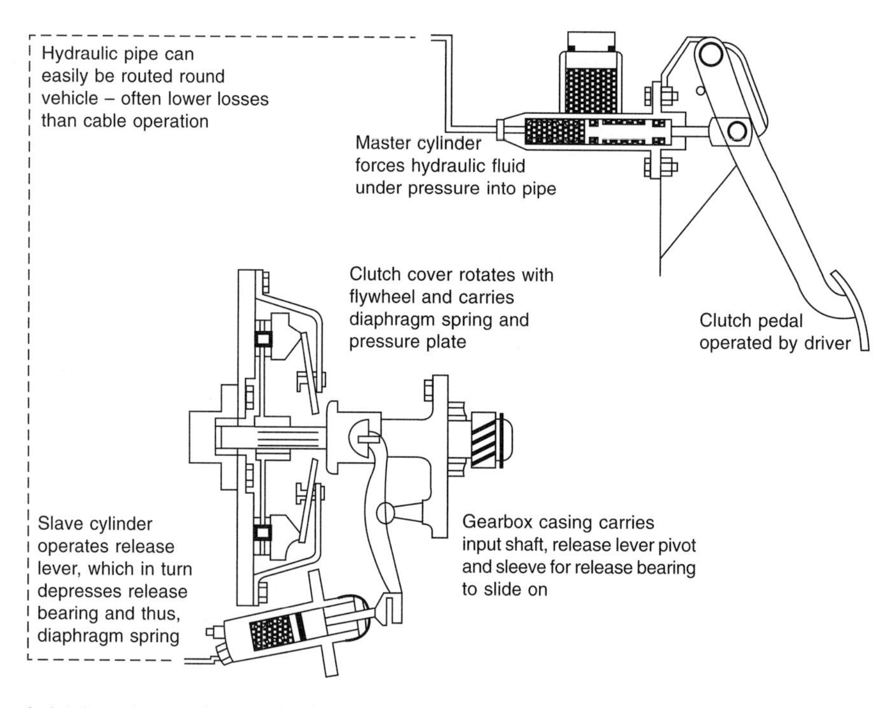

Fig. 5.1-15 Diagram of clutch and actuation mechanism.

automated the clutch so that no pedal was required. At start-up, they operate as an automatic with the control system actuating the clutch to achieve a start from rest when the accelerator pedal is depressed. During gear changes, the clutch is operated in response to movement of the gear lever.

Consideration of the mechanics of the automated manual systems suggests that it may be difficult for these systems to replace the conventional automatic. The fundamental point is that the automated manual systems need to disconnect the drive from the engine to the transmission in order to achieve a gear change. With conventional automatics only a small reduction in the engine power is required to achieve a smooth transition form one gear to another because of the action of the torque converter. There are, however, twin-clutch designs of transmission, which overcome this limitation by providing two parallel torque paths through the transmission where a gearchange simply switches from one path to another and engages one clutch rather than the other. This can be done without reducing the engine output (a 'hot shift'). This has been used in the past by large automotive gearboxes, but could be extended to the car market.

In the commercial market, there are a number of manufacturers now producing automated manual transmissions for trucks. Whereas these developments have needed the driver to indicate the gear selection in the past, the latest developments have the intelligence to completely automate the gearchange. In heavy commercial vehicles, this may need to include missing some gears, especially when unladen, so the control software required is not trivial.

5.1.4 The AT

The concept of an AT offers considerable advantages to vehicle drivers since they can be relieved of the burden of selecting the right gear ratio. This burden, both mental and physical, has become more significant with increasing traffic congestion.

Any reduction in driver fatigue and increased opportunity for the driver to concentrate on other aspects of vehicle control must contribute to increased safety and a reduction in road traffic accidents. There are also benefits in terms of economy and emissions if an automated system can make a better selection of ratio than a non-expert driver does. There are several alternative solutions to achieve this automation including automated layshaft transmissions (described above), CVT (described in the next section) and the 'conventional' AT described here.

The term 'automatic transmission' is used to refer to a combination of torque converter with a ratio change section that is based on epicyclic gearsets. The use of these components can be traced back to the early days of automotive developments, and in a recognizable combination to the middle of the last century. Yet, it is an area that is still seeing extremely rapid development today. The success of this combination lies in the simplicity of the torque converter as a device that inherently has ideal characteristics to start a vehicle from rest, and the opportunity that epicyclic gear sets provide to give relatively easy and controllable changes between ratios.

The controllability of these devices has allowed automatics to be developed with the good shift quality necessary to satisfy the driver's expectations for a gear change. Somehow, drivers of conventional manual-shift vehicles are always more critical in judging the gear change of another driver rather than their own where a misjudged shift can be more easily forgiven. In just the same way, they are more discerning in judging the quality of an automated gear change and, thus, high standards are required. In the past, these have been virtually impossible to achieve from automated manual gearboxes. This situation is, however, changing with the greater use and sophistication of electronic controls.

The downside of an AT in comparison with a manual gearbox alternative is greater cost, greater weight, larger size and lower efficiency. It has thus been used most in larger cars where these penalties are less significant and the driveability advantages most appreciated. This may well account for the large proportion of ATs used in the USA (approaching 90%) in comparison with Europe (around 20%). However, all these disadvantages have acted to maintain the pressure for development of the AT leading to modern designs that achieve a greater number of gear ratios within the same or even a reduced space envelope.

5.1.4.1 The jatco JF506E – a state-of-the-art transmission

An example of today's typical five-speed automatic suitable for a medium to large size front-wheel-drive vehicle, where packaging and space constraints are severe, is shown as a sectioned view in Fig. 5.1-16. The input shaft drives first into a three-element torque converter (top right) directly to a pair of epicyclic gears that give a four-speed change section. A further fifth-speed change is obtained on the secondary shaft with a final drive ratio providing the connection to the differential. The output drive shafts can then be taken either side of the differential shown at the bottom of the section. This layout can also be easily extended to give a four-wheel drive output with an additional gear section. The main components that make up this transmission are described in more detail below and the combined operation in Section 5.1.4.4.

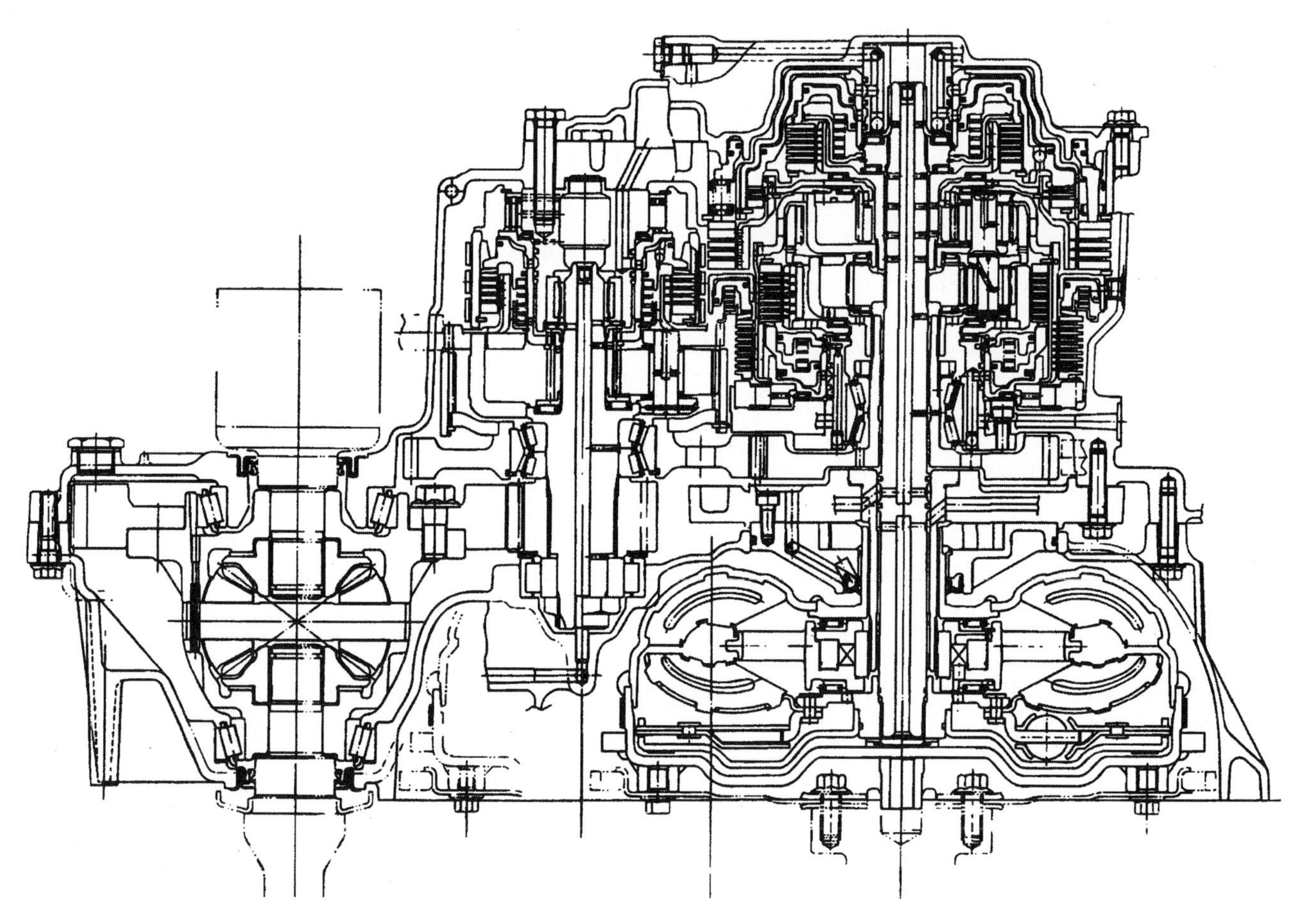

Fig. 5.1-16 Sectioned view of Jatco JF506E (courtesy of Jatco).

This transmission is electronically controlled by a control unit interfaced with other vehicle systems, including engine management, via a CAN link. The electronic control signals are passed to solenoid valves that apply hydraulic pressure to control clutches that select the required ratio. The programmed control strategy takes inputs from speed and temperature sensors to respond to the driver's demand. This demand comes principally from the accelerator pedal position but is modified by brake application and both the gear selector (D, 4, 3, 2) and a pattern selector (drive, sport and snow).

5.1.4.2 The hydrokinetic torque converter

Hydrokinetic drives involve the transfer of power through the 'kinetic energy' or velocity head of a fluid. In such devices an impeller element creates the flow kinetic energy and a turbine element recovers the energy producing a torque output. There are two main types of hydrokinetic devices: fluid coupling and fluid converter. Both these families provide an automatic adjustment of ratio (input speed for a given output speed and load) and an infinite ratio capability that makes them highly appropriate as a 'starting device'. Their features include: stepless variation in torque and speed without external control, vibration isolation, shock load absorption, low maintenance and virtually wear-free operation. Disadvantages include efficiency, design limitations and great difficulty to control precisely.

The term torque converter is used here to describe the converter coupling as most frequently used in automotive applications. This is also known as a Trilok converter. It is so called because, in a part of its operating range, it gives a torque multiplication (behaving as a converter) and in the remainder, it behaves as a coupling with a 1:1 torque ratio.

The basic equation defining the fluid torque acting on impeller or turbine is:

$$T = C\omega^2 D^5$$

where T = torque transferred

C = capacity constant

ω = rotational speed

D = diameter

The capacity factor C, is independent on the detailed geometry (blade angles, etc.), fluid density and viscosity and, most importantly, it varies with speed ratio.

5.1.4.2a Fluid coupling

Fluid couplings contain only two rotating elements – impeller and turbine – within a toroidal casing as shown in Fig. 5.1-17. Both these elements have radial vanes and the cavity is filled with hydraulic fluid. The impeller and casing are driven by the input, and fluid trapped between the rotating vanes must also rotate and this in turn causes flow outwards to the largest diameter as a result of centrifugal action. This outward radial fluid flow is directed by the curvature of the impeller shroud back to the turbine section where the rotational component of velocity gives a torque reaction on the turbine blades as the fluid flow direction is changed. The fluid returns towards the centre line of the assembly and re-enters the impeller at a smaller diameter.

Since there are only two elements, there must always be an equal and opposite torque reaction; thus input torque T_i, must balance output torque T_o:

$$T_o = T_i$$

Vanes in simple fluid couplings are radial and hence it can be an almost symmetrical device where the impeller and turbine functions can be reversed and torque transmitted in the reverse direction. However, it is also possible to use curved vanes that give asymmetry and a higher torque capacity in one sense. The transmitted torque will depend on the relative speed of the impeller and turbine. It will reduce to zero if they are rotating at the same speed and will reverse if the turbine rotates faster than the impeller. The relative speed may be expressed either as a speed ratio (ω_o/ω_i) or by a relative slip s, defined by:

$$s = \frac{(\omega_i - \omega_o)}{\omega_i} = 1 - \frac{\omega_o}{\omega_i}$$

The power transmission efficiency η, is also related to speed ratio as follows:

$$\eta = \frac{\text{power out}}{\text{power in}} = \frac{T_o\omega_o}{T_i\omega_i} = \frac{\omega_o}{\omega_i}$$

The efficiency characteristics are thus a linear function of speed ratio as shown. However, as the speed ratio approaches unity the torque transfer capability will reduce and the flow losses mean that the torque transfer falls rapidly to zero. This occurs in the region where slip is 2–5% (speed ratio 0.95–0.98), depending on the internal clearances within the coupling.

5.1.4.2b Fluid converter

The converter is like the coupling in having a turbine and impeller but, in addition, uses a third vane element called a reactor or stator that does not rotate. To prevent it from rotating, it is connected via a tube concentric with the turbine output shaft to an internal part of the gearbox casing such as a bearing housing. The stator vanes redirect the flow as in Fig. 5.1-18 and *add* to the torque provided by the engine input to give a multiplying effect on the output torque (despite the apparent sequence implied by the flow path). The torque balance then becomes:

$$T_o = T_i + T_s$$

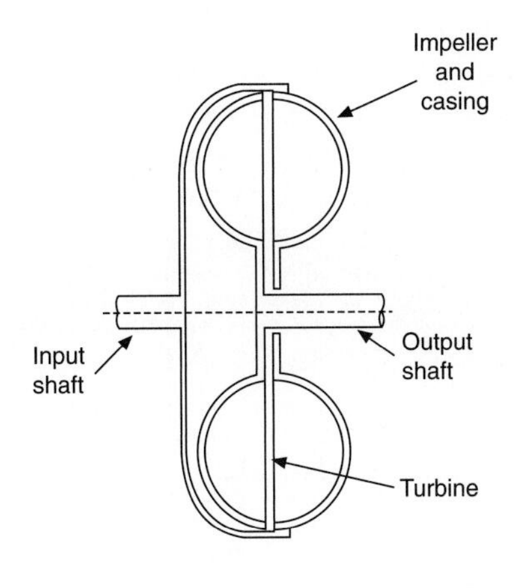

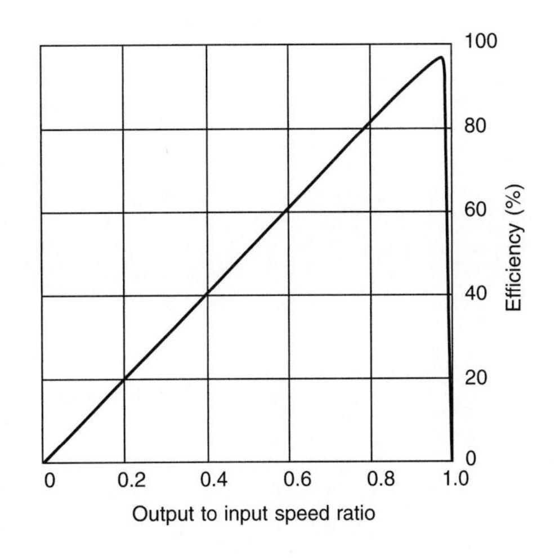

Fig. 5.1-17 Fluid coupling and characteristics.

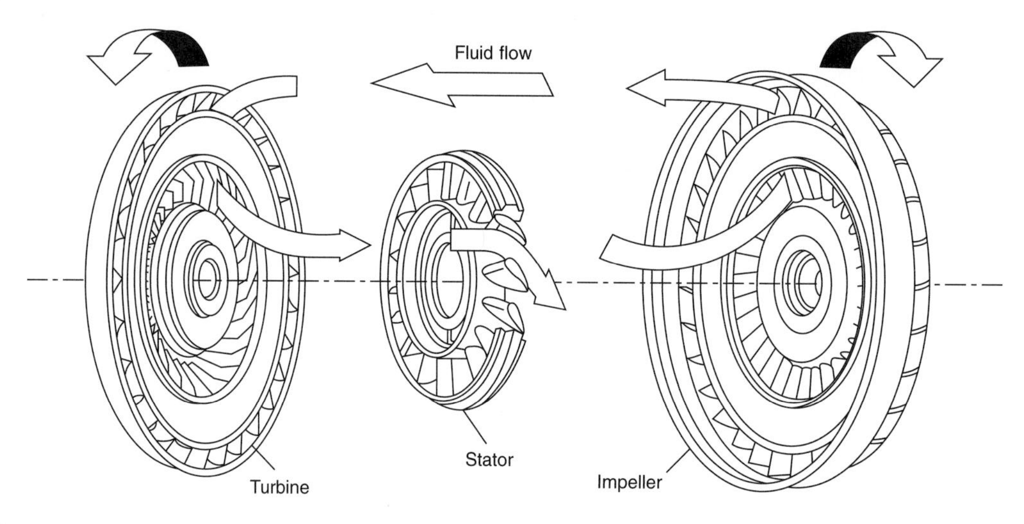

Fig. 5.1-18 Torque converter 3D with flow path.

The efficiency is:

$$\eta = \frac{\text{power out}}{\text{power in}} = \frac{T_o\omega_o}{T_i\omega_i}$$

The blade angles of all three elements are curved to give the easiest flow path at a so-called 'design point'. This point usually represents a peak efficiency with respect to speed ratio as shown in Fig. 5.1-19. At other conditions, additional losses occur as flow meets the vanes at 'awkward' angles, giving rise to shock losses. The blade curvature means that the converter is not symmetrical and will not transmit torque effectively in the reverse sense (negligible engine braking). There is a compromise in design between achieving a high torque ratio at stall (zero output speed) but at the expense of efficiency. It is possible to achieve torque ratios of 5:1, but these days fuel efficiency has become increasingly important and automotive converters tend to operate around 2:1.

Fig. 5.1-19 also shows that beyond the point of maximum efficiency, the torque ratio tends below unity. This region is not attractive from an automotive viewpoint with reducing efficiency and hence the developments of the converter-coupling operation described in the next section.

5.1.4.2c Torque converter

The term torque converter is sometimes applied to the basic fluid converter described above but it is used here to describe the device that combines both converter and coupling operation. This combination is typically used in automotive applications and sometimes called a Trilok converter. The design is based closely on that of the fluid converter but with the addition of an overrun clutch (Heisler, 1989) connecting the reactor (stator) to its fixed reference frame. This prevents the reactor from rotating in one direction but will allow it to rotate freely in the other. Operation can be visualized during an acceleration sequence with an increasing vehicle speed when the operation initially follows that of the fluid converter. The reactor will be locked until a speed ratio is reached where the input and output torques are equal, and consequently the reactor torque has reduced to zero. In converter operation, any further increase in speed ratio above this would give a reduced torque ratio that can only occur by

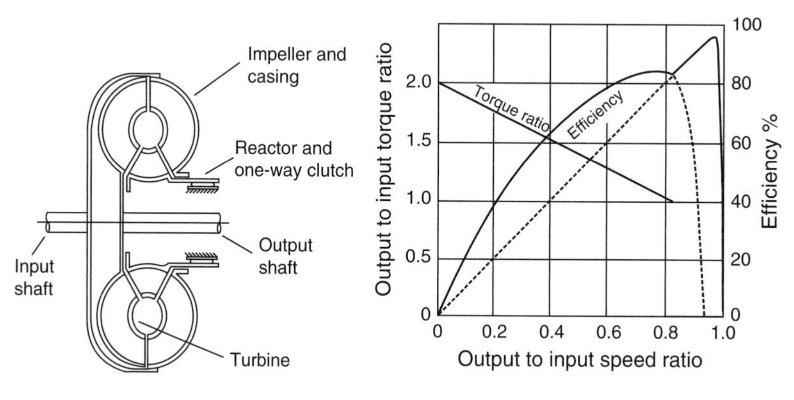

Fig. 5.1-19 Torque converter and characteristics.

a reversal of torque on the reactor (since the three component torques must still be in balance). In a converter coupling, this cannot be reacted by the overrun clutch and the reactor will free wheel. Above this speed ratio, the assembly behaves as a two-element device and operates as a fluid coupling. This gives the combined characteristic shown by the full line in Fig. 5.1-19 with increasing efficiency until the operating limits are reached as with a fluid coupling.

Further improvements in efficiency can be obtained if a lock-up clutch is used to mechanically lock the impeller and casing to the turbine and hence directly connect input and output shafts. This should only take place when the speed ratio is near unity, and needs to be controlled gradually in order to prevent any driveline shock that might be felt by the driver. This action can be actuated hydraulically as required by the transmission controller.

5.1.4.3 The epicyclic gear set – the key component in the AT

The epicyclic or planetary gear contains three sets of concentric gears meshing at two diameters as shown in Fig. 5.1-20. These are connected to three external shafts distributing the transmitted torque between them. These three comprise the sun gear shaft, the annulus gear shaft and the carrier. The planet gears rotate about pinions mounted on the carrier, and this part of the assembly can be considered to behave as a single component. In ATs, epicyclic gears are usually operated with one of these three shafts locked to the gearbox casing or frame leaving the remaining two to act as input and output shafts. Another alternative is for any two of the components to be locked together and the whole unit rotates as one, and although apparently trivial it is a convenient and common option.

An example of how the device works can be envisaged with the carrier shaft locked, leaving the sun and annulus free to rotate connected via the rotation of the planets about their now fixed centres. The peripheral speed of any of the planet gears must be the same at both the contact radius of the sun and the radius of the annulus. Thus, the number of gear teeth on each must determine the relative speed of the sun and annulus, and, since the tooth pitch or module must be the same, their relative diameters. They will of course rotate in opposite directions. This ratio is referred to as the fundamental ratio i for the epicyclic gear and:

$$i = -\frac{\omega_s}{\omega_a} = \frac{t_a}{t_s} = \frac{D_a}{D_s}$$

where t is number of teeth and subscripts a and s refer to annulus and sun respectively

Consideration of the torques acting on the gear teeth at the two meshing diameters indicates that these will be given by the inverse of the speed relationship. Also, the tooth forces on the planet will be the same at both meshing interfaces but the highest tooth stresses will occur where there are fewer teeth to share this load. This must occur at the planet sun mesh and will occur on whichever is the smaller diameter of the two. Thus, in

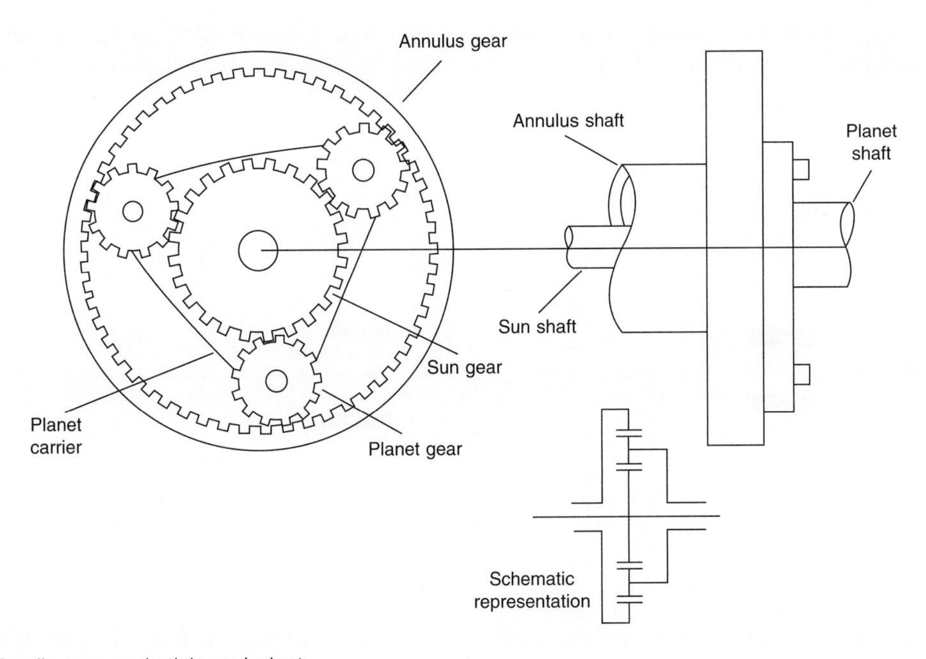

Fig. 5.1-20 An epicyclic gear and stick equivalent.

sizing the gears for a required torque capacity, this mesh region will indicate the limiting case.

Examination of the shaft speeds with other components locked allows the derivation of the overall kinematic relation for an epicyclic set as:

$$\omega_s = (1+i)\omega_c - i\omega_a$$

There are limits in ratio that can be sensibly achieved with an epicyclic gear arising from a combination of physical packaging, tooth and pinion strength and the need to have a whole number of teeth on all components. This gives values for the fundamental ratio generally in the range:

$$2 \le i \le 4$$

The ease with which ratio changes can be implemented by a simple braking or clutching action can be illustrated by looking in more detail at a single epicyclic train. Fig. 5.1-21 shows an epicyclic gear with the carrier and annulus shafts used as the input and output. In Fig. 5.1-21(a), the sun gear is locked to the carrier and the whole assembly rotates as a single unit with a speed ratio of one. In Fig. 5.1-21(b), the sun gear is held stationery by the clutch and for a fundamental ratio $i = 2$, the annulus shaft will rotate at:

$$\omega_a = \left(\frac{1+i}{i}\right)\omega_c = 1.5\,\omega_c$$

The change in ratio from 1 to 1.5 is a typical 'step' between gears. The relative rates at which the release and engagement actions take place on clutches and brakes can be controlled to give very smooth transitions. The constructional details of typical clutches can be found in Heisler (1989). The ways that sets of epicyclic gears can be combined to give different overall ratios are described in the following sections.

5.1.4.4 JF 506E AT operation

The combined operation of these components can be seen by a more detailed examination of the JF 506E transmission introduced above. Fig. 5.1-22 shows a schematic representation of the transmission with different gears selected. For clarity, the main components are only represented on one side of the shaft centreline, effectively half of the transmission. The main torque and hence power flow path is shown by the heavy lines. There are two epicyclic gear sets on the primary shaft, labelled A and B, and a further reduction epicyclic on the secondary shaft.

In all gears the drive passes through the torque converter to the primary shaft with the stator free wheel becoming active at higher speed ratios. The drive from the primary shaft comes from a path that is connected to both the annulus of epicyclic A and carrier of epicyclic B. This passes to the secondary shaft via transfer gearing onto the annulus of the reduction epicyclic set. The output then passes from the carrier of this epicyclic to the final drive and hence to the output shaft. In the first four gears and reverse, the sun gear of the reduction epicyclic is locked by the reduction brake. This gives a carrier rotation in the same sense as the annulus but with a reduction ratio of 0.8.

First gear – drive passes through epicyclic B with the annulus locked to frame through a hydraulic clutch and a free wheel clutch. The carrier is driven by the sun at about a third of the primary shaft speed and in the same direction.

Second gear – in this gear, both of the primary shaft epicyclic play a part to give a combined ratio on the output to the transfer gear. The sun of epicyclic A is now locked and the torque reaction on the annulus of epicyclic B allows it to rotate with the carrier of epicyclic A in the free wheel sense of the overrun clutch. The

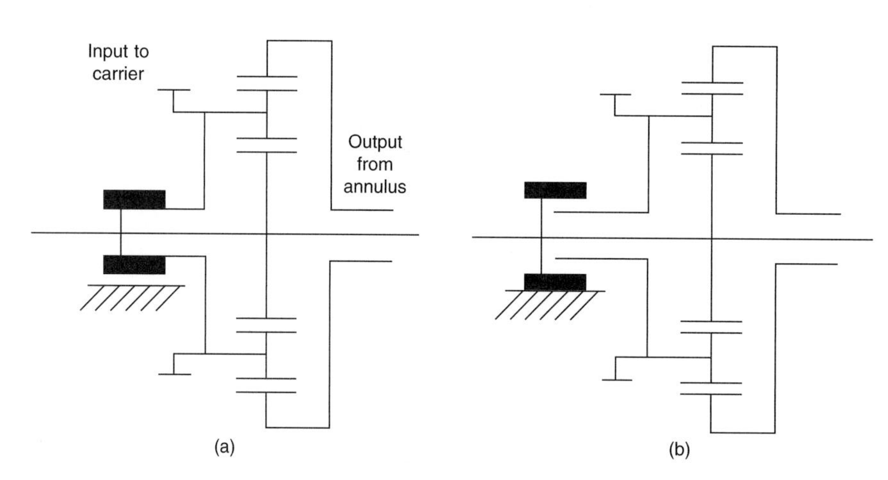

Fig. 5.1-21 Single epicyclic sun locked to carrier and sun locked to frame. (a) Sun shaft locked to carrier shaft, (b) sun shaft held stationary.

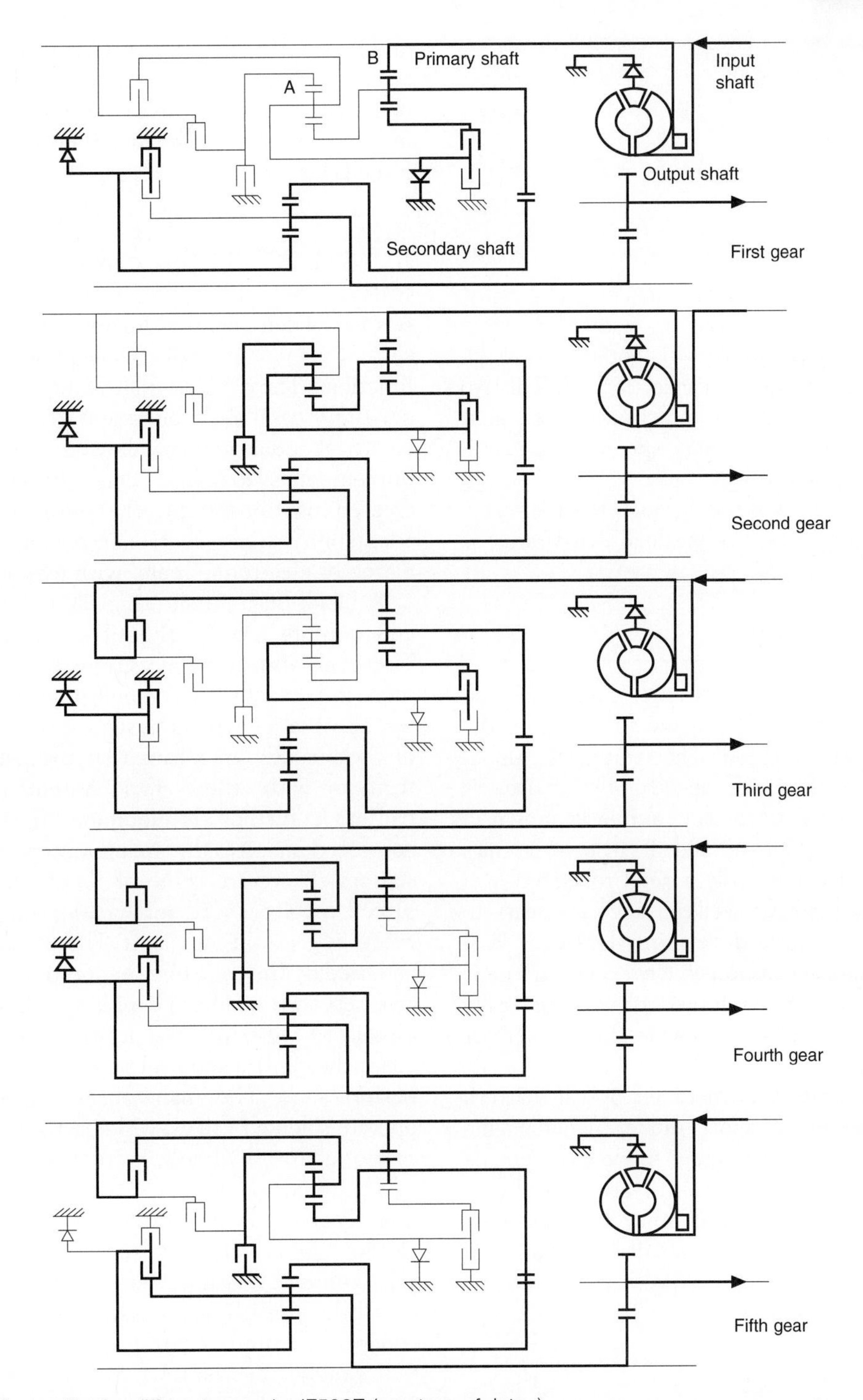

Fig. 5.1-22 Power flow paths for different gears in JF506E (courtesy of Jatco).

combined motion increases the output to about a half of the primary speed.

Third gear – the annulus and the sun of epicyclic B are now locked together (via the carrier of A) and all of the components rotate at the same speed.

Fourth gear – drive comes entirely from epicyclic A with the sun gear locked. The annulus of this gear is driven by the carrier and rotates in the same sense at about 1.5 times the primary speed.

Fifth gear – the path from input to the secondary is the same as fourth gear but now the reduction brake is dis-engaged and the carrier and sun locked together. Again, this means that the whole reduction epicyclic set will rotate as a single unit.

Table 5.1-1 Typical ratio set

Gear	Ratio	Gear	Ratio
1	3.474	4	0.854
2	1.948	5	0.685
3	1.247	Reverse	2.714

Reverse gear (not shown) – drive is obtained through the sun of epicyclic A with the carrier locked. This rotates the annulus in the opposite direction to the sun at a value just below half the primary speed. This is still subject to the reduction epicyclic ratio.

A set of manufacturer's ratios through the main section of the transmission (excluding final drive ratio) is given in Table 5.1-1.

5.1.4.5 Shift strategy

The basic ratio selection depends on a pre-determined shift strategy but there are also many subtleties in the way that changes are executed. The strategy is fundamentally a function of vehicle speed and the driver's accelerator demand and a typical example is shown in Fig. 5.1-23. This shows shift-up and shift-down settings for each gear. There is obviously a need for some hysteresis in the change points between any two gears to prevent hunting phenomena developing. However, it is also obvious that for a given vehicle speed in many sections of the map a heavy-footed driver can easily invoke up and down shifts by moving the accelerator pedal.

This basic shift strategy is modified by the pattern selection for the driver. In 'sport' mode, higher engine speeds will be used, whereas in 'snow' mode lower engine speeds will be used. There are also driver-selected gear hold positions (D, 4, 3, 2) that will limit the highest gear to be selected. In addition, there are failure modes of

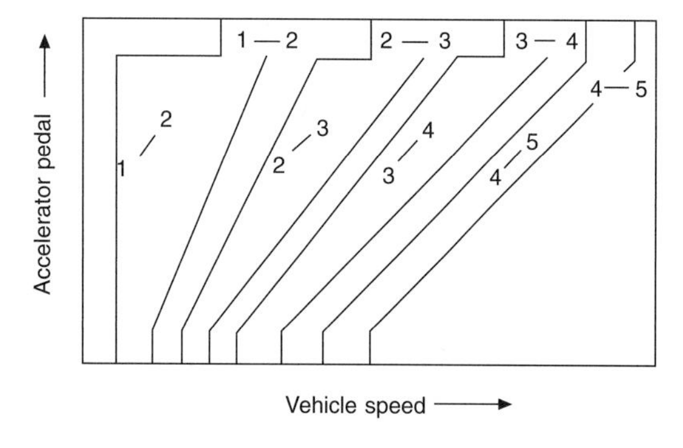

Fig.5.1-23 Shift strategy (courtesy of Jatco).

operation that may either limit the gears that can be selected or hold a particular gear in order to provide safe operation. These are detected and appropriate actions invoked by the Automatic Transmission Control Unit (ATCU).

5.1.4.6 ATCU the controller

ATs have been operated for many years with a hydraulic system providing both the logic and control actuating functions. The development of digital controllers and the necessary parallel development of low-cost sensors, and electrical actuators have allowed the control logic to be implemented digitally. This allows considerably increased functionality together with greater flexibility and adaptability of the controller, overall the classic combination of electronic 'brain' with hydraulic 'muscle'.

A block diagram showing the basic controller inputs and outputs for the JF506E is shown in Fig. 5.1-24. There are also interconnections through the CAN bus, most importantly to not only the engine management, but also brakes and the instrument pack. The flexibility of the bus system means that the path is open for integration with other vehicle systems (e.g. traction control) as a further development. In addition to driver control demands, the main inputs come from speed sensors. There are three of these, and they have been placed to give speed information on both sides of all epicyclic gears during a shift. These speeds are monitored and used by the controller to fine-tune shift operations to give best shift quality. There is a control memory feature associated with this that learns about the hardware to adapt the control actions to give a consistently good shift performance. The main outputs from the controller operate solenoid valves, which activate clutches for each gear or control hydraulic system pressure.

The solenoids for gear change have an on/off action to directly open, or close, a hydraulic valve. When the valve is closed, pilot pressure is raised at one end of a sliding spool valve to move it against a spring return. Such a spool will usually be multi-functional and open or close a number of connections simultaneously. These may supply flow to raise pressure on a clutch or drain flow to decrease clutch pressure, which is of course what is required in many gear change operations. They can also be used to provide other flow connections that may help in a sequence of operations or to inhibit actions like connecting the reverse gear when moving forward. The multi-function feature can allow a reduction in the number of components used and hence lower the cost.

The solenoids for pressure control also open and close a single directly acted valve. However, they are operated more rapidly to give alternate connection between a supply and a drain line. The proportion of time spent

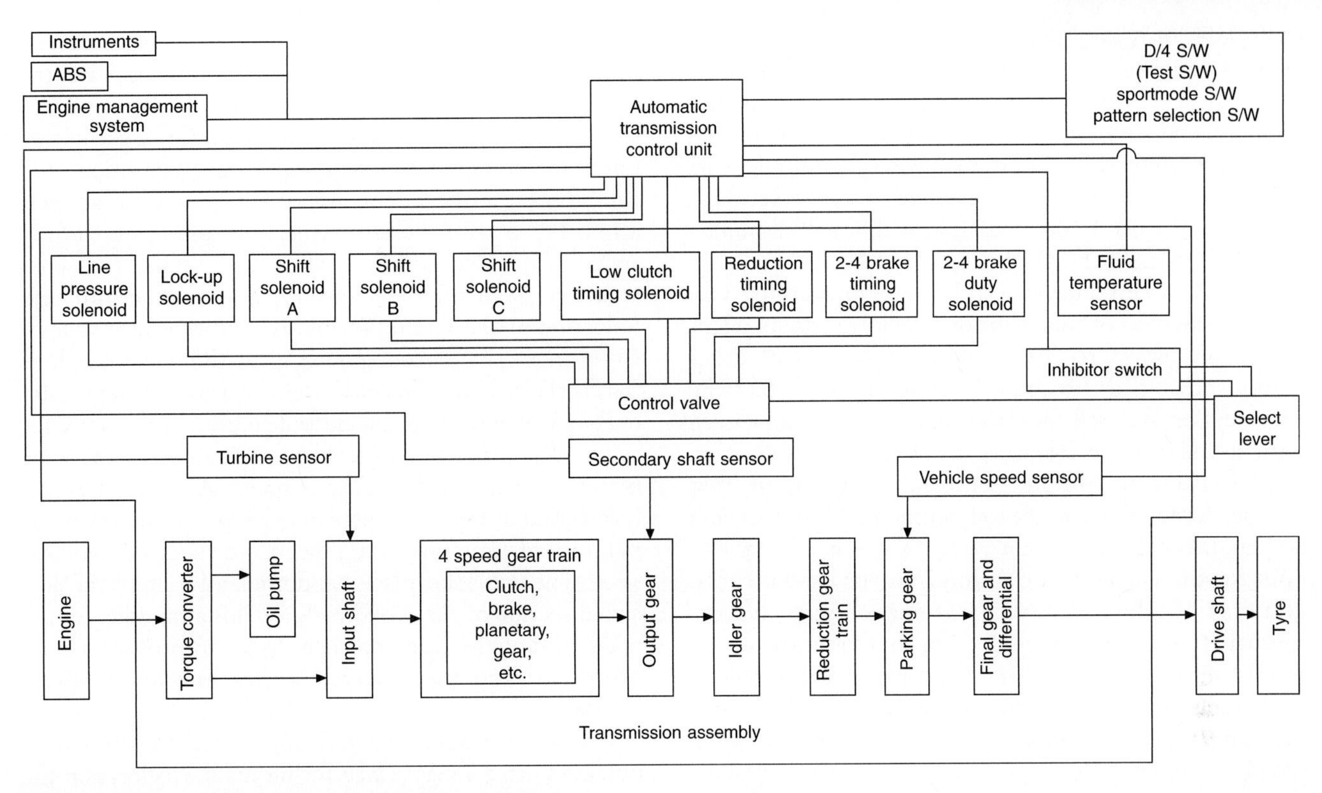

Fig. 5.1-24 Control block diagram (courtesy of Jatco).

open, relative to that closed, gives a mean pressure level established between a minimum drain pressure and a maximum supply pressure. The open/close timing is obtained directly from a pulse width modulated (PWM) output from the ATCU. This transmission uses five on/off solenoids and four modulated solenoids that are used to control the hydraulic system line pressure and some individual clutch pressures.

The hydraulic system is supplied from a fixed capacity positive displacement pump. The pump outlet pressure in all ATs can have a significant effect on system efficiency, since the power absorbed by the pump is directly proportional to pressure, as well as speed. It is thus kept as low as possible, but sufficient for the clutches to grip with a suitable margin for safety. Fig. 5.1-25 shows the type of scheduling of system pressure with accelerator pedal position. This control can be considered typical, but the mean level is changed with a number of conditions, such as starting from rest; or for acceleration then a higher mean level is set. If steady cruise conditions are detected by low-accelerator pedal movements, then a cruise schedule will be adopted that reduces these levels in a process called 'cut-back' as shown. Reduced levels are also adopted during a gearshift operation to allow some controlled slip to take place across the clutches during the transition, and hence give a good shift quality. Also, during gearshifts the ATCU requests the engine management system for a short period of reduced engine torque via the CAN link. There are also occasions when higher pressures can be selected when the driver indicates a need for engine braking by selecting a downshift at relatively high speeds. It should also be remembered that the hydraulic system provides a cooling function by removing heat from high-temperature regions, as well as lubrication, and the actuation functions described here.

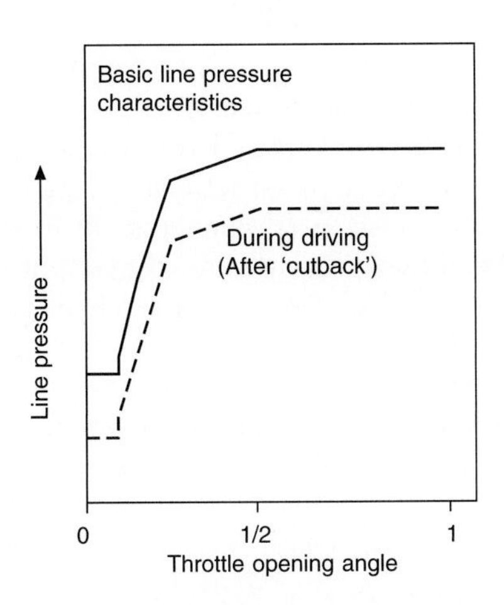

Fig. 5.1-25 Line pressure control strategy (courtesy of Jatco).

Torque converter lock-up is also controlled by the ATCU. This locks the torque converter impeller to the turbine in regions of high-speed ratio. Again, there is some hysteresis on this action so that, once engaged, it does not drop out too quickly. The lock-up clutch is activated by a pressure-modulating type of solenoid to give a smooth and steady transition between states. The lock-up feature is normally only available in 4th and 5th gears but can be extended to both 2nd and 3rd gears if a high fluid temperature is detected, since this will eliminate any heat input via the torque converter losses.

The ATCU intelligence can be used in a similar way for the provision of this additional lock-up action. Monitoring of system responses allows detection of other failure states, and this can lead to action to protect the transmission, but still allow the vehicle to be driven, even with a reduced performance as described in Section 5.1.4.5. There are also low-temperature conditions that can be detected from the oil temperature sensor and actions taken to try to encourage a quicker warm-up, both for the engine and transmission. For example, 5th gear may be inhibited at very low temperatures and the normal torque converter lock-up inhibited even at moderate temperatures. These features are all designed to enhance the overall system performance as well as refining the contribution mode by the transmission itself.

5.1.5 Continuously variable transmissions

The overwhelming majority of transmissions in road going vehicles is either manual or conventional automatic in design. These transmissions use meshing gears that give discrete ratio steps between engine and the vehicle speed. However, alternative designs exist that can transmit power and simultaneously give a stepless change of ratio; in other words a CVT. Strictly speaking, a CVT is a transmission that will allow an input to output ratio to change, continuously without any steps, in a range between two finite limits. An extension of this idea is a transmission that also allows a zero output speed to be included within this operating range. This can be considered as an IVT*. Since all vehicles need to come to rest then, any transmission that inherently has an IVT characteristic is at an advantage. The use of the term CVT is used here in the more generic sense to include both types.

5.1.5.1 The rationale for the CVT

Transmissions that vary ratio continuously can be controlled automatically, and hence have the same advantages in terms of driving comfort as the more conventional ATs that were described above. In addition, they have the ability to vary the engine speed independently of vehicle speed. This brings the opportunity to optimize the engine operating point for any required output conditions and can offer either best economy or best power for acceleration. Fig. 5.1-26 shows a typical petrol engine torque/speed map with contours of brake specific fuel consumption, and lines of constant power superimposed. Selection of the minimum fuel consumption for successively increasing output power gives the 'economy line' that is also shown. This indicates the most efficient point of engine operation for any output conditions and a CVT is required in order to enable these engine conditions to be matched to the vehicle output speed. A study of power levels used in typical vehicle operation indicates that the majority of time only low powers are required and that operation will take place predominantly along the low speed section of the economy line. This also indicates the need for a wide ratio spread in order to provide a good overdrive ratio that allows the low engine speeds to be used.

There are, in addition, performance benefits that can be obtained using a CVT since for ultimate acceleration the engine can be operated at its maximum power when required. In a vehicle with a discrete ratio transmission, the engine can only develop maximum power once for any ratio and there will be an interruption in power transmitted when the ratios are changed. Porsche showed (Kraxner *et al.*, 1999) that a CVT vehicle would out accelerate the same vehicle with a conventional manual transmission. It was further emphasized that it would only be the skilled drivers who could achieve the best from a manual transmission anyway.

Such transmissions have been under development since the early days of automobiles, but have only recently begun to establish a place in the commercial market. Why is this, if there are potential benefits in terms of driving comfort, economy and performance? Different factors have contributed to this unpopularity at various times, and have included reliability, weight and cost. However, the main downside is the lower efficiency of all CVTs relative to geared transmissions. Losses in geared systems occur as a frictional torque required to rotate the transmission components. If an ideal speed ratio is defined for a gear pair then the output torque will be less than the ideal torque as a result of these losses, but the output speed will be the same as the ideal speed. However, all CVTs also incur some speed 'slip' between input and output as well as torque losses, and both output torque and speed will be less than the ideal. Since gears

* The idea of 'infinitely' can be easily accommodated when it is appreciated that trasmission ratios are usually exressed in the form of input to output as a SPEED ratio, and that an infinite speed ratio is obtained when the output speed is zero.

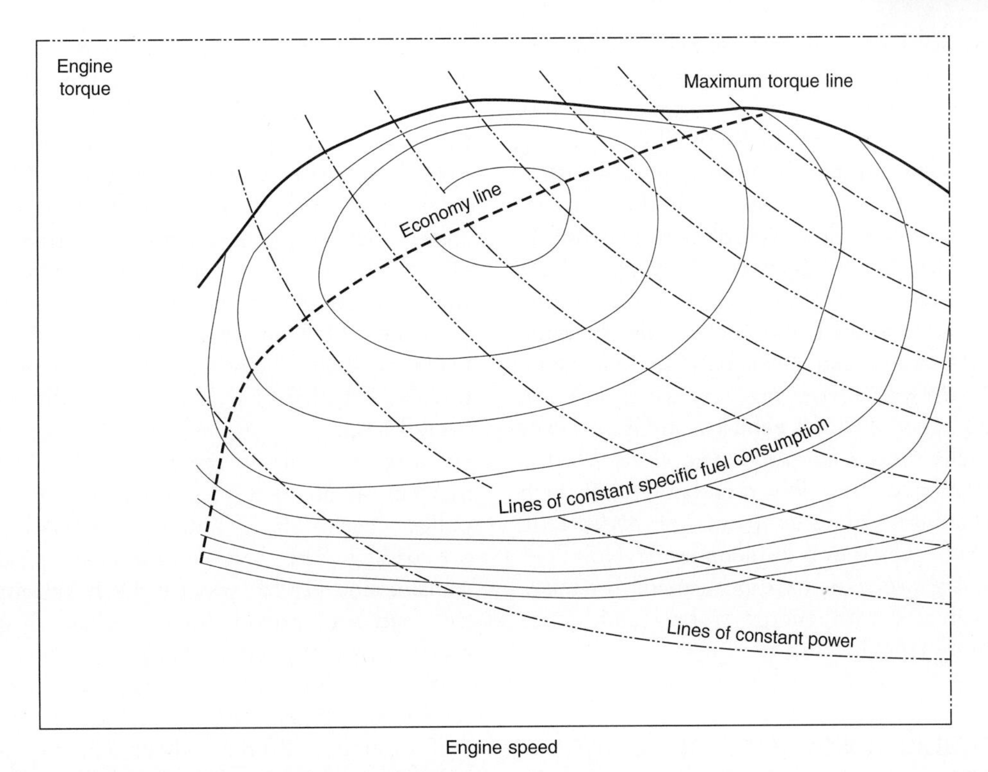

Fig. 5.1-26 A typical petrol engine torque speed characteristic.

are very efficient, it only takes a small additional loss for the effect to become significant.

The other factor is the way that the CVT powertrain is controlled and the resulting driver's feel or perception of the vehicle behaviour in response to accelerator pedal demands. This characteristic is called driveability, and is very much conditioned by the individual driver's previous experience and expectations. People more used to vehicles with manual transmissions are not as comfortable when driving a vehicle with any AT, because they feel they are less in control. There is also a significant compromise required between vehicle driveability and the possibility of achieving the desired economy line operation. A study of the engine map in Fig. 5.1-26 shows that in most regions there is very little extra torque available between the economy line and the maximum torque curve. When a driver demands acceleration in this area then the system has to use this excess torque to first accelerate the engine before a significantly higher power can be transferred to the driving wheels. This gives a delay between a driver pressing the accelerator and the vehicle responding, this is also accompanied by the sound of the engine speeding up but nothing apparently happening. Many drivers find this disconcerting; particularly those used to manual transmissions. Road test reports have made comments about 'the rubber band effect' and '... the car's initial sluggish response ...'.

Most vehicles fitted with CVTs have been with smaller engine sizes, generally below 1.5 litre, and not those noted as sporty cars even in standard form. In this category, the performance margins are smaller and most manufacturers have not controlled the engine to rigidly follow the economy line and have compromised this to allow a larger torque margin for acceleration. The effect of transmission efficiency is also more significant in this category with a greater parasitic power lost. The result has been that most early production CVT vehicles have neither given better economy nor performance relative to vehicles with manual transmissions. However, a more valid comparison is with a conventional AT and here the CVT is the winner for both economy and performance. More recently larger engined vehicles like the Audi A6 have shown that CVT is not just a viable alternative to ATs but can deliver the improvements in both economy and performance even relative to the manual version (Audi's own figures). What is more, even road test reports in motor magazines have begun to report the virtues of CVTs as being realized '... a transmission that for once actually enhances the car's performance' (*Autocar*, 2000).

5.1.5.2 Hydraulic transmissions

There are three main categories of transmission mechanism that can be used to provide the power and speed range for vehicle use: hydraulic systems, variable radius pulleys and traction drives. The first of these, described in this section, can be further sub-divided into categories of hydrostatic transmissions. Both of these are capable of giving a zero output speed and hence give IVT operation. The hydrokinetic drives have been described above in Section 5.1.4, and, as discussed above, torque converters are widely used in road going vehicles of all types.

Hydrostatic drives also rely on fluid flow to transmit power but it is the pressure level in the fluid that is significant rather than the flow velocity. A hydrostatic transmission comprises a pump unit supplying a motor unit, and both these are of the, so-called, positive displacement type. One at least must be variable capacity, usually this is the pump, and this is used to control the overall ratio. Hydrostatic drives are very widely used in agricultural and other off-road vehicles but have never been commercially used in automotive applications. It is generally the efficiency at low powers and potentially high noise levels that are the weak points of hydrostatic transmissions. However, they are still under development and, even recently, proposals have been made for buses and delivery vehicles with energy storage and hence a hybrid capability (Ifield website).

5.1.5.3 Variable pulley variator designs

The idea of a variable pulley system is a logical extension of a conventional V-belt fixed ratio drive. Fig. 5.1-27 shows the principle, where there are two pairs of conical pulley sheaves and a fixed length V-belt. One half of each pulley pair is moveable and their movement is synchronized. This

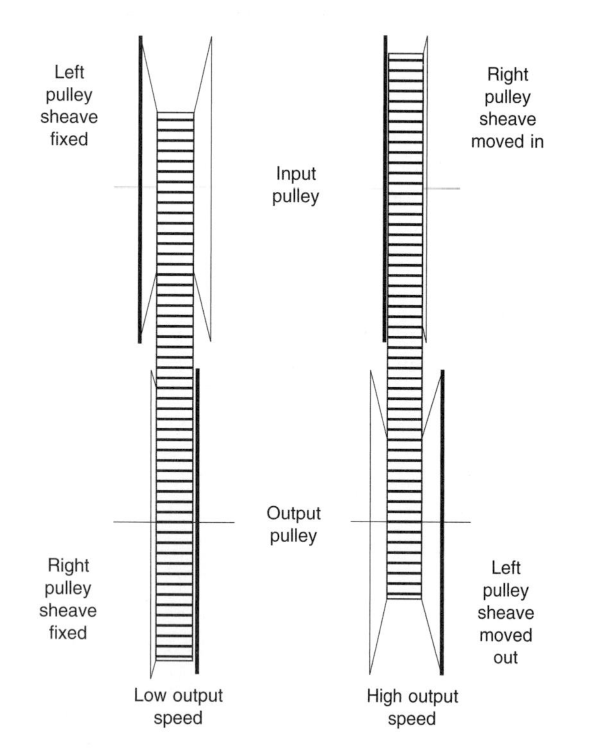

Fig. 5.1-27 Variable pulley drive concept.

allows the belt-rolling radius to be changed such that a relatively low output speed is obtained when its radius at the output is large. The output speed can be increased as this radius is made smaller and the rolling radius on the input is increased. As in a conventional V-belt, there is a frictional force between the pulley sheaves and the angled belt face that provides the transfer of tractive effort. There must be sufficient normal force between them to prevent gross slipping and the clamping forces that hold the pulley sheaves together also produce a tension load in the belt. The relative magnitude of these clamping forces can then also be used to control the belt position and overall ratio.

Early proposals, in the 1920s (Gott, 1991) relied on mechanical linkages to provide the movement of the pulley sheaves and maintain their relative position and pre-loading. The first commercially produced vehicle transmission (1958) was the DAF Variomatic and here the load was provided by pre-loaded springs in the pulleys, and the ratio change by a centrifugal effect working against these springs. The belt was fabric reinforced rubber, and two belts were used in parallel each driving a different wheel. This design was limited in its torque capacity to only smaller vehicles, and both wear and efficiency limitations restricted its broader application. Simple rubber belt designs have continued to evolve and are still widely used in two-wheel and some small city vehicle applications. Most modern variable pulley systems use steel belts to provide the power capability needed for automotive applications in a suitably compact package. The design and manufacture of the metal belts provides the heart of modern systems. The belt must have sufficient strength and rigidity to transmit the driving loads and yet also be flexible to keep the minimum rolling radius as small as possible. This has been achieved with two designs that also have significantly higher efficiency than the earlier rubber versions.

The first metal belt system was introduced into the market place in 1987 as the Transmatic, by Van Doorne's Transmissie (VDT) of the Netherlands. This transmission is the successor to the Variomatic but shares little more than the variable conical pulley concept with its predecessor. This is still the only automotive CVT in quantity production despite the recent launch of vehicles with both an alternative belt design (Audi) and a traction drive (Nissan). Over 3 million VDT units have been supplied for car use and current production of the belt is 500 000 units per year. Ford, Fiat, Honda, Mitsubishi, Nissan, Rover, Subaru, Toyota and Volvo have used it. Recent vehicles to use the this system have engine capacities in the 2-L range, representing a significant advance on the early small and super-mini (sub-compact) use. The capability for operating in extremes was shown by the successful demonstration of a version fitted in

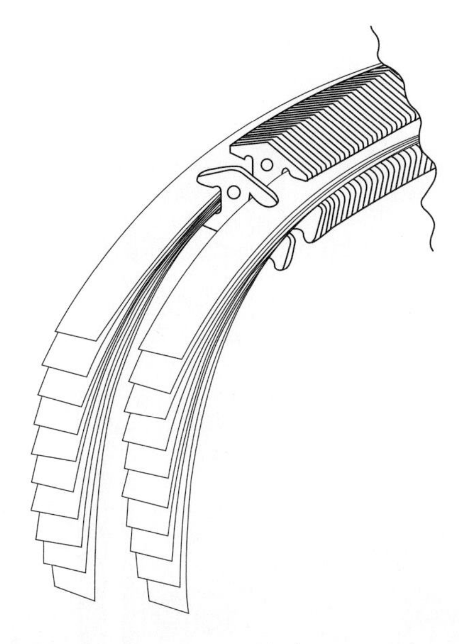

Fig. 5.1-28 The Van Doorne metal belt construction.

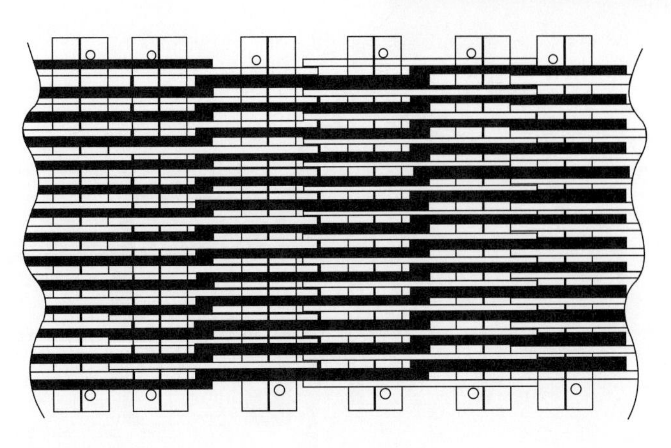

Fig. 5.1-29 The LuK-PIV chain construction.

a Williams–Renault Formula 1 car before this technology was banned by the FIA in 1994.

The construction of the belt used in the VDT system is shown in Fig. 5.1-28. The belt is assembled from two packs of flexible steel bands with a set of individual segment blocks retained by these bands. A typical design for an 85 kW, 165 Nm capability has 12 bands in each pack with about 430 segments in a belt that has a free length of 680 mm. The angled surface of the segments, below the shoulder, provides the contact surface with the pulley sheaves. Since the bands are free to slide relative to the segments and it is not possible for the segments to transfer load in tension, the only remaining mechanism is for the segments to transfer load in compression. Thus this design is often called a 'push belt'. Since the surface area between segments is relatively large the working levels of compression stress are low. When the bands have been placed in tension by the pulley clamping forces they will stretch and gaps open up between the segments on the non-compression side of the belt. Many changes in detail design of the belt, particularly the segments, have been made since its introduction, and given improved performance and reduced manufacturing cost.

Another variable pulley system is based around the LuK-PIV chain as the flexible belt element. This is constructed in a way that is more reminiscent of conventional roller chain as used in motor and pedal cycles and is shown in Fig. 5.1-29. Adjacent pairs of rocker pins are connected through a number of link plates, each set offset relative to the next set of link plates. The link lengths are not all identical and these give a staggered pitch for the pins that reduces acoustic noise. Some of the links near the outer edges of the chain are thicker to increase stiffness and reduce overall distortion under high load. The extended ends of the rocker pins act as the contact faces to the pulley sheaves. The ends of the pins have a crowned face rather than being flat, and this reduces the sensitivity to any misalignment but also reduces the active contact area.

5.1.5.4 Variable pulley transmissions

The variable pulley variators form the basis of a complete transmission system that also includes a hydraulic system, additional gearing and a starting device. A typical transmission based on the VDT push belt is shown in Fig. 5.1-30, and is similar to that used by many major manufacturers. Until recently, all production variants used a clutch as a starting device, most frequently a wet plate clutch as in the figure, although some low-power designs have used an electromagnetic powder clutch. Audi have continued with this in their A6 2.4 Litre Multitronic, but the current trend is towards the use of a torque converter in place of the clutch. This gives a wider useable ratio range and has the start from rest and acceleration feel of a torque converter, and hence gives a better driveability feel for the transmission. However, it is less efficient and increases the torque in the pulley section, which will also increase the losses. The transmission shown also includes transfer and final drive gearing, including the differential appropriate for a front-wheel-drive vehicle. In the region of the clutch pack there is also an epicyclic gear that is used to give a forward and reverse shift by engaging the appropriate clutch.

The hydraulic system provides similar functions to the more conventional ATs including lubrication, cooling and control. Both chain and belt systems use hydraulic pressure applied to pistons that are a part of the moving pulley sheaves. The supply is obtained from a pump, located towards the right of the transmission in Fig. 5.1-30, and permanently driven from the engine via a quill shaft. This preferentially supplies the control system that will also

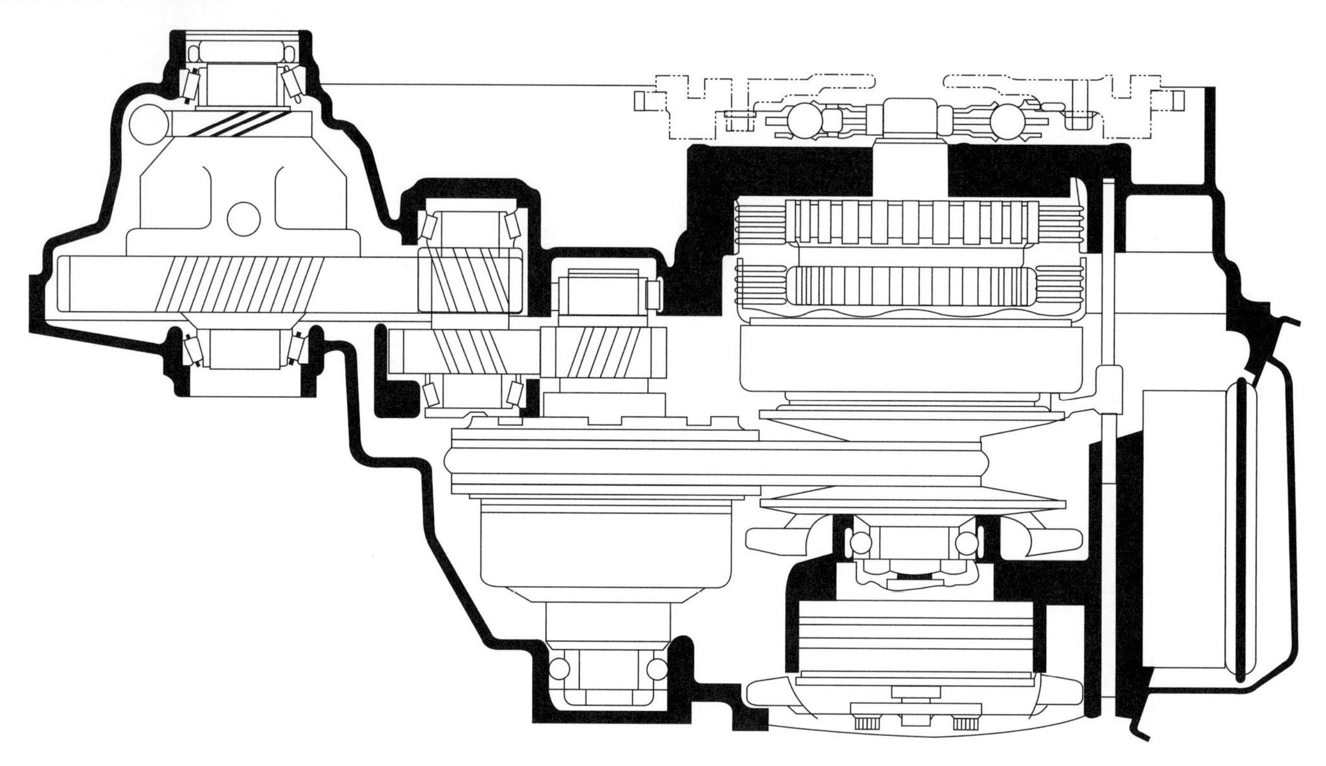

Fig. 5.1-30 A complete push belt transmission.

determine the maximum operating pressure. Since it is better to keep lower operating pressures to improve efficiency, this will be controlled to the minimum necessary to provide the pulley clamping forces to prevent belt slip. Some designs use a single piston that provides both the base clamping force and ratio control by modulating the relative pressures on input and output pulleys. Others use a double-piston design that separates the ratio and clamping functions. Although introducing mechanical complication, this provides the opportunity to reduce clamping pressures particularly in low throttle overdrive situations. This must be accompanied by a device that can raise the pressure rapidly, if the torque suddenly changes (as might occur if one wheel loses grip momentarily on a patch of ice).

The first versions of belt CVTs used hydraulic logic to control the transmission and schedule the ratio and system pressure with operating conditions. This has now changed, and all of the designs now being launched include electrohydraulic control for the same reasons of versatility and flexibility that were described above for ATs. A comparison carried out by LuK (Faust and Linnenbruegger, 1998) showed that a modern 5 speed AT required approximately 6 solenoids with 20 valves in comparison with their CVT design that needed only 3 solenoids and 9 valves.

The only control strategy that has been adopted until recently has been that of a full automatic with some driver selected inputs to hold low ratios to allow an increase in engine braking. As with conventional Ats, input signals come from driver inputs of accelerator and brake, as well as engine and output speeds. In all cases the transmission is scheduled to keep engine speeds as low as possible within the constraints and compromises of driveability and emissions. This gives the best fuel economy. However, several new systems (e.g. Rover/MG, Audi and Nissan) have offered an additional option of a manual transmission emulation. This gives the driver the option to select between six pre-defined ratio settings to give the feel of a manual gearbox. The resulting ratio change is rapid and smooth and of course since there is no clutch there is no interruption in drive to the wheels. If accompanied by electric switches on the steering wheel it also follows the pattern established for Formula 1 and rally cars!

5.1.5.5 Traction drive designs

The other main mechanical alternative to pulley belt systems are those based on the idea of changing the axis of a rotating element between input and output discs. This concept is shown in Fig. 5.1-31 by a single rotating roller that is held in friction contact between the two discs that have the same horizontal axis of rotation. In Fig. 5.1-31(a) the axis of rotation of the roller is angled to the left and contacts the input disc at a smaller radius than the output disc, hence giving a reduced output speed relative to input. Fig. 5.1-31(b) shows the roller axis rotated to the right giving the output disc an increase in speed. Drive is transferred via the contact patch between the roller and the two discs through a thin

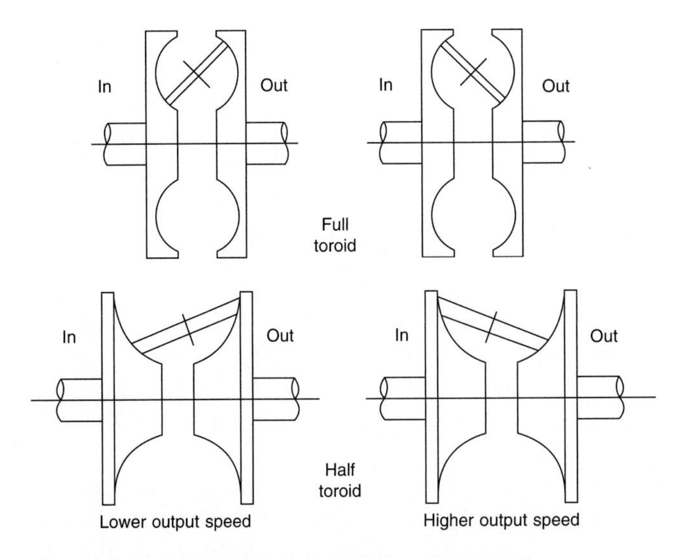

Fig. 5.1-31 The toroidal drive concept (full and half).

elastohydrodynamic liquid film, typically under 0.5 μm. This requires high loads holding the components together to give high contact patch pressures. The resulting high pressure in turn gives a significant increase in viscosity of the fluid in this region.

The cavity between the discs is toroidal in shape and gives this name to variators based on this principle. There are in addition several other practical designs that include the Kopp variator for industrial use and the Milner ball drive (Akehurst *et al.*, 2001) used in lower power applications. The figure shows a full- and a half-toroidal cavity and both these form the basis for transmission designs. The power roller in the half toroid design is pushed outwards by the application of the clamping forces between the discs necessary to provide the surface contact pressures. The full toroid retains the power rollers within the cavity without generating these side forces but experiences a spin loss in the contact surface. This spin loss occurs because the surface of the power roller and the corresponding disc surface cannot be travelling at the same speeds throughout the contact patch.

In these traction drives both the materials used in the contact surfaces and the lubricating or traction fluid are important in giving good reliability and efficiency, ultimately the effectiveness of the transmission. High-grade bearing steels with very low impurity levels have been shown by NSK to give the best life in their half-toroidal variator. Special fluids have been developed by a number of companies including Monsanto, and more recently Shell. These have given good 'traction' coefficients in the order of 0.1 at high pressure whilst retaining good lubrication and low friction under more normal bearing conditions. It has also been demonstrated that cleanliness of the fluid is important in reducing the wear of the rolling components and hence better filtration than normal is required.

5.1.5.6 Toroidal transmissions

The first production toroidal transmission was lunched in the autumn of 1999, fitted in the Nissan Cedric/Gloria but only available in the Japanese market. This vehicle is fitted with a 3 litre engine and the transmission is rated to an input torque capacity of 370 Nm. The variator section has dual toroidal cavities with two power rollers in each, as shown in Fig. 5.1-32. The necessary contact surface loading is obtained through a combination of a preloaded spring establishing a minimum value and a loading cam that gives an additional component that increases with transmitted torque. This transmission also has a torque converter as the starting device and includes forward and reverse gearing via an epicyclic section. The torque converter has a maximum conversion ratio of 1.98 and the variator section has a ratio range of 4.4. The transmission is reputed to give a 10% improvement in fuel consumption relative to a conventional AT.

The implementation of a full-toroidal device in a transmission has been proposed by Torotrak and is currently

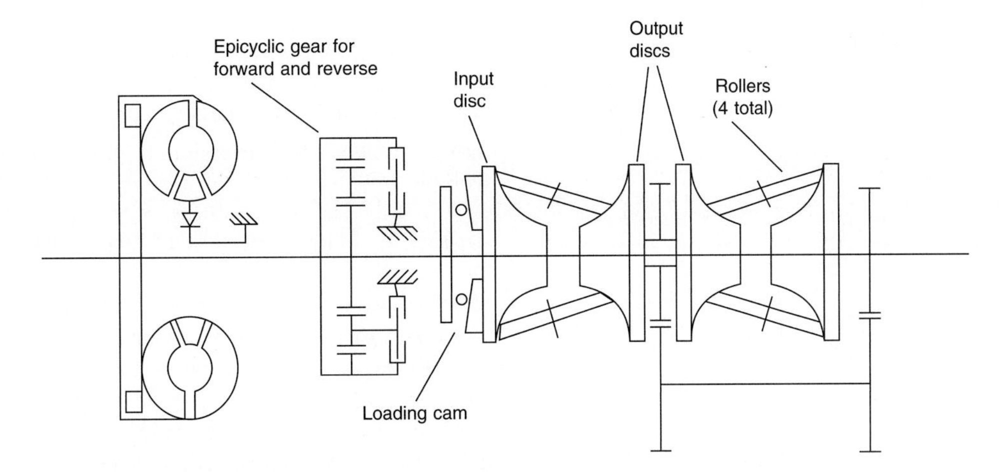

Fig. 5.1-32 A dual-cavity half-toroidal transmission.

under development for demonstration in sports utility vehicles. This is also a dual cavity with three rollers in each, a front rear layout and rated for 5 Litre engines producing around 450 Nm torque. In prototype form it has also been fitted to the 2 Litre Ford Mondeo in a front-wheel-drive configuration. This transmission is notable for the use of the split path principle to provide an IVT characteristic from a CVT variator.

The layout for the transmission is shown schematically in Fig. 5.1-33 and shows two potential drive paths connecting the engine to the output, referred to as low and high regime. One path from the engine is via the variator and the sun of the epicyclic gear, and the second transferred from the engine directly to the planet carrier of the epicyclic gear via the low regime clutch. The output is taken from the annulus of the epicyclic gear and is a summation of the two inputs. A study of the epicyclic equation in Section 5.1.4.6 will show that it if the sun and carrier are driven at the right relative speeds, i.e. the sun is driven at about 3 times the carrier speed, then the resulting annulus speed will be zero. This will result in a zero output speed whilst the engine and internal components are still rotating and is called a geared neutral. If the sun speed is reduced relative to the carrier then the annulus and output will increase in speed in the same sense as the carrier. If, however the sun speed is increased relative to the carrier then the annulus will be driven in a negative sense relative to the carrier. Hence this configuration not only gives the transmission an effective starting device but also a reverse gear capability without any additional components.

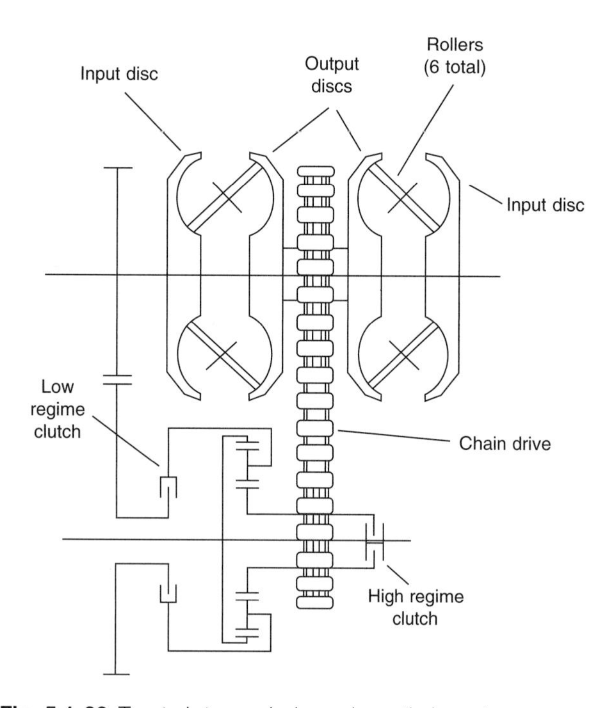

Fig. 5.1-33 Torotrak transmission schematic layout.

Unfortunately it cannot then also meet the high overdrive ratio that is necessary for good economy and the range needs to be extended through what is called a mode change. In high regime the variator output is locked directly to the annulus and transmission output shafts by a multi-plate clutch. At the same time the second clutch disconnects the transfer drive input to the carrier of the epicyclic. The change between high and low regimes has been designed to take place when all three epicyclic components are rotating at the same speed, and corresponds to one end of the variator range. The full range of variator ratio is thus available in both high and low regimes, giving an overdrive capability of 100 km/h per 1000 rev/min.

A hydraulic system is again used for control purposes in both toroidal transmissions. The roller mechanisms transmit torque only when there is a steer effect trying to get them to move to new position and hence change ratio. Applying a load to the roller's mounting produces this steer effect, and the consequent reactions at the roller circumference produce a torque between the two discs. In both implementations the rollers are moved and loaded hydraulically but their control strategies are different. In the half-toroid the rollers are positioned in a closed loop hydromechanical system that sets a transmission ratio. In the full toroid the load is used directly to control the transmission torque rather than its ratio. It is this feature that makes control of the transmission around the geared neutral effective since a zero demanded output torque will give this condition directly. In both cases there is an electronic management system integrated with other vehicle systems and with electrohydraulic control inputs to the transmissions. Scheduling the control can again allow for a variety of different strategies including the discrete speed emulation, described above.

5.1.6 Application issues for transmissions

This section picks up on some of the important issues for transmissions that are not covered in the above sections. These include a number of design considerations required when a transmission is chosen for a particular vehicle.

5.1.6.1 The operating environment

The environment that any of the transmission units operate in can be a very major consideration when installing in the vehicle. It should be pointed out that any vehicle manufacturer will have their own standards to which they would expect a transmission to conform. For this reason, the material here can only ever be an approximate guide.

Firstly, considering the temperature constraints. Many transmission units are installed under the bonnet (hood), alongside, and attached to, the engine. The influence of the engine, cooling pack and exhaust system, together with any air ducting, can lead to extremely high air temperatures around the transmission. This air temperature can certainly reach similar temperatures to the required temperature of the transmission oil. Unless the transmission has cooling oil piped to a remote cooler or heat exchanger, the unit must rely on convection to remove any excess heat. This is obviously only possible if there is air of a lower temperature than the oil passing over the outside of the casing.

Ideally, the transmission oil temperature will warm up quickly on start-up of the vehicle, particularly if the ambient temperatures are near or below freezing. This will enable the transmission to operate efficiently for as much of the time as possible. The transmission oil will obviously start near to the ambient air temperature and at low temperatures can be extremely viscous. This can be a severe problem with automatic and CVT transmissions particularly as the lubricant also acts as a hydraulic control fluid so fluidity is important.

At the other operating extreme we will want to prevent the oil getting too hot and compromising the durability of the oil itself or the components in the unit. Maximum operating temperatures are often around 140 or 150 °C, but it should be noted that these temperatures require a modern specification lubricant. Many older formulations will start to degrade much above 110 °C. It can sometimes be very difficult to control the lubricant to a reasonable upper operating limit when the vehicle is operating in high ambient temperatures and working hard. For example when towing a trailer with ambients around 50 °C in the Gulf States.

From the above discussion, it can be seen that the control of the lubricant temperature is not a trivial problem and many companies are working towards management of the transmission in much the same way as the engine. That is warming the unit up quickly with waste heat from the engine but then controlling the upper temperature, perhaps to a temperature below that of the engine coolant. Simply linking the cooling systems of the engine and gearbox is not necessarily the answer. Part of this development also entails careful consideration of the lubricant itself.

There is not the space here to discuss the choice of lubricant in any depth. It is perhaps, sufficient to highlight some of the main considerations.

- Cold-temperature fluidity – Does the lubricant have to act as a hydraulic fluid? If so, how cold is the transmission expected to operate, and how critical is immediate pressure or oil flow from the pump?
- High-temperature capability – What is the highest temperature the transmission can see in service? Remember it is very difficult to specify 'excursions' to high temperatures, as it is almost impossible to predict the duration and frequency over the life of a vehicle
- Viscosity grade and load carrying additives – What grade of oil is required to provide adequate gear and bearing life? In certain cases this is as dependent on the action of the additives as on the base oil.
- Air release and foaming – This tendency of the oil can be vital in certain cases. If the oil behaves badly, it can lead to oil being ejected from the breather system.
- Corrosion inhibition – Components within the transmission do get exposed to moisture, mostly from condensation. Because of this, the oil needs to prevent the surfaces from corroding, particularly while the transmission is stood for a long duration.
- Friction modification – The level of friction, which occurs at the rubbing surface of some parts, needs to be controlled within certain limits. An example of this is in automatic gearbox clutch packs, where too high a friction could cause problems as significant as too low a value.
- Viscosity modification – The viscosity grade of the base oil used may not be quite what is required. Additives can be used to improve the rate at which viscosity changes with temperature, providing that the oil is thicker at high temperatures to protect the running parts, while retaining the required fluidity at low temperatures.

Finally, other aspects of the operating environment also need to be considered. The corrosion of the outside of the transmission case is rarely too much of a concern, but if ferrous materials are used then they may need to be protected by painting. Similarly, immersion in water is unlikely to cause a significant problem, but for off road vehicles, this needs to be considered along with possible damage from rough terrain or sand etc. Tests on the transmission will be used to validate the design and could include salt spray, wading, and various types of off road driving. The environment within the engine bay may also require the outer transmission parts to resist degradation from fuels, antifreeze, brake fluid, etc.

5.1.6.2 Efficiency

Before continuing with this section, an important point to appreciate is that the efficiency (or losses) of a transmission unit can be considered in three ways:

- The load (torque) related efficiency, this occurs largely as a result of the friction losses at the gear

mesh and can be considered in the form of a percentage efficiency or loss figure, i.e. a loss of 3% would be an efficiency of 97%, whichever is the most convenient.
- The parasitic losses can be considered to be independent of the applied torque, these losses can be considered to be 'drag torques' and take the form of a resistance within the transmission.
- The slip losses that may occur in transmission elements which do not involve a fixed gear ratio. Where the drive is transmitted by gear pairs, the input/output speed ratio is obviously fixed by the tooth numbers on the gears. Where the drive is transmitted by another means, the output speed is not necessarily a fixed ratio to the input.

The overall efficiency of a transmission unit needs to take all three aspects into account to arrive at an overall efficiency figure.

Although a number of authors in the literature consider all the losses in a transmission together, others have considered either the load related or parasitic losses in isolation. In common with this approach we will consider the losses separately. Merritt (1971) uses the definitions of oil churning losses and 'tooth friction' when discussing gear losses. He and other authors have chosen to elaborate on the friction losses at the gear mesh and assume the parasitic losses are small by comparison. Test work completed by the authors has demonstrated that this is not the case with the dip lubricated transmission units typical of those used in the automotive industry. The parasitic losses can be significant and need to be considered in a wide variety of transmission units. This is particularly the case when the gearbox is operated from a cold start where the lubricant is at or near the ambient air temperature. In winter, this could obviously be well below 0 °C.

In automatic units, the losses associated with the oil pump are often the largest cause of parasitic loss. When the gearbox requires a high oil pressure the torque required to drive the pump can be significant proportion of the torque being transmitted. An example of this is a belt CVT operating at its low ratio at low vehicle speeds. The belt system requires high pressure but the transmitted torque is low due to the low road load so the pump load can be very significant.

Most transmission units use rolling element bearings rather than plain bearings. As with gears, bearings will have load related friction losses and parasitic losses due to the oil movement and windage of the rollers and cage. For prediction and comparison purposes, the bearing loss can be treated as part of the gear system losses. It can be assumed these are small, and either constant, or behave in a similar way as the oil churning or gear mesh losses relative to the 'control' variables of speed, torque and viscosity. Many authors consider the bearing losses to be an order of magnitude smaller than the corresponding gear losses.

This split of load related, parasitic and slip losses is important for other related areas of work. The load related, parasitic, and slip losses need to be treated separately in performance prediction and simulation work. Also an overall indication of the losses can often be derived for the transmission from simple parasitic loss testing at zero output torque (i.e. with the output disconnected so no absorbing dynamometer is required).

The efficiency of a transmission unit is particularly important during two operating conditions of the vehicle:

- Cold start, 'gentle' drive cycles, urban driving, test cycles, etc. The parasitic losses have an impact on fuel economy, as they are significant compared to the drive torque required by the vehicle.
- Arduous use, high speed, towing, etc. The friction (load related) losses are roughly proportional to the torque transmitted and can cause very high heat output from the unit. This, in turn, can lead to high oil temperatures and even oil breakdown or component failure due to insufficient operating oil film.

In summary, these loss mechanisms can be described in terms of the three categories discussed above:

Load-related losses:

- Friction losses at the gear tooth mesh point
- Load-related bearing friction losses

Parasitic losses:

- Oil churning where gears and shafts dip in the oil bath or foamed oil
- Oil displacement at the point where the gear teeth enter the mesh point
- Windage losses where gears operate in air or oil mist
- Oil seal drag
- Oil pump drag
- Parasitic losses in bearings due to oil displacement (and windage) within the bearings
- Drag in clutch packs in autos and CVTs (those not engaged)

Slip losses:

- Slip in the contact zone where drive is transmitted by friction (i.e. belt–pulley contact in a CVT)
- Slip that occurs in a fluid drive such as a torque converter.

In most non-pumped automotive transmissions the large proportion of the load related and parasitic losses come from the gear friction losses and the oil churning losses respectively. The pump losses must always be considered if the transmission has a high-pressure hydraulic circuit. In belt and toroidal CVT/IVT transmissions the slip losses can also be significant. As discussed in Section 5.1.4 the speed ratio (indicating slip) is always considered in torque converters.

5.1.6.3 Other transmission components

No review of transmissions would be complete without at least a brief mention of some of the other components that can be found within the driveline system of the vehicle:

- differentials
- breather systems
- the gearchange mechanism.

Differentials are used in order to allow the left and right hand wheels on any one axle to rotate at different speeds. This is essential on road going vehicles as the two wheels take different paths when the vehicle goes round a corner, hence travel different distances and the inner wheel will rotate slower than the outer. This effect can also be seen between the front and rear wheels of a four wheel drive vehicle so a centre differential is required between the front and rear axles.

As a result of the above requirement, differentials are part of the gearbox assembly on front-wheel-drive vehicles as the drive is taken directly to the wheels by two separate, output driveshafts. On rear and four wheel drive vehicles the differential is invariably part of a separate axle assembly. It should be noted that axle assemblies also contain part of the gearing required to reduce the speed of the rotating parts from engine speed to wheel speed. This is often referred to as the final drive and typically on a passenger car would be a reduction of around 4:1. Within an axle this would be achieved using a bevel gear pair which also turns the drive through 90°. The differential assembly then rotates with the output gear of this pair and allows the wheels to rotate as described above. It can be seen that the assembly often referred to as the 'differential' on a vehicle actually serves other purposes as well.

Breather systems are one part of the transmission system often ignored by many vehicle engineers, but can prove to be difficult to design. As the transmission warms up and cools down the air inside needs to be able to expand and contract. The breather allows this to happen without the air inside becoming pressurized, as this could push oil out past the oil seals, joints, etc. The problem with developing these is finding a position on the transmission and a design of breather, which allows the air to move in and out of the transmission without allowing water in or oil out. Considering that:

- oil inside the transmission can be extremely aerated
- there is often no part of the gearcase that does not have oil splashed on the inside
- the gearbox is often partly or wholly submerged during wading through flooded roads. It can be readily seen that the breather not only has an important task, but can prove difficult to engineer.

Sealing of the gearcase on the transmission is required to prevent oil leaking on to the ground and the subsequent loss of oil. The breather system has been considered above and is one part of the whole system. The other areas are where the shafts enter and exit the unit, the joints in the casing and the integrity of the casing itself. Proprietary 'rubbing' lip oil seals are used on most shafts (also called dynamic shaft seals). The temperature requirement of the application means that most are made from a synthetic material which has a thin lip contacting the rotating part. Important design aspects of these parts include the surface hardness, surface roughness and adequate oil supply to the lip; the contact area needs to be lubricated to avoid excessive wear. The manufacturers of these seals provide good literature to allow them to be applied correctly. To complete the sealing of the casing, the joints between the housings need to be sealed carefully, usually using silicone or RTV sealant applied at the point of assembly. Gaskets can also be used. The casings themselves can be a source of leaks if the castings used are allowed to be porous. This can occur in the casting process and needs to be avoided by both good design and manufacturing practice.

The gearchange on both manual and automatic transmissions require a mechanical connection between the driver's controls and the transmission unit. For manual gearboxes this is obviously used to shift the gearchange mechanism within the gearbox. On automatics it can sometimes do little more than engage the parking mechanism within the 'box'. With older ATs though, the shift mechanism can be mechanically linked to the internal hydraulics and used to directly influence the gearchange. The requirement to mechanically link the two is reducing because of the impact of electronic control. In practice, the difficulty of linking the controls (gearlever) with the transmission unit is much to do with the relative position of the two and the interface at the gearbox end. It goes without saying that rear or mid-engined installations can be difficult, as the connections on the power unit often face the rear of the vehicle. For manual transmissions, the use of solid rod connections is often favoured so the driver gets a more direct feel of the gearchange, although cable systems can provide a very good solution. Either solution will have two parts to it; one to transmit the axial movement (i.e. in and out of gear), and the other to transmit the 'cross-gate' movement. The latter is where the gearlever is moved from the 1st/2nd position of the 'gate' over towards the 3rd/4th (usually the centre) and the 5th gear positions. Any solution to the gearchange will have to allow both movements to be transmitted to the mechanism in the gearbox.

References

Akehurst, S. Brace, C.J., Vaughan, N.D. and Milner, P.J. (2001). 'Performance investigation of a novel rolling traction CVT' SAE Int. Congress and Expo, Detroit, USA, March, 2001.

Autocar (2000). 6th September.

Faust, H. and Linnenbruegger, A. (1998). CVT Development at LuK, LuK 6th Symposium 1998, pp. 157–179.

Gott, P.G. (1991). *Changing Gears: The Development of the Automotive Transmission*. SAE ISBN 1 56091 099 2.

Heisler, H. (1989). *Advanced Vehicle Technology.* Butterworth-Heineman ISBN 0 7131 3660.

Ifield website – www.ifieldshep.com

Kraxner D., Baur, P., Petersmann, J. and Seidel, W. (1999). CVTip in Sports Cars, CVT '99, International Congress on CV Power Transmissions, Eindhoven, Netherlands, Sept. 1999, pp. 21–26.

Liebrand, N. (1999). Future Potential for CVT Technology, CVT '96, International Conference on CV Power Transmission, Yokohama, Japan, Sept. 1996.

Merritt, H.E. (1971). *Gear Engineering*, Pitman, ISBN 0 273 42977 9 – out of print but you should be able to find it in good technical/college libraries.

Socin and Walters (1968). *Manual Transmission Synchronizers*. SAE 680008.

Takahsashi, M., Kido, R., Nonaka, K., Takayama, M. and Fujii, T. (1999) Design and Development of a Dry Hybrid Belt for CVT Vehicles, CVT '99, International Congress on CV Power Transmissions, Eindhoven, Netherlands, Sept. 1999, pp. 254–259.

Further reading

Heisler, H. (1989) *Advanced Vehicle Technology.* Butterworth-Heineman ISBN 0-7131-3660. An excellent base text that describes operation and gives details of many of the components used in vehicles. There is a significant section on transmissions and the book is very well illustrated with many line drawings. Included in the transmission material is design and function detail of many of the components and sub systems found within the transmission system.

Vaughan, N.D. and Simner, D. (2001). *Automotive Transmissions and Drivelines*. Butterworth-Heinemann. A text book aimed at degree and postgraduate levels that covers the material in this chapter in much more detail and also expands to include other driveline components. Many aspects of basic analysis are included with worked examples and a number of detailed case studies of typical transmission designs.

Gott, P.G. (1991). *Changing Gears: The Development of the Automotive Transmission*. SAE 1991 ISBN 1-56091-099-2 A fascinating historical view of transmission developments from the earliest days of automobile engineering. There is more emphasis placed on American manufacturers and the automatic transmission but the coverage is much wider, definitively comprehensive and embraces developments globally. Highly recommended.

Bibliography – other worthwhile references

Bosch Automotive Handbook. Good ready reference on many automotive subjects, regularly updated with new editions published.

Wong, J.Y. (1993). *Theory of Ground Vehicles*. Wiley, ISBN 0–471–52496–4

Bearing manufacturers – Most of the large manufacturers produce catalogues with extensive and useful reference material as an introduction. Particularly recommended are SKF, Timken, and NSK/RHP.

Further detail design guidance and information is often available from the component suppliers. The oil companies and clutch suppliers are good examples of this.

Section Six

Electric vehicles

Chapter 6.1

6.1

Battery/fuel-cell EV design packages

John Fenton and Ron Hodkinson

6.1.1 Introduction

The rapidly developing technology of EV design precludes the description of a definitive universal package because the substantial forces which shape the EV market tend to cause quite sudden major changes in direction by the key players, and there are a number of different EV categories with different packages. For passenger cars, it seems that the converted standard IC-engine driven car may be giving way to a more specifically designed package either for fuel-cell electric or hybrid drive. While the volume builders may lean towards the retention of standard platform and body shell, it seems likely that the more specialist builder will try and fill the niches for particular market segments such as the compact city car. It is thus very important to view the EV in the wider perspective of its market and the wider transportation system of which it might become a part.

Because electric drive has a long history, quite a large number of different configurations have already been tried, albeit mostly only for particular concept designs. As many established automotive engineers, brought up in the IC-engine era, now face the real possibility of fuel-cell driven production vehicles, the fundamentals of electric traction and the experience gained by past EV builders are now of real interest to those contemplating a move to that sector. A review of the current 'state of play' in sole electric drive and associated energy storage systems is thus provided, while hybrid drive and fuel-cell applications will be considered in Chapter 7.1.

6.1.2 Electric batteries

According to battery maker, Exide, the state of development of different battery systems by different suppliers puts the foreseeable time availability for the principal battery contenders, relative to the company's particular sphere of interest, lead–acid – as in Fig. 6.1-1a.

6.1.2.1 Advanced lead–acid

The lead–acid battery is attractive for its comparatively low cost and an existing infrastructure for charging, servicing and recyclable disposal. A number of special high energy versions have been devised such as that shown at (b), due to researchers at the University of Idaho. This battery module has three cells, each having a stack of double-lugged plates separated by microporous glass mats. High specific power is obtained by using narrow plates with dual current collecting lugs and a 1:4 height to width aspect ratio. Grid resistance is thus reduced by shortening conductor lengths and specific energy is improved by plates that are thinner than conventional ones. They have higher active mass utilization at discharge rates appropriate to EV use. At an operating temperature of 110 °F specific energy was 35.4 Wh/kg and specific power 200 W/kg. Over 600 discharge cycles were performed in tests without any serious deterioration in performance. The table at (c) lists the main parameters of the battery. The US company Unique Mobility Inc. have compared advanced lead–acid batteries with other proposed systems. In

Lightweight Electric/Hybrid Vehicle Design; ISBN: 9780750650922
Copyright © 2001 John Fenton and Ron Hodkinson; All rights of reproduction, in any form, reserved.

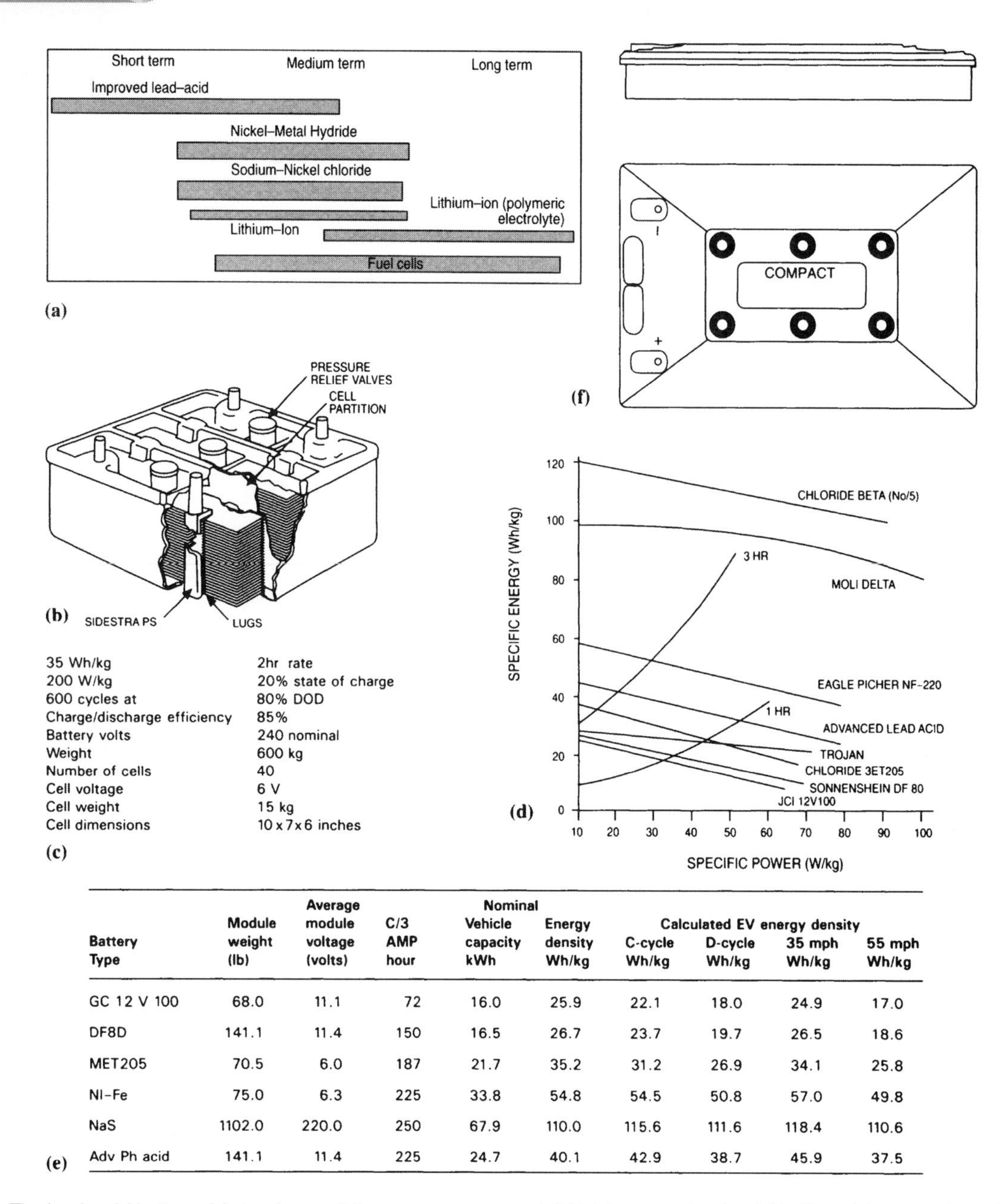

Battery Type	Module weight (lb)	Average module voltage (volts)	C/3 AMP hour	Nominal Vehicle capacity kWh	Energy density Wh/kg	Calculated EV energy density C-cycle Wh/kg	D-cycle Wh/kg	35 mph Wh/kg	55 mph Wh/kg
GC 12 V 100	68.0	11.1	72	16.0	25.9	22.1	18.0	24.9	17.0
DF8D	141.1	11.4	150	16.5	26.7	23.7	19.7	26.5	18.6
MET205	70.5	6.0	187	21.7	35.2	31.2	26.9	34.1	25.8
NI-Fe	75.0	6.3	225	33.8	54.8	54.5	50.8	57.0	49.8
NaS	1102.0	220.0	250	67.9	110.0	115.6	111.6	118.4	110.6
Adv Ph acid	141.1	11.4	225	24.7	40.1	42.9	38.7	45.9	37.5

(e)

Fig. 6.1-1 The lead–acid battery: (a) development time spans compared; (b) high energy lead–acid battery; (c) parameters of H-E battery; (d) battery characteristics; (e) energy-storage comparisons.

carrying out trials on an advanced EV-conversion of a Chrysler Minivan the company obtained the comparisons shown at (d). The graphs also show the extent to which the specific energy content of batteries is reduced as specific power output is increased. Trojan and Chloride 3ET205 are commercial wet acid batteries while the Sonnenshein DF80 and JCI 12 V 100 are gelled electrolyte maintenance-free units which involve an energy density penalty. The Eagle pitcher battery is a nickel–iron one taking energy density up to 50 Wh/kg at the 3 hour rate. The Beta and Delta units are sodium–sulphur batteries offering nominal energy density of 110 Wh/kg. Unique Mobility listed the characteristics of the batteries as at (e).

Exide's semi-bipolar technology has both high electrical performance and shape flexibility. The very low internal resistance allows high specific peak power rates and the electrode design permits ready changes in current capacity. The flat shape of the battery aids vehicle installation. The battery is assembled in a way which allows reduced need for internal connections between cells and a lightweight grid. Coated plates are stacked horizontally into the battery box. Performance is 3.9 Ah/kg and 7.4 Ah/dm^3 and shape profile is at (f).

6.1.2.2 Sodium–sulphur

For the sodium–sulphur battery, Fig. 6.1-2, as used in the Ford Ecostar, the cathode of the cell is liquid sodium immersed in which is a current collector of beta-alumina. This is surrounded by a sulphur anode in contact with the outer case. The cells are inside a battery box containing a heater to maintain them at their operating temperature of 300–350 °C. This is electrically powered and contained within the charge circuit. When discharging, internal resistance produces sufficient heat for the electrode but some 24 hours are required to reach running temperature from cold. In a typical EV application 100 cells would be connected in series to obtain 100 V and give a battery of 300 Ah, 60 kWh. In use, batteries would typically be charged nightly to bring them up to voltage after daily discharge and to keep the electrode molten. A typical battery installation of chloride cells is seen at (a) with expected vehicle performance, compared with lead–acid, shown at (b). The chloride cells are based on an electropheritic process while those from the Asea Brown Boveri Company, used in Ecostar, are made by isostatic pressing.

The ABB cell is seen at (c); the electronic current flowing through the external load resistor during discharge corresponds to a flow of sodium ions through the electrolyte from the sodium side to the sulphur side. Voltage is from 1.78 to 2.08 V according to the degree of discharge involved. A cell with a capacity of 45 Ah has a diameter of 35 mm and length of 230 mm. Its internal resistance is 7 milli ohms and 384 cells of this type can be installed in a battery of 0.25 litre volume. An example produced by ABB has external dimensions $1.42 \times 0.485 \times 0.36$ metres. The cells account for 55% of the total weight of 265 kg. By connecting the cells in four parallel strings of 96, the battery has an open circuit voltage of 170–200 V and a capacity of 180 Ah.

The electrical energy which can be drawn from the battery is shown at (d) as a function of the (constant) discharge power. With a complete discharge in 2 hours, energy content is 32 kWh, corresponding to a density of 120 Wh/kg. Associated discharge efficiency is 92%. Complete discharge at constant power is possible in a minimum of 1 hour, and an 80% discharge in less than three-quarters of an hour. The graph at (e) shows that the battery can cope with a load of up to two-thirds of the no-load voltage for a few minutes. This corresponds to a rating of about 50 kW or 188 W/kg. The portion of the

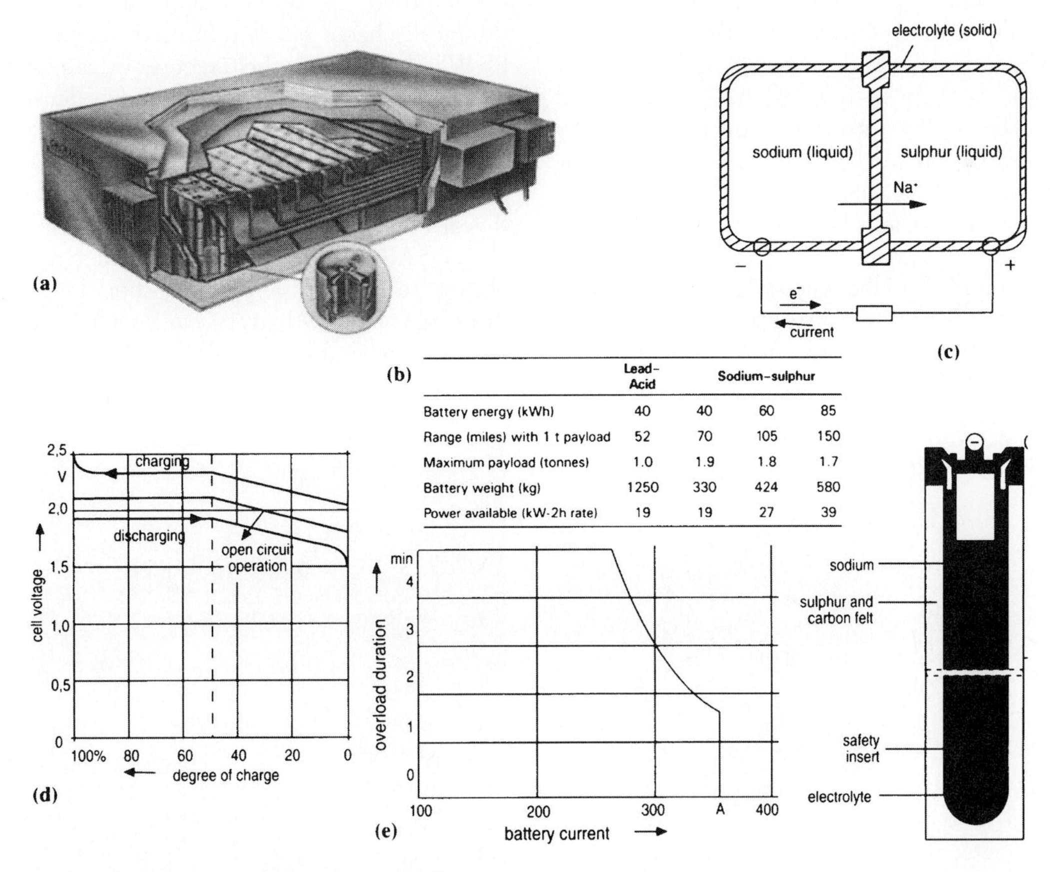

(b)	Lead-Acid	Sodium-sulphur		
Battery energy (kWh)	40	40	60	85
Range (miles) with 1 t payload	52	70	105	150
Maximum payload (tonnes)	1.0	1.9	1.8	1.7
Battery weight (kg)	1250	330	424	580
Power available (kW-2h rate)	19	19	27	39

Fig. 6.1-2 Sodium–sulphur battery: (a) battery assembly; (b) projected performances; (c) ABB s-s cell; (d) energy capability; (e) overload capability.

heat loss not removed by the cooling system which is incorporated into the battery is stored in the heated-up cells – and covers losses up to 30 hours. Additional heat must be supplied for longer standstill periods either from the electric mains or from the battery itself. Effective, vacuum-type, thermal insulation maintains the power loss at just 80 W so that when fully charged it can maintain its temperature for 16 days. In order to maintain the battery in a state of readiness, the battery must be held above a minimum temperature and it takes about 4 – 10 hours to heat up the battery from cold – but a limit of 30 freeze–thaw cycles is prescribed. Life expectancy of the battery otherwise is 10 years and 1000 full discharge cycles, corresponding to an EV road distance of 200 000 km.

6.1.2.3 Nickel–metal hydride

As recently specified as an option on GM's EV1, the nickel–metal hydride alkaline battery, Fig. 6.1-3, was seen as a mid-term solution by the US Advanced Battery Consortium of companies set up to progress battery development. According to the German Varta company, it shares with nickel–cadmium cells the robustness necessary for EV operation; it can charge up quickly and has high cycle stability. The nickel–metal hydride however, is superior, in its specifications relative to vehicle use, with specific energy and power some 20% higher and in volumetric terms 40% higher. Unpressurized hydrogen is taken up by a metallic alloy and its energy then discharged by electrochemical oxidation. The raw material costs are still signalling a relatively high cost but its superiority to lead–acid is likely to ensure its place as its associated control system costs are lower than those of sodium sulphur. Specific energy is 50–60 Wh/kg, energy density 150–210 Wh/litre, maximum power more than 300 W/kg; 80% charge time is 15 minutes and more than 2000 charge/discharge cycles can be sustained.

The negative electrode is a hydrogen energy-storage alloy while nickel hydroxide is the positive electrode. An optimum design would have weight around 300 kg, and capacity of 15 kWh, with life of 2000 discharge cycles. For buses Varta have devised a mobile charging station, in cooperation with Neoplan, which will allow round-the-clock operation of fleets. This removes the need for fixed sites and allows battery charging and changing to be carried out by the bus driver in a few minutes. The mobile station is based on a demountable container which can be unloaded by a conventional truck. Trials have shown that a bus covering a daily total distance of 75 miles on a three-mile-long route needs to stop at the station after eight journeys. Discharged batteries are changed semiautomatically on roller-belt arms, by a hand-held console.

6.1.2.4 Sodium chloride/nickel

Sodium chloride (common salt) and nickel in combination with a ceramic electrolyte are used in the ZEBRA battery, Fig. 6.1-4, under development by Beta Research (AEG and AAC) and Siemens. During charging the salt is decomposed to sodium and nickel chloride while during discharge salt is reformed. Its energy density of 90 Wh/kg exceeded the target set by the USA Advanced Battery Consortium (80 Wh/kg energy density, to achieve 100 miles range under any conditions and 150 W/kg peak power density to achieve adequate acceleration) and can achieve 1200 cycles in EV operation, equivalent to an 8 year life, and has a recharge time of less than 6 hours. The USABC power to energy ratio target of 1.5 was chosen to avoid disappointing short-range high power discharge of a ZEV battery and for a hybrid vehicle a different ratio would be chosen.

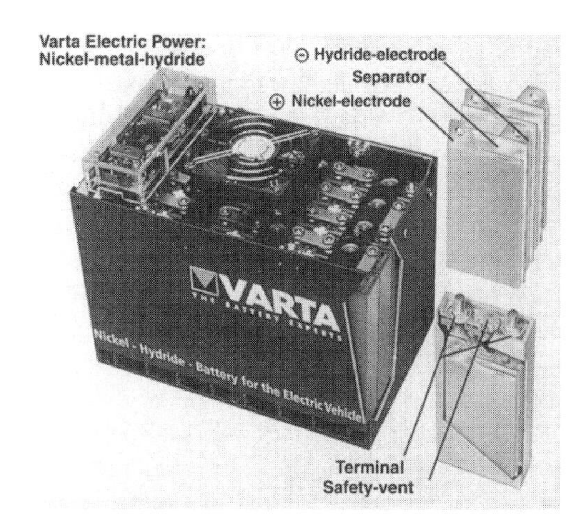

Fig. 6.1-3 Nickel–metal hydride battery.

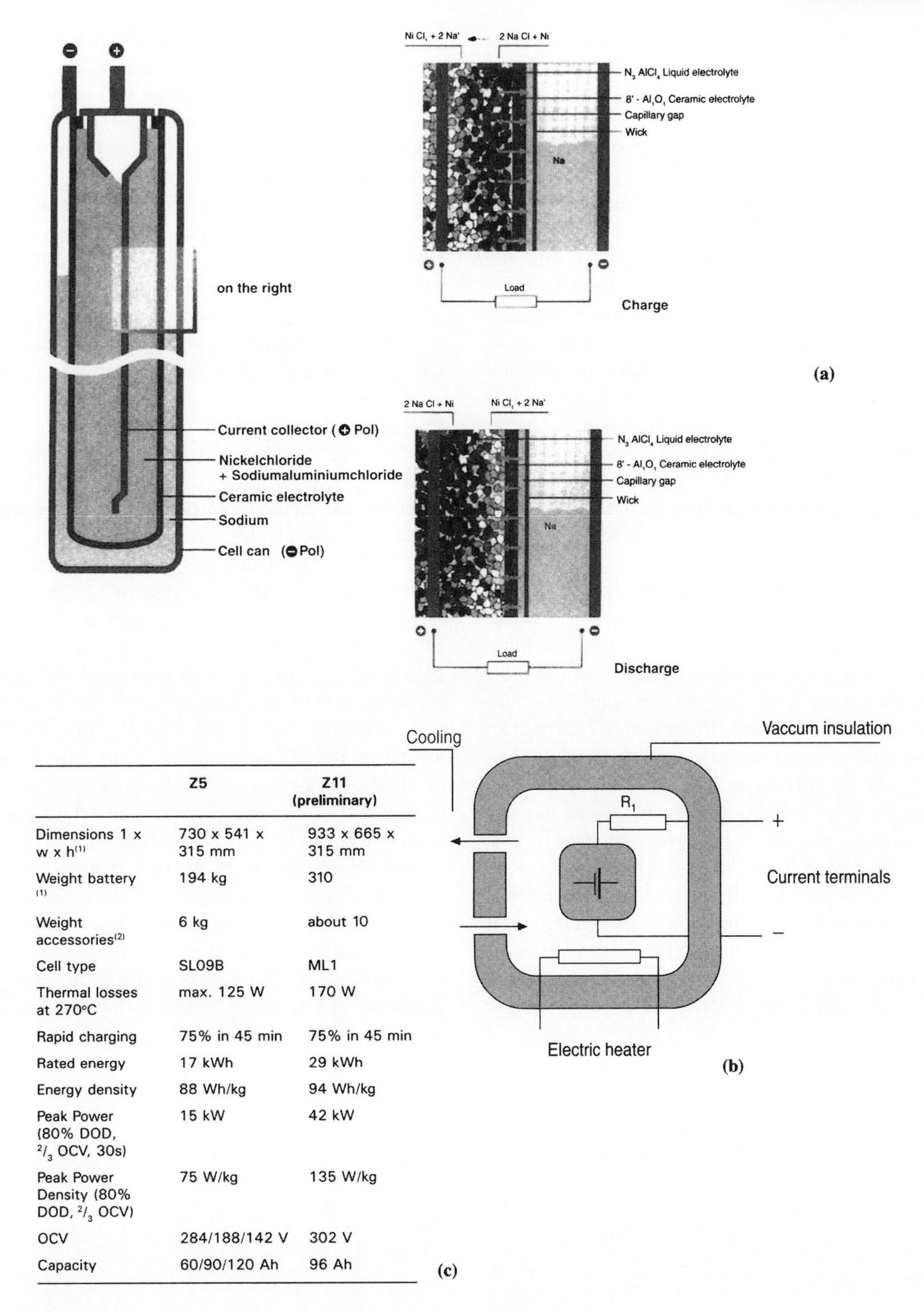

	Z5	Z11 (preliminary)
Dimensions 1 x w x h(1)	730 x 541 x 315 mm	933 x 665 x 315 mm
Weight battery (1)	194 kg	310
Weight accessories(2)	6 kg	about 10
Cell type	SL09B	ML1
Thermal losses at 270°C	max. 125 W	170 W
Rapid charging	75% in 45 min	75% in 45 min
Rated energy	17 kWh	29 kWh
Energy density	88 Wh/kg	94 Wh/kg
Peak Power (80% DOD, $^2/_3$ OCV, 30s)	15 kW	42 kW
Peak Power Density (80% DOD, $^2/_3$ OCV)	75 W/kg	135 W/kg
OCV	284/188/142 V	302 V
Capacity	60/90/120 Ah	96 Ah

Fig. 6.1-4 ZEBRA battery: (a) cell; (b) cell-box; (c) performance comparison.

Each cell is enclosed in a robust steel case with electrodes separated by a β-ceramic partition which conducts sodium ions but acts as a barrier to electrons, (a). The melt of sodium/aluminium chloride conducts sodium ions between the inner ceramic wall and into the porous solid $Ni/NiCl_2$ electrode. As a result, the total material content is involved in the cell reaction. Apart from the main reversible cell reaction there are no side reactions so that the coulometric efficiency of the cell is 100%. The completely maintenance-free cells are hermetically sealed using a thermal compression bond (TCB) ceramic/metal seal.

The cell type SL09B presently produced in the pilot production line has an open-circuit voltage of 2.58 V at 300 °C with a very low temperature coefficient of 3×10^{-4} V/K, a capacity of 30 Ah and an internal resistance that varies between 12 and 25 mW, dependent on temperature, current and rate of discharge. This variation is because, during the charging and discharging process, the electrochemical reaction zone moves from the inner surface of the β-ceramic electrolyte into the solid electrode. During this process the length of the sodium ion path and the current-density in the reaction zone increases and so the internal resistance increases. In principle this effect is used to enable a stable operation of parallel connected strings of cells. But from the vehicle point of view the available power which is directly related to the internal cell resistance should not depend on the battery charge status. The redesigned cell type ML1 is a good compromise between these two requirements. The battery is operated at an internal temperature range of 270–350 °C.

The cells are contained in a completely sealed, double walled and vacuum-insulated battery box as shown at (b). The gap between the inner and outer box is filled with a special thermal insulation material which supports atmospheric pressure and thus enables a rectangular box design to be utilized. In a vacuum better than 1.10^{-1} mbar this material has a heat conductivity as low as 0.006 W/mK. By this means the battery box outside temperature is only 5–10 °C above the ambient temperature, dependent on air convection conditions. Cooling systems have been designed, built and tested using air cooling as well as a liquid cooling. The latter is a system in which high temperature oil is circulated through heat exchangers in the battery with an oil/water heat exchanger outside the battery. By this means heat from the battery can be used for heating the passenger room of the vehicle.

In the ML1 cell, internal resistance is reduced to increase power. The resistance contribution of the cathode is due to a combination of the ion conduction between the inner surface of the β-aluminium ceramic with the reaction zone (80%) and electric conduction between the reaction zone and the cathode current collector (20%). The ML1 has a cloverleaf section shape ceramic to enlarge its surface area over the normal circular section, with resultant two-fold reduction in cathode thickness and 20% reduction in resistance. Based on this form of cell construction a new, Z11, battery has been produced with properties compared with the standard design as shown by the table at (c) and the battery is under development for series production.

6.1.2.5 Solar cells

According to Siemens, solar technology is a probable solution for Third World tropical countries. Solar modules are available from the company to supply 12 V, 100 Ah batteries from a 50 W solar module. The company recently installed a system on the Cape Verde Islands with a collective power output of 550 kW at each of five island sites. Even in Bavaria, the village of Flanitzhutte, which has an average 1700 hours annual sunshine period, has severed its links with the national grid with the installation of 840 solar modules, with a total area of 360 square metres, to provide peak power of 40 kW. Maintenance-free batteries provide a cushion.

The technology of solar cells, Fig. 6.1-5, has been given a recent boost by the Swiss Federal Institute of Technology who claim to have outperformed nature in the efficiency of conversion of sunlight to electricity even under diffuse light conditions. The cell has a rough surface of titanium dioxide semiconductor material and is 8% efficient in full sunlight rising to 12% in diffuse daylight. For more conventional cells, such as those making a Lucas solar panel, these are available in modules

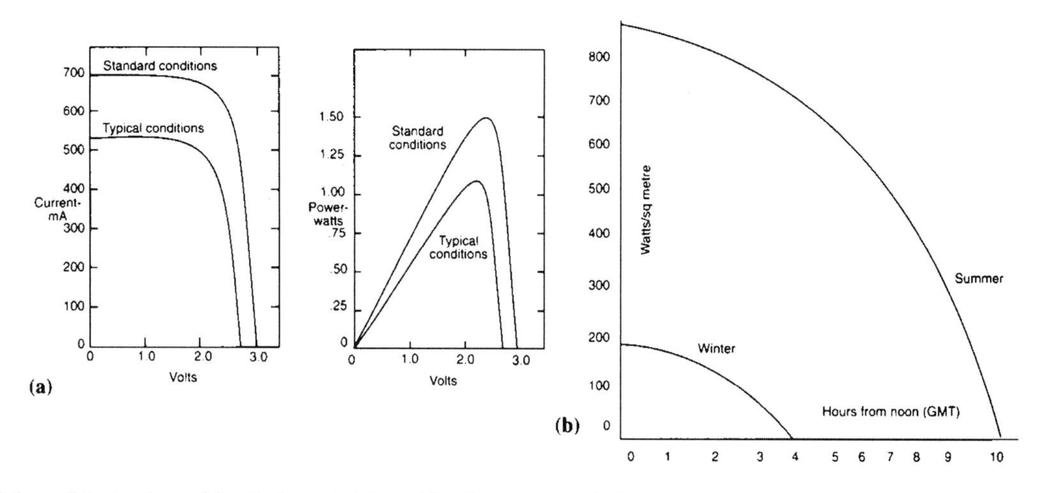

Fig. 6.1-5 Solar cell technology: (a) cell characteristics; (b) solar energy variation.

of five connected in series to give maximum output of 1.3 watts (0.6 A at 2.2 V). Some ten modules might be used in a solar panel giving 13 watts output in summer conditions. Power vs voltage and current vs voltage are shown at (a) for so-called 'standard' and 'typical' operating conditions. 100 mW/cm^2 solar intensity, 0 °C cell temperature at sea level defines the standard conditions against 80 mW/cm^2 and 25 °C which represent 'typical' conditions at which power output per cell drops to 1 W. Temperature coefficients for modules are 0.45% change in power output per 1 °C rise in temperature, relative to 0 °C; cell temperatures will be 20 °C above ambient at 100 mW/cm^2 incident light intensity. Variation of solar energy at 52° north latitude, assuming a clear atmosphere, is shown at (b). On this basis the smallest one person car with a speed of 15 mph and a weight of 300 lb with driver would require 250 W or 50 ft^2 (4.65 m^2) of 5% efficient solar panel – falling to 12.5 ft^2 (1.18 m^2) with the latest technology cells. A 100 Wh sealed nickel–cadmium battery would be fitted to the vehicle for charging by the solar panel while parked.

The future, of course, lies with the further development of advanced cell systems such as those by United Solar Systems in the USA. Their approach is to deposit six layers of amorphous silicon (two identical n-i-p cells) onto rolls of stainless steel sheet. The 4 ft^2 (0.37 m^2) panels are currently 6.2% efficient and made up of layers over an aluminium/zinc oxide back reflector. The push to yet higher efficiencies comes from the layer cake construction of different band-gap energy cells, each cell absorbing a different part of the solar spectrum. Researchers recently obtained 10% efficiency in a 12 in^2 (0.09 m^2) module.

Rapid thermal processing (RTP) techniques are said to be halving the time normally taken to produce silicon solar cells, while retaining an 18% energy conversion efficiency from sunlight. Researchers at Georgia Institute of Technology have demonstrated RTP processing involving a 3 minute thermal diffusion, as against the current commercial process taking 3 hours. An EC study has also shown that mass production of solar cells could bring substantial benefits and that a £350 million plant investment could produce enough panels to produce 500 MW annually and cut the generating cost from 64 p/kWh to 13p.

6.1.2.6 Lithium-ion

A high energy battery receiving considerable attention is the lithium-ion cell unit, the development of which has been described by Nissan and Sony engineers[1] who point out that because of the high cell voltage, relatively few cells are required and better battery management is thus obtained. Accurate detection of battery state-of-charge is possible based on voltage measurement. In the battery system developed, Fig. 6.1-6, cell controllers and a battery controller work together to calculate battery power, and remaining capacity, and convey the results to the vehicle control unit. Charging current bypass circuits are also controlled on a cell-to-cell basis. Maximizing lifetime performance of an EV battery is seen by the authors to be as important as energy density level. Each module of the battery system has a thermistor to detect temperature and signal the controllers to activate cooling fans as necessary.

Nissan are reported to be launching the Ultra EV in 1999 with lithium–ion batteries; the car is said to return a 120 mile range per charge. Even further into the future lithium–polymer batteries are reported to be capable of giving 300 mile ranges.

6.1.2.7 Supercapacitors

According to researchers at NEC Corp.[2], the supercapacitor, Fig. 6.1-7, will be an important contributor to the energy efficient hybrid vehicle, the absence of chemical reaction allowing a durable means of obtaining high

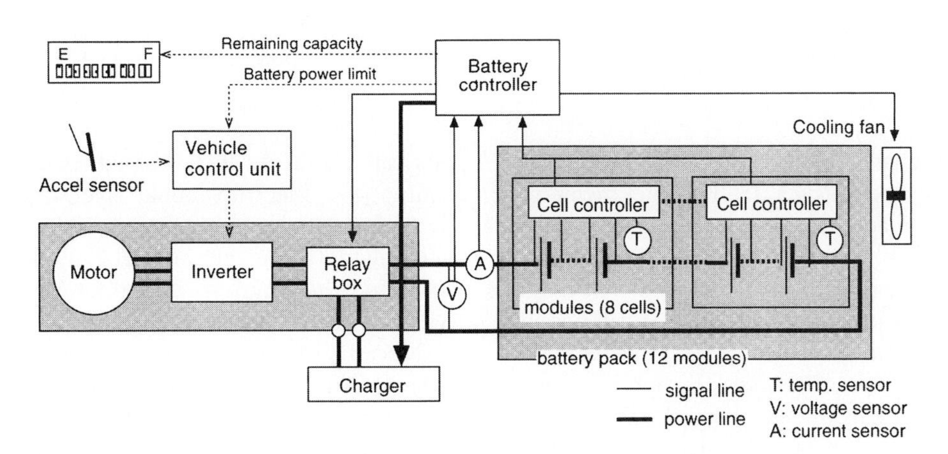

Fig. 6.1-6 Overall system configuration.

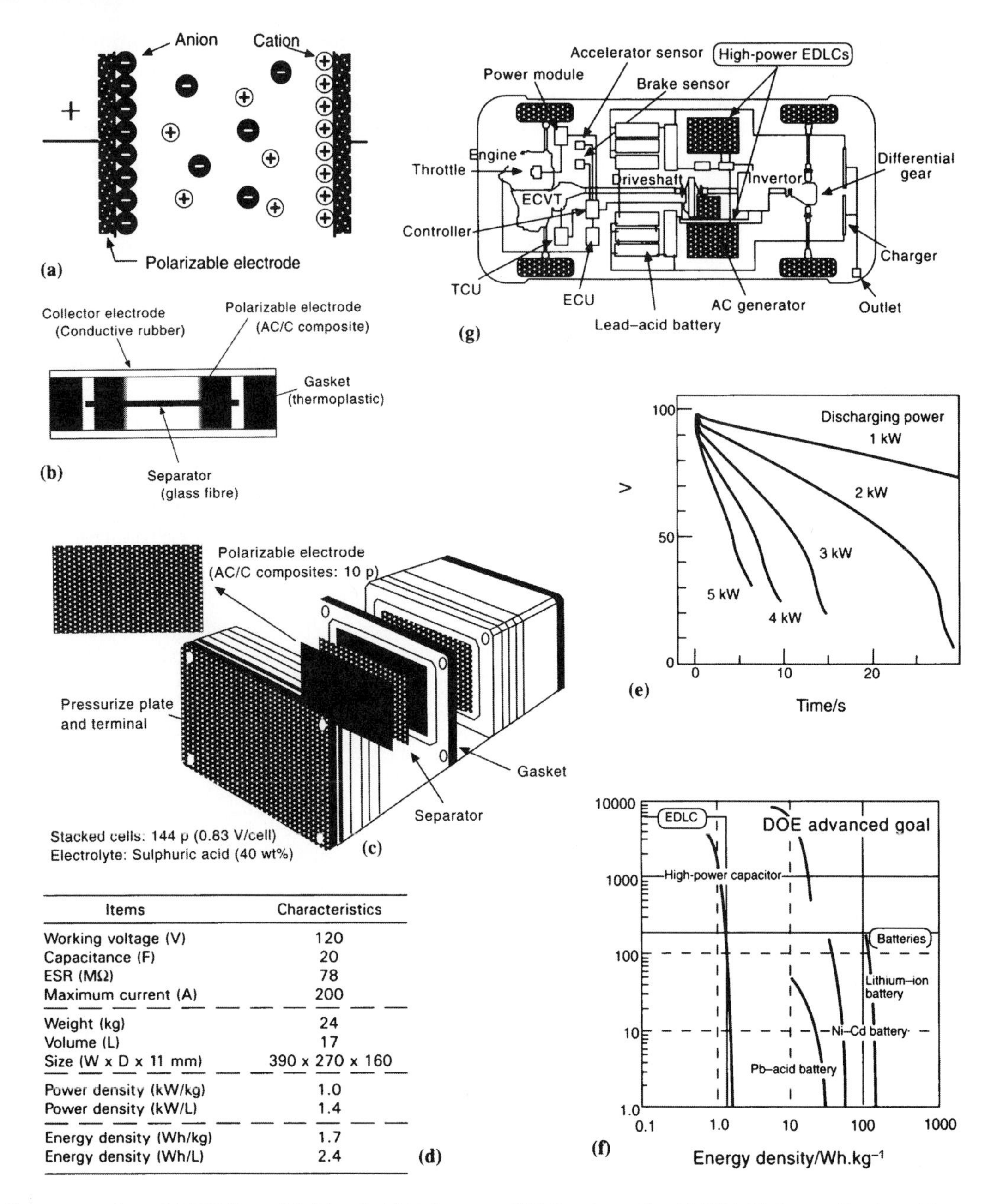

Items	Characteristics
Working voltage (V)	120
Capacitance (F)	20
ESR (MΩ)	78
Maximum current (A)	200
Weight (kg)	24
Volume (L)	17
Size (W x D x 11 mm)	390 x 270 x 160
Power density (kW/kg)	1.0
Power density (kW/L)	1.4
Energy density (Wh/kg)	1.7
Energy density (Wh/L)	2.4

Fig. 6.1-7 Supercapacitors: (a) EDLC model; (b) cell; (c) high-power EDLC schematic; (d) HP EDLC specification; (e) constant power discharge characteristics; (f) power density, *y*-axis in W/kg, vs energy density for high-power EDLC; (g) ELCAPA configuration.

energy charge/discharge cycles. Tests have shown for multi-stop vehicle operations a 25–30% fuel saving was obtained in a compact hybrid vehicle fitted with regenerative braking. While energy density of existing, non-automotive, supercapacitors is only about 10% of that of lead–acid batteries, the authors explain, it is still possible to compensate for some of the weak points of conventional batteries. For effective power assist in hybrids, supercapacitors need a working voltage of over 100 V, alongside low equivalent series resistance and high energy density. The authors have produced 120 V units operating at 24 kW fabricated from newly developed activated carbon/carbon composites. Electric double layer capacitors (EDLCs) depend on the layering between electrode surface and electrolyte, (a) showing an EDLC model. Because energy is stored in physical adsorption/desorption of ions, without chemical reaction, good life is obtained. The active carbon electrodes usually have a specific surface area over 1000 m^2/g and double-layer capacitance is some 20–30 $\mu F/cm^2$ (activated carbon has capacitance over 200–300 F/g). The EDLC has two double layers in series, so it is possible to obtain 50–70 °F using a gram of activated carbon. Working voltage is about 1.2 V and storable energy is thus 50 J/g or 14 Wh/kg.

The view at (b) shows a cell cross-section, the conductive rubber having 0.2 S/cm conductivity and thickness of 20 microns. The sulphuric acid electrolyte has conductivity of 0.7 S/cm. The view at (c) shows the high power EDLC suitable for a hybrid vehicle, and the table at (d) its specification. Plate size is $68 \times 48 \times 1\ mm^3$ and the weight 2.5 g, a pair having 300 F capacity. The view at (e) shows constant power discharge characteristics and (f) compares the EDLC's energy density with that of other batteries. Fuji Industries' ELCAPA hybrid vehicle, (g), uses two EDLCs (of 40 F total capacity) in parallel with lead–acid batteries. The stored energy can accelerate the vehicle to 50 kph in a few seconds and energy is recharged during regenerative braking. When high energy batteries are used alongside the supercapacitors, the authors predict that full competitive road performance will be obtainable.

6.1.2.8 Flywheel energy storage

Flywheel energy storage systems for use in vehicle propulsion has reached application in the light tram vehicle. They have also featured in pilot-production vehicles such as the Chrysler Patriot hybrid-drive racing car concept. Here, flywheel energy storage is used in conjunction with a gas turbine prime-mover engine, Fig. 6.1-8. The drive was developed by Satcon Technologies in the USA to deliver 370 kW via an electric motor drive to the road wheels. A turbine alternator unit is also incorporated which provides high frequency current generation from an electrical machine on a common shaft with the gas turbine. The flywheel is integral with a motor/generator and contained in a protective housing affording an internal vacuum environment. The 57 kg unit rotates at 60 000 rpm and provides 4.3 kW of electrical energy. The flywheel is a gimbal-mounted carbon-fibre composite unit sitting in a carbo-fibre protective housing. In conjunction with its motor/generator it acts as a load leveller, taking in power in periods of low demand on the vehicle and contributing power for hill climbing or high acceleration performance demands.

European research work into flywheel storage systems includes that reported by Van der Graaf at the Technical University of Eindhoven[3]. Rather than using continuously variable transmission ratio between flywheel and driveline, a two-mode system is involved in this work. A slip coupling is used up to vehicle speeds of 13 km/h, when CVT comes in and upshifts when engine and flywheel speed fall simultaneously. At 55 km/h the drive is transferred from the first to the second sheave of the CVT variator, the engine simultaneously being linked to the first sheave. Thus a series hybrid drive exists at lower speeds and a parallel hybrid one at higher speeds. The 19 kg 390 mm diameter composite-fibre flywheel has energy content of 180 kW and rotates up to 19 000 rpm.

6.1.3 Battery car conversion technology

For OEM conversions of production petrol-engined vehicles in the decades up to the 1970s, and up to the present day for aftermarket conversions, are typified by those used by many members of the UK Battery Vehicle Society and documented by Prigmore *et al.*[4] Such conversions rely on basic lead–acid batteries available at motor factors for replacement starter batteries. A ton of such batteries, at traction power loading of 10–15 kW/ton, stores little more than 20 kWh. Affordable motors and transmissions for this market sector have some 70% efficiency, to give only 14 kWh available at the wheels.

6.1.3.1 Conversion case study

The level-ground range of the vehicle can be expressed in terms of an equivalent gradient *1:h*, representing

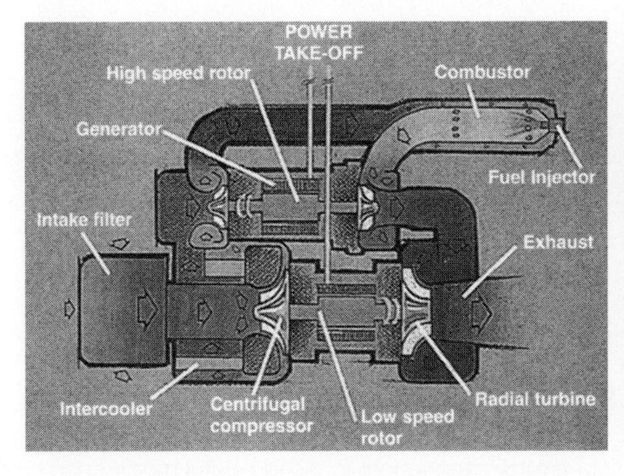

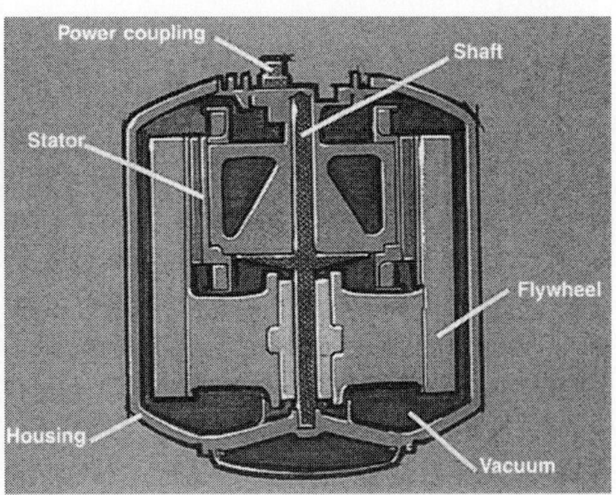

Fig. 6.1-8 Chrysler Patriot flywheel energy storage system: left, turbine: right, flywheel.

rolling resistance, such that a resistance of 100 kgf/tonne is equivalent to a gradient of 1:10. If the fraction of the total vehicle weight contributed by the battery is f_b then range is given by $\{(14 \times 3600)/(9.81 \times 1000)\}f_b h$. Pessimistically h is about 30 at 50 km/h and if f_b is 0.4 then cruising range would be about 60–65 km. This of course is reduced by frequent acceleration and braking.

Series-wound DC motors, Fig. 6.1-9, have been chosen for low cost conversions because of their advantageous torque/speed characteristics, seen at (a), given relatively low expected road speeds. Series winding the field and armature, the same current is carried by both and as it increases in magnitude so does the magnetic flux and the torque increases more than proportionally with current. Rotation of the armature creates a back-EMF in opposition to the applied voltage because the wires at the edge of the armature are moving across the field flux. Motors are designed to equalize applied and back-EMFs at operational speed. This will be low when field current is high and vice versa. The speed/current curve can be made to move up the x-axis by reducing the field current to a fixed fraction of the armature current (0.5–0.7), with the help of the field diverter resistance shown but

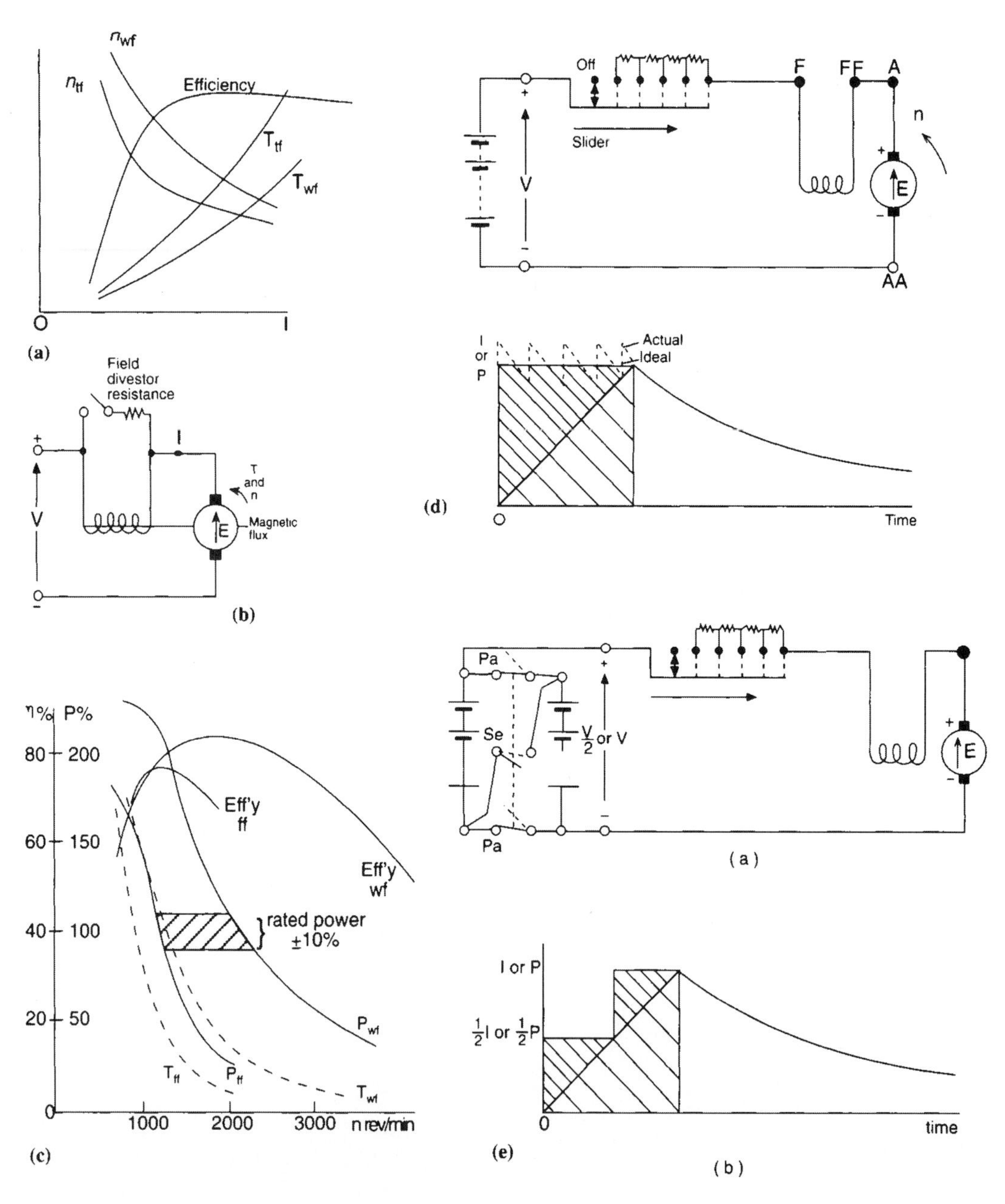

Fig. 6.1-9 Series-wound DC motor: (a) motor characteristics; (b) field diverter resistance; (c) speed-base motor characteristics; (d) rheostatic control; (e) parallel/series battery control.

the torque for a given armature current is, of course, reduced, see (b). The efficiency of the motor is low at low speeds, in overcoming armature inertia, and again at high speeds as heating of the windings absorbs input power. Motors can thus be more highly rated by the provision of cooling fans. Average power in service should in general be arranged at 0.8 of the rated power and the transmission gear ratio be such that the motor is loaded to no more than its rated power for level-ground cruising. The motor characteristics shown at (c) are obtained by replotting the conventional characteristics on a speed base. The wide range of speeds available (up to 2:1) are around rated power and show how full field can be used for uphill running while weak field is used on the level enabling speed reduction to compensate for torque increase in limiting battery power requirements for negotiating gradients.

With little or no back-EMF to limit current at starting, resistance is added to keep the current down to a safe level, as at (d). The current is maintained at the required accelerating value, perhaps 2–4 times rated current. The starting resistance is reduced as the motor gains speed so as to keep the accelerating current constant to the point where the starting resistance is zero, at the 'full voltage point'. Thereafter a small increase in speed causes gradual reduction in current to the steady running value. As the current is supplied from the battery at constant voltage, the current curve can be rescaled as a power curve to a common time base, as at (e). The shaded area then gives energy taken during controlled acceleration with the heavily shaded portion showing the energy wasted in resistance. So rheostatic acceleration has an ideal efficiency of about 50% up to full voltage. This form of control is thus in order for vehicle operation involving, say, twice daily regular runs under cruise conditions but unwise for normal car applications.

6.1.3.2 Motor control alternatives

Alternatives such as parallel/series (two-voltage) rheostatic control, or weak field control, can be better for certain applications, but the more elaborate thyristor, chopper, control of motor with respect to battery (Fig. 6.1-10) is preferred for maintaining efficiencies with drivers less used to electric drive, particularly in city-centre conditions. It involves repetitive on–off switching of the battery to the motor circuit and if the switch is on for a third of the time, the mean motor voltage is a third of the supply voltage (16 V for a 48 V battery), and so on, such that no starting resistance is needed. Effective chopper operation requires an inductive load and it may be necessary to add such load to the inherent field inductance. Because an inductive circuit opposes change in current then motor current rises relatively slowly during 'on' periods and similarly falls slowly during 'off' periods, provided it has a path through which to flow. The latter is provided by the 'flywheel' diode FD, a rectifier placed across the motor to oppose normal voltage. During chopper operation, current i_b flows in pulses from battery to motor while current i_m flows continuously through the motor. Electronic timing circuits control the switching of the thyristors, (a).

Single ratio drives from motor to driveline are not suitable for hilly terrain, despite the torque/speed characteristic, as the motor would have to be geared too low to avoid gradient overloading and thus be inefficient at cruise. A 5:1 CVT drive is preferred so that the motor can be kept at its rated power under different operating conditions. There is also a case for dispensing with the weight of a conventional final drive axle and differential gear by using two, say 3 kW, motors one for each driven wheel.

The behaviour of lead–acid batteries, (b), is such that in the discharged condition lead sulphate is the active material for both cell-plates which stand in dilute sulphuric acid at 1.1 specific gravity. During charging the positive plate material is converted to lead peroxide while that of the negative plate is converted into lead, as seen at (c). The sulphuric acid becomes more concentrated in the process and rises to SG = 1.5 when fully charged, the cells then developing over 2 volts. In discharge the acid is diluted by the reverse process. While thin plates with large surface area are intended for batteries with high discharge rates, such as starter batteries, the expansion process of the active material increases in volume by three times during discharge and the active material of very thin plates becomes friable in numerous charge/discharge cycles, and a short life results. Normal cells, (b), comprise interleaved plates with porous plastic separators; there is one more negative than positive plates, reducing the tendency to buckle on rapid discharge. Expensive traction batteries have tubular plates in some cases with strong plastic tubes as separators to keep the active material in place. Discharge rates of less than half the nominal battery capacity in amphours are necessary to preserve the active material over a reasonable life-span, but short bursts at up to twice the nominal rate are allowable. The graphs at (d) permit more precise assessments of range than the simple formula at the beginning of the section which assumes heavy discharge causes battery capacity to be reduced by 70–80% of normal, 25 kWh becoming 20.

When charging the gassing of plates must be considered, caused by the rise in cell voltage which causes part of the current to electrolyse the water in the electrolyte to hydrogen. Gassing commences at about 75% full charge. At this point, after 3–4 hours of charging at

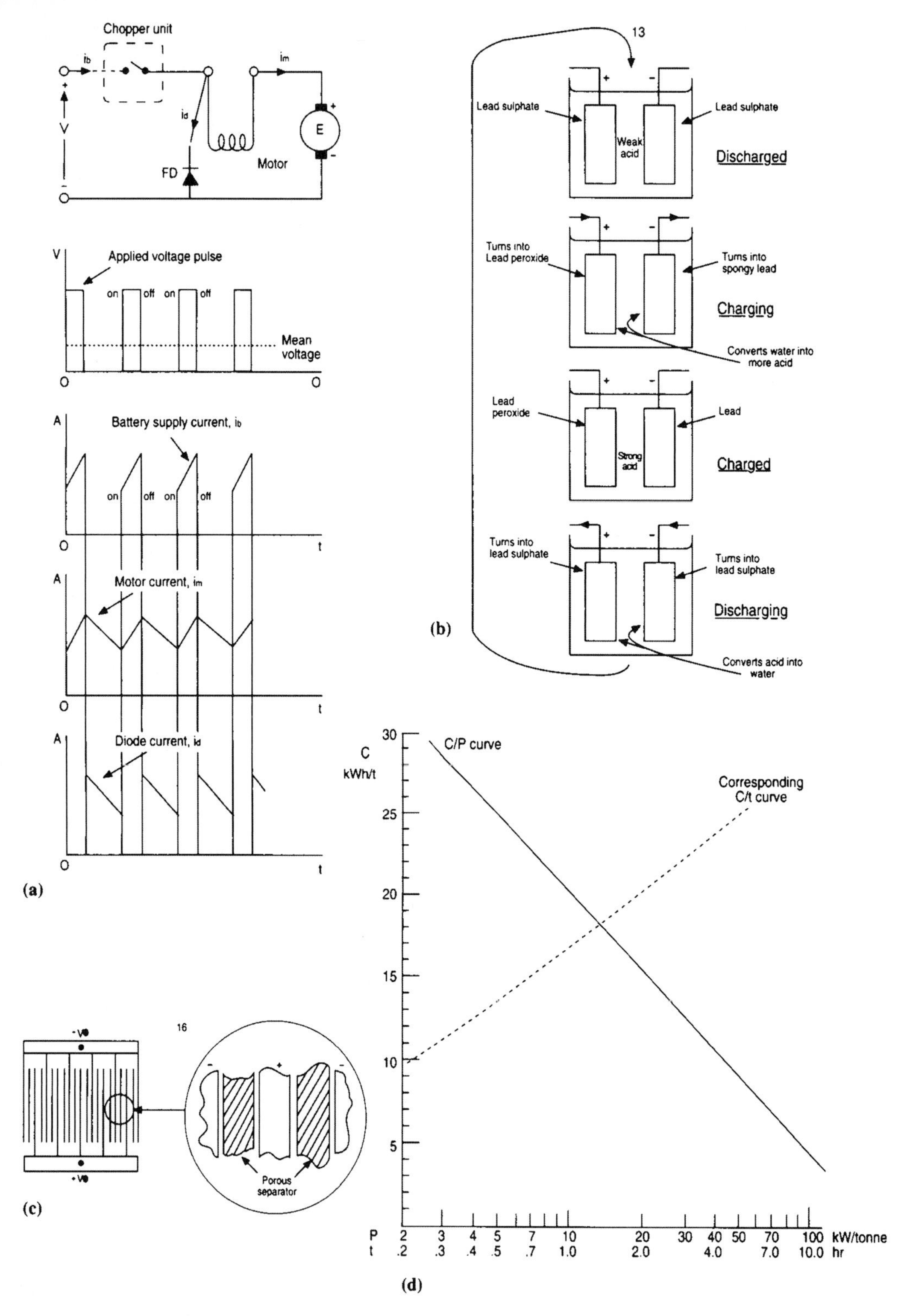

Fig. 6.1-10 Motor control and battery: (a) chopper circuit: (b) battery charge–discharge cycle; (c) cell arrangement; (d) battery time-of-discharge curves.

1/15th battery capacity, the rate should be decreased to 1/20th and carried on until 2.6 volts are shown at the cells. To ensure near-complete removal of the sulphate a periodic 'soak charge' should be provided for several hours until peak voltage remains steady at 2.6–2.8 V, with all cells gassing freely and with constant specific gravity. Such a charge should be followed by topping up with distilled water.

6.1.4 EV development history

According to pioneer UK EV developer and producer Geoffrey Hardings,[5] the Lucas programme was a major event in the renaissance of the electric vehicle. He set up a new Lucas Industries facility to develop battery EVs in 1974 because, as a major transport operator, he had asked Lucas to join him in an approach to a UK government department for some financial assistance to build a battery electric bus which would operate on a route between railway stations in Manchester. The reasons for his interest in this project were twofold. First, there was a major problem with the reliability of many of the diesel buses at that time and he wanted to find out whether electric buses would live up to the attributes of good reliability and minimal maintenance that had been afforded to EVs for many years. Second, a world shortage of oil at that time was causing an apparent continuous and alarming increase in the price.

Having subsequently joined Lucas and set up the new company, he was responsible for building the electric bus in question and providing technical support when it entered service. The bus – the performance of which was comparable with diesel buses, except for range – operated successfully for some years and was popular with both passengers and drivers. On the other hand, it was not popular with schedulers because its restricted range (about 70 km in city service) added yet more limitation to its uses, particularly at weekends. Nevertheless, much was learnt from the in-service operation of this vehicle which proved to be remarkably reliable. He then obtained agreement within Lucas that the battery EV most likely to succeed at that time was a 1-tonne payload van because it would be possible, with relatively minor changes to production vans, to modify the drive to battery electric without reducing either the payload volume or the weight, Fig. 6.1-11.

The converted Bedford vehicles underwent a significant testing programme on that company's test track, and were in fact built on the company's ICE van production line, interspersed between petrol- and diesel-powered versions of CF vans. This method of production was the first of its kind. Some hundreds of Bedford vans and a smaller number of Freight Rover vans were built and sold, all with a working range in excess of 80 km in city traffic, a payload of just under 1 tonne, an acceleration of 0–50 km in 13 s, a maximum speed of 85 km/h, and a battery design life of 4 years. The vehicles had, for that time, sophisticated electronic controllers and DC/DC converters, as well as oilfired heating and demisting systems. Lucas designed and constructed the chargers and battery-watering systems. Some were sold in the USA as the GM Griffon, (a), and it was estimated that collectively their total service had exceeded 32 million km, and even today a few are still operating.

The Lucas Chloride converted Bedford CF van had two-pedal control and a simple selector for forward/reverse. Most of the vehicle's braking was regenerative and batteries were of the tubular cell lead–acid type. Thirty-six monobloc units of 6 volts were used – connected in series to give a 216V, 188 Ah pack. The rear-mounted traction motor drove the wheels through a primary reduction unit coupled to a conventional rear axle, via a prop-shaft. Measured performance of the monobloc is shown at (b); an energy density of 34 Wh/kg was involved, at the 5 hour rate, and a 4 year service life was claimed. The motor used was a separately excited type in order to allow the electronics maximum flexibility in determining the power curve. It weighed 15 kg and had a controlled output of 40 kW; working speed was 6100 rpm corresponding to a vehicle speed of about 60 mph. The motor control system used an electronic bypass to leave the main thyristor uncommutated during field control. The latter uses power transistors which handle up to 25 A.

Within the Lucas development programme, which at one time employed close to 100 personnel, some work on HEVs was undertaken and one five-seat passenger car was designed and built. This utilized an electric Bedford drive system and could be operated either as a series hybrid or a parallel hybrid. The car had a maximum speed of 130 kph, and a pure-electric range of about 70 km. The Lucas Chloride hybrid, (c), has engine (3) driving through the motor (1) but midships positioning of the batteries (4) with on-board charger (5) at the rear. Clutches are shown at (6) while (7) and (8) are alternator and control unit. This used Reliant's 848 cc engine developing 30 kW alongside a 50 kW Lucas CAV traction motor. The 216 V battery set had capacity of 100 amp-hour on a 5 hour rate. Maximum speed in electric drive of 120 km/h rises to 137 km/h in combined mode, (d).

6.1.4.1 Electric vehicle development 1974–1998

In considering the changes which have taken place in the quarter century since the start of the Lucas project, Harding argues that the developments which have taken place in electric cars are not as great as had been hoped and expected. Some hybrids, he considers, are effectively ICEVs with an electric drive which assists when required. A major problem with HEVs has been their cost, which is exacerbated by having two drive systems in one vehicle. Fortunately, the automotive industry is so good at meeting challenges of this nature that who can say what can be achieved? However, it is claimed that micro-turbines together with their associated generators and accessories can be produced cheaply, mainly because they have a very low component count. These turbines

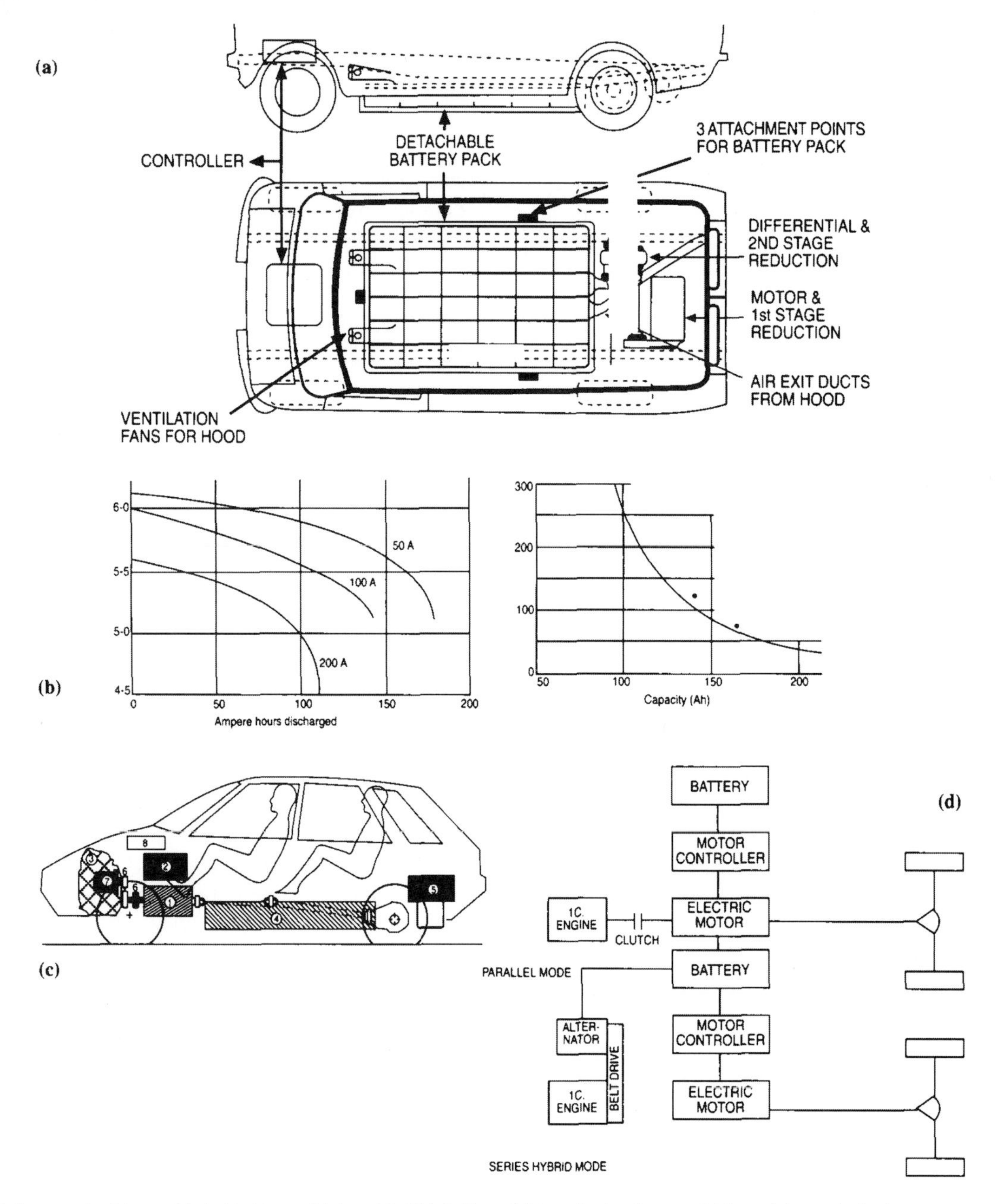

Fig. 6.1-11 Lucas electric and hybrid drive vehicles: (a) GM Griffon; (b) terminal volts per 6 V module and discharge current in amps; (c) Lucas Chloride hybrid car; (d) bi-mode drive system.

are capable of operating on a wide variety of fuels and are considered to produce a very low level of pollutants, but with one or two exceptions, such as Volvo and Chrysler, these claims have not been subjected to any extensive field testing. If what is claimed proves to be true, then such vehicles would be expected to play a large part in the transport scene in the new millennium.

At present, the great hope for the future, he believes, is the fuel cell. Hydrogen is the preferred fuel for fuel cells but its storage presents a problem. One of the ways of overcoming this problem is to convert a liquid fuel, such as methanol, into hydrogen. This was done in the 5 kW unit made by the Shell Oil Company as long ago as 1964. The unit was installed in the world's first fuel-cell powered car. Shell also produced a 300 W nett cell in 1965 which converted methanol directly into electricity, so it is not the case that this technology is new. The principal problem at the time this work was carried out was the cost of the unit. Although a number of fuel-cell powered cars have been built recently by automobile manufacturers, the only vehicle so far offered for sale is the Zevco London taxi which was launched in London in July 1998. The propulsion system is a hybrid arrangement: a battery drives the vehicle and is recharged by a 5 kW fuel cell. The vehicle uses bottled hydrogen as fuel and has a service range of 145 km, and a performance similar to its diesel counterpart. This design works well because the stop–start nature of the traffic provides time for the low output of the fuel cell to replenish the energy drawn from the battery during previous spells of vehicle motion. At a later date, this type of taxi may be fitted with a cryogenic hydrogen-storage system, perhaps placed between the two layers of a sandwich-floor construction of the vehicle. With such an arrangement, it is expected that the fuel cell would be refuelled with very cold liquid hydrogen in minutes and, thereby, would extend the vehicle's range dramatically, but only in stop–start traffic.

Harding opines that what the world really needs is vehicles fitted with fast-response, high-output fuel cells together with on-board clean reformers which would enable a liquid fuel to be turned into hydrogen on vehicles. Initially, the most likely liquid fuel would seem to be methanol, but arranging for methanol to be widely available would necessitate some large changes in infrastructure. If all this is possible, then refuelling vehicles with liquid fuel would be, in principle, little or no different from today. The eventual aim is said, by those developing high-output fuel cells, to be the development of reformers which can produce hydrogen from gasoline. In this case, only the current gasoline infrastructure would be required. Interest and investment in fuel cells is increasing, and the joint arrangements between the Canadian fuel cell company Ballard and motor industry giants Mercedes and Ford would appear to be an almost irresistible force on a course aimed at solving some daunting problems. The Ballard unit is a proton exchange membrane (PEM) fuel cell and amongst early examples of road vehicles fitted with this are buses in the USA. Quite apart from the technical problems still to be resolved, the problem of cost is very great.

6.1.5 Contemporary electric car technology

According to Sir Clive Sinclair, whose abortive efforts to market an electric tricycle have led him to concentrate on economical bicycle conversions, peak efficiencies of 90% are available with EVs for converting electricity into tractive energy – and that attainable electrical generating

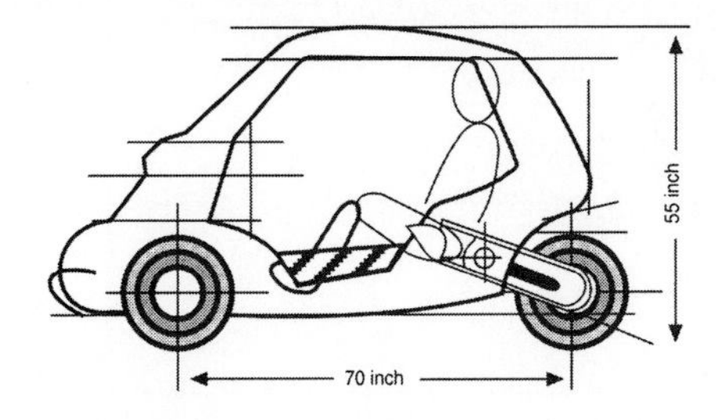

Fig. 6.1-12 Sinclair C 10 proposal.

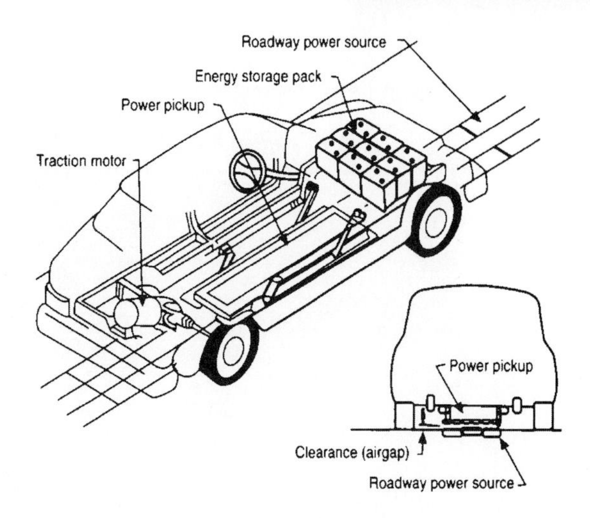

Fig. 6.1-13 Road-induced electricity.

efficiencies of over 50% meant a 45% fuel conversion efficiency could be obtained compared with 30% for the petrol engine. His C10 proposal shown in Fig. 6.1-12 must mean his faith in the future of the electric car is still maintained.

There are other initiatives, too, such as the desire to make motorway driving under very high density peak traffic conditions less dangerous and less tiring. This is generating fresh interest in reserved lanes for vehicle guidance systems. Where these additionally provide roadway-induced powering, Fig. 6.1-13, as described by researchers from the Lawrence Livermore National Laboratory[6], a case for a car to suit relatively long-distant commuters can be made. The success of trials on GM's Impact electric car have so far pointed to the very considerable importance of light weight, good aerodynamics and low rolling resistance but the electrical breakthrough has come in the electronics technology of the DC/alternating current (AC) converter. Ford, too, have had very promising prototype results from their Ecostar 1 car-derived van, using a transistorized DC to AC inverter.

6.1.5.1 Honda 'EV'

The state of the art in pilot-production electric cars is typified by Honda's nickel–metal hydride battery driven electric car, Fig. 6.1-14; it has been given the name 'EV' and claims twice the range obtainable with comparable lead–acid batteried cars. The car is not a conversion of an ICE model and has 95% new componentry. It is a 3-door, 4-seater with battery pack in a separated compartment between the floor. The pack comprises 24 × 12 V batteries and rests between virtually straight underframe longitudinal members running front to rear for maximum crash protection. The motor is a brushless DC type with rare earth high strength magnets and is said to give 96% efficiency. There is a fixed ratio transmission with parking lock. Maximum torque is 275 Nm, available from 0 to

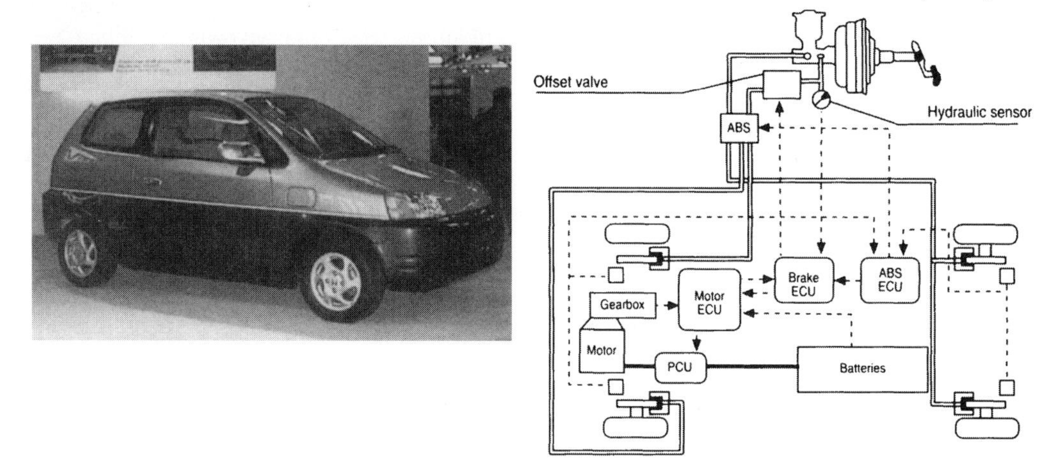

Fig. 6.1-14 Honda 'EV' electric car and Honda regenerative braking/coasting system.

1700 rpm, the speed at which a maximum power of 49 kW is developed and remains constant up to 8750 rpm. The under-bonnet power control unit comprises management and motor ECUs, power driver, junction board, 12 V DC/DC converter, air-conditioning inverter and 110/220 V on board charger. Its aluminium container is liquid cooled in a system shared with motor and batteries. The controller uses IGBT switching devices in a PWM system. A phase control system involves both advanced angle control and field weakening to optimize operation in both urban and motorway conditions. A heat-pump climate control system has an inverter-controlled compressor with a remote-control facility to permit pre-cooling or pre-heating of the cabin prior to driving. Energy recovery is carried out in both braking and 'throttle-off' coasting modes. Low rolling-drag tyres are inflated to 300 kPa and are said to have just 57% the resistance of conventional tyres. A 'power-save' feature automatically reduces peak power when battery state-of-charge drops below 15%. An instrument display shows range and battery-charge state as a biaxial graph with clearly marked segments which even respond to throttle pedal depression. The car has a 125 mile urban range to the FUDS standard while top speed is 80 mph. Recharge time is 8 hours from 20% to fully charged.

6.1.5.2 General motors 'EVI'

The latest generation GM EVI (Fig. 6.1-15) is a purpose-built electric vehicle which offers two battery technologies: an advanced, high capacity lead–acid, and an optional nickel–metal hydride. The EV1 is currently available at selected GM Saturn retailers and is powered by a 137 (102 kW), 3 phase AC induction motor and uses a single

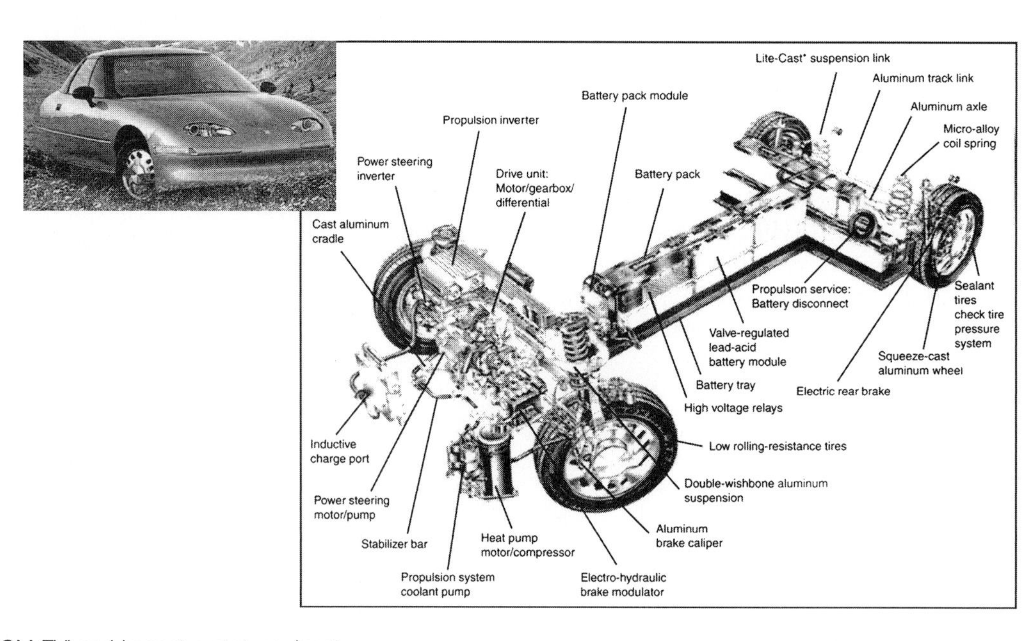

Fig. 6.1-15 GM EVI and Impact prototype, inset.

speed dual reduction planetary gear set with a ratio of 10.946:1. The second generation propulsion system has an improved drive unit, battery pack, power electronics, 6.6 kW charger, and heating and thermal control module. Now, 26 valve-regulated, high capacity, lead–acid (PbA) batteries, 12 V each, are the standard for the EV1 battery pack and offer greater range and longer life. An optional nickel–metal hydride battery pack is also available for the Gen II model. This technology nearly doubles the range over the first generation battery and offers improved battery life as well. The EV1 with the high capacity lead-acid pack has an estimated real world driving range of 55 to 95 miles, depending on terrain, driving habits and temperature; range with the nickel–metal hydride pack is even greater. Again, depending on terrain, driving habits, temperature and humidity, estimated real world driving range will vary from 75 to 130 miles, while only 10% of power is needed to maintain 100 km/h cruising speed, because of the low drag, now aided by Michelin 175/65R14 Proxima tyres mounted on squeeze-cast aluminium alloy wheels.

The 1990 Impact prototype from which the EV1 was developed had one or two more exotic features which could not be carried through to the production derivative but claimed an urban range of 125 miles on lead–acid batteries. In the Impact the 32 10 volt lead–acid batteries weighed 395 kg, some 30% of the car's kerb weight, housed into a central tunnel fared into the smooth underpanel and claimed to have a life of 18 500 miles. The Impact weighed 1 tonne and accelerated from 0 to 100 kph in 8 seconds, maximum power of the motor being 85 kW. The vehicle had 165/65R14 Goodyear low-drag tyres running at 4.5 bar. Two 3 phase induction motors were used, each of 42.5 kW at 6600 rpm; each can develop 64 Nm of constant torque from 0 to 6000 rpm, important in achieving 50–100 km/h acceleration in 4.6 seconds. Maximum current supply to each motor was 159 A, maximum voltage 400 V and frequency range 0–500 Hz. The battery charger was integrated into the regulator and charging current is 50 A for the 42.5 Ah lead–acid batteries, which could at 1990 prices be replaced for about £1000.

The EV1 can be charged safely in all weather conditions with inductive charging. Using a 220 volt charger, charging from 0% to 100% for the new lead–acid pack takes up to 5.5–6 hours. Charging for the nickel–metal hydride pack, which stores more energy, is 6–8 hours. Braking is accomplished by using a blended combination of front hydraulic disk, and rear electrically applied drum brakes and the electric propulsion motor. During braking, the electric motor generates electricity (regenerative) which is then used to partially recharge the battery pack. The aluminium alloy structure weighs 290 pounds and is less than 10% of the total vehicle weight. The exterior composite body panels are dent and corrosion resistant and are made from SMC and RIM polymers. The EV1 is claimed to be the most aerodynamic production vehicle on the road today, with a 0.19 drag coefficient and 'tear drop' shape in plan view, the rear wheels being 9 inches closer together than the front wheels. The EV 1 has an electronically regulated top speed of 80 mph. It comes with traction control, cruise control, anti-lock brakes, airbags, power windows, power door locks and power outside mirrors, AM/FM CD/cassette and also a tyre inflation monitor system.

6.1.5.3 AC drives

An interesting variant on the AC-motored theme, Fig. 6.1-16, is the use of a two-speed transaxle gearbox which reduces the otherwise required weight of the high speed motor and its associated inverter. A system developed by Eaton Corporation is shown at (a) and has a 4 kW battery charger incorporated into the inverter. A 3 phase induction motor operates at 12 500 rpm – the speed being unconstrained by slip-ring commutator systems. A block diagram of the arrangement is at (b) and is based on an induction motor with 18.6 kW 1 hour rating – and base speed of 5640 rpm on a 192 V battery pack. The pulse width modulated inverter employs 100 A transistors. The view at (c) shows the controller drive system functions in association with the inverter. In an AC induction motor, current is applied to the stator windings and then induced into the windings of the rotor. Motor torque is developed by the interaction of rotor currents with the magnetic field in the air gap between rotor and stator. When the rotor is overdriven by coasting of the vehicle, say, it acts as a generator. Three phase winding of the stator armature suits motors of EV size; the rotor windings comprise conducting 'bars' short-circuited at either end to form a 'cage'. Rotation speed of the magnetic field in the air gap is known as the synchronous speed which is a function of the supply-current frequency and the number of stator poles. The running speed is related to synchronous speed by the 'slip'.

If two alternators were connected in parallel, and one was driven externally, the second would take current from the first and run as a 'synchronous motor' at a speed depending on the ratio of each machine's number of poles. While it is a high efficiency machine which runs at constant speed for all normal loads, it requires constant current for the rotor poles; it is not self-starting and will stop if overloaded enough for the rotor to slip too far behind the rotating stator-field. Normally, the synchronous motor is similar in construction to an induction motor but has no short-circuited rotor – which may be of the DC-excited, permanent-magnet or reluctance type.

The view at (d) shows a stator winding for a 2 pole 3 phase induction motor in diagrammatic form. If supply current frequency is f_s, then stator field speed is f_s/p for

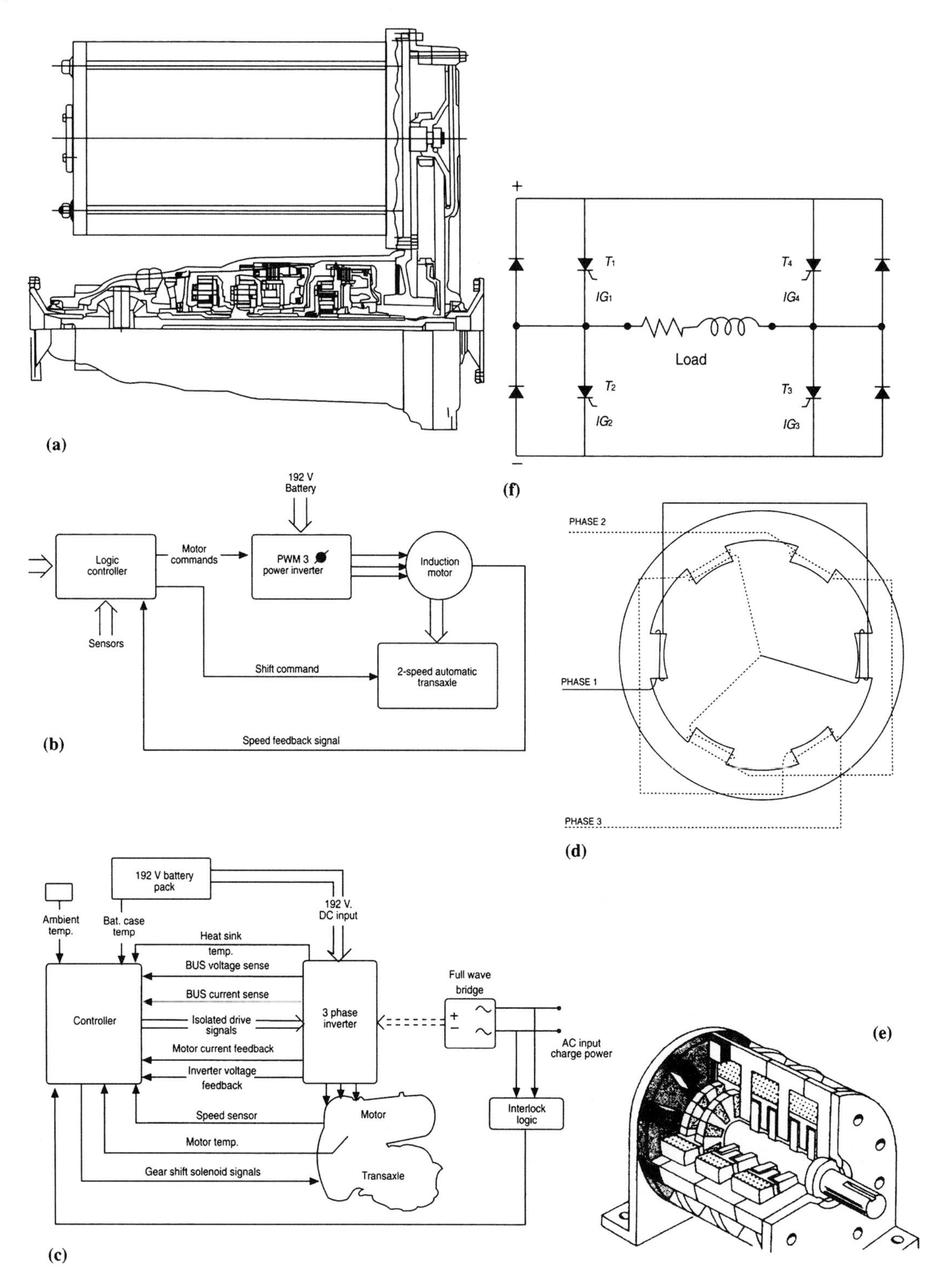

Fig. 6.1-16 AC drive systems: (a) transaxle motor; (b) controller arrangement; (c) controller drive system; (d) 3 phase induction motor stator; (e) air-gap type reluctance motor; (f) induction motor controller.

number of poles p. Rotor current frequency $f_r = sf_p$ where s is the slip. Power supplied to a 3 phase motor can be expressed as $EI(3)^{1/2}f_s n$ where n is efficiency. When a synchronous motor has no exciting voltage on the rotor it is termed a 'reluctance' motor which has very simple construction and, when used with power transistors, can be applied as a variable speed drive. Axial air-gap versions are possible as at (e); such an electronically commutated motor can operate with a DC source by periodic reversal of the rotor polarities.

The general form of control, with DC link inverter, for induction motors is shown diagrammatically at (f). The thyristors of the inverter are generally switched so as to route current through the stator winding as though it were connected to a 3 phase AC. Frequency can be varied by timing of pulses to the thyristors and is typically 5–100 Hz, to give a speed range of almost twice synchronous speed. According to engineers from Chloride EV Systems Division,[7] devices for building such an inverter are not yet available for current levels associated with EV traction. This situation may have already changed in the interim, however.

6.1.5.4 Ford E-KA: Lithium-ion battery power

European Ford claim the first production-car use of lithium-ion batteries in a road vehicle by a major player in the industry. The high energy density and power-to-weight ratio of these storage units puts a prototype electric version of the small Ka hatchback, Fig. 6.1-17, on a par with the petrol-engined model in driving performance. Top speed is said to be 82 mph with acceleration to 62 mph in 12.7 seconds, although range between charges is still only 95 miles, but is extendible to 125 miles with a constant speed of 48 mph.

Until now Li-ion batteries have been used mainly in small consumer electronic products like notebook computers, cellular phones, baby monitors and smoke detectors. Output per cell of 3.6 volts is some three times that of nickel–cadmium and nickel–metalhydride that they widely replace. They also retain full charge regardless of usage, can be recharged from zero to full capacity in 6 house with over 3000 repeated charge/discharge cycles, and are immune from the so-called 'memory effect' suffered by Ni–Cads. The basic Li-ion chemistry was initially redeveloped for automotive use by the French company SAFTSA, a leading battery manufacturer. Their advanced technology was then adapted to the e-Ka by Ford's Research Centre in Aachen, Germany, with financial assistance for the project from the German Ministry for Education and Research. The battery pack consists of 180 cells with 28 kwh rating and weighs only 280 kg (615 lb). This is 30% the weight of the power equivalent in lead–acid batteries, and substantially less than comparable Ni–Cads and Ni–MHs. In terms of volume the Ni-ion has approximately half the bulk of all the other three.

Batteries are divided into three individual sealed 'troughs', each with 30 modules containing six cells. One trough is located in the engine compartment, with the other two on either side of the back axle. Nominal output of 315 V DC is transformed by a solid-state inverter to 3 phase AC for the traction motor. Heat generated by the internal resistance of these second-generation Li-ion batteries is dissipated by one of two fluid cooling systems. A second independent system cools the drivetrain, with a 65 kW (88 bhp) asynchronous motor followed by a fixed-ratio transmission driving the front wheels. Torque rating is 190 Nm (140 lb ft).

Performance of the e-Ka is enhanced by a 45 kg (100 lb) weight reduction to counter the battery load. The roof and hood are of aluminium sandwich construction with a thermoplastic filling, while the front brake callipers with ceramics discs, rear drums, wheel rims and back axle are all in light alloy. Electric power steering supplied by Delphi provides further weight saving, where an electronic control module regulates the assistance needed to minimize battery demand. Brake servo and ABS system are also electric.

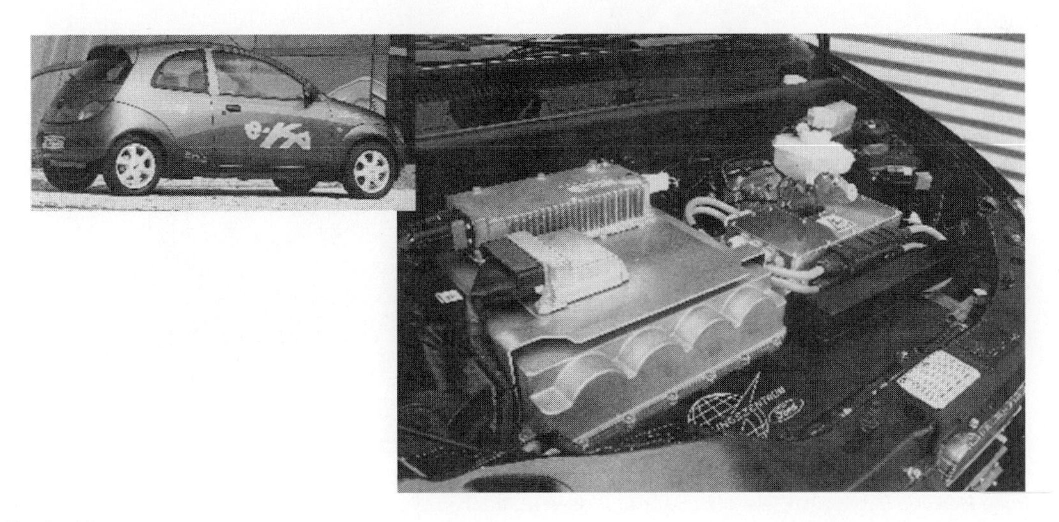

Fig. 6.1-17 Ford e-Ka.

6.1.6 Electric van and truck design

6.1.6.1 Goods van to fleet car conversion

Europe's largest maker of EVs is Peugeot-Citroen whose Berlingo Dynavolt, Fig. 6.1-18, sets out to maximize the benefits of electric vehicles in a fleet car. It has a range extender in the form of an auxiliary generating system which does not quite make the vehicle a hybrid in the conventional sense. The generator feeds current into the traction motor rather than into the battery pack. The generator engine is a 16 ps, 500 cc Lombardini running on LPG which drives a Dynalto-style starter generator unit developing 8 kW at 3300 rpm, to supplement the supply from the 4 kW Saft Ni–Cad batteries. Company designed software controls the cut-in of the generator according to range requirements. The range is 80 km, which can be extended to 260 km with generator assistance. Series production was imminent as we went to press.

6.1.6.2 Ford EXT II

An early key US initiative in AC drives was the Ford EV project EXT11, Fig. 6.1-19, which has been exploring the use of an AC drive motor in the Aerostar Minivan, seen at (a). A sodium–sulphur battery was employed and a single-shaft propulsion system. A battery with the following specification was involved: nominal voltage 200 V; minimum voltage at 60 kW, 135 V; 50 kW capacity on FUDS cycling; 60 kW maximum power (20 seconds) and 35 kW continuous power (40 minutes). Much of the technology has since been carried over to the Ford Ecostar, described later. Dimensions of the battery were 1520 × 1065 × 460 mm and it was based on the use of small (10 Ah) cells connected in 4-cell series strings with parallel 8 V banks arranged to provide the required capacity. A voltage of 200 required 96 cells in series (each had a voltage of 2.076) and a parallel arrangement of 30 cells gave the 300 Ah capacity; overall 2880 cells are used and their discharge performance is shown at (b). Battery internal resistance was 30 milliohms with an appropriate thermal management system under development. The cells rested on a heater plate which incorporates a 710 W element used particularly in the initial warm-up. An associated air cooling system can dissipate 6 kW. Temperature must be maintained above 300 °C to achieve an internal resistance value that will allow sensible current flow.

The US General Electric Company was also a partner in the programme, specifically on the AC drive system, at (c). This comprised a 50 bhp induction motor within a two-speed transaxle, a liquid-cooled transistor inverter being used to convert the 200 battery volts into variable-voltage 3 phase AC, (d). The induction motor was a 2 pole design with just under 13 cm stack length–within a stator diameter just under 23 cm. Calculated stall torque was 104 Nm and the transition between constant torque and constant power operation occurred at 3800 rpm. Design speed was 9000 rpm and absolute maximum 12 000 rpm. Full power was developed at a line current of 244.5 A. Control strategy relating inverter to motor is shown at (e).

Ford produced the transaxle on an in-house basis having coaxial motor and drive, (f). The motor rotor is hollow to allow one of the axle shafts to pass through it from the bevel-gear differential. The installation in a test

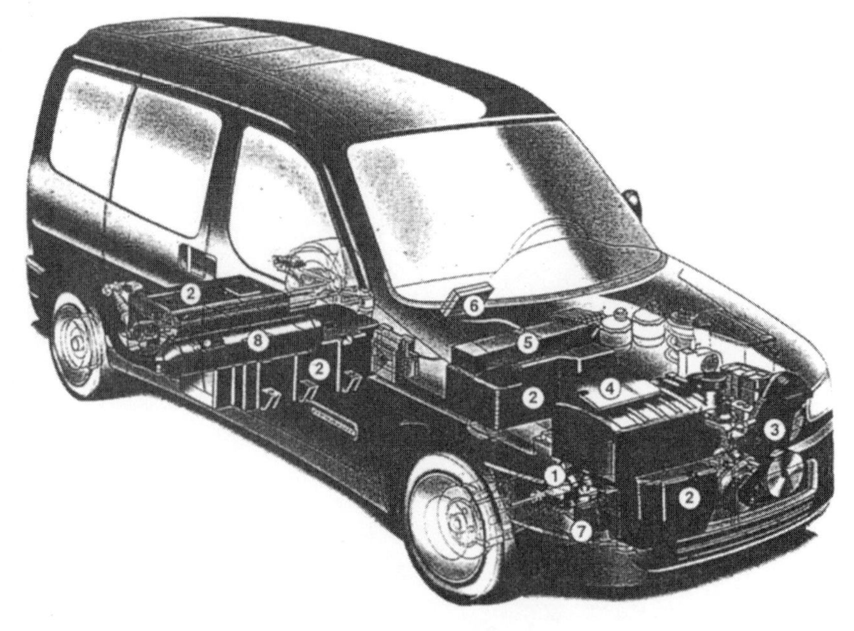

Fig. 6.1-18 Citroen Berlingo Dynavolt: 1. electric motor and drive; 2. traction battery pack; 3. generator set; 4. motor controller; 5. generator controller; 6. drive programme selector; 7. LPG regulator; 8. LPG storage tank.

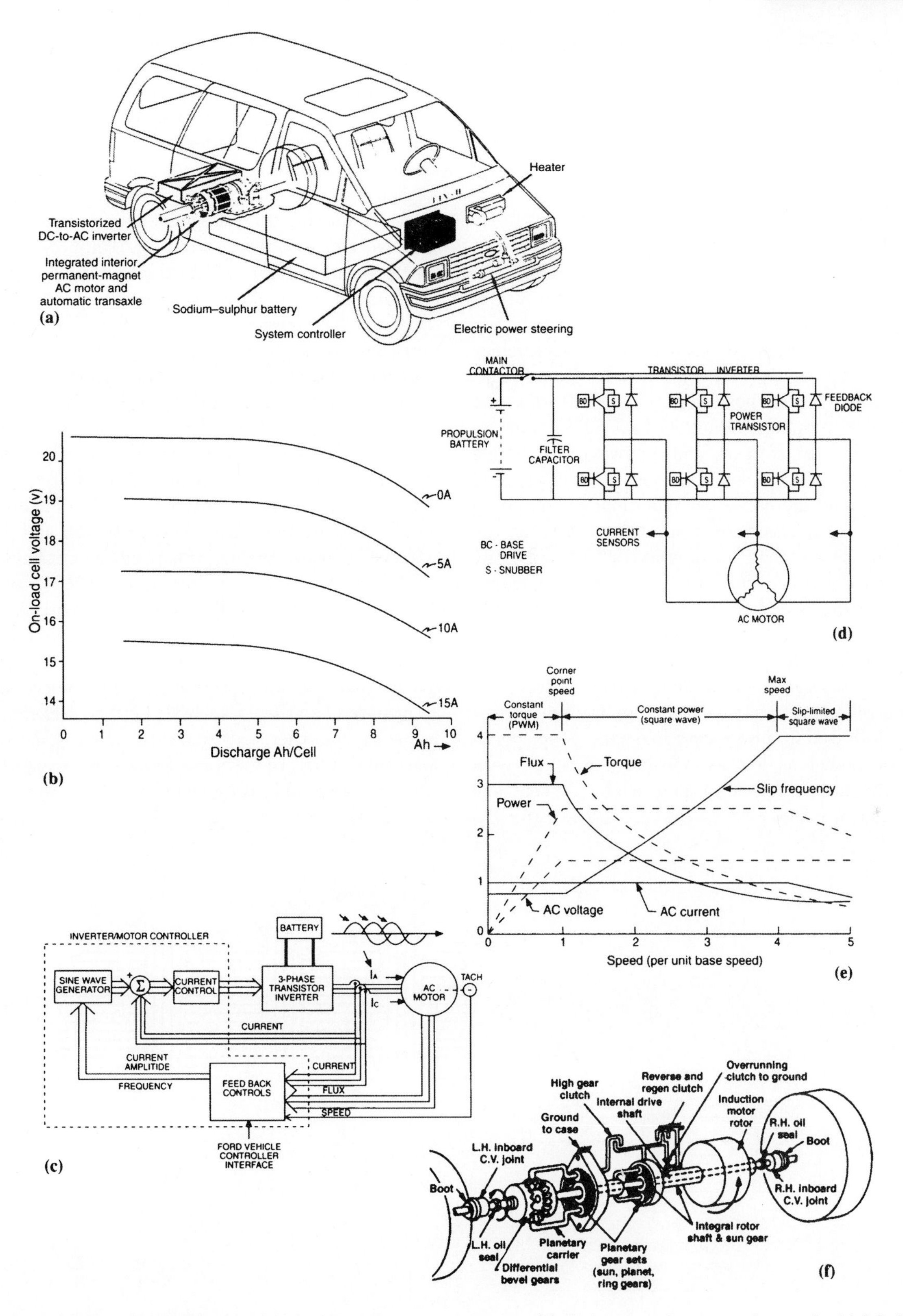

Fig. 6.1-19 Ford EXT 11 EV project: (a) AC-driven Ford Aerostar conversion; (b) discharge performance of s-s cells; (c) AC drive; (d) transistorized inverter; (e) AC control strategy; (f) transaxle drive with coaxial motor.

saloon car achieved 30% gradient ability, 0–50 mph acceleration in less than 20 seconds, 60 mph top speed and 0.25 kWh/mile energy consumption. Gear ratios were 15.52 and 10.15:1, and rearward speed was obtained by electrically reversing the motor; in 'neutral' the motor was electrically disconnected.

6.1.6.3 UK EVA practice for CVS

In its manual of good practice for battery electric vehicles the Electric Vehicle Association lays down some useful ground rules for conceptual design of road-going electric trucks. Exploiting the obvious benefits of EV technology is the first consideration. Thus an ultra-low floor walk-through cab is a real possibility when batteries and motors can be mounted remotely. Lack of fuelling requirement, ease of start-up and getaway–also driving simplicity of two-pedal control without gearshifting – all these factors lend themselves to operations, such as busy city-centre deliveries where a substantial part of the driver's time is spent in off-loading and order-taking. Any aspect of vehicle design which minimizes the driving task thus maximizes his or her other workload duties. Successful builders of electric trucks are thus, say the EVA, specialists in assembling bought-in systems and components. Required expertise is in tailoring a motor/battery/speed-controller package to a given application. Principles to be followed in this process are keeping top speeds and motor power as low as possible consistent with fulfilling the task; using controllers which prevent any unnecessary acceleration once the vehicle has reached running speed in stop–start work; using generously rated motors rather than overspecifying battery capacity–the latter because large batteries cost more, displace payload and waste energy providing tractive force required for their extra weight.

Whereas over a decade ago the rule of thumb applied that 1 tonne of batteries plus one of vehicle and one of payload could be transported at 20 mph over a 20 mile range including 200 stop–starts per charge, now 40/40 to 50/50 mph/miles range is feasible by careful design. Belted radial tyres can now be run at 50% above normal inflation pressures. High voltage series motors of 72 V and above have working efficiencies of 85–90% over a 3:1 speed range and electronic controllers cut peak acceleration currents and avoid resistive losses – to extend ranges by 15–20% over resistance controllers. A further 10% range increase is possible if high frequency controllers (15 000 Hz) are used–also much lower internal-resistance batteries reduce voltage fall-off and the use of DC/DC converters means that cells do not have to work so deeply to power auxiliaries, the EVA explain. The percentage of GVW allocated to the battery is a key parameter which should be kept below 35%. Lightweight lead–acid batteries giving 15–25% greater specific capacity than the BS 2550 standard traction cell should be considered, as they also have better discharge characteristics, but there is usually a trade-off in equivalent percentage life reductions. Separately excited motors and regenerative braking give further bonuses.

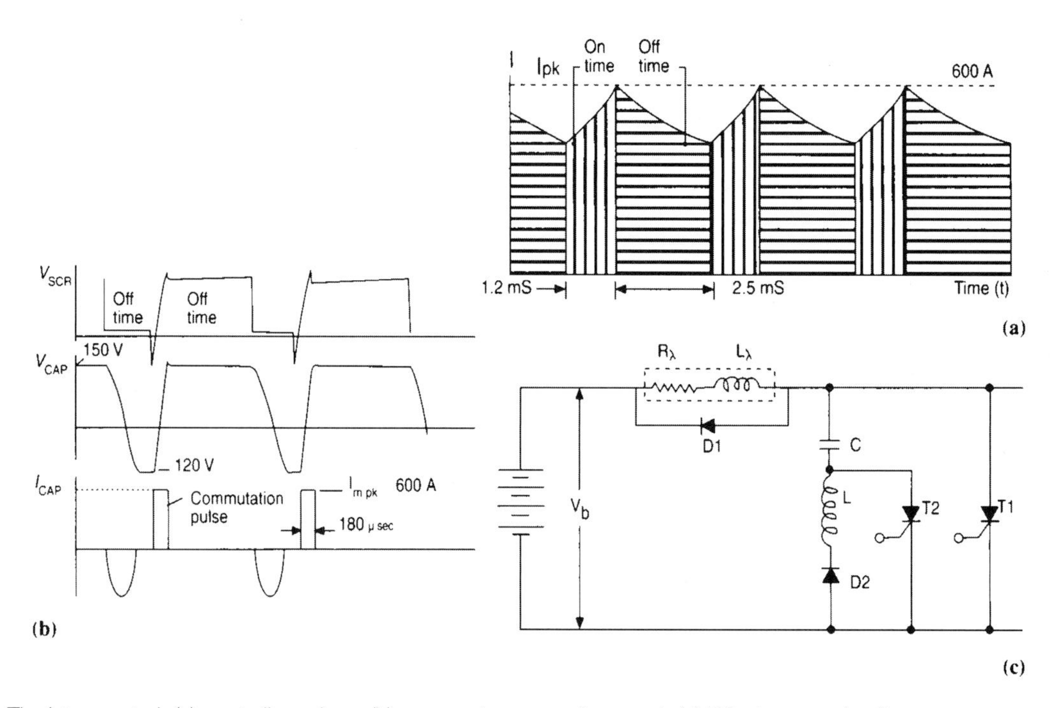

Fig. 6.1-20 Thyristor control: (a) controller pulses: (b) energy storage and reversal: (c) DC chopper circuit.

6.1.6.4 Thyristor control

Thyristor control, Fig. 6.1-20, has been an important advance and units such as the controller by Sevcon are suitable for 24–72 V working with a current limit rating of 500 A for 5 minutes. Continuous rating at any frequency would be 160 A and a bank of eight 50 mF, 160 V commutation capacitors are employed. Overall dimensions are 430 × 301 × 147 mm and weight is 14 kg. Like other units of this type, it supplies the motor with fixed width pulses at a variable repetition frequency to provide stepless control over the average voltage supplied to the motor. The view at (a) shows the typical pulse width of 12 ms, giving a frequency range from 10 to 750 Hz and a good motor current form factor with minimal heating effect commensurate with some iron losses. Ratio of on-time to the total period would be 30–50% while the commutation circuit which switches the main conducting thyristor operates every 2.5–3.5 milliseconds. The graph shows motor consumption at current limit; ripple is 170 while average motor current of 500 A compares with battery current of 165 A. At (b) is seen the energy being stored and reversed each cycle by the main commutation capacitor.

A typical DC chopper circuit for thyristor drive is shown at (c). Thyristor T1 is the 'on' one and T2 the 'off'; inductive load of a series motor is shown as R1 and L1 while Dl is a so-called 'freewheel' diode. When capacitor C is charged to a voltage *V*b in a direction opposing the battery voltage, T1 and T2 are off and load current *I*l is flowing through R1, L1 and D. A pulse is applied to the gate of T1 to turn it on; Dl becomes reverse-biased and load current *I*l begins to flow through R1, L1 and T1. This causes a short-circuit, effectively, across CL and D2, creating a tuned circuit. Its resonance drives a current through T2, D2, L and C, sinusoidally rising to a maximum then decaying to zero, at which point C is effectively charged to V2 in the reverse direction.

When the current attempts the second half of its cycle, D2 prevents any reversal and the reverse charge on C is caught by D2, the 'catching' diode. At time t_{on} a pulse applied to the gate of T2 turns it on, applying reverse voltage on C across T1 – which rapidly turns off and diverts load current through R1, L1, C, L and T2. C is thus charged at a linear rate by *I*l passing through it. When capacitor voltage again approaches *V*b, opposing battery voltage, load current begins to flow through R1, L1 and D1 again, current through T2 dropping below the holding value so turning off T2–completing the cycle ready for reinitiation by pulsing the gate of T2 again. The circuit thus represents a rapid on/off switch which enables effective DC output voltage to be variable according to:

$$[t_{on}/(t_{on} + t_{off})]Vb$$

and output from the device is a train of rectangular voltage pulses. The steady current related to such a train is obtained by connecting Dl across the motor load. Output voltage can thus be controlled by altering t_{off}/t_{on} while holding $t_{on} + t_{off}$ constant – a constant frequency variable mark/space chopper – or altering $1/(t_{on} + t_{off})$ while holding t_{on} constant – a variable frequency constant mark chopper.

6.1.6.5 Ford ecostar

An electric vehicle with equivalent performance to an IC-engined one, and yet based on a conventional Escort van platform, Fig. 6.1-23, has been achieved by Ford with the use of a sodium–sulphur battery. The battery is enclosed in a hermetically sealed metal casing, which contains sodium and sulphur on each side of a porous ceramic separator. In order to react on a molecular level, the sodium and sulphur must be free to flow through the separator in liquid form. This requires maintaining an operating temperature of between 290 and 350° C. The Ecostar's battery pack is formed of a double-skinned, stainless steel case, with a glass fibre insulated cavity which has been evacuated to form a vacuum flask. Inside the case, 480 individual sodium–sulphur cells are embedded in a sand matrix to prevent movement and contain the elements in the event of a severe impact.

Each cell is encased in a corrosion-resistant aluminium cylinder and produces about 2 volts. They are linked to provide a 330 volt power supply and a steady-load current of 80 amps. Power of up to 50 kW is available for short periods. The battery pack is mounted under the vehicle floor, ahead of the rear axle. The vehicle is fitted with an intelligent on-board charging system that automatically adjusts to the supply voltage from the mains. It can be used for overnight recharging and temperature control of the sodium–sulphur battery. A full recharge takes between 5 and 7 hours. In the event of electrical malfunction, the battery pack is isolated by replacing external 200 amp fuses at the battery connection box. The batteries are also equipped with internal fuses that isolate the electrical supply from the vehicle if there is any kind of internal overload.

The Ecostar, Fig. 6.1-21, is powered by a high speed, 3 phase AC motor, rated at 56 kW (75 bhp). It is directly coupled to the transaxle by a single-speed planetary reduction gear set. The motor operates at 330 volts, provided by the sodium–sulphur battery pack. Compared to equivalent DC systems, the AC motor works at a higher rate of efficiency, is smaller, lighter, lower in cost, easier to cool, more reliable and generally more durable. Maximum speed is 13 500 rpm and maximum torque 193 Nm at zero rpm. Regenerative braking comes

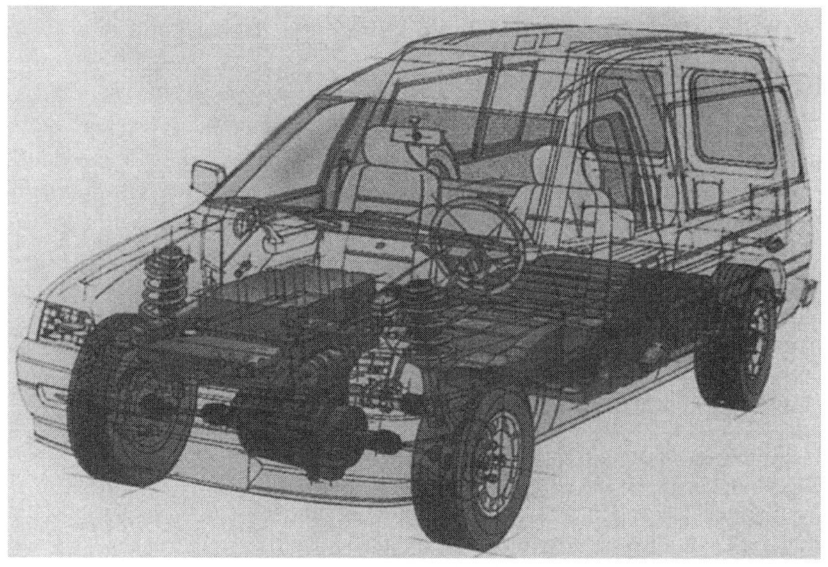

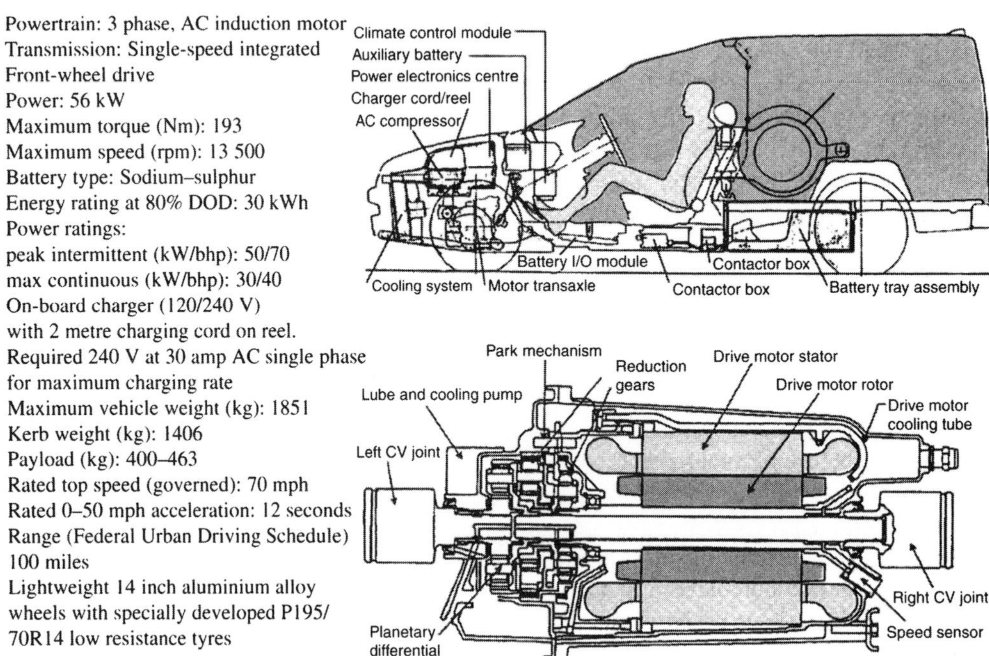

Fig. 6.1-21 Ecostar package, motor and specification.

from the motor acting as a generator, so recharging the battery.

The supply from the sodium–sulphur battery is fed along special heavy-duty cables to the power electronics centre (PEC), which is housed in the engine compartment. The candy-striped, red-on-black cables are strikingly marked to avoid confusion with any other wiring in the van. Encased in aluminium, the PEC incorporates the battery charging electronics and inverters that convert the 330 volt DC supply to AC power. It also includes a transformer which enables a 12 volt auxiliary battery to be recharged from the high voltage traction batteries. The electrical supply to the PEC is connected and isolated by power relays inside a contactor box, controlled by the ignition key and the electronic modules.

On-board microprocessors are linked by a multiplex database that allows synchronous, highspeed communication between all the vehicle's systems. The vehicle system controller (VSC) acts as the user/vehicle interface and is operated by electrical signals from the accelerator pedal. This 'drive-by-wire' system has no mechanical connection to the speed controller. It is supplemented by a battery controller that monitors the sodium–sulphur operating temperature, the state of charge and recharging. The control system also incorporates a diagnostic data recorder which stores information from all the on-board electronic systems. This locates any operational malfunctions quickly and precisely.

A fault detection system (known as the power protection centre (PPC)), has also been built into the battery controller to monitor continuously the main electrical functions. Every 4 seconds it checks for internal and external leakage between the high-voltage system, the vehicle chassis and the battery case. Should a leak be detected in any wire, a warning light tells the driver that service action should be taken. The vehicle can then be driven safely for a short distance so that repairs can be made. If leakage is detected from both battery leads, the vehicle system cuts off the power to the motor and illuminates a red warning light. The auxiliary power supply is maintained to operate the battery cooling system. An inertia switch is also fitted, which is activated in the event of a vehicle collision and isolates power from the main battery pack. Auxiliary power to the battery coolant pump is also cut off to reduce the risk of hot fluids escaping. The vehicle incorporates a small amount of 'creep' whereby slight brake pressure is required to prevent it from moving forwards (or backwards when in reverse gear). This results in easier manoeuvring and smoother transmission of power. Auxiliary vehicle systems are powered by a standard automotive 12 volt lead–acid battery, the exception being the electrically driven cabin air conditioning system, which is powered directly from the main sodium–sulphur traction batteries, via a special AC inverter in the PEC module.

The climate control unit handles both the air conditioning and a highly efficient 4.5 kW ceramic element PTC heater. The heater elements are made from barium titanate with a multi-layer metallic coating on each side, impregnated with special chemical additives. The material has low resistance at low temperature for a very fast warm-up, while at higher temperatures the power supply is automatically regulated to save electrical energy. A strip of solar panels across the top of the windscreen supply power to a supplementary extractor fan that ventilates the cabin when the vehicle is parked in direct sunlight. This relieves the load on the air conditioning system when the journey is resumed. Some lightweight materials have been used in the Ecostar to offset the 350 kg weight of the battery pack.

Elimination of the clutch, torque converter and additional gearing is supplemented by a magnesium transmission casing, aluminium alloy wheels, air conditioning compressor and power electronics housing. Plastic composite materials have been used for the rear suspension springs, load floor and rear bulkhead. Use of these materials has helped keep the Ecostar's kerb weight to between 1338 and 1452 kg, which is 25% heavier than a standard diesel-powered van. The vehicle also has a useful load carrying capacity of up to 463 kg and retains similar load space dimensions to the standard Escort van. The Ecostar has been developed for optimum performance in urban conditions, where it is expected to be driven most frequently. Its top speed is restricted to 70 mph, whilst 0–50 mph takes approximately 12 seconds. Because of the high torque at low speed, acceleration from standing to 30 mph is quicker than diesel and petrol driven vehicles of the same size. The average vehicle range between charges to date has been 94 miles, with a maximum recorded range of 155 miles.

6.1.6.6 Bradshaw envirovan

DC drive is used on the higher payload capacity purpose-built Bradshaw Envirovan. Fig. 6.1-22 shows the Envirovan built in conjunction with US collaborator Taylor-Dunn. This can carry 1500 lb on a 3.55 square metre platform at speeds up to 32.5 mph and is aimed specifically at city deliveries. The vehicle relies on 12.6 volt deep-cycle,

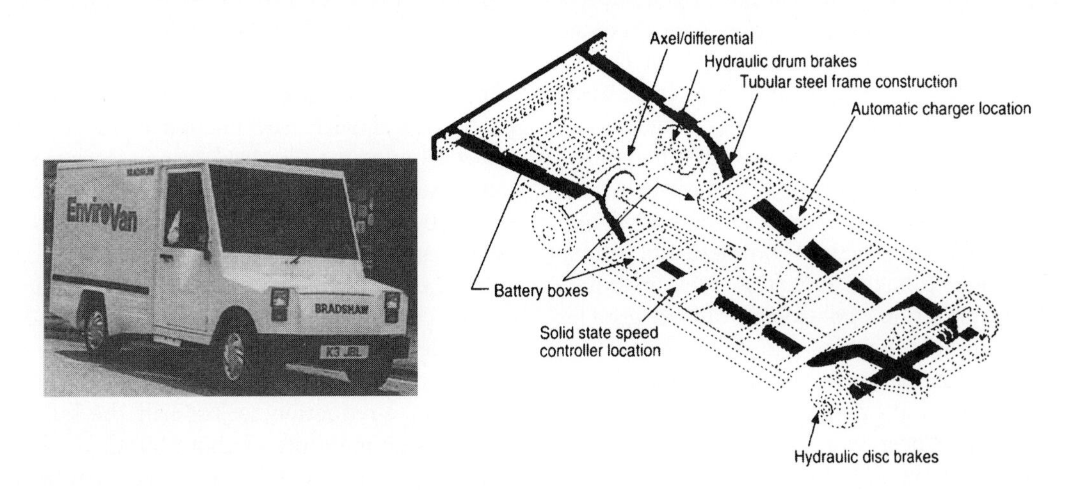

Fig. 6.1-22 Bradshaw Envirovan.

rechargeable lead–acid batteries for a total of 72 V. All accessories, such as internal lights, windscreen wipers, and gauges, run off the 72 V system through a DC/DC converter, which steps the power down to 12 V, so that all batteries discharge equally. This distributes power requirements evenly across all 12 batteries and prevents one or two of the batteries from draining prematurely. A battery warning indicator shows the current percentage of battery power available, with a visual warning when battery charge is below 20%. An on-board battery charger, featured on the Envirovan, can be used to recharge the battery packs simply by plugging it into any standard 240 V AC power socket. The entire 72 V system requires approximately 9.5 hours to fully charge the batteries from an 80% state of discharge (20% remaining charge). The battery pack provides approximately 1000 recharging cycles before replacement is required. Battery packs are available for less than £1000 which equates to less than 2p per mile. Recharging costs add an additional 1 p per mile giving a total cost per mile of 3p. A 20 bhp General Electric motor has been designed for the Envirovan. A range of 8 hours/150 miles is available for the vehicle which measures 4.21 metres long × 1.65 wide. It can accelerate to 25 mph in 6 seconds and its controller can generate up to 28 bhp for quick response.

6.1.7 Fuel-cell powered vehicles

6.1.7.1 General motors Zafira projects

GM and its Opel subsidiary had aimed at a compact fuel-cell driven vehicle, Fig. 6.1-23. By 2010, up to 10% of total sales are expected to be taken by this category. The efficiency of cells tested by the company is over 60% and CO_2 emissions, produced during the reformation of methanol to obtain hydrogen, are about half that of an equivalent powered IC engine. Fuel cells have already been successfully exploited in power generation, at Westervoort in the Netherlands, and experimental versions have been shown to successfully power laptop computers. According to GM, in principle four basic fuels are suitable: sulphur-free modified gasoline, a synthetic fuel, methanol or pure hydrogen. Modified gasoline is preferred because of the existing distribution infrastructure but CO_2 emission in reforming is higher than with methanol. Synthetic fuel and methanol can be obtained from some primary energy sources including natural gas. Transportation and storage of hydrogen is still at the development stage for commercial viability, Liquefying by low temperature and/or pressure being seen as the only means of on-vehicle storage.

GM engineers have been working on a fuel-cell drive version of the Zafira van (a) in which electric motor, battery and controller are accommodated in the former engine compartment (b). The 'cold combustion' of the fuel-cell reaction, hydrogen combining with oxygen to form water, takes place at 80–90° C and a single cell develops 0.6–0.8 V. Sufficient cells are combined to power a 50 kW asynchronous motor driving the front wheels through a fixed gear reduction. The cell comprises fuel anode, electrolyte and oxygen cathode. Protons migrate through the electrolyte towards the cathode, to form water, and in doing so produce electric current. Prospects for operating efficiencies above 60% are in view, pending successful waste heat utilization and optimization of gas paths within the system. The reforming process involved in producing hydrogen from the fuel involves no special safety measures for handling methanol and the long-term goal is to produce no more than 90 g/km of CO_2. In the final version it is hoped to miniaturize the reformer, which now takes up most of the load space, (c), and part of the passenger area, so that it also fits within the former engine compartment. Rate of production of hydrogen in the reformer, and rate of current production in the fuel cell, both have to be accelerated to obtain acceptable throttle response times–the flow diagram is seen at (d). The 20 second start-up time also has to be reduced to 2 seconds, while tolerating outside temperatures of −30°C.

GM Opel was reportedly working in the jointly operated Global Alternative Propulsion Centre (GAPC) on a version of its fuel-celled MPV which is now seen as close to a production design. A 55 kW (75 hp) 3 phase synchronous traction motor drives the front wheels through fixed gearing, with the complete electromechanical package weighing only 68 kg (150 lb). With a maximum torque of 251 Nm (181 lb ft) at all times it accelerates the Zafira to 100 km/h (62 mph) in 16 seconds, and gives a top speed of 140 km/h (85 mph). Range is about 400 km (240 miles).

In contrast to the earlier vehicle fuelled by a chemical hydride system for on-board hydrogen storage, this car uses liquid hydrogen. Up to 75 litres (20 gallons) is stored at a temperature of −253° C, just short of absolute zero, in a stainless steel cylinder 1 metre (39 in) long and 400 mm (15.7 in) in diameter. This cryostat is lined with special fibre glass matting said to provide insulating properties equal to several metres of polystyrene. It is stowed under the elevated rear passenger seat, and has been shown to withstand an impact force of up to 30 g. Crash behaviour in several computer simulations also been tested.

Fuel cells as well as the drive motor are in the normal engine compartment. In the 6 months since mid-2000 the 'stack' generating electricity by the reaction of hydrogen and oxygen now consists of a block of 195 single fuel cells, a reduction to just half the bulk. Running at a process temperature of about 80° C, it has a maximum output of 80 kW. Cold-start tests at ambient temperatures down to −40° C have been successfully conducted.

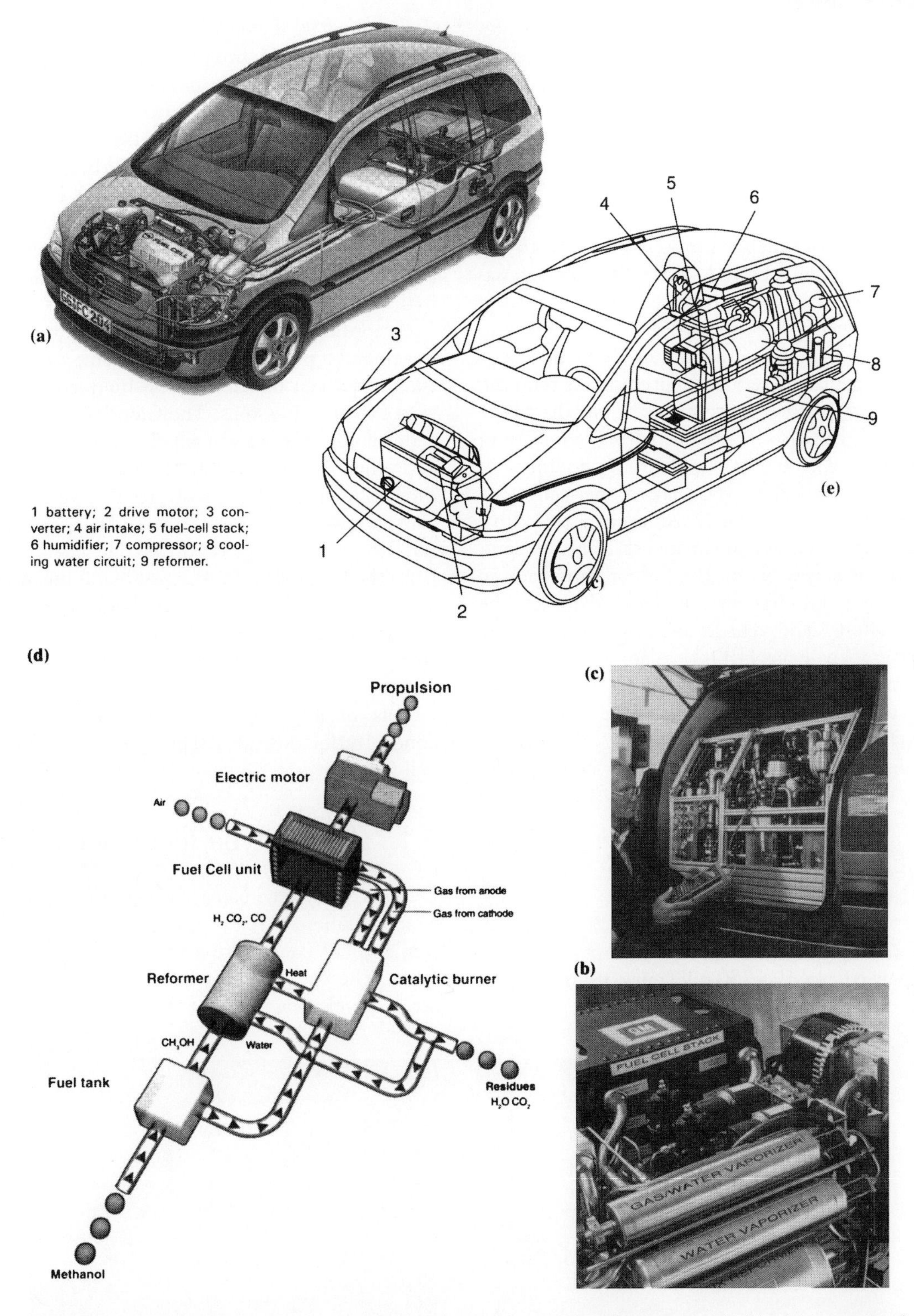

Fig. 6.1-23 GM fuel-cell developments: (a) Zafira conversion package; (b) under-bonnet power-pack; (c) reformer and cells; (d) flow diagram; (e) latest package with on-board hydrogen storage.

GAPC has created strong alliances with several major petroleum companies to investigate the creation of the national infrastructures needed to support a reasonable number of hydrogen-fuelled vehicles once they reach the market. Fuel cost is another critical factor. Although hydrogen is readily available on a commercial basis from various industrial processes, its cost in terms of energy density presents a real problem for the many automakers who research both fuel cells and direct combustion.

According to one calculation based on current market prices, the energy content of hydrogen generated by electrolysis using solar radiation with photovoltaic cells equals gasoline at roughly $10 a gallon.

6.1.7.2 Ford P2000

Mounting most of the fuel-cell installation beneath the vehicle floor has been achieved on Ford's FC5, seen as a static display in 1999, with the result of space for five passengers in a medium-sized package. Their aim is to achieve an efficiency twice that of an IC engine. The company points out that very little alteration is required to a petrol-distributing infrastructure to distribute methanol which can also be obtained from a variety of biomass sources. Oxygen is supplied in the form of compressed air and fed to the Ballard fuel-cell stack alongside reformed hydrogen. Ford use an AC drive motor, requiring conversion of the fuel cell's DC output. Even the boot is accessible on the 5-door hatchback so much miniaturization has already been done to the propulsion system. The vehicle also uses an advanced lighting system involving HID headlamps, with fibre-optic transmission of light in low beam, and tail-lights using high efficiency LED blade manifold optics. The company's running P2000 demonstrator, Fig. 6.1-24, uses fuel in the form of pure gaseous hydrogen in a system developed with Proton Energy Systems.

6.1.7.3 Liquid hydrogen or fuel reformation

Renault and five European partners have produced a Laguna conversion with a 250 mile range using fuel-cell propulsion. The 135 cell stack produces 30 kW at a voltage of 90 V, which is transformed up to 250 V for powering the synchronous electric motor, at a 92% transformer efficiency and 90–92% motor efficiency. Nickel–metal hydride batteries are used to start up the fuel-cell auxiliary systems and for braking energy regeneration. Some 8 kg of liquid hydrogen is stored in an on-board cryogenic container, (a), at −253° C to achieve the excellent range. Renault insist that an on-board reformer would emit only 15% less CO_2 than an IC engine against the 50% reduction they obtain by on-board liquid hydrogen storage, Fig. 6.1-25.

According to Arthur D. Little consultants, who have developed a petrol reforming system, a fuel-cell vehicle thus fitted can realize 80 mpg fuel economy with near zero exhaust emissions. The Cambridge subsidiary Epyx is developing the system which can also reform methanol and ethanol. It uses hybrid partial oxidation and carbon monoxide clean-up technologies to give it a claimed advantage over existing reformers. The view at (b) shows how the fuel is first vaporized (1) using waste energy from the fuel cell and vaporized fuel is burnt with a small amount of air in a partial oxidation reactor (2) which produces CO and O_2. Sulphur compounds are removed from the fuel (3) and a catalytic reactor (4) is used with steam to turn the CO into H_2 and CO_2. The remaining CO is burnt over the catalyst (5) to reduce CO_2 concentration down to 10 ppm before passing to the fuel cell (6).

6.1.7.4 Prototype fuel-cell car

Daimler-Chrysler's Necar IV, Fig. 6.1-26, is based on the Mercedes-Benz A-class car and exploits that vehicle's duplex floor construction to mount key propulsion systems. The fuel cell is a Ballard PEM type, 400 in the stack, developing 55 kW at the wheels to give a top speed

Fig. 6.1-24 Ford P2000 fuel cell platform with two 35 kW Ballard stacks.

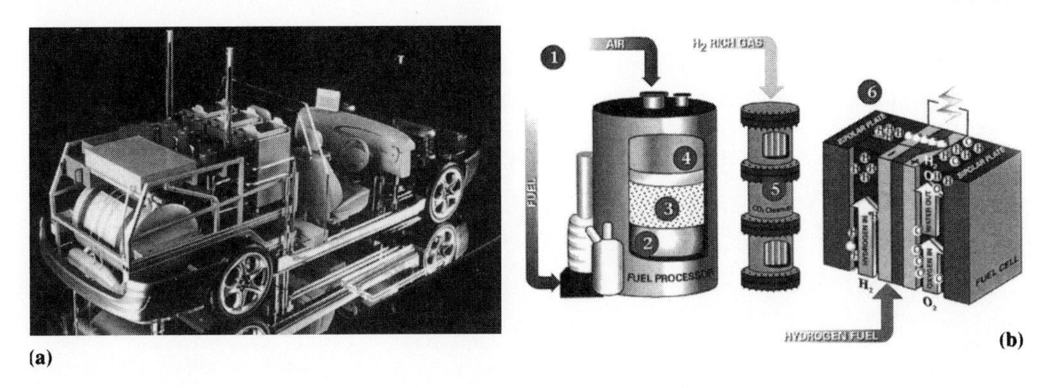

Fig. 6.1-25 Liquid hydrogen or reformed fuel: (a) Renault cryogenic storage: (b) Arthur D. Little reformer.

of 145 km/h and a range of 450 km. Fuel consumption is equivalent to 88 mpg and torque response to throttle movement is virtually instantaneous. While the first prototype weighs 1580 kg, the target weight is 1320 kg, just 150 kg above the standard A-class. Tank to wheel efficiency is quoted as 40% now, with 88% in prospect for a vehicle with a reformer instead of compressed hydrogen. The American Methanol Institute is predicting 2 million thus-fitted cars on the road by 2010 and 35 million by 2020.

In a summer 1999 interview Ballard chief Firoz Rasul put the cost of electricity produced by fuel cells as $500/kW so that car power plants between 50 and 200 kW amount to $25–1 00 000. PEM cells operate at 80° C and employ just a thin plastic sheet as their electrolyte. The sheet can tolerate modest pressure differentials across it, which can increase power density. Ballard's breakthrough in power density came in 1995 with the design of a stack which produced 1000 watts/litre, ten times the 1990 state of the art. Cell energy conversion efficiency, from chemical energy to electricity, is about 50% and the cell does not 'discharge' in the manner of a conventional storage battery. Electrodes are made from porous carbon separated by the porous ion-conduction electrolyte membrane. It is both an electron insulator and proton conductor and is impermeable to gas. A catalyst is integrated between each electrode and the membrane while flow field plates are placed on each side of the membrane/electrode assembly. These have channels formed in their surface through which the reactants flow. The plates are bi-polar in a stack, forming the anode of one cell and the cathode of the adjacent one. The catalyst causes the hydrogen atoms to dissociate into protons and electrons. The protons are carried through to the cathode and the free electrons conducted as a usable current.

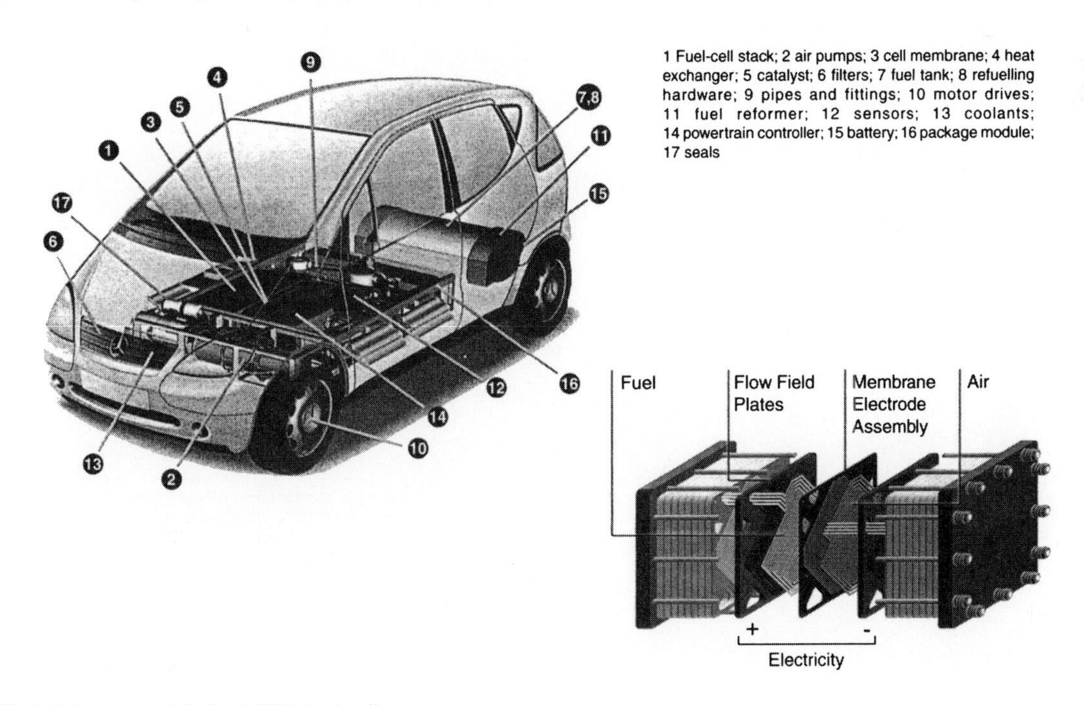

Fig. 6.1-26 D-C Necar and Ballard PEM fuel cell.

References

1. Origuchi *et al.*, Development of a lithium–ion battery system for EVs, SAE paper 970238
2. Saito *et al.*, Super capacitor for energy recycling hybrid vehicle, Convergence 96 proceedings
3. Van der Graaf, R., EAEC paper 87031
4. Prigmore *et al.*, *Battery car conversions*. Battery Vehicle Society, 1978
5. Harding. G., Electric vehicles in the next millennium, *Journal of Power Sources*, 3335, 1999
6. Huettl *et al.*, *Transport Technology USA*. 1996
7. SAE paper 900578, 1990

Further reading

Smith & Alley, *Electrical circuits, an introduction*, Cambridge, 1992

Copus, A., DC traction motors for electric vehicles, *Electric vehicles for Europe*. EVA conference report, 1991

EVA *manual*, Electric Vehicle Association of GB Ltd

Unnewehr and Nasar, *Electric vehicle technology*. Wiley, 1982

Huettl *et al.*, *Transport Technology USA*, 1996

Argonne National Laboratory authors, SAE publication: *Alternative Transportation Problems*, 1996

Strategies in electric and hybrid vehicle design, SAE publication SP-1156, 1996

(ed.) Dorgham, M., *Electric and hybrid vehicles*, Interscience Enterprises, 1982

Electric vehicle technology, MIRA seminar report, 1992

Battery electric and hybrid vehicles, IMechE seminar report, 1992

(ed.) Lovering, D., *Fuel cells*, Elsevier 1989

The urban transport industries report, Campden, 1993

The MIRA electric vehicle forecast, 1992

Niewenhuis *et al.*, *The green car guide*, Merlin, 1992

Combustion engines and hybrid vehicles, IMechE, 1998

Section Seven

Hybrid vehicles

Chapter 7.1

7.1

Hybrid vehicle design

John Fenton and Ron Hodkinson

7.1.1 Introduction

The hybrid-drive concept appears in many forms depending on the mix of energy sources and propulsion systems used on the vehicle. The term can be used for drives taking energy from two separate energy sources, for series or parallel drive configurations or any combination of these. Here the layout and development of systems for cars and buses are described in terms of drive configuration and package-design case studies of recent-year introductions.

7.1.1.1 The hybrid vehicle

This solution is considered by coauthor Ron Hodkinson to be a short-term remedy to the pollution problem. It has two forms, parallel and series hybrid, which he illustrates in Fig. 7.1-1. Conventionally, parallel hybrids are used in lower power electric vehicles where both drives can be operated in parallel to enhance high power performance. Series hybrids are used in high power systems. Typically, a gas turbine drives a turbo-alternator to feed electricity into the electric drive. It is this type of drive that would be used on trucks between 150 kW and 1000 kW. In pollution and fuel economy terms, hybrid technology should be able to deliver two-thirds fuel consumption and one-third noxious emission levels of IC engined vehicles. This technology would just about maintain the overall emissions status quo in 10 years overall. If hybrid vehicles were used on battery only in cities, this would have a major impact on local pollution levels.

7.1.2 Hybrid-drive prospects

A neat description of the problems of hybrid-drive vehicles has come out of the results of the 3 year HYZEM research programme undertaken by European manufacturers (Fig. 7.1-2). According to Rover participants[1], controlled comparisons of different hybrid-drive configurations, using verified simulation tools, are able to highlight the profitable fields of development needed to arrive at a fully competitive hybrid-drive vehicle

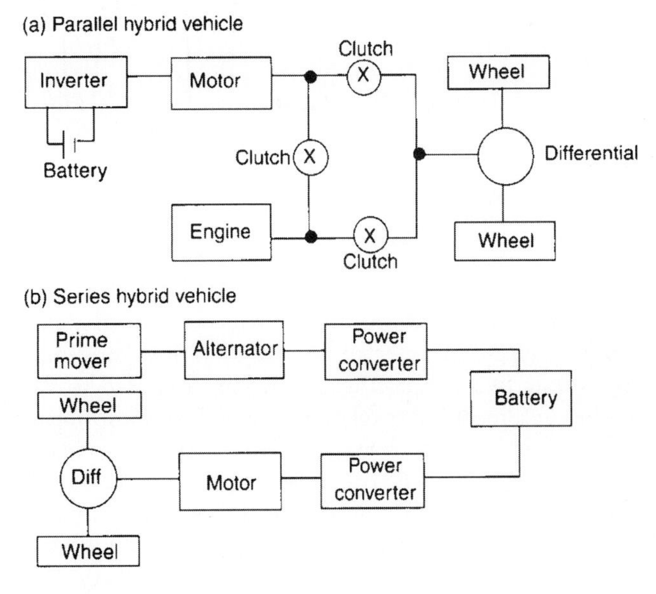

Fig. 7.1-1 Types of hybrid drive.

Lightweight Electric/Hybrid Vehicle Design; ISBN: 9780750650922
Copyright © 2001 John Fenton and Ron Hodkinson; All rights of reproduction, in any form, reserved.

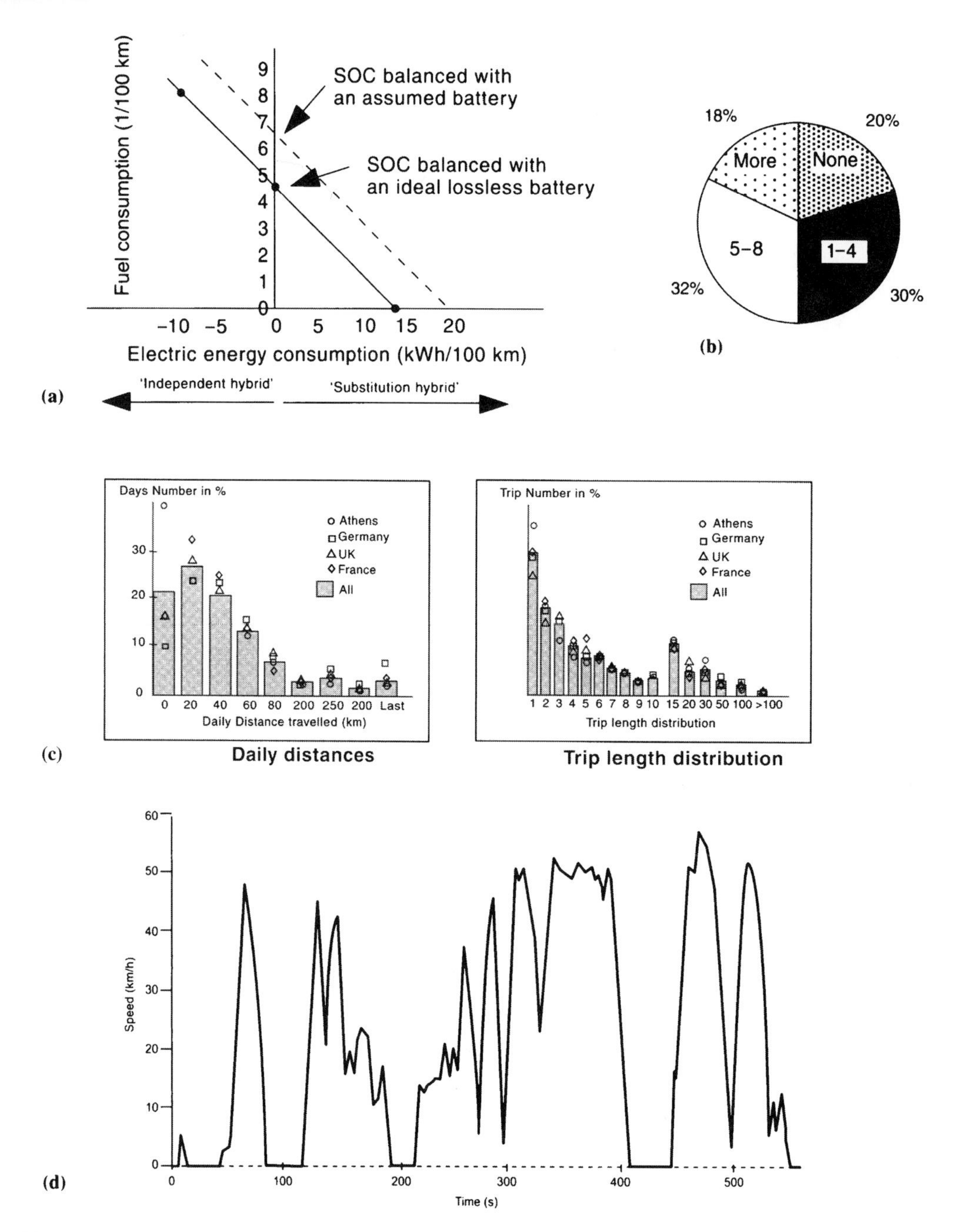

Fig. 7.1-2 HYZEM research programme: (a) characterizing a hybrid powertrain; (b) use of vehicle per day; (c) daily distances and trip lengths; (d) synthetic urban drive cycle.

and demonstrate, in quantitative terms, the tradeoff between emissions, electrical energy, and fuel consumption. Only two standard test points are required to describe the almost linear relationship: fuel consumption at point of no overall change in battery state of charge (SOC) and point of electrical consumption over the same cycle in pure electric mode. A linear characteristic representing an ideal lossless battery can also be added to the graph, to show the potential for battery development, as at (a).

Confirmation was also given to such empirical assessments that parallel hybrids give particularly good fuel economy because of the inherent efficiency of transferring energy direct to the wheels as against the series hybrids' relatively inefficient energy conversion from mechanical to electrical drive. The need for a battery which can cope with much more frequent charge/discharge cycles than one for a pure electric-drive vehicle was also confirmed. Although electric energy capability requirement is less stringent, a need to reduce weight is paramount in overcoming the problem of the redundant drive in hybrid designs.

A useful analysis of over 10 000 car journeys throughout Europe was undertaken for a better understanding of 'mission profile' for the driving cycles involved. Cars were found to be used typically between one and eight times per day, as at (b), and total daily distances travelled were mostly less than 55 km. Some 13% of trips, (c), were less than 500 metres, showing that we are in danger of becoming like the Americans who drive even to visit their next door neighbours! Even more useful velocity and acceleration profiles were obtained, by data recoding at 1 Hz frequency, so that valuable synthetic drive cycles were obtained such as the urban driving one shown at (d).

7.1.2.1 Map-controlled drive management

BMW researchers[2] have shown the possibility of challenging the fuel consumption levels of conventional cars with parallel hybrid levels, by using map-controlled drive management, Fig. 7.1-3. The two-shaft system used by the company, seen at (a), uses a rod-shaped asynchronous motor, by Siemens, fitted parallel to the crankshaft beneath the intake manifold of the 4-cylinder engine, driving the tooth-belt drive system as seen at (b): overall specification compared with the 518i production car from which it is derived is shown at (c). The vehicle still has top speed of 180 kph (100 kph in electric mode) and a range of 500 km; relative performance of the battery options is shown at (d). Electric servo pumps for steering and braking systems are specified for the hybrid vehicle and a cooling system for the electric motor is incorporated. The motor is energized by the battery via a 13.8 V/50 A DC/DC converter. The key electronic control unit links with the main systems of the vehicle as seen at (e).

To implement the driving modes of either hybrid, electric or IC engine the operating strategy is broken down into tasks processed parallel to one another by the CPU, to control and monitor engine, motor, battery and electric clutch. The mode task determines which traction condition is appropriate, balancing the inputs from the power sources; the performance/output task controls power flow within the total system; the battery task controls battery charging. According to accelerator/braking pedal inputs, the monitoring unit transfers the power target required by the driver to the CPU where the optimal operating point for both drive units is calculated in a continuous, iterative process. The graphs at (f) give an example of three iterations for charge efficiency, also determined by the CPU, based on current charge level of the battery.

7.1.2.2 Justifying hybrid drive

Studies carried out at the General Research Corporation in California, where legislation on zero emission vehicles is hotly contested, have shown that the 160 km range electric car could electrify some 80% of urban travel based on the average range requirements of city households, (a). It is unlikely, however, that a driver would take trips such that the full range of electric cars could be totally used before switching to the IC engine car for the remainder of the day's travel. This does not arise with a hybrid car whose entire electric range could be utilized before switching and it has been estimated that with similar electric range such a vehicle would cover 96% of urban travel requirements. In two or more car households, the second (and more) car could meet 100% of urban demand, if of the hybrid-drive type (Fig. 7.1-4).

Because of the system complexities of hybrid-drive vehicles, computer techniques have been developed to optimize the operating strategies. Ford researchers[3], as well as studying series and parallel systems, have also examined the combined series/parallel one shown at (b). The complexity of the analysis is shown by the fact that in one system, having four clutches, there are 16 possible configurations depending on state of engagement. They also differentiated between types with and without wall-plug re-energization of the batteries between trips.

7.1.2.3 Mixed hybrid-drive configurations

Coauthor Ron Hodkinson argues that while initially parallel and series hybrid-drive configurations were seen as possible contenders (parallel for small vehicles and series for larger ones) it has been found in building 'real world' vehicles that a mixture of the two is needed. For cars a mainly parallel layout is required with a small series element. The latter is required in case the vehicle becomes stationary for a long time in a traffic jam to make sure the traction battery always remains charged to sustain the 'hotel loads' (air conditioning etc.) on the vehicle's electrical system. Cars like the Toyota Prius have 3–4 kW series capability but detail configuration of the system as a whole is just a matter of cost vs performance. Generally the most economical solution for passenger cars is with front wheel drive and a conventional differential/final-drive gearbox driven by a single electric motor. No change-speed gearbox is required, where the motor can give constant power over a 4:1 speed range, but reduction gearing is required to match 13 500 rpm typical motor speed with some 800 rpm roadwheel speed. This is usually in the form of a two-stage

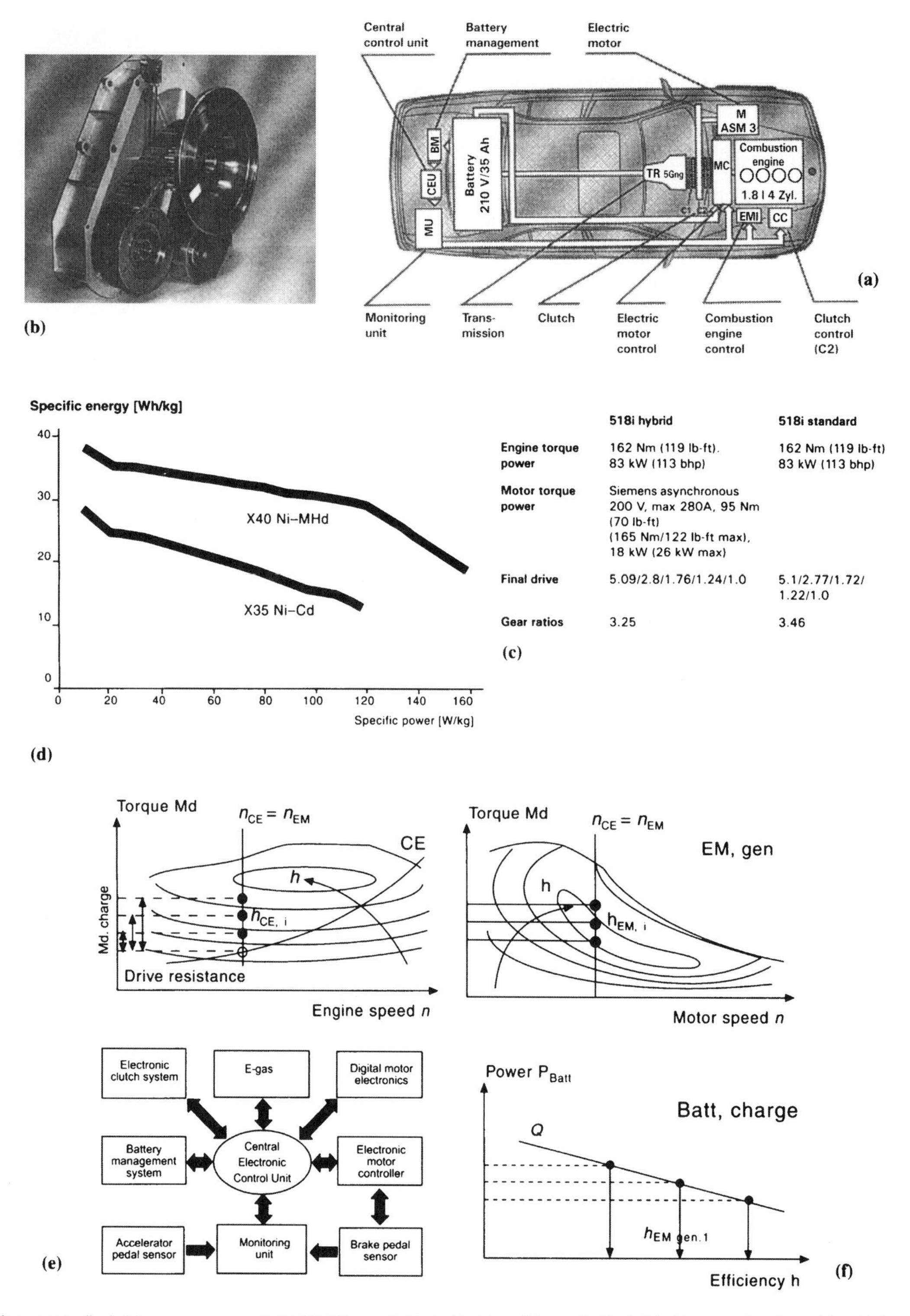

	518i hybrid	518i standard
Engine torque power	162 Nm (119 lb-ft). 83 kW (113 bhp)	162 Nm (119 lb-ft) 83 kW (113 bhp)
Motor torque power	Siemens asynchronous 200 V, max 280A, 95 Nm (70 lb-ft) (165 Nm/122 lb-ft max), 18 kW (26 kW max)	
Final drive	5.09/2.8/1.76/1.24/1.0	5.1/2.77/1.72/ 1.22/1.0
Gear ratios	3.25	3.46

Fig. 7.1-3 Map-controlled drive management: (a) BMW parallel hybrid drive; (b) parallel hybrid-drive mechanism; (c) vehicle specification; (d) ragone diagram for the two battery systems; (e) vehicle management; (f) optimized recharge strategy.

reduction by epicyclic gear trains, the first down to 4000 rpm, and the final drive gearing providing the second stage–typically two stages of 3–4:1 are involved. A change-speed gearbox only becomes necessary in simple lightweight vehicles using brushed DC motors and Curtis controllers. Weight can be saved by using a motor of one-quarter the normal torque capacity and multiplying up the torque via the gearbox.

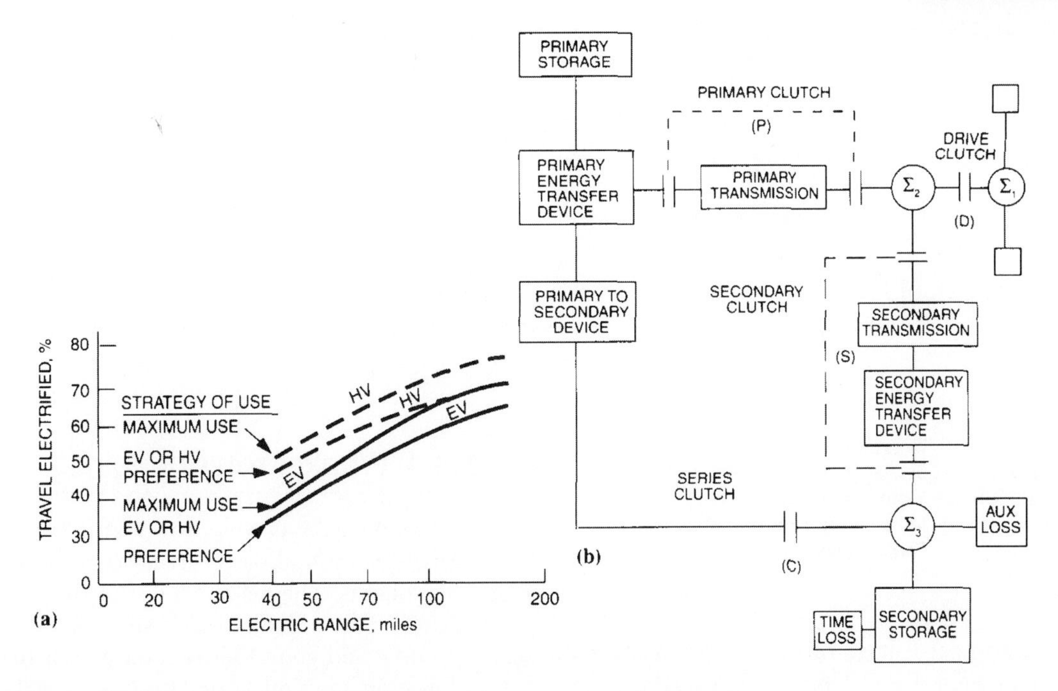

Fig. 7.1-4 Justifying the hybrid: (a) EV traffic potential; (b) combined series–parallel mode.

7.1.3 Hybrid technology case studies

7.1.3.1 The hybrid electric solution for small cars

Ron Hodkinson[4] points out that US President Bill Clinton's initiative for the American family car sets the target that, in 2003, cars will run for 100 miles on one US gallon of unleaded gasoline. The objectives are reduced fuel consumption, reduced imported oil dependency, and reduced pollution to improve air quality. Can it be done? The answer is yes. Work carried out on GM's ultra-lightweight car programme involved composite structure to achieve a body weight of 450 kg, drag coefficient $C_d < 0.2$ by means of streamlining the underside of the car, reduced frontal area ($<1.5\ m^2$), conventional drive train with a 30 bhp two-stroke orbital engine and low-rolling-resistance tyres. Overall, this leads to a vehicle weight of 750 kg, a 400 kg payload at a top speed of 80 mph and an acceleration of 0–60 mph in 20 seconds. This illustrates the dilemma. Reduce the engine size to improve the fuel consumption, and the acceleration performance is sacrificed.

7.1.3.2 Hybrid power pack, a better solution

In the long term we may use electric vehicles using flywheel storage or fuel cells. Until these systems are available the best answer is to use a hybrid-drive line consisting of a small battery, a 45 kW electric drive, and a 22.5 kW engine. This solution would increase the vehicle weight from 750 kg to 860 kg, but it would now accelerate from 0 to 60 mph in 8 seconds. In addition, the vehicle will have automatic transmission with regenerative braking, could operate in electric-only mode with a 30 mile range for use in zero emission mode in city centres, and could be recharged from a wall socket or charging point if desired.

The Polaron subsidiary, Nelco, worked with Wychwood Engineering and Midwest Aero Engines on a parallel hybrid replacement for a front-wheel drive train in family cars and delivery vans, Fig. 7.1-5.

7.1.3.3 Rotary engine with PM motor, the mechanical outlines

The drive line is a marriage of two techniques: a permanent-magnet brushless DC motor and a Wankel two-stroke engine. The electric motor provides instant acceleration with 45 kW of power available from 1500 to 6000 rpm, on this design. A permanent-magnet design is used because it is lightweight, highly efficient, and results in an economical inverter. The concept is to exploit the machine characteristics using vector control. At low speeds, the permanent magnets provide the motor field. At high speeds, the field is weakened by introducing a reactive *Id* component at right angles to the torque-producing component *Iq*. The control objective is to

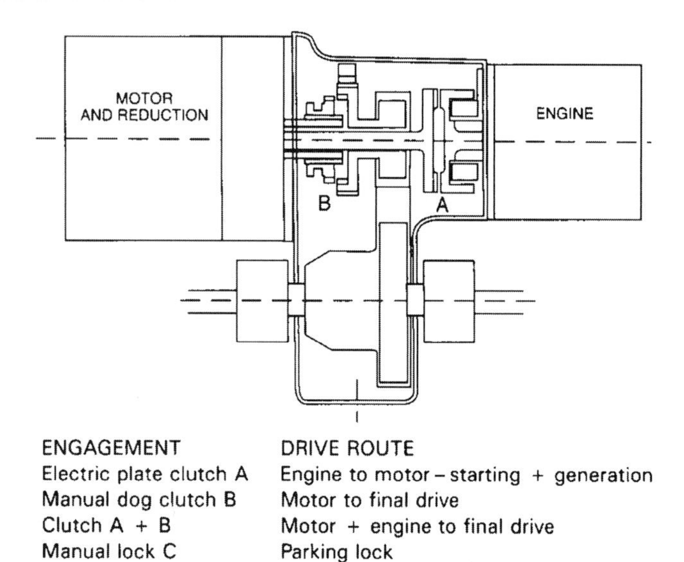

Fig. 7.1-5 The hybrid power unit.

maintain the terminal voltage of the motor constant in the high speed region, Fig. 7.1-6.

The system efficiency is achieved by using a dual-mode control system for the inverter: (i) at low speeds the inverter operates in the current-source mode; the current in the motor windings is pulse-width modulated; (ii) when the rectified motor voltage exceeds that of the DC link, the inverter changes to voltage-source control. Since the system operates with field weakening, the machine has a leading power factor, consequently there is virtually no switching loss in the inverter transistors if square-wave excitation is used. Since the motor has 30% impedance at full load, the impedance at the 5th/7th harmonic is 150/210%, and consequently there are few harmonics in the current. As a result, low-saturation insulation gate bipolar transistors (IGBTs) can be used, and they switch at less than 2 kHz with high efficiency in the cruise mode and low RF interference.

A key benefit of the PM brushless DC motor is the wide band constant-power curve. Consequently there is no need for gear changing for high-torque operation, and the motor gives high efficiency and low rotor heating both on the flat and in hilly terrain. The motor inverter and battery are oil cooled to ensure compact dimensions. The battery uses lead/tinfoil plates to achieve low internal resistance and is thermally managed to ensure charge equalization in the cells. A chopper is used to give a stabilized 300 V DC link. The outstanding feature of the lead–acid battery is the peak power capability of 50 kW for 2 minutes in a weight of 170 kg, Fig. 7.1-7(a).

7.1.3.4 Wankel rotary engine

The Wankel engine is well proven at 300 cc size, having full Air Registration Board certification for use in drones and microlights. The main benefits of the Wankel engine are: light weight and '6 cylinder' smoothness; flat torque/speed curve and good fuel economy with fuel injection very low emissions; on natural gas it is possible to comply without a catalytic converter. On unleaded petrol, two electrically preheated catalytic converters are used. Another advantage is multi-fuel capability. In the control scheme, the electric drive runs continuously. The Wankel engine switches on and off. At speeds above 60 mph, the engine will run continuously. The objective is to avoid discharging the battery by more than 30%. In this way we can obtain 11 000 cycles or 100 000 miles on a small battery. Consequently the economics of this scheme make sense.

In mass production (>100 000 systems) the additional cost per car of this system would be £2500. We believe this cost could be recouped, in fuel savings and reduced maintenance, by a 3 year first user, Fig. 7.1-7(b). The problem is to break through the dichotomy of the electric-vehicle market. Market logic says start with

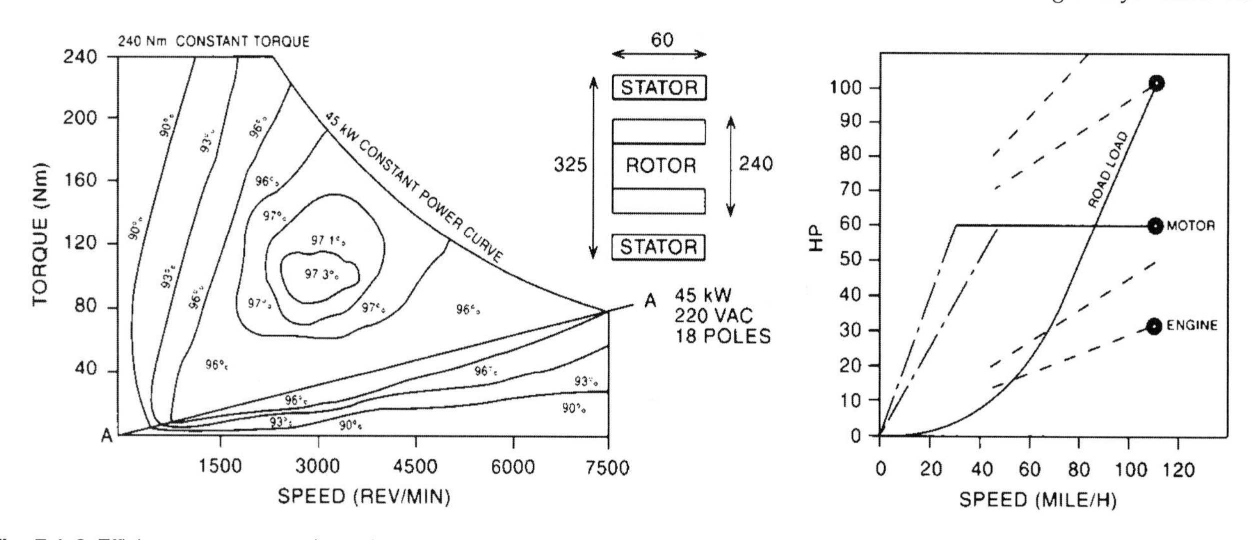

Fig. 7.1-6 Efficiency map, torque/speed curves and load matching.

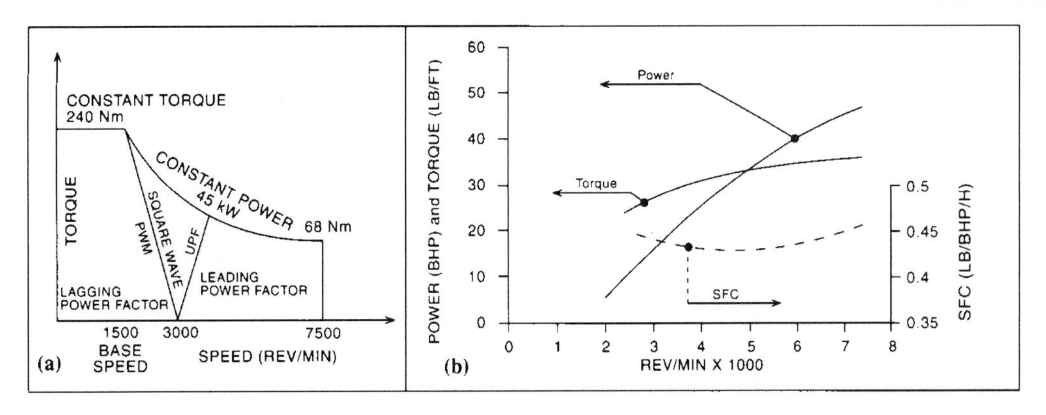

Fig. 7.1-7 System operation states (a), with performance curves of 300 cc fuel-injection Wankel engine (b).

luxury vehicles and work down. The environment requires an impact on the mass market, that is the 'repmobile', as soon as possible. At least we now have a system that can meet the environmental demands at a reasonable cost, and which the market will be prepared to buy.

7.1.3.5 Hybrid passenger cars

A recent method of construction for permanent magnet motors (Fig. 7.1-8), by Fichtel & Sachs and Magnet Motor GmbH, exploits the relationship that specific motor torque, on the basis of weight and bulk, is proportional to the product of magnetic field in the air gap, radius of the air gap squared and the axial length of the motor. The requirement for maximum air gap focuses on the construction of the outside of the rotor. On the inside tangentially magnetized permanent magnets are fitted to the circumference. Shown inset in (a), trapezoidal iron conductors are seen between the rare-earth element magnets which are also trapezoidal in section. These collect magnetic field and divert it to the stator. The laminations of the stator are arranged radially and wound in individual coils which are connected individually or in groups, series or parallel, to the single phase power electronic DC/AC converters.

The latter are supplied with power from a DC link circuit and commutate the coil current at the amplitude required by the rotor angle as detected by remote sensors. The converters only supply the section of the motor assigned to them and thus work independently. They are made as 4 quadrant controllers and have IGBT switching. These are described as Multiple Electronic Permanent (MEP) magnet motors and are made by the Magnet Motor Co. With liquid cooling, the specific performance of these MEP machines is significantly increased over conventional EV motors as seen by the table at (b) and the corresponding characteristic curves at (c). Both companies are jointly engaged in further development for volume production with target performances seen in the tables at (d).

A test vehicle, Fig. 7.1-9, based on an Audi 100 Quattro of 100 kW has been constructed and the drive configuration chosen is seen at (a). The four MEP motors have a nominal performance of 25 kW each and are direct connected to each road wheel without the need for mechanical reduction. The tandem configuration of the motors has the advantages of an electrical torque apportioning differential; the motor is part of the vehicle sprung mass and relatively long drive shafts can be used. The generator is also of MEP construction and direct-flange-connected to the IC engine. Power electronics are used to provide power through a DC link circuit to the four motors.

A drive-by-wire arrangement is involved in the IC-engine throttle control, an ECU matching engine speed to generator output. The view at (b) shows the speed/torque map of the engine set to a low SFC value. The ECU also controls the commutation of the MEP machines. The drive configuration allows the engine to operate at constant speed and the wheel speed reductions are also controlled by software in the ECU and thus different driving programs can be instituted. Handling control is also affected by the software in that different torque distribution to the road wheel can be programmed. The MEP generator also acts as a starter motor for the IC engine.

Test results and simulated performance are seen at (c). The simulation has also been used to test the effect of further developments such as the removal of the IC engine flywheel, redesign of the Audi platform to better exploit electric propulsion and the achievement of the target MEP machine performances previously listed. These estimated performances are seen at (d). A 10% fuel consumption is achieved for road performance equivalent to the standard car and work is under way to design ULEV versions incorporating on-board storage batteries.

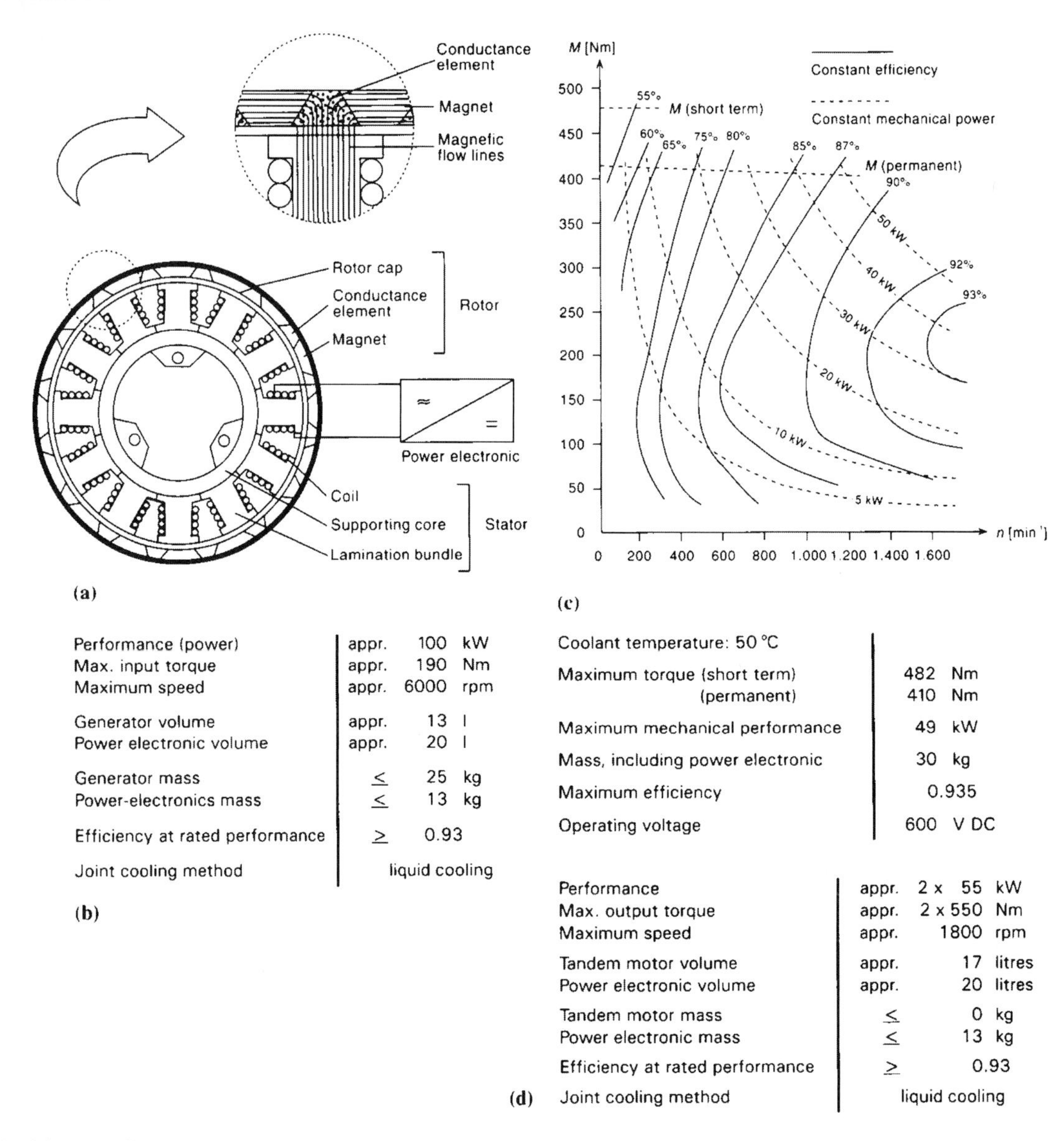

Performance (power)	appr.	100 kW
Max. input torque	appr.	190 Nm
Maximum speed	appr.	6000 rpm
Generator volume	appr.	13 l
Power electronic volume	appr.	20 l
Generator mass	≤	25 kg
Power-electronics mass	≤	13 kg
Efficiency at rated performance	≥	0.93
Joint cooling method		liquid cooling

(b)

Coolant temperature: 50 °C	
Maximum torque (short term)	482 Nm
(permanent)	410 Nm
Maximum mechanical performance	49 kW
Mass, including power electronic	30 kg
Maximum efficiency	0.935
Operating voltage	600 V DC

Performance	appr.	2 x 55 kW
Max. output torque	appr.	2 x 550 Nm
Maximum speed	appr.	1800 rpm
Tandem motor volume	appr.	17 litres
Power electronic volume	appr.	20 litres
Tandem motor mass	≤	0 kg
Power electronic mass	≤	13 kg
Efficiency at rated performance	≥	0.93
Joint cooling method		liquid cooling

(d)

Fig. 7.1-8 Advanced PM motor system: (a) trapezoidal iron conductors; (b) motor performance; (c) motor curve; (d) target performances.

7.1.3.6 Taxi hybrid drive

Based on a Range Rover, an EC project involving Rover, gas turbine maker OPRA, Athens Technical University, Hawker Batteries and Renault was carried out at Imperial College so as to provide a hybrid powertrain for a European taxi, Fig. 7.1-10, and in particular the development of a turbogenerator. A system involving a gas turbine and high speed generator together with a small battery pack for power storage and zero-emission city driving in a series configuration has been proposed by the EC and the turbogenerator work is reported by Pullen and Etemad [5], (a).

Its proponents point to the unpleasant side-effects of traffic congestion with conventional vehicles in that reduced vehicle speeds cause the engine to run more than necessary for each journey: the average operating point for the engine is further away from its design point and greater energy wastage is involved in braking. In urban usage, the average power consumption of the vehicle is quite low since average vehicle speeds are low. Typical power consumption figures are shown at (b) with only 5 kW being sufficient to drive the vehicle at a speed of 60 km/h, generally above the average vehicle speed for European cities. However, large, short duration power demands must be met to provide sufficient acceleration. Since the rate of energy consumption is low in the urban regime, this energy can be provided by means of a lead–acid battery and give an acceptable range of 50–70 km. Such a battery must, however, be capable of meeting the peak power demands for acceleration without overheating. High specific energy 'advanced' lead–acid batteries are hence required.

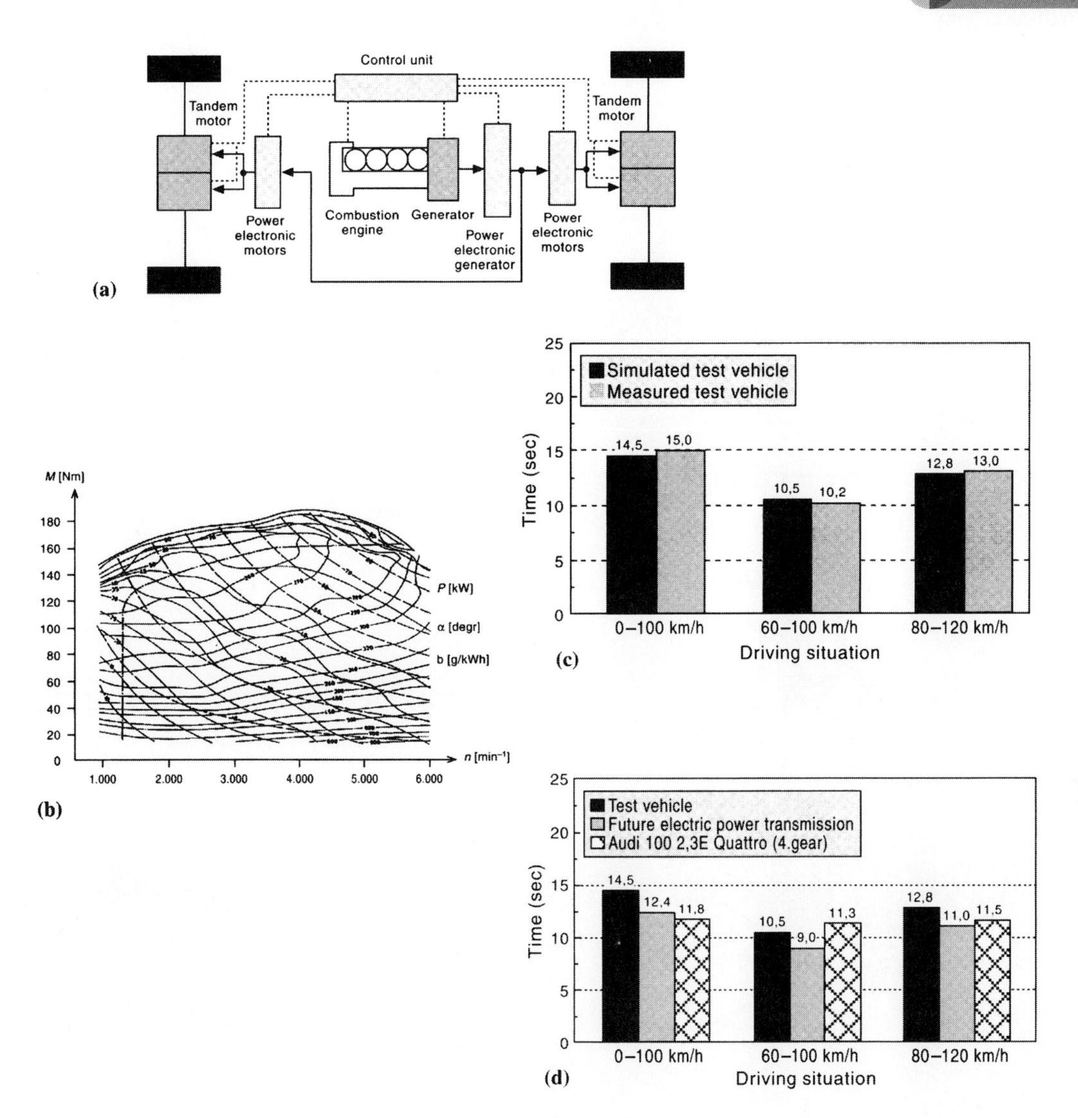

Fig. 7.1-9 Audi Quattro hybrid: (a) package; (b) speed/torque map; (c) simulated performance test; (d) estimated future performance.

While the vehicle will operate at zero emissions in the above mode of operation, if the journey length is greater than the range, the gas turbine engine must be operated to allow the batteries to be recharged. Again, 'advanced' lead–acid batteries are required to allow the full power of the gas turbine to be absorbed, hence preventing it from operating in a part load condition. Although the vehicle will not be ZEV in this mode, the composite emissions will still meet the ultra-low emission vehicle (ULEV) requirements due to the inherent low emission characteristics of the gas turbine engine. Combustion in the gas turbine is steady state as compared to the intermittent combustion in IC engines and is always lean burn. The potential for further reduction of small gas turbine emissions is also very good and has been demonstrated in much larger machines in the industrial sector. In highway usage, operation requires a much greater rate of energy consumption due to the increased average vehicle speeds. Typically 30–50 kW is required for passenger vehicles operating at motorway speeds.

It is currently not possible to store sufficient energy in a battery or any other accumulator such as a flywheel to give anywhere near an acceptable range at such a discharge rate. The gas turbine must hence operate continuously although not necessarily at constant power. Generally, the power output of the gas turbine will be matched to the vehicle power demand although some form of smoothing of the power demand will be needed to avoid rapid cycling of the gas turbine engine. Any excess power generated or power deficit will be handled by absorbing this or taking this from the battery. Since the power demand is high and continuous, the engine must be able to produce such power continuously. If an IC engine generator set was to be used instead of the gas turbine and high speed generator, the resulting machinery would be too voluminous and heavy for the series

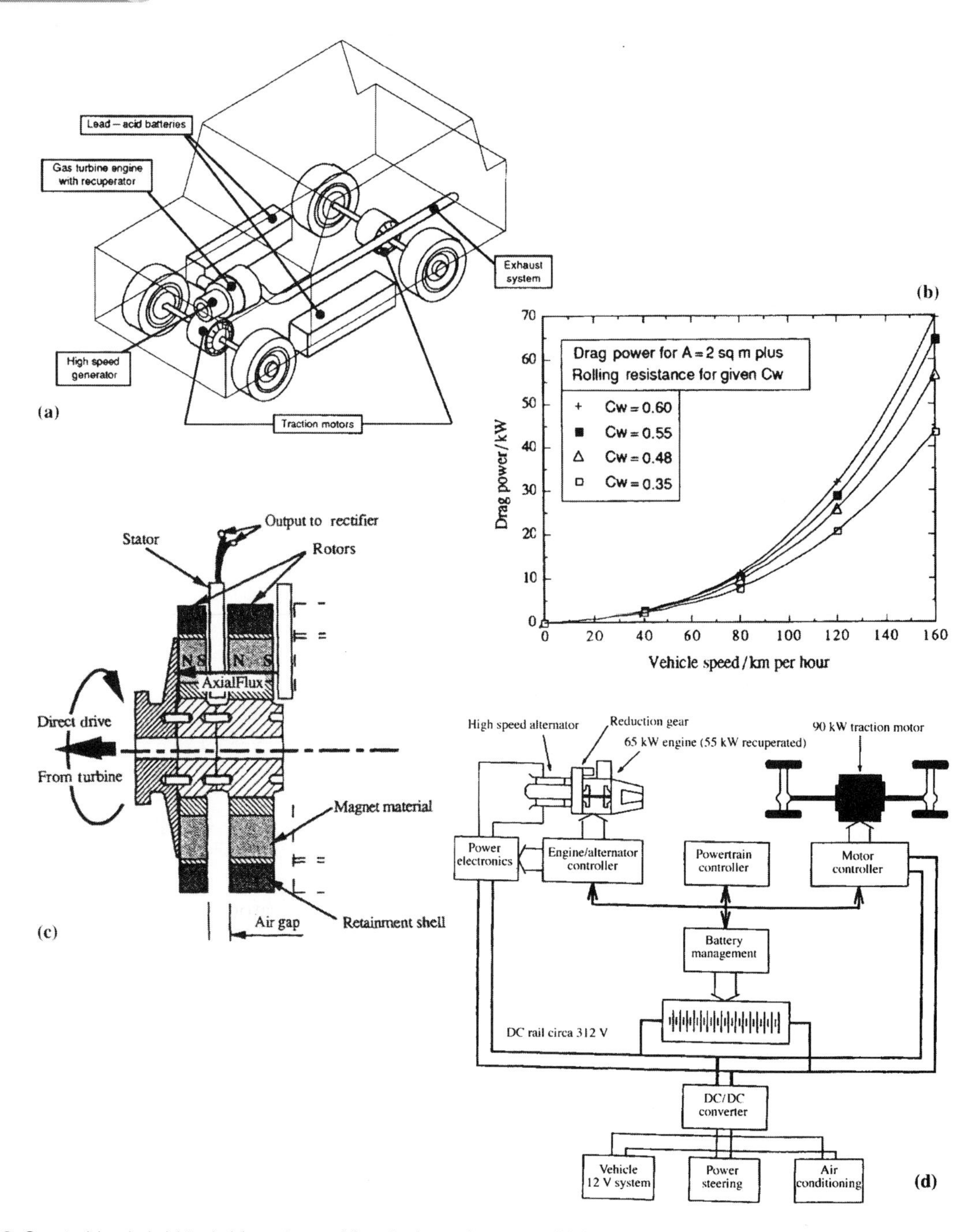

Fig. 7.1-10 Gas turbine hybrid taxi: (a) package; (b) typical steady-state vehicle power consumption; (c) high speed disc alternator; (d) system diagram of the TETLEI vehicle.

hybrid concept. This eliminates the feasibility of a small 'battery charger' IC engine which is often proposed for hybrid vehicles.

A key requirement of the researchers' proposal is to produce a gas turbine at a price corresponding to an IC engine while providing low fuel consumption over the medium to full power range. Tests are now under way at Imperial College on gas turbines of 30 and 50 kW; these are simple cycle machines but a recuperator is envisaged for the larger engine. The generator envisaged is a disc alternator which provides a large surface area for cooling from a relatively low volume machine, (c). The flux path is axial and is returned at the rotor ends by high strength steel keeper discs, the retainment ring being of carbon fibre. A key factor is that the rotor magnet retaining ring is not between the magnets and the stator; the flux within the stator is consequently higher and the diameter of the rotor can be greater. This results in a compact shaft length suitable for high speeds. An induction motor used with the system is considered to be the most reliable of

AC types, which are generally lighter than DC ones, and 94% overall efficiency could be expected, the researchers maintain. It is suggested that one motor per drive wheel be used with differential speed action being geared to the position of the steering wheel and an anti-spin system is envisaged.

The Rover Group has been involved in the TETLEI Euro-taxi project and K. Lillie, with Warwick University co-workers[6], has also gone into print on the gas turbine series hybrid concept. The taxi will be based on the latest Range Rover and the series hybrid mode allows the gas turbine to be decoupled from the wheels so as to operate at its optimum speed and load and avoid the classic limitation of this type of power unit, poor fuel economy at light loads, poor dynamic response and a high rotational speed required, over 60 000 rpm. A sophisticated control system is required to run the turbine in on–off mode according to power demand. One of the vehicle schematics under consideration is seen at (d) which will be computer modelled to assess its effectiveness. The researchers argue that the development of a valid simulation requires that a number of factors are fully considered. As all of the data stems from an initial calculation of the battery current, it is important that this value is accurate. Small variations in the internal resistance of the battery can cause large variations in the rail voltage (square of variation). It is important to have an accurate model of the battery which considers both the variation of internal resistance and open-circuit voltage of the battery at different states of charge. Modelling the drive cycle at the two extremes of battery operation (80% and 20% discharge) gives a good indication of the range of values over which the voltage and currents in the system may vary for specification purposes. The next step is to introduce a more realistic battery model and practical limits on the power sourced from regenerative braking. During electrical regenerative braking, there is a practical upper limit to the voltage permitted across a battery in order to avoid 'gassing'. This restriction may be overcome by the use of an additional form of power sink if the alternator is to continue operating at a fixed load point.

7.1.3.7 Dual hybrid system

Japanese researchers[7] from Equos Research have described the dual system of hybrid drive which differs from the more familiar series–parallel drives and their combinations. It allows free control of the IC engine while keeping mechanical connection between it and the drive wheels; a compact transaxle design integrates the two electric drive motors, to simplify the conversion of conventional vehicles, and use of the generator as a motor in combination permits flexible adaptation to driving conditions. Essentially, the 'split' drive system divides the output from the engine using a planetary gear, Fig. 7.1-11.

Instead of using a switching system, between series and parallel drive, (a), the split system acts as a series and parallel system at all times, the planetary gear dividing the drive between the series path of engine to generator and parallel path of engine to drive wheels. As parallel-path engine speed increases in proportion to vehicle speed, output energy from the engine also increases with vehicle speed, as is normally required. At high speeds most of the engine output is supplied by the parallel path and a smaller generator for the series system can therefore be employed.

The dual system, (b), is an optimized arrangement of the split system and thus far has been applied to a Toyota Corolla with an all-up weight of 1345 kg, involving a 660 cc engine adapted to drive-by-wire throttle control and giving 90–100 kph cruising speed. In the Toyota the dual system engine is mounted, for front wheel drive, onto a transaxle (c) of just 359 mm overall length, which is shorter than the production automatic transmission installation and 30 kg lighter than the engine/transaxle assembly of the standard model. The transaxle is of four shaft configuration, with compactness achieved by mounting motor and engine on separate shafts, each having optimized gear reduction, of 4.19 overall for the engine and 7.99 for the motor. The planetary splitter gear has a carrier connected to the engine and ring gear to the output shaft; it also acts as a speed increasing and torque reducing device for the motor/generator, with a 3.21 reduction ratio.

The motor/generator is a brushless 8 pole DC machine with 6 kW output, having generator brake and planetary gear installed within the coil ends for compactness. The generator functions as a starting device, clutch and form of CVT. The 40 kW traction motor is a 4 pole brushless motor which functions as a torque levelling device of the parallel hybrid system. Under low-load cruising conditions, the system uses parallel hybrid mode with the brake engaged, preventing the motor/generator from causing energy conversion losses. Brake cooling oil is also used for cooling the motor coils. Twenty-four lead–acid batteries are used, of 25 Ah capacity each, to give a total voltage of 288 V, the type being Cyclon-25C VRLA. Overall control strategy is seen at (d).

7.1.3.8 Flywheel addition to hybrid drive

According to Thoolen[8], the problem of providing peak power for acceleration, and recuperation of braking energy, in an efficient hybrid-drive vehicle can be overcome with an electromechanical accumulator. Such systems also admirably suit multiple stop–start vehicles such as city buses by using flywheel and electric power transmission.

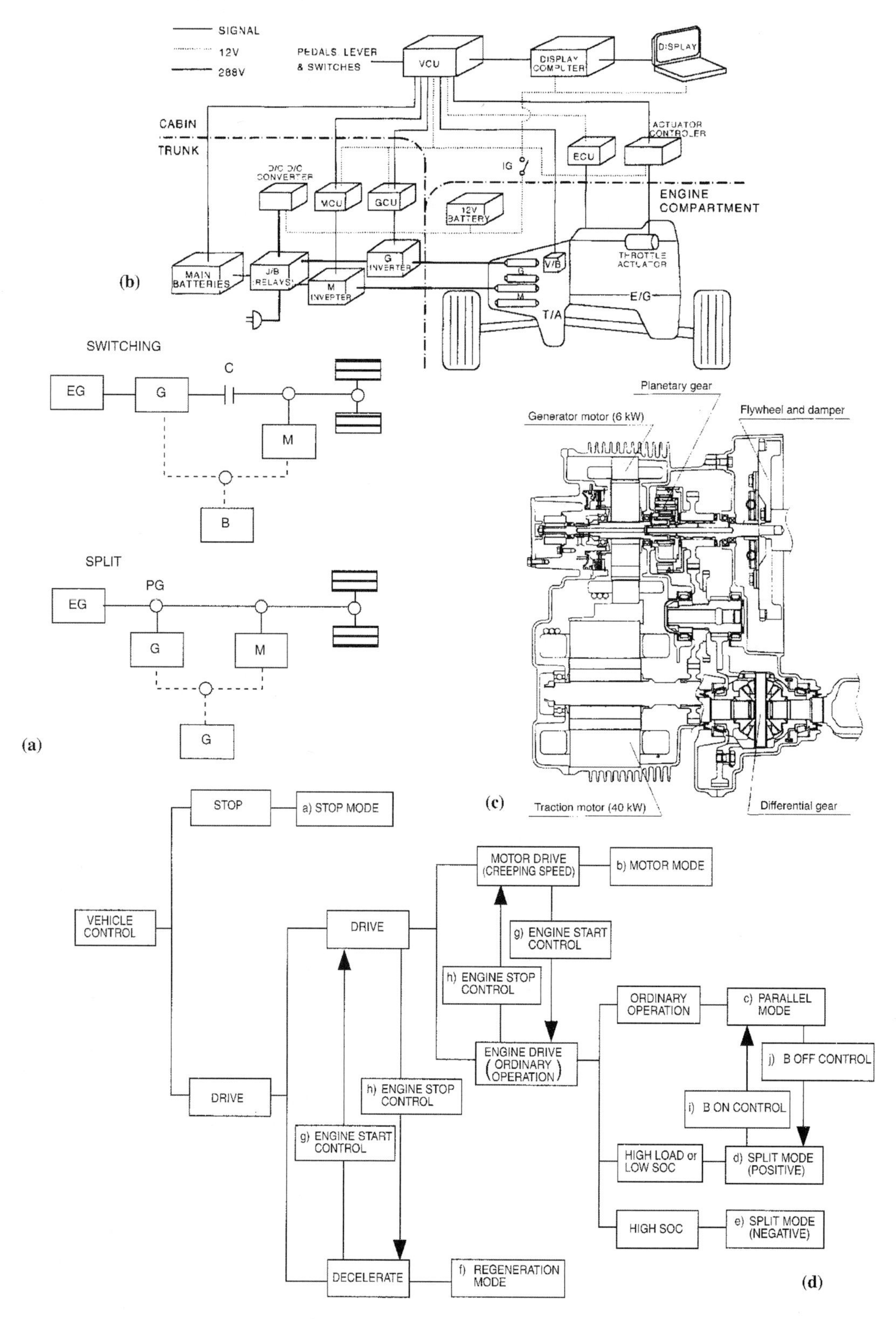

Fig. 7.1-11 Dual system: (a) comparison of switching and split series/parallel hybrid drives, full line is mechanical connection and dotted line electrical; (b) hybrid system; (c) transaxle configuration; (d) control strategy).

In the Emafer concept, Fig. 7.1-12, a flywheel motor/ generator unit is controlled by a power electronic converter. The flywheel (a) is of advanced composites construction and the motor/ generator of the synchronous permanent-magnet type while the converter uses high frequency power switches. The flywheel is comprised of four discs of tangentially wound prestressed fibre composite, designed to achieve a modularity of energy capacity as well as improved failure protection under the 100 000 *g* loading. The motor is of the exterior type with rotor outside the stator. High speed operation is possible as the rotor is merely a steel cylinder with permanent magnets on its inside. All windings are contained in the stator, having hollow journals at its ends for feeding electrical energy, cooling and lubrication fluids. Carefully designed supports for the high speed ball bearings allows the rotor to run 'overcritical' without serious vibration modes. Bearings, rotor and stator are vacuum-enclosed for reducing windage losses and for safety reasons. The containment is cardanically suspended to avoid gyroscopic effects.

The power converter, (b), controls exchange of electrical power between the 3 phase terminals of the motor/generator and the DC load. Of the current source inverter (CSI) type, it comprises a full bridge with six semiconductor switches and GTO thyristors. The latter are driven by measurements by the CSI of rotor position, DC voltage and current. When used as a sole driving source the Emafer is charged at bus stops by overhead supply contacts. In a hybrid-drive line, an IC engine on board, with generator, supplies the average power demand, with the Emafer taking care of fluctuations about the average, the flywheel extracting or applying power according to braking or accelerating mode.

7.1.4 Series-production hybrid-drive cars

During the early stages of introducing hybrid vehicles into the urban scene, state or local authorities may well offer direct and indirect financial inducements to get these 'clean vehicles' into areas that suffer from atmospheric pollution by motor transport. Now Toyota are manufacturing their Prius, Fig. 7.1-13, at the rate of 1000 a month and these cars are selling well in Japan. The Japanese Government, in a deliberate effort to curb urban pollution in Japan, is subsidizing the manufacture and sales drive by a variety of tax concessions, including one that directly benefits the user/operator of the hybrid saloon. The deal with the Japanese Government has enabled Toyota to offer these cars as a competitive package, when taking these taxation inducements into account. Toyota have found a technical solution, which, in engineering terms, is both ingenious and realistic. The company have made use of various new technologies to reduce the weight of the vehicle and its major components and systems. For example, the rolling resistance of the tyres has also been minimized, which reduces power demand by about 5–8%.

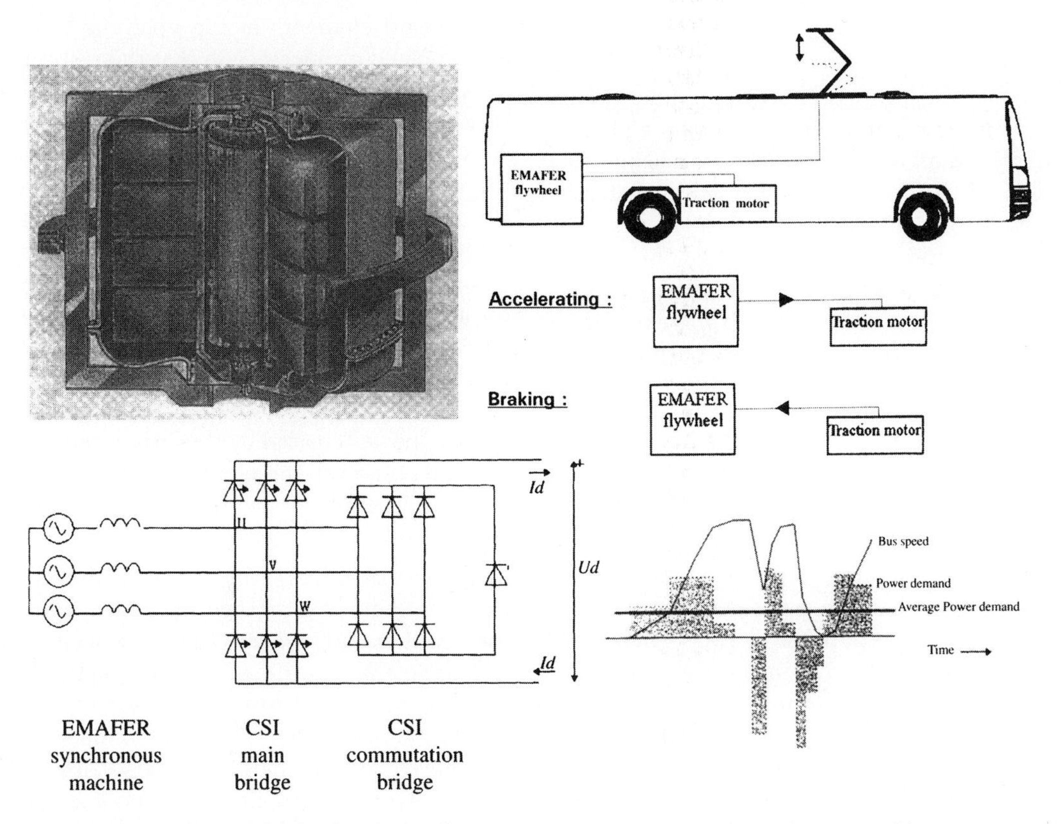

Fig. 7.1-12 Flywheel motor/generator: (a) Emafer flywheel/generator construction and electronic control; (b) power converter: shaded areas show power the Emafer has to supply/extract for accelerating or braking.

(a)

(b)

Fig. 7.1-13 Toyota Prius and its THS inverter.

7.1.4.1 Toyota prius systems

The Toyota Hybrid System (THS), (a), has two motive power sources, which are selectively engaged, depending on driving conditions: (1) A 1.5 litre petrol engine, developing 42.5 kW at 4000 rpm and a peak torque of 102 Nm at 4000 rpm; (2) A battery-powered permanent magnet synchronous electric motor with a maximum output of 30 kW over the speed range of 940–2000 rpm and peak torque of 305 Nm from standstill to 940 rpm. The petrol engine is the hybrid's main power source. It is a 1.5 litre DOHC 16 valve, 13.5:1 compression ratio, engine with Variable Valve Timing: Intelligent (VVT-I, a continuously variable valve mechanism) and electronic fuel injection, using the highly heat-efficient Miller cycle, that, in turn, is a further development of the high expansion Atkinson cycle. In this cycle the expansion continues for longer than in the conventional 4-stroke engine, thereby extracting more of the thermal energy of the burning gases than can be achieved in either a 2-stroke or 4-stroke engine of conventional design. Engine revolutions are restricted to 4000 rpm maximum, and the engine is electronically controlled to run always within a relatively narrow band of engine speed and load, corresponding to optimum fuel efficiency. Toyota claim that this sacrifice of a wide span of engine speed is more than made up for by the greater flexibility of the epicyclic drive system and the power split between the two motive-power units (Fig. 7.1-14).

THS functions as a continuously variable transmission and combines power from the petrol engine and the electric motor, to give smooth power delivery with little lag between the driver depressing the accelerator pedal and vehicle response. The innovative features of the Prius are in the design details of the power sources and the power split device in the hybrid transmission that allocates power from the petrol engine either directly to the vehicle's front wheels or to the electric generator. The power-split device, (b), employs a planetary gear system, which can steplessly effect the optimum power flow to suit the driving conditions encountered at any one moment. One of the output shafts of the power-split device is linked to the electric generator, while the other is linked to the electric motor and road wheels. The complex transmission system (c), which also includes a reduction gear, is electronically controlled, with the power flow allocation constantly being reviewed by the special control unit. This means that the information, which has been gathered by a number of key sensors, is compared with the target values encoded in the ECU, the system's brain. This ECU ensures that the appropriate elements in the epicyclic transmission are being braked or released, so that the respective speed of the petrol engine, the electric generator, and the electric motor are held within the optimum performance band. This power flow allocation split will depend on whether the car is being driven at a steady rate, accelerated or slowing down. The distribution of the petrol engine's power, which is so regulated that it will generally operate mainly in its optimum fuel efficiency band, the high torque zone, is determined by such factors as throttle opening, vehicle speed, and state of battery charge. The portion that is used to turn the wheels is balanced against that which is used to generate electric power. Electric power created by the largish generator may then be used to power the electric motor, to help drive the vehicle. There are a number of systems operating conditions.

In 'normal driving' the engine power is divided into two power-flow paths by the power-split device, one route will directly power the road wheels, and the other will drive the electric generator. Electric current from the generator may be used to power the electric motor, to assist in driving the road wheels. Electric current may also flow into the traction battery pack, to top up its charge. The power-split electronic control system determines the ratio of power flow to these outlets in such a manner that optimum fuel efficiency and responsive

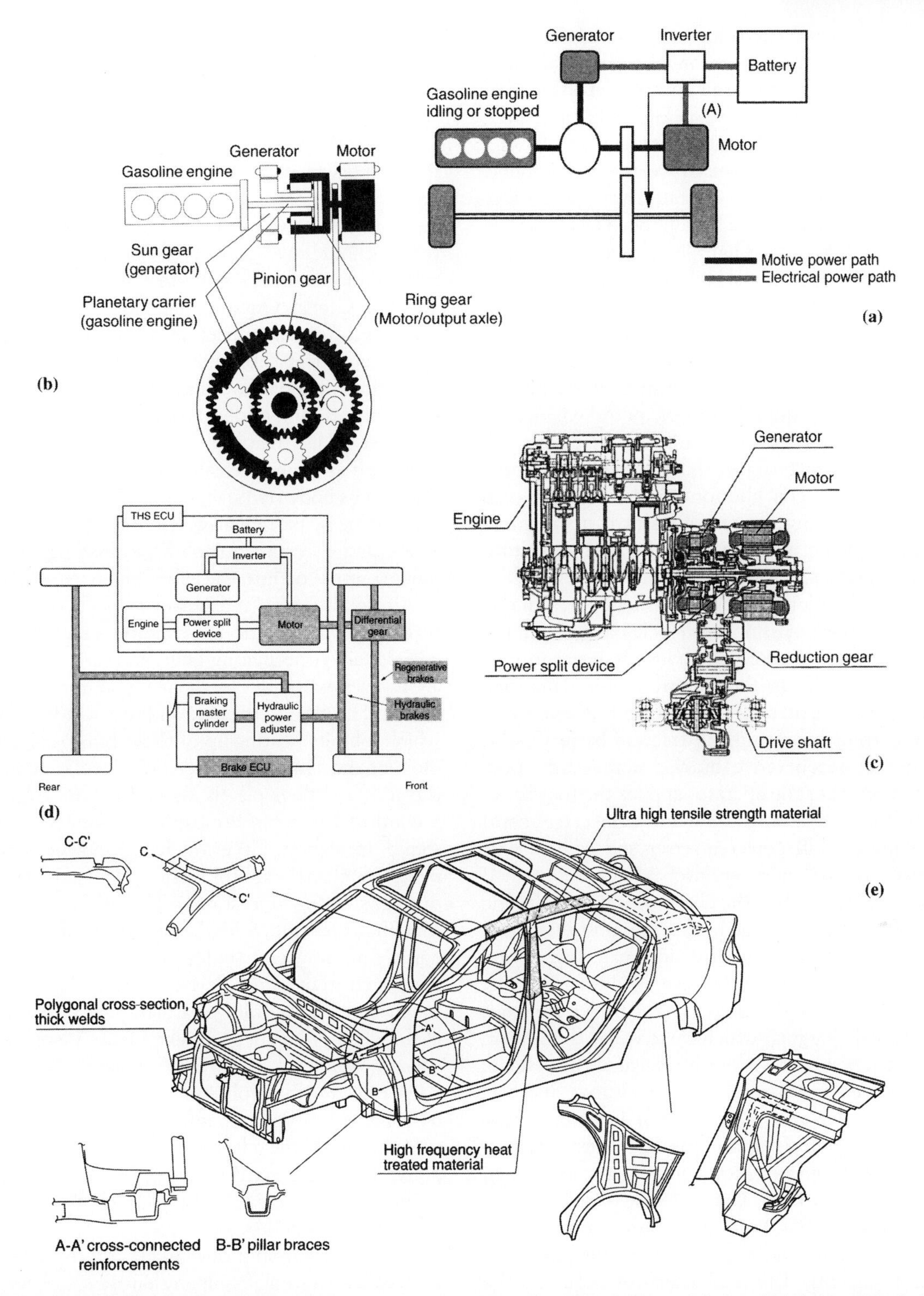

Fig. 7.1-14 Prius systems: (a) THS schematic; (b) power-split device; (c) engine and hybrid transmission; (d) hybrid system in full-throttle acceleration; (e) lightweight structure detail.

driveability are maintained at all times. The battery pack is made up of 40 individual nickel–metal hydride batteries and has a relatively small capacity of only 7.5 Ah, which would not give it much of a range in driving the vehicle in an all-electric mode.

During 'full throttle acceleration' drive mode, (d), power is also supplied from the battery to augment the drive power supplied by the petrol engine. Such a power boost, though adequate for overtaking and short bursts of speed, can generally not be maintained for extended periods of high speed motoring. The vehicle is being promoted as a car that produces only half the amount of CO_2 of conventionally powered compact-size cars and only one-tenth of the amount of HC, CO, and NO_x permitted under current Japanese emission regulations. Despite the vehicle having a kerb weight of 1.5 tonne, Toyota claim that Prius will accelerate from standstill to 400 m in 19.4 seconds and reach a top speed of 160 km/h.

In 'starting from rest and light load' mode (moving at low speed or descending a slight gradient) the electric motor drives the vehicle and the petrol engine is stopped. The high torque characteristic of the electric motor helps to get the car moving and will sustain it during low load demand slow speed progress in urban centres. Should additional power be required from the petrol engine, the computer control system will ensure that the engine will play its part, either by charging the traction battery pack or by some direct contribution to driving the road wheels. But when coming to rest at traffic lights, the fuel supply to the engine is cut off and the engine is automatically stopped.

During 'deceleration and braking' mode, the kinetic energy of the moving mass of the vehicle passes from the road wheels through the epicyclic transmission gearing of the power-split device to the electric motor. This then acts as an electric generator, delivering this energy as a charging current to top up the traction battery pack. This feature of regenerative braking comes into play, regardless of whether the operator applies the foot brake or relies on engine braking to slow down the car. A complex but compact full power inverter and control unit ensures that the traction battery pack is being maintained at a constant charge. When the charge is low, the electric generator routes power to the battery. In most instances, this energy will come from the internal combustion engine rather than the energy recovered during braking. The system has been so designed that the batteries do not require external charging, which means that there is no practical restriction to the operating range of the vehicle.

Toyota have compensated for the dual drive and battery weight penalty with a number of ingenious measures: since the petrol engine has been restricted to a maximum of 4000 rpm, key components have been pared down to save weight. Compared with an engine of comparable size but 5600 rpm maximum speed, the internal dynamic loadings on many of the moving parts are halved. Consequently there is scope for reducing the dimensions of, for instance, crankshaft journals and also pistons, which have remarkably short skirts; the overall effect of paring down of individual components is reduced weight of the built-up assembly. Overall length is only 4.28 m, but the car provides an interior space equal to that of many medium-class cars, by having a relatively long wheelbase of 2.55 m. a 1.7 m wide body, and short overhangs front and rear. The Prius boot has a reasonable capacity, thanks to the newly developed rear suspension which has no internal protrusions into the luggage compartment. The slanted, short bonnet covers the transversely mounted and very compact THS power train assembly. With a height of 1.49 m, the car stands taller than others in its class. The blending of the three box layout into a good aerodynamic shape has resulted in a drag factor of $C_d = 0.30$. Considerable weight saving, (e), without any sacrifice in passive safety, has been achieved in the body-in-white. The platform is based on Toyota's Global Outstanding Assessment (GOA) concept study of an impact-absorbing body and high integrity occupant cabin design, developed to meet 1999 US safety standards. Ribs made of energy-absorbing materials are embedded inside the pillars and roof side rails. GOA also features strong cross members, several produced in higher-tensile-strength sheet steel, linking the various body frame elements. These provide strength and stiffness, particularly in potential collision damage zones, and also spread the impact loading, thereby minimizing intrusion into the cabin, the occupant safety cell.

Air conditioning and power-assisted steering are featured. The automatic air conditioner creates a double layer of air, recirculating only internal air around the leg areas, even when the fresh air intake mode has been selected. The glass in the side and rear windows is of a type which inhibits heating up of the cabin space, by blocking most of the sun's ultraviolet rays. Insulating materials in the roof and floor panels also contribute to maintaining a comfortable cabin atmosphere. They also offer good sound insulation. Steering has a power-assist system using an electric motor, which consumes power only during steering operations. The front suspension has MacPherson struts with L-arms for locating their lower ends. In the semi-trailing-arm rear suspension, the combined coil spring and hydraulic damper units are much shorter. Their lower attachment is to a trailing arm each, which, in turn, is attached to an innovative type of torsion beam, of an inverted channel section. It incorporates toe-control links, to improve handling stability and the double-layer anti-vibration mounts joining the suspension to the chassis suppress much of the road noise. Passenger comfort is appropriate to a car which retails at around 27–30% above a comparable, but conventionally propelled model.

For the power-split device, Fig. 7.1-15, which is a key part of the system, company engineers[9] have provided the diagram at (a) to show how the engine, generator and motor operate under different conditions. At A level with the vehicle at rest, the engine, generator and motor are also at rest; on engine start-up the generator produces electricity acting as a starter to start the engine as well as operating the motor causing the vehicle to move off as at B. For normal driving the engine supplies enough power and there is no need for the generation of electricity, C.

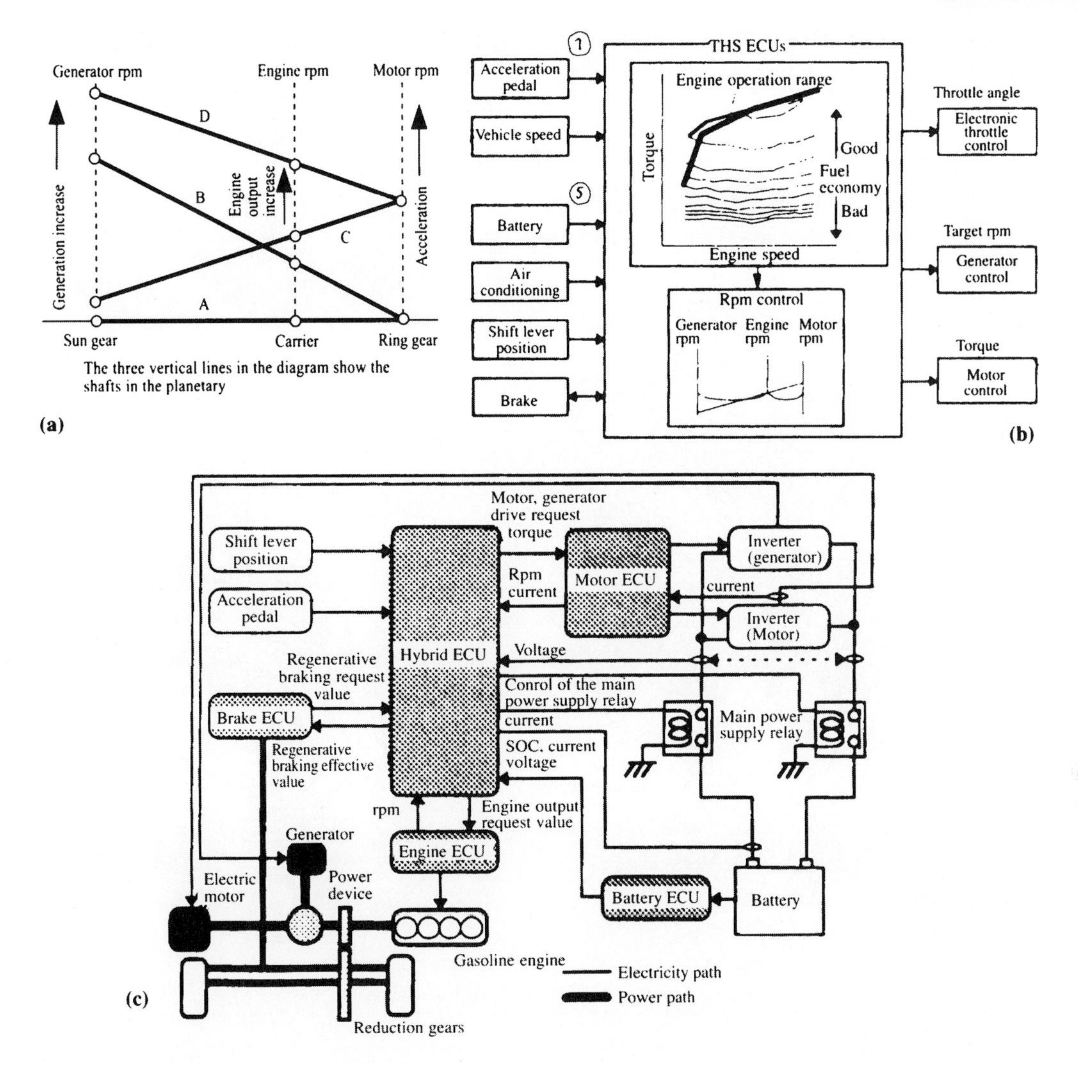

Fig. 7.1-15 Power-split control: (a) power interaction diagram; (b) THS control system; (c) ECU schematic.

As the vehicle accelerates from the cruise condition, generator output increases and the motor sends extra power to the drive shaft for assisting acceleration, D. The system can change engine speed by controlling generator speed; some of the engine output goes to the motor via the generator as extra acceleration power and there is no need for a conventional transmission. The control system schematic for the vehicle is at (b), the THS calculates desired and existing operating conditions and controls the vehicle systems accordingly, in real time.

The ECU keeps the engine operating in a predetermined high torque to maximize fuel economy. The corresponding schematic for the ECU is at (c). It is made up of five separate ECUs for the major vehicle systems. The hybrid ECU controls overall drive force by calculating engine output, motor torque and generator drive torque, based on accelerator and shift position. Request values sent out are received by other ECUs; the motor one controls the generator inverters to output a 3 phase DC current for desired torque; the engine ECU controls the electronic throttle in accordance with requested output; the braking ECU coordinates braking effort of motor regeneration and mechanical brakes; the battery ECU controls charge rate.

Toyota claims that Prius has achieved a remarkably low fuel consumption of 28 km/litre (79.5 mpg or 3.57 litre/100 km) on the 10/15 mode standard Japanese driving cycle.

7.1.4.2 Recent addition to production hybrid vehicles

Honda's Insight hybrid-drive car, Fig. 7.1-16, uses the company's Integrated Motor Assist (IMA) hybrid system, comprising high efficiency petrol engine, electric motor and lightweight 5-speed manual transmission, in combination with a lightweight and aerodynamic aluminium body, seen at (a), to provide acceleration of 0–62 mph in 12 seconds and a top speed of 112 mph, without compromising fuel economy of 83 mpg (3.41/100 km) and 80 g/km CO_2 (EUDC)

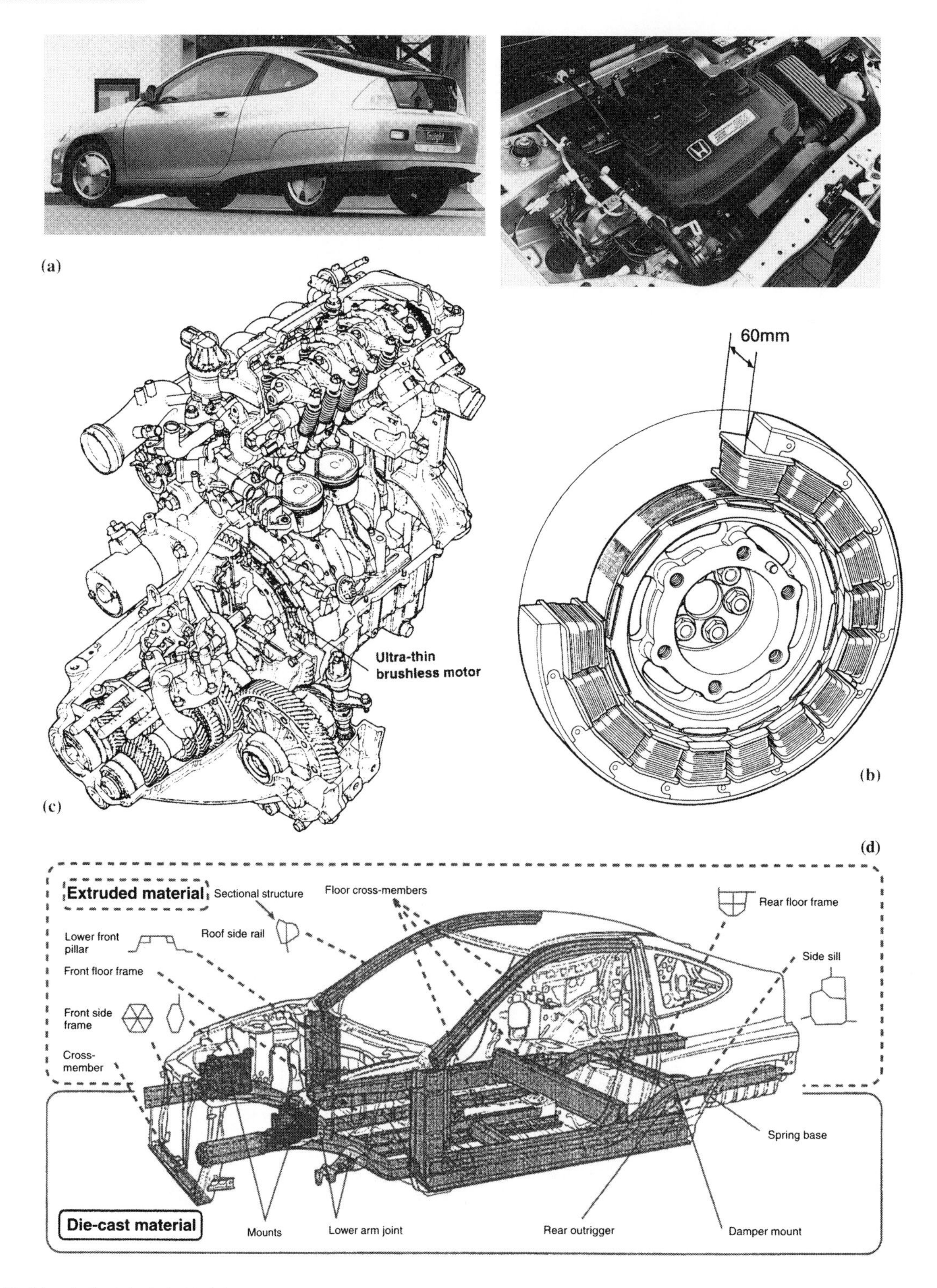

Fig. 7.1-16 Honda Insight hybrid: (a) aerodynamic tailed body and underbonnet power unit; (b) motor; (c) motor installation; (d) body structure.

emission. The car is claimed to have the world's lightest 1.0 litre, 3-cylinder petrol engine, which uses lean burn technology, low friction characteristics and lightweight materials in combination with a new lean burn compatible NO_x catalyst.

The electrical drive consists of an ultra-thin (60 mm) brushless motor, (b), directly connected to the crankshaft, (c), 144 V nickel–metal hydride (Ni–MH) batteries (weighing just 20 kg) and an electronic Power Control Unit (PCU). The electric motor draws power from the batteries during acceleration (so-called motor assist) to boost engine performance to the level of a 1.5 litre petrol engine as well as acting as a generator during deceleration to recharge the batteries. As a result engine output is increased from a high 50 to 56 kW with motor assist, but it is low speed torque that mainly benefits, boosting a non-assist 91 Nm at 4800 rpm to 113 Nm at 1500 rpm.

A new type of lightweight aluminium body, (d), offers a high level of rigidity and advanced safety performance. It is a combination of extruded, stamped and die cast aluminium components and body weight is said to be 40% less than a comparable steel body. All outer panels are aluminium except for the front wings and rear wheel skirt which are made from recyclable abs/nylon composite. Total kerb weight is 835 kg (850 kg including air conditioning). Aerodynamic characteristics include a streamlined nose, a low height and long tapered roof, narrow rear track, low drag grille, aluminium aero wheels, rear wheel skirt, a flat underside, and a tail designed to reduce the area of air separation. Insight also uses low-rolling-resistance tyres that have been designed to provide good handling, ride comfort and road noise characteristics. All these features give the Insight an aerodynamic drag coefficient of 0.25.

Further fuel savings are provided by an auto idle stop system. In simple terms, the engine cuts out as the car is brought to a standstill, and restarting is achieved by dipping the clutch and placing the car in gear. In combination, Honda calculates that weight reduction measures, aerodynamics and reduction of rolling resistance contribute to approximately 35% of the increase in fuel efficiency, and the IMA system a further 65% compared to a 1.5 litre Civic. Further features include ABS, electric power steering, dual air bags, AM/FM stereo cassette, power windows and mirrors, power door locks with keyless entry, automatic air conditioning and an anti-theft immobilizer.

The battery system is designed to avoid overcharging or complete emptying and in the unlikely event of motor failure, the Insight will run on the petrol engine alone. At the front, the suspension consists of struts, with an aluminium forged knuckle and lower arm, anti-roll bar linked to the dampers, and light aluminium cast wheels; while at the rear, a light and compact suspension features a twist beam with variable cross-section, and trailing arms with bushes having a toe-control function. Electric power steering, optimized for feel and feedback, has been used to make further fuel savings. It features a centre takeoff and aluminium forged tie-rod.

The company argue that in conventional petrol/electric hybrid systems, the vehicle is powered by the electric motor alone at low speeds. At higher vehicle speeds, or when recharging is required, engine torque is directed to the driven wheels or used to drive a generator. Such systems require complex control mechanisms, large capacity batteries, as well as a separate motor and generator. Honda chose instead a system in which the motor is linked directly to the engine, assisting it during acceleration for a reduction in consumption and acting as a generator during deceleration. When cruising, there is no assistance and lean burn keeps fuel consumption to a minimum. A very wide, flat torque curve is achieved through the benefits of VTEC technology at high engine speeds and the substantial boost provided by the electric motor at low and mid-range engine speeds. This approach allows for superior fuel efficiency and excellent driving performance over a wide range of driving situations.

The key to the engine operating at exceptionally low air/fuel ratios is rapid combustion of the mixture, since combustion time increases as the mixture becomes leaner. By adopting a new swirl port to enhance the turbulence of the mixture in the cylinder, a compact combustion chamber and a high compression ratio are achieved. The design is an evolution of the conventional VTEC-E mechanism where swirls are generated by almost closing one of the pair of inlet valves. In the new design, the inlet ports are set up in a more vertical direction to generate more powerful swirls flowing into the cylinder. This has been made possible with a new VTEC mechanism. Rather than inlet and exhaust rocker arms carried on separate rocker shafts, the Insight features just one rocker shaft with the included angle of the valves narrowed from 46° to 30°, allowing the high swirl port shape and the compact combustion chamber to be realized. Conventional lean burn engines, with their oxygen rich exhaust gases, mean reducing NO_x emissions is technologically difficult. The Insight's improvement in combustion efficiency goes some way towards solving the problem. However, a newly developed catalytic converter containing additives able to absorb NO_x, provides an elegant solution to the problem. During lean-burn driving, NO_x is directly absorbed; it is later reduced to harmless nitrogen in stoichiometric driving conditions. The system also helps to boost fuel efficiency, since it allows a widening of the lean burn range and therefore improved efficiency. Emissions performance is further improved by an exhaust manifold-integrated cylinder head. Rather than a conventional arrangement of an independent exhaust port for each cylinder, the ports are combined into one in the cylinder head structure. Considerable weight reduction is the result, but just as important, the small radiation area minimizes heat loss, enabling quick activation of the catalytic converter.

New technologies have reduced the overall friction of the engine by 38% compared with a conventional 1.5 litre engine. Among the measures adopted are roller type rocker arms, adapted to the single cam VTEC mechanism, providing a 70% reduction in friction losses.

A special 'micro dimple' surface treatment of the piston skirt improves the retention of the oil film between the piston and the cylinder reducing friction by approximately 30%, in conjunction with offset cylinders and low tensile piston rings. By using case hardening for significantly increased strength, slimmer connecting rods have been adopted, achieving a reduction in weight of 30%. A newly developed magnesium alloy, with a high degree of heat resistance, has been used for the engine sump in place of aluminium alloy, giving a 35% weight reduction. Other weight saving technology includes: a thin sleeve block, the new VTEC cylinder head, bracketless ancillary equipment, a magnesium PCU case, and an increase in plastic parts (intake manifold, cylinder head cover, water pump pulley).

The ultra-thin brushless motor of 10 kW output sandwiched between the engine and transmission has a central rotor manufactured using the lost wax method, to give a precise shape and high strength, which achieves a 20% weight reduction. For the rotor magnet, improvements to the neodymium sintered magnet used in the Honda EV Plus mean an improvement in the magnetic flux density or torque ratio by 8%, while improved heat resistance has made a cooling system unnecessary. In order to create a thin motor, a split stator with compact salient-pole field winding and centralized bus ring forms a very simple structure allowing a width of 60 mm, 40% thinner than if conventional technologies were used. The Ni–MH battery pack installed at the rear of the car is held in a compact cylindrical pack. A series connection of 120 cells each with 1.2 V provides a voltage of 144 V. Ni–MH batteries are said by the company to offer stable output characteristics regardless of the charging condition, as well as excellent durability. The PCU, mounted alongside the battery pack, provides precision control of the motor assist and battery regeneration functions, as well as the supply of electricity to the standard 12 V battery through a DC/DC converter. The inverter which drives the motor, and is the most important element in the PCU, consists of a compact 3 phase integrated type switching module.

The weight target was a body-in-white of 150 kg, or half that of the Civic 3 door, the closest comparable sized Honda model; compared with the Civic it was reduced to 47%, yet torsional rigidity is up by 38%, and bending rigidity up by 13%. Hexagonal extruded aluminium cross-sections are used for the front side frame, bringing a weight saving of 37% while also attaining high energy absorbing characteristics compared to a conventional steel frame. The side sill and roof side rail which contribute considerably to the overall body rigidity, although simpler in cross-section, achieve 47% and 53% weight reductions respectively. A new manufacturing technique, 'three-dimensional bending forming', provides a degree of freedom in design, and a reduction in the number of parts required has been adopted, for example to produce the roof side rails. Widely different sections and the need for high rigidity called for the special jointing method involving die cast aluminium, which permits a high degree of shaping and flexibility in joining different shaped sections. However, in the case of the rear outrigger, where structural frames meet from three directions and which serves as the installation area for the suspension frame, an alternative was required. Its deep box-like shape means that if it were formed with the conventional die cast method its wall thickness would become too thick and too heavy. So the thixo-cast method was used, said to be a first in body frame construction. This involves pouring aluminium in a half solidified rather than molten state to create a uniform and fine metal structure allowing a 22% thinner wall thickness, 20% higher strength, and a 20% weight reduction. In comparison with the NSX, the Insight uses 15% fewer body parts and 24% fewer welding spots to give weight and productivity savings.

7.1.5 Hybrid passenger and goods vehicles

7.5.1 Hybrid-drive buses

Passenger service vehicles have been the first to use hybrid drives on a commercial scale, usually employing a series layout. In a series hybrid configuration, part of the traction energy is converted into electrical energy, and then into mechanical energy, and part flows to the wheels directly via a mechanical transmission. It is argued that this configuration can potentially offer higher overall efficiency. In a series layout, all the IC engine energy is converted into electrical energy and then into mechanical energy. Such a configuration could offer advantages where the electric motor is designed as a very high efficiency unit regardless of the load upon it. Furthermore, the ability to use pure electrical transmission allows for flexibility, in system management, in optimizing engine operating conditions and reducing noise output. It also allows greater freedom in mechanical packaging, including electric motors direct driving the wheels, and for the incorporation of future fuel-cell technology.

This was the thinking behind the choice of a series system by Fiat in their pioneering hybrid bus, Fig. 7.1-17, a flow diagram for the system being shown at (a). The DC compound motor used, with separately excited field, had interpoles and compensating windings to expand the load range of maximum efficiency operation both as a motor and generator. Armature current was varied, by means of a thyristor chopper, from zero to base speed,

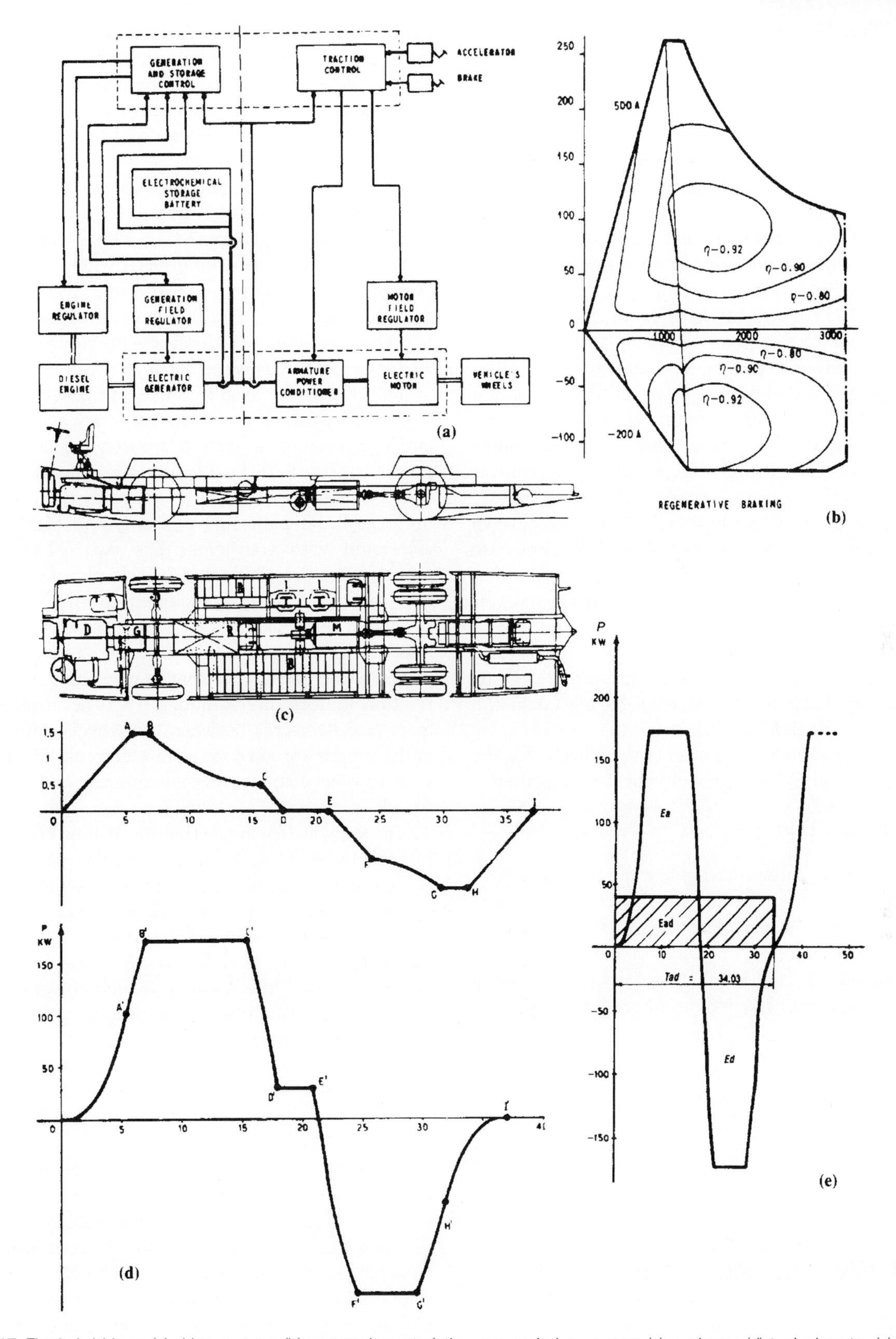

Fig. 7.1-17 Fiat hybrid bus: (a) drive system; (b) motor characteristics: power in kw vs rpm; (c) package; (d) typical route; (e) heaviest duty cycle.

above which field current was varied by an independent chopper. The reversible chopper allowed regenerative braking, down to zero speed. The motor had characteristics as seen at (b) and was direct coupled to the rear-axle driving head. Its nominal voltage was 600 V, continuous power 90 kW, maximum power 180 kW and field control ratio at continuous power 1150/3200 rpm. Efficiency at continuous power in field control was over 90%. The 600 V generating unit comprised a diesel generator set with power rating of 56 kW and maximum power 78 kW. Battery storage comprised 50 12 V lead–acid cells of 135 Ah capacity at a 5 hour discharge rate, their weight of 1930 kg corresponding to 12% of the GVW. Vehicle layout was as seen at (c); in tests the vehicle recorded diesel consumption of 32.3 kg/100 km compared with 37.8 for a conventional vehicle, with battery SOC found to be the same at the beginning and end of the tests. Range in purely electric drive was 30 km of city driving from 100% to 20% battery SOC.

During the design stages Fiat examined a typical town route between two termini as seen at (d), on a time base of seconds. It was also established that the maximum acceptable acceleration for standing passengers was 1.5 m/sec^2 on level ground and 0.27 at a gradient. The subtended area in the power diagram gave energy required between stops, the negative portion representing energy flowing back to the batteries having taken the various system efficiencies into account.

For E_s the total energy required at the wheels, E_m the engine energy and E_r the mains electrical energy, then

$$E_s = (\eta_g E_m + E_r)\eta_b \eta_t \eta_o$$

where efficiencies subscripted g, b, t and o refer to generator, battery, motors/controllers and transmission respectively, with their product the overall efficiency h. Then for total duration of daily service T_s, terminus turnaround time T_c and number of daily runs between termini N, the power required from the engine is:

$$P_m = (E_s - \eta E_r)/[\eta_g \eta \{T_s - (N-1)T_c\}]$$

The total daily energy is calculated for a typical route of length L_p, divided into N_t segments of length L_t. The heaviest cycle is shown at (e), in which $Ea = 584.6$ Wh; $Ed = 442$ Wh. $Ead = \eta_t' \eta_s \eta Ed = 337.6$ Wh and energy spent per run $Ep = Ead(Lp/Lad) = 12\ 250$ Wh.

7.1.5.2 CNG-electric hybrid

Smaller buses have been built with pure electric and alternative forms of hybrid drive. An interesting project by Unique Mobility in North America put a CNG-electric hybrid system into a 25 ft (7.62 m), 24 passenger vehicle (Fig. 7.1-18). Here the compactness and locational flexibility of the hybrid-drive elements meant that considerable gains could be obtained in packaging the vehicle occupants. Using high power-density permanent-magnet motors driving the rear wheels allowed a particularly low floor of 12 in (305 mm) from the ground.

The CNG-engine generator provided steady state power and was augmented by storage batteries to supply the power required above that base level while recharging of the batteries would take place when the power requirement fell below the base level. CNG tanks were roof mounted while batteries were positioned over the rear wheel wells and inside the engine compartment. The 11 tonne GVW bus is seen at (a).

The 90 bhp gas engine drove the generator through a flywheel-positioned step-up planetary gear set and an engine management system allowed engine speed and power to vary with load conditions. The rate at which speed was increased was minimized by the controller in order to avoid poor fuel economy and high emissions associated with transients. The two 70 kW traction motors were provided with a single planetary reduction gear of 2.77:1, directly coupled to the drive wheels through a secondary set of 5.2:1 included in the wheel hub to give an attainable speed of 55 mph. Rear suspension was an independent trailing arm system with traction motors direct mounted to the arms, so as to maximize floor area between the wheels. Motor differential speeds for cornering are electronically controlled with reference to steering wheel angle and road wheel speed.

The view at (b) shows the power flow charts for different modes of operation. In the first, on IC engine power only, a speed of 37 mph was achieved. On IC engine and battery power, higher speeds were possible and a reserve was available for gradients and acceleration; in the final mode of regenerative braking with the IC engine operating, the engine provided power only to the accessories and that from braking was fed into the storage batteries. The latter were used primarily for supplying accelerative power and were 12 V units with 160 Ah capacity. Two series strings of 15 batteries were connected in parallel to yield 180 V and 320 Ah total capacity.

Another pioneering series of hybrid buses has been the Daimler-Benz OE 305 city bus conversions (Fig. 7.1-19), some 20 of the first type were evaluated in trials in German cities in the early 1980s. Electric drive in the city centre and diesel drive in the suburbs was the mode of operation. Seen at (a), the set-up was electric motor (1) , air compressor and power-steering pump (2), motor fan (3), diesel engine and generator (4), battery-fan (5), electronic controller (6), traction batteries (7) and battery cooling unit (8). Range was 30–45 miles on batteries alone and 190 miles as a hybrid diesel combination.

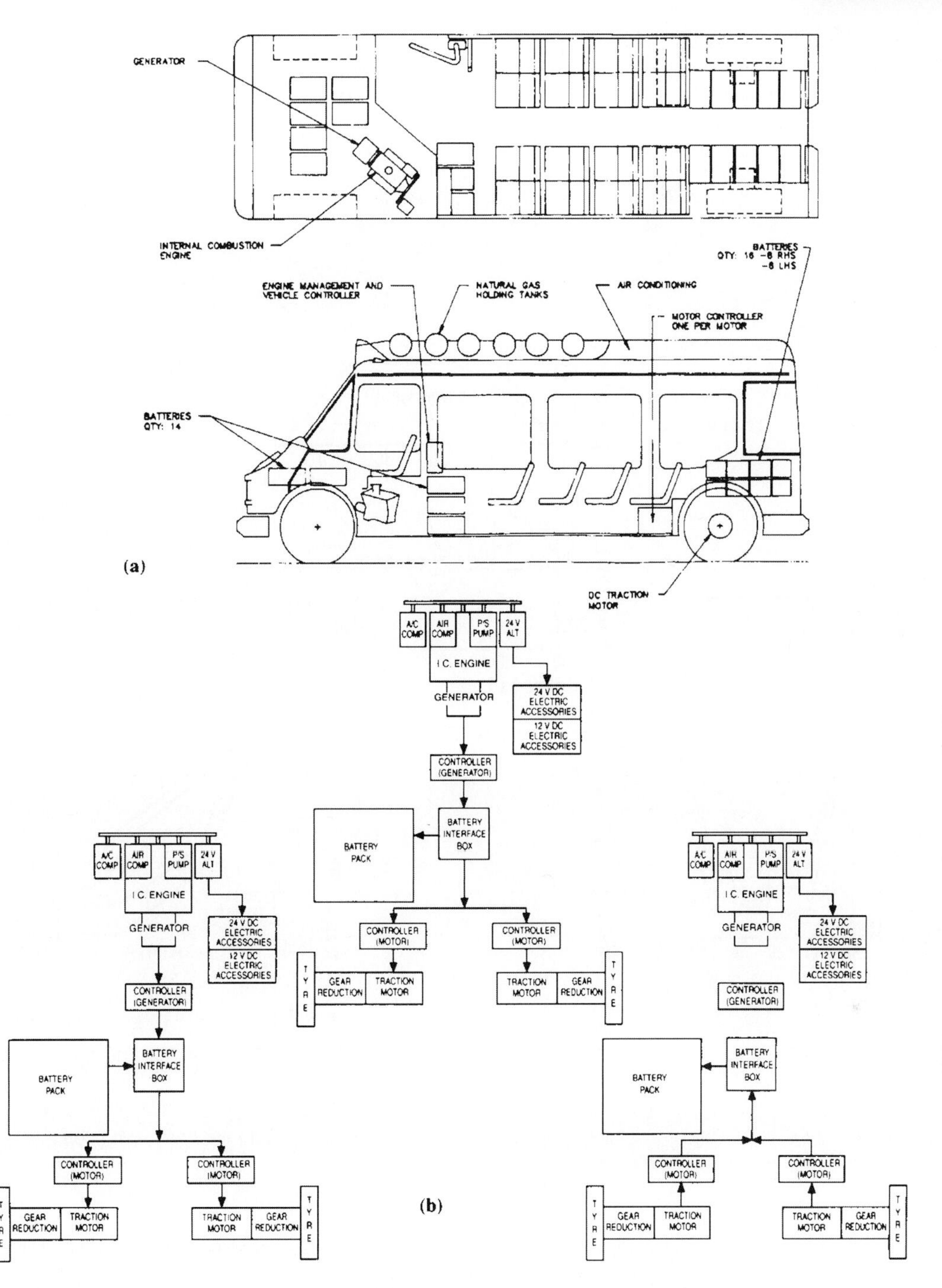

Fig. 7.1-18 Unique Mobility small hybrid bus: (a) Unique Mobility midibus; (b) power flow charts.

The 100 passenger vehicle had a maximum speed of 43 mph and the 19 tonne GVW vehicle had batteries weighing 3.5 tonnes. The motor could develop up to 200 bhp while the diesel engine was rated at 100 bhp. The five 275 Ah batteries operated at 360 V.

At the same period a Daimler-Benz 305 was also converted to flywheel hybrid operation in a study which involved MAN and Berlin University, too. The team calculated the flywheel storage energy requirements of a city bus to be 750 kW for absorbing the kinetic energy of the vehicle at top speed and a 220 lb flywheel was chosen, with a 1500 kW total energy to allow a reserve, which suffered 2 kW power loss at 12 000 rpm. For a typical urban operating cycle, characteristics were plotted, from the starting point of the bus stationary with diesel engine at idle and flywheel charged up from previous operation, for the configuration shown at (b).

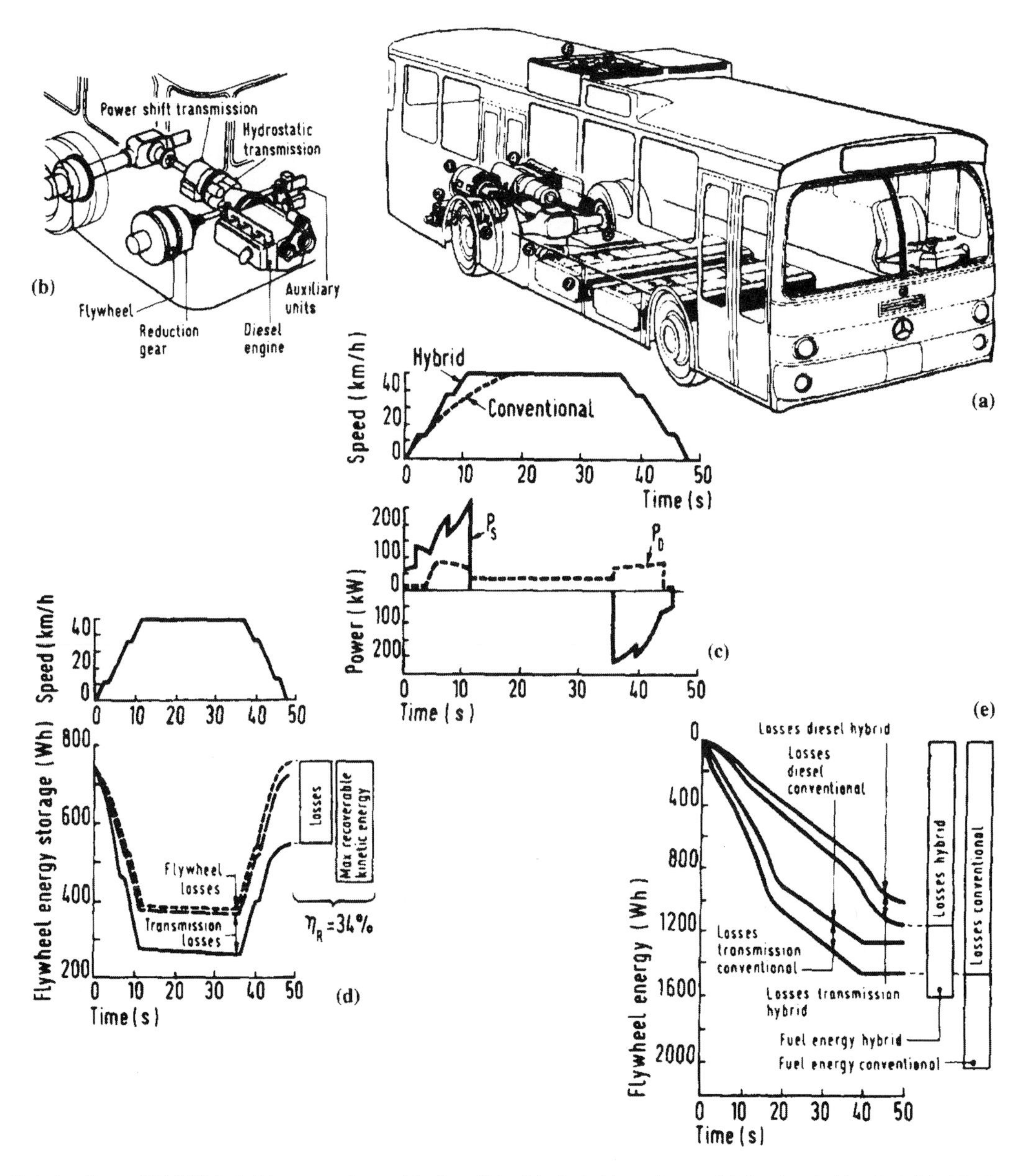

Fig. 7.1-19 Daimler-Benz OE 303 hybrid conversions: (a) diesel/electric hybrid package; (b) flywheel drive hybrid; (c) characteristics of flywheel hybrid; (d) flywheel losses; (e) diesel losses.

During acceleration the mechanical gear stage is automatically shifted as the vari-drive hydraulic transmission changes speed. This results in the almost constant slope (full line) in the curves shown at (c) compared with a conventional bus (dotted line). Initially the flywheel alone is used, then the diesel is brought in during deceleration, as seen in the bottom half of the figure. Flywheel power P_s is 260 kW immediately before the constant-power cruising phase, during which the diesel drives. During deceleration the flywheel is recharged and its power is 200 kW, corresponding diesel power being shown by P_d. Flywheel energy content and losses are seen at (d); the inertia losses are replaced by diesel energy. The diesel losses are seen at (e) together with the transmission losses between engine and drive wheels, compared with those for a conventional vehicle.

7.1.5.3 Advanced hybrid bus

A joint venture between MAN and Voith has resulted in the NL 202 DE low floor concept city bus, Fig. 7.1-20, designed to carry 98 passengers at a maximum speed of 70 kph, (a). No steps are involved at any of the entrances which lead directly to a completely level deck height of between 317 and 340 mm. The rear-mounted horizontally positioned diesel engine allows fitment of a bench seat at the rear of the bus; it drives a generator with only electrical connection to the Voith

(a)

	TFM wheel motor	TFM generator
Power	57 kW	135 kW
Rated speed	735 rpm	1750 rpm
Approx. max. speed	2500 rpm	200 rpm
Max. fundamental frequency of stator	1350 Hz	–
Rated torque	740 Nm	740 Nm
Approx. max. torque	1050 Nm	740 Nm
Approx. torque conversion	1:3	
Power/weight ratio	1.8 kg/kW	0.9 kg/kW

(d)

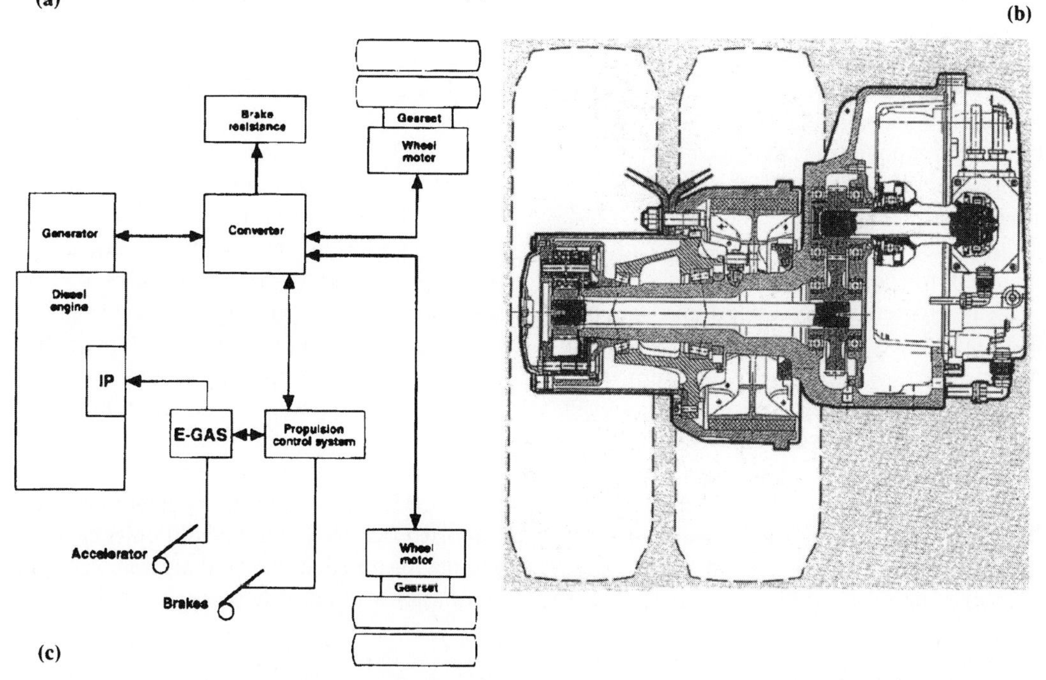

(b)

(c)

Fig. 7.1-20 MAN/Voith concept city bus: (a) low floor package; (b) wheel motor; (c) controller; (d) drive characteristics.

transverse-flux wheel motors which drive the wheels through two-stage hub-reduction gearsets, (b). The diesel is rated at 127 kW and the generator at 135 kW; the controller, (c), is of the IGBT converter type and also developed by Voith. It provides a differential action to the wheel motors on cornering. Permanent-magnet synchronous wheel motors are rated at 57 kW and have a maximum speed of 2500 rev/min; see table at (d). The bus is 12 metres long and has water cooling for its generator, converters and wheel motors. As well as providing virtually jerk-free acceleration, the drive system is seen by MAN as providing the possibility of four-wheel drive on future articulated buses to improve traction and stability in slippery road conditions. The term transverse flux motor refers to the means used to guide the magnetic flux in the stator; this is new to inverter-supplied PM types and involves a novel collector configuration. Double-sided magnetic force generation is also new and involves a patented double air gap construction having high idling inductances and force densities up to 120 kN m/m^2, with relatively low losses. A new control process permits operation of the motor in a field-weakening type mode, in spite of PM excitation. The generator is almost identical in concept but involves no field weakening. Each has concentric construction of permanent magnets, rotor/stator soft-iron elements and stator winding; see below. Armature elements are U-shaped cut strip-wound core sections, embedded in the ring-shaped supporting structures of inner and outer stators. Each core surrounds the windings and forms a stator pole with its cut surfaces facing the rotor. The latter is pot-shaped and positioned between poles of the outer and inner stators. In the stator pole region it comprises magnet and soft-iron element while in the winding region a ring of GRP serves as the connecting element. The inverters supply the motors with sinusoidal currents and voltages until the nominal operating point is reached; operating frequency is 10 kHz. In field-weakening mode the induced voltage exceeds intermediate circuit voltage and only square wave voltages are supplied to the motor. Power output then remains constant and the operating frequency equals the fundamental motor frequency. A large speed ratio, 1.5:1, is thus possible.

7.1.5.4 Advanced hybrid truck

Mitsubishi have been prominent in hybrid truck manufacture and have recently developed a heavier, municipal, version of the light hybrid truck launched in 1995. Because added cost limited market acceptance of the lighter, Canter-based, hybrid the decision to build a heavier municipal version, Fig. 7.1-21, was taken on the grounds of low noise, and greatly reduced emissions, which made the vehicle attractive for city-centre operation, a lift-platform version being particularly popular. The hydraulic pump for operating auxiliaries such as a lift platform is electric motor driven, with the benefit of near silent operation.

Series hybrid mode, (a) was chosen first, because the engine is used solely for power generation and so can be operated in a peak efficiency speed band and secondly, since the engine is isolated from the drive system, it results in a simpler and more flexible drive-system layout with greater freedom for hydraulic equipment mounting. Two electric motors are involved. Shown at (b) are typical operational modes of the truck: when the battery has high SOC the vehicle operates exclusively in battery mode. At less than 65% SOC the power-generating engine starts and hybrid mode is invoked; when 70% SOC is achieved again the vehicle reverts to battery operation. Provision is also made to inhibit hybrid operation until 30% SOC is reached so silent and zero-emission night-time, or in-tunnel, operation is made possible. In hybrid mode SOC is maintained at 65–70%, at which point the generated power, from the generator, and the regenerative power, from the motor, provide sufficient charging.

The overall package layout is shown at (c); dimensions are 5.78 m long × 1.88 m wide × 3.35 m high, with a wheelbase of 2.5 m. Gross vehicle weight is 6.965 tonnes and tyre size 205/85R16. While an elevating platform vehicle normally requires counterweighting, in this case the mass of the dual drive suffices with just modest additions. The central positioning of the battery above the chassis frame was chosen to optimize weight distribution and avoid the weight of cantilever frames. The motors, of 55 kW, are of the induction type and each develop 150 Nm at 3500 rpm, rated voltage being 288 V. The simple two speed transmission has a PTO for driving the hydraulic pump and reverse motion is achieved by altering the rotation of the electric motor. The generator has a maximum output of 30 kW at 3500 rpm; it operates at 220–360 V and weighs 70 kg.

The petrol engine is a 16 valve unit of 1834 cc which has a 1.935:drive gear to the generator. Lead–acid traction batteries are employed, 24 units each weighing 25 kg and having 65 Ah capacity at a 5 hour rate. The company's estimations of unit efficiency are shown at (d).

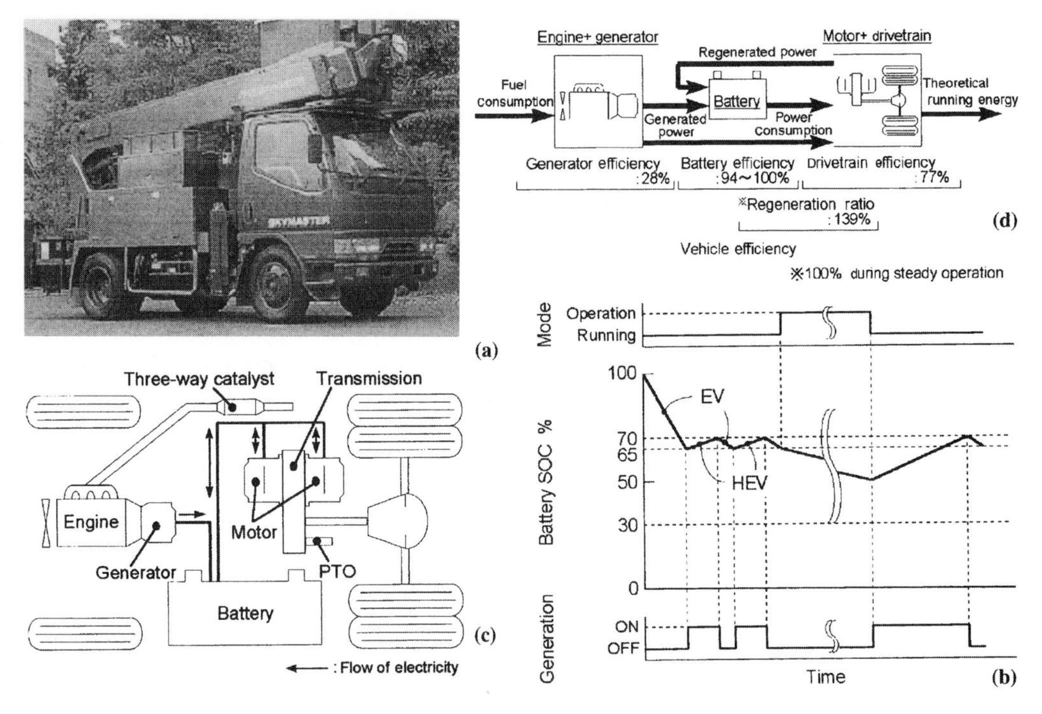

Fig. 7.1-21 Mitsubishi Canter-based hybrid municipal truck: (a) Complete package; (b) hybrid drive; (c) operating modes; (d) unit efficiencies.

Section Eight

Suspensions

Chapter 8.1

8.1

Types of suspension and drive

Jornsen Reimpell, Helmut Stoll and Jurgen Betzler

This chapter deals with the principles relating to drives and suspensions.

8.1.1 General characteristics of wheel suspensions

The suspension of modern vehicles need to satisfy a number of requirements whose aims partly conflict because of different operating conditions (loaded/unloaded, acceleration/braking, level/uneven road, straight running/cornering).

The forces and moments that operate in the wheel contact area must be directed into the body. The kingpin offset and disturbing force lever arm in the case of the longitudinal forces, the castor offset in the case of the lateral forces, and the radial load moment arm in the case of the vertical forces are important elements whose effects interact as a result of, for example, the angle of the steering axis.

Sufficient vertical spring travel, possibly combined with the horizontal movement of the wheel away from an uneven area of the road (kinematic wheel) is required for reasons of *ride comfort*. The recession suspension should also be compliant for the purpose of reducing the rolling stiffness of the tyres and short-stroke movements in a longitudinal direction resulting from the road surface (longitudinal compliance, Fig. 8.1-1), but without affecting the development of lateral wheel forces and hence steering precision, for which the most rigid wheel suspension is required. This requirement is undermined as a result of the necessary flexibility that results from disturbing wheel movements generated by longitudinal forces arising from driving and braking operations.

For the purpose of ensuring the optimum *handling characteristics* of the vehicle in a steady state as well as in a transient state, the wheels must be in a defined position with respect to the road surface for the purpose of generating the necessary lateral forces. The build-up and size of the lateral wheel forces are determined by specific toe-in and camber changes of the wheels depending on the jounce and movement of the body as a result of the axle kinematics (roll steer) and operative forces (compliance steer). This makes it possible for specific operating conditions such as load and traction to be taken into consideration. By establishing the relevant geometry and kinematics of the axle, it is also possible to prevent the undesirable diving or lifting of the body during braking or accelerating and to ensure that the vehicle does not exhibit any tendency to oversteer and displays predictable transition behaviour for the driver.

Other requirements are:

- independent movement of each of the wheels on an axle (not guaranteed in the case of rigid axles);
- small, unsprung masses of the suspension in order to keep wheel load fluctuation as low as possible (important for driving safety);
- the introduction of wheel forces into the body in a manner favourable to the flow of forces;
- the necessary room and expenditure for construction purposes, bearing in mind the necessary tolerances with regard to geometry and stability;
- ease of use;
- behaviour with regard to the passive safety of passengers and other road users;
- costs.

Automotive Chassis: Engineering Principles; ISBN: 9780750650540
Copyright © 2001 Elsevier Ltd. All rights of reproduction, in any form, reserved.

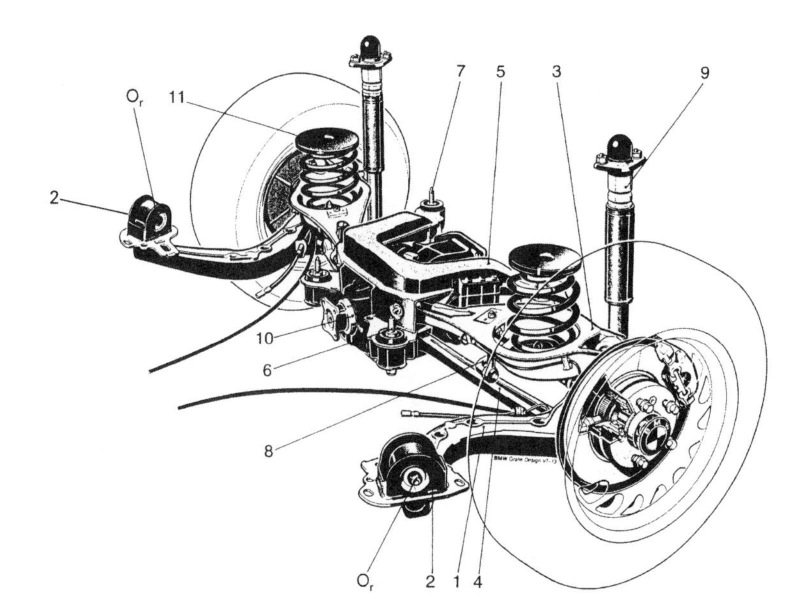

Fig. 8.1-1 A multi-link rear axle – a type of suspension system which is progressively replacing the semi-trailing arm axle, and consists of at least one trailing arm on each side. This arm is guided by two (or even three) transverse control arms (Figs. 8.1-62 and 8.1-77). The trailing arm simultaneously serves as a wheel hub carrier and (on four-wheel steering) allows the minor angle movements required to steer the rear wheels. The main advantages are, however, its good kinematic and elastokinematic characteristics. BMW calls the design shown in the illustration and fitted in the 3-series (1997) a 'central arm axle'. The trailing arms 1 are made from GGG40 cast iron; they absorb all longitudinal forces and braking moments as well as transfering them via the points 2 – the centres of which also form the radius arm axes – on the body. The lateral forces generated at the centre of tyre contact are absorbed at the subframe 5, which is fastened to the body with four rubber bushes (items 6 and 7) via the transverse control arms 3 and 4. The upper arms 3 carry the minibloc springs 11 and the joints of the anti-roll bar 8. Consequently, this is the place where the majority of the vertical forces are transferred between the axle and the body.
The shock absorbers, which carry the additional polyurethane springs 9 at the top, are fastened in a good position behind the axle centre at the ends of the trailing arms. For reasons of noise, the differential 10 is attached elastically to the subframe 5 at three points (with two rubber bearings at the front and one hydro bearing at the back). When viewed from the top and the back, the transverse control arms are positioned at an angle so that, together with the differing rubber hardness of the bearings at points 2, they achieve the desired elastokinematic characteristics. These are:

- toe-in under braking forces;
- lateral force compliance understeer during cornering;
- prevention of torque steer effects (see Section 10.1.10.4);
- lane change and straight running stability.

For reasons of space, the front eyes 2 are pressed into parts 1 and bolted to the attachment bracket. Elongated holes are also provided in this part so toe-in can be set. In the case of the E46 model series (from 1998 onwards), the upper transverse arm is made of aluminium for reasons of weight (reduction of unsprung masses).

The requirements with regard to the steerability of an axle and the possible transmission of driving torque essentially determine the design of the axis.

Vehicle suspensions can be divided into rigid axles (with a rigid connection of the wheels to an axle), independent wheel suspensions in which the wheels are suspended independently of each other, and semi-rigid axles, a form of axle that combines the characteristics of rigid axles and independent wheel suspensions.

On all rigid axles (Fig. 8.1-23), the axle beam casing also moves over the entire spring travel. Consequently, the space that has to be provided above this reduces the boot at the rear and makes it more difficult to house the spare wheel. At the front, the axle casing would be located under the engine, and to achieve sufficient jounce travel the engine would have to be raised or moved further back. For this reason, rigid front axles are found only on commercial vehicles and four-wheel-drive, general-purpose passenger cars (Figs. 8.1-3 and 8.1-4).

With regard to independent wheel suspensions, it should be noted that the design possibilities with regard to the satisfaction of the above requirements and the need to find a design which is suitable for the load paths, increase with the number of wheel control elements (links) with a corresponding increase in their planes of articulation. In particular, independent wheel suspensions include:

- Longitudinal link and semi-trailing arm axles (Figs. 8.1-13 and 8.1-15), which require hardly any overhead room and consequently permit a wide luggage space with a level floor, but which can have considerable diagonal springing.

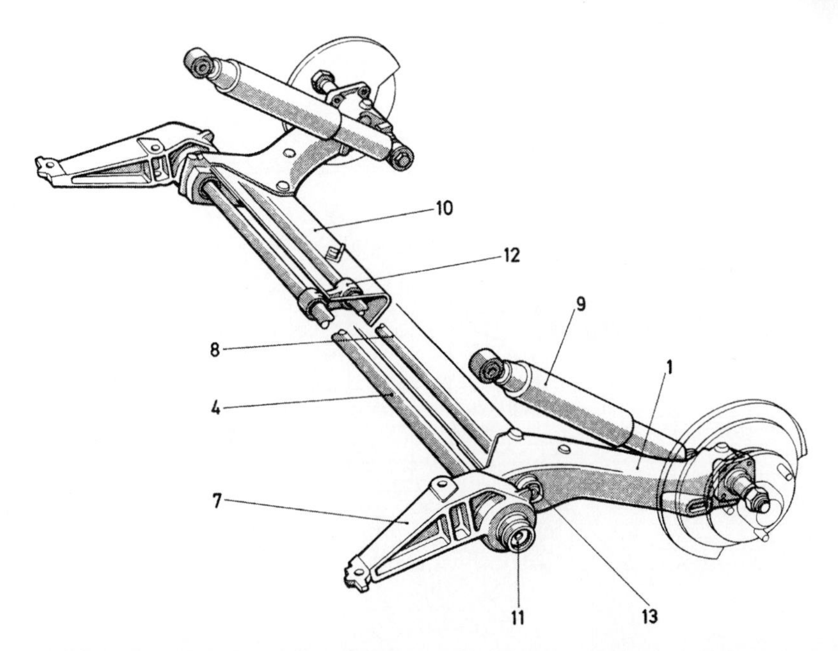

Fig. 8.1-2 An extremely compact four-bar twist beam axle by Renault, with two torsion bar springs both for the left and right axle sides (items 4 and 8). The V-shape profile of the cross-member 10 has arms of different lengths, is resistant to bending but less torsionally stiff and absorbs all moments generated by vertical, lateral and braking forces. It also partially replaces the anti-roll bar. At 23.4 mm, the rear bars 8 are thicker than the front ones (Ø 20.8 mm, item 4). On the outside, part 8 grips into the trailing links 1 with the serrated profile 13 and on the inside they grip into the connector 12. When the wheels reach full bump, a pure torque is generated in part 12, which transmits it to the front bars 4, subjecting them to torsion. On the outside (as shown in Fig. 8.1-63) the bars with the serrated profile 11 grip into the mounting brackets 7 to which the rotating trailing links are attached. The pivots also represent a favourably positioned pitch centre O_r. The mounting brackets (and therefore the whole axle) are fixed to the floor pan with only four screws.
On parallel springing, all four bars work, whereas on reciprocal springing, the connector 12 remains inactive and only the thick rear bars 8 and the cross-member 10 are subject to torsion.
The layout of the bars means soft body springing and high roll stability can be achieved, leading to a reduction of the body roll pitch during cornering.
To create a wide boot without side encroachments, the pressurized monotube shock absorbers 9 are inclined to the front and therefore are able to transmit forces upwards to the side members of the floor pan.

- Wheel controlling suspension and shock-absorber struts (Figs. 8.1-8 and 8.1-57), which certainly occupy much space in terms of height, but which require little space at the side and in the middle of the vehicle (can be used for the engine or axle drive) and determine the steering angle (then also called McPherson suspension struts).
- Double wishbone suspensions (Fig. 8.1-7).
- Multi-link suspensions (Figs. 8.1-1, 8.1-18 and 8.1-19), which can have up to five guide links per wheel and which offer the greatest design scope with regard to the geometric definition of the kingpin offset, pneumatic trail, kinematic behaviour with regard to toe-in, camber and track changes, braking/starting torque behaviour and elastokinematic properties.

In the case of twist-beam axles (Figs. 8.1-2, 8.1-31 and 8.1-58), both sides of the wheels are connected by means of a flexurally rigid, but torsionally flexible beam. On the whole, these axles save a great deal of space and are cheap, but offer limited potential for the achievement of kinematic and elastokinematic balance because of the functional duality of the function in the components and require the existence of adequate clearance in the region of the connecting beam. They are mainly used as a form of rear-wheel suspension in front-wheel-drive vehicles up to the middle class and, occasionally, the upper middle class, for example, the Audi A6, and some high-capacity cars.

8.1.2 Independent wheel suspensions – general

8.1.2.1 Requirements

The chassis of a passenger car must be able to handle the engine power installed. Ever-improving acceleration,

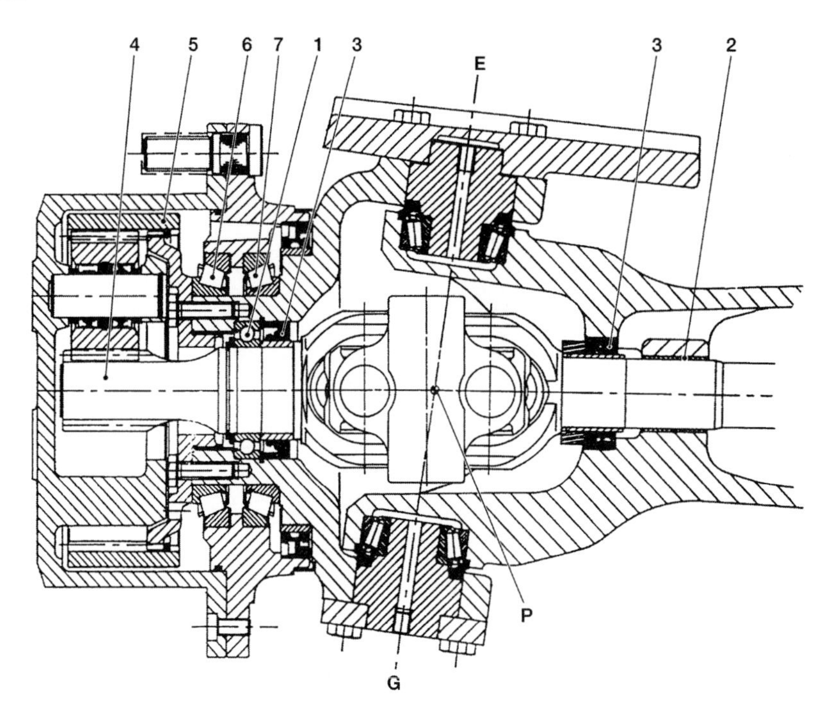

Fig. 8.1-3 Driven, rigid steering axle with dual joint made by the company GKN – Birfield AG for four-wheel-drive special-purpose vehicles, tractors and construction machinery.
The dual joint is centred over the bearings 1 and 2 in the region of the fork carriers; these are protected against fouling by the radial sealing rings 3. Bearing 1 serves as a fixed bearing and bearing 2 as a movable bearing. The drive shaft 4 is also a sun gear for the planetary gear with the internal-geared wheel 5. Vertical, lateral and longitudinal forces are transmitted by both tapered-roller bearings 6 and 7. Steering takes place about the steering axis EG.

higher peak and cornering speeds, and deceleration lead to significantly increased requirements for safer chassis. Independent wheel suspensions follow this trend. Their main advantages are:

- little space requirement;
- a kinematic and/or elastokinematic toe-in change, tending towards understeering is possible;
- easier steerability with existing drive;
- low weight;
- no mutual wheel influence.

The last two characteristics are important for good road-holding, especially on bends with an uneven road surface.

Transverse arms and trailing arms ensure the desired kinematic behaviour of the rebounding and jouncing wheels and also transfer the wheel loadings to the body (Fig. 8.1-5). Lateral forces also generate a moment which, with unfavourable link arrangement, has the disadvantage of reinforcing the roll of the body during cornering. The suspension control arms require bushes that yield under load and can also influence the springing. This effect is either reinforced by twisting the rubber parts in the bearing elements, or the friction increases due to the parts rubbing together (Fig. 8.1-11), and the driving comfort decreases.

The wheels incline with the body (Fig. 8.1-6). The wheel on the outside of the bend, which has to absorb most of the lateral force, goes into a positive camber and the inner wheel into a negative camber, which reduces the lateral grip of the tyres. To avoid this, the kinematic change of camber needs to be adjusted to take account of this behaviour and the body roll in the bend should be kept as small as possible. This can be achieved with harder springs, additional anti-roll bars or a body roll centre located high up in the vehicle.

8.1.2.2 Double wishbone suspensions

The last two characteristics above are most easily achieved using a double wishbone suspension (Fig. 8.1-7). This consists of two transverse links (control arms) either side of the vehicle, which are mounted to rotate on the frame, suspension subframe or body and, in the case of the front axle, are connected on the outside to the steering knuckle or swivel heads via ball joints. The greater the effective distance *c* between the transverse links (Fig. 8.1-5), the smaller the forces in the

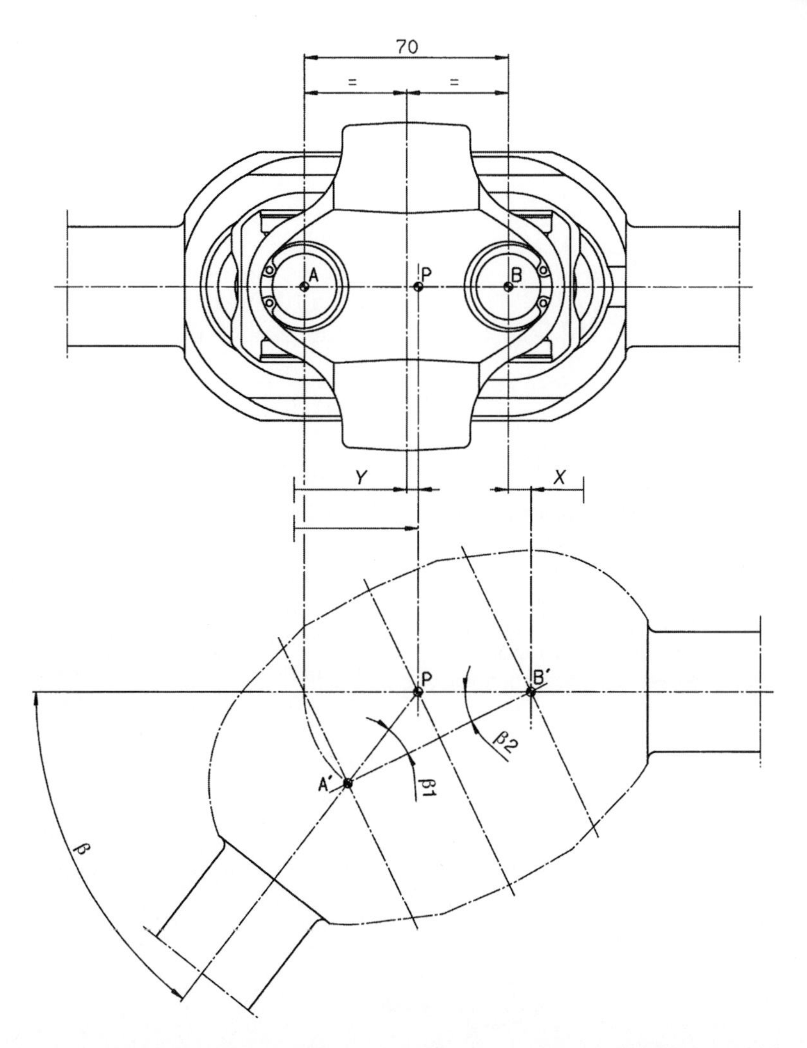

Fig. 8.1-4 Top view of the dual joint (Fig. 8.1-3). The wheel end of the axle is turned about point P in the middle of the steering pivot during steering. The individual joints are constrained at points A and B so that point A is displaced to position A′, P is displaced to P′ and B is displaced along the drive axle by the distance *X* to B′. In order to assimilate the variable bending angle β resulting from the longitudinal displacement of point B, the mid-point of the joint P is displaced by the distance *Y*. The adjustment value *Y* depends on the distance between the joints and the steering angle at which constant velocity is to exist. Where large steering angles can be reached (up to 60°), there should be constant velocity at the maximum steering angle.
The adjustment value *Y* and the longitudinal displacement *X* should be taken into consideration in the design of the axle.

suspension control arms and their mountings become, i.e. component deformation is smaller and wheel control more precise.

The main advantages of the double wishbone suspension are its kinematic possibilities. The positions of the suspension control arms relative to one another–in other words the size of the angles α and β – can determine both the height of the body roll centre and the pitch pole. Moreover, the different wishbone lengths can influence the angle movements of the compressing and rebounding wheels, i.e. the change of camber and, irrespective of this, to a certain extent also the track width change. With shorter upper suspension control arms the compressing wheels go into negative camber and the rebounding wheels into positive. This

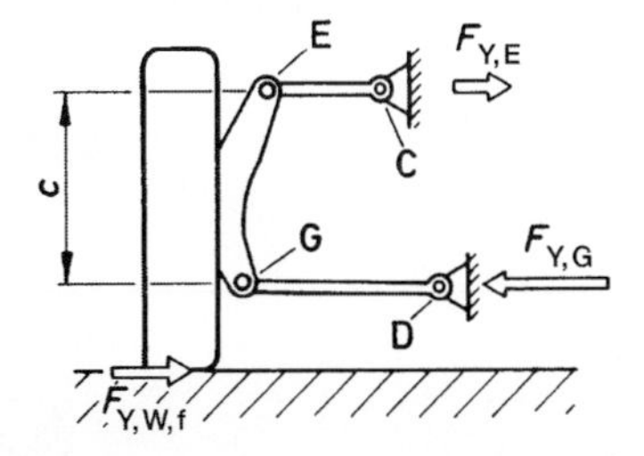

Fig. 8.1-5 On front independent wheel suspensions, the lateral cornering force $F_{Y,W,f}$ causes the reaction forces $F_{Y,E}$ and $F_{Y,G}$ in the links joining the axle with the body. Moments are generated on both the outside and the inside of the bend and these adversely affect the roll pitch of the body. The effective distance *c* between points E and G on a double wishbone suspension should be as large as possible to achieve small forces in the body and link bearings and to limit the deformation of the rubber elements fitted.

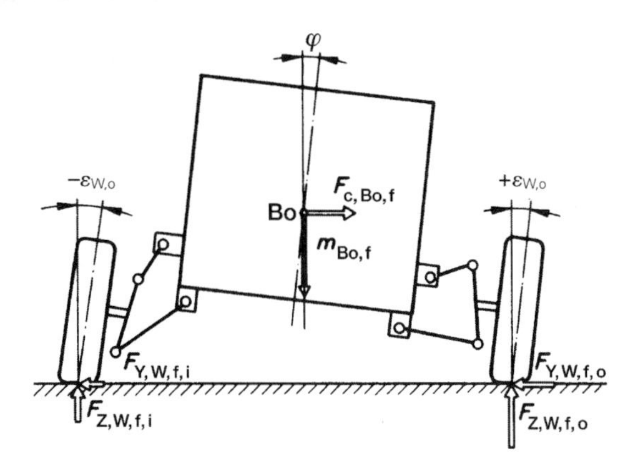

Fig. 8.1-6 If the body inclines by the angle φ during cornering, the outer independently suspended wheel takes on a positive camber $\varepsilon_{W,o}$ and the inner wheel takes on a negative camber $\varepsilon_{W,i}$. The ability of the tyres to transfer the lateral forces $F_{Y,W,f,o}$ or $F_{Y,W,f,i}$ decreases causing a greater required slip angle Equation 10.1.16, $m_{Bo,f}$ is the proportion of the weight of the body over the front axle and $F_{c,Bo,f}$ the centrifugal force acting at the level of the centre of gravity Bo. One wheel rebounds and the other bumps, i.e. this vehicle has 'reciprocal springing', that is:

$$F_{Z,W,f,o} = F_{Z,W,f} + \Delta F_{Z,W,f}$$
$$F_{Z,W,f,i} = F_{Z,W,f} - \Delta F_{Z,W,f}$$

counteracts the change of camber caused by the roll pitch of the body (Fig. 8.1-6). The vehicle pitch pole O is located behind the wheels on the front axle and in front of the wheels on the rear axle. If O_r can be located over the wheel centre, it produces not only a better anti-dive mechanism, but also reduces the squat on the driven rear axles (or lift on the front axles). These are also the reasons why the double wishbone suspension is used as the rear axle on more and more passenger cars, irrespective of the type of drive, and why it is progressively replacing the semi-trailing link axle (Figs. 8.1-1, 8.1-62 and 8.1-77).

8.1.2.3 McPherson struts and strut dampers

The McPherson strut is a further development of double wishbone suspension. The upper transverse link is replaced by a pivot point on the wheel house panel, which takes the end of the piston rod and the coil spring. Forces from all directions are concentrated at this point and these cause bending stress in the piston rod. To avoid detrimental elastic camber and caster changes, the normal rod diameter of 11 mm (in the shock absorber) must be increased to at least 18 mm. With a piston

Fig. 8.1-7 Front axle on the VW light commercial vehicle Lt 28 to 35 with an opposed steering square. A cross-member serves as a subframe and is screwed to the frame from below. Springs, bump/rebound-travel stops, shock absorbers and both pairs of control arms are supported at this force centre. Only the anti-roll bar, steering gear, idler arm and the tie-rods of the lower control arms are fastened to the longitudinal members of the frame. The rods have longitudinally elastic rubber bushings at the front that absorb the dynamic rolling hardness of the radial tyres and reduce lift on uneven road surfaces.

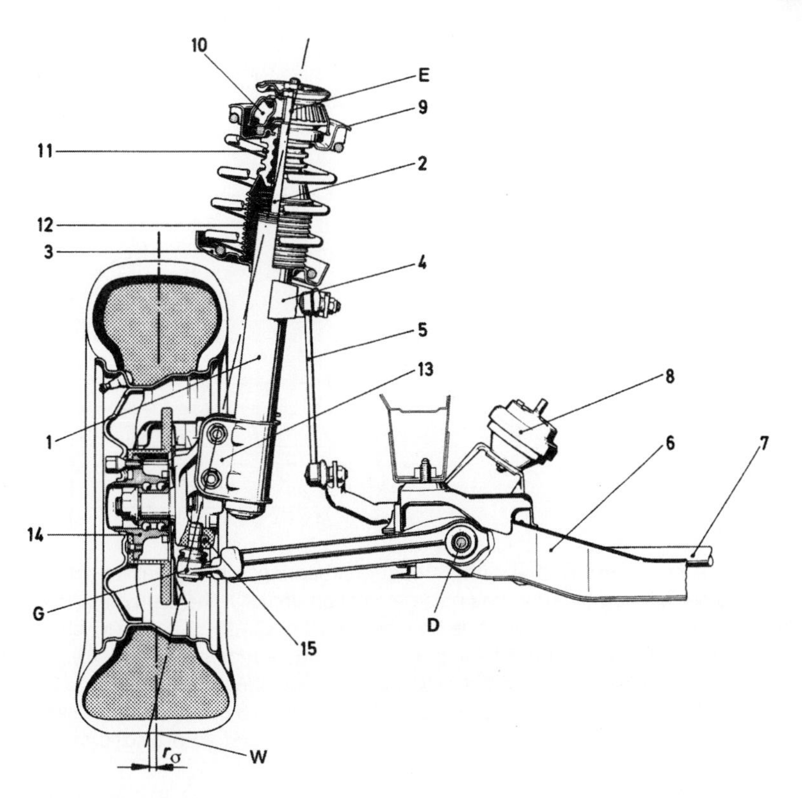

Fig. 8.1-8 Rear view of the left-hand side of the McPherson front axle on the Opel Omega (1999) with negative kingpin offset at ground (scrub radius) r_σ and pendulum-linked anti-roll bar. The coil spring is offset from the McPherson strut to decrease friction between piston rod 2 and the rod guide. Part 2 and the upper spring seat 9 are fixed to the inner wheel house panel via the decoupled strut mount 10.
The additional elastomer spring 11 is joined to seat 9 from the inside, and on the underside it carries the dust boot 12, which contacts the spring seat 3 and protects the chrome-plated piston rod 2. When the wheel bottoms out, the elastomer spring rests on the cap of the supporting tube 1. Brackets 4 and 13 are welded to part 1, on which the upper ball joint of the anti-roll bar rod 5 is fastened from inside. Bracket 13 takes the steering knuckle in between the U-shaped side arms.
The upper hole of bracket 13 has been designed as an elongated hole so that the camber can be set precisely at the factory.
A second-generation double-row angular (contact) ball bearing (item 14) controls the wheel.
The ball pivot of the guiding joint G is joined to the steering knuckle by means of clamping forces. The transverse screw 15 grips into a ring groove of the joint bolt and prevents it from slipping out in the event of the screw loosening.
The subframe 6 is fixed to the body. In addition to the transverse control arms, it also takes the engine mounts 8 and the back of the anti-roll bar 7. The drop centre rim is asymmetrical to allow negative wheel offset (not shown) at ground (scrub radius) (Figs. 10.1-10, 10.1-11 and 10.1-23).

diameter of usually 30 mm or 32 mm the damper works on the twin-tube system and can be non-pressurized or pressurized.

The main advantage of the McPherson strut is that all the parts providing the suspension and wheel control can be combined into one assembly. As can be seen in Fig. 8.1-8, this includes:

- the spring seat 3 to take the underside of the coil spring;
- the auxiliary spring 11 or a bump stop;
- the rebound-travel stop;
- the underslung anti-roll bar 7 via rod 5;
- the steering knuckle.

The steering knuckle can be welded, brazed or bolted firmly to the outer tube (Fig. 8.1-56). Further advantages are:

- lower forces in the body-side mounting points E and D due to a large effective distance c (Fig. 8.1-5);
- short distance b between points G and N;
- long spring travel;
- three bearing positions no longer needed;
- better design options on the front crumple zone;
- space at the side permitting a wide engine compartment; which
- makes it easy to fit transverse engines (Fig. 8.1-50).

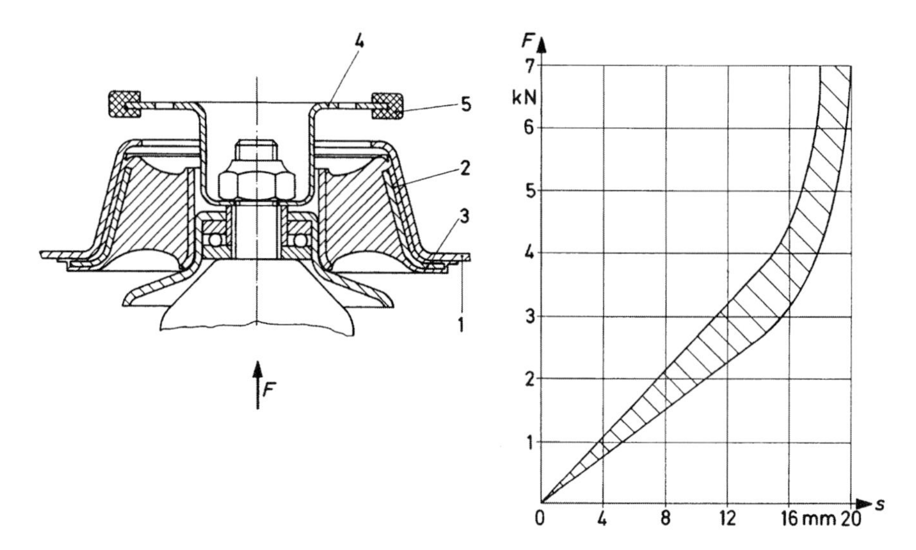

Fig. 8.1-9 McPherson strut mount on the VW Golf III with a thrust ball bearing, which permits the rotary movement of the McPherson strut whereas the rubber anchorage improves noise insulation. Initially, the deflection curve remains linear and then becomes highly progressive in the main work area, which is between 3 kN and 4 kN. The graph shows the scatter. Springing and damping forces are absorbed together so the support bearing is not decoupled (as in Fig. 8.1-10).
In the car final assembly line the complete strut mount is pressed into a conical sheet metal insert on the wheel house inside panel 1. The rubber layer 2 on the outside of the bearing ensures a firm seat and the edge 3 gives the necessary hold in the vertical direction. The rubber ring 5 clamped on plate 4 operates when the wheel rebounds fully and so provides the necessary security (figure: Lemförder Fahrwerktechnik AG).

Nowadays, design measures have ensured that the advantages are not outweighed by the inevitable disadvantages on the front axle. These disadvantages are:

- Less favourable kinematic characteristics.
- Introduction of forces and vibrations into the inner wheel house panel and therefore into a relatively elastic area of the front end of the vehicle.
- It is more difficult to insulate against road noise – an upper strut mount is necessary (Fig. 8.1-9), which should be as decoupled as possible (Fig. 8.1-10, item 10 in Fig. 8.1-8 and item 6 in Fig. 8.1-56).
- The friction between piston rod and guide impairs the springing effect; it can be reduced by shortening distance b (Fig. 8.1-11).
- In the case of high-mounted rack and pinion steering, long tie rods and, consequently, more expensive steering systems are required (Figs. 8.1-57 and 9.1-1); in addition, there is the unfavourable introduction of tie-rod forces in the middle of the shock-absorbing strut (see Section 9.1.2.4) plus additional steering elasticity.
- Greater sensitivity of the front axle to tyre imbalance and radial runout (see Section 10.1.5).
- Greater clearance height requirement.
- Sometimes the space between the tyres and the damping element (Fig. 8.1-41) is very limited.

This final constraint, however, is only important on front-wheel-drive vehicles as it may cause problems with fitting snow chains. On non-driven wheels, at most the lack of space prevents wider tyres being fitted. If such tyres are absolutely necessary, disc-type wheels with a smaller wheel offset e are needed and these lead to a detrimentally larger positive or smaller negative kingpin offset at ground r_o (Fig. 10.1-8).

McPherson struts have become widely used as front axles, but they are also fitted as the rear suspension on front-wheel-drive vehicles (e.g. Ford Mondeo sedan). The vehicle tail, which has been raised for aerodynamic reasons, allows a larger bearing span between the piston rod guide and piston. On the rear axle (Fig. 8.1-12):

- The upper strut mount is no longer necessary, as no steering movements occur.
- Longer cross-members, which reach almost to the vehicle centre, can be used, producing better camber and track width change and a body roll centre that sinks less under load.
- The outer points of the braces can be drawn a long way into the wheel to achieve a shorter distance b.
- The boot can be dropped and, in the case of damper struts, also widened.
- However, rubber stiffness and the corresponding distance of the braces on the hub carriers (points 6 and 14 in Fig. 8.1-12) are needed to ensure that there is no unintentional elastic self-steer.

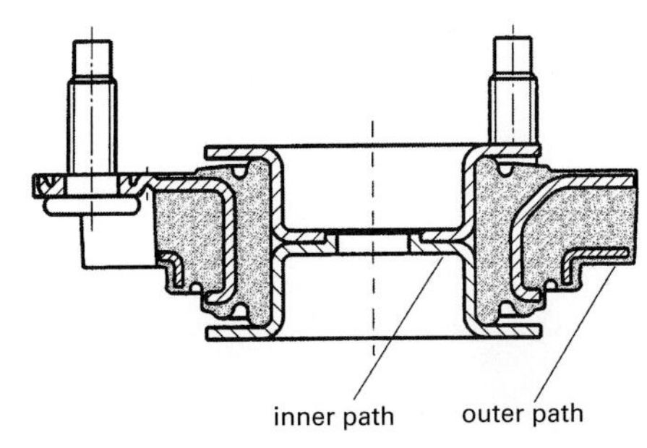

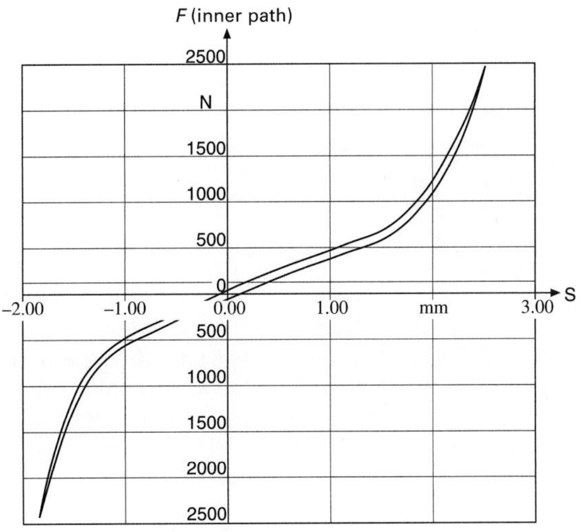

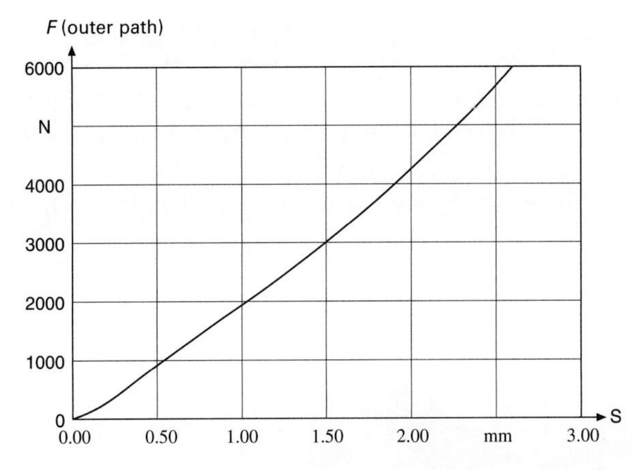

Fig. 8.1-10 The dual path top mount support of the Ford Focus (1998) manufactured by ContiTech Formteile GmbH. The body spring and shock-absorber forces are introduced into the body along two paths with variable rigidity. In this way, it is possible to design the shock-absorber bearing (inner element) in the region of small amplitudes with little rigidity and thus achieve good insulation from vibration and noise as well as improve the roll behaviour of the body. With larger forces of approximately 700 N and above, progression cams, which increase the rigidity of the bearing, come into play. A continuous transition between the two levels of rigidity is important for reasons of comfort. The bearing must have a high level of rigidity in a transverse direction in order to ensure that unwanted displacements and hence changes in wheel position do not occur. The forces of the body springs are directed along the outer path, which has a considerably higher level of rigidity.

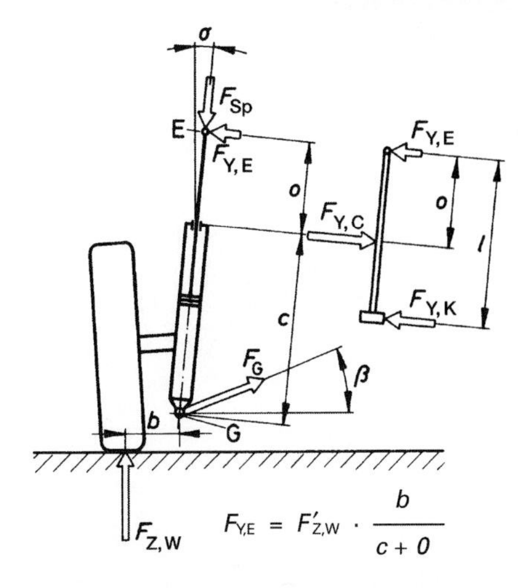

Fig. 8.1-11 If lateral force $F_{Z,W}$ moves lever arm *b* round guiding joint G, the lateral force F_{Sp} continually acts in the body-side fixing point E of the McPherson strut as a result of the force $F_{Y,E}$. This generates the reaction forces $F_{Y,C}$ and $F_{Y,K}$ on the piston rod guide and piston. This is $F_{Y,C} + F_{Y,E} = F_{Y,K}$ and the greater this force becomes, the further the frictional force F_{fr} increases in the piston rod guide and the greater the change in vertical force needed for it to rip away.
As the piston has a large diameter and also slides in shock-absorber fluid, lateral force $F_{Y,K}$ plays only a subordinate role. $F_{Y,K}$ can be reduced by offsetting the springs at an angle and shortening the distance *b* (see Fig 8.1-56).

8.1.2.4 Rear axle trailing-arm suspension

This suspension – also known as a crank axle – consists of a control arm lying longitudinally in the driving direction and mounted to rotate on a suspension subframe or on the body on both sides of the vehicle (Figs. 8.1-13 and 8.1-63). The control arm has to withstand forces in all directions, and is therefore highly subject to bending and torsional stress (Fig. 8.1-14). Moreover, no camber and toe-in changes are caused by vertical and lateral forces.

The trailing-arm axle is relatively simple and is popular on front-wheel drive vehicles. It offers the advantage that the car body floor pan can be flat and the fuel tank and/or spare wheel can be positioned between the suspension control arms. If the pivot axes lie parallel to the floor, the bump and rebound-travel wheels undergo no track width, camber or toe-in change, and the wheel base simply shortens slightly. If torsion springs are applied, the length of the control arm can be used to influence the progressivity of the springing to achieve better vibration behaviour under load. The control arm pivots also provide the radius-arm axis O; i.e. during braking the tail end is drawn down at this point.

The tendency to oversteer as a result of the deformation of the link (arm) when subject to a lateral

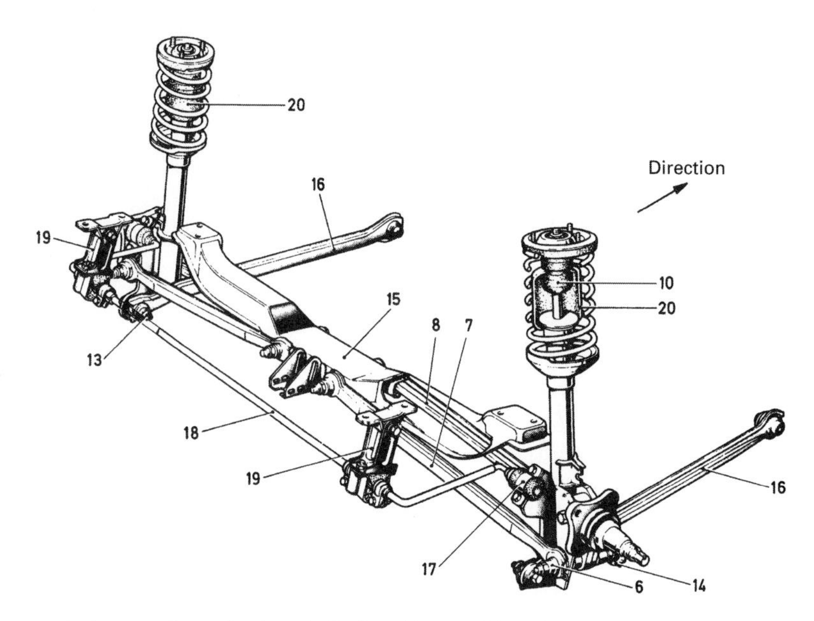

Fig. 8.1-12 The McPherson strut rear axle on the Lancia Delta with equal length transverse links of profiled steel trunnion-mounted close to the centre the cross-members 7 and 8. As large a distance as possible is needed between points 6 and 14 on the wheel hub carrier to ensure unimpaired straight running. The fixing points 13 of the longitudinal links 16 are behind the wheel centre, exactly like mounting points 17 of the anti-roll bar 18. The back of the anti-roll bar is flexibly joined to the body via tabs 19. The additional springs 10 attached to the top of the McPherson struts are covered by the dust tube 20. The cross-member 15 helps to fix the assembly to the body. An important criterion for dimensioning the control arm 16 is reverse drive against an obstruction.

force, the roll centre at floor level the extremely small possibility of a kinematic and elastokinematic effect on the position of the wheels and the inclination of the wheels during cornering consistent with the inclination of the body outwards (unwanted positive camber) are disadvantages.

8.1.2.5 Semi-trailing-arm rear axles

This is a special type of trailing-arm axle, which is fitted mainly in rear-wheel-and four-wheel-drive passenger cars, but which is also found on front-wheel-drive vehicles (Fig. 8.1-15). Seen from the top (Fig. 8.1-16), the

Fig. 8.1-13 Trailing-arm rear suspension of the Mercedes-Benz A class (1997). In order to minimize the amount of room required, the coil spring and monotube gas-pressure shock absorber are directly supported by the chassis subframe. The connecting tube is stress optimized oval shaped in order to withstand the high bending moments from longitudinal and lateral wheel forces which occur in the course of driving. The torsion-bar stabilizer proceeds directly from the shock-absorber attachment for reasons of weight and ease of assembly. When establishing the spring/shock-absorber properties, the line along which the forces act and which is altered by the lift of the wheel is to be taken into consideration, as a disadvantageous load-path can occur with jounce. The two front subframes are hydraulically damped in order to achieve a good level of comfort (hydromounts). The chassis subframe can make minor elastokinematic control movements. When designing subframe mounts, it is necessary to ensure that they retain their defined properties with regard to strength and geometry even with unfavourable conditions of use (e.g. low temperatures) and for a sufficiently long period of time, because variations in the configuration have a direct effect on vehicle performance. The longitudinal arms which run on tapered-roller bearings and which are subject to both flexural as well as torsional stress are designed in the form of a parallelogram linkage. In this way, the inherent disadvantage of a trailing arm axle – unwanted toe-in as a result of the deformation of the link when subject to a lateral force – is reduced by 75%, according to works specifications.

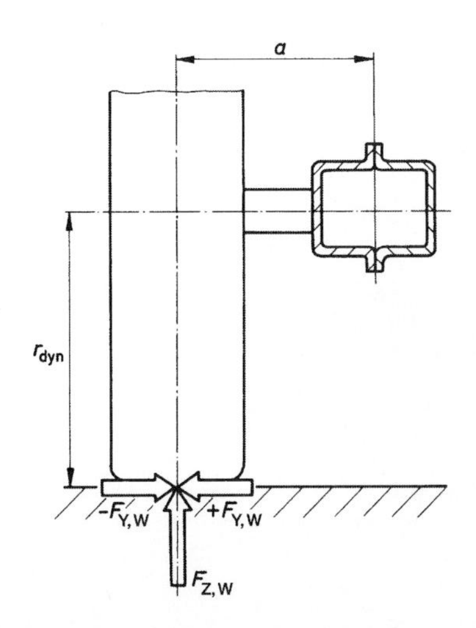

Fig. 8.1-14 On rear axle trailing-link suspensions, the vertical force $F_{Z,W}$ together with the lateral forces $F_{Y,W}$ cause bending and torsional stress, making a corresponding (hollow) profile, e.g. a closed box profile necessary. A force from inside causes the largest torsional moment:

$$T = F_{Z,W} \times a + F_{Y,W} \times r_{dyn}$$

control arm axis of rotation $\overline{EG}$ is diagonally positioned at an angle $\alpha = 10\text{–}25°$, and from the rear an angle $\beta \leq 5°$ can still be achieved. When the wheels bump and rebound-travel they cause spatial movement, so the drive shafts need two joints per side with angular mobility and length compensation (Fig. 8.1-17). The horizontal and vertical angles determine the roll steer properties.

When the control arm is a certain length, the following kinematic characteristics can be positively affected by angles α and β:

- height of the roll centre;
- position of the radius-arm axis;
- change of camber;
- toe-in change;

Camber and toe-in changes increase the bigger the angles α and β: semi-trailing axles have an elastokinematic tendency to oversteering.

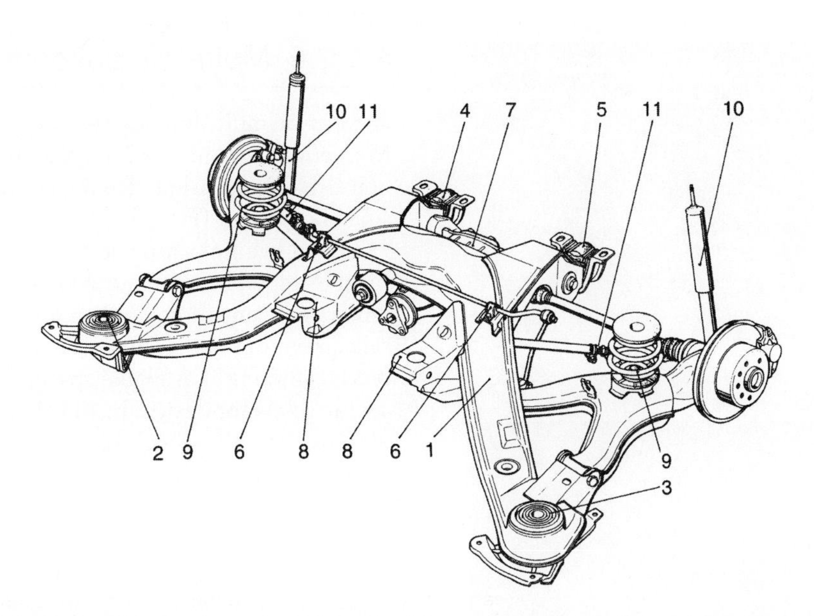

Fig. 8.1-15 Tilted-(Multiple) Staft Steering Rear Axle of the Opel Omega (1999), a further development of the tilted shaft steering axle. The differential casing of the rear-axle drive is above three elastic bearings, noise-isolated, connected with subframe (1), and this subframe is again, with four specially developed elastomer bearings on the installation (pos. 2–5). On top of part seated are the bearings (6) for the back of the stabilizer. Both of the extension arms (8) take up the inner bearings of the tilted shafts, which carry the barrel-shaped helical springs (9). In order to get a flat bottom of the luggage trunk, they were transferred to the front of the axle drive shafts. The transmission i_{Sp} (wheel to spring, becomes thereby with 1.5 comparatively large. The shock absorbers (10) are seated behind the centre of the axle, the transmission is with $i_D = 0.86$ favourable.
The angle of sweep of the tilted shafts amounts to alpha = 10° and the Dachwinkel, assume roof or top angle beta = 1°35′. Both of these angles change dynamically under the influence of the additional tilted shaft (11). These support the sideforces, coming from the wheel carriers directly against the subframe (1). They raise the lateral stability of the vehicle, and provide an absolute neutral elastic steering under side-forces and also, that in driving mode, favourable toe-in alterations appear during spring deflection, and also under load.

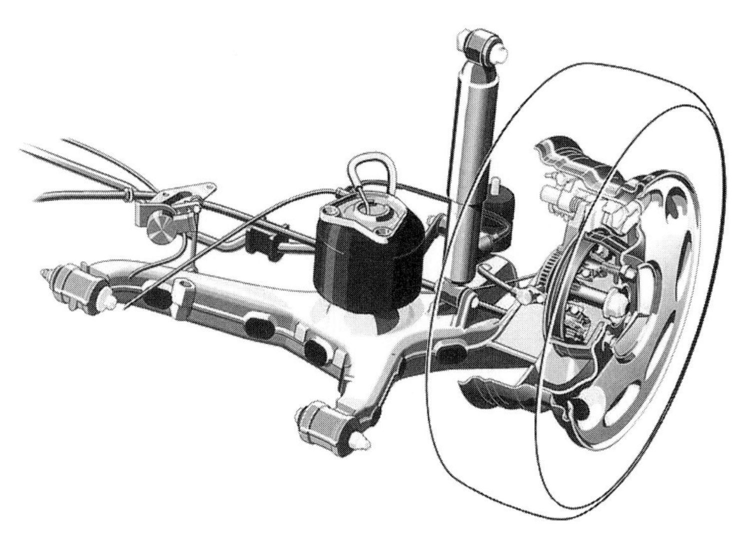

Fig. 8.1-16 Flat, non-driven air-suspended semi-trailing-arm rear axle of the Mercedes-Benz V class, whose driven front axle with spring-and-shock absorber strut has conventional coil springs. The air-spring bellows are supplied by an electrically powered compressor. The individual wheel adjustment permits the lowering or lifting of the vehicle as well as a constant vehicle height, regardless of – even onesided – loading. It is also possible to counteract body tilt during cornering. The damping properties of the shock absorbers are affected by spring bellow pressure depending on the load. The short rolling lobe air-spring elements make a low load floor possible; its rolling movement during compression and rebound results in self-cleaning. In the case of semi-trailing arm axles, roll understeer of the rear axle can be achieved by means of a negative vertical angle of pivot-axis inclination the kinematic toe-in alteration is also reduced.

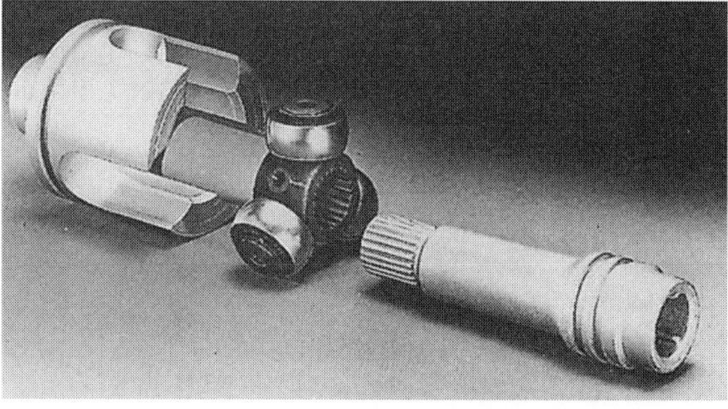

Fig. 8.1-17 Constant velocity sliding joints by GKN Automotive. In front-drive vehicles, considerable articulation angles of the drive axles occur, sometimes even during straight running, as a result of the installation situation, short propshafts and lifting movements of the body due to torque steer effects. These result in force and moment non-conformities and losses which lead to unwanted vibration. The full-load sliding ball joint (top, also see Fig. 8.1-53) permits bending angles of up to 22° and displacements of up to 45 mm. Forces are transmitted by means of six balls that run on intersecting tracks. In the rubber – metal tripod sliding joint (bottom), three rollers on needle bearings run in cylindrically machined tracks. With bending angles of up to 25° and displacements of up to 55 mm, these joints run particularly smoothly and hence quietly.

8.1.2.6 Multi-link suspension

A form of multi-link suspension was first developed by Mercedes-Benz in 1982 for the 190 series. Driven and non-driven multi-link front and rear suspensions have since been used (Figs. 8.1-1, 8.1-18, 8.1-19 and 8.1-44).

Up to five links are used to control wheel forces and torque depending on the geometry, kinematics, elasto-kinematics and force application of the axle. As the arrangement of links is almost a matter of choice depending on the amount of available space, there is extraordinarily a wide scope for design. In addition to the known benefits of independent wheel suspensions, with the relevant configuration the front and rear systems also offer the following advantages:

- Free and independent establishment of the kingpin offset, disturbing force and torque developed by the radial load.

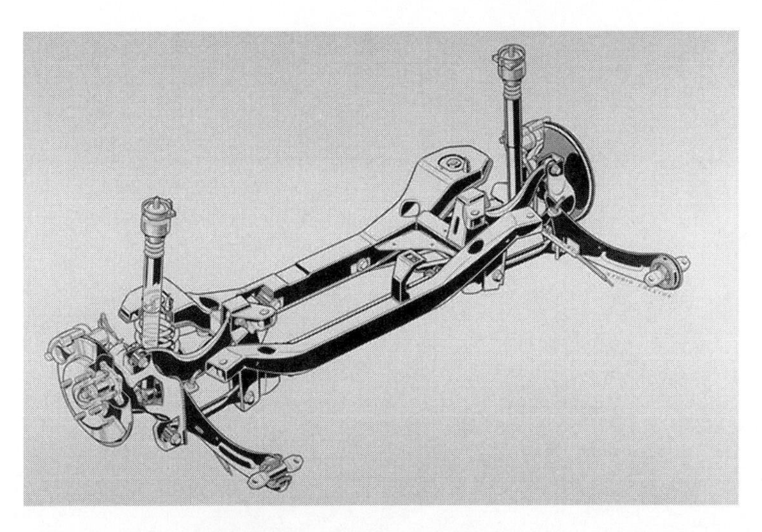

Fig. 8.1-18 Multi-link suspension of Ford Werke AG. Derived from the Mondeo Turnier model series, multi-link suspension is used by Ford for the first time in the Focus models (1998) in the segment of C class vehicles. This is called the 'control sword axle' after the shape of the longitudinal link. As there are five load paths available here, instead of the two that exist in twist-beam axles and trailing arm axles, there is great potential for improvement with regard to the adjustment of riding comfort, driving safety and noise and vibration insulation. As a result of a very elastic front arm bush, the high level of longitudinal flexibility necessary for riding comfort is achieved. At the same time, very rigid and accurate wheel control for increased driving safety is ensured by the transverse link, even at the stability limit. The longitudinal link is subject to torsional stress during wheel lift and to buckling stress when reversing. By using moulded parts, it was possible to reduce the unsprung masses by 3.5 kg per wheel.

- Considerable opportunities for balancing the pitching movements of vehicles during braking and acceleration (up to more than 100% anti-dive, anti-lift and anti-squat possible).
- Advantageous wheel control with regard to toe-in, camber and track width behaviour from the point of view of tyre force build-up, and tyre wear as a function of jounce with almost free definition of the roll centre and hence a very good possibility of balancing the self-steering properties.
- Wide scope for design with regard to elastokinematic compensation from the point of view of (a) specific elastokinematic toe-in changes under lateral and longitudinal forces and (b) longitudinal elasticity with a view to riding comfort (high running wheel comfort) with accurate wheel control.

As a result of the more open design, the wheel forces can be optimally controlled, i.e. without superposition, and introduced into the bodywork in an advantageous way with wide distances between the supports.

The disadvantages are:

- increased expenditure as a result of the high number of links and bearings;
- higher production and assembly costs;
- the possibility of kinematic overcorrection of the axle resulting in necessary deformation of the bearings during vertical or longitudinal movements;
- greater sensitivity to wear of the link bearings;
- high requirements with regard to the observation of tolerances relating to geometry and rigidity.

8.1.3 Rigid and semi-rigid crank axles

8.1.3.1 Rigid axles

Rigid axles (Fig. 8.1-20) can have a whole series of disadvantages that are a consideration in passenger cars, but which can be accepted in commercial vehicles:

- Mutual wheel influence (Fig. 8.1-21).
- The space requirement above the beam corresponding to the spring bump travel.
- Limited potential for kinematic and elastokinematic fine-tuning.
- Weight – if the differential is located in the axle casing (Fig. 8.1-20), it produces a tendency for wheel hop to occur on bumpy roads.
- The wheel load changes during traction (Fig. 8.1-22) and (particularly on twin tyres) there is a poor support base b_{Sp} for the body, which can only be improved following costly design work (Fig. 8.1-42).

The effective distance b_{Sp} of the springs is generally less than the tracking width b_r, so the projected spring

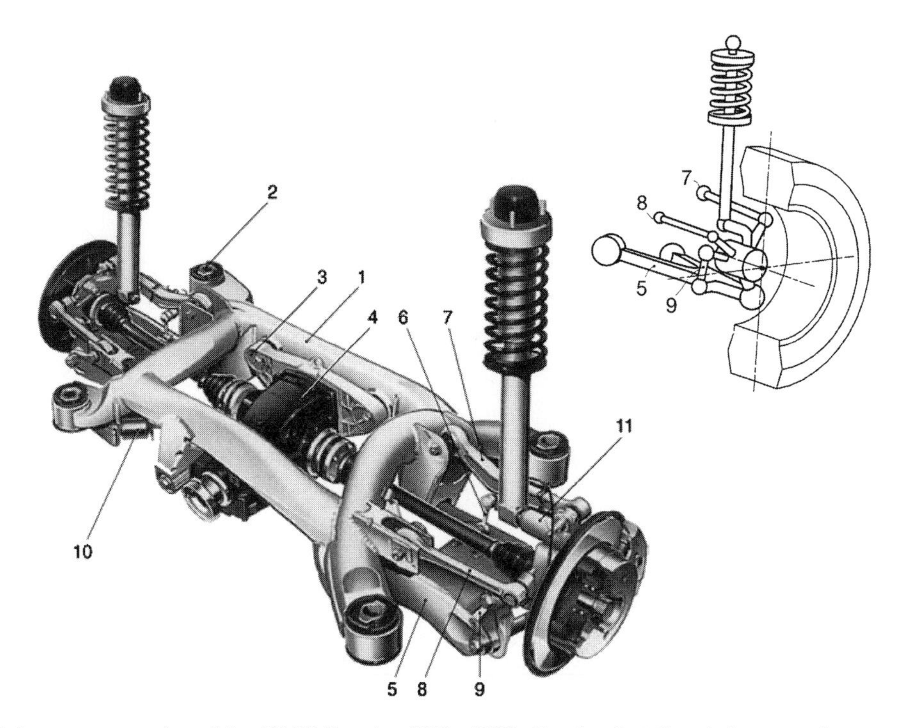

Fig. 8.1-19 Multi-link rear suspension of the BMW 5 series (E39, 1996). For the first time in large-scale car production, mainly aluminium is used for the suspension system derived from the geometry of the BMW 7 series.
The subframe (rear-axle support) (1), produced from welded aluminium tubes, is attached to the bodywork by means of four large rubber mounts (2). These are soft in a longitudinal direction for the purposes of riding comfort and noise insulation and rigid in a transverse direction to achieve accurate wheel control. The differential gear also has compliant mounts (3). The wheel carrier is mounted on a U-shaped arm (5) at the bottom and on the transverse link (7) and inclined guide link (8) at the top. As a result of this inclined position, an instantaneous centre is produced between the transverse link and guide link outside the vehicle which leads to the desired brake understeer during cornering and the elastokinematic compensation of deformation of the rubber bearings and components. The driving and braking torque of the wheel carrier (11) is borne by the 'integral' link (9) on the swinging arm (5), which is subject to additional torsional stress as a result. This design makes it possible to ensure longitudinally elastic control of the swinging arm on the guide bearing (10) for reasons of comfort, without braking or driving torque twisting the guide bearings as would be the case with torque borne by pairs of longitudinal links. The stabilizer behind presses on the swinging arm (5) by means of the stabilizer link (6), whereas the twin-tube gas-pressure shock absorber, whose outer tube is also made of aluminium, and the suspension springs provide a favourably large spring base attached directly to the wheel carrier (11). For reasons of weight, the wheel discs are also made of aluminium plate. The wheel carrier is made of shell cast aluminium. The rear axle of the station wagon BMW Tourer is largely similar in design. However, the shock absorber extends from the U-shaped swinging arm in order to allow for a wide and low loading area.

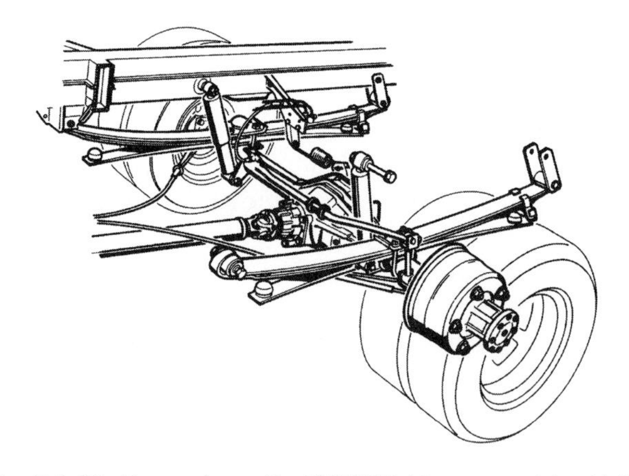

Fig. 8.1-20 Rear axle on the VW LT light commercial vehicle. The long, parabola-shaped rolled-out, dual leaf springs cushion the frame well and are progressive. The rubber buffers of the support springs come into play when the vehicle is laden. Spring travel is limited by the compression stops located over the spring centres, which are supported on the side-members. The spring leaves are prevented from shifting against one another by the spring clips located behind them, which open downwards (see also Fig. 8.1.68).
The anti-roll bar is fixed outside the axle casing. The benefits of this can be seen in Fig. 8.1.23. The shock absorbers, however, are unfortunately located a long way to the inside and are also angled forwards so that they can be fixed to the frame side-members.

rate c_φ is lower (Fig. 8.1-23). As can be seen in Fig. 8.1-61, the springs, and/or suspension dampers, for this reason should be mounted as far apart as possible.

The centrifugal force ($F_{c,Bo}$, Fig. 8.1-6) acting on the body's centre of gravity during cornering increases the roll pitch where there is a rigid axle.

Thanks to highly developed suspension parts and the appropriate design of the springing and damping, it has

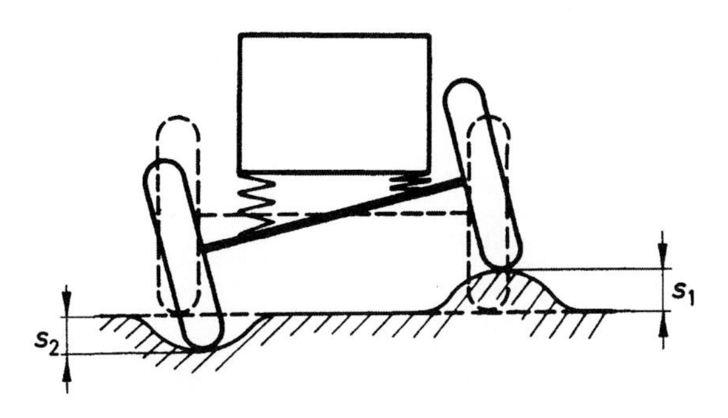

Fig. 8.1.21 Mutual influence of the two wheels of a rigid axle when travelling along a road with pot-holes, shown as 'mutually opposed springing'. One wheel extends along the path s_2 and the other compresses along the path s_1.

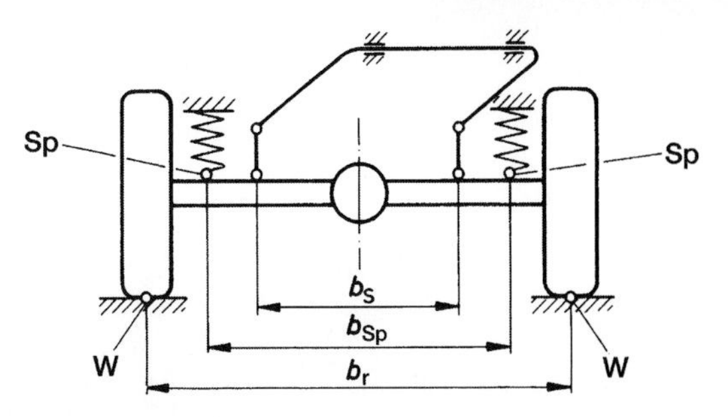

Fig. 8.1-23 When considering the roll pitch of the body with the rigid axle the distances b_{Sp} (of the springs F) and b_S (of the anti-roll bar linkage points) are included in the calculation of the transfer with mutually opposed springing. i_φ is squared to give the rate c_φ:

$$i_\varphi = b_r/b_{Sp} \quad \text{and} \quad c_\varphi = c_r i_\varphi^2$$

The greater the ratio, the less the roll reaction applied by the body, i.e. the springs and anti-roll bar arms should be fixed as far out as possible on the rigid axle casing.

been possible to improve the behaviour of rigid drive axles. Nevertheless, they are no longer found in standard-design passenger cars, but only on four-wheel-drive and special all-terrain vehicles (Figs. 8.1-43 and 8.1-68).

Because of its weight, the driven rigid axle is outperformed on uneven roads (and especially on bends) by independent wheel suspension, although the deficiency in road-holding can be partly overcome with pressurized mono-tube dampers. These are more expensive, but on the compressive stroke, the valve characteristic can be set to be harder without a perceptible loss of comfort. With this, a responsive damping force is already opposing the compressing wheels. This is the simplest and perhaps also the most economic way of overcoming the main disadvantage of rigid axles.

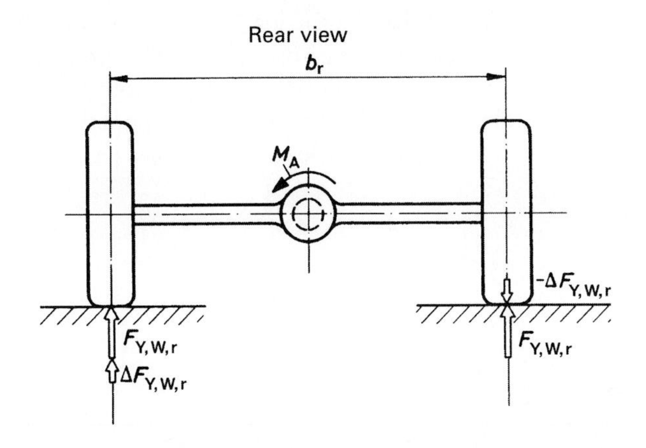

Fig. 8.1-22 If the differential is located in the body of the rigid axle, the driving torque M_A coming from the engine is absorbed at the centres of tyre contact, resulting in changes to vertical force $\pm\Delta F_{y,w,r}$.
In the example, M_A would place an additional load on the left rear wheel ($F_{Y,W,r} + \Delta F_{y,w,r}$) and reduce the vertical force ($F_{Y,W,r} - \Delta F_{y,w,r}$) on the right one.
On a right-hand bend the right wheel could spin prematurely, leading to a loss in lateral force in the entire axle and the car tail suddenly breaking away (Fig. 10.1-37).

In contrast to standard-design vehicles, the use of the rigid rear axle in front-wheel-drive vehicles has advantages rather than disadvantages (Fig. 8.1-24). The rigid rear axle weighs no more than a comparable independent wheel suspension and also gives the option of raising the body roll centre (which is better for this type of drive). Further advantages, including those for driven axles, are:

- they are simple and economical to manufacture;
- there are no changes to track width, toe-in and camber on full bump/rebound-travel, thus giving
- low tyre wear and sure-footed road holding;
- there is no change to wheel camber when the body rolls during cornering (Fig. 8.1-6), therefore there is constant lateral force transmission of tyres;
- the absorption of lateral force moment $M_Y = F_{T,X}\, h_{Ro,r}$ by a transverse link, which can be placed at almost any height (e.g. Panhard rod, Fig. 8.1-25);
- optimal force transfer due to large spring track width b_{Sp}
- the lateral force compliance steering can be tuned towards under- or over-steering (Fig. 8.1-29).

There are many options for attaching a rigid axle rear suspension beneath the body or chassis frame. Longitudinal leaf springs are often used as a single suspension control arm, which is both supporting and springing at the same time, as these can absorb forces in all three directions as well as drive-off and braking moments (Fig. 8.1-26). This economical type of rear suspension also has the advantage that the load area on lorries and

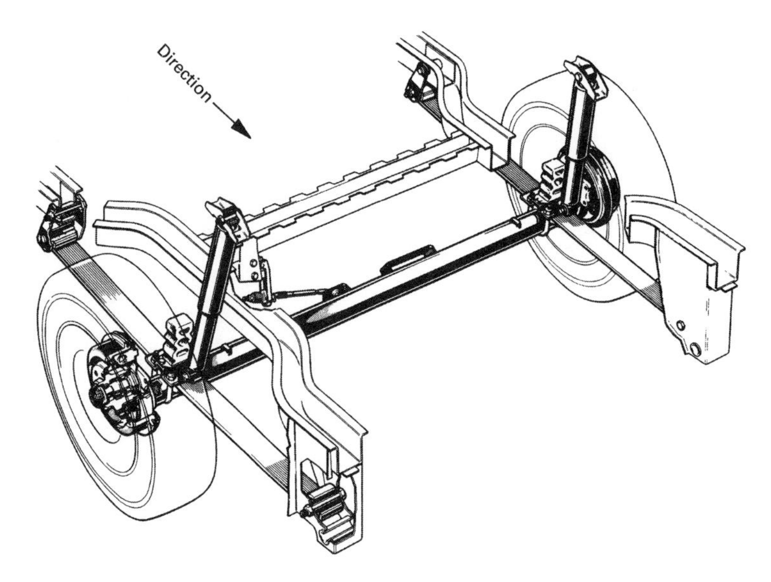

Fig. 8.1-24 The rear axle on a Ford Escort Express delivery vehicle. Single leaf springs carry the axle and support the body well at four points. The shock absorbers (fitted vertically) are located close to the wheel, made possible by slim wheel-carriers/hub units. The additional elastomer springs sit over the axle tube and act on the side members of the body when at full bump.

the body of passenger cars can be supported in two places at the back: at the level of the rear seat and under the boot (Fig. 8.1-27). This reduces the stress on the rear end of the car body when the boot is heavily laden, and also the stress on the lorry frame under full load (Fig. 8.1-20).

The longitudinal leaf springs can be fitted inclined, with the advantage that during cornering the rigid rear axle (viewed from above) is at a small angle to the vehicle longitudinal axis (Fig. 8.1-28). To be precise, the side of the wheel base on the outside of the bend shortens somewhat, while the side on the inside of the bend lengthens by the same amount. The rear axle steers into the bend and, in other words, it is forced to self-steer towards 'roll-understeering' (Fig. 8.1-29).

This measure can, of course, have an adverse effect when the vehicle is travelling on bad roads, but it does prevent the standard passenger car's tendency to oversteer when cornering. Even driven rigid axles exhibit – more or less irrespective of the type of suspension – a tendency towards the load alteration (torque steering) effect, but not to the same extent as semi-trailing link suspensions. Details can be found in Section 10.1.12.2.

On front-wheel-drive vehicles, the wheels of the trailing axle can take on a negative camber. This improves the lateral grip somewhat, but does not promote perfect tyre wear. This is also possible on the compound crank suspension (a suspension-type halfway between a rigid

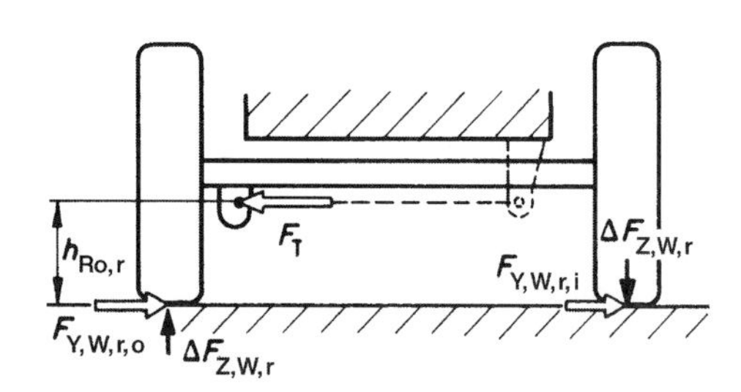

Fig. 8.1-25 On rigid axles the axle body absorbs the bending moments which arise as a result of lateral forces. Only the force F_T occurs between the suspension and the body, and its size corresponds to the lateral forces $F_{Y,W,r,o}$ and $F_{Y,W,r,i}$. On a horizontal Panhard rod, the distance $h_{Ro,r}$ is also the height of the body roll centre. The higher this is above ground, the greater the wheel force change $\pm\Delta F_{z,wr}$.

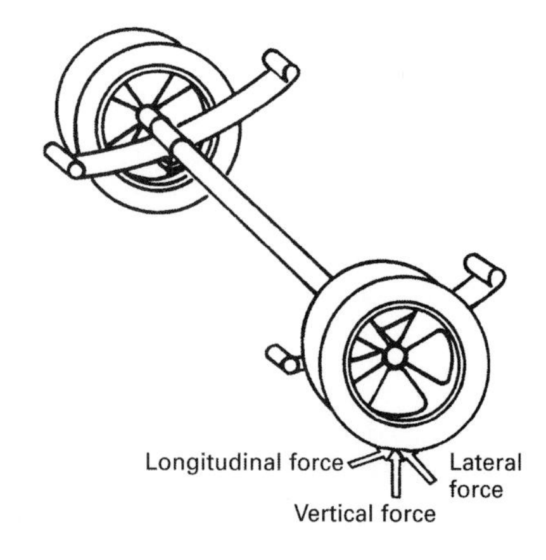

Fig. 8.1-26 Longitudinal leaf springs can absorb both forces in all directions and the drive-off, braking and lateral force moment.

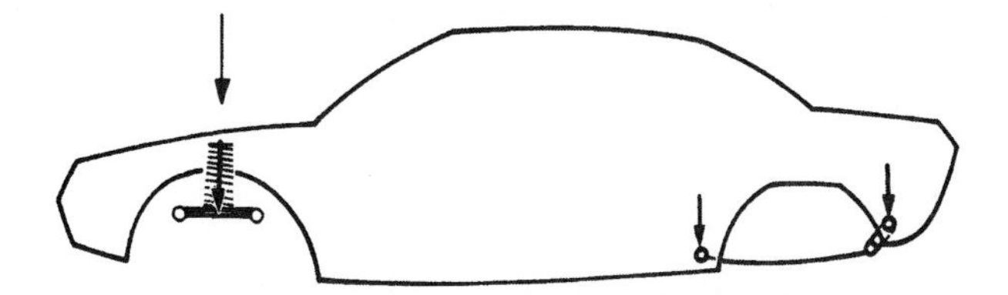

Fig. 8.1-27 Longitudinal rear leaf springs support the body of a car in two places – under the back seats and under the boot – with the advantage of reduced bodywork stress.

axle and independent wheel suspension) which, up to now, has been fitted only on front-wheel drive vehicles. Details are given in Fig. 8.1-2 and Section 8.1.6.4.1.

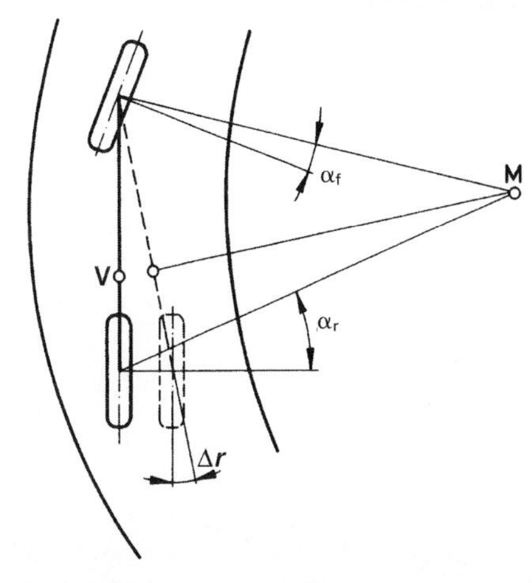

Fig. 8.1-29 If a rigid rear axle steers with the angle Δr towards understeer, the tail moves out less in the bend and the driver has the impression of more neutral behaviour. Moreover, there is increased safety when changing lanes quickly at speed. The same occurs if the outside wheel of an independent wheel suspension goes into toe-in and the inside wheel goes into toe-out.

8.1.3.2 Semi rigid crank axles

The compound crank suspension could be described as the new rear axle design of the 1970s (Figs. 8.1-2 and 8.1-30) and it is still used in today's small-and medium-sized front-wheel-drive vehicles. It consists of two trailing arms that are welded to a twistable cross-member and fixed to the body via trailing links. This member absorbs all vertical and lateral force moments and, because of its offset to the wheel centre, must be less torsionally stiff and function simultaneously as an anti-roll bar. The axle has numerous advantages and is therefore found on a number of passenger cars which have come onto the market.

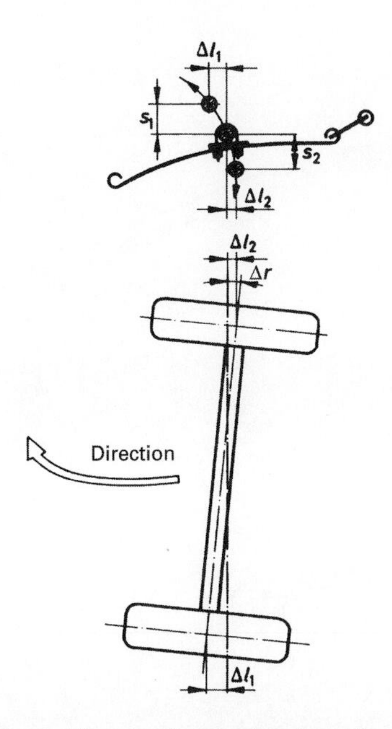

Fig. 8.1-28 Angled longitudinal leaf springs fixed lower to the body at the front than at the back cause the rigid rear axle to self-steer towards understeering (so-called roll pitch understeering). Where there is body roll, the wheel on the outside of the bend, which is compressing along the path s_1, is forced to accommodate a shortening of the wheel base $\Delta l1$, whilst the wheel on the inside of the bend, which is extending by s_2, is forced to accommodate a lengthening of the wheel base by $\Delta l2$. The axle is displaced at the steering angle Δr.

Fig. 8.1-30 Twist-beam suspension of the VW Golf IV (1997), VW Bora (1999) and Audi A3 (1996). The rubber–metal bearings of the axle body are set at 25° to the transverse suspension of the vehicle in order to improve the self-steering properties of the suspension together with the rigidity of the bearings which varies in three directions in space. Compared with the previous model, it was possible to reduce unwanted lateral-force toe-out steer resulting from link deformation by 30% to approximately 1 mm per 500 N of lateral force. Fig. 8.1-72 shows the four-wheel-drive version of the VW Golf IV.

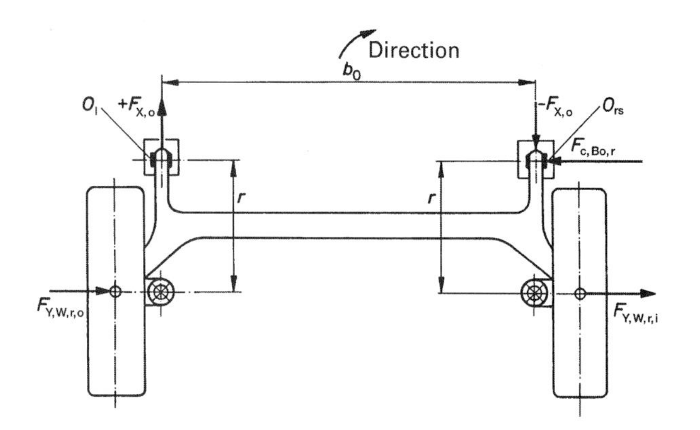

Fig. 8.1-31 The lateral forces $F_{Y,W,r,o}$ and $F_{y,W,r,i}$ occurring at the centres of tyre contact during cornering are absorbed at the bearing points O_l and O_{rs}. This results in a moment $M_y = (F_{Y,W,r,o} + F_{Y,W,r,i}) X r = F_{X,o}\, b_O$ which (depending on the elasticity of the rubber bearing) can cause 'lateral force oversteering'. The longer the control arms (distance r) and the closer the points O_l and O_{rs} (distance b_O), the greater the longitudinal forces $\pm F_{XO}$.

From an installation point of view:

- the whole axle is easy to assemble and dismantle;
- it needs little space;
- a spring damper unit or the shock absorber and springs are easy to fit;
- no need for any control arms and rods; and thus
- only few components to handle.

From a suspension point of view:

- there is a favourable wheel to spring damper ratio;
- there are only two bearing points O_l and O_{rs}, which hardly affect the springing (Fig. 8.1-31);
- low weight of the unsprung masses; and
- the cross-member can also function as an anti-roll bar.

From a kinematic point of view:

- there is negligible toe-in and track width change on reciprocal and parallel springing;
- there is a low change of camber under lateral forces;
- there is low load-dependent body roll understeering of the whole axle; and
- good radius-arm axis locations O_l and O_{rs} (Fig. 8.1-31), which reduce tail-lift during braking.

The disadvantages are:

- a tendency to lateral force oversteer due to control arm deformation;

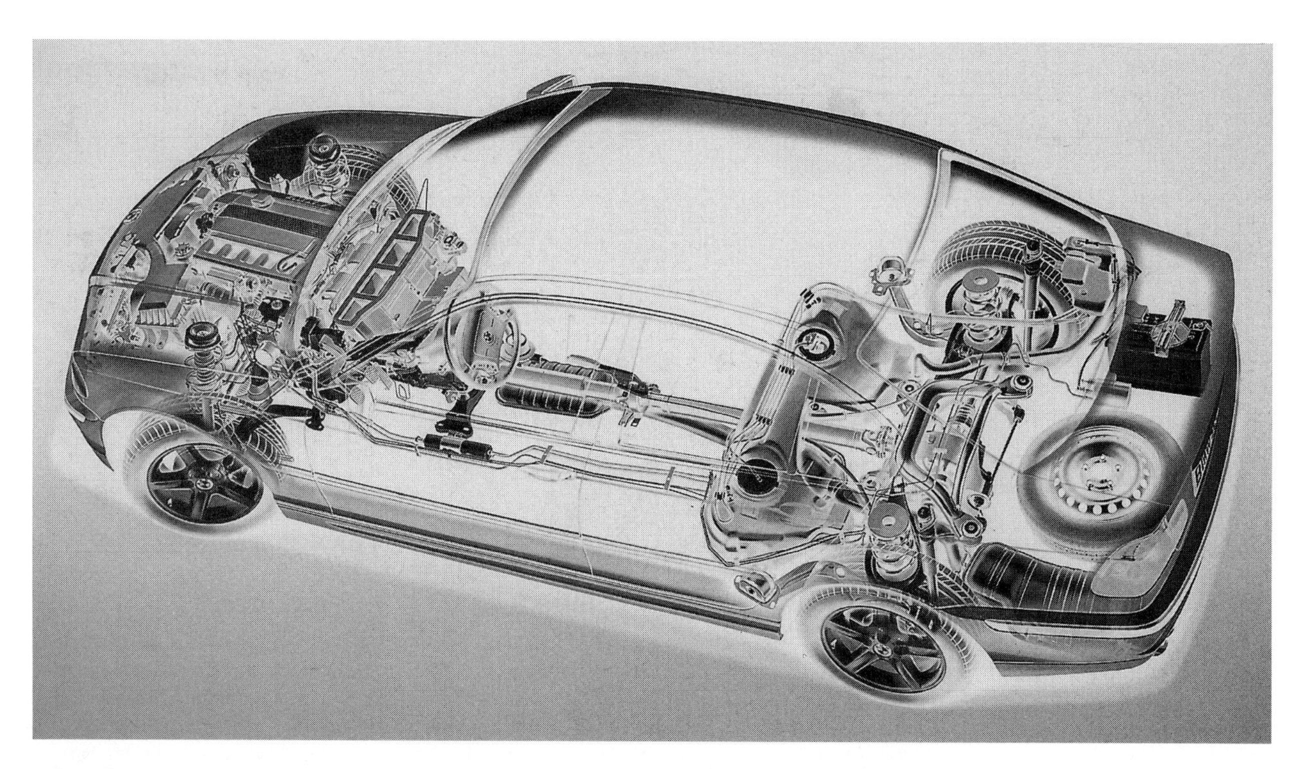

Fig. 8.1-32 Front-mounted engine, rear-mounted drive (BMW 3 series E46, 1998). The manual transmission is flange-mounted on the engine, which is longitudinally positioned over the front axle. The rear-axle differential is driven by means of a propshaft. The fuel tank is situated in front of the rear axle for safety in case of an accident. The battery was placed in the boot in order to achieve a balanced 50:50 axle-load distribution. Fig. 8.1-1 shows the rear axle in detail.

Fig. 8.1-33 Chevrolet Corvette (1998). In order to achieve balanced axle-load distribution, a more rigid overall system (necessary on account of the greater flexibility of the plastic bodywork) and more leg room, the gearbox is integrated with the rear-axle differential. Compared with standard drives, the cardan shaft turns higher (with engine speed) but is subject to correspondingly less torque. The front and rear axles have plastic (fibreglass) transverse leaf springs. Compared with the previous model, unwanted vibration, particularly on an uneven road surface, is reduced as a result of the shorter length of the wheel spindles of 63 mm and the small steering-axle angle of 8.8°. Owing to the combination of a castor angle of 6.5° with a castor trail of 36 mm (previous model: 5.9°, 45 mm), a good compromise is achieved between high lateral rigidity of the axle and good feedback properties.

- torsion and shear stress in the cross-member;
- high stress in the weld seams; which means
- the permissible rear axle load is limited in terms of strength;
- the limited kinematic and elastokinematic opportunities for determining the wheel position;
- the establishment of the position of the instantaneous centre by means of the axle kinematics and rigidity of the twist-beam axle;
- the mutual effect on the wheel;
- the difficult decoupling of the vibration and noise caused by the road surface; and

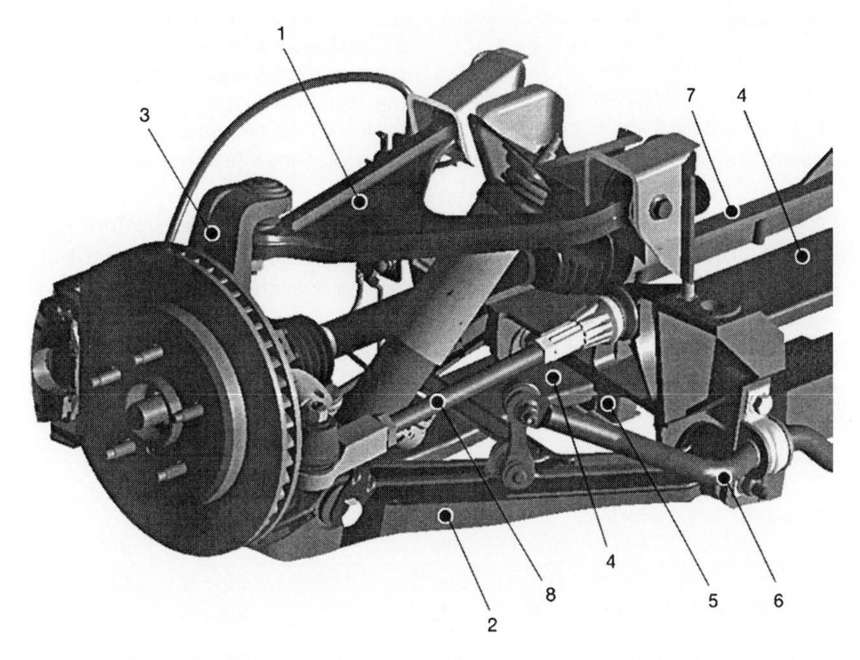

Fig. 8.1-34 Rear axle (left side of wheel) of the Chevrolet Corvette (1998). Links 1, 2 and wheel carrier 3 of the multi-link suspension are made from aluminium in order to reduce the unsprung masses. The plastic leaf spring 4 is mounted at two places on the right and left sides of the body (5) so that it also helps to make the body more resistant to roll. Roll spring stiffness is further increased by stabilizer 6. This is attached to subframe 7, which is also made of aluminium. The design of the wheel carrier 3 on the front and rear axles is the same, but not the wheel links 1 and 2. The toe-in control of the rear axle is exercised very stiffly and precisely, via tie rod 8.

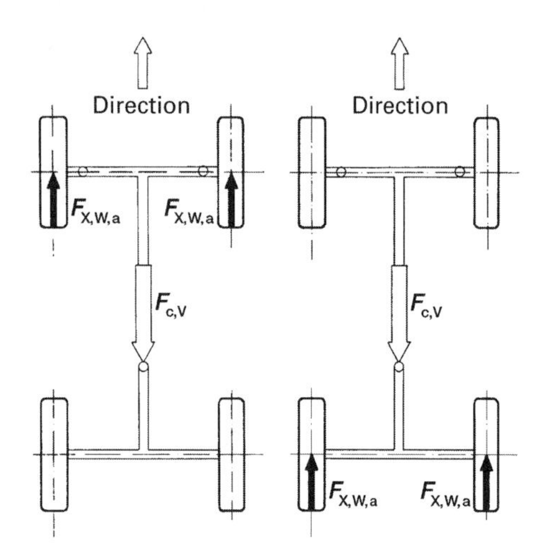

Fig. 8.1-35 On a front-wheel drive (left) the vehicle is pulled. The result is a more stable relationship between the driving forces $F_{X,W,a}$ and the inertia force $F_{c,V}$ Conversely, in the case of driven rear wheels an unstable condition is theoretically evident; front axle settings ensure the necessary stabilization.

- the considerable need for stability of the bodywork in the region of those points on the front bearings at which complex, superposed forces have to be transmitted.

8.1.4 Front-mounted engine, rear-mounted drive

In passenger cars and estate cars, the engine is approximately in the centre of the front axle and the rear wheels are driven (Fig. 8.1-32). To put more weight on the rear axle and obtain a more balanced weight distribution, Alfa Romeo, Porsche (928, 968 models) and Volvo integrated the manual transmission with the differential. This is also the case with the Chevrolet Corvette sports car (1998; Figs. 8.1-33 and 8.1-34). With the exception of light commercial vehicles, all lorries have the engine at the front or centrally between the front and rear axles together with rear-wheel-drive vehicles. The long load area gives hardly any other option. Articulated lorries, where a major part of the trailer weight – the trailer hitch load – is carried over the rear wheels, have the same configuration. On buses, however, the passengers are spread evenly throughout the whole interior of the vehicle, which is why there are models with front, central and rear engines.

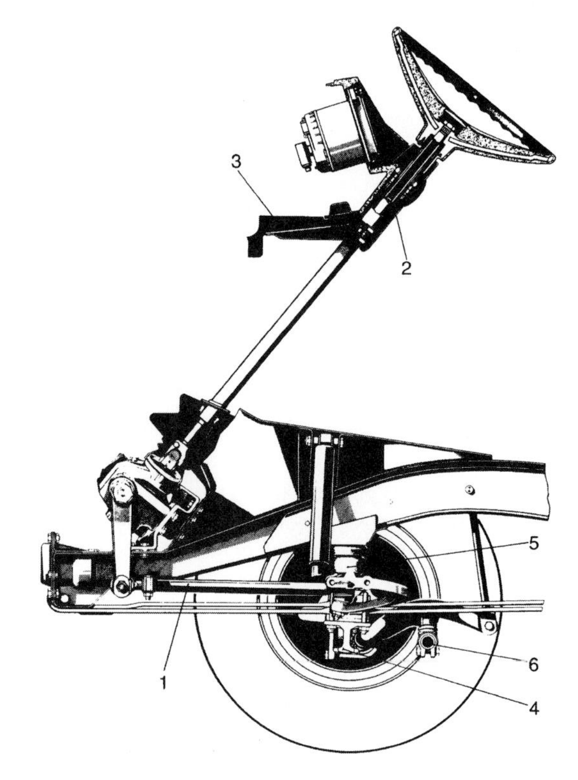

Fig. 8.1-37 The front rigid axle on the Mercedes-Benz light commercial vehicle of the 207 D/308 series with recirculating ball steering gear and steering rod 1 parallel to the two-layer parabolic spring. This rod has to be slightly shorter than the front side of the spring, so that both parts take on the same motion curve when the axle bottoms out (see also Fig. 9.1-6). The brace 3, running from the steering column jacket 2 to the body, bends on impact.
The T-shaped axle casing 4, which is cranked downwards and to which the springs are fastened, can be seen in the section. The elastomer spring 5 sits on the longitudinal member of the frame and the two front wheels are joined by the tie rod 6. The safety steering wheel has additional padding.

	Front-wheel drive		*Rear-wheel drive*		*Rear engine*	
	front	*rear*	*front*	*rear*	*front*	*rear*
Empty	61	39	50	50	40	60
2 passengers at the front	60	40	50	50	42	58
4 passengers	55	45	47	53	40	60
5 passengers and luggage	49	51	44	56	41	59

Fig. 8.1-36 Average proportional axle load distribution based on drive type and loading condition. With the standard-design saloon, when the vehicle is fully laden, the driven rear wheels have to carry the largest load. With the front-wheel drive, however, with only two persons in the vehicle, the front wheels bear the greater load.

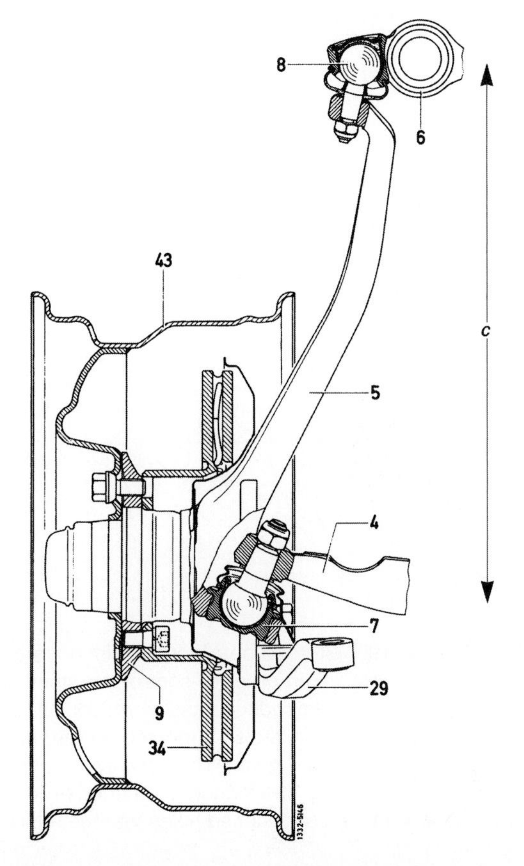

Fig. 8.1-38 Front hub carrier (steering knuckle) on the Mercedes-Benz S class (W40, 1997) with a large effective distance *c* (see also Fig. 8.1.4). The upper transverse control arm 6 forms the casing for the ball pivot of the guiding joint, whereas the lower supporting joint 7 is pressed into the hub carrier 5. The ventilated brake disc 34 (dished inwards), the wheel hub 9, the double-hump rim 43 with asymmetrical drop centre and the space for the brake caliper (not included in the picture) are clearly shown.

8.1.4.1 Advantages and disadvantages of the front-mounted engine, rear-mounted drive design

The standard design has a series of advantages on passenger cars and estate cars:

- There is hardly any restriction on engine length, making it particularly suitable for more powerful vehicles (in other words, for engines with 8–12 cylinders).
- There is low load on the engine mounting, as only the maximum engine torque times the conversion of the lowest gear without differential transmission has to be absorbed.
- Insulation of engine noise is relatively easy.
- Under full load, most of the vehicle mass is on the driven rear axle (important for estate cars and trailers (Fig. 8.1-36).
- A long exhaust system with good silencing and catalytic converter configuration.
- Good front crumple zone, together with the 'submarining' power plant unit, i.e. one that goes underneath the floor panel during frontal collision.
- Simple and varied front axle designs are possible irrespective of drive forces.
- More even tyre wear thanks to function distribution of steering/drive.
- Uncomplicated gear shift mechanism.
- Optimum gearbox efficiency in direct gear because no force-transmitting bevel gear is in action.
- Sufficient space for housing the steering system in the case of a recirculating ball steering gear.
- Good cooling because the engine and radiator are at the front; a power-saving fan can be fitted.
- Effective heating due to short hot-air and water paths.

The following disadvantages mean that, in recent years, only a few saloon cars under 2 l engine displacement have been launched internationally using this design, and performance cars also featured the front-mounted design:

- Unstable straight-running ability (Fig. 8.1-35), which can be fully corrected by special front suspension geometry settings, appropriate rear axle design and suitable tyres.
- The driven rear axle is slightly loaded when there are only two persons in the vehicle, leading to poor traction behaviour in wet and wintry road conditions – linked to the risk of the rear wheels spinning, particularly when tight bends are being negotiated at speed. This can be improved by setting the unladen axle load distribution at 50%/50% which, however, is not always possible (Fig. 8.1-36). It can be prevented by means of drive-slip control.
- A tendency towards the torque steer effect (Fig. 10.1-53) and, therefore,
- complex rear independent wheel suspension with chassis subframe, differential gear case and axle drive causing
- restrictions in boot size
- The need for a propshaft between the manual gearbox and differential (Fig. 8.1-32) and, therefore,
- a tunnel in the floor pan is inevitable, plus an unfavourable interior to vehicle–length ratio.

8.1.4.2 Non-driven front axles

The standard design for passenger cars that have come onto the market in recent years have McPherson struts on the front axle, as well as double wishbone or multi-link

Fig. 8.1-39 Multi-link front suspension of the Mercedes-Benz model W220 (S class, 1998). Based on a double wishbone axle, two individual links (tension strut and spring link) are used instead of the lower transverse link in order to control the steering axle nearer to the middle of the wheel. As a result, the kingpin offset and disturbing force lever arm are reduced and vibrations are caused by tyre imbalances and brake-force fluctuations is consequently minimized. Crash performance is also improved by the more open design. The air-spring struts with integrated shock absorber proceed directly from the spring link. The laterally rigid rack and pinion steering in front of the middle of the wheel leads to the desired elastokinematic understeer effect during cornering owing to the laterally elastic spring link bearings. The manufacturing tolerances are kept so small by means of punched holes that the adjustment of camber and camber angles in production is not necessary.

suspensions. The latter type of suspension is becoming more and more popular because of its low friction levels and kinematic advantages. Even some light commercial vehicles have McPherson struts or double wishbone axles (Fig. 8.1-7). However, like almost all medium-sized and heavy commercial vehicles, most have rigid front axles. In order to be able to situate the engine lower, the axle subframe has to be offset downwards (Fig. 8.1-37).

The front wheels are steerable; to control the steering knuckle 5 (Fig. 8.1-38) on double wishbone suspensions, there are two ball joints that allow mobility in all directions, defined by full bump/rebound-travel of the wheels and the steering angle. The wishbone, which accepts the spring, must be carried on a supporting joint (item 7) in order to be able to transmit the vertical forces. A regular ball joint transferring longitudinal and lateral forces (item 8) is generally sufficient for the second suspension control arm. The greater the distance between the two joint points, the lower the forces in the components. Fig. 8.1-39 shows a front axle with ball joints a long way apart.

The base on McPherson struts is better because it is even longer. Fig. 8.1-40 shows a standard design and Fig. 8.1-8 the details.

The coil spring is offset at an angle to reduce the friction between piston rod 2 and the rod guide. The lower guiding joint (point G) performs the same function as on double wishbones, whereas point E is fixed in the shock tower, which is welded to the wheel house panel. As the wheels reach full bump, piston rod 2 moves in the cylinder tube (which sits in the carrier or outer tube and when there is a steering angle the rod and spring turn in an upper strut mount, which insulates noise and is located at point E (Fig. 8.1-9).

Wheel controlling damper struts do not require such a complex mount. The piston rod turns easily in the damping cylinder (Fig. 8.1-41). Only the rod needs noise insulation. The coil spring sits separately on the lower control arm, which must be joined to the steering knuckle via a supporting joint. The damper is lighter than a shock-absorbing strut and allows a greater bearing span across the damping cylinder, permits a wider, flatter engine compartment (which is more streamlined) and is easier to repair. However, it is likely to be more costly and offsetting the spring from the damper (Figs. 8.1-8 and 8.1-11) may cause slip-stick problems with a loss of ride comfort.

In the case of front-wheel-drive vehicles, there may be a problem in the lack of space between the spring and the drive axle.

8.1.4.3 Driven rear axles

Because of their cost advantages, robustness and ease of repair rigid axles are fitted in practically all commercial

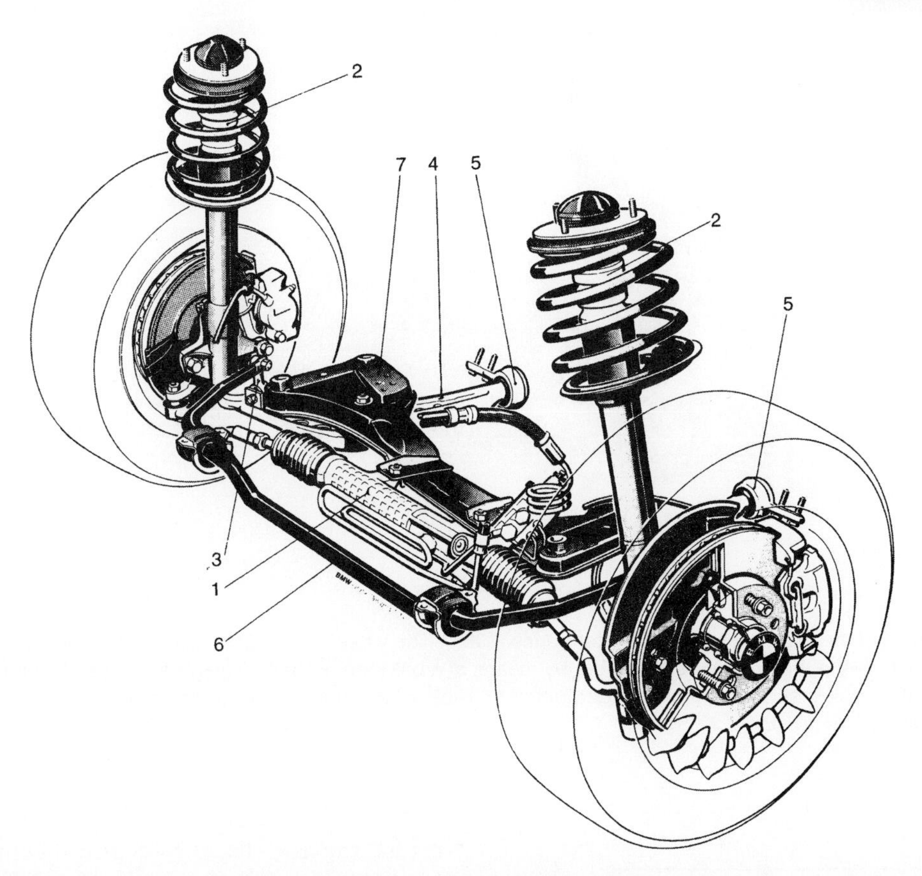

Fig. 8.1-40 Spring strut front axle of the BMW Roadster Z3, which Lemförder Fahrwerktechnik produce in the USA and supply directly to the assembly line there. The additional springs 2 are positioned in the coil springs (Fig. 8.1-11) which are offset at an angle in order to reduce friction. The stabilizer 6 is connected to the lower links by the struts 3.
The cross-member 7 which serves as the subframe takes the hydraulically supported rack and pinion steering 1 at the front and the transverse link 4 on its lower side. The L-shape of the transverse link makes good decoupling of the lateral rigidity and longitudinal elasticity possible: lateral forces are introduced directly into the rigid front bearing, while longitudinal forces produce a rotational movement about the front bearing as a result of the laterally elastic rear bearing 5. These rubber elements ensure a defined lateral springing. The large-diameter internally ventilated brake discs (15″ rim) and the third-generation, two-row angular ball bearings, whose outside ring also acts as a wheel hub, are clearly shown.
The kingpin offset at ground (scrub radius) depends on the tyre width and thus the wheel offset; it is $r_\sigma = +10$ mm on 185/65 R 15 tyres and $r_2 = +5$ mm on 205/60 R 15 tyres.

and off-road vehicles (Fig. 8.1-43) in combination with leaf springs, coil springs or air springing (Figs. 8.1-20 and 8.1-42). They are no longer found in saloons and coupés. In spite of the advantages described in Section 8.1.3, the weight of the axle is noticeable on this type of vehicle.

For independent suspension, the semi-trailing arm axle, shown in Figs. 8.1-15 and 8.1-45, is used as independent wheel suspension in passenger and light commercial vehicles. This suspension has a chassis subframe to which the differential is either fixed or, to a limited degree, elastically joined to give additional noise and vibration insulation. The springs sit on the suspension control arms. This gives a flat, more spacious boot, but with the disadvantage that the forces in all components become higher.

Because of its ride and handling advantages, more and more passenger cars have double wishbone suspension rear axles or so-called multi-link axles (Figs. 8.1-1, 8.1-19, 8.1-34 and 8.1.72).

Most independent wheel suspensions have an easy-to-assemble chassis subframe for better wheel control and noise insulation. However, all configurations (regardless of the design) require drive shafts with length compensation. This is carried out by the sliding constant velocity (CV) joints fitted both at the wheel and the differential. Fig. 8.1-17 shows a section through a joint of this type, and Fig. 8.1-44 shows a typical modern bearing of a driven rear wheel.

8.1.5 Rear and mid engine drive

The rear-mounted power plant consists of the engine and the differential and manual gearbox in one assembly unit,

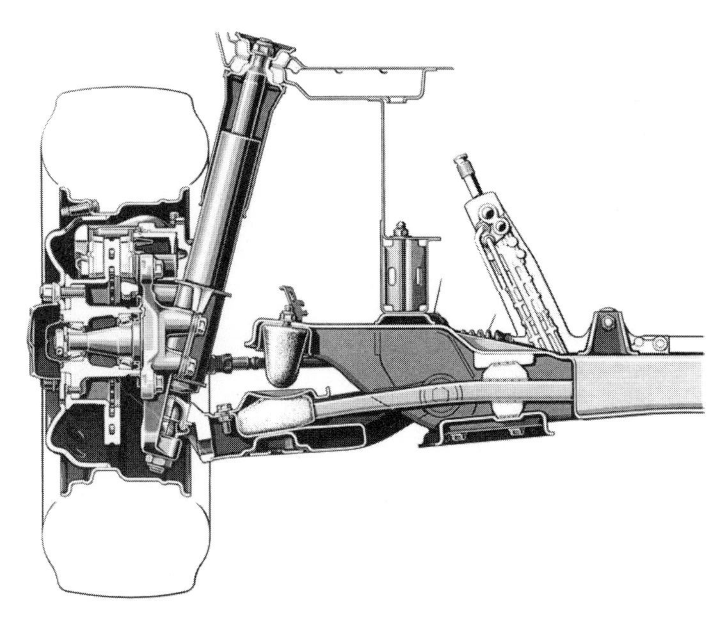

Fig. 8.1-41 Front axle of the Mercedes-Benz Sprinter series (1995). The wheel-controlling strut is screwed on to the wheel carrier, which is, in turn, connected to the lower cross-member by means of a ball joint. Both the vehicle suspension and roll stabilization are ensured by means of a transverse plastic leaf springmounted on rubber elements. Large rubber buffers with progressive rigidity act as additional springs and bump stops.

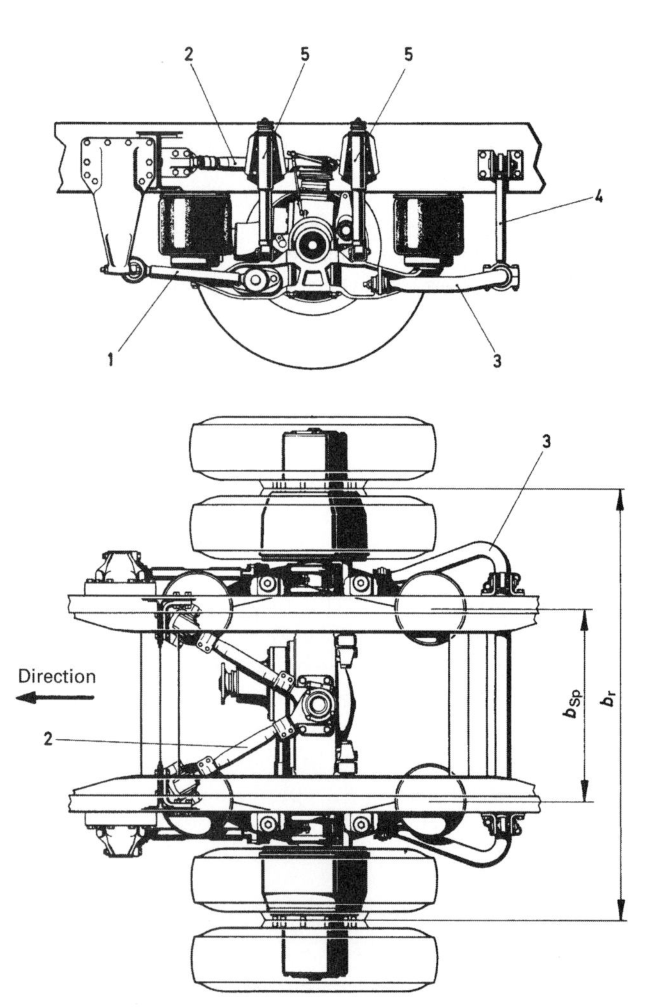

and it drives the rear wheels. The power plant can sit behind the axle (Fig. 8.1-45, rear-mounted engine) or in front of it (Fig. 8.1-46, central engine). This configuration makes it impossible to have a rear seat as the engine occupies this space. The resulting two-seater is only suitable as a sports or rally car.

The disadvantages of rear and central engine drive on passenger cars are:

- moderate straight running abilities (caster offset at ground angles of up to $\tau = 8°$ are factory set);
- sensitivity to side winds;
- indifferent cornering behaviour at the stability limit (central engine);
- oversteering behaviour on bends (rear-mounted engine, see Fig. 10.1-42);

Fig. 8.1-42 Driven rear axle with air springs of the Mercedes-Benz lorry 1017 L to 2219 L 6 × 2. The axle is carried in the longitudinal and lateral directions by the two struts 1 and the upper wishbone type control arm 2. The four spring bellows sit under the longitudinal frame members and, because of the twin tyres, they have a relatively low effective b_{Sp}. The tracking width b_r divided by b_{Sp} yields approximately the ratio $i_\varphi = 2.2$.

To reduce body roll pitch the anti-roll bar 3 was placed behind the axle and is supported on the frame via the rod 4. The four shock absorbers 5 are almost vertical and are positioned close to the wheels to enable roll movements of the body to fade more quickly.

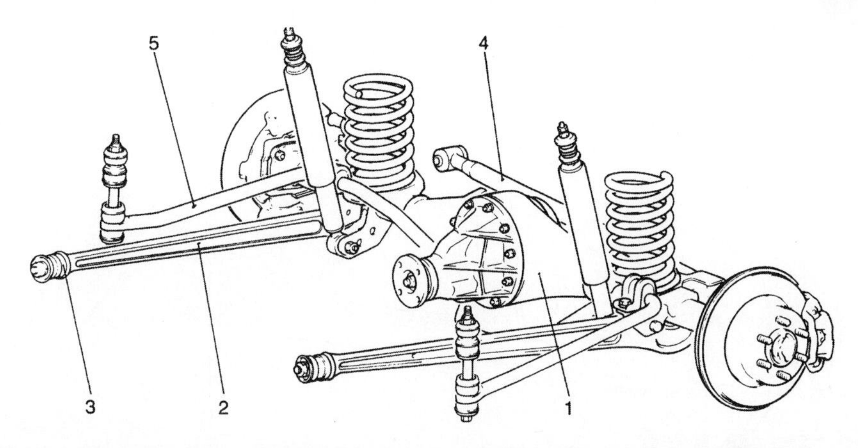

Fig. 8.1-43 The rear axle on the all-terrain, general-purpose passenger car, Mitsubishi Pajero. The rigid axle casing 1 is taken through the longitudinal control arms 2. These absorb the drive-off and braking forces (and the moments which arise) and transmit them to the frame. The rubber mountings 3 in the front fixing points 1, which also represent the vehicle pitch pole O_r, are designed to be longitudinally elastic to keep the road harshness due to the dynamic rolling hardness of the radial tyre away from the body. The Panhard rod 4 absorbs lateral forces. The anti-roll bar 5 is (advantageously) fastened a long way out on the frame (Fig. 8.1-23). The disc brakes, coil springs and almost vertical shock absorbers can be clearly seen.

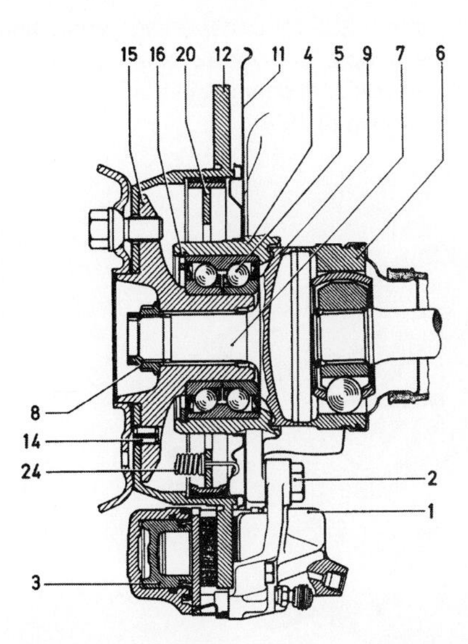

Fig. 8.1-44 Rear axle wheel hub carrier with wheel and brake. The drive shaft 7 is butt-welded to the CV slip joint 6. The drive shaft transmits the driving torque to the wheel hub 15 via a serrated profile. Part 15 is carried by the maintenance-free, two-row angular (contact) ball bearing 5. The one-part outer ring is held in the hub carrier 4 by the snap ring 16.
The seal rings on both sides sit in the permanently lubricated bearing unit. The covering panel 11 (that surrounds the brake disc 12) acts as additional dirt-protection outside, as does collar 9 of the CV joint on the inside. This grips into a cut-out in the wheel hub carrier 4 and creates a cavity. The centrifugal effect of the bell-shaped joint housing prevents ingress of dirt and water. The brake disc 12 is pulled from outside against the flange 15 and fixed by dowel 14 until the wheel is mounted. The jaws 20 of the drum brake acting as a handbrake act on the inside of part 12. At the lower end, the illustration shows the fixed calliper 1 of the disc brake. Two hexagonal bolts (item 2) fix it to the wheel hub carrier 4. Piston 3 and the outer brake pad are shown cut away (illustration: Mercedes-Benz).

- difficult to steer on ice because of low weight on the front wheels;
- uneven tyre wear front to rear (high rear axle load, see Fig. 8.1-36);
- the engine mounting must absorb the engine moment times the total gear ratio;
- the exhaust system is difficult to design because of short paths;
- the engine noise suppression is problematic;
- complex gear shift mechanism;
- long water paths with front radiators (Fig. 8.1-46);
- high radiator performance requirement because of forced air cooling, the electric fan can only be used on the front radiator;
- the heating system has long paths for hot water or warm air;
- the fuel tank is difficult to house in safe zones;
- the boot size is very limited.

In the case of vehicles with a short wheel base and high centre of gravity with the engine on or behind the rear axle, there is a danger that the vehicle will overturn if it is rolling backwards down a steep slope and the parking brake, which acts upon the rear axle, is suddenly applied.

As a result of the logical further development of the kinematics and elasto-kinematics of the axles, Porsche have succeeded in improving straight running as well as cornering in the steady state (vehicles now understeer slightly up to high lateral accelerations) and transient state as well as when subject to torque steer effects. Even in the case of the Boxsters (with mid-engine, see Fig. 8.1-46, since 1996) and 911 (water-cooled since 1997), Porsche are adhering to rear-wheel drive (whereas

Fig. 8.1-45 VW Transporter, a light truck which could be used either as an eight-seater bus or for transporting goods, and which has the optimal axle load distribution of 50%/50% in almost all loading conditions. The double wishbone suspension at the front, the semi-trailing link rear axle and the rack and pinion steering, which is operated via an additional gear set in front, can be seen clearly. To achieve a flat load floor throughout, VW changed the Transporter to front-wheel drive in 1990.

Fig. 8.1-46 The Porsche Boxster (1996) has a water-cooled engine which is longitudinally installed in front of the rear axle. The front axle is designed as a spring strut-type axle. The transverse link is arranged almost in extension of the wheel axle; it is connected to the longitudinal link by a strut bush which is soft for reasons of comfort. This open design and link geometry make it possible to combine a high level of driving precision, a result of rigid wheel control, with riding comfort, owing to the longitudinal elasticity of the axle. At a camber angle of 8°, good straight running results from the large castor displacement of 41 mm. The kingpin offset is −7 mm and the disturbing force lever arm is 83 mm. The pitch centre of the front axle was located near to the road to achieve kinematic wheel recession of the axle, which is important for riding comfort, with the result that braking-torque compensation is only 10%.
The rear axle is also a spring strut-type axle in an open link design; the wheel carrier, hub and bushes as well as the transverse link are the same as those found on the front axle. The open design makes it possible to have an inwardly inclined elastokinematic axis of rotation, so that a stabilizing toe-in position of the rear wheels is produced during braking. The axle can also be designed to understeer when subject to lateral forces.
The main disadvantages of the mid-engine design are apparent from the boot space: only 130 l are available at both the front and back.

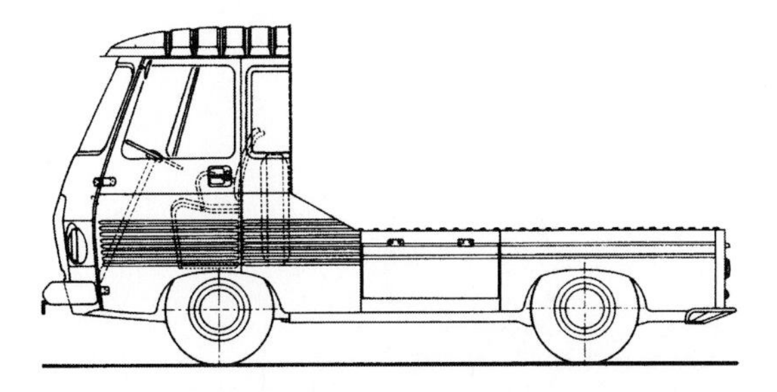

Fig. 8.1-47 The low cargo area on the Peugeot light commercial vehicle J 5/J 7 is achieved due to front-wheel drive and a semi-trailing link axle to the rear (similar to the one in Fig. 8.1-63).

the VW Transporter, Fig. 8.1-45, has not been built since 1991) and, in so doing, obtain the following benefits:

- very agile handling properties as a result of the small yawing moment;
- very good drive-off and climbing capacity, almost irrespective of load;
- a short power flow because the engine, gearbox and differential form one compact unit;
- light steering due to low front axle load;
- good braking force distribution;
- simple front axle design;
- easy engine dismantling (only on rear engine);
- no tunnel or only a small tunnel in the floor pan;
- a small overhang to the front is possible.

8.1.6 Front-wheel drive

The engine, differential and gearbox form one unit, which can sit in front of, over, or behind the front axle. The design is very compact and, unlike the standard design, means that the vehicle can either be around 100–300 mm shorter, or the space for passengers and luggage can be larger. These are probably the main reasons why, worldwide, more and more car manufacturers have gone over to this design. In recent years only a few saloons of up to 2 l capacity without front-wheel drive have come onto the market. Nowadays, front-wheel - drive vehicles are manufactured with V6 and V8 engines and performances in excess of 150 kW.

However, this type of drive is not suitable for commercial vehicles as the rear wheels are highly loaded and the front wheels only slightly. Nevertheless, some light commercial vehicle manufacturers accept this disadvantage so they can lower the load area and offer more space or better loading conditions (Fig. 8.1-47). The propshafts necessary on standard passenger cars would not allow this.

8.1.6.1 Types of design

8.1.6.1.1 Engine mounted longitudinally 'north–south' in front of the axle

In-line or V engines mounted in front of the axle – regardless of the wheelbase – give a high front axle load,

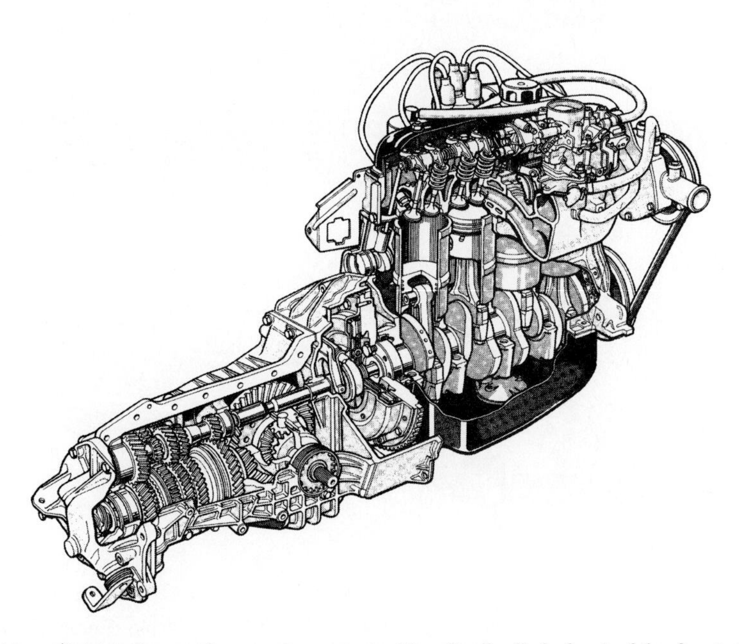

Fig. 8.1-48 In front-wheel drive vehicles the engine can be mounted longitudinally in front of the front axle with the manual gearbox behind. The shaft goes over the transverse differential (illustration: Renault).

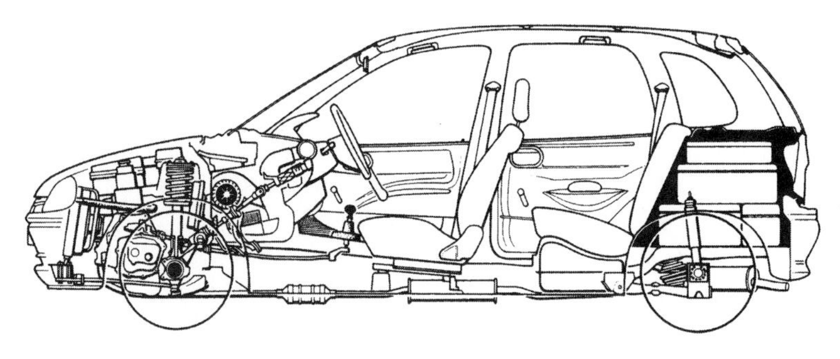

Fig. 8.1-49 Compact power train unit on the Vauxhall Corsa (1997). The engine is transverse mounted with the gearbox on the left. The McPherson front axle and safety steering column can be seen clearly.

whereby the vehicle centre of gravity is pushed a long way forwards (Fig. 8.1-48). Good handling in side winds and good traction, especially in the winter, confirm the merits of a high front axle load, whereas the heavy steering from standing (which can be rectified by power-assisted steering), distinct understeering during cornering and poor braking force distribution would be evidence against it.

This type of design, as opposed to transverse mounting, is preferred in the larger saloons as it allows for relatively large in-line engines. The first vehicles of this type were the Audi 80 and 100. Inclining the in-line engine and placing the radiator beside it means the front overhang length can be reduced. Automatic gearboxes need more space because of the torque converter. This space is readily available with a longitudinally mounted engine.

A disadvantage of longitudinal engines is the unfavourable position of the steering gear: this should be situated over the gearbox. Depending on the axle design, this results in long tie rods with spring strut (McPherson) front axles (Fig. 8.1-57).

8.1.6.1.2 Transverse engine mounted in front of the axle

In spite of the advantage of the short front overhang, only limited space is available between the front wheel housings (Figs. 8.1-49 and 8.1-50). This restriction means that

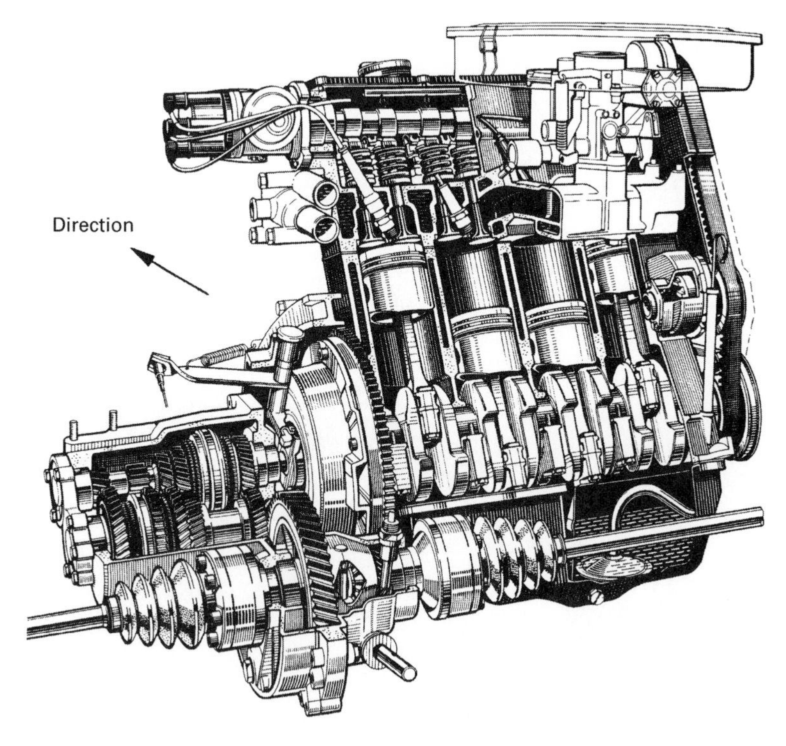

Fig. 8.1-50 Layout of transverse engine, manual gearbox and differential on the VW Polo. Because the arrangement is offset, the axle shaft leading to the left front wheel is shorter than that leading to the right one. The shifter shaft between the two can be seen clearly. The total mechanical efficiency should be around $n \approx 0.9$.

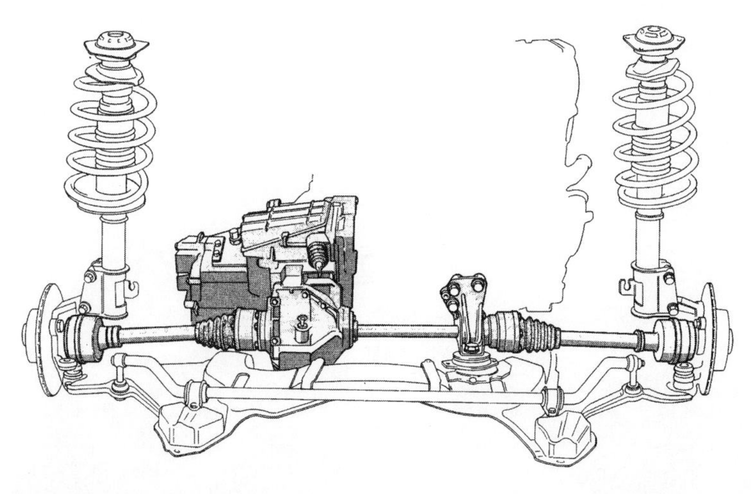

Fig. 8.1-51 Gearbox unit on the Lancia Thema, located beside the transverse engine and between the front axle McPherson struts. Owing to the high engine performance, the design features two equal-length axle shafts joined by an intermediate shaft. There are also internally ventilated disc brakes.

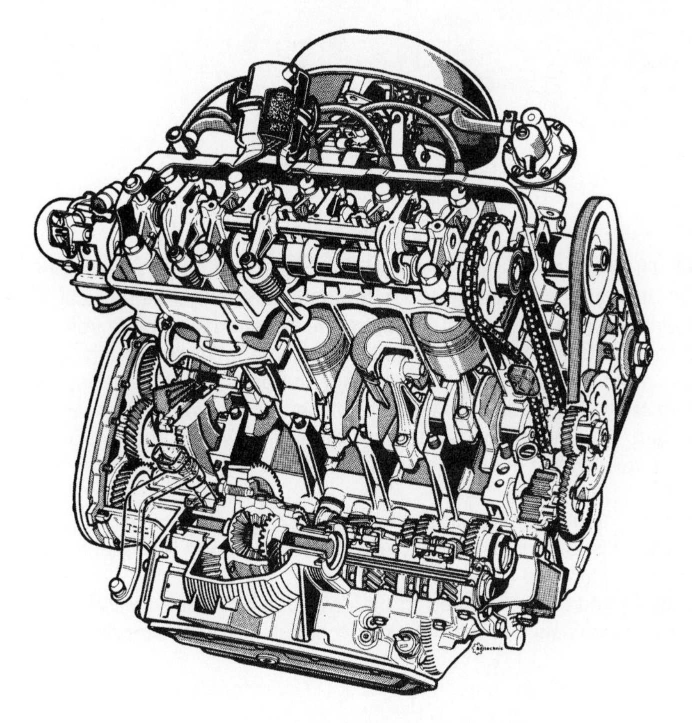

Fig. 8.1-52 Arrangement of the gearbox beneath the motor, which is inclined towards the rear, and the differential gear placed behind it. A single oil-economy undertakes the supply, in this case, of the driving unit, narrow in its design (Works Illustr. Fa Peugeot).

engines larger than an in-line four cylinder or V6 cannot be fitted in a medium-sized passenger car. Transverse, asymmetric mounting of the engine and gearbox may also cause some performance problems. The unequal length of the drive shafts affects the steering. During acceleration the vehicle rises and the drive shafts take on different angular positions, causing uneven moments around the steering axes. The difference between these moments to the left and to the right causes unintentional steering movements resulting in a noticeable pull to one side; drive shafts of equal length are therefore desirable. This also prevents different drilling angles in the drive shaft causing timing differences in drive torque build-up.

The large articulation angle of the short axle shaft can also limit the spring travel of the wheel. To eliminate the adverse effect of unequal length shafts, passenger cars with more powerful engines have an additional bearing next to the engine and an intermediate shaft, the ends of which take one of the two sliding CV joints with angular mobility (Figs. 8.1-51 and 8.1-17). Moreover, 'flexing vibration' of the long drive shaft can occur in the main driving range. Its natural frequency can be shifted by clamping on a suppression weight (Fig. 9.1).

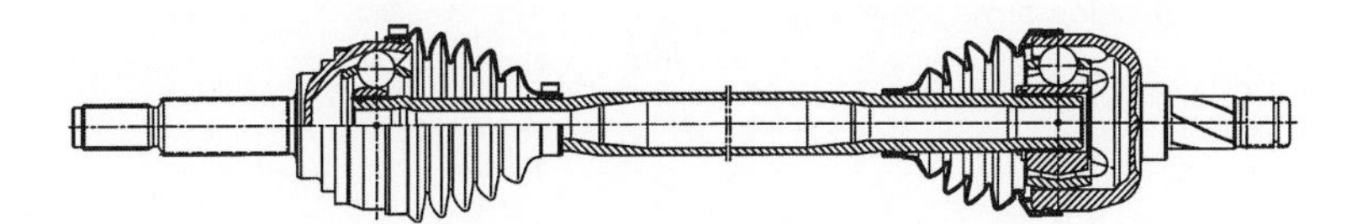

Fig. 8.1-53 Front-wheel output shaft of GKN Automotive. A CV sliding joint is used on the gearbox side and a CV fixed joint is used on the wheel side (Fig. 8.1-17). The maximum bending angles are 22° for the sliding joint and 47° for the fixed joint. For reasons of weight, the sliding joint is placed directly into the differential and fixed axially by a circlip. A central nut secures attachment on the wheel side. The intermediate shaft is designed as a carburized, shaped hollow shaft.

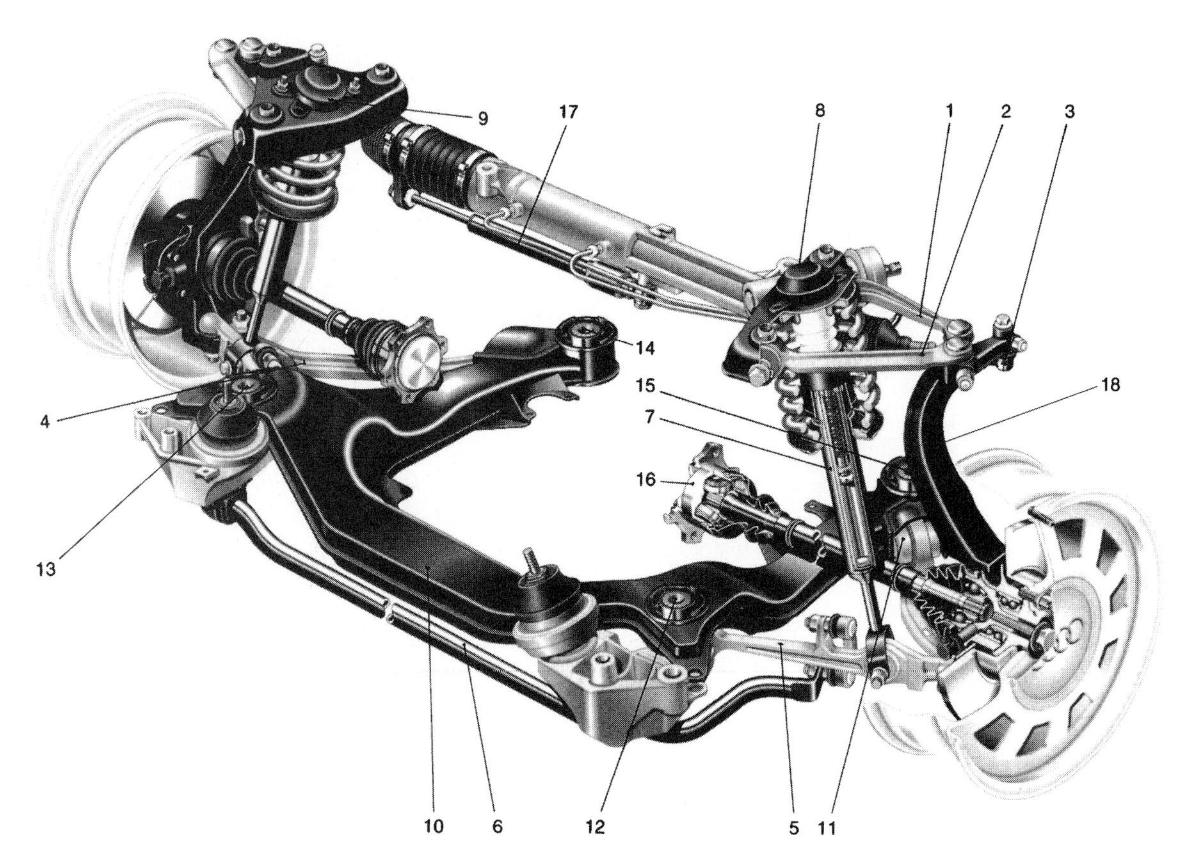

Fig. 8.1-54 Double wishbone front axle assembly of the Audi A4. The Audi A6 of 1997, the Audi A8 (1996) and the VW Passat of 1996 are similar. Four individual transverse 'arms' on each side form what is effectively a double wishbone arrangement which provides lateral and longitudinal wheel location. The two upper members (1 and 2) are attached to the spheroidal graphite iron hollow-section stub-axle post (18) by low-friction ball and socket joints. The track rod (3) provides the steering input through a horizontal extension of this stub-axle post which forms a steering arm. The two lower suspension members consist of the radius arm (4) and the transverse arm (5). This latter must be capable of reacting high loads from the anti-roll bar (6) and spring/damper (7) attachment points. The co-axial spring/damper assembly incorporates a polyurethane rubber bump-stop, as well as the hydro-mechanical tension rod stop. The spring/damper unit (7) and the inner bearings of the upper members (1) and (2) are mounted on the upper suspension bracket.
The inner ends of suspension members (4 and 5) are located by substantial rubber mountings on the inside of the sub-frame (10). The rear mounting (11) is hydraulically damped to absorb any harshness associated with radial tyres. The vehicle body is mounted on four rubber mountings (12 to 15) of specified elasticity to ensure a high standard of ride comfort.
The inner drive shafts are located to the rear of the spring dampers (7) and are connected to the drive-line by 'tripot' flexible couplings (16). The outer ends of the drive shafts transmit the drive to the wheels through double-row angular contact bearings. The inner races of these bearings are integral with the wheel hubs.
The hydraulically assisted steering rack is mounted on the vehicle's scuttle, with the steering damper (17) located on one side of the steering housing, and the other side attached to the steering rack.
The high location of the wheel-joint facilitates space saving and a consequent reduction of the lever-arm forces, and allows the inner valences of the mudguard to be located further outboard.
The advantages of this type of four-link suspension include the location of the points E and G of the paired arms 1 and 2, likewise 4 and 5, which are subjected to outward thrusts resulting from steering input to the steering-arm, which are thereby compressed through $r = 10$ mm. Moreover the high location of the point E (Fig. 8.1-5)–together with the negative steering roll radius $r = -7$ mm – helps to reduce the loads in all components of the front suspension system.
Other design parameters of the suspension arrangement are:

King-pin inclination	$\varepsilon = 30'$
Caster angle	$\tau = +3°50'$
Camber angle	$\sigma = +3°45'$
Caster linear trail	$\eta = +5.5$ mm

Fig. 8.1-55 Double wishbone front suspension on the Honda models, Prelude and Accord, with short upper wishbones with widely spaced bearings, lower transverse control arms and longitudinal rods whose front mounts absorb the dynamic rolling stiffness of the radial tyres. The spring shock absorbers are supported via fork-shaped struts on the transverse control arms and are fixed within the upper link mounts. This point is a good force input node. Despite the fact that the upper wheel carrier joint is located high, which gives favourable wheel kinematics, the suspension is compact and the bonnet can be low to give aerodynamic advantages. The large effective distance *c* between the upper and lower wheel hub carrier joints seen in Fig. 8.1-5 results in low forces in all mounts and therefore less elastic deflection and better wheel control.

8.1.6.2 Advantages and disadvantages of front-wheel drive

Regardless of the engine position (see Fig. 8.1-52), front-wheel drive has numerous advantages:

- there is load on the steered and driven wheels;
- good road-holding, especially on wet roads and in wintry conditions – the car is pulled and not pushed (Fig. 8.1-35);
- good drive-off and sufficient climbing capacity with only few people in the vehicle;
- tendency to understeer in cornering;
- insensitive to side wind;
- although the front axle is loaded due to the weight of the drive unit, the steering is not necessarily heavier (in comparison with standard cars) during driving;
- axle adjustment values are required only to a limited degree for steering alignment;
- simple rear axle design – e.g. compound crank or rigid axles – possible;
- long wheelbase making high ride comfort possible;
- short power flow because the engine, gearbox and differential form a compact unit;
- good engine cooling (radiator in front), and an electric fan can be fitted;
- effective heating due to short paths;
- smooth car floor pan;
- exhaust system with long path (important on cars with catalytic converters);
- a large boot with a favourable crumple zone for rear end crash.

The disadvantages are:

- under full load, poorer drive-off capacity on wet and icy roads and on inclines (Fig 8.1-36);
- with powerful engines, increasing influence on steering;
- engine length limited by available space;
- with high front axle load, high steering ratio or power steering is necessary;
- with high located, dash-panel mounted rack and pinion steering, centre take off tie rods become necessary (Figs. 8.1-57 and 9.1-39) or significant kinematic toe-in change practically inevitable;
- geometrical difficult project definition of a favourable interference force lever arm and a favourable steering roll radius (scrub radius);
- engine gearbox unit renders more difficult the arrangement of the steering package;
- the power plant mounting has to absorb the engine moment times the total gear ratio;
- it is difficult to design the power plant mounting – booming noises, resonant frequencies in conjunction with the suspension, tip in and let off torque effects etc., need to be suppressed;
- with soft mountings, wavy road surfaces excite the power plant to natural frequency oscillation (so-called 'front end shake';
- there is bending stress on the exhaust system from the power plant movements during drive-off and braking (with the engine);
- there is a complex front axle, so inner drive shafts need a sliding CV joint (Fig. 8.1-53);
- the turning and track circle is restricted due to the limited bending angle (up to 50°) of the drive joints;
- high sensitivity in the case of tyre imbalance and non-uniformity on the front wheels;
- higher tyre wear in front, because the highly loaded front wheels are both steered and driven;

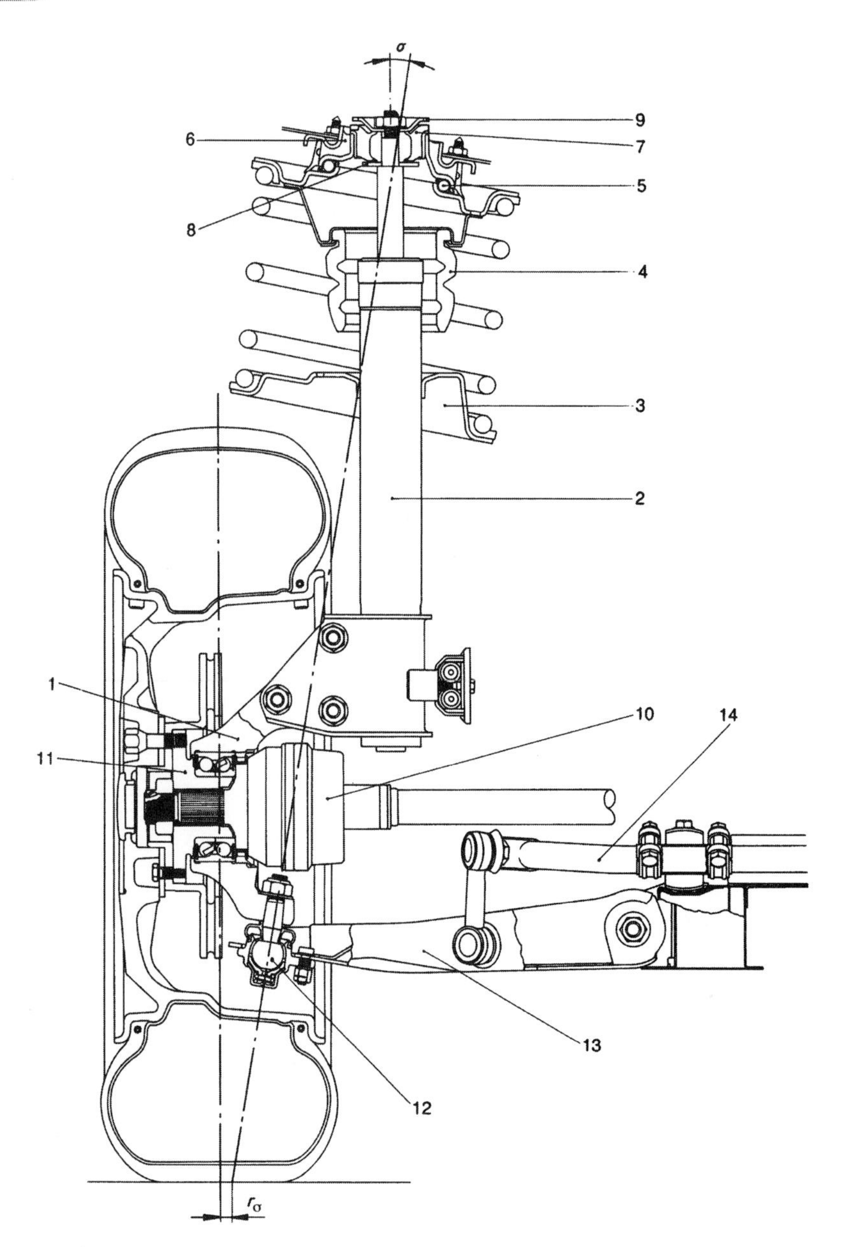

Fig. 8.1-56 Lancia front axle. The McPherson strut consists of the wheel hub carrier 1 and the damping part 2; the two are connected by three screws. The lower spring seat 3 sits firmly on the outer tube and also acts as a buffer for the supplementary spring 4. This surrounds the outer tube 2 giving a longer bearing span (path *l–o*, Fig. 8.1-11). The supporting bearing 5 is arranged diagonally and thus matches the position of the coil spring which is offset to reduce damping friction. The rubber bearing 6 absorbs the spring forces, and the rubber bearing 7 absorbs the forces generated by the damping. Disc 8 acts as a compression buffer and plate 9 acts as a rebound buffer for this elastic bearing. Both parts come into play if the damping forces exceed certain values. The centre of the CV joint 10 lies in the steering axis and the wheel hub 11 fits onto a two-row angular (contact) ball bearing. Guiding joint 12 sits in a cone of the wheel hub carrier 1 and is bolted to the lower transverse control arm 13. Inelastic ball joints provide the connection to the anti-roll bar 14. The steering axis inclination σ between the centre point of the upper strut mount and guiding joint 12 and the (here slightly positive) kingpin offset at ground (scrub radius) r_σ are included.

- poor braking force distribution (about 75% to the front and 25% to the rear);
- complex gear shift mechanism which can also be influenced by power plant movements.

The disadvantage of the decreased climbing performance on wet roads and those with packed snow can be compensated with a drive slip control (ASR) or by shifting the weight to the front axle. On the XM models, Citroën moved the rear axle a long way to the rear resulting in an axle load distribution of about 65% to the front and 35% to the back. The greater the load on the front wheels, the more the car tends to understeer, causing adverse steering angles and heavy

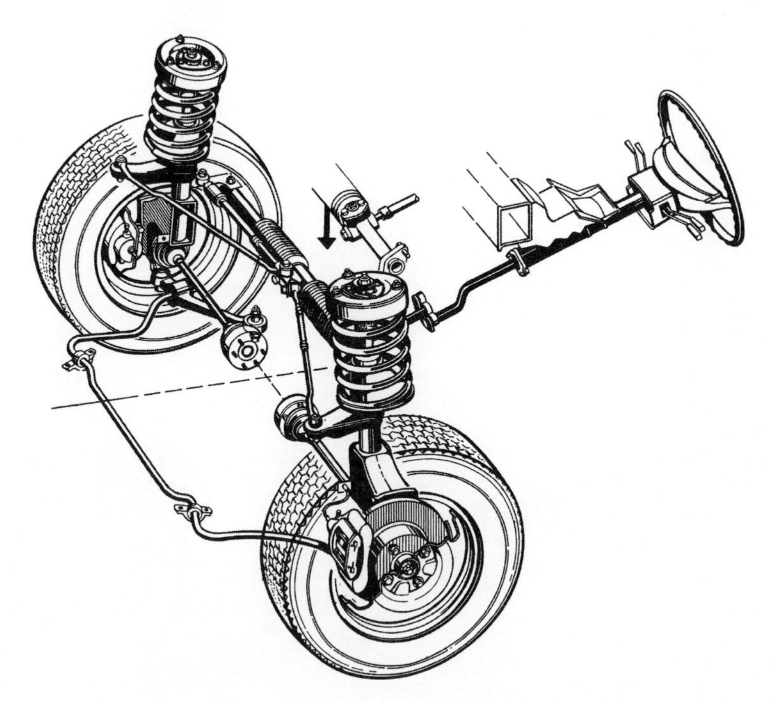

Fig. 8.1-57 Driven McPherson front axle on the Audi 6 (Audi 100, 1991). The dynamic rolling hardness of the radial tyre is absorbed by the rubber bearings, which sit in the lower transverse control arms. The inner sleeves of these bearings take the arms of the anti-roll bar, which also act as a trailing link (classic McPherson construction). To avoid greater toe-in changes when the wheels are at full bump/rebound-travel, centre take-off tie rods are used on the rack and pinion steering higher up and in the centre (Fig. 9.1-39). Together with these rods, the steering damper located on the right is fastened to the end of the steering rack. The engine is mounted longitudinally, which means the drive shafts are of equal length.
The development of the axle since 1997 is shown in Fig. 8.1-54.

steering, which makes power steering mandatory (see Section 9.1.2.5).

8.1.6.3 Driven front axles

The following are fitted as front axles on passenger cars, estate cars and light commercial vehicles:

- double wishbone suspensions;
- multi-link axles;
- McPherson struts, and (only in very few cases);
- damper struts.

On double wishbone suspensions, the drive shafts require free passage in those places where the coil springs are normally located on the lower suspension control arms. This means that the springs must be placed higher up with the disadvantage that (as on McPherson struts) vertical forces are introduced a long way up on the wheel house panel. It is better to leave the springs on the lower suspension control arms and to attach these to the stiffer body area where the upper control arms are fixed. Shock absorbers and springs can be positioned behind the drive shafts (see Fig. 8.1-54) or sit on split braces, which grip round the shafts and are jointed to the lower suspension control arms (Fig. 8.1-55). The axle is flatter and the front end (bonnet contour) can be positioned further down. The upper suspension control arms are relatively short and have mountings that are wide apart. This increases the width of the engine compartment and the spring shock absorber unit can also be taken through the suspension control arms; however, sufficient clearance to the axle shaft is a prerequisite. Due to the slight track width change, the change of camber becomes favourable. Furthermore, the inclination of the control arms provides an advantageous radius arm axis position and anti-dive when braking (see also Fig. 8.1-75).

Most front-wheel drives coming on to the market today have McPherson struts. It was a long time after their use in standard design cars that McPherson struts were used at the front axle on front-wheel drive vehicles. The drive shaft requires passage under the damping part (Fig. 8.1-56). This can lead to a shortening of the effective distance l–o, which is important for the axle (Fig. 8.1-11), with the result that larger transverse forces $F_{Y,C}$ and $F_{Y,K}$ occur on the piston and rod guide and therefore increase friction.

On front-wheel-drive vehicles there is little space available to fit rack and pinion steering. If the vehicle has spring dampers or damper struts, and if the steering gear is housed with short outer take-off tie rods, a toe-in change is almost inevitable (Fig. 9.1-4). A high

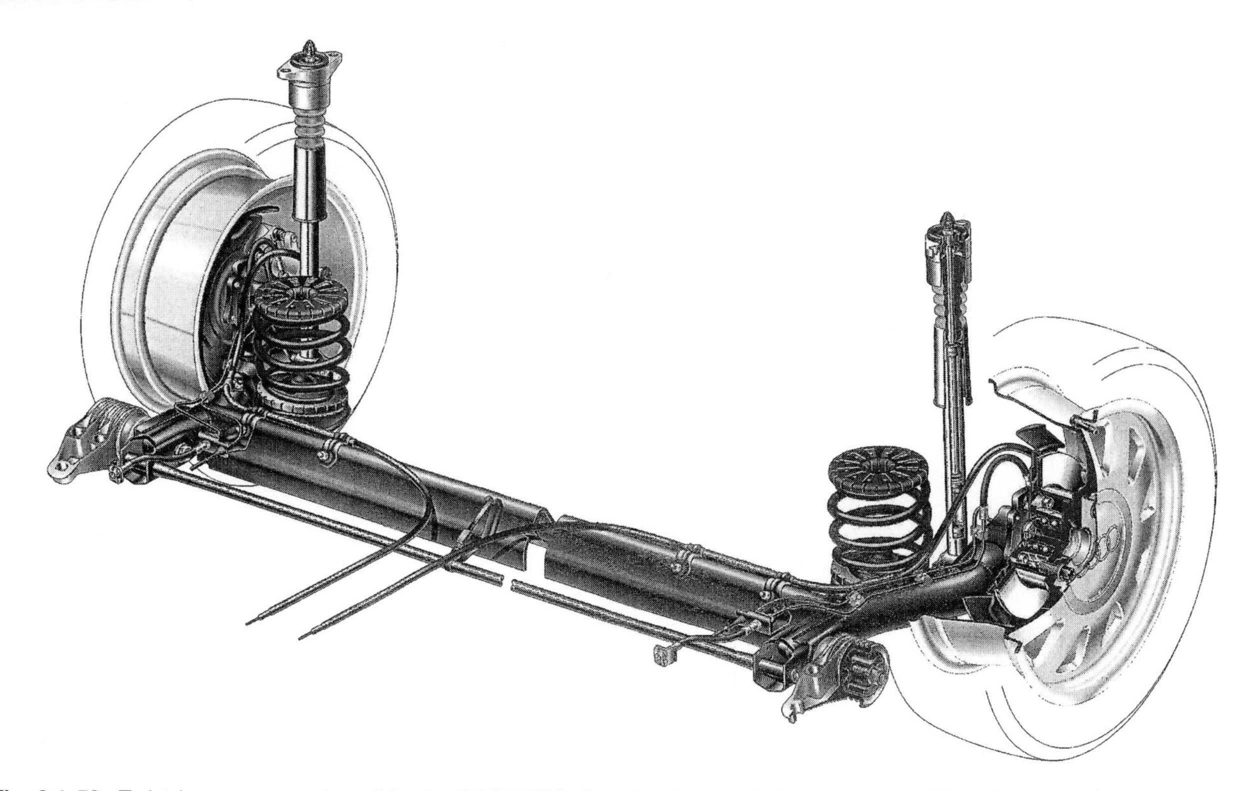

Fig. 8.1-58 Twist-beam suspension of the Audi A6 (1997). An advantageously large support width of the guide on the links – important for force application – was chosen because of the overhung arrangement. The flexurally resistant, but torsionally soft V section profile of the axle is in an upright position in order to ensure that the suspension has roll understeer properties through the high position of the centre of thrust of the profile. The instantaneous centre height is 3.7 mm and the toe-in alteration is 0.21 min/mm. Braking-torque compensation of 73% is reached. The stabilizer situated in front of the axis of rotation increases the lateral rigidity of the axle design, because it accepts tension forces upon the occurrence of lateral wheel forces. The linear coil springs mounted on noise-insulating moulded rubber elements on both sides are separated from the shock absorbers to allow the maximum loading width of the boot as a result of their location under the side rails. The gas-pressure shock absorbers support additional springs made of cellular polyurethane which act softly, through specific rigidity balancing, to avoid uncomfortable changes in stiffness when reaching the limits of spring travel. Owing to the rigid attachment of the shock absorbers to the bodywork, these also work at low amplitudes; so-called 'parasitic' springing resulting from the unwanted flexibility of wheel suspension or bodywork components is thereby reduced.

steering system can readily be attached to the dash panel (Fig. 8.1-57), but a centre take-off is then necessary and the steering system becomes more expensive (Figs. 9.1-1, 9.1-11 and 9.1-39). Moreover, the steering force applied to the strut is approximately halfway between mountings E and G (Fig. 8.1-11). The inevitable, greater yield in the transverse direction increases the steering loss angle and makes the steering less responsive and imprecise.

8.1.6.4 Non-driven rear axles

If rear axles are not driven, use can frequently be made of more simple designs of suspension such as twist-beam or rigid.

8.1.6.4.1 Twist-beam suspension

There are only two load paths available on each side of the wheel in the case of twist-beam axles. As a result of their design (superposed forces in the links, only two load paths), they suffer as a result of the conflicting aims of longitudinal springing – which is necessary for reasons of comfort – and high axle rigidity – which is required for reasons of driving precision and stability. This is particularly noticeable with the loss of comfort resulting from bumpy road surfaces. If the guide bearings of the axle are pivoted, the superposition of longitudinal and lateral forces should particularly be taken into consideration. As a result of the design, twist-beam suspensions exhibit unwanted oversteer when subject to lateral forces as a result of deformation of the swinging arms. In order to reduce the tendency to oversteer, large guide bearings which, as 'toe-in correcting' bearings, permit lateral movements of the whole axle body towards understeer when subject to a lateral force are provided. As the introduction of longitudinal and lateral forces into the body solely occurs by means of the guide bearings, it

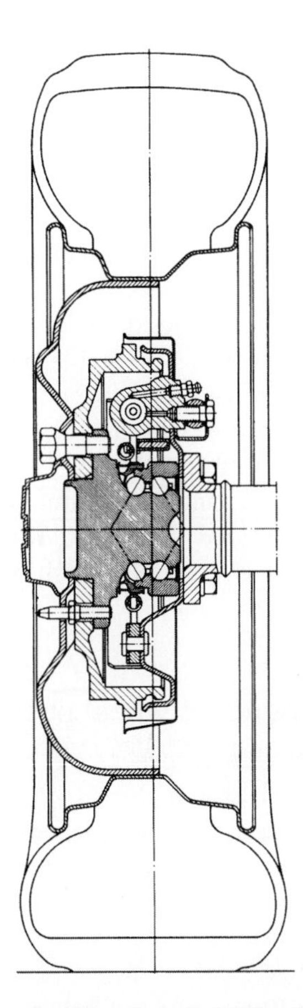

Fig. 8.1-59 Rear+wheel bearing on the Fiat Panda with a third-generation, two-row angular (contact) ball bearing. The wheel hub and inner ball bearing ring are made of one part, and the square outer ring is fixed to the rigid axle casing with four bolts (picture: SKF).

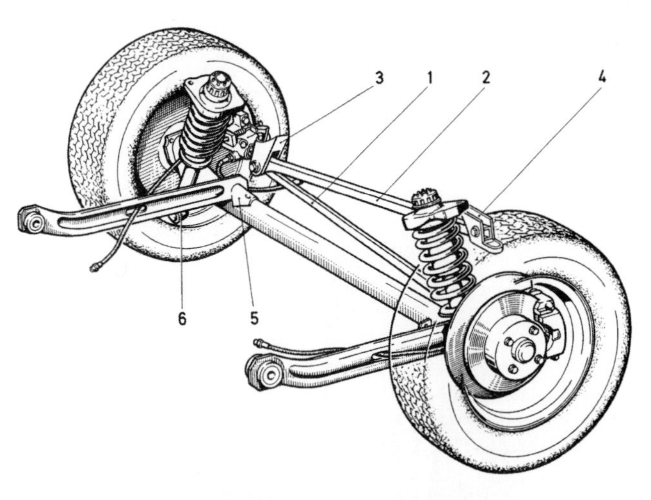

Fig. 8.1-61 Torsion crank axle on the Audi A6 (Audi 100, 1991) with spring dampers fixed a long way out at points 6 and which largely suppress body roll vibrations.
The longitudinal control arms therefore had to be welded further in to the U-profile acting as a cross-member and reinforced by shoe 5. The U-profile is also raised at the side to achieve higher torsional resistance. The anti-roll bar is located inside the U-profile.
Brace 2 distributes the lateral forces coming from the Panhard rod 1 to the two body-side fixing points 3 and 4. Bar 1 is located behind the axle, and the lateral force understeering thus caused could be largely suppressed by the length of the longitudinal control arms. Furthermore, it was possible to increase the comfort and to house an 80 l fuel tank as well as the main muffler in front of the axle.
The only disadvantage is that the link fixing points, and therefore the body roll axis O_r, moves further forward and this reduces the 'anti-dive', and that the suspension requires much space when assembled.

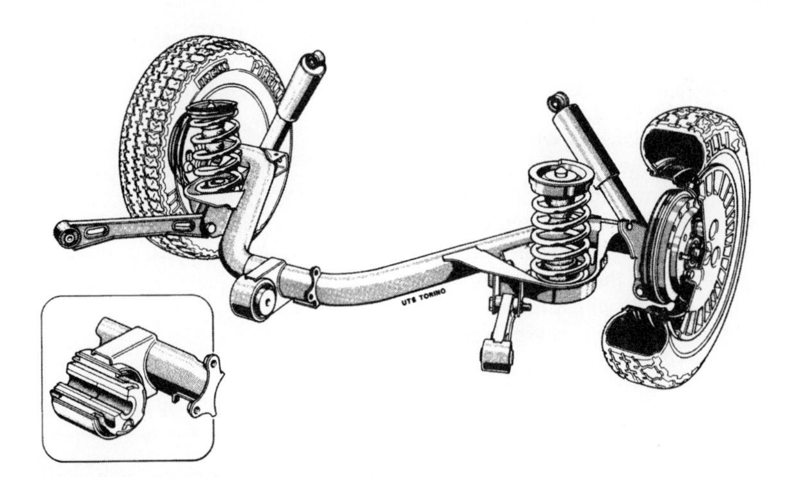

Fig. 8.1-60 'Omega' rear wheel suspension on the Lancia Y 10 and Fiat Panda, a trailing axle with a U-shaped tube, drum brakes, inclined shock absorbers and additional elastomer springs seated inside the low-positioned coil springs. The rubber element in the shaft axle bearing point, shown separately, has cut-outs to achieve the longitudinal elasticity necessary for comfort reasons; the same is true for the front bearings of the two longitudinal trailing links. The middle bearing point is also the body roll axis.
The body roll centre is located in the centre of the axle but is determined by the level of the three mounting points on the body. The lateral forces are absorbed here. The angled position of the longitudinal trailing links is chosen to reduce the lateral force oversteering that would otherwise occur (shown in Fig 8.1-31). The coil springs are located in front of the axle centre and so have to be harder, with the advantage that the body is better supported on bends.

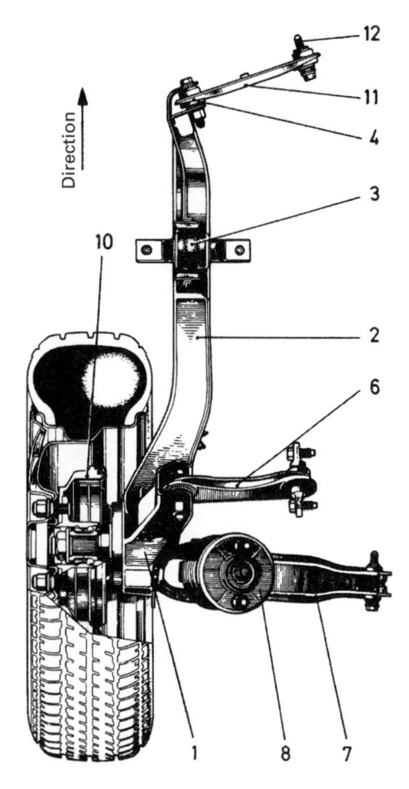

Fig. 8.1-62 Top view of the double wishbone rear axle on the Honda Civic. The trailing arm 2, which is stiff under flexure and torsion, and the wheel hub carrier 1 form a unit and, along with the two widely spaced lower transverse control arms 7 and 11, ensure precise wheel control and prevent unintentional toe-in changes. The rubber bearing in point 3, which represents the so-called 'vehicle roll axis' O_r, provides the real longitudinal wheel control of the axle. The lateral control of wheel carrier 1 is performed by the short upper transverse control arm 6 and the longer lower one 7, which accepts the spring shock absorber 8 in point 9. The length difference in the control arms gives favourable camber and track width change.
During braking, bearing 3 yields in the longitudinal direction and, due to the angled position of the links 11 when viewed from the top, the front point 4 moves inwards and the wheel goes into toe-in. Behaviour during cornering is similar: the axle understeers due to lateral force and body roll (see Figs. 8.1-1 and 8.1-77). The wheel is carried by 'third-generation' angular (contact) ball bearings on which the outside ring is also designed as a wheel hub. In models with smaller engines, brake drums (item 10) are used, which are fixed to the wheel hub.

must be ensured that the structure of the bodywork is very rigid in these places (see Figs. 8.1-30 and 8.1-58).

8.1.6.4.2 Rigid axle

Non-driven rigid axles can be lighter than comparable independent wheel suspensions. Their advantages outweigh the disadvantages because of the almost non-variable track and camber values during drive. Fig. 8.1-24 illustrates an inexpensive yet effective design:

- axle casing in steel tubing;
- suspension on single leaf springs.

The lateral and longitudinal wheel control characteristics are sufficient for passenger cars in the medium to small vehicle range and delivery vehicles. The resultant hard springing is acceptable and may even be necessary because of the load to be moved. The wheel bearing can be simple on such axles (Fig. 8.1-59). Faster, more comfortable vehicles, on the other hand, require coil springs and, for precise axle control, trailing links and a good central guide (Fig. 8.1-60) or Panhard rod. This is generally positioned behind the axle (Fig. 8.1-61).

8.1.6.4.3 Independent wheel suspension

An independent wheel suspension is not necessarily better than a rigid axle in terms of handling properties. The wheels may incline with the body and the lateral grip characteristics of the tyres decrease, and there are hardly any advantages in terms of weight. This suspension usually needs just as much space as a compound crank axle.

Among the various types, McPherson struts (Fig. 8.1-12), semi-trailing or trailing link axles (Figs. 8.1-2, 8.1-13 and 8.1-63) and – having grown in popularity for some years now – double wishbone suspensions, mostly as so-called multi-link axles (Figs. 8.1-1, 8.1-8 and 8.1-62) are all used. The latter are currently the best solution, due to:

- kinematic characteristics;
- elastokinematic behaviour;
- space requirements;
- axle weight;
- the possibility of being able to retrofit the differential on four-wheel drive (Figs. 8.1-77 and 8.1-1; see also Section 8.1.4.3).

8.1.7 Four-wheel drive

In four-wheel drives, either all the wheels of a passenger car or commercial vehicle are continuously – in other words permanently – driven, or one of the two axles is always linked to the engine and the other can be selected manually or automatically. This is made possible by what is known as the 'centre differential lock'. If a middle differential is used to distribute the driving torque between the front and rear axles, the torque distribution

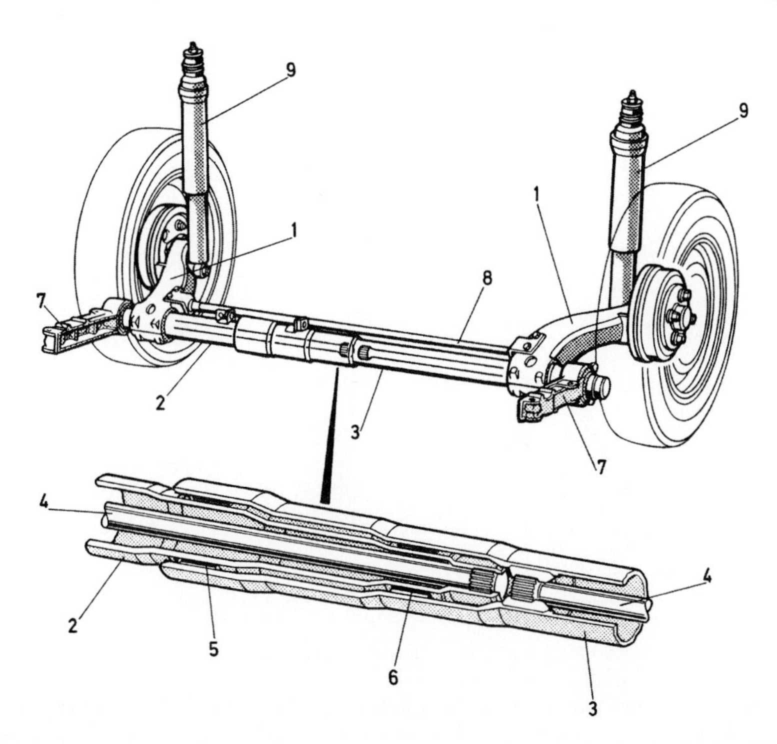

Fig. 8.1-63 Compact trailing arm rear axle, fitted by Renault to less powerful medium-sized vehicles. The short torsion bar springs grip into the guide tubes 2 and 3 in the centre of the vehicle. Parts 2, 3 and 4 are jointly subjected to torsional stresses and so the torsional stiffness of the transverse tubes contributes to the spring rate. On the outside, the cast trailing arms 1 are welded to the transverse tubes, which (pushed into each other) support each other on the torsionally elastic bearings 5 and 6. This creates a sufficiently long bearing basis, which largely prevents camber and toe-in changes when forces are generated.
The entire assembly is fixed by the brackets 7 which permits better force transfer on the body side sill. Guide tubes 2 and 3 are mounted in the brackets and can rotate, as well as the outer sides of the two torsion bars 4. The two arms thus transfer all vertical forces plus the entire springing moment to the body. The anti-roll bar 8 is connected to the two trailing arms via two U-shaped tabs. The two rubber bearings 5 and 6 located between the tubes 2 and 3 also contribute to the stabilizing effect.
The bump and rebound travel stops are fitted into the shock absorber 9. As shown in Fig. 8.1-2, on the newer models, the dampers would be inclined so that they can be fixed to the side members of the floor pan which also leads to more space between the wheel housings.

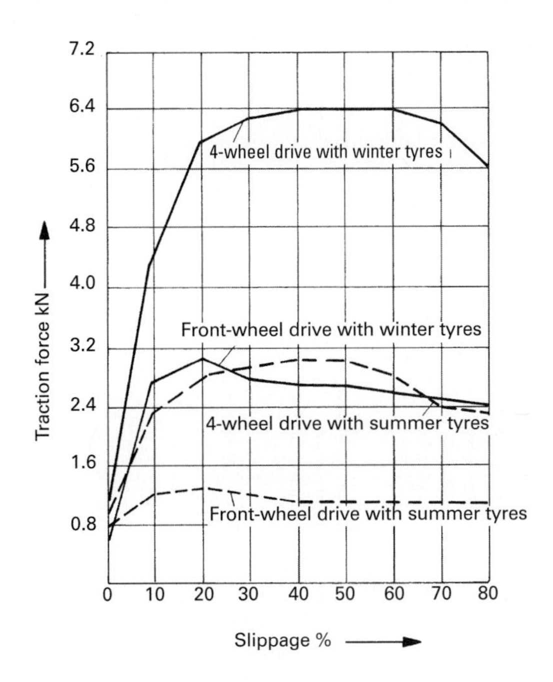

Fig. 8.1-64 With a loaded Vauxhall Cavalier on compacted snow ($\mu_{X,W} = 0.2$) driving forces are measured on the flat as a function of the slip (Fig. 10.1-33). The illustration shows the advantage of four-wheel drive, and the necessity, even with this type of drive, of fitting correct tyres. Regardless of the type of drive, winter tyres also give shorter braking (stopping) distances on these road surface conditions.

can be established on the basis of the axle–load ratios, the design philosophy of the vehicle and the desired handling characteristics. That is why Audi choose a 50%:50% distribution for the V8 Quattro and Mercedes-Benz choose a 50%:50% distribution for M class off-road vehicles, whereas Mercedes-Benz transmits only 35% of the torque to the front axle and as much as 65% to the rear axle in vehicles belonging to the E class.

This section deals with the most current four-wheel-drive designs. In spite of the advantages of four-wheel drive, suitable tyres – as shown in Fig. 8.1-64 – should be fitted in winter.

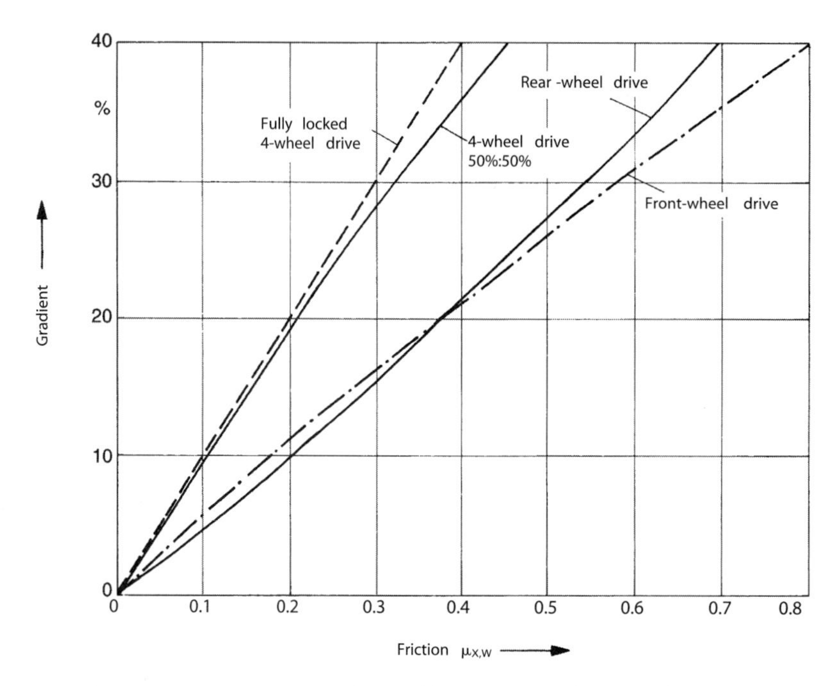

Fig. 8.1-65 Hill-climbing capacity on a homogeneous surface with front, rear-wheel and four-wheel drive, and with locked centre differential and a driving force distribution of 50%/50% on four-wheel drive. Of the cars studied, the front/rear axle load distribution was (Fig. 8.1-36):

front-wheel drive 57%/43%;
rear-wheel drive 51%/49%;
four-wheel drive 52%/48%.

8.1.7.1 Advantages and disadvantages

In summary, the advantages of passenger cars with permanent four-wheel drive over those with only one driven axle are:

- better traction on surfaces in all road conditions, especially in wet and wintry weather (Figs. 8.1-64–8.1-66);
- an increase in the drive-off and climbing capacity regardless of load;
- better acceleration in low gear, especially with high engine performance;
- reduced sensitivity to side wind;
- stability reserves when driving on slush and compacted snow tracks;
- better aquaplaning behaviour;

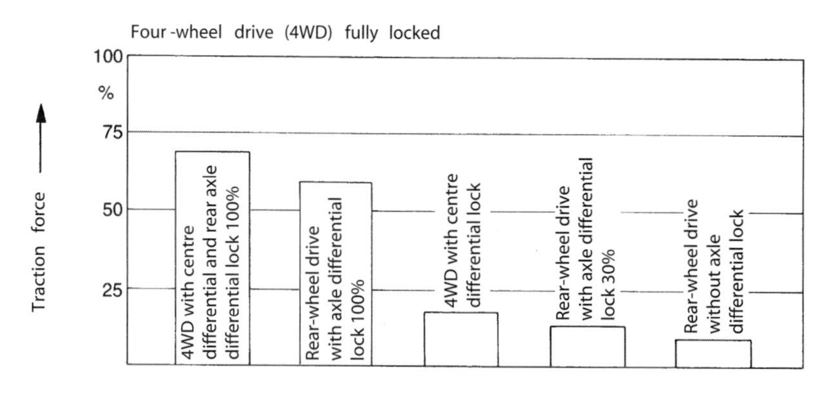

Fig. 8.1-66 Influence of the type of drive and differential lock on the propulsion force with μ split', in other words, a slippery road surface with $\mu_{x,w} = 0.1/0.8$ on one side only. 100% locking of the rear axle differential gives most benefits.
Some car manufacturers offer this option as *ASR* (or *EDS*) or using a hydraulic manual selection clutch. However, only 25% to 40% locking is provided on the multi-disc limited-slip differentials that have usually been fitted on vehicles to date.

Fig. 8.1-67 The Mercedes G all-terrain vehicle, according to DIN 70, a so-called 'all-purpose passenger car', has high ground clearance and short overhangs both front and rear. This, together with the large ramp angles (α_f, α_r,) and the overhang angle β, makes it particularly suitable for off-road driving.

- particularly suitable for towing trailers;
- balanced axle load distribution;
- reduced torque steer effect;
- even tyre wear.

According to EU Directive 70/156/EWG, a 'towed trailer load' of 1.5 times the permissible total weight has been possible for multi-purpose passenger vehicles (four-wheel passenger vehicles) since 1994.

However, the system-dependent, obvious disadvantages given below should not be ignored:

- acquisition costs;
- around 6–10% higher kerb weight of the vehicle;
- generally somewhat lower maximum speed;
- 5–10% increased fuel consumption;
- in some systems, limited or no opportunity for using controlled brake gearing, for instance for anti-locking or ESP systems;
- not always clear cornering behaviour;
- smaller boot compared with front-wheel-drive vehicles.

Predictability of self-steering properties even in variable driving situations, traction, toe-in stability and deceleration behaviour when braking, manoeuvrability, behaviour when reversing and interaction with wheel control systems are the principal characteristics of the

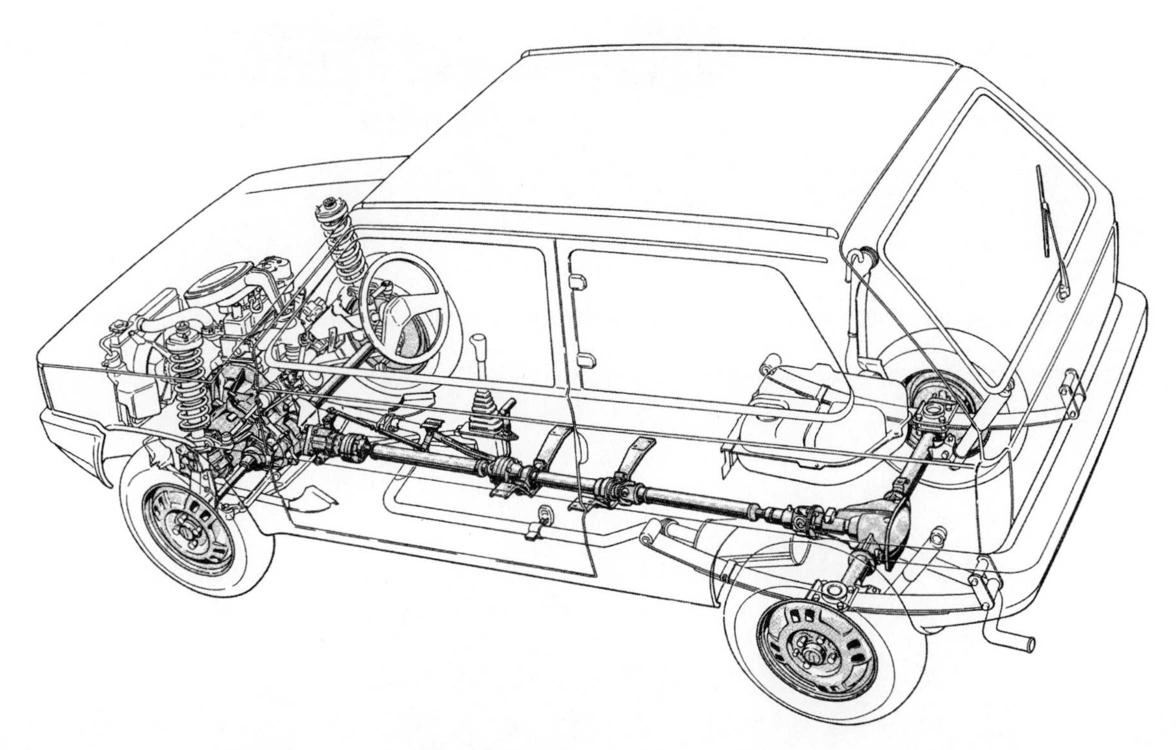

Fig. 8.1-68 The Fiat Panda Treking 4 × 4, a passenger car based on front-wheel drive with transverse engine. The vehicle has McPherson struts at the front and a rigid axle with longitudinal leaf springs at the back. The propshaft leading to it is divided into three to be able to take the rotational movements of the rigid axle around the transverse (*y*) axis during drive-off and braking and to absorb movements of the drive unit. The Fiat Panda is an estate car with the ratio:

$$i_1 = \frac{2159\ \text{mm}}{3689\ \text{mm}} = 0.59.$$

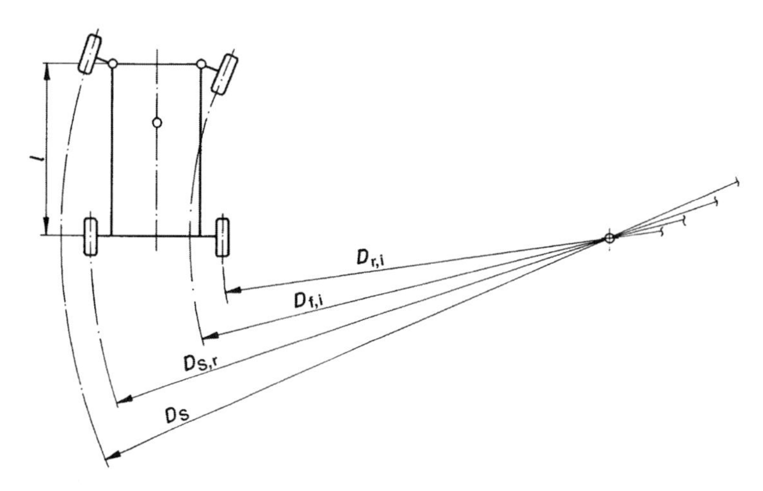

Fig. 8.1-69 The front wheel on the outside of the bend draws the largest arc during slow cornering, the track circle diameter D_S, while the inner wheel draws the considerably smaller arc $D_{f,i}$. This is the reason for the differential in the driven front axle of the front-wheel drive. The bend diameters $D_{S,r}$ and $D_{r,i}$ to the rear are even smaller, so the rolling distance of the two wheels of this axle decreases further and there can be tensions in the drive train if both axles are rigidly connected, a bend is being negotiated and when a dry road surface makes wheel slip more difficult because of high coefficients of friction.

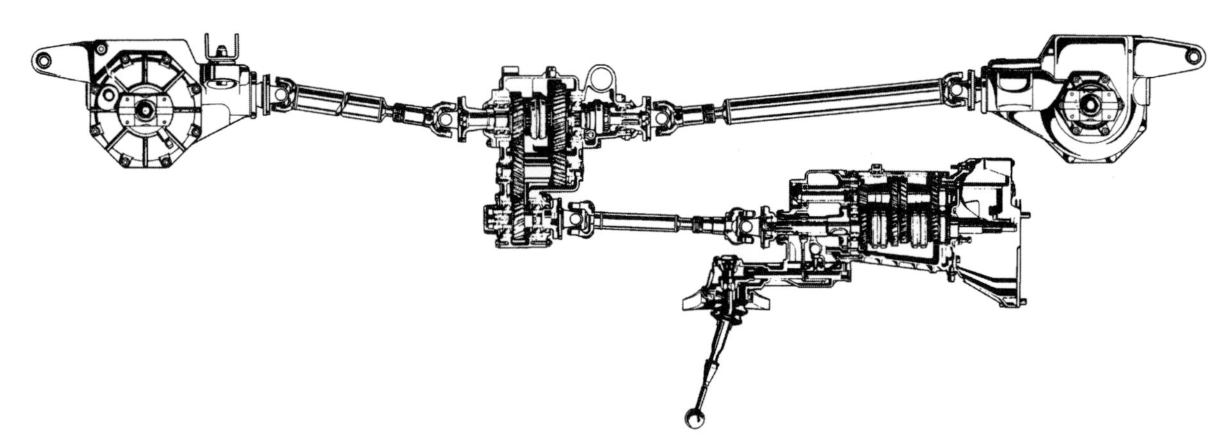

Fig. 8.1-70 Complex power distribution on the Fiat Campagnolo, a four-wheel drive, all-purpose passenger car. The drive moment is transferred from the manual gearbox via a centrally located two-gear power take-off gear to the differentials of the front and rear axles. Efficiency is not likely to be especially good.

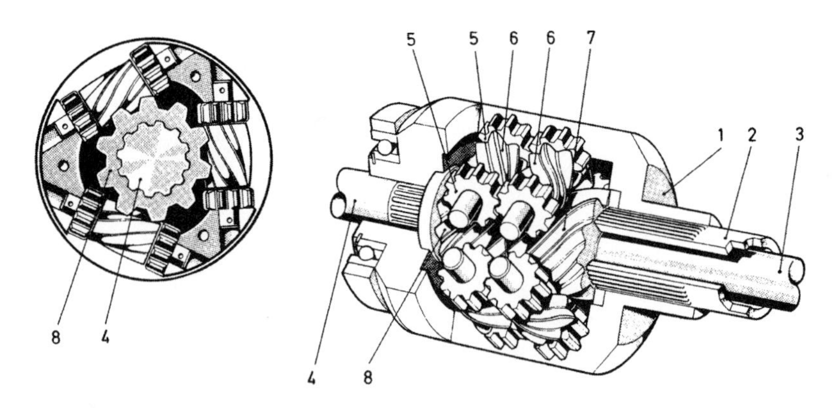

Fig. 8.1-71 Torsen central differential fitted in Quattro models (apart from the TT) by Audi. It consists of two worm gears, which are joined by spur gears and, depending on the traction requirement, can distribute the driving torque up to 75% to the front or rear axle. Under normal driving conditions 50% goes to each axle.

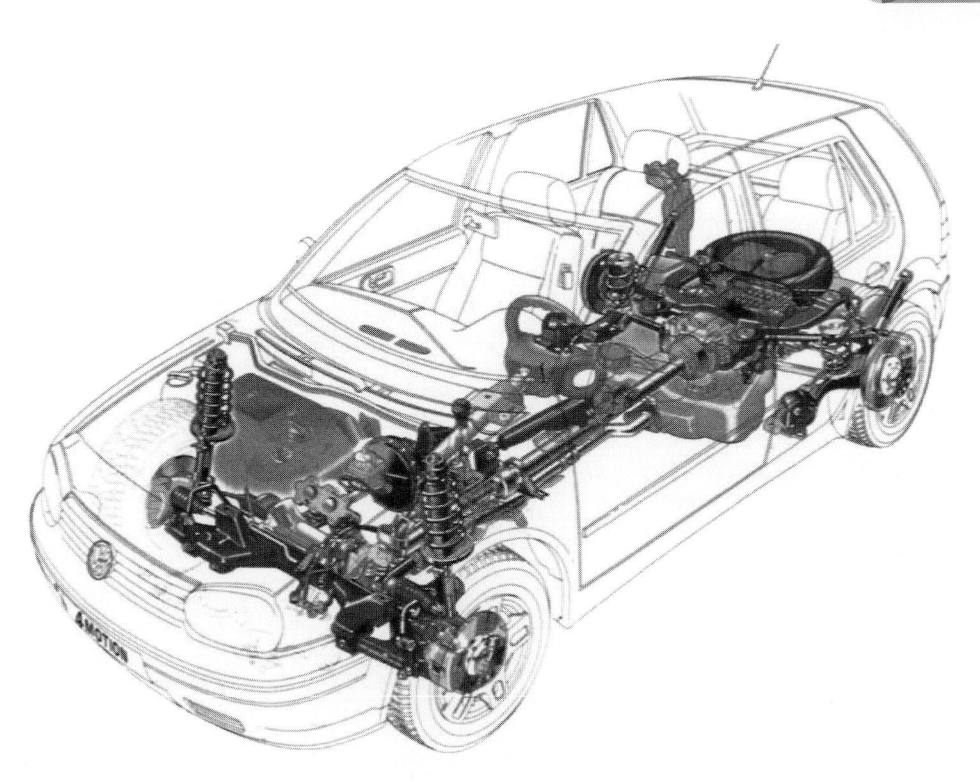

Fig. 8.1-72 Four-wheel-drive Golf 4motion (1998). In the four-wheel-drive vehicle, Volkswagen uses a multi-link suspension consisting of one longitudinal and two transverse links mounted on a subframe. The driving torque is transmitted to the rear axle via a wet multi-disc clutch by the Swedish company Haldex which is flange-mounted on the rear axle drive and runs in oil. This electronically controlled clutch can build up a coupling torque of up to 3200 N m even at small cardan-shaft rotation angles of 45° and can be combined to good effect with brake-power control systems. The drive train of the Audi TT Quattro (1998) built on the same platform is built to almost the same design.

vehicle movement dynamics which are taken into consideration for an assessment of four-wheel-drive systems.

To transmit the available engine torque to all four wheels, interaxle differentials (such as cone, planet or Torsen differentials), which are manually or automatically lockable, or clutches (such as sprag, multi-disc or visco clutches) must be installed on the propshaft between the front and rear axles. Differentials must be present on both drive axles. However, on roads with different coefficients of friction on the left and right wheels, known as 'μ-split', and with traditional differentials, each driven axle can, at most, transmit double the propulsion force of the wheels running on the side with the lower coefficient of friction (μ-low).

Higher driving forces can be achieved with an 'axle differential lock' or controlled wheel brake gearing which creates the need for 'artificial' torque on the spinning wheel. Differential-locking can only be 100% effective on the rear axle as, at the front, there would no longer be problem-free steering control. The lock partially or completely stops equalization of the number of revolutions between the left and right wheel of the respective axle and prevents wheelspin on the μ-low-side.

In passenger cars, automatic locking differentials are used between front and rear axles. These can operate mechanically (multi-disc limited slip differentials (see Torsen differential, Fig. 8.1-71) or based on fluid friction (visco lock, Fig. 8.1-74) and produce a locking degree of usually 25–40%. Higher values severely impede cornering due to the tensions in the power train (Fig. 8.1-69) Nevertheless, up to 80% locking action can be found in motor sport.

The locking action of uncontrolled or slip-dependent differential locks necessitates increased expenditure with the use of brake-power control systems (ABS, ESP). Thus, in the case of the visco lock, a free-wheel clutch is required that is engaged during reversing. Here the advantage of controlled differentials (Haldex clutch, automatically controlled locking differential, see Sections 8.1.7.4 and 8.1.7.5) becomes apparent: They can be used to maximum effect in any operating conditions and with any brake-power control system, because the locking action is produced by an electronically controlled, hydraulically activated multi-disc clutch (Fig. 8.1-67).

Traditional differential locks are increasingly being replaced because of the use of wheel control systems in both front- and rear-wheel drive as well as four-wheel-drive

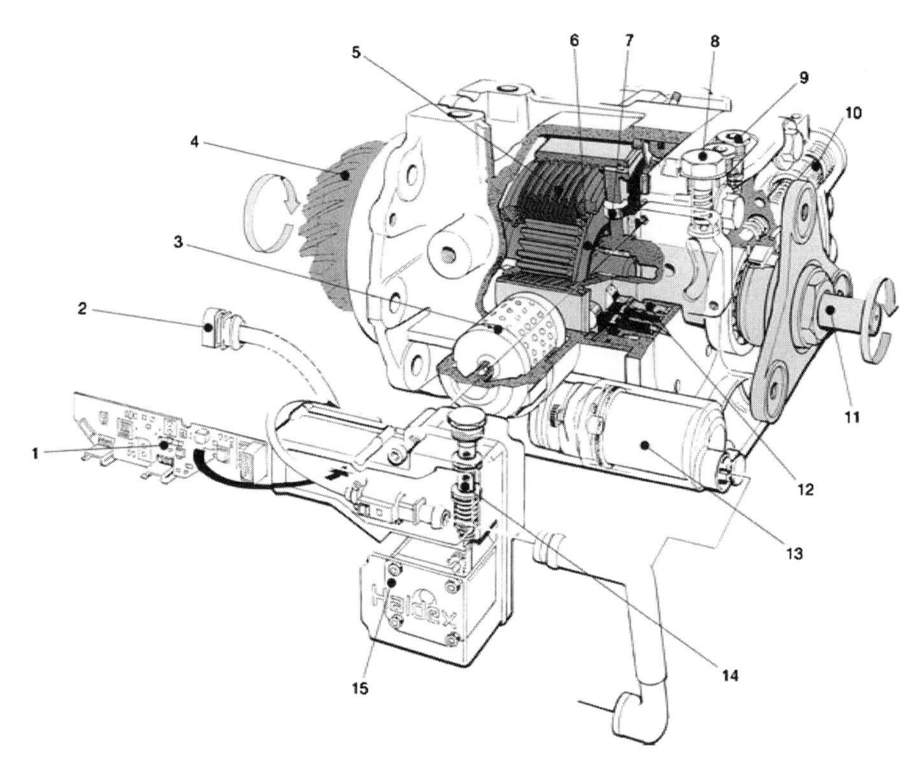

Fig. 8.1-73 Multi-disc clutch of the Swedish company Haldex, used in the Golf 4motion (1998) and Audi TT Quattro (1998). When there is a difference in speed between the front and rear axles, the disc cam 6 on the output shaft activates the working elements of the axial-piston pump 12 by means of the rollers 7. Via the control valve 14, the pressure produced activates the working piston which moves the discs. The torque transmitted is adjusted continuously by the control unit up to the maximum value, depending on the driving situation described by the wheel sensors, the signals from the slip and brake-power control systems, the position of the accelerator pedal, the engine speed etc. The clutch is disengaged when the ABS function is used. 1 electronic control unit, 2 connector vehicle (voltage, CAN, K leads), 3 oil filter, 4 shaft bevel wheel exit (rear axle gearing), 5 lamella, 6 cam plate, 7 coil, 8 relief valve, 9 pressure regulating valve, 10 accumulator, 11 input shaft, 12 axial piston pump, 13 pre-load pump, 14 control valve, 15 intermittent or step motor.

vehicles. In these systems, the wheel speed is measured, usually with the use of ABS sensors. If the speed of a wheel is established, this wheel is retarded by means of the wheel braking device. In the case of the differential, this corresponds to the build-up of torque on the side of the spinning wheel and it can now transmit torque at the higher coefficient of friction up to the adhesion limit of the wheel. Volkswagen AG calls this system electronic differential lock, as front-wheel-drive forces which correspond to those of a driven axle with differential lock and 100% locking action, and which can even be exceeded in intelligent (slip-controlled) systems, such wheel gearing systems produce. The system can ensure that the driving torque that is to be applied to the side with the retarded wheel is equal to the torque on the side with the higher coefficient of friction. This 'lost torque' must be generated by the transmission, on the one hand, and retarded by the wheel brake, on the other, so that loss of engine power and heating of the wheel brakes are produced. The braking temperatures are calculated on the basis of the braking torque and period of application of the brakes. If the temperatures calculated exceed the permissible limits, application of the brakes is discontinued during the front-wheel-drive phase until a calculated cooling of the system has taken place; the transmission then corresponds to that found in a conventional vehicle.

Another possibility for maximum utilization of grip is afforded by traction control systems in which engine power is reduced by means of the throttle, injection and ignition point so that the spinning wheels work in the region of lower slip and consequently higher adhesion. Both systems are used together, even without four-wheel drive. In models of the E and M class with an electronic traction system (ETS), Mercedes-Benz uses electronic locks instead of mechanical differential locks.

8.1.7.2 Four-wheel-drive vehicles with overdrive

In four-wheel-drive vehicles with overdrive the middle differential is not used. The engine torque is distributed

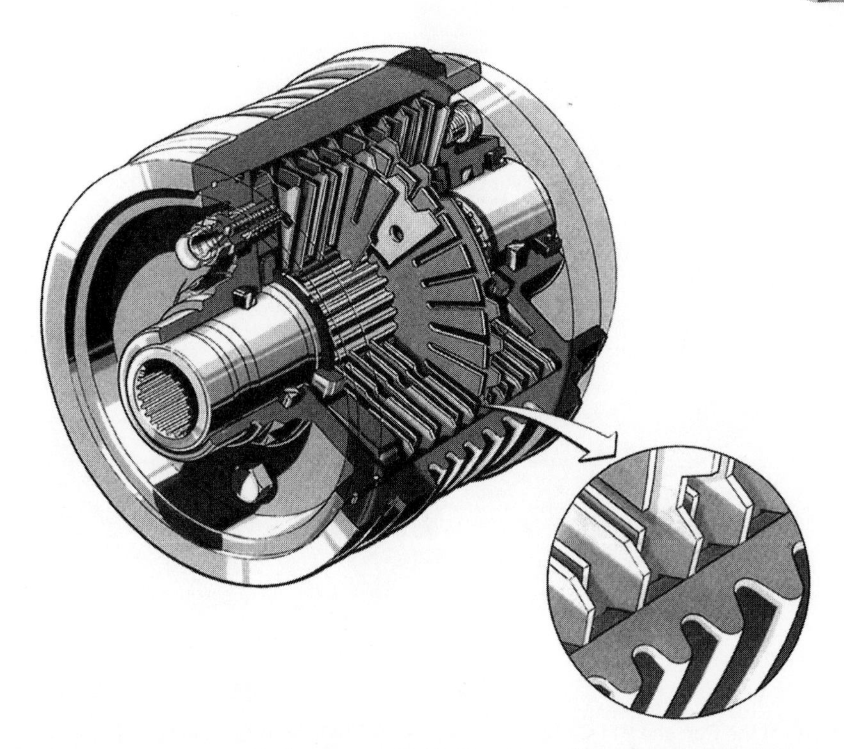

Fig. 8.1-74 Visco clutch with slip-dependent drive moment distribution. Two different packages sit in the closed drum-shaped housing: radially slit steel discs, which are moved by the serrated profile of the hollow shaft, and perforated discs which grip (as can be seen below) into housing keys. The shaft is joined with the differential and the casing with the propshaft going to the rear axle. The discs are arranged in the casing so that a perforated disc alternates with a slit one. The individual parts have no definite spacing but can be slid against one another axially. The whole assembly is filled with viscous silicone fluid and the torque behaviour (therefore the locking effect) can be adjusted via the filling level.
If slip occurs between the front and rear axle, the sets of discs in the clutch rotate relative to one another and shearing forces are transferred via the silicone fluid. These increase with increasing slip and ensure a torque increase in the rear axle. The power consumed in the visco clutch leads to warming and thus to growing inner pressure. This causes an increase in the transferable torque which, under conditions of extreme torque requirement, ultimately leads to an almost slip-free torque transfer (rigid drive). With ABS braking, a free-wheeling device disengages the clutch; the latter must be engaged again when reversing.

to all four wheels by means of a clutch on the propshaft, as required. The clutch can be engaged manually, or automatically in response to slip. With the use of sprag clutches, which are usually engaged manually, the torque is transmitted in a fixed ratio between the front and rear axles; multi-disc or visco clutches permit variable torque distribution. As these systems have essential similarities with permanent four-wheel drive varieties, they are discussed in Section 8.1.7.4.

With sprag-clutch engaged transmissions, the design complexity, and therefore the costs, are lower than on permanent drive. Usually there is no rear axle differential lock, which is important on extremely slippery roads; while this results in price and weight advantages, it does lead to disadvantages in the traction.

Front-wheel drive is suitable as a basic version and the longitudinal engine has advantages here (Fig. 8.1-48). With the transverse engine, the force from the manual gearbox is transmitted via a bevel gear and a divided propshaft, to the rear axle with a differential (Fig. 8.1-68). There is relatively little additional complexity compared with the front-wheel-drive design, even if, on the Fiat Panda (Trecking 4 × 4), there is a weight increase of about 11% (90 kg), not least because of the heavy, driven, rigid axle. It is possible to select rear-wheel drive during a journey using a shift lever that is attached to the prop-shaft tunnel.

Manual selection on the Subaru Justy operates pneumatically at the touch of a button (even while travelling). This vehicle has independent rear-wheel suspension and weighs only 6% more than the basic vehicle with front-wheel drive. Traction is always improved considerably if the driver recognizes the need in time and switches the engine force onto all four wheels. In critical situations, this usually happens too late, and the abrupt change in drive behaviour becomes an additional disadvantage.

Conversely, if the driver forgets to switch to single axle drive on a dry road, tensions occur in the power train during cornering, as the front wheels travel larger arcs

Fig. 8.1-75 Driven front axle of the Porsche 911 Carrera 4 (1996, 1998). The visco clutch is flange mounted directly on the front axle to achieve a better distribution of axle load. With corresponding slip of the rear wheels, up to 40% of the driving torque is transmitted to the front axles. Particular attention was paid during the adjustment of the four-wheel drive to predictable self-steering properties independent of drive distribution and to controllability of the handling characteristics even at the stability limit. Instead of differential locks, specific wheel brake engagements are made in order to retard spinning wheels. Four-wheel drive is integrated into the Porsche Stability Management (PSM), a system for controlling the dynamics of vehicle movement with brake actuation.

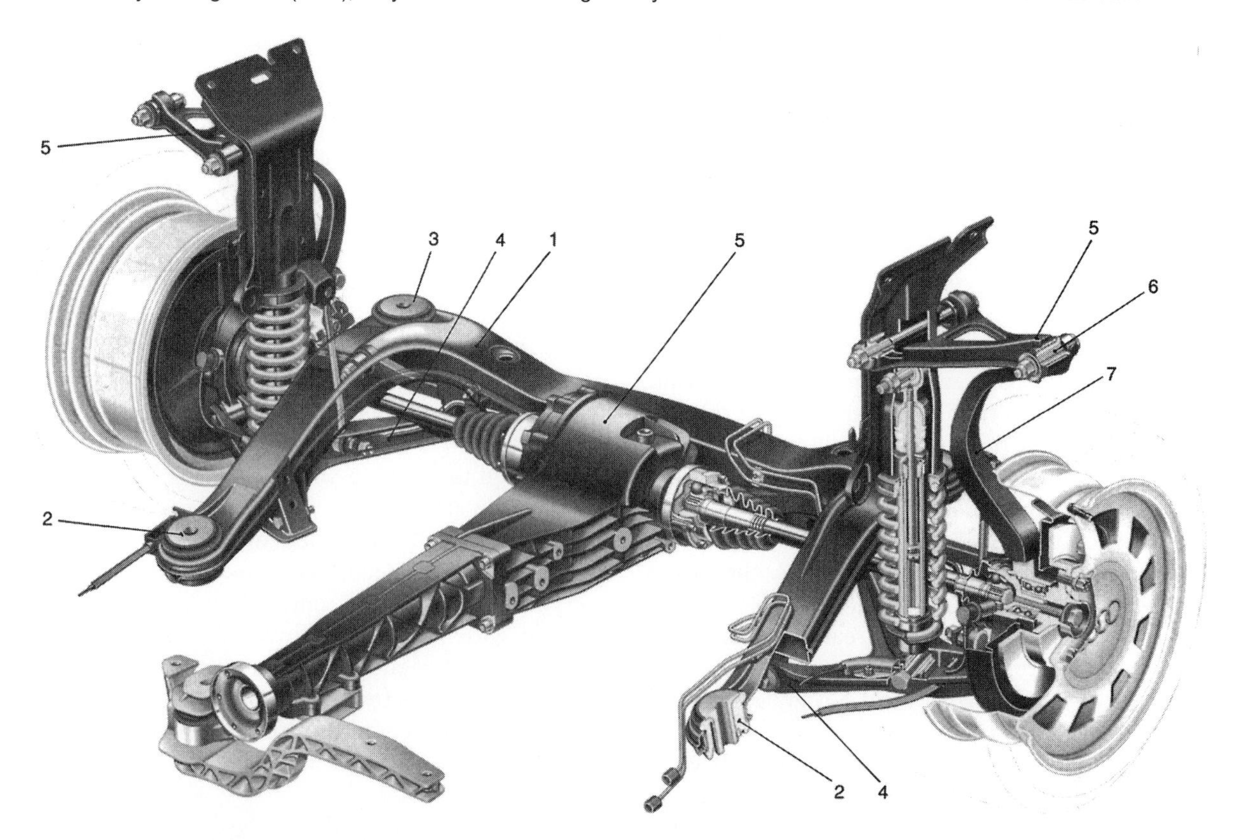

Fig. 8.1-76 Double wishbone rear axle on the Audi A4 Quattro. The suspension subframe 1 is fixed to the body with four widely spaced rubber mountings (items 2 and 3) and houses the differential casing 8 and transverse control arms (items 4 and 5). The springs and shock absorbers are mounted next to the fixings for the upper control arms 7. The location 6 of the wheel hub carrier 5 was raised (long base *c*, Fig. 8.1-4) and drawn outwards. The lower transverse control arm 4 is fixed to part 1 with widely spaced mountings. These measures ensure a wide boot and low forces, making it easier to attain the desired kinematic characteristics.

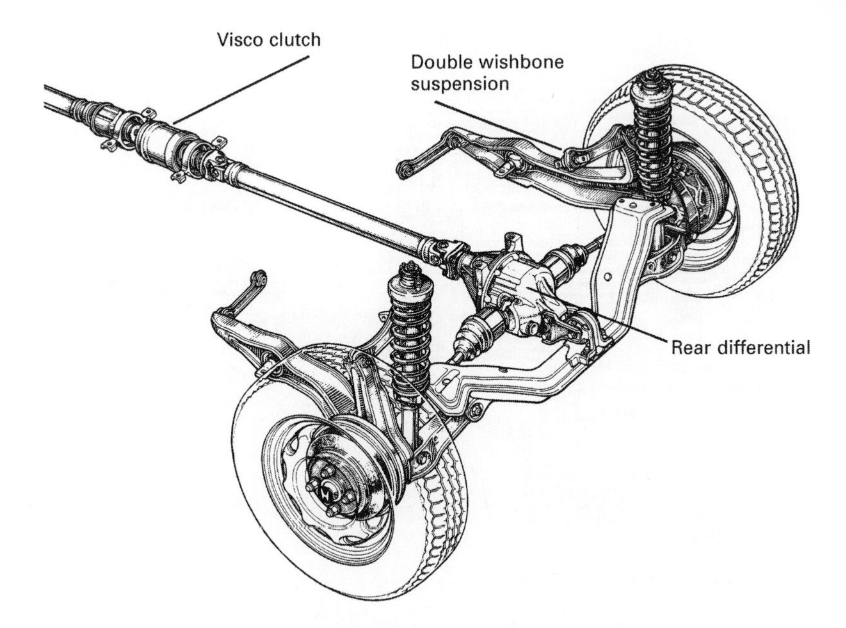

Fig. 8.1-77 Double wishbone rear axle of the Honda Civic Shuttle 4 WD. The visco clutch sits (held by two shaft bearings) in the centre of the divided propshaft. The rear axle differential has been moved forwards and is mounted to the rear on the body via a cross-member. Apart from the different type of wheel bearings and the lower transverse control arm positioned somewhat further back (to make it possible to bring the drive shafts through in front of the spring dampers), the axle corresponds to Fig. 8.1-62 and resembles the suspension shown in Fig. 8.1-1.

than the back ones (Fig. 8.1-69). The tighter the bend, the greater the stress on the power train and the greater the tendency to unwanted tyre slip.

A further problem is the braking stability of these vehicles. If the front axle locks on a wet or wintry road during braking, the rear one is taken with it due to the rigid power train. All four wheels lock simultaneously and the car goes into an uncontrollable skid.

8.1.7.3 Manual selection four-wheel drive on commercial and all-terrain vehicles

The basis for this type of vehicle is the standard design which, because of the larger ground clearance necessary in off-road vehicles (Fig. 8.1-67), has more space available between the engine and front axle differential and

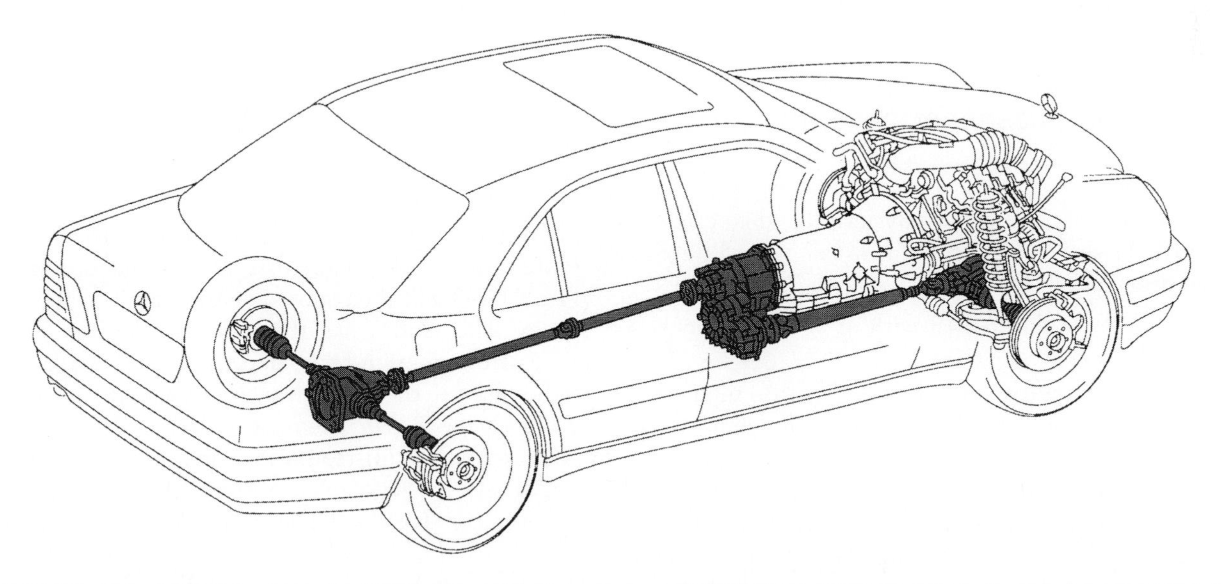

Fig. 8.1-78 Drive train of the four-wheel-drive Mercedes-Benz E class 4MATIC (from 1997). In order to be able to control the drive shafts to the front wheels, an integrated spring-and-shock absorber strut in the shape of a fork on the lower transverse link is used. In the almost identical suspension design of other than off-road varieties, the springs and shock absorbers are separate.

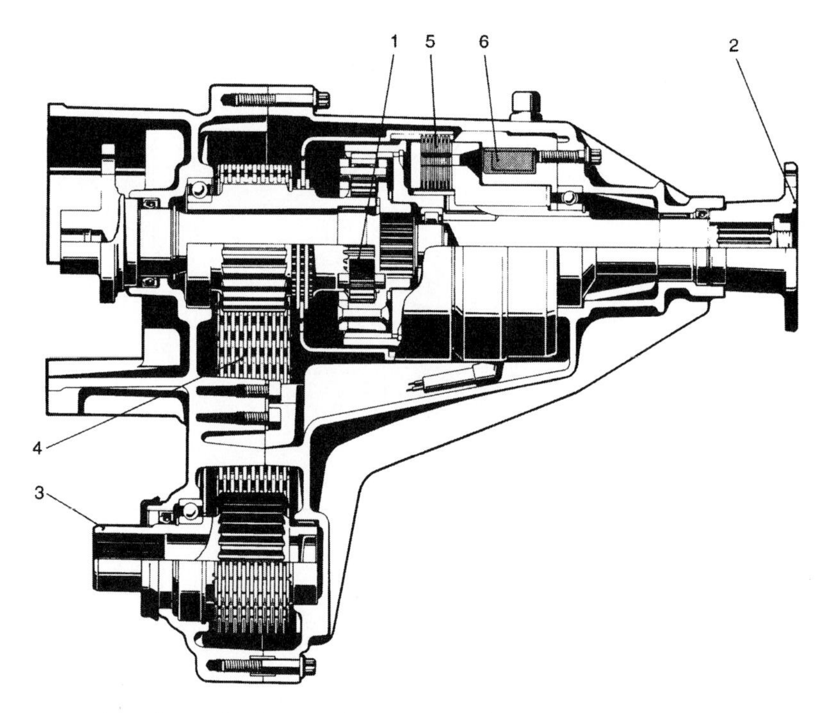

Fig. 8.1-79 The torque coming from the engine is apportioned by the Planet Wheel-Centric Differential 1 in such one, to the rear cardan shaft 2 (64%) and to the front one 3 (36%). The offset to this shaft is bridged-over by the inserted tooth type chain 4. The adaptation of the distribution of driving power is taken over the multiple-disk clutch 5, which is driven (controlled) by the electromagnet. Power Divider A110 of the Fa. ZF (Zahnradfabrik Friedrichshafen).

between the cargo area and the rear axle. Fig. 8.1-70 shows the design details:

- a central power take-off gear with manual selection for the front axle, plus a larger ratio off-road gear, which can be engaged if desired;
- three propshafts;
- complex accommodation of the drive joints if there is a rigid front axle (Fig. 8.1-3).

8.1.7.4 Permanent four-wheel drive; basic passenger car with front-wheel drive

All four wheels are constantly driven; this can be achieved between the front and rear axle with different design principles:

- a bevel centre differential with or without manual lock selection;
- a Torsen centre differential with moment distribution, based on the traction requirement (Fig. 8.1-71);
- a planet gear central differential with fixed moment distribution and additional visco clutch, which automatically takes over the locking function when a difference in the number of revolutions occurs or a magnetic clutch (which is electronically controlled, Fig. 8.1-79);
- electronically controlled multi-disc clutches (Haldex clutch, Fig. 8.1-73);
- a visco clutch in the propshaft power train, which selects the initially undriven axle depending on the tyre slip (Figs. 8.1-72, 8.1-74 and 8.1-75).

Here too, the front-wheel-drive passenger car is suitable as a basic vehicle. In 1979, Audi was the first company to bring out a car with permanent four-wheel drive, the Quattro, and today vehicles with this type of drive are available throughout the entire Audi range. On a longitudinally mounted engine, a Torsen centre differential distributes the moment according to the traction requirement (Fig. 8.1-71). The four-wheel drive increases the weight by around 100 kg.

VW used a visco clutch in the power train (without centre differential) for the first time on the Transporter (Fig. 8.1-74) and then subsequently used it in the Golf syncro. The clutch has the advantage of the engine moment distribution being dependent on the tyre slip. If the slip on the front wheels, which are otherwise driven at the higher moment, increases on a wet or frozen surface or off-road, more drive is applied to the rear wheels. No action on the part of the driver is either necessary or possible. The transverse engine makes a bevel gear in front of the split prop shaft necessary. The visco clutch sits in the rear differential casing and there is also an overrunning clutch, which ensures that the rear wheels are automatically disengaged from the

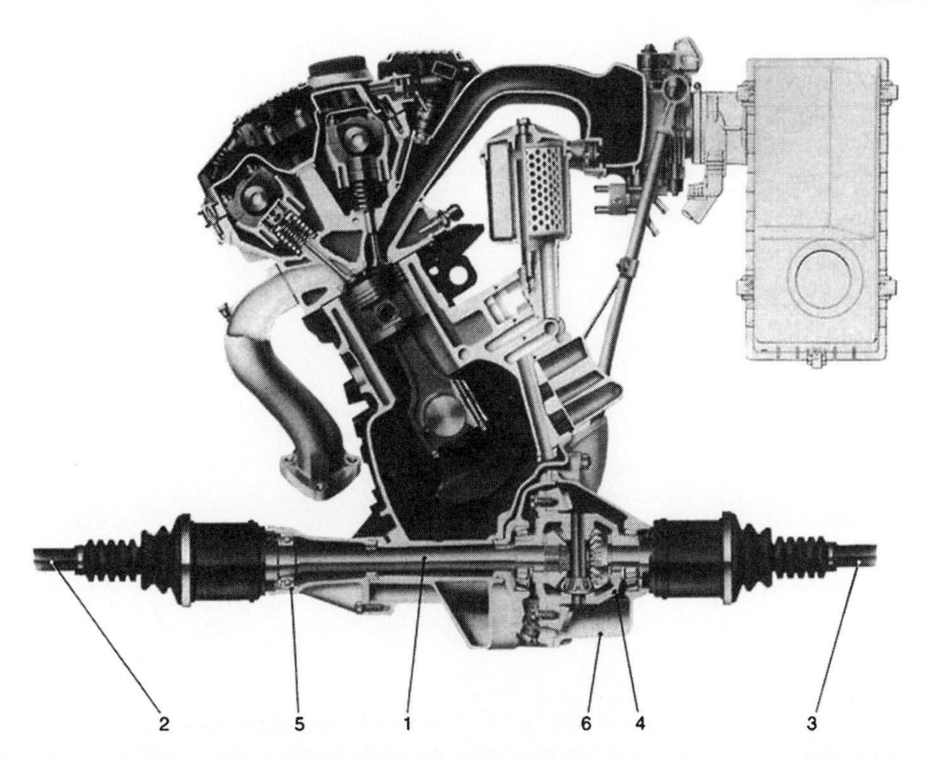

Fig. 8.1-80 Front cross-section view of the engine; and drive axle of a standard four-wheel-drive vehicle (BMW assembly diagram). The basic vehicle has rear-wheel drive and, in order to also be able to drive the front wheels, the front axle power take-off 4 had to be moved into the space of the oil pan. The intermediate shaft 1 bridges the distance to the right inner CV joint and thus ensures drive shafts of equal length to both wheels (items 2 and 3 and Fig. 8.1-51). Part 1 is mounted on one side in the non-lockable differential 4 and on the other side in the outrigger 5. This, and the casing 6, are screwed to the oil pan.

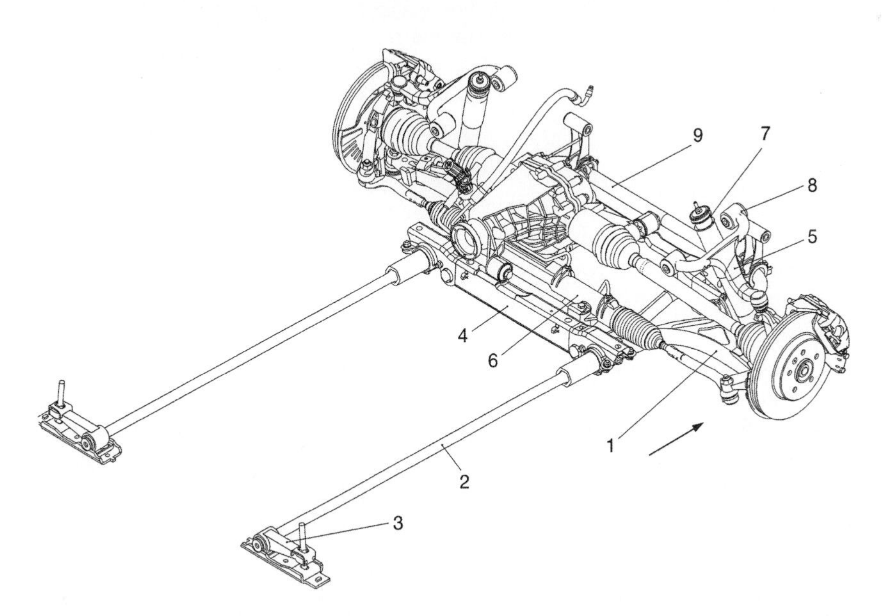

Fig. 8.1-81 Front suspension and drive axle of the Mercedes-Benz off-road vehicle of the M series. In off-road vehicles, rigid axles are mostly used. Instead of these, Mercedes-Benz installs double wishbone suspensions at the front and rear. In this way, the proportion of unsprung masses can be reduced by approximately 66%; driving safety and riding comfort are increased. For space reasons, torsion-bar springs are used for the suspension of the front axle.
1 lower transverse link in the form of a forged steel component because of the introduction of torque by the torsion bars (2) and notch insensitivity off road conditions; 2 torsion bars (spring rate of 50 Nm/degree); 3 vertically adjustable torque support which can be placed in any position in a transverse direction; 4 integral bearers (subframe) attached to the box-type frame by 4 bolts; 5 upper transverse link in the form of a forged aluminium component; 6 rack and pinion power steering, 7 twin-tube shock absorber with integrated rubber bump stop, 8 transverse link mounting points; 9 stabilizer application of force to lower transverse link.

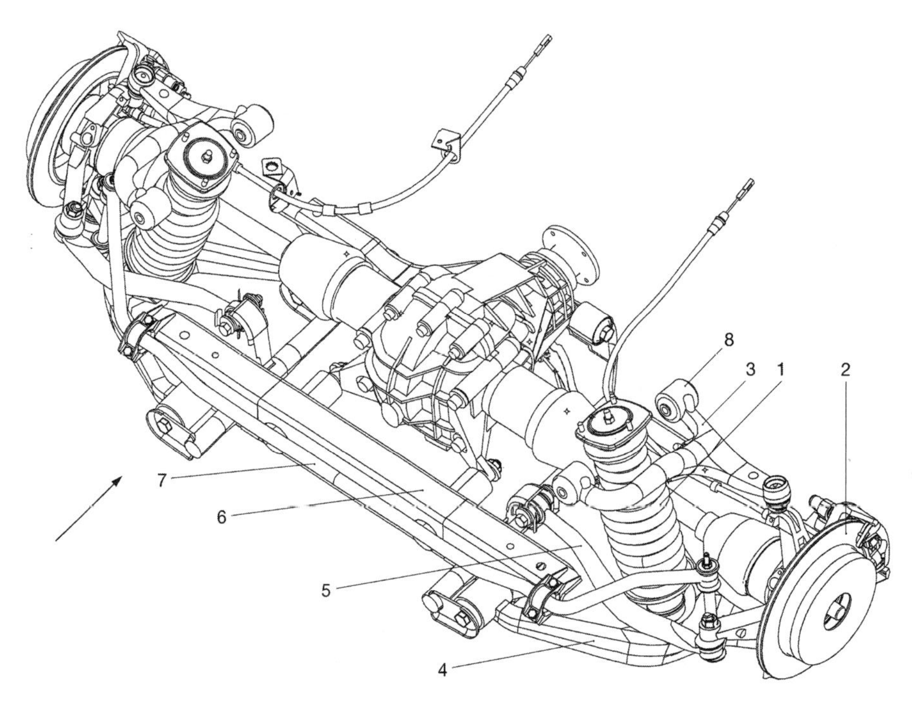

Fig. 8.1-82 Rear axle of the Mercedes-Benz off-road vehicle of the M series. Suspension and damping are ensured by the spring strut (1) whose spring is tapered for reasons of construction space (spring rate gradually increasing from 70 to 140 N/mm), 2 brake disc with integrated drum parking brake, 3 upper transverse link (forged aluminium component), 4 lower transverse link (forged aluminium component), 5 tie rod (forged steel component), 6 integral bearer (subframe), 7 stabilizer, 8 transverse link mounting points. Common characteristics of front and rear axles: camber and castor are adjusted by positioning the transverse link mounting points (8) in long holes during assembly. Technical data: spring travel ±100 mm, kingpin offset −5 mm, disturbing force moment arm 56.7 mm, kingpin inclination 10.5°, camber angle −0.5°, castor for front axle/rear axle 7/−8.5°, castor trail for front axle/rear axle 37/−55 mm, wheel castor trail for front axle/rear axle 5/−4.5 mm, instantaneous centre height for front axle/rear axle 80/119 mm, braking-torque compensation for front axle/rear axle 38/21%, starting-torque compensation for front axle/rear axle −7/3%. The axle concept was designed and developed by Mercedes-Benz. Mass production and assembly is undertaken by Zahnradfabrik Friedrichshafen AG who, via Lemförder Fahrwerktechnik AG, supply the complete subassemblies to the assembly line as required.

drive, on overrun, to guarantee proper braking behaviour. This type of drive is fully ABS compatible. When reverse is engaged, a sliding sleeve is moved, which bridges the overrunning clutch to make it possible to drive backwards.

When selecting their rear axle design, manufacturers choose different paths. Audi fits a double wishbone suspension in the A4 and A6 Quattro (Fig. 8.1-76), Honda uses the requisite centre differential on the double wishbone standard suspension in the Civic Shuttle 4WD (Figs. 8.1-77 and 8.1-62).

8.1.7.5 Permanent four-wheel-drive, basic standard design passenger car

Giving a standard design car four-wheel drive requires larger modifications, greater design complexity and makes the drive less efficient (Fig. 8.1-78). A power take-off gear is required, from which a short propshaft transmits the engine moment to the front differential. The lateral offset must be bridged, for example, with a toothed chain (Fig. 8.1-79). The ground clearance must not be affected and so changes in the engine oil pan are indispensable if the axle drive is to be accommodated (Fig. 8.1-80).

The power take-off gear (Fig. 8.1-79) contains a planet gear centre differential which facilitates a variable force distribution (based on the internal ratio); 36% of the drive moment normally goes to the front and 64% to the rear axle. A multi-disc clutch can also be installed that can lock the differential electromagnetically up to 100%, depending on the torque requirement (front to rear axle). Moreover, there is a further electrohydraulically controlled lock differential in the rear axle which is also up to 100% effective.

The two differentials with variable degrees of lock offer decisive advantages:

- to reach optimal driving stability, they distribute the engine moments during overrun and traction according to the wheel slip on the drive axles;

Motor position	Reduction drive	Drive on	Four-wheel drive			Middle differential locks via	Front axle differential		Rear axle differential		Example
			switched by	slip dependent	perm. by		locking	brake	locking	brake	
longit.	2.05:1	rear	sprag-clutch man.			N/A	n	n	n	n	Opel Frontera
longit.	1.425:1	rear	sprag-clutch man.			N/A	n	n	multi-disc clutch	n	Mitsubishi Pajero
longit.	2.15:1	rear	sprag-clutch man.			N/A	n	n	n	n	Suzuki Jimny Cross Country
longit.	2.43:1	rear	sprag-clutch man.			N/A	n	n	multi-disc clutch	n	Chevrolet Blazer
longit.	2.48:1	rear	multi-disc.	electron.		N/A	n	n	multi-disc clutch	n	Ford Explorer
transv.		front	multi-disc	electron.		N/A	n	n	nn	n	Honda CR-V
longit.	2.72:1	rear		visco		N/A	n	n	multi-disc clutch	n	Jeep Grand Cherokee
transv.		front		visco		N/A	n	n	n	n	Chrysler Voyager 4WD
transv.		front		visco		N/A	n	y	n	y	Land Rover Freelander Discovery
longit.		rear		visco		N/A	n	y	n	y	Porsche Carrera 4
transv.		front		visco		N/A	n	y	multi-disc clutch	n	Volvo V70 R AWD
transv.		front		visco		N/A	n	y	multi-disc clutch	n	VW Multivan Synchro
transv.		front		Haldex multi-disc		N/A	n	y	n	n	Audi TT Quattro, Golf 4motion
longit.		f + r			longit. diff.	claw clutch	n	n	n	n	Daihatsu Terios
longit.	2.64:1	f + r			longit. diff.	N/A	n	y	n	y	Mercedes-Benz M series
longit.	y	f + r			longit. diff.	y	y	n	y	n	Mercedes-Benz G series
longit.		f + r			longit. diff.	N/A	n	y	n	y	Mercedes-Benz E series
longit.		f + r			longit. diff.	N/A	n	y	n	y	BMW Sports Activity Vehicle (SAV, E53)
longit.	1.21:1	f + r			longit. diff.	N/A	n	y	n	y	Land Rover Discovery
	y	f + r			longit. diff.	visco + man.					Toyoto LandCruiser
longit.	1.93:1	f + r			longit. diff.	visco + man.	n	n	sprag-clutch	n	Mitsubishi Pajero
longit.	1.20:1	f + r			longit. diff.	visco lock	n	n	n	n	Subaru Legacy Outback
longit.	1.45:1	f + r			longit. diff.	visco lock	n	n	n	n	Subaru Forester
transv		f + r			Torsen diff.	self-locking	n	y	n	y	Audi A4/A6 Quattro, VW Passat Synchro
transv		f + r			Torsen diff.	self-locking	n	y	n	y	Audi A4/A6 Quattro, VW Passat Synchro

Fig. 8.1-83 Different kinds of four-wheel drive.

- they allow maximum traction without loss of driving stability (Fig. 8.1-66).

The locks are open during normal driving. By including the front axle differential, they make it possible to equalize the number of revolutions between all wheels, so tight bends can be negotiated without stress in the power train and parking presents no problems. If the car is moved with locked differentials and the driver is forced to apply the brakes, the locks are released in a fraction of a second. The system is therefore fully ABS compatible.

In its four-wheel drive vehicles of the E class (Fig. 8.1-78), Mercedes-Benz uses a transfer gear with central differential situated on the gearbox outlet and a front axle gear integrated into the engine-oil pan. The (fixed) driving torque distribution is 35%:65%. Instead of traditional differential locks, the wheel brakes are activated on the spinning wheels as in off-road vehicles of the M class (Figs. 8.1-81 and 8.1-82). This system permits maximum flexibility, its effect not only corresponds to differential locks on front and rear axles as well as on the central differential, but also makes it possible for other functions such as ABS and electronic yaw control (ESP) to be integrated without any problem. Design complexity – and thus cost – is considerable.

8.1.7.6 Summary of different kinds of four-wheel drive

The list in Fig. 8.1-83 shows the increasing use of slip-controlled clutches (visco and Haldex clutches) for the transmission of torque instead of an interaxle differential and the importance of electronic brake application systems which are used instead of lockable differential gears. Modern four-wheel varieties operate without functional restrictions with antilocking, slip and driving stabilization systems.

Section Nine

Steering

Chapter 9.1

9.1

Steering

Jornsen Reimpell, Helmut Stoll and Jurgen Betzler

This chapter gives only the essential aspects of the subject, 'Steering'.

The steering system is type-approved on all new passenger cars and vans coming on to the market; it is governed by the following EC directives.

70/311/EWG 91/662/EWG
74/297/EWG 92/62/EWG

Figs. 9.1-1, 8.1-45, 8.1-56 and 8.1-70 show the complete steering system of a front-wheel-drive passenger vehicle with left-hand steering.

9.1.1 Steering system

9.1.1.1 Requirements

On passenger cars, the driver must select the steering wheel angle to keep deviation from the desired course low. However, there is no definite functional relationship between the turning angle of the steering wheel made by the driver and the change in driving direction, because the correlation of the following is not linear (Fig. 9.1-2):

- turns of the steering wheel;
- alteration of steer angle at the front wheels;
- development of lateral tyre forces;
- alteration of driving direction.

This results from elastic compliance in the components of the chassis. To move a vehicle, the driver must continually adjust the relationship between turning the steering wheel and the alteration in the direction of travel. To do so, the driver will monitor a wealth of information, going far beyond the visual perceptive faculty (visible deviation from desired direction). These factors would include for example, the roll inclination of the body, the feeling of being held steady in the seat (transverse acceleration) and the self-centring torque the driver will feel through the steering wheel. The most important information the driver receives comes via the steering moment or torque which provides him with feedback on the forces acting on the wheels.

It is therefore the job of the steering system to convert the steering wheel angle into as clear a relationship as possible to the steering angle of the wheels and to convey feedback about the vehicle's state of movement back to the steering wheel. This passes on the actuating moment applied by the driver, via the steering column to the steering gear 1 (Fig. 9.1-3) which converts it into pulling forces on one side and pushing forces on the other, these being transferred to the steering arms 3 via the tie rods 2. These are fixed on both sides to the steering knuckles and cause a turning movement until the required steering angle has been reached. Rotation is around the steering axis EG, also called kingpin inclination, pivot or steering rotation axis (Fig. 8.1-3).

9.1.1.2 Steering system on independent wheel suspensions

If the steering gear is of a type employing a rotational movement, i.e. the axes of the meshing parts (screw shaft 4 and nut 5, Fig. 9.1-15) are at an angle of 90° to one another, on independent wheel suspensions, the insides of the tie rods are connected on one side to the pitman arm 4 of the gear and the other to the idler arm 5 (Fig. 9.1-3). As shown in Figs. 9.1-12 and 9.1-36–9.1-38, parts 4 and 5 are connected by the intermediate rod 6. In the case of steering gears, which operate using a shift movement (rack and

Automotive Chassis: Engineering Principles; ISBN: 9780750650540
Copyright © 2001 Elsevier Ltd; All rights of reproduction, in any form, reserved.

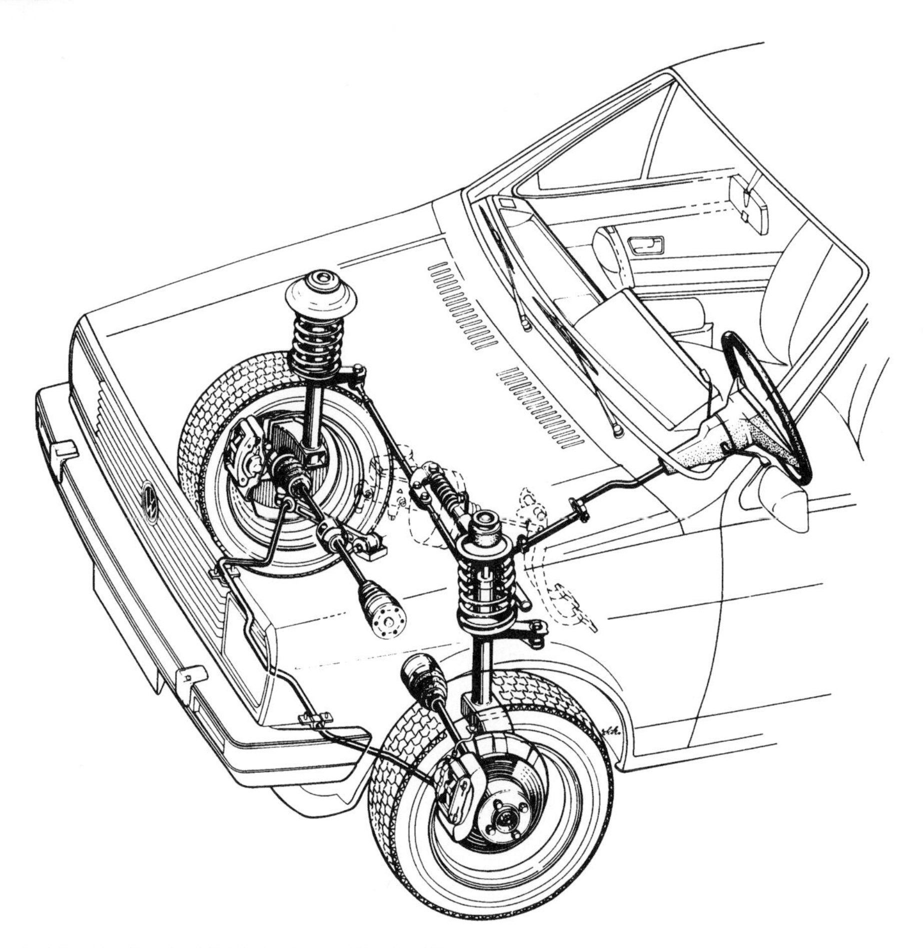

Fig. 9.1-1 Damper strut front axle of a VW Polo (up to 1994) with 'steering gear', long tie rods and a 'sliding clutch' on the steering tube; the end of the tube is stuck onto the pinion gear and fixed with a clamp. The steering arms, which consist of two half shells and point backwards, are welded to the damper strut outer tube. An 'additional weight' (harmonic damper) sits on the longer right drive shaft to damp vibrations. The anti-roll bar carries the lower control arm. To give acceptable ground clearance, the back of it was designed to be higher than the fixing points on the control arms. The virtual pitch axis is therefore in front of the axle and the vehicle's front end is drawn downwards when the brakes are applied.

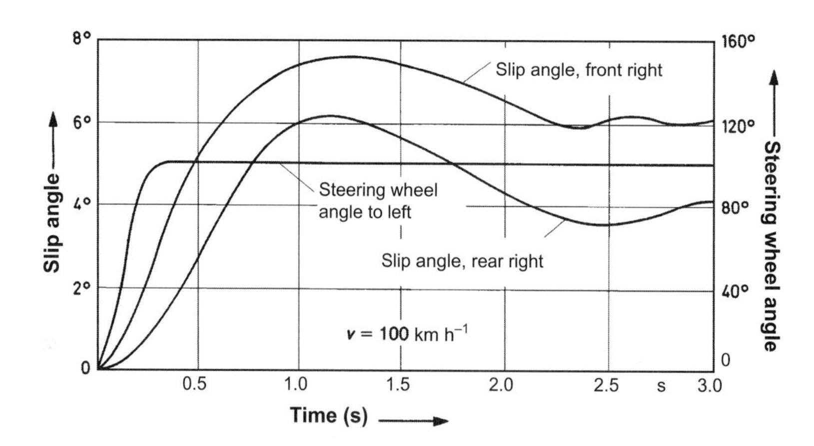

Fig. 9.1-2 Delayed, easily manageable response of the right front wheel when the steering wheel is turned by 100° in 0.2 s, known as step steering input. A slip angle of $\alpha_f \approx 7°$ on both front tyres is generated in this test. The smaller angle α_r on the rear axle, which later increases, is also entered. Throughout the measurement period it is smaller than α_f (x-axis), i.e. the model studied by Mercedes Benz understeers and is therefore easy to handle.

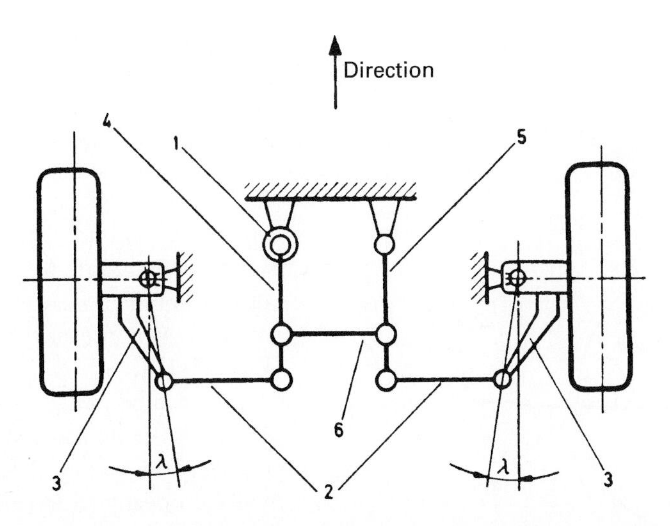

Fig. 9.1-3 Synchronous steering A-bar on the front suspension of a left-hand drive passenger car or light van; on the right-hand drive vehicle, the steering gear is on the other side. The steering arm (3) and the pitman arm (4) rotate in the same direction. The tie rods (2) are fixed to these arms.

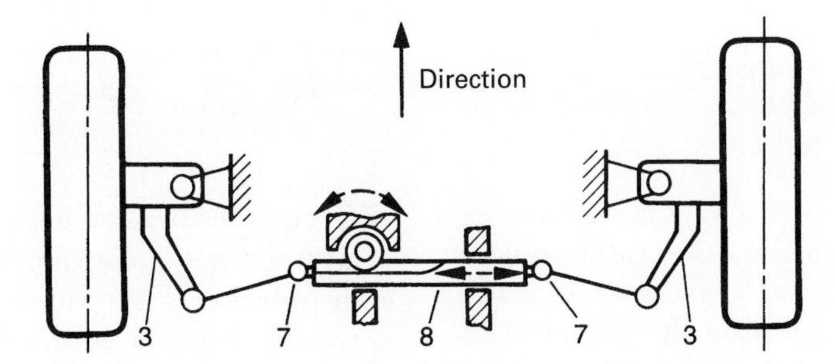

Fig. 9.1-4 Rack and pinion steering with the steering linkage 'triangle' behind the front axle. The spigots of the inner tie rod joints 7 are fixed to the ends of the steering rack 8 and the outside ones to the steering arms 3 (see also Figs. 8.1-40 and 8.1-54).

pinion steering), it is most economical to fix the inner tie rod joints 7 to the ends of the steering rack 8 (Fig. 9.1-4).

9.1.1.3 Steering system on rigid axles

Rack and pinion steering systems are not suitable for steering the wheels on rigid front axles, as the axles move in a longitudinal direction during wheel travel as a result of the sliding-block guide. The resulting undesirable relative movement between wheels and steering gear cause unintended steering movements. Therefore, only steering gears with a rotational movement are used. The intermediate lever 5 sits on the steering knuckle (Fig. 9.1-5). The intermediate rod 6 links the steering knuckle and the pitman arm 4. When the wheels are turned to the left, the rod is subject to tension and turns both wheels simultaneously, whereas when they are turned to the right, part 6 is subject to compression. A single tie rod connects the wheels via the steering arm.

However, on front axles with leaf springs, the pitman arm joint 4, which sits on the steering gear 1, must be disposed in such a manner that when the axle is at full suspension travel, the lower joint 8 describes the same arc 9 as the centre of the front axle housing (Figs. 9.1-6 and 8.1-37). The arc 9 must be similar to the curved path 7,

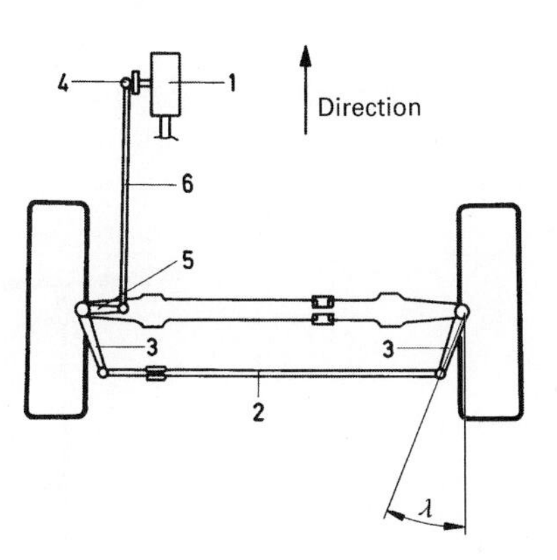

Fig. 9.1-5 On rigid axles, apart from the two steering arms 3, only the tie rod 2, the idler arm 5 and the drag link 6 are needed to steer the wheels. If leaf springs are used to carry the axle, they must be aligned precisely in the longitudinal direction, and lie vertical to the lever 5 when the vehicle is moving in a straight line. Steering arm angle λ is an essential factor in the relationship between the outer and the inner curve steering angles.

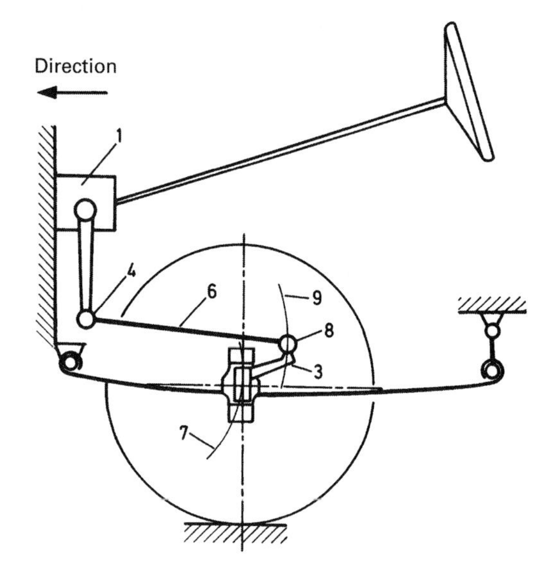

Fig. 9.1-6 Side view of a rigid front axle showing the movement directions 9 and 7 of the drag link and axle housing during bump and rebound-travel. The path of point 7 is determined by the front half of the leaf spring and can be calculated on a spring-balance by measuring the change in length when a load is added to and removed from the spring.

otherwise there is a danger of the wheels experiencing a parallel toe-in alteration when the suspension reaches full travel, i.e. both being turned in the same direction (Fig. 9.1-7). If a rigid axle is laterally controlled by a panhard rod, the steering rod must be parallel to it.

Its construction is similar to that of the intermediate rod of the steering linkage shown in Fig. 9.1-13; length adjustment and ball joints on both sides are necessary.

9.1.2 Rack and pinion steering

9.1.2.1 Advantages and disadvantages

This steering gear with a shift movement is used not only on small and medium-sized passenger cars, but also on heavier and faster vehicles, such as the Audi A8 and Mercedes E and S Class, plus almost all new light van designs with independent front wheel suspension. The advantages over manual recirculating ball steering systems are (see also Section 9.1.3.1):

- simple construction;
- economical and uncomplicated to manufacture;
- easy to operate due to good degree of efficiency;
- contact between steering rack and pinion is free of play and even internal damping is maintained (Fig. 9.1-10);
- tie rods can be joined directly to the steering rack;
- minimal steering elasticity compliance;
- compact (the reason why this type of steering is fitted in all European and Japanese front-wheel-drive vehicles);
- the idler arm (including bearing) and the intermediate rod are no longer needed;
- easy to limit steering rack travel and therefore the steering angle.

The main disadvantages are:

- greater sensitivity to impacts;
- greater stress in the case of tie rod angular forces;
- disturbance of the steering wheel is easier to feel (particularly in front-wheel drivers);
- tie rod length sometimes too short where it is connected at the ends of the rack (side take-off design;
- size of the steering angle dependent on steering rack travel;
- this sometimes requires short steering arms 3 (Fig. 9.1-4) resulting in higher forces in the entire steering system;
- decrease in steering ratio over the steer angle associated with heavy steering during parking if the vehicle does not have power-assisted steering;
- cannot be used on rigid axles.

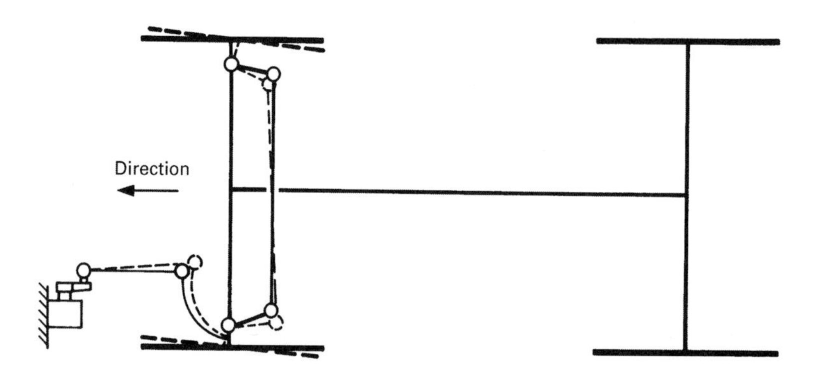

Fig. 9.1-7 If the movement curve 7 of the axle housing and curve 9 of the rear steering rod joint do not match when the body bottoms out, the wheels can turn and therefore an unwanted self-steering effect can occur.

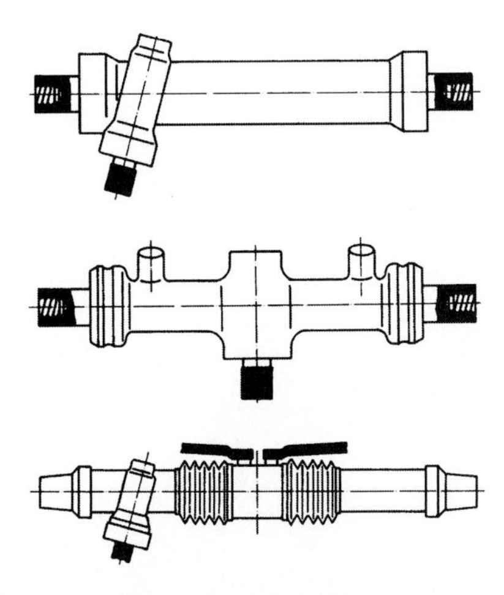

Fig. 9.1-8 The three most common types of rack and pinion steering on left-hand drive passenger cars; right-hand drive vehicles have the pinion gear on the other side on the top and bottom configurations (shown in Fig. 9.1-39). The pinion gear can also be positioned in the centre to obtain longer steering rod travel.

9.1.2.2 Configurations

There are four different configurations of this type of steering gear (Fig. 9.1-8):

Type 1 Pinion gear located outside the vehicle centre (on the left on left-hand drive and on the right on right-hand drive) and tie rod joints screwed into the sides of the steering rack (side take-off).

Type 2 Pinion gear in vehicle centre and tie rods taken off at the sides.

Type 3 Pinion gear to the side and centre take-off, i.e. the tie rods are fixed in the vehicle centre to the steering rack.

Type 4 'Short steering' with off-centre pinion gear and both tie rods fixed to one side of the steering rack (Fig. 9.1-1).

Types 1 and 3 are the solutions generally used, whereas Type 2 was found in some Porsche vehicles, and Type 4 used to be preferred by Audi and VW.

9.1.2.3 Steering gear, manual with side tie rod take-off

Type 1 (Fig. 9.1-8) is the simplest solution, requiring least space; the tie rod joints are fixed to the sides of the steering rack (Fig. 9.1-9), and neither when the wheels are turned, nor when they bottom out does a moment occur that seeks to turn the steering rack around its centre line. It is also possible to align the pinion shaft pointing to the

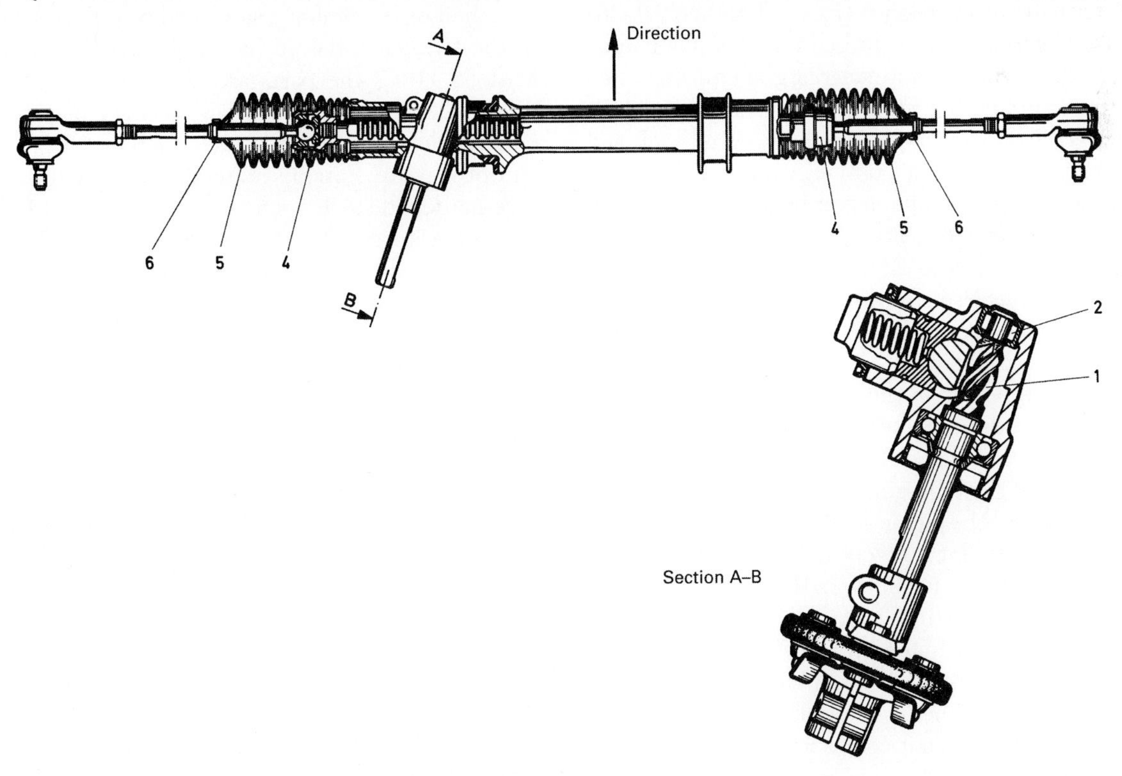

Fig. 9.1-9 Rack and pinion steering on the Vauxhall Corsa (1997). The tie rod axial joints 4 bolted to the side of the steering rack and the sealing gaiters 5 can be seen clearly. To stop them from being carried along when the toe-in is set (which is done by rotating the middle part of the rod) it is necessary to loosen the clamps 6.
The pinion 1 has been given a 'helical cut', due to the high ratio, and is carried from below by the needle bearing 2. The bearing housing has been given a cover plate to facilitate assembly and prevent dirt ingress.

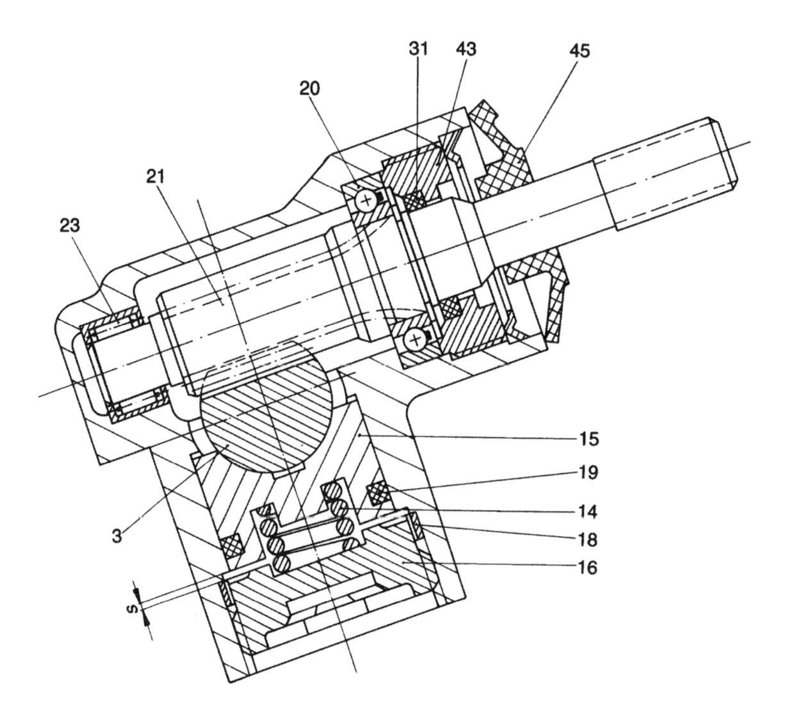

Fig. 9.1-10 Rack- and- pinion steering by ZF; section through pinion gear, bearing and rod guide. The distance ring 18 is used for setting the plays, and the closing screw 16 is tightened against it. The O-ring 19 provides the damping function and prevents rattling noises.

steering tube (Figs. 8.1-57, 9.1-24 and 9.1-29) making it easy to connect the two parts together. Using an intermediate shaft with two joints (Figs. 8.1-49 and 9.1-26) enables the steering column to bend at this point in an accident. In this event the entire steering gear is turned when viewed from the side (i.e. around the y-axis).

Fig. 9.1-10 is a section showing how, on all rack and pinion steering systems, not only can the play between the steering rack and the pinion gear be easily eliminated, but it also adjusts automatically to give the desired damping. The pinion gear 21 is carried by the grooved ball bearing 20; this also absorbs any axial forces. Ingress of dirt and dust are prevented by the seal 31 in a threaded ring 43 and the rubber cap 45. The lower end of the pinion gear is supported in the needle bearing 23.

In a left-hand-drive passenger car or light van, the steering rack 3 is carried on the right by a plastic bearing shell and on the right by guide 15, which presses the steering rack against the pinion gear. On a right-hand-drive vehicle this arrangement is reversed. The half-round outline of the guide 15 does not allow radial movement of the steering rack. To stop it from moving off from the pinion gear, when subject to high steering wheel moments (which would lead to reduced tooth contact), the underside of the guide-bearing 15 is designed as a buffer; when it has moved a distance of $s \leq 0.12$ mm it comes into contact with the screw plug 16.

Depending on the size of the steering system, coil spring 14 has an initial tension force of 0.6–1.0 kN, which is necessary to ensure continuous contact between steering rack and pinion gear and to compensate for any machining imprecision, which might occur when the toothing is being manufactured or the steering rack broached or the pinion gear milled or rolled. The surface of the two parts should have a Rockwell hardness of at least 55 HRC; the parts are not generally post-ground due to the existence of a balance for the play. Induction-hardenable and annealed steels such as Cf 53, 41 Cr 4 and others are suitable materials for the steering rack, case-hardened steels such as 20 MnCr 5, 20 MoCr 4, for example, are suitable for the pinion gear. In order to ensure a good response and feedback of the steering, the frictional forces between guide-bearing 15 and gear rack 3 must be kept as small as possible.

Sealing the steering rack by means of gaiters to the side (Fig. 9.1-9) makes it possible to lubricate them with grease permanently, and lubrication must be provided through a temperature range of $-40°C$ to $+80°C$. It is important to note that if one of the gaiters is damaged, the lubricant can escape, leading to the steering becoming heavier and, in the worst case, even locking. Gaiters should therefore be checked at every service inspection. They are also checked at the German TÜV (Technischer Überwachungs Verein) annual vehicle inspection.

9.1.2.4 Steering gear, manual with centre tie rod take-off

As shown in Figs. 8.1-57 and 9.1-1, and as described in Section 9.1.7.3.2, with McPherson struts and strut dampers the tie rods must be taken off from the centre if

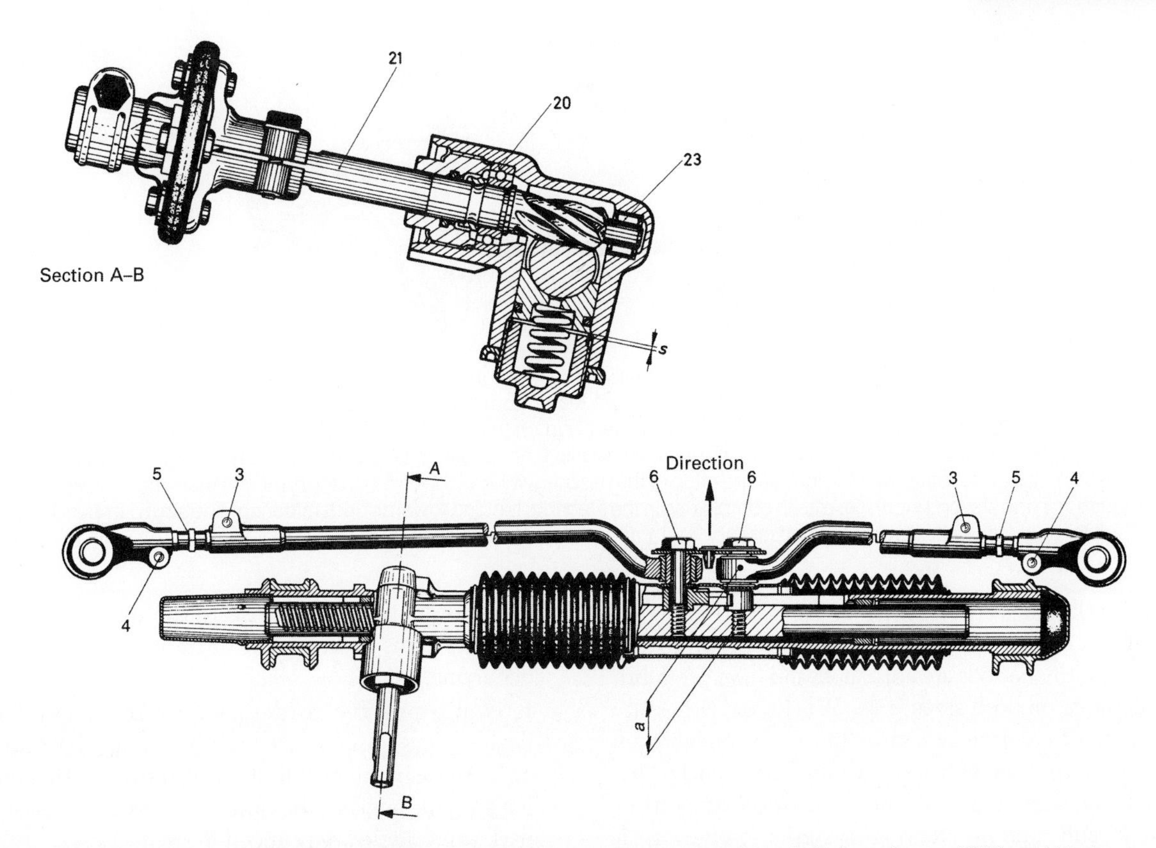

Fig. 9.1-11 Top view of the rack and pinion steering of the front-wheel-drive Opel (Vauxhall) Astra (up to 1997) and Vectra (up to 1996); the steering arms on the McPherson strut point backwards and the steering gear is located relatively high. For this reason the tie rods have to be jointed in the middle and (in order not to come into contact with the gear housing when the wheels are turned) have to be bent. The guide-bearing in the groove of the housing prevents the steering rack from twisting. On the inside, both tie rods have the eye-type joint; the distance *a* to the steering rack centre, which causes a bending moment, and a torque (when the wheels bump and rebound) is also shown. The two bolts 6 gripping into the steering rack are secured.
Once the screws 3 and 4 have been loosened, toe-in to the left and right can be set by turning the connecting part 5. The steering gear has two fixing points on the dashpanel, which are a long way apart and which absorb lateral force moments with minimal flexing. As also shown in Fig. 9.1-10, the pinion is carried by a ball and a needle bearing (positions 20 and 23) and is also pressed onto the steering rack by a helical spring. The illustration shows the possible path *s* of the rack guide. Figs. 9.1-46–9.1-48 show the reason for the length of the tie rods on McPherson struts and strut dampers.

the steering gear has to be located fairly high up. This is because the steering tie rods must thus be very long in order to prevent unwanted steering movements during wheel travel (Fig. 9.1-46).

In such cases, the inner joints are fixed in the centre of the vehicle to the steering rack itself, or to an isolator that is connected to it. The designer must ensure that the steering rack cannot twist when subject to the moments that arise. When the wheels rebound and compress, the tie rods are moved to be at an angle, something which also happens when the wheels are steered. The effective distance *a* between the eye-type joints of the tie rods and the steering rack centre line, shown in Fig. 9.1-11, gives a lever, via which the steering could be twisted. Two guide pieces which slide in a groove in the casing stop this from happening. However, the need to match the fit for the bearing of the steering rack and the guide groove can lead to other problems. If they are too tight, the steering will be heavy, whereas if they are too loose, there is a risk of rattling noises when the vehicle is in motion.

As the steering forces are introduced at a relatively large distance from the bearing points of the steering axle (suspension strut support bearings at the top, ball-and-socket joint on the transverse link at the bottom), elastic (flexural) deformations occur on the suspension strut and shock-absorber strut. As a result, steering precision and response characteristics worsen.

9.1.3 Recirculating ball steering

9.1.3.1 Advantages and disadvantages

Steering gears with a rotating movement are difficult to house in front-wheel-drive passenger cars and, in a standard design vehicle with independent wheel suspension,

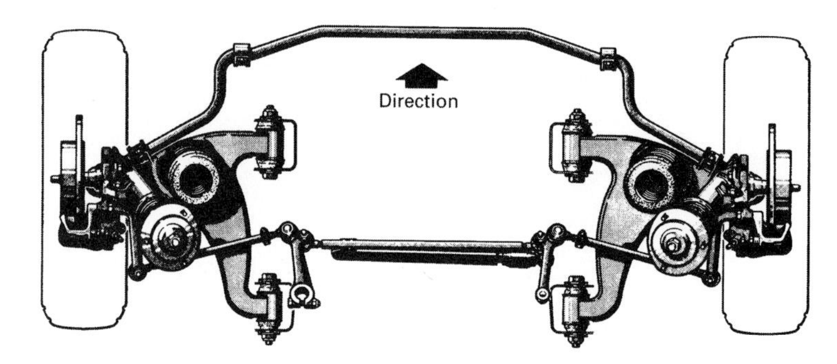

Fig. 9.1-12 Top view of the strut damper front axle on a Mercedes vehicle. The intermediate rod and the tie rods are fixed side by side on the pitman and idler arms and one grips from the top and the other from the bottom into the two levers; the steering square is opposed. The steering damper is supported on the one side at the intermediate rod and on the other side on the suspension subframe.
The anti-roll bar is linked to the lower wishbone type control arms whose inner bearings take large rubber bushings. The defined springing stiffness of these bearings, together with the inclined position of the tie rods (when viewed from the top) means that when the vehicle corners, there is a reduction in the steering input, i.e. elastic compliance in the steering, tending towards understeering. The strut dampers are screwed to the steering knuckles; the negative kingpin offset is $r_\sigma = -14$ mm.

also require the idler arm 5 (see Fig. 9.1-3) and a further intermediate rod, position 6, to connect them to the pitman arm 4; the tie rods are adjustable and have pre-lubricated ball joints on both sides (Figs. 9.1-13 and 9.1-14).

This type of steering system is more complicated on the whole in passenger cars with independently suspended front wheels and is therefore more expensive than rack and pinion steering systems; however, it sometimes has greater steering elasticity, which reduces the responsiveness and steering feel in the on-centre range.

Comparing the two types of configuration (without power-assisted steering) indicates a series of advantages:

- Can be used on rigid axles (Figs. 9.1-5 and 8.1-37).
- Ability to transfer high forces.
- A large wheel input angle possible – the steering gear shaft has a rotation range up to ±45°, which can be further increased by the steering ratio.
- It is therefore possible to use long steering arms.
- This results in only low load to the pitman and intermediate arms in the event of tie rod diagonal forces occurring.
- It is also possible to design tie rods of any length desired, and to have steering kinematics that allow an increase in the overall steering ratio i_S with increase ing steering angles. The operating forces necessary to park the vehicle are reduced in such cases.

9.1.3.2 Steering gear

The input screw shaft 4 (Fig. 9.1-15) has a round thread in which ball bearings run, which carry the steering nut 5 with them when the steering wheel is rotated. The balls which come out of the thread at the top or the bottom (depending on the direction of rotation) are returned through the tube 6. The nut has teeth on one side which mesh with the toothed segment 7 and therefore with the steering output shaft 8. When viewed from the side, the

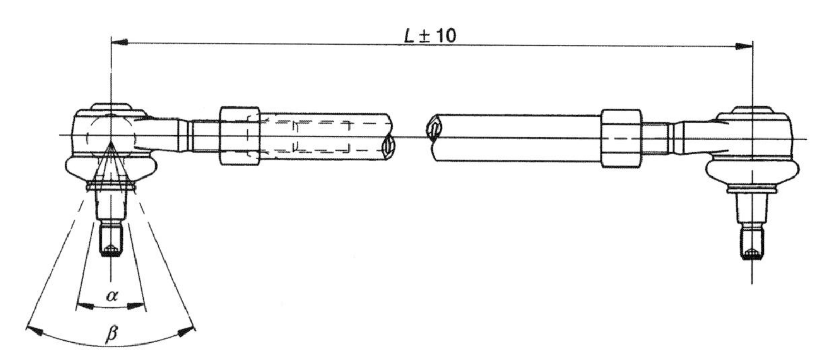

Fig. 9.1-13 Configuration of an adjustable tie rod with pre-lubricated joints and buckling-resistant central tube, the interior of which has a right-hand thread on one side and a left-hand thread on the other. It can usually be continuously adjusted by ±10 mm. When toe-in has been set, the length on the right and left tie-rod may differ, resulting in unequal steering inputs and different size turning circles; for this reason, the central tube should be turned the same amount on the left and right wheel.
The configuration shown in the illustration is used on rigid front axles and as a drag link (illustration: Lemförder Fahwerktechnik).

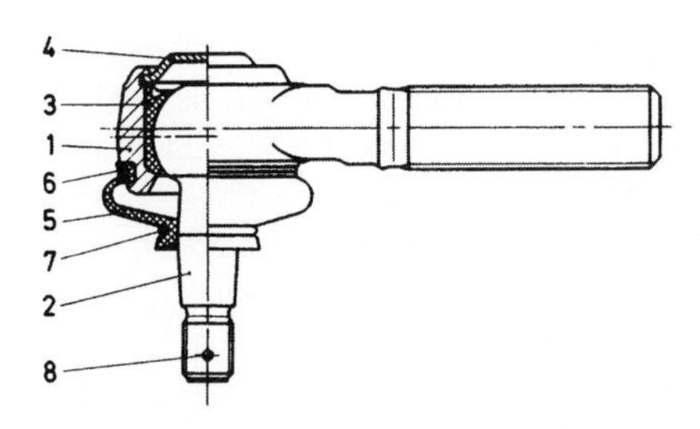

Fig. 9.1-14 Lemförder Fahrwerktechnik pre-lubricated tie rod joint, used on passenger cars and light vans. The joint housing 1 has a fine thread on the shaft (M14 × 1.5 to M22 × 1.5) and is made of annealed steel C35V; surface-hardenable steel 41Cr4V is used for the ball pivot 2.
The actual bearing element – the one-part snap-on shell 3 made from polyacetal (e.g. DELRIN, made by Dupont) – surrounds the ball; the rolled-in panel cover 4 ensures a dirt- and waterproof seal. The polyurethane or rubber sealing gaiter 5 is held against the housing by the tension ring 6. The gaiter has a bead at the bottom (which the second tension ring 7 presses against the spigot) and a sealing lip, which comes into contact with the steering arm.
The ball pivot 2 has the normal 1:10 taper and a split pin hole (position 8). If there is a slit or a hexagonal socket (with which the spigot can be held to stop it twisting), a self-locking nut can be used instead of a slotted castle nut and split pin.

slightly angular arrangement of the gearing can be seen top right. This is necessary for alignment bolt 1 to overcome the play of the wheels when pointing straight ahead, by axial adjustment. If play occurs in the angular ball bearings 2 and 3, the lock-nut must be loosened and the sealing housing cover re-tightened.

Only a few standard design larger saloons can be found on the road with manual recirculating ball steering. For reasons of comfort, newer passenger cars of this type have hydraulic power-assisted steering. The same applies to commercial vehicles; only a few light vans are still fitted with manual configurations as standard and even these are available with power-assisted steering as an option.

9.1.4 Power steering systems

Power steering systems have become more and more widely used in the last few years, due to the increasing front axle loads of vehicles on the one hand and the trend towards vehicles with more agile steering properties and hence direct transmission steering systems on the other. With the exception of some members of the 'sub-compact' class, power steering systems are optionally or automatically included as one of the standard features.

Manual steering systems are used as a basis for power steering systems, with the advantage that the mechanical connection between the steering wheel and the wheel and all the components continues to be maintained with or without the help of the auxiliary power. The steering-wheel torque applied by the driver is

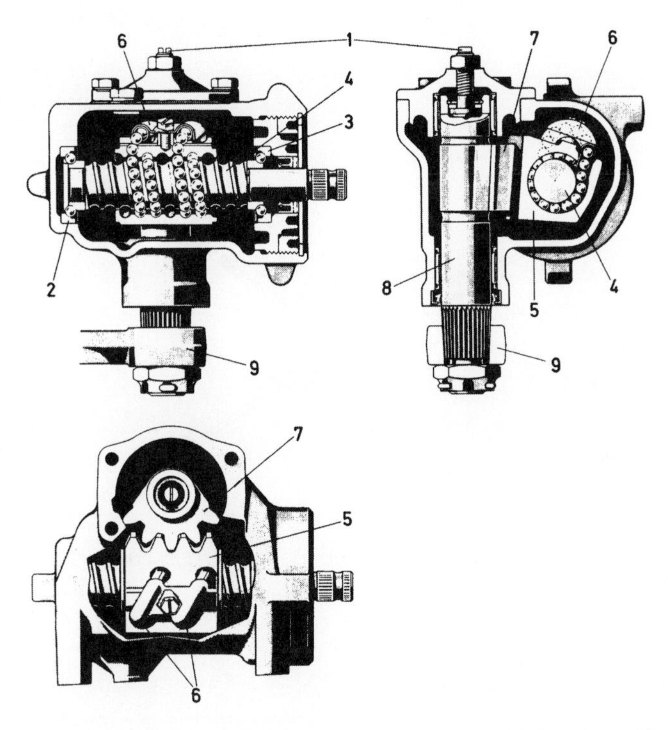

Fig. 9.1-15 Mercedes Benz recirculating ball steering suitable for passenger cars and light vans; today, apart from in a few exceptional cases, this is only fitted as a hydraulic power-assisted version. Pitman arm 9 is mounted onto the tapered toothed profile with a slotted castle nut 11 (Fig. 9.1-24).

detected by a measurement system located in the region of the input shaft of the steering gear or in the steering tube, and additional forces or moments are introduced into the system. This follows a characteristic curve (valve characteristic) or group of curves depending on the height of the steering-wheel torque, if another quantity, e.g. driving speed, is entered as a signal. The steering boost is thereby reduced, with the aim of achieving better road contact at higher speeds.

9.1.4.1 Hydraulic power steering systems

Hydraulic power steering systems are still the most widely used. The method of using oil under pressure to boost the servo is sophisticated and advantageous in terms of cost, space and weight. Sensitivity to movements caused by the road surface and hence the effect of torsional impacts and torsional vibrations passing into the steering wheel is also noticeably reduced, particularly with rack and pinion steering. This can be attributed to the hydraulic self-damping. It might also be the reason why it is possible to dispense with an additional steering shock absorber in most vehicles with hydraulical rack and pinion steering, whereas it is required for the same vehicles with manual steering (see Section 9.1.6).

The oil pump is directly driven by the engine and constantly generates hydraulic power. As hydraulic power steering systems have to be designed in such a way that a sufficient supply volume is available for fast steering movements even at a low engine speed, supply flow limiting valves are required. These limit the supply flow to about 8 l per minute in order to prevent the hydraulic losses which would otherwise occur at higher engine speeds. Depending on the driving assembly and pump design, the additional consumption of fuel can lie between 0.2 and 0.7 l per 100 km.

Assemblies which are added to provide auxiliary power are shown in Fig. 9.1-16, taking the example of the rack and pinion steering used by Opel in the Vectra (1997). The pressure oil required for steering boost is supplied direct to the steering valve 6 located in the pinion housing from vane pump 1 via the high-pressure line 2 and the cooling circuit 3. From here, depending on the direction of rotation of the steering wheel and the corresponding counterforce on the wheels, distribution to the right or left cylinder line takes place (items 7 and 8). Both lead to the working cylinder which is integrated in the steering-gear housing 5. A disc located on the gear rack divides the pressure chamber. Differences in pressure generate the required additional axial force in the gear rack F_{Pi} via the active areas of the disc:

$$F_{Pi} = (p_{hyd,2} - p_{hyd,1})A_{Pi} \qquad (9.1.1)$$

where A_{Pi} is the effective piston surface, here the difference between the disc and gear rack surfaces, and $p_{hyd,1 \text{ or } 2}$ are the pressures acting on the working piston.

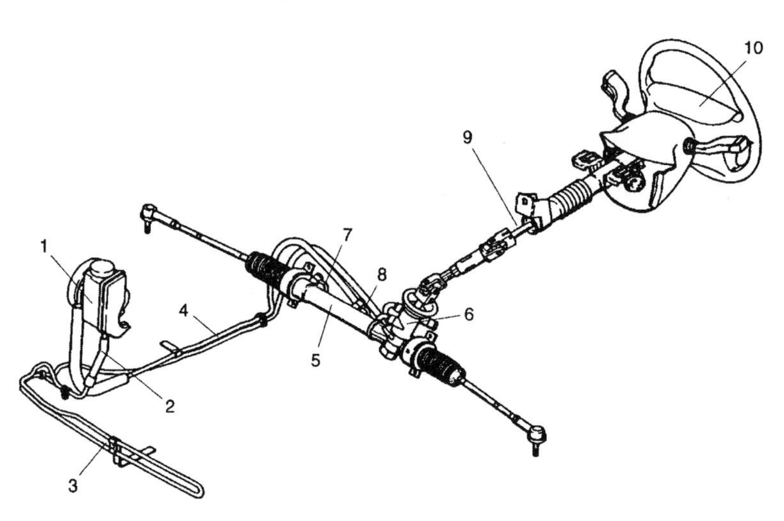

Fig. 9.1-16 Hydraulic power steering system of the Opel Vectra (1997). The individual components are:
1 vane pump, driven by V-belts
2 high-pressure line
3 cooling circuit
4 return line, from the steering valve to the pump
5 steering gear with external drive, attached to the auxiliary frame
6 steering valve
7/8 pressure lines to the working cylinder
9 steering column with intermediate shaft
10 steering wheel with integrated airbag

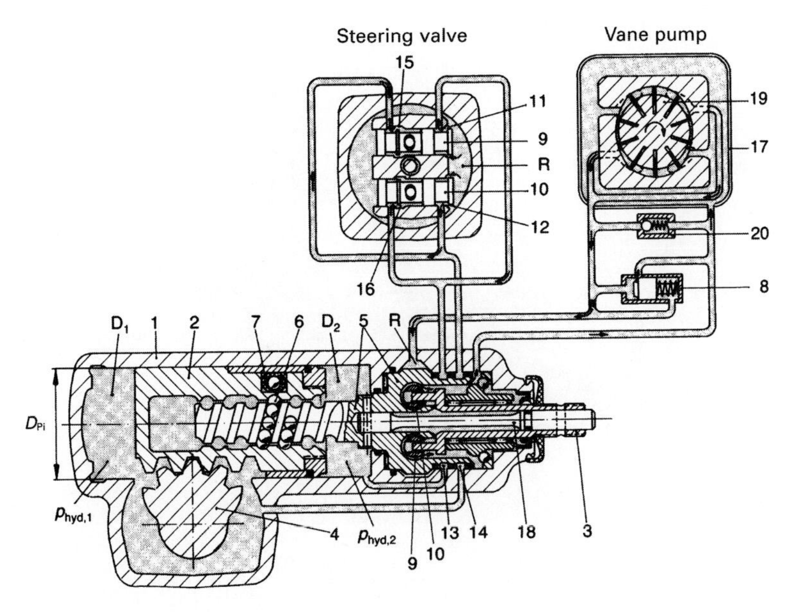

Fig. 9.1-17 Illustration of the principles of the ZF recirculating ball steering in the neutral position (vehicle travelling in a straight line). The steering valve, the working piston and the mechanical gear sit in a common housing. The two valve pistons of the steering valve have been turned out of their operating plane to make the diagram easier to see. The individual parts are:

1	gear housing	9/10	valve piston
2	piston with steering nut	11/12	inlet groove
3	steering spindle connection	13/14	radial groove
4	steering shaft with toothed segment	15/16	return groove
5	steering worm roller with valve body	17	fluid reservoir
6	balls	18	torsion bar
7	recirculation tube	19	hydraulic pump
8	fluid flow limitation valve	20	pressure-limiting valve

In a situation where there is no torque, for example during straight running, the oil flows direct from the steering valve 6 back to the pump 1 via the return line 4.

The method of operation of the steering valve is shown in Fig. 9.1-17, using the example of recirculating-ball steering. In a similar way to rack and pinion steering, it is integrated into the input shaft of the steering gear. As is the case with most hydraulic power steering systems, the measurement of the steering-wheel torque is undertaken with the use of a torsion bar 18. The torsion bar connects the valve housing 5 (part of the steering screw) to the valve pistons 9/10 in a torsionally elastic way. Steering-wheel torque generates torsion of the torsion bar. These valve pistons then move and open radial groove 13 or 14, depending on the direction of rotation. This leads to a difference in pressure between pressure chambers D1 and D2. The resultant axial force on the working piston 2 is calculated using Equation 9.1-2. Because $p_{\text{hyd},2}$ also operates in the interior space of the piston behind the steering screw 5, the surface areas are the same on both sides:

$$F_{\text{Pi}} = p_{\text{hyd},1 \text{ or } 2} A_{\text{Pi}} = p_{\text{hyd},1 \text{ or } 2} \frac{\pi D_{\text{Pi}}^2}{4} \qquad (9.1.2)$$

9.1.4.2 Electro-hydraulic power steering systems

With electro-hydraulic power steering systems, the power-steering pump driven by the engine of the vehicle via V-belts is replaced by an electrically operated pump.

Fig. 9.1-18 shows the arrangement of the system in an Opel Astra (1997). The electrically operated power pack supplies the hydraulic, torsion-bar controlled steering valve with oil. The pump is electronically controlled – when servo boost is not required, the oil supply is reduced.

The supply of energy by electricity cable allows greater flexibility with regard to the position of the power pack. In the example shown, it is located in the immediate vicinity of the steering gear. Compared with the purely hydraulic system, the lines can be made considerably shorter and there is no cooling circuit. The steering gear, power pack and lines are installed as a ready-assembled and tested unit.

To sum up, electro-hydraulic power steering systems offer the following advantages:

- The pressure supply unit (Fig. 9.1-19) can be accommodated in an appropriate location (in relation to space and crash safety considerations).

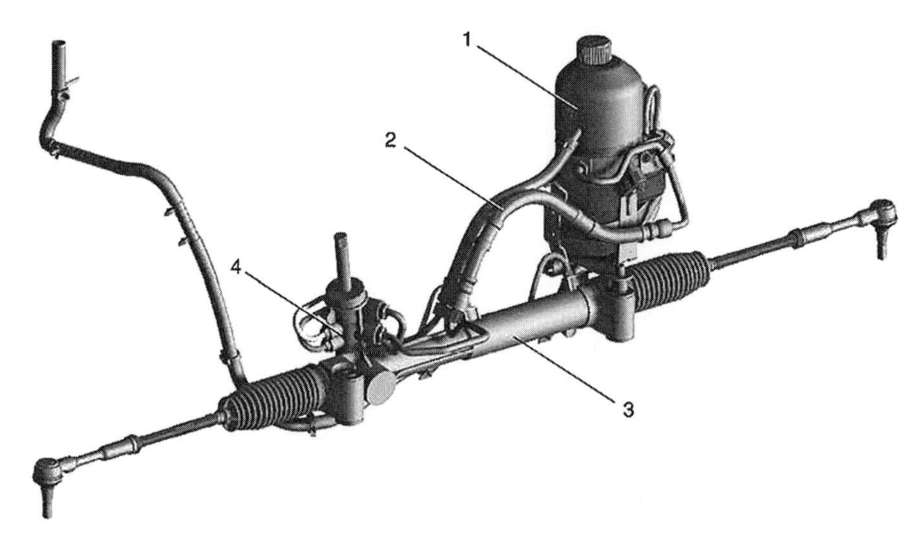

Fig. 9.1-18 Electro-hydraulic power steering system of the Opel Astra (1997). The individual components are:
1 electrically operated power-steering pump with integrated reserve tank ('power pack')
2 pump–steering valve hydraulic lines
3 rack and pinion steering gear with external drive, attached to auxiliary frame
4 steering valve.

- Servo boost is also guaranteed by the electrical pressure supply even when the engine is not running.
- Pressure-controlled systems generate only the amount of oil required for a particular driving situation. Compared with standard power steering systems, energy consumption is reduced to as little as 20%.
- The steering characteristics (nature and amount of steering boost, sensitivity, speed dependency) can be adjusted by the control electronics individually for the particular vehicle.

9.1.4.3 Electrical power steering systems

The bypass of the hydraulic circuit and direct steering boost with the aid of an electric motor has additional advantages in terms of weight and engine bay space compared with electro-hydraulic steering, because of

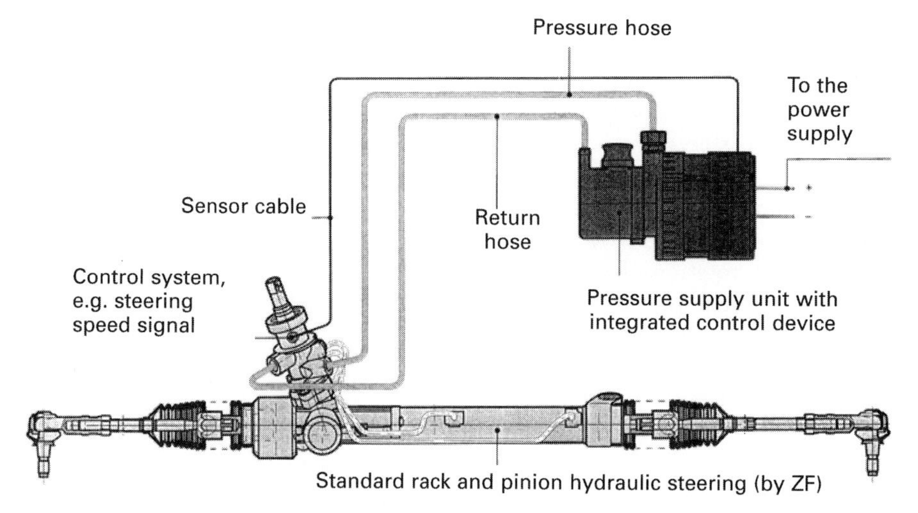

Fig. 9.1-19 Open-centre control system from ZF. The pressure supply unit designed as a modular unit can be fitted with different electric motors (DC motor with or without brushes) and pump fuel feed volumes (1.25–1.75 cm^3 per rpm) depending on its particular function. Oil tanks for horizontal or vertical installation are also available. Operating pressure is up to 120 bar, with a maximum power consumption of 80 A.

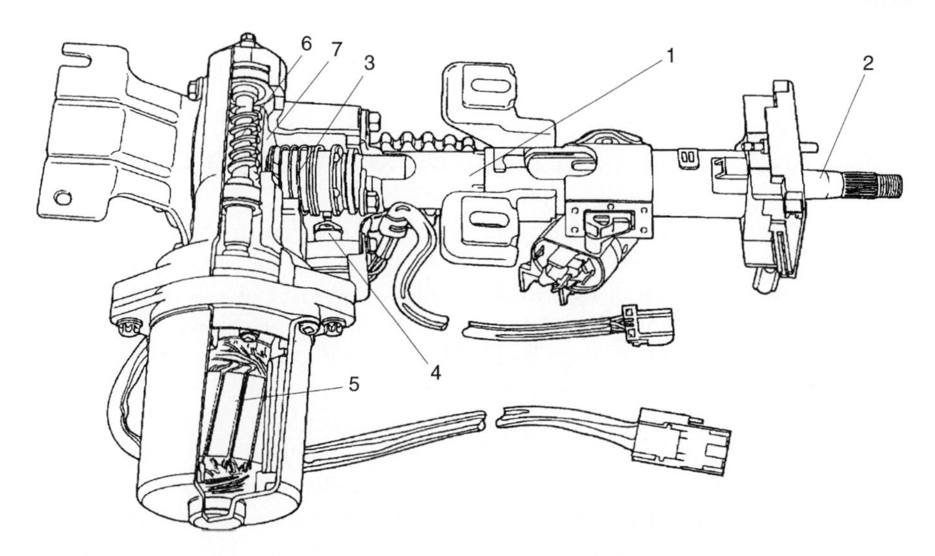

Fig. 9.1-20 Steering column with power-steering assembly of the Opel Corsa (1997). The individual components are:
1 column tube
2 steering tube
3 sliding sleeve with groove
4 rotary potentiometer with tap
5 servomotor
6 drive worm
7 worm gear

the omission of all the hydraulic components. Other advantages are obtained through more variations of the steering boost because of the purely electrical signal processing.

The electrical servo unit can be installed on the steering column (Fig. 9.1-20), pinion (Fig. 9.1-21) or gear rack (Fig. 9.1-22). The steering axle loads and maximum gear rack forces are, depending on the particular

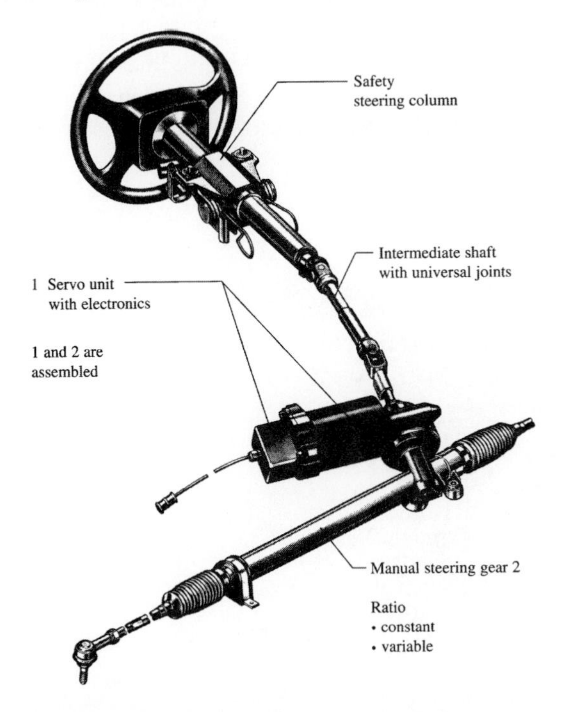

Fig. 9.1-21 Electrical power steering system by ZF. The servo unit acts directly upon the pinion of the rack and pinion steering. Consequently, the amount of stress to which the pinion is subjected increases by the amount of steering boost, compared with a mechanical or hydraulic power steering system.

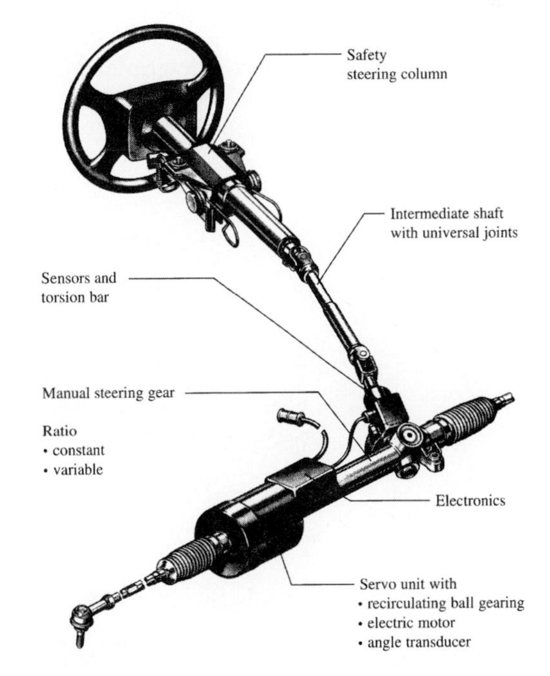

Fig. 9.1-22 Electrical power steering system by ZF. The servo unit acts on the gear rack itself. This system is suitable for high axle loads and steering forces. The maximum current strength is 105 A with a 12 V electric system; with a 42 V system, it is only 35 A.

arrangement, about 650 kg and 6000 N, 850 kg and 8000 N or 1300 kg and 10 000 N.

The systems only have limited power because the current is limited by an operating voltage of 12 V. They are of interest though for smaller vehicles. In this class of vehicles in particular, electric power steering systems show their advantages, not least because of the small amount of energy required. The introduction of the increased voltage of 42 V will make the use of electrical power steering systems and wheel brakes much easier.

Fig. 9.1-23 shows the steering system of the Opel Corsa (1997) with electric power steering. It is a system with steering-tube transmission, i.e. the intermediate spindle transmits the whole of the torque resulting from the steering wheel force and servo boost. Due to the more direct steering transmission, this torque is clearly higher than in a comparable manual steering system, something which must be taken into consideration when deciding on the size of the components which control performance.

In Fig. 9.1-20, the method of operation of the servo assembly (EPAS system by NSK) becomes clear: a plastic worm gear 7 is applied to the steering tube 2. This is engaged by the worm 6, which in its turn is connected to the shaft of the servomotor 5. Steering-wheel torque generates a torsional movement of the torsion bar (concealed by the sliding sleeve 3). The steering tube area is axially grooved above the torsion bar and spindle-shaped below. As the spindle rises, the sliding sleeve makes an axial movement on the steering tube proportional to the torsion of the torsion bar. This axial movement is transmitted to the rotary potentiometer 4 via a tap. Corresponding to a group of curves, the servo boost is determined from the steering-wheel torque and driving speed signals and the servomotor 5 controlled accordingly.

9.1.5 Steering column

In accordance with the German standard DIN 70 023 'nomenclature of vehicle components', the steering column consists of the jacket tube (also known as the outer tube or protective sleeve), which is fixed to the body, and the steering shaft, also called the steering tube. This is only mounted in bearings at the top (or top and bottom, positions 9 and 10 in Fig. 9.1-26) and transfers the steering-wheel moment M_H to the steering gear.

A compliant cardan joint (part 10 in Fig. 9.1-24) can be used to compensate for small angular deviations. This also keeps impacts away from the steering wheel and, at the same time, performs a noise insulation function on hydraulic power-assisted steering. If the steering column does not align with the extension of the pinion gear axis (or the steering screw), an intermediate shaft with two universal joints is necessary (part 6 in Fig. 9.1-26). When universal joints are used, attention should be paid to their transmission properties, which are dependent on their angle of inflexion for steering wheel angle and moment, because a non-linear steering moment above the steering angle, noticeable for the driver, can occur.

The steering tube should be torsionally stiff to keep the steering elasticity low. On the other hand, it should show, together with the jacket tube, a deformation behaviour which is defined in a longitudinal direction, as steering wheel intrusion in case of a head-on crash is to be avoided while the absorption of force necessary for the

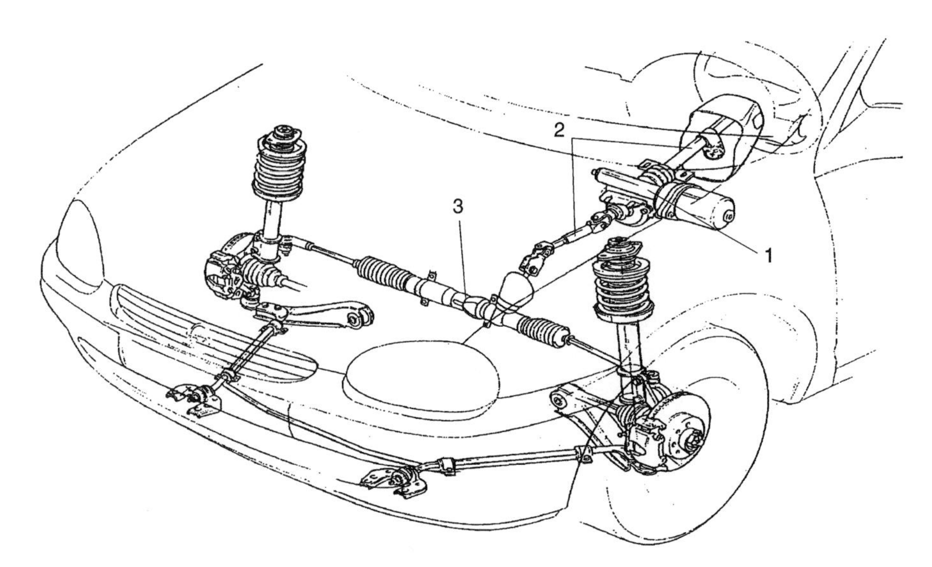

Fig. 9.1-23 Electric power steering system of the Opel Corsa (1997). The individual components are:
1 steering-column assembly
2 steering column with intermediate spindle
3 rack and pinion steering with external drive.

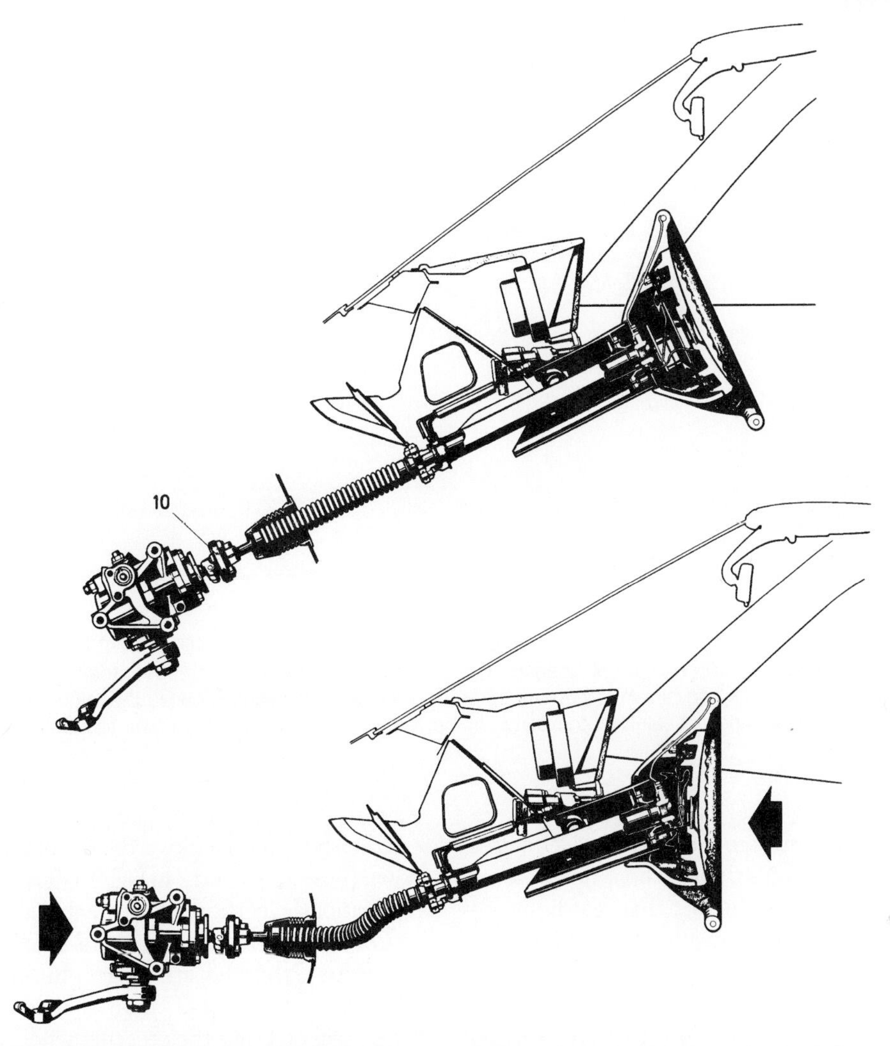

Fig. 9.1-24 Mercedes Benz safety steering tube and dished steering wheel; it is fixed to the recirculating ball steering gear with a compliant 'joint'. The bottom illustration shows the corrugated tube bending out in a head-on crash. The illustration also shows the energy-absorbing deformation of the steering wheel and the flexibility of the steering gear mounting.

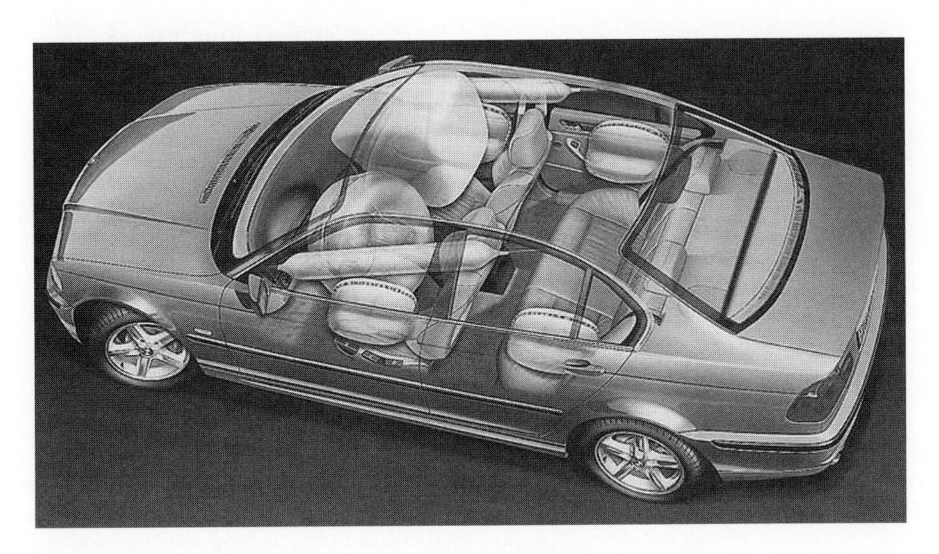

Fig. 9.1-25 BMW passenger car with air bags for the front, sides and head (front and back).

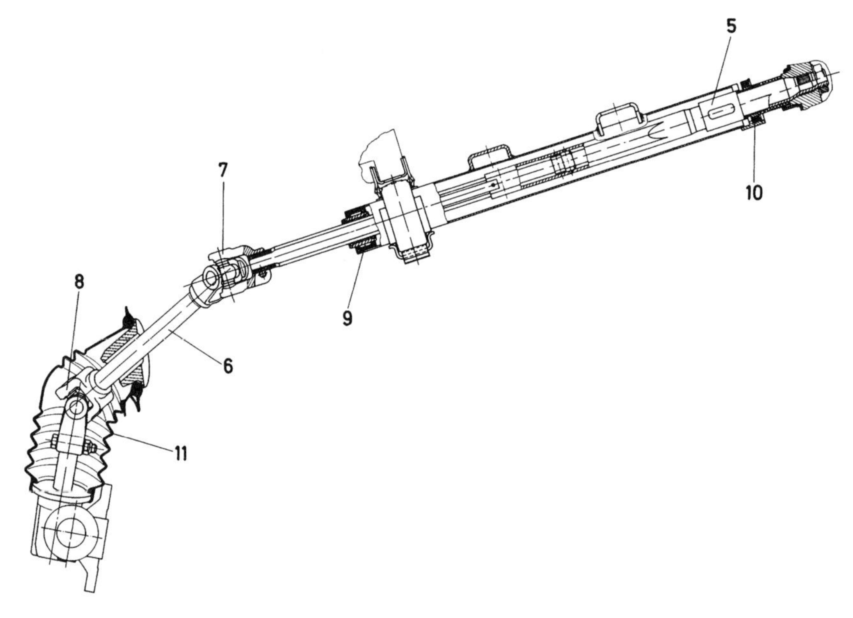

Fig. 9.1-26 Steering column of the VW Golf III and Vento (1996). The collapsible steering tube (Fig. 9.1-27) is carried from the bottom by the needle bearing 9 and through the top by the ball bearing 10 in the jacket tube; the spigot of the steering lock grips into part 5. The almost vertical pinion gear of the rack and pinion steering is linked to the inclined steering tube via the intermediate shaft 6 with the universal joints 7 and 8. The dashpanel is sealed by the gaiter 11 between this and the steering gear (illustration: Lemförder Fahrwerktechnik).

functionality of the airbag (Fig. 9.1-25) must be safeguarded. As there is a requirement in some US states that the airbag should cushion a driver who is not wearing his seatbelt in a crash, despite the fact that seat belts are mandatory, the steering column must be designed for this borderline contingency.

Three types of steering tube configuration meet these requirements with vehicle-specific deformation paths on passenger cars:

- steering tubes with flexible corrugated tube portion (Fig. 9.1-24);
- collapsible (telescopic) steering tubes (Figs. 9.1-27 and 9.1-28);
- detachable steering tubes (Figs. 9.1-1 and 9.1-29).

To increase ride and seating comfort, most automobile manufacturers offer an adjustable steering column, either as standard or as an option. The position of the steering wheel can then be altered backwards and forwards as well as up and down (positions 1 and 2 in Fig. 9.1-30). As can be seen in the illustrations, electrical adjustment is also possible.

On light vans, which have a steering gear in front of the front axle, the steering column is almost vertical (Figs. 8.1-7 and 8.1-37). In a head-on crash the outer tube bracket 1 and the steering wheel skeleton must flex (Fig. 9.1-31).

9.1.6 Steering damper

Steering dampers absorb shocks and torsional vibrations from the steering wheel and prevent the steering wheel over-shooting (also known as free control) on front-wheel-drive vehicles – something which can happen when

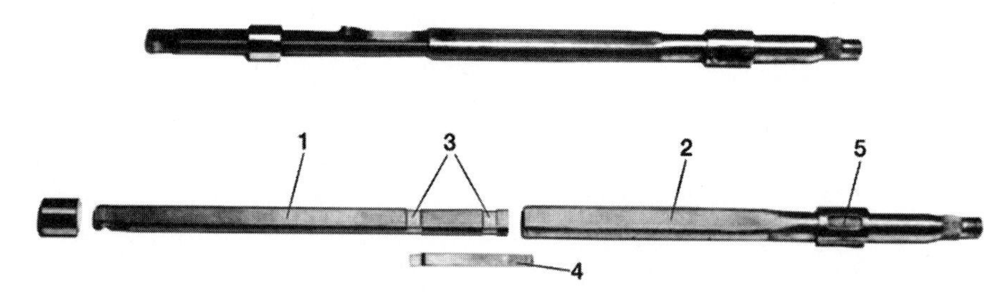

Fig. 9.1-27 Telescopic collapsible steering tubes consist of a lower part 1, which is flattened on the outside, and a hollow part 2, which is flattened on the inside. The two will be fitted together; the two plastic bushes 3 ensure that the assembly does not rattle and that the required shear-off force in the longitudinal direction is met. The tab 4 fixed to part 1 ensures the passage of electric current when the horn is operated. The spigot of the steering wheel lock engages with the welded-on half shells 5 (illustration: Lemförder Fahrwerktechnik).

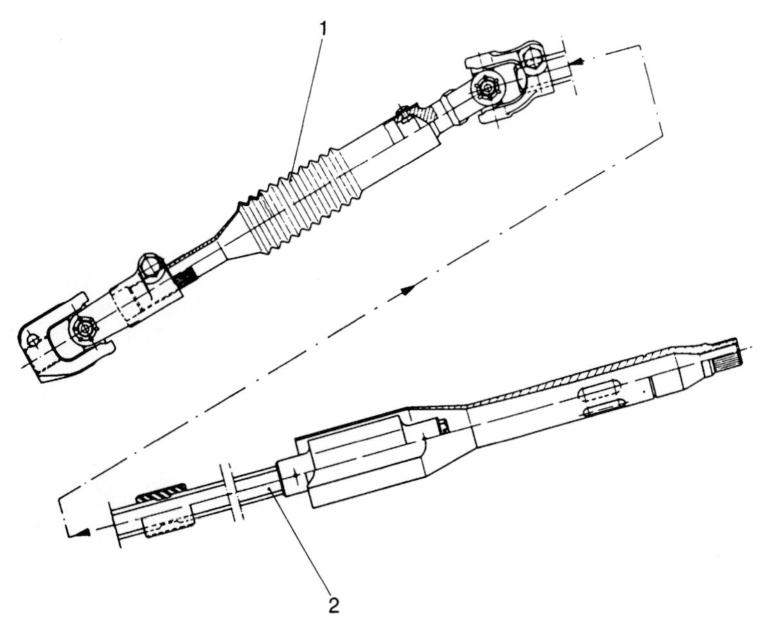

Fig. 9.1-28 Volvo steering column. Both the corrugated tube 1 in the intermediate shaft and the collapsible steering tube 2 meet the safety requirements. To save weight, the universal joints are made of aluminium alloy Al Mg Si 1 F31 (illustration: Lemförder Fahrwerktechnik).

the driver pulls the steering wheel abruptly. The dampers therefore increase ride comfort and driving safety, mainly on manual steering gears. The setting, which generally operates evenly across the whole stroke range, allows sufficiently light steerability but stops uncontrollable wheel vibrations where the front wheels are subjected to uneven lateral and longitudinal vibrational disturbances; in this event the damper generates appropriate forces according to the high piston speeds involved.

The dampers are fitted horizontally. As shown in Fig. 8.1-57, on rack and pinion steering, one side of the damper is fixed to the steering rack via an eye or pin-type joint and the other to the steering housing. On recirculating ball steering systems, the pitman arm on independent wheel suspensions or the intermediate rod can be used as a pivot point (Figs. 8.1-39 and 9.1-12) or the tie rod on rigid axles. As shown in Fig. 9.1-5, this is parallel to the axle housing.

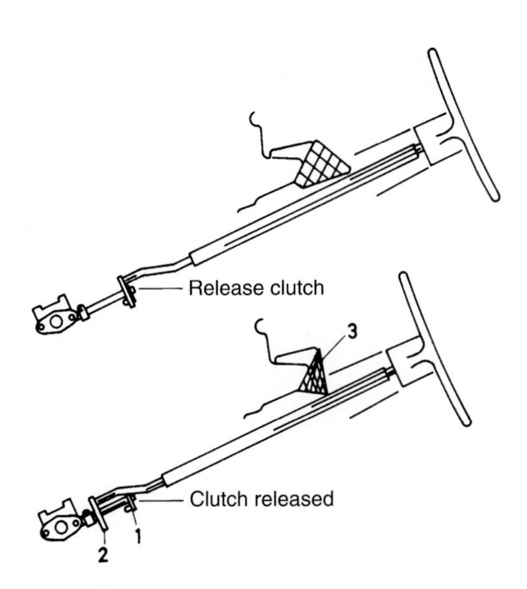

Fig. 9.1-29 'Release clutch' used by VW on steering columns. A half-round plate sits on the short shaft that is linked to the steering pinion gear, and carries the two pins 1 which point downwards. They grip into the two holes of the clutch 2 sitting on the steering tube from the top. The jacket tube is connected to the dashboard via a deformable bracket. As shown in a head-on crash, this part 3 flexes and the pins 1 slide out of part 2.

9.1.7 Steering kinematics

9.1.7.1 Influence of type and position of the steering gear

Calculating the true tie rod length u_0 (Fig. 9.1-32) and the steering arm angle λ (Fig. 9.1-3) creates some difficulties in the case of independent wheel suspensions. The position of the steering column influences the position of the steering gear by the type of rotational movement. If this deviates from the horizontal by the angle ω (Fig. 9.1-33), a steering gear shaft, which is also inclined by the angle ω, becomes necessary. The inner tie rod joint T which sits on the pitman arm, is carried through a three-dimensional arc, influenced by this angle ω when the wheels are turned. However, the outer joint U on the steering knuckle whose steering axis is inclined inwards (Fig. 9.1-34) by the kingpin inclination angle σ and is

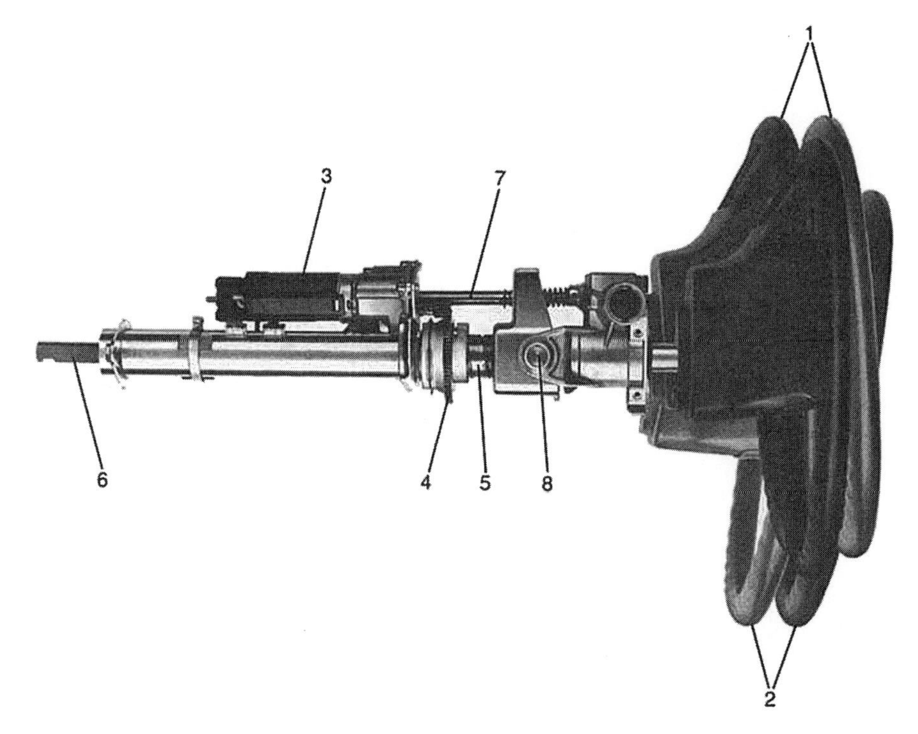

Fig. 9.1-30 Electrically adjustable steering column manufactured by Lemförder Fahrwerktechnik. The electric motor 3 turns a ball nut via the gears 4 and this engages with the grooves 5 of the steering tube and shifts it (position 6) in the longitudinal direction (position 1). To change the height of the steering wheel (position 2), the same unit tips around the pivot 8 by means of the rod 7.

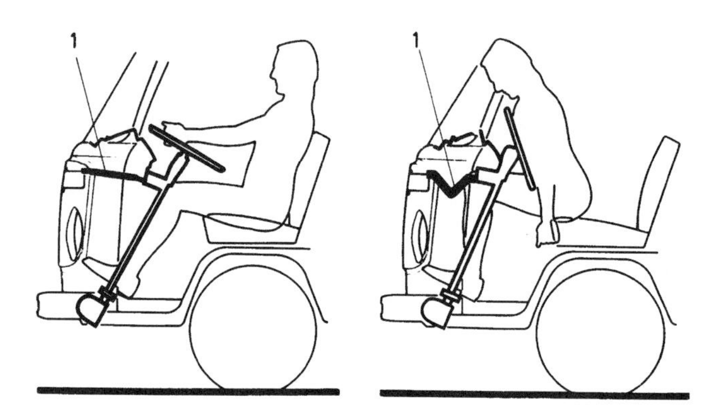

Fig. 9.1-31 The VW Bus Type II has an almost vertical steering column. In a head-on crash, first the steering wheel rim gives and then the retaining strut 1, which is designed so that a given force is needed to make it bend inwards.

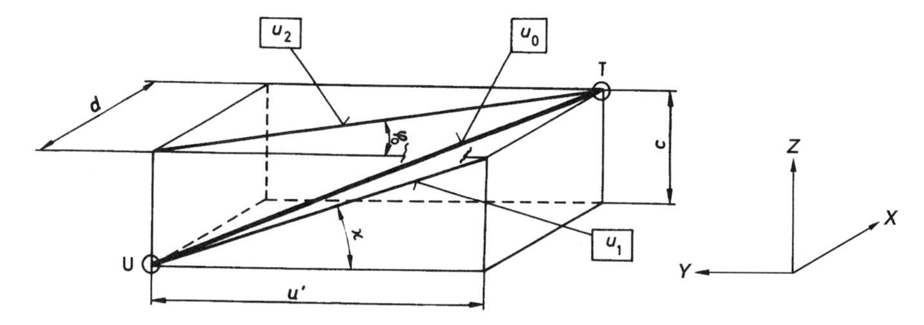

Fig. 9.1-32 On independent wheel suspensions, the tie rod UT is spatially inclined. The path u' (i.e. the lateral distance of points U and T from one another) or the angle κ must be determined when viewed from the rear. From the top view, the distance d or the angle ω_0 is more important; the projected lengths which appear in both views are u_1 and u_2. The true tie rod length is then: $u_0 = (u'^2 + c^2 + d^2)^{1/2}$.

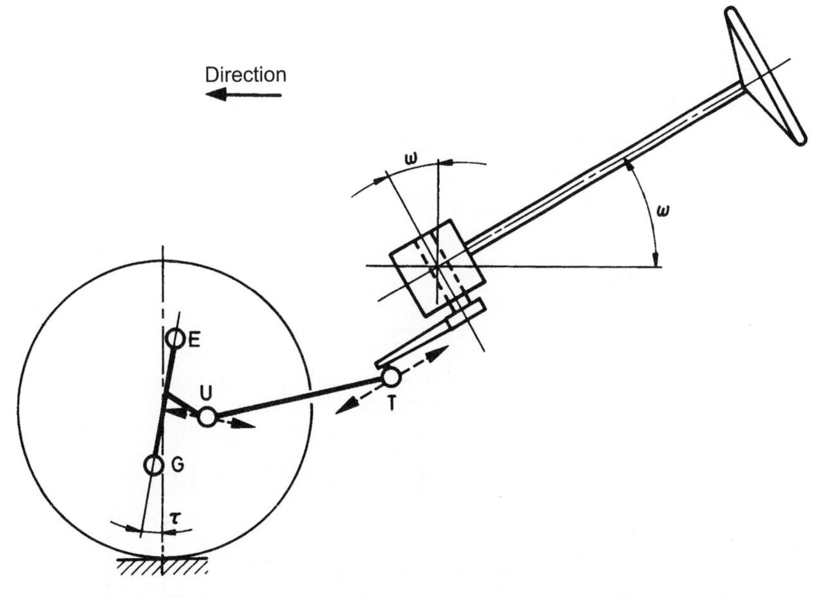

Fig. 9.1-33 The central points of the tie rod joints (T on the inside and U on the outside) change their position relative to one another, based on the wheel travel (vertical and horizontal) on independent wheel suspensions. The reasons for this are the different directions of movement of pitman arm and steering arm. The former depends on the inclined position of the steering gear (angle ω) and that of the point U from the inclination of the steering axis EG, i.e. the kingpin inclination σ and the caster angle τ.

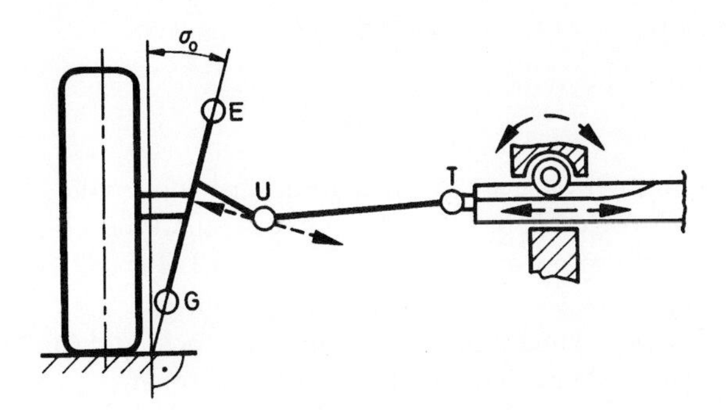

Fig. 9.1-34 When viewed from the rear, the inner tie rod joint T on rack and pinion steering moves parallel to the ground, whereas the outer tie rod joint U moves on an arc running vertical to the steering axis EG. Any caster angle τ must also be considered.

often also inclined backwards by the caster angle τ (shown in Fig. 9.1-33). This joint therefore moves on a completely different three-dimensional arc.

The construction designer's job is to calculate the steering arm angle λ (and possibly also the angle *o* of the pitman arm, Fig. 9.1-37) in such a manner that when the wheels are turned, the specified desired curve produced comes as close as possible. The achievement of the necessary balance is made more difficult still by the movements of the wheel carrier during driving: for example, wheel travel, longitudinal flexibility and vertical springing.

Two curves that are desirable on passenger cars with an initially almost horizontal shape ($\Delta\delta \approx +30'$) and a subsequent rise in the curve to nearly half the nominal value when the wheels are fully turned. The more highly loaded wheel on the outside of the bend can even be turned further in than the inner wheel (and not just parallel to it, $\Delta\delta \approx -30'$); due to the higher slip angle that then has been forced upon it, the tyre is able to transfer higher lateral forces. When the wheels are fully turned, the actual curve should, nevertheless, remain below the nominal curve to achieve a smaller turning circle.

The steering angle δ_o of the wheel on the outside of the bend depends on the angle of the one on the inside of the bend δ_i via the steering difference angle $\Delta\delta$:

$$\Delta\delta = \delta_i - \delta_o \quad \text{(axis of the ordinate)}$$

9.1.7.2 Steering linkage configuration

The main influences on $\Delta\delta$ are the steering arm angle λ, the inclined position of the tie rod when viewed from the top (angle φ_o, Fig. 9.1-32) and the angle *o* of the pitman and idler arms on steering gears with a rotational movement. The tie rod position is determined by where the steering gear can be packaged. The amount of space available is prescribed and limited and the designer is unlikely to be able to change it by more than a little. The task consists of determining the angles λ and *o* by drawing or calculation. Both also depend on the bearing elasticities, which are not always known precisely.

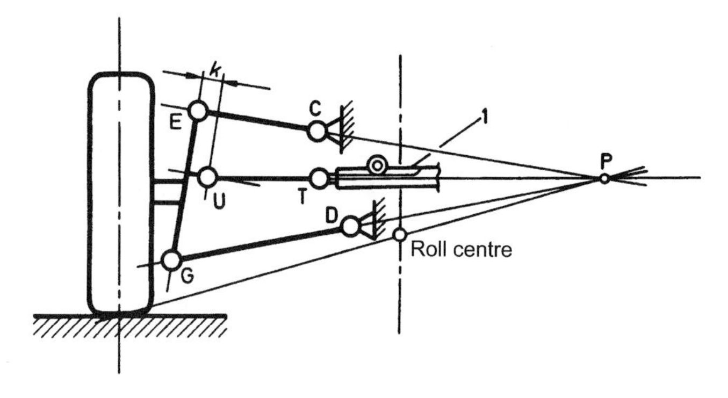

Fig. 9.1-35 Path and movement points necessary for determining the tie rod length and position. The position of the tie rods is given by the connecting line UP (to the pole). The illustration also shows the roll centre Ro.

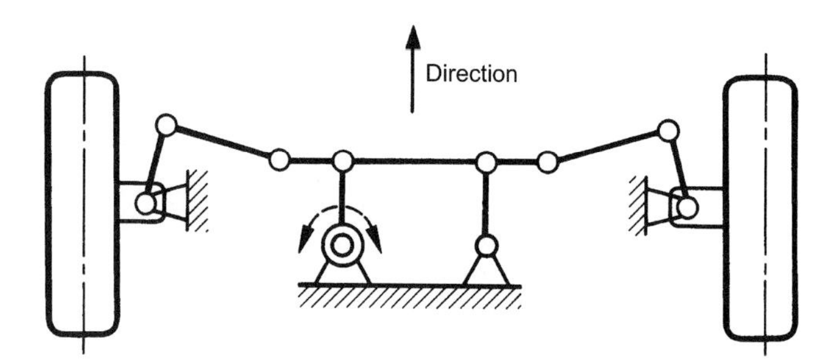

Fig. 9.1-36 'Synchronous' 4-bar linkage with steering arms pointing forwards. The inner joints are fixed to the sides of the intermediate rod.

The configuration of the steering kinematics on rack and pinion steering is comparatively simple; here, it is only necessary to transfer a straight-line lateral shift movement into the three-dimensional movement of the steering knuckle (Fig. 9.1-34). However, the extension of the tie rod UT must point to virtual centre of rotation P (Fig. 9.1-35); this is necessary on all individual wheel suspensions for determining the body roll centre Ro and is therefore known (see Section 9.1.6.3).

On steering gears with a rotational movement, the 4-bar linkage can be either in front or behind the axle and can be opposed or synchronous; Figs. 9.1-3 and 9.1-36–9.1-38 show four different configurations.

From a kinematic point of view, rack and pinion steering systems have a triangular linkage that can either be in front of or behind the axle or even across it. Figs. 9.1-4 and 9.1-39–9.1-41 show the individual options for left- and right-hand drive vehicles and also where the

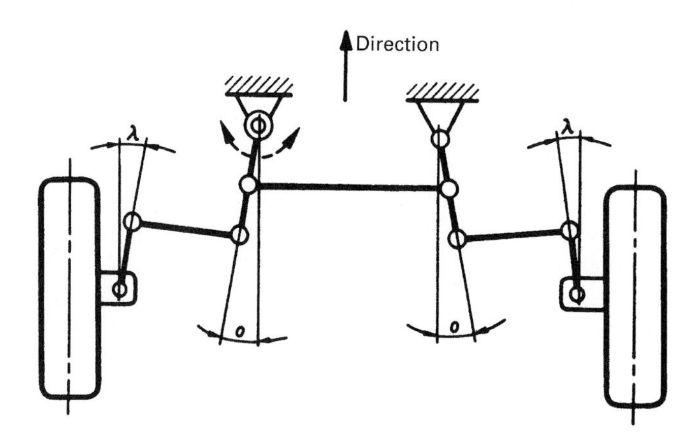

Fig. 9.1-37 'Opposed' 4-bar linkage located in front of the wheel centre. Steering arm and pitman arm rotate in opposite directions towards one another, similar to meshing gears. The tie rods are fixed directly to pitman and idler arms. For kinematic reasons, these can have the pre-angle *o* (see also Fig. 8.1-7).

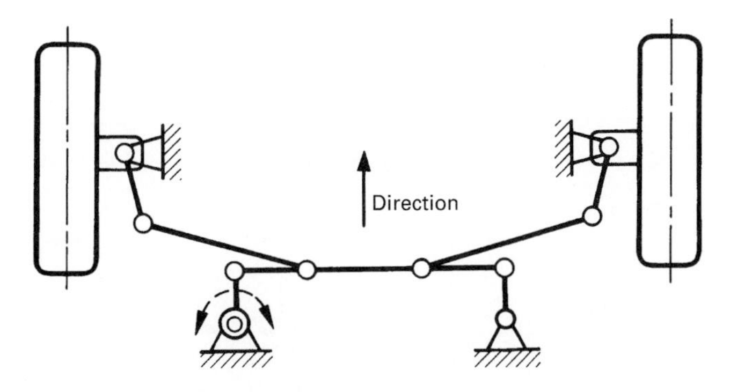

Fig. 9.1-38 'Opposed' 4-bar linkage located behind the wheel centre. The inner tie rod joints can be fixed to the middle part of the intermediate rod or directly to the pitman and idler arm (see Fig. 9.1-12).

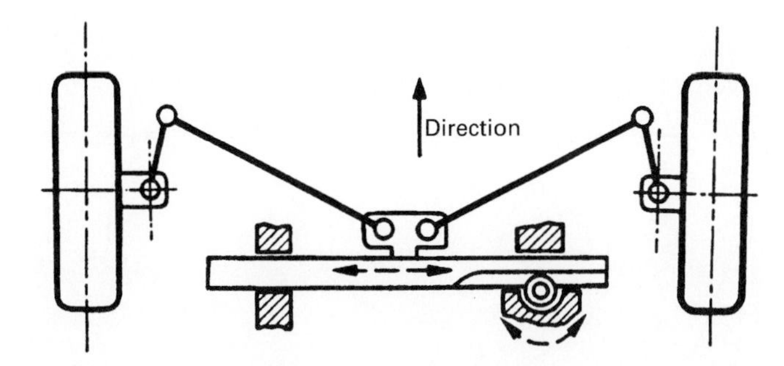

Fig. 9.1-39 The rack-and-pinion steering is behind and above the wheel centre and the steering arms point forward (shown for a right-hand-drive vehicle). For kinematic reasons, the inner tie rod joints are fixed to a central outrigger – known as a central take-off. This type of solution (also shown in Fig. 8.1-57) is necessary on McPherson and strut damper front axles with a high-location steering system as the tie rods have to be very long to avoid unwanted steering angles during jounce.

pinion gear must be located – above or below the steering rack – to make the wheels turn in the direction in which the steering wheel is turned. The steering arms (negative angles λ) which point outwards, shown in Fig. 9.1-41, allow longer tie rods; something which is useful when the inner joints are pivoted on the ends of the steering rack.

The significantly simpler steering kinematics on rigid axles are shown in Figs. 9.1-5–9.1-7.

9.1.7.3 Tie rod length and position

When the wheels compress and rebound as well as in longitudinal movement, there should not be any, or only

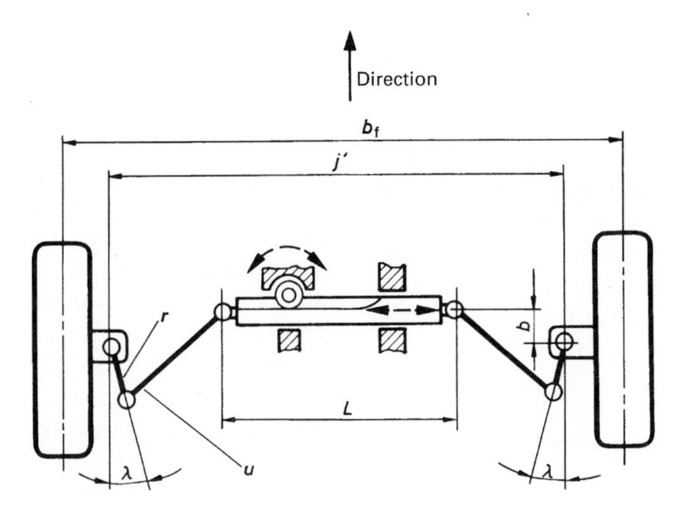

Fig. 9.1-40 The steering is in front of the wheel centre and the triangular linkage behind it, with the inner joints fixed to the ends of the steering rack.

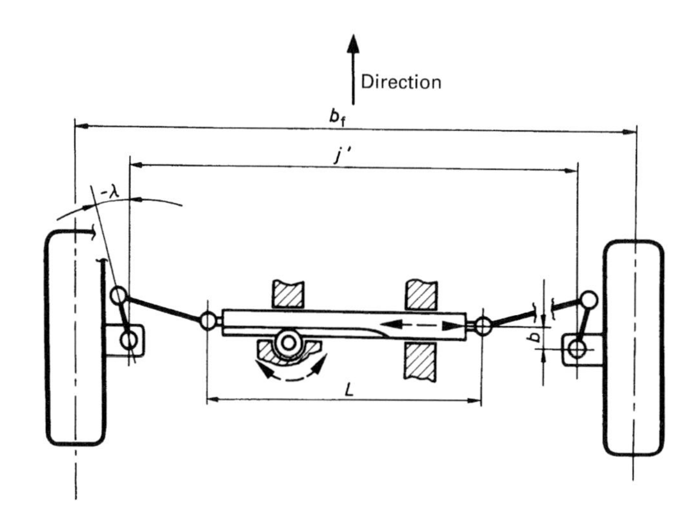

Fig. 9.1-41 Where rack and pinion steering and the steering triangle are shifted in front of the wheel centre, for kinematic reasons the steering arms must point outwards, making longer tie rods possible (see also Fig. 8.1-40).

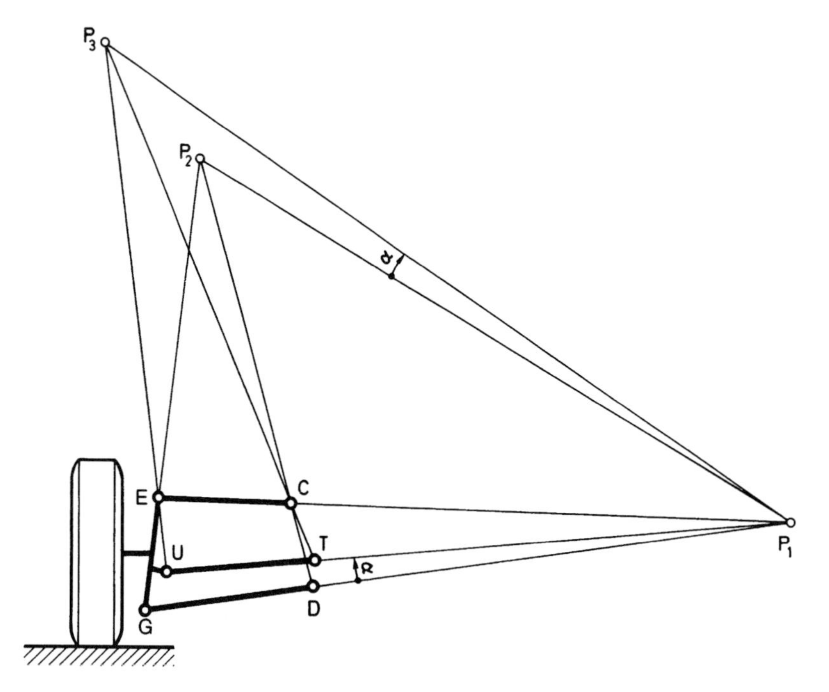

Fig. 9.1-42 Double wishbone suspension with steering arm pointing inwards. The tie rod is above the lower control arm.

a very specific, toe-in alteration; both depend primarily on the tie rods being the correct length and on their position. Various illustrations in an earlier work show the results of incorrect toe-in and the possibility of achieving a roll–steer effect on the front wheels and steer-fight during braking. The elasticity in the steering system or that in the bearings of the steering control arms, is also a contributory factor.

9.1.7.3.1 Double wishbone and multi-link suspensions

There are two ways of determining the central point T of the inner tie rod joint as a function of the assumed position U of the outer joint, the template and 'virtual centre' procedure. Both methods consider one side of the front axle when viewed from the rear (here the left side, Fig. 9.1-42). The projected length u' of the tie rod shown in Fig. 9.1-32 and the angle κ, which determines its position, must be calculated. This must match the line connecting the outer joint U with pole P_1 which is also needed for calculating the roll centre.

Initially, the position of the outer tie rod joint U is unknown when viewed from the rear; to obtain an approximation of this point, the height of the steering gear must be specified (Fig. 9.1-35). The angle λ is assumed so

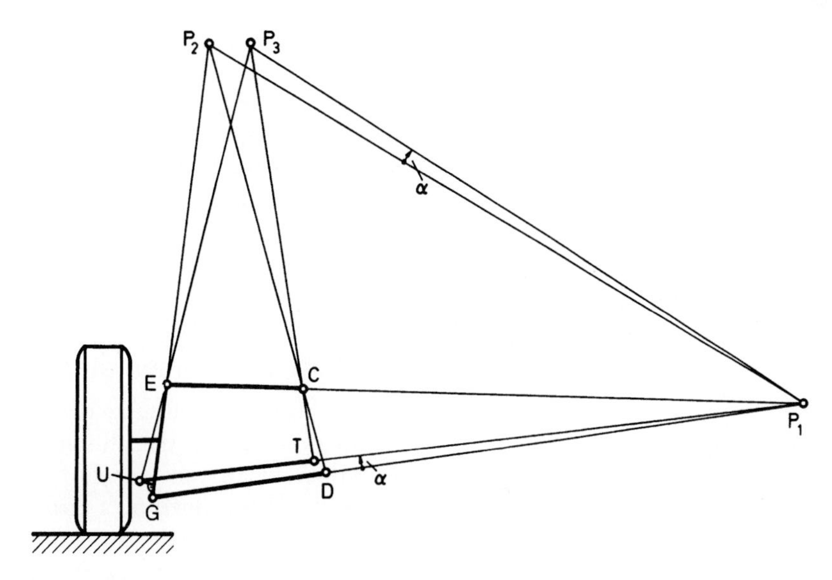

Fig. 9.1-43 In the case of a steering gear located in front of the wheel centre, the centre of the tie rod joint U lies outside the steering axis EG.

that, together with the known steering arm length r the path required for configuring it

$$k = r \sin \lambda \tag{9.1.3}$$

can be calculated (for r and λ see Fig. 9.1-40).

All figures contain point U and the curve of its movement. It only remains to find point T on the connecting line UP. Twould be the centre point of the arc which best covers the path of point U.

It is likely to be simpler and more precise to determine the point T graphically, using virtual centres. First, as shown in Figs. 9.1-42, the virtual-centre at P (marked here as P_1) must be calculated so that it can be connected to U. The extension of the paths EG and DC gives P_2, which is also required and from which a line to P_1 must be drawn. If the path UP_1 is above GD, the angle α enclosed by the two must be moved up to P_1P_2; if UP_1 were to lie below it, the line would have to be moved down. A line drawn from P_1 at the angle α must be made to intersect with the extension of the connecting path UE to give the tie rod virtual-centre P_3. To calculate the desired point T – i.e. the centre of the inner joint – P_3 is connected to C and extended.

The path k (i.e. the distance of point U from the steering axis EG, Fig. 9.1-35 and Equation 9.1-3) is the

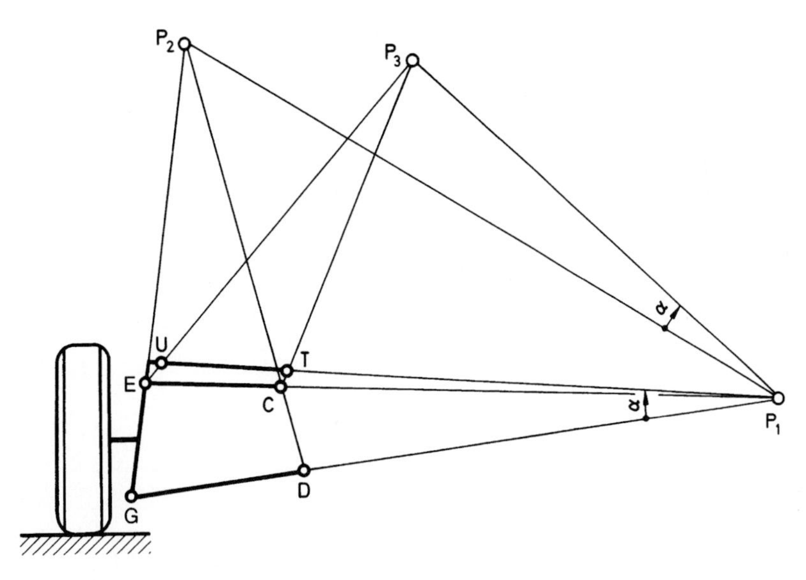

Fig. 9.1-44 A high-location steering gear can involve a tie rod above the upper control arm. The steering arm points backwards and towards the inside in the example.

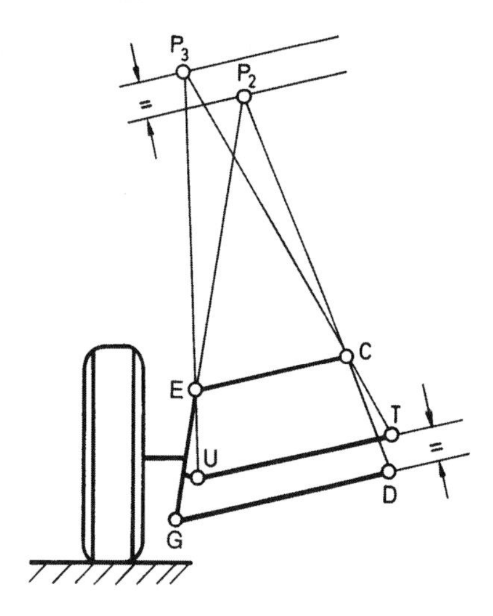

Fig. 9.1-45 Suspension control arms, which are parallel to one another in the design position of the vehicle, have to have a tie rod in the same position.

determining factor for the position of virtual-centre P_3 in the lateral direction. Fig. 9.1-43 shows the case of point U, which lies left of the path EG. This is something that is only possible where the steering gear is located in front of the axle (Fig. 9.1-41). P_3 moves to the right, resulting in an inner link T moving further away from the centre of the vehicle. This is beneficial if it is to be fixed to the end of the steering rod.

A tie rod that is located above the upper suspension control arm (Fig. 9.1-44) causes a large angle α and P_3 that is shifted a long way to the right. Where the control arms are parallel to one another (Figs. 9.1-45), P_1 is at ∞. In such cases, a line parallel to the path GD must be drawn through U and, at the same distance, a further one drawn through the virtual centre P_2. The intersection of this second parallel with the extension of the path UE gives P_3, which must be linked to C to obtain T.

9.1.7.3.2 McPherson struts and strut dampers

When the vehicle is fitted with McPherson struts or strut dampers – due to the alteration in distance between E and G when the wheels compress and rebound – point T is determined by a different method. To obtain pole P_1, a vertical to the centre line of the shock absorber is drawn in the upper mounting point E and made to intersect with the extension of the suspension control arm GD (Fig. 9.1-46); P_1 linked with U gives the position of the tie rod. A line parallel to EP_1 must be drawn through G; the intersection with the extension of ED then gives the second virtual-centre P_2. The angle α, included by the paths EP_1 and UP_1, must be entered downwards to the connection P_1P_2 to obtain P_3 as the intersection of this line with the extension of the path UG. The extension of the connecting line P_3D then gives the central point T of the inner tie rod joint on UP_1.

If, in the case of $\lambda = 0°$, point U is on the steering axis EG which dominates the rotation movement

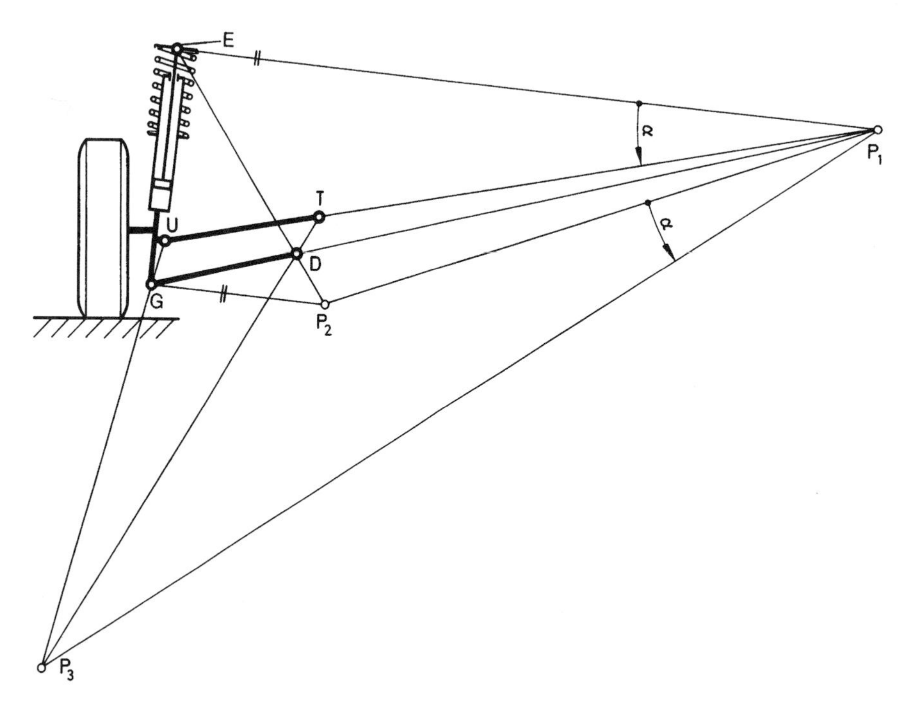

Fig. 9.1-46 On the McPherson strut or strut damper, the tie rod is above the lower control arm; the steering arms point inwards with the result that the outer joint U lies more to the vehicle centre.

(Fig. 9.1-47), P_3 is on the extension of this path. The determining factor for the position of P_1 is the direction of the shift in the damping part of the McPherson strut; for this reason, the vertical in point E must be created on its centre-line (not on the steering axis EG). The important thing in this calculation is the position of point U, i.e. the extension of the connecting line UG downwards. U is shown on the steering axis EG simply for reasons of presentation.

A low mounted tie rod causes the virtual-centre P_3 to move to the right (Fig. 9.1-48) and this then causes a shorter rod. This situation is favourable if the inner joint needs to sit on the ends of the steering rack. The figures clearly show that the higher U, which constitutes the connection between steering arm and tie-rod, is situated, the longer the tie rods must be, i.e. a centre take-off becomes necessary on a high-mounted rack and pinion steering (Figs. 8.1-57, 9.1-11 and 9.1-39).

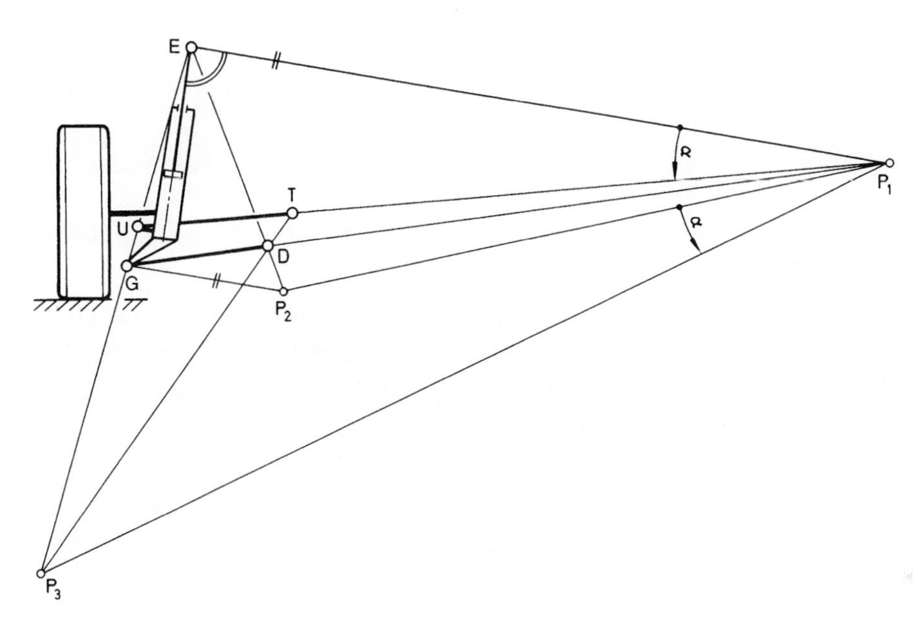

Fig. 9.1-47 On a McPherson strut with the joint G shifted to the wheel, the outer one, U of the tie rod, can lie in the plane of the steering axis (i.e. on the connecting line EG) when viewed from the rear. Extending the path UG is crucial for determining the virtual centre P_3, whereas the direction of movement of the damper, i.e. a vertical on the piston rod in point E, must be the starting point for calculating P_1.

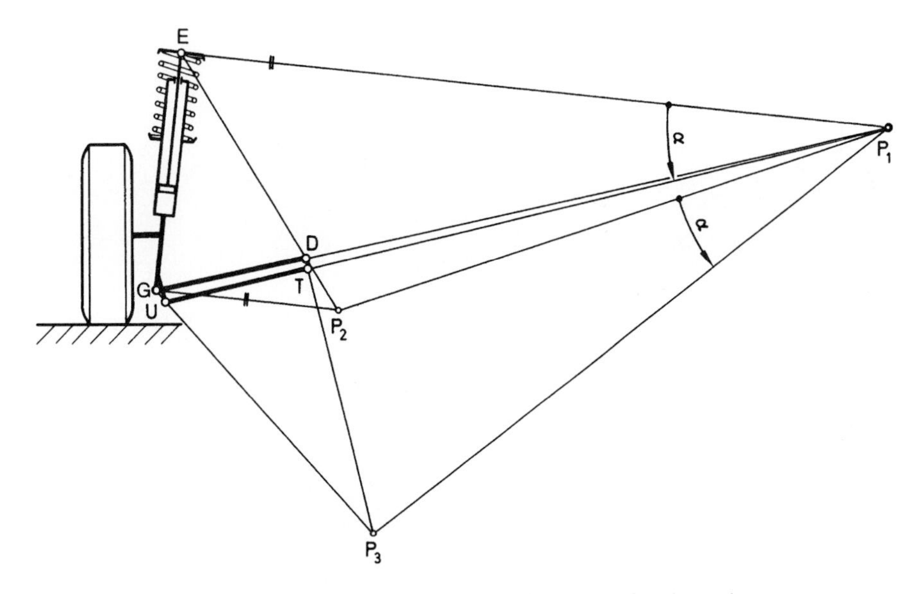

Fig. 9.1-48 The tie rod can also lie under the control arm when the steering arm points inwards.

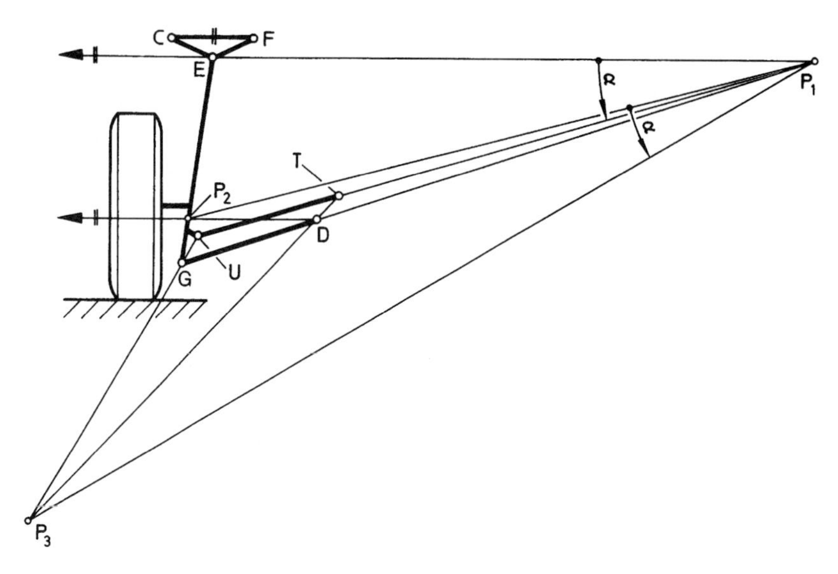

Fig. 9.1-49 Longitudinal transverse axle with the tie rod located above the lower control arm and the steering arm pointing inwards.

9.1.7.3.3 Longitudinal transverse axles

On longitudinal wishbone axles the upper point E moves in a straight line vertical to the steering axis CF and the lower point G on an arc around D (Fig. 9.1-49). To obtain P_1, a parallel to CF must therefore be drawn through E and made to intersect with the control arm extension GD. A parallel to EP_1 laid through point D gives the virtual centre P_2 on the connecting line EG. The angle α enclosed by the paths EP_1 and UP_1 must be drawn downwards to the connecting line P_1P_2 to obtain the virtual centre P_3 as the intersection with the extension of the path UG. P_3 linked with D then gives the centre T of the inner tie rod joint.

9.1.7.3.4 Reaction on the steering arm angle λ

Figs. 9.1-40–9.1-49 indicate that shifting the outer joint U to the side results in a slight alteration in the distance UT. However, this shift is necessary if the angle λ has to be reduced or increased with a given steering arm length *r*. The projected length u' of the tie rod, and therefore also its overall length u_0 (Fig. 9.1-32), changes when viewed from the rear. However, the latter is one of the determining factors for the aspects relating to the steering angles δ_i (inside) and δ_o (outside), i.e. for the actual steering curve. It is, therefore, likely to be essential to check the desired position of point T with the tie rod, which has become longer or shorter.

Section Ten

Tyres

Chapter 10.1

10.1

Tyres and wheels

Jornsen Reimpell, Helmut Stoll and Jurgen Betzler

10.1.1 Tyre requirements

The tyres are crucial functional elements for the transmission of longitudinal, lateral and vertical forces between the vehicle and road. The tyre properties should be as constant as possible and hence predictable by the driver. As well as their static and dynamic force transmission properties, the requirements described below – depending on the intended use of the vehicle – are also to be satisfied.

As tyres significantly affect the handling properties of vehicles, the properties of original tyres – the tyres with which the vehicle is supplied to the customer –are specified by the vehicle manufacturers in conjunction with the tyre manufacturers. However, spare tyres usually differ from the original tyres, despite their similar designation; hence handling characteristics can change. Individual vehicle manufacturers have therefore decided to identify tyres produced in accordance with their specifications by means of a symbol on the sidewall of the tyre or to sell tyres which meet the specifications of original tyres at their manufacturing branches.

10.1.1.1 Interchangeability

All tyres and rims are standardized to guarantee interchangeability, i.e. to guarantee the possibility of using tyres from different manufacturers but with the same designation on one vehicle and to restrict the variety of tyre types worldwide.

Within Europe, standardization is carried out by the European Tyre and Rim Technical Organization or ETRTO, which specifies the following:

- tyre and rim dimensions;
- the code for tyre type and size;
- the load index and speed symbol.

Passenger car tyres are governed by UNO regulation ECE-R 30, commercial vehicles by R 54, spare wheels by R 64, and type approval of tyres on the vehicle by EC directive 92/23/EC.

In the USA the Department of Transportation (or DOT, see item 9 in Fig. 10.1-18) is responsible for the safety standards. The standards relevant here are:

Standard 109	Passenger cars
Standard 119	Motor vehicles other than passenger cars.

The Tire and Rim Association, or TRA for short, is responsible for standardization.

In Australia, binding information is published by the Federal Office of Road Safety, Australian Motor Vehicle Certification Board.

ARD 23	Australian Design Rule 23/01: Passenger car tyres

is the applicable standard.

In Germany, the DIN Standards (Deutsches Institut für Normung) and the WdK Guidelines (Wirtschaftsverband der Deutschen Kautschukindustrie Postfach 900360,

Automotive Chassis: Engineering Principles; ISBN: 9780750650540
Copyright © 2001 Elsevier Ltd; All rights of reproduction, in any form, reserved.

D-60443, Frankfurt am Main) are responsible for specifying tyre data. All bodies recognize the publications of these two organizations.

At the international level, the International Organization for Standardization (ISO) also works in the field of tyre standardization and ISO Standards are translated into many languages.

10.1.1.2 Passenger car requirements

The requirements for tyres on passenger cars and light commercial vehicles can be subdivided into the following six groups:

- driving safety
- handling
- comfort
- service life
- economy
- environmental compatibility.

To ensure driving safety it is essential that the tyre sits firmly on the rim. This is achieved by a special tyre bead design (tyre foot) and the safety rim, which is the only type of rim in use today (Figs. 10.1-5 and 10.1-21). Not only is as great a degree of tyre-on-rim retention as possible required, but the tyre must also be hermetically sealed; on the tubeless tyre this is the function of the inner lining. Its job is to prevent air escaping from the tyre, i.e. it stops the tyre from losing pressure. However, this pressure reduces by around 25–30% per year, which shows how important it is to check the tyre pressure regularly.

In order to guarantee driving safety, the aim is also to ensure that tyres are as insensitive to overloading and as puncture-proof as possible and that they have emergency running properties which make it possible for the driver to bring the vehicle safely to a halt in case of tyre failure.

Handling characteristics include the properties:

- high coefficients of friction in all operating conditions;
- steady build-up of lateral forces without sudden changes;
- good cornering stability;
- direct and immediate response to steering movements;
- guarantee requirement of sustained maximum speed;
- small fluctuations in wheel load.

Riding comfort includes the characteristics:

- good suspension and damping properties (little rolling hardness);
- high smoothness as a result of low radial tyre run-out and imbalances;
- little steering effort required during parking and driving;
- low running noise.

Durability refers to:

- long-term durability
- high-speed stability.

Both are tested on drum test stands and on the road.

Economic efficiency is essentially determined by the following:

- purchase cost;
- mileage (including the possibility of profile regrooving in the case of lorry tyres);
- wear;
- rolling resistance;
- the necessary volume, which determines
- the amount of room required in the wheel houses and spare-wheel well;
- load rating.

Of increasing importance is environmental compatibility, which includes:

- tyre noise;
- raw material and energy consumption during manufacture and disposal;
- possibility of complete remoulding inherent in the construction.

The importance of

- tyre design, profile design and the 'radius–width appearance' must not be neglected either.

10.1.1.3 Commercial vehicle requirements

In principle, the same requirements apply for commercial vehicles as for passenger cars, although the priority of the individual groups changes. After safety, economy is the main consideration for commercial vehicle tyres. The following properties are desirable:

- high mileage and even wear pattern
- low rolling resistance
- good traction
- low tyre weight
- ability to take chains
- remoulding/retreading possibilities.

Compared with passenger car tyres, the rolling resistance of commercial vehicle tyres has a greater influence on fuel consumption (20–30%) and is therefore an important point (Fig. 10.1-32).

10.1.2 Tyre designs

10.1.2.1 Diagonal ply tyres

In industrialized countries, cross-ply tyres are no longer used on passenger cars, either as original tyres or as replacement tyres, unlike areas with very poor roads where the less vulnerable sidewall has certain advantages. The same is true of commercial vehicles and vehicles that tow trailers, and here too radial tyres have swept the board because of their many advantages. Nowadays, cross-ply tyres are used only for:

- temporary use (emergency) spare tyres for passenger cars (due to the low durability requirements at speeds up to 80 or 100 km h^{-1});
- motor cycles (due to the inclination of the wheels against the lateral force);
- racing cars (due to the lower moment of inertia);
- agricultural vehicles (which do not reach high speeds).

Cross-ply tyres consist of the substructure (also known as the tyre carcass, Fig. 10.1-1) which, as the 'supporting framework' has at least two layers of rubberized cord fibres, which have a zenith or bias angle ξ of between 20° and 40° to the centre plane of the tyre (Fig. 10.1-2). Rayon (an artificial silk cord), nylon or even steel cord may be used, depending on the strength requirements. At the tyre feet the ends of the layers are wrapped around the core of the tyre bead on both sides; two wire rings, together with the folded ends of the plies, form the bead. This represents the frictional connection to the rim. The bead must thus provide the permanent seat and transfer drive-off and braking moments to the tyre. On tube-less tyres it must also provide the airtight seal.

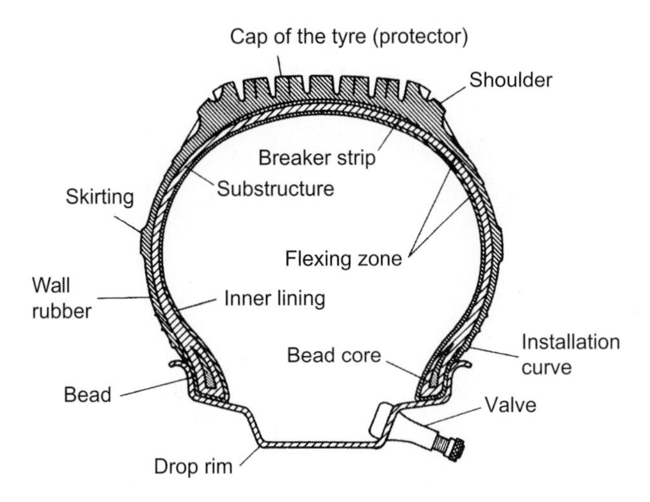

Fig. 10.1-1 Design of a diagonal ply tubeless car tyre with a normal drop rim and pressed-in inflating valve (see also Fig. 10.1-6).

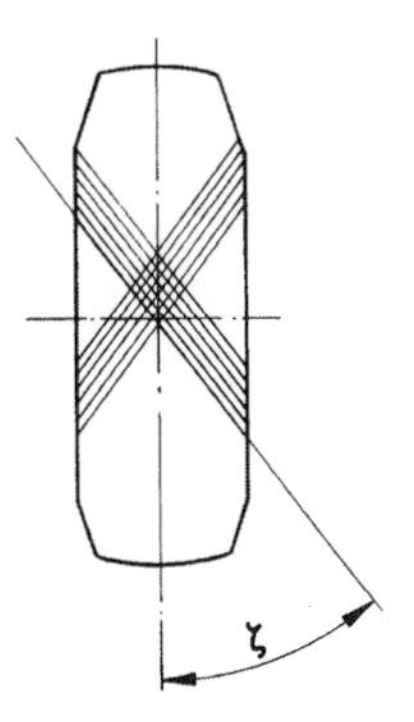

Fig. 10.1-2 The diagonal ply tyre has crossed-bias layers; the zenith angle ξ was 30–40° for passenger cars. The 4 PR design should have two layers in each direction. Smaller angles ξ can be found in racing cars. Rolling resistance, lateral and suspension stiffness are significantly determined by the zenith angle.

The running tread, which is applied to the outer diameter of the substructure, provides the contact to the road and is profiled. Some tyres also have an intermediate structure over the carcass as reinforcement.

At the side, the running tread blends into the shoulder, which connects to the sidewall (also known as the side rubber), and is a layer that protects the substructure. This layer and the shoulders consist of different rubber blends from the running tread because they are barely subjected to wear; they are simply deformed when the tyre rolls. This is known as flexing. Protective mouldings on the sides are designed to prevent the tyre from being damaged through contact with kerbstones. There are also GG grooves, which make it possible to see that the tyre is seated properly on the rim flange.

Cross-ply design and maximum authorized speed are indicated in the tyre marking by a dash (or a letter, Fig. 10.1-12) between the letters for width and rim diameter (both in inches) and a 'PR' (ply rating) suffix. This ply rating refers to the carcass strength and simply indicates the possible number of plies (Fig. 10.1-5). The marking convention is:

5.60-15/4 PR (VW rear-engine passenger car, tyres authorized up to 150 km h^{-1})
7.00-14/8 PR (VW Transporter, tyres authorized up to 150 km h^{-1})
9.00-20/14 PR (reinforced design for a commercial vehicle)

and on the temporary use spare wheel of the VW Golf, which requires a tyre pressure of $p_T = 4.2$ bar and may only be driven at speeds up to 80 km h^{-1} (F symbol)

T 105/70 D 14 38 F

10.1.2.2 Radial ply tyres

The radial ply tyre consists of two bead cores joined together radially via the carcass (Fig. 10.1-3) – hence the

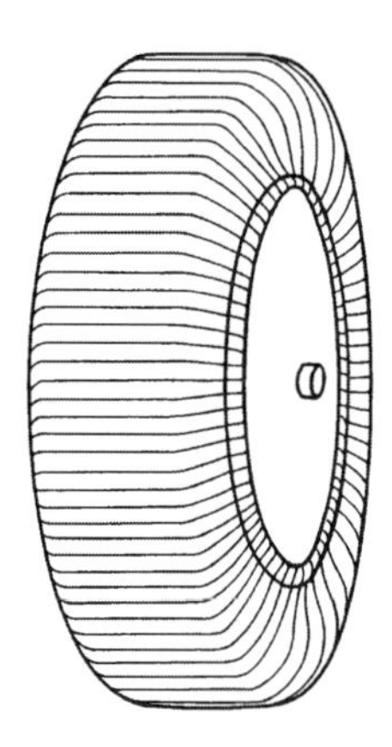

Fig. 10.1-3 Substructure of a radial tyre. The threads have a bias angle between 88° and 90°.

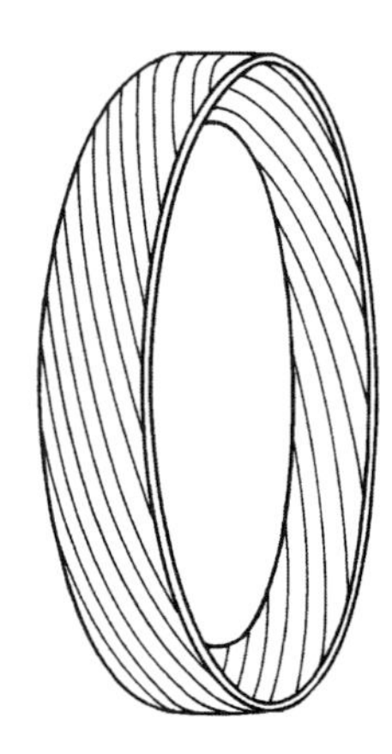

Fig. 10.1-4 The belt of the radial tyre sits on the substructure. The threads are at angles of between 15° and 25° to the plane of the tyre centre.

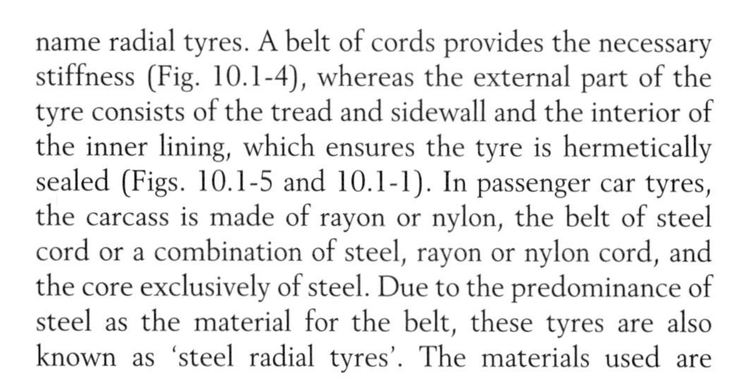

name radial tyres. A belt of cords provides the necessary stiffness (Fig. 10.1-4), whereas the external part of the tyre consists of the tread and sidewall and the interior of the inner lining, which ensures the tyre is hermetically sealed (Figs. 10.1-5 and 10.1-1). In passenger car tyres, the carcass is made of rayon or nylon, the belt of steel cord or a combination of steel, rayon or nylon cord, and the core exclusively of steel. Due to the predominance of steel as the material for the belt, these tyres are also known as 'steel radial tyres'. The materials used are indicated on the sidewall (Fig. 10.1-18, points 7 and 8). In commercial vehicle designs this is particularly important and the carcass may also consist of steel.

The stiff belt causes longitudinal oscillation, which has to be kept away from the body by wheel suspensions with a defined longitudinal compliance, otherwise this would cause an unpleasant droning noise in the body, when on cobbles and poor road surfaces at speeds of less than 80 km h^{-1}. The only other disadvantage is the greater

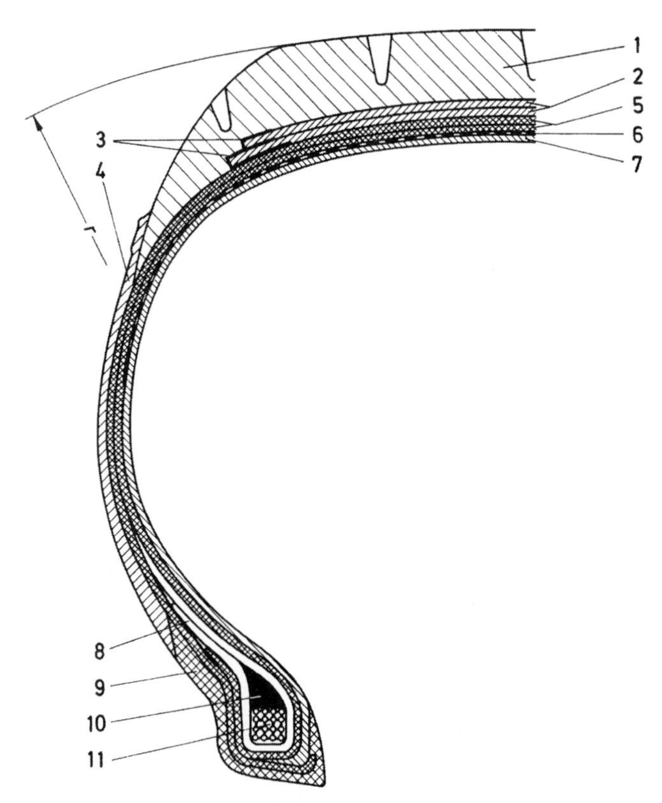

Fig. 10.1-5 Radial design passenger car tyres in speed category T (Fig. 10.1-12); the number of layers and the materials are indicated on the sidewall (see Fig. 10.1-18). The components are: 1 running tread; 2 steel belt; 3 edge protection for the belt, made of rayon or nylon; 4 sidewall; 5 substructure with two layers; 6 cap; 7 inner lining; 8 flipper; 9 bead profile; 10 core profile; 11 bead core.

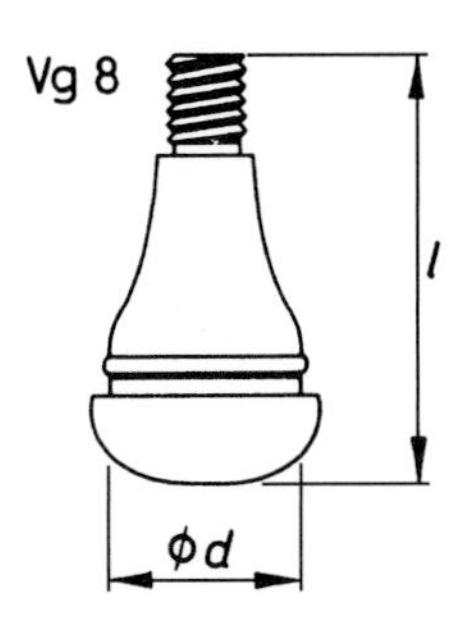

DIN	l	Diameter d
43 GS 11.5	43	15.2
43 GS 16	43	19.5

Fig. 10.1-6 Snap-in rubber valve for tubeless tyres, can be used on rims with the standard valve holes of 11.5 mm and 16 mm diameter. The numerical value 43 gives the total length in mm (dimension *l*). There is also the longer 49 GS 11.5 design.

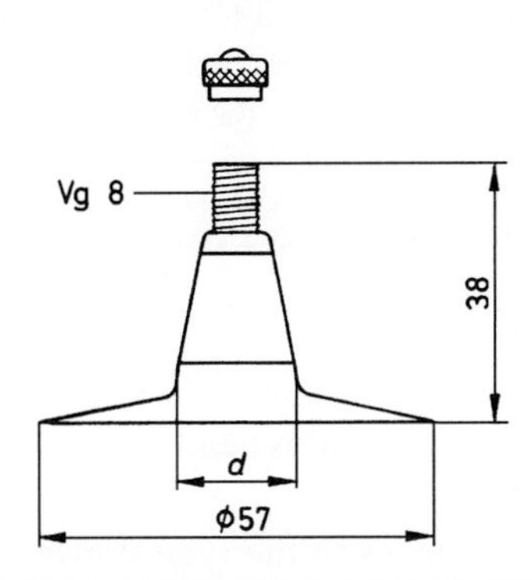

Valve specification	d
38/11.5	11.7
38/16	16.5

Fig. 10.1-7 Rubber valve vulcanized onto tubes. Designations are 38/11.5 or 38/16.

susceptibility of the thinner sidewalls of the tyres to damage compared with diagonal ply tyres. The advantages over cross-ply tyres, which are especially important for today's passenger cars and commercial vehicles, are:

- significantly higher mileage
- greater load capacity at lower component weight
- lower rolling resistance
- better aquaplaning properties
- better wet-braking behaviour
- transferable, greater lateral forces at the same tyre pressure
- greater ride comfort when travelling at high speeds on motorways and trunk roads.

10.1.2.3 Tubeless or tubed

In passenger cars, the tubeless tyre has almost completely ousted the tubed tyre. The main reasons are that the tubeless tyre is:

- easier and faster to fit
- the inner lining is able to self-seal small incisions in the tyre.

In tubeless tyres the inner lining performs the function of the tube, i.e. it prevents air escaping from the tyre. As it forms a unit with the carcass and (unlike the tube) is not under tensional stress, if the tyre is damaged the incision does not increase in size, rapidly causing loss of pressure and failure of the tyre. The use of tubeless tyres is linked to two conditions:

- safety contour on the rim (Fig. 10.1-21)
- its air-tightness.

Because this is not yet guaranteed worldwide, tubed tyres continue to be fitted in some countries. When choosing the tube, attention should be paid to ensuring the correct type for the tyre. If the tube is too big it will crease, and if it is too small it will be overstretched, both of which reduce durability. In order to avoid confusion, the tyres carry the following marking on the sidewall:

tubeless (Fig. 10.1-18, point 3)
tubed or tube type.

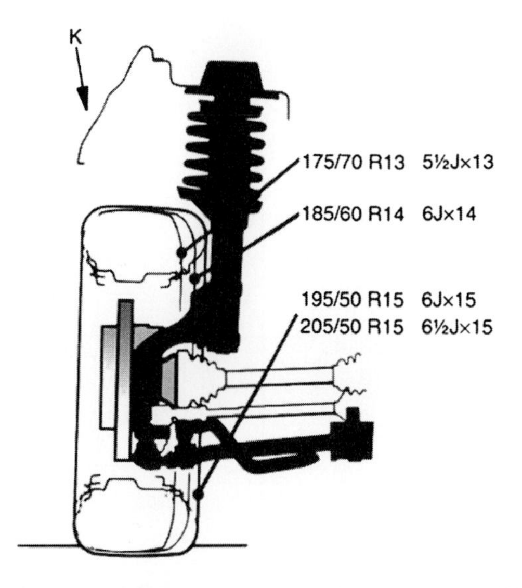

Fig. 10.1-8 Tyre sizes and associated rims used on the VW Golf III. All tyres fit flush up to the outer edge of the wing (wheel house outer panel) K. To achieve this, differing wheel offsets (depth of dishing) *e* are used on disc-type wheels with the advantage of a more negative rolling radius r_σ on wider tyres. A disadvantage then is that snow chains can no longer be fitted and steering sensitivity changes very slightly.

Valves are needed for inflating the tyre and maintaining the required pressure. Various designs are available for tubeless and tubed tyres (Figs. 10.1-6 and 10.1-7). The most widely used valve is the so-called 'snap-in valve'. It comprises a metal foot valve body vulcanized into a rubber sheath, which provides the seal in the rim hole (Fig. 10.1-20). The functionality is achieved by a valve insert, while a cap closes the valve and protects it against ingress of dirt.

At high speeds, the valve can be subjected to bending stress and loss of air can occur. Hub caps and support areas on alloy wheels can help to alleviate this (see Fig. 10.1-24).

10.1.2.4 Height-to-width ratio

The height-to-width ratio *H/W* – also known as the 'profile' (high or low) – influences the tyre properties and affects how much space the wheel requires (Fig. 10.1-8). As shown in Fig. 10.1-9, the narrower tyres with a *H/W* ratio = 0.70 have a reduced tread and therefore good aquaplaning behaviour (Fig. 10.1-35). Wide designs make it possible to have a larger diameter rim and bigger brake discs (Fig. 10.1-10) and can also transmit higher lateral and longitudinal forces.

W is the cross-sectional width of the new tyre (Fig. 10.1-11); the height *H* can easily be calculated from the rim diameter given in inches and the outside diameter of the tyre OD_T. The values OD_T and *W* are to be taken from the new tyre mounted onto a measuring rim at a measuring tyre pressure of 1.8 bar or 2.3 bar on V-, W- or ZR tyres, Fig. 10.1-15):

$$H = 0.5(OD_T - d) \qquad (10.1.1)$$

$$1'' = 1\text{ in} = 25.4\text{ mm} \qquad (10.1.1a)$$

The 175/65 R 14 82 H tyre mounted on the measuring rim 5J × 14 can be taken as an example:

$$OD_T = 584\text{ mm},\ d = 14 \times 25.4 = 356\text{ mm and } W = 177\text{ mm}$$

$$H/W = [0.5 \times (OD_T - d)]/W = 114/177 = 0.644$$

The cross-section ratio is rounded to two digits and given as a percentage. We talk of 'series', and here the ratio profile is 65% as shown in the tyre marking – in other words it is a 65 series tyre. A wider rim, e.g. 6J × 14 would give a smaller percentage.

10.1.2.5 Tyre dimensions and markings

10.1.2.5.1 Designations for passenger cars up to 270 km h^{-1}

The standards manual of the European Tire and Rim Technical Organization (ETRTO) includes all tyres for passenger cars and delivery vehicles up to 270 km h^{-1} and specifies the following data:

- tyre width in mm
- height-to-width ratio as a percentage
- code for tyre design
- rim diameter in inches or mm
- operational identification, comprising load index; LI (carrying capacity index) and speed symbol GSY.

The following applies to the type shown in Fig. 10.1-15:

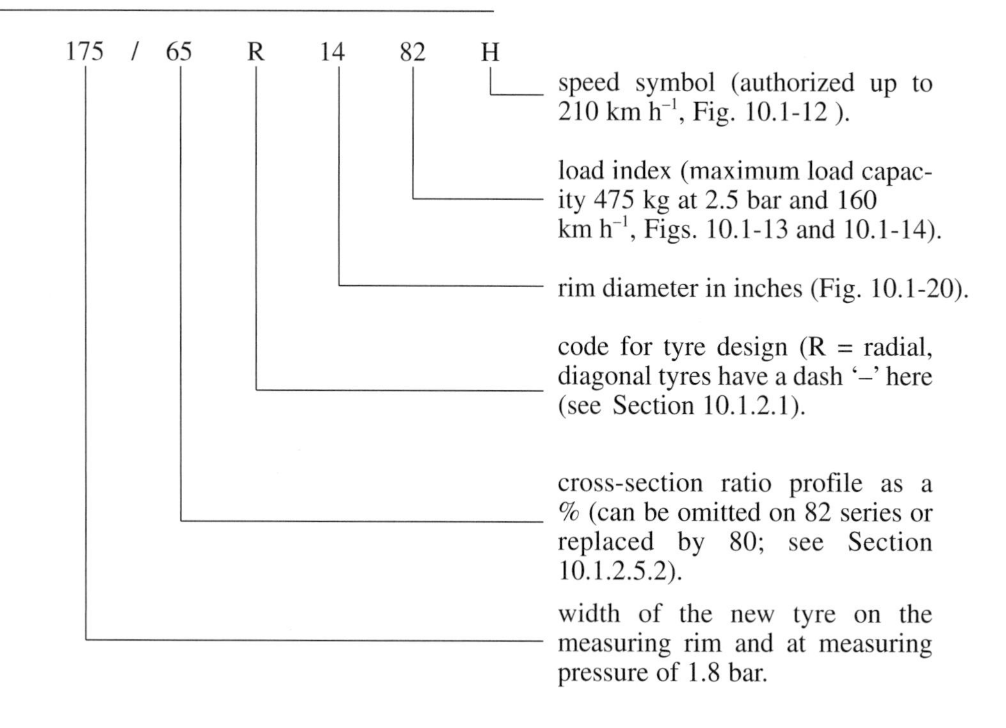

Fig. 10.1-9 If they have the same outside diameter and load capacity the four tyre sizes used on medium-sized passenger cars are interchangeable. The series 65, 55 and 45 wide tyres each allow a 1" larger rim (and therefore larger brake discs). The different widths and lengths of the tyre contact patch, known as 'tyre print', are clearly shown (Fig. 10.1-19), as are the different designs of the standard road profile and the asymmetric design of the sports profile (see also Section 10.1.2.10). The 65 series is intended for commercial vehicles, and the 60, 55 and 45 series for sports cars. (Illustration: Continental; see also Fig. 10.1-19.)

The old markings can still be found on individual tyres:

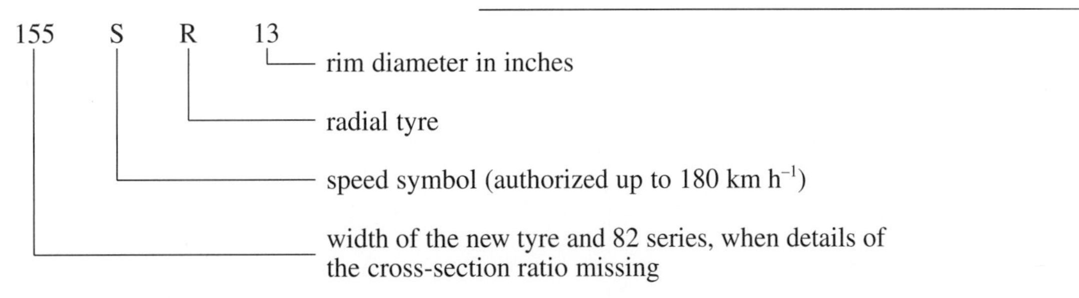

10.1.2.5.2 Designations of US tyres and discontinued sizes for passenger cars

Tyres manufactured in the USA and other non-European countries may also bear a 'P' for passenger car (see Fig. 10.1-17) and a reference to the cross-section ratio:

P 155/80 R 13 79 S

The old system applied up until 1992 for tyres which were authorized for speeds of over $V = 210$ km h^{-1} (or 240 km h^{-1}, Fig. 10.1-12); the size used by Porsche on the 928 S can be used as an example:

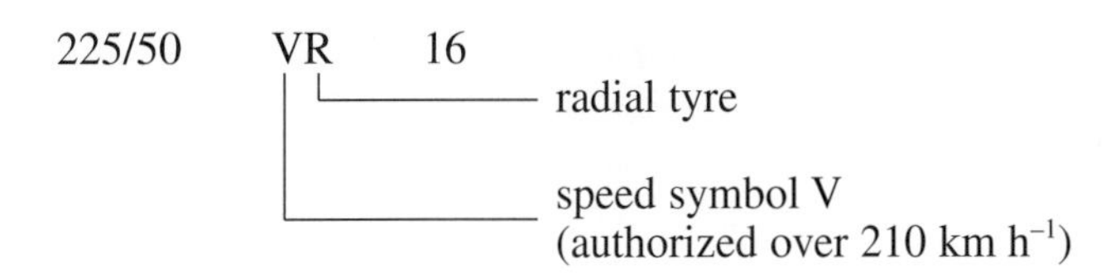

Wheel rim diameter in inches	12	13	14	15	16	17
Brake disc outer diameter in mm	221	256	278	308	330	360
Brake drum inner diameter in mm	200	230	250	280	300	325

Fig. 10.1-10 The flatter the tyre, i.e. the larger the rim diameter *d* (Fig. 10.1-1) in comparison with the outside diameter OD_T, the larger the brake discs or drums that can be accommodated, with the advantage of a better braking capacity and less tendency to fade. An asymmetric well-base rim is favourable (Figs. 8.1-8 and 10.1-11).

The following should be noted for VR tyres:

- over 210 km h^{-1} and up to 220 km h inclusive, the load may only be 90% of the otherwise authorized value;
- over 220 km h^{-1} the carrying capacity reduces by at least 5% per 10 km h^{-1} speed increment.

10.1.2.5.3 Designation of light commercial vehicle tyres

Tyres for light commercial vehicles have a reinforced substructure compared with those for passenger cars (Fig. 10.1-5), so they can take higher pressures, which means they have a higher load capacity. The suffix 'C'

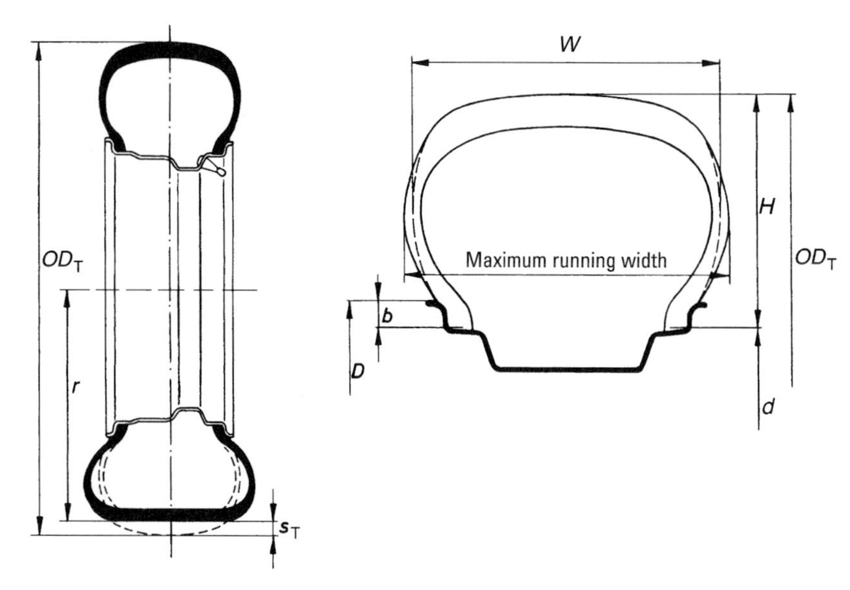

Fig. 10.1-11 Tyre dimensions specified in standards and directives. *B* is the cross-section width of the new tyre; the tread moulding (as can be seen in Fig. 10.1-1) is not included in the dimension. For clearances, the maximum running width with the respective rim must be taken into consideration, as should the snow chain contour for driven axles. The tyre radius, dependent on the speed, is designated *r* (see Section 10.1.2.8). Pictured on the left is an asymmetrical well-base rim, which creates more space for the brake caliper and allows a larger brake disc (Fig. 10.1-10).

v_{max} *in km/h^{-1}*	*Speed symbol*	*Identification*
80	F	
130	M	
150	P	
160	Q	
170	R	
180	S	
190	T	
210	H	
240	V	
270	W	
300	Y	
over 210	—	VR (old system)
over 240	—	ZR (old system)

Fig. 10.1-12 Standardized speed categories for radial tyres, expressed by means of a speed symbol and – in the case of discontinued sizes – by means of the former speed marking. Sizes marked VR or ZR may be used up to maximum speeds specified by the tyre manufacturer. The symbols F and M are intended for emergency (temporary use) spare wheels.

Load index	Wheel load capacity in kg with tyre pressure measured in bars										
	1.5	1.6	1.7	1.8	1.9	2.0	2.1	2.2	2.3	2.4	2.5
69	215	225	240	250	260	270	285	295	305	315	325
70	225	235	245	260	270	280	290	300	315	325	335
71	230	240	255	265	275	290	300	310	325	335	345
72	235	250	260	275	285	295	310	320	330	345	355
73	245	255	270	280	295	305	315	330	340	355	365
74	250	260	275	290	300	315	325	340	350	365	375
75	255	270	285	300	310	325	335	350	360	375	387
76	265	280	295	310	320	335	350	360	375	385	400
77	275	290	305	315	330	345	360	370	385	400	412
78	280	295	310	325	340	355	370	385	400	410	425
79	290	305	320	335	350	365	380	395	410	425	437
80	300	315	330	345	360	375	390	405	420	435	450
81	305	325	340	355	370	385	400	415	430	445	462
82	315	330	350	365	380	395	415	430	445	460	475
83	325	340	360	375	390	405	425	440	455	470	487
84	330	350	365	385	400	420	435	450	470	485	500
85	340	360	380	395	415	430	450	465	480	500	515
86	350	370	390	410	425	445	460	480	495	515	530
87	360	380	400	420	440	455	475	490	510	525	545
88	370	390	410	430	450	470	485	505	525	540	560
89	385	405	425	445	465	485	505	525	545	560	580
90	400	420	440	460	480	500	520	540	560	580	600
91	410	430	450	475	495	515	535	555	575	595	615
92	420	440	465	485	505	525	550	570	590	610	630
93	430	455	475	500	520	545	565	585	610	630	650
94	445	470	490	515	540	560	585	605	625	650	670
95	460	485	505	530	555	575	600	625	645	670	690
96	470	495	520	545	570	595	620	640	665	685	710
97	485	510	535	560	585	610	635	660	685	705	730
98	500	525	550	575	600	625	650	675	700	725	750
99	515	540	570	595	620	650	675	700	725	750	775
100	530	560	590	615	640	670	695	720	750	775	800

Fig. 10.1-13 Load capacity/air pressure category specified in the directives. The load capacity on the left – also known as 'load index' (LI) – applies for all passenger cars up to the speed symbol W; they relate to the minimum load capacity values up to 160 km h^{-1} at tyre pressure 2.5 bar (see Section 10.1.2.6). Further criteria, such as maximum speed, handling etc., are important for the tyre pressures to be used on the vehicle. For LI values above 100, further load increases are in 25-kg increments:

LI = 101 corresponds to 825 kg,
LI = 102 corresponds to 850 kg etc. to
LI = 108 corresponds to 1000 kg.

followed by information on the carcass strength (6, 8 or 10 PR) used to indicate suitability for use on light commercial vehicles, or the word 'reinforced' simply appeared at the end of the marking. The current marking (as for passenger cars) retains the speed symbol as well as the load index which, behind the slash, gives the reduced load capacity on twin tyres. Compared with the previous marking, the new system is as follows:

Former	Current
–	205/65 R 15 98 S (Fig. 10.1-15)
185 SR 14	185 R 14 90 S
185 SR 14 reinforced	185 R 14 94 R
185 R 14 C 6 PR	185 R 14 99/97 M
185 R 14 C 8 PR	185 R 14 102/100 M

The 185 R 14 tyre is passenger car size, which is also fitted to light commercial vehicles.

10.1.2.5.4 Tyre dimensions

Figure 10.1-15 shows the important data for determining tyre size:

- size marking;
- authorized rims and measuring rim;
- tyre dimensions: width and outside diameter new and maximum during running;
- static rolling radius (Fig. 10.1-11);
- rolling circumference (at 60 km h^{-1}, Fig 10.1-16, see also Section 10.1.2.8);
- load capacity coefficient (load index LI, Fig. 10.1-13);
- tyre load capacity at 2.5 bar and up to 160 km h^{-1} (see Section 10.1.2.6).

Top speed of car ($km\ h^{-1}$)	Tyre load capacity (%)		
	Speed symbol		
	V	W	Y Tyres
210	100	100	100
220	97	100	100
230	94	100	100
240	91	100	100
250	–	95	100
260	–	90	100
270	–	85	100
280	–	–	95
290	–	–	90
300	–	–	85

Fig. 10.1-14 The tyre load capacity shown in the ETRTO standards manual in the form of the load index LI is valid for V tyres up to vehicle speeds of 210 $km\ h^{-1}$; for W tyres up to 240 $km\ h^{-1}$ and for Y tyres up to 270 $km\ h^{-1}$. At higher speeds, lower percentages of the load capacity must be incurred; for VR and ZR tyres, which are no longer made, these values were determined by vehicle and tyre manufacturers.

10.1.2.6 Tyre load capacities and inflation pressures

The authorized axle loads $m_{V,f,max}$ and $m_{V,r,max}$, and the maximum speed v_{max} of the vehicle, determine the minimum tyre pressure. However, the required tyre pressure may be higher to achieve optimum vehicle handling (see also Section 10.1.10.3.5 and Fig. 10.1-44).

10.1.2.6.1 Tyre load capacity designation

The load capacities indicated in the load index (item 6, Fig. 10.1-18) are the maximum loads per tyre permitted for all tyres up to the speed symbol 'H'. They are valid up to speeds of 210 $km\ h^{-1}$ for tyres marked 'V' and up to 240 $km\ h^{-1}$ for those marked 'R' 'W' or 'ZR'. For vehicles with a higher top speed, the load capacity has to be reduced accordingly.

Consequently, for tyres with speed symbol 'V', at a maximum speed of 240 $km\ h^{-1}$ the load capacity is only 91% of the limit value (Fig. 10.1-14). Tyres designated 'W' on the sidewall are only authorized up to 85% at 270 $km\ h^{-1}$. In both cases, the load capacity values between 210 $km\ h^{-1}$ ('V' tyre) and 240 $km\ h^{-1}$ ('W' tyre) and the maximum speed must be determined by linear interpolation.

For higher speeds (ZR tyres), the interpolation applies to the 240–270 $km\ h^{-1}$ speed range. At higher speeds, the load capacity as well as the inflating pressure will be agreed between the car and tyre manufacturers. However, this approval does not necessarily apply to tyres which are specially produced for the US market and which bear the additional marking 'P' (Fig. 10.1-17 and Section 10.1.2.5.2).

10.1.2.6.2 Tyre pressure determination

For tyres with speed symbols 'R' to 'V' and standard road tyres the minimum pressures set out in the tables and corresponding with load capacities are valid up to 160 $km\ h^{-1}$ (see Fig. 10.1-15 and Section 10.1.1.1).

Special operating conditions, the design of the vehicle or wheel suspension and expected handling properties can all be reasons for higher pressure specification by the vehicle manufacturer.

Further, for speeds up to 210 $km\ h^{-1}$ the linear increase of basic pressure has to be by 0.3 bar (i.e. by 0.1 bar per $\Delta v = 17\ km\ h^{-1}$) and at speeds above 210 $km\ h^{-1}$ the tyre load capacity has to be reduced in accordance with item 10.1.2.6.1. If the tyre load is lower than the maximum load capacity, a lower additional safety pressure can be used in consultation with the tyre manufacturer.

For tyres with the speed symbol 'W', the pressures in Fig. 10.1-13 apply up to 190 km^{-1}. After this it has to be increased by 0.1 bar for every 10 $km\ h^{-1}$ up to 240 $km\ h^{-1}$. For higher speeds, the load capacity must be reduced (see Section 10.1.2.6.1).

On vehicles, pressure should be tested on cold tyres, i.e. these must be adjusted to the ambient temperature. If the tyre pressure is set in a warm area in winter there will be an excessive pressure drop when the vehicle is taken outside.

On M & S winter tyres, it has long been recommended that inflation pressures be increased by 0.2 bar compared with standard tyres. Newer brands of tyre no longer require this adjustment.

10.1.2.6.3 Influence of wheel camber

Wheel camber angles ϵ_W considerably influence tyre performance and service life. The camber angle should therefore not exceed 4° even in full wheel jounce condition. For angles above ±2°, the loadability of the tyres reduces at

$\varepsilon_W > 2°$ to 3° to 95%

$\varepsilon_W > 3°$ to 4° to 95%

Tyre size	Dimensions of new tyre: Measuring rim	Width of cross-section	Outer diameter	Permissible rims according to DIN 7817 and DIN 7824	Max. width	Manufacturer's measurements: Max. outer diameter[4]	Static radius ±2.0%	Circumference +1.5% −2.5%	Load index (LI)	Wheel load capacity[5]
155/65 R 13	4.50 B × 13	157	532	4.00 B × 13[1]	158	540	244	1625	73	365
				4.50 B × 13[1]	164					
				5.00 B × 13[1]	169					
				5.50 B × 13[1]	174					
155/65 R 14	4½ J × 14	157	558	4 J × 14[2]	158	566	257	1700	74	375
				4½ J × 14[2]	164					
				5 J × 14[2]	169					
				5½ J × 14[2]	174					
165/65 R 13	5.00 B × 13	170	544	4.50 B × 13[1]	171	533	248	1660	76	400
				5.00 B × 13[1]	176					
				5.50 B × 13[1]	182					
				6.00 B × 13[1,3]	187					
165/65 R 14	5 J × 14	170	570	4½ J × 14[2]	171	579	261	1740	78	425
				5 J × 14[2]	176					
				5½ J × 14[2]	182					
				6 J × 14	187					
175/65 R 13	5.00 B × 13	177	558	5.00 B × 13[1]	184	567	254	1700	80	450
				5.50 B × 13[1]	189					
				6.00 B × 13[1,3]	194					
175/65 R 14	5 J × 13	177	584	5 J × 14[2]	184	593	267	1780	82	475
				5½ J × 14[2]	189					
				6 J × 14	194					
175/65 R 15	5 J × 15	177	609	5 J × 15[2]	184	618	279	1855	83	487
				5½ J × 15[2]	189					
				6 J × 15	194					
185/65 R 13	5.50 B × 14	189	570	5.50 B × 13[1]	191	580	259	1740	84	500
				5.50 B × 13[1]	197					
				6.00 B × 13[1,3]	202					
				6½ J × 13	207					
185/65 R 14	5½ J × 14	189	596	5 J × 14	191	606	272	1820	86	530
				5½ J × 14	197					
				6 J × 14	202					
				6½ J × 14	207					

Fig. 10.1-15 Radial 65 series tyres, sizes, new and running dimensions, authorized rims and load capacity values (related to maximum 160 km h^{-1} and 2.5 bar); the necessary increase in pressures at higher speeds can be taken from Section 10.1.2.5.6. The tyre dimensions apply to tyres of a normal and increased load capacity design (see Section 10.1.2.5.3) and to all speed symbols and the speed marking ZR.

185/65 R 15	5½ J × 15	189	621	5 J × 15	191	631	284	1895	88	560
				5½ J × 15	197					
				6 J × 15	202					
				6½ J × 15	207					
195/65 R 14	6 J × 14	201	610	5½ J × 14	204	620	277	1860	89	580
				6 J × 14	209					
				6½ J × 14	215					
				7 J × 14	220					
195/65 R 15	6 J × 15	201	635	5½ J × 15	204	645	290	1935	91	615
				6 J × 15	209					
				6½ J × 15	215					
				7 J × 15	220					
205/65 R 14	6 J × 14	209	622	5½ J × 14	212	633	282	1895	91	615
				6 J × 14	217					
				6½ J × 14	222					
				7 J × 14	227					
				7½ J × 14	233					
205/65 R 15	6 J × 15	209	647	5½ J × 15	212	658	294	1975	94[6]	670
				6 J × 15	217					
				6½ J × 15	222					
				7 J × 15	227					
				7½ J × 15	233					
215/65 R 15	6½ J × 15	221	661	6 J × 15	225	672	300	2015	96[7]	710
				6½ J × 15	230					
				7 J × 15	235					
				7½ J × 15	240					
215/65 R 16	6½ J × 16	221	686	6 J × 16	225	697	312	2090	98	750
				6½ J × 16	230					
				7 J × 16	235					
				7½ J × 16	240					
225/65 R 15	6½ J × 15	228	673	6 J × 15	232	685	304	2055	99	775
				6½ J × 15	237					
				7 J × 15	242					
				7½ J × 15	248					
				8 J × 15	253					

[1] Instead of wheel rims with the identification letter B, same-sized rims with the identification letter J may be used. For example 5½ J × 13 instead of 5.50 B × 13. (See Section 10.1.3.2.)
[2] Instead of wheel rims with the identification letter J, same-sized rims with the identification letter B may be used. For example 4.50 B × 14 instead of 4½ J × 14.
[3] The wheel rims without identification letters mentioned in the table are expected to be identified with DIN 7824 Part 1.
[4] The outer diameter of wheels with M & S – tread can be up to 1% bigger than the standard tread.
[5] Maximum in kg at 2.5 bar.
[6] Reinforced model, 750 kg at 3.0 bar (LI 98).
[7] Reinforced model, 800 kg at 3.0 bar (LI 100).

Fig. 10.1-15 Continued.

Intermediate values have to be interpolated. Compensation can be achieved by increasing the inflation pressure. The values are as follows:

Camber angle	2°20′	2°40′	3°	3°20′	3°40′	4°
Pressure increase	2.1%	4.3%	6.6%	9.0%	11.5%	14.1%

Taking all the influences into account, such as top speed, wheel camber and axle load, the minimum tyre pressure required can be calculated for each tyre category (size and speed symbol). Formulas are shown in the 'WdK 99' guidelines from the Wirtschaftsverband der Deutschen Kautschukindustrie.

10.1.2.6.4 Tyre pressure limit values

Tyre pressure limit values should be adhered to. These values are

Q and T tyres	3.2 bar
H to W and ZR tyres	3.5 bar
M & S tyres (Q and T tyres)	3.5 bar

10.1.2.7 Tyre sidewall markings

All tyres used in Europe should be marked in accordance with the ETRTO standards (see Section 10.1.1.1).

In the USA, Japan and Australia, additional markings are required to indicate the design of the tyre and its characteristics. The characters must also bear the import sizes – the reason why these can be found on all tyres manufactured in Europe (Fig. 10.1-18).

10.1.2.8 Rolling circumference and driving speed

The driving speed is:

$$v = 0.006(1 - S_{X,W,a})\frac{C_{R,dyn} \times n_M}{i_D \times i_G}\ (\mathrm{km/h}) \qquad (10.1.1b)$$

This includes:

$S_{X,W,a}$ the absolute traction slip (Equation 10.1.4f)
$C_{R,dyn}$ the dynamic rolling circumference in m (Equation 10.1.1d)
n_M the engine speed in rpm
i_D the ratio in the axle drive (differential)
i_G the ratio of the gear engaged

The following can be assumed for slip $S_{X,W,a}$:

1st gear	0.08	4th gear	0.035
2nd gear	0.065	5th gear	0.02
3rd gear	0.05		

According to DIN 75020 Part 5, the rolling circumference C_R given in the tyre tables relates to 60 km/h and operating pressure of 1.8 bar. At lower speeds it goes down to $C_{R,stat}$:

$$C_{R,stat} = r_{stat} 2\pi \qquad (10.1.1c)$$

The values for r_{stat} are also given in the tables. At higher speeds, C_R increases due to the increasing centrifugal force. The dynamic rolling circumference $C_{R,dyn}$ at speeds over 60 km h^{-1} can be determined using the speed factor k_v. Figure 10.1-16 shows the details for k_v as a percentage, increasing by increments of 30 km h^{-1}. Intermediate values must be interpolated. The circumference would then be:

$$C_{R,dyn} = C_R(1 + 0.01 \times k_v)(\mathrm{mm}) \qquad (10.1.1d)$$

The dynamic rolling radius can be calculated from $C_{R,dyn}$ as

$$r_{dyn} = C_R/2\pi$$

or, at speeds of more than 60 km h^{-1},

$$r_{dyn} = C_{R,dyn}/2\pi \qquad (10.1.2)$$

Taking as an example the tyre 175/65 R 14 82 H at $v = 200$ km h^{-1} (Fig. 10.1-15) gives:

$$k_{v180} = 0.7\% \quad \text{and} \quad k_{v210} = 1.1\%$$

and interpolation gives:

$$k_{v200} = 0.007 + 0.0027 = 0.0097$$
$$k_{v200} = 0.97\%$$

The rolling circumference C_R taken from Fig. 10.1-15, according to Equation 10.1.1d, gives

$$C_{R,dyn200} = 1780 \times (1 + 0.0097) = 1797\ \mathrm{mm}$$

and thus the dynamic radius in accordance with Equation 10.1.2 is:

$$r_{dyn60} = 283\ \mathrm{mm} \quad \text{and} \quad r_{dyn200} = 286\ \mathrm{mm}$$

The outside diameter (construction measure) is

$$OD_T = 584\ \mathrm{mm} \quad \text{and} \quad \text{thus } OD_T/2 = 292\ \mathrm{mm}$$

v (km h^{-1})	60	90	120	150	180	210	240
Factor k_v (%)	–	+0.1	+0.2	+0.4	+0.7	+1.1	+1.6
Deviation Δk_v (%)	–	±0.1	±0.2	±0.4	±0.7	±1.1	±1.6

Fig. 10.1-16 Factor k_v, which expresses the speed dependence of the rolling circumference of passenger vehicle radial tyres above 60 km h^{-1} as a percentage. The permissible tolerances Δk_v have to be added (see Section 10.1.2.8), all taken from the German WDK Guideline 107, page 1.

a value which shows the extent to which the tyre becomes upright when the vehicle is being driven: r_{dyn} is only 9 mm or 6 mm less than $OD_T/2$.

10.1.2.9 Influence of the tyre on the speedometer

The speedometer is designed to show slightly more than, and under no circumstances less than, the actual speed. Tyres influence the degree of advance, whereby the following play a role:

- the degree of wear
- the tolerances of the rolling circumference
- the profile design
- associated slip.

The EC Council directive 75/443, in force since 1991, specifies an almost linear advance Δv,

$$+\Delta v \leq 0.1 \times v + 4\ (\text{km h}^{-1}) \tag{10.1.2a}$$

On vehicles registered from 1991 onwards the values displayed may only be as follows:

Actual speed (km h^{-1})	30	60	120	180	240
Max displayed value (km h^{-1})	37	70	136	202	268

As Fig. 10.1-15 indicates, at 60 km h^{-1} the rolling circumference C_R has a tolerance range of $\Delta C_R = +1.5\%$ to -2.5%, and according to Fig. 10.1-16 with a speed factor of k_v, deviations of up to $\Delta k_v = \pm 1.6\%$ are possible. When related to the dynamic rolling circumference $C_{R,dyn}$ (Equation 10.1.1d), the following tolerance limits (rounded to the nearest figure) may prevail and result in the displayed values when only the minus tolerances are considered, and if the speedometer has the maximum authorized advance:

Actual speed (km h^{-1})	60	120	180	240
Possible overall tolerance (%)	+1.5 −2.5	+1.7 −2.7	+2.2 −3.2	+3.1 −4.1
Max display value at minus tolerance (km h^{-1})	72	140	208	279

The slip should be added directly to this, which in direct gear amounts to around 2% (see Equations 10.1.1b and 10.1.4f), in other words

$$S_{X,W,a} = 0.02$$

If the manufacturer fully utilizes the advance specified in Equation 10.1.2a, it is possible that although the speedometer indicates 140 km h^{-1}, the vehicle is only moving at 120 km h^{-1}. This occurs, in particular, when the tyres are worn:

3 mm wear gives an advance of around 1%

Tyres with an M & S winter profile can, however, have a 1% larger outside diameter so that the profile can be deeper (Fig. 10.1-15, note 5 and Fig. 10.1-19). They would therefore reduce the degree by which the speedometer is advanced if the tyres are not yet worn. The same applies where the positive tolerances given in the above table are used. In this instance, it is also possible that even a very precise speedometer could display too low a speed.

10.1.2.10 Tyre profiles

The design of tyre profiles (Fig. 10.1-19) depends on the intended use, taking into account the parameters of height-to-width ratio, construction and mixture and design. The aquaplaning properties are improved by increasing the negative proportion (light places in the tyre impression, Fig. 10.1-9). The shoulder region with its transverse water-drainage grooves is particularly important for its properties in a lateral direction and the middle region with straight longitudinal grooves is important

Fig. 10.1-17 ZR tyres manufactured specially for the American market and marked with a 'P' do not meet the European standard and are therefore not authorized here (photograph: Dunlop factory).

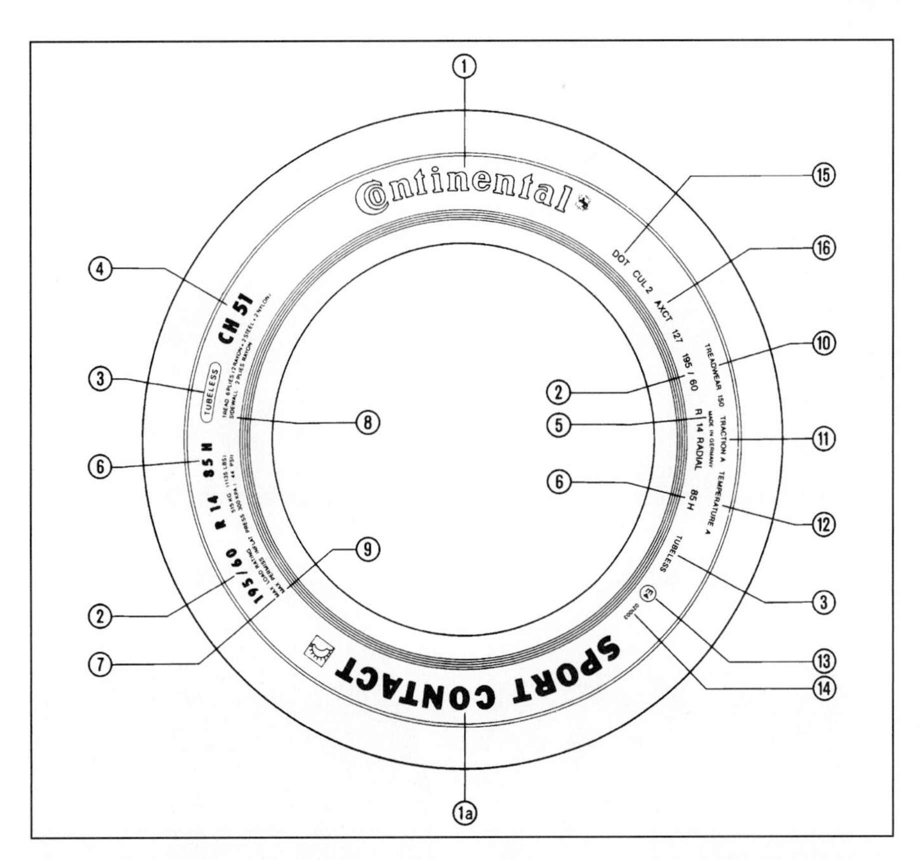

Fig. 10.1-18 Explanation of the marking on the sidewall of a tyre manufactured by Pneumatiques Kléber SA:

Legal and industry standard markings on the sidewalls of tyres according to:
FMVSS and CIR 104
UTQG (USA)
CSA Standard (Canada)
ADR 23B (Australia)
ECE–R30 (Europe)

1 Manufacturer (brand)
1a Product name
2 Size marking
195 = nominal tyre width in mm
60 = height–width ratio (60%)
radial type construction
14 rim diameter in inches
3 Tubeless
4 Trade code
5 Country of manufacture
6 Load capacity index (LI)
7 Maximum load capacity for the USA
8 Tread: under the tread are 6 plies carcass rayon, 2 plies steel belt, 2 plies nylon)
Sidewall: the substructure consists of 2 plies rayon
9 Maximum tyre pressure for the USA
10, 11, 12 USA: manufacturer's guarantee of compliance with the Uniform Tire Quality Grade (UTQG) which specifies: 10 tread wear: relative life expectancy compared with US-specific standard test values; 11 traction: A, B, C = braking performance on wet surfaces 12 temperature resistance: A, B or C = temperature resistance at higher test stand speeds; C fulfills the legal requirement in the USA
13 E 4 = tyre fulfills the ECE R30 value requirements
4 = country in which approval was carried out
(4 = The Netherlands)
14 identity number according to ECE R-30
15 DOT = tyre fulfills the requirements according to FMVSS 109 (DOT = Department of Transportation)
16 Manufacturer's code:
CU = factory (Continental)
L2 = tyre size
AXCT = model
127 = date of manufacture: production week 12, 1987

for its properties in a longitudinal direction. An asymmetrical profile design ('sports' profile) is chosen for wide tyres, tread lugs in the outside shoulder, which are subject to greater stress during cornering, can be designed to be more rigid. By adjusting the correct balance between profile rigidity and belt rigidity, it must be ensured that no conical forces are produced. Profiled bands around the middle region increase noise reduction and improve the steering response properties and, via the increase in circular rigidity, the brake-response properties.

Winter tyre profiles are improved, in terms of their force transmission properties in the wet, snow and ice, by a higher negative profile component, transverse grooves and a large number of sipes. Directional profiles (TS770) can be used to increase water dispersal, the longitudinal force coefficient and self-cleaning by means of transverse grooves which run diagonally outwards. Noise control is improved by variation in block length, sipes cut up to under the groove base or ventilation grooves running around the tyre.

Fig. 10.1-19 Designs of Continental tyre. (Top) Summer tyre (tyre foot prints, see Fig. 10.1-9) EcoContact EP (size 185/65 R14T) and Sport Contact (size 205/55 R16W). (Below) Winter tyre WinterContact TS760 (size 185/65 R14T) and WinterContact TS770 (size 235/60 R16H).

10.1.3 Wheels

10.1.3.1 Concepts

Tyres are differentiated according to the loads to be carried, the possible maximum speed of the vehicle, and whether a tubed or tubeless tyre is driven. In the case of a tubeless tyre, the air-tightness of the rim is extremely important. The wheel also plays a role as a 'styling element'. It must permit good brake ventilation and a secure connection to the hub flange. Figure 10.1-20 shows a passenger car rim fitted with a tubeless tyre.

10.1.3.2 Rims for passenger cars, light commercial vehicles and trailers

For these types of vehicle only well-base rims are provided. The dimensions of the smallest size, at 12″ and 13″ diameter and rim width up to 5.0″, are contained in the standard DIN 7824. The designation for a standard rim, suitable for the 145 R 13 tyre (Fig. 10.1-1) for example is:

DIN 7824 – drop base rim 4.00 B × 13

This type of rim used on passenger cars up to around 66 kW (90 PS) has only a 14-mm high rim flange and is identified with the letter B. The DIN standard can generally be dropped.

In order to make it possible to fit bigger brakes (Fig. 10.1-10), more powerful vehicles have larger diameter rims as follows:

- series production passenger cars: 14–17″ rims
- sports cars: –18″ rims.

The J rim flange applied here is used on rims from 13″ upwards and is 17.3 mm high. The rim base can (as shown in Fig. 10.1-1) be arranged symmetrically or shifted outwards. The rim diameter, which is larger on

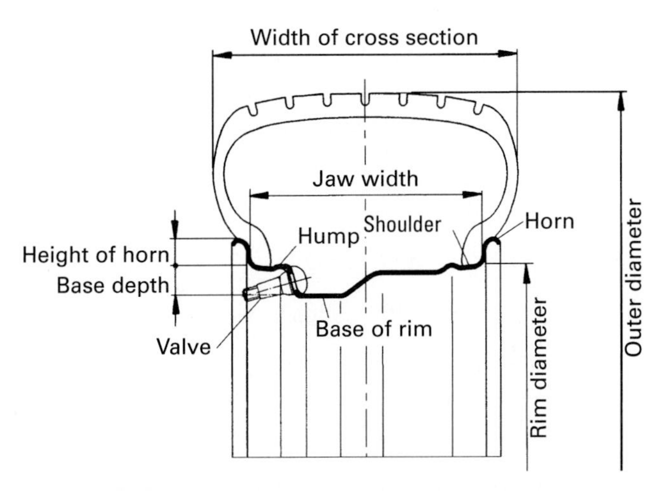

Fig. 10.1-20 Series 55 wide tyre designs, mounted on a double hump rim with the inflating valve shown in Fig. 10.1-6. The actual rim consists of the following:

- rim horns, which form the lateral seat for the tyre bead (the distance between the two rims is the jaw width a);
- rim shoulders, the seat of the beads, generally inclined at 5° ± 1° to the centre where the force transfer occurs around the circumference (Fig. 10.1-5);
- well base (also known as the inner base), designed as a drop rim to allow tyre fitting, and mostly shifted to the outside (diagram: Hayes Lemmerz).

the inside, creates more space for the brake (Figs. 8.1-8, 8.1-56, 10.1-10, 10.1-11 and 10.1-20). DIN 7817 specifies the rim widths from 3½″ to 8½″. The definition of a normal asymmetrical rim with a 5″ width, J rim flange and 14″ diameter is:

DIN 7817 drop base rim – 5 J × 14

The symmetrical design is identified by the suffix ‘S’. The standards also contain precise details on the design and position of the valve hole (see also Figs. 10.1-20 and 10.1-24).

C tyres for light commercial vehicles require a broader shoulder (22 mm instead of 19.8 mm), which can be referred to by adding the letters LT (light truck) at the end of the marking:

DIN 7817 drop base rim – $5\frac{1}{2}$ J × 15 – LT

There is a preference worldwide for using tubeless radial tyres on passenger cars and light commercial vehicles. Where these tyres are used, it is essential to have a ‘safety contour’ at least on the outer rim shoulder. This stops air suddenly escaping if the vehicle is cornering at reduced tyre pressure.

The three different contours mainly used are (Fig. 10.1-21):

Hump (H, previously H1)
Flat-hump (FH, previously FHA)
Contre Pente (CP)

Sheets 2 and 3 of DIN 7817 specify the dimensions of the first two designs. The ‘hump’ runs around the rim, which is rounded in H designs, whereas a flat hump rim is simply given a small radius towards the tyre foot. The fact that the bead sits firmly between the hump and rim flange is advantageous on both contours. An arrangement on both the outside and inside also prevents the tyre feet sliding into the drop bases in the event of all the air escaping from the tyre when travelling at low speeds, which could otherwise cause the vehicle to swerve. The disadvantage of hump rims is that changing the tyre is difficult and requires special tools.

A French design, intended only for passenger car rims, is the ‘Contre Pente’ rim, known as the CP for short. This has an inclined shoulder towards the rim base, which for rim widths between 4″ and 6″ is provided on one or both sides.

For years, the rims of most passenger cars have had safety shoulders on both sides, either a double hump (Figs. 10.1-20 and 10.1-24) or the sharp-edged flat-hump on the outside and the rounder design on the inside (Fig. 10.1-23). The desired contour must be specified in the rim designation. Figure 10.1-22 gives the possible combinations and abbreviations which must appear after the rim diameter data. A complete designation for an asymmetrical rim would then be as follows:

10.1.3.3 Wheels for passenger cars, light commercial vehicles and trailers

Most passenger cars and light commercial vehicles are fitted with sheet metal disc wheels, because these are economic, have high stress limits and can be readily serviced. They consist of a rim and a welded-on wheel disc (also known as an attachment face, Fig. 10.1-23). Cold-formable sheet metal, or band steel with a high elongation, can be used (e.g. RSt37-2 to European standard 20) depending on the wheel load, in thicknesses from 1.8 to 4.0 mm for the rim and 3.0 to 6.5 mm for the attachment faces.

There is a direct correlation between wheel offset e and ‘kingpin offset at ground’ r_σ; the more positive

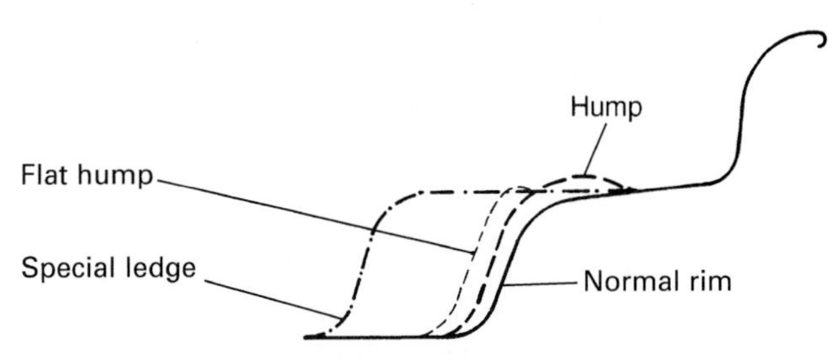

Fig. 10.1-21 Standard rim and contours of the safety shoulders which can be used on passenger cars and light commercial vehicles.

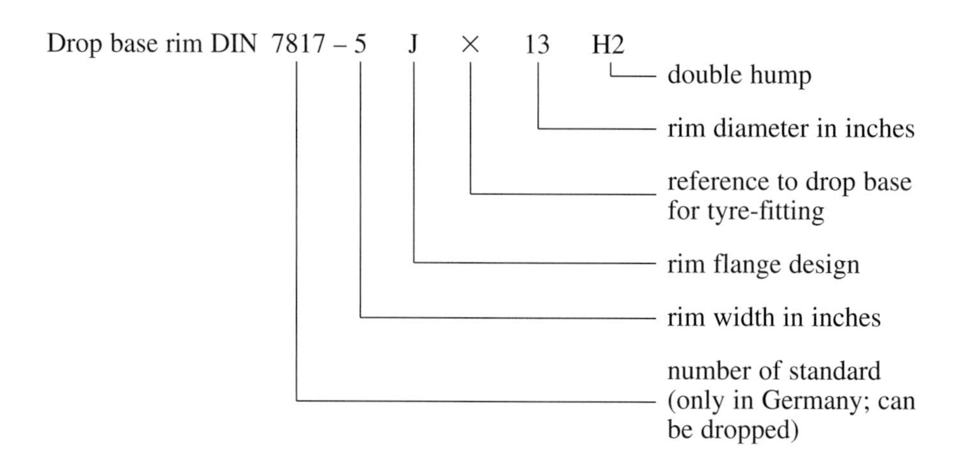

Denomination	Nature of safety shoulder		Identification letters
	Outside of rim	*Inside of rim*	
One-sided hump	Hump	Normal	H
Double hump	Hump	Hump	H2
One-sided flat hump	Flat hump	Normal	FH[1]
Double-sided flat hump	Flat hump	Flat hump	FH2[1]
Combination hump	Flat hump	Hump	CH[2]

[1] In place of the identification letters FH the identification letters FHA were also permitted.
[2] In place of the identification letters CH the identification letters FH1-H were also permitted.

Fig. 10.1-22 Marking of the various safety shoulders when used only on the outside of the rim or on both the inside and outside. Normal means there is no safety contour (Fig. 10.1-1). Further details are contained in standard DIN 7817.

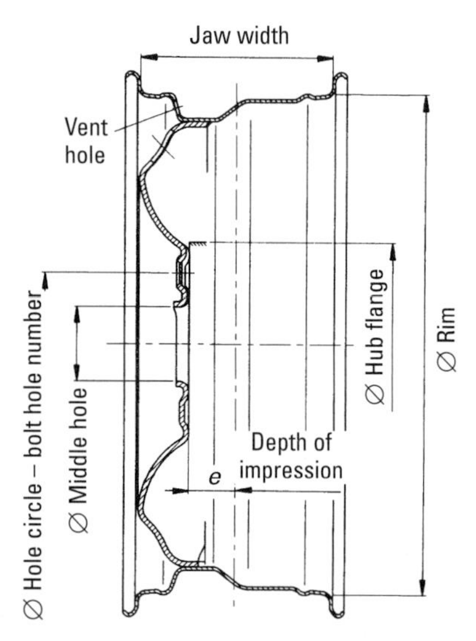

Fig. 10.1-23 The sheet metal disc-type wheel used in series production vehicles consists of a rim and disc. To avoid fatigue fractures, the wheel hub flange diameter should be greater than the dish contact surface. Wheel offset e (depth of impression) and kingpin offset at ground r_σ are directly correlated. A change in e can lead to an increase or a reduction in r_σ. The dome-shaped dish leading to the negative kingpin offset at ground is clearly shown (diagram: Hayes Lemmerz).

r_σ, the smaller can be the depth dimension *e*. However, a negative kingpin offset $- r_\sigma$, especially on front-wheel drive, results in a significant depth *e* and severe bowing of the attachment faces (as can be seen in Figs. 10.1-8, 10.1-23 and 10.1-25).

The wheel disc can be perforated to save weight and achieve better brake cooling. Despite the fact that they cost almost four times as much as sheet metal designs, alloy wheels are becoming increasingly popular (Figs. 8.1-56 and 10.1-24). Their advantages are:

- lower masses;
- extensive styling options; and therefore
- better appearance;
- processing allows precise centring and limitation of the radial and lateral runout (see Section 10.1.5);
- good heat transfer for brake-cooling.

Often incorrectly called aluminium rims, alloy wheels are mainly manufactured using low-pressure chill casting, occasionally forging or aluminium plate, and generally consist of aluminium alloys with a silicon content (which are sometimes heat hardenable), such as GK-Al Si 11 Mg, GK-Al Si 7 Mg T (T = tempered after casting) etc.

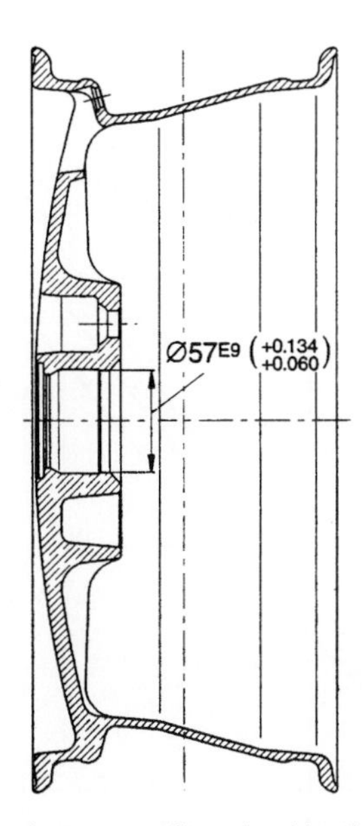

Fig. 10.1-24 Hayes Lemmerz alloy wheel for the Audi 80, made of the aluminium alloy GK-Al Si 7 Mg wa. The wheel has a double-hump rim (H2) and middle centring and is fixed with four spherical collar bolts. The different wall thicknesses, which are important for the strength, the shape of the bolt hole, the different shape of the drop-rim and the position of the valve hole are clearly shown. At high speeds the snap-fit valve (Fig. 10.1-6) is pressed outwards by the centrifugal force and supported below the rim base.

Regardless of the material, the wheels must be stamped with a marking containing the most important data (Fig. 10.1-25).

10.1.3.4 Wheel mountings

Many strength requirements are placed on the wheel disc sitting in the rim (or the wheel spider on alloy wheels); it has to absorb vertical, lateral and longitudinal forces coming from the road and transfer them to the wheel hub via the fixing bolts.

The important thing here is that the contact area of the attachment faces, known as the 'mirror', should sit evenly and, for passenger cars, that the hub flange should have a slightly larger diameter (Fig. 10.1-23), otherwise it is possible that the outer edge of the hub will dig into the contact area, with a loss of torque on the bolts. The notch effect can also cause a fatigue fracture leading to an accident.

The number of holes and their circle diameter are important in this context. This should be as large as possible to introduce less force into the flange and fixing bolts. If the brake discs are placed onto the wheel hub from the outside – which is easier from a fitting point of view – it is difficult to create a hole larger than 100 mm on 13″ wheels, and

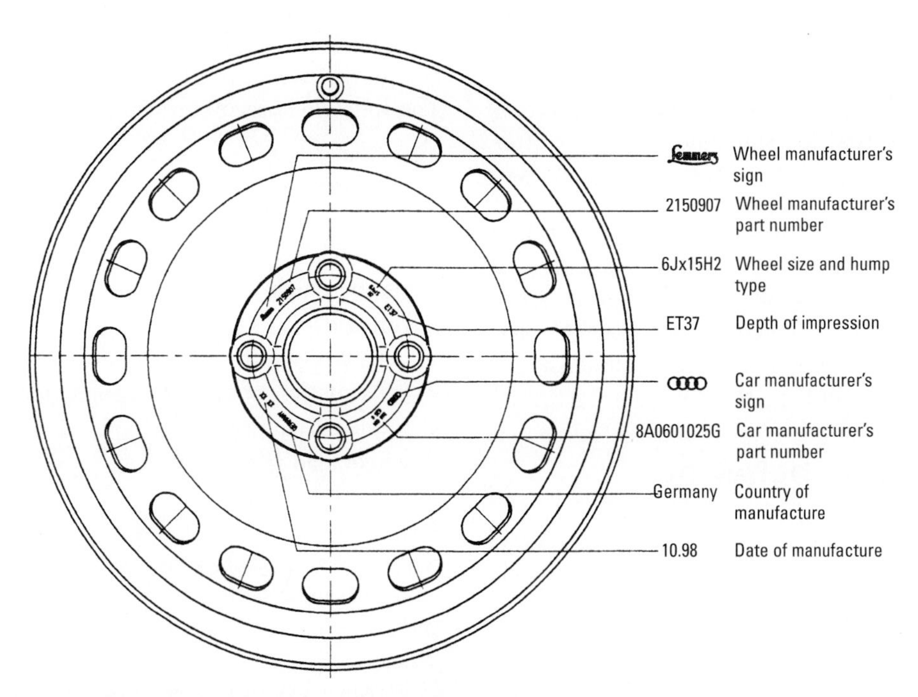

Fig. 10.1-25 Double-hump sheet metal disc-type wheel with openings for cooling the brakes. Also pictured is the stamp in accordance with the German standard DIN 7829, indicating manufacturer code, rim type and date of manufacture (week or month and year). Also specified is the wheel offset (ET37) and, in the case of special wheels with their own ABE (General operating approval), the allocation number of the KBA, the German Federal Vehicle Licensing Office. If there is not much space the stamp may be found on the inside of the dish. The date of manufacture also points to when the vehicle was manufactured (diagram: Hayes Lemmerz).

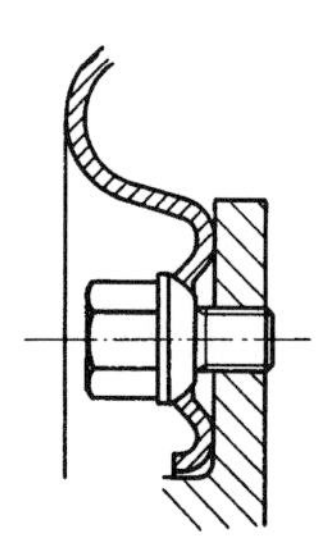

Fig. 10.1-26 Depression design with special springing characteristics on a passenger car sheet metal disc-type wheel. The wheel can be centred using the fixing bolts or by fitting into the toleranced hole (Fig. 10.1-24).

using a 14″ or 15″ wheel should make for the best compromise (Figs. 8.1-1, 8.1-41, 8.1-44 and 10.1-10). German standard DIN 74361 contains further details.

The brake disc can also be fixed to the wheel hub from the inside (Fig. 8.1-38). However, the disadvantage of this is that the hub has to be removed before the disc can be changed. This is easy on the non-driven axle, but time-consuming on the driven axle (see Section 10.1.5). This brief look shows that even the brakes play a role in the problems of fixing wheels.

Nowadays, wheels are almost always fixed with four or five metric M12 × 1.5 or M14 × 1.5 DIN 74361 spherical collar bolts. The high friction between the spherical collar and the stud hole prevents the bolts from coming loose while the vehicle is in motion. For this reason, some car manufacturers keep the contact surface free of paint. On sheet metal disc wheels with attachment faces up to 6.5 mm thick, the spring action of the hole surround (Fig. 10.1-26) is an additional safety feature, which also reduces the stress on the wheel bolts as a result of its design elasticity. Sheet metal rings are often inserted in the alloy wheels to withstand high stresses underneath the bolt head.

Generally, the spherical collar nuts also do the job of centring the wheels on the hub. Hub centring has become increasingly popular because of a possible hub or radial run-out and the associated steering vibrations. A toleranced collar placed on the hub fits into the dimensioned hole which can be seen in Fig. 10.1-24.

10.1.4 Springing behaviour

The static tyre spring rate c_T – frequently also known as spring stiffness or (in the case of a linear curve) spring constant – is the quotient of the change in vertical force $\Delta F_{Z,W}$ in newtons and the resultant change Δs_T – the compression in mm within a load capacity range corresponding to the tyre pressure p_T (Fig. 10.1-27; see also Section 10.1.2.5.4):

$$c_T = \Delta F_{Z,W}/\Delta s_T \text{ (N/mm)} \qquad (10.1.3)$$

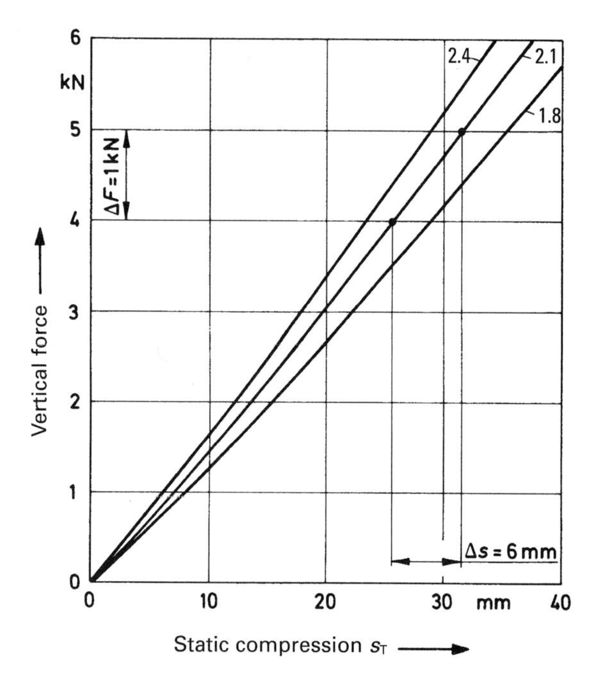

Fig. 10.1-27 The static tyre spring rate c_T is the quotient of the force and the deflection travel shown on the radial tyre 175/70 R 13 80 S at p_T = 1.8 bar, 2.1 bar and 2.4 bar; the example shown gives:
$c_T = \frac{\Delta F_{Z,W}}{\Delta s_T} = \frac{1000\text{N}}{6\text{ mm}} = 167$ N/mm

The parameter c_T forms part of the vibration and damping calculation and has a critical influence on the wheel load impact factor (see Section 9.1.1). The stiffer the tyre, the higher the damping must be set and the greater the stress experienced by the chassis components. The following parameters influence the spring rate:

- vertical force
- tyre pressure
- driving speed
- slip angle
- camber angle
- rim width
- height-to-width ratio
- construction of tyre (bias angle, material)
- tyre wear and tear
- wheel load frequency.

As can be seen in Fig. 10.1-27, apart from in the low load range, the spring rate is independent of the load. A linear increase can be seen as the speed increases (Figs. 10.1-16 and 10.1-28), which persists even when the tyre pressure changes.

During cornering, the force $F_{Y,W}$ shifts the belt in a lateral direction, and so it tips relative to the wheel plane. This leads to a highly asymmetrical distribution of pressure and (as can be seen from Fig. 10.1-28) to a reduction in the spring rate as the slip angles increase.

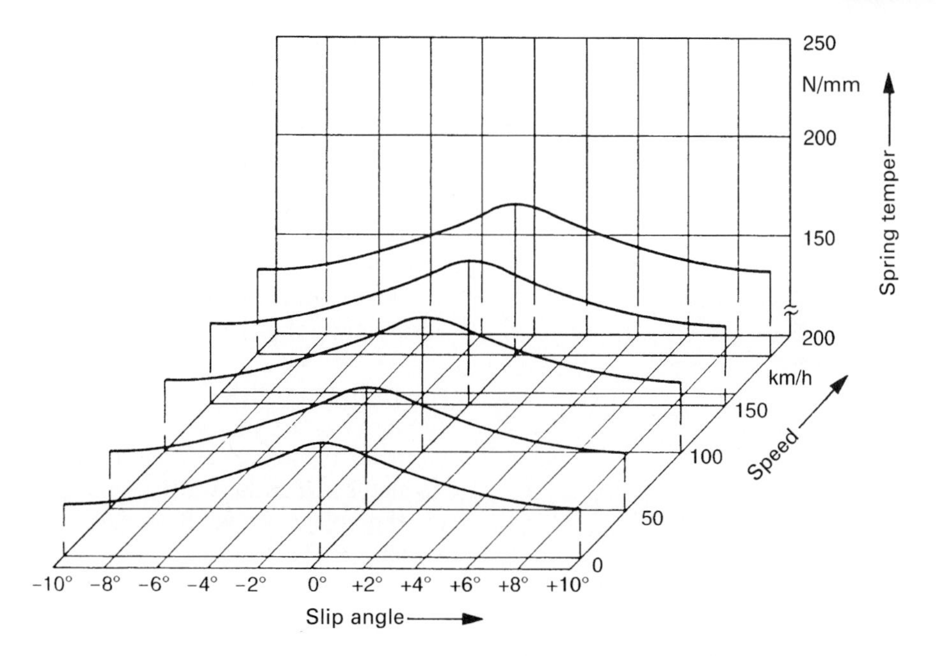

Fig. 10.1-28 Tyre springing rate as a function of slip angle and road speed, measured on a radial tyre 185/70 R 13 86 S at $p_T = 2.1$ bar. Speed increases the springing rate as the belt stands up due to the centrifugal force. However, the slip angle makes it softer because the belt is pushed away to the side and the shoulders take over part of the springing effect.

10.1.5 Non-uniformity

The tyre consists of a number of individual parts, e.g. carcass layers, belt layers, running tread, sidewall stock and inner lining, which – put together on a tyre rolling machine – give the tyre blank (Fig. 10.1-5). In the area where it is put together, variations in thickness and stiffness occur, which can lead to non-uniformity.

Owing to the irregularities caused during manufacture, the following occur around the circumference and width of the tyre:

- thickness variations
- mass variations
- stiffness variations.

These cause various effects when the tyre rolls:

- imbalance
- radial tyre runout
- lateral tyre runout
- variation in vertical and/or radial force
- lateral force variations
- longitudinal force variation
- ply steer (angle) force
- conicity force.

Imbalance U occurs when an uneven distribution of mass and the resulting centrifugal forces are not equalized. Because the uneven distribution occurs not only around the circumference, but also laterally, we have to differentiate between static and dynamic imbalance (Fig. 10.1-29). This is calculated in size and direction on balancing machines and eliminated with balancing weights on the rim bead outside and inside the wheel.

Radial and lateral runout are the geometrical variations in the running tread and the sidewalls. They are measured with distance sensors on a tyre-uniformity machine. The German WdK Guideline 109 contains full details.

The most important of the three force variations is the radial force variation. For greater clarity, it is shown on the model in Fig. 10.1-30, where the tyre consists of different springs whose rates fluctuate between c_1 and c_8. The resulting phenomena should be indicated on the

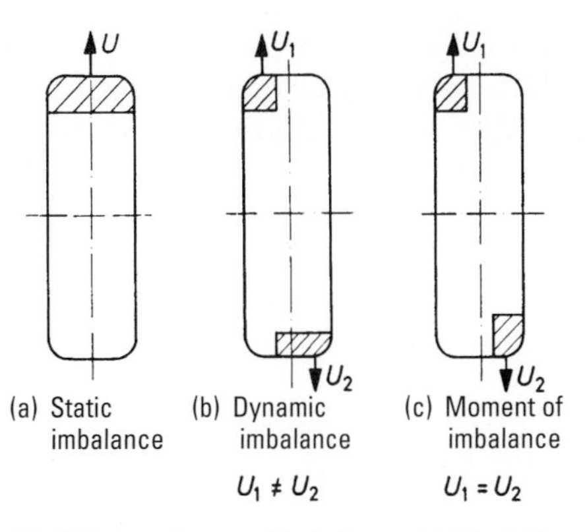

Fig. 10.1-29 Different forms of imbalance U: (a) static, (b) dynamic. The imbalance is equalized in (c).

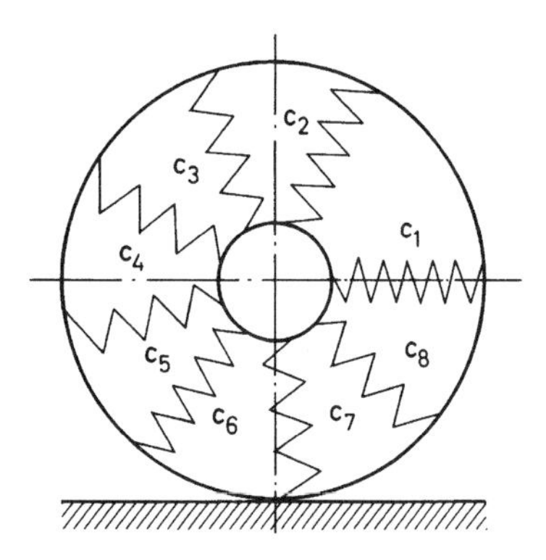

Fig. 10.1-30 The tyre spring rate can fluctuate depending on the manufacturing process, shown as c_1 to c_8.

175 R 14 88 S steel radial tyre, loaded at $F_{Z,W} = 4.5$ kN and pressurized to $p_T = 1.9$ bar. Assuming this had a mean spring rate $c_T = 186$ N m^{-1}, which fluctuates by ±5%, the upper limit would be $c_{T,max} = 195$ N mm^{-1} and the lower limit would be $c_{T,min} = 177$ N mm^{-1}. Under vertical force $F_{Z,W} = 4.5$ kN = 4500 N the tyre would, according to Equation 10.1.3a, have as its smallest jounce travel:

$$s_{T,min} = \frac{F_{Z,W}}{C_{T,max}} = \frac{4500}{195}; \quad s_{T,min} = 23.1 \text{ mm} \tag{10.1.3a}$$

and

$$s_{T,max} = 25.4 \text{ mm}$$

as the greatest travel. The difference is:

$$\Delta s_T = s_{T,max} - s_{T,min} = 2.3 \text{ mm}$$

This difference in the dynamic rolling radius of $\Delta s_T = 2.3$ mm would cause variations in vertical force $\Delta F_{z,w}$, which nevertheless is still smaller than the friction in the wheel suspension bearings. At a speed of perhaps 120 km/h and travelling on a completely smooth road surface, this would nevertheless lead to vibration that would be particularly noticeable on the front axle.

The vehicle used as an example should have a body spring rate of $c_f = 15$ N/mm per front axle side. The travel Δs_T would then give a vertical force difference of:

$$\Delta F_{Z,W,f} = c_f \, \Delta s_T = 15 \times 2.3; \quad \Delta F_{Z,W,f} = 34.5 \text{ N}$$

The friction per front axle side is, however, not generally below

$$F_{fr} = \pm 100 \text{ N}$$

so it can only be overcome if greater variations in vertical force occur as a result of non-uniformity in the road surface. The more softly sprung the vehicle, the more the variations in radial force in the tyre make themselves felt.

The lateral force variations of the tyre influence the straight-running ability of the vehicle. Even with a tyre that is running straight, i.e. where the slip angle is zero, lateral forces occur, which also depend on the direction of travel.

The variations in longitudinal force that occur must be absorbed on the chassis side by the rubber bearings.

The ply steer force dependent on the rolling angle results from the belt design because of the lateral drift of the tyre contact area as a consequence of flat spotting. In contrast, the conicity force, resulting from a change in diameter across the width of the tyre, is not dependent on the rolling angle. Both forces disturb the straight running of the vehicle.

10.1.6 Rolling resistance

10.1.6.1 Rolling resistance in straight-line driving

Rolling resistance is a result of energy loss in the tyre, which can be traced back to the deformation of the area of tyre contact and the damping properties of the rubber. These lead to the transformation of mechanical into thermal energy, contributing to warming of the tyre.

Sixty to 70% of the rolling resistance is generated in the running tread (Fig. 10.1-5) and its level is mainly dependent on the rubber mixture. Low damping running tread mixtures improve the rolling resistance, but at the same time reduce the coefficient of friction on a wet road surface. It can be said that the ratio is approximately 1:1, which means a 10% reduction in the rolling resistance leads to a 10% longer braking distance on a wet road surface. The use of new combinations of materials in the running tread (use of silica) has led to partial reduction of the conflict between these aims.

Rolling resistance is either expressed as a rolling resistance force F_R or as the rolling resistance factor k_R – also known as the coefficient of rolling resistance:

$$F_R = k_R \times F_{Z,W}(\text{N}) \tag{10.1.4}$$

The factor k_R is important for calculating the driving performance diagram and depends on the vertical force $F_{Z,W}$ and the tyre pressure p_T. Figure 10.1-31 shows the theoretical k_R curve of tyres of different speed classes as a function of the speed. Although the coefficient of rolling friction of the T tyre increases disproportionally from around 120 km h^{-1}, this increase does not occur in H and V tyres until 160 to 170 km h^{-1}. The reason for

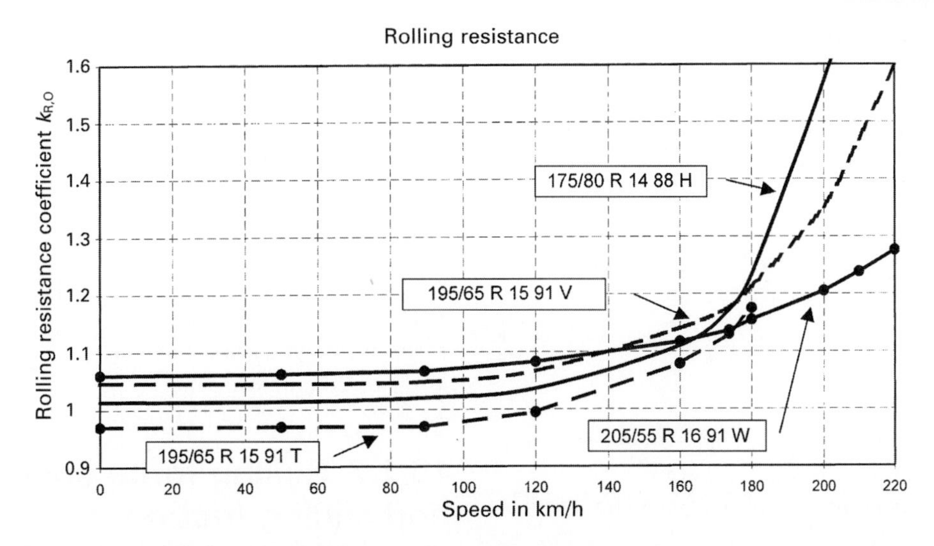

Fig. 10.1-31 Rolling resistance coefficients $\kappa_{R,0}$, average values of radial tyres as a function of the speed, measured on a drum test rig. Tyres authorized up to 210 km h^{-1} have a lower rolling resistance below 160 km h^{-1} (than the V and W designs), whilst the value rises sharply above this speed (measurements: Continental). Asphalted roads cause $k_{R,0}$ to increase by around 20% as k_R and rough concrete to at least 30%. The ratios i_R are then 1.2 or 1.3 to 1.4 and the actual value of k_R is: $K_R = i_R \times K_{R,0}$ (2.4a)

this behaviour is the shape of the rolling hump that occurs at different speeds depending on the speed class, and is dependent on the stiffness of the belt, in other words on its design. The lower k_R values for the T tyres result from the usually poorer wet skidding behaviour of this speed class.

The difference is due to the different design emphases during development of the tyres. The design priorities for H, V and W tyres are high-speed road holding and good wet skidding and aquaplaning behaviour, whereas T tyres are designed more for economy, i.e. lower rolling resistance (which plays an important role at lower speeds and influences urban driving fuel consumption, Fig. 10.1-32) and long service life.

10.1.6.2 Rolling resistance during cornering

Rolling resistance can change dramatically during cornering; its value depends on the speed and the rolling radius R, in other words on $\mu_{Y,W}$ (see Equations 10.1.9 and 10.1.11 and Fig. 10.1-43) and α_f or r. The rolling resistance $k_{R,co}$, which is included in some calculations, comprises the coefficient k_R for straight running and the increase Δk_R:

$$k_{R,co} = k_R + \Delta k_R$$
$$\Delta k_R \approx \mu_{Y,W} \times \sin \alpha \qquad (10.1.4b)$$

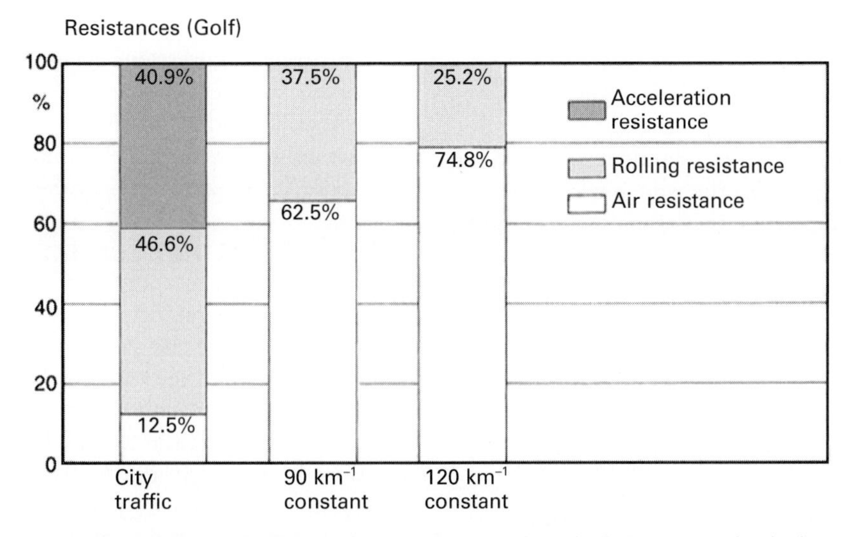

Fig. 10.1-32 In town and when the vehicle is travelling at low speeds on rural roads, fuel consumption is determined up to 40% by the rolling resistance, whereas at higher speeds the air drag is the determining factor see Section 10.1.1 and Section 10.1.2. The figure shows a study carried out by VW on the Golf.

The following data can provide an example:

Front axle force $F_{Z,V,f} = 7\text{kN}$; $\mu_{Y,W} = 0.7$ (asphalted road)
Tyres 155 R 13 78 S $p_T = 1.8$ bar, $v \leq 120$ km h

In accordance with Equation 10.1.11 related to one wheel:

$$F_{Y,W,f} = \mu_{Y,W} F_{Z,W,f} = \mu_{Y,W} F_{Z,V,f}/2 = 0.7 \times 3.5 \text{ kN}$$
$$F_{Y,W,f} = 2.45 \text{ kN}$$

The slip angle read off at $F_{Y,W,f}$ in Fig. 10.1-44 is 4° and corresponds to the values in Fig. 10.1-43.

However, the dynamic wheel load transfer seen in Fig. 10.1-5 plays a role during cornering, leading to a greater slip angle on the wheel on the outside of the curve (and thus also on the inner wheel), than resulted from test rig measurements. On '82' series tyres, α is about 5°, in accordance with Fig. 10.1-38:

$$\alpha \approx 7\,\mu_{Y,W} \tag{10.1.4c}$$

With sin 5° in accordance with Equation 10.1.4b there is an increase of

$$\Delta k_R \approx 0.7 \times 0.087 = 0.061$$

Assuming a value of $k_{R,0} = 0.012$, in accordance with Equation 10.1.4a, on asphalted road

$$k_R = i_R\, k_{R,0} = 1.2 \times 0.010 = 0.012$$

and therefore the rolling resistance during cornering is

$$k_{R,co} = 0.012 + 0.061 \approx 0.073$$

In the case of the understeering vehicles (Fig. 10.1-41) $k_{R,co}$ increases as a result of the additional steering input and – if the wheels are driven – μ_{rsl} should be inserted for $\mu_{Y,W}$ (see Equation 10.1.18); the slip angle increases further. '65 Series' tyres, on the other hand, require a smaller steering input and thus make the vehicle easier to handle:

$$\alpha = 3 \times \mu_{Y,W} \tag{10.1.4d}$$

10.1.6.3 Other influencing variables

The rolling resistance increases in certain situations:

- in the case of a large negative or positive camber (the influence can be ignored up to ±2°);
- due to a change to track width;
- in the case of deviations in zero toe-in around 1% per $\delta = 10'$ or $v = 1$ mm;
- on uneven ground.

In general it can be said that the ratio i_R (see Fig. 10.1-31) will take the following values:

- around 1.5 on cobbles
- around 3 on potholed roads
- around 4 on compacted sand
- up to 20 on loose sand.

10.1.7 Rolling force coefficients and sliding friction

10.1.7.1 Slip

If a tyre transfers drive or braking forces, a relative movement occurs between the road and tyre, i.e. the rolling speed of the wheel is greater or less than the vehicle speed (see Equation 10.1.1b). The ratio of the two speeds goes almost to ∞ when the wheel is spinning, and is 0 when it locks. Slip is usually given as a percentage. The following equation applies during braking:

$$S_{X,W,b} = \frac{\text{vehicle speed} - \text{circumferential speed of wheel}}{\text{vehicle speed}}$$

$$S_{X,W,b} = \frac{v - v_W}{v} \times 100\ (\%) \tag{10.1.4e}$$

Drive slip is governed by:

$$S_{X,W,a} = \frac{v_W - v}{v_W} \times 100\ (\%) \tag{10.1.4f}$$

The different expressions have the advantage that, in both cases where the wheel is spinning or locked, the value is 100% and is positive.

Further details can be found in Section 10.1.2.8, Section 8.1.2 Chapter 8.1 and Section 10.1.2.

10.1.7.2 Friction coefficients and factors

The higher the braking force or traction to be transmitted, the greater the slip becomes. Depending on the road condition, the transferable longitudinal force reaches its highest value between 10% and 30% slip and then reduces until the wheel locks (100% slip). The quotient from longitudinal force F_x and vertical force $F_{Z,W}$ is the coefficient of friction, also known as the circumferential force coefficient

$$\mu_{X,W} = F_{X,W}/F_{Z,W} \tag{10.1.5}$$

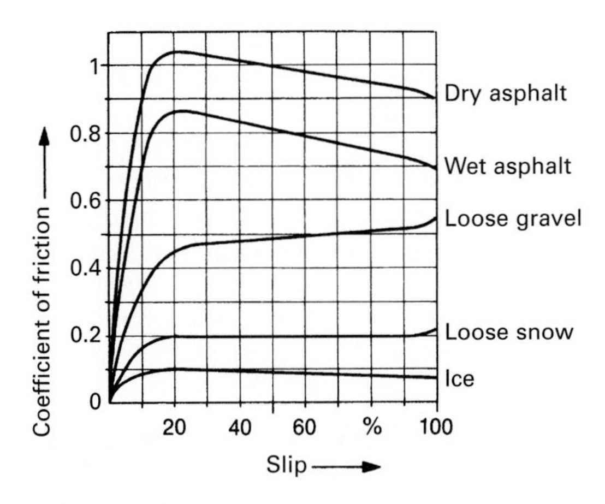

Fig. 10.1-33 Coefficient of friction $\mu_{X,W}$ of a summer tyre with 80–90% deep profile, measured at around 60 km/h and shown in relation to the slip on road surfaces in different conditions (see also Fig. 8.1-64). Wide tyres in the '65 series' and below have the greatest friction at around 10% slip, which is important for the ABS function.

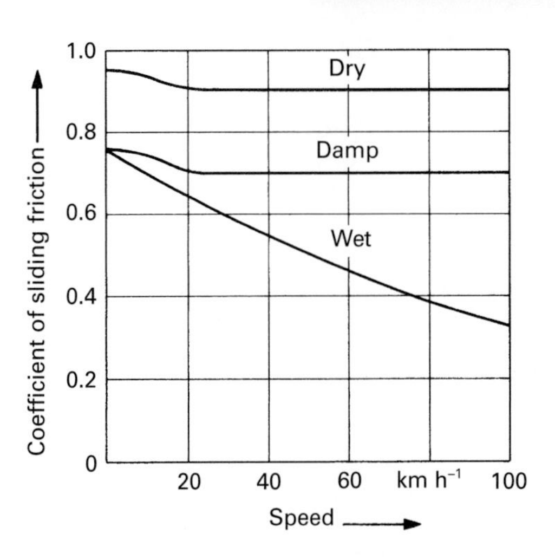

Fig. 10.1-34 Dependency of the coefficient of sliding friction $\mu_{X,W,lo}$ on speed on different road conditions.

when it relates to the maximum value, and the coefficient of sliding friction, also called sliding friction factor

$$\mu_{X,W,lo} = F_{X,W}/F_{Z,W} \qquad (10.1.5a)$$

when it is the minimal value (100% slip) (Fig. 10.1-33). F_x is designated $F_{X,W,b}$ during braking and $F_{X,W,a}$ during traction.

In all cases $\mu_{x,w}$ is greater than $\mu_{x,w,lo}$; in general it can be said that

$$\text{on a dry road } \mu_{x,w} \approx 1.2\ \mu_{x,w,lo} \qquad (10.1.6)$$

$$\text{on a wet road } \mu_{x,w} \approx 1.3\ \mu_{x,w,lo} \qquad (10.1.6a)$$

10.1.7.3 Road influences

10.1.7.3.1 Dry and wet roads

On a dry road, the coefficient of friction is relatively independent of the speed (Fig. 10.1-34), but a slight increase can be determined below 20 km/h. The reason lies in the transition from dynamic to static rolling radius (see the example in Section 10.1.2.5.4) and is therefore linked to an increasing area of tyre contact. At speeds a little over zero, on a rough surface, a toothing cogging effect can occur, which causes a further increase in the coefficient of friction, then:

$$\mu_{x,w} \geq 1.3 \qquad (10.1.6b)$$

When the road is wet, the coefficient of friction reduces, but is still independent of the speed. This situation changes as the amount of water increases and also with shallower profile depth. The water can no longer be moved out of the profile grooves and the μ value falls as speed increases.

10.1.7.3.2 Aquaplaning

The higher the water level, the greater the risk of aquaplaning. Three principal factors influence when this occurs:

- road
- tyres
- speed.

With regard to the road, the water level is the critical factor (Fig. 10.1-35). As the level rises, there is a disproportionate increase in the tendency towards aquaplaning. When the level is low, the road surface continues to play a role because the coarseness of the surface absorbs a large part of the volume of water and carries it to the edge of the road. Following rainfall, the water levels on roads are generally up to 2 mm; greater depths can also be found where it has been raining for a long time, during storms or in puddles.

On the tyre, the tread depth has the greatest influence (Fig. 10.1-47). There can be up to a 25 km h^{-1} difference in speed between a full tread and the legal minimum tread depth of 1.4 mm. High tyre pressure and low running surface radius r (Fig. 10.1-5) lead to the area of contact becoming narrower, giving the advantage of improved aquaplaning behaviour as the distribution of ground pressure becomes more even (Fig. 10.1-9). Lower tyre pressure and contours with larger radii make aquaplaning more likely; this also applies to wider tyres (Fig. 10.1-19) particularly when tread depths are low. However, the greatest influence by far is the speed, especially when the water level increases and tread depths are low. This is why reducing speed is the best way to lessen the risk of aquaplaning, and is a decision drivers can make for themselves.

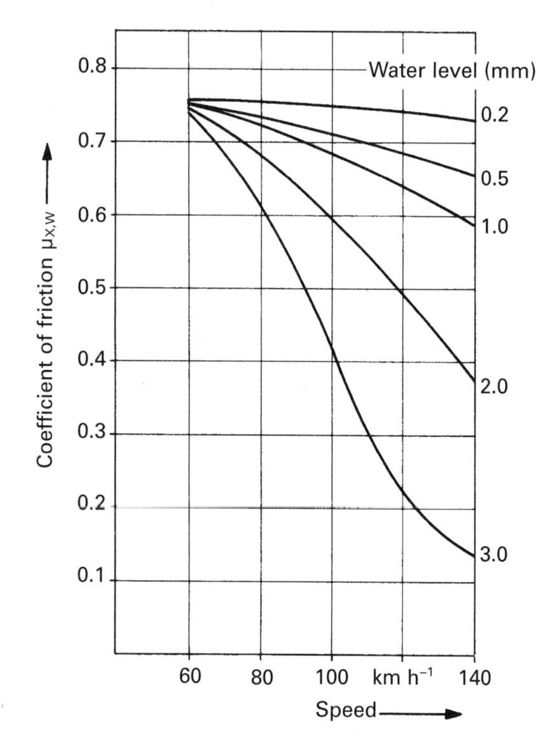

Fig. 10.1-35 Coefficients of friction $\mu_{X,W}$ of a summer tyre with an 8-mm deep profile dependent on speed at different water levels. Hardly any influence can be detected under 60 km h^{-1}; at higher speeds and 3-mm water depth, the curve shows a lowering of $\mu_{X,W}$ which indicates the aquaplaning effect.

10.1.7.3.3 Snow and ice

Similar to aquaplaning, low coefficients of friction occur on icy roads, although these are highly dependent on the temperature of the ice. At close to 0°C, special conditions occur; compression of the surface can lead to the formation of water which has a lubricating effect and reduces the coefficient of friction to $\mu_{x,w} \leq 0.08$ (Fig. 10.1-36). At −25°C, a temperature that is by no means rare in the Nordic countries, values of around $\mu_{x,w} = 0.6$ can be reached. At low temperatures, coefficients of friction and sliding friction are further apart:

$$\mu_{X,W} \sim 2\mu_{X,W,lo} \qquad (10.1.7)$$

10.1.8 Lateral force and friction coefficients

10.1.8.1 Lateral forces, slip angle and coefficient of friction

Lateral forces on a rolling tyre can be caused by the tyre rolling diagonal to the direction of travel (so-called slip), the tendency of a tyre to move from its position vertical to the road, camber or conical effects. The build-up of lateral forces as a result of slip will be discussed next.

If a disturbing force $F_{c,V}$ acts at the centre of gravity of the vehicle (e.g. a wind or side negative lift force), lateral wheel forces $F_{Y,W,f,o}$, $F_{Y,W,f,i}$, $F_{Y,W,r,o}$ and $F_{Y,W,r,i}$ are needed to balance the forces (Fig. 10.1-37). To build up these forces, the vehicle must alter its direction of travel about the angle α, the slip angle. The size of the slip angle depends on the force transmission properties of the tyre and the disturbing force (Fig. 10.1-38).

When cornering, the interference force should be equal to the centrifugal force $F_{c,V}$, which results from the speed

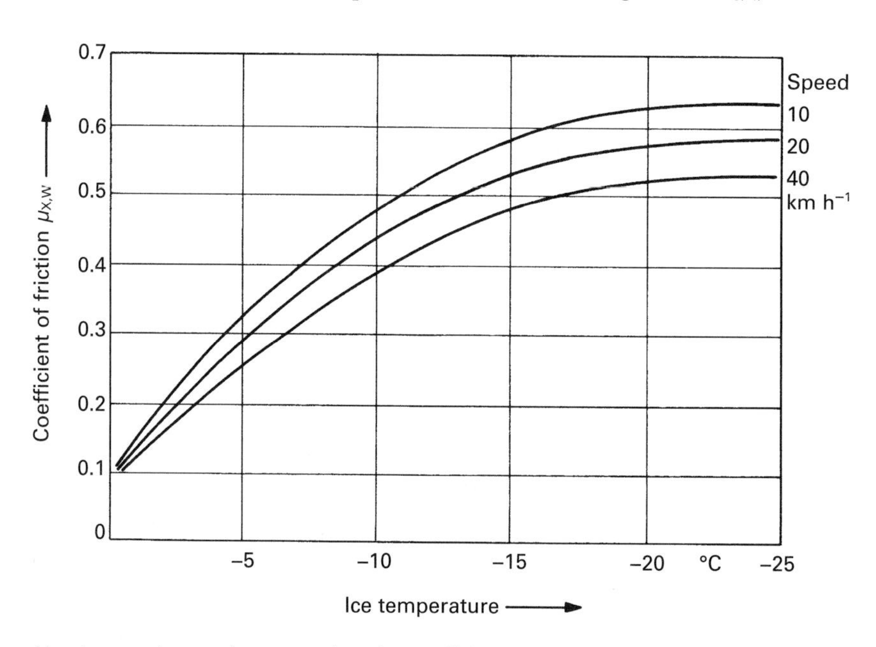

Fig. 10.1-36 Influence of ice temperature and car speed on the coefficient of friction $\mu_{X,W}$ of an 82 series winter tyre; the extremely low values at 0 °C can be seen clearly.

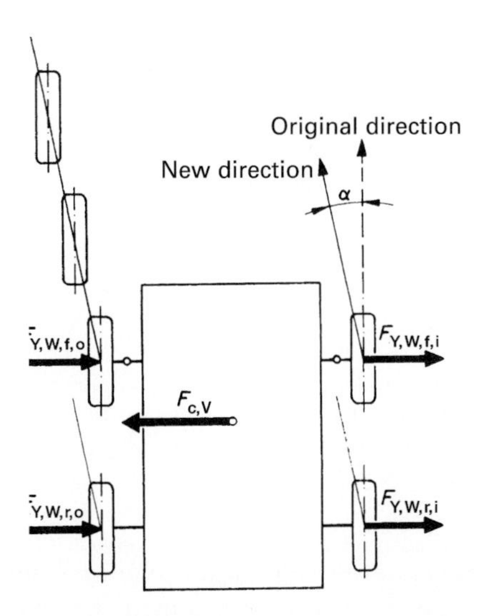

Fig. 10.1-37 Tyres are only able to transfer a lateral force $F_{Y,V}$ acting on the vehicle if they are rolling at an angle to the vehicle. Regardless of whether these are $F_{Y,V}$ or the centrifugal force $F_{C,Y}$ during cornering, the lateral forces $F_{Y,W}$ should be regarded as being perpendicular to the wheel centre plane.

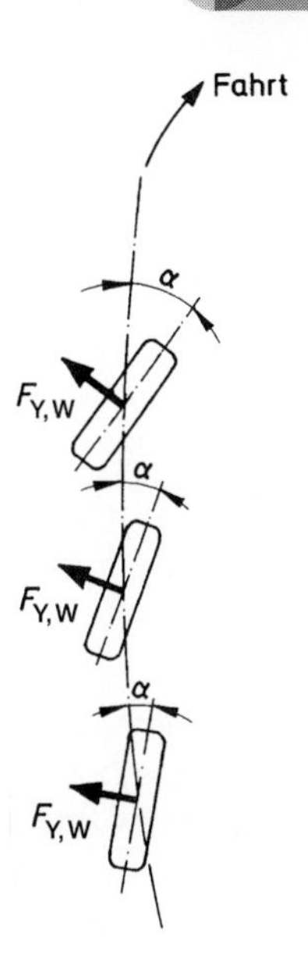

Fig. 10.1-39 Increasing lateral forces $F_{Y,W}$ during cornering caused by the centrifugal force $F_{c,V}$ leads to increasing slip angles α.

v in m/s and the radius of the bend R in m, on which the vehicle centre of gravity V (Fig. 10.1-29a) moves. With the total weight $m_{v,t}$ of the vehicle the equation is:

$$F_{c,V} = m_{V,t} \times v^2/R = m_{V,t} \times a_y = F_{Y,V}(\text{N}) \quad (10.1.8)$$

The centrifugal or disturbance force is just as large as the lateral forces on the wheels (Fig. 10.1-37):

$$F_{Y,V} = F_{Y,W,f,o} + F_{Y,W,f,i} + F_{Y,W,r,o} + F_{Y,W,r,i} = \sum F_{Y,W} \quad (10.1.8a)$$

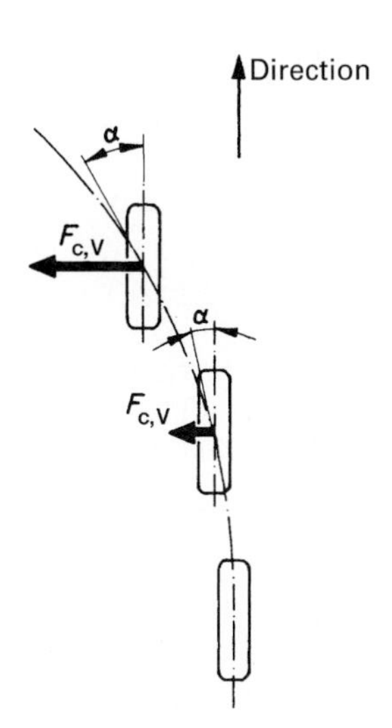

Fig. 10.1-38 The higher the lateral force $F_{Y,W}$, the greater the tyre slip angle α.

and

$$\sum F_{Y,W} = \mu_{Y,W} \times \sum F_{Z,W} = \mu_{Y,W} \times F_{Z,V,t}$$

Together the two equations give:

$$\mu_{Y,W} F_{Z,V,t} = \mu_{Y,W} \times m_{V,t}\, g = m_{V,t} \times a_y \quad (10.1.9)$$

and

$$\mu_{Y,W} = g/a_y$$

The coefficient of friction $\mu_{Y,W}$ is not dependent on the radius of the curve and driving speed and is therefore more suitable for calculating cornering behaviour.

The faster the vehicle negotiates a bend, the higher the coefficient of friction used and the greater the slip angles (Fig. 10.1-39).

10.1.8.2 Self-steering properties of vehicles

The self-steering properties of a vehicle describe the lateral force and hence slip angle ratios produced during steady-state cornering (radius and driving speed constant; no external disturbances). In the case of an understeering vehicle, a larger slip angle is required on the front axle than at the rear axle ($\alpha_f > \alpha_r$, Fig. 10.1-41). During cornering with an increase in lateral acceleration, the driver must force the vehicle into the bend by increasing the steering angle (see Fig. 10.1-2). If the necessary slip angles on the front and rear axles are the same

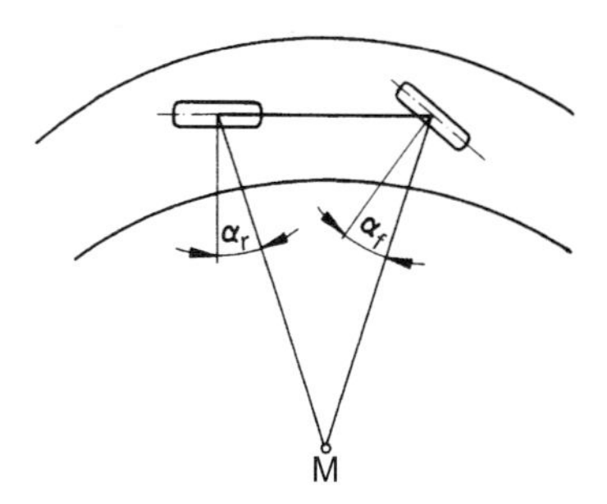

Fig. 10.1-40 If, during cornering, $\alpha_f \sim \alpha_r$, the handling of a vehicle can be described as neutral.

($\alpha_f = \alpha_r$, Fig. 10.1-40), one speaks of neutral handling characteristics. Over-steering behaviour is present if the tail of the vehicle moves outwards during cornering and the slip angle on the rear axle is greater than on the front axle ($\alpha_f < \alpha_r$, Fig. 10.1-42). The driver must respond to this by reducing the steering angle.

As understeering behaviour is consistent with the expectations and experience of the driver, it is this which needs to be aimed for. In normal driving conditions (anti-skid roadway, lateral acceleration of less than 6 m/s), all vehicles, therefore, are now designed to understeer. With increasing lateral acceleration, the under-steering behaviour should be as linear as possible and then, also as a warning to the driver that the stability limit is about to be reached, increase progressively. If the handling characteristics change to oversteer at the stability limit, for instance with very high acceleration, this is an unpredictable driving situation which the untrained driver can only control with difficulty. For active riding safety, the predictability of self-steering properties in all kinds of conditions (vehicle loading, the distribution of driving torque in four-wheel drive vehicles, different coefficients of friction, acceleration or braking procedures, changes in tyre pressure, etc.) is of paramount importance.

For a simplified representation of the relationships described, the so-called single-track model is used, in which the wheels of the vehicle are drawn together in the middle of the vehicle, without taking into account the height of the centre of gravity (flat model).

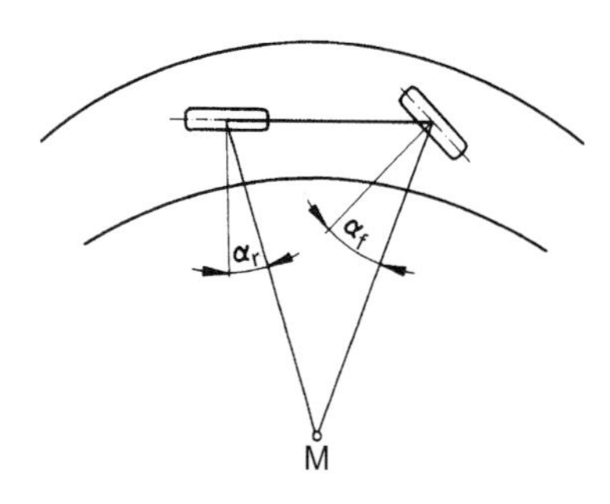

Fig. 10.1-41 If there is a greater slip angle α_f on the front wheels than α_r on the rear, the vehicle understeers.

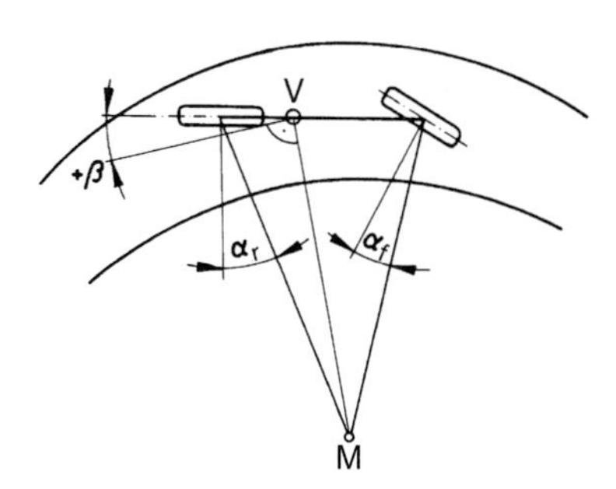

Fig. 10.1-42 If there is a greater slip angle α_r on the rear wheels than on the front (α_f), the vehicle oversteers. The positive angle describes the angle between the vehicle longitudinal axis and its speed at the centre of gravity.

Since in greater bend radii the average steering angle δ_m is less than 5°, it can be assumed that the sine and radius values of the angle are equal, and the angles δ_o and δ_i correspond to this:

$$\sin \delta_m \approx \delta_m \approx \delta_o \approx \delta_i (\text{rad})$$

It is now possible to determine the relationship between steering angle, turning circle diameter D_S (Fig. 8.1-69) and slip angles at a constant cornering speed:

$$\delta_m = \frac{2 \times l}{D_S} + \alpha_f - \alpha_r \tag{10.1.10}$$

The Kingpin offset at ground r_σ is so negligable in comparision to Ds that it can be ignored.

10.1.8.3 Coefficients of friction and slip

To determine the cornering behaviour, the chassis engineer needs the lateral forces (or the coefficient of friction) based on the slip angle and the parameters:

- vertical force (or wheel load) in the centre of tyre contact
- tyre pressure
- wheel camber
- tyre type.

The measurements are generally taken on test rigs, up to slip angles of $\alpha = 10°$. The drum surface with its friction values of $\mu_0 = 0.8$–0.9 sets limits here, and larger angles hardly give increasing lateral coefficients of friction:

$$\mu_{Y,W} = F_{Y,W}/F_{Z,W} \tag{10.1.11}$$

Conditions on the road are very different from those on the test rig; the type of road surface and its condition play a role here. As can be seen in Fig. 10.1-43, the coefficient

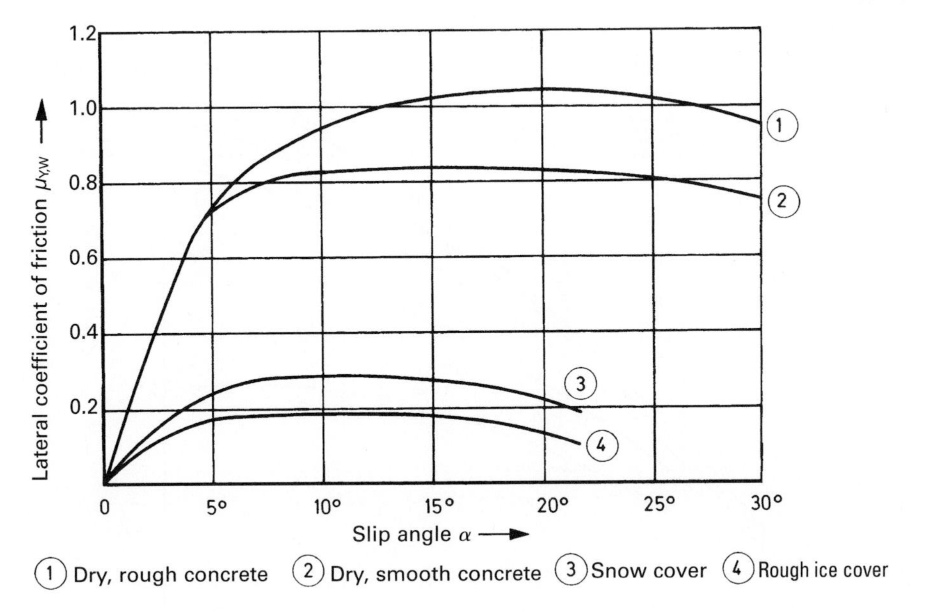

Fig. 10.1-43 Lateral coefficients of friction $\mu_{Y,W}$ as a function of slip angle and road condition, shown for an '82 series' summer tyre with around 90% deep profile. The ice temperature is around −4 °C. The vertical force $F_{Z,W}$ was kept constant during the measurements to obtain the dimensionless values of $\mu_{Y,W}$. The maximum at $\alpha = 20°$ on a very skid-resistant road can be seen clearly. The further $\mu_{Y,W}$ sinks, the further it moves towards smaller angles.

of friction on rough, dry concrete increases to $\alpha = 20°$ and then falls. In precisely the same way as with the longitudinal force the slip $S_{Y,W}$ (in the lateral direction) is also taken into consideration; this is as a percentage of the sine of the slip angle times 100:

$$S_{Y,W} = \sin\alpha \times 100\ (\%) \quad (10.1.12)$$

In conjunction with the drum value $\alpha = 10°$, this would give a slip of $S_{Y,W} = 17\%$, and on the street at $\alpha = 20°$ slip values of up to $S_{Y,W} = 34\%$. If the tyre is further twisted to $\alpha = 90°$, it slides at an angle of 90° to the direction of travel; sin α would then be equal to one and $S_{Y,W} = 100\%$. The coefficient of friction then becomes the coefficient of lateral sliding friction $\mu_{Y,W,lo}$, which on average is around 30% lower:

$$\mu_{Y,W,lo} \approx 0.7 \times \mu_{Y,W} \quad (10.1.13)$$

In contrast to dry concrete (as also shown in Fig. 10.1-43) on asphalt and, in particular on wet and icy road surfaces, no further increase in the lateral cornering forces can be determined above $\alpha = 10°$ (i.e. $S_{Y,W} \approx 17\%$).

10.1.8.4 Lateral cornering force properties on dry road

Figure 10.1-44 shows the usual way in which a measurement is carried out for a series 82 tyre. The lateral force appears as a function of the vertical force in kilo newtons and the slip angle α serves as a parameter.

A second possibility can be seen in Fig. 10.1-45; here, for the corresponding series 70 tyre, $\mu_{Y,W} = F_{Y,W}/F_{Z,W}$ is plotted against α and $F_{Z,W}$ serves as a parameter. The degree of curvature of the graphs in both figures shows that slope at any point changes as a function of $F_{Z,W}$ or $\mu_{Y,W}$. The maximum occurs with large angles and small vertical forces. A less stressed tyre in relation to its load capacity therefore permits greater coefficients of friction and higher cornering speeds than one whose capacity is fully used.

This result, which has been used for a long time in racing and sports cars, has also become popular in modern cars, A mid-range standard car can be taken as an example. The car manufacturer specifies $p_T = 2.2$ bar/2.5 bar under full load for the front and rear wheels 185/65 R 15 88H. At these pressures, the load capacity, in accordance with Figs. 10.1-13 and 10.1-15, is:

front 505 kg and rear 560 kg

and the wheel load (divided by two) results:

front 375 kg and rear 425 kg

As described in Section 10.1.2.6, at speeds up to 210 km h^{-1} (H tyres), an increase in tyre pressure of 0.3 bar is necessary or there is only a correspondingly lower load capacity. This then is, with $p_T = 1.9$ bar at the front or 2.2 bar at the back,

450 kg and 505 kg

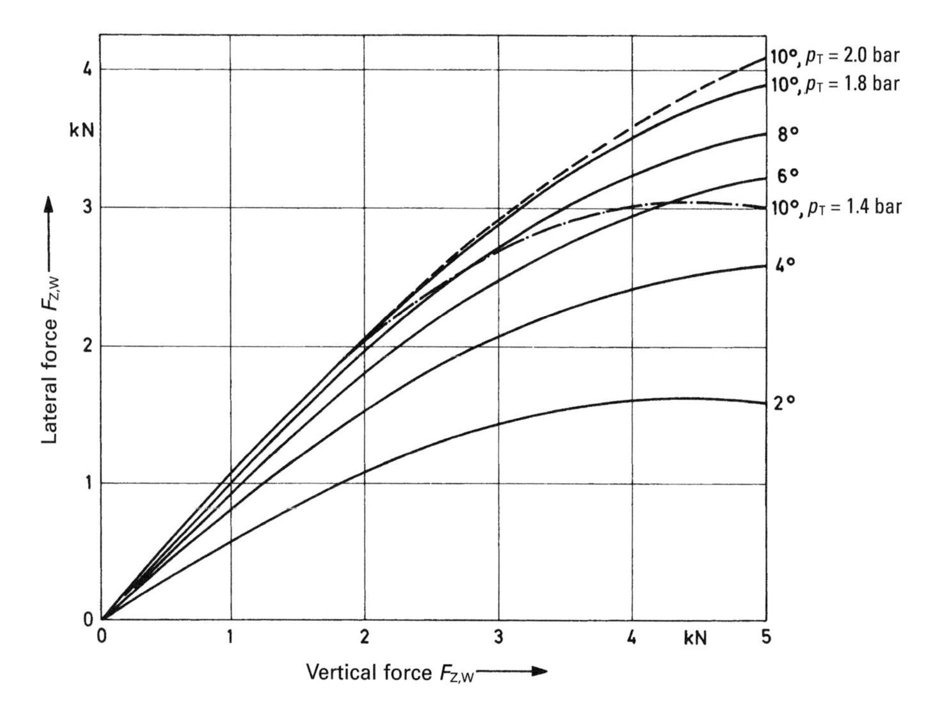

Fig. 10.1-44 Lateral cornering forces of the 155 R 13 78 S '82 series' steel radial tyre, measured on a dry drum at $p_T = 1.8$ bar. The load capacity at this pressure is around 360 kg, corresponding to a vertical force $F_{Z,W} = 3.53$ kN. Also shown are the forces at $\alpha = 10°$ and $P_T = 1.4$ bar and 2.0 bar to indicate the influence of the tyre pressure on the lateral cornering properties.

Thus, the actual load factor k_m at 210 km/h becomes:

$$\begin{aligned} \text{front } k_{m,f} &= (375/450) \times 100 = 83\% \\ \text{back } k_{m,r} &= (425/505) \times 100 = 84\% \end{aligned} \quad (10.1.14)$$

10.1.8.5 Influencing variables

10.1.8.5.1 Cross-section ratio *H/W*

The 185/65 R 15 88H size used as an example in the previous section is a 65 series wide tyre; the 15″ diameter also allows a good sized brake disc diameter (Fig. 10.1-10).

In contrast to the 82 series standard tyre, the sizes of the 70 series and wide tyres ($H/W = 0.65$ and below) generate higher lateral cornering forces at the same slip angles (Figs. 10.1-9, 10.1-45 and 10.1-46). As can be seen in Fig. 8.1-6, these, as $F_{Y,W,o} = \mu_{Y,w} (F_{Z,W} + \Delta F_{Z,W})$, are all the greater, the faster the vehicle takes a bend.

10.1.8.5.2 Road condition

The force transmission ratios between the tyres and road are determined by the state of the road (see construction, surface roughness and condition; Figs. 10.1-43 and 10.1-47).

10.1.8.5.3 Track width change

The track width change that exists, in particular on independent wheel suspensions, causes undesirable lateral forces at the centres of tyre contact on both wheels when the vehicle is moving unimpeded in a straight line. This effect is magnified by an increase in slip rigidity, as, for example, in wide tyres.

10.1.8.5.4 Variations in vertical force

During cornering, vertical force variations $\pm \Delta F_{Z,W}$ in the centre of tyre contact cause a reduction in the transferable lateral forces $F_{Y,W}$ as the tyre requires a certain amount of time and distance for the build-up of lateral forces. The loss of lateral force $\Delta F_{Y,W,4}$ depends on the effectiveness of the shock absorbers, the tyre pressure p_T (which can enhance the 'springing' of the wheels) and the type of wheel suspension link mountings. Further influences are wheel load and driving speed. To calculate cornering behaviour, an average loss of lateral force $\Delta F_{Y,W,4}$ due to variations in vertical force and dependent only on tyre design and slip angle α, should be considered:

$$\Delta F_{Y,W,4} \approx 40 \text{ N per degree } \alpha \quad (10.1.15)$$

10.1.8.5.5 Camber change

Wheels that incline with the body during cornering have a similar, detrimental influence on the transferability of lateral forces. As can be seen from Fig. 8.1-6, positive angle ($+\varepsilon_W$) camber changes occur on the outside of the

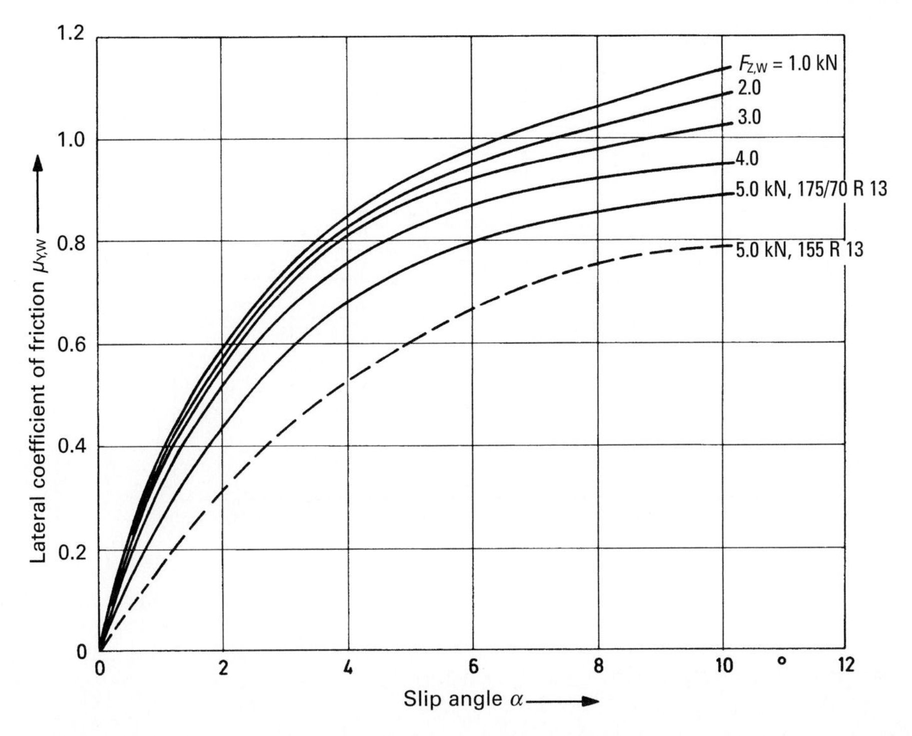

Fig. 10.1-45 Lateral coefficients of friction $\mu_{Y,W}$ as a function of the slip angle α and the vertical force $F_{Z,W}$, measured on a dry drum on a 175/70 R 13 82 S tyre at $p_T = 2.0$ bar. The tyre, which has been inflated in such a manner, carries 395 kg or $F_{Z,W} = 3.87$ kN. In order to indicate the influence of the cross section on the transferable lateral forces the 82 series 155 R 13 78 S tyre was also included.

bend and negative angles ($-E_W$) on the inside of the bend as a consequence of the body roll. The lateral forces are directed to the centre point of the bend. If a wheel is 'cambered' against this, in other words inclined at the top towards the outside of the bend, the possibility of transferring lateral forces reduces; on

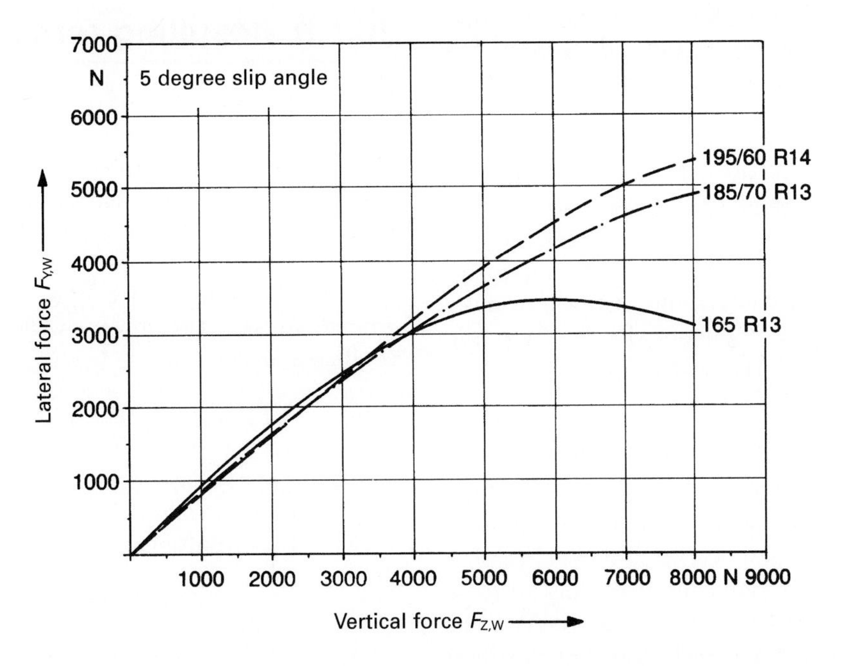

Fig. 10.1-46 Lateral force $F_{Y,W}$ dependent on vertical force $F_{Z,W}$ and tyre sizes of different *H/W* ratios: 165 R 13 82 H, 185/70 R 13 85 H and 195/60 R 14 85 H. Up to $F_{Z,W} = 4000$ N the curves are more or less the same, but at higher loads the more favourable lateral cornering properties of the wide tyre are evident.

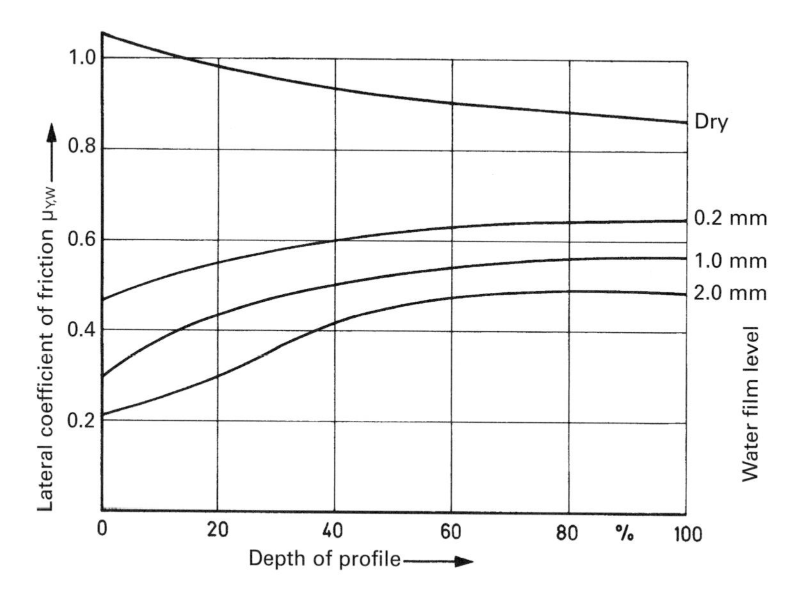

Fig. 10.1-47 Possible lateral friction coefficients $\mu_{Y,W}$ of a steel radial tyre 155 R 13 78 S depending on the depth of the tyre profile as a percentage (starting from 8 mm = 100%) at $p_T = 1.8$ bar, $\alpha = 10°$, $v = 60$ km/h and varying water film levels in mm. The improved grip of the treadless tyre on a dry road can be seen clearly as can its significantly poorer grip in the wet; a fact which also applies to the coefficient of friction in the longitudinal direction (see Section 10.1.7.2).

a dry road surface, depending on the tyre size, the change is

$$\Delta F_{Y,W,3} = 40\text{ N to }70\text{ N per degree of camber} \tag{10.1.16}$$

To counteract this, a greater slip angle must occur and greater steering input becomes necessary for the front wheels. This makes the vehicle understeer more (Fig. 10.1-41) and appear less easy to handle. Furthermore, the steering aligning moment also increases. If this effect occurs on the rear axles – as is the case with longitudinal link axles (Fig. 10.1-14) – the vehicle has a tendency to oversteer. Negative camber $-\varepsilon_W$ on the outside of the bend and positive $+\varepsilon_W$ on the inside would have exactly the opposite effect. Wheels set in this manner would increase the lateral forces that can be absorbed by the amount stated previously for $\Delta F_{Y,W,3}$ and cause a reduction in the tyre slip angle.

10.1.8.5.6 Lateral force due to camber

Wheels according to the body roll inclined towards the outside edge of the bend (Fig. 8.1-6) try to roll outwards against the steering direction, so that additional camber forces are required in the tyre contact patches to force the wheels in the desired steering direction. As these camber forces act in the same direction as the centrifugal force $F_{c,Bo\text{ or }V}$ in the case described, greater lateral slip forces $F_{Y,W,f,o}$, $F_{Y,W,f,i}$, $F_{Y,W,r,o}$ and $F_{Y,W,r,i}$ and hence greater slip angles must be applied to maintain the balance of forces on the part of the tyres.

The average force $F_{\varepsilon W}$ with the standard camber values for individual wheel suspensions on a dry road are (see Section 10.1.2.3):

$$F_{\varepsilon_W} \approx F_{Z,W} \times \sin \varepsilon_W \tag{10.1.17}$$

10.1.9 Resulting force coefficient

Rolling resistance increases when negotiating a bend (see Equation 10.1.4a), and the vehicle would decelerate if an increased traction force $F_{X,W,A}$ did not create the equilibrium needed to retain the cornering speed selected. $F_{X,W,A}$ is dependent on a series of factors and the type of drive system (front- or rear-wheel drive); on single-axle drive (see Sections 8.1.4–8.1.6), the traction force on the ground stresses the force coefficient of friction (the coefficient of)

$$\mu_{X,W} = F_{X,W,A,f,\text{ or }r}/F_{Z,V,f\text{ or }r} \tag{10.1.15}$$

and thus greater slip angles at the driven wheels. With given values for cornering speed and radius (see Equation 10.1.8) the resulting force coefficient μ_{rsl} can be determined:

$$\mu_{rsl} = (\mu_{Y,W}^2 + \mu_{X,W}^2)^{\frac{1}{2}} \tag{10.1.18}$$

μ_{rsl} cannot be exceeded because the level depends on the road's surface and the condition.

When braking on a bend, additional longitudinal forces $F_{X,W,b}$ occur on all wheels, and act against the direction of travel. In this case Equation 10.1.18 also applies.

On standard vehicles and front-wheel drives, the front wheels take 70–80% of the braking force and the rear wheels only 20–30%. This means that the slip angles increase on both axles, but more at the front than the rear and the vehicle tends to understeer (Fig. 10.1-41). If the wheels of an axle lock, the friction becomes sliding friction and the vehicle pushes with this pair of wheels towards the outside of the bend.

Taking into consideration the maximum possible values in the longitudinal and lateral direction of the road – known respectively as $\mu_{X,W,max}$ and $\mu_{X,W,min}$ – the increasing force coefficient can be calculated:

$$\mu_{X,W} = \mu_{X,W,max}\left[1 - \left(\frac{\mu_{Y,W}}{\mu_{Y,W,max}}\right)^2\right]^{\frac{1}{2}} \qquad (10.1.19)$$

Consider as an example a braking process on a dry road at 100 km/h on a bend with $R = 156$ m. Using Equation 10.1.9 the calculation gives $\mu_{Y,W} = 0.5$.

Figure 10.1-48 shows a measurement on the tyre in question where the greatest coefficient of friction in the lateral direction at $F_{Z,W} = 2490$ N, $\varepsilon_W = 10\%$ and $\alpha = 4°$ (see Equation 10.1.11) amounts to

$$\begin{aligned}\mu_{Y,W,max} &= F_{Y,W}/F_{Z,W}\\ &= 2850/2940\ (\mathrm{N/N})\\ \mu_{Y,W,max} &= 0.97\end{aligned}$$

In the longitudinal direction the possible braking force $F_{X,W,b} = 3130$ N is at $\alpha = 0°$ and therefore (see Equation 10.1.5),

$$\mu_{X,W,max} = F_{X,W,b}/F_{Z,W} = 3130/2940(\mathrm{N/N}) = 1.06$$

and

$$\mu_{X,W} = 1.06\left[1 - \left(\frac{0.5}{0.97}\right)^2\right]^{\frac{1}{2}} = 0.91$$

The lateral forces that the tyre can absorb during braking can also be calculated:

$$\mu_{Y,W} = \mu_{Y,W,max}\left[1 - \left(\frac{\mu_{X,W}}{\mu_{X,W,max}}\right)^2\right]^{\frac{1}{2}} \qquad (10.1.19a)$$

$\mu_{X,W} = 0.7$ should be given. The lateral force coefficient (which can be used) is:

$$\mu_{Y,W} = 0.97\left[1 - \left(\frac{0.7}{1.06}\right)^2\right]^{\frac{1}{2}} = 0.73$$

At $S_{X,W,b} = 10\%$ and $\alpha = 4°$ the transferable lateral force is:

$$F_{Y,W} = \mu_{Y,W} \times F_{Z,W} = 0.73 \times 2940 = 2146\ \mathrm{N}$$

and the available braking force is:

$$F_{X,W,b} = \mu_{X,W} \times F_{Z,M} = 0.7 \times 2940 = 2058\ \mathrm{N}$$

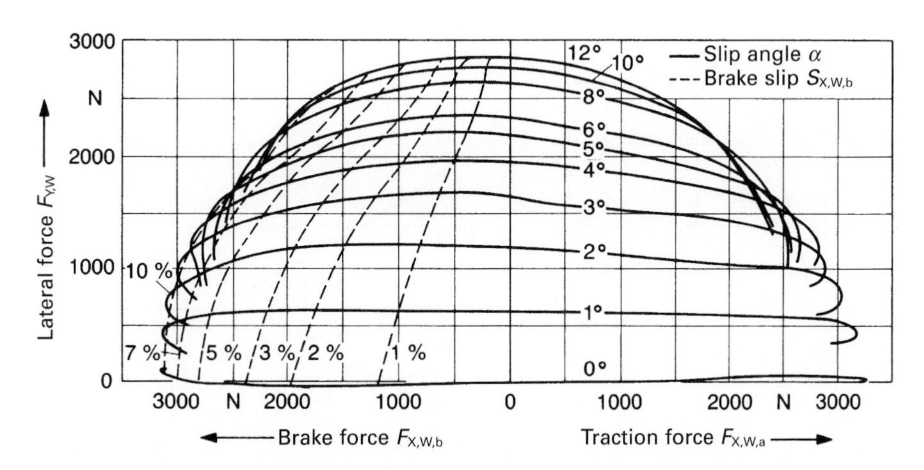

Fig. 10.1-48 Tyre-tangential lateral force performance characteristics with slip angles and brake slip as parameters. The study was carried out on a 18565 R 14 86 S radial tyre loaded at 300 kg at $p_T = 1.5$ bar. The shape of the curves indicates that, with increasing longitudinal forces, those which can be absorbed laterally reduce. At 1.5 bar, the tyre carries a weight of 350 kg, i.e. it is only operating at 86% capacity.

10.1.10 Tyre self-aligning torque and caster offset

10.1.10.1 Tyre self-aligning torque in general

The focal point of the force of the tyre contact patch lies behind the middle of the wheel because of its load- and lateral-force-related deformation. As a result, the point of application of the lateral force alters by the amount $r_{\tau,T}$, known as the caster offset, and comes to lie behind the centre of the wheel (Fig. 10.1-49). On the front wheels, the lateral cornering force $F_{Y,W,f}$ together with $r_{\tau,T}$ (as the force lever) gives the self-aligning moment $M_{Z,T,Y}$ which superimposes the kinematic alignment torque and seeks to bring the input wheels back to a straight position.

The self-aligning torque, lateral force and slip angle are measured in one process on the test rig. $M_{Z,T,Y}$ is plotted as a function of the slip angle (Fig. 10.1-49), the vertical force $F_{Z,W}$ serves as a parameter. The higher $F_{Z,W}$, the greater the self-alignment and, just like the lateral force, the moment increases to a maximum and then falls again. $M_{Z,T,Y,max}$ is, however, already at $\alpha \approx 4°$ (as can be seen in Fig. 10.1-43) and not, on a dry road, at $\alpha \geq 10°$.

10.1.10.2 Caster offset

Caster offset, $r_{\tau,T}$, is included in practically all calculations of the self-aligning moment during cornering. The length of this can easily be calculated from the lateral force and moment:

$$r_{\tau,T} = M_{Z,T,Y}/F_{Y,W} \text{ (m)} \tag{10.1.20}$$

This requires two images, one which represents $F_{Y,W} = f(F_{Z,W}$ and $\alpha)$ or $\mu_{Y,W} = f(F_{Z,W}$ and $\alpha)$, and another with $M_{Z,T,Y} = f(F_{Z,W}$ and $\alpha)$. The values of the 175/70R 13 82 S steel radial tyre shown in Figs. 10.1-45 and 10.1-49 and measured at $p_T = 2.0$ bar serve as an example. At $\alpha = 2°$ and $F_{Z,W} = 5.0$ kN the coefficient of friction $\mu_{y,w} = 0.44$ and therefore:

$$F_{Y,W} = \mu_{Y,W} \times F_{Z,W} = 0.44 \times 5.0 = 2.2 \text{ kN} = 2200 \text{ N}$$

At the same angle and with the same wheel force, the self-aligning torque is $M_{Z,T,Y} = 95$ N m and therefore:

$$r_{\tau,T} = M_{Z,T,Y}/F_{Y,W} = 95/2200 = 0.043 \text{ m} = 43 \text{ mm}$$

Figure 10.1-50 shows the caster (caster offset trail) calculated in this manner. Higher lateral forces necessitate

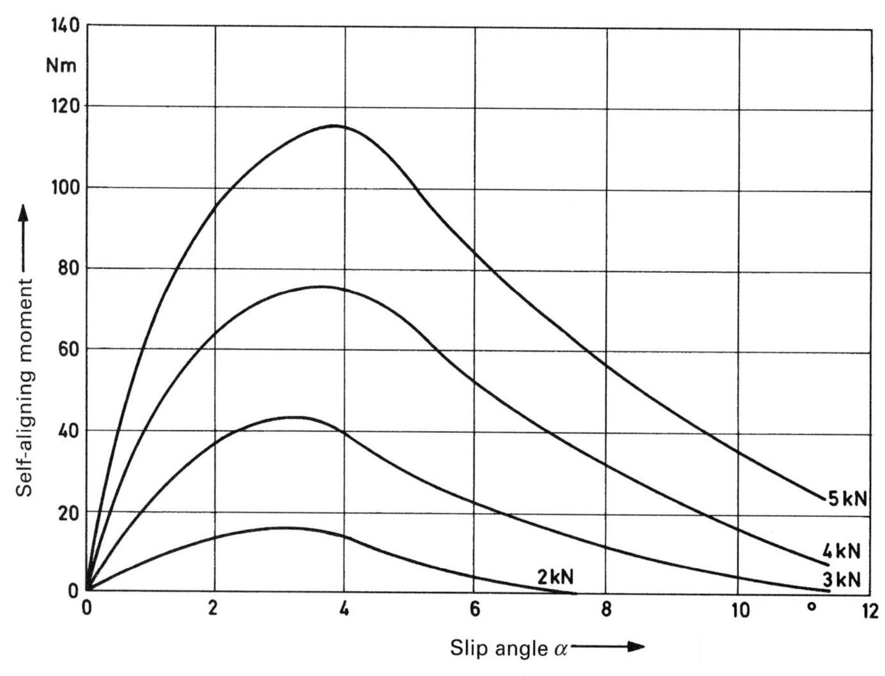

Fig. 10.1-49 Self-aligning torques of a 175/70 R 13 82 S steel radial tyre measured on a dry drum as a function of the slip angle at $p_T = 2.0$ bar. The vertical force $F_{Z,W}$ in kilonewtons is used as a parameter. The torques increase sharply at low angles, reach a maximum at $\alpha = 3$–4° and then reduce slowly. As the cornering speed increases, the tyre self-aligning torque decreases, while the kinematically determined torque increases.

greater slip angles, and the latter result in smaller self-aligning moments and a reduced caster offset. The explanation for this fact is that, at low slip angles, only the tyre profile is deformed at the area of contact. The point of application of the lateral force can therefore move further back, unlike large angles where, principally, the carcass is deformed. High vertical wheel forces cause the tyre to be severely compressed and therefore an increase both in the area of tyre contact and also in the caster offset occur.

10.1.10.3 Influences on the front wheels

The tyre self-aligning torque is one of the causes for the steering forces during cornering; its level depends on various factors.

10.1.10.3.1 Dry roads

The self-aligning torque is usually measured on a roller test bench with the drum allowing a coefficient of friction of $\mu_0 = 0.8$–0.9 between its surface and the tyre. If the resultant self-aligning torque on the open road is required, it is possible to approximate the value $M_{Z,T,Y,\mu}$ using a correction factor:

$$k_\mu = \mu_{Y,W}/\mu_0 \qquad (10.1.21)$$

A cement block with $\mu_{Y,W} \sim 1.05$ (Fig. 10.1-43) and the 175/70 R 13 82 S radial tyre can be used as an example. In accordance with Fig. 10.1-49,

$$M_{Z,T,Y} = 40\ \text{N m with } F_{Z,W} = 3\ \text{kN and } \alpha = 4^\circ$$

As a correction factor this gives

$$k_\mu = \mu\frac{\text{road}}{\text{roller}} = \frac{\mu_{Y,W}}{\mu_0} = \frac{1.05}{0.80} = 1.31$$

and thus

$$M_{Z,T,Y,\mu} = k_\mu \times M_{Z,T,Y} = 1.31 \times 40 = 52.4\ \text{Nm}$$

10.1.10.3.2 Wet roads

Provided that k_μ is independent of tyre construction and profile, the approximate value for a wet road can also be determined. In accordance with Fig. 10.1-47, with 1 mm of water on the surface and full profile depth the $\mu_{Y,W}$ value reduces from 0.86 to 0.55. Owing to the reduced coefficient of friction, only a smaller value $M_{Z,T,Y,\mu}$, can be assumed; in other words,

$$k_\mu = \mu_{Y,W}\frac{\text{wet}}{\text{roller}} = \frac{0.55}{0.86} = 0.64, \text{ and}$$
$$M_{Z,T,Y,\mu} = 0.64 \times 40\ Nm = 25.6\ Nm$$

A greater water film thickness may cause the coefficient of friction to reduce but the self-aligning moment increases and the water turns the wheel back into the straight position. Furthermore, the self-aligning maximum shifts towards smaller slip angles when the road is wet.

10.1.10.3.3 Icy roads

Only with greater vertical forces and small slip angles is the smoothness of the ice able to deform the area of tyre contact and generate an extremely small moment, which is nevertheless sufficient to align the tyre. Low front-axle loads or greater angles α arising as a result of steering corrections would result in a negative moment $-M_{Z,T,Y}$ (in other words in a 'further steering input' of the tyres). The wheel loads at the front, which were only low, were already a problem on rear-engine passenger vehicles.

10.1.10.3.4 Longitudinal forces

Traction forces increase the self-aligning torque; the equation for one wheel is

$$M_{Z,W,a} = F_{Y,W} \cdot r_{\tau,T} + F_{X,W,a} \cdot r_T = F_{z,W}(\mu_{Y,W} \cdot r_{\tau,T} + \mu_{X,W} \cdot r_T) \qquad (10.1.22)$$

During braking the moment fades and reduces to such an extent that it even becomes negative and seeks to input the wheels further. The formula for one wheel is

$$M_{Z,W,b} = F_{Y,W} \cdot r_{\tau,T} - F_{X,W,b} \cdot r_T = F_{Z,W}(\mu_{Y,W} \cdot r_{\tau,T} - \mu_{X,W} \cdot r_T) \qquad (10.1.23)$$

10.1.10.3.5 Tyre pressure

When the tyre pressure is increased the self-aligning torque reduces by 6–8% per 0.1 bar, and increases accordingly when the pressure reduces, by 9–12% per 0.1 bar

A reduction in pressure of, for example, 0.5 bar could thus result in over a 50% increase in the moment, a value which the driver would actually be able to feel.

10.1.10.3.6 Further influences

The following have only a slight influence:

- positive camber values increase the torque slightly, whereas negative ones reduce it;
- $M_{Z,T,Y}$ falls as speeds increase because the centrifugal force tensions the steel belt which becomes more difficult to deform (Fig. 10.1-16);
- widening the wheel rim width slightly reduces self-alignment.

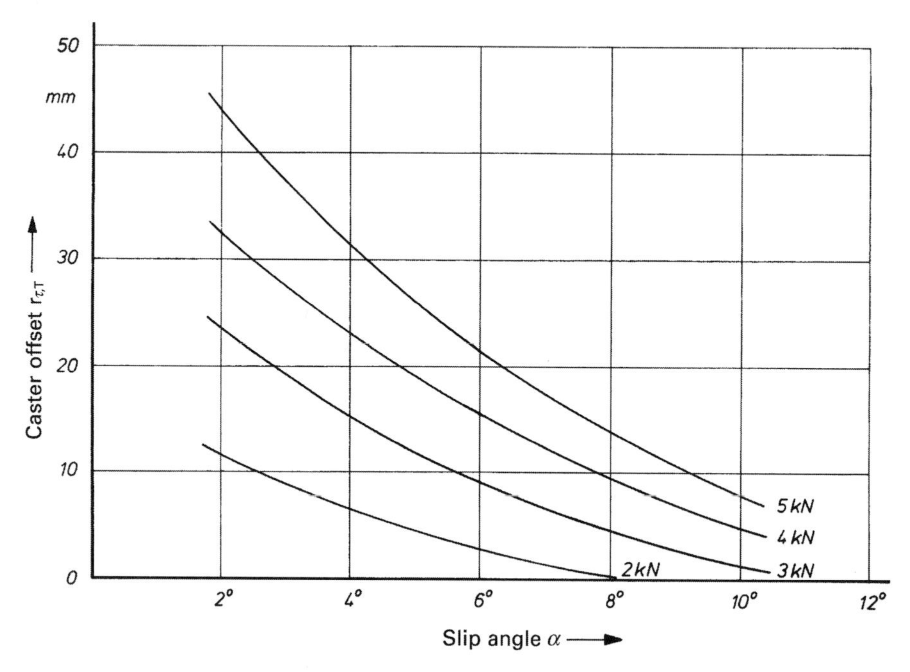

Fig. 10.1-50 Caster offset of tyre $r_{\tau,T}$ calculated from Figs. 10.1-45 and 10.1-49 for 175/70 R 13 82 S steel radial tyres at $p_T = 2.0$ bar. The higher the vertical force $F_{Z,W}$ (in kN) and the smaller the angle α, the longer is $r_{\tau,T}$.

10.1.11 Tyre overturning moment and displacement of point of application of force

A tyre which runs subject to lateral forces on the tyre contact patch is subject to deformation; there is a lateral displacement between the point of application of the normal force (wheel load; Fig. 10.1-49) and the centre plane of the wheel. Figure 10.1-51 shows the lateral drift of the normal (wheel load) point of application which is dependent on the size of the tyre, the lateral force and the camber angle and to a large extent on the construction of the tyre. Low section tyres with a small height-to-width ratio and a high level of sidewall rigidity exhibit greater lateral displacement. The roll-over resistance of the vehicle is considerably reduced, as there is a decrease in the distance between the point

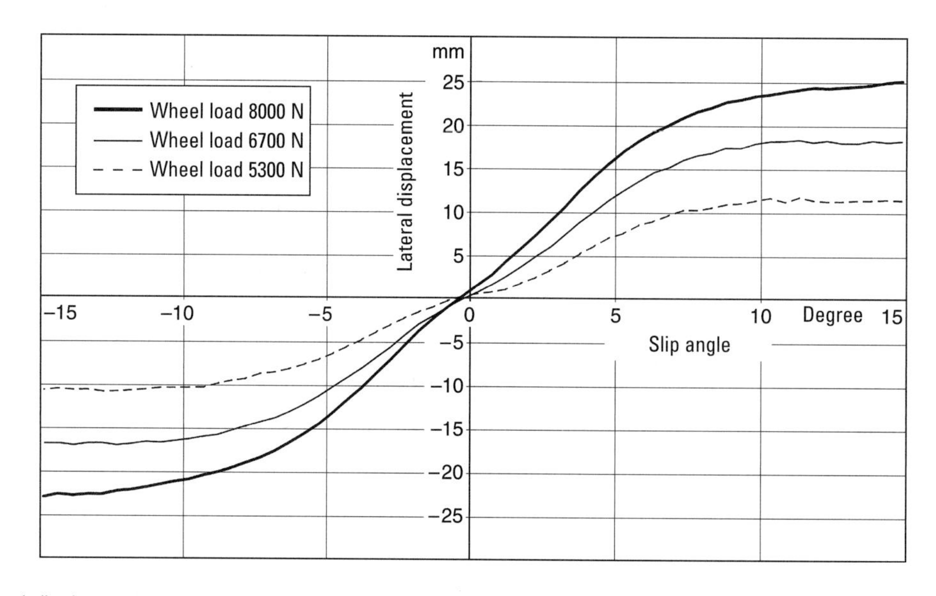

Fig. 10.1-51 Lateral displacement of normal (wheel load) point of application depending on slip angle and wheel load; measurements by Continental on a tyre of type 205/65 R 15 94 V ContiEcoContact CP.

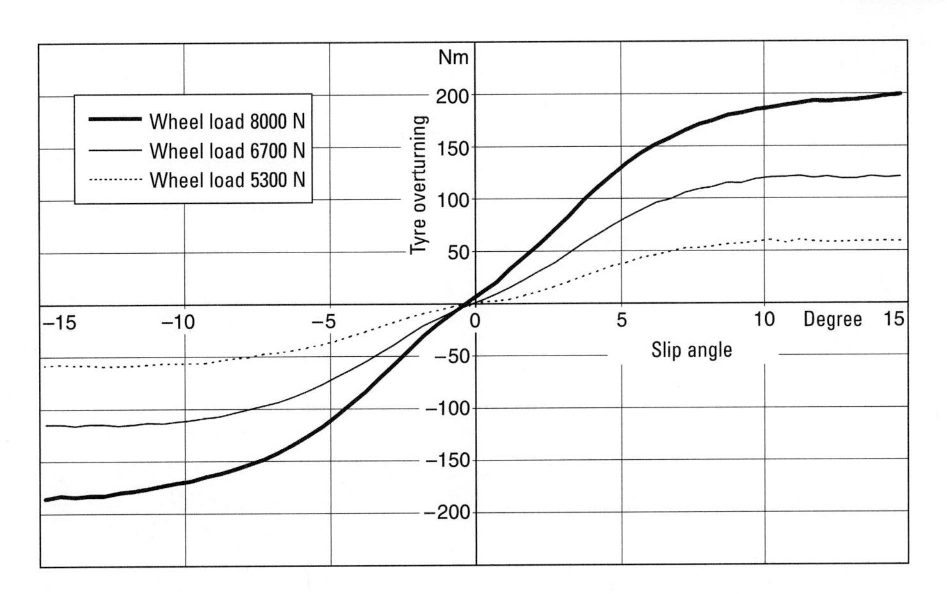

Fig. 10.1-52 Tyre overturning moments $M_{X,T,\alpha}$ on the wheel as a result of the buildup of lateral forces at different slip angles and wheel loads $F_{Z,W}$; measurements by Continental on a tyre of type 205/65 R 15 94 V ContiEcoContact CP.

of contact of the wheel and the centre of gravity of the vehicle.

This displacement results in the emergence of tyre overturning moments $M_{X,T,\alpha}$ about the longitudinal axis of the tyre (Fig. 10.1-52).

Both the lateral displacement of the point of application of the normal force and the tyre overturning moments must be taken into account when considering the overturning behaviour of vehicles, as they can considerably reduce rollover resistance, if, for example, a vehicle has a high centre of gravity and a small track dimension.

10.1.12 Torque steer effects

Torque steer effects, i.e. changes in longitudinal forces during cornering, are an important criterion for the definition of transient handling characteristics. The torque

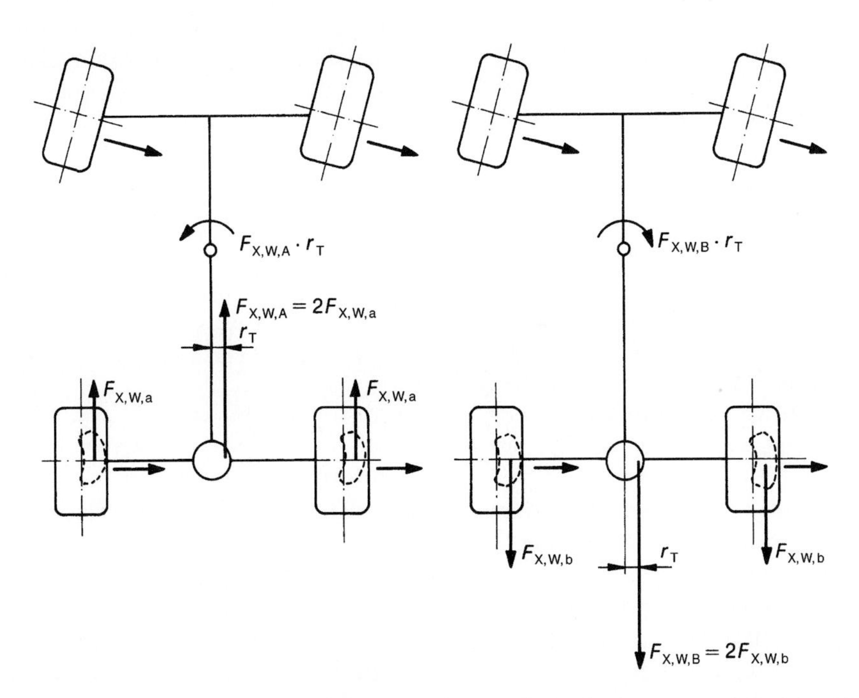

Fig. 10.1-53 The deformation of the tyre contact area during cornering results in aligning torque of the lateral forces which is further intensified by tractive forces and produces an understeering yawing moment. If there is a change in load, the braking forces produce an oversteering yawing moment.

steer effects depend on the size of the change in the longitudinal force, the adherence potential between the tyres and the road, the tyres and the kinematic and elastokinematic chassis design.

10.1.12.1 Torque steer effects as a result of changes in normal force

Torque steer effects usually occur during cornering when a driver has to slow down on a wrongly assessed bend by reducing the amount of acceleration or applying the brake.

The reaction force acting at the centre of gravity of the vehicle causes an increase in front axle load with a simultaneous reduction in the load on the rear axle. At an initially unchanged slip angle, the distribution of lateral forces changes as a result. If the force coefficient relating to the simultaneous transfer of longitudinal and transverse forces is sufficient, e.g. in the case of torque steer effects owing to reduction in acceleration or gentle braking (cf. Fig. 10.1-48), the increased lateral force corresponding to the increase in normal force on the front axle results in a yawing moment which allows the vehicle to turn into the bend.

If the adhesion potential is exceeded as a result of fierce braking or a low force coefficient, the tyres are no longer able to build up the necessary lateral forces. This results in an over- or understeering vehicle response depending on the specific case, be it a loss of lateral force on the front axle or rear axle or both.

10.1.12.2 Torque steer effects resulting from tyre aligning torque

The lateral displacement of the tyre contact area as a result of lateral forces leads to longitudinal forces being applied outside the centre plane of the wheel (Fig. 10.1-53).

This effect causes an increase in tyre aligning torque in driven wheels. In rear-wheel drive vehicles, this torque has an understeering effect with tractive forces, whereas it has an oversteering effect where there is a change in braking power.

In front-wheel drive vehicles, the resultant tractive force vector applies about lever arm $l_f \times \sin \delta_f$ offset from the centre of gravity of the vehicle (Fig. 10.1-54), so that an oversteering yawing moment is produced during driving which alters with application of a braking force to a (small) understeering yawing moment.

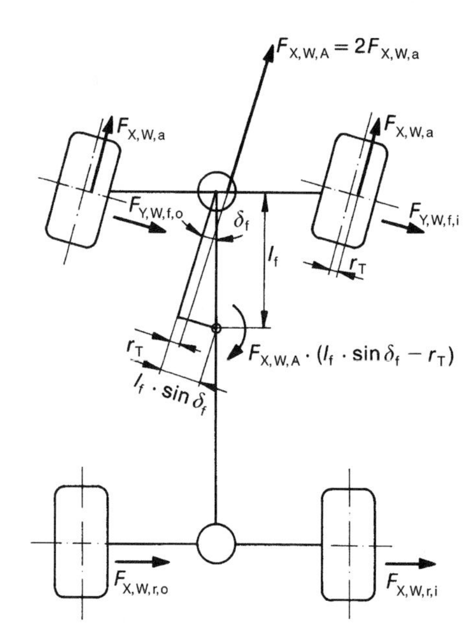

Fig. 10.1-54 With front-wheel drive, an oversteering yawing moment is produced, because the resultant tractive force vector is applied about lever arm l_f X sin δ_f displaced to the centre of gravity of the vehicle.

10.1.12.3 Effect of kinematics and elastokinematics

An attempt is made to keep the torque steer effects of a vehicle low by means of specific chassis design. The above-mentioned changes in forces produce bump and rebound travel movements on the axles. The results, depending on the design of the chassis, in kinematic and elastokinematic toe-in and camber changes which can be used to compensate for unwanted changes in lateral forces, particularly in the case of multi-link suspensions. With unfavourable axle design and construction, there is, however, also the possibility of an increase in the torque steer effects.

LI = 101 corresponds to 825 kg,
LI = 102 corresponds to 850 kg etc. to
LI = 108 corresponds to 1000 kg.

- rim horns, which form the lateral seat for the tyre bead (the distance between the two rims is the jaw width a);
- rim shoulders, the seat of the beads, generally inclined at 5° ± 1° to the centre where the force transfer occurs around the circumference (Fig. 10.1-5);
- well base (also known as the inner base), designed as a drop rim to allow tyre fitting, and mostly shifted to the outside (diagram: Hayes Lemmerz).

Section Eleven

Handling

Chapter 11.1

11.1

Tyre characteristics and vehicle handling and stability

Hans Pacejka

11.1.1 Introduction

This chapter is meant to serve as an introduction to vehicle dynamics with emphasis on the influence of tyre properties. Steady-state cornering behaviour of simple automobile models and the transient motion after small and large steering inputs and other disturbances will be discussed. The effects of various shape factors of tyre characteristics (cf. Fig. 11.1-1) on vehicle handling properties will be analysed. The slope of the side force F_y vs slip angle α near the origin (the cornering or side slip stiffness) is the determining parameter for the basic linear handling and stability behaviour of automobiles. The possible offset of the tyre characteristics with respect to their origins may be responsible for the occurrence of the so-called tyre-pull phenomenon. The further non-linear shape of the side (or cornering) force characteristic governs the handling and stability properties of the vehicle at higher lateral accelerations. The load dependency of the curves, notably the non-linear relationship of cornering stiffness with tyre normal load has a considerable effect on the handling characteristic of the car. For the (quasi) steady-state handling analysis simple single-track (two-wheel) vehicle models will be used. Front- and rear-axle effective side force characteristics are introduced to represent effects that result from suspension and steering system design factors such as steering compliance, roll steer and lateral load transfer. Also the effect of possibly applied (moderate) braking and driving forces may be incorporated in the effective characteristics. Large braking forces may result in wheel lock and possibly large deviations from the undisturbed path. The application of the handling and stability theory to the dynamics of heavy trucks will also be briefly dealt with in the present chapter. Special attention will be

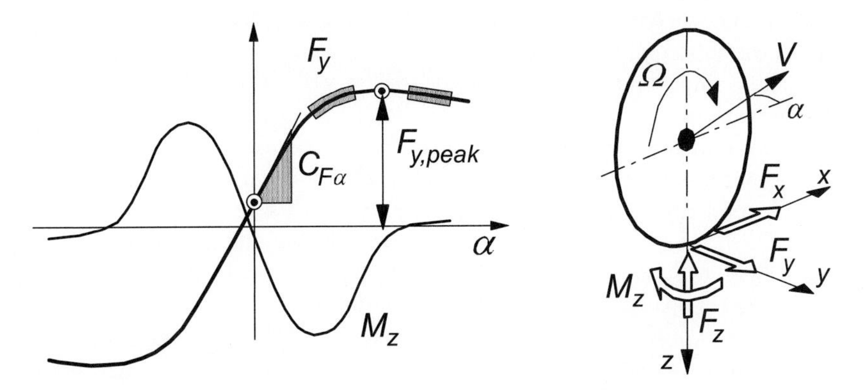

Fig. 11.1-1 Characteristic shape factors (indicated by points and shaded areas) of tyre or axle characteristics that may influence vehicle handling and stability properties. Slip angle and force and moment positive directions.

Tyre and Vehicle Dynamics; ISBN: 9780750669184
Copyright © 2005 Hans Pacejka. All rights of reproduction, in any form, reserved.

given to the phenomenon of oscillatory instability that may show up with the car trailer combination.

When the wavelength of an oscillatory motion of the vehicle that may arise from road unevenness, brake torque fluctuations, wheel unbalance or instability (shimmy), is smaller than say 5 m, a non-steady-state or transient description of tyre response is needed to properly analyse the phenomenon. Applications demonstrate the use of transient and oscillatory tyre models and provide insight into the vehicle dynamics involved.

11.1.2 Tyre and axle characteristics

Tyre characteristics are of crucial importance for the dynamic behaviour of the road vehicle. In this section an introduction is given to the basic aspects of the force and moment generating properties of the pneumatic tyre. Both the pure and combined slip characteristics of the tyre are discussed and typical features presented. Finally, the so-called effective axle characteristics are derived from the individual tyre characteristics and the relevant properties of the suspension and steering system.

11.1.2.1 Introduction to tyre characteristics

The upright wheel rolling freely, that is without applying a driving torque, over a flat level road surface along a straight line at zero side slip, may be defined as the starting situation with all components of slip equal to zero. A relatively small pulling force is needed to overcome the tyre rolling resistance and a side force and (self) aligning torque may occur as a result of the not completely symmetric structure of the tyre. When the wheel motion deviates from this by definition zero-slip condition, wheel slip occurs that is accompanied by a build-up of additional tyre deformation and possibly partial sliding in the contact patch. As a result, (additional) horizontal forces and the aligning torque are generated. The mechanism responsible for this is treated in detail in the subsequent chapters. For now, we will suffice with some important experimental observations and define the various slip quantities that serve as inputs into the tyre system and the moment and forces that are the output quantities (positive directions according to Fig. 11.1-1). Several alternative definitions are in use as well.

For the freely rolling wheel the forward speed V_x (longitudinal component of the total velocity vector V of the wheel centre) and the angular speed of revolution Ω_o can be taken from measurements. By dividing these two quantities the so-called effective rolling radius r_e is obtained:

$$r_e = \frac{V_x}{\Omega_o} \tag{11.1.1}$$

Although the effective radius may be defined also for a braked or driven wheel, we restrict the definition to the case of free rolling. When a torque is applied about the wheel spin axis a longitudinal slip arises that is defined as follows:

$$\kappa = -\frac{V_x - r_e\Omega}{V_x} = -\frac{\Omega_o - \Omega}{\Omega_o} \tag{11.1.2}$$

The sign is taken such that for a positive κ a positive longitudinal force F_x arises, that is: a driving force. In that case, the wheel angular velocity Ω is increased with respect to Ω_o and consequently $\Omega > \Omega_o = V_x/r_e$. During braking, the fore and aft slip becomes negative. At wheel lock, obviously, $\kappa = -1$. At driving on slippery roads, κ may attain very large values. To limit the slip to a maximum equal to one, in some texts the longitudinal slip is defined differently in the driving range of slip: in the denominator of (11.1.2) Ω_o is replaced by Ω. This will not be done in the present text.

Lateral wheel slip is defined as the ratio of the lateral and the forward velocity of the wheel. This corresponds to minus the tangent of the slip angle α (Fig. 11.1-1). Again, the sign of α has been chosen such that the side force becomes positive at positive slip angle:

$$\tan\alpha = -\frac{V_y}{V_x} \tag{11.1.3}$$

The third and last slip quantity is the so-called spin which is due to rotation of the wheel about an axis normal to the road. Both the yaw rate resulting in path curvature when α remains zero, and the wheel camber or inclination angle γ of the wheel plane about the x axis contribute to the spin. The camber angle is defined positive when looking from behind the wheel is tilted to the right. The forces F_x and F_y and the aligning torque M_z are results of the input slip. They are functions of the slip components and the wheel load. For steady-state rectilinear motions we have in general:

$$F_x = F_x(\kappa,\alpha,\gamma,F_z),\ F_y = F_y(\kappa,\alpha,\gamma,F_z),$$
$$M_z = M_z(\kappa,\alpha,\gamma,F_z) \tag{11.1.4}$$

The vertical load F_z may be considered as a given quantity that results from the normal deflection of the tyre. The functions can be obtained from measurements for a given speed of travel and road and environmental conditions.

Fig. 11.1-1 shows the adopted system of axes (x, y, z) with associated positive directions of velocities and forces and moments. The exception is the vertical force F_z acting from road to tyre. For practical reasons, this force is defined to be positive in the upward direction and thus equal to the normal load of the tyre. Also Ω (not provided with a y subscript) is defined positive with respect to the negative y axis. Note, that the axes system is in accordance with SAE standards (SAE J670e 1976). The sign of the slip angle, however, is chosen opposite with respect to the SAE definition.

In Fig. 11.1-2 typical pure lateral ($\kappa = 0$) and longitudinal ($\alpha = 0$) slip characteristics have been depicted together with a number of combined slip curves. The camber angle γ was kept equal to zero. We define pure slip to be the situation when either longitudinal or lateral slip occurs in isolation. The figure indicates that a drop in force arises when the other slip component is added. The resulting situation is designated as combined slip. The decrease in force can be simply explained by realising that the total horizontal frictional force F cannot exceed the maximum value (radius of 'friction circle') which is dictated by the current friction coefficient and normal load. The diagrams include the situation when the brake slip ratio has finally attained the value 100% ($\kappa = -1$) which corresponds to wheel lock.

The slopes of the pure slip curves at vanishing slip are defined as the longitudinal and lateral slip stiffnesses respectively. The longitudinal slip stiffness is designated as $C_{F\kappa}$. The lateral slip or cornering stiffness of the tyre, denoted with $C_{F\alpha}$, is one of the most important property parameters of the tyre and is crucial for the vehicle's handling and stability performance. The slope of minus the aligning torque vs slip angle curve (Fig. 11.1-1) at zero slip angle is termed as the aligning stiffness and is denoted with $C_{M\alpha}$. The ratio of minus the aligning torque and the side force is the pneumatic trail t. This length is the distance behind the contact centre (projection of wheel centre onto the ground in wheel plane direction) to the point where the resulting lateral force acts. The linearised force and moment characteristics (valid at small levels of slip) can be represented by the following expressions in which the effect of camber has been included:

$$\begin{aligned} F_x &= C_{F\kappa}\kappa \\ F_y &= C_{F\alpha}\alpha + C_{F\gamma}\gamma \\ M_z &= -C_{M\alpha}\alpha + C_{M\gamma}\gamma \end{aligned} \qquad (11.1.5)$$

These equations have been arranged in such a way that all the coefficients (the force and moment slip and camber stiffnesses) become positive quantities.

It is of interest to note that the order of magnitude of the tyre cornering stiffness ranges from about 6 to about 30 times the vertical wheel load when the cornering stiffness is expressed as force per radian. The lower value holds for the older bias-ply tyre construction and the larger value for modern racing tyres. The longitudinal slip stiffness has been typically found to be about 50% larger than the cornering stiffness. The pneumatic trail is approximately equal to a quarter of the contact patch length. The dry friction coefficient usually equals ca. 0.9, on very sharp surfaces and on clean glass ca. 1.6; racing tyres may reach 1.5 to 2.

For the side force which is the more important quantity in connection with automobile handling properties, a number of interesting diagrams have been presented in Fig. 11.1-3. These characteristics are typical for truck and car tyres and are based on experiments conducted at the University of Michigan Transportation Research Institute (UMTRI, formerly HSRI). The car tyre cornering stiffness data stem from newer findings. It is seen that the cornering stiffness changes in a less than proportional fashion with the normal wheel load. The maximum normalised side force $F_{y,peak}/F_z$ appears to decrease with increasing wheel load. Marked differences in level and slope occur for the car and truck tyre curves also when normalised with respect to the rated or nominal load. The cornering force vs slip angle characteristic shown at different speeds and road conditions indicate that the slope at zero slip angle is not or hardly affected by the level of speed and by the condition wet or dry. The peak force level shows only little variation if the road is dry. On a wet road a more pronounced peak occurs and the peak level drops significantly with increasing speed.

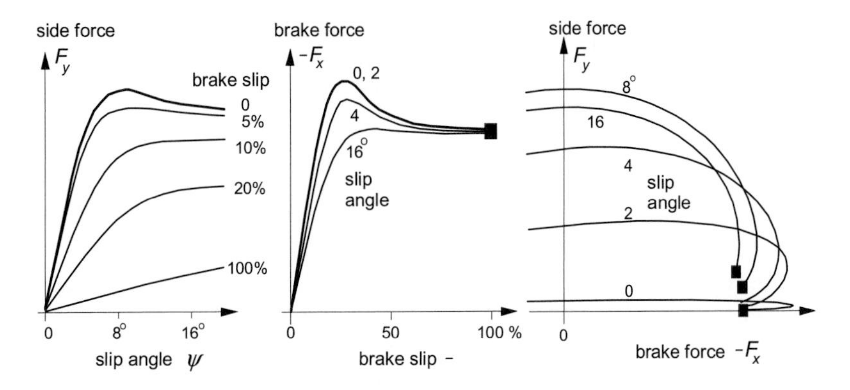

Fig. 11.1-2 Combined side force and brake force characteristics.

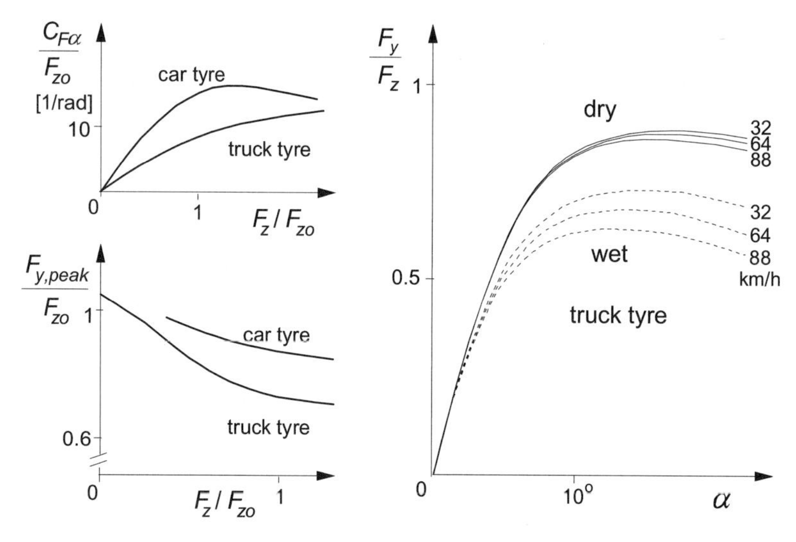

Fig. 11.1-3 Typical characteristics for the normalised cornering stiffness, peak side force and side force vs normalised vertical load and slip angle, respectively. F_{zo} is the rated load.

Curves which exhibit a shape like the side force characteristics of Fig. 11.1-3 can be represented by a mathematical formula that has become known by the name *'Magic Formula'*. The basic expressions for the side force and the cornering stiffness are:

$$F_y = D \sin[C \arctan\{B\alpha - E(B\alpha - \arctan(B\alpha))\}]$$

with stiffness factor

$$B = C_{F\alpha}/(CD)$$

peak factor

$$D = \mu F_z \ (= F_{y,peak}) \tag{11.1.6}$$

and cornering stiffness

$$C_{F\alpha} \ (= BCD) = c_1 \sin\{2 \arctan(F_z/c_2)\}$$

The shape factors C and E as well as the parameters c_1 and c_2 and the friction coefficient μ (possibly depending on the vertical load and speed) may be estimated or determined through regression techniques.

11.1.2.2 Effective axle cornering characteristics

For the basic analysis of (quasi) steady-state turning behaviour a simple two-wheel vehicle model may be used successfully. Effects of suspension and steering system kinematics and compliances such as steer compliance, body roll and also load transfer may be taken into account by using effective axle characteristics. The restriction to (quasi) steady state becomes clear when we realise that for transient or oscillatory motions, exhibiting yaw and roll accelerations and differences in phase, variables like roll angle and load transfer can no longer be written as direct algebraic functions of one of the lateral axle forces (front or rear). Consequently, we should drop the simple method of incorporating the effects of a finite centre of gravity height if the frequency of input signals such as the steering wheel angle cannot be considered small relative to the body roll natural frequency. Since the natural frequency of the wheel suspension and steering systems are relatively high, the restriction to steady-state motions becomes less critical in case of the inclusion of e.g. steering compliance in the effective characteristic. Chiesa and Rinonapoli were among the first to employ effective axle characteristics or 'working curves' as these were referred to by them. Vågstedt determined these curves experimentally.

Before assessing the complete non-linear effective axle characteristics we will first direct our attention to the derivation of the effective cornering stiffnesses which are used in the simple linear two-wheel model. For these to be determined, a more comprehensive vehicle model has to be defined.

Fig. 11.1-4 depicts a vehicle model with three degrees of freedom. The forward velocity u may be kept constant. As motion variables we define the lateral velocity v of reference point A, the yaw velocity r and the roll angle φ. A moving axes system (A,x,y,z) has been introduced. The x axis points forwards and lies both in the ground plane and in the plane normal to the ground that passes through the so-called roll axis. The y axis points to the right and the z axis points downwards. This latter axis passes through the centre of gravity when the roll angle is equal to zero. In this way the location of the point of reference A has been defined. The longitudinal distance to the front axle is a and the distance to the rear axle is b.

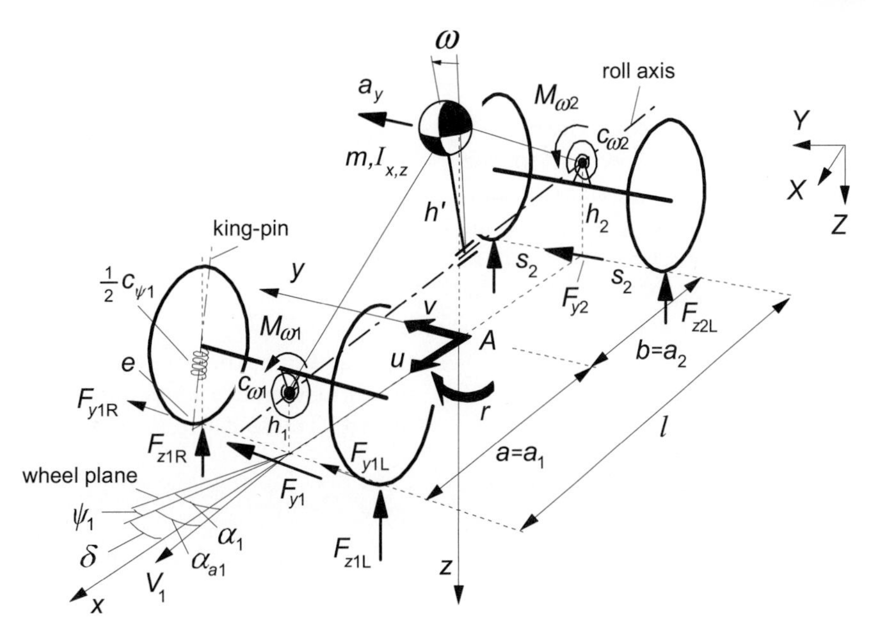

Fig. 11.1-4 Vehicle model showing three degrees of freedom: lateral, yaw and roll.

The sum of the two distances is the wheel base l. For convenience we may write: $a = a_1$ and $b = a_2$.

In a curve, the vehicle body rolls about the roll axis. The location and attitude of this virtual axis is defined by the heights $h_{1,2}$ of the front and rear roll centres. The roll axis is assessed by considering the body motion with respect to the four contact centres of the wheels on the ground under the action of an external lateral force that acts on the centre of gravity. Due to the symmetry of the vehicle configuration and the linearisation of the model these locations can be considered as fixed. The roll centre locations are governed by suspension kinematics and possibly suspension lateral compliances. The torsional springs depicted in the figure represent the front and rear roll stiffnesses $c_{\varphi 1,2}$ which result from suspension springs and anti-roll bars.

The fore and aft position of the centre of gravity of the body is defined by a and b; its height follows from the distance h' to the roll axis. The body mass is denoted by m and the moments of inertia with respect to the centre of mass and horizontal and vertical axes by I_x, I_z and I_{xz}. These latter quantities will be needed in a later phase when the differential equations of motion are established. The unsprung masses will be neglected or they may be included as point masses attached to the roll axis and thus make them part of the sprung mass, that is, the vehicle body.

Furthermore, the model features torsional springs around the steering axes. The king-pin is positioned at a small caster angle that gives rise to the caster length e as indicated in the drawing. The total steering torsional stiffness, left plus right, is denoted by $c_{\psi 1}$.

Effective axle cornering stiffness

Linear analysis, valid for relatively small levels of lateral accelerations allows the use of approximate tyre characteristics represented by just the slopes at zero slip. We will first derive the effective axle cornering stiffness that may be used under these conditions. The effects of load transfer, body roll, steer compliance, side force steer and initial camber and toe angles will be included in the ultimate expression for the effective axle cornering stiffness.

The linear expressions for the side force and the aligning torque acting on a tyre have been given by Eqs.(11.1.5). The coefficients appearing in these expressions are functions of the vertical load. For small variations with respect to the average value (designated with subscript o) we write for the cornering and camber force stiffnesses the linearised expressions:

$$\begin{aligned} C_{F\alpha} &= C_{F\alpha o} + \zeta_\alpha \Delta F_z \\ C_{F\gamma} &= C_{F\gamma o} + \zeta_\gamma \Delta F_z \end{aligned} \qquad (11.1.7)$$

where the increment of the wheel vertical load is denoted by ΔF_z and the slopes of the coefficient vs load curves at $F_z = F_{zo}$ are represented by $\zeta_{\alpha,\gamma}$.

When the vehicle moves steadily around a circular path, a centripetal acceleration a_y occurs and a centrifugal force $K = ma_y$ can be said to act on the vehicle body at the centre of gravity in the opposite direction. The body roll angle φ, that is assumed to be small, is calculated by dividing the moment about the roll axis by the apparent roll stiffness which is reduced with the term mgh' due to the additional moment $mgh'\varphi$:

$$\varphi = \frac{-ma_y h'}{c_{\varphi 1} + c_{\varphi 2} - mgh'} \qquad (11.1.8)$$

The total moment about the roll axis is distributed over the front and rear axles in proportion to the front and rear roll stiffnesses. The load transfer ΔF_{zi} from the inner to the outer wheels that occurs at axle i (=1 or 2) in a steady-state cornering motion with centripetal acceleration a_y follows from the formula:

$$\Delta F_{zi} = \sigma_i m a_y \qquad (11.1.9)$$

with the load transfer coefficient of axle i

$$\sigma_i = \frac{1}{2s_i}\left(\frac{c_{\varphi i}}{c_{\varphi 1} + c_{\varphi 2} - mgh'}h' + \frac{l - a_i}{l}h_i\right) \qquad (11.1.10)$$

The attitude angle of the roll axis with respect to the horizontal is considered small. In the formula, s_i denotes half the track width, h' is the distance from the centre of gravity to the roll axis and $a_1 = a$ and $a_2 = b$. The resulting vertical loads at axle i for the left (L) and right (R) wheels become after considering the left and right increments in load:

$$\begin{aligned} \Delta F_{ziL} &= \Delta F_{zi}, & \Delta F_{ziR} &= -\Delta F_{zi} \\ F_{ziL} &= \tfrac{1}{2}F_{zi} + \Delta F_{zi}, & F_{ziR} &= \tfrac{1}{2}F_{zi} - \Delta F_{zi} \end{aligned} \qquad (11.1.11)$$

The wheels at the front axle are steered about the king-pins with the angle δ. This angle relates directly to the imposed steering wheel angle δ_{stw} through the steering ratio n_{st}, that is:

$$\delta = \frac{\delta_{stw}}{n_{st}} \qquad (11.1.12)$$

In addition to this imposed steer angle the wheels may show a steer angle and a camber angle induced by body roll through suspension kinematics. The functional relationships with the roll angle may be linearised. For axle i we define:

$$\begin{aligned} \psi_{ri} &= \varepsilon_i \varphi \\ \gamma_{ri} &= \tau_i \varphi \end{aligned} \qquad (11.1.13)$$

Steer compliance gives rise to an additional steer angle due to the external torque that acts about the king-pin (steering axis). For the pair of front wheels this torque results from the side force (of course also from the here not considered driving or braking forces) that exerts a moment about the king-pin through the moment arm which is composed of the caster length e and the pneumatic trail t_1. With the total steering stiffness $c_{\psi 1}$ felt about the king-pins with the steering wheel held fixed, the additional steer angle becomes when for simplicity the influence of camber on the pneumatic trail is disregarded:

$$\psi_{c1} = -\frac{F_{y1}(e + t_1)}{C\Psi} \qquad (11.1.14)$$

In addition, the side force (but also the fore and aft force) may induce a steer angle due to suspension compliance. The so-called side force steer reads:

$$\psi_{sfi} = c_{sfi} F_{yi} \qquad (11.1.15)$$

For the front axle, we should separate the influences of moment steer and side force steer. For this reason, side force steer at the front is defined to occur as a result of the side force acting in a point on the king-pin axis.

Beside the wheel angles indicated above, the wheels may have been given initial angles that already exist at straight ahead running. These are the toe angle ψ_o (positive pointing outwards) and the initial camber angle γ_o (positive: leaning outwards). For the left and right wheels we have the initial angles:

$$\begin{aligned} \psi_{iLo} &= -\psi_{io}, & \psi_{iRo} &= \psi_{io} \\ \gamma_{iLo} &= -\gamma_{io}, & \gamma_{iRo} &= \gamma_{io} \end{aligned} \qquad (11.1.16)$$

Adding all relevant contributions (11.1.12) to (11.1.16) together yields the total steer angle for each of the wheels.

The effective cornering stiffness of an axle $C_{eff,i}$ is now defined as the ratio of the axle side force and the virtual slip angle. This angle is defined as the angle between the direction of motion of the centre of the axle i (actually at road level) when the vehicle velocity would be very low and approaches zero (then also $F_{yi} \to 0$) and the direction of motion at the actual speed considered. The virtual slip angle of the front axle has been indicated in Fig. 11.1-4 and is designated as α_{a1}. We have in general:

$$C_{eff,i} = \frac{F_{yi}}{\alpha_{ai}} \qquad (11.1.17)$$

The axle side forces in the steady-state turn can be derived by considering the lateral force and moment equilibrium of the vehicle:

$$F_{yi} = \frac{l - a_i}{l} m a_y \qquad (11.1.18)$$

The axle side force is the sum of the left and right individual tyre side forces. We have:

$$\begin{aligned} F_{yiL} &= (½C_{F\alpha i} + \zeta_{\alpha i}\Delta F_{zi})(\alpha_i - \psi_{io}) \\ &\quad + (½C_{F\gamma i} + \zeta_{\gamma i}\Delta F_{zi})(\gamma_i - \gamma_{io}) \\ F_{yiR} &= (½C_{F\alpha i} - \zeta_{\alpha i}\Delta F_{zi})(\alpha_i + \psi_{io}) \\ &\quad + (½C_{F\gamma i} - \zeta_{\gamma i}\Delta F_{zi})(\gamma_i + \gamma_{io}) \end{aligned} \qquad (11.1.19)$$

where the average wheel slip angle α_i indicated in the figure is:

$$\alpha_i = \alpha_{ai} + \psi_i \qquad (11.1.20)$$

and the average additional steer angle and the average camber angle are:

$$\begin{aligned} \psi_i &= \psi_{ri} + \psi_{ci} + \psi_{sfi} \\ \gamma_i &= \gamma_{ri} \end{aligned} \qquad (11.1.21)$$

The unknown quantity is the virtual slip angle α_{ai} which can be determined for a given lateral acceleration a_y. Next, we use the Eqs. (11.1.8), (11.1.9), (11.1.13)–(11.1.15) and (11.1.18), substitute the resulting expressions (11.1.21) and (11.1.20) in (11.1.19) and add up these two equations. The result is a relationship between the axle slip angle α_{ai} and the axle side force F_{yi}. We obtain for the slip angle of axle i:

$$\begin{aligned} \alpha_{ai} &= \frac{F_{yi}}{C_{eff,i}} \\ &= \frac{F_{yi}}{C_{F\alpha i}}\left(1 + \frac{l(\varepsilon_i C_{F\alpha i} + \tau_i C_{F\gamma i})h'}{(l - a_i)(c_{\varphi 1} + c_{\varphi 2} - mgh')} \right. \\ &\quad + \frac{C_{F\alpha i}(e_i + t_i)}{C_{\psi i}} - C_{F\alpha i}c_{sfi} \\ &\quad \left. + \frac{2l\sigma_i}{l - a_i}(\zeta_{\alpha i}\psi_{io} + \zeta_{\gamma i}\gamma_{io}) \right) \end{aligned} \qquad (11.1.22)$$

The coefficient of F_{yi} constitutes the effective axle cornering compliance, which is the inverse of the effective axle cornering stiffness (11.1.17). The quantitative effect of each of the suspension, steering and tyre factors included can be easily assessed. The subscript i refers to the complete axle. Consequently, the cornering and camber stiffnesses appearing in this expression are the sum of the stiffnesses of the left and right tyre:

$$\begin{aligned} C_{F\alpha i} &= C_{F\alpha iL} + C_{F\alpha iR} = C_{F\alpha iLo} + C_{F\alpha iRo} \\ C_{F\gamma i} &= C_{F\gamma iL} + C_{F\gamma iR} = C_{F\gamma iLo} + C_{F\gamma iRo} \end{aligned} \qquad (11.1.23)$$

in which (11.1.7) and (11.1.11) have been taken into account. The load transfer coefficient σ_i follows from Eq.(11.1.10). Expression (11.1.22) shows that the influence of lateral load transfer only occurs if initially, at straight ahead running, side forces are already present through the introduction of e.g. opposite steer and camber angles. If these angles are absent, the influence of load transfer is purely non-linear and is only felt at higher levels of lateral accelerations. In the next subsection, this non-linear effect will be incorporated in the effective axle characteristic.

Effective non-linear axle characteristics

To illustrate the method of effective axle characteristics we will first discuss the determination of the effective characteristic of a front axle showing steering compliance. The steering wheel is held fixed. Due to tyre side forces and self-aligning torques (left and right) distortions will arise resulting in an incremental steer angle ψ_{c1} of the front wheels (ψ_{c1} will be negative in Fig. 11.1-5 for the case of just steer compliance). Since load transfer is not considered in this example, the situation at the left and right wheels are identical (initial toe and camber angles being disregarded). The front tyre slip angle is denoted with α_1. The 'virtual' slip angle of the axle is denoted with α_{a1} and equals (cf. Fig. 11.1-5):

$$\alpha_{a1} = \alpha_1 - \psi_{c1} \qquad (11.1.24)$$

where both α_1 and ψ_{c1} are related with F_{y1} and M_{z1}. The subscript 1 refers to the front axle and thus to the pair of tyres. Consequently, F_{y1} and M_{z1} denote the sum of the left and right tyre side forces and moments. The objective is, to find the function $F_{y1}(\alpha_{a1})$ which is the effective front axle characteristic. Fig. 11.1-6 shows a graphical approach. According to Eq.(11.1.24) the points on the $F_{y1}(\alpha_{a1})$ curve must be shifted horizontally over a length ψ_{c1} to obtain the sought $F_{y1}(\alpha_{a1})$. The slope of the curve at the origin corresponds to the effective axle cornering stiffness found in the preceding subsection. Although the changes with respect to the original characteristic may be small, they can still be of considerable importance since it is the difference of slip angles front and rear which largely determines the vehicle's handling behaviour.

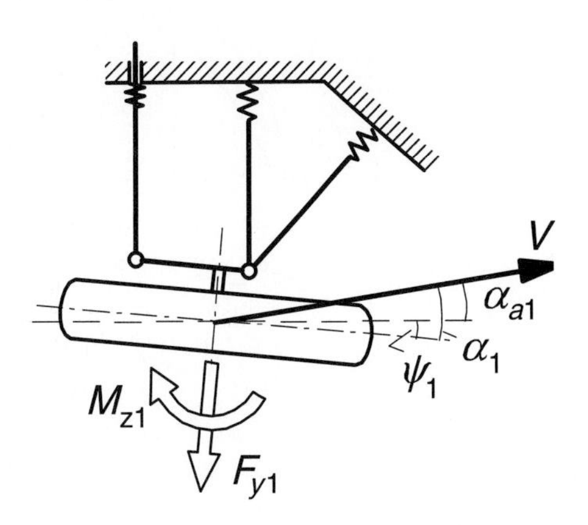

Fig. 11.1-5 Wheel suspension and steering compliance resulting in additional steer angle ψ_1.

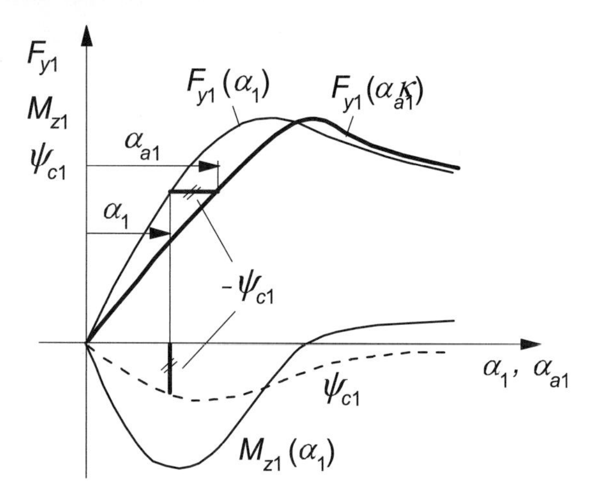

Fig. 11.1-6 Effective front axle characteristic $F_{y1}(\alpha_{a1})$ influenced by steering compliance.

The effective axle characteristic for the case of roll steer can be easily established by subtracting ψ_{ri} from α_i. Instead of using the linear relationships (11.1.8) and (11.1.13) non-linear curves may be adopted, possibly obtained from measurements. For the case of roll camber, the situation becomes more complex. At a given axle side force the roll angle and the associated camber angle can be found. The cornering characteristic of the pair of tyres at that camber angle is needed to find the slip angle belonging to the side force considered.

Load transfer is another example that is less easy to handle. In Fig. 11.1-7 a three-dimensional graph is presented for the variation of the side force of an individual tyre as a function of the slip angle and of the vertical load. The former at a given load and the latter at a given slip angle. The diagram illustrates that at load transfer the outer tyre exhibiting a larger load will generate a larger side force than the inner tyre. Because of the non-linear degressive F_y vs F_z curve, however, the average side force will be smaller than the original value it had in the absence of load transfer. The graph indicates that an increase $\Delta\alpha$ of the slip angle would be needed to compensate for the adverse effect of load transfer. The lower diagram gives a typical example of the change in characteristic as a result of load transfer. At the origin the slope is not affected but at larger slip angles an increasingly lower derivative appears to occur. The peak diminishes and may even disappear completely. The way to determine the resulting characteristic is the subject of the next exercise.

Exercise 11.1.1. Construction of effective axle characteristic at load transfer

For a series of tyre vertical loads F_z the characteristics of the two tyres mounted on, say, the front axle of an automobile are given. In addition, it is known how the load transfer ΔF_z at the front axle depends on the centrifugal force K ($= mgF_{y1}/F_{z1} = mg\,F_{y2}/F_{z2}$) acting at the centre of gravity. From this data the resulting cornering characteristic of the axle considered (at steady-state cornering) can be determined.

1. Find the resulting characteristic of one axle from the set of individual tyre characteristics at different tyre loads F_z and the load transfer characteristic (both shown in Fig. 11.1-8).

   Hint: First draw in the lower diagram the axle characteristics for values of ΔF_z = 1000, 2000, 3000 and 4000 N and then determine which point on each of these curves is valid considering the load transfer characteristic (left-hand diagram). Then, draw the resulting axle characteristic.

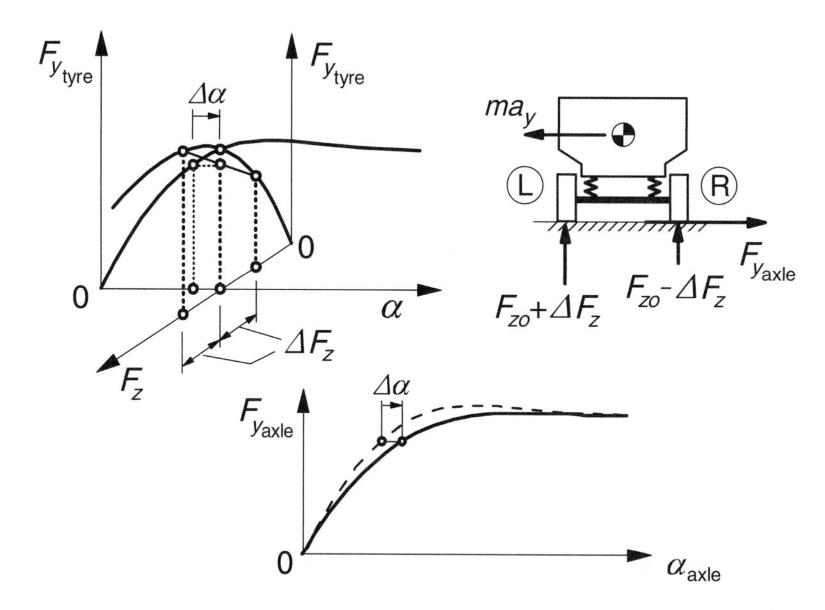

Fig. 11.1-7 The influence of load transfer on the resulting axle characteristic.

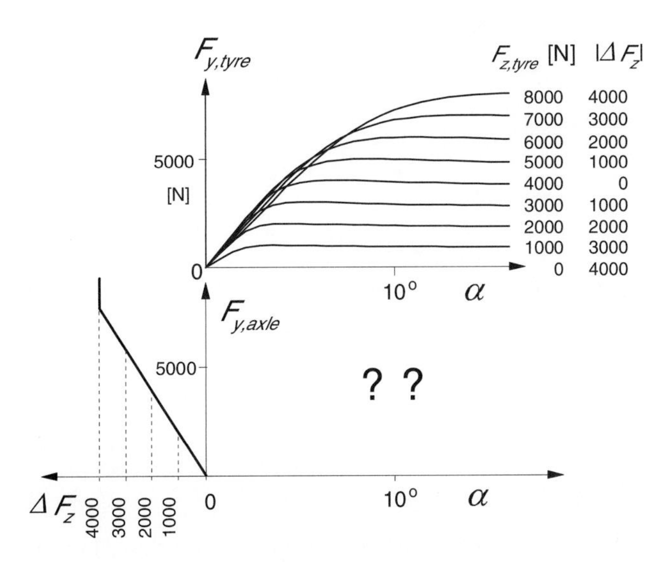

Fig. 11.1-8 The construction of the resulting axle cornering characteristics at load transfer (Exercise 11.1.1).

It may be helpful to employ the Magic Formula (11.1.6) and the parameters shown below:

side force: $F_y = D \sin[C \arctan\{B\alpha - E(B\alpha - \arctan(B\alpha))\}]$

with factors: $B = C_{F\alpha}/(CD)$, $C = 1.3$, $D = \mu F_z$, $E = -3$, with $\mu = 1$

cornering stiffness: $C_{F\alpha} = c_1 \sin[2 \arctan\{F_z/c_2\}]$

with parameters: $c_1 = 60000$ [N/rad], $c_2 = 4000$ [N]

In addition, we have given for the load transfer: $\Delta F_z = 0.52 F_{y,axle}$ (up to lift-off of the inner tyre, after which the other axle may take over to accommodate the increased total load transfer).

2. Draw the individual curves of F_{yL} and F_{yR} (for the left and right tyre) as a function of α which appear to arise under the load transfer condition considered here.
3. Finally, plot these forces as a function of the vertical load F_z (ranging from 0 to 8000 N). Note the variation of the lateral force of an individual (left or right) tyre in this same range of vertical load which may be covered in a left and in a right-hand turn at increasing speed of travel until (and possibly beyond) the moment that one of the wheels (the inner wheel) lifts from the ground.

11.1.3 Vehicle handling and stability

In this section attention is paid to the more fundamental aspects of vehicle horizontal motions. Instead of discussing results of computer simulations of complicated vehicle models we rather take the simplest possible model of an automobile that runs at constant speed over an even horizontal road and thereby gain considerable insight into the basic aspects of vehicle handling and stability. Important early work on the linear theory of vehicle handling and stability has been published by Riekert and Schunck; Whitcomb and Milliken; and Segel. Pevsner studied the non-linear steady-state cornering behaviour at larger lateral accelerations and introduced the handling diagram. One of the first more complete vehicle model studies has been conducted by Pacejka and by Radt and Pacejka.

The derivation of the equations of motion for the three-degree-of-freedom model of Fig. 11.1-4 will be treated first after which the simple model with two degrees of freedom is considered and analysed. This analysis comprises the steady-state response to steering input and the stability of the resulting motion. Also, the frequency response to steering fluctuations and external disturbances will be discussed, first for the linear vehicle model and subsequently for the non-linear model where large lateral accelerations and disturbances are introduced.

The simple model to be employed in the analysis is presented in Fig. 11.1-9. The track width has been neglected with respect to the radius of the cornering motion which allows the use of a two-wheel vehicle model. The steer and slip angles will be restricted to relatively small values. Then, the variation of the geometry may be regarded to remain linear, that is: $\cos\alpha \approx 1$ and $\sin\alpha \approx \alpha$ and similarly for the steer angle δ. Moreover, the driving force required to keep the speed constant is assumed to remain small with respect to the lateral tyre force. Considering combined slip curves like those shown in Fig. 11.1-2 (right), we may draw the conclusion that the influence of F_x on F_y may be neglected in that case.

In principle, a model as shown in Fig. 11.1-9 lacks body roll and load transfer. Therefore, the theory is actually limited to cases where the roll moment remains small, that is at low friction between tyre and road or a low centre of gravity relative to the track width. This restriction may be overcome by using the effective axle characteristics in which the effects of body roll and load transfer have been included while still adhering to the simple (rigid) two-wheel vehicle model. As has been mentioned before, this is only permissible when the frequency of the imposed steer angle variations remains small with respect to the roll natural frequency. Similarly, as has been demonstrated in the preceding section, effects of other factors like compliance in the steering system and suspension mounts may be accounted for.

The speed of travel is considered to be constant. However, the theory may approximately hold also for quasi-steady-state situations for instance at moderate braking or driving. The influence of the fore-and-aft

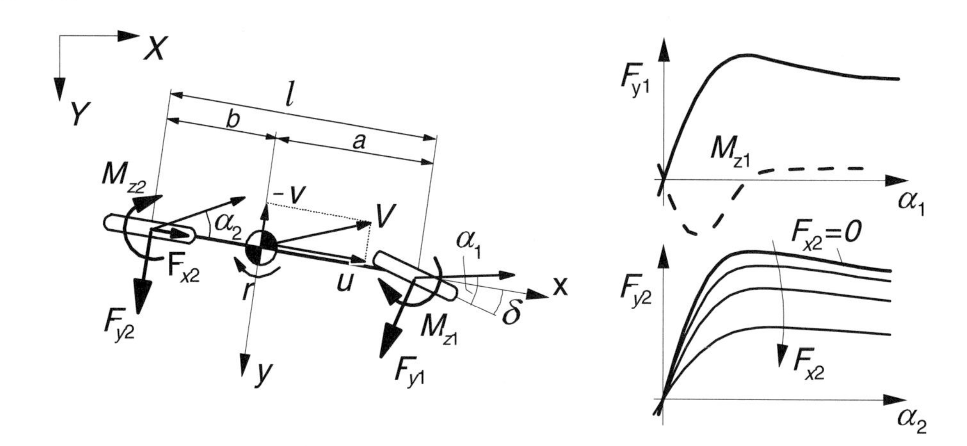

Fig. 11.1-9 Simple car model with side force characteristics for front and rear (driven) axle.

force F_x on the tyre or axle cornering force vs slip angle characteristic (F_y, α) may then be regarded (cf. Fig. 11.1-9). The forces F_{y1} and F_{x1} and the moment M_{z1} are defined to act upon the single front wheel and similarly we define F_{y2} etc. for the rear wheel.

11.1.3.1 Differential equations for plane vehicle motions

In this section, the differential equations for the three-degree-of-freedom vehicle model of Fig. 11.1-4 will be derived. In first instance, the fore and aft motion will also be left free to vary. The resulting set of equations of motion may be of interest for the reader to further study the vehicle's dynamic response at somewhat higher frequencies where the roll dynamics of the vehicle body may become of importance. From these equations, the equations for the simple two-degree-of-freedom model of Fig. 11.1-9 used in the subsequent section can be easily assessed. In Subsection 11.1.3.6 the equations for the car with trailer will be established. The possible instability of the motion will be studied.

We will employ Lagrange's equations to derive the equations of motion. For a system with n degrees of freedom n (generalised) coordinates q_i are selected which are sufficient to completely describe the motion while possible kinematic constraints remain satisfied. The moving system possesses kinetic energy T and potential energy U. External generalised forces Q_i associated with the generalised coordinates q_i may act on the system and do work W. Internal forces acting from dampers to the system structure may be regarded as external forces taking part in the total work W. The equation of Lagrange for coordinate q_i reads:

$$\frac{d}{dt}\frac{\partial T}{\partial \dot{q}_i} - \frac{\partial T}{\partial q_i} + \frac{\partial U}{\partial q_i} = Q_i \qquad (11.1.25)$$

The system depicted in Fig. 11.1-4 and described in the preceding subsection performs a motion over a flat level road. Proper coordinates are the Cartesian coordinates X and Y of reference point A, the yaw angle ψ of the moving x axis with respect to the inertial X axis and finally the roll angle φ about the roll axis. For motions near the X axis and thus small yaw angles, Eq.(11.1.25) is adequate to derive the equations of motion. For cases where ψ may attain large values, e.g. when moving along a circular path, it is preferred to use modified equations where the velocities u, v and r of the moving axes system are used as generalised motion variables in addition to the coordinate φ. The relations between the two sets of variables are (the dots referring to differentiation with respect to time):

$$\begin{aligned} u &= \dot{X}\cos\psi + \dot{Y}\sin\psi \\ v &= -\dot{X}\sin\psi + \dot{Y}\cos\psi \\ r &= \dot{\psi} \end{aligned} \qquad (11.1.26)$$

The kinetic energy can be expressed in terms of u, v and r. Preparation of the first terms of Eq.(11.1.25) for the coordinates X, Y and ψ yields:

$$\begin{aligned} \frac{\partial T}{\partial \dot{X}} &= \frac{\partial T}{\partial u}\frac{\partial u}{\partial \dot{X}} + \frac{\partial T}{\partial v}\frac{\partial v}{\partial \dot{X}} = \frac{\partial T}{\partial u}\cos\psi - \frac{\partial T}{\partial v}\sin\psi \\ \frac{\partial T}{\partial \dot{Y}} &= \frac{\partial T}{\partial u}\frac{\partial u}{\partial \dot{Y}} + \frac{\partial T}{\partial v}\frac{\partial v}{\partial \dot{Y}} = \frac{\partial T}{\partial u}\sin\psi + \frac{\partial T}{\partial v}\cos\psi \\ \frac{\partial T}{\partial \dot{\psi}} &= \frac{\partial T}{\partial r} \\ \frac{\partial T}{\partial \psi} &= \frac{\partial T}{\partial u}v - \frac{\partial T}{\partial v}u \end{aligned} \qquad (11.1.27)$$

The yaw angle ψ may now be eliminated by multiplying the final equations for X and Y successively with $\cos\psi$ and $\sin\psi$ and subsequently adding and subtracting them. The resulting equations represent the equilibrium in the x and y (or u and v) directions, respectively.

We obtain the following set of modified Lagrangean equations for the first three variables u, v and r and subsequently for the remaining real coordinates (for our system only φ):

$$\begin{aligned}
&\frac{d}{dt}\frac{\partial T}{\partial u} - r\frac{\partial T}{\partial v} = Q_u\\
&\frac{d}{dt}\frac{\partial T}{\partial v} + r\frac{\partial T}{\partial u} = Q_v\\
&\frac{d}{dt}\frac{\partial T}{\partial r} - v\frac{\partial T}{\partial u} + u\frac{\partial T}{\partial v} = Q_r\\
&\frac{d}{dt}\frac{\partial T}{\partial \dot\varphi} - \frac{\partial T}{\partial \varphi} + \frac{\partial U}{\partial \varphi} = Q_\varphi
\end{aligned} \qquad (11.1.28)$$

The generalised forces are found from the virtual work:

$$\delta W = \sum_{j=1}^{4} Q_j\, \delta q_j \qquad (11.1.29)$$

with q_j referring to the quasi coordinates x and y and the coordinates ψ and φ. Note that x and y can not be found from integrating u and v. For that reason the term 'quasi' coordinate is used. For the vehicle model we find for the virtual work as a result of the virtual displacements δx, δy, $\delta\psi$ and $\delta\varphi$:

$$\begin{aligned}\delta W &= \sum F_x\delta x + \sum F_y \delta y\\ &+ \sum M_z \delta\psi + \sum M_\varphi \delta\varphi\end{aligned} \qquad (11.1.30)$$

where apparently

$$\begin{aligned}
Q_u &= \sum F_x = F_{x1}\cos\delta - F_{y1}\sin\delta + F_{x2}\\
Q_v &= \sum F_y = F_{x1}\sin\delta + F_{y1}\cos\delta + F_{y2}\\
Q_r &= \sum M_z = aF_{x1}\sin\delta + aF_{y1}\cos\delta + M_{z1}\\
&\qquad - bF_{y2} + M_{z2}\\
Q_\varphi &= \sum M_\varphi = -(k_{\varphi1} + k_{\varphi2})\dot\varphi
\end{aligned} \qquad (11.1.31)$$

The longitudinal forces are assumed to be the same at the left and right wheels and the effect of additional steer angles ψ_i are neglected here. Shock absorbers in the wheel suspensions are represented by the resulting linear moments about the roll axes with damping coefficients $k_{\varphi i}$ at the front and rear axles.

With the roll angle φ and the roll axis inclination angle $\theta_r \approx (h_2 - h_1)/l$ assumed small, the kinetic energy becomes:

$$\begin{aligned}
T &= \tfrac{1}{2}m\{(u - h'\varphi r)^2 + (v + h'\dot\varphi)^2\}\\
&+ \tfrac{1}{2}I_x\dot\varphi^2 + \tfrac{1}{2}I_y(\varphi r)^2\\
&+ \tfrac{1}{2}I_z(r^2 - \varphi^2 r^2 + 2\theta_r r\dot\varphi) - I_{xz}r\dot\varphi
\end{aligned} \qquad (11.1.32)$$

The potential energy U is built up in the suspension springs (including the radial tyre compliances) and through the height of the centre of gravity. We have, again for small angles:

$$U = \tfrac{1}{2}(c_{\varphi1} + c_{\varphi2})\varphi^2 - \tfrac{1}{2}mgh'\varphi^2 \qquad (11.1.33)$$

The equations of motion are finally established by using the expressions (11.1.31), (11.1.32) and (11.1.33) in the Eqs. (11.1.28). The equations will be linearised in the assumedly small angles φ and δ. For the variables u, v, r and φ we obtain successively:

$$\begin{aligned}&m(\dot u - rv - h'\varphi\dot r - 2h'r\dot\varphi)\\ &\quad = F_{x1} - F_{y1}\delta + F_{x2}\end{aligned} \qquad (11.1.34a)$$

$$\begin{aligned}&m(\dot v + ru - h'\ddot\varphi + h'r^2\varphi)\\ &\quad = F_{x1}\delta + F_{y1} + F_{y2}\end{aligned} \qquad (11.1.34b)$$

$$\begin{aligned}&I_z\dot r + (I_z\theta_r - I_{xz})\ddot\varphi - mh'(\dot u - rv)\varphi\\ &\quad = aF_{x1}\delta + aF_{y1} + M_{z1} - bF_{y2} + M_{z2}\end{aligned} \qquad (11.1.34c)$$

$$\begin{aligned}&(I_x + mh'^2)\ddot\varphi + mh'(\dot v + ru) + (I_z\theta_r - I_{xz})\dot r\\ &\quad - (mh'^2 + I_y - I_z)r^2\varphi + (k_{\varphi1} + k_{\varphi2})\dot\varphi\\ &\quad + (c_{\varphi1} + c_{\varphi2} - mgh')\varphi = 0\end{aligned} \qquad (11.1.34d)$$

Note that the small additional roll and compliance steer angles ψ_i have been neglected in the assessment of the force components. The tyre side forces depend on the slip and camber angles front and rear and on the tyre vertical loads. We may need to take the effect of combined slip into account. The longitudinal forces are either given as a result of brake effort or imposed propulsion torque or they depend on the wheel longitudinal slip which follows from the wheel speed of revolution requiring four additional wheel rotational degrees of freedom. The first equation (11.1.34a) may be used to compute the propulsion force needed to keep the forward speed constant.

The vertical loads and more specifically the load transfer can be obtained by considering the moment equilibrium of the front and rear axle about the respective roll centres. For this, the roll moments $M_{\varphi i}$ (cf. Fig. 11.1-4) resulting from suspension springs and dampers as appear in Eq.(11.1.34d) through the terms with subscript 1 and 2 respectively, and the axle side forces appearing in Eq.(11.1.34b) are to be regarded. For a linear model the load transfer can be neglected if initial (left/right opposite) wheel angles are disregarded. We have at steady-state (effect of damping vanishes):

$$\Delta F_{zi} = \frac{-c_{\varphi i}\varphi + F_{yi}h_i}{2s_i} \qquad (11.1.35)$$

The front and rear slip angles follow from the lateral velocities of the wheel axles and the wheel steer angles with respect to the moving longitudinal x axis. The longitudinal velocities of the wheel axles may be regarded the same left and right and equal to the vehicle longitudinal speed u. This is allowed when $s_i|r|\ll u$. Then the expressions for the assumedly small slip angles read:

$$\alpha_1 = \delta + \psi_1 - \frac{v + ar - e\dot{\delta}}{u}$$
$$\alpha_2 = \psi_2 - \frac{v - br}{u} \tag{11.1.36}$$

The additional roll and compliance steer angles ψ_i and the wheel camber angles γ_i are obtained from Eq.(11.1.21) with (11.1.13)–(11.1.15) or corresponding non-linear expressions. Initial wheel angles are assumed to be equal to zero. The influence of the steer angle velocity appearing in the expression for the front slip angle is relatively small and may be disregarded. The small products of the caster length e and the time rate of change of ψ_i have been neglected in the above expressions.

Eqs. (11.1.34) may be further linearised by assuming that all the deviations from the rectilinear motion are small. This allows the neglection of all products of variable quantities which vanish when the vehicle moves straight ahead. The side forces and moments are then written as in Eq.(11.1.5) with the subscripts $i = 1$ or 2 provided. If the moment due to camber is neglected and the pneumatic trail is introduced in the aligning torque we have:

$$F_{yi} = F_{y\alpha i} + F_{y\gamma i} = C_{F\alpha i}\alpha_i + C_{F\gamma i}\gamma_i$$
$$M_{zi} = M_{z\alpha i} = -C_{M\alpha i}\alpha_i = -t_i F_{y\alpha i} = -t_i C_{F\alpha i}\alpha_i \tag{11.1.37}$$

The three linear equations of motion for the system of Fig. 11.1-4 with the forward speed u kept constant finally turn out to read expressed solely in terms of the three motion variables v, r and φ:

$$m(\dot{v} + ur + h'\ddot{\varphi}) = C_{F\alpha 1}\{(1 + c_{sc1})(u\delta + e\dot{\delta} - v - ar)/u + c_{sr1}\varphi\} + C_{F\alpha 2}\{(1 + c_{sc2})(-v + br)/u + c_{sr2}\varphi\} + (C_{F\gamma 1}\tau_1 + C_{F\gamma 2}\tau_2)\varphi \tag{11.1.38a}$$

$$I_z\dot{r} + (I_z\theta_r - I_{xz})\ddot{\varphi} = (a - t_1)C_{F\alpha 1}\{(1 + c_{sc1})(u\delta + e\dot{\delta} - v - ar)/u + c_{sr1}\varphi\} - (b + t_2)C_{F\alpha 2}\{(1 + c_{sc2})(-v + br)/u + c_{sr2}\varphi\} + (aC_{F\gamma 1}\tau_1 - bC_{F\gamma 2}\tau_2)\varphi \tag{11.1.38b}$$

$$(I_x + mh'^2)\ddot{\varphi} + mh'(\dot{v} + ur) + (I_z\theta_r - I_{xz})\dot{r} + (k_{\varphi 1} + k_{\varphi 2})\dot{\varphi} + (c_{\varphi 1} + c_{\varphi 2} - mgh')\varphi = 0 \tag{11.1.38c}$$

In these equations the additional steer angles ψ_i have been eliminated by using expressions (11.1.21) with (11.1.13) – (11.1.15). Furthermore, the resulting compliance steer and roll steer coefficients for $i = 1$ or 2 have been introduced:

$$c_{sci} = \frac{\left(c_{sfi} - \frac{e_i + t_i}{c_{\psi i}}\right)C_{F\alpha i}}{1 - \left(c_{sfi} - \frac{e_i + t_i}{c_{\psi i}}\right)C_{F\alpha i}},$$
$$c_{sri} = \frac{\varepsilon_i + \tau_i\left(c_{sfi} - \frac{e_i + t_i}{c_{\psi i}}\right)C_{F\gamma i}}{1 - \left(c_{sfi} - \frac{e_i + t_i}{c_{\psi i}}\right)C_{F\alpha i}} \tag{11.1.39}$$

where the steer stiffness at the rear $c_{\psi 2}$ may be taken equal to infinity. Furthermore, we have the roll axis inclination angle:

$$\theta_r = \frac{h_2 - h_1}{l} \tag{11.1.40}$$

The relaxation length denoted by σ_i is an important parameter that controls the lag of the response of the side force to the input slip angle. For the Laplace transformed version of the Eqs. (11.1.38) with the Laplace variable s representing differentiation with respect to time, we may introduce tyre lag by replacing the slip angle α_i by the filtered transient slip angle. This may be accomplished by replacing the cornering stiffnesses $C_{F\alpha i}$ appearing in (11.1.38) and (11.1.39) by the 'transient stiffnesses':

$$C_{F\alpha i} \rightarrow \frac{C_{F\alpha i}}{1 + s\sigma_i/u} \tag{11.1.41}$$

A similar procedure may be followed to include the tyre transient response to wheel camber variations. The relaxation length concerned is about equal to the one used for the response to side slip variations. At nominal vertical load the relaxation length is of the order of magnitude of the wheel radius. A more precise model of the aligning torque may be introduced by using a transient pneumatic trail with a similar replacement as indicated by (11.1.41) but with a much smaller relaxation length approximately equal to half the contact length of the tyre.

11.1.3.2 Linear analysis of the two-degree-of-freedom model

From the Eqs. (11.1.34b) and (11.1.34c) the reduced set of equations for the two-degree-of-freedom model can be derived immediately. The roll angle φ and its derivative are set equal to zero and furthermore, we will assume the forward speed u ($\approx V$) to remain constant and neglect the influence of the lateral component of the longitudinal forces F_{xi}. The equations of motion of the simple model of Fig. 11.1-9 for v and r now read:

$$m(\dot{v}+ur) = F_{y1}+F_{y2} \tag{11.1.42a}$$

$$I\dot{r} = aF_{y1}-bF_{y2} \tag{11.1.42b}$$

with v denoting the lateral velocity of the centre of gravity and r the yaw velocity. The symbol m stands for the vehicle mass and I ($=I_z$) denotes the moment of inertia about the vertical axis through the centre of gravity. For the matter of simplicity, the rearward shifts of the points of application of the forces F_{y1} and F_{y2} over a length equal to the pneumatic trail t_1 and t_2, respectively (that is the aligning torques), have been disregarded. Later, we come back to this. The side forces are functions of the respective slip angles:

$$F_{y1} = F_{y1}(\alpha_1) \quad \text{and} \quad F_{y2} = F_{y2}(\alpha_2) \tag{11.1.43}$$

and the slip angles are expressed by:

$$\alpha_1 = \delta-\frac{1}{u}(v+ar) \quad \text{and} \quad \alpha_2 = -\frac{1}{u}(v-br) \tag{11.1.44}$$

neglecting the effect of the time rate of change of the steer angle appearing in Eq.(11.1.36). For relatively low-frequency motions the effective axle characteristics or effective cornering stiffnesses according to Eqs.(11.1.17) and (11.1.22) may be employed.

When only small deviations with respect to the undisturbed straight-ahead motion are considered, the slip angles may be assumed to remain small enough to allow linearisation of the cornering characteristics. For the side force the relationship with the slip angle reduces to the linear equation:

$$F_{yi} = C_i\alpha_i = C_{F\alpha i}\alpha_i \tag{11.1.45}$$

where C_i denotes the cornering stiffness. This can be replaced by the symbol $C_{F\alpha i}$ which may be preferred in more general cases where also camber and aligning stiffnesses play a role.

The two linear first-order differential equations now read:

$$\begin{aligned} m\dot{v}+\frac{1}{u}(C_1+C_2)v+\left\{mu+\frac{1}{u}(aC_1-bC_2)\right\}r &= C_1\delta \\ I\dot{r}+\frac{1}{u}(a^2C_1+b^2C_2)r+\frac{1}{u}(aC_1-bC_2)v &= aC_1\delta \end{aligned} \tag{11.1.46}$$

After elimination of the lateral velocity v we obtain the second-order differential equation for the yaw rate r:

$$\begin{aligned} &Imu\ddot{r}+\{I(C_1+C_2)+m(a^2C_1+b^2C_2)\}\dot{r} \\ &+\frac{1}{u}\{C_1C_2l^2-mu^2(aC_1-bC_2)\}r \\ &= muaC_1\dot{\delta}+C_1C_2l\delta \end{aligned} \tag{11.1.47}$$

Here, as before, the dots refer to differentiation with respect to time, δ is the steer angle of the front wheel and $l\,(=a+b)$ represents the wheel base. The equations may be simplified by introducing the following quantities:

$$\begin{aligned} C &= C_1+C_2 \\ Cs &= C_1a-C_2b \\ Cq^2 &= C_1a^2+C_2b^2 \\ mk^2 &= I \end{aligned} \tag{11.1.48}$$

Here, C denotes the total cornering stiffness of the vehicle, s is the distance from the centre of gravity to the so-called neutral steer point S (Fig. 11.1-11), q is a length corresponding to an average moment arm and k is the radius of gyration. Eqs.(11.1.46) and (11.1.47) now reduce to:

$$\begin{aligned} m(\dot{v}+ur)+\frac{C}{u}v+\frac{Cs}{u}r &= C_1\delta \\ mk^2\dot{r}+\frac{cq^2}{u}r+\frac{Cs}{u}v &= C_1a\delta \end{aligned} \tag{11.1.49}$$

and with v eliminated:

$$\begin{aligned} &m^2k^2u^2\ddot{r}+mC(q^2+k^2)u\dot{r}+(C_1C_2l^2-mu^2Cs)r \\ &= mu^2aC_1\dot{\delta}+uC_1C_2l\delta \end{aligned} \tag{11.1.50}$$

The neutral steer point S is defined as the point on the longitudinal axis of the vehicle where an external side force can be applied without changing the vehicle's yaw angle. If the force acts in front of the neutral steer point, the vehicle is expected to yaw in the direction of the force; if behind, then against the force. The point is of interest when discussing the steering characteristics and stability.

11.1.3.2.1 Linear steady-state cornering solutions

We are interested in the path curvature ($1/R$) that results from a constant steer angle δ at a given constant speed of travel V. Since we have at steady state:

$$\frac{1}{R} = \frac{r}{V} \approx \frac{r}{u} \tag{11.1.51}$$

the expression for the path curvature becomes using (11.1.47) with u replaced by V and the time derivatives omitted:

$$\frac{1}{R} = \frac{C_1C_2l}{C_1C_2l^2-mV^2(aC_1-bC_2)}\delta \tag{11.1.52}$$

By taking the inverse, the expression for the steer angle required to negotiate a curve with a given radius R is obtained:

$$\delta = \frac{1}{R}\left(l - mV^2 \frac{aC_1 - bC_2}{lC_1C_2}\right) \tag{11.1.53}$$

It is convenient to introduce the so-called understeer coefficient or gradient η. For our model, this quantity is defined as:

$$\eta = -\frac{mg}{l}\frac{aC_1 - bC_2}{C_1C_2} = -\frac{s}{l}\frac{mgC}{C_1C_2} \tag{11.1.54}$$

with g denoting the acceleration due to gravity. After having defined the lateral acceleration which in the present linear analysis equals the centripetal acceleration:

$$a_y = Vr = \frac{V^2}{R} \tag{11.1.55}$$

Eq.(11.1.53) can be written in the more convenient form as:

$$\delta = \frac{l}{R}\left(1 + \eta\frac{V^2}{gl}\right) = \frac{l}{R} + \eta\frac{a_y}{g} \tag{11.1.56}$$

The meaning of understeer vs oversteer becomes clear when the steer angle is plotted against the centripetal acceleration while the radius R is kept constant. In Fig. 11.1-10 (left-hand diagram) this is done for three types of vehicles showing understeer, neutral steer and oversteer. Apparently, for an understeered vehicle, the steer angle needs to be increased when the vehicle is going to run at a higher speed. At neutral steer the steer angle can be kept constant while at oversteer a reduction in steer angle is needed when the speed of travel is increased and at the same time a constant turning radius is maintained.

According to Eq.(11.1.56) the steer angle changes sign when for an oversteered car the speed increases beyond the critical speed that is expressed by:

$$V_{crit} = \sqrt{\frac{gl}{-\eta}} \quad (\eta < 0) \tag{11.1.57}$$

As will be shown later, the motion becomes unstable when the critical speed is surpassed. Apparently, this can only happen when the vehicle shows oversteer.

For an understeered car a counterpart has been defined which is the so-called characteristic speed. It is the speed where the steer angle required to maintain the same curvature increases to twice the angle needed at speeds approaching zero. We may also say that at the characteristic speed the path curvature response to steer angle has decreased to half its value at very low speed. Also interesting is the fact that at the characteristic speed the yaw rate response to steer angle r/δ reaches a maximum (the proof of which is left to the reader). We have for the characteristic velocity:

$$V_{char} = \sqrt{\frac{gl}{\eta}} \quad (\eta > 0) \tag{11.1.58}$$

Expression (11.1.54) for the understeer gradient η is simplified when the following expressions for the front and rear axle loads are used:

$$F_{z1} = \frac{b}{l}mg \quad \text{and} \quad F_{z2} = \frac{a}{l}mg \tag{11.1.59}$$

We obtain:

$$\eta = \frac{F_{z1}}{C_1} - \frac{F_{z2}}{C_2} \tag{11.1.60}$$

which says that a vehicle exhibits an understeer nature when the relative cornering compliance of the tyres at the front is larger than at the rear. It is important to note that in (11.1.59) and (11.1.60) the quantities $F_{z1,2}$ denote the vertical axle loads that occur at stand-still and thus represent the mass distribution of the vehicle. Changes of these loads due to aerodynamic down forces

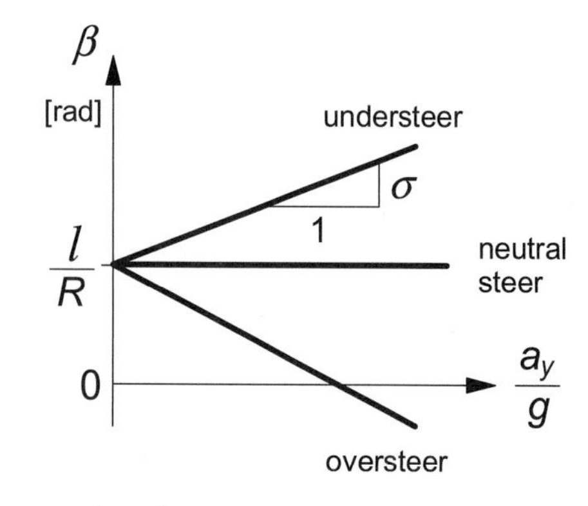

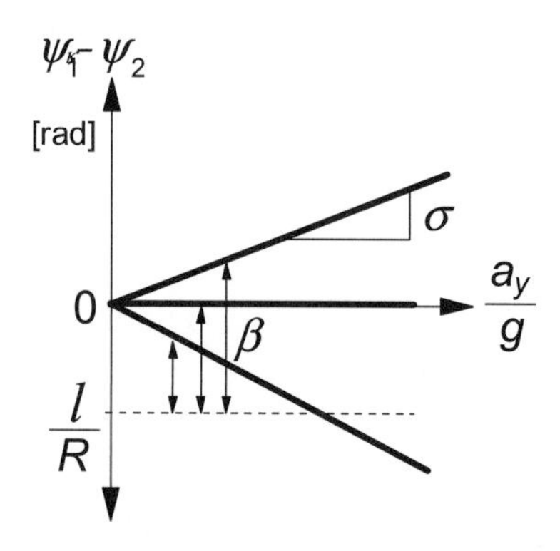

Fig. 11.1-10 The steer angle vs lateral acceleration at constant path curvature (left graph). The difference in slip angle versus lateral acceleration and the required steer angle at a given path curvature (right graph). The understeer gradient η.

and fore and aft load transfer at braking or driving should not be introduced in expression (11.1.60).

In the same diagram the difference in slip angle front and rear may be indicated. We find for the side forces:

$$F_{y1} = \frac{b}{l} m a_y = F_{z1} \frac{a_y}{g}, \quad F_{y2} = \frac{a}{l} m a_y = F_{z2} \frac{a_y}{g} \tag{11.1.61}$$

and hence for the slip angles:

$$\alpha_1 = \frac{F_{z1}}{C_1} \frac{a_y}{g}, \quad \alpha_2 = \frac{F_{z2}}{C_2} \frac{a_y}{g} \tag{11.1.62}$$

The difference now reads when considering the relation (11.1.59) as:

$$\alpha_1 - \alpha_2 = \eta \frac{a_y}{g} \tag{11.1.63}$$

Apparently, the sign of this difference is dictated by the understeer coefficient. Consequently, it may be stated that according to the linear model an understeered vehicle ($\eta > 0$) moves in a curve with slip angles larger at the front than at the rear ($\alpha_1 > \alpha_2$). For a neutrally steered vehicle the angles remain the same ($\alpha_1 = \alpha_2$) and with an oversteered car the rear slip angles are bigger ($\alpha_2 > \alpha_1$). As is shown by the expressions (11.1.54), the signs of η and s are different. Consequently, as one might expect when the centrifugal force is considered as the external force, a vehicle acts oversteered when the neutral steer point lies in front of the centre of gravity and understeered when S lies behind the c.g.. As we will see later on, the actual non-linear vehicle may change its steering character when the lateral acceleration increases. It appears then that the difference in slip angle is no longer directly related to the understeer gradient.

Consideration of Eq.(11.1.56) reveals that in the left-hand graph of Fig. 11.1-10 the difference in slip angle can be measured along the ordinate starting from the value l/R. It is of interest to convert the diagram into the graph shown on the right-hand side of Fig. 11.1-10 with ordinate equal to the difference in slip angle. In that way, the diagram becomes more flexible because the value of the curvature l/R may be selected afterwards. The horizontal dotted line is then shifted vertically according to the value of the relative curvature l/R considered. The distance to the handling line represents the magnitude of the steer angle.

Fig. 11.1-11 depicts the resulting steady-state cornering motion. The vehicle side slip angle β has been indicated. It is of interest to note that at low speed this angle is negative for right-hand turns. Beyond a certain value of speed the tyre slip angles have become sufficiently large and the vehicle slip angle changes into positive values. In Exercise 11.1.2 the slip angle β will be used.

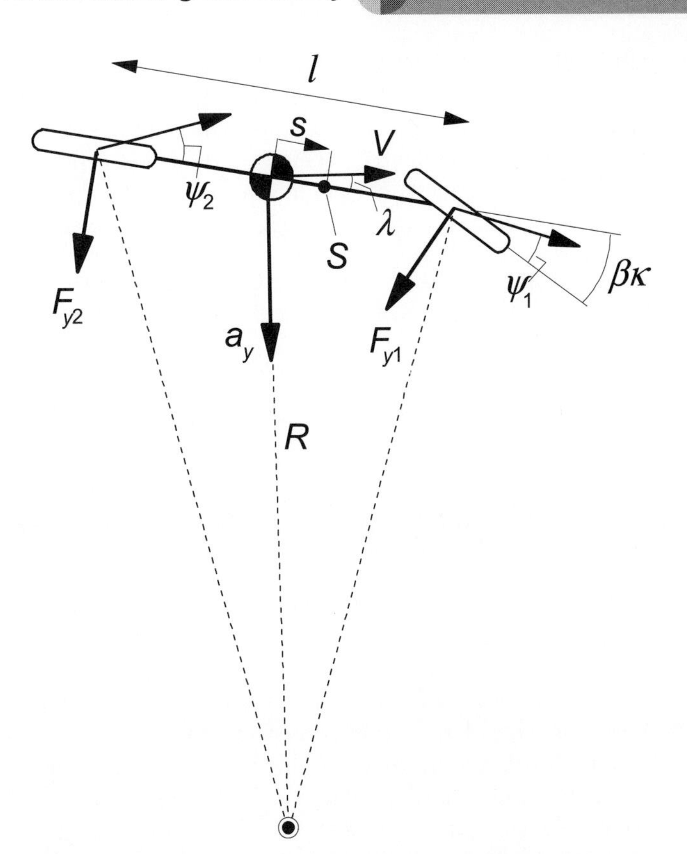

Fig. 11.1-11 Two-wheel vehicle model in a cornering manoeuvre.

11.1.3.2.2 Influence of the pneumatic trail

The direct influence of the pneumatic trails t_i may not be negligible. In reality, the tyre side forces act a small distance behind the contact centres. As a consequence, the neutral steer point should also be considered to be located at a distance approximately equal to the average value of the pneumatic trails, more to the rear, which means actually more understeer. The correct values of the position s of the neutral steer point and of the understeer coefficient η can be found by using the effective axle distances $a' = a - t_1$, $b' = b + t_2$ and $l' = a' + b'$ in the Eqs.(11.1.48) and (11.1.59) instead of the original quantities a, b and l.

Stability of the motion

Stability of the steady-state circular motion can be examined by considering the differential equation (11.1.47) or (11.1.50). The steer angle is kept constant so that the equation gets the form:

$$a_0 \ddot{r} + a_1 \dot{r} + a_2 r = b_1 \delta \tag{11.1.64}$$

For this second-order differential equation stability is assured when all coefficients a_i are positive. Only the last coefficient a_2 may become negative which corresponds to divergent instability (spin-out without oscillations). As already indicated, this will indeed occur when for an oversteered vehicle the critical speed (11.1.57) is exceeded. The condition for stability reads:

$$a_2 = C_1C_2l^2\left(1+\eta\frac{V^2}{gl}\right) = C_1C_2l^2\left(\frac{\delta}{l/R}\right)_{ss} > 0 \tag{11.1.65}$$

with the subscript *ss* referring to steady-state conditions, or

$$V < V_{crit} = \sqrt{\frac{gl}{-\eta}} \quad (\eta < 0) \tag{11.1.66}$$

The next section will further analyse the dynamic nature of the stable and unstable motions.

It is of importance to note that when the condition of an automobile subjected to driving or braking forces is considered, the cornering stiffnesses front and rear will change due to the associated fore and aft axle load transfer and the resulting state of combined slip. In expression (11.1.60) for the understeer coefficient η the quantities F_{zi} represent the static vertical axle loads obtained through Eqs.(11.1.59) and are to remain unchanged! In Subsection 11.1.3.4 the effect of longitudinal forces on vehicle stability will be further analysed.

Free linear motions

To study the nature of the free motion after a small disturbance in terms of natural frequency and damping, the eigenvalues, that is the roots of the characteristic equation of the linear second-order system, are to be assessed. The characteristic equation of the system described by the Eqs.(11.1.49) or (11.1.50) reads after using the relation (11.1.54) between s and η:

$$m^2k^2V^2\lambda^2 + mC(q^2+k^2)V\lambda + C_1C_2l^2\left(1+\frac{\eta}{gl}V^2\right) = 0 \tag{11.1.67}$$

For a single mass-damper-spring system shown in Fig. 11.1-13 with r the mass displacement, δ the forced displacement of the support, M the mass, D the sum of the two damping coefficients D_1 and D_2 and K the sum of the two spring stiffnesses K_1 and K_2 a differential equation similar in structure to Eq.(11.1.50) arises:

$$M\ddot{r} + D\dot{r} + Kr = D_1\dot{\delta} + K_1\delta \tag{11.1.68}$$

and the corresponding characteristic equation:

$$M\lambda^2 + D\lambda + K = 0 \tag{11.1.69}$$

When an oversteered car exceeds its critical speed, the last term of (11.1.67) becomes negative which apparently corresponds with a negative stiffness K. An inverted pendulum is an example of a second-order system with negative last coefficient showing monotonous (diverging) instability.

The roots λ of Eq.(11.1.67) may have loci in the complex plane as shown in Fig. 11.1-12. For positive values of the cornering stiffnesses only the last coefficient of the characteristic equation can become negative which is responsible for the limited types of eigenvalues that can occur. As we will see in Subsection 11.1.3.3, possible negative slopes beyond the peak of the non-linear axle characteristics may give rise to other types of unstable motions connected with two positive real roots or two conjugated complex roots with a positive real part. For the linear vehicle model we may have two real roots in the oversteer case and a pair of complex roots in the understeer case, except at low speeds where the understeered vehicle can show a pair of real negative roots.

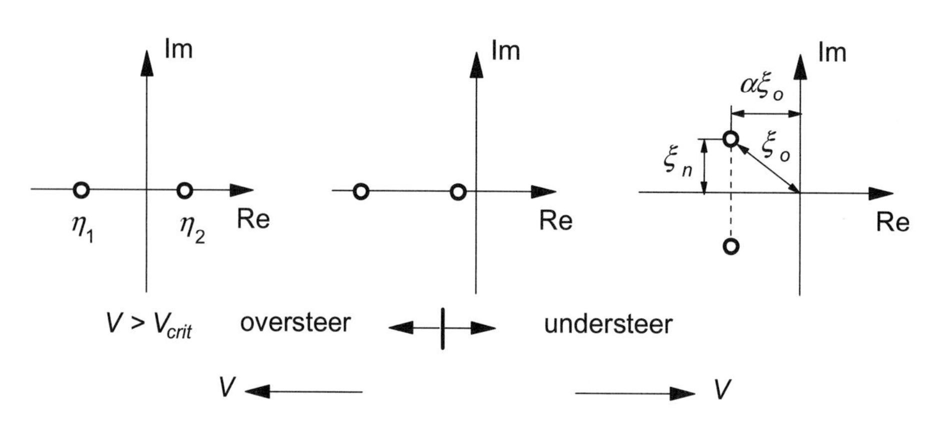

Fig. 11.1-12 Possible eigenvalues for the over and understeered car at lower and higher speeds.

As indicated in the figure, the complex root is characterised by the natural frequency ω_o of the undamped system ($D = 0$), the damping ratio ζ and the resulting actual natural frequency ω_n. Expressions for these quantities in terms of the model parameters are rather complex. However, if we take into account that in normal cases $|s| \ll l$ and $q \approx k \approx \frac{1}{2}\, l$ we may simplify these expressions and find the following useful formulae:

The natural frequency of the undamped system:

$$\omega_o^2 = \frac{K}{M} \approx \left(\frac{C}{mV}\right)^2 \cdot \left(1 + \frac{\eta}{gl}V^2\right) \tag{11.1.70}$$

The damping ratio:

$$\zeta = \frac{D}{2M\omega_o} \approx \frac{1}{\sqrt{1 + \frac{\eta}{gl}V^2}} \tag{11.1.71}$$

The natural frequency:

$$\omega_n^2 = \omega_o^2(1 - \zeta^2) \approx \left(\frac{C}{m}\right)^2 \frac{\eta}{gl} \tag{11.1.72}$$

The influence of parameters has been indicated in Fig. 11.1-13. An arrow pointing upwards represents an increase of the quantity in the same column of the matrix.

The yaw rate response to a step change in steer angle is typified by the rise time t_r indicated in Fig. 11.1-14 and expressed in terms of the parameters as follows:

$$t_r = \frac{r_{ss}}{\left(\frac{\partial r}{\partial t}\right)_{t=0}} = \frac{mk^2V}{aC_1 l\left(1 + \frac{\eta}{gl}V^2\right)} = \frac{mk^2V}{\frac{a}{l}\left\{C_1 l^2 + \left(b - a\frac{C_1}{C_2}\right)mV^2\right\}} \tag{11.1.73}$$

which expression may be readily obtained with the aid of Eqs.(11.1.46) and (11.1.47).

The parameter influence has been indicated in the figure. The results correspond qualitatively well with the 90% response times found in vehicle model simulation studies. A remarkable result is that for an understeered automobile the response time is smaller than for an oversteered car.

Forced linear vibrations

The conversion of the equations of motion (11.1.46) into the standard state space representation is useful when the linear system properties are the subject of investigation. The system at hand is of the second order and hence possesses two state variables for which we choose: v and r. The system is subjected to a single input signal: the steer angle δ. Various variables may be of interest to analyse the vehicle's response to steering input oscillations. The following quantities are selected to illustrate the method and to study the dynamic behaviour of the vehicle: the lateral acceleration a_y of the centre of gravity of the vehicle, the yaw rate r and the vehicle slip angle β defined at the centre of gravity. In matrix notation, the equation becomes:

$$\begin{aligned} \dot{\boldsymbol{x}} &= \boldsymbol{Ax} + \boldsymbol{Bu} \\ \boldsymbol{y} &= \boldsymbol{Cx} + \boldsymbol{Du} \end{aligned} \tag{11.1.74}$$

with

$$\dot{\boldsymbol{x}} = \begin{pmatrix} \dot{v} \\ \dot{r} \end{pmatrix}, \quad \boldsymbol{u} = \delta, \qquad \boldsymbol{y} = \begin{pmatrix} a_y \\ r \\ \beta \end{pmatrix} = \begin{pmatrix} \dot{v} + Vr \\ r \\ -v/V \end{pmatrix} \tag{11.1.75}$$

and

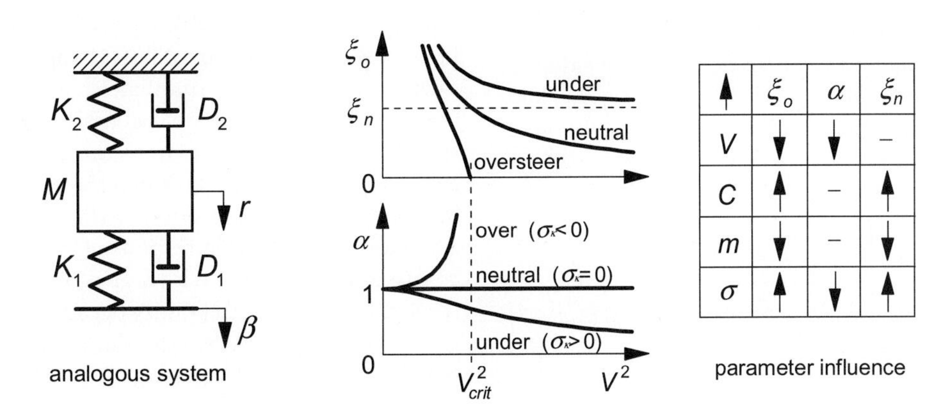

▲	ξ_o	α	ξ_n
V	▼	▼	–
C	▲	–	▲
m	▼	–	▼
σ	▲	▼	▲

Fig. 11.1-13 The influence of parameters on natural frequency and damping.

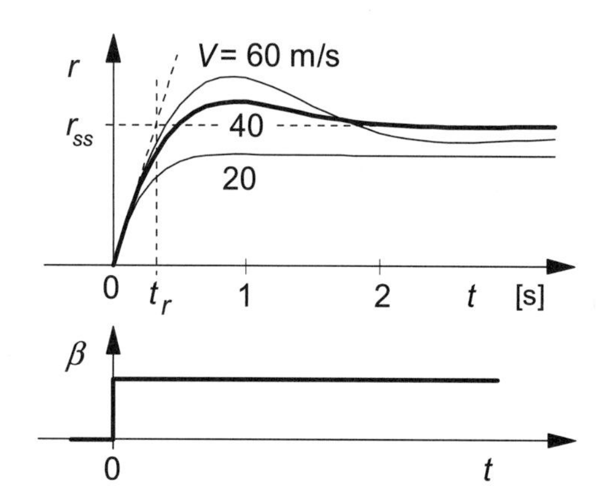

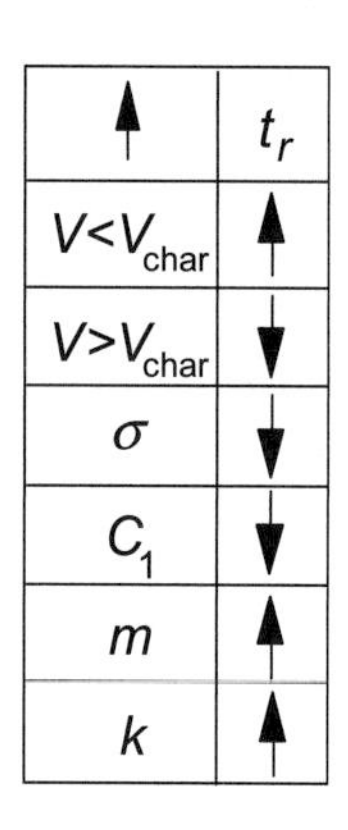

Fig. 11.1-14 Step response of yaw rate to steer angle. Parameters according to Table 11.1-1. Parameter influence on the rise time t_r.

$$A = -\begin{pmatrix} \frac{C}{mV} & V + \frac{Cs}{mV} \\ \frac{Cs}{mk^2V} & \frac{Cq^2}{mk^2V} \end{pmatrix}, \quad B = \begin{pmatrix} \frac{C_1}{m} \\ \frac{C_1 a}{mk^2} \end{pmatrix}$$

$$C = -\begin{pmatrix} \frac{C}{mV} & \frac{Cs}{mV} \\ 0 & -1 \\ 1/V & 0 \end{pmatrix}, \quad D = \begin{pmatrix} \frac{C_1}{m} \\ 0 \\ 0 \end{pmatrix} \tag{11.1.76}$$

The frequency response functions have been computed using Matlab software. Fig. 11.1-15 presents the amplitude and phase response functions for each of the three output quantities and at three different values of speed of travel. The values of the chosen model parameters and a number of characteristic quantities have been listed in Table 11.1-1.

Explicit expressions of the frequency response functions in terms of model parameters are helpful to understand and predict the characteristic aspects of these functions which may be established by means of computations or possibly through full scale experiments.

From the differential equation (11.1.50) the frequency response function is easily derived. Considering the quantities formulated by (11.1.70) and (11.1.71) and the steady-state response $(r/\delta)_{ss} = (V/R)/\delta$ obtained from (11.1.56) we find:

$$\frac{r}{\delta}(j\omega) = \left(\frac{r}{\delta}\right)_{ss} \cdot \frac{1 + \frac{mVa}{C_2 l} j\omega}{1 - \left(\frac{\omega}{\omega_o}\right)^2 + 2\zeta\left(\frac{j\omega}{\omega_o}\right)};$$
$$\left(\frac{r}{\delta}\right)_{ss} = \frac{V/l}{1 + \frac{\eta}{gl}V^2} \tag{11.1.77}$$

Similarly, the formula for the response of lateral acceleration a_y can be derived:

$$\frac{a_y}{\delta}(j\omega) = \left(\frac{a_y}{\delta}\right)_{ss} \cdot \frac{1 - \frac{mk^2}{C_2 l}\omega^2 + \frac{b}{V} j\omega}{1 - \left(\frac{\omega}{\omega_o}\right)^2 + 2\zeta\left(\frac{j\omega}{\omega_o}\right)};$$
$$\left(\frac{a_y}{\delta}\right)_{ss} = \frac{V^2/l}{1 + \frac{\eta}{gl}V^2} \tag{11.1.78}$$

and for the slip angle β:

$$\frac{\beta}{\delta}(j\omega) = \left(\frac{\beta}{\delta}\right)_{ss} \cdot \frac{1 - \frac{mk^2V}{amV^2 - bC_2 l} j\omega}{1 - \left(\frac{\omega}{\omega_o}\right)^2 + 2\zeta\left(\frac{j\omega}{\omega_o}\right)};$$
$$\left(\frac{\beta}{\delta}\right)_{ss} = -\frac{b}{l}\frac{1 - \frac{a}{b}\frac{m}{C_2 l}V^2}{1 + \frac{\eta}{gl}V^2} \tag{11.1.79}$$

By considering Eq.(11.1.77) it can now be explained that for instance at higher frequencies the system exhibits features of a first-order system: because of the $j\omega$ term in the numerator the yaw rate amplitude response tends to a decay at a 6 dB per octave rate (when plotted in log–log scale) and the phase lag approaches 90°. The phase increase at low frequencies and higher speeds is due to the presence of the speed V in that same term. At speeds beyond approximately the characteristic speed, the corresponding (last) term in the denominator has less influence on the initial slope of the phase characteristic. The lateral acceleration response (11.1.78) shown in the centre graph of Fig. 11.1-15 gives a finite amplitude at frequencies tending to infinity because of the presence of ω^2 in the numerator. For the same reason, the phase lag goes back to zero at large frequencies. The side slip phase response tends to –270° (at larger speeds) which is due to the negative coefficient of $j\omega$ in the numerator of

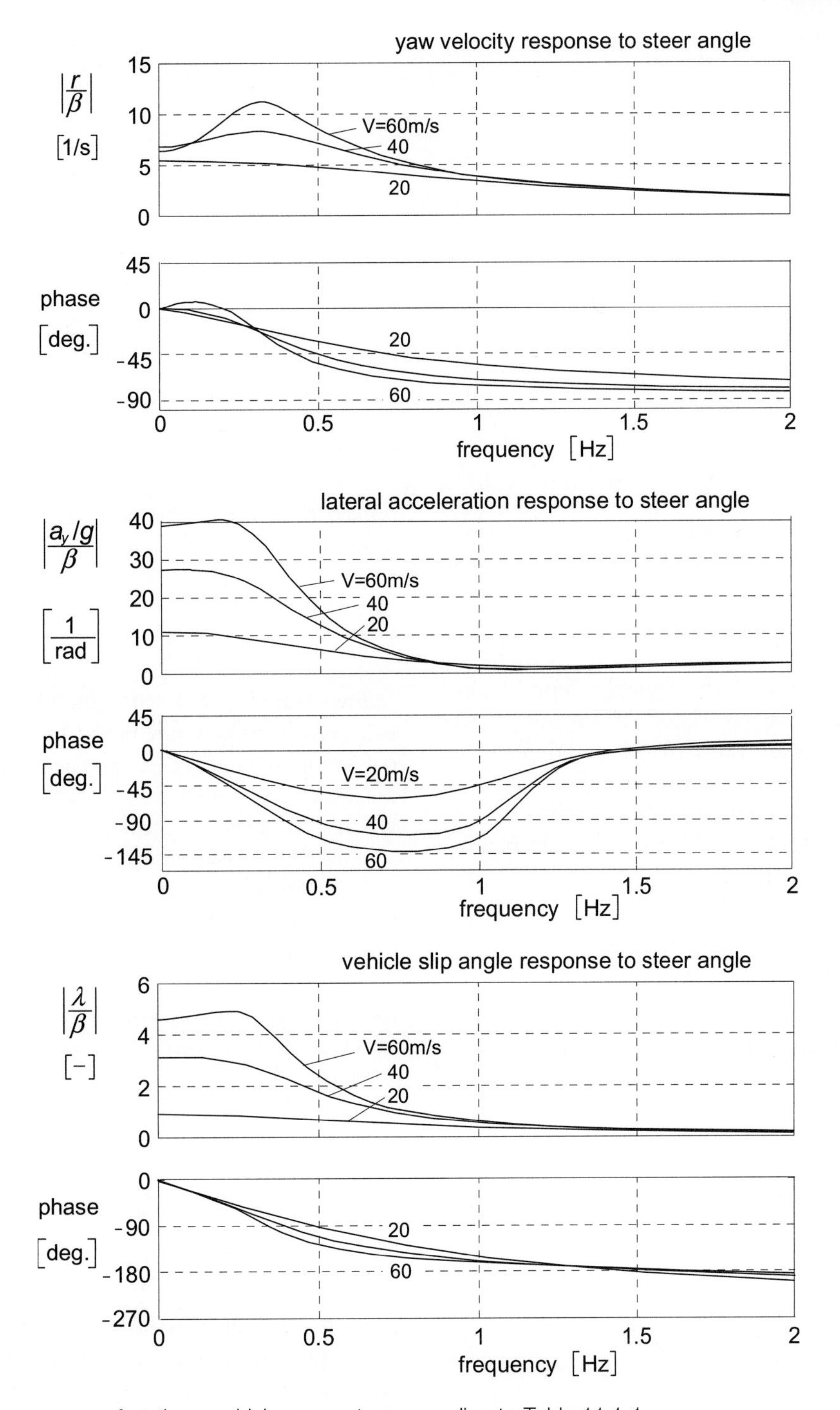

Fig. 11.1-15 Frequency response functions; vehicle parameters according to Table 11.1-1.

(11.1.79). This in contrast to that coefficient of the yaw rate response (11.1.77).

It is of interest to see that the steady-state slip angle response, indicated in (11.1.79), changes sign at a certain speed V. At low speeds where the tyre slip angles are still very small, the vehicle slip angle obviously is negative for positive steer angle (considering positive directions as adopted in Fig. 11.1-11). At larger velocities the tyre slip angles increase and as a result, β changes into the positive direction.

Exercise 11.1.2. Four-wheel steer, condition that the vehicle slip angle vanishes

Consider the vehicle model of Fig. 11.1-16. Both the front and the rear wheels can be steered. The objective is to have a vehicle moving with a slip angle β remaining

Table 11.1-1 Parameter values and typifying quantities

Parameters		Derived typifying quantities					
a	1.4 m	l	3 m	V [m/s]	20	40	60
b	1.6 m	F_{z1}	8371 N	ω_o [rad/s]	4.17	2.6	2.21
C_1	60000 N/rad	F_{z2}	7325 N	ζ [-]	0.9	0.7	0.57
C_2	60000 N/rad	q	1.503 m	ω_n [rad/s]	1.8	1.8	1.82
m	1600 kg	s	−0.1 m	t_r [s]	0.23	0.3	0.27
k	1.5 m	η	0.0174 rad (~1° extra steer/g lateral accel.)				

equal to zero. In practice, this may be done to improve handling qualities of the automobile (reduces to first-order system!) and to avoid excessive side slipping motions of the rear axle in lane change manoeuvres. Adapt the equations of motion (11.1.46) and assess the required relationship between the steer angles δ_1 and δ_2. Do this in terms of the transfer function between δ_2 and δ_1 and the associated differential equation. Find the steady-state ratio $(\delta_2/\delta_1)_{ss}$ and plot this as a function of the speed V. Show also the frequency response function $\delta_2/\delta_1(j\omega)$ for the amplitude and phase at a speed $V = 30$ m/s. Use the vehicle parameters supplied in Table 11.1-1.

11.1.3.3 Non-linear steady-state cornering solutions

From Eqs.(11.1.42) and (11.1.59) with the same restrictions as stated below Eq.(11.1.60), the following force balance equations can be derived (follows also from Eqs.(11.1.61)) (the effect of the pneumatic trails will be dealt with later on):

$$\frac{F_{y1}}{F_{z1}} = \frac{F_{y2}}{F_{z2}} = \frac{a_y}{g}\left(= \frac{K}{mg}\right) \quad (11.1.80)$$

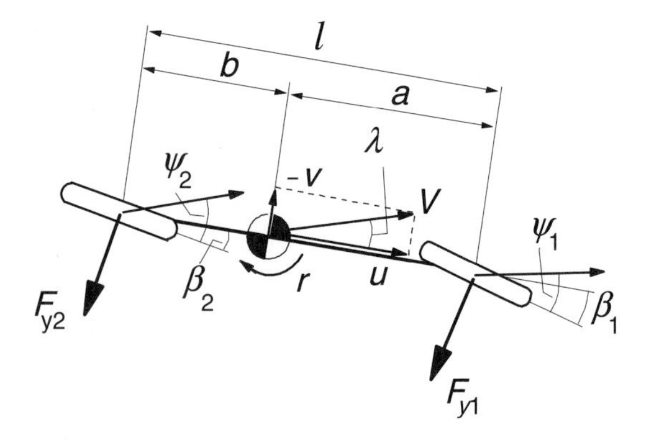

Fig. 11.1-16 'Four-wheel' steering to make slip angle $\beta = 0$ (Exercise 11.1.2).

where $K = ma_y$ represents the centrifugal force. The kinematic relationship

$$\delta - (\alpha_1 - \alpha_2) = \frac{l}{R} \quad (11.1.81)$$

follows from Eqs.(11.1.44) and (11.1.51). In Fig. 11.1-11 the vehicle model has been depicted in a steady-state cornering manoeuvre. It can easily be observed from this diagram that relation (11.1.81) holds approximately when the angles are small.

The ratio of the side force and vertical load as shown in (11.1.80) plotted as a function of the slip angle may be termed as the normalised tyre or axle characteristic. These characteristics subtracted horizontally from each other produce the 'handling curve'. Considering the equalities (11.1.80) the ordinate may be replaced by a_y/g. The resulting diagram with abscissa $\alpha_1 - \alpha_2$ is the non-linear version of the right-hand diagram of Fig. 11.1-10 (rotated 90° anti-clockwise). The diagram may be completed by attaching the graph that shows for a series of speeds V the relationship between lateral acceleration (in g units) a_y/g and the relative path curvature l/R according to Eq.(11.1.55).

Fig. 11.1-17 shows the normalised axle characteristics and the completed handling diagram. The handling curve consists of a main branch and two side lobes. The different portions of the curves have been coded to indicate the corresponding parts of the original normalised axle characteristics they originate from. Near the origin the system may be approximated by a linear model. Consequently, the slope of the handling curve in the origin with respect to the vertical axis is equal to the under steer coefficient η. In contrast to the straight handling line of the linear system (Fig. 11.1-10), the non-linear system shows a curved line. The slope changes along the curve which means that the degree of understeer changes with increasing lateral acceleration. The diagram of Fig. 11.1-17 shows that the vehicle considered changes from understeer to oversteer. We define:

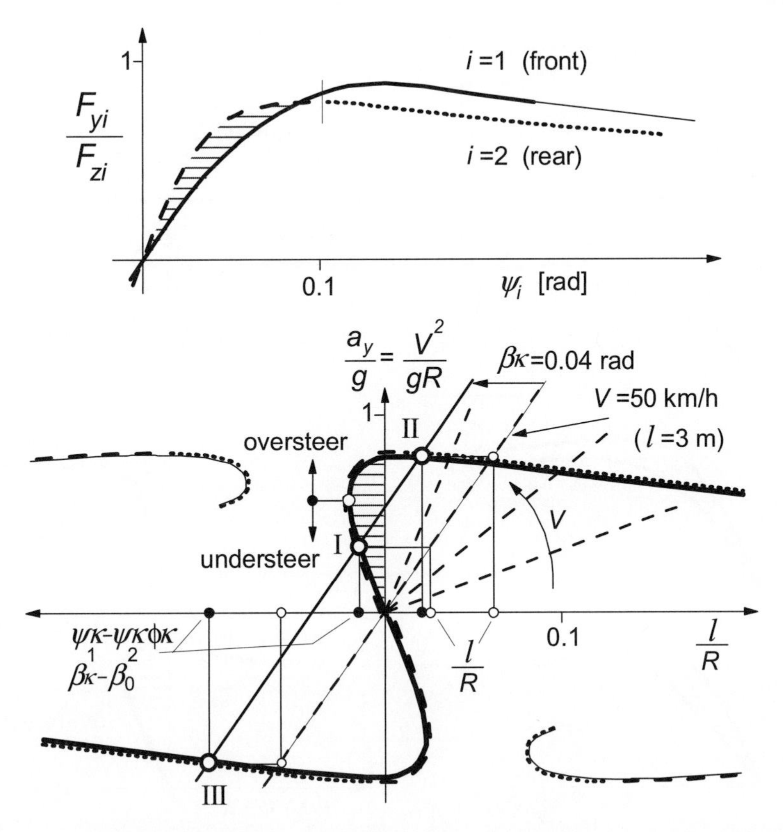

Fig. 11.1-17 Handling diagram resulting from normalised tyre characteristics. Equilibrium points I, II and III (steady turns) of which only I is stable, arise for speed $V = 50$ km/h and steer angle $\delta = 0.04$ rad. From the different line types the manner in which the curves are obtained from the upper diagram may be retrieved.

$$\begin{aligned} &\textit{understeer} \text{ if: } \left(\frac{\partial \delta}{\partial V}\right)_R > 0 \\ &\textit{oversteer} \text{ if: } \left(\frac{\partial \delta}{\partial V}\right)_R < 0 \end{aligned} \qquad (11.1.82)$$

The family of straight lines represents the relationship between acceleration and curvature at different levels of speed. The speed line belonging to $V = 50$ km/h has been indicated (wheel base $l = 3$ m). This line is shifted to the left over a distance equal to the steer angle $\delta = 0.04$ rad and three points of intersection with the handling curve arise. These points I, II and III indicate the possible equilibrium conditions at the chosen speed and steer angle. The connected values of the relative path curvature l/R can be found by going back to the speed line. As will be shown further on, only point I refers to a stable cornering motion. In points II and III ($R < 0$!) the motion is unstable.

At a given speed V, a certain steer angle δ is needed to negotiate a circular path with given radius R. The steer angle required can be read directly from the handling diagram. The steer angle needed to negotiate the same curve at very low speed ($V \to 0$) tends to l/R. This steer angle is denoted with δ_0. Consequently, the abscissa of the handling curve $\alpha_1 - \alpha_2$ may as well be replaced by $\delta - \delta_0$. This opens the possibility to determine the handling curve with the aid of simple experimental means, i.e. measuring the steering wheel input (reduced to equivalent road wheel steer angle by means of the steering ratio, which method automatically includes steering compliance effects) at various speeds running over the same circular path.

Subtracting normalised characteristics may give rise to very differently shaped handling curves only by slightly modifying the original characteristics. As Fig. 11.1-17 shows, apart from the main branch passing through the origin, isolated branches may occur. These are associated with at least one of the decaying ends of the pair of normalised tyre characteristics.

In Fig. 11.1-18 a set of four possible combinations of axle characteristics have been depicted together with the resulting handling curves. This collection of characteristics shows that the nature of steering behaviour is entirely governed by the normalised axle characteristics and in particular their relative shape with respect to each other.

The way in which we can use the handling diagram is presented in Fig. 11.1-19. The speed of travel may be kept constant and the lateral acceleration is increased by running over a spiral path with decreasing radius. The required variation of the steer angle follows from the

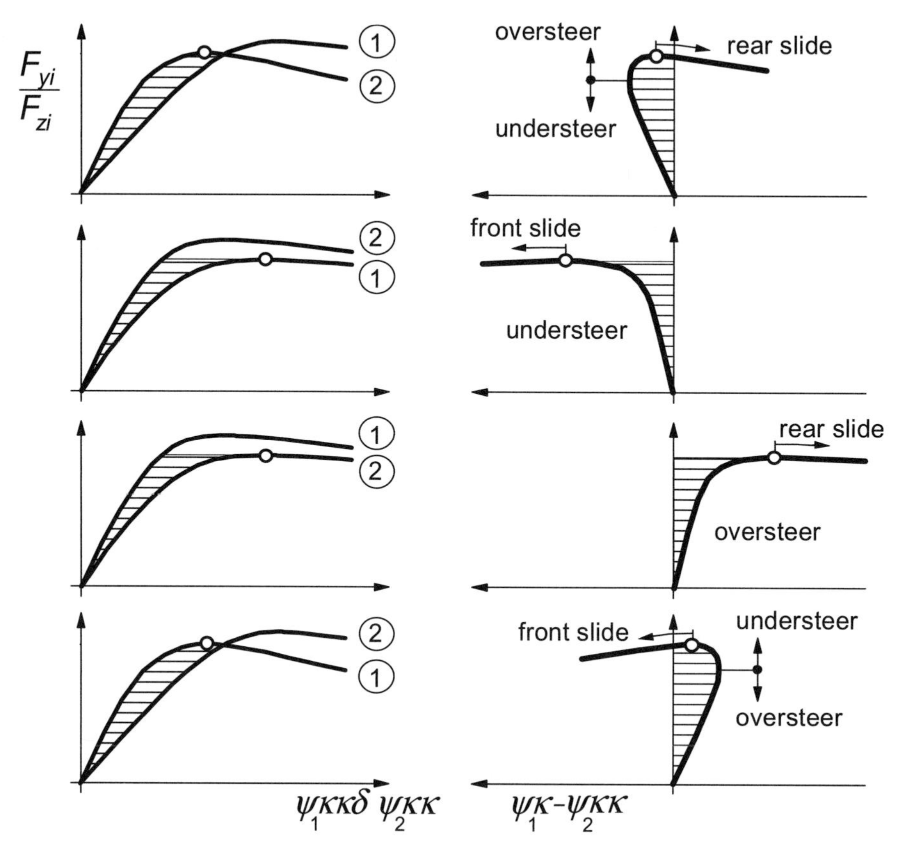

Fig. 11.1-18 A number of handling curves arising from the pairs of normalised tyre characteristics are shown on the left. Only the main branch of the handling curve has been drawn (1: front, 2: rear).

distance between the handling curve and the speed line. Similarly we can observe what happens when the path curvature is kept constant and the speed is increased.

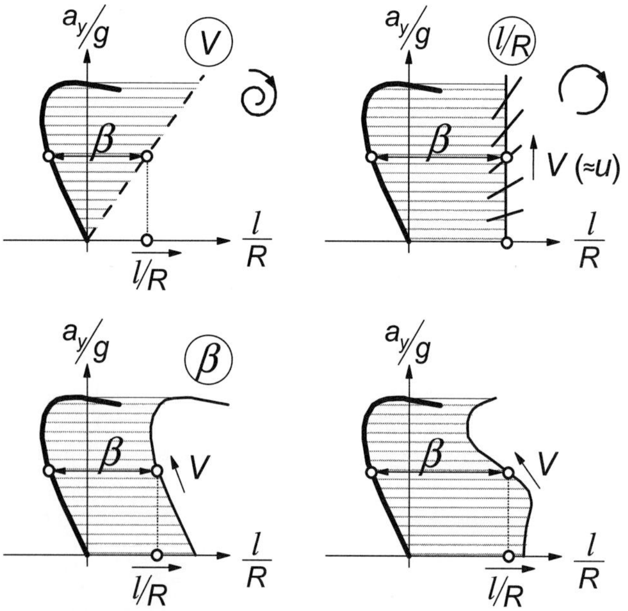

Fig. 11.1-19 Types of quasi-steady-state manoeuvres.

Also, the resulting variation of the curvature at a constant steer angle and increasing speed can be found. More general cases of quasi-steady-state motions may be studied as well.

Stability of the motion at large lateral accelerations

The non-linear set of Eqs.(11.1.42 – 11.1.44) may be linearised around the point of operation, that is one of the equilibrium states indicated above. The resulting second-order differential equation has a structure similar to Eq.(11.1.64) or (11.1.47) but with the variables replaced by their small variations with respect to the steady-state condition considered. Analysis of the coefficients of the characteristic equation reveals if stability exists. Also the nature of stability (monotonous, oscillatory) follows from these coefficients. This is reflected by the type of singular points (node, spiral, saddle) representing the equilibrium solutions in the phase plane as treated in the next section.

It now turns out that not only the last coefficient can become negative but also the second coefficient a_1. Instead of the cornering stiffnesses C defined in the origin of the tyre cornering characteristics, the slope of the normalised characteristics at a given level of a_y/g becomes now of importance. We define:

$$\Phi_i = \frac{1}{F_{zi}} \frac{\partial F_{yi}}{\partial \alpha_i} \quad (i = 1,2) \tag{11.1.83}$$

The conditions for stability, that is: second and last coefficient of equation comparable with Eq.(11.1.47) must be positive, read after having introduced the radius of gyration k ($k^2 = I/m$):

$$(k^2 + a^2)\Phi_1 + (k^2 + b^2)\Phi_2 > 0 \tag{11.1.84}$$

$$\Phi_1\Phi_2\left(\frac{\partial \delta}{\partial 1/R}\right)_V > 0 \tag{11.1.85}$$

The subscript V refers to the condition of differentiation with V kept constant, that is while staying on the speed line of Fig. 11.1-17. The first condition (11.1.84) may be violated when we deal with tyre characteristics showing a peak in side force and a downwards sloping further part of the characteristic. The second condition corresponds to condition (11.1.65) for the linear model. Accordingly, instability is expected to occur beyond the point where the steer angle reaches a maximum while the speed is kept constant. This, obviously, can only occur in the oversteer range of operation. In the handling diagram the stability boundary can be assessed by finding the tangent to the handling curve that runs parallel to the speed line considered.

In the upper diagram of Fig. 11.1-20 the stability boundary, that holds for the right part of the diagram (a_y vs l/R), has been drawn for the system of Fig. 11.1-17 that changes from initial understeer to oversteer. In the middle diagram a number of shifted V-lines, each for a different steer angle δ, has been indicated. In each case the points of intersection represent possible steady-state solutions. The highest point represents an unstable solution as the corresponding point on the speed line lies in the unstable area. When the steer angle is increased the two points of intersections move towards each other. It turns out that for this type of handling curve a range of δ values exists without intersections with the positive half of the curve. The fact that both right-hand turn solutions may vanish has serious implications which follows from the phase plot. At increased steer angle, however, new solutions may show up. At first, these solutions appear to be unstable, but at rather large steer angles of more than about 0.2 rad we find again stable solutions. These occur on the isolated branch where α_2 is small and α_1 is large. Apparently, we find that the vehicle that increases its speed while running at a constant turning radius will first cross the stability boundary and may then recover its stability by turning the steering wheel to a relatively large angle. In the diagram the left part of the isolated branch is reached where stable spirals appear to occur. This phenomenon may correspond to similar experiences in the racing practice.

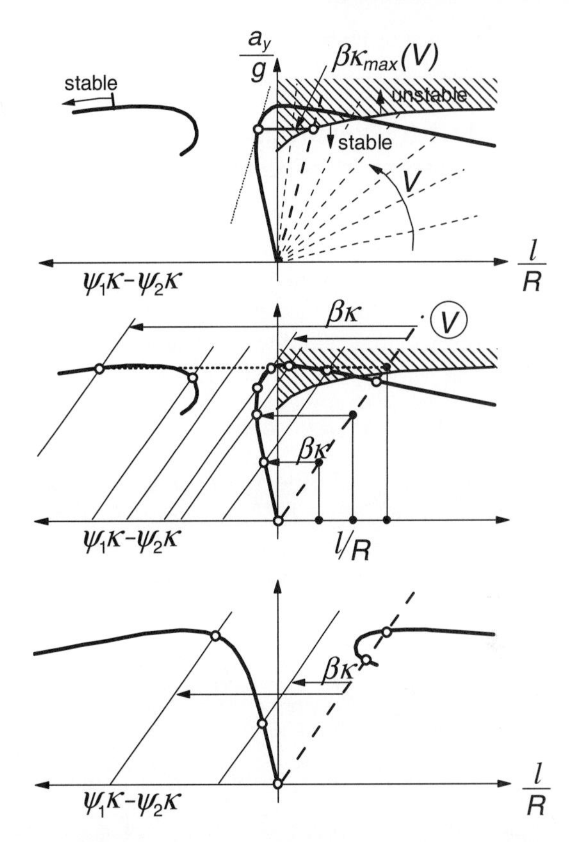

Fig. 11.1-20 Construction of stability boundary (upper diagram, from Fig. 11.1-17). On the isolated branch, a stable range may occur (large steer angle as indicated in middle diagram). The lower diagram shows the case with complete understeer featuring a stable main branch.

The lower diagram depicts the handling curve for a car that remains understeered throughout the lateral acceleration range. Everywhere the steady-state cornering motion remains stable. Up to the maximum of the curve the tangents slope to the left and cannot run parallel to a speed line. Beyond the peak, however, we can find a speed line parallel to the tangent, but at the same time one of the slopes (Φ_1) of the normalised axle characteristics starts to show a negative sign so that condition (11.1.85) is still satisfied. Similarly, the limit oversteer vehicle of the upper graph remains unstable beyond the peak. On the isolated part of the handling curve of the lower diagram the motion remains unstable. It will be clear that the isolated branches vanish when we deal with axle characteristics that do not show a peak and decaying part of the curve.

It may seem that the establishment of unstable solutions has no particular value. It will become clear, however, that the existence and the location of both stable and unstable singular points play an important role in shaping the trajectories in the phase-plane. Also, the nature of stability or instability in the singular points are of importance.

Exercise 11.1.3. Construction of the complete handling diagram from pairs of axle characteristics

We consider three sets of hypothetical axle characteristics (a, b and c) shown in the graph of Fig. 11.1-21. The dimensions of the vehicle model are: $a = b = \frac{1}{2}l = 1.5$ m. For the tyres we may employ axle characteristics described by the *Magic Formula* (11.1.6):

$$F_y = D \sin[C \arctan\{B\alpha - E(B\alpha - \arctan(B\alpha))\}]$$

We define: the peak side force $D = \mu F_z$ and the cornering stiffness $C_{F\alpha} = BCD = C_{F\alpha}F_z$ so that $B = C_{F\alpha}/(C\mu)$. For the six tyre/axle configurations the parameter values have been given in the table below.

Axle	Case	μ	$C_{F\alpha}$	C	E
Front	a,b	0.8	8	1.2	–2
	c	0.78	8	1.3	–2
Rear	a	0.9	11	1.2	–2
	b	0.9	6	1.2	–2
	c	0.65	11	1.5	–1

Determine for each of the three combinations (two dry, one wet):

1. The handling curve (cf. Fig. 11.1-17).
2. The complete handling diagram (cf. Fig. 11.1-17).
3. The portion of the curves where the vehicle shows an oversteer nature.

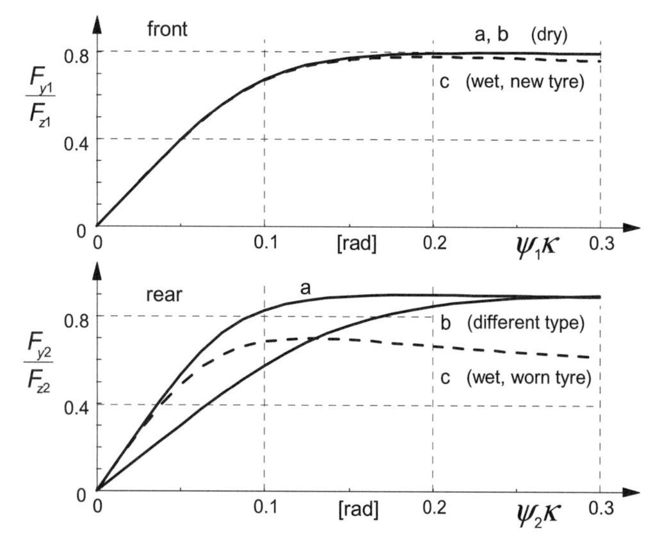

Fig. 11.1-21 Three sets of hypothetical axle cornering characteristics (Exercise 11.1.3).

4. The stability boundary (associated with these oversteer ranges) in the (a_y/g vs l/R) diagram (= right-hand side of the handling diagram) (cf. Fig. 11.1-20).
5. Indicate in the diagram (or in a separate graph):
   - **a.** the course of the steer angle δ required to negotiate a curve with radius $R = 60$ m as a function of the speed V. If applicable, indicate the stability boundary, that is the critical speed V_{crit}, belonging to this radius.
   - **b.** the course of steer angle δ as a function of relative path curvature l/R at a fixed speed $V = 72$ km/h and if applicable assess the critical radius R_{crit}.

For the vehicle systems considered so far a unique handling curve appears to suffice to describe the steady-state turning behaviour. Cases may occur, however, where more curves are needed, one for each velocity. A simple example is the situation when the car runs over a wet surface where the tyre characteristics change considerably with speed. Also, as a result of the down forces acting on e.g. the body of a racing car, the tyre loads increase with speed. Consequently, the tyre characteristics change accordingly which requires an adaptation of the handling curve.

A more difficult and fundamentally different situation occurs when the vehicle is equipped with a third axle. Also in this case multiple handling curves arise. A tandem rear axle configuration of a heavy truck for example, strongly opposes movement along a curved track. The slip angles of the two rear axles are different so that a counteracting torque arises. This torque gets larger when the turning radius becomes smaller. This may for instance occur at a given level of lateral acceleration. When at this level the speed becomes lower, the curvature must become larger and the opposing torque will increase which entails an increased front steer angle to generate a larger side force needed to balance the vehicle. This increased steer angle goes on top of the steer angle which was already larger because of the increased l/R. Here, l is the average wheel base. Consequently, in the handling diagram, the points on the handling curve belonging to the lower speed lie more to the left.

Assessment of the influence of the pneumatic trail on the handling curve

So far the direct influence of the pneumatic trails have not been taken into account. As with the linear analysis we may do this by considering the effective axle positions

$$a' = a - t_1, \quad b' = b + t_2 \quad \text{and} \quad l' = a' + b' \tag{11.1.86}$$

The difficulty we have to face now is the fact that these pneumatic trails t_i will vary with the respective slip angles. We have if the residual torques are neglected:

$$t_i(\alpha_i) = -\frac{M_{zi}(\alpha_i)}{F_{yi}(\alpha_i)} \tag{11.1.87}$$

Introducing the effective axle loads

$$F'_{z1} = \frac{b'}{l'}mg, \quad F'_{z2} = \frac{a'}{l'}mg \tag{11.1.88}$$

yields for the lateral force balance instead of (11.1.80):

$$\frac{F_{y1}}{F'_{z1}} = \frac{F_{y2}}{F'_{z2}} = \frac{a_y}{g} \tag{11.1.89}$$

or after some rearrangements:

$$\frac{a'}{a}\frac{F_{y1}}{F_{z1}} = \frac{b'}{b}\frac{F_{y2}}{F'_{z2}} = Q\frac{a_y}{g} \tag{11.1.90}$$

with

$$Q = \frac{l}{l'}\frac{a'b'}{ab} \approx 1 \tag{11.1.91}$$

The corrected normalised side force characteristics as indicated in (11.1.90) can be computed beforehand and drawn as functions of the slip angles and the normal procedure to assess the handling curve can be followed. This can be done by taking the very good approximation $Q = 1$ or we might select a level of Qa_y/g then assess the values of the slip angles that belong to that level of the corrected normalised side forces and compute Q according to (11.1.91) and from that the correct value of a_y/g.

Large deviations with respect to the steady-state motion

The variables r and v may be considered as the two state variables of the second-order non-linear system represented by the Eqs. (11.1.42). Through computer numerical integration the response to a given arbitrary variation of the steer angle can be easily obtained. For motions with constant steer angle δ (possibly after a step change), the system is autonomous and the phase-plane representation may be used to find the solution. For that, we proceed by eliminating the time from Eqs.(11.1.42). The result is a first-order non-linear equation (using $k^2 = I/m$):

$$\frac{dv}{dr} = k^2\frac{F_{y1} + F_{y2} - mVr}{aF_{y1} - bF_{y2}} \tag{11.1.92}$$

Since F_{y1} and F_{y2} are functions of α_1 and α_2 it may be easier to take α_1 and α_2 as the state variables. With (11.1.44) we obtain:

$$\frac{d\alpha_2}{d\alpha_1} = \frac{dv/dr - b}{dv/dr + a} \tag{11.1.93}$$

a which becomes with (11.1.92):

$$\frac{d\alpha_2}{d\alpha_1} = \frac{F_{y2}(\alpha_2)/F_{z2} - (\delta - \alpha_1 + \alpha_2)V^2/gl}{F_{y1}(\alpha_1)/F_{z1} - (\delta - \alpha_1 + \alpha_2)V^2/gl} \tag{11.1.94}$$

For the sake of simplicity we have assumed $I/m = k^2 = ab$.

By using Eq.(11.1.94) the trajectories (solution curves) can be constructed in the (α_1, α_2) plane. The isocline method turns out to be straightforward and simple to employ. The pattern of the trajectories is strongly influenced by the so-called singular points. In these points, the motion finds an equilibrium. In the singular points the motion is stationary and consequently, the differentials of the state variables vanish.

From the handling diagram, K/mg and l/R are readily obtained for given combinations of V and δ. Used in combination with the normalised tyre characteristics F_{y1}/F_{z1} and F_{y2}/F_{z2} the values of α_1 and α_2 are found, which form the coordinates of the singular points. The manner in which a stable turn is approached and from what collection of initial conditions such a motion can or cannot be attained may be studied in the phase-plane. One of the more interesting results of such an investigation is the determination of the boundaries of the domain of attraction in case such a domain with finite dimensions exists. The size of the domain may give indications as to the so-called stability in the large. In other words the question may be answered: does the vehicle return to its original steady-state condition after a disturbance and to what degree does this depend on the magnitude and point of application of the disturbance impulse?

For the construction of the trajectories we draw isoclines in the (α_1, α_2) plane. These isoclines are governed by Eq.(11.1.94) with slope $d\alpha_2/d\alpha_1$ kept constant. The following three isoclines may already provide sufficient information to draw estimated courses of the trajectories. We have for $k^2 = ab$:

vertical intercepts ($d\alpha_2/d\alpha_1 \to \infty$):

$$\alpha_2 = \frac{gl}{V^2}\frac{F_{y1}(\alpha_1)}{F_{z1}} + \alpha_1 - \delta \tag{11.1.95}$$

horizontal intercepts ($d\alpha_2/d\alpha_1 \to 0$):

$$\alpha_1 = -\frac{gl}{V^2}\frac{F_{y2}(\alpha_2)}{F_{z2}} + \alpha_2 + \delta \tag{11.1.96}$$

intercepts under 45° $(d\alpha_2/d\alpha_1 = 1)$:

$$\frac{F_{y1}(\alpha_1)}{F_{z1}} = \frac{F_{y2}(\alpha_2)}{F_{z2}} \tag{11.1.97}$$

Fig. 11.1-22 illustrates the way these isoclines are constructed. The system of Fig. 11.1-17 with $k = a = b$, $\delta = 0.04$ rad and $V = 50$ km/h has been considered. Note that the normalised tyre characteristics appear in the left-hand diagram for the construction of the isoclines. The three points of intersection of the isoclines are the singular points. They correspond to the points I, II and III of Fig. 11.1-17. The stable point is a focus (spiral) point with a complex pair of solutions of the characteristic equation with a negative real part. The two unstable points are of the saddle type corresponding to a real pair of solutions, one of which is positive. The direction in which the motion follows the trajectories is still a question to be examined. Also for this purpose the alternative set of axes with r and v as coordinates (multiplied with a factor) has been introduced in the diagram after using the relations (11.1.44).

From the original equations (11.1.42) it can be found that the isocline (11.1.97) forms the boundary between areas with $\dot{r} > 0$ and $\dot{r} < 0$ (indicated in Fig. 11.1-22). Now it is easy to ascertain the direction along the trajectories. We note that the system exhibits a bounded domain of attraction. The boundaries are called separatrices. Once outside the domain, the motion finds itself in an unstable situation. Remains the disturbance limited so that resulting initial conditions of the state variables stay within the boundaries, then ultimately the steady-state condition is reached again.

For systems with normalised characteristics showing everywhere a positive slope, a handling curve arises that consists of only the main branch through the origin. If the rear axle characteristic (at least in the end) is higher than the front axle characteristic, the vehicle will show (at least in the limit) an understeer nature and unstable singular points cannot occur. This at least if for the case of initial oversteer the speed remains under the critical speed. In such cases, the domain of attraction is theoretically unbounded so that for all initial conditions ultimately the stable equilibrium is attained. The domain of Fig. 11.1-22 appears to be open on two sides which means that initial conditions, in a certain range of (r/v) values, do not require to be limited in order to reach the stable point. Obviously, disturbance impulses acting in front of the centre of gravity may give rise to such combinations of initial conditions.

In Figs. 11.1-23 and 11.1-24 the influence of an increase in steer angle δ on the stability margin (distance between stable point and separatrix) has been shown for the two vehicles considered in Fig. 11.1-20. The system of Fig. 11.1-23 is clearly much more sensitive. An increase in δ (but also an increase in speed V) reduces the stability margin until it is totally vanished as soon as the two singular points merge (also the corresponding points I and II on the handling curve of Fig. 11.1-17) and the domain breaks open. As a result, all trajectories starting above the lower separatrix tend to leave the area. This can only be stopped by either quickly reducing the steer angle or enlarging δ to around 0.2 rad or more. The latter situation appears to be stable again (focus) as has been stated before. For the understeered vehicle of Fig. 11.1-24 stability is practically always ensured.

For a further appreciation of the phase diagram it is of interest to determine the new initial state (r_o, v_o) after the action of a lateral impulse to the vehicle (cf. Fig. 11.1-25). For an impulse S acting at a distance x

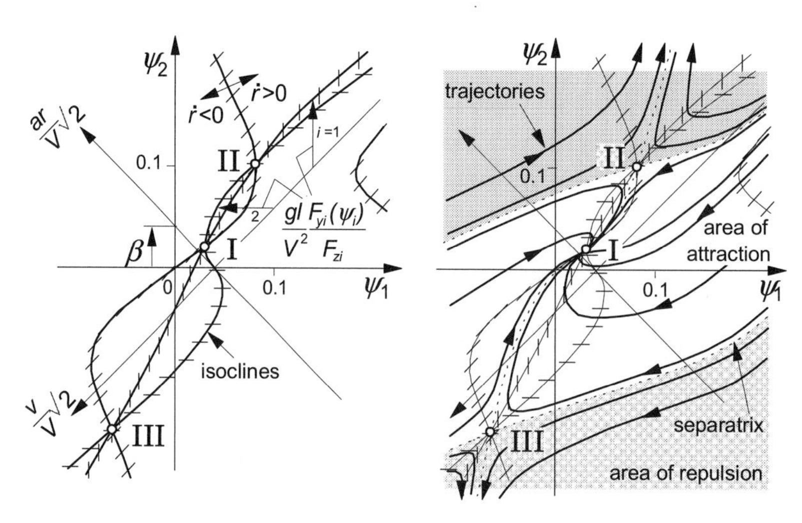

Fig. 11.1-22 Isoclines for the construction of trajectories in the phase-plane. Also shown: the three singular points I, II and III (cf. Fig. 11.1-17) and the separatrices constituting the boundary of the domain of attraction. Point I represents the stable cornering motion at steer angle δ.

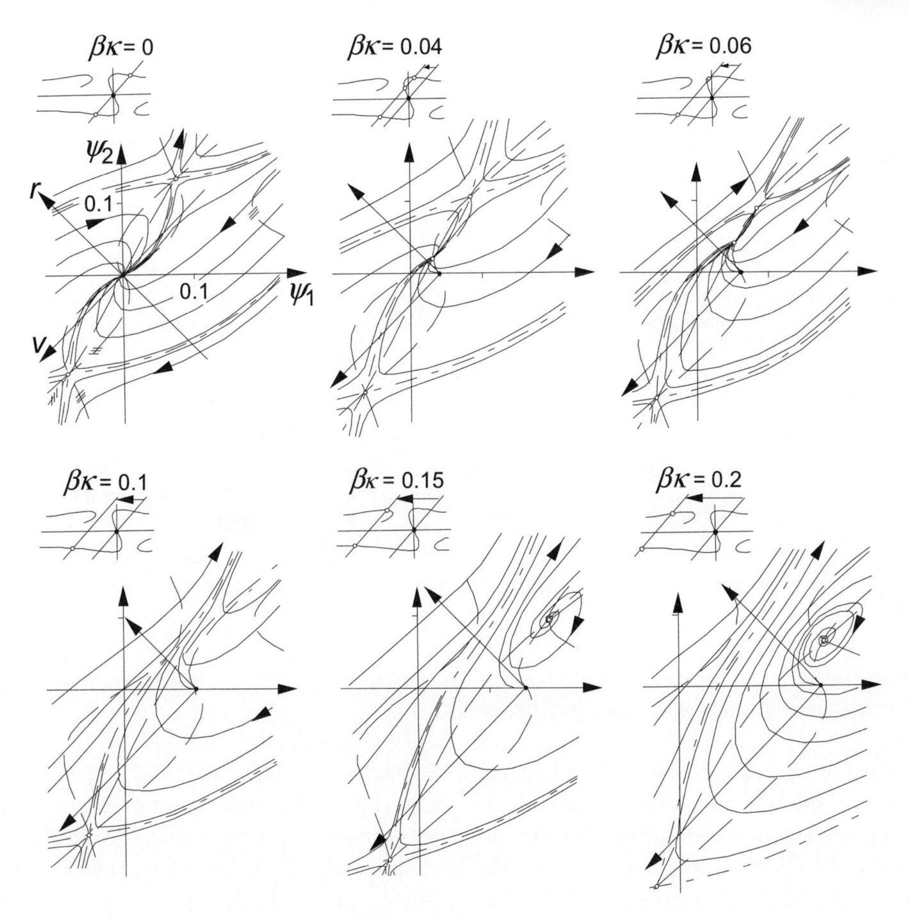

Fig. 11.1-23 Influence of steering on the stability margin (system of Fig. 11.1-20 (top)).

in front of the centre of gravity the increase in r and v becomes:

$$\Delta r = \frac{Sx}{I}, \quad \Delta v = \frac{S}{m} \tag{11.1.98}$$

which results in the direction

$$\frac{a\Delta r}{\Delta v} = \frac{x}{b}\frac{ab}{k^2} \tag{11.1.99}$$

The figure shows the change in state vector for different points of application and direction of the impulse S ($k^2 = I/m = ab$). Evidently, an impulse acting at the rear (in outward direction) constitutes the most dangerous disturbance. On the other hand, an impulse acting in front of the centre of gravity about half way from the front axle does not appear to be able to get the new starting point outside of the domain of attraction no matter the intensity of the impulse.

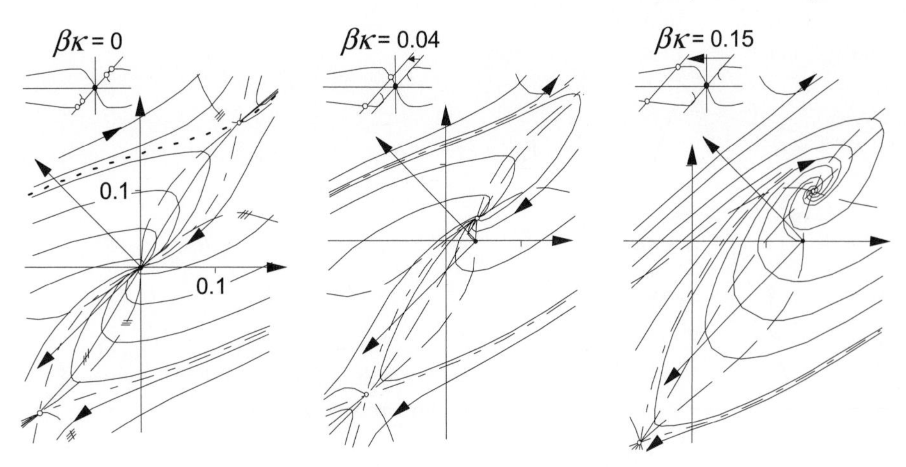

Fig. 11.1-24 Influence of steering on the stability margin (system of Fig. 11.1-20 (bottom)).

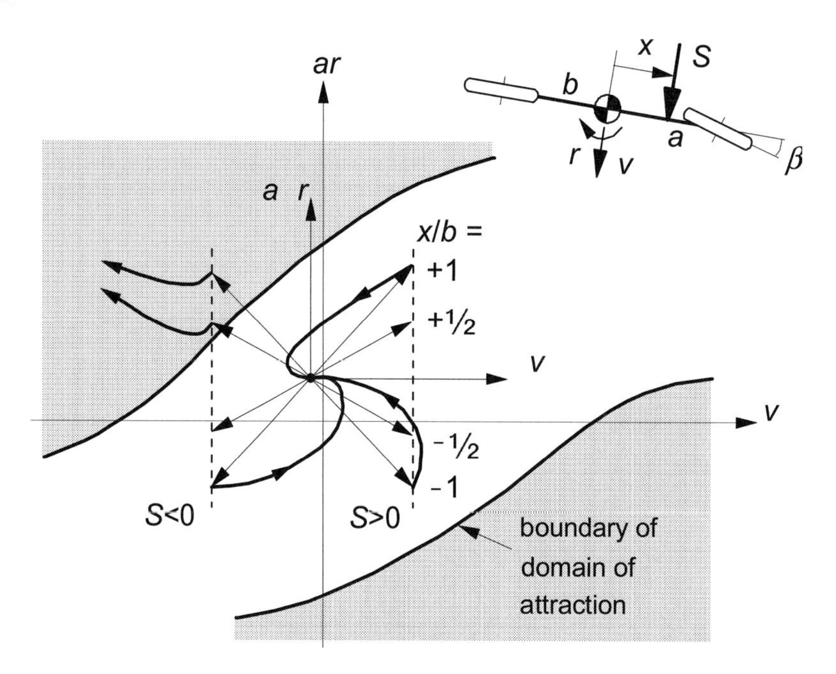

Fig. 11.1-25 Large disturbance in a curve. New initial state vector (Δv, Δr) after the action of a lateral impulse S. Once outside the domain of attraction the motion becomes unstable and may get out of control.

When the slip angles become larger, the forward speed u may no longer be considered as a constant quantity. Then, the system is described by a third-order set of equations. In the paper by Pacejka, the solutions for the simple automobile model have been presented also for yaw angles >90°.

11.1.3.4 The vehicle at braking or driving

When the vehicle is subjected to longitudinal forces that may result from braking or driving actions possibly to compensate for longitudinal wind drag forces or down or upward slopes, fore and aft load transfer will arise (Fig. 11.1-26). The resulting change in tyre normal loads causes the cornering stiffnesses and the peak side forces of the front and rear axles to change. Since, as we assume here, the fore and aft position of the centre of gravity is not affected (no relative car body motion), we may expect a change in handling behaviour indicated by a rise or drop of the understeer gradient. In addition, the longitudinal driving or braking forces give rise to a state of combined slip, thereby affecting the side force in a way as shown in Fig. 11.1-2.

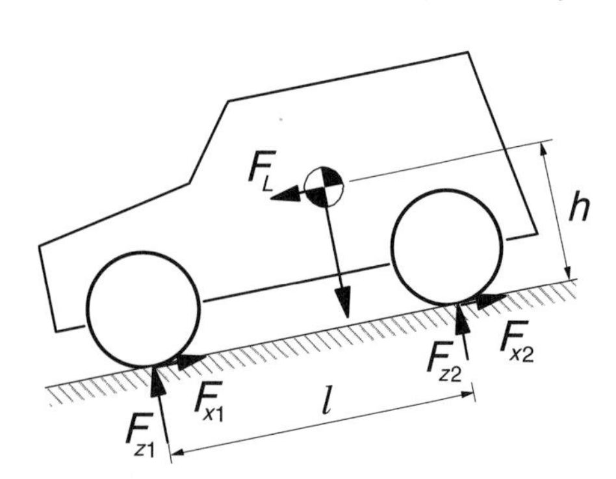

Fig. 11.1-26 The automobile subjected to longitudinal forces and the resulting load transfer.

For moderate driving or braking forces the influence of these forces on the side force F_y is relatively small and may be neglected for this occasion. This means that, for now, the cornering stiffness may be considered to be dependent on the normal load only. The upper left diagram of Fig. 11.1-3 depicts typical variations of the cornering stiffness with vertical load.

The load transfer from the rear axle to the front axle that results from a forward longitudinal force F_L acting at the centre of gravity at a height h above the road surface (F_L possibly corresponding to the inertial force at braking) becomes:

$$\Delta F_z = \frac{h}{l} F_L \tag{11.1.100}$$

The understeer gradient reads according to Eq.(11.1.60):

$$\eta = \frac{F_{z1o}}{C_1(F_{z1})} - \frac{F_{z2o}}{C_2(F_{z2})} \tag{11.1.101}$$

The static axle loads F_{zio} ($i = 1$ or 2) are calculated according to Eq.(11.1.59), while the actual loads F_{zi} front and rear become:

$$F_{z1} = F_{z1o} + \Delta F_z, \quad F_{z2} = F_{z2o} - \Delta F_z \tag{11.1.102}$$

At moderate braking with deceleration $-a_x = F_L/m$ the load transfer remains small and we may use the linearised approximation of the variation of cornering stiffness with vertical load:

$$C_i = C_{io} + \zeta_{\alpha i}\Delta F_{zi} \quad \text{with } \zeta_{\alpha i} = \left(\frac{\partial C_i}{\partial F_{zi}}\right)_{F_{zio}} \tag{11.1.103}$$

The understeer gradient (11.1.101) can now be expressed in terms of the longitudinal acceleration a_x (which might be: minus the forward component of the acceleration due to gravity parallel to the road). We obtain:

$$\eta = \eta_o + \lambda\frac{a_x}{g} \tag{11.1.104}$$

with the determining factor λ approximately expressed as:

$$\lambda = \zeta_{\alpha 1}\frac{h}{b}\left(\frac{F_{z1o}}{C_{1o}}\right)^2 + \zeta_{\alpha 2}\frac{h}{a}\left(\frac{F_{z2o}}{C_{2o}}\right)^2 \tag{11.1.105}$$

and η_o denoting the original value not including the effect of longitudinal forces. Obviously, since $\zeta_{\alpha 1,2}$ is usually positive, negative longitudinal accelerations a_x, corresponding to braking, will result in a decrease of the degree of understeer.

To illustrate the magnitude of the effect we use the parameter values given in Table 11.1-1 (above Eq.(11.1.77)) and add the c.g. height $h = 0.6$ m and the cornering stiffness vs load gradients $\zeta_{\alpha i} = 0.5C_{io}/F_{zio}$. The resulting factor appears to take the value $\lambda = 0.052$. This constitutes an increase of η equal to $0.052a_x/g$. Apparently, the effect of a_x on the understeer gradient is considerable when regarding the original value $\eta_o = 0.0174$.

As illustrated in Fig. 11.1-9 the peak side force will be diminished if a longitudinal driving or braking force is transmitted by the tyre. This will have an impact on the resulting handling diagram in the higher range of lateral acceleration. The resulting situation may be represented by the second and third diagrams of Fig. 11.1-18 corresponding to braking (or driving) at the front or rear respectively. The problem becomes considerably more complex when we realise that at the front wheels the components of the longitudinal forces perpendicular to the x axis of the vehicle are to be taken into account. Obviously, we find that at braking of the front wheels these components will counteract the cornering effect of the side forces and thus will make the car more understeer. The opposite occurs when these wheels are driven (more oversteer).

At hard braking, possibly up to wheel lock, stability and steerability may deteriorate severely.

11.1.3.5 The moment method

Possible steady-state cornering conditions, stable or unstable, have been portrayed in the handling diagram of Fig. 11.1-17. In Fig. 11.1-22 motions tending to or departing from these steady-state conditions have been depicted. These motions are considered to occur after a sudden change in steer angle. The potential available to deviate from the steady turn depends on the margin of the front and rear side forces to increase in magnitude. For each point on the handling curve it is possible to assess the degree of manoeuvrability in terms of the moment that can be generated by the tyre side forces about the vehicle centre of gravity. Note that at the steady-state equilibrium condition the tyre side forces are balanced with the centrifugal force and the moment equals zero.

In general, the handling curve holds for a given speed of travel. That is so, when e.g. the aerodynamic down forces are essential in the analysis. In Fig. 11.1-27 a diagram has been presented that is designated as the Milliken Moment Method (*MMM*) diagram and is computed for a speed of 60 mph. The force-moment concept was originally proposed by W.F. Milliken in 1952 and thereafter continuously further developed by the Cornell Aeronautical Laboratory staff and by Milliken Research Associates. A detailed description is given in Milliken's book.

The graph shows curves of the resulting tyre moment N vs the resulting tyre side force Y in non-dimensional form. The resulting force and moment result from the individual side forces and act from ground to vehicle. For greater accuracy, one may take the effect of the pneumatic trails into consideration. Two sets of curves have been plotted: one set for constant values of the vehicle side slip angle β with the steering wheel angle δ_{stw} as parameter and the other set for contant steer angle and varying slip angle. Along the horizontal axis the moment is zero and we have the steady-state equilibrium cornering situation that corresponds with the handling curve. It is observed that for the constant speed considered in the diagram, the steer angle increases when the total side force Y or lateral acceleration a_y is chosen larger which indicates that the motion remains stable. At the limit (near number 2) the maximum steady-state lateral acceleration is attained. At that point the ability to generate a positive moment is exhausted. Only a negative moment may still be developed by the car that tends to straighten the curve that is being negotiated. As we have seen in Fig. 11.1-18, second diagram, there is still some side force margin at the rear tyre which can be used to increase the lateral acceleration in a transient fashion. At the same time, however, the car yaws outwards because the associated moment is negative (cf. diagram near number 8). How to get at points below the equilibrium point near the

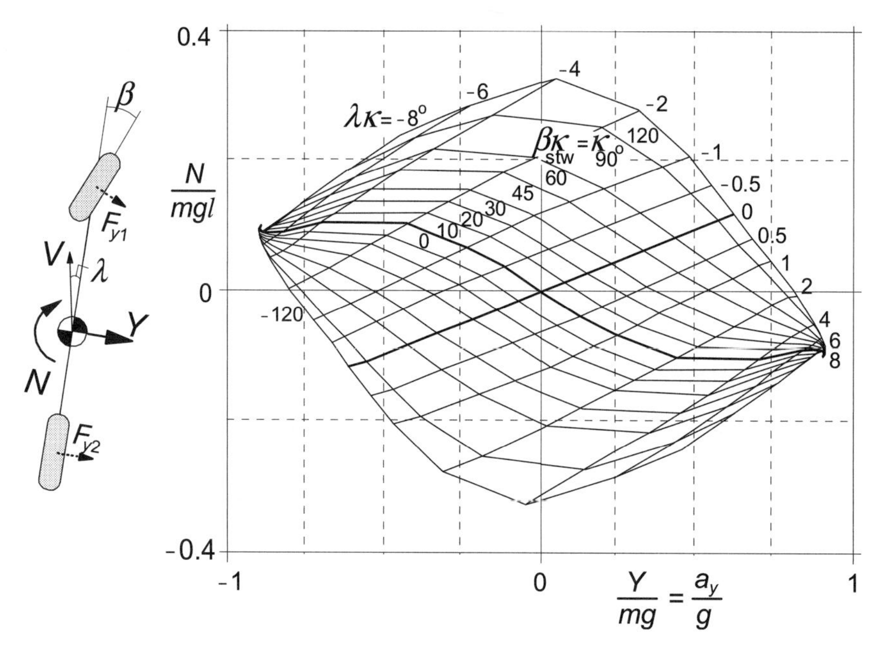

Fig. 11.1-27 The *MMM* diagram portraying the car's potential manoeuvring capacity.

number 2 is a problem. Rear wheel steering is an obvious theoretical option. In that way, the vehicle slip angle β and front steer angle δ can remain unchanged while the rear steer angle produce the desired rear tyre slip angle. Of course, the diagram needs to be adapted in case of rear wheel steering. Another more practical solution would be to bring the vehicle in the desired attitude ($\beta \rightarrow 8°$) by briefly inducing large brake or drive slip at the rear that lowers the cornering force and lets the car swing to the desired slip angle while at the same time the steering wheel is turned backwards to even negative values.

The *MMM* diagram, which is actually a Gough plot established for the whole car at different steer angles, may be assessed experimentally either through outdoor or indoor experiments. On the proving ground a vehicle may be attached at the side of a heavy truck or railway vehicle and set at different slip angles while the force and moment are being measured (tethered testing). Fig. 11.1-28 depicts the remarkable laboratory MMM test machine. This MTS Flat-Trac Roadway Simulator™ uses four flat belts which can be steered and driven independently. The car is constrained in its centre of gravity but is free to roll and pitch.

11.1.3.6 The car-trailer combination

In this section we will discuss the role of the tyre in connection with the dynamic behaviour of a car that tows a trailer. More specifically, we will study the

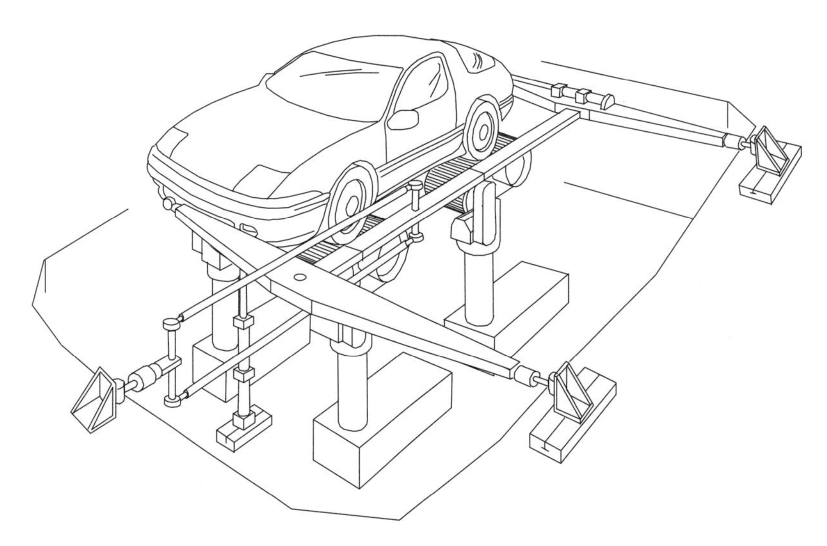

Fig. 11.1-28 The MTS Flat-Trac Roadway Simulator™, Milliken (1995).

possible unstable motions that may show up with such a combination. Linear differential equations are sufficient to analyse the stability of the straight ahead motion. We will again employ Lagrange's equations to set up the equations of motion. The original equations (11.1.25) may be employed because the yaw angle is assumed to remain small. The generalised coordinates Y, ψ and θ are used to describe the car's lateral position and the yaw angles of car and trailer, respectively. The forward speed dX/dt ($\approx V \approx u$) is considered to be constant. Fig. 11.1-29 gives a top view of the system with three degrees of freedom. The alternative set of three variables v, r and the articulation angle φ and the vehicle velocity V (a parameter) which are not connected to the inertial axes system $(0, X, Y)$ has been indicated as well and will be employed later on. The kinetic energy for this system becomes, if we neglect all the terms of the second order of magnitude (products of variables):

$$T = \tfrac{1}{2}m(\dot{X}^2 + \dot{Y}^2) + \tfrac{1}{2}I\dot{\psi}^2 + \tfrac{1}{2}m_c\{\dot{X}^2 + (\dot{Y} - h\dot{\psi} - f\dot{\theta})^2\} + \tfrac{1}{2}I_c\dot{\theta}^2 \quad (11.1.106)$$

The potential energy remains zero:

$$U = 0 \quad (11.1.107)$$

and the virtual work done by the external road contact forces acting on the three axles reads:

$$\delta W = F_{y1}\delta(Y + a\psi) + F_{y2}\delta(Y - b\psi) + F_{y3}\delta(Y - h\psi - g\theta) \quad (11.1.108)$$

With the use of the Eqs. (11.1.25) and (11.1.29) the following equations of motion are established for the generalised coordinates Y, ψ and θ:

$$(m + m_c)\ddot{Y} - m_c(h\ddot{\psi} + f\ddot{\theta}) = F_{y1} + F_{y2} + F_{y3} \quad (11.1.109)$$

$$(I_c + m_c f^2)\ddot{\theta} - m_c f(\ddot{Y} - h\ddot{\psi}) = -gF_{y3} \quad (11.1.110)$$

$$(I + m_c h^2)\ddot{\psi} - m_c h(\ddot{Y} - f\ddot{\theta}) = aF_{y1} - bF_{y2} - hF_{y3} \quad (11.1.111)$$

This constitutes a system of the sixth order. By introducing the velocities v and r the order can be reduced to four. In addition, the angle of articulation φ will be used. We have the relations:

$$\dot{Y} = V\psi + v, \quad \dot{\psi} = r, \quad \theta = \psi - \varphi \quad (11.1.112)$$

and with these the equations for v, r and φ:

$$(m + m_c)(\dot{v} + Vr) - m_c\{(h + f)\dot{r} - f\ddot{\varphi}\} = F_{y1} + F_{y2} + F_{y3} \quad (11.1.113)$$

$$\{I + m_c h(h + f)\}\dot{r} - m_c h(\dot{v} + Vr + f\ddot{\varphi}) = aF_{y1} - bF_{y2} - hF_{y3} \quad (11.1.114)$$

$$(I_c + m_c f^2)(\ddot{\varphi} - \dot{r}) + m_c f(\dot{v} + Vr - h\dot{r}) = gF_{y3} \quad (11.1.115)$$

The right-hand members are still to be expressed in terms of the motion variables. With the axle cornering stiffnesses C_1, C_2 and C_3 we have:

$$F_{y1} = C_1\alpha_1 = -C_1\frac{v + ar}{V}$$
$$F_{y2} = C_2\alpha_2 = -C_2\frac{v - br}{V}$$
$$F_{y3} = C_3\alpha_3 = -C_3\left(\frac{v - hr - g(r - \dot{\varphi})}{V} + \varphi\right) \quad (11.1.116)$$

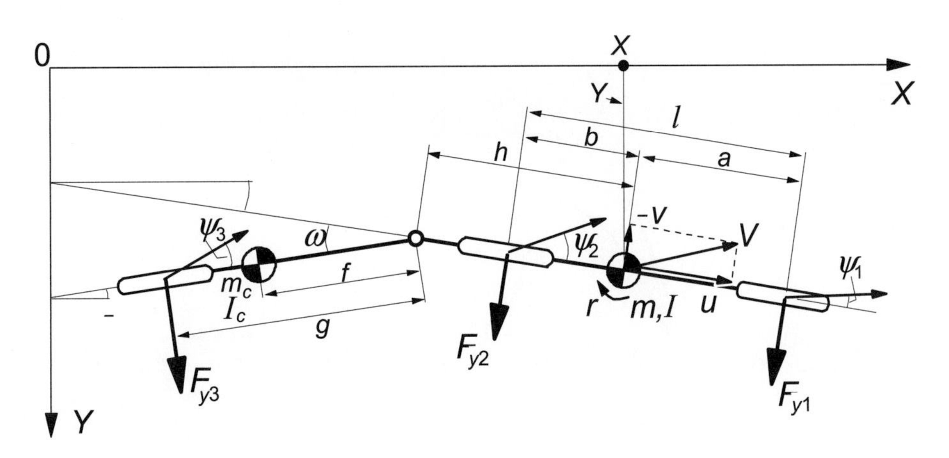

Fig. 11.1-29 Single-track model of car trailer combination.

From the resulting set of linear differential equations the characteristic equation may be derived which is of the fourth degree. Its general structure is:

$$a_0 s^4 + a_1 s^3 + a_2 s^2 + a_3 s + a_4 = 0 \qquad (11.1.117)$$

The stability of the system can be investigated by considering the real parts of the roots of this equation or we might employ the criterium for stability according to Routh–Hurwitz. According to this criterion the system of order n is stable when all the coefficients a_i are positive and the Hurwitz determinants H_{n-1}, H_{n-3} etc. are positive. For our fourth-order system the complete criterion for stability reads:

$$\begin{aligned} H_3 &= \begin{bmatrix} a_1 & a_0 & 0 \\ a_3 & a_2 & a_1 \\ 0 & a_4 & a_3 \end{bmatrix} \\ &= a_1\, a_2\, a_3 - a_1^2\, a_4 - a_0\, a_3^2 > 0 \\ a_i > 0 &\quad \text{for } i = 0, 1, \ldots, 4 \end{aligned} \qquad (11.1.118)$$

In Fig. 11.1-30, the boundaries of stability have been presented in the caravan axle cornering stiffness vs speed parameter plane. The three curves belong to the three different sets of parameters for the position f of the caravan's centre of gravity and the caravan's mass m_c as indicated in the figure. An important result is that a lower cornering stiffness promotes oscillatory instability: the critical speed beyond which instability occurs decreases. Furthermore, it appears from the diagram that moving the caravan's centre of gravity forward (f smaller) stabilises the system which is reflected by the larger critical speed. A heavier caravan (m_c larger) appears to be bad for stability. Furthermore, it has been found that a larger draw bar length g is favourable for stability.

It turns out that a second type of instability may show up. This occurs when the portion of the weight of the caravan supported by the coupling point becomes too large. This extra weight is felt by the towing vehicle and makes it more oversteer. The critical speed associated with this phenomenon is indicated in the diagram by the vertical lines. This divergent instability occurs when (starting out from a stable condition) the last coefficient becomes negative, that is $a_n = a_4 < 0$.

The oscillatory instability connected with the 'snaking' phenomenon arises as soon as (from a stable condition) the second highest Hurwitz determinant becomes negative, $H_{n-1} = H_3 < 0$ (then also $H_n < 0$). When the critical speed is surpassed self-excited oscillations are created which shows an amplitude that in the actual non-linear case does not appear to limit itself. This is in contrast to the case of the wheel shimmy phenomenon. The cause of the unlimited snaking oscillation is that with increasing amplitudes also the slip angle increases which lowers the average cornering stiffness as a consequence of the degressively non-linear cornering force characteristic. From the diagram we found that this will make the situation increasingly worse. As has been seen from full vehicle/caravan model simulations, the whole combination will finally overturn. Another effect of this reduction of the average cornering stiffness is that when the vehicle moves at a speed lower than the critical speed, the originally stable straight ahead motion may become unstable if through the action of an external disturbance (side wind gust) the slip angle of the caravan axle becomes too large (surpassing of the associated unstable

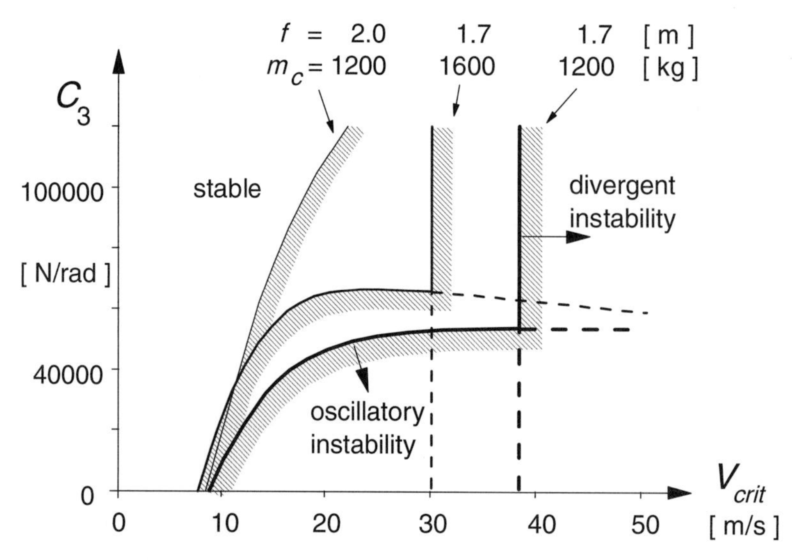

Fig. 11.1-30 Stability boundaries for the car caravan combination in the caravan cornering stiffness vs critical speed diagram. Vehicle parameters according to Table 11.1-1, in addition: $h = 2$ m, $g = 2$ m, $k_c = 1.5$ m ($I_c = m_c k_c^2$), cf. Fig. 11.1-29.

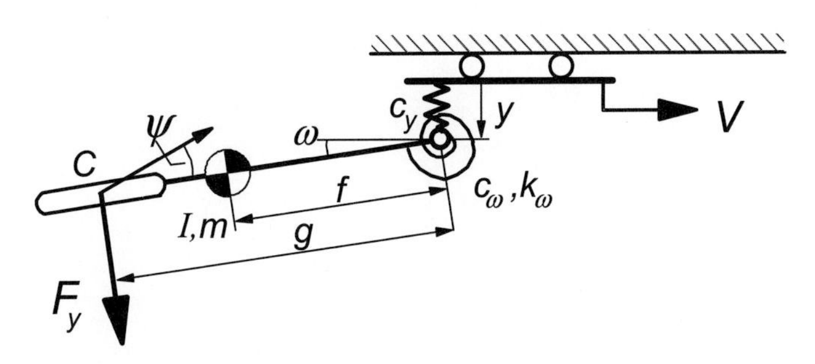

Fig. 11.1-31 On the stability of a trailer (Exercise 11.1.4).

limit-cycle). This is an unfortunate, possibly dangerous situation!

Exercise 11.1.4. Stability of a trailer

Consider the trailer of Fig. 11.1-31 that is towed by a heavy steadily moving vehicle at a forward speed V along a straight line. The trailer is connected to the vehicle by means of a hinge. The attachment point shows a lateral flexibility that is represented by the lateral spring with stiffness c_y. Furthermore, a yaw torsional spring and damper are provided with coefficients c_φ and k_φ.

Derive the equations of motion of this system with generalised coordinates y and φ. Assume small displacements so that the equations can be kept linear. The damping couple $k_\varphi \dot{\varphi}$ may be considered as an external moment acting on the trailer or we may use the dissipation function $D = \frac{1}{2} k_\varphi \dot{\varphi}^2$ and add $+\partial D/\partial \dot{q}_i$ to the left-hand side of Lag range's equation (11.1.25). Obviously, the introduction of this extra term will be beneficial particularly when the system to be modelled is more complex.

Assess the condition for stability for this fourth-order system. Simplify the system by putting $g = f$ and $c_\varphi = k_\varphi = 0$. Now find the explicit conditional statement for the cornering stiffness C.

11.1.3.7 Vehicle dynamics at more complex tyre slip conditions

So far, relatively simple vehicle dynamics problems have been studied in which the basic steady-state cornering force vs slip angle characteristic plays the dominant role. The situation becomes more complex when matters like combined slip at hard braking, wheel camber, tyre transient and vibrational properties and e.g. obstacle crossings are to be considered.

Section Twelve

Brakes

Chapter 12.1

12.1

Braking systems

Julian Happian-Smith

The aim of this chapter is to:

- Aid the designer to understand the legal requirements of braking systems;
- Understand the basic requirements for braking systems to be successful;
- Understand the design process for achieving an efficient braking system;
- Appreciate the material requirements for efficient braking systems;
- Understand current developments in braking control systems.

12.1.1 Introduction

The safe and reliable use of a road vehicle necessitates the continual adjustment of its speed and distance in response to change in traffic conditions. This requirement is met in part by the braking system, the design of which plays a key role in ensuring a particular vehicle is suitable for a given application. This is achieved through the design of a system that makes as efficient use as possible of the finite amount of traction available between the tyre and the road over the entire range of operating conditions that are likely to be encountered by the vehicle during normal operation.

The purpose of this chapter is to introduce the reader to the basic mechanics associated with the deceleration behaviour of a road vehicle and provide insight to the many issues that must be addressed when selecting the brake rotor and friction materials. A complete coverage is not feasible within the confines of a single chapter and so a set of references and additional reading is provided at its end that points the interested reader to further sources of information.

The chapter commences with a review of the function of a brake system together with an outline of the principal components and their possible configurations. The subject of legislation is reviewed and its importance as a tool to aid the designer of a brake system is highlighted. Straight forward kinematic and kinetic analyses are used to address the fundamentals of the braking problem as a precursor to the analysis of brake proportioning, adhesion utilization and other related issues. A case study is built into this section of the chapter that illustrates the application of the theory and so reinforces understanding. The selection of appropriate materials from which to manufacture the friction pair is reviewed and problems linked to thermo-mechanical behaviour highlighted. The chapter concludes with a brief summary of more advanced topics, often linked to modern chassis control, that integrate the braking system with other chassis systems.

12.1.1.1 The functions and conditions of use of a brake system

In order to understand the behaviour of a braking system it is useful to define three separate functions that must be fulfilled at all times:

(a) The braking system must decelerate a vehicle in a controlled and repeatable fashion and when appropriate cause the vehicle to stop.

(b) The braking system should permit the vehicle to maintain a constant speed when travelling downhill.

(c) The braking system must hold the vehicle stationary when on a flat or on a gradient.

Introduction to Modern Vehicle Design; ISBN: 9780750650441
Copyright © 2001 Julian Happian-Smith. All rights of reproduction, in any form, reserved.

When simply stated, as above, the importance of the role played by the brakes/braking system in controlling the vehicle motion is grossly understated. Consideration of the diverse conditions under which the brakes must operate leads to a better appreciation of their role. These include, but are not limited to, the following:

- slippery wet and dry roads.
- rough or smooth road;
- split friction surfaces;
- straight line braking or when braking on a curve;
- wet or dry brakes;
- new or worn linings;
- laden or unladen vehicle;
- vehicle pulling a trailer or caravan;
- frequent or infrequent applications of short or lengthy duration;
- high or low rates of deceleration;
- skilled or unskilled drivers.

Clearly the brakes, together with the steering components and tyres, represent the most important accident avoidance systems present on a motor vehicle which must reliably operate under various conditions. The effectiveness of any braking system is, however, limited by the amount of traction available at the tyre–road interface.

12.1.1.2 System design methodology

The primary functions of a brake system, listed above, must be fulfilled at all times. In the event of a system failure, the same functions must also be performed albeit with a reduced efficiency. Consequently the braking system of a typical passenger car comprises a service brake for normal braking, a secondary/emergency brake used in the event of a service brake failure and a parking brake. Current practice permits service brake components to be used in the secondary/parking brake systems.

Irrespective of the detail design considerations all brake systems divide into the following subsystems:

(1) *Energy source*
This includes all those components which generate, store or release energy required by the braking system. In standard passenger cars muscular pedal effort, applied by the driver, in combination with a vacuum boost system comprise the energy source. In the event of a boost failure, the driver can still apply the brakes by muscular effort alone. Alternative sources of energy include power braking systems, surge brakes, drop weight brakes, electric and spring brakes.

(2) *Modulation system*
This embraces those elements of the brake system which are used to control the level of braking effort applied to each brake. Included in this system are the driver, pressure limiting/modulating values and, if fitted, anti-lock braking systems (ABSs).

(3) *Transmission system*
The components through which energy travels to the wheel brakes comprise the transmission system. Brake lines (rigid tubes) and brake hoses (flexible tubes) are used in hydraulic and air brake systems. Mechanical brakes make use of rods, levers, cams and cables to transmit energy. The parking brake of a car quite often makes use of a mechanical transmission system.

(4) *Foundation brakes*
These assemblies generate the forces that oppose the motion of the vehicle and in doing so convert the kinetic energy associated with the longitudinal motion of the vehicle into heat.

There are four main stages involved in the design of a brake system. The first, and perhaps most fundamental stage, is the choice of brake force distribution between the axles of the vehicle. This is primarily a function of the vehicle dimensions and its weight distribution. Next is the design of the transmission system and this activity embraces the sizing of the master cylinder together with the front and rear wheel cylinders. Additional components, such as special valves that modulate the hydraulic pressure applied to each wheel are physically accounted for at this stage. The foundation brakes form the focus of the third stage of the process. As well as being able to react the applied loads and torques, the foundation brakes must be endowed with adequate thermal performance, wear and noise characteristics. The last phase in the process results in the incorporation of the pedal assembly and vacuum boost system into the brake system. To accomplish this design task, the engineer requires access to several fundamental vehicle parameters. These include:

- laden and unladen vehicle mass;
- static weight distribution when laden and unladen;
- wheelbase;
- height of centre of gravity when laden and unladen;
- maximum vehicle speed;
- tyre and rim size;
- vehicle function;
- braking standards.

It is essential to recognize that each of the preceding stages are closely linked and that the final design will take many iterations to realize. Thus any formal methodology must be designed so as not to compromise the overall system quality that could result from design changes at the component level. By way of example, a reduction in package space could lead to smaller diameter wheel brakes having to be fitted to the vehicle. This will change the brake force distribution unless checked, by say resizing the wheel cylinders, and in the worst case this could lead to premature wheel lock and a violation of the governing legislation.

12.1.1.3 Brake system components and configurations

The principal components put together comprise a conventional braking system that is outlined below together with possible brake system layouts. The discussion of the components begins with the pedal assembly and moves through the brake system finishing with the foundation or wheel brakes.

12.1.1.3.1 Pedal assembly

A brake pedal consists of an arm, pad and pivot attachments. The majority of passenger cars make use of hanging pedals. A linkage is connected to the pedal and this transmits both force and movement to the master cylinder.

12.1.1.3.2 Brake booster

The brake booster serves to amplify the foot pressure generated when the brake pedal is depressed. This has the effect of reducing the manual effort required for actuation. Boosters are invariably combined with the master cylinder assembly. A vacuum booster employs the negative pressure generated in the intake manifold of a spark ignition engine, whereas a hydraulic booster relies upon the existence of a hydraulic energy source and typically finds application in vehicles powered by diesel engines that generate only a minimal amount of intake vacuum.

12.1.1.3.3 Master cylinder

The master cylinder essentially initiates and controls the process of braking. The governing regulations demand that passenger vehicles be equipped with two separate braking circuits and this is satisfied by the so-called tandem master cylinder. A tandem master cylinder has two pistons housed within a single bore. Each section of the unit acts as a single cylinder and the piston closest to the brake pedal is called the primary piston whilst the other is called the secondary piston. Thus, if a leak develops within the primary circuit, the primary piston moves forward until it bottoms against the secondary piston. The push rod force is transmitted directly to the secondary piston through piston-to-piston contact, thus allowing the secondary piston to pressurize the secondary circuit. Conversely, if the secondary circuit develops a leak then the secondary piston moves forward until it stops against the end of the master cylinder bore. This then allows trapped fluid between the two pistons to become pressurized and so the primary circuit remains operative.

12.1.1.3.4 Regulating valves

The dynamics of the braking process gives rise to need for some means of reducing the magnitude of the brake force generated at the rear of a vehicle under the action of increasing rates of deceleration. This need arises form the load transfer that takes place from rear to front during any braking event. This function is realized through the incorporation of some form of brake pressure regulating valve into the rear brake circuit. The exact nature of the valve depends upon the detail design but they fall into three generic types;

- Load sensitive pressure regulating valve: Valves of this type are fitted to vehicles that experience large in-service changes in axle load. The valve is anchored to the vehicle body and is also connected to the rear suspension through a mechanical linkage. This permits the valve to sense the relative displacement between the body and suspension and adjust the valve performance to effect control over the rear line pressure and so enable the rear brakes to compensate for the change in axle load.
- Pressure-sensitive pressure regulating valve: Otherwise known as a pressure limiter, this type of valve isolates the rear brake circuit when the line pressure exceeds a predetermined value. They find application on vehicles that are characterized by a low centre of gravity and a limited cargo volume.
- Deceleration-sensitive pressure regulating valve: This class of valve finds wide application. The actuation point is determined by the rate of deceleration of the vehicle and this is typically of the order of 0.3*g*.
  A benefit of this type of valve is that it does provide for a degree of load-sensitive operation as the overall deceleration of the vehicle is the function of the vehicle weight and the line pressure. They are also sensitive to braking on a slope. Mathematical models of this class of valve are developed later in the text and their influence on the performance of a brake system is demonstrated.

12.1.1.3.5 Foundation brakes

Foundation, or wheel brakes, divide into two distinct classes, namely disc (axial) and drum (radial) brakes. Modern vehicles are invariably fitted with disc units on the front axle and there is a growing tendency to fit similar units to the rear axle. If drum brakes are fitted to

1 Brake circuit 1
2 Brake circuit 2
II variant X variant

Figure 12.1-1 Common brake system layouts.

the rear axle then these are typically of the Simplex type which employs a leading and trailing shoe configuration to generate the required brake torque. The torque output of this type of drum brake is not sensitive to change in vehicle direction. On vehicles fitted entirely with disc brakes, then a small drum unit is often employed to act as a parking brake on the rear axle of the vehicle. Issues surrounding the selection of the materials used to manufacture both discs and drums together with their friction material partners are discussed in more detail later in the text.

12.1.1.3.6 Brake system layouts

Legislative requirements demand a dual circuit transmission system to be installed on all road vehicles. Of the five possible configurations, two have become standard and these are known as the II and X variants shown in Figure 12.1-1. The II design is characterized by separate circuits for both the front and rear axles whilst in the X configuration, each circuit actuates one wheel at the front and the diagonally opposed rear wheel. The II design is often found on vehicles that are rear heavy and the X layout has application on vehicles that are front heavy.

12.1.2 Legislation

Without exception, motorized road vehicles, whether cars, buses or lorries, represent a potentially lethal hazard to other road users and pedestrians. Also the rise in 'green thinking' during the past decade has led to serious consideration of the impact of road vehicles on the environment in which we all must live. It is one of the many responsibilities of a government to ensure that all road vehicles are as safe as possible and that any adverse effects of the vehicle on the environment are minimized. This task is achieved through legislation which, in so far as the brakes are concerned, primarily sets the minimum standards for the performance of the systems and their components that combine to arrest the motion of a vehicle in a controlled manner. A design engineer has to take into account many factors associated with the mechanics of braking when designing a new brake system. In addition to these elements, conformity to the legislative requirements of the country or countries in which the vehicle is to operate is absolutely essential. Thus a working knowledge of the content and scope of such documents forms a very important part of the brake engineer's database of information.

There are many arguments both for and against legislation. However, despite some inevitable drawbacks standards and legislation form a necessary and desirable part of today's society and they are here to stay. Poorly written legislative documents may smother initiative and restrict technical progress by enforcing unrealistic standards and by failing to recognize the advance of technology. The phrasing of the documents is somewhat complex which can lead to difficulties in understanding their content. This is, to a large extent, unavoidable because they are legal documents and they must attempt to cover all eventualities, prevent ambiguity and close any loopholes. By default, proof of compliance with a national or international standard generates an overhead which is transferred to the consumer as an added cost. It is essential that this process does not inhibit either new or small manufacturers from entering the marketplace. Finally, national legislation can be used as an economic weapon (termed a technical barrier to trade), particularly by those countries operating an approval system, to protect their industry from worldwide intrusion into the local market. The job of the importer is made very difficult through the use of standards and test procedures that favour the home industry or by withholding interpretations, by allowing the test authority the use of subjective judgement and by showing a lack of co-operation in test scheduling and in the issue of approval documents.

Technical standards and legal requirements need to be kept under review in order to force quality upwards. They must set realistic standards for new and in-service vehicles that result in real improvements in safety and environmental protection. Bearing in mind the necessary legal constraints, they should be straightforward to understand and interpret as well as be universally acceptable to encourage free trade and so prevent the production of trade barriers. Standards and legislation must also be applicable to all types of vehicle and should not preclude innovative design by being so inflexible as to limit technical advance; ideally they must actively encourage the use of new technology. Given time, companies incorporate the formal test procedures in their design programmes and develop close working relationships with the national approval bodies. In principle this leads to improved export performance since approval obtained in the country of manufacture is automatically valid for all others. Also the clearly defined ground rules quite often act as an aid to product development.

With regard to the braking system, legislation first appeared in the form of the Motor Cars Order of 1904. Since this time, the range and complexity of the vehicles that populate the road network has markedly increased. Inevitably this has been accompanied by a similar

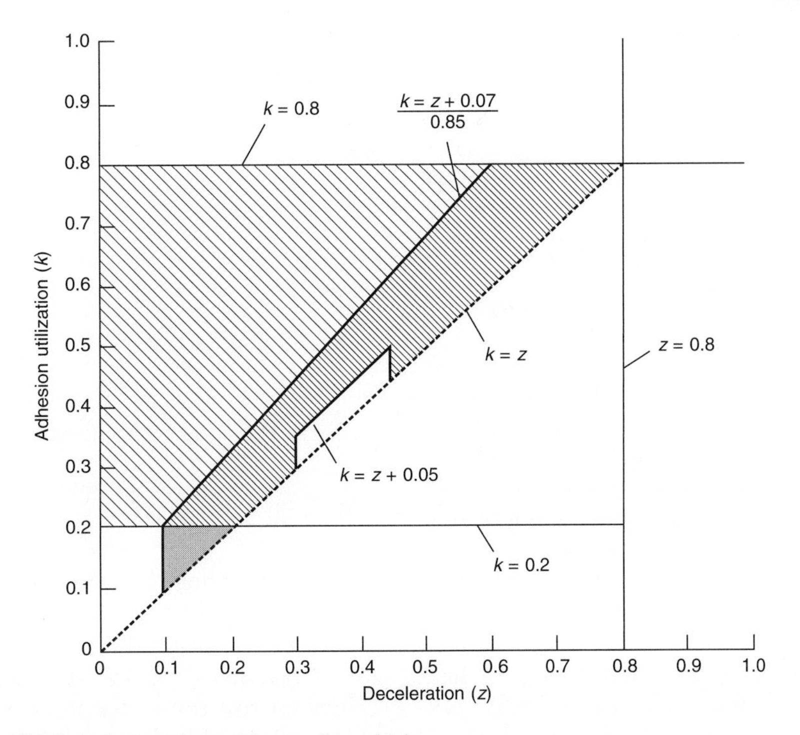

Figure 12.1-2 Adhesion utilization diagram for a category M_1 vehicle.

increase in the size and complexity of the regulations pertaining to braking. The first major change away from the self-certification process came when the UK commenced along the road of Type Approval, favoured by continental Europe, by the incorporation of the Economic Commission for Europe (ECE) Regulations. These voluntary regulations attempt to harmonize vehicle legislation and they provide for the reciprocal acceptance and notification of vehicle systems and component approval. The relevant braking legislation is contained in Regulation 13. The next major change occurred when the UK became a member of the Common Market in which the acceptance of EEC Directives is binding on all member states. Member states are not allowed to impose more stringent standards than those contained in the Directives. They are, however, free to demand additional standards with regard to matters not covered by EEC legislation. EEC Directives differ from the corresponding ECE Regulations in one major aspect: Approvals issued by one member must be accepted by all others. The objective of either is, however, the same, namely the harmonization of differing technical requirements. In 1978, it became mandatory in the UK for all new cars to be type approved to the EEC Braking Directives. A similar exemption was granted in 1980 for vehicles approved to ECE Regulation 13. The current EEC Directive on braking is Directive 71/320/EEC as last amended by Directive 91/422/EEC.

Of the many requirements laid down in the EEC regulations, perhaps the single most important aspect that must be satisfied by a road vehicle relates to its use of available tyre–ground adhesion. Manufacturers have to submit adhesion utilization curves that demonstrate compliance with the limits defined in figures IA and IB of EEC Directive 71/320 Annex II. The adhesion utilization diagram that refers to a category M_1 vehicle (passenger vehicle with seating capacity up to eight including the driver) is shown in Figure 12.1-2, and is derived from that contained within EEC Council Directive 71/320/EEC as last amended by 91/422/EEC.

The Directive uses the letter k to indicate adhesion utilization and z for deceleration and it states that for all categories of vehicle for values of adhesion utilization between 0.2 and 0.8,

$$z \geq 0.1 + 0.85(k - 0.2)$$

For category M_1 vehicles, the adhesion utilization curve of the front axle must be greater than that of the rear for all load cases and values of deceleration between $0.15g$ and $0.8g$. Between deceleration levels of $0.3g$ and $0.45g$, an inversion of the adhesion utilization curves is allowed provided the rear axle adhesion curve does not exceed the line defined by $k = z$ by more than 0.05. The above provisions are applicable within the area defined by the lines $k = 0.8$ and $z = 0.8$.

Compliance of the braking system to the constraints defined in Figure 12.1-2 ensures that the rear wheels do not lock in preference to the front wheels and that the proportion of braking effort exerted at the front of the vehicle is limited so that the braking system does not become too inefficient. A detailed interpretation of this requirement is outlined later in the text.

12.1.3 The fundamentals of braking

12.1.3.1 Kinematics of a braking vehicle

12.1.3.1.1 Kinematic analysis of a braking vehicle

The distance travelled by a vehicle when braking, either during a stop or snub (when the final velocity is non-zero) is a basic measure of the effectiveness of a brake system. Before addressing issues coupled to the forces which act on a vehicle during a braking manoeuvre it is worthwhile to first consider the kinematic behaviour of the vehicle. A straightforward kinematic analysis assuming straight line (one-dimensional) motion and constant deceleration provides a ready indication of stopping distance. Predictions of stopping distance made by this analysis find use in accident reconstruction.

With reference to Figure 12.1-3, the total distance travelled is made up of two parts. In part 1, the total distance travelled by the vehicle moving with constant velocity U is:

$$S_1 = Ut_1 \tag{12.1.1}$$

In part 2 the vehicle is decelerated at a constant rate until such time as the vehicle comes to rest. The distance travelled is:

$$S_2 = \frac{Ut_2}{2} = \frac{U^2}{2a} \tag{12.1.2}$$

Thus, the total stopping distance is simply

$$S_i = S_1 + S_2 = Ut_1 + \frac{U^2}{2a} \tag{12.1.3}$$

The preceding analysis assumes the vehicle deceleration is achieved instantaneously and is sustained for the duration of the stop. No account is taken of driver reaction time, initial system response time, deceleration rise time, change in deceleration during the period of actual braking and, if applicable, release time. These factors are now defined with reference to

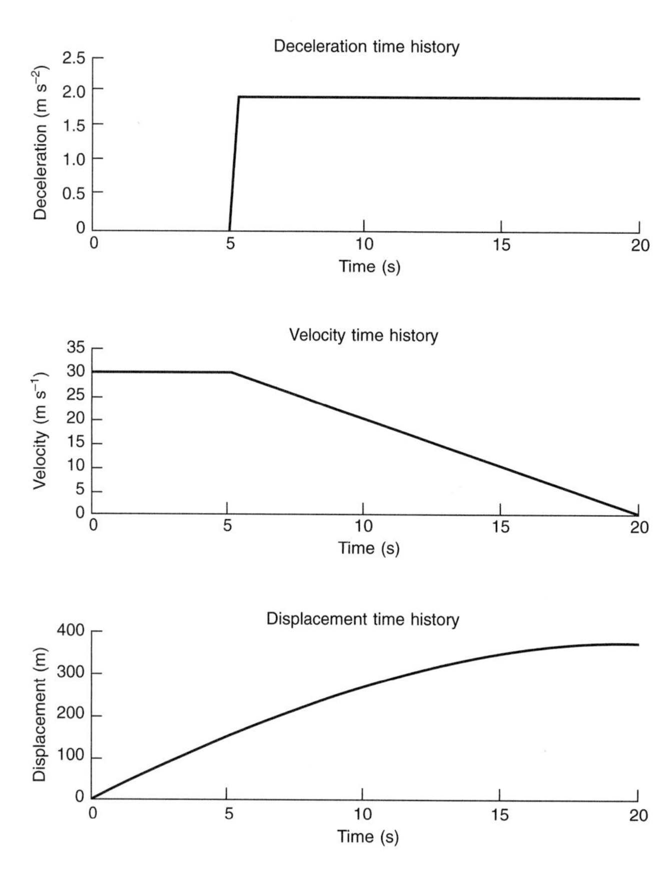

Figure 12.1-3 Kinematics of a simple stop.

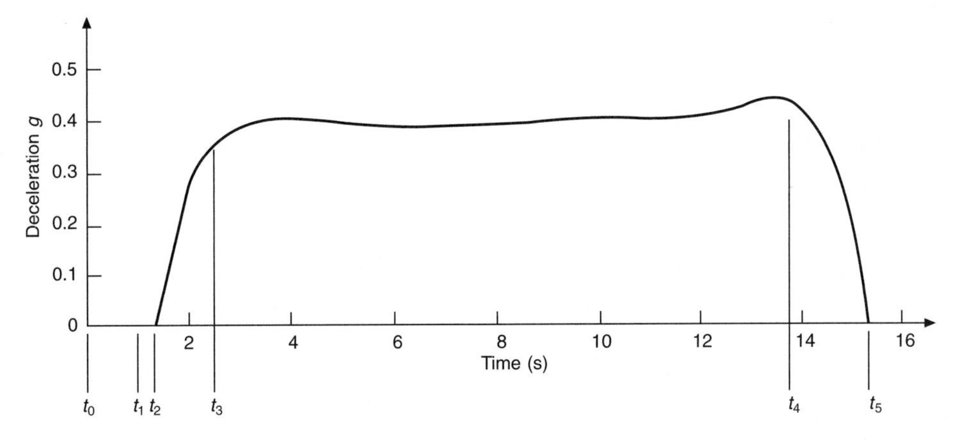

Figure 12.1-4 Typical measured deceleration time history.

Figure 12.1-4, which illustrates the characteristics of a typical measured deceleration time history.

12.1.3.1.2 Driver reaction time

The time taken by the driver to respond to the danger, formulate an avoidance strategy and physically move his/her right foot from the accelerator to the brake, $(t_1 - t_0)$.

12.1.3.1.3 Initial system response time

The time taken from the point at which the brake pedal begins to move to the time at which a braking force is generated at the tyre road interface, $(t_2 - t_1)$.

12.1.3.1.4 Deceleration rise time

The elapsed time for the deceleration to reach the value, determined by the driver, for the stop in question, $(t_3 - t_2)$.

12.1.3.1.5 Braking time

The time taken from the point at which fully developed braking is reached to the time at which the vehicle stops or brake release begins, $(t_4 - t_3)$.

12.1.3.1.6 Release time

The elapse time between the point at which brake release starts to the time at which brake force generation ceases, $(t_5 - t_4)$.

The stopping time and distance are measured from the time t_0 to the time t_4 or t_5, if appropriate, whilst the braking time and distance is measured from time t_1. Inclusion of these factors modifies the simple kinematic analysis with the result that the predicted stopping times and distances are increased.

Additional deceleration forces, such as those arising from engine drag, aerodynamic drag, rolling resistance and gravity, are not taken into account in the following analysis, which is again built around one-dimensional particle kinematics. With reference to Figure 12.1-5, the following expressions for the distance travelled during each stage of the stop can be derived. Note that as a stop is the subject of the analysis, the effect of release time $(t_5 - t_4)$ is ignored.

Distance travelled during reaction time S_1:

$$S_1 = U(t_1 - t_0) \tag{12.1.4}$$

Distance travelled during the initial system response time S_2:

$$S_2 = U(t_2 - t_1) \tag{12.1.5}$$

The distance travelled during the deceleration rise time S_3, assuming this to be linear, can be shown to be:

$$S_3 = U(t_3 - t_2) - \frac{a_f(t_3 - t_2)^2}{6} \tag{12.1.6}$$

where a_f is the value of the mean fully developed deceleration. The distance travelled during the course of braking, S_4, under the deceleration of a_f, which is assumed to remain constant, can be expressed as:

$$S_4 = \frac{1}{2a_f}\left[U^2 + \frac{a_f^2(t_3 - t_2)^2}{4} - Ua_f(t_3 - t_2)\right] \tag{12.1.7}$$

Thus, the total stopping distance, S_s, is simply

$$S_s = \sum_{i=1}^{4} S_i \tag{12.1.8}$$

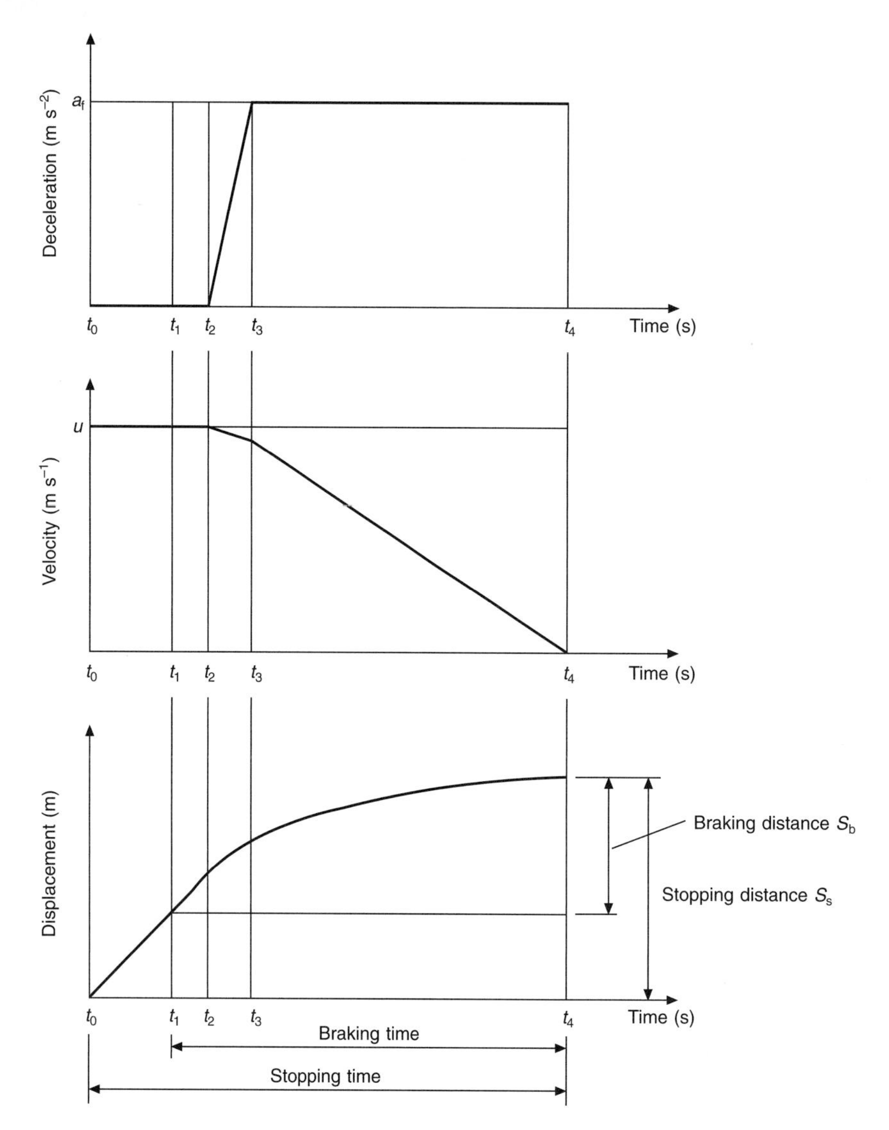

Figure 12.1-5 Four-stage stop simulation.

and the braking distance, S_b, is:

$$S_b = \sum_{i=2}^{4} S_i \tag{12.1.9}$$

12.1.3.2 Kinetics of a braking vehicle

A general equation for braking performance can be easily derived through application of Newton's second law to a simplified free-body diagram of a vehicle in the direction of its travel (Figure 12.1-6). Assuming x is positive in the direction of travel, then:

$$\sum F_x = M\ddot{x} \tag{12.1.10}$$

and so

$$-T_f - T_r - D - P\sin\theta = M\ddot{x} \tag{12.1.11}$$

where

M = Vehicle mass
P = Vehicle weight
g = Acceleration due to gravity
T_f = Front axle braking force
T_r = Rear axle braking force
R_f = Front axle load
R_r = Rear axle load
θ = Angle of incline

Note that the front and rear braking force terms, T_f and T_r, represent the sum of all the effects that combine to generate the forces which act between the front and rear axles and ground. These include the torque generated by the brakes together with rolling resistance effects, bearing friction and drive train drag.

If an additional variable for linear deceleration, d, is defined such that

$$d = -\ddot{x} \tag{12.1.12}$$

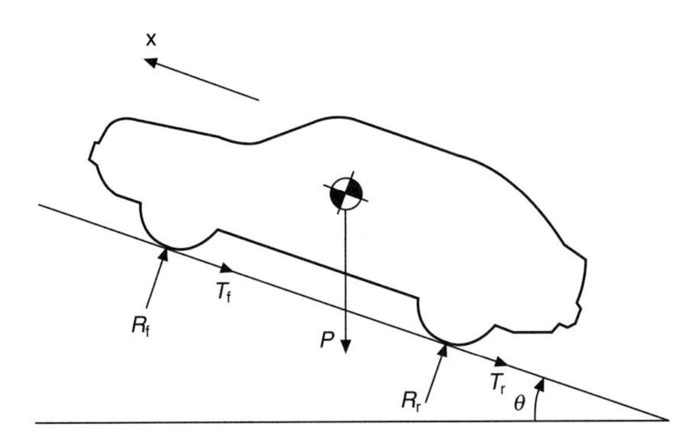

Figure 12.1-6 Free body diagram of a braking vehicle.

then equation 12.1.11 becomes:

$$Md = T_f + T_r + D + P\sin\theta = T \quad (12.1.13)$$

in which T is the sum of all those forces that contribute to the overall braking effort.

By considering the case of constant deceleration, straightforward and fundamental relationships can be derived that yield an appreciation of the physics which governs all braking events.

From equation 12.1.13, the linear deceleration of a vehicle can be expressed as:

$$d = \frac{T}{M} = -\frac{dv}{dt} \quad (12.1.14)$$

in which v is the forward velocity of the vehicle. Since the deceleration is assumed constant, then the total brake force is also constant and so equation 12.1.14 can be integrated with respect to time between the limits of the initial velocity, v_0, to the final velocity, v_f, to determine the duration of the braking event, t_b.

On rearranging, equation 12.1.14 becomes:

$$\int_{v_0}^{v_f} dv = -\frac{T}{M}\int_0^{t_b} dt \quad (12.1.15)$$

which leads to

$$v_0 - v_f = \frac{T}{M}t_b \quad (12.1.16)$$

The fact that velocity and displacement are related by $v = dx/dt$ permits an expression for stopping distance to be derived from equation 12.1.14 through substitution for dt and integration between v_0 and v_f as before. On rearranging, equation 12.1.14 becomes:

$$\frac{T}{M}\int_{x_0}^{x_f} dx = -\int_{v_0}^{v_f} v dv \quad (12.1.17)$$

which leads to

$$\frac{T}{M}(x_f - x_0) = \frac{Tx}{M} = \frac{v_0^2 - v_f^2}{2} \quad (12.1.18)$$

where x is the distance travelled during the brake application.

When considering a stop, the final velocity v_f is zero and so the stopping distance x is, from equation 12.1.18, given by:

$$x = \frac{Mv_0^2}{2T} \quad (12.1.19)$$

and the time, t_b, taken to stop the vehicle is, from equation 12.1.16,

$$t_b = \frac{Mv_0}{T} = \frac{v_0}{d} \quad (12.1.20)$$

Thus, from equation 12.1.19 the distance required to stop the vehicle is proportional to the square of the initial velocity and, from equation 12.1.20, the time taken to stop the vehicle is proportional to the initial velocity.

To achieve maximum deceleration and hence minimum stopping distance on a given road surface, each axle must simultaneously be on the verge of lock. If this is so then, if z is the deceleration as a proportion of g, $z = \frac{d}{g}$, and the brake force T is equal to the product of the vehicle weight and the coefficient of tyre–ground adhesion, P_μ, then from equation 12.1.13,

$$P_z = T = P_\mu \quad (12.1.21)$$

from which it can be deduced that

$$z = \mu \quad (12.1.22)$$

This represents a limiting case in which it is clear that the maximum deceleration cannot exceed the value of tyre–ground adhesion. A deceleration in excess of $1g$ therefore implies that the tyre–ground adhesion has a value greater than unity and this is quite realizable with certain types of tyre compound.

As already indicated, the primary source of retardation force arises from the foundation brake. Secondary forces which contribute to the overall braking performance include:

- Rolling resistance, expressed by a coefficient of rolling resistance. The total rolling resistance is independent of the load distribution between axles and the force is typically equivalent to a nominal $0.01g$ deceleration.
- Aerodynamic drag which depends on dynamic pressure and is proportional to the square of the vehicle speed. It is negligible at low speeds, however, aerodynamic drag

may account for a force equivalent to $0.03g$ when travelling at high speed.

- Gradient makes either a positive (uphill) or negative (downhill) contribution to the total braking force experienced by a vehicle. This force is simply the component of the total vehicle weight acting in the plane of the road.
- Drivetrain drag may either help or hinder the braking performance of a vehicle. If the vehicle is decelerating faster than the components of the drivetrain would slow down under their own friction then a proportion of the brake torque generated by the wheel brakes must be used to decelerate the rotating elements within the drivetrain. Thus, the inertia of the elements of the drivetrain effectively adds to the mass of the vehicle and so should be considered in any rigorous brake design programme. Conversely, the drivetrain drag may be sufficient to decelerate the rotating elements and so contribute to the overall vehicle braking effort and this is often the case during braking manoeuvres involving a low rate of deceleration.

12.1.3.3 Tyre–road friction

The brake force, F_b, which acts at the interface between a single wheel and the road is related to the brake torque, T_b, by the relationship:

$$F_b = \frac{T_b}{r} \tag{12.1.23}$$

where r is the radius of the wheel. The brake force on a vehicle can be predicted using equation 12.1.23 as long as all the wheels are rolling. The brake force F_b cannot increase without bound as it is limited by the extent of the friction coupling between the tyre and the road.

The friction coupling that gives rise to the brake force characteristic reflects the combination of tyre and road surface materials together with the condition of the surface. The best conditions occur on dry, clean road surfaces on which the brake force coefficient, defined as the ratio of brake force to vertical load, μ_b, can reach values between 0.8 and unity. Conversely, icy surfaces reflect the poorest conditions and on ice the brake force coefficient can lie between 0.05 and 0.1. On wet surfaces or on roads contaminated by dirt, the brake force coefficient typically spans the range 0.2–0.65.

Hysteresis and adhesion are the two mechanisms responsible for friction coupling. Surface adhesion comes about from the intermolecular bonds which exist between the rubber and the aggregate in the road surface. Hysteresis, on the other hand, represents an energy loss in the rubber as it deforms when sliding over the aggregate. Each of these mechanisms relies on a small amount of slip taking place at the tyre–road interface and so brake force and slip co-exist. The longitudinal slip of the tyre is defined as a ratio:

$$\text{slip} = \frac{\text{slip velocity in contact patch}}{\text{forward velocity}} = \frac{v - \omega r}{v} \tag{12.1.24}$$

where v is the forward velocity of the vehicle, ω is the angular velocity of the wheel and r the wheel radius.

Useful information can be obtained by plotting brake force coefficient against slip (Figure 12.1-7). During straight line braking, no lateral forces are generated which means that all of the force that is potentially available within the tyre–ground contact patch can be used to decelerate the vehicle. The uppermost characteristic illustrates the brake coefficient derived from both the adhesive and hysteretic mechanisms and it increases linearly with increase in slip up to around 20% slip. On dry roads, the adhesion component dominates the production of friction coupling. The peak coefficient, denoted by μ_p, defines the maximum braking force that can be obtained for a given tyre–road friction pair. At higher values of slip this coefficient decreases to its lowest value of μ_s at 100% slip, which represents the full lock condition. The maximum brake force, corresponding to μ_p, is a theoretical maximum as the system becomes unstable at this point. Once a wheel is decelerated to the point at which μ_p is achieved, any disturbance about this point results in an excess of brake torque that causes the wheel to decelerate further. This leads to an increase in slip and this in turn reduces the brake force leading to a rapid deceleration to the full lock

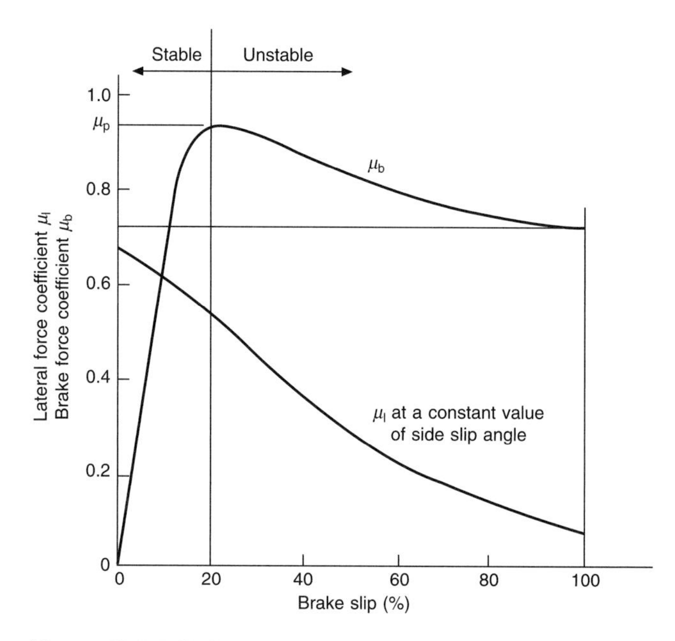

Figure 12.1-7 Brake force against wheel slip.

condition. It is worthwhile to note that ABSs make use of this phenomenon.

The negotiation of a bend requires a vehicle to develop a lateral force in the tyre–ground plane through the deformation of the tyre carcass brought about by a slip angle. Lateral forces, characterized by the coefficient μ_l, being a function of slip angle, and longitudinal forces, characterized by μ_b thus co-exist and compete for the finite amount of force that is available within the contact patch. A typical lateral force coefficient is shown on Figure 12.1-7 for a given value of slip angle. It has a maximum value when the brake slip is zero and falls with increase in brake slip. The minimum value occurs when the wheel has locked and is unable to generate further lateral force.

12.1.4 Brake proportioning and adhesion utilization

The vertical loads carried by the front and rear wheels of a rigid, two axle vehicle are not, in general, equal. In order to efficiently utilize the available tyre–road adhesion the braking effort must be apportioned between the front and rear of the vehicle in an intelligent and controlled fashion. Failure to do so could result in any of the following:

- A vehicle being unable to generate the necessary deceleration for a given pedal pressure.
- Front axle lock, in which the vehicle remains stable yet suffers from a loss of steering control.
- Rear axle lock that causes the vehicle to become unstable.

The terms front wheel lock and rear wheel lock can be alternatively labelled front axle over brakes and rear axle over brakes, respectively. In this section factors influencing the fore–aft axle loads are identified and their effect on braking examined, concentrating on vehicles that have a fixed brake ratio. Devices used to modify the braking ratio are discussed in the latter part of this section and their effect on braking performance evaluated.

The theory developed in Sections 12.1.4.2–12.1.4.4 and 12.1.4.8 is applied to the analysis of the braking requirements of a fictitious two axle road vehicle. This vehicle is assumed to be unladen and is described by the parameters given in Table 12.1-1.

12.1.4.1 Static analysis

A simple representation of a two axle road vehicle is shown in Figure 12.1-8, which has a mass, M, concentrated at the centre of gravity. The centre of gravity is assumed to lie on the longitudinal centreline of the vehicle and the road surface is flat with no camber. As a consequence of this, the loads on the two wheels mounted on each

Table 12.1-1 Prototype vehicle parameters

Parameter		Symbol	Units	Value
Mass		M	kg	980
Wheel base		l	m	2.45
Height of centre of gravity above ground		h	m	0.47
Static axle loads (% of total)	Front	F_f	–	65
	Rear	F_r	–	35
Fixed brake ratio	Front	X_f	–	75
$R = \frac{X_f}{X_r}(\%)$	Rear	X_r	–	25

axle are equal and so the following analyses treat axle loading rather than the individual wheels.

The vertical load due to the vehicle mass is simply:

$$P = Mg \tag{12.1.25}$$

where g is the acceleration due to gravity.

Taking moments about the rear tyre–road contact gives:

$$F_f = \frac{Pb}{l} \tag{12.1.26}$$

and moments about the front tyre–road contact provides a value of the vertical load acting at the rear axle given by:

$$F_r = \frac{Pa}{l} \tag{12.1.27}$$

The influence of the fore–aft location of the centre of gravity on the vertical wheel loads, due to change in loading conditions is readily apparent from equations 12.1.26 and 12.1.27.

The maximum payload of a passenger car forms only a fraction of the unladen vehicle mass and the inherent space restrictions limit the extent to which the centre of gravity can move. For a light rigid truck, such as a transit type van or pickup, there exists considerable scope for

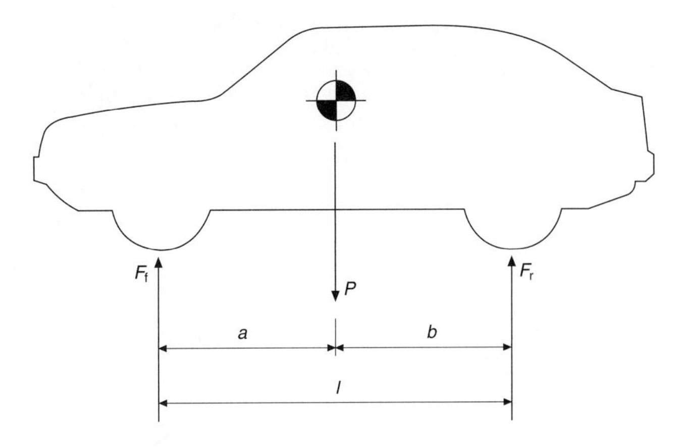

Figure 12.1-8 Static axle loads.

varying the payload and so the change in mass from unladen to laden can be severe. The distribution and size of the payload, which also governs the location of the centre of gravity, may change on a daily basis or throughout a delivery cycle in which the payload is discharged in stages.

12.1.4.2 Braking with a constant brake ratio

The object of this analysis is to show how braking affects the vertical loads carried by the front and rear axles. This in turn leads to a way of determining the maximum deceleration attainable by a vehicle under specified conditions that does not result in axle lock.

If the front and rear axles are to be on the point of locking, then the braking forces T_f and T_r acting at each axle must be in proportion to the vertical loads being carried, R_f and R_r. The magnitude of the braking force generated by each axle, up to the point at which it locks, is a function of the design of the braking system.

Changes in load transfer between the front the rear axles occur during braking and so a variable brake effort ratio is required to provide ideal braking. The situation is in reality complicated by the following:

- Change in vehicle weight.
- Change in weight distribution.
- The effect of gradients (positive and negative).
- Cornering, in which some proportion of the total force present at the tyre–ground interface is used to generate lateral forces.
- Varying road surfaces and weather conditions.
- Split friction surfaces where the coefficient of adhesion changes from port to starboard.

Consider the case for a vehicle with a fixed brake ratio, $R = \frac{x_f}{x_r}$, on a flat road that has a uniform coefficient of tyre–ground adhesion μ. In the analysis that follows, the governing equations of motion for the decelerating vehicle are derived by the direct application of Newton's second law to the free body diagram of the vehicle rather than through D'Alembert's method which is adopted in the EEC Council Directive 71/320/EEC. This permits easy extension of the model to embrace additional degrees of freedom or to take account of the presence of a trailer. The free body diagram of the decelerating vehicle is shown in Figure 12.1-9.

In the x direction

$$\begin{aligned} M\ddot{x} &= \sum F_x \\ &= -D - T_f - T_r - T_{fr} - T_{rr} \end{aligned} \qquad (12.1.28)$$

which through combination with equation 12.1.12 becomes:

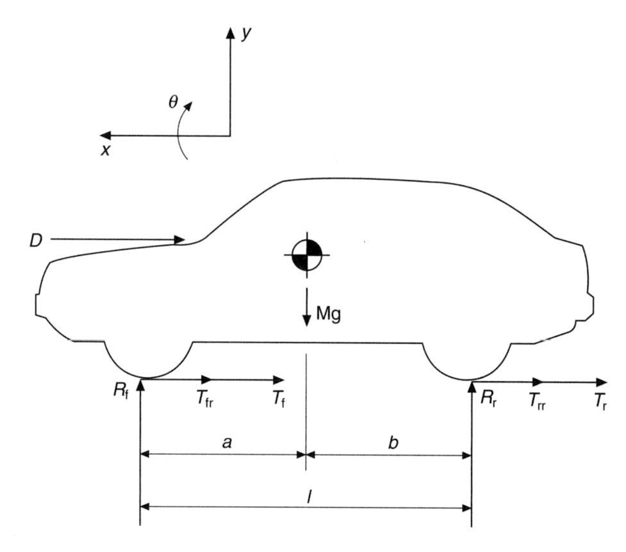

Figure 12.1-9 Free body diagram of the decelerating vehicle.

If the aerodynamic drag and rolling resistance forces are assumed to be negligible, then equation 12.1.29 reduces to:

$$Md = D + T_f + T_r + T_{fr} + T_{rr} \qquad (12.1.29)$$

$$Md = T_f + T_r \qquad (12.1.30)$$

By defining z to be the vehicle deceleration as a proportion of g:

$$z = \frac{d}{g} \qquad (12.1.31)$$

then equation 12.1.30 takes the form:

$$Mgz = T_f + T_r = Pz \qquad (12.1.32)$$

In the vertical direction

$$M\ddot{y} = \sum Fy = R_r + R_f - Mg = 0 \qquad (12.1.33)$$

as $y = 0\ \text{ms}^{-2}$, whilst in the θ direction, taking moments about the centre of gravity of the vehicle leads to:

$$\begin{aligned} I\ddot{\theta} \\ = \sum M_{cg} = R_f a - R_r b - T_f h - T_r h = 0 \end{aligned} \qquad (12.1.34)$$

as $\ddot{\theta} = 0\ \text{rad s}^{-2}$.

Manipulation of equations 12.1.33 and 12.1.34 results in the following expressions for the front and rear dynamic axle loads:

$$R_f = \frac{Mgb}{l} + \frac{h}{l}(T_f + T_r) \qquad (12.1.35)$$

$$R_r = \frac{Mga}{l} - \frac{h}{l}(T_f + T_r) \qquad (12.1.36)$$

which can be combined with equation 12.1.32 and the static axle loads, equations 12.1.26 and 12.1.27, to give

$$R_f = F_f + \frac{Pzh}{l} \tag{12.1.37}$$

$$R_r = F_r - \frac{Pzh}{l} \tag{12.1.38}$$

The above are in accord with those given in the EEC Directive and they show that a change in axle load in favour of the front axle occurs during a braking manoeuvre. In order for each axle to be simultaneously on the verge of locking, the brake force generated at each axle must be in direct proportion to the vertical axle load. This means that to fully utilize the available tyre–ground adhesion, the braking system must support an infinitely variable brake ratio.

Consider first the case of a vehicle in which the brake ratio is fixed. If the ratio has been set so that the front axle locks in preference to the rear, then the brake force generated at the front axle when about to lock is given by:

$$\begin{aligned} T_f &= \mu R_f \\ &= \mu\left(F_f + \frac{Pzh}{l}\right) \end{aligned} \tag{12.1.39}$$

During the same braking event, the rear axle is also generating a brake force that has not exceeded its limiting value and this is found by considering the vehicle brake ratio:

$$R = \frac{x_f}{x_r} = \frac{T_f}{T_r} \tag{12.1.40}$$

from which

$$T_r = T_f\frac{x_r}{x_f} = \mu\left(F_f + \frac{Pzh}{l}\right)\frac{x_r}{x_f} \tag{12.1.41}$$

leading to a total brake force of:

$$\begin{aligned} T &= Pz = T_f + T_r \\ &= \mu\left(F_f + \frac{Pzh}{l}\right) + \mu\left(F_f + \frac{Pzh}{l}\right)\frac{x_r}{x_f} \end{aligned} \tag{12.1.42}$$

which reduces to:

$$T = Pz = \mu\left(F_f + \frac{Pzh}{l}\right)\frac{1}{x_f} \tag{12.1.43}$$

This equation can be rearranged to yield the maximum value of deceleration as a proportion of g as:

$$z = \frac{\mu F_f}{P(lx_f - \mu h)} \tag{12.1 44}$$

If, however, the ratio has been set so that the rear axle locks in preference to the front, then the brake force generated at the rear axle when about to lock is given by:

$$\begin{aligned} T_r &= \mu R_r \\ &= \mu\left(F_r - \frac{Pzh}{l}\right) \end{aligned} \tag{12.1.45}$$

In this case, the brake force that is generated at the front axle is not necessarily the limiting value and its magnitude is found from the brake ratio as:

$$\begin{aligned} T_f &= T_r\frac{x_f}{x_r} \\ &= \mu\left(F_r - \frac{Pzh}{l}\right)\frac{x_f}{x_r} \end{aligned} \tag{12.1.46}$$

which leads to a total brake force of:

$$\begin{aligned} T &= Pz = T_f + T_r \\ &= \mu\left(F_r - \frac{Pzh}{l}\right)\frac{x_f}{x_r} + \mu\left(F_r - \frac{Pzh}{l}\right) \\ &= \mu\left(F_r - \frac{Pzh}{l}\right)\frac{1}{x_r} \end{aligned} \tag{12.1.47}$$

and this can be solved for the deceleration as a proportion of g to be:

$$z = \frac{l\mu Fr}{P(lx_r + \mu h)} \tag{12.1.48}$$

Direct solution of equations 12.1.44 and 12.1.48 for z is straightforward, however, greater insight to the mechanics of the braking process can be gained through the following graphical solution that deals with each axle in turn.

The adhesion force acting between the front tyres and ground depends upon the ratio of the tangential forces at the front and rear wheels due to the brake torques. It is therefore linked to the fixed brake ratio and so the front adhesion force, T_f, as a proportion of the total is given by:

$$\begin{aligned} T_f &= x_f T \\ &= x_f Pz \end{aligned} \tag{12.1.49}$$

which when normalized to the vehicle weight, P, becomes:

$$\frac{T_f}{P} = x_f z \tag{12.1.50}$$

Equation 12.1.50 is shown in Figure 12.1-10 labelled as T_f/P. The total available braking force at the front of the vehicle is, from equation 12.1.43,

$$Tx_f = \mu\left(F_f + \frac{Pzh}{l}\right) \quad (12.1.51)$$

which when normalized to the vehicle weight becomes:

$$\frac{Tx_f}{P} = \frac{\mu}{P}\left(F_f + \frac{Pzh}{l}\right) \quad (12.1.52)$$

This represents the maximum braking force, expressed as a proportion of the total vehicle weight, that could be sustained between the front tyres and road surface for a given set of vehicle parameters. It is shown in Figure 12.1-10 as the line Tx_f/P.

Application of the same procedure to the rear axle of the vehicle results in the following normalized expression for the rear adhesion force T_r:

$$\frac{T_r}{P} = X_r z \quad (12.1.53)$$

and this is labelled T_r/P in Figure 12.1-10. Similarly, the total available braking force at the rear axle, normalized to the vehicle weight, is, from equation 12.1.47:

$$\frac{Tx_r}{P} = \frac{\mu}{P}\left(F_r \frac{Pzh}{l}\right) \quad (12.1.54)$$

and this is the line Tx_r/P in Figure 12.1-10.

The point of intersection of lines T_f/P and Tx_f/P, labelled a, represents the solution to equation 12.1.44 and the point of intersection of lines T_r/P and Tx_r/P, labelled b, is the solution to equation 12.1.48. The data used to generate Figure 12.1-10 are that of the prototype vehicle defined in Section 12.1.4.4. The vehicle is assumed to be braking on a road that has a tyre–ground adhesion coefficient of unity. The lower value of deceleration, b, is that value of deceleration that would first give rise to wheel lock for the given value of tyre–ground adhesion. In this instance, the rear axle will lock first and the tyres will be unable to generate the braking force required by the rear brakes at higher levels of deceleration. Direct solution of equation 12.1.48 leads to a limiting value of deceleration of $z = 0.79\,g$.

12.1.4.3 Braking efficiency

The efficiency with which a brake system uses the available tyre–ground adhesion, η, can be conveniently defined as the ratio of the deceleration, z, to the tyre–ground adhesion coefficient, μ, that is:

$$\eta = \frac{z}{\mu} \quad (12.1.55)$$

There are two expressions for η; one for the case in which the front axle is about to lock and the other for the case in which the rear axle is about to lock. To determine which is applicable recall that:

$$\eta \le 1.0 \quad (12.1.56)$$

For the case of front axle lock, η can be written using equation 12.1.44 as:

$$\eta = \frac{z}{\mu} = \frac{\dfrac{l\mu F_f}{P(lx_f - \mu h)}}{\mu} = \frac{F_f}{P\left(x_f - \dfrac{\mu h}{l}\right)} \quad (12.1.57)$$

For the rear axle lock case, application of equation 12.1.48 results in the second expression for efficiency

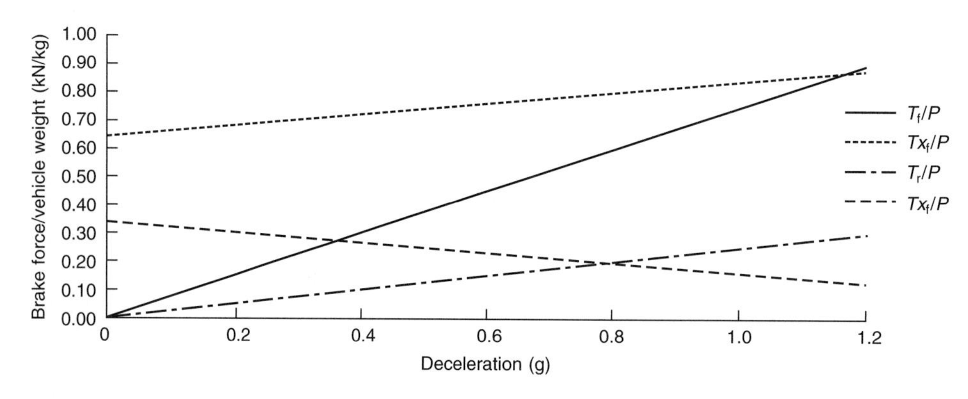

Figure 12.1-10 Normalized brake force against deceleration.

$$\eta = \frac{z}{\mu} = \frac{\dfrac{l\mu F_r}{P(lx_r + \mu h)}}{\mu} = \frac{F_r}{P\left(x_r + \dfrac{\mu h}{l}\right)} \qquad (12.1.58)$$

A measure of the efficiency with which a vehicle, having a particular brake design with fixed brake ratio, performs over a variety of road surfaces can be shown by a graph of efficiency, η, against tyre–ground adhesion coefficient, μ, Figure 12.1-11. This has been generated using the prototype vehicle data. Both axles are on the verge of lock and the system is 100% efficient when the vehicle is braked on a road surface that has a tyre–ground adhesion coefficient of 0.52 indicated by the point a. On road surfaces below this value of adhesion, the vehicle is limited by front axle lock, equation 12.1.57, and the efficiency falls to a minimum of 87%, whilst on road surfaces with an adhesion coefficient greater than 0.52 the vehicle is limited by rear axle lock, equation 12.1.58, and this falls to a minimum of 79%. Data presented above the line defining 100% efficiency has no physical meaning and can be ignored.

An alternative means of presenting efficiency data can be achieved by plotting deceleration, z, against tyre–ground adhesion, μ, for the cases of front and rear axle lock defined by equations 12.1.44 and 12.1.48, respectively. This method of presentation has the advantage that the brake engineer can obtain a comparison of possible deceleration levels attainable on different road surfaces along with a measure of the system efficiency. In this case, the efficiency is the gradient of the curve drawn on the deceleration–adhesion space. A line with unit gradient represents optimum performance.

Recall that for front wheel lock,

$$z = \frac{l\mu F_f}{P(lx_f - \mu h)} \qquad (12.1.59)$$

and for rear wheel lock,

$$z = \frac{l\mu F_r}{P(lx_r + \mu h)} \qquad (12.1.60)$$

The two curves that define the limiting deceleration of the front and rear axles, derived from equations 12.1.59 and 12.1.60, respectively, are shown in Figure 12.1-12 for the prototype vehicle. They intersect the optimum line at the point a which indicates 100% efficiency. Meaningful information is taken from those portions of the curves that lie below the optimum line and it is clear that the vehicle is governed by rear axle lock on road surfaces with a high tyre–ground adhesion coefficient and that the system is least efficient in this area.

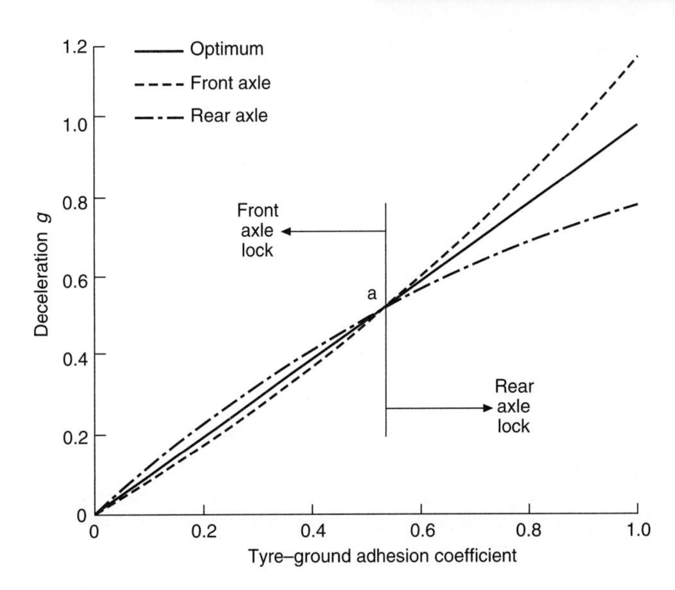

Figure 12.1-11 Deceleration as a function of tyre–ground adhesion.

12.1.4.4 Adhesion utilization

Adhesion utilization, f, is the theoretical coefficient of adhesion that would be required to act at the tyre–road interface of a given axle for a particular value of deceleration. It is therefore the minimum value of tyre–ground adhesion required to sustain a given deceleration and is defined as the ratio of the braking force to the vertical axle load during braking.

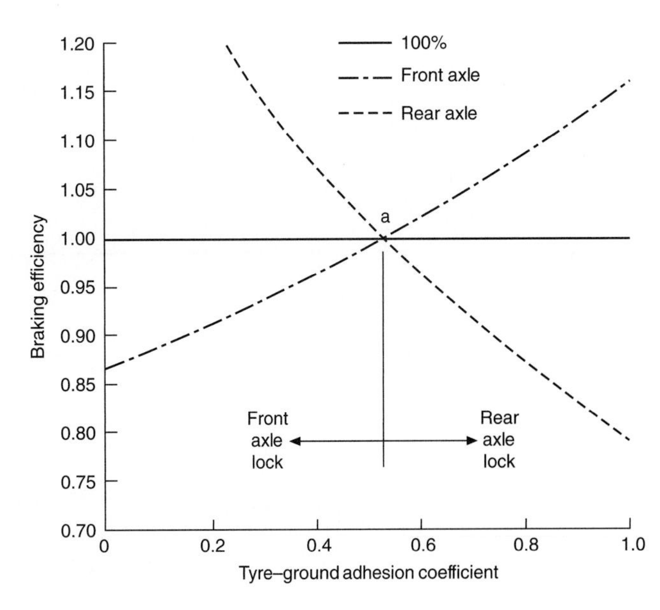

Figure 12.1-12 Efficiency as a function of tyre–ground adhesion.

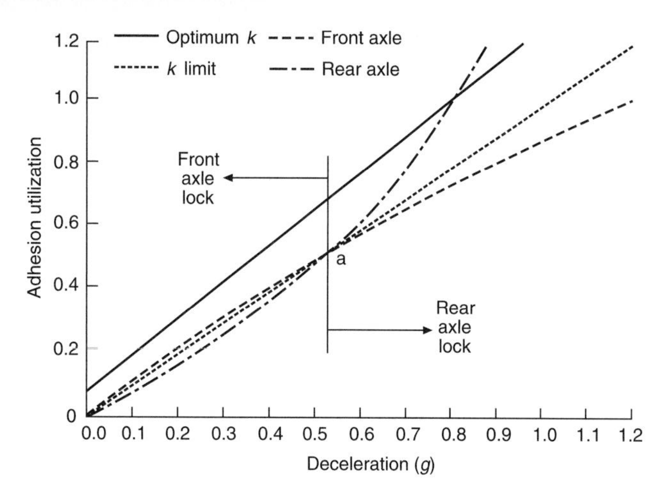

Figure 12.1-13a Adhesion utilization, datum prototype vehicle.

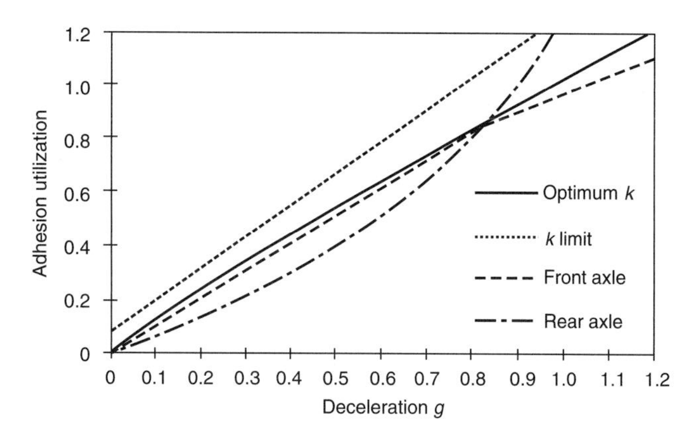

Figure 12.1-13b Adhesion utilization, modified prototype vehicle.

For the front of the vehicle the adhesion utilization is defined by:

$$f_f = \frac{T_f}{R_f} \qquad (12.1.61)$$

The vertical axle load is defined by equation 12.1.37 and the front axle brake force, expressed as a proportion of the total is $x_f Pz$, leads to:

$$f_f = \frac{x_f Pz}{F_f + \dfrac{Pzh}{l}} \qquad (12.1.62)$$

Similarly, for the rear of the vehicle:

$$f_r = \frac{T_r}{R_r} \qquad (12.1.63)$$

The vertical axle load is defined by equation 12.1.38 and the rear axle brake force, expressed as a proportion of the total is $x_r Pz$, leads to:

$$f_r = \frac{x_r Pz}{F_r - \dfrac{Pzh}{l}} \qquad (12.1.64)$$

Using the data that describe the prototype vehicle leads to Figure 12.1-13a. The optimum line has unit gradient and defines the ideal adhesion utilization characteristic in which the brake system remains 100% efficient over all possible values of deceleration. The upper limit on allowable adhesion utilization, defined in the *EEC Braking Directive*, Section 12.1.2, is shown for reference purposes. The remaining two lines define the axle adhesion

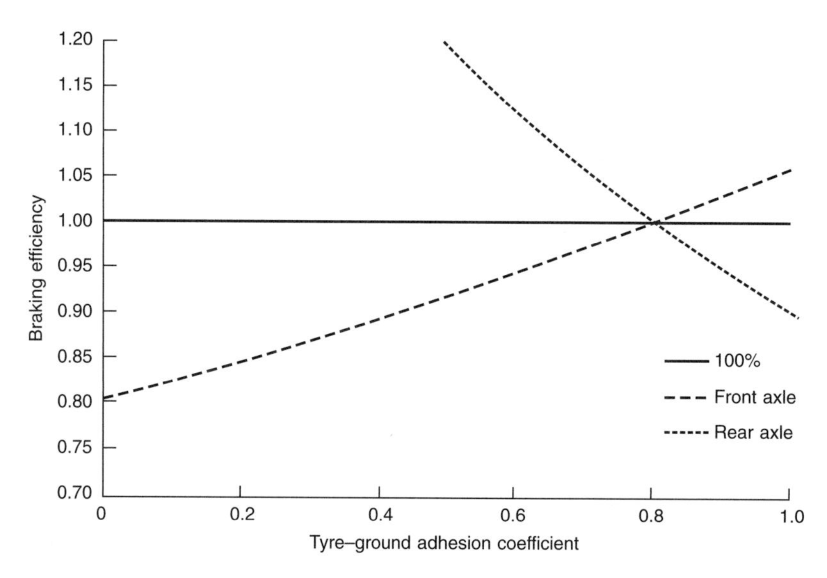

Figure 12.1-13c Brake system efficiency of modified prototype vehicle.

characteristics for the vehicle. The point labelled ***a***, at which the curves cross, intersect the optimum line of adhesion utilization indicating that at this value of deceleration both axles are on the verge of lock. The axle having the highest adhesion utilization coefficient for a given value of deceleration is that which limits the braking performance of the vehicle and, in this case, braking is limited by front axle lock up to a deceleration of 0.52*g*. Thereafter, braking is limited by rear axle lock. It is also possible to find from this diagram the maximum deceleration for a given coefficient of adhesion utilization.

Comparison of the adhesion utilization diagram derived for the prototype vehicle with the legislative requirements outlined in Section 12.1.2 shows that the vehicle brake system does not meet the minimum standard, as the front axle adhesion curve does not lie above that of the rear axle for all values of deceleration between 0.15*g* and 0.8*g*. This can be remedied by changing the fixed brake ratio in favour of the rear axle and this causes the point ***a*** to move up the optimum adhesion line. The limiting deceleration is set at 0.8*g* which leads to a new fixed brake ratio of $\frac{x_f}{x_r} = \frac{0.803}{0.197}$. This in turn results in the modified adhesion diagram shown in Figure 12.1-13b, which this satisfies the legislative requirements. The modified vehicle is governed by front axle lock up to a deceleration level of 0.8*g*, achieved at the expense of the overall brake system efficiency (Figure 12.1-13c).

12.1.4.5 Wheel locking

The role of tyre–ground friction and the dependency of the brake force coefficient on the degree of longitudinal slip has been outlined in Section 12.1.3.3. From 0% to approximately 20% longitudinal slip, the magnitude of the brake force coefficient increases in a roughly linear fashion to its maximum value, μ_p, at 20% longitudinal slip. Further increase, due to increase in applied brake torque, causes the wheel to decelerate rapidly to a condition of full lock and the brake force coefficient takes a value of μ_s at 100% longitudinal slip. The ratio of μ_p/μ_s depends upon the nature of the road surface in question and it takes its highest value under wet or icy conditions.

This leads to a possible scenario in which a vehicle is capable of generating its maximum braking potential when one axle is locked and the second is on the verge of lock. This contrasts with the generally accepted idea that maximum deceleration occurs when the first axle is about to lock and is dependent upon the vehicle weight distribution and the fixed braking ratio.

If the front axle is locked and the rear axle is about to lock then the total brake force is given by:

$$Pz = \mu_s\left(F_f + \frac{Pzh}{l}\right) + \mu_p\left(F_r - \frac{Pzh}{l}\right) \qquad (12.1.65)$$

Similarly, if the rear axle is locked and the front axle is on the verge of lock, then the total brake force is:

$$Pz = \mu_p\left(F_f + \frac{Pzh}{l}\right) + \mu_s\left(F_r - \frac{Pzh}{l}\right) \qquad (12.1.66)$$

12.1.4.6 Effect of axle lock on vehicle stability

When an axle locks, there is reduced friction in both the longitudinal and lateral directions and so the ability of the

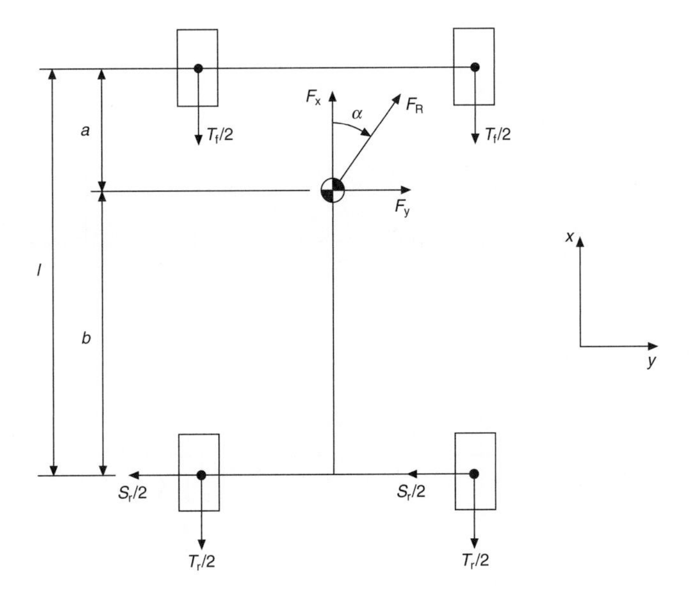

Figure 12.1-14 Front axle lock.

vehicle to generate the lateral forces required to maintain directional control and stability is severely impaired.

Irregularities in the road surface or lateral forces can cause the vehicle to deviate from its direction of travel. The nature of the ensuing motion, which is rotational about the vehicle vertical axis, depends on which axle has locked together with the vehicle speed, tyre–ground friction coefficient, yaw moment of inertia of the vehicle body and the vehicle dimensions. By considering the two cases of front and rear axle lock it is possible to derive useful insight into the stability problem:

12.1.4.6.1 Front axle lock

Any disturbance in the lateral direction due to gradient, sidewind or left to right brake imbalance produces a side force F_y that acts through the centre of gravity of the vehicle, as shown in Figure 12.1-14.

The resultant force F_R that is due to the inertia force F_x caused by the braking event and the lateral force F_y gives rise to a slip angle α. This slip angle represents the difference between the longitudinal axis of the vehicle and the direction in which the vehicle centre of gravity is moving. The lateral force F_y must be balanced by the side forces generated in the tyre–ground contact patches. As the front axle is locked, no side force is generated by the front wheels and the resulting side force is developed solely by the still rolling rear wheels. This gives rise to a total moment of $S_r b$. This yaw moment has a stabilizing effect since it causes the longitudinal axis of the vehicle to align with the direction of travel, thereby reducing the initial slip angle α. Thus, when the front axle is locked, the vehicle is unable to respond to any steering inputs and so its forward motion continues in a straight line.

12.1.4.6.2 Rear axle lock

Assume now that the fixed brake ratio associated with the same vehicle has been changed such that the rear axle locks in preference to the front as depicted in Figure 12.1-15. If the vehicle is subject to the same lateral disturbance, then this can only be reacted by a side force generated between the front wheels and ground and the resulting moment about the vehicle centre of gravity has a magnitude of $S_f a$. In contrast, this yaw moment now has a destabilizing effect as it causes the longitudinal axis of the vehicle to move away from the direction of travel, thereby increasing the vehicle slip angle α. This in turn leads to a rise in lateral force at the front of the vehicle causing an increase in yaw acceleration.

It is thus preferable, from a safety point of view, for the front axle to lock in preference to the rear as this is a stable condition and the driver is able to regain directional control of the vehicle simply by releasing the brakes. If the rear axle has locked and the vehicle has begun to spin, driver reaction must be rapid if control of the situation is to be regained.

In a collision situation, a frontal impact, linked to front axle lock, will usually result in less serious occupant injury than the possible side impact that could well be associated with the uncontrolled yawing of the vehicle that results from rear axle lock.

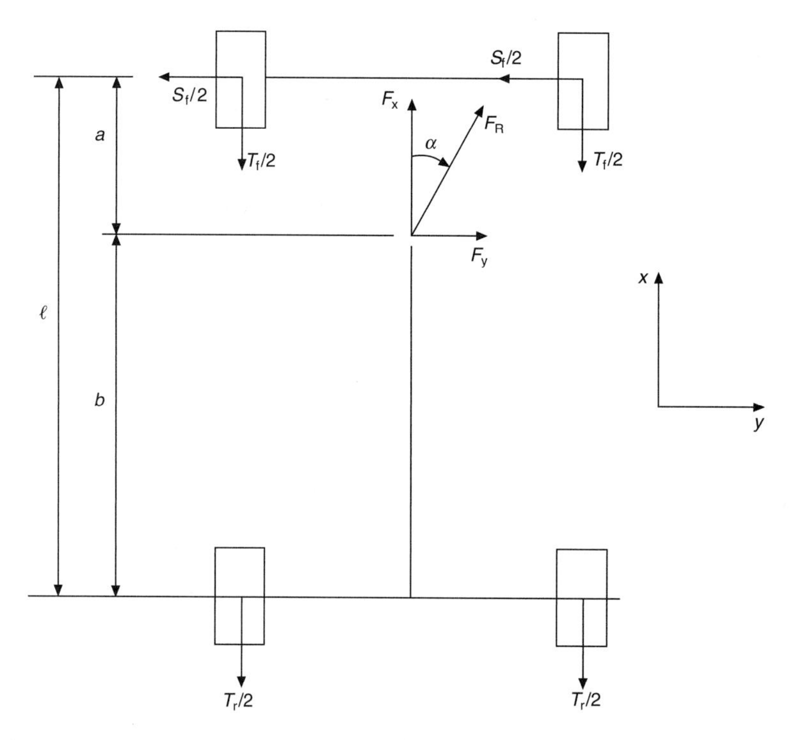

Figure 12.1-15 Rear axle lock.

It is therefore feasible to apply the preceding ideas to the formulation of a fixed brake ratio that will invariably lead to front axle lock and this is commonly applied to the design of brake systems found on passenger vehicles. The fixed brake ratio is chosen such that for the unladen case both front and rear axles are on the verge of lock when the vehicle undertakes a $1g$ stop on a road surface that has a tyre–ground adhesion coefficient of unity. Under such conditions, the brake ratio is equal to:

$$\frac{x_f}{x_r} = \frac{F_f + \frac{Ph}{l}}{F_r - \frac{Ph}{l}} \qquad (12.1.67)$$

and on all surfaces where the tyre–ground adhesion is less than unity, the braking will be limited by front axle lock.

The effect of axle lock on vehicle stability may also be assessed through the formal derivation of the equation of motion associated with the yawing of the vehicle. Analysis of the same cases of axle lock leads to identical conclusions regarding the behaviour of the vehicle with the added benefit that measures of yaw acceleration, velocity and displacement can be deduced.

12.1.4.7 Pitch motion of the vehicle body under braking

The transfer of load from the rear to the front axle that takes place during a braking event will cause the vehicle body to rotate about its lateral axis. This pitching motion also results in a change in the height of the vehicle centre of gravity. Both of these quantities can be determined as a function of vehicle deceleration using the notation in Figure 12.1-16. The following analysis assumes the vehicle body to be rigid and that the front and rear suspension spring rates, k_f and k_r, are linear. The spring rates used are the axle rates.

The opposed spring forces generated during a braking event are equal to the load transfer that takes place and so are equal to

$$y_r \pm \frac{Pzh}{l}$$

and this causes the vehicle to go down at the front and move upwards at the rear as shown in Figure 12.1-16. Thus, on the assumption of linear springing, the compression travel at the front is:

$$y_f = \frac{\frac{Pzh}{l}}{k_f} \qquad (12.1.68)$$

and the corresponding travel at the rear is:

$$y_r = \frac{\frac{Pzh}{l}}{k_r} \qquad (12.1.69)$$

The pitch angle, θ, in degrees, adopted by the vehicle body is therefore given by:

$$\theta = \left(\frac{y_f + y_r}{l}\right) \times \frac{360}{2\pi} \qquad (12.1.70)$$

Vertical and longitudinal movement of the vehicle body centre of gravity occurs as a result of the body pitch motion and this in turn causes a small change in the overall centre of gravity of the vehicle. The extent of movement of the vehicle body centre of gravity, initially located a distance a_b from the front axle at a height h_b above ground, depends upon its location within the

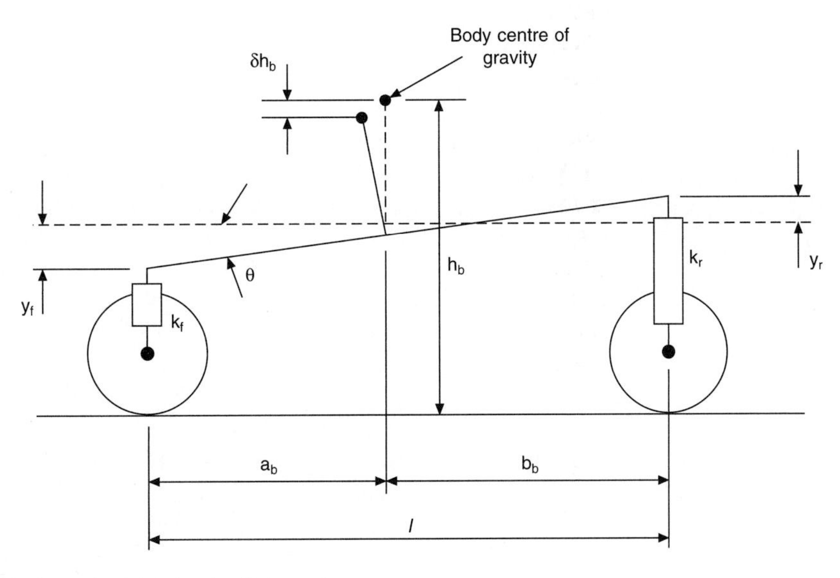

Figure 12.1-16 Determination of vehicle body pitch angle.

structure, the suspension rates and the rate of deceleration. An indication of the extent of this movement can be seen in Figure 12.1-16. Under severe braking conditions, the vertical displacement, δh_b, of the vehicle body centre of gravity equates to approximately 5% of its original height. A detailed account of the relevant theory can be found in Reimpell and Stoll (1996) and from this the change in height is given by:

$$\delta h_b = -y_f \frac{F_{bf}}{F_b} + y_r \frac{F_{br}}{F_b} \quad (12.1.71)$$

where

$$F_{bf} = F_{sf} + F_{af} \quad (12.1.72)$$

$$F_{br} = F_{sr} + F_{ar} \quad (12.1.73)$$

$$F_b = F_{bf} + F_{br} \quad (12.1.74)$$

in which F_b is the vehicle body weight, $F_{bf,r}$ are the brake reaction loads applied to the front and rear of the vehicle body, $F_{af,r}$ are the unsprung weights of the front and rear axles and $F_{sf,r}$ are the front and rear axle loads. If the loads due to the unsprung axle masses are ignored then a corresponding expression for the change in the height of the overall centre of gravity of the vehicle, δ_h, can be found using:

$$\delta_h = -y_f \frac{F_{sf}}{P} + y_r \frac{F_{sr}}{P} \quad (12.1.75)$$

in which P is the total vehicle weight.

12.1.4.8 Braking with a variable braking ratio

If a vehicle is to achieve maximum retardation, equal to the value of the tyre–ground adhesion coefficient, equation 12.1.22, then the brake system must be designed with a continuously variable brake ratio. This must be equal to the ratio of the dynamic load distribution between the front and rear for all values of deceleration. Thus the variable brake ratio, R_v, is defined as:

$$R_v = \frac{x_{fv}}{x_{rv}} = \frac{R_f}{R_r} = \frac{F_f + \frac{Pzh}{l}}{F_r - \frac{Pzh}{l}} \quad (12.1.76)$$

from which it can be shown that:

$$x_{fv} = \frac{F_f}{P} + \frac{zh}{l} \quad (12.1.77)$$

and

$$x_{rv} = \frac{F_r}{P} - \frac{zh}{l} \quad (12.1.78)$$

A situation giving rise to the need for a variable braking ratio might result from a given vehicle design in which the maximum deceleration using a fixed braking ratio is too low. In practice the introduction of a regulating valve into the braking system helps to optimize the braking efficiency over a wide range of operating conditions. Although such devices do not permit a continuously variable braking ratio, they do offer a means of improving the overall braking performance. Mathematical models of deceleration sensitive pressure regulating valves are now derived.

12.1.4.8.1 Deceleration-sensitive pressure limiting valve

A typical valve design is shown in Figure 12.1-17. At a predetermined deceleration, determined by the mass of the ball and the angle of installation, the inertial force acting on the ball causes it to roll up the valve body and close the valve thereby isolating the rear brakes. These valves are gradient sensitive but do act in a favourable manner. On a rising slope the valve closes at higher levels of deceleration allowing the rear brakes to contribute more to the total braking effort, whilst on a falling slope the rear brakes are isolated sooner reflecting the load transfer to the front of the vehicle caused by the gradient.

The effect on performance brought about by the inclusion of a regulating valve in the rear brake line can be assessed by deriving equations which define the brake ratio for all possible values of deceleration. These may then be used in the equations for efficiency and adhesion utilization, derived earlier, which quantify the brake system performance. In the following analysis it is assumed the valve isolates the line to the rear brakes when the vehicle deceleration has reached a certain value of deceleration, z_v. Note that the mechanism through which cut-off is achieved depends upon the chosen valve type and this determines the actual value of z_v.

Figure 12.1-18 shows a typical front to rear brake force characteristic. For all values of deceleration less than z_v, the brake force is apportioned between the front and rear axles in the fixed ratio R. Once the deceleration has exceeded z_v, the line pressure to the rear brakes is held constant and so they can no longer generate

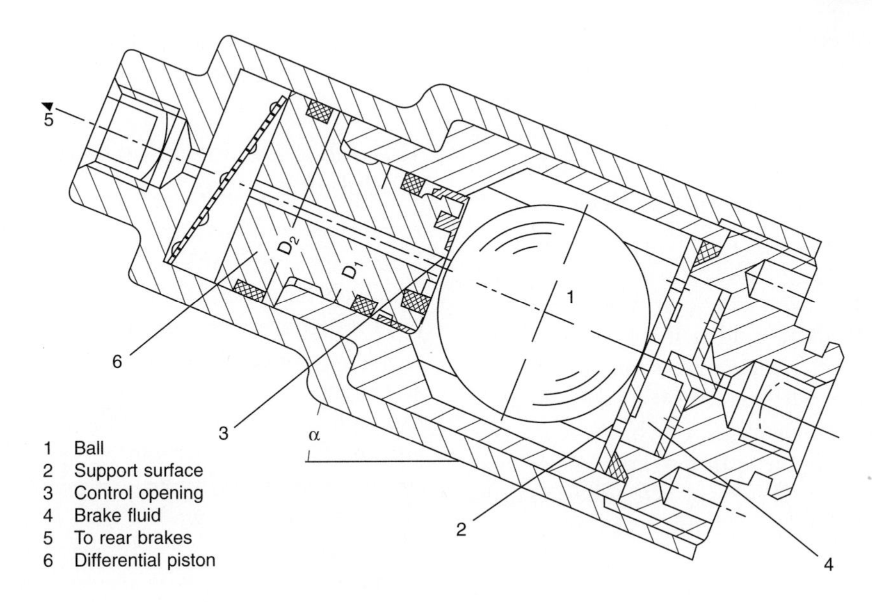

Figure 12.1-17 Deceleration-sensitive pressure limiting valve (Limpert, 1992).

additional braking force. Consequently the brake ratio changes from its original value.

In region 1, for $z \le z_v$, the proportion of braking effort at the rear of the vehicle is:

$$x_{rv} = x_r \tag{12.1.79}$$

The proportion of braking effort at the front of the vehicle is therefore:

$$\begin{aligned} x_{fv} &= 1 - x_{rv} \\ &= 1 - x_r = x_f \end{aligned} \tag{12.1.80}$$

and so the brake ratio R_v is:

$$R_v = \frac{x_{fv}}{x_{rv}} = \frac{x_f}{x_r} \tag{12.1 81}$$

In region 2, $z > z_v$, the valve actuates and isolates the rear brakes from any further increase in line pressure and

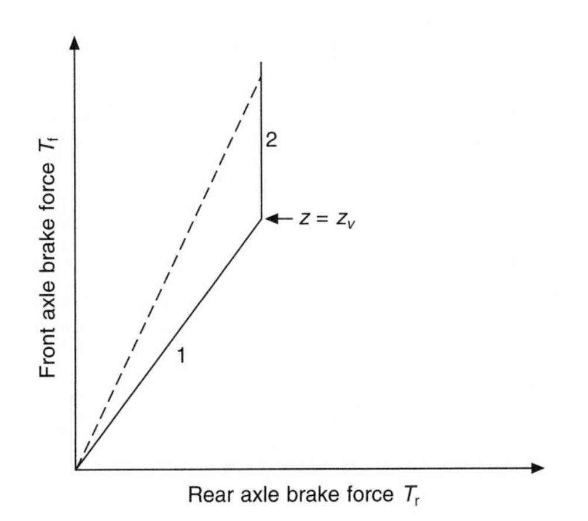

Figure 12.1-18 Typical limiting valve brake force distribution.

the rear brakes can no longer generate additional braking force. As the front brakes are able to respond to further increase in line pressure then the rate of deceleration can increase above z_v and the brake ratio changes, being equal to the slope of the dashed line. When in region 2, the brake force at the rear, T_r, is constant and is:

$$T_r = Pz_v x_r \tag{12.1.82}$$

Simultaneously, the brake force acting at the front of the vehicle, T_f, increases and is equal to the difference between the total brake force, Pz, and that sustained at the rear axle

$$T_f = Pz - Pz_v x_r \tag{12.1.83}$$

Thus, in region 2, the brake ratio is defined by:

$$R_v = \frac{T_f}{T_r} = \frac{Pzx_{fv}}{Pzx_{rv}} = \frac{Pz - Pz_v x_r}{Pz_v x_r} \tag{12.1.84}$$

from which

$$x_{fv} = \frac{z - z_v x_r}{z} \tag{12.1.85}$$

and

$$x_{rv} = \frac{z_v x_r}{z} \tag{12.1.86}$$

The incorporation of such a valve into the brake system of the prototype vehicle results in improved adhesion utilization and efficiency. Biasing the fixed brake ratio in favour of rear axle lock improves front axle adhesion up to the point of lock. This can be set, through the fixed brake ratio, to lie within the deceleration range of 0.35g to 0.45g. The deceleration-sensitive pressure limiting valve is chosen to actuate at the point of lock and this results in an adhesion utilization diagram that has the form shown in Figure 12.1-19a. The fixed brake ratio has been changed to $R = \frac{x_f}{x_r} = \frac{0.73}{0.27}$ which results in a critical deceleration of 0.4 g and the valve is assumed to actuate at this level of deceleration. With reference to Figure 12.1-19a, the vehicle is now governed by front axle lock over all values of deceleration as the front axle adhesion lies above that of the rear. The brake system now makes much better use of the available adhesion when executing low/moderate g stops as, in comparison to Figure 12.1-13b, the front axle adhesion now lies closer to the optimum line. However at higher rates of deceleration, the front axle adhesion utilization has reduced. The beneficial effect of the valve on efficiency can be seen through comparison of Figure 12.1-19b (valve fitted) to Figure 12.1-13c (no valve).

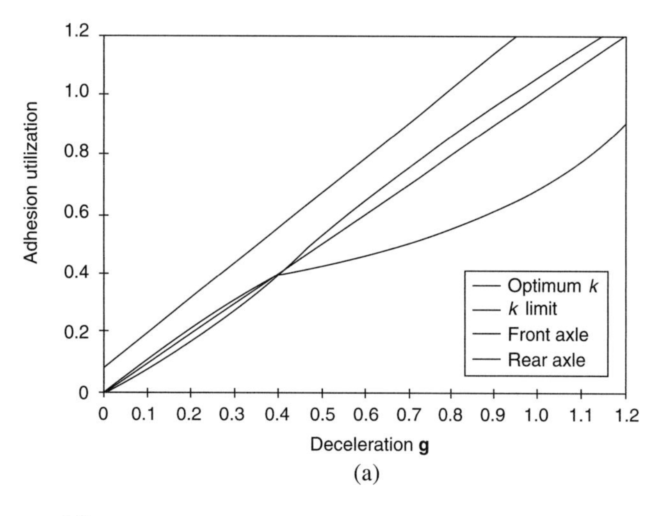

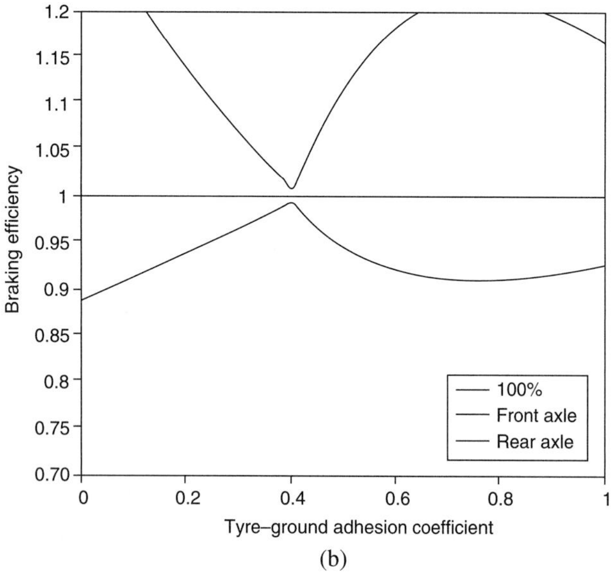

Figure 12.1-19 (a) Adhesion utilization, modified vehicle fitted with deceleration-sensitive pressure, (b) brake system efficiency, modified vehicle fitted with deceleration sensitive.

12.1.4.8.2 Deceleration-sensitive pressure modulating valve

A pressure modulating valve, or reducer valve, differs from a pressure limiting valve as once the activation point has been exceeded, they do not isolate the rear brakes but for higher pressures the rear brake pressure increases at a lower rate than that of the front brakes. The main advantages of this type of valve are that the rear pressure can be increased even after the front brakes have locked and the front and rear line pressures lie close to the optimum values. A typical pressure modulating valve characteristic is shown in Figure 12.1-20.

The effect on system performance of a pressure modulating valve may be assessed in a similar fashion to that of a limiting value. As before, it is assumed that the valve is actuated once the vehicle deceleration has exceeded a certain value of deceleration, z_v, and the exact value of z_V is determined by the mechanics of the valve.

In region 1, the vehicle deceleration is less than z_V and so the vehicle brakes in accordance with the fixed brake ratio assigned to the system,

$$x_{rv} = x_r \tag{12.1.87}$$

$$x_{fv'} = x_f \tag{12.1.88}$$

and

$$R_v = \frac{x_{fv}}{x_{rv}} = \frac{x_f}{x_r} \tag{12.1 89}$$

In region 2, the vehicle deceleration is greater than z_V and the valve has actuated. The front and rear brake

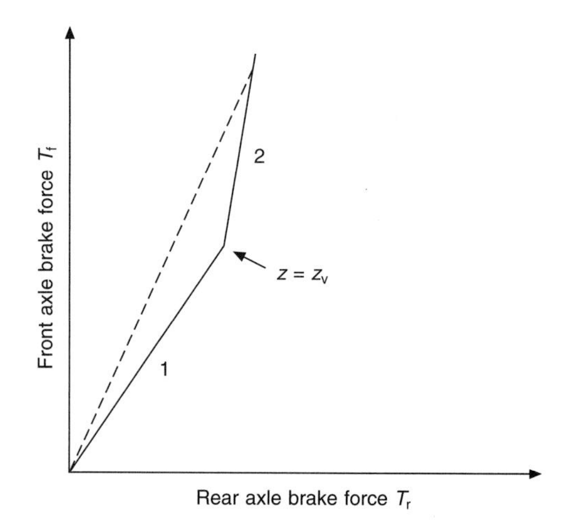

Figure 12.1-20 Typical modulating valve brake force distribution.

forces increase in accordance with the slope of the valve characteristic and this causes the overall vehicle brake ratio to vary, being equal to the slope of the dashed line. If the slope of the brake force characteristic in region 2 is defined as $\frac{x_{f2}}{x_{r2}}$ then the brake force at the rear axle, T_r, is:

$$T_r = Pz_v x_r + (Pz - Pz_v)x_{r2} \tag{12.1.90}$$

and the brake force at the front axle is:

$$\begin{aligned} T_f &= T - T_r \\ &= Pz - (Pz_v x_r + (Pz - Pz_v)x_{r2}) \end{aligned} \tag{12.1.91}$$

Thus, the overall brake ratio, defined by the slope of the dashed line, is:

$$R_v = \frac{T_f}{T_r} = \frac{Pzx_{fv}}{Pzx_{rv}} = \frac{Pz - (Pz_v x_r + (Pz - Pz_v)x_{r2})}{Pz_v x_r + (Pz - Pz_v)x^2} \tag{12.1.92}$$

from which

$$x_{fv} = \frac{z - z_v x_r - (z - z_v)x_{r2}}{z} \tag{12.1.93}$$

and

$$x_{rv} = \frac{z_v x_r + (z - z_v)x_{r2}}{z} \tag{12.1.94}$$

The substitution of the modulation valve for the limiting valve into the brake system of the prototype vehicle enables improvements to be made to the adhesion utilization at high rates of deceleration. This is due to the ability of the valve to increase the line pressure to the rear brakes at a reduced rate. By appropriate choice of the slope of the brake force characteristic in region 2, the front axle adhesion curve can be forced to move closer to the optimum. In this example, setting the ratio for region 2 to be $\frac{x_{f2}}{x_{r2}} = \frac{0.88}{0.12}$ causes the front and rear adhesion curves to cross at a deceleration of $0.8g$ and gives rise to the adhesion utilization diagram of Figure 12.1-21a.

The adhesion behaviour is identical to that shown in Figure 12.1-19a up until valve actuation. Thereafter the front axle adhesion converges on that of the optimum at $0.8g$. Decelerations greater than $0.8g$ lead to rear axle lock but this is strictly admissible according to the adhesion utilization requirement specified in the governing EEC braking directive. The fact that the front axle adhesion now deviates little from the optimum illustrates the positive advantage that can be gained through the introduction of a bias valve into the brake system. Comparison of Figures 12.1-11, 12.1-13c, 12.1-19b and 12.1-21b illustrates this point by showing the progressive refinement of the brake system efficiency during the design process.

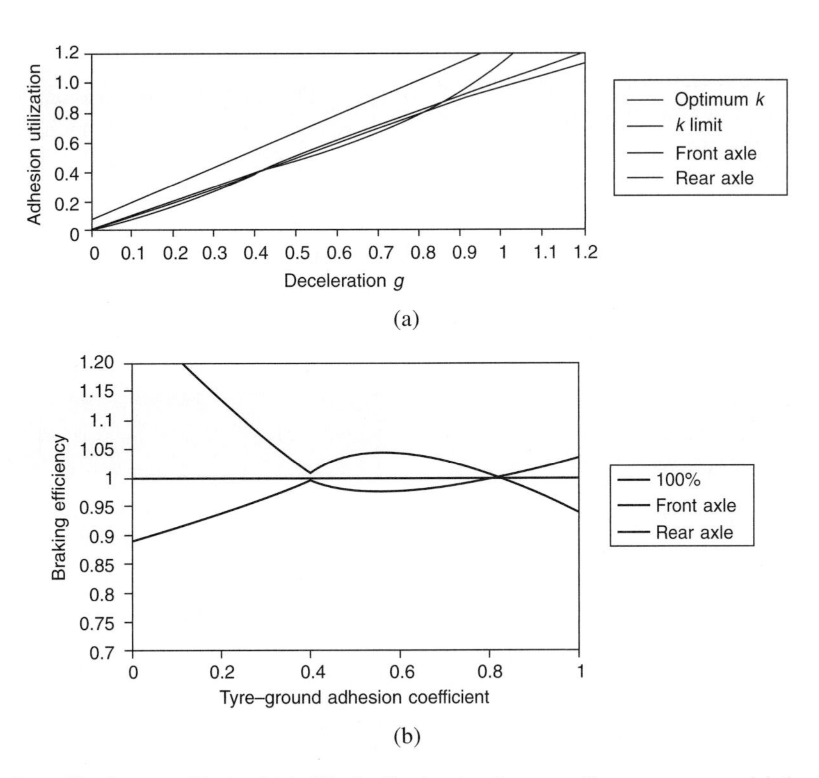

Figure 12.1-21 (a) Adhesion utilization, modified vehicle fitted with deceleration-sensitive pressure modulating valve, (b) brake system efficiency, modified vehicle fitted with deceleration sensitive.

12.1.5 Materials design

12.1.5.1 Materials requirements for braking systems

In any conventional foundation brake, the relative rotation of the so-called 'friction pair' under the action of the brake system activating force is responsible for generating the frictional retarding torque required to slow the vehicle. Most friction pairs consist of a hard, usually metallic, rotating component and a relatively compliant 'friction' material in the form of a brake pad or shoe. The materials requirements for the rotating and stationary components of the friction pair are therefore quite different as discussed below.

Any rotor material must be sufficiently stiff and strong to be able to transmit the frictional torque to the hub without excessive deformation or risk of failure. However, the stresses arising from thermal effects are much higher than purely mechanical stresses and are more likely to give concerns over disc integrity. Thus the rotor material should have high volumetric heat capacity ($\rho \cdot c_P$) and good thermal conductivity (k) in order to absorb and transmit the heat generated at the friction interface without excessive temperature rise. Furthermore the maximum operating temperature (MOT) of the material should be sufficiently greater than the maximum expected temperature rise to ensure integrity of the rotor even under the most severe braking conditions. Ideally the rotor material should have a low coefficient of thermal expansion (α) to minimize thermal distortions such as 'coning' of a disc. It should also have low density (ρ) to minimize the unsprung mass of the vehicle. It should be resistant to wear since generally it is far easier and cheaper to replace the friction pads or shoes than the rotor itself. Finally, and most importantly, the rotor should be cheap and easy to manufacture.

The brake pad or shoe represents the stationary part of the foundation brake assembly. Normally a proprietary composite friction material is bonded to a steel backing plate or shoe platform. The primary function of the friction material is generally considered to be the production of a stable and predictable coefficient of friction to enable reliable and efficient braking of the vehicle over a wide range of conditions. In fact, it is the combined tribological characteristics of both rotor and stator materials (i.e. the 'friction pair') which are responsible for the generation of the frictional torque. As for the rotor, the friction material must have sufficient structural integrity to resist the mechanical and thermal stresses. This is particularly important for the bond between the friction material itself and the steel structure which supports it, as a complete failure here could have disastrous consequences. The friction material should have a relatively high MOT to prevent thermal degradation of the surface although, due to the nature of its composition, the MOT of the pad material will always be lower than that of the disc. A low conductivity for the pad or shoe material is desirable to minimize conduction of heat to other components of the system, in particular to the hydraulic fluid. The material should be reasonably wear resistant but not excessively so since wear can be beneficial in promoting a uniform contact pressure distribution and preventing 'hot spotting'. Likewise the elastic modulus of the material should be relatively low to give good conformity with a roughened or thermally distorted rotor surface. Finally, as for the rotor, the friction material should be cheap and easy to manufacture.

The friction material selected to meet the above requirements is invariably a complex composite consisting of a variety of fibres, particles and fillers bonded together in a polymeric matrix such as phenolic resin. For many years, asbestos fibres were an important element of friction materials due to their excellent thermal and friction properties. For health and safety reasons, asbestos has now largely been replaced by other less harmful fibres, e.g. Kevlar. The exact composition of any friction material must be tailored to the application and knowledge of the formulation is proprietary to the supplier.

12.1.5.2 Cast iron rotor metallurgy

The overwhelming majority of rotors for conventional automotive brakes is manufactured from grey cast iron (GI). This material, also known as flake graphite iron, is cheap and easy to cast and machine in high volumes. It has good volumetric heat capacity due mainly to its relatively high density, and reasonable conductivity due largely to the presence of the graphite (or carbon) flakes. The coefficient of thermal expansion is relatively low and the material has an MOT well in excess of 700 °C (but note that martensitic transformations at high temperatures can

Table 12.1-2 Tensile strength and conductivity of some common cast irons

Grade	Min. tensile strength (MPa)	Thermal conductivity at 300 °C (W/m K)
400/18 SG*	400	36.2
250 GI	250	45.4
200 GI	200	48.1
150 GI	150	50.5

*Spherical graphite iron.

lead to hot judder problems). Although the compressive strength is good, the tensile strength is relatively low and the material is brittle and prone to microcracking in tension. As the proportion of flake graphite in GI is increased, the tensile strength reduces but thermal conductivity increases as shown in Table 12.1-2. Note that spheroidal graphite iron (SG) has a higher tensile strength than GI but a much reduced conductivity which explains why it is rarely used for brake rotors.

Currently GI grades used for disc brakes fall into two categories reflecting two different design philosophies (MacNaughton and Krosnar, 1998):

1. Medium carbon GI (e.g. Grade 220)

These irons are used for small diameter discs such as on small- and medium-sized passenger cars. Such discs will run hot under extreme conditions, and good strength and thermal crack resistance at high temperatures are therefore required.

2. High carbon GI (e.g. Grade 150)

These grades tend to be used for larger vehicles where space constraints are not as content limited. Discs are larger and, with the improved conductivity due to the high carbon, will run cooler. Strength retention at high temperature is therefore not as critical and manufacturability improves with the higher carbon content.

Alloying elements can be applied to all grades of cast iron with the general effect of improving strength but at expense of thermal properties and manufacturability. The most commonly used elements and their effects are as follows:

- chromium increases strength by stabilizing pearlitic matrix at high temperatures (preventing martensitic transformations) but tends to promote formation of bainitic structures which cause casting/machining difficulties and can reduce pad life;
- molybdenum similar to chromium;
- copper increases strength without causing manufacturing difficulties;
- nickel as for copper but more expensive;
- titanium reported to influence friction performance but rarely used at significant levels.

12.1.5.3 Alternative rotor materials

Although GI is a cheap material with good thermal properties and strength retention at high temperature, its density is high and, because section thickness must be maintained for both manufacturability and performance, cast iron rotors are heavy. Currently there are significant incentives to reduce rotor weights in order to: (a) reduce emissions by improving the overall fuel consumption of the vehicle, and (b) aid refinement and limit damage to roads by reducing the unsprung mass. Thus much effort

Table 12.1-3 Physical properties of three candidate disc materials

Disc material	ρ kg m^{-3}	c_P J kg^{-1} K^{-1}	$\rho \cdot c_P$ kJ m^{-3} K^{-1}	k Wm^{-1} K^{-1}	$\alpha \times 10^{-6}$ K^{-1}
High carbon cast iron	7150	438	3132	50	10
Generic 20% SiC-reinforced Al MMC	2800	800	2240	180	17.5
Carbon–carbon composite	1750	1000	1750	40–150	0.7

has been directed at investigating light-weight alternatives to cast iron. Two such alternatives which have received serious attention are aluminium metal matrix composites (MMCs) and carbon–carbon composites, typical properties for each of which are displayed in Table 12.1-3. together with corresponding properties for a high carbon cast iron (Grieve, *et al.* 1995).

Aluminium MMCs normally incorporate 10–30% by volume silicon carbide particle reinforcement within a silicon-containing alloy matrix. The resulting composite has much lower density than cast iron and much improved conductivity. Thus the thermal diffusivity ($k/\rho \cdot c_P$) is much higher which opens the possibility of lighter discs running cooler by being able to rapidly conduct heat away from the friction interface. However, aluminium MMCs have a low MOT (c. 500 °C) and there are serious consequences if this MOT is exceeded since complete surface disruption may then occur leading to extremely rapid pad wear. Ideally higher reinforcement contents or alternative reinforcing materials (e.g. alumina) should be used to increase the MOT but the former causes severe casting difficulties whilst alumina reinforcement results in poorer thermal properties.

It can be seen from Table 12.1-3 that carbon–carbon composites have an even lower density than aluminium MMCs and can have a conductivity almost as high. Their MOT is also very high, raising the possibility of using thin rotors which run much hotter and lose heat by radiation as well as by conduction/convection. Also the very low coefficient of thermal expansion of carbon minimizes thermal distortions. Thus there is the potential for very significant weight savings with carbon–carbon composite discs. However, the material has a poor low temperature friction performance and moreover is currently much more expensive than metallic alternatives. Hence, it is likely to remain confined to high performance race car applications for the foreseeable future.

When considering alternative materials or designs for disc brakes, reference can be made to the so-called

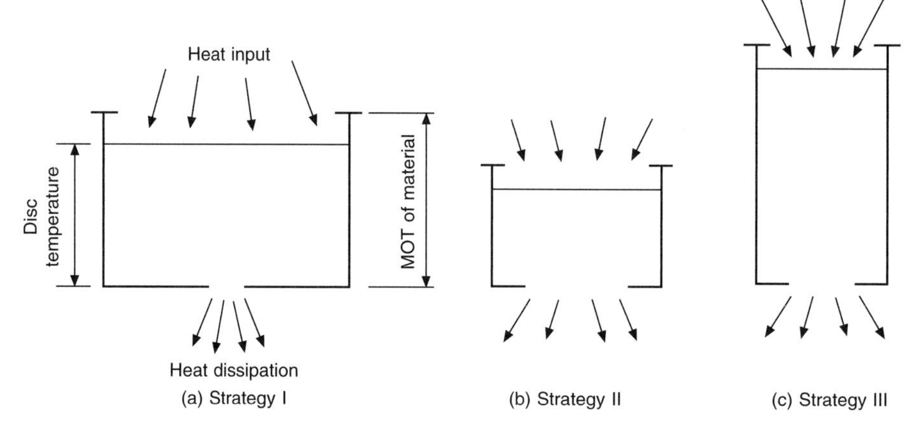

Figure 12.1-22 The 'bucket-and-hole' analogy.

'bucket-and-hole' analogy in which the rate of water flow into the bucket is taken to represent the heat flow into the disc and the height of the water level in the bucket represents the maximum temperature of the disc surface. A hole in the bucket represents the ability of the disc to lose heat to the surroundings. The volume of the bucket is therefore the heat capacity of the disc whilst the height is the MOT of the disc material. The question then is how close does the level of water in the bucket get to overflow!

Consideration of the 'bucket-and-hole' analogy and with reference to the typical material properties of Table 12.1-3, three distinct strategies for brake rotor materials can be identified (Grieve *et al.*, 1996):

Strategy I

Large diameter and relatively deep bucket with small hole (see Figure 12.1-22a). This implies a high volumetric heat capacity to store heat during braking and a relatively high MOT but only moderate conductivity of heat away from the rubbing surfaces. Current GI discs represent such a system but some steels may also meet these criteria.

Strategy II

Smaller diameter and relatively shallow bucket but large hole (see Figure 12.1-22b). This implies smaller volumetric heat capacity and a relatively low MOT. Hence, it is important to have high conductivity to transfer heat to other parts of the rotor and then the surroundings in order to prevent temperature build-up at the rubbing surfaces. Aluminium MMC may meet these criteria but recent research (Grieve *et al.*, 1998) suggests that this can only be successfully achieved for currently available MMCs if the brake rotor is redesigned to increase its thermal mass and cooling capability. Other materials that may be successful with appropriate development include high reinforcement content MMCs and coated alloy discs but again there are manufacturing, integrity and cost issues to be resolved.

Strategy III

Even smaller diameter bucket but much deeper with moderately sized hole (see Figure 12.1-22c). This implies a material with high MOT which can be allowed to run much hotter than current designs and so lose significant amounts of heat by radiation as well as more moderate amounts by conduction/convection. Carbon–carbon composites are a possibility here but, as mentioned above, these are currently too expensive for mass produced vehicles. High-temperature steels with good strength retention at temperatures well in excess of 1000 °C may also be candidate materials under this heading. Such discs could perhaps be made much thinner and without vents, and therefore also save significant weight. However, there would be concerns over compatible friction materials and heat transfer to other components in the underbody wheel arch area if discs were allowed to run much hotter than is currently the practice with cast iron.

12.1.5.4 Disc materials/design evaluation

Ultimately, any new brake material or design must be validated by experimental trials on actual vehicles to allow accurately for model-specific parameters such as the effect of body trim on rotor cooling. However, much can be learnt about potential new rotor materials or designs by numerical simulations of critical brake tests using finite element (FE) analysis. Such techniques require the rotor and/or stator geometry to be broken down into a number of small non-overlapping regions known as elements which are assumed to be connected to one another at certain points known as nodes. A 2D axisymmetric FE idealization can be used as a first approximation but, for more accurate simulation of the heat flow and stresses, a 3D model is desirable such as the 10° segment model of a brake disc and hub shown

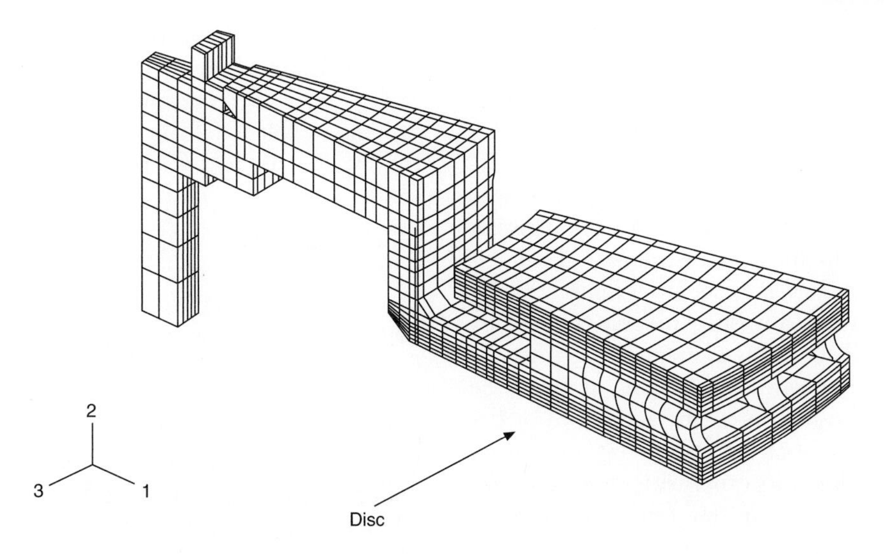

Figure 12.1-23 Finite element model of 10° segment of vented disc and hub.

in Figure 12.1-23. Note that in order to accurately simulate the heat loss from the rotor, it is sometimes necessary to include the wheel and other components in the model.

The heat input to the system is estimated from theoretical consideration and applied over the rubbing surface. The heat loss to the surrounding is specified by convective and sometimes radiative heat transfer conditions along relevant boundaries of the model. The temperatures predicted by a thermal analysis can be used as input conditions to a structural analysis in order to predict thermal deformations and stresses. If the pad is included in the model, the contact pressure distribution (and hence the distribution of heat input) can be estimated leading to the possibility of a fully coupled thermal-structural analysis (Brooks *et al.*, 1994).

In addition to details of geometry and material properties, accurate date on heat loss to other components and to the atmosphere are vital to allow accurate predictions of rotor temperatures using FE methods. Such data can be generated by conducting the so-called 'cooling tests' on actual vehicles fitted with representative brake rotors carrying rubbing or embedded thermocouples. The rotor surface is first heated to a predetermined temperature by dragging the brakes and then allowed to cool whilst the vehicle is driven at constant velocity. By comparing the experimental rate of cooling with that predicted by the FE simulation for different boundary conditions, optimized heat transfer coefficients can be derived which are then assumed to apply for different rotor materials and factored for the varying air stream velocity under different test conditions.

Two very different vehicle brake tests are often simulated to critically examine the maximum temperatures and integrity of new rotor materials or designs: (i) a long slow Alpine descent during which the brakes are dragged and the vehicle is subsequently left to stand at the end of the descent; (ii) a repeated high speed autobahn stop with the rotor allowed to cool only moderately between stops. The former test determines the ability of the design to limit temperature build-up in the rotor by heat transfer to the atmosphere whilst the high-speed repeated stop examines the ability of the rotor material to withstand repeated thermal cycling and the ability of the friction pair to resist 'fade' under these severe conditions.

Friction performance cannot easily be predicted by the FE approach and there remains a requirement for dynamometer testing to determine the fade-and-wear characteristics of every new friction pair. The dynamometer can either be a full-scale device or a small sample rig in which the geometry and loading conditions are scaled to give an accurate representation of the actual brake. These tests will not only give data on friction performance over a wide range of conditions but can also be used to determine the MOT of the pad and rotor materials by progressively increasing the temperature at the rubbing interface until some form of failure occurs (Grieve *et al.*, 1996).

12.1.6 Advanced topics

12.1.6.1 Driver behaviour

The driver of a vehicle plays a key role during any braking event since his/her reactions to external stimuli have a direct bearing on his/her ability to maintain complete control over the vehicle trajectory and deceleration rate. A knowledge of how the driver interacts with these external stimuli and the way in which the vehicle responds to the control signals generated by the driver is vital to the future development of safe road transport systems.

Many experimental studies, including Newcomb (1981), Newcomb and Spurr (1974), Mortimer (1976) and Spurr (1972), have been undertaken that have led to improved understanding of driver behaviour during braking. These have focused on the study of limb dynamics, pedal effort, braking kinematics and response to external stimuli such as obstacles and road signs. This has given rise to the development of mathematical models that embody a representation of the driver into a model of the vehicle dynamics. Any such model, typified by McLean *et al.* (1976), contains elements that describe the dynamics of the vehicle, the braking system, the neuro-muscular system and force characteristics of the driver and finally the motion detection system/sensory characteristics of the driver together with feedback loops as appropriate to the model in question. The adaptive nature of the driver that is captured in such models requires enhancement but simulation of vehicle braking performance with the driver can yield deceleration characteristics that match closely those from experiment.

12.1.6.2 Brake by wire

The driver behind brake-by-wire systems has arisen from the ongoing development of modern braking systems such as anti-lock and traction control systems (TCSs) along with the need to effect their seamless integration within the overall chassis control strategy. There are two strategies currently receiving attention.

The first utilizes a conventional hydraulically actuated braking system, that includes the brake fluid, brake lines and conventional actuators, together with a significant number of electro-hydraulic components (Jonner *et al.*, 1996).

The second relies upon a full electro-mechanical system (Bill, 1991; Maron *et al.*, 1997; Schenk *et al.*, 1995) in which the brake force is generated directly by electro-mechanical foundation brake actuators. The electro-mechanical system potentially requires little maintenance due to the removal of the hydraulic fluid as the means of energy transmission and this conveniently combines with a reduction in the amount of hardware demanded by the brake system which in turn leads to an overall weight reduction. Such systems may also contribute towards the enhancement of passenger safety as the location of the pedal assembly within the vehicle can be optimized so that the likelihood of lower leg injury is minimized during impact events. As with all advanced control systems, it is the control unit, its associated software and the array of sensors that combine to define the overall effectiveness of the system. The controller must operate in closed-loop fashion, be able to take into account the in-use variation of the system parameters and fail safe.

12.1.6.3 Anti-lock braking systems

Under normal braking conditions, the driver of a vehicle makes use of the linear portion of the brake slip vs brake force characteristic (Figure 12.1-7). The brake force coefficient, μ, builds from zero in the free rolling state to a maximum, μ_p, at around 20% slip and within this region the wheel is both stable and controllable. When braking under extreme conditions the driver may demand a brake torque that is greater than that which is capable of being reacted by the wheel. This results in a torque imbalance that causes the wheel slip to increase and the wheel rapidly decelerates to the full lock condition and in this state, the brake force coefficient is approximately $0.7\,\mu_p$. If the front wheels have locked, then steering control is lost and if rear wheel lock takes place then the vehicle becomes unstable. Simultaneously, the ability of the vehicle to generate side force markedly reduces (Figure 12.1-7), and this explains why limiting wheel slip, thereby avoiding wheel lock, is more critical for steering and directional stability of the car than for stopping distance alone.

The purpose of ABS is to control the rate at which individual wheels accelerate and decelerate through the regulation of the line pressure applied to each foundation brake. The control signals, generated by the controller and applied to the brake pressure modulating unit, are derived from the analysis of the outputs taken from wheel speed sensors. Thus, when active, the ABS makes optimum use of the available friction between the tyres and the road surface.

12.1.6.4 Traction control systems

Traction control systems aim (TCSs) to control and maintain vehicle stability during acceleration manoeuvres, by, for example, preventing wheel spin when accelerating on a low friction surface or on a steep up-grade. This is achieved by the optimization of individual wheel torques through the control of some combination of fuel mixture, ignition and driven wheel brake torque. TCSs are able to utilize components used in ABS and integration of the two systems is becoming commonplace.

12.1.7 References and further reading

Automotive brake systems. (1995). Pub. Robert Bosch GmbH. Distributed by SAE. ISBN 1-56091-708-3.

Bill, K. (1991). *Investigations on the behaviour of electrically actuated friction brakes for passenger cars*. EAEC 91021.

Brake handbook. (1981) (2nd edition). Alfred Teves GmbH.

Brake technology and ABS/TCS systems. (1999). Pub SAE. SP-1413ISBN 0-7680-0345-8.

Brooks, P.C., Barton, D.C., Crolla, D.A., Lang, A.M. and Schafer, D.R. (1994). A study of disc brake judder using a fully coupled thermo-mechanical finite element model. Proceedings FISITA 94 Conference, Paper No 945042, Beijing, October.

Grieve, D., Barton, D.C., Crolla, D.A., Buckingham, J.T. and Chapman, J. (1995). Investigation of light weight materials for brake rotor applications. IoM Conference on Materials for Lean Weight Vehicles, University of Warwick, November.

Grieve, D., Barton, D. C., Crolla, D. A., Buckingham, J. T., & Chapman, J. (1996). Investigation of light weight materials for brake rotor applications. In Barton D.C. (Ed.), *Advances in Automotive Braking Technology*. PEP.

Grieve, D., Barton, D. C., Crolla, D. A., Chapman, J., & Buckingham, J. T. (1998). *Design of a light weight automotive disc brake using finite element and Taguchi techniques*. In: *Proc IMechE Part D*. 245–254.

Jonner, W.-D., Winner, H., Dreilich, L., & Schunck, E. (1996). *Electrohydraulic brake system – the first approach to brake-by-wire technology*. Detroit: SAE 960991.

Limpert, R. (1992). *Brake design and safety*. SAE. ISBN 1-56091-261-8.

MacNaughton, M. P., & Krosnar, J. G. (1998). Cast iron – a disc brake material of the future? In D. C. Barton, & M. J. Haigh (Eds.), *Automotive Braking: Recent Developments and Future Trends* PEP.

Maron, C., Dieckmann, T., Hauck, S., & Prinzler, H. (1997). *Electromechanical brake system: Actuator control development system*. Detroit: SAE 970814.

McLean, D., Newcomb, T. P., & Spurr, R. T. (1976). Simulation of driver behaviour during braking. *Proc IMechE Conference on Braking of Road vehicles*. paper C41/76.

Mortimer, R. G. (1976). Implications of some characteristics of drivers for brake system performance. *Proc ImechE Conf Braking of Road Vehicles*187–195, Loughborough.

Newcomb, T. P. (1981). *Driver behaviour during braking*. SAE/IMechE Exchange Lecture 1981. SAE 810832.

Newcomb, T. P., & Spurr, R. T. (1967). *Braking of Road Vehicles*. Chapman and Hall Ltd. Out of print.

Newcomb, T. P., & Spurr, R. T. (1974). *Aspects of driver behaviour during braking*. XV Congress FISITA. paper A. 1.4.

Puhn, F. (1995). *Brake handbook*. HP. Books. ISBN 0-89526-232-8.

Reimpell, J., & Stoll, H. (1996). *The Automotive Chassis: Engineering Principles*. Arnold.

Schenk, O. E., Wells, R. L., & Miller, I. E. (1995). *Intelligent braking for current and future vehicles*. Detroit: SAE 950762.

Spurr, R.T. (1972). Driver behaviour during braking. Proc. Symp. on Psychological Aspects of Driver Behaviour, paper 1-2, Holland.

Section Thirteen

Vehicle control systems

Chapter 13.1

13.1

Vehicle motion control

William Ribbens

13.1.1 Introduction

The term *vehicle motion* refers to its translation along and rotation about all three axes (i.e., longitudinal, lateral, and vertical). By the term *longitudinal axis*, we mean the axis that is parallel to the ground (vehicle at rest) along the length of the car. The lateral axis is orthogonal to the longitudinal axis and is also parallel to the ground (vehicle at rest). The vertical axis is orthogonal to both the longitudinal and lateral axes.

Rotations of the vehicle around these three axes correspond to angular displacement of the car body in roll, yaw, and pitch. *Roll* refers to angular displacement about the longitudinal axis; *yaw* refers to angular displacement about the vertical axis; and *pitch* refers to angular displacement about the lateral axis.

Electronic controls have been recently developed with the capability to regulate the motion along and about all three axes. Individual car models employ various selected combinations of these controls. This chapter discusses motion control electronics beginning with control of motion along the longitudinal axis in the form of a cruise control system.

The forces that influence vehicle motion along the longitudinal axis include the powertrain (including, in selected models, traction control), the brakes, the aerodynamic drag, and tire-rolling resistance, as well as the influence of gravity when the car is moving on a road with a nonzero inclination (or grade). In a traditional cruise control system, the tractive force due to the powertrain is balanced against the total drag forces to maintain a constant speed. In an ACC system, brakes are also automatically applied as required to maintain speed when going down a hill of sufficiently steep grade.

13.1.2 Typical cruise control system

Automotive cruise control is an excellent example of the type of electronic feedback control system. Recall that the components of a control system include the plant, or system being controlled, and a sensor for measuring the plant variable being regulated. It also includes an electronic control system that receives inputs in the form of the desired value of the regulated variable and the measured value of that variable from the sensor. The control system generates an error signal constituting the difference between the desired and actual values of this variable. It then generates an output from this error signal that drives an electromechanical actuator. The actuator controls the input to the plant in such a way that the regulated plant variable is moved toward the desired value.

In the case of a cruise control, the variable being regulated is the vehicle speed. The driver manually sets the car speed at the desired value via the accelerator pedal. Upon reaching the desired speed the driver activates a momentary contact switch that sets that speed as the command input to the control system. From that point on, the cruise control system maintains the desired speed automatically by operating the throttle via a throttle actuator.

Under normal driving circumstances, the total drag forces acting on the vehicle are such that a net positive traction force (from the powertrain) is required to maintain a constant vehicle speed. However, when the car is on a downward sloping road of sufficient grade, constant vehicle speed requires a negative tractive force that the powertrain cannot deliver. In this case, the car will accelerate unless brakes are applied. For our initial

Understanding Automotive Electronics; ISBN: 9780750675994
Copyright © 2003 Elsevier Ltd; All rights of reproduction, in any form, reserved.

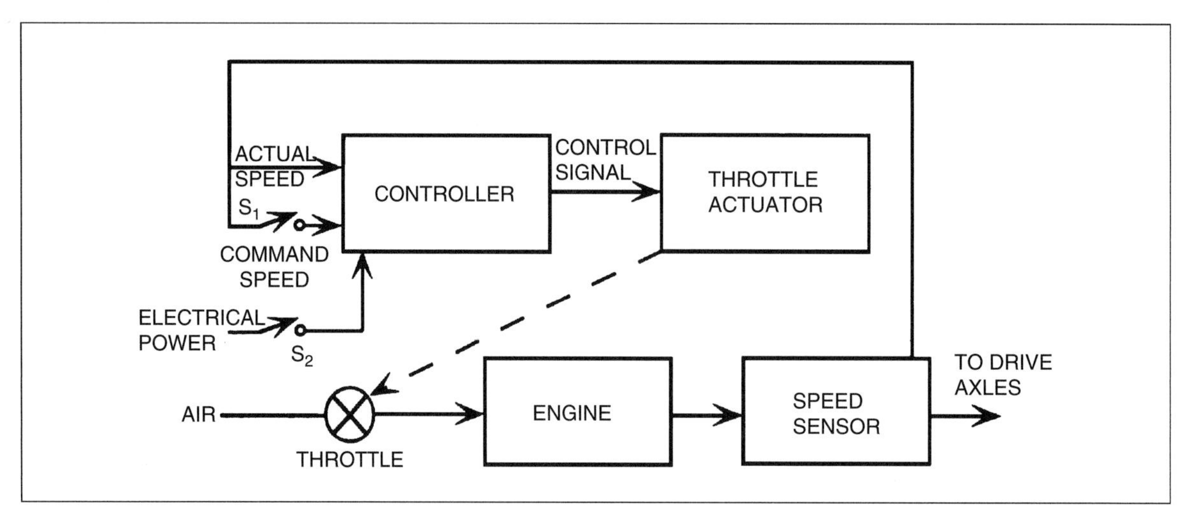

Fig. 13.1-1 Cruise control configuration.

discussion, we assume this latter condition does not occur and that no braking is required.

The plant being controlled consists of the powertrain (i.e., engine and drivetrain), which drives the vehicle through the drive axles and wheels. As described above, the load on this plant includes friction and aerodynamic drag as well as a portion of the vehicle weight when the car is going up and down hills.

The configuration for a typical automotive cruise control is shown in Fig. 13.1-1. The momentary contact (pushbutton) switch that sets the command speed is denoted S_1 in Fig. 13.1-1. Also shown in this figure is a disable switch that completely disengages the cruise control system from the power supply such that throttle control reverts back to the accelerator pedal. This switch is denoted S_2 in Fig. 13.1-1 and is a safety feature. In an actual cruise control system the disable function can be activated in a variety of ways, including the master power switch for the cruise control system, and a brake pedal-activated switch that disables the cruise control any time that the brake pedal is moved from its rest position. The throttle actuator opens and closes the throttle in response to the error between the desired and actual speed. Whenever the actual speed is less than the desired speed the throttle opening is increased by the actuator, which increases vehicle speed until the error is zero, at which point the throttle opening remains fixed until either a disturbance occurs or the driver calls for a new desired speed.

A block diagram of a cruise control system is shown in Fig. 13.1-2. In the cruise control depicted in this figure, a proportional integral (PI) control strategy has been assumed. However, there are many cruise control systems still on the road today with proportional (P) controllers. Nevertheless, the PI controller is representative of good design for such a control system since it can reduce speed errors due to disturbances (such as hills) to zero. In this strategy an error e is formed by

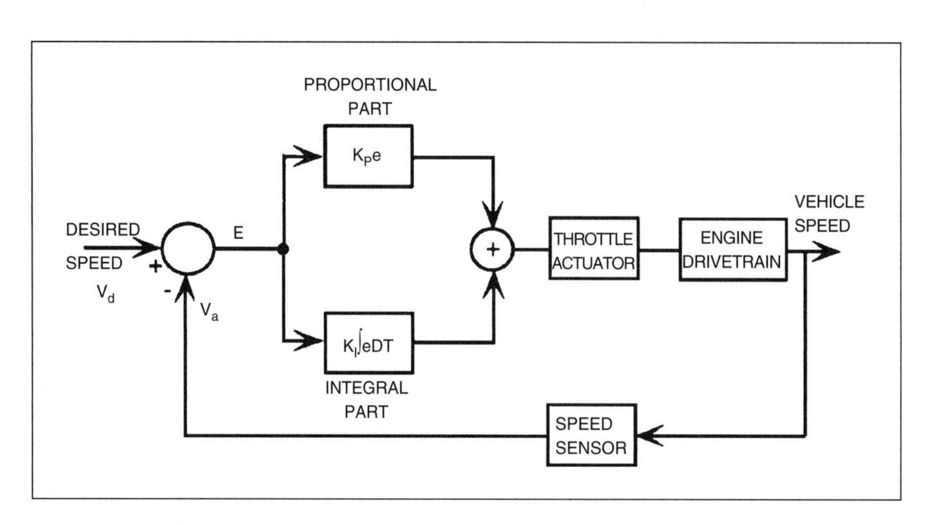

Fig. 13.1-2 Cruise control block diagram.

subtracting (electronically) the actual speed V_a from the desired speed V_d:

$$e = V_d - V_a$$

The controller then electronically generates the actuator signal by combining a term proportional to the error ($K_P e$) and a term proportional to the integral of the error:

$$K_I \int e\, dt$$

The actuator signal u is a combination of these two terms:

$$u = K_P e + K_I \int e\, dt$$

The throttle opening is proportional to the value of this actuator signal.

Operation of the system can be understood by first considering the operation of a proportional controller (i.e., imagine that the integral term is not present for the sake of this preliminary discussion). We assume that the driver has reached the desired speed (say, 60 mph) and activated the speed set switch. If the car is traveling on a level road at the desired speed, then the error is zero and the throttle remains at a fixed position.

If the car were then to enter a long hill with a steady positive slope (i.e., a hill going up) while the throttle is set at the cruise position for level road, the engine will produce less power than required to maintain that speed on the hill. The hill represents a disturbance to the cruise control system. The vehicle speed will decrease, thereby introducing an error to the control system. This error, in turn, results in an increase in the signal to the actuator, causing an increase in engine power. This increased power results in an increase in speed. However, in a proportional control system the speed error is not reduced to zero since a nonzero error is required so that the engine will produce enough power to balance the increased load of the disturbance (i.e., the hill).

The speed response to the disturbance is shown in Fig. 13.1-3a. When the disturbance occurs, the speed drops off and the control system reacts immediately to increase power. However, a certain amount of time is required for the car to accelerate toward the desired speed. As time progresses, the speed reaches a steady

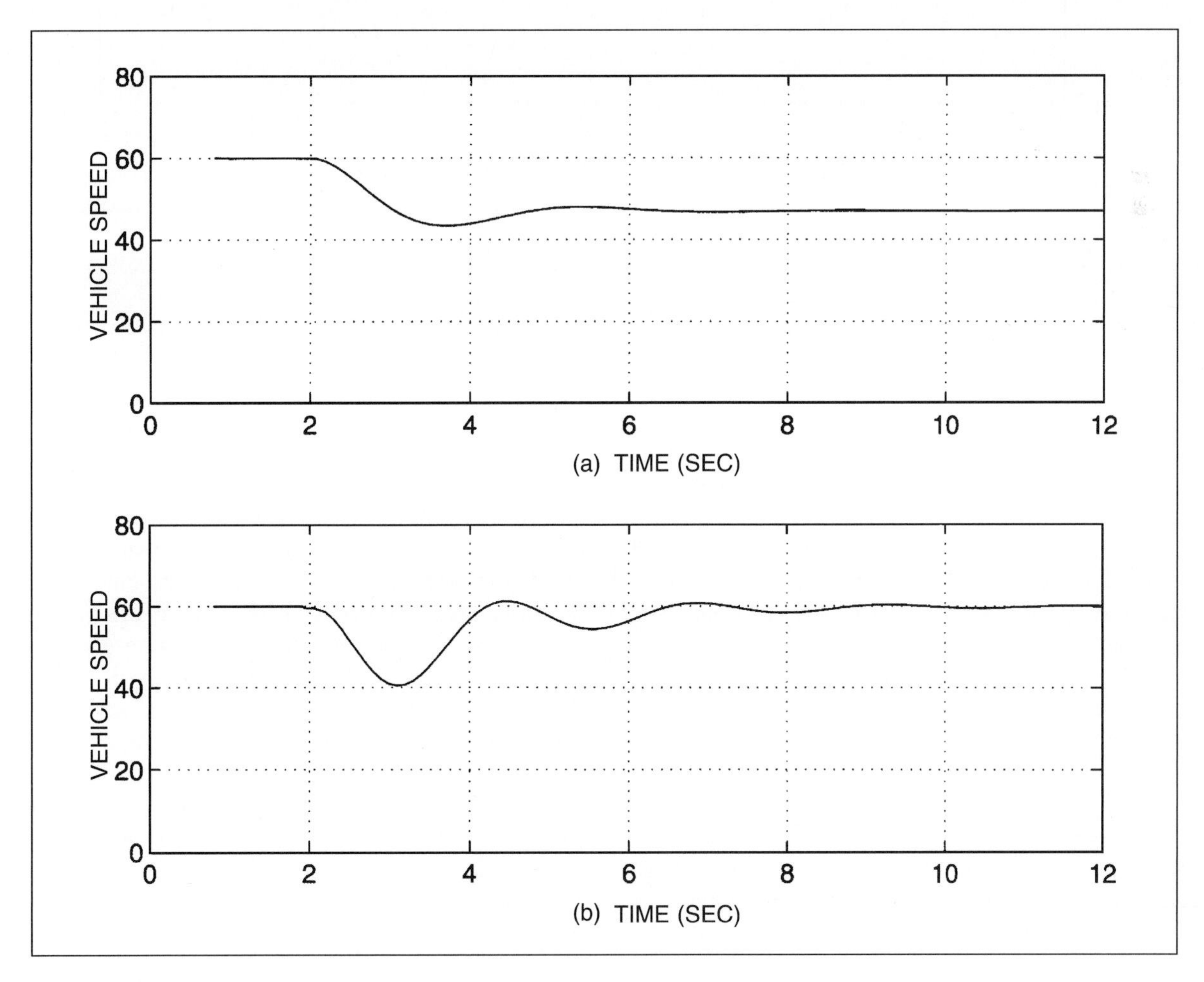

Fig. 13.1-3 Cruise control speed performance.

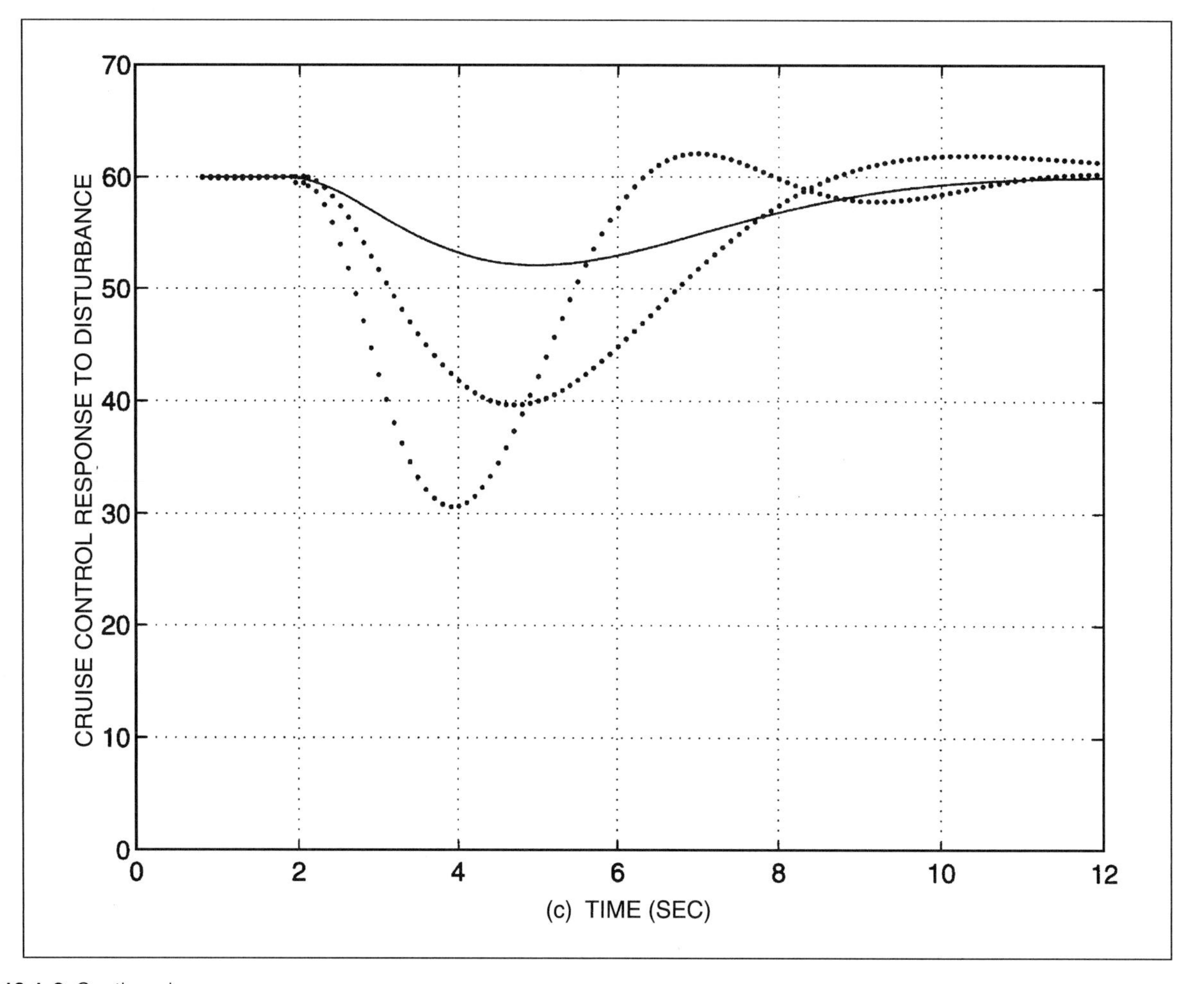

Fig. 13.1-3 Continued

value that is less than the desired speed, thereby accounting for the steady error (e_s) depicted in Fig. 13.1-3a (i.e., the final speed is less than the starting 60 mph).

If we now consider a PI control system, we will see that the steady error when integrated produces an ever-increasing output from the integrator. This increasing output causes the actuator to increase further, with a resulting speed increase. In this case the actuator output will increase until the error is reduced to zero. The response of the cruise control with PI control is shown in Fig. 13.1-3b.

The response characteristics of a PI controller depend strongly on the choice of the gain parameters K_P and K_I. It is possible to select values for these parameters to increase the speed of the system response to disturbance. If the speed increases too rapidly, however, overshoot will occur and the actual speed will oscillate around the desired speed. The amplitude of oscillations decreases by an amount determined by a parameter called the *damping ratio*. The damping ratio that produces the fastest response without overshoot is called *critical damping*. A damping ratio less than critically damped is said to be *underdamped*, and one greater than critically damped is said to be *overdamped*.

13.1.2.1 Speed response curves

The curves of Fig. 13.1-3c show the response of a cruise control system with a PI control strategy to a sudden disturbance. These curves are all for the same car cruising initially at 60 mph along a level road and encountering an upsloping hill. The only difference in the response of these curves is the controller gain parameters.

Consider, first, the curve that initially drops to about 30 mph and then increases, overshooting the desired speed and oscillating above and below the desired speed until it eventually decays to the desired 60 mph. This curve has a relatively low damping ratio as determined by the controller parameters K_P and K_I and takes more time to come to the final steady value.

Next, consider the curve that drops initially to about 40 mph, then increases with a small overshoot and decays to the desired speed. The numerical value for this damping ratio is about 0.7, whereas the first curve had a damping ratio of about 0.4. Finally, consider the solid curve of Fig. 13.1-3c. This curve corresponds to critical damping. This situation involves the most rapid response of the car to a disturbance, with no overshoot.

The importance of these performance curves is that they demonstrate how the performance of a cruise control system is affected by the controller gains. These gains are simply parameters that are contained in the control system. They determine the relationship between the error, the integral of the error, and the actuator control signal.

Usually a control system designer attempts to balance the proportional and integral control gains so that the system is optimally damped. However, because of system characteristics, in many cases it is impossible, impractical, or inefficient to achieve the optimal time response and therefore another response is chosen. The control system should make the engine drive force react quickly and accurately to the command speed, but should not overtax the engine in the process. Therefore, the system designer chooses the control electronics that provide the following system qualities:

1. Quick response
2. Relative stability
3. Small steady-state error
4. Optimization of the control effort required

13.1.2.2 Digital cruise control

The explanation of the operation of cruise control thus far has been based on a continuous-time formulation of the problem. This formulation correctly describes the concept for cruise control regardless of whether the implementation is by analog or digital electronics. Cruise control is now mostly implemented digitally using a microprocessor-based computer. For such a system, proportional and integral control computations are performed numerically in the computer. A block diagram for a typical digital cruise control is shown in Fig. 13.1-4. The vehicle speed sensor (described later in this chapter) is digital. When the car reaches the desired speed, S_d, the driver activates the speed set switch. At this time, the output of the vehicle speed sensor is transferred to a storage register.

The computer continuously reads the actual vehicle speed, S_a, and generates an error, e_n, at the sample time, t_n (n is an integer). $e_n = S_d - S_a$ at time t_n. A control signal, d, is computed that has the following form:

$$d_n = K_P e_n + K_I \sum_{m=1}^{M} e_{n-m}$$

(*Note*: The symbol $\sum$ in this equation means to add the M previously calculated errors to the present error.) This sum, which is computed in the cruise control computer, is then multiplied by the integral gain K_I and added to the most recent error multiplied by the proportional gain K_P to form the control signal.

This control signal is actually the duty cycle of a square wave (V_c) that is applied to the throttle actuator (as explained later). The throttle opening increases or decreases as d increases or decreases due to the action of the throttle actuator.

The operation of the cruise control system can be further understood by examining the vehicle speed

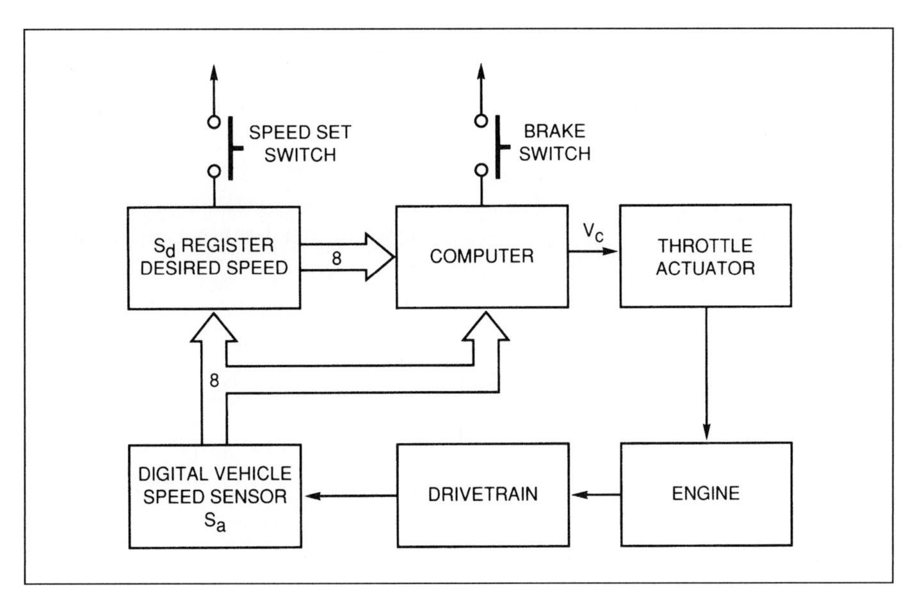

Fig. 13.1-4 Digital cruise control system.

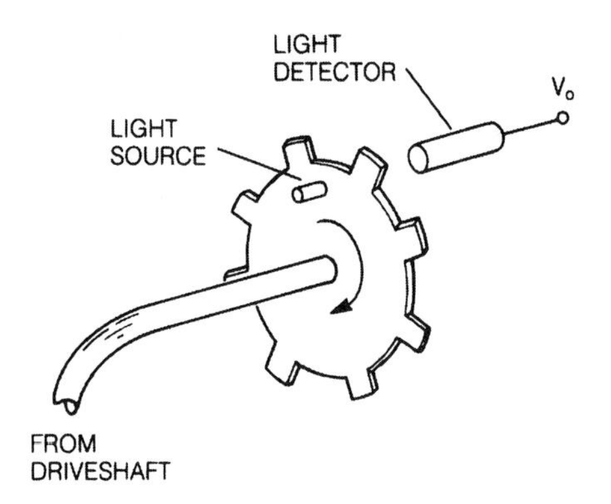

Fig. 13.1-5a Digital speed sensor.

sensor and the actuator in detail. Fig. 13.1-5a is a sketch of a sensor suitable for vehicle speed measurement.

In a typical vehicle speed measurement system, the vehicle speed information is mechanically coupled to the speed sensor by a flexible cable coming from the driveshaft, which rotates at an angular speed proportional to vehicle speed. A speed sensor driven by this cable generates a pulsed electrical signal (Fig. 13.1-5b) that is processed by the computer to obtain a digital measurement of speed.

A speed sensor can be implemented magnetically or optically. For the purposes of this discussion. For the hypothetical optical sensor, a flexible cable drives a slotted disk that rotates between a light source and a light detector. The placement of the source, disk, and detector is such that the slotted disk interrupts or passes the light from source to detector, depending on whether a slot is in the line of sight from source to detector. The light detector produces an output voltage whenever a pulse of light from the light source passes through a slot to the detector. The number of pulses generated per second is proportional to the number of slots in the disk and the vehicle speed:

$$f = NSK$$

where

f is the frequency in pulses per second
N is the number of slots in the sensor disk
S is the vehicle speed
K is the proportionality constant that accounts for differential gear ratio and wheel size

It should be noted that either a magnetic or an optical speed sensor generates a pulse train such as described here.

The output pulses are passed through a sample gate to a digital counter (Fig. 13.1-6). The gate is an electronic switch that either passes the pulses to the counter or does not pass them, depending on whether the switch is closed or open. The time interval during which the gate is closed is precisely controlled by the computer. The digital counter counts the number of pulses from the light detector during time t that the gate is open. The number of pulses P that is counted by the digital counter is given by:

$$P = tNSK$$

That is, the number P is proportional to vehicle speed S. The electrical signal in the binary counter is in a digital format that is suitable for reading by the cruise control computer.

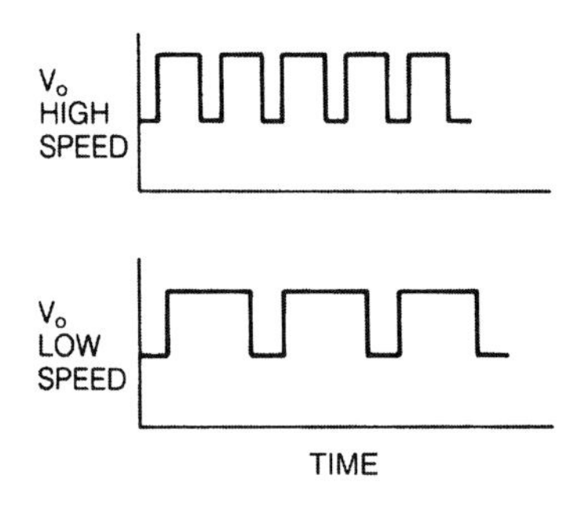

Fig. 13.1-5b Digital speed sensor.

13.1.2.3 Throttle actuator

The throttle actuator is an electromechanical device that, in response to an electrical input from the controller, moves the throttle through some appropriate mechanical linkage. Two relatively common throttle actuators operate either from manifold vacuum or with a stepper

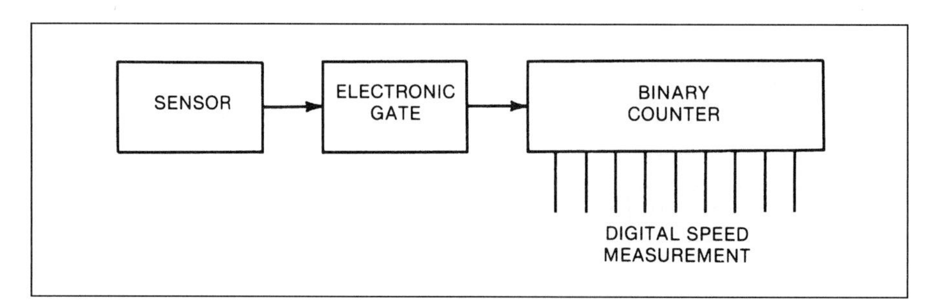

Fig. 13.1-6 Digital speed measurement system.

motor. The stepper motor implementation operates similarly to the idle speed control actuator described in Chapter 4.1. The throttle opening is either increased or decreased by the stepper motor in response to the sequences of pulses sent to the two windings depending on the relative phase of the two sets of pulses.

The throttle actuator that is operated by manifold vacuum through a solenoid valve is similar to that used for the EGR valve described in Chapter 4.1 and further explained later in this chapter. During cruise control operation the throttle position is set automatically by the throttle actuator in response to the actuator signal generated in the control system. This type of manifold-vacuum-operated actuator is illustrated in Fig. 13.1-7.

A pneumatic piston arrangement is driven from the intake manifold vacuum. The piston-connecting rod assembly is attached to the throttle lever. There is also a spring attached to the lever. If there is no force applied by the piston, the spring pulls the throttle closed. When an actuator input signal energizes the electromagnet in the control solenoid, the pressure control valve is pulled down and changes the actuator cylinder pressure by providing a path to manifold pressure. Manifold pressure is lower than atmospheric pressure, so the actuator cylinder pressure quickly drops, causing the piston to pull against the throttle lever to open the throttle.

The force exerted by the piston is varied by changing the average pressure in the cylinder chamber. This is done by rapidly switching the pressure control valve between the outside air port, which provides atmospheric pressure, and the manifold pressure port, the pressure of which is lower than atmospheric pressure. In one implementation of a throttle actuator, the actuator control signal V_c is a variable-duty-cycle type of signal like that discussed for the fuel injector actuator. A high V_c signal energizes the electromagnet; a low V_c signal deenergizes the electromagnet. Switching back and forth between the two pressure sources causes the average pressure in the chamber to be somewhere between the low manifold pressure and outside atmospheric pressure. This average pressure and, consequently, the piston force are proportional to the duty cycle of the valve control signal V_c. The duty cycle is in turn proportional to the control signal d (explained above) that is computed from the sampled error signal e_n.

This type of duty-cycle-controlled throttle actuator is ideally suited for use in digital control systems. If used in an analog control system, the analog control signal must first be converted to a duty-cycle control signal. The same frequency response considerations apply to the throttle actuator as to the speed sensor. In fact, with both in the closed-loop control system, each contributes to the total system phase shift and gain.

13.1.3 Cruise control electronics

Cruise control can be implemented electronically in various ways, including with a microcontroller with special-purpose digital electronics or with analog electronics. It can also be implemented (in proportional control strategy alone) with an electromechanical speed governor.

The physical configuration for a digital, microprocessor-based cruise control is depicted in Fig. 13.1-8. A system such as is depicted in Fig. 13.1-8 is often called a *microcontroller* since it is implemented with a microprocessor operating under program control. The actual

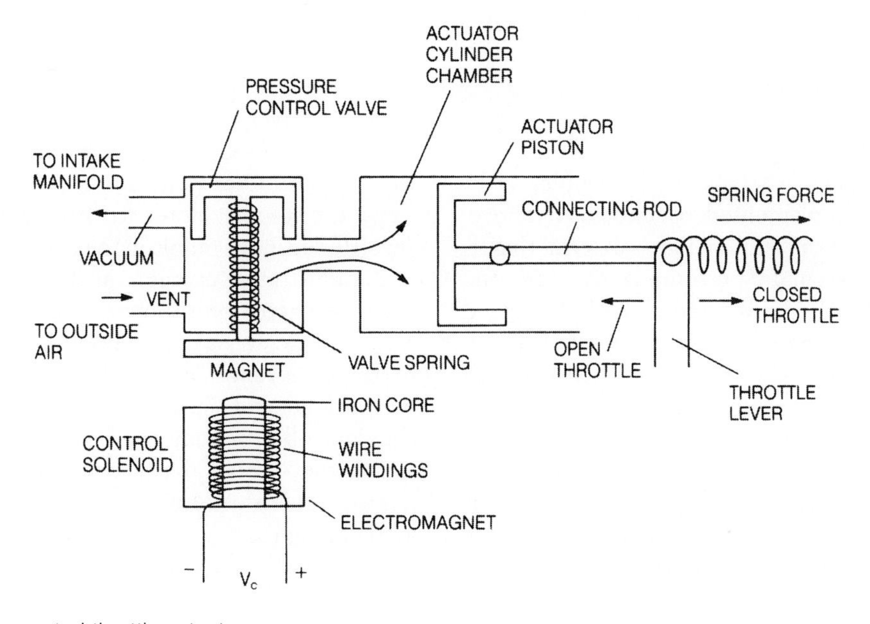

Fig. 13.1-7 Vacuum-operated throttle actuator.

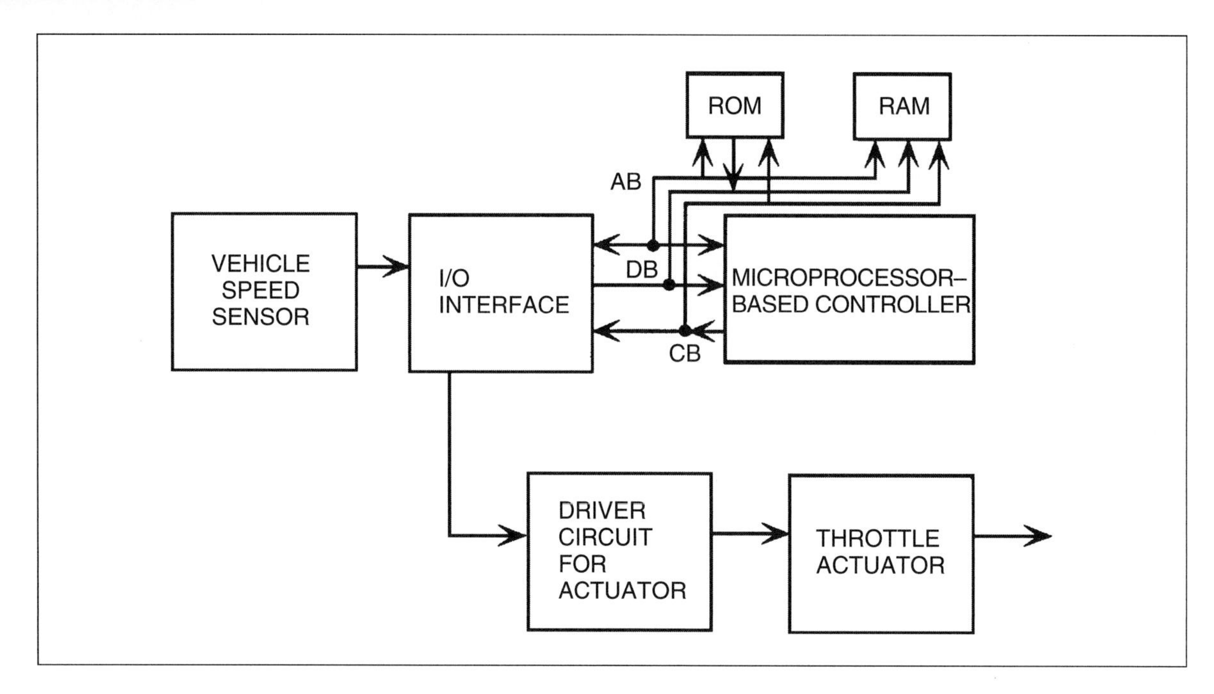

Fig. 13.1-8 Digital cruise control configuration.

program that causes the various calculations to be performed is stored in read-only memory (ROM). Typically, the ROM also stores parameters that are critical to the correct calculations. Normally a relatively small-capacity RAM memory is provided to store the command speed and to store any temporary calculation results. Input from the speed sensor and output to the throttle actuator are handled by the I/O interface (normally an integrated circuit that is a companion to the microprocessor). The output from the controller (i.e., the control signal) is sent via the I/O (on one of its output ports) to the so-called driver electronics. The latter electronics receives this control signal and generates a signal of the correct format and power level to operate the actuator (as explained below).

A microprocessor-based cruise control system performs all of the required control law computations digitally under program control. For example, a PI control strategy is implemented as explained above, with a proportional term and an integral term that is formed by a summation. In performing this task the controller continuously receives samples of the speed error e_n, and where n is a counting index ($n = 1, 2, 3, 4, \ldots$). This sampling occurs at a sufficiently high rate to be able to adjust the control signal to the actuator in time to compensate for changes in operating condition or to disturbances. At each sample the controller reads the most recent error. As explained earlier, that error is multiplied by a constant K_P that is called the proportional gain, yielding the proportional term in the control law. It also computes the sum of a number of previous error samples (the exact sum is chosen by the control system designer in accordance with the desired steady-state error). Then this sum is multiplied by a constant K_I and added to the proportional term, yielding the control signal.

The control signal at this point is simply a number that is stored in a memory location in the digital controller. The use of this number by the electronic circuitry that drives the throttle actuator to regulate vehicle speed depends on the configuration of the particular control system and on the actuator used by that system.

13.1.3.1 Stepper motor-based actuator

For example, in the case of a stepper motor actuator, the actuator driver electronics reads this number and then generates a sequence of pulses to the pair of windings on the stepper motor (with the correct relative phasing) to cause the stepper motor to either advance or retard the throttle setting as required to bring the error toward zero.

An illustrative example of driver circuitry for a stepper motor actuator is shown in Fig. 13.1-9. The basic idea for this circuitry is to continuously drive the stepper motor to advance or retard the throttle in accordance with the control signal that is stored in memory. Just as the controller periodically updates the actuator control signal, the stepper motor driver electronics continually adjusts the throttle by an amount determined by the actuator signal.

This signal is, in effect, a signed number (i.e., a positive or negative numerical value). A sign bit indicates the direction of the throttle movement (advance or retard).

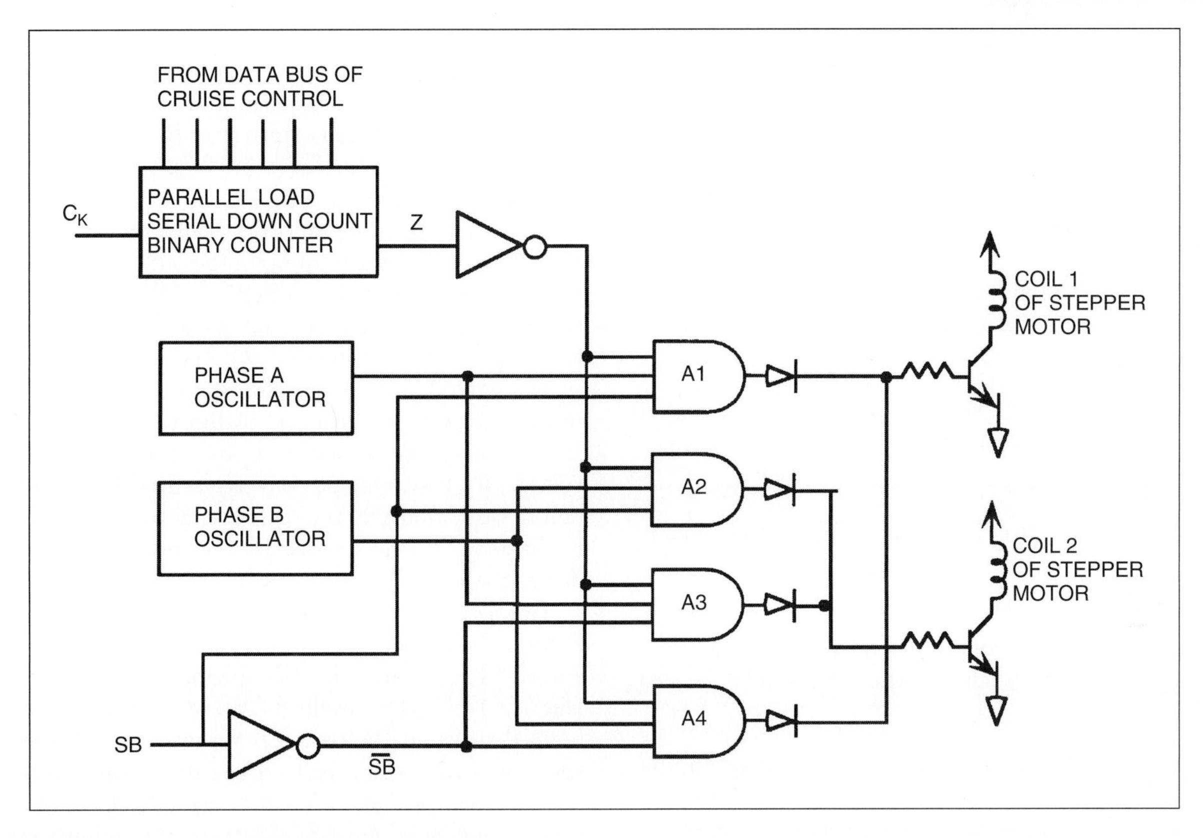

Fig. 13.1-9 Stepper motor actuator for cruise control.

The numerical value determines the amount of advance or retard.

The magnitude of the actuator signal (in binary format) is loaded into a parallel load serial down-count binary counter. The direction of movement is in the form of the sign bit (SB of Fig. 13.1-9). The stepper motor is activated by a pair of quadrature phase signals (i.e., signals that are a quarter of a cycle out of phase) coming from a pair of oscillators. To advance the throttle, phase A signal is applied to coil 1 and phase B to coil 2. To retard the throttle these phases are each switched to the opposite coil. The amount of movement in either direction is determined by the number of cycles of A and B, one step for each cycle.

The number of cycles of these two phases is controlled by a logical signal (Z in Fig. 13.1-9). This logical signal is switched high, enabling a pair of AND gates (from the set A1, A2, A3, A4). The length of time that it is switched high determines the number of cycles and corresponds to the number of steps of the motor.

The logical variable Z corresponds to the contents of the binary counter being zero. As long as Z is not zero, a pair of AND gates (A1 and A3, or A2 and A4) is enabled, permitting phase A and phase B signals to be sent to the stepper motor. The pair of gates enabled is determined by the sign bit. When the sign bit is high, A1 and A3 are enabled and the stepper motor advances the throttle as long as Z is not zero. Similarly, when the sign bit is low, A2 and A4 are enabled and the stepper motor retards the throttle.

To control the number of steps, the controller loads a binary value into the binary counter. With the contents not zero the appropriate pair of AND gates is enabled. When loaded with data, the binary counter counts down at the frequency of a clock (C_K in Fig. 13.1-9). When the countdown reaches zero, the gates are disabled and the stepper motor stops moving.

The time required to count down to zero is determined by the numerical value loaded into the binary counter. By loading signed binary numbers into the binary counter, the cruise controller regulates the amount and direction of movement of the stepper motor and thereby the corresponding movement of the throttle.

13.1.3.2 Vacuum-operated actuator

The driver electronics for a cruise control based on a vacuum-operated system generates a variable-duty-cycle signal. In this type of system, the duty cycle at any time is proportional to the control signal. For example, if at any given instant a large positive error exists between the command and actual signal, then a relatively large control signal will be generated. This control signal will cause the driver electronics to produce a large duty-cycle signal to operate the solenoid so that most of the time the actuator cylinder chamber is nearly at manifold vacuum level. Consequently, the piston will move against the restoring spring and cause the throttle opening to increase. As a result, the engine will produce more power and will accelerate the vehicle until its speed matches the command speed.

It should be emphasized that, regardless of the actuator type used, a microprocessor-based cruise control system will:

1. Read the command speed.
2. Measure actual vehicle speed.
3. Compute an error (error = command – actual).
4. Compute a control signal using P, PI, or PID control law.
5. Send the control signal to the driver electronics.
6. Cause driver electronics to send a signal to the throttle actuator such that the error will be reduced.

An example of electronics for a cruise control system that is basically analog is shown in Fig. 13.1-10. Notice that the system uses four operational amplifiers (op amps) and that each op amp is used for a specific purpose. Op amp 1 is used as an error amplifier. The output of op amp 1 is proportional to the difference between the command speed and the actual speed. The error signal is then used as an input to op amps 2 and 3. Op amp 2 is a proportional amplifier with a gain of $K_P = -R_2/R_1$. Notice that R_1 is variable so that the proportional amplifier gain can be adjusted. Op amp 3 is an integrator with a gain of $K_I = -1/R_3C$. Resistor R_3 is variable to permit adjustment of the gain. The op amp causes a current to flow into capacitor C that is equal to the current flowing into R_3. The voltage across R_3 is the error amplifier output voltage, V_e. The current in R_3 is found from Ohm's law to be

$$I = \frac{V_e}{R_3}$$

which is identical to the current flowing into the capacitor. If the error signal V_e is constant, the current I will be constant and the voltage across the capacitor will steadily change at a rate proportional to the current flow. That is, the capacitor voltage is proportional to the integral of the error signal:

$$V_I = -\frac{1}{R_3C}\int V_e\,dt$$

The output of the integral amplifier, V_I, increases or decreases with time depending on whether V_e is above or below zero volts. The voltage V_I is steady or unchanging only when the error is exactly zero; this is why the integral gain block in the diagram in Fig. 13.1-10a can reduce the system steady-state error to zero. Even a small error (e.g., due to a disturbance) causes V_I to change to correct for the error.

The outputs of the proportional and integral amplifiers are added using a summing amplifier, op amp 4. The summing amplifier adds voltages V_P and V_I and inverts the resulting sum. The inversion is necessary because both the proportional and integral amplifiers invert their input signals while providing amplification. Inverting the sum restores the correct sense, or polarity, to the control signal.

The summing amplifier op amp produces an analog voltage, V_s, that must be converted to a duty-cycle signal before it can drive the throttle actuator. A voltage-to-duty-cycle converter is used whose output directly drives the throttle actuator solenoid.

Two switches, S_1 and S_2, are shown in Fig. 13.1-10a. Switch S_1 is operated by the driver to set the desired speed. It signals the sample-and-hold electronics (Fig. 13.1-10b) to sample the present vehicle speed and hold that value. Voltage V_I, representing the vehicle speed at which the driver wishes to set the cruise controller, is sampled and it charges capacitor C. A very high input impedance amplifier detects the voltage on the capacitor without causing the charge on the capacitor to "leak" off. The output from this amplifier is a voltage, V_s, proportional to the command speed that is sent to the error amplifier.

Switch S_2 (Fig. 13.1-10a) is used to disable the speed controller by interrupting the control signal to the throttle actuator. Switch S_2 disables the system whenever the ignition is turned off, the controller is turned off, or the brake pedal is pressed. The controller is switched on when the driver presses the speed set switch S_1.

For safety reasons, the brake turnoff is often performed in two ways. As just mentioned, pressing the brake pedal turns off or disables the electronic control. In certain cruise control configurations that use a vacuum-operated throttle actuator, the brake pedal also mechanically opens a separate valve that is located in a hose connected to the throttle actuator cylinder. When the valve is opened by depression of the brake pedal, it allows outside air to flow into the throttle actuator cylinder so that the throttle plate instantly snaps closed. The valve is shut off whenever the brake pedal is in its inactive position. This ensures a fast and complete shutdown of the speed control system whenever the driver presses the brake pedal.

13.1.3.3 Advanced cruise control

The cruise control system previously described is adequate for maintaining constant speed, provided that any

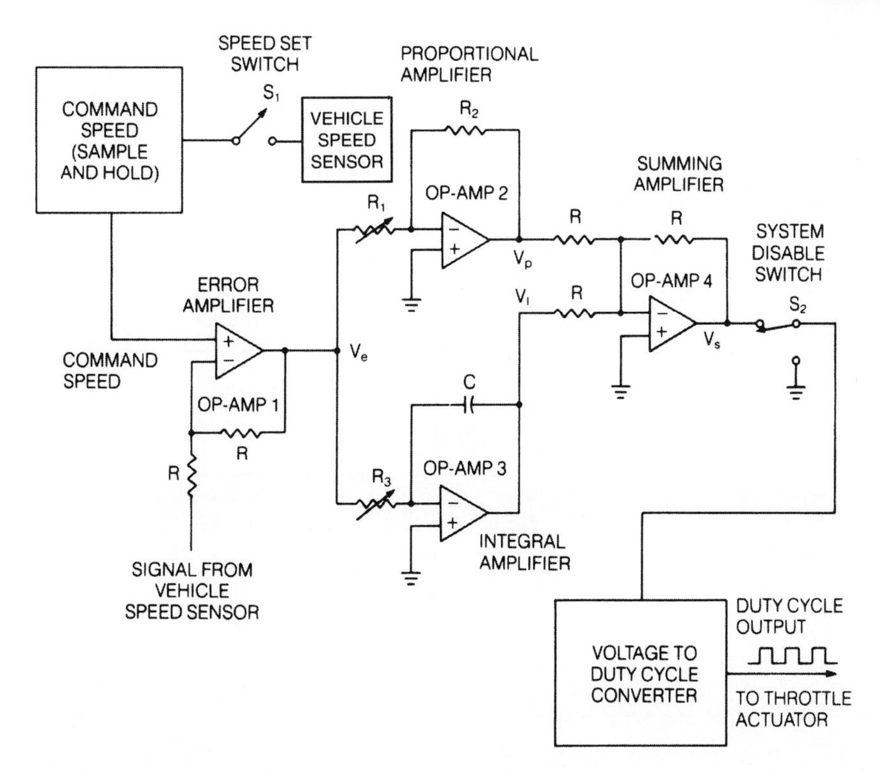

Fig. 13.1-10a Cruise control electronics (analog).

required deceleration can be achieved by a throttle reduction (i.e., reduced engine power). The engine has limited braking capability with a closed throttle, and this braking in combination with aerodynamic drag and tire-rolling resistance may not provide sufficient deceleration to maintain the set speed. For example, a car entering a long, relatively steep downgrade may accelerate due to gravity even with the throttle closed.

For this driving condition, vehicle speed can be maintained only by application of the brakes. For cars equipped with a conventional cruise control system, the driver has to apply braking to hold speed.

An ACC system has a means of automatic brake application whenever deceleration with throttle input alone is inadequate. A somewhat simplified block diagram of an ACC is shown in Fig. 13.1-11 emphasizing the automatic braking portion.

This system consists of a conventional brake system with master cylinder wheel cylinders, vacuum boost (power brakes), and various brake lines. Fig. 13.1-11 shows only a single-wheel cylinder, although there are four in actual practice. In addition, proportioning valves are present to regulate the front/rear brake force ratio.

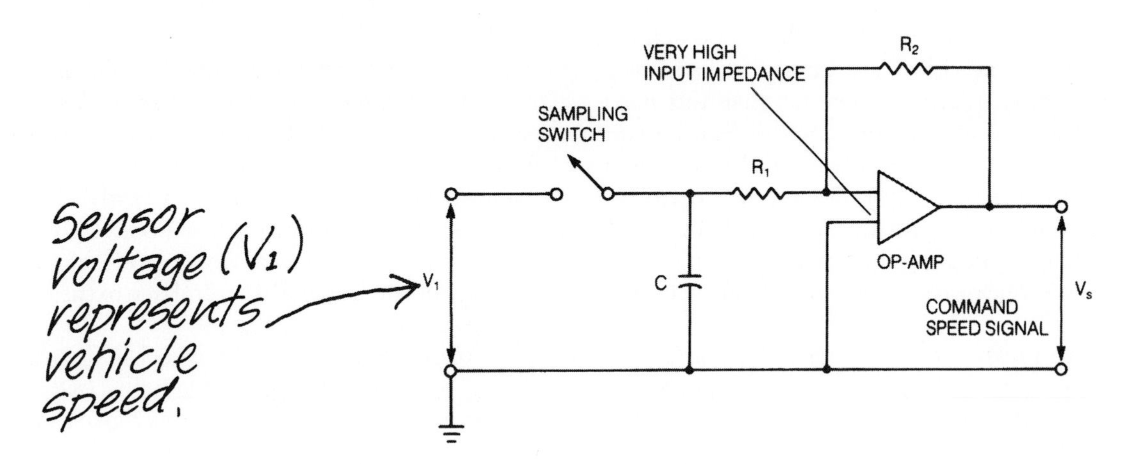

Fig. 13.1-10b Typical sample-and-hold circuit.

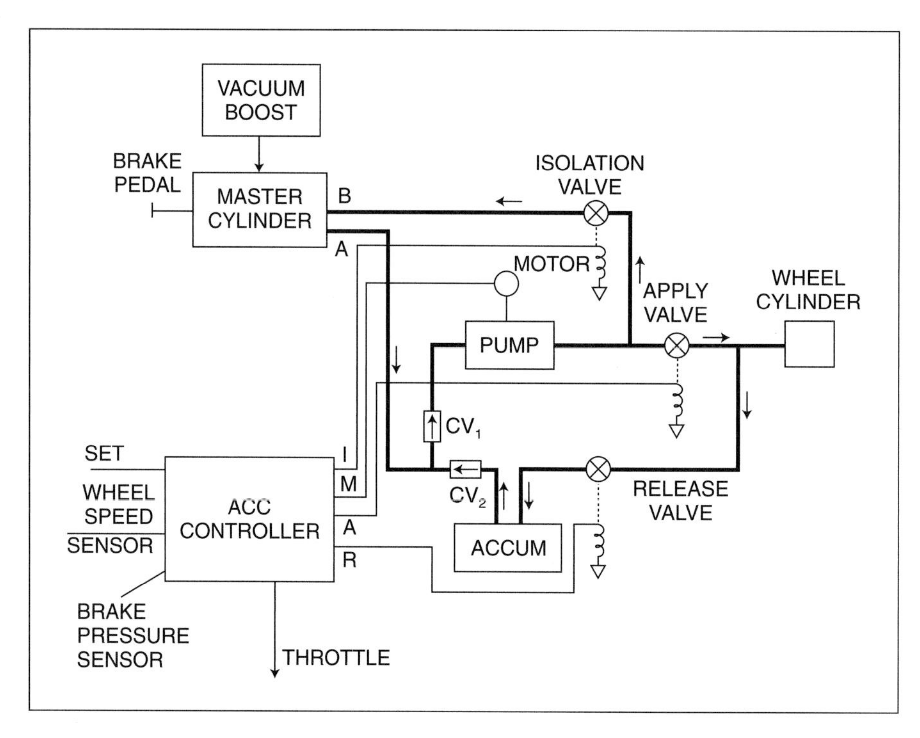

Fig. 13.1-11 ACC emphasizing the automatic braking portion.

In normal driving, the system functions like a conventional brake system. As the driver applies braking force through the brake pedal to the master cylinder, brake fluid (under pressure) flows out of port and through a brake line to the junction of check valves CV_1 and CV_2. Check valve CV_2 blocks brake fluid, whereas CV_1 permits flow through a pump assembly P and then through the apply valve (which is open) to the wheel cylinder(s), thereby applying brakes.

In cruise control mode, the ACC controller regulates the throttle (as explained above for a conventional cruise control) as well as the brake system via electrical output signals and in response to inputs, including the vehicle speed sensor and set cruise speed switch. The ACC system functions as described above until the maximum available deceleration with closed throttle is inadequate. Whenever there is greater deceleration than this maximum valve, the ACC applies brakes automatically. In this automatic brake mode, an electrical signal is sent from the M (i.e., motor) output, causing the pump to send more brake fluid (under pressure) through the apply valve (maintained open) to the wheel cylinder. At the same time, the release valve remains closed such that brakes are applied.

The braking pressure can be regulated by varying the isolation valve, thereby bleeding some brake fluid back to the master cylinder. By activating isolation valves separately to the four wheels, brake proportioning can be achieved. Brake release can be accomplished by sending signals from the ACC to close the apply valve and open the release valve.

Another potential future application for automatic braking involves separate brake pressure applied individually to all four wheels. This independent brake application can be employed for improved handling when both braking and steering are active (e.g., braking on curves).

A further application of the ACC involves maintaining a constant headway (separation) behind another vehicle on the road.

13.1.4 Antilock braking system

One of the most readily accepted applications of electronics in automobiles has been the antilock brake system (ABS). ABS is a safety-related feature that assists the driver in deceleration of the vehicle in poor or marginal braking conditions (e.g., wet or icy roads). In such conditions, panic braking by the driver (in non-ABS-equipped cars) results in reduced braking effectiveness and, typically, loss of directional control due to the tendency of the wheels to lock.

In ABS-equipped cars, the wheel is prevented from locking by a mechanism that automatically regulates braking force to an optimum for any given low-friction condition. The physical configuration for an ABS is shown in Fig. 13.1-12. In addition to the normal brake

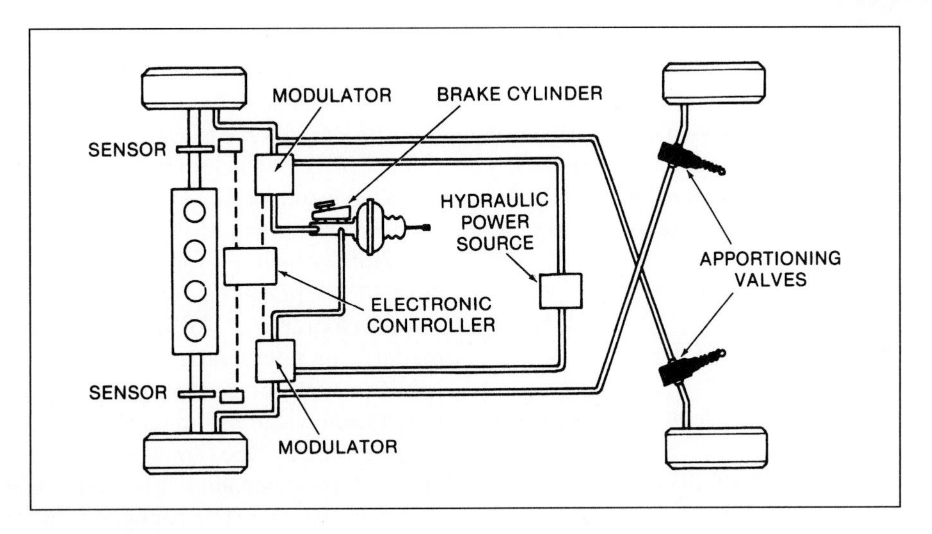

Fig. 13.1-12 Antilock braking system.

components, including brake pedal, master cylinder, vacuum boost, wheel cylinders, calipers/disks, and brake lines, this system has a set of angular speed sensors at each wheel, an electronic control module, and a hydraulic brake pressure modulator (regulator).

In order to understand the ABS operation, it is first necessary to understand the physical mechanism of wheel lock and vehicle skid that can occur during braking. Fig. 13.1-13 illustrates the forces applied to the wheel by the road during braking.

The car is traveling at a speed U and the wheels are rotating at an angular speed w where

$$w = \frac{\pi \mathrm{RPM}}{30}$$

and where RPM is the wheel revolutions per minute. When the wheel is rolling (no applied brakes),

$$U = Rw$$

where R is the tire radius. When the brake pedal is depressed, the calipers are forced by hydraulic pressure against the disk. This force acts as a torque T_b in opposition to the wheel rotation. The actual force that decelerates the car is shown as F_b in Fig. 13.1-13. The

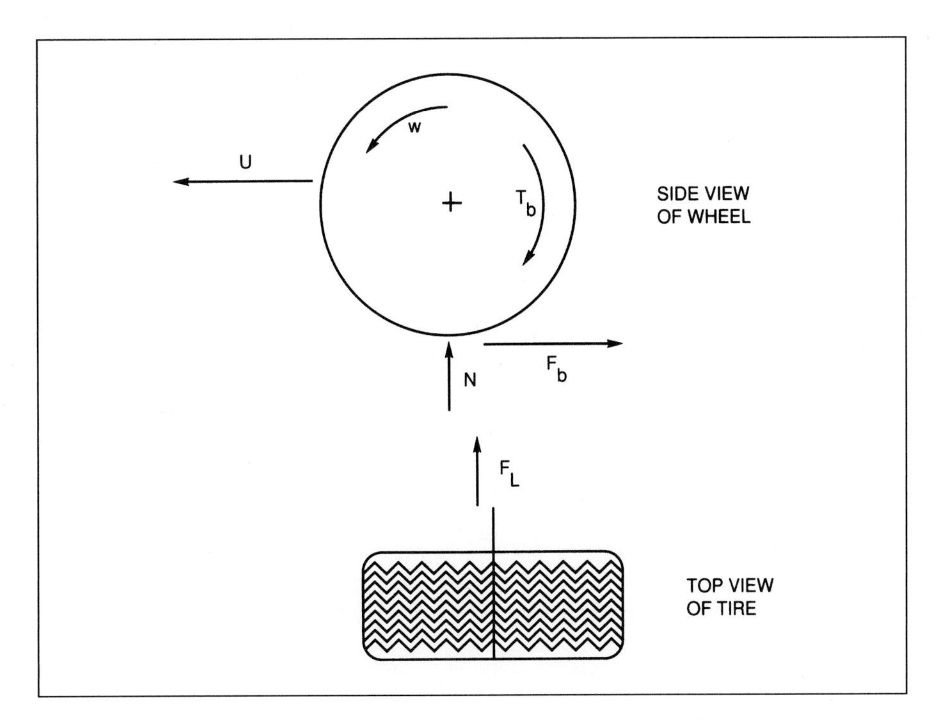

Fig. 13.1-13 Forces during braking.

lateral force that maintains directional control of the car is shown as F_L in Fig. 13.1-13.

The wheel angular speed begins to decrease, causing a difference between the vehicle speed U and the tire speed over the road (i.e., wR). In effect, the tire slips relative to the road surface. The amount of slip (S) determines the braking force and lateral force. The slip, as a percentage of car speed, is given by

$$S = \frac{U - wR}{U} \times 100\%$$

Note: A rolling tire has slip $S = 0$, and a fully locked tire has $S = 100\%$.

The braking and lateral forces are proportional to the normal force (from the weight of the car) acting on the tire/road interface (N in Fig. 13.1-13) and the friction coefficients for braking force (F_b) and lateral force (F_L):

$$F_b = N\mu_b$$
$$F_L = N\mu_L$$

where

μ_b is the braking friction coefficient
μ_L is the lateral friction coefficient

These coefficients depend markedly on slip, as shown in Fig. 13.1-14. The solid curves are for a dry road and the dashed curves for a wet or icy road. As brake pedal force is increased from zero, slip increases from zero. For increasing slip, μ_b increases to $S = S_o$. Further increase in slip actually decreases μ_b, thereby reducing braking effectiveness.

On the other hand, μ_L decreases steadily with increasing S such that for fully locked wheels the lateral force has its lowest value. For wet or icy roads, μ_L at $S = 100\%$ is so low that the lateral force is insufficient to maintain directional control of the vehicle. However, directional control can often be maintained even in poor braking conditions if slip is optimally controlled. This is essentially the function of the ABS, which performs an operation equivalent to pumping the brakes (as done by experienced drivers before the development of ABS). In ABS-equipped cars under marginal or poor braking conditions, the driver simply applies a steady brake force and the system adjusts tire slip to optimum value automatically.

In a typical ABS configuration, control over slip is effected by controlling the brake line pressure under electronic control. The configuration for ABS is shown in Fig. 13.1-12. This ABS regulates or modulates brake pressure to maintain slip as near to optimum as possible (e.g., at S_o in Fig. 13.1-14). The operation of this ABS is based on estimating the torque T_w applied to the wheel at the road surface by the braking force F_b:

$$T_w = RF_b$$

In opposition to this torque is the braking torque T_b applied to the disk by the calipers in response to brake pressure P:

$$T_b = k_b P$$

where k_b is a constant for the given brakes.

The difference between these two torques acts to decelerate the wheel. In accordance with basic Newtonian mechanics, the wheel torque T_w is related to

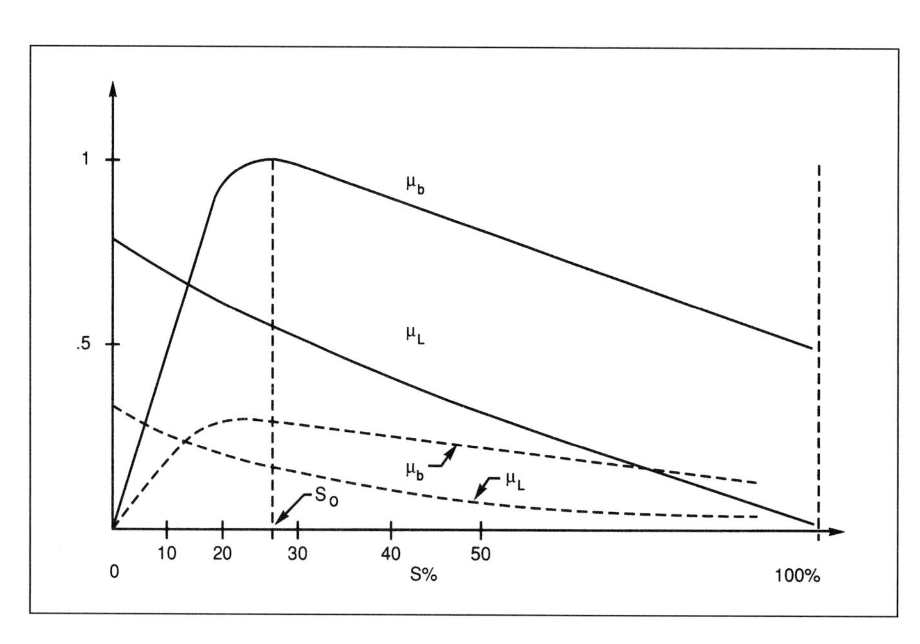

Fig. 13.1-14 Braking coefficients versus tire slip (solid curves for dry road, dashed curves for wet or icy road).

braking torque and wheel deceleration by the following equation:

$$T_w = T_b + I_w \dot{w}$$

where I_w is the wheel moment of inertia and $\dot{w}$ is the wheel deceleration (dw/dt, i.e., the rate of change of wheel speed).

During heavy braking under marginal conditions, sufficient braking force is applied to cause wheel lock-up (in the absence of ABS control). We assume such heavy braking for the following discussion of the ABS. As brake pressure is applied, T_b increases and w decreases, causing slip to increase. The wheel torque is proportional to μ_b, which reaches a peak at slip S_o. Consequently, the wheel torque reaches a maximum value (assuming sufficient brake force is applied) at this level of slip.

Fig. 13.1-15 is a sketch of wheel torque versus slip illustrating the peak T_w. After the peak wheel torque is sensed electronically, the electronic control system commands that brake pressure be reduced (via the brake pressure modulator). This point is indicated in Fig. 13.1-15 as the limit point of slip for the ABS. As the brake pressure is reduced, slip is reduced and the wheel torque again passes through a maximum.

The wheel torque reaches a value below the peak on the low slip side and at this point brake pressure is again increased. The system will continue to cycle, maintaining slip near the optimal value as long as the brakes are applied and the braking conditions lead to wheel lock-up.

The mechanism for modulating brake pressure is illustrated in Fig. 13.1-16. The numbers in Fig. 13.1-16a refer to the following:

1. Applied master cylinder pressure
2. Bypass brake fluid
3. Normally open solenoid valve
4. EMB braking action
5. DC motor pack
6. ESB braking
7. Gear assembly
8. Ball screw
9. Check valve unseated
10. Outlet to brake cylinders
11. Piston

The numbers in Fig. 13.1-16b refer to the following:

1. Trapped bypass brake fluid
2. Solenoid valve activated
3. EMB action released
4. DC motor pack
5. ESB braking action released
6. Gear assembly
7. Ball screw
8. Check valve seated
9. Applied master cylinder pressure

Under normal braking, brake pressure from the master cylinder passes without reduction through the passageways associated with check valve 9 and solenoid valve 3 in Fig. 13.1-16a.

Whenever the wheel slip limit is reached, the solenoid valve is closed and the piston (11) retracts, closing the check valve. This action effectively isolates the brake cylinders from the master cylinder, and brake line pressure is controlled by the position of piston 11. This piston retracts, lowering the brake pressure sufficiently so that slip falls below S_o. At this point, the control system detects low T_w and the piston moves up, thereby increasing

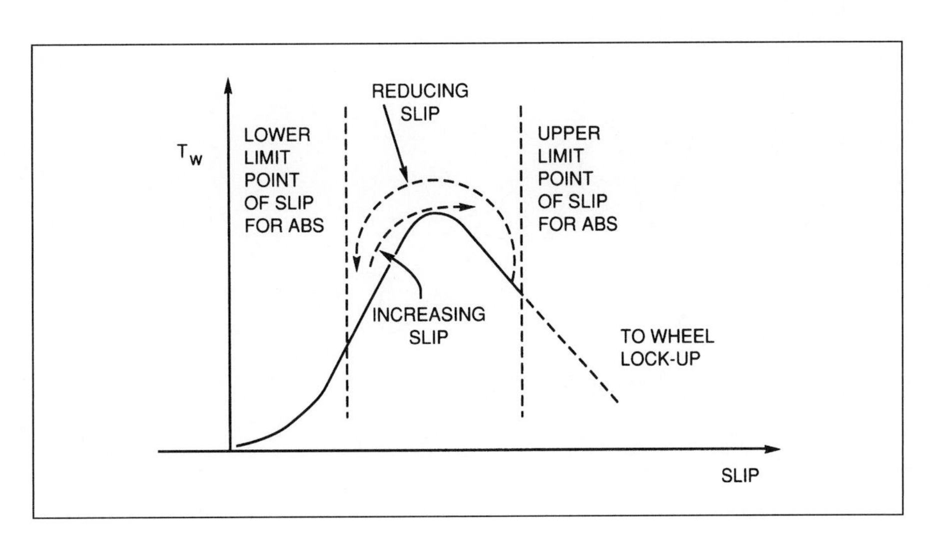

Fig. 13.1-15 Wheel torque versus slip.

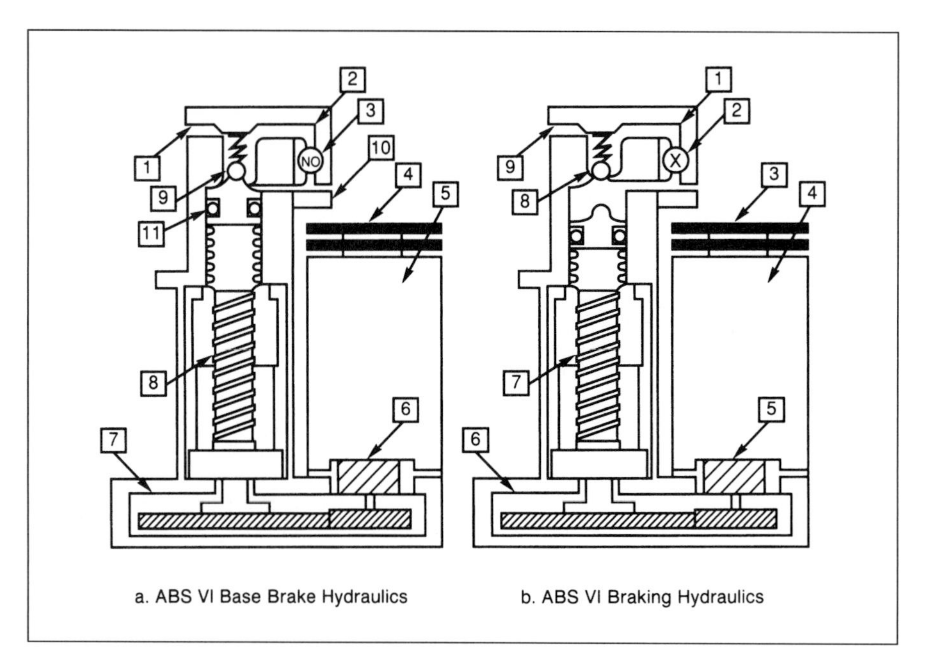

Fig. 13.1-16 Brake pressure modulating mechanism.

brake line pressure. The ABS system will continue to cycle until the vehicle has stopped, the braking conditions are normal, or the driver removes the brake pressure from the master cylinder.

In the latter case, the operation of the brake pressure modulator restores normal braking function. For example, should the driver release the brake pedal, then the pressure at the inlet (1) is reduced. At this point, the check valve (9) opens and brake line pressure is also removed. The solenoid valve opens and the piston returns to its normal position (fully up) such that the check valve is held open.

Fig. 13.1-17 illustrates the braking during an ABS action. In this illustration, the vehicle is initially traveling at 55 mph and the brakes are applied as indicated by the rising brake pressure. The wheel speed begins to drop until the slip limit is reached. At this point, the ABS reduces brake pressure and the wheel speed increases. With the high applied brake pressure, the wheels again tend toward lock-up and ABS reduces

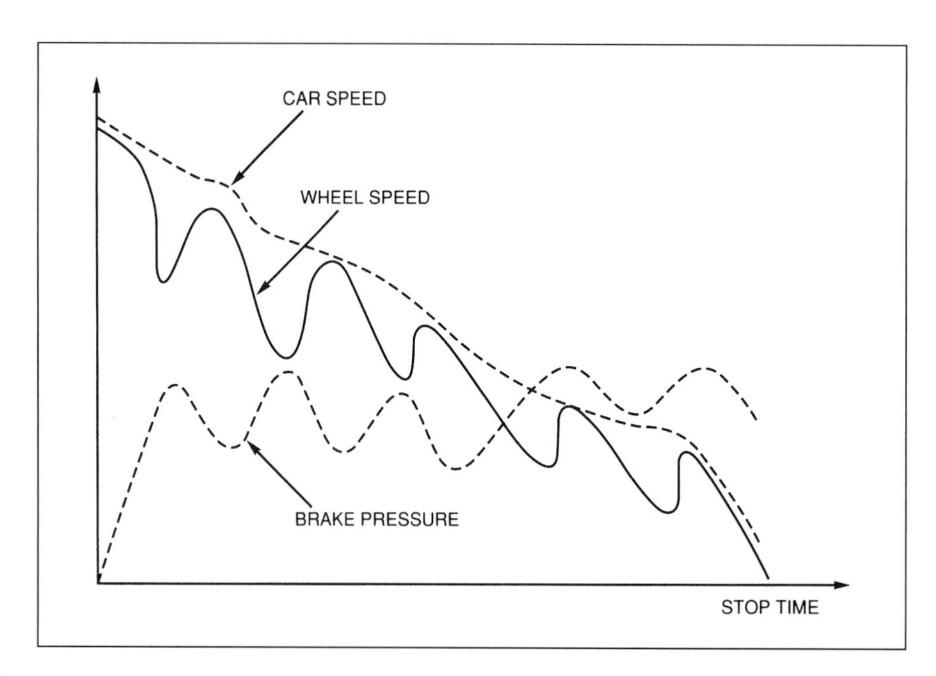

Fig. 13.1-17 ABS braking action.

brake pressure. The cycle continues until the vehicle is stopped.

It should be noted that by maintaining slip near S_0, the maximum deceleration is achieved for a given set of conditions. Some reduction in lateral force occurs from its maximum value by maintaining slip near S_0. However, in most cases the lateral force is large enough to maintain directional control.

In some ABSs, the slip oscillations are shifted below S_0, sacrificing some braking effectiveness to enhance directional control. This can be accomplished by adjusting the upper and lower slip limits.

13.1.4.1 Tire-slip controller

Another benefit of the ABS is that the brake pressure modulator can be used for tire-slip control. Tire slip is effective in moving the car forward just as it is in braking. Under normal driving circumstances, the slip that was defined previously for braking is negative. That is, the tire is actually moving at a speed that is greater than for a purely rolling tire. In fact, the traction force is proportional to slip.

For wet or icy roads, the friction coefficient can become very low and excessive slip can develop. In extreme cases, one of the driving wheels may be on ice or in snow while the other is on a dry (or drier) surface. Because of the action of the differential, the low-friction tire will spin and relatively little torque will be applied to the dry-wheel side. In such circumstances, it may be difficult for the driver to move the car even though one wheel is on a relatively good friction surface.

The difficulty can be overcome by applying a braking force to the free spinning wheel. In this case, the differential action is such that torque is applied to the relatively dry wheel surface and the car can be moved. In the example ABS, such braking force can be applied to the free spinning wheel by the hydraulic brake pressure modulator (assuming a separate modulator for each drive wheel). Control of this modulator is based on measurements of the speed of the two drive wheels. Of course, the ABS already incorporates wheel speed measurements, as discussed previously.

The ABS electronics have the capability to perform comparisons of these two wheel speeds and to determine that braking is required of one drive wheel to prevent wheel spin.

Antilock braking can also be achieved with electrohydraulic brakes. An electrohydraulic brake system was described in the section of this chapter devoted to ACC.

Recall that for ACC a motor-driven pump supplied brake fluid through a solenoid-operated apply valve to the wheel cylinder. In the case of ABS, the driver supplies the pressurized brake fluid instead of the motor-driven pump. For ACC application of the brakes, the apply and isolation valves independently regulate the braking to each of the four wheels.

For ABS applications, the braking pressure is regulated by alternately opening and closing the apply and release valves. These valves are operated by output signals from the ABS controller in accordance with an algorithm applied to wheel speed measurements as described above, which attempts to maintain slip near a value corresponding to the maximum friction coefficient.

13.1.5 Electronic suspension system

Automotive suspension systems consist of springs, shock absorbers, and various linkages to connect the wheel assembly to the car frame. The purpose of the suspension system is to isolate the car body motion as much as possible from wheel motion due to rough road input; and the performance of the suspension system is strongly influenced by the damping parameter of the shock absorber.

The two primary subjective performance measures are ride and handling. *Ride* refers to the motion of the car body in response to road bumps or irregularities. *Handling* refers to how well the car body responds to dynamic vehicle motion such as cornering or hard braking.

Generally speaking, ride is improved by lowering the shock absorber damping, whereas handling is improved by increasing this damping. In traditional suspension design, the damping parameter is fixed and is chosen to achieve a compromise between ride and handling (i.e., an intermediate value for shock absorber damping is chosen).

In electronically controlled suspension systems, this damping can be varied depending on driving conditions and road roughness characteristics. That is, the suspension system adapts to inputs to maintain the best possible ride subject to handling constraints that are associated with safety.

There are two major classes of electronic suspension control systems: active and semiactive. The semiactive suspension system is purely dissipative (i.e., power is absorbed by the shock absorber under control of a microcontroller). In this system, the shock absorber damping is regulated to absorb the power of the wheel motion in accordance with the driving conditions.

In an active suspension system, power is added to the suspension system via a hydraulic or pneumatic power source. At the time of the writing of this book, commercial suspension systems are primarily semiactive. The active suspension system is just beginning to appear in production vehicles. In this chapter, we explain the semiactive system first, then the active one.

The primary purpose of the semiactive suspension system is to provide a good ride for as much of the time as possible without sacrificing handling. Good ride is achieved if the car's body is isolated as much as possible from the road. A semiactive suspension controls the shock absorber damping to achieve the best possible ride.

In addition to providing isolation of the sprung mass (i.e., car body and contents), the suspension system has another major function. It must also dynamically maintain the tire normal force as the unsprung mass (wheel assembly) travels up and down due to road roughness. Recall from the discussion of antilock braking that cornering forces depend on normal tire force. Of course in the long-term time average, the normal forces will total the vehicle weight plus any inertial forces due to acceleration, deceleration, or cornering.

However, as the car travels over the road, the unsprung mass moves up and down in response to road input. This motion causes a variation in normal force, with a corresponding variation in potential cornering or braking forces. For example, while driving on a rough curved road, there is a potential loss of steering or braking effectiveness if the suspension system does not have good damping characteristics.

Fig. 13.1-18 illustrates typical tire normal force variation as a function of frequency of excitation for a fixed-amplitude, variable-frequency sinusoidal excitation. The solid curve is the response for a relatively low-damping-coefficient shock absorber and the dashed curve is the response for a relatively high damping coefficient.

In Fig. 13.1-18, the ordinate is the ratio of amplitude of force variation to the average normal load (i.e., due to weight). There are two relative peaks in this response. The lower peak is approximately 1–2 Hz and is generally associated with spring/sprung mass oscillation. The second peak, which is in the general region of 12–15 Hz, is resonance of the spring/unsprung mass combination.

Generally speaking, for any given fixed suspension system, ride and handling cannot both be optimized simultaneously. A car with a good ride is one in which the sprung mass motion/acceleration due to rough road input is minimized. In particular, the sprung mass motion in the frequency region from about 2 to 8 Hz is most important for good subjective ride. Good ride is achieved for relatively low damping (low D in Fig. 13.1-18).

For low damping, the unsprung mass moves relatively freely due to road input while the sprung mass motion remains relatively low. Note from Fig. 13.1-18 that this low damping results in relatively high variation in normal force, particularly near the two peak frequencies. That is, low damping results in relatively poor handling characteristics.

With respect to the four frequency regions of Fig. 13.1-18, the following generally desired suspension damping characteristics can be identified.

Another major input to the vehicle that affects handling is steering input that causes maneuvers parallel to the road surface (e.g., cornering). Whenever the car is executing such maneuvers, there is a lateral acceleration. This acceleration acting through the center of gravity causes the vehicle to roll in a direction opposite to the maneuver.

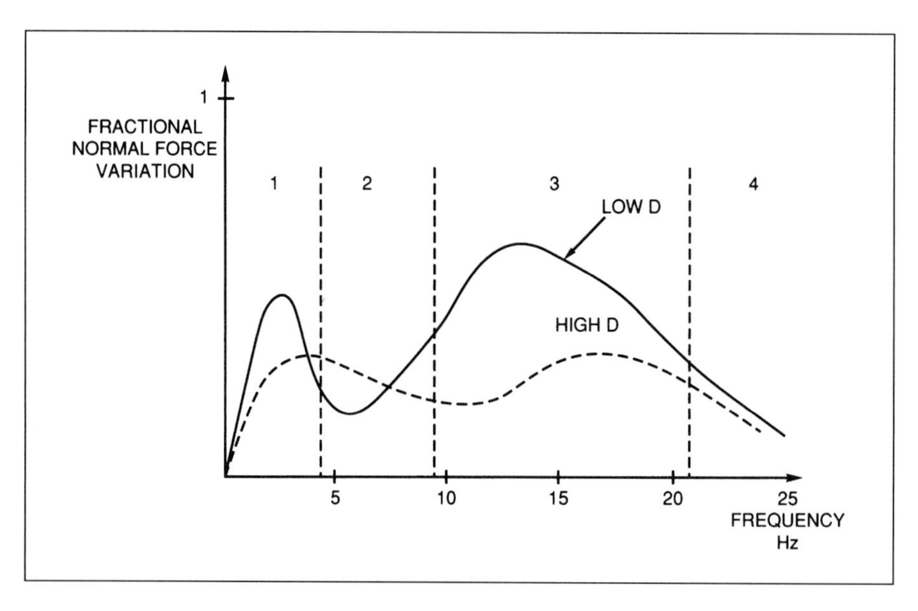

Fig. 13.1-18 Tire force variation.

Region	Frequency (Hz)	Damping
1: Sprung mass mode	1–2	High
2: Intermediate ride	2–8	Low
3: Unsprung mass resonance	8–20	High
4: Harshness	>20	Low

Car handling generally improves if the amount of roll for any given maneuver is reduced. The rolling rate for a given car and maneuver is improved if spring rate and shock absorber damping are increased. Although the semiactive control system regulates only the damping, handling is improved by increasing this damping as lateral acceleration increases.

Lateral acceleration A_L is proportional to vehicle speed and input steering angle:

$$A_L = kVq_s$$

where

V is the speed of the car

q_s is the steering angle

The dynamics of a spring/mass/damping system, identifying resonant frequency and critical damping (D_c) is

$$D_c = 2\sqrt{KM}$$

For good ride, the damping should be as low as possible. However, from practical design considerations, the minimum damping is generally in the region of $0.1 < D/D_c < 0.2$. For optimum handling, the damping is in the region of $0.6 < D/D_c < 0.8$.

Technology has been developed permitting the damping characteristics of shock absorber/strut assembly to be varied electrically, which in turn permits the ride/handling characteristics to be varied while the car is in motion. Under normal steady-cruise conditions, damping is electrically set low, yielding a good ride. However, under dynamic maneuvering conditions (e.g., cornering), the damping is set high to yield good handling. Generally speaking, high damping reduces vehicle roll in response to cornering or turning maneuvers, and it tends to maintain tire force on the road for increased cornering forces. Variable damping suspension systems can improve safety, particularly for vehicles with a relatively high center of gravity (e.g., SUVs).

The damping of a suspension system is determined by the viscosity of the fluid in the shock absorber/strut and by the size of the aperture through which the fluid flows as the wheel moves relative to the car body.

The earliest active or semiactive suspension systems employed variable aperture. One scheme for achieving variable damping is to switch between two aperture sizes using a solenoid. Another scheme varies aperture size continuously with a motor-driven mechanism.

Although there are many potential control strategies for regulating shock absorber damping, we consider first switched damping as in our example. In such a system, the shock absorber damping is switched to the higher value whenever lateral acceleration exceeds a predetermined threshold. Fig. 13.1-19 illustrates such a system in which the threshold for switching to firm damping (i.e., higher damping) is 0.35 g.

The variation in shock absorber damping is achieved by varying the aperture in the oil passage through the piston. In practical semiactive suspension systems, there are two means used to vary this aperture size—a solenoid-operated bypass valve and a motor-driven variable-orifice valve (Fig. 13.1-20). Fig. 13.1-21 is an illustration of the force/relative velocity characteristics of a shock absorber having a solenoid-switched aperture.

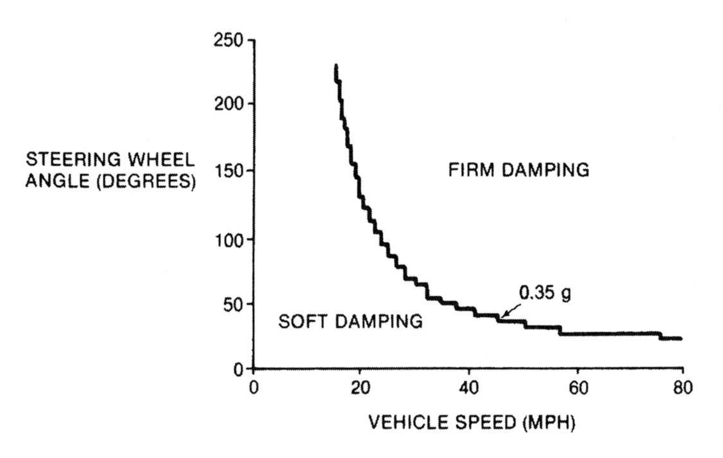

Fig. 13.1-19 Switching threshold versus speed and steering inputs.

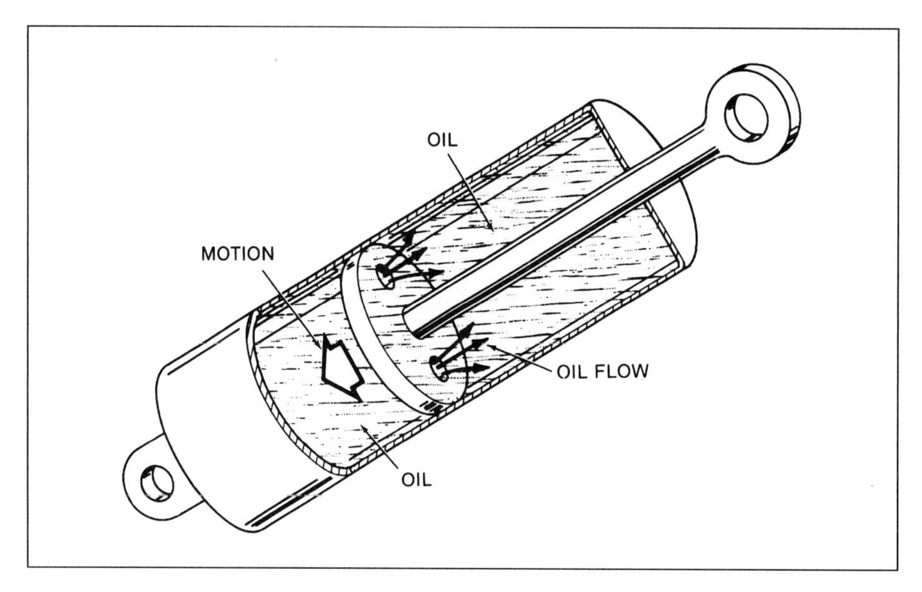

Fig. 13.1-20 Adjustable shock absorber.

13.1.5.1 Variable damping via variable strut fluid viscosity

Variable suspension damping is also achieved with a fixed aperture and variable fluid viscosity. The fluid for such a system consists of a synthetic hydrocarbon with suspended iron particles and is called a magneto-rheological fluid (MR). An electromagnet is positioned such that a magnetic field is created whose strength is proportional to current through the coil. This magnetic field passes through the MR fluid. In the absence of the magnetic field, the iron particles are randomly distributed and the MR fluid has relatively low viscosity corresponding to low damping. As the magnetic field is increased from zero, the

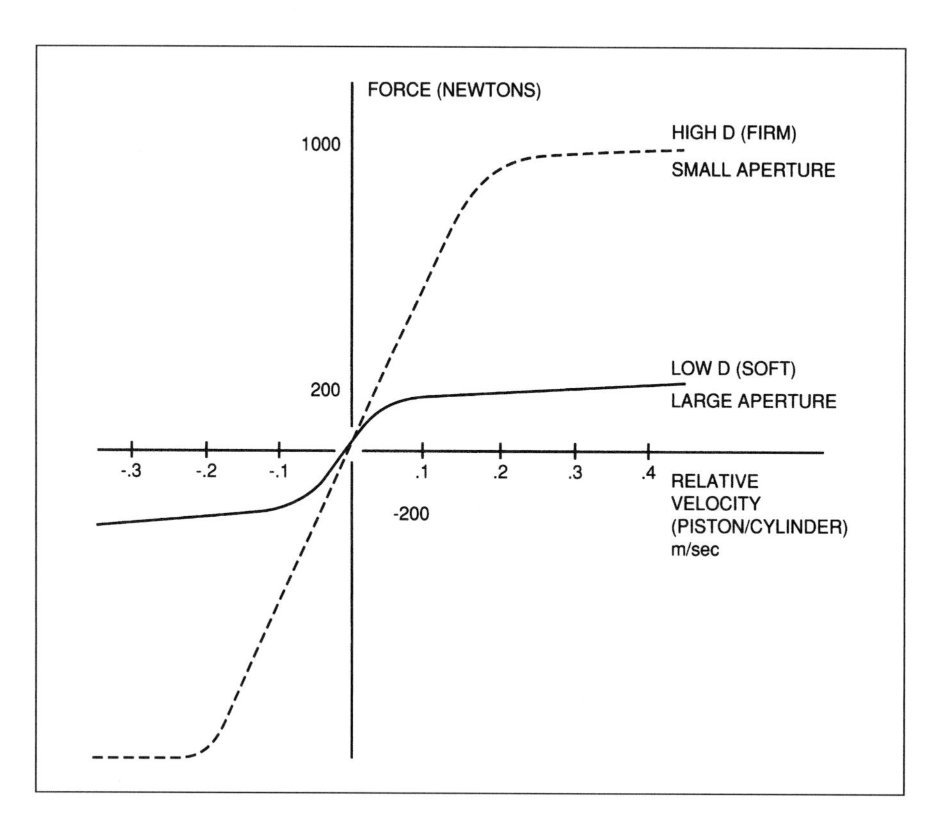

Fig. 13.1-21 Force versus relative velocity of a solenoid-switched aperture shock absorber.

iron particles begin to align with the field, and the viscosity increases in proportion to the strength of the field (which is proportional to the current through the electromagnet coil). That is, the damping of the associated shock absorber/strut varies continuously with the electromagnet coil current.

13.1.5.2 Variable spring rate

The frequency response characteristics of a suspension system are influenced by the springs as well as the shock absorber damping. Conventional steel springs (i.e., coil or leaf) have a fixed spring rate (i.e., force-deflection characteristics). The vehicle height above the ground is determined by vehicle weight, which in turn depends on loading (i.e., passengers, cargo, and fuel). Some vehicles, having electronically controlled suspension, are also equipped with pneumatic springs as a replacement for steel springs. A pneumatic spring consists of a rubber bladder mounted in an assembly and filled with air under pressure. This mechanism is commonly called an air suspension system. The spring rate for such pneumatic springs is proportional to the pressure in the bladder. A motor-driven pump is provided that varies the pressure in the bladder, yielding a variable spring rate suspension. In conjunction with a suitable control system, the pneumatic springs can automatically adjust the vehicle height to accommodate various vehicle loadings.

13.1.5.3 Electronic suspension control system

The control system for a typical electronic suspension system is depicted in the block diagram of Fig. 13.1-22. The control system configuration in Fig. 13.1-22 is generic and not necessarily representative of the system for any production car. This system includes sensors for measuring vehicle speed; steering input (i.e., angular deflection of steered wheels); relative displacement of the wheel assembly and car body/chassis; lateral acceleration; and yaw rate. The outputs are electrical signals to the shock absorber/strut actuators and to the motor/compressor that pressurizes the pneumatic springs (if applicable). The actuators can be solenoid-operated (switched) orifices or motor-driven variable orifices or electromagnets for RH fluid-type variable viscosity struts.

The control system typically is in the form of a microcontroller or microprocessor-based digital controller. The inputs from each sensor are sampled, converted to digital format, and stored in memory. The

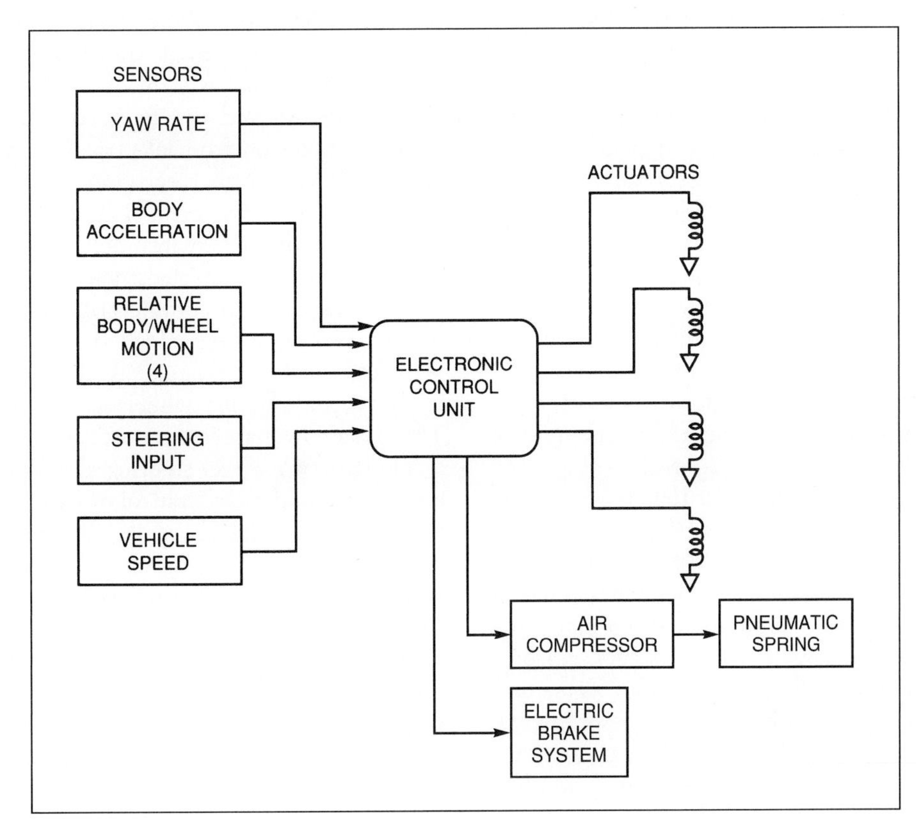

Fig. 13.1-22 Electronic suspension system.

body acceleration measurement can be used to evaluate ride quality. The controller does this by computing a weighted average of the spectrum of the acceleration. The relative body/wheel motion can be used to estimate tire normal force, and damping is then adjusted to try to optimize this normal force.

The yaw rate sensor provides data which in relationship to vehicle speed and steering input measurements can be used to evaluate cornering performance. In certain vehicles, these measurements combine in an algorithm that is used to activate the electrohydraulic brakes.

Under program control in accordance with the control strategy, the electronic control system generates output electrical signals to the various actuators. The variable damping actuators vary either the oil passage orifice or the RH fluid viscosity independently at each wheel to obtain the desired damping for that wheel.

There are many possible control strategies and many of these are actually used in production vehicles. For the purposes of this book, it is perhaps most beneficial to present a representative control strategy that typifies features of a number of actual production systems.

The important inputs to the vehicle suspension control system come from road-roughness-induced forces and inertial forces (due, for example, to cornering or maneuvering), steering inputs, and vehicle speed. In our hypothetical simplified control strategy these inputs are considered separately. When driving along a nominally straight road with small steering inputs, the road input is dominant. In this case, the control is based on the spectral content (frequency region) of the relative motion. The controller (under program control) calculates the spectrum of the relative velocity of the sprung and unsprung mass at each wheel (from the corresponding sensor's data). Whenever the weighted amplitude of the spectrum near the peak frequencies exceeds a threshold, damping is increased, yielding a firmer ride and improved handling. Otherwise, damping is kept low (soft suspension).

If in addition the vehicle is equipped with an accelerometer (usually located in the car body near the center of gravity) and with motor-driven variable-aperture shock absorbers, then an additional control strategy is possible. In this latter control strategy, the shock absorber apertures are adjusted to minimize sprung mass acceleration in the 2–8-Hz frequency region, thereby providing optimum ride control. However, at all times, the damping is adjusted to control unsprung mass motion to maintain wheel normal force variation at acceptably low levels for safety reasons. Whenever a relatively large steering input is sensed (sometimes in conjunction with yaw rate measurement), such as during a cornering maneuver, then the control strategy switches to the smaller aperture, yielding a "stiffer" suspension and improved handling. In particular, the combination of cornering on a relatively rough road calls for damping that optimizes tire normal force, thereby maximizing cornering forces.

13.1.6 Electronic steering control

The steering effort required of the driver to overcome restoring torque generally decreases with vehicle speed and increases with steering angle. Traditionally, the steering effort required by the driver has been reduced by incorporating a hydraulic power steering system in the vehicle. Whenever there is a steering input from the driver, hydraulic pressure from an engine-driven pump is applied to a hydraulic cylinder that boosts the steering effort of the driver.

Typically, the effort available from the pump increases with engine speed (i.e., with vehicle speed), whereas the required effort decreases. It would be desirable to reduce steering boost as vehicle speed increases. Such a feature is incorporated into a power steering system featuring electronic controls. An electronically controlled power steering system adjusts steering boost adaptively to driving conditions. Using electronic control of power steering, the available boost is reduced by controlling a pressure relief valve on the power steering pump.

An alternative power steering scheme utilizes a special electric motor to provide the boost required instead of the hydraulic boost. Electric boost power steering has several advantages over traditional hydraulic power steering. Electronic control of electric boost systems is straightforward and can be accomplished without any energy conversion from electrical power to mechanical actuation. Moreover, electronic control offers very sophisticated adaptive control in which the system can adapt to the driving environment.

An example of an electronically controlled steering system that has had commercial production is for four-wheel steering systems (4WS). In the 4WS-equipped vehicles, the front wheels are directly linked mechanically to the steering wheel, as in traditional vehicles. There is a power steering boost for the front wheels as in a standard two-wheel steering system. The rear wheels are steered under the control of a microcontroller via an actuator. Fig. 13.1-23 is an illustration of the 4WS configuration.

In this illustration, the front wheels are steered to a steering angle δ_f by the driver's steering wheel input. A sensor (S) measures the steering angle and another sensor (U) gives the vehicle speed. The microcontroller (C) determines the desired rear steering angle δ_r under program control as a function of speed and front steering angle.

The details of the control strategy are proprietary and not available for this book. However, it is within the

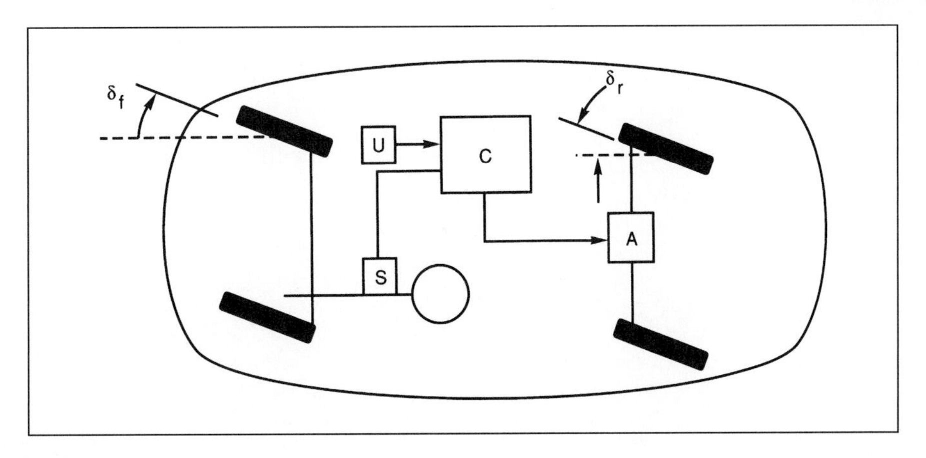

Fig. 13.1-23 4WS Configuration.

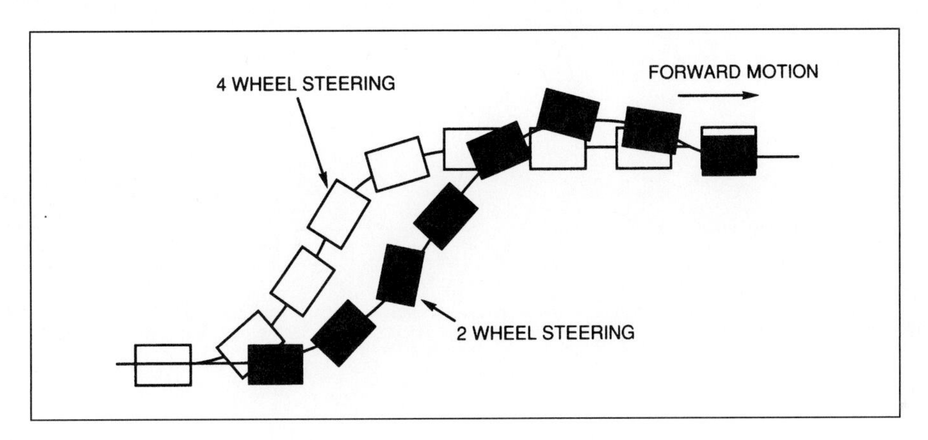

Fig. 13.1-24 Lane Change Maneuver.

scope of this book to describe a representative example control strategy as follows.

For speeds below 10 mph, the rear steering angle is in the opposite direction to the front steering angle. This control strategy has the effect of decreasing the car's turning radius by as much as 30% from the value it has for front wheel steering only. Consequently, the maneuvering ability of the car at low speeds is enhanced (e.g., for parking).

At intermediate speeds (e.g., 11 mph $< U <$ 30 mph), the steering might be front wheel only. At higher speeds (including highway cruise), the front and rear wheels are steered in the same direction. At least one automaker has an interesting strategy for higher speeds (e.g., at highway cruise speed). In this strategy, the rear wheels turn in the opposite direction to the front wheels for a very short period (on the order of one second) and then turn in the same direction as the front wheels. This strategy has a beneficial effect on maneuvers such as lane changes on the highway. Fig. 13.1-24 illustrates the lane change for front wheel steering and for this latter 4WS strategy, in which the same front steering angle was used. Notice that the 4WS strategy yields a lane change in a shorter distance and avoids the overshoot common in a standard-steering vehicle.

Turning the wheels in the same direction at cruising speeds has another benefit for a vehicle towing a trailer. When front and rear wheels turn in the same direction, the angle between the car and trailer axes is less than it is for front wheel steering only. The reduction in this angle means that the lateral force applied to the rear wheels by the trailer in curves is less than that for front wheel only steering. This lateral force reduction improves the stability of the car or truck/trailer combination relative to front steering only.

Section Fourteen

Intelligent transport systems

Chapter 14.1

14.1

Global positioning technology

Ljubo Vlacic and M. Parent

There is no doubt that one of the most important enabling technologies in the intelligent vehicle space is the global positioning system (GPS). Without the ability to accurately determine a vehicle's position on demand, there would be no way to cost-effectively implement autonomous or server-based vehicle navigation, nor would the ability to deliver customized, location-based services to the vehicle be possible.

This chapter will provide a brief overview of the GPS, and how it can be leveraged in intelligent vehicle applications. This chapter begins with a section describing the history of space-based positioning projects that have led to the current GPS, followed by a detailed description of the system as it exists and operates today. This is followed by a discussion of the science behind the GPS, and the techniques and components required to accurately and cost-effectively determine a user's position. The chapter concludes with some example applications where GPS is being used in the intelligent vehicle and related spaces, as well as future services that will be made possible because of GPS-based positioning capabilities.

14.1.1 History of GPS

Long before the development of the GPS in use today, the concept of time transfer and positioning via signals from space was being researched around the world. These costly research projects were mainly sponsored by government agencies, to address their long-standing need to improve techniques for quickly and accurately positioning military vehicles and personnel on or above the battlefield. Troops and vehicles of centuries past relied on maps, charts, the stars and various electronic devices to find their location; however, with each improved method of determining position came inherent limitations. Boundaries and landmarks change with the passage of time, making mapping a continual, time-consuming task. Positioning via the stars has long been a necessity for mariners, but accurate time keeping and clear skies are at times elusive. Until the deployment of today's GPS, the ultimate solution did not exist – an 'always on, always available' system for determining an exact position anywhere on the globe.

The constellation of satellites being used for global positioning today has it roots in the satellite positioning and time transfer systems of the early 1960s. Like many successful endeavours, the GPS was conceived from building blocks of other programmes such as the Navy Navigation Satellite System (NNSS, or Transit), Timation and Project 621B. It is worthwhile to have a brief understanding of these predecessors of GPS in order fully to understand and appreciate the complexity of space-based radio navigation.

Transit was conceived to provide positioning capabilities for the US submarine fleet, and originally deployed in 1964. While Transit proved to be a tremendous success in demonstrating the concept of radio navigation from space, the system was inherently inaccurate and required long periods of satellite observation in order to provide a user with enough information to calculate a position. Periods of observation in excess of 90 minutes were not uncommon, which limited the system's effectiveness for positioning a submarine at sea, since extended surface time could leave the vessel vulnerable. In its simplest form, Transit consisted of a small constellation of satellites broadcasting signals at 150 MHz and 400 MHz. The Doppler shift of these signals as measured by observers at sea, coupled with the known positions of the satellites in space, was sufficient to provide range

Intelligent Vehicle Technologies; ISBN: 9780750650939
Copyright © 2001 Ljubo Vlacic and M. Parent. All rights of reproduction, in any form, reserved.

measurements to the satellites, enabling the user to compute their position in two dimensions. Since all Transit satellites broadcast their signals at the same frequencies, the potential for interference allowed for only a small number of satellites. It was this limited number of satellites that necessitated the long periods of data collection, reducing the overall effectiveness of the system. This system was finally decommissioned in 1996.

In order to overcome the limitations of signal interference inherent in Transit and thereby increase the availability and effectiveness of satellite observation, an alternate technique for signal broadcast was necessary. US Air Force Project 621B, also begun in the late 1960s, demonstrated the use of pseudorandom noise (PRN) to encode a useful satellite ranging signal. PRN code sequences are relatively easy to generate, and by carefully choosing PRN codes which are nearly orthogonal to one another, multiple satellites can broadcast ranging signals on the same frequency simultaneously without interfering with one another. This simple concept forms the fundamental basis for GPS satellite ranging, and for the future implementation of the Wide Area Augmentation System (WAAS), which will be discussed later in this chapter.

The US Navy's Timation satellite system, initially launched in 1967, was also in full swing by the early 1970s. Timation satellites carried payloads with atomic time standards used for time keeping and time transfer applications. This enabled a receiver to use the signal broadcast by each Timation satellite to measure the distance to that satellite by measuring the time it took the signal to reach the receiver. Timation provided a key proof of concept and a foundation building block for the GPS, because without accurate time standards, the current GPS would not be possible.

In 1973, building on the success and knowledge gained from Transit, Timation and Project 621B, and with inputs and support from multiple branches of the military, the US Department of Defense (DoD) launched the Joint Program for GPS. Thus, the NAVSTAR GPS project was born.

14.1.2 The NAVSTAR GPS system

The NAVSTAR project was conceived as an excellent way to provide satellite navigation capabilities for a wide variety of military and civilian applications, and it has been doing so quite effectively since full operational capability (FOC) was declared in 1995. Building on previous satellite technology, the initial GPS satellites were launched between 1978 and 1985. These so-called Block I satellites were used to demonstrate the feasibility of the GPS concepts. Subsequent production models included Block II, Block IIA and Block IIR, each designed with improved capabilities, longer service life and at a lower cost. The next-generation models, known as Block IIF, are now being designed for launch in 2002.

This system, which currently consists of 28 fully operational satellites, cost an estimated $10 billion to deploy. The constellation is maintained and managed by the US Air Force Space Command from five monitoring sites around the world, at an annual cost of between $300 million and $500 million.

14.1.2.1 GPS system characteristics

The 28 satellites in the GPS are deployed in six orbital planes, each spaced 60° apart and inclined 55° relative to the equatorial plane. The orbit of each satellite (space vehicle, or SV) has an approximate radius of 20 200 km, resulting in an orbital period of slightly less than 12 hours. The system design ensures users worldwide should be able to observe a minimum of five satellites, and more likely six to eight satellites, at any given time, provided they have an unobstructed view of the sky. This is important because users with no knowledge of their position or accurate time require a minimum of four satellites to determine what is commonly known as a position, velocity and time solution, or PVT. The PVT data consists of latitude, longitude, altitude, velocity, and corrections to the GPS receiver clock.

The GPS satellites continuously broadcast information on two frequencies, referred to as L_1 and L_2, at 1575.42 MHz and 1227.6 MHz, respectively. The L_1 frequency is used to broadcast the navigation signal for non-military applications, called the Standard Positioning Service (SPS). Because the original design called for the SPS signal to be a lower resolution signal, it is modulated with a PRN code referred to as the Coarse Acquisition (C/A) code. For the purposes of reserving the highest accuracy potential for military users, the DoD may also impose intentional satellite clock and orbital errors to degrade achievable civilian positioning capabilities. This intentional performance degradation is commonly known as Selective Availability (S/A). For US military and other DoD-approved applications, a more accurate navigation signal known as the Precise Positioning Service (PPS) is broadcast on both the L_1 and L_2 frequencies. The PPS, in addition to the C/A code available on L_1, includes a more accurate signal-modulated with a code known as the Precise code (P-code) if unencrypted, and as the P(Y)-code if encrypted. Authorized users who have access to the PPS can derive more accurate positioning information from the L_1 and L_2 signals. Refer to Table 14.1-1 for a list of the original positioning and timing accuracy goals of the SPS and PPS services.

On 1 May, 2000, US President Bill Clinton announced the cessation of the S/A, which immediately resulted in

Table 14.1-1 Original navigation signal accuracy targets for SPS and PPS

	Horizontal accuracy	Vertical accuracy	Timing accuracy
Standard positioning service	100 metres	156 metres	340 ns
Precise positioning service	22 metres	27.7 metres	200 ns

Note: By design, all accuracies are statistically achievable 95% of the time.

greatly increased positioning accuracy for non-military GPS applications. The cessation of S/A should allow users of the SPS a level of accuracy similar to those using the PPS. Within the first week of the discontinuation of S/A, positioning accuracies within 10 metres were already being reported, without any upgrade to the GPS receivers being used.

14.1.2.2 The navigation message

The navigation message broadcast by every GPS satellite contains a variety of information used by each GPS receiver to calculate a PVT solution. The information in this message includes time of signal transmission, clock correction and ephemeris data for the specific SV, and an extensive amount of almanac and additional status and health information on all of the satellites in the GPS.

Each SV repeatedly broadcasts a navigation message that is 12.5 minutes in length, and consists of 25 1500-bit data frames transmitted at 50 bits per second. A single data frame is composed of five 300-bit subframes, each containing different status or data information for the receiver, preceded by two 30-bit words with SV-specific telemetry and handover information. The first three subframes, containing clock correction and ephemeris data relevant to the specific SV, are refreshed as necessary for each data frame transmitted during the navigation message broadcast. The almanac and other data transmitted in the final two subframes are longer data segments, relevant to the entire GPS, requiring the full 25 data frames to be broadcast completely. Below is a brief description of the contents of each subframe. For an illustration of the complete Navigation Message, refer to Fig. 14.1-1.

14.1.2.2.1 Clock correction subframe

The clock correction subframe, the first subframe transmitted in the navigation message data frame, contains the GPS week number, accuracy and health information specific to the transmitting SV, and various clock correction parameters relating to overall system time, such as clock offset and drift.

14.1.2.2.2 SV ephemeris subframes

The second and third subframes of the navigation message contain ephemeris data. This data provides the GPS receiver with precise orbit information and correction parameters about the transmitting SV that the receiver uses accurately to calculate the satellite's current position in space. This information, in turn, is used with the clock information to calculate the range to the SV. Included in the ephemeris subframes are telemetry parameters specific to the transmitting SV, such as correction factors to the radius of orbit, angle of inclination, and argument of latitude, as well as the square root of the semi-major axis of rotation, the eccentricity of the orbit of the SV, and the reference time that the ephemeris data was uploaded to the SV.

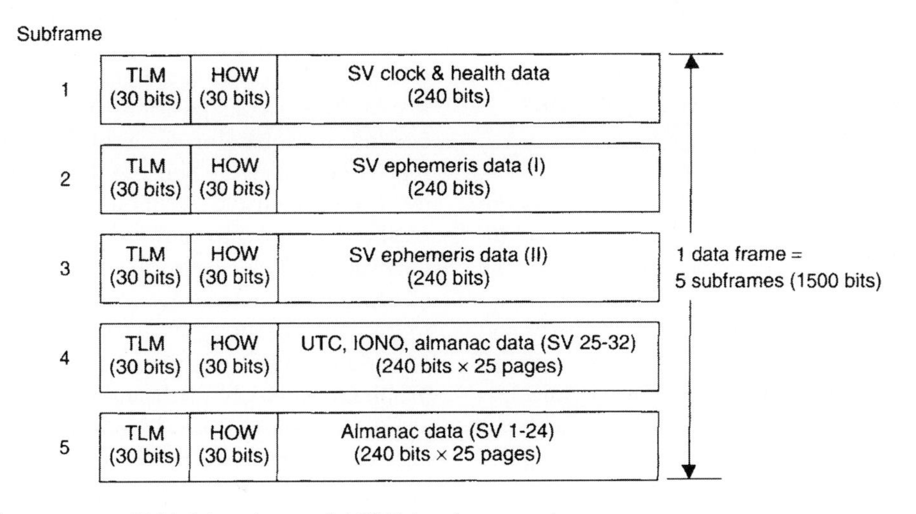

Fig. 14.1-1 Navigation message. TLM, telemetry word; HOW, handover word.

14.1.2.2.3 Almanac and support data subframes

Subframes four and five of the navigation message data frame contain comprehensive almanac data for the entire GPS constellation, along with delay parameters that the receivers use for approximating phase delay of the transmitted signal through the ionosphere, and correction factors to correlate GPS and Universal Time Coordinated (UTC).

The almanac data contains orbit and health information on all of the satellites in the GPS constellation. GPS receivers use this information to speed up the acquisition of SV signal transmissions. The almanac data in subframe four contains health and status information on the operational satellites numbered 25 through 32, along with ionospheric and UTC data. The almanac data in subframe five contains health and status information on the operational satellites in the GPS numbered 1 through 24.

For a more detailed description of the information contained in the Navigation Message, refer to the ICD-GPS-200c specification, which is available from the US Coast Guard Navigation Center.

14.1.3 Fundamentals of satellite-based positioning

To understand the true value and cost of the positioning capabilities of the GPS, it is important for the user to have a basic understanding of the science behind positioning, and the types of components and techniques that may be used to calculate accurate positions. This section divides this discussion into three main areas: the basic science behind GPS; the different unassisted and assisted position calculation techniques that may be used, depending upon the needs of the specific application; and the hardware and software components necessary for calculating a position.

14.1.3.1 The basic science of global positioning

The design of the GPS makes it an all-weather system whereby users are not limited by cloud cover or inclement weather. Broadcasting on two frequencies, the GPS provides sufficient information for users to determine their PVT with a high degree of accuracy and reliability. As mentioned previously, frequency L_1 is generally regarded as the civilian frequency while frequency L_2 is primarily used for military applications. Applications and positioning techniques in this chapter will focus on GPS receiver technology capable of tracking L_1 only, as cost and security issues typically preclude most users from taking full advantage of both GPS frequencies. Without a complete knowledge of the encrypted L_2 frequency, only mathematical exercises enable high accuracy applications of GPS such as surveying to take advantage of any information provided by L_2.

14.1.3.1.1 Position calculation

The fundamental technique for determining position with the GPS is based on a basic range measurement made between the user and each GPS satellite observed. These ranges are actually measured as the GPS signal time of travel from the satellite to the observer's position. These time measurements may be converted to ranges simply by multiplying each measurement by the speed of light; however, since most GPS receiver internal clocks are incapable of keeping time with sufficient accuracy to allow accurate ranging, the mathematical PVT solution must solve for errors in the receiver clock at the time each observation of a satellite is made. Satellite ranges are commonly called pseudoranges to include this receiver clock error and a variety of other errors inherent in using GPS. These receiver clock errors are included as one component in a least squares calculation, which is used to solve for position using a technique called trilateration.

To calculate the values for PVT, the concept of triangulation in two-dimensions as is commonly practised in determining the location of an earthquake epicentre is extended into three-dimensions, with the ranges from the satellites prescribing the radius of a sphere (see Fig. 14.1-2). This technique is known as trilateration, since it uses ranges to calculate position, whereas triangulation uses angular measurements. If a sphere centred on the satellites' position in space is hypothetically created with the range from the user to each satellite as its radius, the intersection of three of these spheres may be used to determine a user's two-dimensional position. While it may seem counterintuitive that ranges to three satellites will allow for only a two-dimensional position, in fact one observation is needed to solve for each of latitude, longitude and receiver clock error. Thus, to determine a user's position in three-dimensions a minimum of four satellites is required, in order to solve for altitude, as well as latitude, longitude and clock error.

Once pseudoranges have been determined to three or more SVs, the user's PVT can be calculated by solving N simultaneous equations as a classic least squares problem, where N is the number of satellite pseudoranges measured. The relationship between the receiver and each satellite's position can best be written by extending the Pythagorean Theorem as illustrated in equation 14.1.1, where i is the number of each satellite

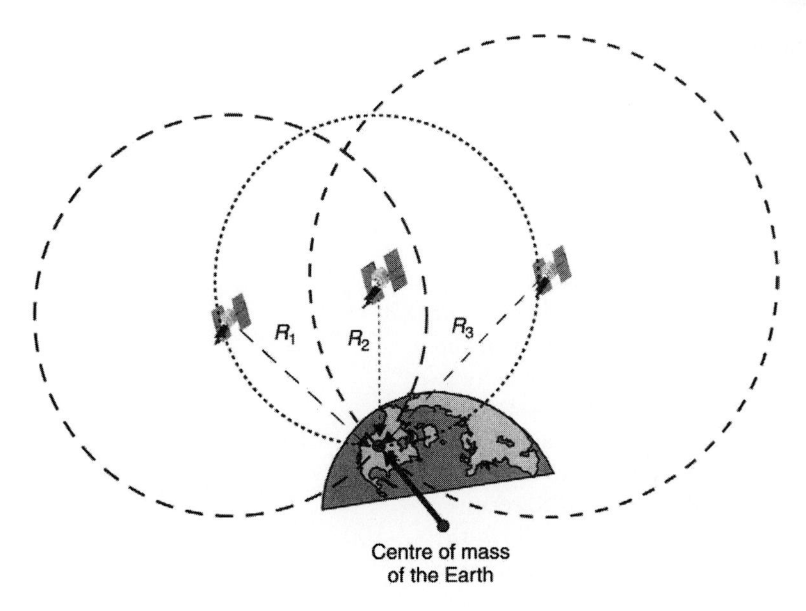

Fig. 14.1-2 3-Dimensional trilateration of GPS satellites.

detected (3–N), $\{x_i, y_i, z_i\}$ is the known position of each satellite i, R_i is the pseudorange measurement for each satellite i, and b is the receiver clock error:

$$R_i = \sqrt{(x_i - x)^2 + (y_i - y)^2 + (z_i - z)^2} - b \tag{14.1.1}$$

While a three-dimensional PVT calculation may be made using pseudoranges from four satellites, improved accuracy can be achieved if five or more are used, as the redundancy can help reduce the effects of position and receiver clock errors in the calculation.

14.1.3.1.2 Coordinate systems

The coordinate frame used by the GPS to map a satellite's position, and thus a receiver's position, is based on the World Geodetic System 1984 (WGS 84). This coordinate reference frame is an Earth-centred, Earth-fixed (ECEF) Cartesian coordinate system, for which curvilinear coordinates (latitude, longitude, height above a reference surface) have also been defined, based on a reference ellipsoid, to allow easier plotting of a user's position on a traditional map. This coordinate frame, or datum, is the standard reference used for calculating position with the GPS. However, many regional and local maps based on datums developed from different ground-based surveys are also in use today, whose coordinates may differ substantially from WGS 84. Simple mathematical transformations can be used to convert calculated positions between WGS 84 and these regional datums, provided they meet certain minimum criteria for the mapping of their longitude, latitude and local horizontal and vertical references. At last count, more than 100 regional or local geodetic datums were in use for positioning applications in addition to WGS 84.

14.1.3.2 Positioning techniques

Several different techniques have been developed for using the GPS to pinpoint a user's position, and to refine that positioning information though a combination of GPS-derived data and additional signals from a variety of sources. Some of the more popular techniques, such as autonomous positioning, differential positioning and server-assisted positioning, are briefly described below.

14.1.3.2.1 Autonomous GPS positioning

Autonomous positioning, also known as single-point positioning, is the most popular positioning technique used today. It is the technique that is commonly thought of when a reference to using the GPS to determine the location of a person, object or address is made. In basic terms, autonomous positioning is the practice of using a single GPS receiver to acquire and track all visible GPS satellites, and calculate a PVT solution. Depending upon the capabilities of the system being used and the number of satellites in view, a user's latitude, longitude, altitude and velocity may be determined. As mentioned earlier, until May of 2000 this technique was limited in its accuracy for commercial GPS receivers. However, with the discontinuation of S/A this technique may now be used to determine a user's location with a degree of accuracy and precision that was previously available only to privileged users.

14.1.3.2.2 Differential GPS positioning

The use of differential GPS (DGPS) has become popular among GPS users requiring accuracies not previously achievable with single-point positioning. DGPS effectively eliminated the intentional errors of S/A, as well as errors introduced as the satellite broadcasts pass through the ionosphere and troposphere.

Unlike autonomous positioning, DGPS uses two GPS receivers to calculate PVT, one placed at a fixed point with known coordinates (known as the master site), and a second (referred to here as the mobile unit) which can be located anywhere in the vicinity of the master site where an accurate position is desired. For example, the master site could be located on a hill or along the coastline, and the mobile unit could be a GPS receiver mounted in a moving vehicle. This would allow the master site to have a clear view of the maximum number of satellites possible, ensuring that pseudorange corrections for satellites being tracked by the mobile unit in the vicinity would be available.

The master site tracks as many visible satellites as possible, and processes that data to derive the difference between the position calculated based on the SV broadcasts and the known position of the master site. This error between the known position and the calculated position is translated into errors in the pseudorange for each tracked satellite, from which corrections to the measured distance to each satellite are derived. These pseudorange corrections may then be applied to the pseudoranges measured by the mobile unit, effectively eliminating the affects of SA and other timing errors in the received signals (see Fig. 14.1-3).

Corrections to measured pseudoranges at the master site are considered equally applicable to both receivers with minimum error as long as the mobile unit is less than 100 km from the master site. This assumption is valid because the distance at which the GPS satellites are orbiting the earth is so much greater than the distance between the master site and the mobile unit that both receivers can effectively be considered to be at the same location relative to their distance from each SV. Therefore, the errors in the pseudorange calculated for a particular satellite by the mobile unit are effectively the same as errors in the same pseudorange at the master site (i.e. the tangent of the angle between the master site and second receiver is negligible (see Fig 14.1-4)).

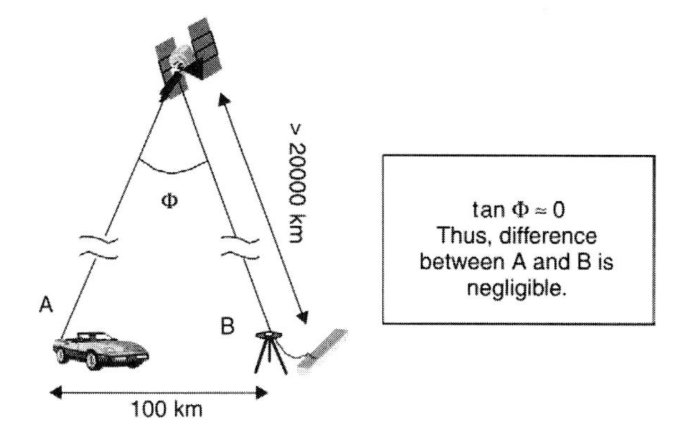

Fig. 14.1-4 Pseudorange correction in DGPS (not to scale).

Of course, to calculate a position using DGPS, a mobile unit must establish communication with a master site broadcasting DGPS correction information. One source is the US Coast Guard, which operates a series of DGPS master sites that broadcast DGPS corrections across approximately 70 per cent of the continental US, including all coastal areas. Alternatively, a GPS receiver that has wireless communication capabilities, such as one that is integrated into an intelligent

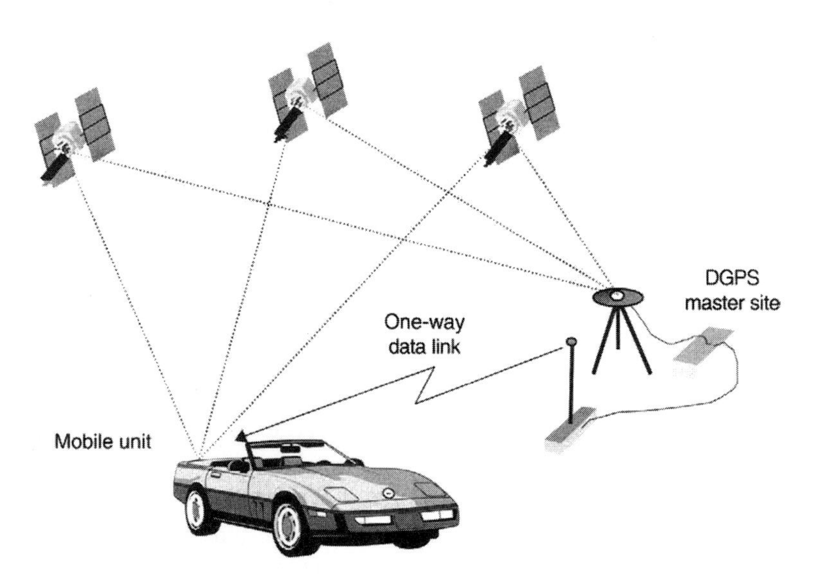

Fig. 14.1-3 Differential GPS positioning.

vehicle, may be able to access DGPS correction data on the Internet, or have it delivered on a subscription basis from a private differential correction service provider.

With the discontinuation of S/A, using the DGPS positioning technique will still provide enhanced positioning accuracy, since other timing errors are inherent in the SV broadcasts that DGPS may help correct. However, these much smaller improvements in accuracy may no longer offset the additional cost of receiving and processing the DGPS correction information for many applications.

14.1.3.2.3 Inverse differential GPS positioning

Inverse differential GPS (IDGPS) is a variant of DGPS in which a central location collects the standard GPS positioning information from one or more mobile units, and then refines that positioning data locally using DGPS techniques. With IDGPS, a central computing centre applies DGPS correction factors to the positions transmitted from each receiver, tracking to a high degree of accuracy the location of each mobile unit, even though each mobile unit only has access to positioning data from a standard GPS receiver (see Fig. 14.1-5).

This technique can be more cost-effective in some ways than standard DGPS, since there is no requirement that each mobile unit be DGPS-enabled, and only the central site must have access to the DGPS correction data. However, there is an additional cost for each mobile device, since each unit must have a means of communicating position data back to the central computer for refinement. For applications such as delivery fleet management or mass transit, IDGPS may be an ideal technique for maintaining highly accurate position data for each vehicle at a central dispatch facility, since the communication channel is already available, and the relative cost of refining the positioning information for each mobile unit at the central location is minimal. Of course, with the discontinuation of S/A, DGPS refinement may no longer be necessary for many of these applications.

14.1.3.2.4 Server-assisted GPS positioning

Server-assisted GPS is a positioning technique that can be used to achieve highly accurate positioning in obstructed environments. This technique requires a special infrastructure that includes a location server, a reference receiver in the mobile unit, and a two-way communication link between the two, and is best suited for applications where location information needs to be available on demand, or only on an infrequent basis, and the processing power available in the mobile unit for calculating position is minimal.

In a server-assisted GPS system, the location server transmits satellite information to the mobile unit, providing the reference receiver with a list of satellites that are currently in view. The mobile unit uses this satellite view information to collect a snapshot of transmitted data from the relevant satellites, and from this calculates the pseudorange information. This effectively eliminates the time and processing power required for satellite discovery and acquisition. Also, because the reference receiver is provided with the satellite view, the sensitivity of the mobile unit can be greatly improved, enabling operation inside buildings or in other places where an obstructed view will reduce the capabilities of an autonomous GPS receiver.

Once the reference receiver has calculated the pseudoranges for the list of satellites provided by the location server, the mobile unit transmits this information back to the location server, where the final PVT solution is

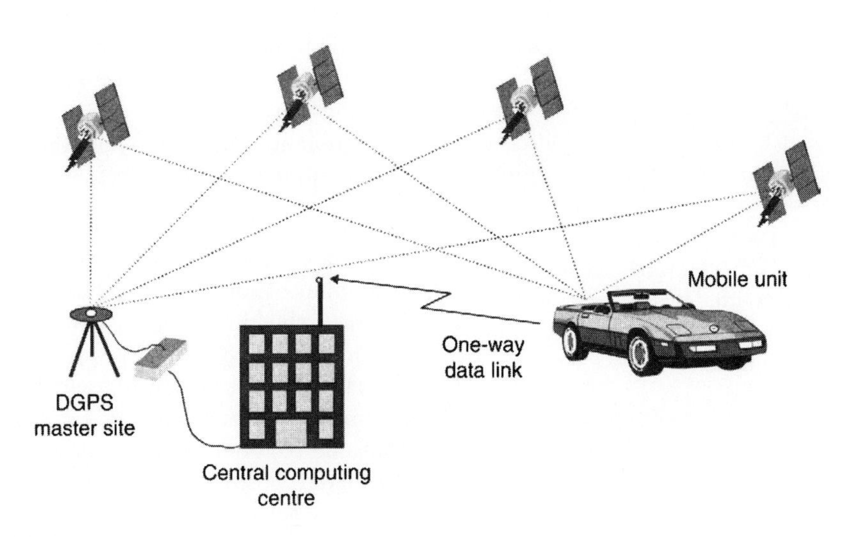

Fig. 14.1-5 IDGPS positioning.

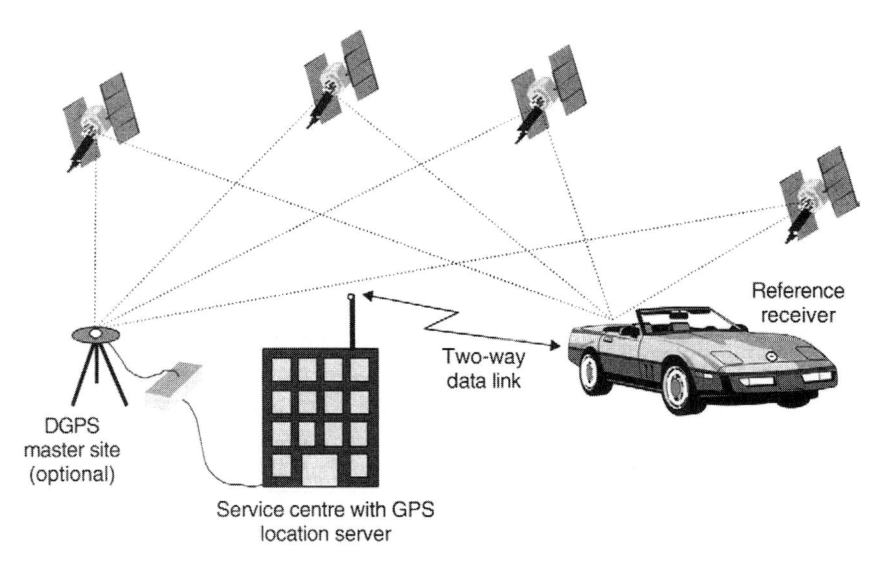

Fig. 14.1-6 Server-assisted GPS positioning.

calculated. The location server then transmits this final position information back to the mobile device as needed. Because the final position data is calculated at the location server, some of the key benefits of DGPS can also be leveraged to improve the accuracy of the position calculation. An illustration of the relationship between the reference receiver and the location server in a server-assisted GPS system can be seen in Fig. 14.1-6.

14.1.3.2.5 Enhanced client-assisted GPS positioning

The enhanced client-assisted GPS positioning technique is a hybrid between autonomous GPS and server-assisted GPS. This type of solution is similar to the server-assisted GPS, with the location server providing the mobile unit with a list of visible satellites on demand. However, in an enhanced client-assisted system, the mobile unit does the complete PVT calculation rather than sending pseudorange information back to the location server.

This technique essentially requires the same processing power and capabilities as an autonomous GPS solution, in addition to a communication link between the mobile unit and the location server. However, the amount of time required to complete the PVT calculation is much less than with an autonomous GPS solution, because of the satellite view information provided by the location server, and fewer exchanges with the location server are required than with a server-assisted solution.

14.1.3.2.6 Dead reckoning

Dead reckoning (DR) is a technique used in conjunction with other GPS-based positioning solutions to maintain an estimate of position during periods when there is poor or no access to the GPS satellite broadcasts. DR is used primarily to enhance navigation applications, since maintaining an accurate position in real time is crucial to the performance of a navigation system, and there may be times during a trip when the GPS-derived position may be intermittent, or not available at all. These GPS outages can be caused by a variety of environmental and terrain features. Examples of areas where GPS coverage could be interrupted include:

- **tunnels** through mountains or in urban areas, which prevent signal reception
- **urban canyons**, such as downtown areas populated by tall buildings, which can result in either blocked signals, or multipath errors caused by signal reflection
- **heavy foliage**, where overhanging trees or bushes block reception of the signal broadcasts
- **interference/jamming**, which can be caused by either harmonics of commercial radio transmissions, or by transmissions specifically designed to interfere with the reception of the satellite broadcasts for security reasons
- **system malfunction**, where the GPS receiver itself is functioning intermittently.

When a positioning data outage of this sort is encountered, a system that is DR-enabled will monitor inputs from one or more additional sensors in order to continue to track the direction, distance and speed the unit is moving. The system will process that data starting from the last known position fix, which will enable it to keep a running estimate of its position. The system will continue to monitor these sensor inputs and update its estimated position until the GPS receiver can again obtain an accurate position fix. At this point, the system updates its position with the satellite-based data.

For example, in an intelligent vehicle with an autonomous navigation system, the GPS receiver normally calculates the position data used by the navigation algorithm to determine the progress of the vehicle along the desired path. However, when driving in some environments, the GPS receiver may have trouble maintaining a continuous satellite lock, resulting in intermittent periods where the vehicle's position cannot be determined based on valid satellite data. In situations like this, DR is used to 'fill in the gaps', providing a method for estimating the current position based on the vehicle's movements since the last known positioning fix.

A variety of input sensors can be used to provide DR capability. In the intelligent vehicle example, several different sensor inputs can be made available to the navigation system to assist in DR calculation. The types of sensors that could be used to enable DR in a vehicle system include:

- **magnetic compass**, which can provide a continuous, coarse-grained indication of the direction in which the vehicle is moving
- **gyroscope**, which can be used to detect the angular movement of the vehicle
- **speedometer**, which can provide the current speed of the vehicle
- **odometer**, which can provide continuous data on the elapsed distance
- **wheel speed sensors**, such as Hall-effect or variable reluctance sensors (VRSs), which can provide fine-grained vehicle speed information
- **accelerometers**, which can detect changes in the velocity of the vehicle.

Many of these sensors are already widely used in vehicles for other applications. Accelerometers are being used today in impact detection (airbag) systems; wheel speed sensors are being used in traction-control and anti-lock braking systems; and of course the trip meters available today in many cars use inputs from the speedometer, odometer and compass to calculate distance travelled, distance remaining and fuel economy.

Systems that leverage inputs from remote vehicle sensors to enable DR can certainly provide more consistent positioning information under some circumstances than may be possible with a single-point GPS receiver. However, depending upon the mix of sensor inputs used, the accuracy of the resulting position data may vary. Some of these sensors are more accurate than others, and most are subject to a variety of environmental, alignment and computational errors that can result in faulty readings. Some vendors of DR-enabled positioning systems have been exploring methods of reducing the effects of these errors. The development of self-correcting algorithms and self-diagnosing sensors may help reduce the impact that sensor errors can have on these systems in the future.

14.1.3.2.7 Additional GPS augmentation techniques

Additional techniques are being developed for increasing the accuracy of the positioning information derived from the GPS for certain applications. One technique, which has been developed by the US Federal Aviation Administration (FAA), uses transmissions from communication satellites to improve the positioning accuracy of GPS receivers in aircraft. This technique, known as the WAAS, uses a network of wide area ground reference stations (WRSs) and two wide area master stations (WMSs) to calculate pseudorange correction factors for each SV, as well as to monitor the operational health of each SV. This information is uplinked to communication satellites in geostationary earth orbit (GEO), which transmit the information on the L_1 frequency, along with additional ranging signals. This system has improved the positioning accuracy of GPS receivers on board aircraft to within 7 metres horizontally and vertically, allowing the system to be used by aircraft for Category I precision approaches. A Category I system is intended to provide an aircraft operating in poor weather conditions with safe vertical guidance to a height of not less than 200 feet with runway visibility of at least 1800 feet.

Another method for improving positioning accuracy is known as carrier-phase GPS. This is a technique where the number of cycles of the carrier frequency between the SV and the receiver is measured, in order to calculate a highly accurate pseudorange. Because of the much shorter wavelength of the carrier signal relative to the code signal, positioning accuracies of a few millimetres are possible using carrier-phase GPS techniques. In order to make a carrier-phase measurement, standard code-phase GPS techniques must first be used to calculate a pseudorange to within a few metres, since it would not be possible to derive a pseudorange using only the fixed carrier frequency. Once an initial pseudorange is calculated, a carrier-phase measurement can then be used to improve its accuracy by determining which carrier frequency cycle marks the beginning of each timing pulse. Of course, receivers that can perform carrier-phase measurements will bear additional hardware and software costs to achieve these improved accuracies.

14.1.4 GPS receiver technology

In order to design and build a GPS receiver, the developer must understand the basic functional blocks that comprise the device, and the underlying hardware and software necessary to implement the desired capabilities. The sections below describe the main functional blocks of a GPS receiver, and the types of solutions that are either available today or in development to provide that functionality.

14.1.4.1 GPS receiver components

GPS receivers are composed of three primary components: the antenna, which receives the radio frequency (RF) broadcasts from the satellites; the downconverter, which converts the RF signal into an intermediate frequency (IF) signal; and the baseband processor or correlator, which uses the IF signal to acquire, track, and receive the navigation message broadcast from each SV in view of the receiver. In most systems, the output of the correlator is then processed by a microprocessor (MPU) or microcontroller (MCU), which converts the raw data output from the correlator into the positioning information which can be understood by a user or another application.

The sections below provide an overview of the three key components of a GPS receiver, describing in generic terms the functionality and capabilities typically found in these systems. As the capabilities of the MPU or MCU needed to process the correlator output is largely dependent on the needs of the applications and the particular GPS chip set being considered, MPU/MCU requirements and capabilities are not discussed here.

14.1.4.1.1 Antennas

As with most RF applications, important performance characteristics to be considered when selecting the antenna for a GPS receiver include impedance, bandwidth, axial ratio, standing wave ratio, gain pattern, ground plane, and tolerance to moisture and temperature. In addition, the relatively weak signal transmitted by GPS satellites is right-hand circularly polarized (RHCP). Therefore, to achieve the maximum signal strength the polarization of the receiving antenna must match the polarization of the transmitted satellite signal. This restriction limits the types of antennas that can be used. Some of the more common antennas used for GPS applications include:

- **microstrip**, or **patch**, antennas are the most popular antenna because of their simple, rugged construction and low profile, but the antenna gain tends to roll-off near the horizon. This makes it more difficult to acquire SVs near the horizon, but it also makes the antenna less sensitive to multipath signals. This type of antenna can be used in single or dual frequency receivers.
- **helix-style** antennas have a relatively high profile compared to the other antennas, maintaining good gain near to the horizon. This can provide easier acquisition of SVs lower on the horizon, but also makes it more sensitive to multipath signals that can contribute to receiver error. The spiral helix antenna is used in dual-frequency receivers, while the quadrifilar helix antenna is used in single frequency systems.
- **monopole** and **dipole** antennas are low cost, single frequency antennas with simple construction and relatively small elements.

Systems with an antenna that is separate from the receiver unit, such as a GPS receiver installed in a vehicle with a trunk-mounted antenna, often use an active antenna which includes a low noise pre-amplifier integrated into the antenna housing. These amplifiers, which boost the very weak received signal, typically have gains ranging from 20 dB to 36 dB. Active antennas are connected to the receiver via a coax cable, using a variety of connectors, including MMCX, MCX, BNC, Type N, SMA, SMB, and TNC. Systems that have the antenna integrated directly into the receiver unit (such as a handheld GPS device) use passive antennas, which do not include the integrated pre-amplifier.

The demand for the integration of positioning technology into smaller devices is challenging antenna development. The industry is already pushing for smaller antennas for applications such as a wristwatch with integrated GPS, which is smaller than most patch antennas available today. Another demand is for dual-purpose antennas that do double duty in wireless communication devices, such as in a mobile telephone with an integrated GPS receiver. Inevitably, the future will bring smaller and more flexible antennas for GPS applications.

14.1.4.1.2 Downconverter

The function of the downconverter is to step down each GPS satellite signal from its broadcast RF frequency to an IF signal that can be output to the base-band processor. The signal from each SV in view of the antenna (active or passive) is filtered and amplified by a low noise pre-amplifier, which sets the overall noise of the system, and rejects out of band interference. The output of this pre-amplifier is input into the downconverter, where the conversion to the IF signal is typically made in two stages. The two-stage mixer is clocked by a fixed-frequency phase-locked loop controlled by an external reference oscillator that provides frequency and time references for the downconverter and base-band processor.

The mixer outputs, which are composed of in-phase (I) and quadraphase (Q) signals, are amplified again and latched as the IF input to the base-band processor to be used for satellite acquisition and tracking. To enable the baseband processor to account for frequency variation over temperature, an integrated temperature sensor is often included in the downconverter circuit.

The downconverter in a GPS receiver is often susceptible to performance degradation from external RF interference from both narrowband and wideband sources. Common sources of narrowband interference include transmitter harmonics from Citizens Band (CB) radios and AM and FM transmitters. Sources of wideband interference can include broadcast frequency harmonics from microwave and television transmitters. In mobile GPS applications such as in intelligent vehicle systems, the GPS receiver will often encounter this type of interference, and must rely on the antenna and downconverter design to attenuate the effects.

14.1.4.1.3 Correlator/data processor

The correlator component in a GPS receiver performs the high-speed digital signal processing functions on the IF signal necessary to acquire and track each SV in view of the antenna. The IF signal received by the correlator from the downconverter is first integrated to enhance the signal, then the correlator performs further demodulation and despreading to extract each individual SV signal being received. Each signal is then multiplied by a stored replica of the C/A signal from the satellite being received, known as the Gold code for that satellite. The timing of this replica signal is adjusted relative to the received signal until the exact time delay is determined. This adjustment period to calculate the time delay between the local clock and the SV signal is defined as the acquisition mode. Once this time delay is determined, that SV signal is then considered acquired, or locked.

After acquisition is achieved, the receiver transitions into tracking mode, where the PRN is removed. Thereafter, only small adjustments must be made to the local reference clock to maintain correlation of the signal. At this point, the extraction of the satellite timing and ephemeris data from the navigation message is done. This raw data and the known pseudoranges are then used to calculate the location of the GPS receiver. This information is then displayed for the user, or otherwise made available to other applications, either through an external port (for remote applications) or through a software API (for integrated applications).

In the past, GPS correlators were designed with a single channel, which was multiplexed between each SV signal being received. This resulted in a very slow process for calculating a position solution. Today, systems come with up to 12 channels, allowing the correlator to process multiple SV signals in parallel, achieving a position solution in a fraction of the time. Also, while the correlator functionality is sometimes performed in software using a high-performance digital signal processor (DSP), the real-time processing requirements and repetitive high rate signals involved make a hardware correlator solution ideal, from both a cost and throughput standpoint.

14.1.4.2 GPS receiver solutions

When access to the GPS first became available for military and commercial use, only a few companies had the technology and expertise to develop reliable, accurate GPS receivers. Application developers who needed GPS services would simply purchase a board level solution from a GPS supplier, and integrate it into their design.

More recently, the demand for putting GPS capabilities into customized packaging has grown dramatically. To meet that demand a variety of solutions are now available, ranging from traditional board-level solutions that connect to an application via a serial interface, to integrated circuit (IC) chip sets, which application developers can embed directly into their designs. The sections below will give a brief overview of the types of solutions available on the market today.

14.1.4.2.1 System level solutions

The first commercially available GPS receivers were designed as either standalone units with connectors for power, an antenna, and a serial interface to a computer or other device, or as more basic board-level solutions, which could be integrated into an application enclosure, but which still required an external antenna connection and serial network interface. These units were entirely self-contained, with the RF interface, downconverter and baseband processing done entirely independent of the application. With this type of solution, the PVT information was transmitted out of the serial port, to be displayed or used as appropriate depending upon the application. In some cases, the user could provide some configuration data to the system, such as the choice of a local datum, and in that way 'customize' the resulting positioning information for their needs. This type of solution is still widely available, and for many applications provides a cost-effective way of adding GPS positioning or timing services to an existing design.

One variation of the board-level solution that is becoming more popular today is to supply the RF section of the GPS receiver, including the discrete RF interface and downconverter, as a self-contained module, along with a standalone correlator ASIC or an MCU with an integrated correlator and software to perform the baseband processing of the IF signal. Typically, the RF section of a GPS receiver is the most challenging portion of the design because of the sensitivity to component layout and extraneous signals, and many of the RF circuits that exist today were designed with a combination of technical know-how and trial-and-error experience that few application developers can afford. By comparison, designing the hardware layout for the baseband processor and interface to the RF module is a relatively minor task, which is what has made this an attractive solution for application developers who want to integrate GPS into their designs, but cannot afford the cost or space necessary for a board-level solution.

14.1.4.2.2 IC chip set solutions

For those developers that have the skill (or want the challenge) of designing the entire GPS receiver circuit into their application, several semiconductor manufacturers now offer GPS chip set solutions. These chip sets, offered with either complete or partial reference designs and control software, enable the designer to integrate GPS into an application at the lowest possible cost, while also conserving power, board space and system resources. However, this high level of integration is achieved at the expense of doing the RF and IF circuit layout and software integration in-house, which can take significant resources and effort.

The custom chip sets used for the original GPS receivers often had up to seven ICs, including the external memory chips, amplifiers, downconverter, correlator ASIC and system processor, in addition to a variety of discrete components. Continuous advances in the performance and integration level of MCUs have greatly increased the performance of the newer GPS chip sets while reducing the power consumption and physical size of the complete system. System-on-a-Chip (SoC) technology has resulted in the integration of the GPS correlator directly onto the MCU, along with embedded RAM, ROM and FLASH memory. In some cases, this increased level of integration has reduced the device count down to a mere two ICs and a handful of discrete components, further decreasing the cost and development effort required.

Even more recently, high-performance RISC MCUs have begun showing up in low-cost GPS chip set solutions. These powerful processors have many more MIPS available for GPS computations, which in turn increases the overall performance and reliability of the GPS solution. This level of computational power is making it possible to execute DR or WAAS algorithms on the same processor as the GPS algorithms, further improving the accuracy of the positioning solution at little or no increase in chip set cost.

A block diagram illustrating the primary components of a GPS receiver as described in the previous sections is pictured in Fig. 14.1-7. This diagram illustrates all of the functional blocks required by a basic GPS system, including an active antenna, a downconverter with an integrated temperature sensor, and a correlator integrated onto a basic MCU along with the additional MCU peripherals required to perform a basic tracking loop routine and calculate a PVT solution.

14.1.4.2.3 Development tools

The development tools available for GPS application design vary depending on the complexity of the target system and the GPS solution being used. Most GPS solution vendors offer software tool suites that allow a developer to communicate with the GPS receiver through the serial port of a personal computer. These software tools typically use messages compatible with the standard National Marine Electronic Association (NMEA) format, but many vendors also offer their own customized sets of messages and message formats.

The more advanced development tools, available for some GPS chip sets, are intended to help the application developer integrate their software with the GPS tracking software running on the same MCU. Because of the hard real-time constraints typical of GPS software implementations, the most efficient way to enable the smooth integration of the GPS tracking loop with the application software is through a clearly defined software API. With a standard interface to the GPS software and the necessary development/debugger tools to support it, an application developer can easily configure the GPS receiver software, enabling access to the appropriate PVT information by the application as needed. For an illustration of the basic software architecture of

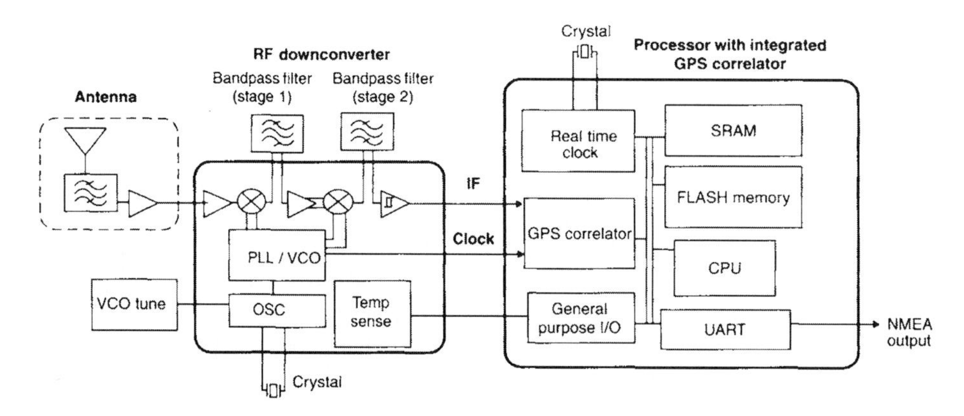

Fig. 14.1-7 Functional block diagram of GPS receiver.

a GPS-enabled application running on a single MCU that is supported by this type of tool suite, refer to Fig. 14.1-8.

14.1.4.3 Performance considerations

There are many parameters used by the industry to assess the performance of a GPS receiver, and to evaluate the relative performance of comparable receivers. The most common parameters being used to evaluate GPS receiver performance include positioning and timing accuracy, time-to-first-fix (TTFF), reacquisition time, and receiver sensitivity.

14.1.4.3.1 Positioning and timing accuracy

The most obvious of these parameters is positioning accuracy – how accurate are the positions calculated by an autonomous GPS receiver, based on the number of satellites that can be seen by that receiver? This is typically measured by performing a mapmatching test, where positions calculated by a receiver for landmarks on a map are compared to their known positions. This is a standard test that is often used to compare the accuracy of multiple GPS receivers simultaneously. When the S/A feature was still enabled, the accuracy of the SPS signal served as the baseline for positioning accuracy for commercial GPS receivers. With the discontinuation of S/A, the accuracy of autonomous GPS receivers has increased significantly, but as of this writing there is no new accepted baseline for the measurement of post-S/A receivers, except perhaps for the PPS signal accuracy.

A performance parameter closely related to positioning accuracy is timing accuracy – how close to UTC is the time calculated by the GPS receiver. This parameter essentially measures the deviation of the calculated time from UTC as maintained by the US Naval Observatory. However, since the accuracy of the time component of the SPS signal with the S/A feature enabled was within 340 ns, this is obviously a test requiring sophisticated time measurement equipment to perform. The timing accuracy of most GPS receivers is more than adequate for commercial applications such as intelligent vehicle systems.

14.1.4.3.2 Time-to-first-fix

TTFF is the measure of the time required for a receiver to acquire satellite signals and calculate a position. The three variants of a TTFF measurement, which depend upon the condition of the GPS receiver when the TTFF is measured, are referred to as hot start, warm start, and cold start. These TTFF measurements include the amount of time it takes the GPS receiver to acquire and lock each satellite signal, calculate the pseudorange for each satellite, and calculate a position fix.

Hot start occurs when a GPS receiver has recent versions of almanac and satellite ephemeris data, and current time, date and position information. This condition might occur when a receiver has gone into a power-conserving stand-by mode due to application requirements. In this situation, most receivers should be able to acquire a position fix within 15 seconds.

Warm start occurs when a GPS receiver is powered on after having been off or out of signal range for several hours to several weeks. In this condition, the receiver has an estimate of time and date information, and a recent copy of satellite almanac data, but no valid satellite ephemeris data. In this state, a receiver can begin tracking satellites immediately, but must still receive updated ephemeris data from each satellite, which is only valid for approximately four hours. Under these conditions, most receivers should be able to acquire a position fix within 45 seconds.

Cold start occurs when a GPS receiver has inaccurate date and time information, no satellite ephemeris data, and no almanac data, or data which is significantly out of date. In this state, the receiver must perform a search for the available satellite signals, and can take 90 seconds or more to acquire a position fix. This condition is encountered when the GPS receiver is powered up for the

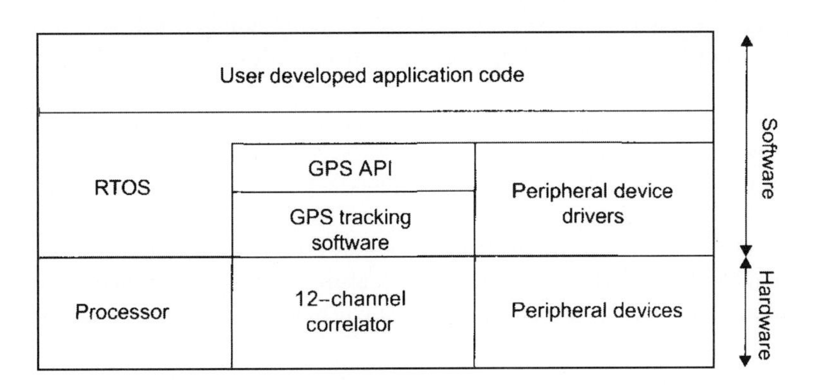

Fig. 14.1-8 GPS-enabled application software architecture.

first time after leaving the factory, or in other situations where the device has not been powered up or used for long periods of time.

Many GPS receivers allow the user to enter time, date and even current position information, which can reduce the TTFF in a cold start situation down close to that of a warm start.

14.1.4.3.3 Reacquisition time

Reacquisition time is the amount of time required by a GPS receiver to regain a position fix when the satellite signal is temporarily disrupted due to a loss of visibility of one or more satellites. This condition can occur when the receiver is operating in areas of dense foliage, or in urban canyons, or anywhere that the satellite views may be intermittently blocked. Most GPS receivers should have reacquisition times of five seconds or less. This is an important parameter for assessing the capability of GPS receivers for intelligent vehicle applications, since navigation systems must routinely operate in locations, such as downtown areas, where reception can be intermittent.

14.1.4.3.4 Receiver sensitivity

The sensitivity to satellite transmissions is another measure of performance of a GPS receiver. This is basically an assessment of how many satellites a receiver can detect under varying conditions. Because operating conditions for GPS receivers in intelligent vehicle applications can range from high elevations with an unobstructed view of the sky to locations inside or between buildings where reception can be more difficult, it is important for the application developer to understand under what conditions the GPS receiver can detect the SV signals, and under what conditions alternate methods of positioning must be relied upon.

14.1.5 Applications for GPS technology

There are a variety of uses for GPS technology today, from basic positioning applications which might provide a traveller with their current location, speed, and direction to their destination, to highly complex applications where the user's position information is feed into a system that provides location-specific features and services tailored for that user. What follows is some examples of how GPS technology is being used to enhance the capabilities of intelligent vehicle platforms. The initial examples illustrate some of the more traditional positioning applications, such as basic location and autonomous navigation systems, which are already seeing widespread use today. This is followed by examples describing how GPS-derived positioning information is being used to provide location-based services in vehicles today, and how the richness and complexity of those services will increase in the near future.

14.1.5.1 Basic positioning applications

The most basic applications for the positioning capabilities provided by the GPS are those providing the user/operator with information regarding their current location, where they have been, and more recently, where they are going. These systems include today's wide assortment of handheld devices, which provide the user with their current location, speed and elevation, as well as track their most recent movements. More complex devices integrate navigation capabilities, which map the user's position information onto a map database, and use that information to provide the user with text-based or graphical directions to their destination. Each type of system is described in more detail in the following sections.

14.1.5.1.1 Handheld GPS system

A handheld GPS device, while not strictly an intelligent vehicle system, provides a good example of an application utilizing the most basic positioning services provided by GPS technology. These units range from less than one hundred dollars to a few thousand dollars, depending upon their features and capabilities. These devices are battery operated, have a small LCD display and basic, menu-driven user interface. Most of the newer models can detect at least 6–8 satellites concurrently, with many now offering the ability to track up to 12 satellites at once. The more complex units also provide a variety of configuration options, such as the choice of localized datums or different information display formats. Because these devices are small enough to be easily carried in a briefcase or purse, users can take them as a positioning aid when travelling, whether on foot, or by private or public transport. The basic functional blocks of a handheld GPS receiver are illustrated in Fig. 14.1-9.

All of these devices provide the user with their current location information, usually in the form of a latitude and longitude reading. Most also provide the user with additional positioning information, including the current local time, elevation above sea level, and velocity, provided enough satellites can be detected by the device. Many of today's handheld devices also provide some tracking services, enabling the user to store the location of points they have previously reached, allowing them to return easily to their starting location. This can be a very useful feature to those travelling in unfamiliar areas, whether in a wilderness area or in an unfamiliar town or city.

Some of today's newer handheld systems now have larger displays and removable memory devices that enable the unit to plot graphically the user's current

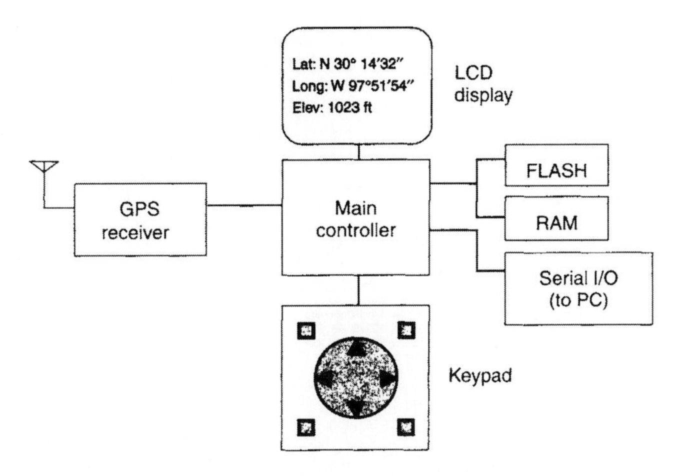

Fig. 14.1-9 Block diagram of basic handheld GPS device.

location onto a map of the local area. These devices may also provide the ability to enter in destination information, so the user can more easily understand where they are in relation to where they want to be. However, most of these devices fall short of being true navigation systems, since they do not provide any assistance to the user in reaching their destination. Instead, they simply give the user a more complete picture of their current location.

14.1.5.1.2 Autonomous navigation systems

True navigation, which provides the user with detailed instructions on how to reach a specific destination, is one of the fastest-growing areas in intelligent vehicle technology. Navigation devices utilize map-matching and best-path algorithms, along with user-defined filtering, to allow the user to choose between the fastest or most direct route to a desired destination. Some systems even allow the user to indicate specific routes to be avoided. The map databases used all provide basic mapping information (streets, major landmarks, etc.), but can also include points of interest and/or helpful location information (restaurants, etc.), depending upon their level of detail and how often they are revised.

Autonomous navigation devices range from in-dash units that are small enough to fit into a 1-DIN slot, to multi-component systems with CD-ROM changers and large multi-plane colour displays. The price of these systems can vary from a few hundred dollars to several thousand dollars, depending upon the complexity and capabilities. These systems utilize position information from an integrated GPS receiver, along with map database information provided from a CD-ROM or memory cartridge, to determine the user's current geographical location on the map. The smaller, in-dash units typically have a limited ability to display the user's position graphically, instead indicating the current location using a text description, such as the current address or location relative to a near-by landmark. Systems supporting a larger display can graphically indicate the user's current position superimposed on a map of the surrounding area. Also, because these systems are typically mounted in the dashboard, displacing the existing vehicle entertainment system, many of them include entertainment functionality such as an AM and FM stereo tuner or audio CD player. The more advanced systems with direct interconnections into the vehicle may also include HVAC system controls or other vehicle-specific comfort and convenience controls, although this is usually limited to systems installed by the vehicle manufacturer or dealer. An illustration of the functional blocks of an autonomous navigation system with an integrated GPS receiver can be seen in Fig. 14.1-10.

To determine the appropriate travelling instructions with one of these devices, the user enters the desired destination using a menu-driven system via a hardware or software keypad, depending upon the system. Some systems also support voice-based destination entry using basic voice-recognition technology. While these voice-driven systems are becoming more sophisticated, much progress is still necessary to improve them to the point where non-technical users are satisfied with their accuracy and reliability. The methods in which the directions are communicated to the user also depend upon the complexity of the system. Navigation systems with limited displays may use simple graphics combined with text to indicate the directions in a turn-by-turn manner. Some systems may combine these graphical turn-by-turn instructions with spoken instructions, using text-to-speech technology. Systems with larger displays can indicate the current position and immediate directions on a map of the immediate area, as well as the desired destination, once it is within the boundaries of the current map being displayed.

The value of GPS technology to these systems is obvious. GPS provides the essential positioning elements of location and speed necessary to make dynamic navigation possible. However, the occasional difficulties in

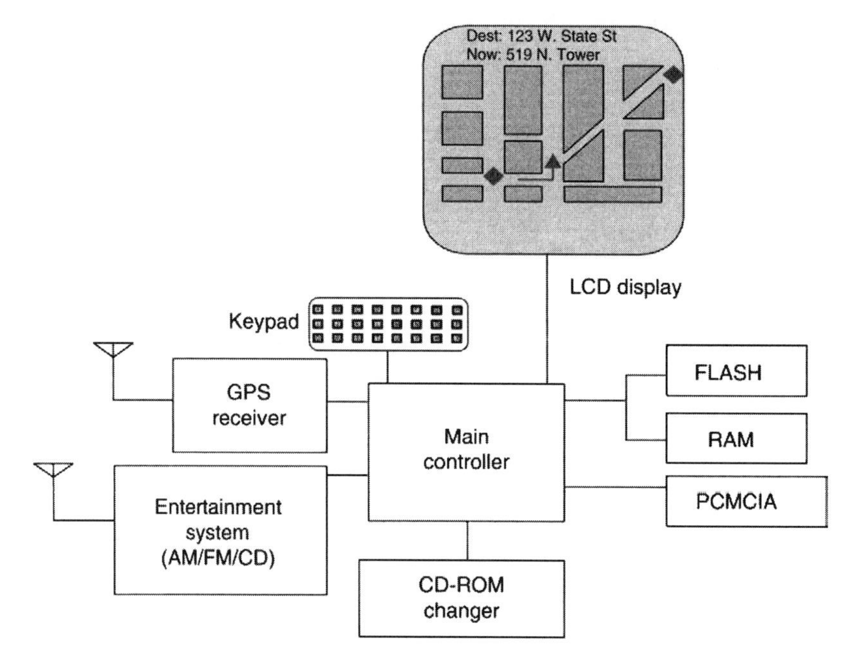

Fig. 14.1-10 Block diagram of basic GPS-enabled navigation system.

maintaining a GPS position lock, particularly in 'urban canyon' areas such as in the downtown districts of big cities, often require the use of additional techniques to maintain the accuracy of the user's location and movements between position locks. These DR techniques, described earlier in this chapter, include the use of internal gyroscopes or accelerometers to track the movement of the vehicle between the times that a solid position fix can be obtained by the GPS receiver. Vehicle speed and direction information, often obtained directly from the vehicle's internal communication network, can also be used to enhanced DR capabilities, although the capabilities for the input and processing of this type of data arc typically only found in systems installed by the vehicle manufacturers and dealers.

14.1.5.2 Location-based services

The delivery of location-based services is really the next phase in the evolution of intelligent vehicle systems. These services, which use GPS technology to pinpoint the user's current position, can then use that information to provide location-specific services to the user, such as relevant points of interest, or the nearest locations where a desired service or product may be available.

14.1.5.2.1 Current location-based services

The most common types of location-based services available in intelligent vehicles today are emergency and concierge/assist services. These services are accessed using a system combining GPS and wireless communication technology with a very basic user interface. This provides the vehicle operator with an on-demand wireless voice link to a call centre staffed 24 hours a day. At the time the wireless connection is initiated, the co-ordinates of the vehicle are transmitted to the call centre, indicating the exact position of the vehicle. This allows the call centre to provide timely and appropriate services relevant to the location of the customer. These services are available today in multiple vehicle models from several manufacturers, and will likely become standard features in the near future on many vehicle lines.

The most common emergency services being offered today include the notification of emergency response personnel in the case of an accident, and the notification of automotive service personnel in the case of a vehicle malfunction. When one of these events occurs, the appropriate local authorities are vectored to the exact position of the vehicle by the call centre, using the uploaded GPS positioning data to pinpoint the location of the vehicle. In some systems, the contact with the call centre can be made automatically if the system detects that an incident has occurred, upon the deployment of an airbag, for example. Other systems rely on a vehicle occupant to initiate the contact, even in the case of an accident. In the case of a vehicle malfunction, most systems today require the vehicle operator to initiate the call to the service centre.

The other class of location-based services currently being offered which rely on this combination of GPS and communication technologies are concierge/assist services. Examples of the services available include: getting directions to a desired destination ('Help, I'm lost, I need to get to . . .') , getting recommendations on a local point of interest ('We are hungry and don't know the area, can you tell us where a nearby restaurant is?'), and remote vehicle services, such as the remote unlocking of the car doors, or recovering a vehicle which has been stolen. Some providers even offer such highly personalized services as helping their customers purchase tickets for local events like plays or concerts.

All of the above services are available today in one form or another. In the future, the providers of location-based services will take advantage of the data capabilities of newer communication technologies to greatly expand their services. This will result in more advanced vehicle and user services, examples of which are illustrated below.

14.1.5.2.2 Future location-based services

In the near future, location-based services available in the vehicle will begin expanding beyond operator-assisted services to include wireless data-oriented services, as well. The wireless communication technology to support these services is already available today due to the accelerating roll-out of digital cellular, which is already in use in Europe and Japan, and is growing rapidly in the US. The current digital standards are still somewhat limited in their ability to support data services, but the roll-out of the next generation of digital wireless communication technology will make data services much more widely available, and will provide significantly improved bandwidth for the delivery of digital content to the user. Digital communication, along with more advanced positioning technology, will enable a wide variety of advanced location-based services for deployment in intelligent vehicles. Examples of these more advanced services include:

- **Server-based navigation systems**, which allow the user to dynamically download the most up-to-date, and therefore accurate, map information of the area in which they are travelling. This will help to ensure that the user gets the most current directions and points of interest possible without having to maintain a subscription with a map database provider. It will also reduce the cost of the navigation system itself, since a costly memory storage subsystem would not be required to access the map database.
- **Dynamic traffic routing and management services**, which will enable navigation systems to take into account current traffic conditions when calculating directions. This will allow travellers to easily route around congested areas, getting them to their destination faster and helping to prevent additional congestion. One approach to this is for the travellers' systems to periodically provide their position information to a central server, which uses the positioning deltas to map traffic flow. This data can then be fed back to each system to provide real-time traffic movement updates.
- **Location-based marketing services**, which can be used by local providers of goods and services to target advertising to travellers who are entering their local service area. For example, if a traveller is looking for a restaurant near their current location, a request including their position information and food types of interest could be submitted, which would return directions to nearby restaurants of the desired type. A restaurant could even include with the directions a coupon for a meal specially to encourage the traveller to visit that establishment.

These are just a few examples of the types of location-based services that will become available in intelligent vehicles as these technologies mature. Of course, the development of the technology to support these advanced location-based services also raises a variety of privacy issues. While the positioning information of individuals can be very valuable to merchants with goods or services to sell, many individuals may consider this information to be very personal and private, and wish to limit its distribution. Therefore the protection and methods of distribution of this information will very likely be the subject of intense debate between merchants and privacy advocates, and may ultimately result in legislation regarding how and to whom that information is disseminated.

14.1.6 Conclusion

The GPS was hailed as a technological success soon after it became fully operational in 1995. With the continual improvements the system is undergoing, and in particular with the discontinuation of the S/A feature in May 2000, many more commercial, military and space applications will be able to derive benefit from this system's services and capabilities in the future. Intelligent vehicle applications will be one of the biggest beneficiaries of these improvements in service.

This chapter has presented an overview of the GPS, including the history of satellite-based positioning, the basic system architecture, the science and mathematics used for determining location, an overview of the components and solutions which are available for use in GPS-enabled applications, and some examples of current and future applications which will utilize the positioning services made possible by the GPS. It is hoped that the reader has gained a basic understanding of the system architecture and requirements, and the impact on applications that require the use of GPS services.

For more detailed information about the science and technology behind the GPS, please refer to the publications listed under further reading.

Further reading

Kaplan, E. (ed.) (1996). *Understanding GPS: Principles and Applications*. Norwood: MA: Artech House.

Parkinson, B. and Spilker, J. (eds) (1996). Global Positioning System: Theory and Applications Volume I. *Progress in Astronautics and Aeronautics*, Vol. 163. Washington DC: American Institute of Astronautics and Aeronautics, Inc.

Farrell, J. and Barth, M. (1999). *The Global Positioning System & Inertial Navigation*. New York, NY, McGraw-Hill.

Enge, P. and Misra, P. (eds) (1999). Special Issue on Global Positioning. *Proceedings of the IEEE*, **87**, No. 1.

Anonymous (1995). *Global Positioning System Standard Positioning Service Signal Specification*. US Department of Defense, 2nd edition.

Chapter 14.2

14.2

Decisional architecture

Ljubo Vlacic and M. Parent

14.2.1 Introduction

Autonomy in general and motion autonomy in particular has been a long standing issue in robotics. In the late 1960s and early, 1970s, Shakey (Nilsson, 1984) was one of the first robots able to move and perform simple tasks autonomously. Ever since, many authors have proposed control architectures to endow robot systems with various autonomous capabilities. Some of these architectures are reviewed in Section 14.2.2 and compared to the one presented in Section 14.2.3. These approaches differ in several ways, however it is clear that the control structure of an autonomous robot placed in a dynamic and partially known environment must have both *deliberative* and *reactive* capabilities. In other words, the robot should be able to decide which actions to carry out according to its goal and current situation; it should also be able to take into account events (expected or not) in a timely manner.

The control architecture presented in this chapter aims at meeting these two requirements. It is designed to endow a car-like vehicle moving on the road network with motion autonomy. It was initially developed within the framework of the French Praxitèle programme aimed at the development of a new urban transportation system based on a fleet of electric vehicles with autonomous motion capabilities (Daviet and Parent, 1996); more recent work on this topic is done within the framework of the LaRA (Automated Road) French project. The road network is a complex environment; it is partially known and highly dynamic with moving obstacles (other vehicles, pedestrians, etc.) whose future behaviour is not known in advance. However the road network is a structured environment with motion rules (the highway code) and it is possible to take advantage of these features in order to design a control architecture that is efficient, robust and flexible.

This chapter is organized as follows: in Section 14.2.2, an overview of the existing approaches to implementing a control architecture is presented and discussed; this section describes the three main classes of existing approaches (deliberative, reactive and hybrid), along with their main implementation alternatives and their respective advantages and drawbacks.

Section 14.2.3 describes the hybrid control architecture developed by Sharp. The rationale of the architecture and its main features are overviewed in Section 14.2.3.1. This section introduces the key concept of the *sensor-based manoeuvre,* i.e. general templates that encode the knowledge of how a specific motion task is to be performed. The models of the experimental car-like vehicles that are used throughout the chapter are then described in Section 14.2.3.2. Afterwards the concept of the *sensor-based manoeuvre* is explored in Section 14.2.3.3 and three types of manoeuvres are presented in detail (Sections 14.2.3.4, 14.2.3.5 and 14.2.3.6). These manoeuvres have been implemented and successfully tested on our experimental vehicles; the results of these experiments are finally presented in the Section 14.2.4.

One important component of the architecture is the *motion planner* whose purpose is to determine the trajectory leading the vehicle to its goal. Motion planning for car-like vehicles in dynamic environments remains an open problem and a practical solution to this intricate problem is presented in Section 14.2.5.

Intelligent Vehicle Technologies; ISBN: 9780750650939
Copyright © 2001 Ljubo Vlacic and M. Parent. All rights of reproduction, in any form, reserved.

14.2.2 Robot control architectures and motion autonomy

14.2.2.1 Definitions and taxonomy

The development of robot control architectures constitutes for engineers and scientists one of the most challenging frameworks for integrating and testing *intelligent* systems, inspired from attributes of living beings such as perception, interaction and reasoning. Robot control architectures are rather understood in terms of software architectures, and consequently are closer to domains related to computer science and control engineering. A basic definition for a robot control architecture can be found in Arkin (1998): 'Robotic architecture is the discipline devoted to the design of highly specific and individual robots from a collection of common software building blocks.'

The state of the art in this domain includes a large number of approaches, sometimes guided by research work on ethology and cognitive sciences. One of the most challenging domains for testing and evaluating these approaches, particularly when real-time constraints have to be verified, is mobile robotics. This is why most of the significant contributions in this research field come from work on mobile robot and autonomous guided vehicles. The next sections outline the state of the art in mobile robot architectures, using a commonly agreed taxonomy. This taxonomy is based on three main paradigms on which a large number of control architectures have been developed:

- *The deliberative paradigm.* In this approach, the system uses a model of the world – an *a priori* known model, or a model reconstructed from sensory data – in order to plan the actions that the robots have to execute. This approach leads to a sequential decomposition of the whole process, and to highly hierarchical systems.
- *The reactive paradigm.* This approach is based on a tight coupling between sensors and actuators, for continuously producing the required controls. This approach usually relies on a decomposition of the system into elementary behaviours which can be combined and executed concurrently.
- *The hybrid paradigm.* This approach consists in combining the deliberative and reactive paradigms, in order to try to exploit the advantages of the two previous approaches. Most of the current approaches are of this type.

14.2.2.2 Deliberative architectures

This approach relies on traditional paradigms of artificial intelligence. It tries to implement a simplified view of human reasoning, and it is often referred as the 'sense-model-plan-act' (SMPA) scheme. In practice, this concept is implemented into robotic control systems using a hierarchical architecture made of three main components: perception (which includes sensing and modeling functions), decision and action.

- *Perception.* Considered as a key feature in a robotic system, the *perception function* may be seen as the first basic function of a deliberative architecture. The main purpose of this first stage of the control process is to construct a model of the environment from sensory data and *a priori* knowledge (e.g. topological or a grid-based models). This model is subsequently used for planning robot actions and for checking that the robot actions have correctly been executed. However, world reconstruction from sensory data is a complete active research domain, having already motivated a great number of research works and approaches; this research domain is still open.
- *Decision.* The second processing phase of a deliberative architecture is referred as the *decision* module. It consists in 'reasoning' about the task model and the environment model, in order to decide what is the more appropriate sequence of actions to execute. In practice, this reasoning phase is often implemented as an *off-line motion planning* task. This is why motion planning has been a very active research domain for about 20 years.
- *Action.* The last processing phase of a deliberative architecture is to control the robot actuators in order to execute the planned actions. Recent research work in this domain addresses robust control techniques and sensor-based control approaches.

The first robot control architectures reported in the literature are based on such an approach. In particular, this type of architecture has been used for controlling the first mobile robot having a partial autonomy: the robot Shakey (Nilsson, 1984). This robot, designed at the beginning of the 1970s at the Stanford Research Institute, used a video-camera as a sensor and was theoretically able to move in a highly constrained environment. Its reasoning capabilities were derived from problem-solving techniques developed in the field of artificial intelligence. The typical tasks that could be achieved by Shakey consisted in finding a known object (i.e. an object described by its shape and its colour) in a room, and in pushing this object up to a given point. Unfortunately, each simple movement of Shakey required more than one hour of external computing, and it had a strong probability of failure at execution time.

The architecture of the Stanford Cart developed at the University of Carnegie-Mellon (Moravec, 1983) is also representative of the deliberative paradigm. In this approach a 3D-vision system provided the robot with the positions of the objects located in its environment. Then, the motion planner generated a collision-free path allowing the robot to reach the desired placement. Finally, the system controlled the robot actuators in order to move it along the planned path, for a distance of about 1 metre. Unfortunately, the complete SMPA process had to be carried out after each motion of this type until the goal had been reached, and each iteration required to wait for about 15 minutes (mainly because of the image processing time).

The main serious problems related to purely deliberative architectures are the following:

- The main drawback of this type of approach relies in its intrinsic incapacity to cope with unpredicted events (mainly because of the large reaction time which is required for processing the whole (SMPA) cycle). Consequently, it is almost impossible to take into account dynamic objects or obstacles detected while the robot is moving. The main reason for these limitations comes from the slowness of the modelling and planning phases, which cannot be carried out in real time, even when they are implemented on large computers external to the robot.
- The second difficulty is related to the modelling phase itself, which is in charge of reconstructing a model of the robot environment from sensory data. Indeed, this problem in its whole generality represents a complete research domain which is still open (even if impressive results have already been obtained by researchers in the held of computer vision). It is well known, that relating sensory data to real objects is a difficult task because of the noisy, inaccurate and often spread nature of the information to process. This difficulty is related to the fact that there is a great difference between *sensing* and *perception*.
- The intrinsic differences which exist between the model and the real world introduce strong uncertainties on the positions/orientations of the robot and of the obstacles. Taking into account these uncertainties is obviously necessary for obtaining a robust system. Unfortunately, this requirement makes the planning phase much more complex, and poses several modelling and algorithmic problems which are still open.

Consequently, the use of such an approach seems to be limited to the case of a robot evolving in a static, strongly constrained and *a priori* known environment. This is why the purely deliberative approach is not used anymore in recent robot control architectures (even if it has raised key research issues that are still actively studied). However, it should be stressed that the deliberative approach has a significant advantage: it includes the property to apply high-level reasoning capabilities at the planning phase, and consequently to be potentially able to cope with complex missions combining several goals and task constraints. As it will be shown in Section 14.2.2.3, this property does not hold when applying purely reactive approaches.

14.2.2.3 Reactive architectures

14.2.2.3.1 The basic idea

Because of the above-mentioned strong limitations of the purely deliberative architectures, some researchers have developed in the 1980s a new approach inspired by ethology and entomology. Nowadays, we know that many live beings, in particular the insects, have very few capacities for 'modelling' and 'reasoning'. Despite this, they are able to achieve quite complex tasks; to an external observer, they exhibit a global behaviour which seems to be the result of intelligence. On the basis of this observation, some researchers have proposed an alternative to deliberative architectures consisting in making use of purely 'reactive behaviours'. This approach basically consists in removing the 'modelling' and 'planning' phases from the decisional loop, and in trying to produce 'intelligent behaviours' driven by sensory data: the robot *reacts intelligently* to what it *senses*. Brooks (1990) justifies the use of such an approach, by claiming that the 'best model of the world is the world itself'.

The implementation of such an approach is based upon the combination of several elementary modules implementing very simple (reactive) behaviours. Such approaches are often referred as *behavioural based architectures:* the observed behaviour of the robot is the result of the combination of some various elementary behaviours; it *emerges* from the interaction of the involved elementary behaviours with themselves and with the environment. Each elementary behaviour (e.g. avoiding an obstacle or the heading to a goal) performs a close coupling between the sensors and the effectors of the robot. The intrinsic low complexity of the involved processing, along with the parallel structure of the behaviours, leads to a high speed execution property.

Note: The previous type of system is usually referred as a 'reactive control system'. However, one can find two different definitions of reactivity in the literature: for peoples in the fields of computer science and robotics, a reactive system is 'a system able to react continuously to a physical environment, at a speed determined by this environment' (Harel and Pnueli, 1985); for peoples in the field of cognitive sciences, an agent is said to be reactive 'if it does not have an explicit representation of the world' (Ferber, 1995). Even if they look different, these two definitions are not contradictory, since they imply that the involved controllers react directly to the stimuli coming from the physical world; hence avoiding the construction of a 'non-correct' model of the world, and enabling high execution speeds. The next subsections illustrate some typical implementations of this type of architecture.

14.2.2.3.2 Subsumption architecture

R. Brooks is one of the pioneers in the domain of reactive architectures for robots, and his work on this topic is certainly the most known in our scientific community (Brooks, 1990). His behavioural architecture is based on a vertical decomposition into several *levels of competence* (see Fig. 14.2-1). Each of these levels represents an independent behaviour, receiving data from sensors and acting on the robot actuators. The first implementation of this architecture has been realized using augmented finite state machine (AFSM) models, a communication mechanism based on messages, and an inhibition/suppression mechanism for modifying dynamically the stimulus or response signals of some active modules.

It should be noticed that such an architecture does not have any central decisional kernel for selecting the required behaviours: this choice is continuously done at execution time using the structure of the implemented layers and communication network.

Eight levels of competence have been defined by Brooks as a guide for his work (see Brooks (1990) for more details): the lower levels of competence consist in (1) 'avoiding obstacles', (2) 'wandering aimlessly around' and (3) 'exploring the world'; the higher levels consist in (7) 'formulating and executing plans', and (8) 'reasoning about the behaviour of objects and modifying plans accordingly'. This approach has been used to develop several small autonomous robots at MIT. However, only the first three levels of competence have been implemented.

Such an architecture exhibit some interesting properties underlined by Brooks: (1) *a short response time* which provides the robot with the capacity to move in a dynamic environment, (2) *a robust controller* which can potentially work even if one of the modules does not work correctly (this is due to the parallel structure and the relative independence of the behaviours), (3) *an incremental structure* which potentially allows the implementation of new higher levels of competence and (4) *a simple implementation* which drastically reduces the production cost and makes miniaturization possible (see for instance the proposal of Brooks for exploration robots (Brooks and Flynn, 1989)).

However, the experience has proven that such an architecture can hardly be fully implemented (except the lower levels of competence), and consequently it exhibits strong limitations coming from its lack of high-level reasoning. Indeed, even if this capacity is theoretically present in the highest levels of competence, these levels have never been implemented and the feasibility still remains doubtful. Consequently, only simple behaviours can be implemented in the current state of the art.

14.2.2.3.3 Other reactive architectures

The supposed potentialities of the behavioural approach have motivated various developments and theories.

As has been previously mentioned, the subsumption architecture uses a hierarchy of behaviours, in which only one module at a time 'controls' the robot (by possibly integrating the results of some other behaviours located lower in the hierarchy). Another possible approach, as the one proposed by Anderson and Donath (1990), consists in combining the controls simultaneously 'recommended' by several behaviours. In such an approach, each reactive module implements a simple 'reflex behaviour' having no memory capacity. Such a type of behaviour has been defined by McFarland (1987): 'a reflex behaviour is the simplest form of reaction to an external stimulation; stimuli such as a sudden change in the level of illumination or a contact arising at a given point of the body, generate an automatic, involuntary and stereotyped response.' In the approach of Anderson and Donath, the involved modules generate artificial potential fields whose composition is used to determine the motion direction to be followed by the robot (by using the classical gradient technique). The set of the behaviours to be considered at a given time is selected according to criteria taking into the account the

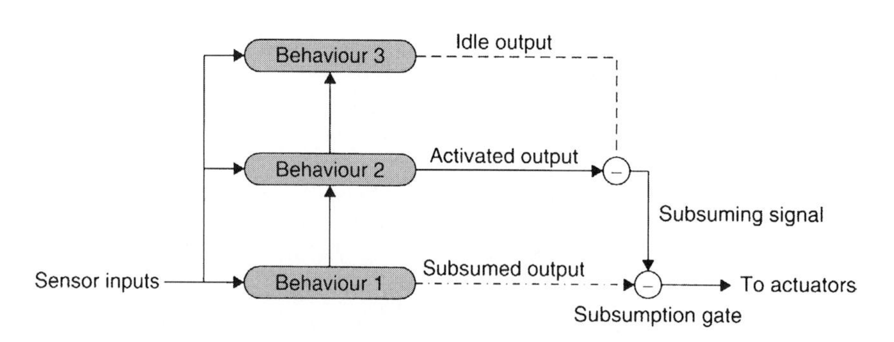

Fig. 14.2-1 Principle of the subsumption architecture (Brooks).

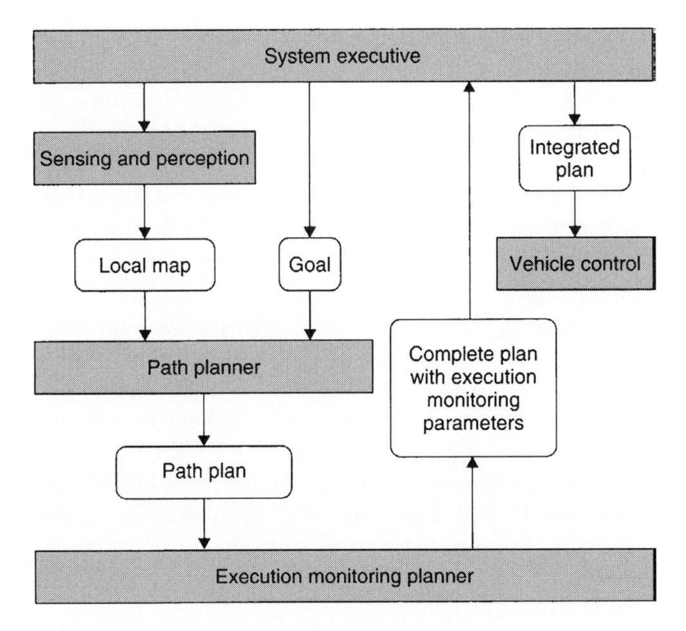

Fig. 14.2-2 JPL exploratory robot architecture (Gat).

task to be achieved and the characteristics of the environment. In a similar way, Rosenblatt and Payton (1989) have proposed to use an artificial neural network to combine the controls recommended by the selected behaviours; however, their system has been developed using an empirical approach, and the way the controls have to be appropriately combined is rather difficult to define.

In order to overcome the previous difficulty, some authors have tried to use training techniques for realizing the selection mechanism. For instance, Humphrys (1995) has proposed to apply a learning phase to each individual behavioural module, in order to construct a function for evaluating the 'quality' of the proposed actions; then this function is used on-line to select the best ranked behaviours. Lin (1993) developed a similar approach by using Q-learning techniques for implementing the training phase.

Although some improvements have been proposed for the behaviours selection and composition mechanisms, all these approaches exhibit the same general limitations as the subsumption architecture (see Section 14.2.2.2).

14.2.2.4 Hybrid architectures

14.2.2.4.1 How to hybridize?

The purely reactive and purely deliberative approaches represent two extremes that many authors naturally tried to combine. The objective is to try to preserve the potential high-level reasoning capacity of the deliberative approaches, while ensuring the robustness and the short response time of reactive approaches. Such approaches are referred to as *hybrid architectures*. However, several types of hybrid architecture can be distinguished depending on the way the deliberative and reactive components have been combined. A commonly used classification (with some minor variations) in the literature consists in considering three main types of approaches:

- *Deliberative-based hybrid approaches.* In such approaches, the *planning function* has a predominant role (i.e. motion is driven by planning using the SMPA paradigm), and the *reactive functions* are only added for dealing with some exceptions. The related elementary reactive actions can either be integrated as part of the architecture, or inserted into the generated motion plans by the planner.
- *Reactive-based hybrid approaches.* In this type of approach, the robot motions are executed under the supervision of the reactive functions, and the planning functions are mainly used as a 'guidance resource' for the reactive component of the system.
- *Three-layered hybrid approaches.* Most of the current developments on robot control architectures are based on this type of approach. Such an approach may be seen as the implementation of an 'adaptable planning-reacting scheme'. The basic idea consists in adding an intermediate layer for appropriately interfacing the deliberative and reactive functions. Then, one can consider that this type of architecture relies on the three following basic logical functions: *planning, sequencing* and *reacting.* In such a paradigm, the behaviours of the reactive layer are conditionally instantiated by the sequencing module, according to some sensing conditions and constraints defined by the planning module.

14.2.2.4.2 Deliberative-based hybrid architectures

JPL exploratory robot architecture (Gat) This architecture has been proposed by Gat *et al.* (1990) at Jet Propulsion Laboratory (JPL), for providing a planet exploratory mobile robot with some partial autonomy capabilities (since such a robot cannot be directly teleoperated from the earth because of communication delays). As shown in Fig. 14.2-2, this architecture is based on four main modules: the Perception Module, the Path Planner, the Execution Monitoring Planner and the System Executive.

The main task of the Perception Module is to build a local map of the environment, using sensory data and global data provided by the orbiter. Then, this map is used by the Path Planner and by the Execution Monitoring Planner to respectively plan a collision-free path (of roughly 10 metres) and a complete plan including execution monitoring parameters. This motion plan is obtained by simulating the displacement of the robot along the planned collision-free path, and by anticipating the possible failures and the sensory data to monitor. The parameterized motion plan is finally used by the System Executive to monitor the execution of the robot task. In practice, only some predefined and simple 'Reflex Actions' can be introduced into the motion plan (e.g. stopping the robot and moving back to a safe position). Such an architecture basically applies the sequential SMPA paradigm, while executing some predefined reflex manoeuvres when dangerous situations have been detected. This approach represents to some extent the minimum level of integration of a reactive component into a deliberative architecture.

Payton's architecture This architecture (Payton, 1986) is based on a hierarchical decomposition, in which each layer is characterized by a type of sensory data processing (modelling). As shown in Fig. 14.2-3, this architecture is composed of a layered perception system and of four main decisional modules: (1) the 'Mission planner' defines a sequence of geographical goals to reach along with their associated motion constraints; (2) the 'Map-based planner' uses the global world model to generate paths connecting the previous geographical goals (the response time of this planner is of a few minutes); (3) the 'Local planner' determines the details of the motions which are required for moving the robot along the planned paths (the response time of this planner is of a few seconds); (4) the 'Reflexive planner' controls in real time the execution of the motion task.

From the implementation point of view, this architecture has been developed using expert agents communicating between them using a blackboard technique. Using this approach, the activity of a particular module can theoretically be controlled by a higher layer through the selection of the expert agents to be activated.

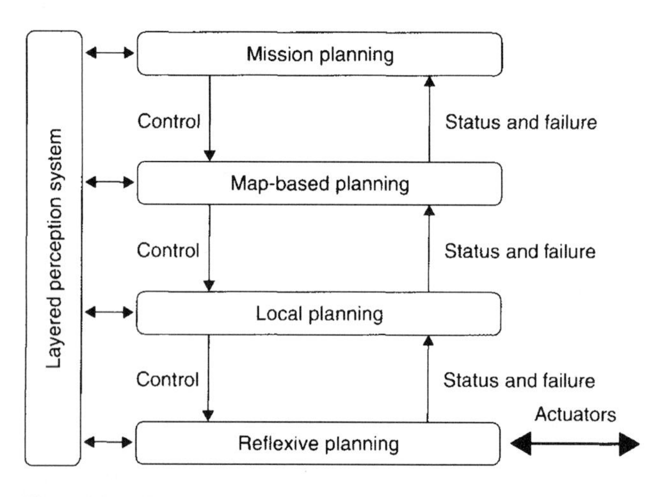

Fig. 14.2-3 Payton's architecture.

However, only the implementation of some expert agents belonging to the 'Reflexive planner' (e.g. follow a wall, or avoid an obstacle) has been deeply described. In this implementation, the related reflex behaviours are associated with some virtual sensors in charge of providing a specialized information (e.g. obstacle detection, object recognition, localization). Then, the activation of the appropriate reflex behaviours is done using a blackboard technique and some predefined priorities.

Quite complex missions have been planned and executed with a significant level of reactivity using this approach. The main limitations of the system come from both the limited communication mechanism existing between the different layers, and the predefined combination of behaviours. Later on, Payton *et al.* (1990) and Rosenblatt (1997) have improved the reactivity of the system by using a distributed control arbitration technique, allowing them to combine controls coming from both the reactive behaviours and the planning layers.

Task control architecture (TCA; Simmons) The TCA proposed by Simmons (1994) represents a new alternative to traditional hierarchical approaches. This architecture is composed of an arbitrary number of specialized modules, communicating through messages with a central management module. The specialized modules carry out the tasks which are specific to the robot to control, whereas the central management module supervises the functioning of the whole system and controls the routeing of the messages between the various modules; messages can be used for an information request, for sending a command, or for asking for a task decomposition to the planners. The TCA architecture makes use of a hierarchical representation of the tasks/sub-tasks relationships (called the 'task tree') for maintaining an internal representation of the robot task to execute.

Fig. 14.2-4 shows how the TCA architecture has been implemented for controlling the Ambler legged robot. However, this implementation put the emphasis onto the planning functions (gait planner, footfall planner, etc.), and reduces the reactivity to the processing of some exceptions (for stabilizing the robot). The main drawback of this architecture relies on the centralized processing schema and its associated communication mechanism, which often implies rather long response times incompatible with fast robots. This is why the author has also implemented additional (i.e. apart from the TCA architecture) some 'emergency reflexes' for quickly stabilizing the Ambler robot when a problem arose.

14.2.2.4.3 Reactive-based hybrid architectures

AuRA architecture (Arkin) The AuRA architecture proposed by Arkin (1987; 1989; 1990) is mainly based on the concepts of 'motor schema' and 'perceptive schema', which are used for describing the links existing between

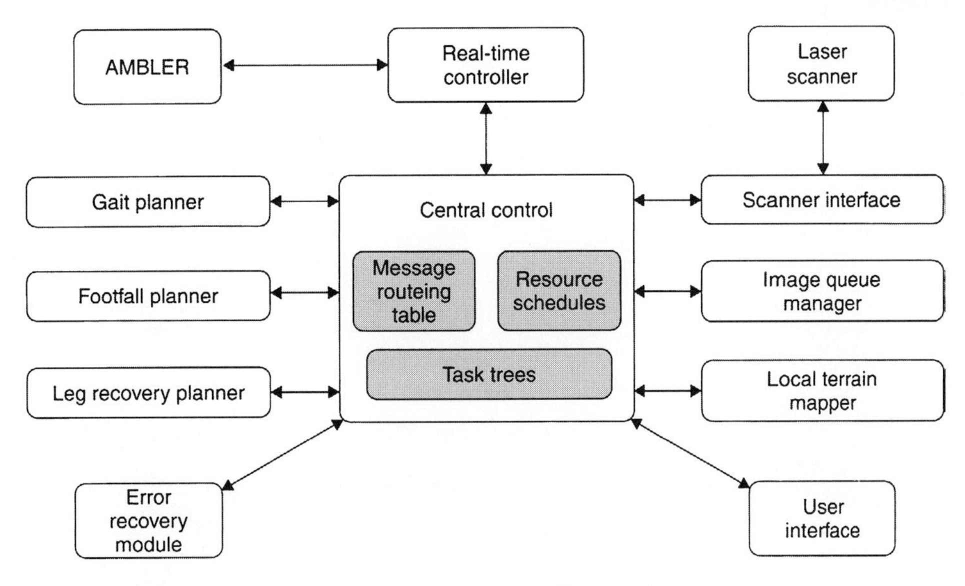

Fig. 14.2-4 TCA architecture implemented for the AMBLER legged robot (Simmons).

action and perception (like in reactive approaches). A motor schema specifies a generic behaviour which can be instantiated under some conditions, for generating a particular type of robot motion (e.g. moving along a straight line, moving towards a goal position, or avoiding a given obstacle); each motor schema is associated to a perceptive schema (*action-oriented perception*) in charge of providing the required information.

The AuRA architecture is mainly composed of two components: a hierarchical component in charge of the modelling and planning tasks, and a reactive component inhabited by the motor and perceptual schemas (see Fig. 14.2-5). The hierarchical component is composed of three classical layers: the *Mission planner* which generates a sequence of sub-goals to achieve, the *spatial reasoner* (*or navigator*) which constructs executable paths using cartographic data stored in a long-term memory, and the *plan sequencer* (*or pilot*) which selects and instantiates the appropriate behaviours. The reactive component makes use of a vector field approach to combine the movements proposed by the activated motor schemas, and to generate the required controls, in practice, the deliberative part of the system mainly produces way-points and associated behaviours; it is reactivated only when a fatal failure has been detected (no more motion or timeout). This approach does not allow the processing of more complex missions combining several manoeuvres; it also suffers from the classical drawback of reactive approaches: the combination of behaviours generates motions which can hardly been predicted, and conflicts may appear when 'opposite' behaviours have to be considered.

Symbolic, Subsumption, Servo (SSS) architecture (Connell) The SSS architecture proposed by Connell (1992) is composed of the three layers (see Fig. 14.2-6), associated with three levels of discretization of the robot state space and of the time. The lower layer (Servo) operates using continuous space and time representations for controlling the robot and the sensing functions. The intermediate layer (Subsumption) works using a continuous time representation and a discrete state space model for generating specialized behaviours (e.g. wall following, or crossing a door). The higher layer (Symbolic) operates under discrete space and time representations for selecting the behaviours to apply according to

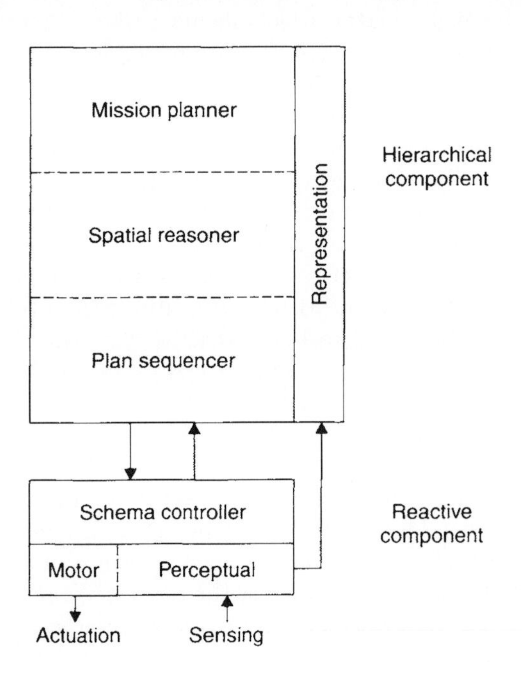

Fig. 14.2-5 AuRA architecture (Arkin).

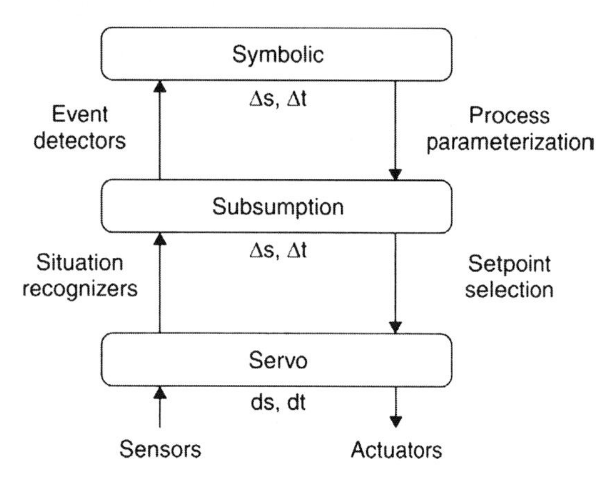

Fig. 14.2-6 SSS architecture (Connel).

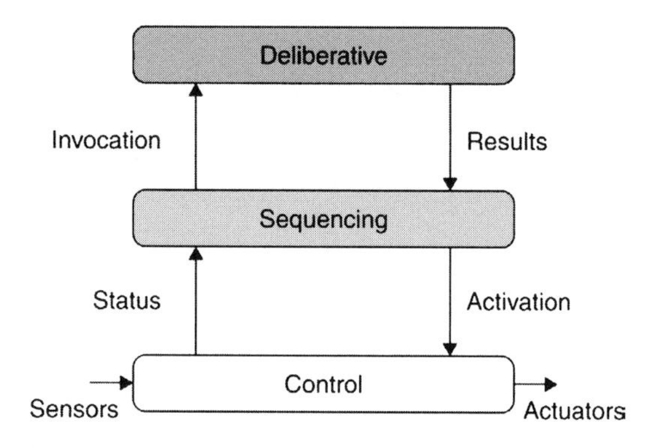

Fig. 14.2-7 ATLANTIS architecture (Gat).

the task to achieve and to the arising events; this layer makes use of a map of the environment containing 'landmarks' connected by paths (straight line segments).

In practice, this architecture mainly behaves as an improved subsumption architecture, having a more elaborated mechanism for selecting or inhibiting behaviours (the role of the symbolic layer is mainly to construct a contingency table indicating under which conditions the behaviours of the subsumption layer have to be activated).

14.2.2.4.4 Three-layered hybrid architectures

ATLANTIS (Gat) and 3T (Bonasso) architectures The ATLANTIS (Gat, 1992; 1997) and 3T (Bonasso *et al.*, 1996) architectures are both based on three main layers: (1) the higher layer which includes the deliberative functions (planners), (2) the intermediate layer whose purpose is to manage the various sequences of actions to execute and (3) the lower layer which includes the reactive control mechanisms. These three functional layers are generally represented in all the contemporary mobile robot control architectures. The intermediate layer, which can be seen as an 'advanced interface' between the deliberative and reactive components, has an important role to play for appropriately integrating high-level task representations and real-time reactive capabilities. The three previous layers are respectively called *Deliberator, Sequencer* and *Controller* by Gat (Fig. 14.2-7) and *Planner, Sequencer* and *Skill Manager* by Bonasso (Fig. 14.2-8).

In these architectures, the lower layer gathers together several functions implementing simple behaviours such as wall following or obstacle avoidance; it constitutes a library *of skills,* used at the request of the higher levels. The intermediate layer (the *Sequencer)* selects and parameterizes the set of behaviours to apply in the current state of the robot task; in order to authorize the concurrent execution of several alternative action plans (the appropriate solution being chosen at execution time according to some identified internal an external events), this layer makes use of the 'conditional sequencing' principle (Gat, 1997). Using this approach, the role of the higher layer is to produce action plans for 'guiding' the robot movements, rather than generating a single sequence of actions. In the 3T architecture, the action plans are previously sent to the *Sequencer,* whereas they are produced in response to the requests from the *Sequencer* in the ATLANTIS system.

The implementation of these three basic layers varies from one system to another. The way the *Sequencer* is implemented may obviously have strong consequences to the scope and the robustness of the whole system. In the 3T and ATLANTIS architectures, this layer has respectively been implemented using the Reactive Action Packages (RAP) and ESL languages. These two languages have similar characteristics, and they both rely on the same basic principle (Noreils, 1990): 'Rather than trying to build algorithms that never fail (which is impossible when dealing with real robots), it is better to build algorithms that never fail to detect a failure'; Gat (1997) calls such a type of failure, a '*cognizant failure*'. The RAP (Firby, 1989) language, basically allows the specification of a set of procedures (sequences of actions) which have to be activated when some predefined conditions are verified. A particular RAP specifies the different strategies which are known to achieve a given goal according to some contextual conditions; in this description, any strategy may in turn be specified using some other RAPs. Then, the appropriate RAPs are successively activated according to an agenda constructed and updated by an interpreter (Fig. 14.2-8).

HILARE architecture (LAAS) In the HILARE architecture from LAAS[1] (Alami *et al.* 1998), the three

[1] Laboratoire d'Analyse et d'Architecture des Systèmes, Toulouse.

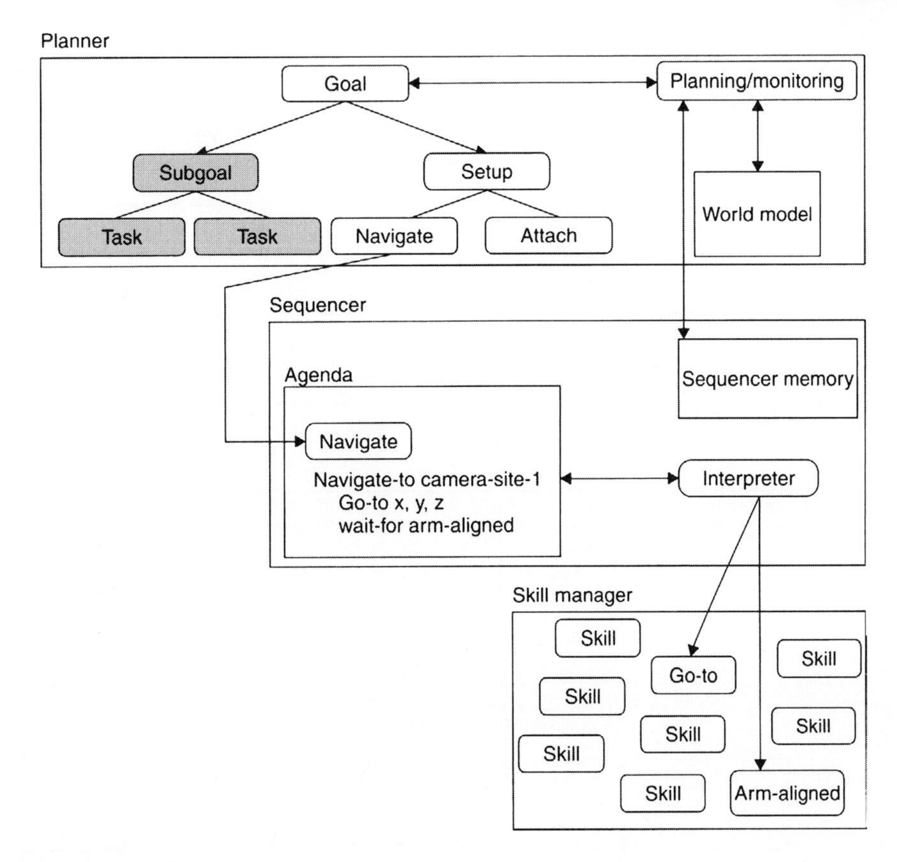

Fig. 14.2-8 3T architecture (Bonasso).

previous layers are roughly represented by the *Decisional, Execution Control* and *Functional layers* (Fig. 14.2-9); this architecture is completed by a *Logical robot layer* whose purpose is to interface the system with the physical resources (in order to increase the portability).

The *Decisional layer* is in charge of processing all the requests requiring a global knowledge of the task and of the execution context (i.e. planning and decision-making). In order to be compatible with the constraints of the reactive functions of the architecture, this layer has been divided into two components having different response times: the *Plan supervisor,* which generates appropriate action plans from a description of the tasks to achieve, and the *Task supervisor,* which is in charge of supervising the work of the intermediate layer and of producing (when it is needed) the task refinements. The action plans that are produced include the required modalities of execution (mainly events to monitor and their associated reflex actions).

The *Execution control layer* is in charge of executing in a reactive way the action plans produced by the *Decisional layer.* For that purpose, it continuously selects, parameterizes, and activates the appropriate modules of the *Functional layer.* This is done using an automation automatically generated from a set of logical rules.

The *Functional layer* is composed of all the functions which are required for processing the perception and robot actions (sensory data processing, events observers, motion controls, etc.). These functions have been encapsulated into modules communicating between themselves and with the *Execution control layer* using a client/server protocol. Then, a particular module is activated through a request (including the involved execution parameters) sent by the *Execution control layer* or by an other module; this module remains active until the task has been completed or a failure has been detected. From the implementation point of view, this layer is organized as a network of interacting modules.

Sharp architecture (INRIA) The Sharp architecture developed at INRIA[2] (Laugier *et al.*, 1998) exhibits the same basic functional structures as the previously described three-layered systems (i.e. the Deliberator, the Sequencer and the Controller). The main objective of this architecture is to take into account the particular constraints and properties of a car-like vehicle, in order to

[2] Institut National de Recherche en Informatique et Automatique.

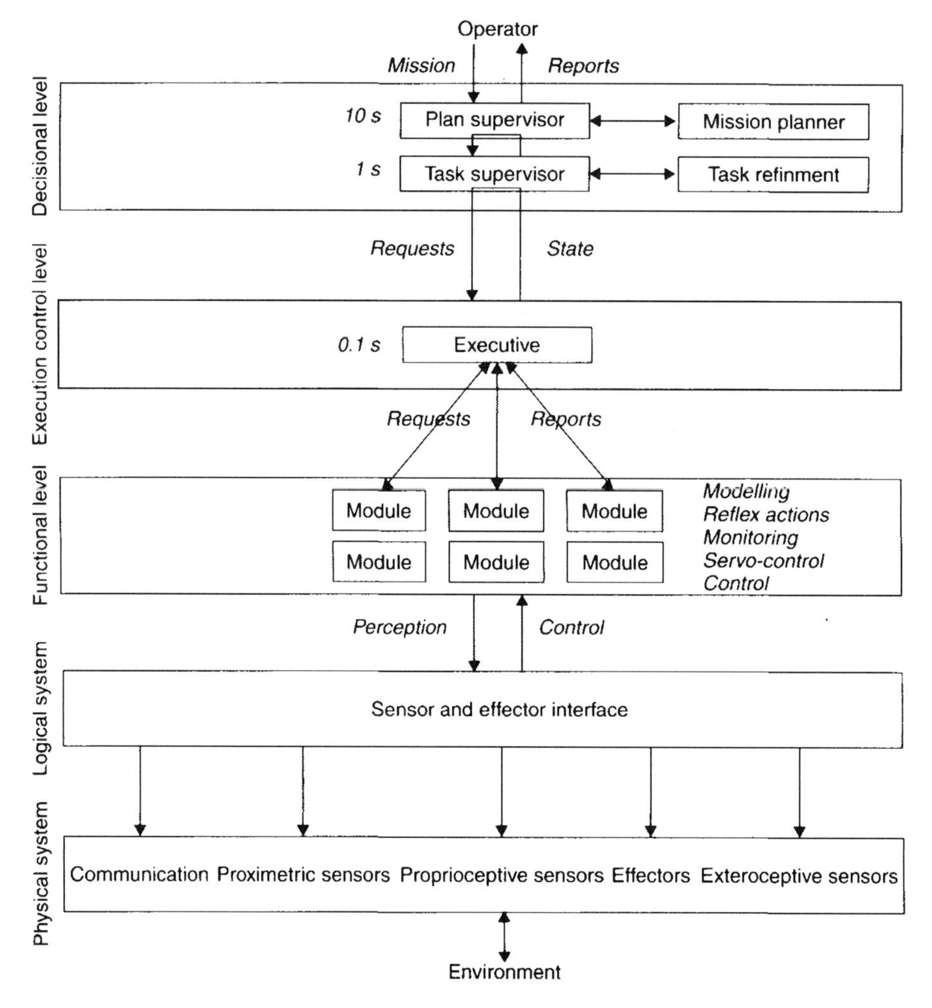

Fig. 14.2-9 HILARE architecture (LAAS).

be able to *efficiently and safely* control the motions of a car moving on the road network; another constraint is to obtain *smooth motions*, i.e. to appropriately control velocities and accelerations.

In the Sharp architecture, the three previous functional layers are respectively represented by the *Planner*, the *Mission Scheduler* and the *Motion Controller* (see Section 14.2.3.1 for more details). However, the efficiency and robustness of the system has been improved by introducing a new concept for constructing on-line the action plans: the concept of *'sensor-based manoeuvre'* (SBM) which can be seen as *meta-skill*. Using this approach, the *Mission Scheduler* can efficiently construct motion plans (i.e. with the required response time) by combining previously planned trajectories with appropriate instantiations of these generic skills (SBM). This approach has been both motivated and justified by the fact that in the considered application (i.e. an intelligent vehicle moving on the road network), the number of different types of manoeuvres to execute is finite and not too large. In particular, it is useless to fully re-plan a similar sequence of actions (i.e. which mainly differ from one to the other by the execution parameters) each time that the vehicle is going to execute an overtaking or a parallel parking manoeuvre. This approach is described in more detail in Section 14.2.3.

14.2.2.5 Conclusion

Nowadays, the three-layer paradigm represents a consensus about the conception of a hybrid control architecture. Yet, the level of abstraction, of reactivity, and of capacity of these layers are still a design problem and a motivation for many research works. This is probably due to the fact that the nature of the integration of deliberative and reactive features is not yet well understood; it is also probably due to the fact that no universal solution exists and that a particular type of control architecture responds to a particular niche of application. The control architecture presented in Section 14.2.3 has specially been designed and implemented for the intelligent vehicle application.

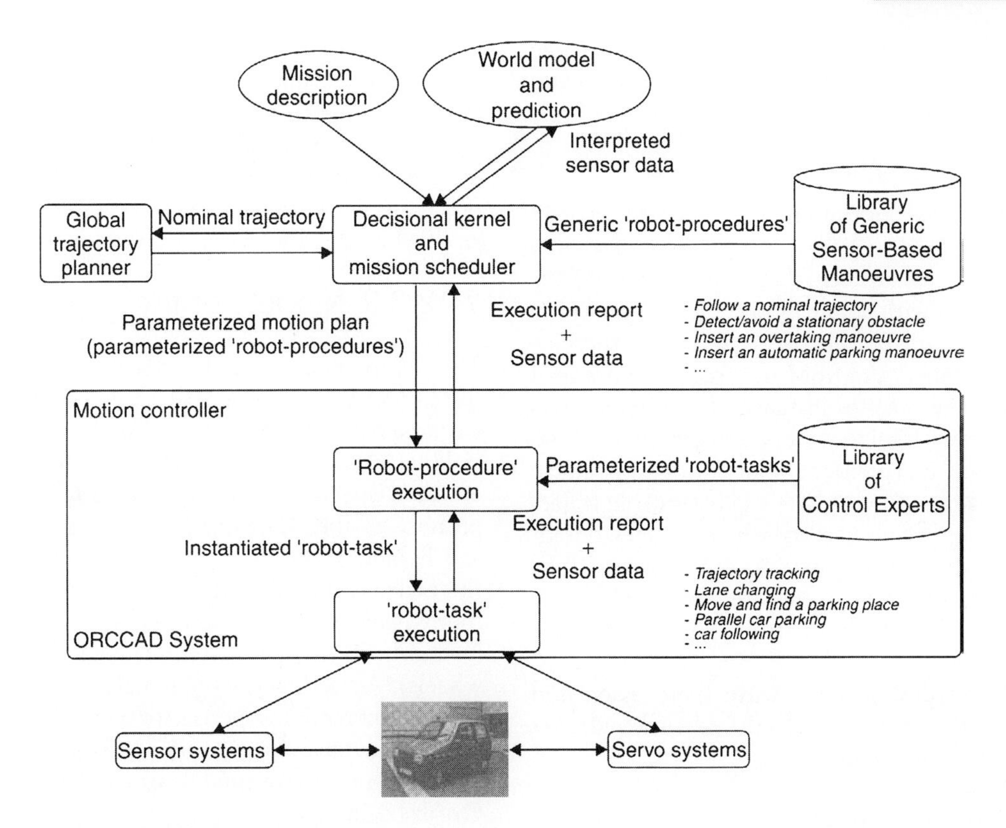

Fig. 14.2-10 Sharp control and decisional architecture.

14.2.3 Sharp control and decisional architecture for autonomous vehicles

14.2.3.1 Overview of the sharp architecture

The Sharp control and decisional architecture is depicted in Fig. 14.2-10. As has previously been mentioned, this architecture is a three-layer-based hybrid architecture that has been specially designed and implemented for controlling an autonomous road vehicle. This architecture takes advantages of the particular features of the considered application domain (and in particular the fact that the road network is a structured environment with motion rules), for making use of appropriate constructions (the *SBMs*) for improving the efficiency, the robustness and the flexibility of the system. The SBM concept is a key concept of our control and decisional architecture; it is derived from the artificial intelligence paradigm of *script* (Rich and Knight, 1983). A script is a general template that encodes procedural knowledge of how a specific type of task is to be performed. A script is fitted to a specific task through the instantiation of variable parameters in the template; these parameters can come from a variety of sources (*a priori* knowledge, sensor data, output of other modules, etc.). Script parameters fill in the details of the script steps and allow it to deal easily with the current task conditions.

The introduction of SBM was motivated by the observation that the kind of motion task that a vehicle has to perform can usually be described as a series of simple steps (a script). An SBM is a script, it combines *control* and *sensing skills*. Skills are elementary functions with real-time abilities: sensing skills are functions processing sensor data whereas control skills are control programs (open or closed loop) that generate the appropriate commands for the vehicle. Control skills may use data provided directly by the sensors or by the sensing skills.

As it has already been mentioned, the idea of combining basic real-time skills to build a plan in order to perform a given task can be found in some other robot control architectures (see Section 14.2.2); this approach permits the authors to obtain robust, flexible and reactive behaviours. From the conceptual point of view, an SBM can be seen as 'meta-skill, which encapsulates high-level expert human knowledge and heuristics about how to perform a specific motion task (see Section 14.2.3.3). Accordingly they allow a reduction in the planning effort required to address a given motion task, thus improving the overall response-time of the system, while retaining the good properties of a skill-based architectures, i.e. robustness, flexibility and reactivity.

Our control and decisional architecture features three main components, the *Mission scheduler*, the *Motion planner* and the *Motion controller*, which are described below.

14.2.3.1.1 Mission scheduler

When given a mission description, e.g. 'park at location l', the *Mission scheduler* generates a *parameterized motion plan (PMP)* which is an ordered set of generic SBMs possibly completed with nominal trajectories. The SBMs are selected from an SBM library, according to the current execution context. An SBM may require a nominal trajectory (as is the case for instance of the 'follow trajectory' SBM). A nominal trajectory is a continuous time-ordered sequence of (position, velocity) of the vehicle that represents a theoretically safe and executable trajectory, i.e. a collision free trajectory which satisfies the kinematic and dynamic constraints of the vehicle. When they are needed, such trajectories are computed by the *Motion planner*, under the request of the *Mission scheduler*.

The involved SBMs, along with their associated nominal trajectories, are passed to the *Motion controller* for their reactive executions.

14.2.3.1.2 Motion planner

The *Motion planner* is in charge of generating collision-free trajectories which satisfy the kinematic and dynamic constraints of the vehicle. Such trajectories are computed using:

- an *a priori known* or *acquired model* of the vehicle environment,
- the current *sensor data*, e.g. position and velocity of the moving obstacles,
- a *world prediction* that gives the most likely behaviours of the moving obstacles.

Motion planning is detailed in the Section 14.2.5.

14.2.3.1.3 Motion controller

The goal of the *motion controller* is to execute in a reactive way the current SBM of the PMP. For that purpose, the current SBM is instantiated according to the current execution context, i.e. the variable parameters of the SBM are set by using the *a priori* known or sensed information available at the time, e.g. road curvature, available lateral and longitudinal space, velocity and acceleration bounds, distance to an obstacle. As mentioned above, an SBM combines control and sensing skills that are either parameterized control programs or sensor data processing functions. It is up to the *Motion controller* to control and coordinate the execution of the different skills required. The sequence of *control skills* that is executed for a given SBM is determined by the events detected by the *sensor skills*. When an event that cannot be handled by the current SBM happens (e.g. the intrusion of an unexpected obstacle which cannot be avoided using the current *control skills*), the *Motion controller* reports a failure to the *mission scheduler* which updates the current PMP either by applying a re-planning procedure (time permitting), or by selecting in real-time an SBM adapted to the new situation.

14.2.3.2 Models of the vehicles

The Sharp control and decisional architecture has been tested on two experimental vehicles with slightly different kinematic characteristics. The first one is a commercial *Ligier* electrical vehicle (Fig. 14.2-11(a)). The second one is a special prototype especially designed for the purpose of the *Automated Public Car* project (Laugier and Parent, 1999) (Fig. 14.2-11(b)). The kinematics of the Ligier is that of a regular car whereas the Cycab has four wheels that can be steered (a steering angle ϕ on the front wheels induces a steering angle $-k\phi$ on the rear wheels). Accordingly, its kinematics is slightly different.

The kinematic properties of a car-like vehicle are explored in detail in Section 14.2.5. From a control point of view, the respective models of the Ligier (left) and the Cycab (right) are:

$$\begin{cases} \dot{x} = v\cos(\theta+\phi) \\ \dot{y} = v\cos(\theta+\phi) \\ \dot{\theta} = \dfrac{v}{L}\sin\phi \end{cases} \qquad \begin{cases} \dot{x} = v\cos(\theta+\phi) \\ \dot{y} = v\cos(\theta+\phi) \\ \dot{\theta} = v\dfrac{\sin(\phi+k\phi)}{L\cos(k\phi)} \end{cases} \tag{14.2.1}$$

where x and y are the coordinates of the front axle midpoint, θ is the orientation of the vehicle and L is the wheel base. The controls are ϕ the steering angle and v the velocity of the front wheels.

14.2.3.3 Concept of SBM

As has previously been mentioned, our control and decisional architecture strongly relies upon the concept of SBM for providing the system with the required

(a)

(b)

Fig. 14.2-11 The *Ligier* vehicle (a), and the *Cycab* (b).

reactivity and robustness properties, while being able to generate smart motion controls for the vehicle. At a given time instant, the vehicle is carrying out a particular SBM that has been instantiated to fit the current execution context (see Section 14.2.3.1). SBMs are general templates encoding the knowledge of how a given motion task is to be performed. They combine real-time functions, control and sensing skills, that are either control programs or sensor data processing functions. From the practical point of view, a SBM can be seen as a specialized controller which generates *safe and smooth motions* for executing in a reactive way a given type of manoeuvre (i.e. by combining some predefined sensory modalities and controls).

In the sequel, we will use two particular types of SBM for illustrating this concept and for showing how it works in practice: the 'trajectory following' SBM, and the 'parallel parking' SBM. These two types of SBM have been developed and integrated in our control and decisional architecture; they have also been implemented and successfully tested on a real automatic vehicle; the results of these experiments are presented in Section 14.2.4. The Orccad tool (Simon *et al*, 1993) has been selected to implement both SBMs and *skills:* robot procedures (in the Orccad formalism) are used to encode SBM's, while 'robot-tasks' encode *skills*; robot procedures and robot tasks can both be represented as finite automata or transition diagrams. The 'trajectory following' and 'parallel parking' SBMs are depicted in Fig. 14.2-12 as transition diagrams. The control skills are represented by square boxes, e.g. 'find parking place', whereas the sensing skills appear as predicates attached to the arcs of the diagram, e.g. 'parking place detected', or conditional statements, e.g. 'obstacle overtaken?'. The control skills are used to control the motions of the vehicle and to activate the selected sensors; the task of the sensing skills is to evaluate the involved perception-based predicates or conditional statements.

The next two sections describe how the two SBMs illustrated in Fig. 14.2-12 operate. Section 14.2.3.6 presents an other type of SBM involving a specialized sensing device.

14.2.3.4 Reactive trajectory following

14.2.3.4.1 Outline of the SBM

The purpose of the trajectory following SBM is to allow the vehicle to follow a given nominal trajectory as closely as possible, while reacting appropriately to any unforeseen obstacle obstructing the way of the vehicle. Whenever such an obstacle is detected, the nominal trajectory is locally modified in real time, in order to avoid the collision. This local modification of the trajectory is done in order to satisfy a set of different motion constraints: collision avoidance, time constraints,

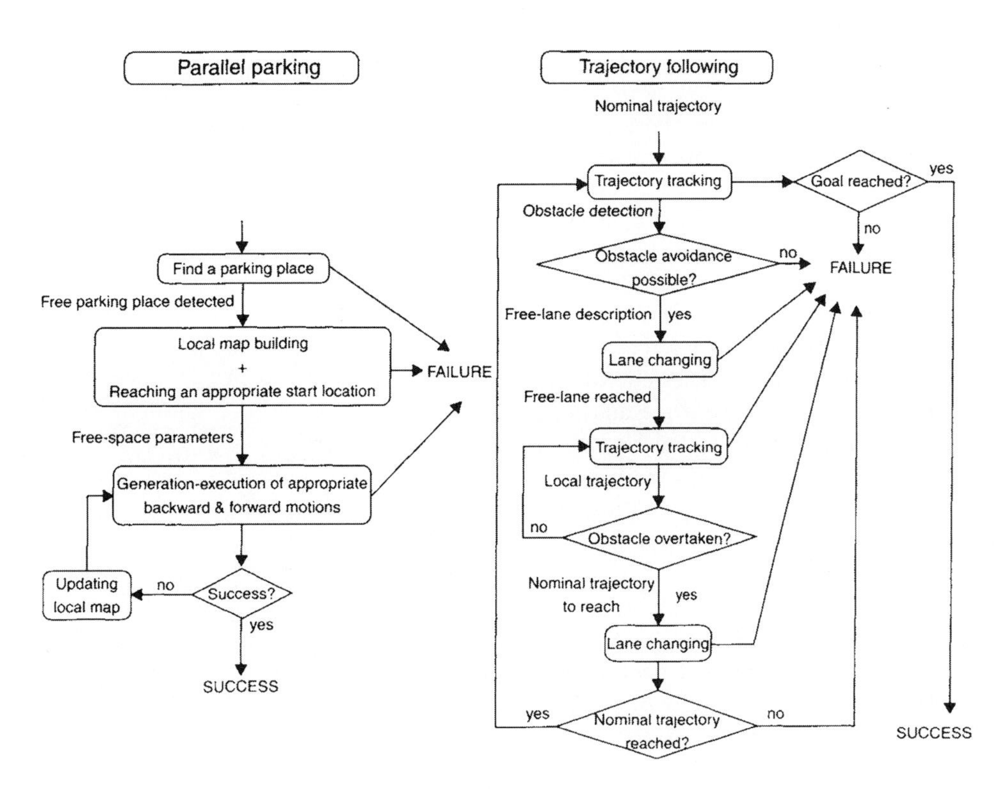

Fig. 14.2-12 The 'parallel parking' and 'trajectory following' SBM.

kinematic and dynamic constraints of the vehicle. In a previous approach, a fuzzy controller combining different basic behaviours (trajectory tracking, obstacle avoidance, etc.) was used to perform trajectory following (Garnier and Fraichard, 1996). However this approach proved unsatisfactory: it yields oscillating behaviours, and does not guarantee that all the aforementioned constraints are always satisfied.

The trajectory following SBM makes use of *smooth local trajectories* to avoid the detected obstacles. These local trajectories allow the vehicle to move away from the obstructed nominal trajectory, and to catch up this nominal trajectory when the (stationary or moving) obstacle has been overtaken. All the local trajectories verify the motion constraints. This SBM relies upon two control skills, *trajectory tracking* and *lane changing* (see Fig. 14.2-12), that are detailed now.

14.2.3.4.2 Trajectory tracking

The purpose of this control skill is to issue the control commands that will allow the vehicle to track a given nominal trajectory. Several control methods for nonholonomic robots have been proposed in the literature. The method described in Kanayama *et al.* (1991) ensures stable tracking of a feasible trajectory by a car-like robot. It has been selected for its simplicity and efficiency. The vehicle's control commands are of the following form:

$$\dot{\theta} = \dot{\theta}_{ref} + v_{R.ref}(k_y y_e + k_\theta \sin \theta_e) \qquad (14.2.2)$$

$$v_R = v_{R.ref} \cos \theta_e + k_x x_e, \qquad (14.2.3)$$

where $q_e = (x_e, y_e, \theta_e)^T$ represents the error between the reference configuration q_{ref} and the current configuration q of the vehicle ($q_e = q_{ref} - q$), $\dot{\theta}_{ref}$ and $v_{R.ref}$ are the reference velocities, $v_R = v \cos\phi$ is the rear axle midpoint velocity, k_x, k_y, k_θ are positive constants (the reader is referred to Kanayama *et al.* (1991) for full details about this control scheme).

When the reference trajectory is considered as too far from the current vehicle configuration (i.e. out of the range of validity of the error parameters of the Kanayama control law), a smooth local trajectory is generated and tracked in order to appropriately catch up the reference trajectory (Fig. 14.2-13). These local trajectories are generated using second degree polynomial functions.

14.2.3.4.3 Lane changing

This control skill is applied to execute a lane changing manoeuvre. The lane changing is carried out by generating and tracking an appropriate smooth local trajectory. Let T be the nominal trajectory to track, d_T be the distance between T and the middle line of the free lane to reach, s_T be the curvilinear distance along T between the vehicle and the obstacle (or the selected end point for the lane change), and $s = s_t$ be the curvilinear abscissa along T since the starting point of the lane change (see Fig. 14.2-14).

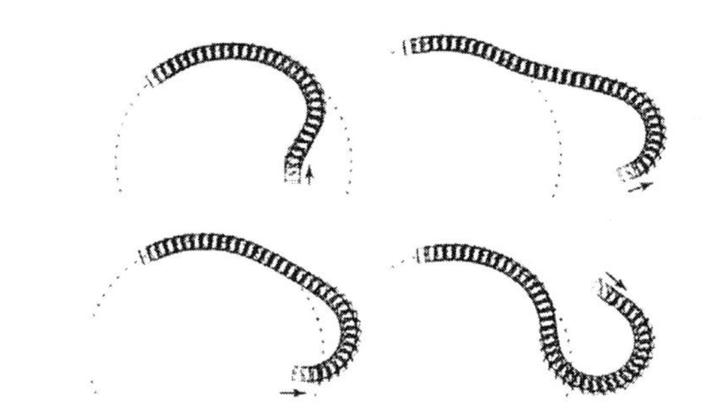

Fig. 14.2-13 Examples of local 'catching up' trajectories.

A feasible smooth trajectory for executing a lane change can be obtained using the following quintic polynomial (see Nelson (1989)):

$$d(s) = d_T\left(10\left(\frac{s}{s_T}\right)^3 - 15\left(\frac{s}{s_T}\right)^4 + 6\left(\frac{s}{s_T}\right)^5\right). \qquad (14.2.4)$$

In this approach, the distance d_T is supposed to be known beforehand. Then, the minimal value required for s_T can be estimated as follows:

$$s_{T.\min} = \frac{\pi\sqrt{kd_T}}{2C_{\max}}, \qquad (14.2.5)$$

where $C_{\max}$ stands for the maximum allowed curvature:

$$C_{\max} = \mathbf{min}\left\{\frac{\tan(\phi_{\max})}{L}, \frac{Y_{\max}}{V^2_{R.ref}}\right\}, \qquad (14.2.6)$$

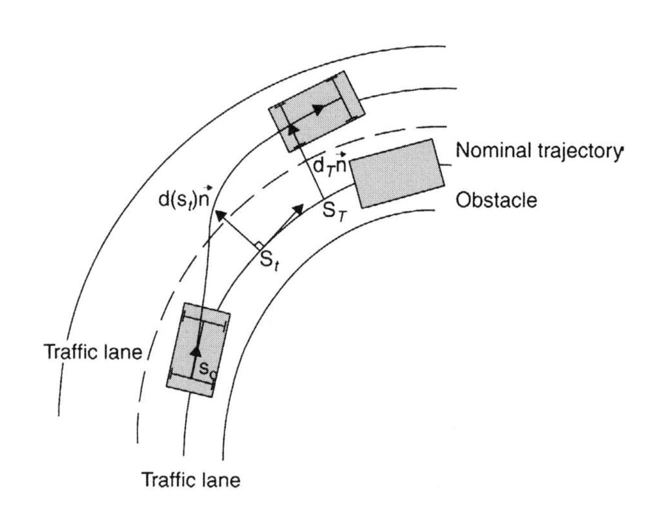

Fig. 14.2-14 Generation of smooth local trajectories to avoid an obstacle.

Y_{max} is the maximum allowed lateral acceleration, and $k > 1$ is an empirical constant, e.g. $k = 1.17$ in our experiments.

At each time t from the starting time T_0, the reference position p_{ref} is translated along the vector $d(S_t).\vec{n}$, where $\vec{n}$ represents the unit normal vector to the nominal velocity vector along $\mathcal{T}$; the reference orientation θ_{ref} is converted into $\theta_{ref} + \arctan(\partial d/\partial s(s_t))$, and the reference velocity $V_{R,ref}$ obtained using the following equation:

$$V_{R.ref}(t) = \frac{dist(p_{ref}(t), p_{ref}(t + \Delta t))}{\Delta t}, \qquad (14.2.7)$$

where *dist* stands for the Euclidean distance. As shown in Fig. 14.2-12, this type of control skill can also be used to avoid a stationary obstacle, or to overtake another vehicle. As soon as the obstacle has been detected by the vehicle, a value $s_{T,\min}$ is computed according to Equation 14.2.5 and compared with the distance between the vehicle and the obstacle. The result of this computation is used to decide which behaviour to apply: avoid the obstacle, slow down or stop. In this approach, an obstacle avoidance or overtaking manoeuvre consists of a lane changing manoeuvre towards a collision-free 'virtual' parallel trajectory (see Fig. 14.2-14). The lane changing skill operates the following way:

1. Generate a smooth local trajectory τ_1 which connects τ with a collision-free local trajectory τ_2 'parallel' to $\mathcal{T}$ (τ_2 is obtained by translating appropriately the involved piece of $\mathcal{T}$).
2. Track τ_1 and τ_2 until the obstacle has been overtaken.
3. Generate a smooth local trajectory τ_3 which connects τ_2 with $\mathcal{T}$, and track τ_3.

14.2.3.5 Parallel parking

The purpose of the parallel parking SBM is to automatically park the vehicle within an unknown parking area. This SBM comprises three main steps (Fig. 14.2-12): (1) localizing a free parking place, (2) reaching an appropriate start location with respect to the selected parking place and (3) performing the parallel parking manoeuvre using iterative backward and forward motions until the vehicle is parked.

14.2.3.5.1 Finding a parking place

During this step, the vehicle moves slowly along the traffic lane and uses its range sensors to build a local map of the environment and detect obstacles. The local map is used to determine whether free parking space is available to park the vehicle. If an obstacle is detected during the motion of the vehicle, another SBM e.g. the trajectory following SBM is activated for avoiding this obstacle.

14.2.3.5.2 Reaching an appropriate start location

A typical situation at the beginning of a parallel parking manoeuvre is depicted in Fig. 14.2-15. The autonomous vehicle A1 is in the traffic lane. The parking lane with parked vehicles B1, B2 and a parking place between them is on the right-hand side of A1. L1 and L2 are respectively the length and width of A1, and D1 and D2 are the distances available for longitudinal and lateral displacements of A1 within the place. D3 and D4 are the longitudinal and lateral displacements of the corner A13 of A1 relative to the corner B24 of B2.

Distances D1, D2, D3 and D4 are computed from data obtained by the sensor systems. The length (D1—D3) and width (D2—D4) of the free parking place are compared with the length L1 and width L2 of A1 in order to determine whether the parking place is sufficiently large.

14.2.3.5.3 Performing the parking manoeuvre

During parallel parking, iterative low-speed backward and forward motions with coordinated control of the steering angle and locomotion velocity are performed to produce a lateral displacement of the vehicle into the parking place. The number of such motions depends on the distances D1, D2, D3, D4 and the necessary parking depth (that depends on the width L2 of the vehicle A1). The start and end orientations of the vehicle are the same for each iterative motion.

For the i-th iterative motion (but omitting the index 'i'), let the start coordinates of the vehicle be $x_0 = x(0)$, $y_0 = y(0), \theta_0 = \theta(0)$ and the end coordinates be $x_T = x(T)$, $y_T = y(T), \theta_T = \theta(T)$, where T is duration of the motion. The 'parallel parking' condition means that

$$\theta_0 - \delta_0 < \theta_T < \theta_0 + \delta_\theta, \qquad (14.2.8)$$

where $\delta_\theta > 0$ is a small admissible error in orientation of the vehicle.

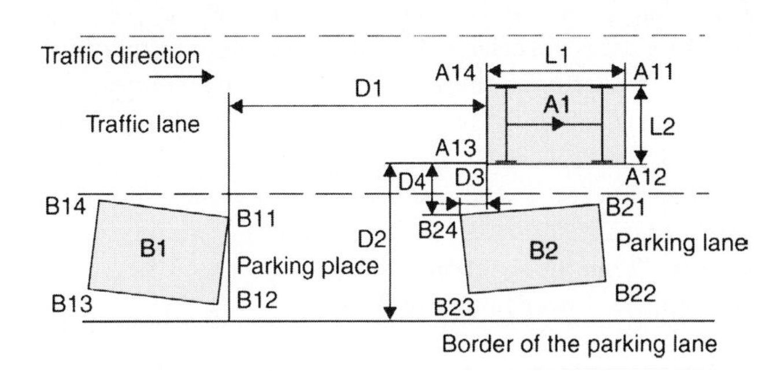

Fig. 14.2-15 Situation at the beginning of a parallel parking manoeuvre.

The following control commands of the steering angle ϕ and locomotion velocity v provide the parallel parking manoeuvre (Paromtchik and Laugier, 1996b):

$$\phi(t) = \phi_{\max} k_\phi A(t), \quad 0 \le t \le T, \tag{14.2.9}$$

$$v(t) = v_{\max} k_v B(t), \quad 0 \le t \le T, \tag{14.2.10}$$

where $\phi_{\max} > 0$ and $v_{\max} > 0$ the admissible magnitudes of the steering angle and locomotion velocity respectively, $k_\phi = \pm 1$ corresponds to a right side (+1) or left side (−1) parking place relative to the traffic lane, $k_v = \pm 1$ corresponds to forward (+1) or backward (−1) motion. Therefore,

$$A(t) = \begin{cases} 1, & 0 \le t < t', \\ \cos\frac{\pi(t-t')}{T^*}, & t' \le t \le T - t', \\ -1, & T - t' < t \le T, \end{cases} \tag{14.2.11}$$

$$B(t) = 0.5\left(1 - \cos\frac{4\pi t}{T}\right), \quad 0 \le t \le T, \tag{14.2.12}$$

where $t' = \frac{T-T^*}{2}$, $T^* < T$. The shape of the type of paths that corresponds to the controls (Equations 14.2.11 and 14.2.12) is shown in Fig. 14.2-16.

The commands (Equations 14.2.9 and 14.2.10) are open-loop in the (x, y, θ)-coordinates. The steering wheel servo-system and locomotion servo-system must execute the commands (Equations 14.2.9 and 14.2.10), in order to provide the desired (x, y)-path and orientation θ of the vehicle. The resulting accuracy of the motion in the (x, y, θ)-coordinates depends on the accuracy of these servo-systems. Possible errors are compensated by subsequent iterative motions.

For each pair of successive motions $(i, i+1)$, the coefficient k_v in Equation 14.2.10 has to satisfy the equation $k_{v,i+1} = -k_{v,i}$ that alternates between forward and backward directions. Between successive motions, when the velocity is null, the steering wheels turn to the opposite side in order to obtain a suitable steering angle $\phi_{\max}$ or $-\phi_{\max}$ to start the next iterative motion.

In this way, the form of the commands in Equations 14.2.9 and 14.2.10 is defined by Equations 14.2.11 and 14.2.12 respectively. In order to evaluate Equations 14.2.9–14.2.12 for the parallel parking manoeuvre, the durations T^* and T, the magnitudes $\phi_{\max}$ and $v_{\max}$ must be known.

The value of T^* is lower-bounded by the kinematic and dynamic constraints of the steering wheel servo-system. When the control command (Equation 14.2.9) is applied, the lower bound of T^* is:

$$T^*_{\min} = \pi \max\left\{\frac{\phi_{\max}}{\dot{\phi}_{\max}}, \sqrt{\frac{\phi_{\max}}{\ddot{\phi}_{\max}}}\right\}, \tag{14.2.13}$$

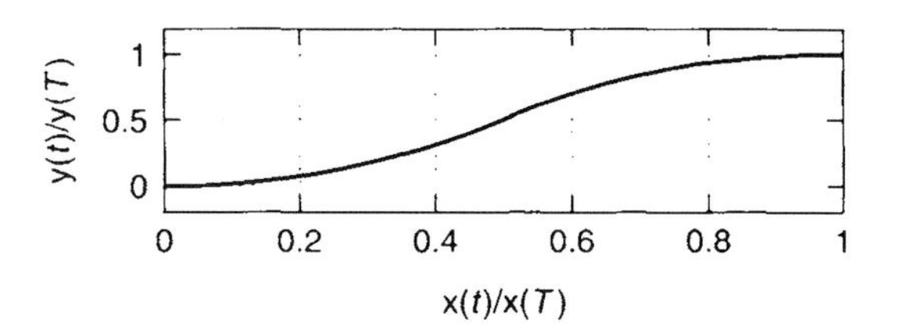

Fig. 14.2-16 Shape of a parallel forward/backward motion.

where $\dot{\phi}_{\max}$ and $\ddot{\phi}_{\max}$ are the maximal admissible steering rate and acceleration respectively for the steering wheel servo-system. The value of $T^*_{\min}$ gives duration of the full turn of the steering wheels from $-\phi_{\max}$ to $\phi_{\max}$ or vice versa, i.e. one can choose $T^* = T^*_{\min}$.

The value of T is lower-bounded by the constraints on the velocity $v_{\max}$ and acceleration $\dot{v}_{\max}$ and by the condition $T^* < T$. When the control command (14.2.10) is applied, the lower bound of T is:

$$T_{\min} = \max\left\{\frac{2\pi v'(D1)}{\dot{v}_{\max}}, T^*\right\}, \tag{14.2.14}$$

where $v'(D1) \le v_{\max}$, empirically obtained function, serves to provide a smooth motion of the vehicle when the available distance D1 is small.

The computation of T and $\phi_{\max}$ aims to obtain the maximal values such that the following 'longitudinal' and 'lateral' conditions are still satisfied:

$$|(x_T - x_0)\cos\theta_0 + (y_T - y_0)\sin\theta_0| \langle D1, \tag{14.2.15}$$

$$|(x_0 - x_T)\sin\theta_0 + (y_T - y_0)\cos\theta_0| \langle D2. \tag{14.2.16}$$

Using the maximal values of T and $\phi_{\max}$ assures that the longitudinal and, especially, lateral displacement of the vehicle is maximal within the available free parking space. The computation is carried out on the basis of the model (Equation 14.2.1) when the commands (Equations 14.2.9 and 14.2.10) are applied. In this computation, the value of $v_{\max}$ must correspond to a safety requirement for parking manoeuvres, e.g. $v_{\max} = 0.75$ m/s was found empirically.

At each iteration i the parallel parking algorithm is summarized as follows:

1. Obtain available longitudinal and lateral displacements D1 and D2 respectively by processing the sensor data.
2. Search for maximal values T and $\phi_{\max}$ by evaluating the model (Equation 14.2.1) with controls (Equations 14.2.9 and 14.2.10) so that the conditions (Equations 14.2.15 and 14.2.16) are still satisfied.
3. Steer the vehicle by controls (Equations 14.2.9 and 14.2.10) while processing the range data for collision avoidance.

4. Obtain the vehicle's location relative to environmental objects at the parking place. If the 'parked' location is reached, stop; otherwise, go to step 1.

When the vehicle A1 moves backwards into the parking place from the start location shown in Fig. 14.2-15, the corner A12 (front right corner of the vehicle) must not collide with the corner B24 (front left corner of the place). The start location must ensure that the subsequent motions will be collision-free with objects limiting the parking place. To obtain a convenient start location, the vehicle has to stop at a distance D3 that will ensure a desired minimal safety distance D5 between the vehicle and the nearest corner of the parking place during the subsequent backward motion. The relation between the distance D1, D2, D3, D4 and D5 is described by a function $\mathcal{F}(D1, D2, D3, D4, D5) = 0$.

This function can not be expressed in closed form, but it can be estimated for a given type of vehicle by using the model (Equation 14.2.1) when the commands (Equations 14.2.9 and 14.2.10) are applied. The computations are carried out off-line and the results are stored in a look-up table which is used on-line, to obtain an estimate of D3 corresponding to a desired minimal safety distance D5 for given D1, D2 and D4 (Paromtchik and Laugier, 1996a). When the necessary parking 'depth' has been reached, clearance between the vehicle and the parked ones is provided, i.e. the vehicle moves forwards or backwards so as to be in the middle of the parking place between the two parked vehicles.

14.2.3.6 Platooning

The platooning SBM allows the controlled vehicle to automatically follow an other vehicle (this leading vehicle can either have been moved autonomously, or driven by a human driver). This SBM takes as input the current (velocity, position, orientation) parameters of the vehicle to control,[3] and it generates in real-time the required lateral and longitudinal controls. The platooning SBM operates in two phases (Daviet and Parent, 1996): (1) determining the relative velocity and position/orientation parameters, and (2) generating the required longitudinal and lateral controls.

14.2.3.6.1 Determining the state parameters

The assessment of the velocity and of the position/orientation parameters of the leading vehicle has to be performed at a rate consistent with the servo-loop frequency (50 Hz in practice). In our implementation, these parameters are evaluated using a linear camera (equipped with appropriate optical lenses) located in the automatic vehicle, and an infrared target located at the rear side of the leading vehicle (see Section 14.2.4). The position/orientation parameters are represented by the longitudinal and lateral distances DX and DY between the two vehicles, and by the angle $D\psi$ between the main axes of the two vehicles; the velocity parameter is obtained by derivating the position parameters.

14.2.3.6.2 Generating the required controls

Following the leading vehicle is performed by controlling, at the servo-loop frequency, the acceleration/deceleration of the automated vehicle along with the angular velocity of its steering wheel.

As for the *longitudinal control,* the basic idea is to set a linear relation between the distance and the speed of the two vehicles:

$$X_l - X_f = d_{\min} + hV_f \qquad (14.2.17)$$

where X_l, X_f, and V_f are respectively the position of the leading vehicle, the position of the following vehicle, and the velocity of the following vehicle, $d_{\min}$ is the minimum distance between the two vehicles, and h is a time constant ($d_{\min} = 1$ m and $h = 0.35$ s in the reported experiments). This approach has led us to make use of the following controller (see Daviet and Parent (1996)) for more details:

$$A_f = C_v \Delta V + C_p(\Delta X - hV_f - d_{\min}) \qquad (14.2.18)$$

where A_f is the acceleration of the following vehicle, $\Delta V = V_l - V_f$, and $\Delta X = X_l - X_f$: the control gains C_p and C_v have been chosen as follows: $C_v = 1/h$ and $C_p = \min(1/h, A_{\max}/V_f)$. The fact that the position gain factor is variable allows the controller to take into account the acceleration saturation and to deal with large initial errors (since C_p decreases when the speed increases).

As for the *lateral control* we have applied a simple approach based onto the classical 'tractor model'. This approach leads the controller to always set the orientation of the steering wheel in a direction parallel to the orientation of the leading vehicle. This approach generate stable behaviours, but it leads the following vehicle to weakly cut the turns (this might be a problem for controlling a platoon of several vehicles in a constrained area).

In a more recent work, we have slightly modified the longitudinal and lateral controls in order to avoid the

[3] The (velocity, position, orientation) parameters of the following vehicle are computed in real-time from the sensory data; they are expressed relatively to the leading vehicle reference frame.

above-mentioned problem and to increase the robustness of the 'target tracking' behaviour. The chosen approach mainly consists in coupling the controller with an on-line 'local trajectory generator' which tries to continuously evaluate the sequences of states – i.e. the (position, orientation, velocity) parameters – of the leading vehicle on a short time interval (instead of only using an instantaneous approach). This approach allows us to still control the motions of the following vehicle, when the target has been lost for a short time period. Current work deals with the processing of exceptions and of the car entrance and exit procedures.

14.2.4 Experimental results

14.2.4.1 Experimental vehicles

The approach described in this chapter has been implemented and tested on our experimental automatic vehicles (a modified Ligier electric car, and the Cycab electric vehicle designed and developed at INRIA (Baille *et al.*, 1999). These vehicles are equipped with the main following capabilities:

1. a *sensor unit* to measure relative distances between the vehicle and environmental objects,
2. a *servo unit* to control the steering angle and the locomotion velocity,
3. a *control unit* that processes data from the sensor and servo units in order to 'drive' the vehicle by issuing appropriate servo commands.

These vehicles can either be manually driven, or they can move autonomously using the *control unit;* this unit is based on a Motorola VME162-CPU board and a transputer network for the Ligier, and on a distributed control architecture implemented using a CAN bus and microcontrollers for the Cycab (Baille *et al.*, 1999). The *sensor unit* of the vehicle makes use of a belt of ultrasonic range sensors (Polaroid 9000) and of a linear CCD-camera. The *servo unit* consists of a steering wheel servo-system, a locomotion servo-system for forward and backward motions, and a braking servo-system to slow down and stop the vehicle. The steering wheel servo-system is equipped with a direct current motor and an optical encoder to measure the steering angle; the locomotion servo-system is equipped with an asynchronous motor and two optical encoders located onto the rear wheels (for odometry data); the Ligier vehicle is also equipped with a hydraulic braking servo-system.

The ultrasonic range sensors used in the described experiments have a measurement range of 0.5–10.0 m, and a sampling rate is 60 ms. The sensors are activated sequentially in order to make more robust measurements in the different regions defined by the vehicle i.e. four sensors are emitting/receiving signals at each time step for sensing each side of the car). The minimal number of ultrasonic sensors required by the parallel parking SBM is eight: three for looking in the forward direction, two located on each side of the vehicle and one for looking in the backward direction. This ultrasonic sensor system is intended to test the control algorithms for low-speed motion only; a more complex sensor system (e.g. a combination of vision and ultrasonic sensors) should be used to ensure reliable operation in a more dynamic environment. This is the purpose of current work.

The CCD-camera has a resolution of 2048 pixels, and it operates at a rate of 1000 Hz; it is equipped with a cylindrical lens and an infrared polarized filter. This device operates in relation with an infrared target for localizing the leading vehicle in the platooning SBM; this target is made of three pulsing sets of LED organized along vertical lines, as shown in Fig. 14.2-22.

The *motion controller* of our control and decisional architecture monitors the current steering angle, locomotion velocity, travelled distance, coordinates of the vehicle and range data from the environment, calculates appropriate local trajectories and issues the required servo commands. It has been implemented using the Orccad software tools (Simon *et al.* 1993) running on a workstation; the compiled code is transmitted via Ethernet to the *control unit* operating under the VxWorks real-time operating system.

11.4.2 Experimental run of the trajectory following manoeuvre

An experimental run of the trajectory following SBM with obstacle avoidance on a circular road (roundabout) is shown in Fig. 14.2-17. In this experiment, the Ligier vehicle follows a nominal trajectory along the curved traffic lane, and it finds on its way another vehicle moving at a lower velocity (see Fig. 14.2-17(a). When the moving obstacle is detected, a local trajectory for a right lane change is generated by the system, and the Ligier performs the lane changing manoeuvre, as illustrated in Fig. 14.2-17(b). Afterwards, the Ligier moves along a trajectory parallel to its nominal trajectory, and a left lane change is performed as soon as the obstacle has been overtaken (Fig. 14.2-17(c)). Finally the Ligier catches up its nominal trajectory, as illustrated in Fig. 14.2-17(d)).

The corresponding motion of the vehicle is depicted in Fig. 14.2-18(a). The steering and velocity controls applied during this manoeuvre are shown in Fig. 14.2-18(b) and Fig. 14.2-18(c). It can be noticed in this example that the velocity of the vehicle has increased when moving along the local 'parallel' trajectory (Fig. 14.2-18(c)); this is due to the fact that the vehicle has to satisfy the time constraints associated with its nominal trajectory.

14.2.4.3 Experimental run of the parallel parking manoeuvre

An experimental run of the parallel parking SBM in a street is shown in Fig. 14.2-19. This manoeuvre can be carried out in environments including moving obstacles, e.g. pedestrians or some other vehicles (see the video Paromtchik and Laugier (1997)). In this experiment, the Ligier was manually driven to a position near the parking place, the driver started the autonomous parking mode and left the vehicle. Then, the Ligier moved forward autonomously in order to localize the parking place, obtained a convenient start location and performed a parallel parking manoeuvre. When, during this motion a pedestrian crosses the street in a dangerous proximity to the vehicle, as shown in Fig. 14.2-19(a), this moving obstacle is detected, the Ligier slows down and stops to avoid the collision. When the way is free, the Ligier continues its forward motion. Range data are used to detect the parking space. A decision to carry out the parking manoeuvre is made and a convenient start position for the initial backward movement is obtained, as shown in Fig. 14.2-19(b). Then, the Ligier moves backwards into the parking space, as shown in Fig. 14.2-19(c). During this backward motion, the front human-driven vehicle starts to move backwards, reducing the length of the bay. The change in the environment is detected and taken into account. The range data shows that the necessary 'depth' in the bay has not been reached, so further iterative motions are carried out until it has been reached. Then, the Ligier moves to the middle between the rear and front vehicles, as shown in Fig. 14.2-19(d). The parallel parking manoeuvre is completed.

The corresponding motion of the vehicle is depicted in Fig. 14.2-20(a) where the motion of the corners of the vehicle and the midpoint of the rear wheel axle are plotted. The control commands (Equations 14.2.9 and 14.2.10) for parallel parking into a parking place situated at the right side of the vehicle are shown in Fig. 14.2-20(b) and (c) respectively. The length of the vehicle is L1 = 2.5 m, the width is L2 = 1.4 m, and the wheelbase is L = 1.785 m. The available distances are D1 = 4.9 m, D2 = 2.7 m relative to the start location of the vehicle.

The lateral distance D4 = 0.6 m was measured by the sensor unit. The longitudinal distance D3 = 0.8 m was estimated so as to ensure the minimal safety distance D5 = 0.2 m. In this case, five iterative motions are performed to park the vehicle. As seen in Fig. 14.2-20, the duration T of the iterative motions, magnitudes of the steering angle ϕ_{max} and locomotion velocity v_{max} correspond to the available displacements D1 and D2 within

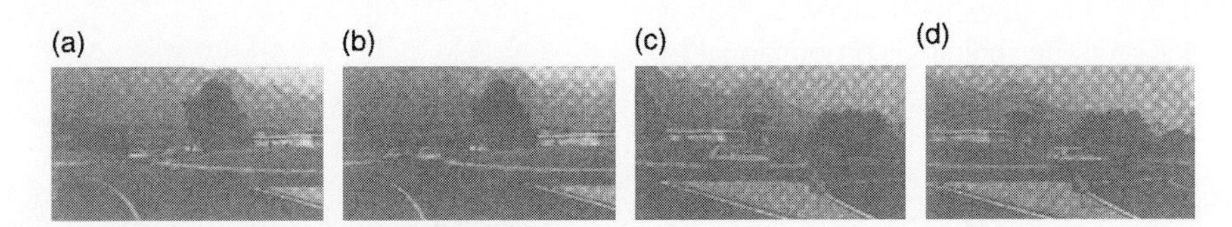

Fig. 14.2-17 Snapshots of trajectory following with obstacle avoidance in a roundabout: (a) following the nominal trajectory, (b) lane changing to the right and overtaking, (c) lane changing to the left, (d) catching up with the nominal trajectory.

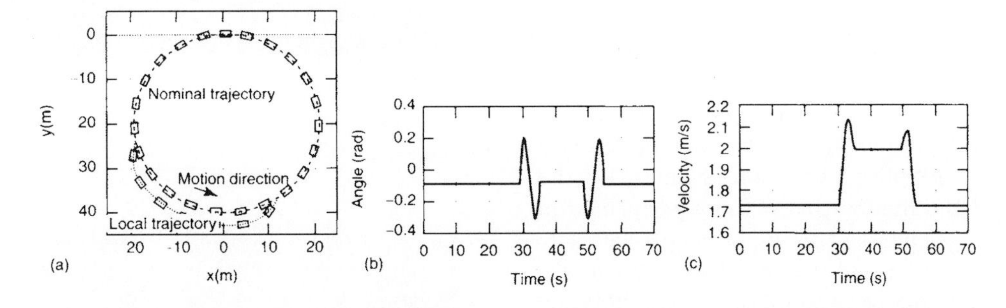

Fig. 14.2-18 Motion and control commands in the "roundabout" scenario: (a) motion, (b) steering angle and (c) velocity controls applied.

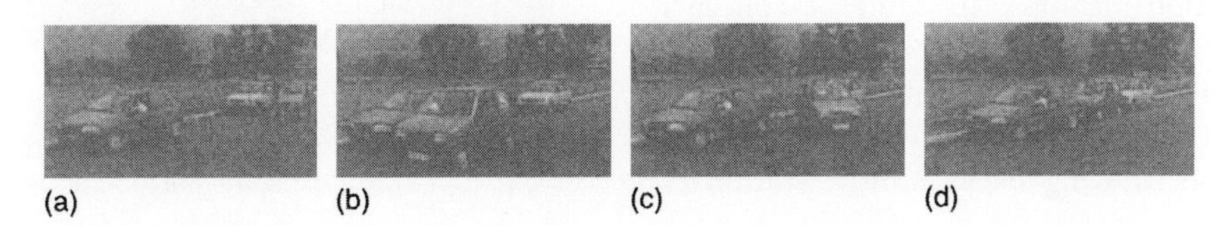

Fig. 14.2-19 Snapshots of parallel parking: (a) localizing a free parking place, (b) selecting an appropriate start location, (c) performing a backward parking motion; (d) completing the parallel parking.

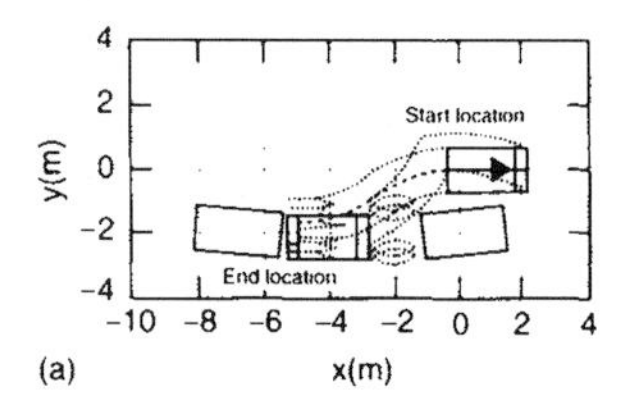

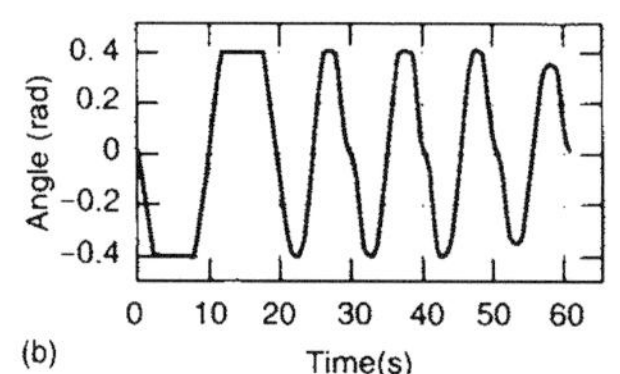

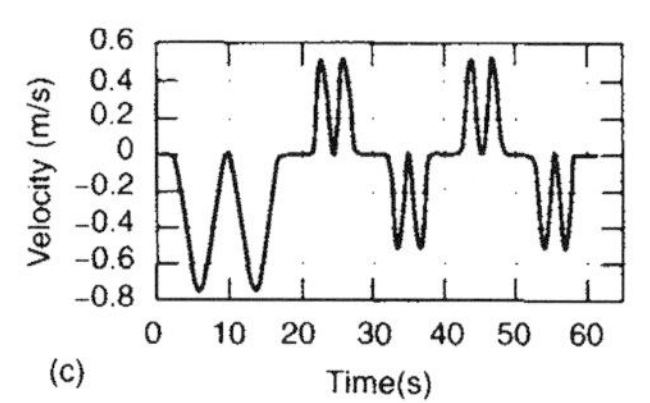

Fig. 14.2-20 Motion and control commands in the parallel parking scenario: (a) motion, (b) steering angle and (c) velocity controls applied.

the parking place (e.g. the values of T, ϕ_{max} and v_{max} differ for the first and last iterative motion).

14.2.4.4 Experimental run of the platooning manoeuvre

An experimental run of the platooning SBM in a street is shown in Fig. 14.2-21. The linear camera and the infrared target is shown in Fig. 14.2-22. During the execution of a platooning manoeuvre, the linear camera operates at a frequency of 1000 Hz for providing the relative position/orientation parameters of the two vehicles; the accuracy of the measurement has been estimated at a value of 1 mm for a distance of 10 m. It has experimentally been shown that the system is robust according to various lighting and light reflecting conditions (thanks to the camera characteristics, to the pulsing infrared target, and to the used filters). Experiments have been conducted at speeds up to 60 km/h, with decelerations up to 2 m/s^2. The distance between the vehicles is proportional to the speed (see Section 14.2.3.6), with a gap of 0.3 s.

14.2.5 Motion planning for car-like vehicles

14.2.5.1 Introduction

The purpose of every robot is to perform actions in its workspace (grasping and mating parts, moving around to explore or survey, etc.). Carrying out a given action usually implies that a motion be made by the robot hence the importance, in robotics, of motion planning, i.e. the determination of the motion that is to be performed in order to achieve a given task. This importance is naturally reflected in the number and variety of research works that have dealt with motion planning in the past 30 years.

Latombe's (1991) book is undoubtedly the reference book for robot motion planning. Its table of contents reveals the importance of what Latombe refers to as the *basic motion planning problem*. Six out of ten chapters are dedicated to this problem, which is to plan a collision-free path for a robot moving freely amidst stationary obstacles. The basic motion planning problem is readily illustrated with the concept of *configuration space* that was introduced in robotics in the late 1970s by Udupa (1977) and Lozano-Perez and Wesley (1979a). The *configuration* of a robot is a set of independent parameters representing the position and orientation of every part of the robot. In its configuration space, a robot is represented as a point, stationary obstacles are represented as forbidden regions[4] and motion planning between a start and a goal configuration is reduced to finding a path, i.e. a continuous sequence of configurations, that avoids the forbidden regions.

The basic motion planning problem is essentially geometric, it deals with collision avoidance of stationary obstacles and it computes a path, i.e. a geometric curve in

Fig. 14.2-21 A platoon of two vehicles: a leader Ligier and a following Cycab.

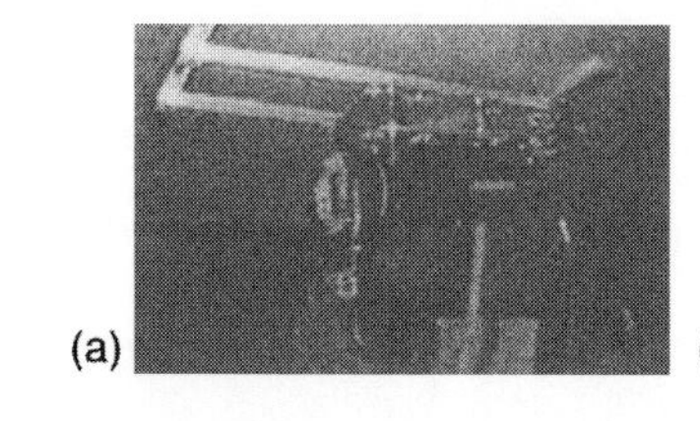

Fig. 14.2-22 Experimental setup for platooning: (a) the linear camera, (b) the first experimental infrared target.

[4] The set of configurations yielding a collision between the robot and the obstacle.

the configuration space. However there is much more to motion planning than that, especially when the robot considered is a car-like vehicle. For a start, such a vehicle cannot move freely: it is subject to *nonholonomic constraints* that restrict its motion capabilities (a car cannot make sidewise motions for instance). Then it usually moves in a workspace that contains *moving obstacles* (they should be avoided too!). In addition to that, it may also be necessary to take into account *dynamic constraints*, e.g. bounded accelerations, that further affect the vehicle's motion capabilities (they cannot be ignored when the vehicle is moving fast).

In summary, physical and temporal constraints as well as geometrical ones must be considered when planning the motions of a car-like vehicle. Such additional constraint yields extensions to the basic motion planning problem that raise new problems and further complicate motion planning. In this case, the output of motion planning is a trajectory, i.e. a path parameterized by time. Trajectory planning with its time dimension permits to take into account time-dependent constraints such as moving obstacles and the dynamic constraints of the vehicle.

The rest of this section explores how to deal with these kinds of constraints. It reviews the main approaches that have been developed in order to deal with all or part of the constraints considered (Section 14.2.5.2). Then it introduces a method that, unlike most of the methods developed before, attempts to take into account all the aforementioned constraints simultaneously; it is based upon the concept of *state-time space* (Section 14.2.5.3). For the sake of clarity, this general method is presented in the case of a car-like vehicle moving along a given path (Sections 14.2.5.4 and 14.2.5.5). Such a path is collision-free and respects the vehicle's nonholonomic constraints. The particular problem of planning a nonholonomic path is explored afterwards (Section 14.2.5.6).

14.2.5.2 Main approaches to trajectory planning

14.2.5.2.1 Nonholonomic constraints

Path planning with nonholonomic constraints is a research field in itself and has motivated a large number of research works in the past 15 years. The review of the relevant literature is made in Section 14.2.5.6, which is dedicated to this topic.

14.2.5.2.2 Dynamic constraints

There are several results for time-optimal trajectory planning for Cartesian robots subject to bounds on their velocity and acceleration (Canny *et al.*, 1990; Ó'Dúnlaing, 1987). Besides optimal control theory provides some exact results in the case of robots with full dynamics and moving along a given path (Shiller and Dubowsky, 1985; Shiller and Lu, 1990). Using these results, some authors have described methods that compute a local time-optimal trajectory (Shiller and Dubowsky, 1989; Shiller and Chen, 1990). The key idea of these works is to formulate the problem as a two-stage optimization process: optimal motion time along a given path is used as a cost function for a local path optimization (hence local time-optimality). However the difficulty of the general problem and the need for practical algorithms led some authors to develop approximate methods. Their basic principle is to define a grid which is searched in order to find a near-time-optimal solution. Such grids are defined either in the workspace (Shiller and Dubowsky, 1988), the configuration space (Sahar and Hollerbach, 1985), or the state space of the robot (Canny *et al.*, 1988; Donald and Xavier, 1990; Jacobs *et al.*, 1989).

14.2.5.2.3 Moving obstacles

A general approach that deals with moving obstacles is the configuration-time space approach which consists of adding the time dimension to the robot's configuration space (Erdmann and Lozano-Perez, 1987). The robot maps in this configuration-time space to a point moving among stationary obstacles. Accordingly the different approaches developed in order to solve the path planning problem in the configuration space can be adapted in order to deal with the specificity of the time dimension and used (see Latombe (1991)). Among the existing works are those based upon extensions of the visibility graph (Erdmann and Lozano-Perez, 1987; Fujimura and Samet, 1990; Reif and Sharir, 1985) and those based upon cell decomposition (Fujimura and Samet, 1989; Shih *et al.*, 1990).

Few research works take into account moving obstacles and dynamic constraints simultaneously, and they usually do so with far too simplifying assumptions e.g. Fujimura and Samet (1989) and Ó'Dúnlaing (1987). More recently (Fiorini and Shiller, 1996), has presented a two-stage algorithm that computes a local time-optimal trajectory for a manipulator arm with full dynamics and moving in a dynamic workspace: the solution is computed by first generating a collision-free path using the concept of velocity obstacle, and then by optimizing it thanks to dynamic optimization.

14.2.5.3 Trajectory planning and state-time space

State-time space is a tool to formulate problems of trajectory planning in dynamic workspaces. In this respect, it is similar to the concept of configuration space (Lozano-Perez and Wesley, 1979b) which is a tool to formulate path planning problems. State-time space permits to study the different aspects of dynamic trajectory planning, i.e.

moving obstacles and dynamic constraints, in a unified way. It seems from two concepts which have been used before in order to deal respectively with moving obstacles and dynamic constraints, namely the concepts of *configuration-time space* (Erdmann and Lozano-Perez, 1987), and *state space*, i.e. the space of the configuration parameters and their derivatives. Merging these two concepts leads naturally to state-time space, i.e. the state space augmented of the time dimension. In this framework, the constraints imposed by both the moving obstacles and the dynamic constraints can be represented by static forbidden regions of state-time space. In addition, a trajectory maps to a curve in state-time space hence trajectory planning in dynamic workspaces simply consists in finding a curve in state-time space, i.e. a continuous sequence of state-times between the current state of the robot and a goal state. Such a curve must obviously respect additional constraints due to the fact that time is irreversible and that velocity and acceleration constraints translate to geometric constraints on the slope and the curvature along the time dimension. However it is possible to extend previous methods for path planning in configuration space in order to solve the problem at hand.

In particular, a method to solve trajectory planning in dynamic workspaces problems when cast in the state-time space framework is presented (Section 14.2.5.5). It is derived from a method originally presented in Canny *et al.* (1988), and extended to take into account the time dimension of the state-time space. It follows the paradigm of near-time-optimization: the search for the solution trajectory is performed over a restricted set of *canonical trajectories* hence the near-time-optimality of the solution. These canonical trajectories are defined as having piecewise constant acceleration that changes its value at given times. Moreover, the acceleration is selected so as to be either minimum, null or maximum (bang controls). Under these assumptions, it is possible to transform the problem of finding the time-optimal canonical trajectory to finding the shortest path in a directed graph embedded in the state-time space.

14.2.5.4 Case study

State-time space was first introduced in Fraichard and Laugier (1992) to plan the motion of a point robot subject to simple velocity and acceleration bounds and moving along a given path amidst moving obstacles. Later, a mobile robot subject to full dynamic constraints was considered; first, in the case of a one-dimensional motion along a given path Fraichard (1993), and then, in the case of a two-dimensional motion on a planar surface (Fraichard and Scheuer, 1994).

It has been decided to focus herein on the case of a car-like vehicle with full dynamics and moving along a given path. The main reason for this choice is that, in this particular case, the state-time space is three-dimensional thus permitting a clear presentation of the concept of state-time space. Besides, although one-dimensional only, this motion planning problem does feature all the key characteristics of trajectory planning in dynamic workspaces, i.e. full dynamics and moving obstacles, and the concepts presented hereafter can easily be extended to problems of higher dimension (see for instance Fraichard and Scheuer (1994)). Accordingly the method presented here could readily be used within a path-velocity decomposition scheme (Kant and Zucker, 1986) to plan motions along a given path taking into account the vehicle dynamics (as in Shin and McKay (1985)) or moving obstacles (as in Kyriakopoulos and Saridis (1991) or Ó'Dúnlaing (1987)).

In summary, this section addresses trajectory planning for a car-like vehicle A which moves along a given path S on a planar workspace W cluttered up with stationary and moving obstacles. It is assumed that S is collision-free with the stationary obstacles of W and that it is feasible, i.e. that it respects the kinematic constraints that restricts the motion capabilities of A. The problem then is to compute a trajectory for A that follows S, is collision-free with the moving obstacles of W and satisfies the dynamic constraints of A.

14.2.5.4.1 Model of the path S

As mentioned earlier, the car-like vehicle A moves along a given path S which is collision-free with the stationary obstacles of W and which is feasible, i.e. it respects the kinematic constraints of A. Those kinematic constraints are studied in more detail in Section 14.2.5.6. It appears there that good feasible paths for a car-like vehicle should be planar curve made up of straight segments and circular arcs of radius $1/\kappa_{max}$ connected with clothoid arcs. S is defined accordingly. Our main concern is that in planning 'high' speed and forward motions only. S should also be of class C^2. The C^2 property ensures that the path is manoeuvre-free and that A can follow it without having to stop to change its direction. Assuming that A moves along S, it is possible to reduce a configuration of A to the single variable s which represents the distance travelled along S.

14.2.5.4.2 Model of the vehicle A

In this section, we start by presenting the dynamic model of A that is used.[5] Then we describe the dynamic constraints that are taken into account.

[5] This model is the two-dimensional instance of the model presented in Shiller and Chen (1990).

Dynamic model of $\mathcal{A}\mathcal{A}$ is modelled as a rigid body supported by four wheels with rigid suspensions. Without loss of generality, it is assumed that the $\vec{t}$ axis of the frame attached to $\mathcal{A}$ coincides with the unit vector tangent to the path S at point R (Fig. 14.2-23). The $\vec{b}$ axis points in the positive direction normal to the plane. The $\vec{n}$ axis is chosen so that $(\vec{t}, \vec{n}, \vec{b})$ is right-handed. Note that the line of the radius of curvature at point R coincides with $\vec{n}$.

The motion of $\mathcal{A}$ along S obeys Newtonian dynamics. The external forces acting on $\mathcal{A}$ are the gravity force $\vec{G}$ and the ground reaction $\vec{R}$ which can be decomposed into their perpendicular components:

$$\vec{G} = -mg\vec{b} \quad (14.2.19)$$

$$\vec{R} = R_t\vec{t} + R_n\vec{n} + R_b\vec{b} \quad (14.2.20)$$

where m is the mass of $\mathcal{A}$ and g the gravity constant. The equation of motion of $\mathcal{A}$ can be expressed in terms of the tangential velocity $\dot{s}$ and the tangential acceleration $\ddot{s}$ namely:

$$\vec{G} + \vec{R} = m\ddot{s}\vec{t} + mK_s\dot{s}^2\vec{n}$$

where K_s, is the signed curvature of the path at position s (κ_s is positive if the radial direction coincides with $\vec{n}$ and negative otherwise, $-\kappa_{max} \le \kappa_s \le K_{max}$). Using Equations 14.2.19 and 14.2.20, this equation can be rewritten in the following set of equations:

$$R_t = m\ddot{s} \quad (14.2.21)$$

$$R_n = m\kappa_s\dot{s}^2 \quad (14.2.22)$$

$$R_b = mg \quad (14.2.23)$$

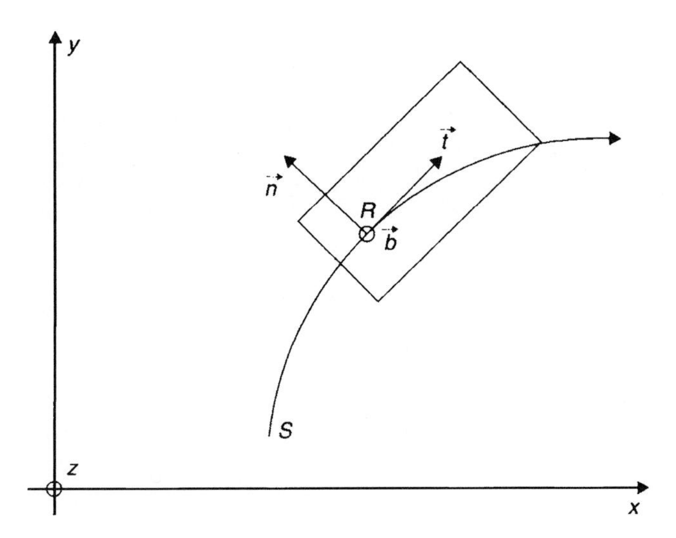

Fig. 14.2-23 The frame attached to $\mathcal{A}$.

Equations 14.2.21 to 14.2.23 represent the forces required to maintain the velocity $\dot{s}$ and the acceleration $\ddot{s}$ of A at a given position s along the path. Although simple, this model is rich enough in the sense that the constraints associated are truly dynamic (they lead to state-dependence of the set of allowable accelerations).

Dynamic constraints of $\mathcal{A}$ Three dynamic constraints are taken into account (engine force, sliding and velocity constraints). They are presented in the next three sections. Afterwards they are transformed into constraints on the tangential velocity $\dot{s}$ and the tangential acceleration $\ddot{s}$.

14.2.5.4.2.1 Engine force constraint

When the vehicle is moving, the torque applied by the engine on the wheels translates into a planar force F whose direction is $\vec{t}$ and whose modulus is $m\ddot{s}$. This force is bounded by the maximum (resp. minimum) equivalent engine force:

$$F_{min} \le F \le F_{max} \quad (14.2.24)$$

These bounds are assumed to be constant and independent of the speed.

14.2.5.4.2.2 Sliding constraint

The component of $\vec{R}$ in the plane $\vec{t} \times \vec{n}$ represents the friction that is applied from the ground to the wheels. This friction is constrained by the following relation:

$$\sqrt{R_t^2 + R_n^2} \le \mu R_b \quad (14.2.25)$$

where μ is the friction coefficient between the wheels and the ground. If this constraint is violated then A will slide off the path.

14.2.5.4.2.3 Velocity constraint

Our main constraint being in planning forward motions, the velocity $\dot{s}$ is constrained by the following relation:

$$0 \le \dot{s} \le \dot{s}_{max} \quad (14.2.26)$$

where $\dot{s}_{max}$ is the highest velocity allowed.

14.2.5.4.2.4 Tangential acceleration constraints

The engine force constraint (Equation 14.2.24) yields the following feasible acceleration range:

$$\frac{F_{min}}{m} \le \ddot{s} \le \frac{F_{max}}{m} \quad (14.2.27)$$

Besides substituting Equations 14.2.21, 14.2.22 and 14.2.23 in Equation 14.2.25 and solving it for $\ddot{s}$ yields the following relation which expresses the feasible acceleration range due to the sliding constraint:

$$-\sqrt{\mu^2 g^2 - \kappa_s^2\,\dot{s}^4} \le \ddot{s} \le \sqrt{\mu^2 g^2 - \kappa_s^2\,\dot{s}^4} \quad (14.2.28)$$

The final feasible acceleration range is therefore given by the intersection of Equations 14.2.27 and 14.2.28:

$$\max\left(\frac{F_{min}}{m}, -\sqrt{\mu^2 g^2 - \kappa_s^2 \dot{s}^4}\right) \leq \ddot{s} \leq \min\left(\frac{F_{max}}{m}, \sqrt{\mu^2 g^2 - \kappa_s^2 \dot{s}^4}\right)$$

14.2.5.4.2.5 Tangential velocity constraints

Velocity $\dot{s}$ must respect Equation 14.2.26. In addition, the argument under the square roots in Equation 14.2.28 should be positive. When $\kappa_s \neq 0, \dot{s}$ must respect the following constraint:

$$-\sqrt{\frac{\mu g}{|\kappa_s|}} \leq \dot{s} \leq \sqrt{\frac{\mu g}{|\kappa_s|}} \quad (14.2.30)$$

The final feasible velocity range is therefore given by the intersection of Equations 14.2.26 and 14.2.30:

$$0 \leq \dot{s} \leq \min\left(\dot{s}_{max}, \sqrt{\frac{\mu g}{|\kappa_s|}}\right) \quad (14.2.31)$$

The latter constraint can be expressed as a set of forbidden states, i.e. points of the $s \times \dot{s}$ plane. Let $\mathcal{TV}$ be this set of states, it is defined as:

$$\mathcal{TV} = \left\{(s, \dot{s}) \middle| 0 > \dot{s} \text{ or } \dot{s} > \min\left(\dot{s}_{max}, \sqrt{\frac{\mu g}{|\kappa_s|}}\right)\right\}$$

14.2.5.4.3 Moving obstacles

A moves in a workspace $\mathcal{W} \in \mathbb{R}^2$ which is cluttered up with stationary and moving obstacles. The path $\mathcal{S}$ being collision-free with the stationary obstacles, only the moving obstacles have to be considered when it comes to planning A' s trajectory.

Let $\mathcal{B}_i, i \in \{1, \ldots, b\}$, be the set of moving obstacles. Let $\mathcal{B}_i(t)$ denote the region of W occupied by $\mathcal{B}_i$ at time t and $A(s)$ the region of W occupied by A at position s along $\mathcal{S}$. If, at time t, A is at position s and if there is an obstacle $\mathcal{B}_i$ such that $\mathcal{B}_i(t)$ intersects $A(s)$ then a collision occurs between A and $\mathcal{B}_i$. Accordingly the constraints imposed by the moving obstacles on A's motion can be represented by a set of forbidden points of the $s \times t$ plane. Let $\mathcal{TB}$ be this set of forbidden points, it is defined as:

$$\mathcal{TB} = \{(s,t) | \exists i \in \{1, \ldots, b\}, \mathcal{A}(s) \cap \mathcal{B}_i(t) \neq \emptyset\}$$

14.2.5.4.4 State-time space of $\mathcal{A}$

As mentioned earlier, the configuration of A is reduced to the single variable s which represents the distance travelled along S. A state of A is therefore represented by a pair $(s, \dot{s}) \in |0, s_{max}| \times |0, \dot{s}_{max}|$ where s_{max} is the arc-length of $\mathcal{S}$.

A **state-time** of A is defined by adding the time dimension to a state hence it is represented by a triple $(s, \dot{s}, t) \in |0, s_{max}| \times |0, \dot{s}_{max}| \times |0, \infty)$. The set of every state-time is the **state-time** space of A; it is denoted by $\mathcal{ST}$.

A state-time is admissible if it does not violate the no-collision and velocity constraints presented earlier. Before defining an admissible state-time formally, let us define $\mathcal{TB}'$, the set of state-times which entail a collision between A and a moving obstacle. $\mathcal{TB}'$ is simply derived from $\mathcal{TB}$:

$$\mathcal{TB}' = \{(s, \dot{s}, t) | \exists i \in \{1, \ldots, b\}, \mathcal{A}(s) \cap B_i(t) \neq \emptyset\}$$

Similarly we define $\mathcal{TV}'$, the set of state-times which violate the velocity constraint Equation 14.2.31. $\mathcal{TV}'$ is simply derived from $\mathcal{TV}$:

$$\mathcal{TV}' = \left\{(s, \dot{s}, t) \middle| 0 > \dot{s} \text{ or } \dot{s} > \min\left(\dot{s}_{max}, \sqrt{\frac{\mu g}{|K_s|}}\right)\right\}$$

Accordingly a state-time q is **admissible** if and only if:

$$q \in \mathcal{ST} / (\mathcal{TB}' \cup \mathcal{TV}')$$

where $E \backslash F$ denotes the complement of F in E. The set of every admissible state-time is the **admissible state-time space** of A, it is denoted by AST and defined as:

$$\mathcal{AST} = \mathcal{ST} / (\mathcal{TB}' \cup \mathcal{TV}')$$

Fig. 14.2-24 depicts the state-time space of $\mathcal{A}$ in a simple case where there is only one moving obstacle which crosses $\mathcal{S}$.

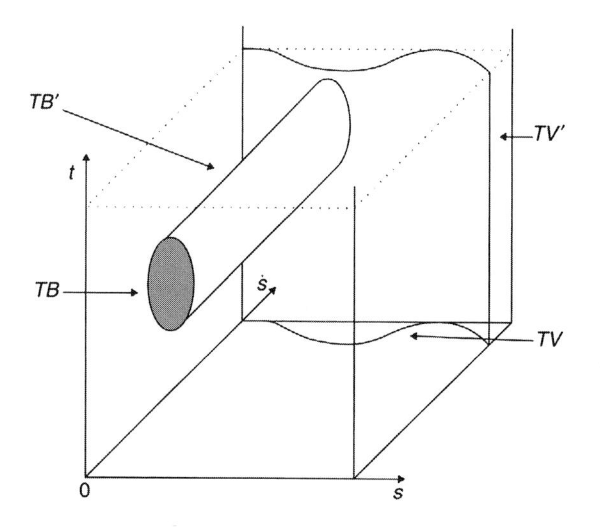

Fig. 14.2-24 $\mathcal{ST}$; the state-time space of $\mathcal{A}$.

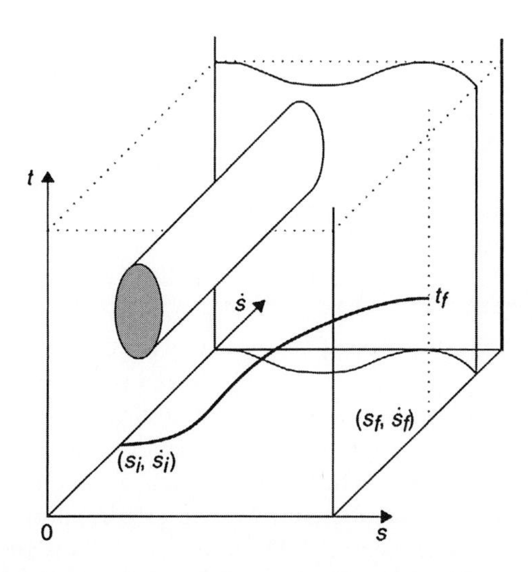

Fig. 14.2-25 A trajectory between $(s_i, \dot{s}_i)$ and $(s_f, \dot{s}_f)$.

In this framework, a *trajectory* Γ for $\mathcal{A}$ between an initial state $(s_i, \dot{s}_i)$ and a final state $(s_f, \dot{s}_f)$ can be represented by a curve of $\mathcal{ST}$, i.e. a continuous sequence of state-times between the initial state-time $(s_i, \dot{s}_i, 0)$ and a final state-time $(s_f, \dot{s}_f, t_f)$. t_f is the duration of the trajectory Γ. The acceleration profile of Γ is a continuous map $\ddot{s} : |0, t_f| \rightarrow \mathbb{R}$. $\ddot{s}(t)$ represents the acceleration which is applied to A at time t. Note that the velocity $\dot{s}$ and position $\dot{s}$ of A along $\mathcal{S}$ are respectively defined as the first and second integral of $\ddot{s}$ subject to an initial position and velocity. In order to be feasible Γ has to verify the different constraints presented in the previous sections, i.e. it must be collision-free with the moving obstacles and respect Equations 14.2.29 and 14.2.31. Fig. 14.2-25 depicts an example of trajectory between $(s_i, \dot{s}_i)$ and $(s_f, \dot{s}_f)$.

14.2.5.4.5 Statement of the problem

Finally, we can formally state the problem which is to be solved. Let $(s_i, \dot{s}_i)$ be the state of $\mathcal{A}$ and $(s_f, \dot{s}_f)$ its goal state. A trajectory $\Gamma : |0, 1| \rightarrow ST$ is a solution to the problem at hand if and only if:

1. $\Gamma(0) = (s_i, \dot{s}_i, 0)$ and $\Gamma(1) = (s_f, \dot{s}_f, t_f)$
2. $\Gamma \subset \mathcal{AST}$
3. Γ's acceleration profile respects Equation 14.2.29.

Naturally, we are interested in finding a time-optimal trajectory, i.e. a trajectory such that t_f should be minimal.

14.2.5.5 Solution algorithm

14.2.5.5.1 Outline of the approach

The method that we have developed in order to solve the problem at hand, i.e. to find a curve Γ of the state-time space $\mathcal{ST}$ which respect the various constraints presented in the previous section, was initially motivated by the work described in Canny *et al.* (1988). For reasons which will be discussed later in Section 14.2.5.5, we follow the paradigm of near-time-optimization, i.e. instead of trying to find out the exact time-optimal trajectory between an initial and a final state, we compute an approximate time-optimal solution by performing the search over a restricted set of *canonical trajectories*. These canonical trajectories are defined as having piecewise constant acceleration $\ddot{s}$ that can only change its value at given times $k\tau$ where τ is a time-step and k some positive integer. Besides $\ddot{s}$ is selected so as to be either minimum, null or maximum. Under these assumptions, it is possible to transform the problem of finding the time-optimal canonical trajectory to finding the shortest path in a directed graph $\mathcal{G}$ embedded in $\mathcal{ST}$. The vertices $\mathcal{G}$ form a regular grid embedded in $\mathcal{ST}$ while the edges corresponds to canonical trajectory segments that each takes time τ. The next sections respectively present the canonical trajectories, the graph $\mathcal{G}$, the search algorithm and experimental results. Finally we discuss the interest of such an approach.

14.2.5.5.2 Canonical trajectories

The definition of the canonical trajectories depends on discretizing time – a time-step τ is chosen – and selecting an acceleration $\ddot{s}$ that respects the acceleration constraint (Equation 14.2.29) and which is either minimum, null or maximum. From a practical point of view, the set of accelerations is discretized – an acceleration-step δ is chosen – and the acceleration applied to A at each time-step, i.e. the minimum, null or maximum one, is selected from this discrete set. As we will see further down, this discretization yields a regular grid in $\mathcal{ST}$.

First let us determine the minimum (resp. maximum) acceleration $\ddot{s}_{\min}$(resp. $\ddot{s}_{\max}$) that can be applied to A. $\ddot{s}_{\min}$ and $\ddot{s}_{\max}$ are derived from Equation 14.2.29 by noting that the acceleration that can be applied to A is maximum (resp. minimum) when the curvature K_s is null, in other words:

$$\ddot{s}_{\min} = \max\left(\frac{F_{\min}}{m}, -\sqrt{\mu^2 g^2}\right)$$

$$\ddot{s}_{\max} = \min\left(\frac{F_{\max}}{m}, -\sqrt{\mu^2 g^2}\right)$$

The interval $[\ddot{s}_{\min}, \ddot{s}_{\max}]$ is therefore the overall range of accelerations that can be applied to $\mathcal{A}$. Now, when $\mathcal{A}$ follows a given path S, at each time instant, it can withstand a range of accelerations that is a subset of $[\ddot{s}_{\min}, \ddot{s}_{\max}]$, this subset is derived from Equation 14.2.29 and depends on the current curvature of S. Given the acceleration step σ, Δ, the overall discrete set of accelerations that can be applied to A is defined as:

$$\Delta = \left\{ i\delta | i \in \mathbb{N}, \left[\frac{\ddot{s}_{\min}}{\delta}\right] \leq i \leq \left[\frac{\ddot{s}_{\max}}{\delta}\right] \right\}$$

Let Γ: $[0, 1] \rightarrow \mathcal{ST}$ be a trajectory and $\ddot{s} : [0, t_f] \rightarrow \Delta$ its acceleration profile. Γ is a **canonical trajectory** if and only if:

- $\ddot{s}$ only changes its value at times $\kappa\tau$ where $\kappa \in \mathbb{N}$, $0 \leq \kappa \leq \lfloor t_f/\tau \rfloor$.
- Let $\ddot{s}^{\kappa\tau}_{\min}$ (resp. $\ddot{s}^{\kappa\tau}_{\max}$) be the minimum (resp. maximum) acceleration allowed w.r.t. the state of A at time $\kappa\tau$. $\ddot{s}(\kappa\tau)$ is chosen from Δ so as to be either null or as close as possible to $\ddot{s}^{\kappa\tau}_{\min}$ and $\ddot{s}^{\kappa\tau}_{\max}$. Thus we have:

$$\ddot{s}(\kappa\tau) \in \left\{ \delta \left[\frac{\ddot{s}^{\kappa\tau}_{\min}}{\delta}\right], 0, \delta \left[\frac{\ddot{s}^{\kappa\tau}_{\max}}{\delta}\right] \right\}$$

As we will see later in this section, $\ddot{s}^{\kappa\tau}_{\min}$ and $\ddot{s}^{\kappa\tau}_{\max}$ are computed so as to ensure that the acceleration constraint (Equation 14.2.29) is respected along the trajectory until the next acceleration change. Note that such a trajectory is very similar to the so-called 'bang–bang' trajectory of the control literature except that, in our case, the acceleration switches occur at regular time intervals.

14.2.5.5.3 State-time graph $\mathcal{G}$

Let q be a state-time, i.e. a point of ST. It is a triple $(s, \dot{s}, t)$. It can equivalently be represented by $q(t) = (s(t), \dot{s}(t))$. Let $q(\kappa\tau) = (s(\kappa\tau), \dot{s}(\kappa\tau))$ be a state-time of A and $q((\kappa + 1)\tau)$ one of the state-times that A can reach by a canonical trajectory of duration τ. $q((\kappa + 1)\tau)$ is obtained by applying an acceleration $\ddot{s} \in \Delta$ to A for the duration τ. Accordingly we have:

$$\dot{s}((K + 1)\tau) = \dot{s}(K\tau) + \ddot{s}\tau \qquad (14.2.32)$$

$$\dot{s}((K + 1)\tau) = s(K\tau) + \dot{s}(K\tau)\tau + \frac{1}{2}\ddot{s}\tau^2 \qquad (14.2.33)$$

By analogy with (Canny *et al.*, 1988), the trajectory between $q(K\tau)$ and $q((K + 1)\tau)$ is called a $(\ddot{s}, \tau)$-**bang**. The state-time $q((K + 1)\tau)$ is reachable from $q(K\tau)$. Obviously a canonical trajectory is made up of a sequence of $(\ddot{s}, \tau)$-bangs.

Let $q(m\tau), m \geq K$, be a state-time reachable from $q(K\tau)$. Assuming that $\dot{s}(K\tau)$ is a multiple of σT, it can be shown that the following relations hold for some integers α_1 and α_2:

$$s(m\tau) = s(K\tau) + \alpha_1 \frac{1}{2} \delta\tau^2$$

$$\dot{s}(m\tau) = \dot{s}(K\tau) + \alpha_2 \delta\tau$$

Thus all state-times reachable from one given state-time by a canonical trajectory lie on a regular grid embedded in ST. This grid has spacings of $\delta\tau^2/2$ in position, of σT in velocity and of τ in time.

Consequently it becomes possible to define a directed graph G embedded in ST. The nodes of G are the grid-points while the edges of G are $(\ddot{s}, \tau)$-bangs between pairs of nodes. G is called the **state-time graph.** Let η be a node in G, the state-times reachable from η by a $(\ddot{s}, \tau)$-bang lie on the grid, they are nodes of G (Fig. 14.2-26). An edge between η and one of its neighbours represents the corresponding $(\ddot{s}, \tau)$-bang. A sequence of edges between two nodes defines a canonical trajectory. The time of such a canonical trajectory is trivially equal to τ times the number of edges in the trajectory. Therefore the shortest path between two nodes (in term of number of edges) is the *time-optimal canonical trajectory* between these nodes.

Let $\mathbf{s} = (s_i, \dot{s}_i)$ be the initial state of A and $\mathbf{g} = (s_f, \dot{s}_f)$ be its goal state. Without loss of generality it is assumed that the corresponding initial state-time $s^* = (s_i, \dot{s}_i, 0)$ and the corresponding set of goal state-times $\mathbf{G}^* = \{(s_f, \dot{s}_f, k\tau) \text{with } k \geq 0\}$ are grid-points. Accordingly searching for a time-optimal canonical trajectory between $\mathbf{s}$ and $\mathbf{g}$ is equivalent to searching a shortest path in G between the node $\mathbf{s}^*$ and a node in $\mathbf{G}^*$.

From a practical point of view, the state-time graph G is embedded in a compact region of ST. More precisely, the time component of the grid-points is upper bounded by a certain value $t_{\max}$ which can be viewed as a time-out. The number of grid-points is therefore finite and so is G. Accordingly the search for the time-optimal canonical trajectory can be done in a finite amount of time.

14.2.5.5.4 Searching the state-time graph

Search algorithm We use an A* algorithm to search $\mathcal{G}$ (Nilsson, 1980). Starting with $\mathbf{s}^*$ as the current node, we

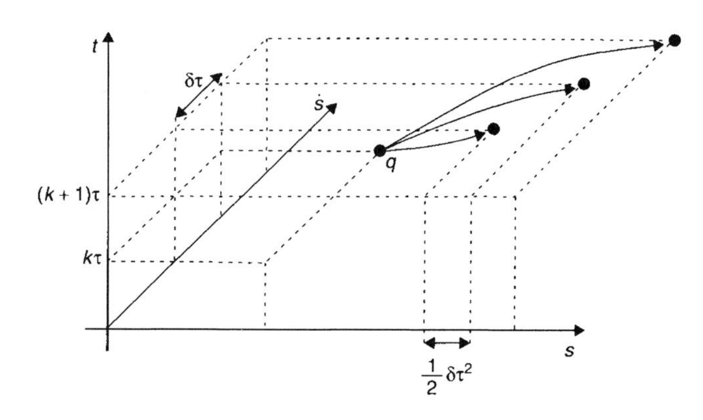

Fig. 14.2-26 G, the graph embedded in $\mathcal{ST}$.

expand this current node, i.e. we determine all its neighbours, then we select the neighbour which is the best according to a given criterion (a cost function) and it becomes the current node. This process is repeated until the goal is reached or until the whole graph has been explored. The time-optimal path is returned using back-pointers. In the next two sections, we detail two key-points of the algorithm, namely the cost function assigned to each node and the node expansion.

Cost function A^* assigns a cost $f(\eta)$ to every node η in G. Since we are looking for a time-optimal path, we have chosen $f(\eta)$ as being the estimate of the time-optimal path in G connecting $\mathbf{s}^*$ to $\mathbf{G}^*$ and passing through η. $f(\eta)$ is classically defined as the sum of two components $g(\eta)$ and $h(\eta)$:

- $g(\eta)$ is the duration of the path between $\mathbf{s}^*$ and η, i.e. the time component of η.
- $h(\eta)$ is the estimate of the time-optimal path between η and an element of $\mathbf{G}^*$, i.e. the amount of time it would take A to reach $\mathbf{g}$ from its current state with a 'bang-coast-bang' acceleration profile, i.e. maximum overall acceleration $\ddot{s}_{max}$, null acceleration and minimum overall acceleration $\ddot{s}_{min}$. When such an acceleration profile does not exist,[6] $h(\eta)$ is set to $+\infty$.

The heuristic function $h(\eta)$ is trivially admissible, thus A^* is guaranteed to generate the time-optimal path whenever it exists (Nilsson, 1980). Moreover the fact that $f(\eta)$ is locally consistent improves the efficiency of the algorithm.

Expansion of a node The neighbours of a given node $\eta = (s, \dot{s}, k\tau)$ are the nodes which can be reached from η by a $(\ddot{s}, \tau)$-bang. As mentioned earlier, $\ddot{s} \in \{\lfloor \ddot{s}^{k\tau}_{min} + \delta \rfloor, 0, \lceil \ddot{s}^{k\tau}_{max} + \delta \rceil\}$. $\ddot{s}^{k\tau}_{min}$ and $\ddot{s}^{k\tau}_{max}$ have to be computed so as to ensure that the acceleration constraint (Equation 14.2.29) is respected along the corresponding $(\ddot{s}, \tau)$-bang. This computation is done in a conservative way. First the farthest position, say s^+, that A can reach from its current state is determined. It is the position reached after $(\ddot{s}_{max}, \tau)$- bang. Then the maximum curvature between s and s^+ is determined and substituted into (Equation 14.2.29) so as to yield the desired acceleration bounds $\ddot{s}^{k\tau}_{min}$ and $\ddot{s}^{k\tau}_{max}$. Finally it remains to check that the $(\ddot{s}, \tau)$-bang associated with each of the candidate neighbours does not violate the velocity and collision avoidance constraints, i.e. that the $(\ddot{s}, \tau)$-bang is included in *AST*.

As we will see now, it is not necessary to compute *AST* to check these two points. Velocity checking is done by using the motion equation of A, while collision checking is performed directly in W. Let us consider a $(\ddot{s}, \tau)$-bang taking place between time instants $k\tau$ and $(k+1)\tau$. $\forall t \in [0, \tau]$ and according to Equations 14.2.32 and 14.2.33, we have:

$$\dot{s}(k\tau + t) = \dot{s}(k\tau) + \ddot{s}t \tag{14.2.34}$$

$$s(k\tau + t) = s(k\tau) + \dot{s}(k\tau)t + \frac{1}{2}\ddot{s}t^2 \tag{14.2.35}$$

Velocity constraint Using Equation 14.2.34, it is straightforward to check that a $(\ddot{s}, \tau)$-bang does not violate the velocity bonds of Equation 14.2.30 stated in Section 14.2.5.4.

Collision avoidance Recall that a $(\ddot{s}, \tau)$-bang between $k\tau$ and $(k+1)\tau$ is collision-free if and only if:

$$\forall t \in [k\tau, (k+1)\tau], \forall i \in \{l, \ldots, b\}, \mathcal{A}(s(t)) \cap \mathcal{B}_i(t) = \emptyset$$

Equation 14.2.35 provides the position of A at every time along the $(\ddot{s}, \tau)$-bang and collision checking can efficiently be performed by computing the intersection between the two planar regions $A(s(t))$ and $B_i(t)$

It might be desirable to add a safety margin to the collision checking procedure so as to incorporate the uncertainty on the motions of A and of the moving obstacles. In this case, the collision avoidance condition becomes:

$$\forall t \in [k\tau, (k+1)\tau], \forall i \in \{1, \ldots, b\},$$
$$\mathcal{G}(\mathcal{A}(s(t)), sm) \cap \mathcal{B}_i(t) = \emptyset$$

where $G(X, sm)$ denotes the planar region X isotropically grown of the safety margin *sm*. The safety margin can integrate both a fixed and a velocity-dependent term, e.g. $sm = c_0 + c_1\dot{s}(t)$ with c_0 and $c_1 \in \mathbb{R}$.

Complexity issues Expanding a node of the graph G can be done efficiently in constant time (recall that the admissible state-time space *AST* is not computed, collision checking is performed directly in the two-dimensional workspace W). The heuristic function used for the A^* search is both admissible and locally consistent (see Section 14.2.5.5), the time complexity of the A^* algorithm is therefore $O(n)$ where n is the number of vertices in G (Farreny and Ghallab, 1987). n is defined as:

$$n = \frac{2s_{max}}{\delta\tau^2}\frac{\dot{s}_{max}}{\delta\tau}\frac{t_{max}}{\tau}$$

The acceleration step and, to a greater extent, the time step are key factors as far as the running time of the algorithm is concerned. Experimental running times and a discussion about the choice of the discretization steps are given later in Section 14.2.5.5.

[6] In this case, η is no longer reachable.

14.2.5.5.5 Implementation and experiments

The algorithm presented earlier has been implemented in C on a Sparc station. Two examples of trajectory planning are depicted in Figs. 14.2-27 and 14.2-28. In each case, there are two windows: a trace window showing the part of the graph which has been explored and a result window displaying the final trajectory. Any such window represents the $s \times t$ plane (the position axis is horizontal while the time axis is vertical; the frame origin is at the upper-left corner). The thick black segments represent the trails left by the moving obstacles and the little dots are points of the underlying grid. Note that the vertical spacing of the dots corresponds to the time-step τ. In both examples, $\mathcal{A}$ starts from position 0 (upper-left corner) with a null velocity, it is to reach position s_{max} (right border) with a null velocity.

The values of the different parameters and discretization steps in these experiments are selected in order to simulate a car-like vehicle moving in the road network: $\dot{s}_{max} = 20\ \mathrm{m/s} - \ddot{s}_{min} = \ddot{s}_{max} = 1\ \mathrm{m/s}^2$. The idea is to plan the motion of the vehicle for the next 500 m ($s_{max} = 500$ m). The time horizon t_{max} is set to 25 s and the obstacles are assumed to keep a constant velocity over the time horizon. For a value of τ set of 0.5 s, the running time is of the order of 1 s.

14.2.5.5.6 Discussion on the proposed solution

As mentioned above in this section, the running time of the search algorithm depends on the size of the graph $\mathcal{G}$ which is to be explored (number of nodes). In turn this size is directly related to the value of the time-step τ–the smaller τ, the higher the number of vertices in $\mathcal{G}$. On the other hand, we intuitively[7] feel that the quality of the solution trajectory is also related to the value of

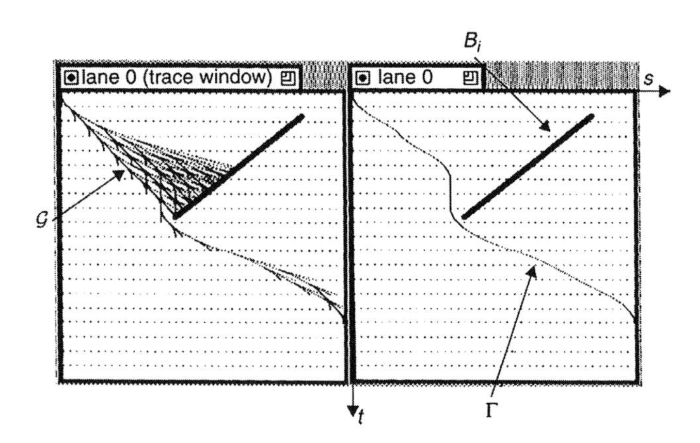

Fig. 14.2-27 Experimental results.

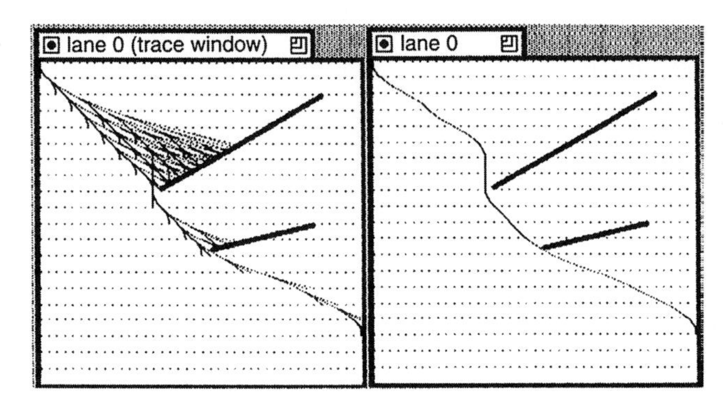

Fig. 14.2-28 Experimental results.

τ – the smaller τ, the better the approximation. Thus it is possible to trade off the computation speed against the quality of the solution.

This property is very important and we would like to advocate this type of approach when dealing with an actual dynamic workspace. In such a workspace, it is usually impossible to have a full *a priori* knowledge of the motion of the moving obstacles. It is more likely that the knowledge that we have of their motions be restricted to a certain time interval, i.e. a time horizon. This time horizon may represent the duration over which an estimation of the motions of the moving obstacles is sound. The main consequence of this assumption is to set an upper bound on the time available to plan the motion of our vehicle (in a highly dynamic workspace, this upper bound may be very low). In this case, an approach such as the one we have presented is most interesting because its average running time can be tuned w.r.t. the time horizon considered.

14.2.5.6 Nonholonomic path planning

Nonholonomy is a classical concept from mechanics that was introduced in robotics by Laumond (1986). A nonholonomic system is subject to non-integrable equations involving the time derivative of its configuration parameters. These equations express constraints in the tangent space of the system at a given configuration, i.e. on the allowable velocities of the system. Nonholonomy usually arises when the system has less control parameters than configuration parameters. A car-like vehicle for instance has three configuration parameters (xy position and orientation) but only two control parameters (acceleration and steering). Thus it cannot change its orientation without also changing its position. As a consequence, any given path in the configuration space is not necessarily admissible which means that,

[7] This intuition is confirmed in Canny *et al.*, (1988) where it is shown that, for a correct choice of τ, any safe trajectory can be approximated to a tolerance ε by a safe canonical trajectory.

even in the absence of obstacles, planning the motion of a nonholonomic system is not straightforward.

In the basic motion planning problem, the existence of a path between two configurations is characterized by the fact that these two configurations lie in the same connected component of the collision-free configuration space of the robot. In other words, a holonomic robot can reach any configuration within the connected component of the configuration space where it is located. This property no longer holds in the presence of nonholonomic constraints. Nonholonomy therefore raises a first problem which is: what is the reachable configuration space? The second problem is of course how to compute an admissible path, i.e. a path that respects the nonholonomic constraints of the robot.

Nonholonomy appears in systems as different as multifingered hands (Murray, 1990), hopping robots (Wang, 1996) or space robots (Nakamura and Mukherjee, 1989). However it concerns primarily wheeled mobile robots and most of the results obtained since 1986 have been obtained for wheeled vehicles such as unicycles, bicycles, two wheel-drive robots, cars, cars with one or several trailers, fire trucks, etc. This section presents the main results regarding path planning for the archetypal nonholonomic system represented by a car-like vehicle. The reader interested to know more about nonholonomy in general is referred to Li and Canny (1992) and Laumond (1998).

This section comprises two parts: the first part considers the 'classical' car-like robot, i.e. the one whose model is equivalent to that of an oriented particle moving in the plane. Henceforth, this car is called the *Reeds and Shepp car.*[8] The Reeds and Shepp car has been extensively studied in the literature and key results have been obtained as far as path planning is concerned. However, as will be seen below, the properties of the Reeds and Shepp car model restricts its applicability, hence the definition of a more complex model for the car. This new car is henceforth called the *Continuous-curvature car,* it is considered in the second part of this section.

14.2.5.6.1 Reeds and Shepp car

As mentioned earlier, the Reeds and Shepp car, or RS car, denotes a car-like vehicle whose mathematical model corresponds to that of an oriented particle moving in the plane. This model is by far the one that has been most widely used. The case where the car can move forward only was first addressed by Dubins (1957) who, among other things, gave a characterization of the shortest paths. Later, Reeds and Shepp (1990) considered the case where the car can change its direction of motion and extended Dubins' results. This section first presents the model of the RS car. Then it summarizes its main properties and overviews the main path planning techniques that were developed.

Model of the RS car Let $\mathcal{A}$ represent a RS car-like robot; it moves on a planar workspace $\mathcal{W} = \mathbb{R}^2$ cluttered up with a set of stationary obstacles $\mathcal{B}_i, i \in \{1,\ldots,b\}$, modelled as forbidden regions of $\mathcal{W}$. A is modelled as a rigid body moving on the plane supported by four wheels making point contact with the ground: two rear wheels and two directional front wheels. It is designed so that the front wheels' axles intersect the rear wheels' axle at a given point C which is the rotation centre of $\mathcal{A}$. It takes three parameters to characterize the position and orientation of $\mathcal{A}$. A configuration of $\mathcal{A}$ is then defined by the triple $q = (x, y, \theta) \in \mathbb{R}^2 \times S^1$ where (x, y) are the coordinates of the rear axle midpoint A and θ the orientation of $\mathcal{A}$ (Fig. 14.2-29).

Under perfect rolling assumption, a wheel moves in a direction normal to its axle. Therefore A must move in a direction normal to the rear wheels' axle and the following constraint holds accordingly (*perfect rolling constraint*):

$$\begin{cases} \dot{x} = v \cos\theta \\ \dot{y} = v \sin\theta \end{cases} \tag{14.2.36}$$

where v is the linear velocity of A, $|v| \le v_{\max}$ ($\mathcal{A}$ moves forward when $v > 0$, stands still when $v = 0$, and moves backward when $v < 0$).

Let ϕ denote the steering angle of $\mathcal{A}$, i.e. the average orientation of the front wheels, and let κ denote the

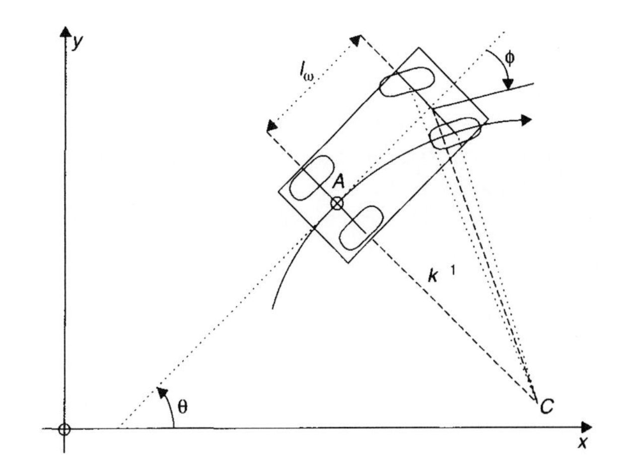

Fig. 14.2-29 A car-like robot.

[8] After Reeds and Shepp (1990) who established its main properties.

curvature of the xy-curve traced by A. κ is the inverse of the distance between C and A: $\kappa = l_w - 1 \tan\phi$, where l_w is the wheelbase of $\mathcal{A}$. Since ϕ is mechanically limited, $|\phi| \leq \phi_{max}$, the following constraint holds (*bounded curvature constraint*):

$$|\kappa| \leq \kappa_{max} = l_w^{-1} \tan\phi_{max} \tag{14.2.37}$$

According to Equation 14.2.36, θ is always tangent to the xy-curve traced by A and its derivative, i.e. the angular velocity ω, therefore satisfies $\dot{\theta} = \omega = vk$. Selecting v and ω as (coupled) control parameters, the model of $\mathcal{A}$ can be described by the following differential system:

$$\begin{pmatrix} \dot{x} \\ \dot{y} \\ \dot{\theta} \end{pmatrix} = \begin{pmatrix} \cos\theta \\ \sin\theta \\ 0 \end{pmatrix} v + \begin{pmatrix} 0 \\ 0 \\ 1 \end{pmatrix} \omega \tag{14.2.38}$$

with $|v| \leq v_{max}, \omega = vK$, and $|\kappa| \leq \kappa_{max}$. Because path planning is generally interested in computing shortest paths, it is furthermore assumed that $|v| = 1$ (thus the time and the arc length of a path are the same). The system (Equation 14.2.38) under the different control constraints defines the Reeds and Shepp car.

Admissible paths for the RS car Let C denote the configuration space of the RS car $\mathcal{A}$:$\mathcal{C} \equiv \mathbb{R}^2 \times [0, 2\pi]$. Let Π denote a path for $\mathcal{A}$, it is a continuous sequence of configurations: $\Pi(t) = (x(t), y(t), \theta(t))$. An admissible path must satisfy both constraints (Equations 14.2.36 and 14.2.37), it is a solution to the differential system (Equation 14.2.38); it is such that:

$$\begin{cases} x(t) = x(0) + \int_0^t v(\tau)\cos\theta(\tau)d\tau \\ y(t) = y(0) + \int_0^t v(\tau)\sin\theta(\tau)d\tau \\ \theta(t) = \theta(0) + \int_0^t \omega(\tau)d\tau \end{cases} \tag{14.2.39}$$

with $|v(\tau)| = 1, \omega(\tau) = v(\tau)\kappa(\tau) \text{and} |\kappa(\tau)| \leq \kappa_{max}$.

Reachable configuration space for the RS car As mentioned earlier, the first question raised by nonholonomy is to determine whether the presence of nonholonomic constraints reduces the set of configurations that the RS car can reach. This question was first answered by Laumond (1986). Through an *ad hoc* geometric reasoning, Laumond established that the RS car could reach any configuration within the same connected component of the collision-free configuration space.

In fact, it turns out that this question is directly related to the controllability of differential systems. The small-time controllability[9] of a differential system implies that the existence of an admissible collision-free path is equivalent to the existence of a collision-free path (Laumond *et al.*, 1998). Using tools from differential geometric control theory, it proved possible to show the small-time controllability of the RS car and therefore to redemonstrate Laumond's result (Barraquand and Latombe, 1990).

In summary, in spite of the presence of nonholonomic constraints, the RS car can reach any configuration within the connected component of the collision-free configuration space where it is located.

Optimal paths for the RS car Once the small-time controllability of the RS car has been established, it is interesting to find out the shortest path between two configurations in the absence of obstacles. Reeds and Shepp (1990) used differential calculus tools to give a first characterization of the shortest paths for the RS car. Later, Boissonnat *et al.* (1991), and Sussmann and Tang (1991) used optimal control theory to refine Reeds and Shepp's result.

The optimal path for the RS car is made up of line segments and circular arcs of radius $1/\kappa_{max}$; it is the shortest among a set of 46 paths that belong to one of the nine following families:

$$\begin{array}{ll} (i) & l^+l^-l^+ \text{ or } r^+r^-r^+ \\ (ii)(iii) & A|AA \text{ or } AA|A \\ (iv) & AA|AA \\ (v) & A|AA|A \\ (vi) & A|ASA|A \\ (vii)(vii) & A|ASA \text{ or } ASA|A \\ (ix) & ASA \end{array} \tag{14.2.40}$$

where A (resp. S) denotes a circular arc (resp. line segment). | denotes a change of direction of motion (a cusp point). A may be replaced by r or l to specify a right (clockwise) or left (counterclockwise) turn. A + or − superscript indicates a forward or backward motion. Fig. 14.2-30 depicts two examples of optimal paths for the RS car.

This result enables definition of what Laumond *et al.* (1998) call a *steering method* for the RS car, i.e. an algorithm that computes an admissible path between two configurations in the absence of obstacles. Let $Steer_{RS}$ denote the steering method returning the optimal path for the RS car, i.e. the shortest path among the set defined by Equation 14.2.40.

Collision-free path planning for the RS car The complete path planning problem for the RS car must consider the constraints imposed both by the nonholonomic

[9] A differential system is *locally controllable* if the set of configurations reachable from any configuration q by an admissible path contains a neighbourhood of q. It is *small-time controllable* if the set of configurations reachable from q before a given time t contains a neighbourhood of q for any t.

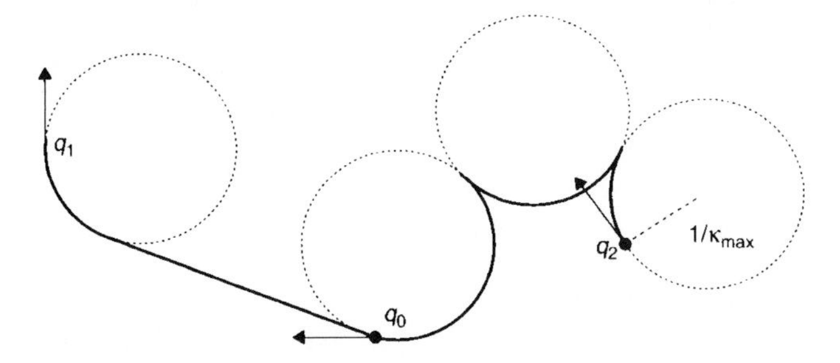

Fig. 14.2-30 Optimal paths for the Reeds and Shepp car.

constraints and the obstacles of the environment. Although the results on the controllability and the optimal paths presented above did not consider the obstacles of the environment, it will be seen further down that they proved useful in the design of path planners for the RS car.

In the past 15 years, several solutions have been proposed to solve the full path planning problem for both the RS car (e.g. Barraquand and Latombe, 1989; Laumond *et al.*, 1989; Pommier, 1991), and the RS car that can move forward only (e.g. Jacobs and Canny, 1989; Fraichard, 1991; Laumond, 1987; Wilfong, 1988). This paragraph focuses on three of them that all share common features. For a start, they are generic, i.e. they can be applied to other types of robotics systems (the first two are completely general, they can deal with holonomic and nonholonomic systems). Then, all of them make use of a steering method such as $Steer_{RS}$. $Steer_{RS}$ computes the optimal path between two configurations for the RS car but other steering methods could be used instead (see the review made in Laumond *et al.* (1998)).

- *Prohahilistic path planning:* this generic path planning scheme was developed by Svestka and Overmars (1998) and Kavraki *et al.* (1996). It operates in two phases:

  – **Learning phase:** build a roadmap reflecting the connectivity of the collision-free configuration space. The roadmap is a graph whose nodes are randomly selected configurations and whose edges are admissible collision-free paths computed with a steering method and a collision checker.

  – **Query phase:** given a start and a goal configurations, use $Steer_{RS}$ to connect them to the roadmap. Search the roadmap for a solution path.
- *Ariadne's Clew algorithm:* this generic path planning scheme developed by Ahuactzin-Larios (1994) is slightly different from the probabilistic path planning approach. It incrementally builds a tree rooted at the start configuration. At each step, a new configuration is added to the tree; it is the farthest configuration reachable from the tree by the steering method considered (optimization tools such as genetic algorithms are used). In parallel, the steering method is used to determine whether the goal configuration is reachable from the new configuration.
- *Holonomic path approximation:* the controllability result on the existence of an admissible collision-free path is at the origin of a two-step nonholonomic path planning scheme that was introduced in Laumond *et al.* (1994):

  Step 1: compute a holonomic[10] collision-free path.

  Step 2: approximate the holonomic path by a sequence of admissible collision-free paths.

Step 2 recursively subdivides the holonomic path and tries to connect the endpoints by using a steering method along with a collision checker.

14.2.5.6.2 Continuous-curvature car

As mentioned earlier, most path planning techniques for the RS car compute paths made up of line segments connected with tangential circular arcs of minimum radius. Reeds and Shepp's (1990) result concerning the shortest path for the RS car is a reason that explains this situation. No doubt that another reason for this situation is that they are easy to deal with from a computational point of view. However the curvature of this type of path is discontinuous: discontinuities occur at the transitions between segments and arcs. The curvature is directly related to the orientation of the front wheels of the car. Accordingly, if a car were to track precisely such a type of path, it would have to stop at each curvature discontinuity so as to reorient its front wheels. It is therefore desirable to plan continuous-curvature paths.[11] To address this issue, a new model for the car-like vehicle is introduced: the Continuous-curvature car, or CC car. This model is presented in the next paragraph. Unlike the RS car, the CC car has been little studied; several results have been obtained however, they are presented afterwards.

Model of the CC *car* Let A now represent a CC car-like robot. As per Boissonnat *et al.* (1994), a configuration of $\mathcal{A}$ is now defined by the quadruple $q = (x, y, \theta, \kappa) \in \mathbb{R}^2 \times \mathbf{S}^1 \times \mathbb{R}$. κ is introduced to characterize the orientation of the front wheels of $\mathcal{A}$.

[10] It does not satisfy the nonholonomic constraints.

[11] As a matter of fact, it is emphasized in De Luca *et al.* (1998) that feedback controllers for car-like robots require this property in order to guarantee the exact reproducibility of a path.

Considering κ as a configuration parameter ensures that it will vary continuously.

$\mathcal{A}$ is subject to the perfect rolling constraint (Equation 14.2.36) and the bounded curvature constraint (Equation 14.2.37). Let σ denote the derivative of: $\kappa : \sigma = \dot{\phi}/\cos^2\phi$. The steering velocity of $\mathcal{A}$ is physically limited, $|\dot{\phi}| \leq \dot{\phi}_{max}$, and the following constraint is added (*bounded curvature derivative constraint*):

$$|\sigma| \leq \sigma_{max} = \dot{\phi}_{max} \tag{14.2.41}$$

Accordingly the model of $\mathcal{A}$ can now be described by the following differential system:

$$\begin{pmatrix} \dot{x} \\ \dot{y} \\ \dot{\theta} \\ \dot{\kappa} \end{pmatrix} = \begin{pmatrix} \cos\theta \\ \sin\theta \\ \kappa \\ 0 \end{pmatrix} v + \begin{pmatrix} 0 \\ 0 \\ 0 \\ 1 \end{pmatrix} \sigma \tag{14.2.42}$$

with $|\kappa| \leq \kappa_{max}, |v| = 1$ and $|\sigma| \leq \sigma_{max}$. The system (Equation 14.2.42) under the different control constraints defines the Continuous-curvature car.

Admissible paths for the CC *car* The configuration space of $\mathcal{A}$ is now 4-dimensional: $\mathcal{C} \equiv \mathbb{R}^2 \times |0.2\pi| \times \mathbb{R}$. A path for $\mathcal{A}$ is a continuous sequence of configurations: $\Pi(t) = (x(t), y(t), \theta(t), \kappa(t))$. An admissible path must satisfy the constraints (Equations 14.2.36, 14.2.37, 14.2.41); it is a solution of the differential system (Equation 14.2.42); it is such that:

$$\begin{cases} x(t) = x(0) + \int_0^t v(\tau)\cos\theta(\tau)d\tau \\ y(t) = x(0) + \int_0^t v(\tau)\sin\theta(\tau)d\tau \\ \theta(t) = \theta(0) + \int_0^t v(\tau)\kappa(\tau)d\tau \\ \kappa(t) = \kappa(0) + \int_0^t \sigma(\tau)\tau \end{cases} \tag{14.2.43}$$

with $|\kappa(\tau)| \leq \kappa_{max}, |v(\tau)| = 1$ and $|\sigma(\tau)| \leq \sigma_{max}$.

Reachable configuration space for the CC *car* The small-time controllability of the CC car has been established in Scheuer and Laugier (1998, Theorem 1). Accordingly, the existence of an admissible collision-free path is equivalent to the existence of a collision-free path. Furthermore, if a path exists between two configurations then an optimal path exists as well (Scheuer and Laugier, 1998, Theorem 2).

Optimal paths for the CC *car* The nature of the optimal paths for the CC car is more difficult to establish than for the RS car. However Scheuer (1998) demonstrates that, for the forward CC car, i.e. the CC car moving forward only, the optimal paths are made up of: (a) line segments, (b) circular arcs of radius $1/\kappa_{max}$, and (c) clothoid arcs[12] of sharpness $\pm\sigma_{max}$. Unfortunately, it appears that, whenever the shortest path includes a line segment, it is irregular and contains an infinite number of clothoid arcs that accumulate towards each endpoint of the segment (Boissonnat *et al.*, 1994). Furthermore, when the distance between two configurations is large enough, the shortest path contains a line segment hence an infinite number of clothoid arcs (Degtiariova-Kostova and Kostov, 1998).

In summary, although the exact nature of the optimal paths for the CC car has not been established yet, it seems reasonable to conjecture that they will (at least) be made up of line segments, circular arcs and clothoid arcs, and that they will be irregular in most cases.

Collision-free path planning for the CC *car* The main path planning techniques presented earlier are generic: they can be used for a wide variety of robotics systems. To some extent, they mainly rely upon the existence of a steering method for the robotics system considered.

As for the CC car and to the best of the authors' knowledge, there is only one steering method that has been

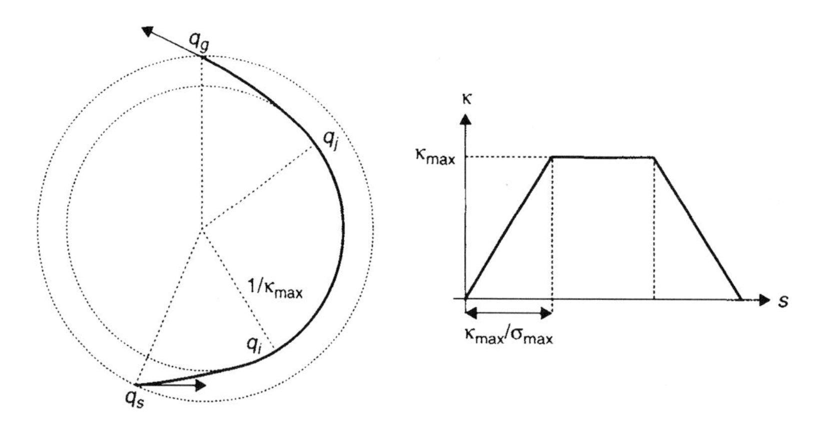

Fig. 14.2-31 A CC turn and its curvature profile.

[12] A clothoid is a curve whose curvature varies linearly with its arc length.

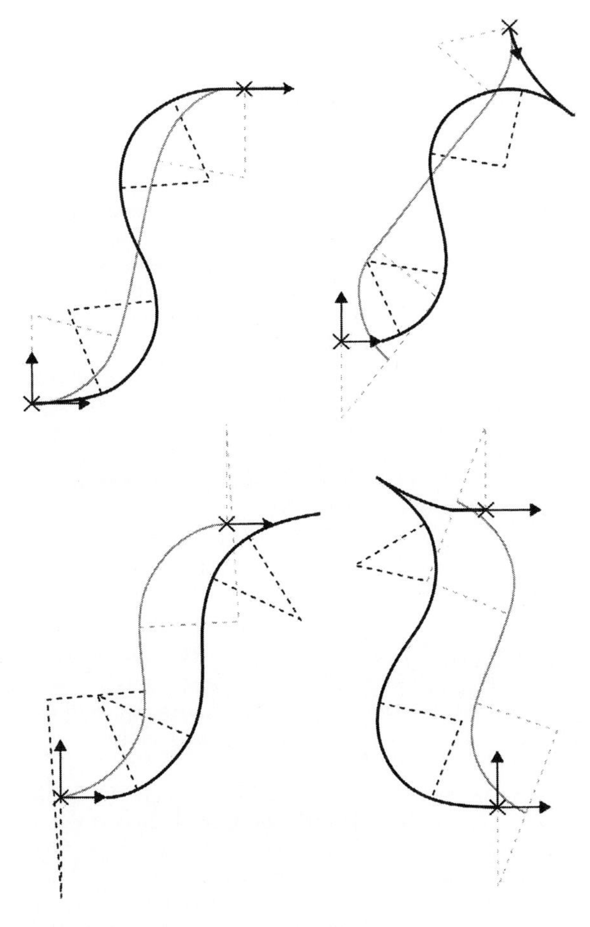

Fig. 14.2-32 RS (in grey) versus CC paths (in black).

developed so far. It has been proposed in Fraichard *et al.* (1999). Since the conjectured irregularity of the optimal paths for the CC car leaves little hope to ever have a steering method for the CC car that finds out the optimal path between two configurations, Fraichard *et al.* have developed a steering method $Steer_{CC}$ that computes admissible paths derived from the optimal paths for the RS car.

The paths computed by $Steer_{CC}$ are similar to those computed by $Steer_{RS}$ but, in order to ensure curvature continuity, the circular arcs are replaced by transitions called CC *turns* whose curvature varies continuously from 0 up and then down back to 0. A CC turn is made up of three parts: (a) a clothoid arc of sharpness $\sigma = \pm\sigma_{max}$ whose curvature varies from 0 to K_{max}, (b) a circular arc of radius $1/\kappa_{max}$ and (c) a clothoid arc of sharpness $-\sigma$ whose curvature varies from K_{max} to 0 (Fig. 14.2-31). The paths computed by $Steer_{CC}$ are not optimal but, based upon the result already established in Scheuer and Laugier (1998) for the forward CC car, it is conjectured that they are suboptimal, i.e. longer than the optimal path of no more than a given constant. This result is yet to be demonstrated however.

Table 14.2-1 RS *vs.* CC paths' length

	Min.	Average	Max.	Deviation
Ratio	1.00253	1.1065	2.45586	0.172188

Table 14.2-2 RS *vs.* CC paths' computation time

RS (1000 paths)	CC (1000 paths)	Average ratio
3.466586 s.	4.483492 s.	1.33

14.2.5.6.3 Reeds and shepp car versus the continuous-curvature car

$Steer_{CC}$ and $Steer_{RS}$ have both been has been implemented and compared. Fig. 14.2-32 illustrates the results obtained. It appears that, for a given pair of (initial, goal) configurations, the resulting RS and CC paths may belong to the same family of path (Fig. 14.2-32, top left), or to different families. CC paths may have the same number of back-up manoeuvres (Fig. 14.2-32, top right), more (Fig. 14.2-32, bottom left) or less (Fig. 14.2-32, bottom right).

Further comparisons were made regarding the respective length of the paths and the time required for their computation. The ratio of CC over RS paths' lengths were computed for 100 pairs of (initial, goal) configurations. The results obtained are summarized in Table 14.2-1. In most cases (82 per cent), CC paths are only about 10 per cent longer than RS paths. Similar experiments were carried out for the computation time. The running time of both $Steer_{CC}$ and $Steer_{RS}$ are of the same order of magnitude

(Table 14.2-2). Given that continuous curvature paths can be tracked with a much greater accuracy by a real car-like vehicle (see the experimental results obtained in Scheuer and Laugier, 1998), the results reported herein demonstrate the interest of CC paths (about the same computation time and same length).

Acknowledgements

This work was partially supported by the Eureka EU-45 European project *Prometheus*, the Inria-lnrets French programme *Praxitèle*, the Inco-Copernicus European project *Multi-agent robot systems for industrial applications in the transport domain* and the French research programme *La route automatisée*.

The authors would like to thank J. Hermosillo, F. Large, E. Gauthier, Ph. Garnier and 1. Paromtchik for their invaluable contributions.

References

Ahuactzin-Larios, J.-M. (1994). Le Fil d'Ariane: Une Methode de Planification Génerale. Application à la Planification Automatique de Trajectoires. Thése de doctoral Inst. Nat. Polytechnique de Grenoble, Grenoble, France.

Alami, R, Chatila, R, Fleury, S, Ghallab, M, and Ingrand, F. (1998). An architecture for autonomy. *International Journal of Robotics Research*, **17**(4), 315–37. Special issue on integrated architectures for robot control and programming.

Anderson, T.L, Donath, M. (1990). Autonomous Robots and Emergent Behavior: A Set of Primitive Behaviors for Mobile Robot Control. In: *Proceedings of the lEEE-RSJ International Workshop on Intelligent Robots and Systems*, **2**, Tsuchiura, Japan, pp. 723–30.

Arkin, R.C. (1987). Motor Schema Based Navigation for a Mobile Robot. In: *Proceedings of the IEEE International Conference on Robotics and Automation*, Raleigh, NC, USA, pp. 264–71.

Arkin, R.C. (1989). Motor schema-based mobile robot navigation. *International Journal of Robotics Research*, **8**; 92–112.

Arkin, R.C. (1990). Integrating Behavioral, Perceptual, and World Knowledge in Reactive Navigation. *Robotics and Autonomous Systems*, **6**, 105–22.

Arkin, R.C. (1998). *Behavior-based robotics – Intelligent robots and autonomous agents*. The MIT Press.

Baille, G, Garnier, Ph, Mathieu, H, and Pissard-Gibollet, R. (1999). Le Cycab de l'Inria Rhêne Alpes. Technical Report 229 Inst. Nat. de Recherche en Informatique et en Automatique. Montbonnot, France.

Barraquand, J, and Latombe, J.-C. (1989). On nonholonomic mobile robots and optimal maneuvering. *Revue d'Intelligence Artificielle*, **3**(2), 77–103.

Barraquand, J, and Latombe, J.-C. (1990). Controllability of Mobile Robots with Kinematic Constraints. Research Report STAN-CS-90–1317. Stanford University, Stanford, CA, USA.

Bobrow, J.E, Dubowsky, S, and Gibson, J.S. (1985). Time-optimal control of robotic manipulators along specified paths. *International Journal of Robotics Research*, **4**(3), 3–17.

Boissonnat, JD, Cérézo, A, and Leblond, J. (1991). Shortest paths of bounded curvature in the plane. Research Report 1503. Inst. Nat. de Recherche en Informatique et en Automatique. Rocquencourt, France.

Boissonnat, JD, Cérézo, A, and Leblond, J. (1994). A note on shortest paths in the plane subject to a constraint on the derivative of the curvature. Research Report 2160. Inst. Nat. de Recherche en Informatique et en Automatique, Rocquencourt, France.

Bonasso, RP, Kortenkamp, D, Miller, DP, and Slack, M. (1996). Experiences with an Architecture for Intelligent, Reactive Agents. *Lecture Notes in Computer Science*, **1037**, 187–202.

Brooks, RA. (1990). A Robust Layered Control System for a Mobile Robot. In: *Readings in Uncertain Reasoning*, G. Shafer and J. Perl (eds). Morgan Kaufmann, pp. 204–13.

Brooks, RA, and Flynn, AM. (1989). Fast, Cheap and Out of Control: A Robot Invasion of the Solar System. *Journal of the British Interplanetary System*, **42**(10), 478–85.

Canny, J, Donald, B, Reif, J, and Xavier, P. (1988). On the complexity of kinodynamic planning. In: *Proceedings of the IEEE Symposium on the Foundations of Computer Sciences*. White Plains, NY, USA, pp. 306–16.

Canny, J, Rege, A and Reif, J. (1990). An exact algorithm for kinodynamic planning in the plane. In: *Proceedings of the ACM Symposium on Computational Geometry*. Berkeley, CA, USA, pp. 271–80.

Connell, J.H. (1992). SSS: A Hybrid Architecture Applied to Robot Navigation. In: *Proceedings of the IEEE International Conference on Robotics and Automation*, vol. 2, Nice, France, pp. 2719–24.

Daviet, P, and Parent, M. (1996). Longitudinal and Lateral servoing of vehicles in a platoon. In: *Proceedings of the IEEE International Symposium on Intelligent Vehicles*, Tokyo, Japan, pp. 41–6.

De Luca, A., Oriolo, G., and Samson, C. (1998). Feedback control of a nonholonomic car-like robot. In: *Robot motion planning and control*, (J.P. Laumond, ed.). *Lecture Notes in Control and Information Science*, **229**, pp. 171–253. Springer.

Degtiariova-Kostova, E., and Kostov, V. (1998). Irregularity of Optimal Trajectories in a Control Problem for a Car-like Robot. Research Report 3411. Inst. Nat. de Recherche en Informatique et en Automatique.

Donald, B, and Xavier, P. (1990). Probably good approximation algorithms for optimal kinodynamic planning for cartesian robots and open-chain manipulators. In: *Proceedings of the ACM Symposium on Computational Geometry*, Berkeley, CA, USA, pp. 290–300.

Dubins, L.E. (1957). On curves of minimal length with a constraint on average curvature, and with prescribed initial and terminal positions and tangents. *American Journal of Mathematics*, **79**,497–517.

Erdmann, M, and Lozano-Perez, T. (1987). On multiple moving objects. *Algorithmica*, **2**, 477–521.

Farreny, H. and Ghallab, M. (1987). *Eléments d'Intelligence Artificielle*. Hermès.

Ferber, J. (1995). *Les systemes multi-agents, vers une intelligence collective*. Intereditions.

Fiorini, P, and Shiller, Z. (1996). Time optimal trajectory planning in dynamic environments. In: *Proceedings of the IEEE International, Conference on Robotics and Automation*, vol. 2, Minneapolis, MN, USA, pp. 1553–8.

Firby, R.J. (1989). Adaptive Execution in Complex Dynamic Domains. PHD thesis YALEU/CSD RR 672 Yale University.

Fraichard, Th. (1991). Smooth trajectory planning for a car in a structured world. In: *Proceedings of the IEEE International Conference on Robotics and Automation*, vol. 1, Sacramento, CA, USA, pp. 318–23.

Fraichard, Th. (1993). Dynamic Trajectory Planning with Dynamic Constraints: a 'State-Time Space' Approach. In: *Proceedings of the IEEE-RSJ International Conference on Intelligent Robots and Systems*, vol. 2, Yokohama, Japan, pp. 1394–1400.

Fraichard, Th, and Laugier, C. (1992). Kinodynamic planning in a structured and time-varying 2D workspace. In *Proceedings of the IEEE International Conference on Robotics and Automation*, Vol. 2, Nice, France, pp. 1500–1505.

Fraichard, Th, and Scheuer, A. (1994). Car-Like Robots and Moving Obstacles. In *Proceedings of the IEEE International Conference on Robotics and Automation*, vol. 1, San Diego, CA, USA, pp. 64–9.

Fraichard, Th, Scheuer, A, and Desvigne, R. (1999). From Reeds and Shepp's to continuous-curvature paths. In *Proceedings of the IEEE International Conference on Advanced Robotics*, Tokyo, Japan, pp. 585–90.

Fujimura, K, and Samet, H. (1989). A hierarchical strategy for path planning among moving obstacles. *IEEE Transactions on Robotics and Automation*, **5**(1), 61–9.

Fujimura, K, and Samet, H. (1990). Motion planning in a dynamic domain. In: *Proceedings of the IEEE International Conference on Robotics and Automation*, Cincinnatti, OH, USA, pp. 324–30.

Garnier, Ph, and Fraichard, Th. (1996). A Fuzzy Motion Controller for a Car-Like Vehicle. In: *Proceedings of the IEEE-RSJ International Conference on Intelligent Robots and Systems*, vol. 3, Osaka, Japan, pp. 1171–8.

Gat, E. (1992). Integrating planning and reacting in a heterogeneous asynchronous architecture for controlling real-world mobile robots. In: *Proceedings of the Tenth National Conference on Artifuial Intelligence*, San Jose, CA, USA, pp. 809–15.

Gat, E. (1997). On Three-Layer Architectures. In *Artificial Intelligence and Mobile Robots*, (D. Kortenkamp, RP. Bonnasso,and R. Murphy, eds). MIT/ AAAl Press.

Gat, E, Slack, M, Miller, D.P, and Firby, RJ. (1990). Path Planning and Execution Monitoring for a Planetary Rover. In: *Proceedings of the IEEE International Conference on Robotics and Automation*, vol. 1, Cincinatti, OH, USA, pp. 20–25.

Harel D, and Pnueli, A. (1985). On the development of reactive systems. In: *Logics and Models of Concurrent Systems*, KR. Apt (ed.). Springer-Verlag, pp. 477–98.

Humphrys, M. (1995). W-learning: Competition among selfish Q-learners. Technical Report 362 University of Cambridge Computer Laboratory. Available at: *ftp://ftp.cl.cam.ac.uk/ papers/20reports/TR36)2-mh2010006-w-learning.0ps.g/*.

Jacobs, P and Canny, J. (1989). Planning smooth paths for mobile robots. In: *Proceedings of the IEEE International Conference on Robotics and Automation*, Scotlsdale, AZ, USA, pp. 2–7.

Jacobs, P, Heinzinger, G, Canny, J, and Paden, B. (1989). Planning guaranteed near-time-optimal trajectories for a manipulator in a cluttered workspace. Research Report ESRC 89–20/RAMP 89–15. Engineering Systems Research Center, University of California, Berkeley, CA, USA.

Kanayama, Y, Kimura, Y, Myazaki, F, and Noguchi, T. (1991). A Stable Tracking Control Method for a Nonholonomic Mobile Robot. In: *Proceedings of the IEEE-RSJ International Workshop on Intelligent Robots and Systems*, vol. 2, Osaka, Japan, pp. 1236–41.

Kant, K., and Zucker, S. (1986). Toward efficient trajectory planning: the path-velocity decomposition. *International Journal of Robotics Research*, **5**(3): 72–89.

Kavraki, L., Svestka, P., Latombe, J,-C. and Overmars, M.H. (1996). Probabilistic roadmaps for path planning in high dimensional configuration spaces. *IEEE Transactions on Robotics and Automation*, **12**, 566–580.

Kyriakopoulos, K.J, and Saridis, G.N. (1991). Collision avoidance of mobile robots in non-stationary environments. In: *Proceedings of the IEEE International Conference on Robotics and Automation*. Sacramento, CA, USA, pp. 904–9.

Latombe, J.-C. (1991). *Robot motion planning*. Kluwer Academic Press.

Laugier, C, and Parent, M. (1999). Towards Motion Autonomy for Future vehicles. In: *Proceedings of the International Symposium on Robotics Research*, Snowbird, USA, Invited paper.

Laugier, C, Fraichard, Th, Paromtchik, I.E, and Garnier, Ph. (1998). Sensor-Based Control Architecture for a Car-Like Vehicle. In *Proceedings of the IEEE-RSJ International Conference on Intelligent Robots and Systems*, vol. 1, Victoria, BC, Canada, pp. 216–22.

Laumond, J.-P. (1986). Feasible trajectories for mobile robots with kinematic and environment constraints. In *Proceedings of the International Conference on Intelligent Autonomous Systems*. Amsterdam, The Netherlands, pp. 346–54.

Laumond, J.-P. (1987). Finding collision-free smooth trajectories for a nonholonomic mobile robot. In: *Proceedings of the International Joint Conference on Artificial Intelligence*, Milan, Italy, pp. 1120–23.

Laumond, J.-P (ed) (1998). Robot motion planning and control. *Lecture Notes in Control and Information Science*, 229, Springer.

Laumond, J.-P, Jacobs, PE, Taix, M, and Murray, RM. (1994). A motion planner for nonholonomic mobile robots. *IEEE Trans! Robotics and Automation*, *10*(5), 577–93.

Laumond, J.-P, Sekhavat, S and Lamiraux, F. (1998). Guidelines in nonholonomic motion planning for mobile robots. In *Robot motion planning and control*, JP. Laumond (ed.). *Lecture Notes in Control and Infonnation Science*, 129, pp. 1–53, Springer.

Laumond, J.-P, Siméon, T, Chatila, R, and Giralt, G. (1989). Trajectory planning and motion control for mobile robts. In: *Geometry and Robotics*, JD. Boissonnat, and JP. Laumond (eds). *Lecture Notes in Computer Science, 391*, pp. 133–49, Springer.

Li, Z, and Canny, J.F. (eds) (1992). Nonholonomic Motion Planning. *The Kluwer International Series in Engineering and Computer Science, 192* Kluwer Academic Press.

Lin, L.J. (1993). Reinforcement learning for robots using neural networks. Technical Report CMU-CS-93–l03 Carnegie Mellon University.

Lozano-Perez, T, and Wesley, M.A. (1979a). An algorithm for planning collision-free paths among polyhedral obstacles. *Communications of the ACM* **22**(10), 560–70.

McFarland, D. (1987). *The Oxford Companion to Animal Behaviour*. Oxford University Press.

Moravec, H.P. (1983). The Stanford Cart and the CMU Rover. *Proceedings of the IEEE*, **71**(7), 872–84.

Murray, R.M. (1990). Robotic control and nonholonomic motion planning. Research Report UCB/ERL M90/l17. Univ. of California at Berkeley, Berkeley, CA, USA.

Nakamura, Y and Mukherjee, R. (1989). Nonholonomic path planning of space robots. In: *Proceedings of the IEEE International Conference on Robotics and Automation*, vol. 2, Scottsdale, AZ, USA, pp. 1050–55.

Nelson, W.L. (1989). Continuous curvature paths for autonomous vehicles. In: *Proceedings of the IEEE International Conference on Robotics and Automation*, vol. 3, Scottsdale, AZ, USA, pp. 1260–64.

Nilsson, N.J. (1980). *Principles of artificial intelligence*. Los Altos, CA, USA, Morgan Kaufmann.

Nilsson, N.J. (1984). Shakey The Robot. Technical note 323 AI Center, SRI International, Menlo Park, CA, USA.

Noreils, F. (1990). Integrating Error Recovery in a Mobile Robot Control System. In: *Proceedings of the IEEE International Conference on Robotics and Automation*, vol. 1, Cincinatti, OH, USA, pp. 396–401.

Ó'Dúnlaing, C. (1987). Motion planning with inertial constraints. *Algorithmica*, **2**, 431–75.

Paromtchik, I.E, and Laugier, C. (1996a). Autonomous Parallel Parking of a Nonholonomic Vehicle. In: *Proceedings of the IEEE Symposium on Intelligent Vehicles*, Tokyo, Japan, pp. 13–18.

Paromtchik, IE, and Laugier, C. (1996b). Motion Generation and Control for Parking an Autonomous Vehicle. In: *Proceedings of the IEEE International Conference on Robotics and Automation*, Minneapolis, MN, USA, pp. 3117–22.

Paromtchik, I.E. and Laugier, Ch. (1997). Automatic Parallel Car Parking. In *Video-Proceedings of the IEEE International Conference on Robotics and Automation*, Albuquerque, NM, USA. Produced by Inst. Nat. de Recherche en Informatique et en Automatique-Unité de Communication et Information Scientifique (3 min.).

Payton, D.W. (1986). An Architecture for Reflexive Autonomous Vehicle Control. In: *Proceedings of the IEEE International Conference on Robotics and Automation*. San Franciso, CA, USA, pp. 1838–45.

Payton, D.W, Rosenblatt, JK, and Keirsey, DM. (1990). Plan Guided Reaction. *IEEE Transactions on Systems, Man and Cybernetics*, **20**(6), 1370–82.

Pommier, E. (1991). Génération de trajectoires pour robot mobile nonholonome par gestion des centres de rotation. These de doctorat Lab. d'Informatique, de Robotique et de Microèlectronique Montpellier, France.

Reeds, J.A, and Shepp, LA. (1990). Optimal paths for a car that goes both forwards and backwards. *Pacific Journal of Mathematics*, **145**(2), 367–93.

Reif, J, and Sharir, M. (1985). Motion planning in the presence of moving obstacles. *Proceedings of the IEEE Symposium on the Foundations of Computer Science*, Portland, OR, USA, pp. 144–54.

Rich, E, and Knight, K. (1983). *Artificial Intelligence*. McGraw-Hill.

Rosenblatt, J.K. (1997). DAMN: A Distributed Architecture for Mobile Navigation. PhD thesis, Carnegie Mellon University.

Rosenblatt, J.K, and Payton, DW. (1989). A Fine-Grained Alternative to the Subsumption Architecture for Mobile Robot Control. In: *Proceedings of the International Joint Conference on Neural Networks*, vol. 2, Washington, DC, pp. 317–23.

Sahar, G, and Hollerbach, JH. (1985). Planning of miniinum-time trajectories for robot arms. In: *Proceedings of the IEEE International Conference on Robotics and Automation*. St Louis, MI, USA, pp. 751–8.

Scheuer, A. (1998). Planification de chemins à courbure continue pour robot mobile nonholonome. Thèse de doctorat Inst. Nat. Polytechnique de Grenoble, Grenoble, France.

Scheuer, A, and Laugier, Ch. (1998). Planning Sub-Optimal and Continuous-Curvature Paths for Car-Like Robots. In *Proceedings of the IEEE-RSJ International Conference on Intelligent Robots and Systems*, vol. 1, Victoria, BC, Canada, pp. 25–31.

Shih, CL, Lee, T.T, and Gruver, WA. (1990). Motion planning with time-varying polyhedral obstacles based on graph search and mathematical programming. In: *Proceedings of the IEEE International Conference on Robotics and Automation*. Cincinnatti, OH, USA, pp. 331–7.

Shiller, Z, and Chen, J.C. (1990). Optimal motion planning of autonomous vehicles in three dimensional terrains. In: *Proceedings of the IEEE International Conference on Robotics and Automation*. Cincinnatti, OH, USA, pp. 198–203.

Shiller, Z, and Lu, H.-H. (1990). Robust computation of path constrained time-optimal motion. In: *Proceedings of the IEEE International Conference on Robotics and Automation*. Cincinnatti, OH, USA, pp. 144–9.

Shiller, Z, and Dubowsky, S. (1985). On the optimal control of robotic manipulators with actuator and end-effector constraints. In: *Proceedings of the IEEE International Conference on Robotics and Automation*. St Louis, MI, USA, pp. 614–20.

Shiller, Z, and Dubowsky, S. (1988). Global time optimal motions of robotic manipulators in the presence of obstacles. In: *Proceedings of the IEEE International Conference on Robotics and Automation*. Philadelphia, PA, USA, pp. 370–75.

Shiller, Z, and Dubowsky, S. (1989). Robot path planning with obstacles, actuator, gripper and pay load constraints. *International Journal of Robotics Research*, 8(6), 3–18.

Shin, K.G, and McKay, N.D. (1985). Minimum-time control of robotic manipulators with geometric path constraints. *IEEE Trans Autom Contr*, 30, 531–41.

Simon, D, Espiau, B, Castillo, K, and Kapellos, K. (1993). Computer-Aided Design of a Generic Robot Controller Handling Reactivity and Real-Time Control Issues. *IEEE Transactions on Control Systems Technology*, 1(4), 213–29.

Simmons, R.G. (1994). Structured Control for Autonomous Robots. *IEEE Transactions on Robotics and Automation*, 10(1), 34–43.

Sussmann, H.J, and Tang, G. (1991). Shortest paths for the Reeds-Shepp car: a worked out example of the use of geometric techniques in nonlinear optimal control. Research Report Sycon-91–10. Rutgers University Center for Systems and Control.

Svestka, P. and Overmars, M.H. (1998). Probabilistic Path Planning. In: *Robot motion planning and control*, JP. Laumond (ed.). *Lecture Notes in Control and Information Science*, 229, pp. 255–304. Springer.

Udupa, S.M. (1977). Collision detection and avoidance in computer-controlled manipulators. In: *Proceedings of the International Joint Conference on Artificial Intelligence*. Cambridge, MA, USA, pp. 737–48.

Wang, Y. (1996). Nonholonomic motion planning: a polynomial fitting approach. In: *Proceedings of the IEEE International Conference on Robotics and Automation*, vol. 3, Minneapolis, MN, USA, pp. 2956–61.

Wilfong, G. (1988). Motion planning for an autonomous vehicle. In: *Proceedings of the IEEE International Conference on Robotics and Automation*. Philadelphia, PA, USA, pp. 529–33.

Section Fifteen

Vehicle modelling

Chapter 15.1

15.1

Modelling and assembly of the full vehicle

Michael Blundell and Damian Harty

15.1.1 Introduction

In this chapter we will address the main systems that must be modelled and assembled to create and simulate the dynamics of the full vehicle system. The term 'full vehicle system' needs to be understood within the context of this textbook. The use of powerful modern multibody systems software allows the modelling and simulation of a range of vehicle subsystems representing the chassis, engine, driveline and body areas of the vehicle. This is illustrated in Fig. 15.1-1 where it can be seen that multibody systems models for each of these areas are integrated to provide a detailed 'literal' representation of the full vehicle. Note that Fig. 15.1-1 includes the modelling of the driver and road as elements of what is considered to constitute a full vehicle system model.

In this chapter we restrict our discussion of 'full vehicle system' modelling to a level appropriate for the simulation of the vehicle dynamics. As such the modelling of the suspension systems, anti-roll bars, steering system, steering inputs, brake system and drive inputs to the road wheels will all be covered. With regard to steering the modelling of the driver inputs will also be described with a range of driver models.

Note at this stage we do not consider the active elements of vehicle control other than to introduce the modelling of ABS for vehicle braking.

For the vehicle dynamics task a starting point involving models of less elaborate construction than that suggested in Fig. 15.1-1 will provide useful insights much earlier in the design process. Provided such models correctly distribute load to each tyre and involve a usefully accurate tyre model, such as the 'Magic Formula', good predictions of the vehicle response for typical proving ground manoeuvres can be obtained.

The treatment that follows in this chapter will discuss a range of options that addresses the representation of the suspension in the full vehicle as either an assembly of linkages or using simpler 'conceptual' models. It is necessary here to start with the discussion of suspension representation in the full vehicle to set the scene for following sections dealing with the modelling of springs in simple suspension models or the derivation of roll stiffness. A case study provided at the end of this chapter will compare the simulated outputs for a simulated vehicle manoeuvre using a range of suspension modelling strategies that are described in Section 15.1.4.

15.1.2 The vehicle body

For the vehicle dynamics task the mass, centre of mass position and mass moments of inertia of the vehicle body require definition within the multibody data set describing the full vehicle. It is important to note that the body mass data may include not only the structural mass of the body-in-white but also the mass of the engine, exhaust system, fuel tank, vehicle interior, driver, passengers and any other payload. A modern CAD system, or the pre-processing capability, for example, in ADAMS/View, can combine all these components to provide the analyst with a single lumped mass.

Figure 15.1-2 shows a detailed representation of a full vehicle model. In a model such as this there are a number of methods that might be used to represent the individual components. Using a model that most closely resembles

The Multibody Systems Approach to Vehicle Dynamics; ISBN: 9780750651127
Copyright © 2004 Michael Blundell and Damian Harty. All rights of reproduction, in any form, reserved.

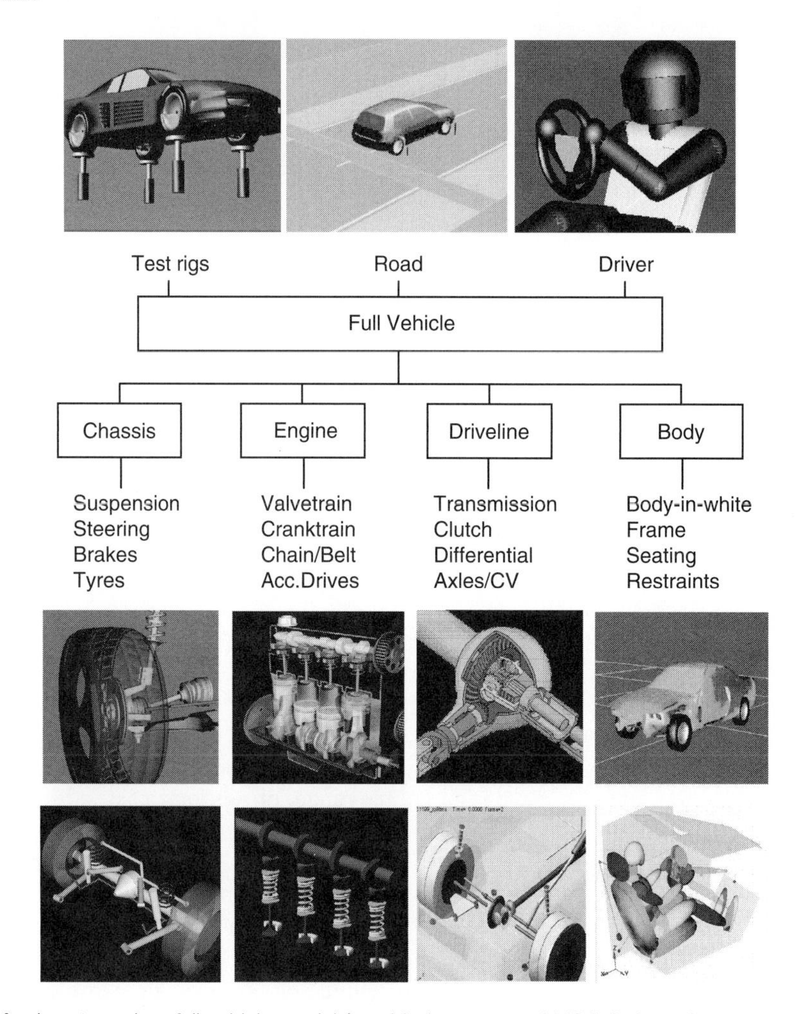

Fig. 15.1-1 Integration of subsystems in a full vehicle model (provided courtesy of MSC.Software).

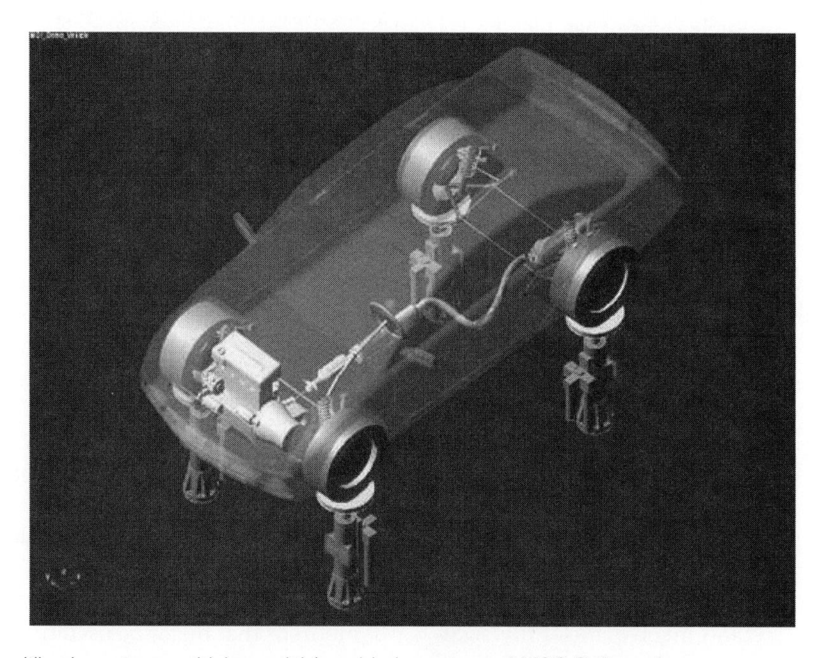

Fig. 15.1-2 A detailed multibody systems vehicle model (provided courtesy of MSC.Software).

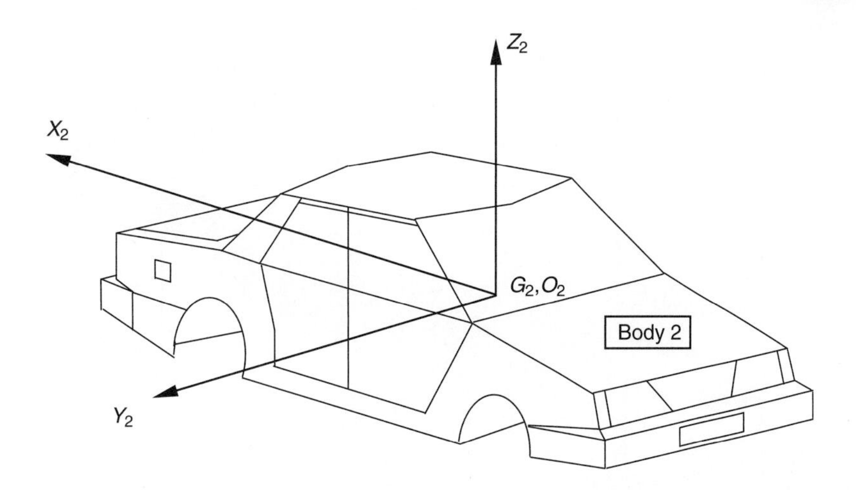

Fig. 15.1-3 Vehicle body reference frame.

the actual vehicle, components such as the engine might, for example, be elastically mounted on the vehicle body using bush elements to represent the engine mounts.

The penalty for this approach will be the addition of 6 degrees of freedom for each mass treated in this way. Alternatively a fix joint may be used to rigidly attach the mass to the vehicle body. Although this would not add degrees of freedom, the model would be less efficient through the introduction of additional equations representing the extra body and the fix joint constraint. The use of fix joint constraints may also introduce high reaction moments that would not exist in the model when using elastic mounts distributed about the mass.

An example of a vehicle body referenced frame O_2 located at the mass centre G_2 for Body 2 is shown in Fig. 15.1-3. For this model the XZ plane is located on the centre line of the vehicle with gravity acting parallel to the negative Z_2 direction. From inspection of Fig. 15.1-3 it can be seen that a value would exist for the I_{xz} cross product of inertia but that I_{xy} and I_{yz} should approximate to zero given the symmetry of the vehicle. In reality there may be some asymmetry that results in a CAD system outputting small values for the I_{xy} and I_{yz} cross products of inertia.

The dynamics of the actual vehicle are greatly influenced by the yaw moment of inertia I_{zz} of the complete vehicle, to which the body and associated masses will make the dominant contribution. A parameter often discussed is the ratio k^2/ab, sometimes referred to as the 'Dynamic Index', where k is the radius of gyration associated with I_{zz} and a and b locate the vehicle mass centre longitudinally relative to the front and rear axles respectively.

The assumption so far has been that the vehicle body is represented as a single rigid body but it is possible to model the torsional stiffness of the vehicle structure if it is felt that this could influence the full vehicle simulations. A simplistic representation of the torsional stiffness of the body may be used where the vehicle body is modelled as two rigid masses, front and rear half body parts, connected by a revolute joint aligned along the longitudinal axis of the vehicle and located at the mass centre. The relative rotation of the two body masses about the axis of the revolute joint is resisted by a torsional spring with a stiffness corresponding to the torsional stiffness of the vehicle body. Typically, the value of torsional stiffness may be obtained using a finite element model of the type shown in Fig. 15.1-4. For efficiency symmetry has been exploited here to model with finite elements only one-half of the vehicle body. This requires the use of anti-symmetry constraints along the centre line of the finite element model, for all nodes on the plane of geometric symmetry, to carry out the asymmetric torsion case.

It is also possible to incorporate a finite element representation of the vehicle body within the multibody system full vehicle model. Despite the capability of modern engineering software to include this level of detail it will be seen from Case study 7 at the end of this chapter that a single lumped mass is an efficient and accurate representation of a relatively stiff modern vehicle body for the simulation of a vehicle handling manoeuvre.

15.1.3 Measured outputs

Before continuing in this chapter to describe the subsystems that describe the full vehicle we need to consider the typical outputs measured on the proving ground and predicted by simulation. An initial treatment is given here to support the following discussion and the case study presented at the end of this chapter. For a full vehicle system simulation the predicted outputs are generally

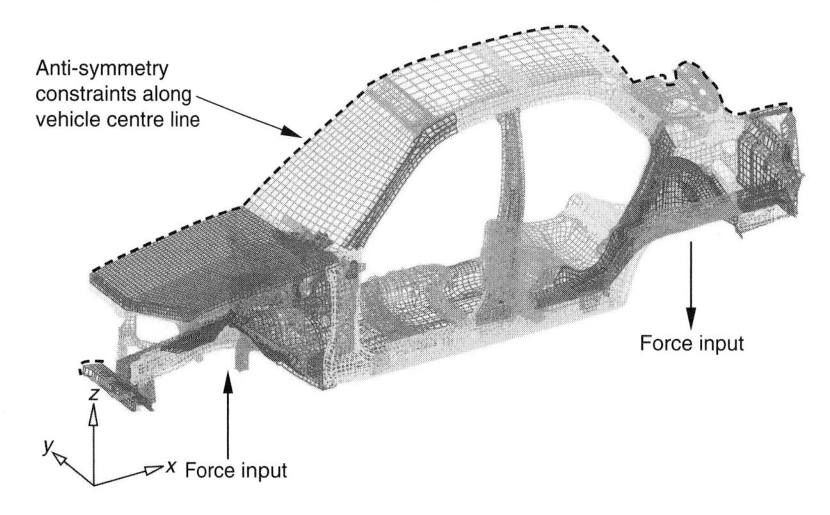

Fig. 15.1-4 Finite element model of body-in-white.

plotted as time history graphs where the outputs are computed in a reference frame fixed in the vehicle body as indicated in Fig. 15.1-5. Typical outputs can include:

(i) Forward velocity
(ii) Lateral acceleration
(iii) Roll angle
(iv) Pitch angle
(v) Yaw rate
(vi) Roll rate

Another measure often determined during test or simulation is the body slip angle, β. This is the angle of the vehicle velocity vector measured from a longitudinal axis through the vehicle as shown in Fig. 15.1-6. The components of velocity of the vehicle mass centre V_x and V_y, measured in vehicle body reference frame, can be used to readily determine this.

15.1.4 Suspension system representation

15.1.4.1 Overview

In this section the representation of the suspension as a component of the full vehicle system model will be considered. As stated the use of powerful multibody systems analysis programs often results in modelling the suspension systems as installed on the actual vehicle.

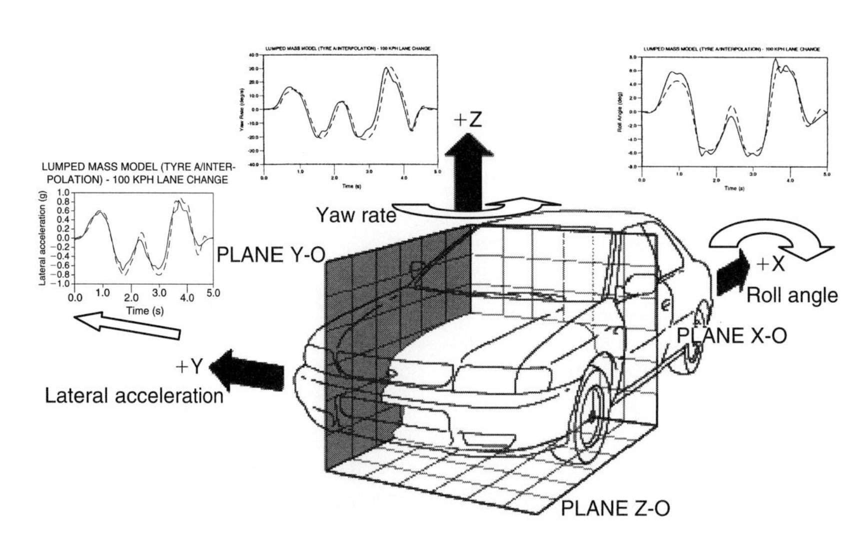

Fig. 15.1-5 Typical lateral responses measured in vehicle co-ordinate frame.

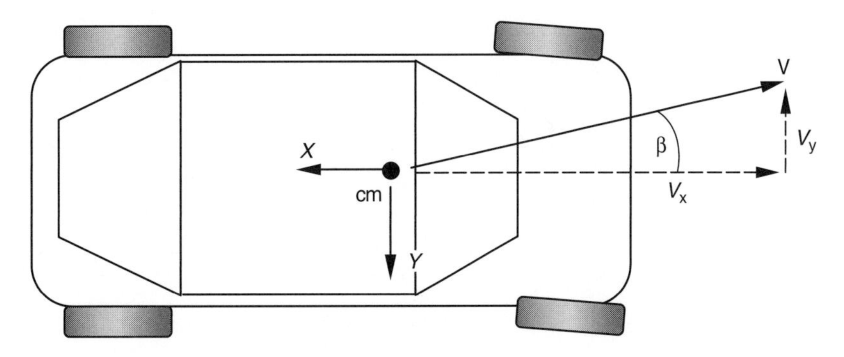

Fig. 15.1-6 Body slip angle.

In the following discussion a vehicle modelled with the suspension represented in this manner is referred to as a 'Linkage model'.

Before the advent of computer simulation classical vehicle dynamicists needed to simplify the modelling of the vehicle to a level where the formulation of the equations of motion was manageable and the solution was amenable with the computational tools available at the time. Such an approach encouraged efficiency with the analyst identifying the modelling issues that were important in representing the problem in hand. The use of modern software need not discourage such an approach. The following sections summarize four vehicle models, one of which is based on modelling the suspension linkages with three other models that use alternative simplified implementations. All four models have been used to simulate a double lane change manoeuvre and are compared in Case study 7 at the end of this chapter. The four models described here involve levels of evolving detail and elaboration and can be summarized as follows:

(i) A *lumped mass model*, where the suspensions are simplified to act as single lumped masses which can only translate in the vertical direction with respect to the vehicle body.

(ii) An *equivalent roll stiffness model*, where the body rotates about a single roll axis that is fixed and aligned through the front and rear roll centres.

(iii) A *swing arm model*, where the suspensions are treated as single swing arms that rotate about a pivot point located at the instant centres for each suspension.

(iv) A *linkage model*, where the suspension linkages and compliant bush connections are modelled in detail in order to recreate as closely as possible the actual assemblies on the vehicle.

15.1.4.2 Lumped mass model

For the lumped mass model the suspension components are considered lumped together to form a single mass. The mass is connected to the vehicle body at the wheel centre by a translational joint that only allows vertical sliding motion with no change in the relative camber angle between the road wheels and the body. The camber angle between the road wheels and the road will therefore be directly related to the roll angle of the vehicle. Spring and damper forces act between the suspensions and the body. Such suspensions have been used on early road vehicles, notably the Lancia Lambda (1908–1927), where it was termed 'sliding pillar'.

The front wheel knuckles are modelled as separate parts connected to the lumped suspension parts by revolute joints. The steering motion required for each manoeuvre is achieved by applying time dependent rotational motion inputs about these joints. Each road wheel is modelled as a part connected to the suspension by a revolute joint. The lumped mass model is shown schematically in Fig. 15.1-7.

15.1.4.3 Equivalent roll stiffness model

This model is developed from the lumped mass model by treating the front and rear suspensions as rigid axles connected to the body by revolute joints. The locations of the joints for the two axles are their respective 'roll centres'. A torsional spring is located at the front and rear roll centres to represent the roll stiffness of the vehicle. The determination of the roll stiffness of the front and rear suspensions required an investigation as described in the following section. The equivalent roll stiffness model is shown schematically in Fig. 15.1-8.

Note that this model shows the historical background to much of the current unclear thinking about roll centres and their influence on vehicle behaviour. With beam axles, as were prevalent in the 1920s, this model is a good equivalent for looking at handling behaviour on flat surfaces and ignoring ride inputs. For independent suspensions where the anti-roll geometry remains relatively consistent with respect to the vehicle and where the roll centres are relatively low (i.e. less than around 100 mm for a typical passenger car) – a fairly typical double wishbone

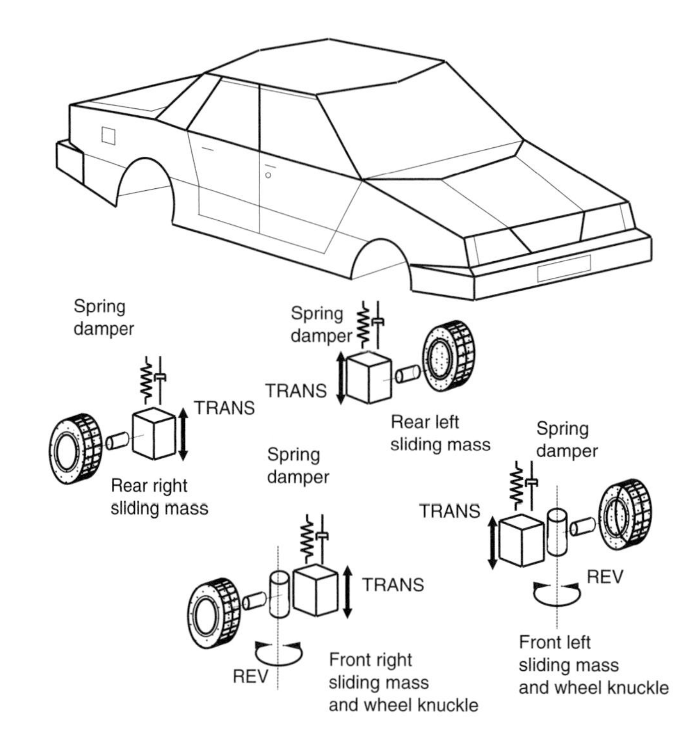

Fig. 15.1-7 Lumped mass model approach.

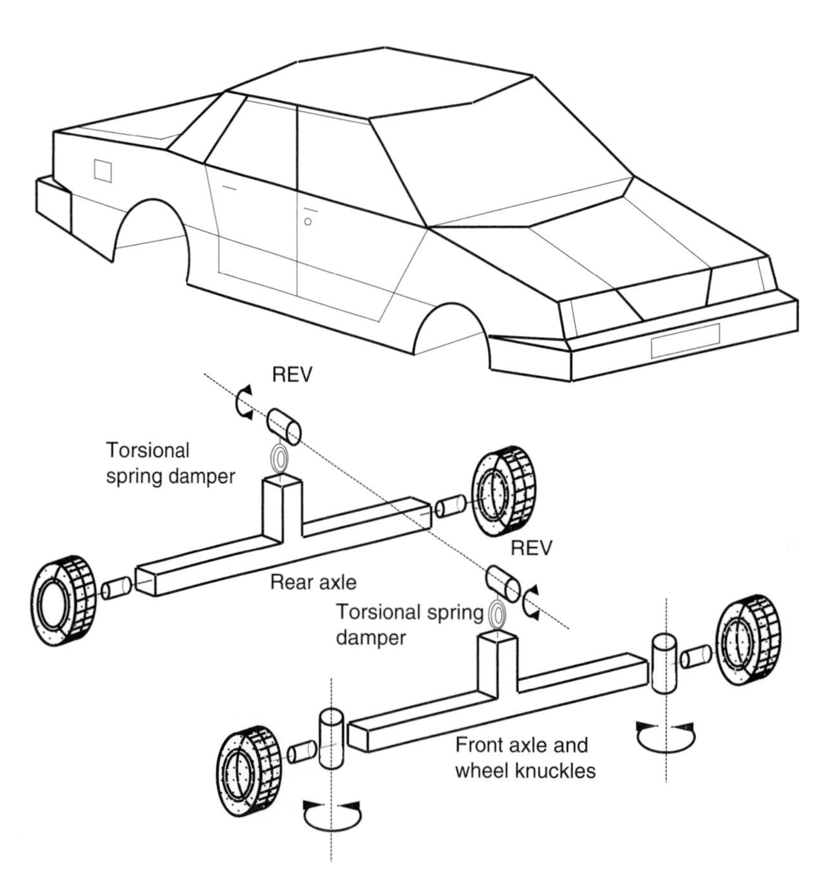

Fig. 15.1-8 Equivalent roll stiffness model approach.

setup, for example – then this approximation can be useful despite its systematic inaccuracy. However, drawing general conclusions from such specific circumstances can be dangerous; vehicles which combine a strut suspension at one end (with very mobile anti-roll geometry) and double wishbone at the other (with relatively constant anti-roll geometry) may not be amenable to such simplifications. With this and all other simplified models, the analyst must consider whether or not the conclusions that are drawn reflect upon the simplification adopted or actually reveal some useful insight. The case study presented at the end of the chapter shows a vehicle that behaves acceptably when modelled in this way.

15.1.4.4 Swing arm model

This model is developed from the equivalent roll stiffness mass model by using revolute joints to allow the suspensions for all four wheels to 'swing' relative to the vehicle body rather than using the suspensions linked on an axle. The revolute joints are located at the instant centres of the actual suspension linkage assembly. These positions are found by modelling the suspensions separately. The swing arm model has an advantage over the roll centre model in that it allows the wheels to change camber angle independently of each other and relative to the vehicle body. The swing arm model is shown schematically in Fig. 15.1-9. Although in the sketch the swing arms are shown with an axis parallel to the vehicle axis this need not be so in general. Also, although in the sketch the swing arms are shown as a 'plausible' mechanical arrangement (i.e. not overlapping) this also need not be so; in general contact between elements is not modelled for vehicle dynamics studies and in general the instant centres are widely spaced and not necessarily within the physical confines of the vehicle body. The swing arm model has the advantage over the equivalent roll stiffness model in that the heave and pitch ride behaviour can be included.

15.1.4.5 Linkage model

The model based on linkages as shown in Fig. 15.1-10 is the model that most closely represents the actual vehicle. This sort of vehicle model is the most common approach adopted by MSC.ADAMS users in the automotive industry often extending the model definition to include full non-linear bush characteristics.

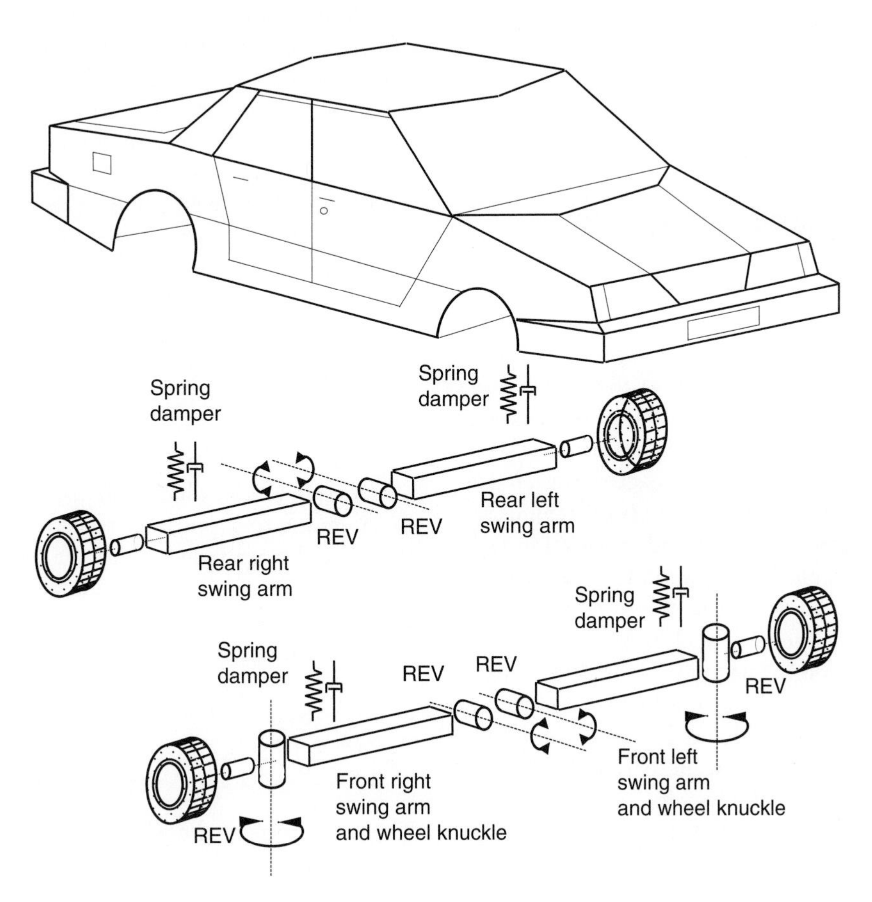

Fig. 15.1-9 Swing arm model approach.

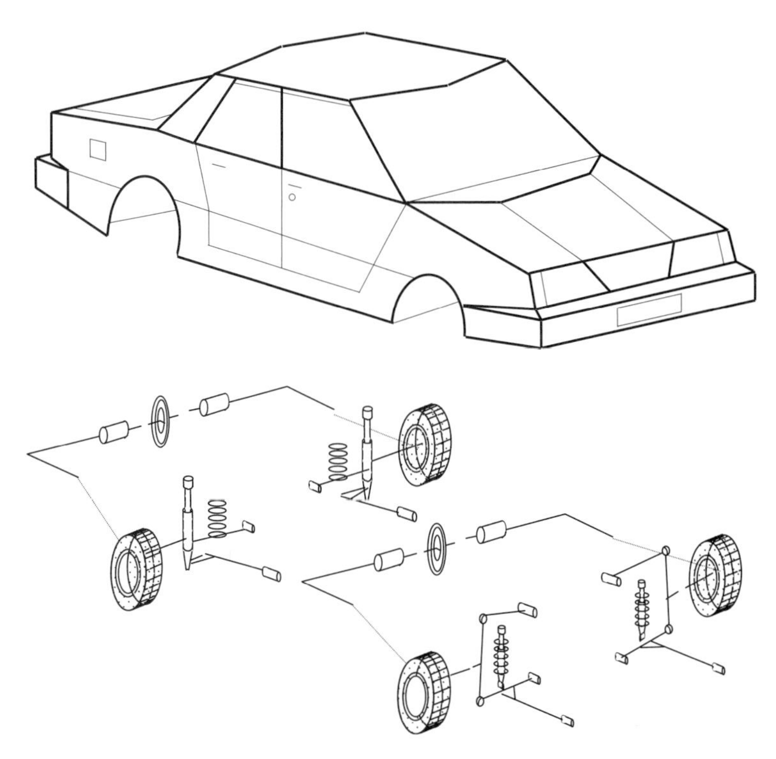

Fig. 15.1-10 Linkage model 'as is' approach.

A simplification of a model based on linkages is to treat the joints as rigid and generate a kinematic representation of the suspension system. A double wishbone arrangement is typical of a suspension system that can be modelled in this way and used for handling simulations.

15.1.4.6 The concept suspension approach

In addition to the four suspension modelling approaches just described another form of suspension model simplification considers an approach where the model contains no elements representing a physical connection between the road wheel and the chassis. Instead the movement of the road wheel with respect to the chassis is described by a functional representation, which describes the wheel centre trajectory and orientation as it moves vertically between full bump and rebound positions. Scapaticci and Minen describe this approach as the implementation of synthetic wheel trajectories. Such a method has been adopted within MSC.ADAMS where the model is referred to as a 'Concept Suspension' and is the basis of many dedicated vehicle dynamics modelling software tools such as Milliken Research Associate's VDMS, MSC's CarSim, University of Michigan's ArcSim, and Leeds University's VDAS. The way in which such a model is applied is summarized in Fig. 15.1-11. In essence the vehicle model containing the concept suspension can be used to investigate the suspension design parameters that can contribute to the delivery of the desired vehicle handling characteristics without modelling of the suspension linkages. In this way, the analyst can gain a clear understanding of the dominant issues affecting some aspect of vehicle dynamics performance. A case study is given in Section 15.1.14 describing the use of a reduced (3 degrees of freedom) linear model to assess the influence of suspension characteristics on straight-line stability. These models belong very firmly in the 'analysis' segment of the overall process diagram.

The functional representation of the model is based on components that describe effects due to kinematics dependent on suspension geometry and also elastic effects due to compliance within the suspension system. A schematic to support an explanation of the function of this model is provided in Fig. 15.1-12.

If we consider first the kinematic effects due to suspension geometry we can see that there are two variables that provide input to the model:

Δz is the change in wheel centre vertical position (wheel travel)
Δv is the change in steering wheel angle

The magnitude of the wheel travel Δz will depend on the deformation of the surface, the load acting vertically through the tyre resulting from weight transfer during

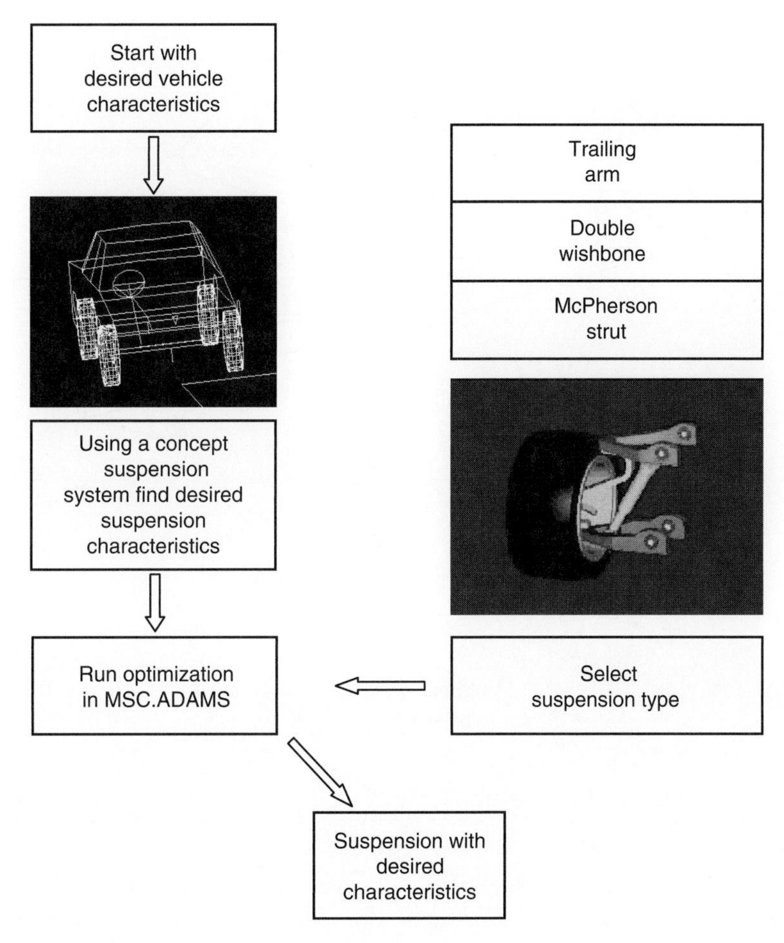

Fig. 15.1-11 Application of a concept suspension model (provided courtesy of MSC.Software).

a simulated manoeuvre and a representation of the suspension stiffness and damping acting through the wheel centre. The magnitude of the change in steering wheel angle Δv will depend on either an open loop fixed time dependent rotational motion input or a closed loop torque input using a controller to feed back vehicle position variables so as to steer the vehicle to follow a predefined path. The modelling of steering inputs is discussed in more detail later in this chapter. The dependent variables that dictate the position and orientation of the road wheel are:

Δx is the change in longitudinal position of the wheel
Δy is the change in lateral position (half-track) of the wheel
$\Delta\delta$ is the change in steer angle (toe in/out) of the wheel
$\Delta\gamma$ is the change in camber angle of the wheel

The functional dependencies that dictate how the suspension moves with respect to the input variables can be obtained through experimental rig measurements, if the vehicle exists and is to be used as a basis for the model, or by performing simulation with suspension models.

The movement of the suspension due to elastic effects is dependent on the forces acting on the wheel. In their 1992 paper Scapaticci and Minen describe the relationship using the equation shown in (15.1.1) where the functional dependencies due to suspension compliance are defined using the matrix F_E:

$$\begin{bmatrix} \Delta x \\ \Delta y \\ \Delta\delta \\ \Delta\gamma \end{bmatrix} = \begin{bmatrix} F_E \end{bmatrix} \begin{bmatrix} Fxt \\ Fxb \\ Fy \\ Mz \end{bmatrix} \tag{15.1.1}$$

and the inputs are the forces acting on the tyre:

Fxt is the longitudinal tractive force
Fxb is the longitudinal braking force
Fy is the lateral force
Mz is the self-aligning moment

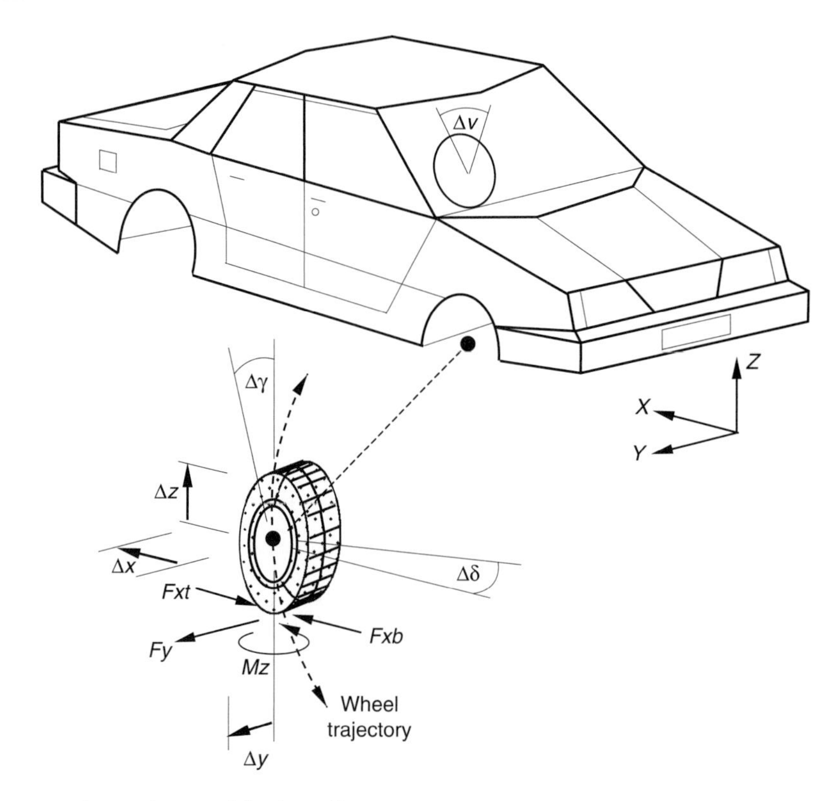

Fig. 15.1-12 Concept suspension system model schematic.

Note that the dimensions of the matrix F_E are such that cross-coupling terms, such as toe change under braking force, can exist. The availability of such data early in the design phase can be difficult but the adoption of such a generalized form allows the user to speculate on such values and thus use the model to set targets for acceptable behaviour.

15.1.5 Modelling of springs and dampers

15.1.5.1 Treatment in simple models

The treatment of road springs and dampers in a vehicle where the suspensions have been modelled using linkages is generally straightforward. A road spring is often modelled as linear but the damper will usually require a non-linear representation. It is also common for the bump travel limiter to be engaged early and to have both stiffness and damping elements to its behaviour; both those aspects may be modelled using the methods discussed here. The choice of whether to combine them with the road spring and damper forces is entirely one of modelling convenience; the authors generally find the ease of debugging and auditing the model is worth the carriage of two not strictly necessary additional force generating terms.

For the simplified modelling approach used in the lumped mass and swing arm models the road springs cannot be directly installed in the vehicle model as with the linkage model. Consider the lumped mass model when compared with the linkage model as shown in Fig. 15.1-13.

Clearly there is a mechanical advantage effect in the linkage model that is not present in the lumped mass vehicle model. At a given roll angle for the lumped mass model the displacement and hence the force in the spring will be too large when compared with the corresponding situation in the linkage model.

For the swing arm model the instant centre about which the suspension pivots is often on the other side of the vehicle. In this case the displacement in the spring is approximately the same as at the wheel and a similar problem occurs as with the lumped mass model. For all three simplified models this problem can be overcome as shown in Fig. 15.1-14 by using an ‘equivalent’ spring which acts at the wheel centre.

As an approximation, ignoring exact suspension geometry, the expression (15.1.2) can be used to represent the stiffness k_w of the equivalent spring at the wheel:

$$\begin{aligned} k_w &= F_w/\delta_w = (l_s/l_w)F_s/(l_w/l_s)\delta_s \\ &= (l_s/l_w)^2 k_s \end{aligned} \qquad (15.1.2)$$

The presence of a square function in the ratio can be considered a combination of both the extra mechanical

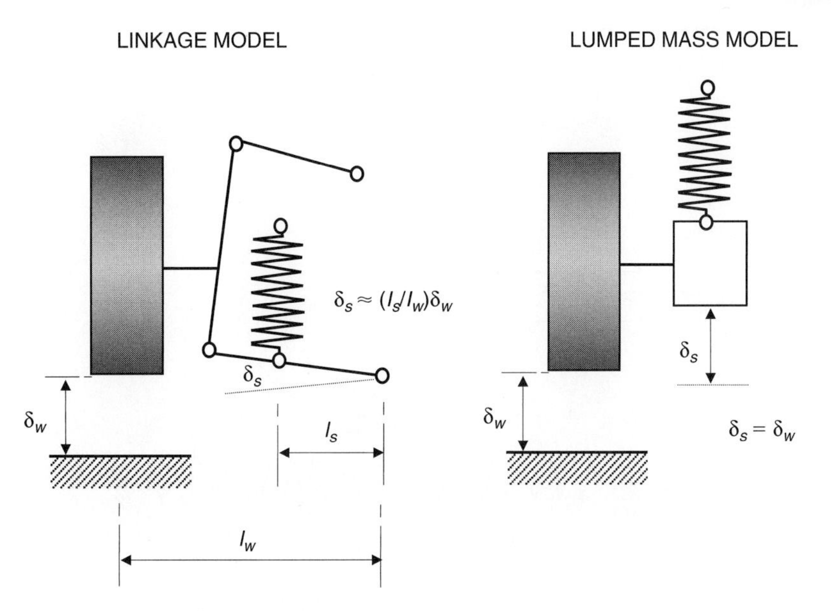

Fig. 15.1-13 Road spring in linkage and lumped mass models.

advantage in moving the definition of spring stiffness to the wheel centre and the extra spring deflection at the wheel centre.

15.1.5.2 Modelling leaf springs

Although the modelling of leaf springs is now rare on passenger cars they are still fitted extensively on light trucks and goods vehicles where they offer the advantage of providing relatively constant rates of stiffness for large variations in load at the axle. The modelling of leaf springs has always been more of a challenge in an MBS environment when compared with the relative simplicity of modelling a coil spring. Several approaches may be adopted the most common of which are shown in Fig. 15.1-15.

Early attempts at modelling leaf springs utilized the simple approach based on equivalent springs to represent the vertical and longitudinal force–displacement characteristic of the leaf spring. On the actual vehicle the leaf springs also contribute to the lateral positioning of the axle, with possible additional support from a panhard rod. Although not shown in Fig. 15.1-15 lateral springs could also be incorporated to represent this.

The next approach is based on modelling the leaf spring as three bodies (SAE 3-link model) interconnected by bushes or revolute joints with an associated torsional stiffness that provides equivalent force–displacement characteristics as found in the actual leaf spring. The last approach shown in Fig. 15.1-15 uses a detailed 'as is' approach representing each of the leaves as a series of distributed lumped masses interconnected by beam

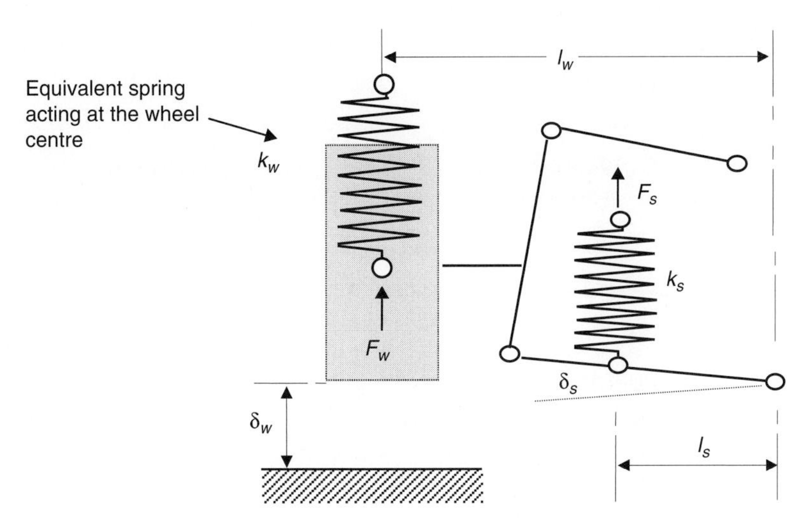

Fig. 15.1-14 Equivalent spring acting at the wheel centre.

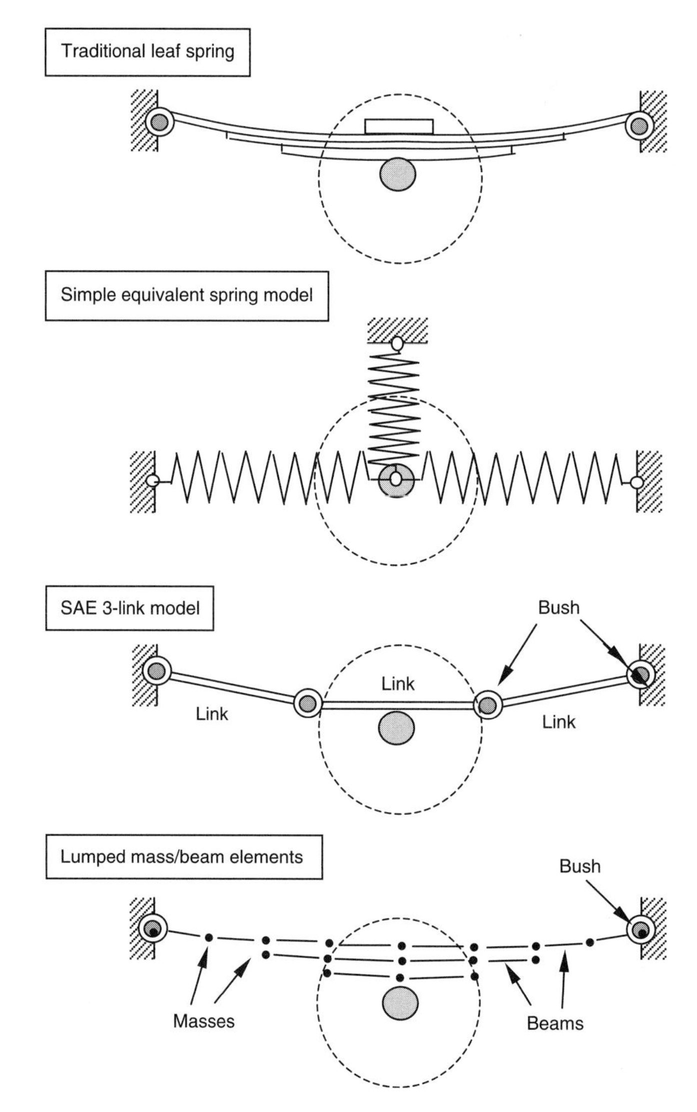

Fig. 15.1-15 Leaf spring modelling strategies.

elements with the correct sectional properties for the leaf. This type of model is also complicated by the need to model the interleaf contact forces between the lumped masses with any associated components of sliding friction.

15.1.6 Anti-roll bars

As shown in Fig. 15.1-16 anti-roll bars may be modelled using two parts connected to the vehicle body by revolute

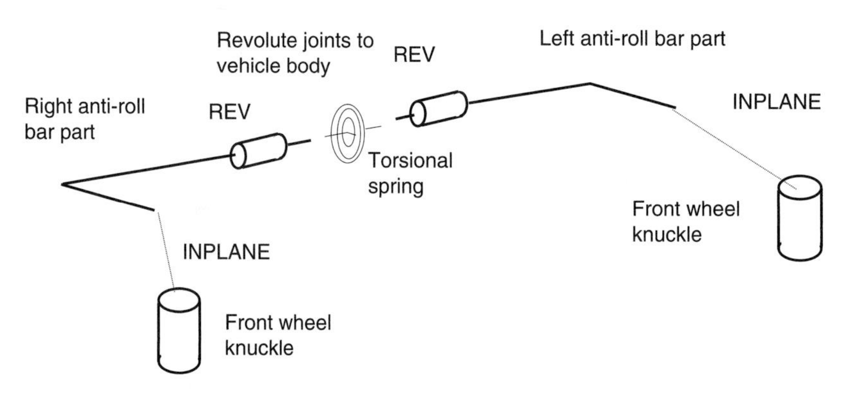

Fig. 15.1-16 Modelling the anti-roll bars using joint primitives.

joints and connected to each other by a torsional spring located on the centre line of the vehicle. In a more detailed model the analyst could include rubber bush elements rather than the revolute joints shown to connect each side of the anti-roll bar to the vehicle. In this case for a cylindrical bush the torsional stiffness of the bush would be zero to allow rotation about the axis, or could have a value associated with the friction in the joint. In this model the connection of the anti-roll bars to the suspension system is not modelled in detail, rather each anti-roll bar part is connected to the suspension using an inplane joint primitive that allows the vertical motion of the suspension to be transferred to the anti-roll bars and hence produce a relative twisting motion between the two sides.

A more detailed approach, shown in Fig. 15.1-17, involves including the drop links to connect each side of the anti-roll bar to the suspension systems. The drop link is connected to the anti-roll bar by a universal joint and is connected to the suspension arm by a spherical joint. This is similar to the modelling of a tie rod where the universal joint is used to constrain the spin of the link about an axis running along its length, this degree of freedom having no influence on the overall behaviour of the model.

The stiffness K_T of the torsional spring can be found directly from fundamental torsion theory for the twisting of bars with a hollow or solid circular cross-section. Assuming here a solid circular bar and units that are consistent with the examples that support this text we have:

$$K_T = \frac{GJ}{L} \tag{15.1.3}$$

where

G is the shear modulus of the anti-roll bar material (N/mm^2)
J is the second moment of area (mm^4)
L is the length of the anti-roll bar (mm)

Note that the length L used in equation (15.1.3) is the length of the bar subject to twisting. For the configuration shown in Fig. 15.1-17 this is the transverse length of the anti-roll bar across the vehicle and does not include the fore–aft lengths of the system that connect to the drop links. These lengths of the bar provide the lever arms to twist the transverse section of bar and are subject to bending rather than torsion. An externally solved FE model could be used to give an equivalent torsional stiffness for a simplified representation such as this.

Given that bending or flexing of the roll bar may have an influence the next modelling refinement of the anti-roll bar system uses finite element beams to interconnect a series of rigid bodies with lumped masses distributed along the length of the bar. Such sophistication becomes necessary to investigate anti-roll bar interactions with steer torque, or anti-roll bar lateral 'walking' problems in the vehicle; in general though, such detail is not required for vehicle behaviour modelling.

Again these joints could be modelled with bushes if needed. Such a model is shown in Fig. 15.1-18 would be to model the drop links with lumped masses and beams if the flexibility of these components needed to be modelled.

The modelling described so far has been for the modelling of the conventional type of anti-roll bar found on road vehicles. Vehicles with active components in the anti-roll bar system might include actuators in place of the drop links or a coupling device connecting the two halves of the system providing variable torsional stiffness at the connection. Space does not permit a description of the modelling of such systems here, but with ever more students becoming involved in motorsport this section will conclude with a description of the type of anti-roll bar model that might be included in a typical student race vehicle. A graphic for the system is shown in Fig. 15.1-19.

The modelling of this system is illustrated in the schematic in Fig. 15.1-20 where it can be seen that the anti-roll bar is installed vertically and is connected to

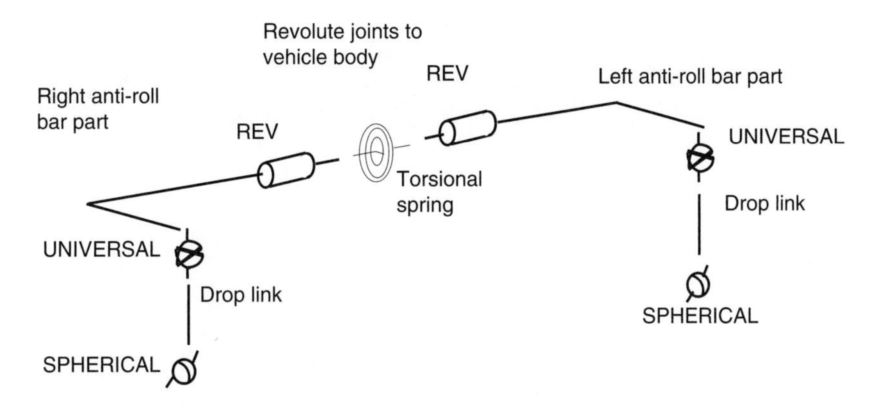

Fig. 15.1-17 Modelling the anti-roll bars using drop links. (This material has been reproduced from the Proceedings of the Institution of Mechanical Engineers, K2 Vol. 213 'The modelling and simulation of vehicle handling. Part 2: vehicle modelling', M.V. Blundell, page 131, by permission of the Council of the Institution of Mechanical Engineers.)

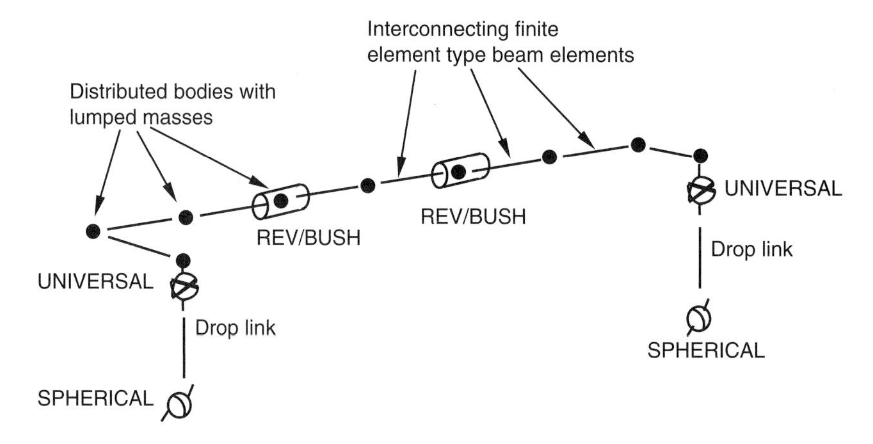

Fig. 15.1-18 Modelling the anti-roll bars using interconnected finite element beams.

the chassis by a revolute joint. The revolute joint allows the anti-roll bar to rock back and forward as the bell cranks rotate during parallel wheel travel but prevents rotation during opposite wheel travel when the body rolls. As the body rolls the torsional stiffness of the anti-roll bar, modelled with the rotational spring damper, resists the pushing motion of one push rod as the suspension moves in bump on one side and the pulling motion as the suspension moves in rebound on the other side. The small spring damper helps to locate the anti-roll bar with respect to the vehicle chassis and adds to the heave stiffness and damping. Alternative linkage designs are possible that allow the use of a translational spring element and hence allow independent control of damping in roll compared to damping in heave. Such 'three spring' systems are common in higher formula motorsports events when allowed by the rules.

15.1.7 Determination of roll stiffness for the equivalent roll stiffness model

In order to develop a full vehicle model based on roll stiffness it is necessary to determine the roll stiffness and damping of the front and rear suspension elements separately. The estimation of roll damping is obtained by assuming an equivalent linear damping and using the

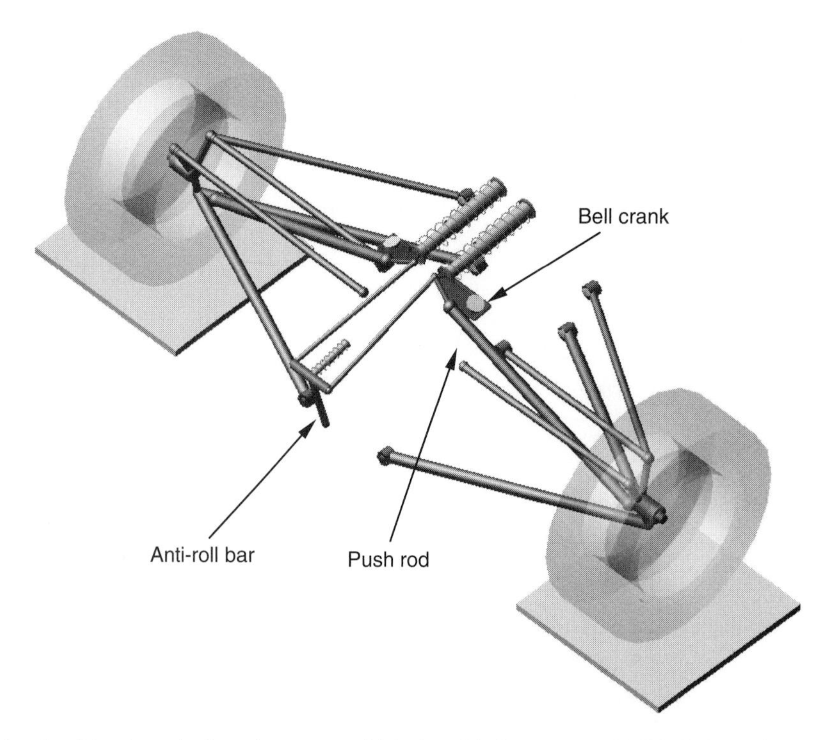

Fig. 15.1-19 Graphic of anti-roll bar in typical student race vehicle (provided courtesy of MSC.Software).

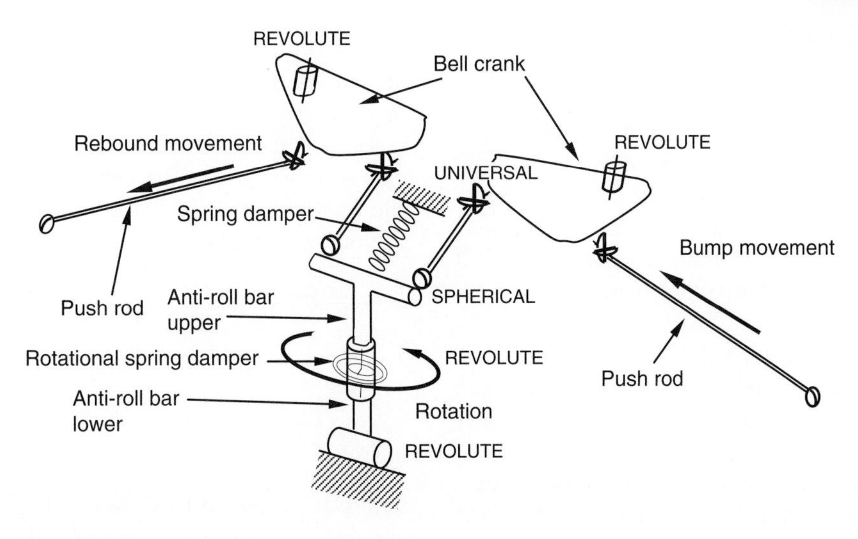

Fig. 15.1-20 Modelling of anti-roll bar mechanism in student race car.

positions of the dampers relative to the roll centres to calculate the required coefficients. If a detailed vehicle model is available the procedure used to find the roll stiffness for the front suspension elements involves the development of a model as shown in Fig. 15.1-21. This model includes the vehicle body, this being constrained to rotate about an axis aligned through the front and rear roll centres. The vehicle body is attached to the ground part by a cylindrical joint located at the front roll centre and aligned with the rear roll centre. The rear roll centre is attached to the ground by a spherical joint in order to prevent the vehicle sliding along the roll axis. A motion input is applied at the cylindrical joint to rotate the body through a given angle. By requesting the resulting torque acting about the axis of the joint it is possible to calculate the roll stiffness associated with the front end of the vehicle. The road wheel parts are not included nor are the tyre properties. The tyre compliance is represented separately by a tyre model and should not be included in the determination of roll stiffness. The wheel centres on either side are constrained to remain in a horizontal plane using inplane joint primitives. Although the damper force elements can be retained in the suspension models they have no contribution to this calculation as the roll stiffness is determined using static analysis. The steering system, although not shown in Fig. 15.1-21, may also be included in the model. If present a motion input is needed to lock the steering in the straight-ahead position during the roll simulation.

For the rear end of the vehicle the approach is essentially the same as for the front end, with in this case a

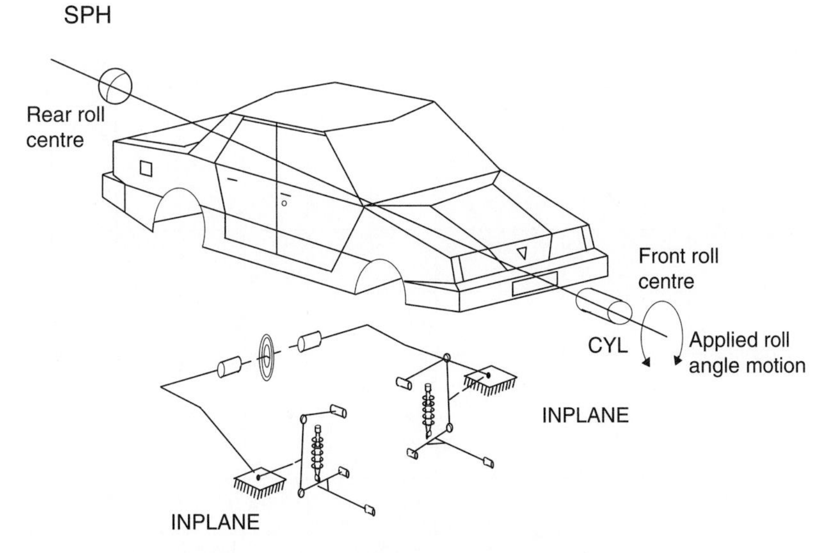

Fig. 15.1-21 Determination of front end roll stiffness. (This material has been reproduced from the Proceedings of the Institution of Mechanical Engineers, K2 Vol. 213 'The modelling and simulation of vehicle handling. Part 2: vehicle modelling', M.V. Blundell, page 127, by permission of the Council of the Institution of Mechanical Engineers.)

cylindrical joint located at the rear roll centre and a spherical joint located at the front roll centre.

For both the front and rear models the vehicle body can be rotated through an appropriate angle either side of the vertical. For the example vehicle used in this text the body was rotated 10 degrees each way. The results for the front end model are plotted in Fig. 15.1-22. The gradient at the origin can be used to obtain the value for roll stiffness used in the equivalent roll stiffness model described earlier.

In the absence of an existing vehicle model that can be used for the analysis described in the preceding section, calculations can be performed to estimate the roll stiffness. In reality this will have contributions from the road springs, anti-roll bars and possibly the suspension bushes. Fig. 15.1-23 provides the basis for a calculation of the road spring contribution for the simplified arrangement shown. In this case the inclination of the road springs is ignored and have a separation across the vehicle given by L_s.

As the vehicle rolls through an angle ϕ the springs on each side are deformed with a displacement δ_s given by

$$\delta_s = \phi L_s/2 \tag{15.1.4}$$

The forces generated in the springs F_s produce an equivalent roll moment M_s given by

$$M_s = F_s L_s = k_s \delta_s L_s = k_s \phi L_s^2/2 \tag{15.1.5}$$

The roll stiffness contribution due to the road springs K_{Ts} at the end of the vehicle under consideration is given by

$$K_{Ts} = M_s/\phi = k_s L_s^2/2 \tag{15.1.6}$$

In a similar manner the contribution to the roll stiffness at one end of the vehicle due to an anti-roll bar can be determined as shown in Fig. 15.1-24.

In this case if the ends of the anti-roll bar are separated by a distance L_r and the vehicle rolls through an angle ϕ, the relative deflection of one end of the anti-roll bar to the other δ_r is given by

$$\delta_r = a\theta = \phi L_r \tag{15.1.7}$$

The angle of twist in the roll bar is given by

$$\theta = \frac{TL_r}{GJ} \tag{15.1.8}$$

where as discussed earlier G is the shear modulus of the anti-roll bar material, J is the polar second moment of area and T is the torque acting about the transverse section of the anti-roll bar. Note that in this analysis we are ignoring the contribution due to bending. The forces acting at the ends of the anti-roll bar F_r produce an equivalent roll moment M_r given by

$$M_r = F_r L_r = TL_r/a = \theta GJ/a = \phi L_r GJ/a^2 \tag{15.1.9}$$

The roll stiffness contribution due to the anti-roll bar K_{Tr} at the end of the vehicle under consideration is given by

$$K_{Tr} = M_r/\phi = L_r GJ/a^2 \tag{15.1.10}$$

The contribution of both the road springs and the anti-roll bar can then be added, ignoring suspension bushes here, to give the roll stiffness K_T:

$$K_T = K_{Ts} + K_{Tr} \tag{15.1.11}$$

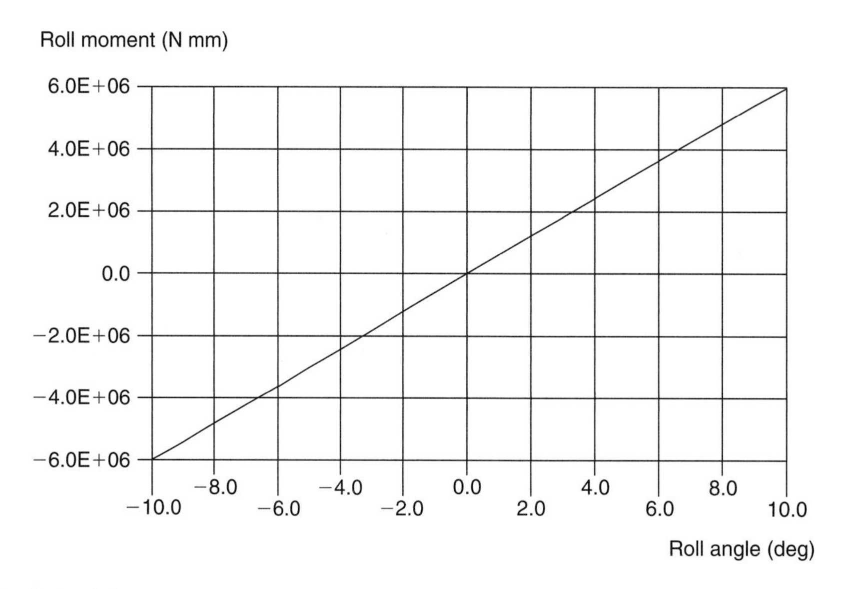

Fig. 15.1-22 Front end roll simulation.

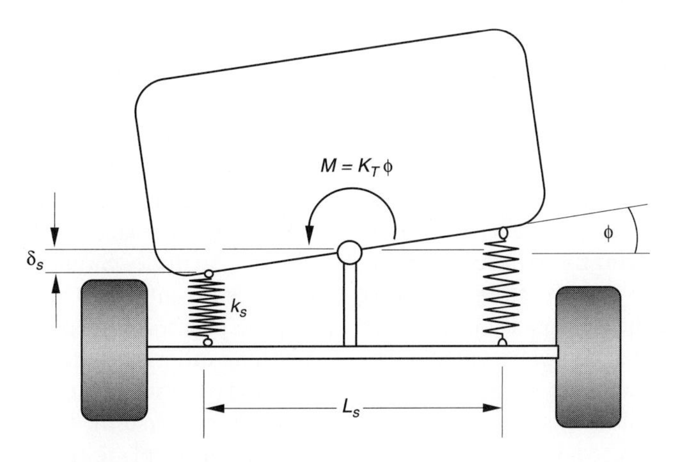

Fig. 15.1-23 Calculation of roll stiffness due to road springs.

Note that current practice in vehicles is to have relatively soft springs and fit stiffer anti-roll bars than was the norm some years ago. If vehicles achieve a large proportion of their roll stiffness from anti-roll bars, the subjective phenomenon of 'roll rock' (also known as 'lateral head toss') becomes problematic. A rule of thumb is that such phenomena begin to emerge when the anti-roll bars form more than about one-third of the overall roll stiffness – in other words if K_{Tr} is greater than 0.5 K_{Ts}.

15.1.8 Aerodynamic effects

Some treatment of aerodynamics is generally given in existing text books (Milliken and Milliken, 1995; Gillespie, 1992) dealing with vehicle dynamics. Other textbooks are dedicated to the subject. The flow of air over the body of a vehicle produces forces and moments acting on the body resulting from the pressure distribution (form) and friction between the air and surface of the body. The forces and moments are considered using a body centred reference frame where longitudinal forces (drag), lateral forces, and vertical forces (lift or down thrust) will arise. The aerodynamic moments will be associated with roll, pitch and yaw rotations about the corresponding axes.

Current practice is generally to ignore aerodynamic forces for the simulation of most proving ground manoeuvres but for some applications and classes of vehicle this is clearly not representative of the vehicle dynamics in the real world, for example winged vehicles. It is often said that for some vehicles of this type the down thrust is so great that this could overcome the weight of a vehicle, allowing it, for example, to drive upside-down through a tunnel, although this has never been demonstrated.

The lack of speed limits on certain autobahns in Germany also means that a vehicle manufacturer selling a high performance vehicle to that market will need to test the vehicle at speeds well over twice the legal UK limit. The possibility of aerodynamic forces at these high speeds destabilizing the vehicle needs to be investigated and where physical testing is to be done, equivalent computer simulation is also desirable. Other effects such as side gusting are also tested for and have been simulated by vehicle dynamicists in the past.

An approach that has been commonly used is to apply forces and moments to the vehicle body using measured results, look-up tables, from wind tunnel testing. As the vehicle speed and the attitude of the body change during the simulation the forces and moments are interpolated

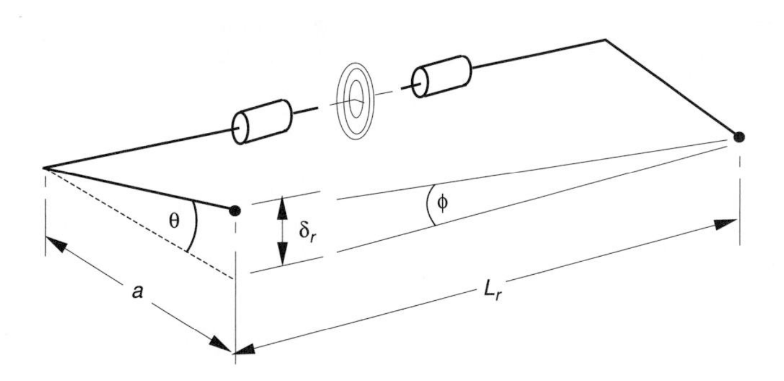

Fig. 15.1-24 Calculation of roll stiffness due to the anti-roll bar.

from the measured data and applied to the vehicle body. A difficulty with such an approach is that the measured results are for steady state in each condition and that transient effects are not included in the simulation. Consideration has been given to the use of a computational fluid dynamics (CFD) program to calculate aerodynamic forces and moments in parallel with (co-simulation) an MBS program solving the vehicle equations of motion. The problem at the current time with this approach is the mismatch in the computation time for both methods. MBS models of a complete vehicle can simulate vehicle handling manoeuvres in seconds, or even real time, whereas complex CFD models can involve simulation times running into days. Current CFD methods also have difficulty with aerodynamic transient effects (e.g. vortex shedding) although an emerging group of 'multi-physics' codes look set to address these problems. Thus there is no realistic prospect of the practical use of transient aerodynamics effects being modelled in the near future. However, genuine transient aerodynamic effects, such as those involved in so-called 'aeroelastic flutter' – an unsteady aerodynamic flow working in sympathy with a structural resonance – are extremely rare in ground vehicles.

In order to introduce readers to the fundamentals consider a starting point where it is intended only to formulate an aerodynamic drag force acting on the vehicle body.

The drag force F_D can be considered to act at a frontal centre of pressure for the vehicle centre of pressure (CP) and have the following formulation:

$$F_D = \frac{1}{2}\frac{\rho V^2 C_D A}{GC} \quad (15.1.12)$$

where

C_D = the aerodynamic drag coefficient

ρ = the density of air

A = the frontal area of the vehicle (projected onto a yz plane)

V = the velocity of the vehicle in the direction of travel

GC = a gravitational constant

The gravitational constant is included in equation (15.1.12) to remind readers that this is a dynamic force. If the model units are SI then GC is equal to 1. If as commonly used the model units for length are mm then GC is equal to 1000. When formulating the aerodynamic drag force it should be considered that the force acts at the CP and that this point generally moves as the vehicle changes attitude. Similarly the drag coefficient C_D and projected frontal area A also change as the body moves. For the position shown in Fig. 15.1-25 it is clear that for anything other than straight-line motion it is going to be necessary to model the forces as components in the body centred axis system. If we consider the vehicle moving only in the xy plane then this is going to require at least the formulation of a longitudinal force Fx, a lateral force Fy and a yawing moment Mz all resolved from the centre of pressure to the body centred axis system, usually located at the mass centre. Wind tunnel testing or computational fluid dynamic analysis is able to yield coefficients for all six possible forces and moments acting on the body, referred back to the mass centre. Note that for passenger vehicles it is typical that the aerodynamic yaw moment is as shown in the figure, i.e. is such to make the vehicle turn away from the wind. For other vehicles this may not be true and individual research on the vehicle in question is needed.

15.1.9 Modelling of vehicle braking

In an earlier work the force and moment generating characteristics of the tyre were discussed and it was shown how the braking force generated at the tyre contact patch depends on the slip ratio as the wheel is braked from a free rolling wheel with a slip ratio of zero to a fully locked wheel where the slip ratio is unity. In this section we are not so much concerned with the tyre, given that we would be using a tyre model interfaced with our full vehicle model to represent this behaviour. Rather we now address the modelling of the mechanisms used to apply a braking torque acting about the spin axis of the road wheel that produces the change in slip ratio and subsequent braking force.

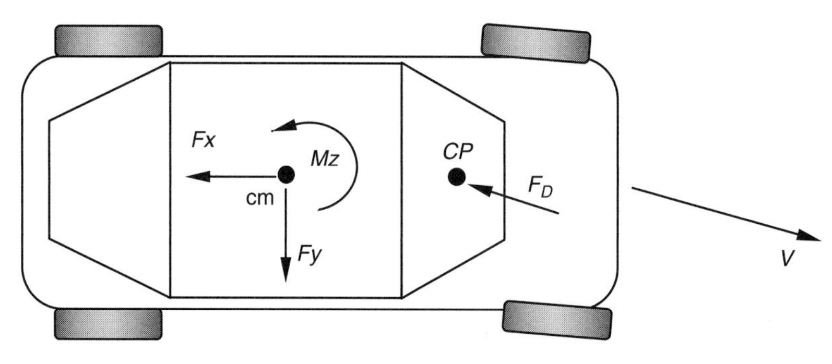

Fig. 15.1-25 Application of aerodynamic drag force.

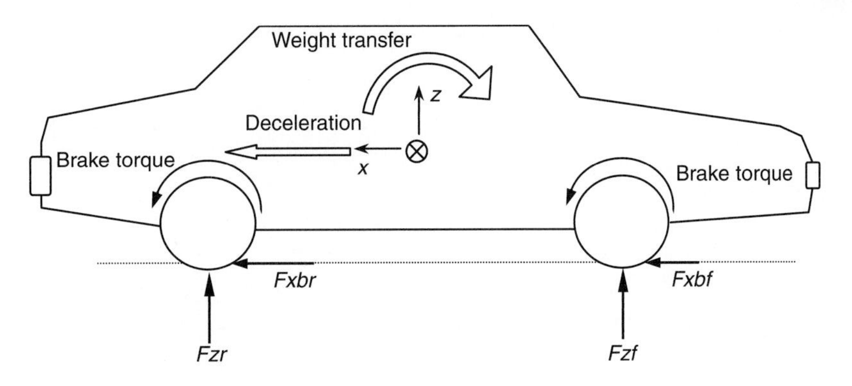

Fig. 15.1-26 Braking of a full vehicle.

Clearly as the vehicle brakes, as shown in Fig. 15.1-26, there is weight transfer from the rear to the front of the vehicle. Given what we know about the tyre behaviour the change in the vertical loads acting through the tyres will influence the braking forces generated. As such the braking model may need to account for real effects such as proportioning the braking pressures to the front and rear wheels or the implementation of anti-lock braking systems (ABS). Before any consideration of this we need to address the mechanism to model a braking torque acting on a single road wheel.

If we consider a basic arrangement the mechanical formulation of a braking torque, based on a known brake pressure, acting on the piston can be derived from Fig. 15.1-27.

The braking torque B_T is given by

$$B_T = n\mu pAR_d \tag{15.1.13}$$

where

n = the number of friction surfaces (pads)
μ = he coefficient of friction between the pads and the disc
p = the brake pressure
A = the brake piston area
R_d = the radius to the centre of the pad

Note that depending on the sophistication of the model the coefficient of friction μ may be constant or defined as a run-time variable as a function of brake rotor temperature. Obtaining such data is usually relatively easy, but the calculation of rotor temperature can be a little more involved. Figure 15.1-28 shows typical specific heat capacity versus temperature characteristics for different brake rotor materials. Brake rotor temperature, T, can be calculated using the expression:

$$T = T_0 + \frac{B_T \omega t - hA_c(T - T_{\text{env}})}{mc} \tag{15.1.14}$$

where

T_0= initial brake rotor temperature (K)
ω = brake rotor spin velocity (rads second^{-1})
t = time (seconds)
h = brake rotor convection coefficient (Wm^{-2}K^{-1})

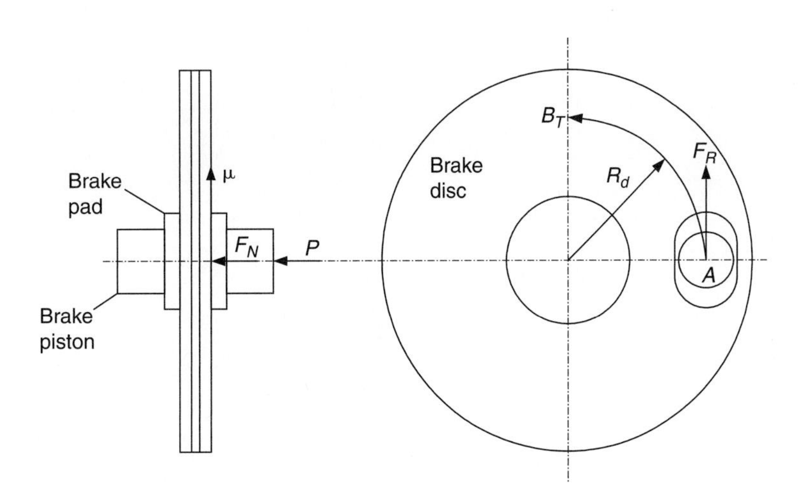

Fig. 15.1-27 Braking mechanism.

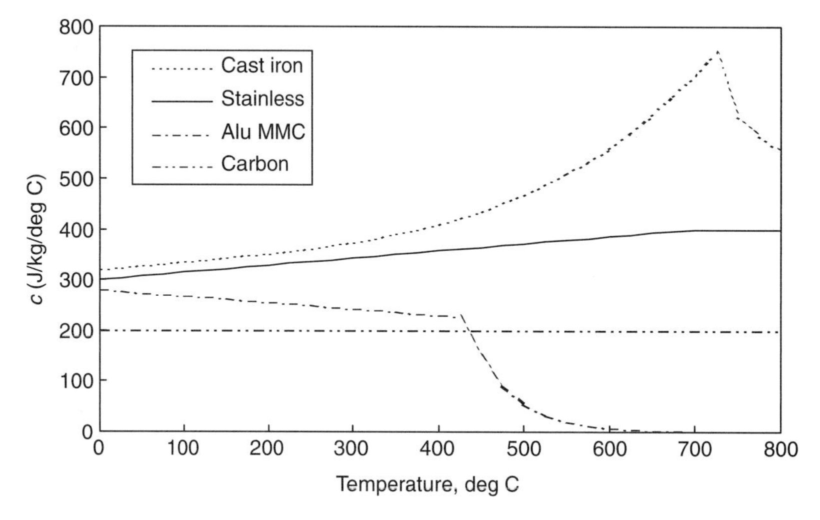

Fig. 15.1-28 Specific heat capacity, c, versus temperature, T.

A_c = convective area of brake disc (m^2)
T_{env} = environmental temperature (K)
m = mass of brake rotor (kg)
c = specific heat capacity of brake rotor ($J\,kg^{-1}\,K^{-1}$)

For the most common brake rotor material, cast iron, the specific heat versus temperature characteristic can be approximated in the working range (0–730°C) by the expression:

$$c = 320 + 0.15T + 1.164 \times 10^{-9}T^4 \quad (15.1.15)$$

Note that in the above expression, temperature T is in centigrade (celsius) and not kelvin. The brake torque and temperature models may be used easily within a multibody system model using a combination of design variables (declared in MSC.ADAMS using the 'variable create' command) and run-time variables (declared in MSC.ADAMS using 'data_element create' variable) as shown in Table 15.1-1 where we are using for the first time here an input format that corresponds to a command language used in MSC.ADAMS. Note the need for an explicit iteration since the temperature depends on the heat capacity and the heat capacity depends on the temperature. When modelling such behaviour in a spreadsheet, it is sufficient to refer to the temperature of the preceding time step. Although this is possible within many multibody system packages, it can be awkward to implement and can also lead to models with some degree of numerical delicacy.

Note also that it is common practice within brake manufacturers to separate the brake energizing event from the brake cooling event for initial design calculations, leading to a systematic overestimation of the temperature during fade/recovery testing. This conservative approach

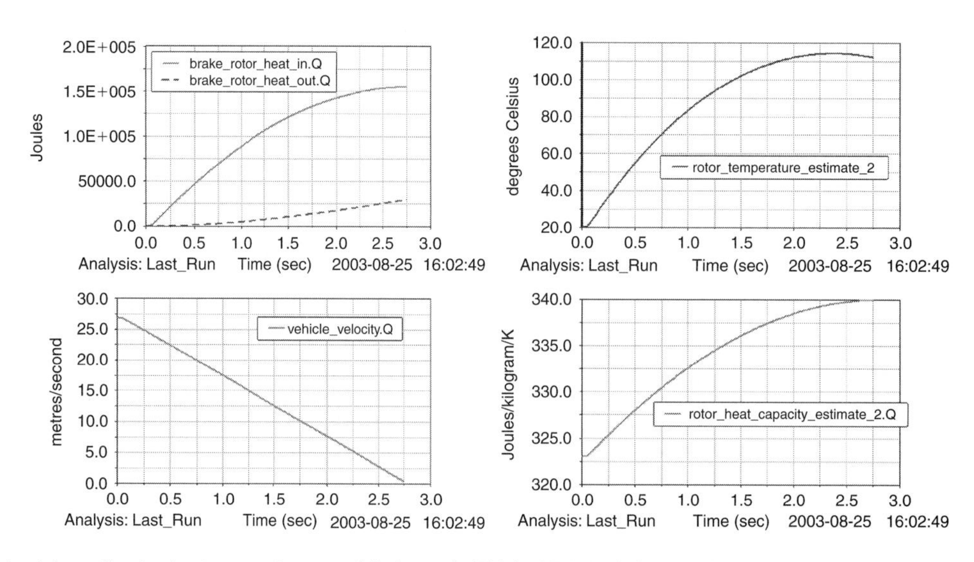

Fig. 15.1-29 Output from the brake temperature model shown in Table 15.1-1 during a 60 mph–0 stop.

Table 15.1-1 A brake rotor temperature model based on brake torque

```
!-------------------- Function definitions --------------------!
!
part create equation differential_equation &
   differential_equation_name = .model_1.brake_heating_integral &
   adams_id = 2 &
   comments = "Brake Heat Input Integral" &
   initial_condition = 0.0 &
   function = "VARVAL(Brake_Torque)*VARVAL(vehicle_velocity)/0.3" &
   implicit = off &
   static_hold = off


data_element create variable &
   adams_id = 102 &
   variable_name = brake_rotor_heat_in &
   function = "DIF(2)"
!
data_element create variable &
   variable_name = "rotor_temperature_kelvin_estimate_1" &
   function = "T_env + VARVAL(brake_rotor_heat_in)/(rotor_mass*350)"
!
data_element create variable &
   variable_name = "rotor_temperature_estimate_1" &
   function = "VARVAL(rotor_temperature_kelvin_estimate_1)-273"
!
data_element create variable &
   variable_name = "rotor_heat_capacity_estimate_2" &
   function = "320 + 0.15*VARVAL(rotor_temperature_estimate_1)", &
              " + 1.164E-9*VARVAL(rotor_temperature_estimate_1)**4"
!
part create equation differential_equation &
   differential_equation_name = .model_1.brake_cooling_integral &
   adams_id = 3 &
   comments = "Brake Heat Rejection Integral" &
   initial_condition = 0.0 &
   function = "hAc*(VARVAL(rotor_temperature_kelvin_estimate_1)-T_env)" &
   implicit = off &
   static_hold = off


data_element create variable &
   adams_id = 103 &
   variable_name = brake_rotor_heat_out &
   function = "DIF(3)"
!
data_element create variable &
   variable_name = "rotor_temperature_kelvin_estimate_2" &
   function = "T_env + ", &
              "(VARVAL(brake_rotor_heat_in) - VARVAL(brake_rotor_heat_out))/", &
              "(rotor_mass*VARVAL(rotor_heat_capacity_estimate_2) + 0.001)"
!
data_element create variable &
   variable_name = "rotor_temperature_estimate_2" &
   function = "VARVAL(rotor_temperature_kelvin_estimate_2)-273"
```

is unsurprising given the consequences of brake system underdesign. The example given is a relatively simple one, with convection characteristics that are independent of vehicle velocity and no variation of brake friction with brake temperature (Fig. 15.1-29). Although in practice these simplifications render the results slightly inaccurate, they are useful when used for comparative purposes – for example, if the brake temperature model is used with an ESP algorithm it can rank control strategies in terms of the energy added to individual brake rotors. Similar modelling is of course possible for other frictional systems within the vehicle, such as drive or transmission clutches. Typical

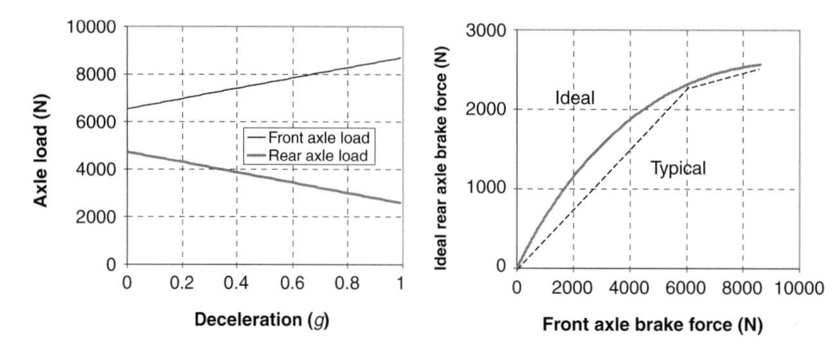

Fig. 15.1-30 Force distribution for ideal and typical braking events.

values of the convection constant *hAc* are around 150 W K^{-1} for a front disc brake installation, around 80 W K^{-1} for a rear brake installation and as low as 20 W K^{-1} for a rear drum brake.

A further key factor in modelling brake performance is the distribution of brake torques around the vehicle. While decelerating, the vertical loads on the axles change due to the fact that the mass centre of the vehicle is above the ground.

It may be presumed that for ideal braking, the longitudinal forces should be distributed according to the vertical forces. Using the above expressions, the graphs in Fig. 15.1-30 can be calculated for vertical axle load versus deceleration. Knowing the total force necessary to decelerate the vehicle it is possible to calculate the horizontal forces for 'ideal' (i.e. matched to vertical load distribution) deceleration. Plotting rear force against front force leads to the characteristic curve shown in Fig. 15.1-30. However, in general it is not possible to arrange for such a distribution of force and so the typical installed force distribution is something like that shown by the dashed line in the figure. Note that the ideal distribution of braking force varies with loading condition and so many vehicles have a brake force distribution that varies with vehicle loading condition. For more detailed information on brake system performance and design, Limpert's 1999 work gives a detailed breakdown of performance characteristics and behaviour, all of which may be incorporated within a multibody system model of the vehicle using an approach similar to that shown in Table 15.1-1 if desired.

Described in some detail in Limpert's work is the function of a vehicle ABS system. The key ingredient of such a system is the ability to control brake pressure in one of three modes, often described as 'hold, dump and pump'. Hold is fairly self-explanatory, the wheel cylinder pressure is maintained regardless of further demanded increases in pressure from the driver's pedal. 'Dump' is a controlled reduction in pressure, usually at a predetermined rate and 'pump' is a controlled increase in pressure, again usually at a predetermined rate.

The main variable is the brake pressure p. In the 1998 work by Ozdalyan a slip control model was initially developed as a precursor to the implementation of an ABS model. This is illustrated in Fig. 15.1-31 where it can be seen that on initial application of the brakes the

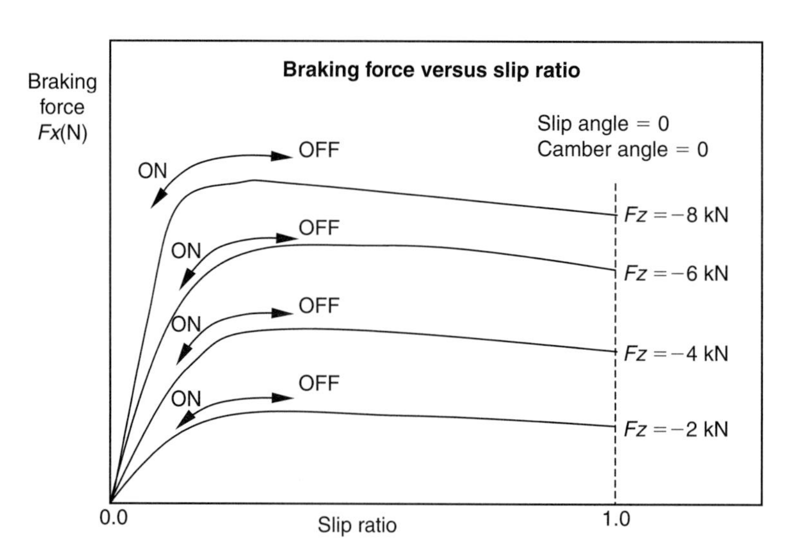

Fig. 15.1-31 Principle of a brake slip control model.

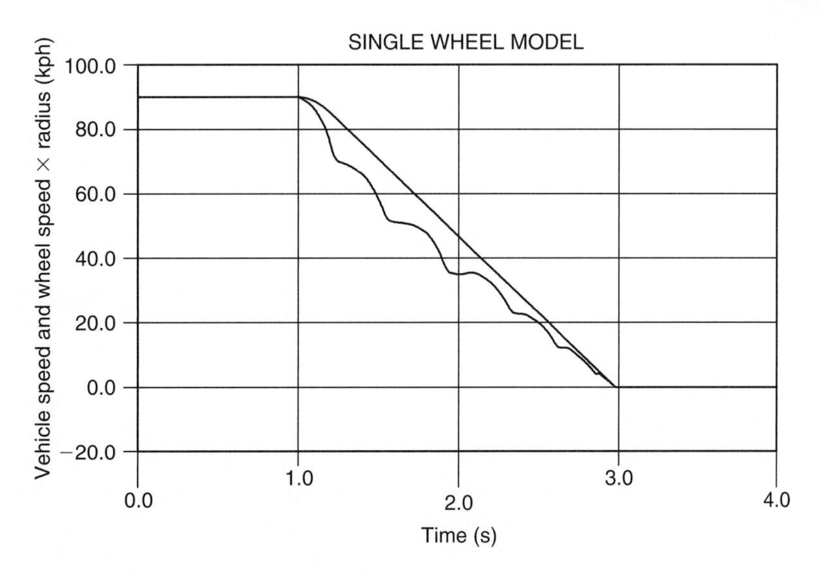

Fig. 15.1-32 Plot of vehicle speed and wheel speed during ABS braking simulation.

brake force rises approximately linearly with slip ratio depending on the wheel load. If the braking is severe the slip ratio increases past the point where the optimum brake force is generated. To prevent the slip ratio increasing further to the point where the wheel is locked an ABS system will then cycle the brake pressure on and off maintaining peak braking performance and a rolling wheel to assist manoeuvres during the braking event.

In this model the brake pressure is found by integrating the rate of change of brake pressure, this having set values for any initial brake application or subsequent application during the ABS cycle phase. Implementation of these changing dump, pump and hold states requires care to ensure no discontinuities in the brake pressure formulation.

The modelling in MBS of more realistic ABS algorithms is more challenging as the forward velocity and hence slip ratio is not directly available for implementation in the model. The implementation of such a model allows the angular velocity of the wheel to be factored with the rolling radius to produce an output commonly referred to as wheel speed by practitioners in this area. A plot of wheel speed is compared with vehicle speed in Fig. 15.1-32 where the typical oscillatory nature of the predicted wheel speed reflects the cycling of the brake pressure during the activation of the ABS model in this vehicle simulation.

15.1.10 Modelling traction

For some simulations it is necessary to maintain the vehicle at a constant velocity. Without some form of driving torque the vehicle will 'drift' through the manoeuvre

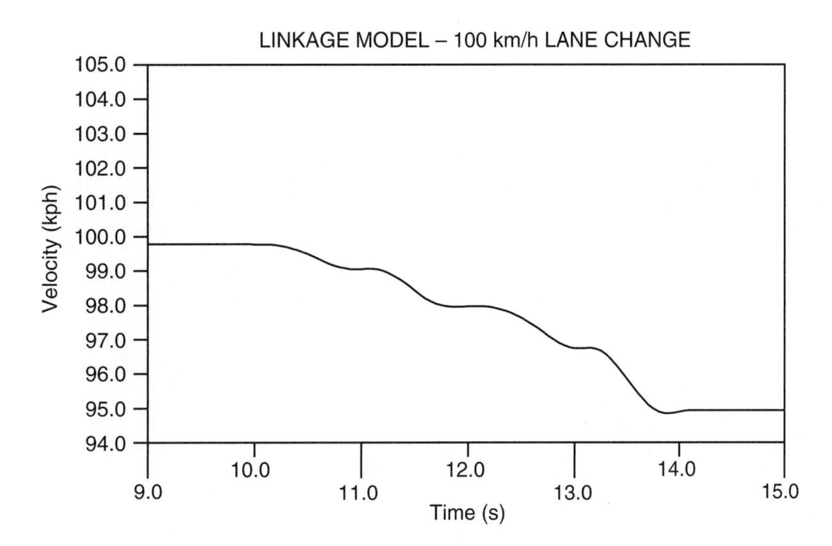

Fig. 15.1-33 Loss in velocity as vehicle 'drifts' through the lane change manoeuvre.

using the momentum available from the velocity defined with the initial conditions for the analysis. Ignoring rolling resistance and aerodynamic drag will reduce losses but the vehicle will still lose momentum during the manoeuvre due to the 'drag' components of tyre cornering forces generated during the manoeuvre. An example is provided in Fig. 15.1-33 where for a vehicle lane change manoeuvre it can be seen that during the 5 seconds taken to complete the manoeuvre the vehicle loses about 5 km/h in the absence of any tractive forces at the tyres.

The emphasis with programs such as ADAMS/Car and ADAMS/Chassis is to include a driveline model as part of the full vehicle as a means to impart torques to the road wheels and hence generate tractive driving forces at the tyres. Space does not permit a detailed consideration of driveline modelling here but as a start a simple method of imparting torque to the driven wheels is shown in Fig. 15.1-34.

The rotation of the front wheels is coupled to the rotation of the dummy transmission part shown in Fig. 15.1-34. The coupler introduces the following constraint equation:

$$s_1 \cdot r_1 + s_2 \cdot r_2 + s_3 \cdot r_3 = 0 \qquad (15.1.16)$$

where s_1, s_2 and s_3 are the scale factors for the three revolute joints and r_1, r_2 and r_3 are the rotations. In this example suffix 1 is for the driven joint and suffixes 2 and 3 are for the front wheel joints. The scale factors used are $s_1 = 1$, $s_2 = 0.5$ and $s_3 = 0.5$ on the basis that 50% of the torque from the driven joint is distributed to each of the wheel joints. This gives a constraint equation linking the rotation of the three joints:

$$r_1 = 0.5r_2 + 0.5r_3 \qquad (15.1.17)$$

Note that this equation is not determinant. For a given input rotation r_1, there are two unknowns r_2 and r_3 but only the single equation. In order to solve r_2 and r_3 this equation must be solved simultaneously with all the other equations representing the motion of the vehicle. This is important particularly during cornering where the inner and outer wheels must be able to rotate at different speeds.

15.1.11 Other driveline components

The control of vehicle speed is significantly easier than the control of vehicle path inside a vehicle dynamics model. In the real vehicle, speed is influenced by the engine torque, brakes and aerodynamic drag. As discussed earlier these are relatively simple devices to represent in a multibody systems model, with the exception of turbochargers and torque converters. Even these latter components can be represented using differential equations of the form:

$$T_{BOOST} = T_2 \cdot \widehat{T}_{BOOST} \qquad (15.1.18)$$

$$\frac{d}{dt}(T_2) = \frac{T_1}{k_2} \cdot (t_{boost} - T_2) \qquad (15.1.19)$$

$$\frac{d}{dt}(T_1) = k_1 \cdot (t_{boost} - T_1) \qquad (15.1.20)$$

where $\widehat{T}_{BOOST}$ is the maximum possible torque available, t_{boost} is the throttle setting to be applied to the boost torque (which may be different to the throttle setting applied to the normally aspirated torque to model the rapid collapse of boost off-throttle) and $k_{1,2}$ are mapped, state dependent values to calibrate the behaviour of the engine (i.e. large delays at low engine speed, reducing delays with rising engine speed). An example of the statements required to model the resulting torque is shown in Table 15.1-2.

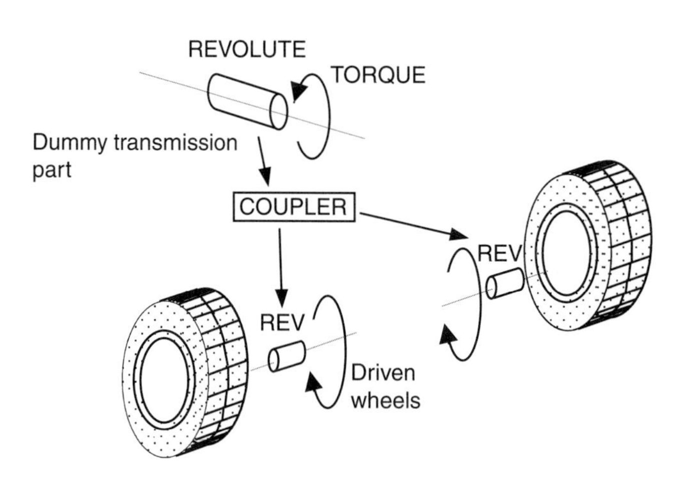

Fig. 15.1-34 Simple drive torque model.

Table 15.1-2 Example MSC.ADAMS command statements for an empricial mean-state turbocharger

```
! -- First First Order Differential Equation --
part create equation differential_equation &
   differential_equation_name = turbo_lag_equation_1 &
   adams_id = 12 &
   comments = "Lag Equation 1 - Explicit" &
   initial_condition = 0.0 &
   function = "varval(K1_now) * ( varval(boost_throttle)*100-DIF(12) )" &
   implicit = off &
data_element create variable &
   variable_name=K2 &
   function="STEP(varval(throttle_derivative),", &
              "-10, 100.0,", &
              " -1, (DIF(12))/varval(K2_divisor_now)", &
              "   )"

! -- Second First Order Differential Equation --
part create equation differential_equation &
   differential_equation_name = turbo_lag_equation_2 &
   adams_id = 13 &
   comments = "Lag Equation 2 - Explicit" &
   function = "varval(K2) * ( varval(boost_throttle)*100-DIF(13) )" &
   implicit = off &
data_element create variable &
   variable_name = boost_torque_scaling &
   function = "DIF(13)/100"

! -- Sum both normally aspirated and turbocharged (delayed) component
data_element create variable &
   variable_name = prop_torque &
   function = "(", &
                " VARVAL(na_engine_torque)*VARVAL(throttle)*1000", &

" +VARVAL(boosted_engine_torque)*VARVAL(boost_torque_scaling)*1000", &
         ")"
```

In this example the variable throttle runs from –0.3 to 1.0 to simulate overrun torque. The variable boost_throttle is a clipped version from 0 to 1.0 since no turbocharger boost is available on overrun. Throttle_derivative is the first time derivative of throttle. All the other variables (varvals) are retrieved from the relevant curves (splines) plotted in Fig. 15.1-35.

The delays inherent in a torque converter are amenable to such modelling techniques using typical torque converter characteristic data in a similar empirical manner.

Once the physical elements of the system are modelled, the task of modelling the driver behaviour is largely similar to that for path following described later. In order to represent, for example, the effect of a driver using the throttle to maintain a steady velocity through a manoeuvre a controller can be developed to generate the torque shown in Fig. 15.1-34.

A simple but workable solution is to model the driving torque T, with the following formulation:

$$T = K^{*}(Vs - Va)^{*}\mathrm{STEP}(\mathrm{Time}, 0, 0, 1, 1) \quad (15.1.21)$$

where

K = a constant which is tuned to stabilize the torque
Vs = the desired velocity for the simulation
Va = the forward velocity of the vehicle, which can be obtained using a system variable

The purpose of the STEP FUNCTION is to define a change of state in the expression that is continuous.

The step function can be used to factor a force function by ramping it on over a set time period. In this case the driving torque is being switched on between time = 0 and time = 1 second. This is important because it is necessary to perform an initial static analysis of the vehicle at time = 0 when $Va = 0$ and the torque must not act.

As can be seen a ‘reference’ (desired) state is needed, an error term is defined by the difference between the current state and the reference state and finally, responses to that in terms of throttle or brake application to adjust the speed back towards the reference value. There are two possible approaches; the simplest provides a speed ‘map’

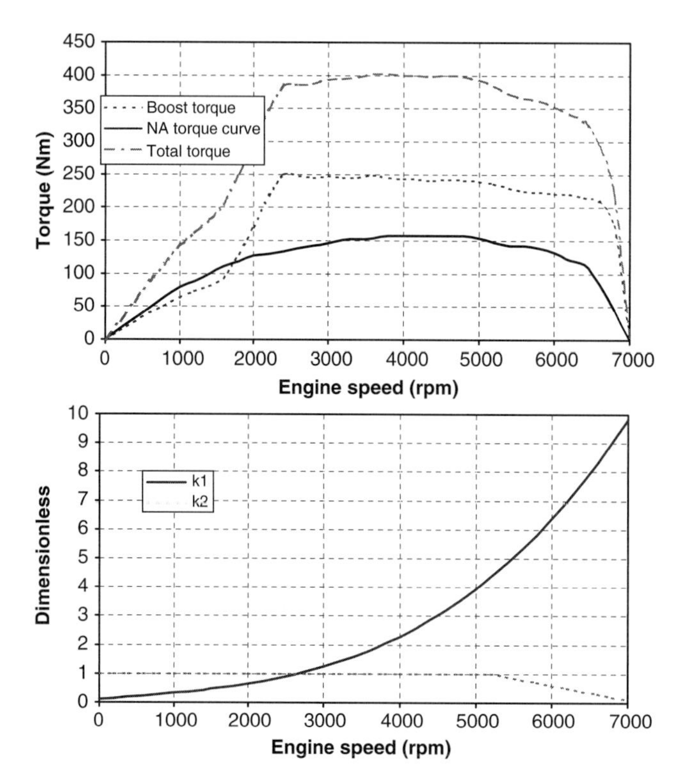

Fig. 15.1-35 Empricial mean-state turbocharger model.

for the track, similar to the curvature map description of it. More elaborately, it is possible to examine the path curvature map locally and decide (through a knowledge of the ultimate capabilities of the vehicle, perhaps) whether or not the current speed is excessive, appropriate or insufficient for the local curvature and use brakes or engine appropriately. For the development of vehicles, open loop throttle or brake inputs may be preferable and are sometimes mandated in defined test manoeuvres, rendering the whole issue of speed control moot.

In many ways the skill of the competition driver lies entirely in this ability to judge speed and adjust it appropriately. It is also a key skill to cultivate for limit handling development and arguably for road driving too, so as not to arrive at hazards too rapidly to maintain control of the vehicle. For this functionality, some form of preview is essential. It is both plausible and reasonable to run a 'here and now at the front axle' model for the path follower and a 'previewing' speed controller within the same model, described in subsequent sections.

15.1.12 The steering system

15.1.12.1 Modelling the steering mechanism

There are a number of steering system configurations available for cars and trucks based on linkages and steering gearboxes. The treatment in the following sections is limited to a traditional rack and pinion system. Space does not permit discussion of the modelling of power steering or steer-by-wire here.

For the simple full vehicle models discussed earlier, such as that modelled with lumped mass suspensions, there are problems when trying to incorporate the steering system. Consider first the arrangement of the steering system on the actual vehicle and the way this can be modelled on the detailed linkage model as shown in Fig. 15.1-36. In this case only the suspension on the right-hand side is shown for clarity.

The steering column is represented as a part connected to the vehicle body by a revolute joint with its axis aligned along the line of the column. The steering inputs required to manoeuvre the vehicle are applied as motion or torque inputs at this joint. The steering rack part is connected to the vehicle body by a translational joint and connected to the tie rod by a universal joint. The translation of the rack is related to the rotation of the steering column by a coupler statement that defines the ratio. An example of a statement that would define the ratio is

```
COUPLER/510502, JOINTS = 501, 502,
TYPE = T:R, SCALES = 8.45D, 1.0
```

In this case joint 501 is the translational joint and 502 is the revolute joint. The coupler statement ensures that

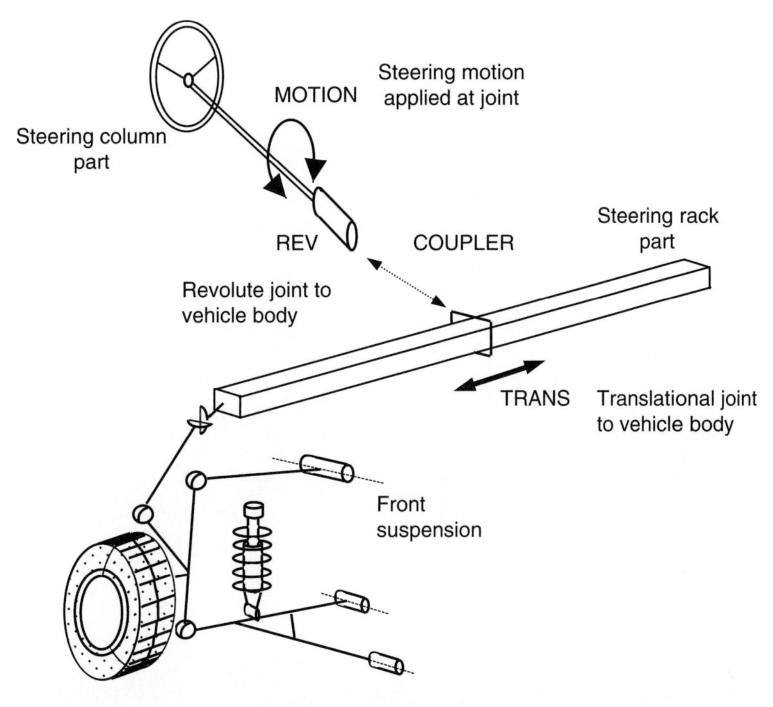

Fig. 15.1-36 Modelling the steering system. (This material has been reproduced from the Proceedings of the Institution of Mechanical Engineers, K2 Vol. 213 'The modelling and simulation of vehicle handling. Part 2: vehicle modelling', M.V. Blundell, page 129, by permission of the Council of the Institution of Mechanical Engineers.)

for every 8.45 degrees of column rotation there will be 1 mm of steering rack travel.

Attempts to incorporate the steering system into the simple models using lumped masses, swing arms and roll stiffness will be met with a problem when connecting the steering rack to the actual suspension part. This is best explained by considering the situation shown in Fig. 15.1-37.

The geometry of the tie rod, essentially the locations of the two ends, is designed with the suspension linkage layout and will work if implemented in an 'as-is' model of the vehicle including all the suspension linkages.

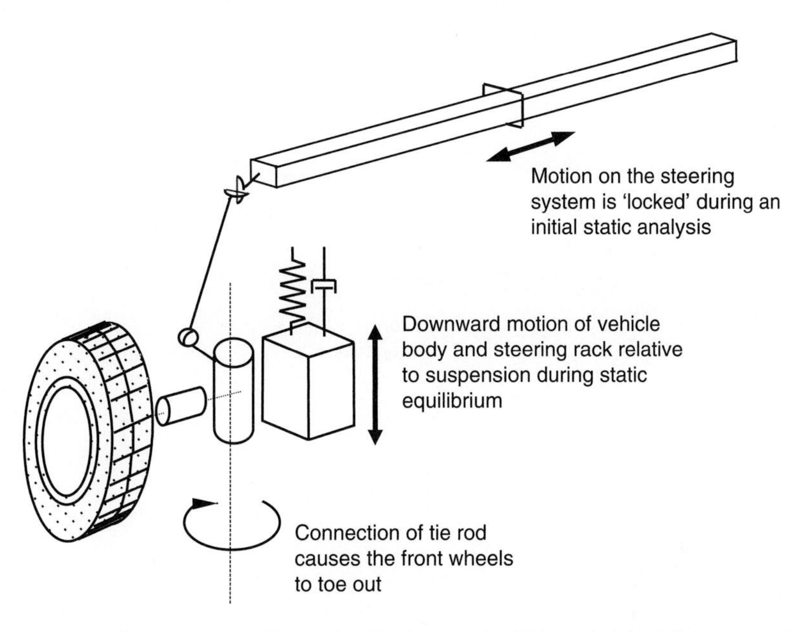

Fig. 15.1-37 Toe change in front wheels at static equilibrium for simple models. (This material has been reproduced from the Proceedings of the Institution of Mechanical Engineers, K2 Vol. 213 'The modelling and simulation of vehicle handling. Part 2: vehicle modelling', M.V. Blundell, page 129, by permission of the Council of the Institution of Mechanical Engineers.)

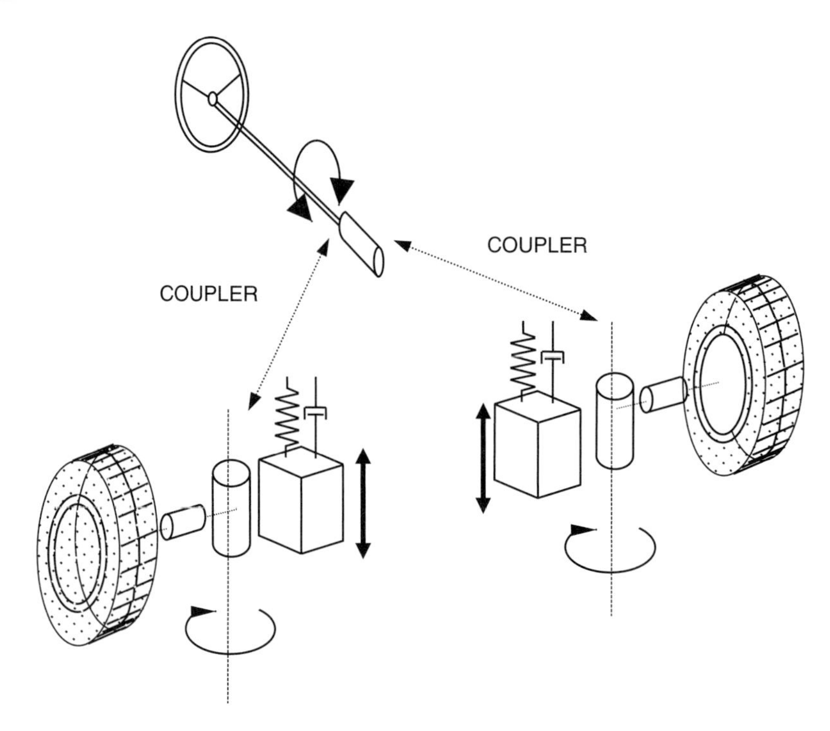

Fig. 15.1-38 Coupled steering system model. (This material has been reproduced from the Proceedings of the Institution of Mechanical Engineers, K2 Vol. 213 'The modelling and simulation of vehicle handling. Part 2: vehicle modelling', M.V. Blundell, page 130, by permission of the Council of the Institution of Mechanical Engineers.)

Physically connecting the tie rod to the simple suspensions does not work. During an initial static analysis of the full vehicle, to settle at kerb height, the rack moves down with the vehicle body relative to the suspension system. This has a pulling effect, or pushing according to the rack position, on the tie rod that causes the front wheels to steer during the initial static analysis. The solution to this is to establish the relationship between the steering column rotation and the steer change in the front wheels and to model this as a direct ratio using two

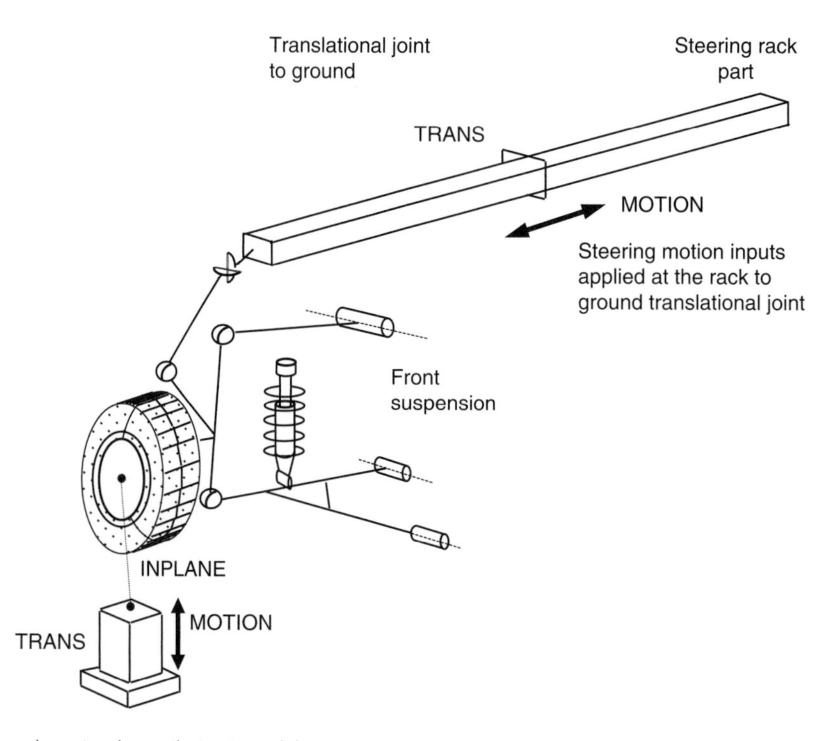

Fig. 15.1-39 Front suspension steering ratio test model.

coupler statements to link the rotation between the steering column and each of the front wheel joints as shown in Fig. 15.1-38.

15.1.12.2 Steering ratio

In order to implement the ratios used in the couplers shown in Fig. 15.1-38 linking the rotation of the steering column with the steer change at the road wheels it is necessary to know the steering ratio. At the start of a vehicle dynamics study the steering ratio can be a model design parameter. In the examples here a ratio of 20 degrees of handwheel rotation to 1 degree of road wheel steer is used. On some vehicles this may be lower and on trucks or commercial vehicles it may be higher. To treat steering ratio as linear is a simplification of the situation on a modern vehicle. For example, the steering ratio may vary between a higher value on centre to a lower value towards the limits of rack travel or vice versa. This would promote a feeling of stability for smaller handwheel movements at higher motorway speeds and assist lower speed car park manoeuvres.

Using the multibody systems approach the steering ratio can be investigated through a separate study carried out using the front suspension system connected to the ground part instead of the vehicle body. The modelling of these two subsystems, with only the suspension on the right side shown, is illustrated in Fig. 15.1-39.

The approach of using a direct ratio to couple the rotation between the steering column and the steer angle of the road wheels is common practice in simpler models but may have other limitations in addition to the treatment of the ratio as linear:

(i) In the real vehicle and the linkage model the ratio between the column rotation and the steer angle at the road wheels would vary as the vehicle rolls and the road wheels move in bump and rebound.

(ii) For either wheel the ratio of toe out or toe in as a ratio of left or right handwheel rotation would not be exactly symmetric.

Modelling the suspension with linkages will capture these effects. Although this may influence the modelling of low speed turning they have little effect for handling manoeuvres with comparatively small steer motions.

With simpler vehicle models, not including suspension linkages, the ratio would need to be functionally dependent on the vertical movement of the suspension and direction of handwheel rotation if the behaviour is to be modelled. It should also be noted that compliance in the steering rack or rotational compliance in the steering column could be incorporated if the analysis dictates this.

In the following example the geometric ratio between the rotation of the steering column and the travel of the rack is already known, so it is possible to apply a motion input at the rack to ground joint that is equivalent to handwheel rotations either side of the straight ahead position. The jack part shown in Fig. 15.1-39 can be used to set the suspension height during a steering test simulation. Typical output is shown in Fig. 15.1-40 where the steering wheel angle is plotted on the *x*-axis and the road wheel angle is plotted on the *y*-axis. The three lines plotted represent the steering ratio test for the suspension in the static (initial model set up here), bump and rebound positions.

Having decided on the suspension modelling strategy and how to manage the relationship between the handwheel rotation and steer change at the road wheels the steering inputs from the driver and the manoeuvre to be performed need to be considered.

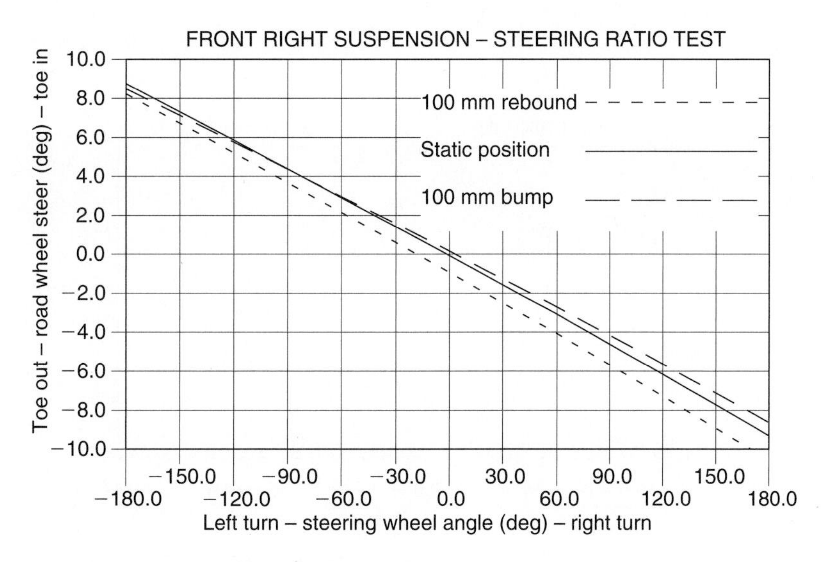

Fig. 15.1-40 Results of steering ratio test for MSC.ADAMS front right suspension model.

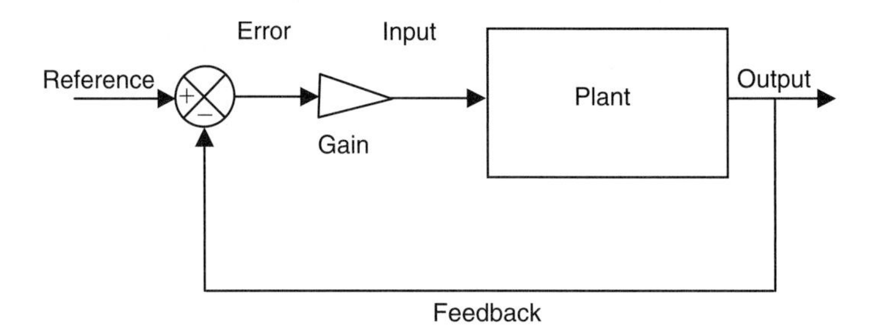

Fig. 15.1-41 An open-loop system, in black, is a subset of a closed-loop system, in grey.

15.1.12.3 Steering inputs for vehicle handling manoeuvres

The modelling of steering inputs suggests for the first time some representation of the driver as part of the full vehicle system model. Any system can be considered to consist of three elements – the 'plant' (the item to be controlled), the input to the plant and the output from the plant (Fig. 15.1-41). Inputs to the system (i.e. handwheel inputs) are referred to as 'open loop' or 'closed loop'. An open loop steering input requires a time dependent rotation to be applied to the part representing a steering column or handwheel in the simulation model. In the absence of these bodies an equivalent translational input can be applied to the joint connecting a rack part to the vehicle body or chassis, assuming a suspension linkage modelling approach has been used. Examples will be given here where the time dependent motion is based on a predetermined function or equation to alter the steering inputs or a series of measured inputs from a vehicle on the proving ground.

Any system can be considered to consist of three elements – the 'plant' (the item to be controlled), the input to the plant and the output from the plant (Fig. 15.1-41).

When a closed-loop controller is added to the system, its goal is to allow the input to the plant to be adjusted so as to produce the desired output. The desired output is referred to as the 'reference' state; a difference between the actual output and the reference is referred to as an 'error' state. The goal of the control system is to drive the error to zero.

We can consider an example of a closed loop steering input that requires a torque to be applied to the handwheel or steering column such that the vehicle will follow a predetermined path during the simulation. A mechanism must be modelled to measure the deviation of the vehicle from the path and process this in a manner that feeds back to the applied steering torque. As the simulation progresses the torque is constantly modified based on the observed path of the vehicle and the desired trajectory. Such an input is referred to as closed loop since the response is observed and fed back to the input, thus closing the control loop.

To return to the open-loop case, we can consider an example of an open loop manoeuvre for a steering input where we want to ramp a steering input of 90 degrees between 1 and 1.5 seconds of simulation time. Using an MSC.ADAMS solver statement the function applied to the steering motion would be:

FUNCTION = STEP(TIME, 1, 0, 1.5, 90D)

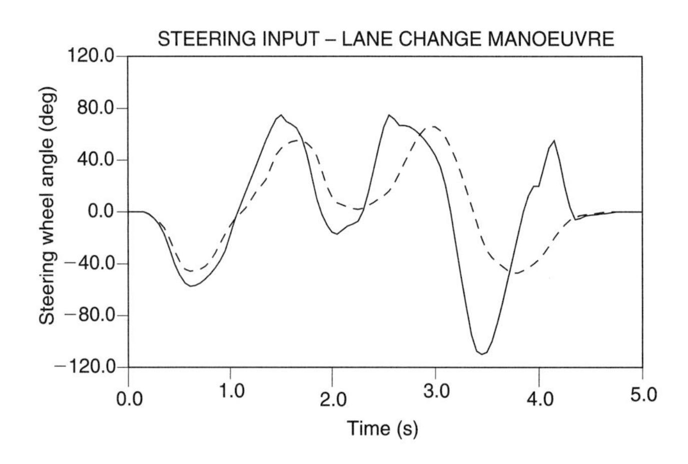

Fig. 15.1-42 Steering input for the lane change manoeuvre at 70 km/h (dashed line) and 100 km/h (solid line).

Table 15.1-3 MSC.ADAMS statements for lane change steering inputs

```
MOTION/502,JOINT=502,ROT
,FUNC=(PI/180)*CUBSPL(TIME,0,1000)

SPLINE/1000
,X=0,1,2,3,4,5,6,7,8,9
,9.1,9.2,9.3,9.4,9.5,9.6,9.7
,9.8,9.9,10,10.1,10.2,10.3,10.4,10.5,10.6,10.7,10.8,10.9,11
,11.1,11.2,11.25,11.3,11.4,11.5,11.6,11.7,11.8,11.9,12,12.1
,12.2,12.3,12.4,12.5,12.6,12.7,12.8,12.9,13,13.1,13.2,13.3
,13.4,13.5,13.6,13.7,13.75,13.8,13.9,14,14.1,14.2,14.3,14.4,14.5
,14.6,14.7,14.8,14.9,15
,Y = 0,0,0,0,0,0,0,0,0,0
,0,0,0,0,0,0,0
,0,0,-5,-17,-40,-55,-57,-52,-43,-30,-5,15,35,55,72,75,70,65,45,10
,-10,-17,-11,-7,15,50,75,67,66,60,50,35,0,-50,-95,-110,-100,-70,-35,0
,20,20,35,55,20,-6,-3,-2,-1,0,0,0,0,0,0
```

In a similar manner if we wanted to apply a sinusoidal steering input with an amplitude of 30 degrees and a frequency of 0.5 Hz we could use:

$$\text{FUNCTION} = 30\text{D} * \text{SIN}(\text{TIME} * 180\text{D})$$

For the lane change manoeuvre described earlier the measured steering wheel angles from a test vehicle can be extracted and input as a set of *XY* pairs, which can be interpolated using a cubic spline fit. A time history plot for the steering inputs is shown in Fig. 15.1-42 for lane change manoeuvres at 70 and 100 km/h.

By way of example the MSC.ADAMS statements which apply the steering motion to the steering column to body revolute joint and the spline data are shown in Table 15.1-3 for a 100 km/h lane change. The *x* values are points in time and the *y* values are the steering inputs in degrees. In the absence of measured data it is possible to construct an open loop single or double lane change manoeuvre using a combination of nested arithmetic IF functions with embedded step functions with some planning and care over syntax. Note that for a fixed steering input a change in vehicle configuration will produce a change in response so that the vehicle fails to follow a path.

For a closed loop steering manoeuvre a torque is applied to the steering column, or a force to the steering rack if the column is not modelled, that will vary during the simulation so as to maintain the vehicle on a predefined path. This requires a steering controller to process feedback of the observed deviation from the path (error) and to modify the torque accordingly as illustrated in Fig. 15.1-43.

15.1.13 Driver behaviour

It becomes inevitable with any form of vehicle dynamics modelling that the interaction of the operator with the vehicle is a source of both input and disturbance. In flight dynamics, the phenomenon of 'PIO' – pilot induced oscillation – is widely known. This occurs when

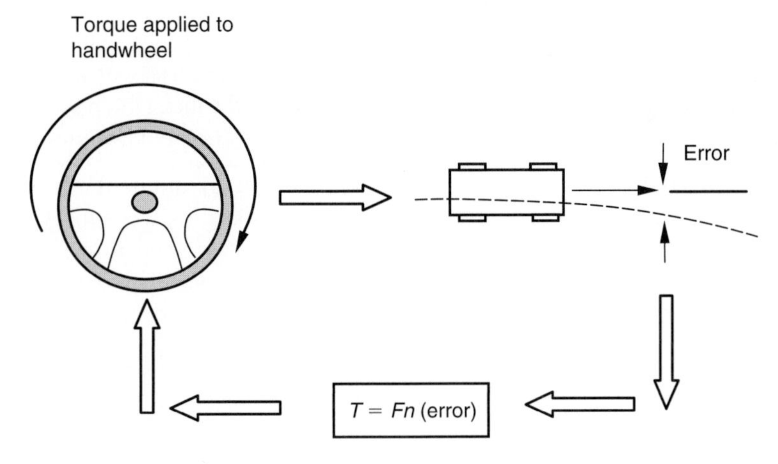

Fig. 15.1-43 Principle of a closed loop steering controller.

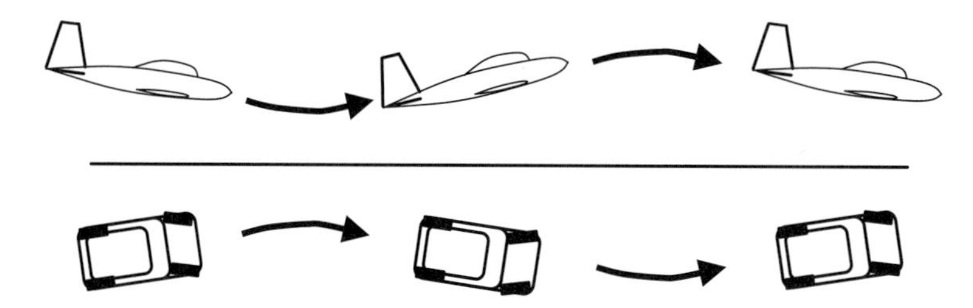

Fig. 15.1-44 Pilot Induced Oscillation (PIO) – not exclusively an aeronautical phenomenon.

inexperienced pilots, working purely visibly and suffering from some anxiety, find their inputs are somewhat excessive and cause the aircraft to, for example, pitch rhythmically instead of holding a constant altitude (Fig. 15.1-44).

PIO is caused when the operator is unable to recognize the effects of small control inputs and therefore increases those inputs, before realizing they were excessive and reversing them through a similar process. It is analogous to the 'excess proportional control gain oscillation' discussed in classical control theory. For road vehicles, drivers most likely to induce PIO in steering tend to be inexperienced or anxious drivers travelling at a speed with which they are uncomfortable. This type of PIO is not to be confused with the typical experience of drivers of skidding vehicles when the initial skid is corrected but the vehicle subsequently 'fishtails' or simply departs in the opposite direction – this is a 'phasing at resonance' control error. The driver fails to apply a 'feed forward' (open loop, knowledge-based) correction in advance of the vehicle's response to compensate for the delay in vehicle response.

PIO also occurs in tractive (i.e. throttle) control inputs and is the reason even experienced drivers are incapable of travelling at a constant speed on highways; perception of changes in following distance is universally poor. If too little attention is spent on the driving task or if insufficient following distance is left, these PIOs become successively amplified by following drivers until the speed variation results in a 'shunt' accident. Radar-based cruise control systems will alleviate this risk but are no substitute for attentive driving while anything less than the whole vehicle fleet is fitted with it.

15.1.13.1 Steering controllers

There are a variety of controller models suitable for modelling driver behaviour in existence. Some, such as ADAMS/Driver™ developed as part of the MSC.ADAMS modelling package, are very complete – others, such as the two-loop feedback control model used by the authors, are simpler. The analyst must consider the needs of the simulation (and the financial constraints of the company) and choose the most appropriate level of modelling to achieve the task at hand. Driver models in general fall into two categories:

(1) **Optimum control models.** Optimum control models use some form of 'penalty function' – a measure used to assess the quality of control achieved. For example, for a vehicle steering model the appropriate variable might be lateral deviation from the intended path. Optimum control models use repeated simulations of the event and numerical optimization methods to 'tune' the parameters for a control system to minimize the value(s) of the penalty function(s) over the duration of the event of interest. For learned events, such as circuit driving, these methods are excellent in producing a prediction of likely driver behaviour. However, some care must be exercised with their use. For road vehicles, drivers are generally unskilled and so the application of modelling techniques in which repeated solutions are used to discover the 'best' way of achieving a manoeuvre may not be appropriate when simulating a manoeuvre that the driver has only one attempt at completing, for example emergency evasive manoeuvres. For race vehicle simulation, some care must also be exercised lest extended calculations result in the proof that the driver can adapt to a remarkable variety of vehicle changes – without any real insight into which will improve performance in a competition environment.

(2) **Moment-by-moment feedback models**[1]. Such models are really a subset of the optimum control models described above – the optimum control method repeatedly uses feedback models in order to discern the best state of tune for the controller.

[1] Feedback models ought to be known as 'instantaneous feedback' models but the word instantaneous has become slightly muddled in recent times. It should be used unambiguously to mean 'existing for a moment in time' but has become sadly confused with 'instant', meaning immediate. Instant feedback would imply the inability to represent transport delays and the like in the controller model, which is incorrect. The use of moment-by-moment is therefore preferred although it introduces confusion with moment in the sense of torque.

When the feedback controllers are used alone, the analyst must set their tuning. Although the absence of 'automated' correlation makes them less appropriate for circuit racing, it also adds clarity in the sense that the parameters, once set, remain constant and so changes in the vehicle behaviour and/or driver inputs can be readily understood.

In general, the driver behaves as the most generic form of loop-closing controller. There are several attractive control technologies represented in the literature and some of their proponents believe they represent a 'one size fits all' solution for the task of applying control to any system. The competing technologies are outlined for comparison:

(i) Logic controller. A logic controller produces output that has only certain possible values. For example, if a driver model were implemented using logic, the logic might be 'if the vehicle is to the left of the intended path, steer right and vice versa'. With a logic controller, the amount of steer is fixed and so any control of the vehicle would be achieved as a series of jerks, oscillating about the intended path. While probably functional it would be unlikely to represent any normal sort of driver.

(ii) PID controller. As stated PID stands for 'Proportional, Integral and Derivative'. The error is used in three ways; used directly, a control effort is applied in proportion to (and opposition to) the error – this is the 'P', proportional, element of the control. The fact that the control effort is in opposition to the error is important, since otherwise the control effort would increase the error instead of reducing it. For this reason, such systems are often referred to as 'negative feedback' systems. The error can also be integrated and differentiated, with control forces applied proportional to the integral and the differential – these are the 'I' and the 'D' terms in the controller. One or more of the terms may not be used at all in any particular controller. An analogy for PID controllers can be found in vehicle suspensions. If the ride height is thought of as the desired output, then individual components of the suspension behave as parts of a control system. The springs produce a force proportional to the change in ride height and the dampers produce a force proportional to the derivative of ride height. Real dampers are often non-linear in performance, and there is nothing to stop non-linear gains being used for any of the control terms. The D term has the effect of introducing damping into the control system. An analogy for the I term is a little harder to come by. The best analogy is that of a self-levelling unit fitted to the suspension, which applies a restoring force related to the length of time the vehicle has been at the wrong ride height and how wrong the ride height is. (This is an imperfect analogy for many reasons but allows the notion to be understood at least.) In real systems, when the output is *nearly* the same as the reference state it is frequently the case that the control forces become too small to influence the system, either because of mechanical hysteresis or sensor resolution or some similar issue. One important measure of the quality of any control system is the accuracy with which it achieves its goals. Such an offset characterizes an inaccurate system; an integral term 'winds up' from a small error until powerful enough to restore the system to the reference state. Thus for classical control, integral terms are important for accuracy. However, since they take some time to act they can introduce delays into the system. In general PID controllers have the advantage that they produce 'continuous' output – that is to say all the derivatives are finite, the output has no steps – which is quite like the behaviour of real people.

(iii) Fuzzy logic. Fuzzy logic was first described in the 1960s but found favour in the 1980s as a fashionable 'new' technology. Notions of 'true' and 'false' govern 'logic' in computer algorithms. Simple control systems assess a set of conditions and make a decision based on whether or not such variables are true or false. Fuzzy logic simply defines 'degrees of truth' by using numbers between 0 and 1 such that the actions taken are some blend of actions that would be taken were something completely true and other actions that would be taken were something completely false. Fuzzy logic is most applicable to control systems where actions taken are dependent on circumstance and where a simple PID controller is unable to produce the correct output in every circumstance. For example, throttle demand in a rear-wheel drive vehicle model might be controlled with a PID controller to balance understeer; however, too much throttle would cause oversteer and some more sophisticated blend of steer and throttle input would be required to retain control under these circumstances.

(iv) Neural networks. Where the system of interest is highly non-linear and a lot of data exists that describes desired outputs of the system for many different combinations of inputs, it is possible to use a neural network to 'learn' the patterns inherently present in the data. A neural network is quite simply a network of devices that is 'neuron-like'. Neurons are the brain's building blocks and are switches with multiple inputs and some threshold to decide when they switch. In general, neural networks are run on transistor devices or in computer simulations. They require a period of 'training' when they learn what settings need to be made for individual neurons in order to produce the required outputs. Once trained, they are extremely rapid in operation since there is very little 'processing' as such, simply a cascade of voltage switching through the transistor network. If the network is implemented as semiconductor transistors then it works at a speed governed only by the latency of the semiconductor medium – extremely fast indeed. Neural networks are extremely useful for controlling highly non-linear systems for which it is too difficult to code a traditional algorithm. However, the requirement for a large amount of data can make the learning exercise a difficult one. Recent advances in the field reduce the need for precise data sets of input and corresponding outputs; input data and 'desirable outcome' definitions allow neural networks to learn how to produce a desirable outcome by identifying patterns in the incoming data. Such networks are extremely slow in comparison to the more traditional types of network during the learning phase. In general, for driver modelling there is little applicability for neural networks at present due to the lack of fully populated data sets with which to teach them. It is also worth commenting that for any input range that was not encountered during the learning phase, the outputs are unknown and may not prove desirable. This latter feature is not dissimilar to real people; drivers who have never experienced a skid are very unlikely to control it at the first attempt.

(v) System identification. System identification is a useful technique, not dissimilar in concept to neural networking. A large amount of data is passed through one of several algorithms that produce an empirical mathematical formulation that will produce outputs like the real thing when given the same set of inputs. The formulation is more mathematical than neural networking and so the resulting equations are amenable to inspection – although the terms and parameters may lack any immediately obvious significance if the system is highly non-linear. System identification methods select the level of mathematical complexity required to represent the system of interest (the 'order' of the model) and generate parameters to tune a generic representation to the specific system of interest. As with neural networks, the representation of the system for inputs that are beyond the bounds of the original inputs (used to identify the model) is undefined. System identification is useful as a generic modelling technique and so has been successfully applied to components such as dampers as well as control system and plant modelling. System identification is generally faster to apply than neural network learning but the finished model cannot work as quickly. The same data set availability problems for neural networking also mean system identification is not currently applicable to driver modelling.

(vi) Adaptive controllers. Adaptive control is a generic term to describe the ability of a control system to react to changes in circumstances. In general, people are adaptive in their behaviour and so it would seem at first glance that adaptive control is an appropriate tool for modelling driver behaviour. Optimum control models, described above, generally use some form of adaptive control to optimize the performance of a given controller architecture to the system being controlled and the task at hand. Adaptation is a problem in real world testing since it obscures real differences in performance; equally it can obscure performance changes and so adaptive modelling of driver behaviour is not preferred except for circuit driving. Several techniques come under the headline of adaptive control; the simplest is to change the control parameters in a predetermined fashion according to the operating regime, an operation referred to as 'gain scheduling'. Gain is the term used for any treatment given to an error state before it is fed to an input – thus the PID controller described above has a P-gain, an I-gain and a D-gain. It might be, for example, that under conditions of opposite lock the P-gain is increased since the driver needs to work quickly to retain control, or under conditions of increasing speed the P-gain is reduced since slower inputs are good for stability at higher speeds. A more complex method is to carry a model of the plant on board in the controller and to use it to better inform some form of gain scheduling, perhaps using information that cannot readily be discerned from on-board instrumentation – such as body slip angle. This is referred to as a Model Reference Adaptive Scheme (MRAS). A further variation on the theme is to use the controller to calculate model parameters using system or parameter identification methods (described above). The control system parameters can be modified based on this information – in effect there is an ongoing redesign of the control system using a classical deterministic method, based on the reference state and the plant characteristics according to the latest estimate. This is referred to as a 'self-tuning-regulator' and is useful for unpredictably varying systems. Finally, a method known as 'dual control' intentionally disturbs the system in order to learn its characteristics, while simultaneously controlling it towards a reference state. In many ways this is similar to a top level rally driver stabbing the brakes in order to assess friction levels while disturbing the overall speed of the vehicle as little as possible; the knowledge gained allows the driver to tune their braking behaviour according to recently learned characteristics. Such behaviour is in marked contrast to circuit drivers, who concentrate on learned braking points and sometimes have difficulty adapting to changing weather conditions. With the exception of the simplest gain scheduling methods, in general adaptive control techniques are unsuitable for the modelling of driver behaviour as part of any practicable process. Once again the variation in simulation output cannot readily be traced to any particular aspect of the system and hence the success or otherwise of an intended modification is difficult to interpret.

In the light of the preceding description, the authors believe a PID controller, with some form of simple gain scheduling, is most appropriate for the modelling of driver behaviour in a multibody system context. The art of implementing a successful model is in selecting the state variables within the model to use with the controller.

15.1.13.2 A path following controller model

The first hurdle to be crossed is the availability of suitable state variables and the use of gain terms to apply to them. Typically in a multibody system model, many more variables are available than in a real vehicle. Within the model, these variables can be the subject of differential equations in order to have available integral and differential terms. Table 15.1-4 shows a portion of a command file from MSC.ADAMS implementing those terms for yaw rate. While it is a working example, no claim is made that is in any sense optimum.

Such variables can usually be manipulated within the model using the programming syntax provided with the code being used. For simulation codes such as MSC.ADAMS, the format of such calculations can appear a little clumsy but this soon disappears with familiarity. The most recent versions of MSC.ADAMS include a 'control toolbox' to facilitate the implementation of PID controllers. For codes such as MATLAB/Simulink the implementation of control systems is arguably easier since they are written with the prime objective of control system modelling. However, the modelling of the vehicle as a plant is more difficult within these systems and so there is an element of swings and roundabouts if choosing between the codes. In general, codes like MSC.ADAMS have a history in very accurate simulation of mechanical systems and can be coerced into representing control systems. Codes like MATLAB and MATLAB/Simulink are the reverse; they have a history in very detailed control system simulation and can be coerced into representing mechanical systems. For this reason, a recent development suggests using each code to perform the tasks at which it is best; this is often referred to as 'co-simulation'. The authors' experiences to date have been universally disappointing for entirely prosaic reasons – the speed of execution is extremely poor and the robustness of the software suppliers in dealing with different releases of each other's product has been somewhat inconsistent. The effort required to persuade the relevant software to work in an area where it is weak is usually made only once and in any case the additional understanding gained is almost always worthwhile for the analyst involved. Until the performance and robustness of the software improve, the authors do not favour co-simulation except for the most detailed software verification exercises.

The next hurdle to be crossed is the representation of the intended behaviour of the vehicle – the 'reference' states. Competition-developed lap simulation tools use a 'track map' based on distance travelled and path curvature. This representation allows the reference path to be of any form at all and allows for circular or crossing paths (e.g. figures of eight) to be represented without the one-to-many mapping difficulties that would be encountered with any sort of y-versus-x mapping. Integrating the longitudinal velocity for the vehicle gives a distance-travelled measure that shows itself to be tolerably robust against drifting within simulation models. Using this measure, the path curvature can be surveyed in the vicinity of the model.

Some authors favour the use of a preview distance for controlling the path of the vehicle, with an error based on lateral deviation from the intended path. However, there is usually a difficulty associated with this since the lateral direction must be defined with respect to the vehicle. (Failure to anchor the reference frame to the vehicle means that portions of the path approaching 90 degrees to the original direction of travel rapidly diverge to large errors.) Projecting a preview line forward of the mass centre based on vehicle centre line is unsatisfactory due to the body slip angle variations mentioned previously. Either the proportional gain must be reduced to avoid 'PIO'-type behaviour, which leads to unsatisfactory behaviour through aggressive avoidance manoeuvres, or else some form of gain scheduling must be applied.

Table 15.1-4 A portion of an MSC.ADAMS command file showing the implementation of differential equations to retrieve and use integral and derivative terms for a state variable

```
! -- Derivative Term - not generally used --
part create equation differential_equation &
   differential_equation_name = .test.yaw_rate_error_equation_1 &
   adams_id = 3 &
   comments = "Yaw Rate Error Equation - Implicit" &
   initial_condition = 0.0 &
   function = "DIF(3)-varval(yaw_rate_error)" &
   implicit = on &
   static_hold = off
data_element create variable &
   variable_name = yaw_rate_error_derivative &
   function = "DIF1(3)"
! -- Integral Term --
part create equation differential_equation &
   differential_equation_name = .test.yaw_rate_error_equation_2 &
   adams_id = 4 &
   comments = "Yaw Rate Error Equation - Explicit" &
   initial_condition = 0.0 &
   function = "varval(yaw_rate_error)" &
   implicit = off &
   static_hold = off
data_element create variable &
   variable_name = yaw_rate_error_integral &
   function = "(DIF(4))"
! Steer input torque in response to path error.
force create direct single_component_force &
      single_component_force_name = yaw_rate_handwheel_torque &
      type_of_freedom = rotational &
      action_only = on &
      i_marker_name = .hand_wheel_column.m_wheel_column &
      j_marker_name = .hand_wheel_column.m_wheel_column &
      function = "(", &
         "   VARVAL(yaw_rate_error)           * VARVAL(yp_gain) ", &
         " + VARVAL(yaw_rate_error_integral)  * VARVAL(yi_gain) ", &
         " + VARVAL(yaw_rate_error_derivative)* VARVAL(yd_gain) ", &
        ")", &
        "* STEP(TIME,0.0,0.0,1.0,1.0)"
```

Alternatively, the preview vector can be adjusted for body slip angle before it is used if oscilliatory behaviour is to be avoided. The length of the preview vector must be adjusted with speed if reasonably consistent behaviour is to be produced. For 'normal' driving this type of model can produce acceptably plausible results but for manoeuvres such as the ISO 3888 Lane Change the behaviour becomes unacceptably oscilliatory particularly after the manoeuvre.

An alternative method, used by the authors with some success for a variety of extreme manoeuvres, is to focus on the behaviour of the front axle. This model fits with one of the author's (Harty) experience of driving at or near the handling limit, particularly on surfaces such as snow where large body slip angles highlight the mechanisms used in the driver's mind. High performance driving coaches rightly concentrate on the use of a 'model' the driver needs in order to retain control in what would otherwise become stressful circumstances of non-linear vehicle behaviour and multiple requirements for control – typically vehicle orientation (body slip angle) and velocity (path control). Useful learning occurs on low grip environments that can be readily transferred across to high grip. In low grip environments, the extreme non-linearity of response of the vehicle can be explored at low speeds and with low stress levels, allowing the driver to piece together a model to be used within their own heads; it is then a matter of practice to transfer the lessons to a high grip environment. The same concepts can be used to explore the behaviour of a driver model within a multibody system environment such as MSC.ADAMS.

The formulation used is described below. All subscripts x and y are in the vehicle reference frame. The

Table 15.1-5 MSC.ADAMS command file sample for 'front axle control' driver model

```
data_element create variable &
   variable_name = ground_plane_velocity &
   function = "(                   (", &
      "VX(m_body_CG,base)**2 +", &
      "VY(m_body_CG,base)**2", &
         ")**0.5 ) / 1000"

data_element create variable &
   variable_name = demanded_yaw_rate &
   function = "varval(ground_plane_velocity) *", &
                  "AKISPL(varval(path_length),0,path_curvature_spline)"

data_element create variable &
   variable_name = beta &
   function="ASIN(VY(m_body_CG,base,m_body_CG)/", &
              "(varval(ground_plane_velocity) + 0.00001))"

data_element create variable &
   variable_name = centacc &
   function = "(VARVAL(latacc)*COS(VARVAL(beta))) +", &
                  "(VARVAL(longacc)*SIN(VARVAL(beta)))"

data_element create variable &
   variable_name = front_axle_no_slip_yaw &
   function = "-(", &
             "VARVAL(centacc) ", &
             "- WDTZ(m_body_CG,base,m_body_CG)", &
             "* DX(m_body_CG,mfr_upright_wheel_centre,m_body_CG)", &
         " )", &
         " /", &
         " ( varval(ground_plane_velocity) + 0.00001 )"

data_element create variable &
   variable_name = yaw_rate_error &
   function = " varval(demanded_yaw_rate) - varval(front_axle_no_slip_yaw)"
```

ground plane velocity V_g is given from the components V_x and V_y using

$$V_g = \sqrt{V_x^2 + V_y^2} \quad (15.1.22)$$

The demanded yaw rate ω_d is found from the forward velocity V_x and path curvature k using

$$\omega_d = V_g k \quad (15.1.23)$$

The body slip angle β is found from the velocities V_y and V_g using

$$\beta = \arcsin\left(\frac{V_y}{V_g}\right) \quad (15.1.24)$$

The centripetal acceleration A^p is given from the components of acceleration A_x and A_y using

$$A^p = A_y \cos(\beta) + A_x \sin(\beta) \quad (15.1.25)$$

The front axle no-slip yaw rate ω_{fNS} is found from the centripetal acceleration A^p, the yaw acceleration α_z, the distance, a, from the mass centre to the front axle and the ground plane velocity V_g using

$$\omega_{fNS} = \frac{A^P - \alpha_z a}{V_g} \quad (15.1.26)$$

The yaw error ω_{err} is then found from the demanded yaw rate ω_d and the front axle no-slip yaw rate ω_{fNS} using

$$\omega_{err} = \omega_d - \omega_{fNS} \quad (15.1.27)$$

The implementation of equations (15.1.22) to (15.1.27) is illustrated, using again an example of the MSC.ADAMS command file format, in Table 15.1-5.

For a variety of events, this formulation produces good driver/vehicle behaviour, representative of real vehicle and driver behaviour (Fig. 15.1-45). The simulated driver and vehicle behaviour for a post-limit turn-in event is

Fig. 15.1-45 Driver and vehicle behaviour for a post-limit turn-in event (photograph courtesy of Don Palmer, www.donpalmer.co.uk).

compared here to a real vehicle. Note the freewheeling analytical model (of a significantly different vehicle) displays greater body slip angle, while the real vehicle displays greater oversteer.

15.1.13.3 Body slip angle control

Skilled drivers, particularly rally drivers, frequently operate at large body slip angles. Colloquially, there is much talk of body slip angles being in excess of 45 degrees but recorded data suggests this is not the case despite appearances. Large body slip angles generally slow progress; although some of the yaw transients are rapid, in general the actual body slip angles are comparatively small (Fig. 15.1-46). In general, drivers greatly overestimate body slip angle subjectively (Fig. 15.1-47).

The steering system on a vehicle has only 20–25 degrees (less on the rally car) of lock and so realistically, control beyond these body slip angles is unlikely without very large amounts of space indeed.

Most drivers are acutely sensitive to the rate of change of body slip angle, albeit they do not always respond correctly to it. Instead a 'threshold' behaviour appears common, with drivers neglecting body slip angle until either the angle becomes large or its rate of change becomes large. For road cars, our goals are to have a road car manage its own body slip angle so as not to put pressure on drivers in an area where in general skill is lacking. For driver modelling purposes, a separate body slip angle control loop is desirable to catch spins but need not be terribly sophisticated since if it is invoked then we have to some extent failed. Such behaviour is desirable in the real vehicle too and is the goal of active intervention systems such as brake-based stability control systems; however, the robust sensing of body slip angle still proves elusive in a cost-effective manner despite its apparent simplicity.

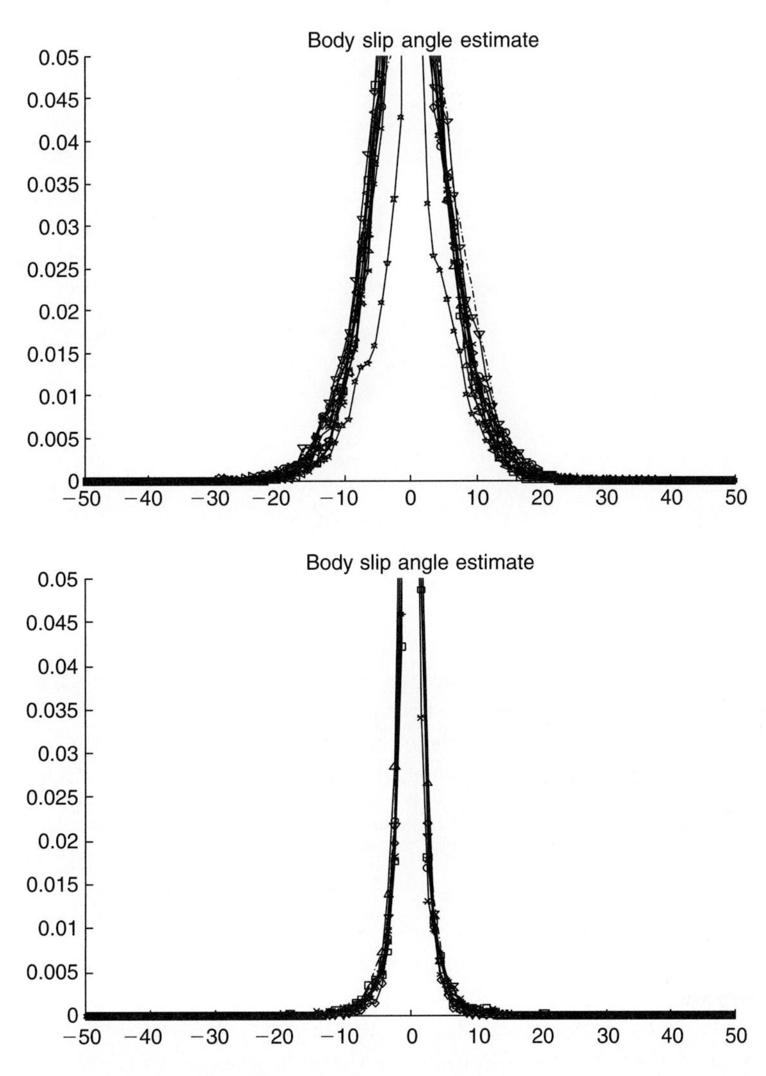

Fig. 15.1-46 Probability density for body slip angle estimates – Greece 2002 (top) and Germany 2002 (bottom) for Petter Solberg, Subaru World Rally Team.

15.1.13.4 Two-loop driver model

For general use, the authors favour a simple and robust two-loop driver model comprising a path follower and spin catcher, with a separate speed control as appropriate to the task at hand. Fig. 15.1-48 shows such a model.

15.1.14 Case study 7 – comparison of full vehicle handling models

As mentioned at the start of this chapter the use of modern multibody systems software provides users with the capability to develop a model of a full vehicle that incorporates all the major vehicle subsystems. Clearly the development of such a model is dependent on the stage of vehicle design and the availability of the data needed to model all the subsystems. For the vehicle dynamics task, however, the automotive engineer will want to carry out simulations before the design has progressed to such an advanced state.

In this case study the level of vehicle modelling detail required to simulate a 'full vehicle' handling manoeuvre will be explored. We will consider a 100 km/h double lane change manoeuvre, as a start. The test procedure for the double lane change manoeuvre is shown schematically in Fig. 15.1-49.

For the simulations performed in the case study the measured steering wheel inputs from a test vehicle have been extracted and applied as a time dependent handwheel rotation (Fig. 15.1-50) as described in Section 15.1.12.3.

To appreciate the use of computer simulations to represent this manoeuvre an example of the superimposed animated wireframe graphical outputs for this simulation is given in Fig. 15.1-51.

In this study the influence of suspension modelling on the accuracy of the simulation outputs is initially

Fig. 15.1-47 Large body slip angles are unavailable to normal drivers except as part of an accident. Subaru WRC, Greece 2002 (courtesy of Prodrive).

discussed based on results obtained using the four vehicle models described in section 15.1.4 and summarized schematically again here in Fig. 15.1-52. The models shown can be thought of as a set of models with evolving levels of elaboration leading to the final linkage model that involves the modelling of the suspension linkages and the bushes.

For each of the vehicle models described here it is possible to estimate the model size in terms of the degrees of freedom in the model and the number of equations that MSC.ADAMS uses to formulate a solution. The calculation of the number of degrees of freedom (DOF) in a system is based on the Greubler equation. It is therefore possible for any of the vehicle models to calculate the degrees of freedom in the model. An example is provided here for the equivalent roll stiffness model where the degrees of freedom can be calculated as follows:

$$
\begin{array}{lrl}
\text{Parts} & 9 \times 6 & = 54 \\
\text{Rev} & 8 \times -5 & = -40 \\
\text{Motion} & 2 \times -1 & = -2 \\
\hline
 & \sum_{DOF} & = 12
\end{array}
$$

In physical terms it is more meaningful to describe these degrees of freedom in relative terms as follows. The vehicle body part has 6 degrees of freedom. The

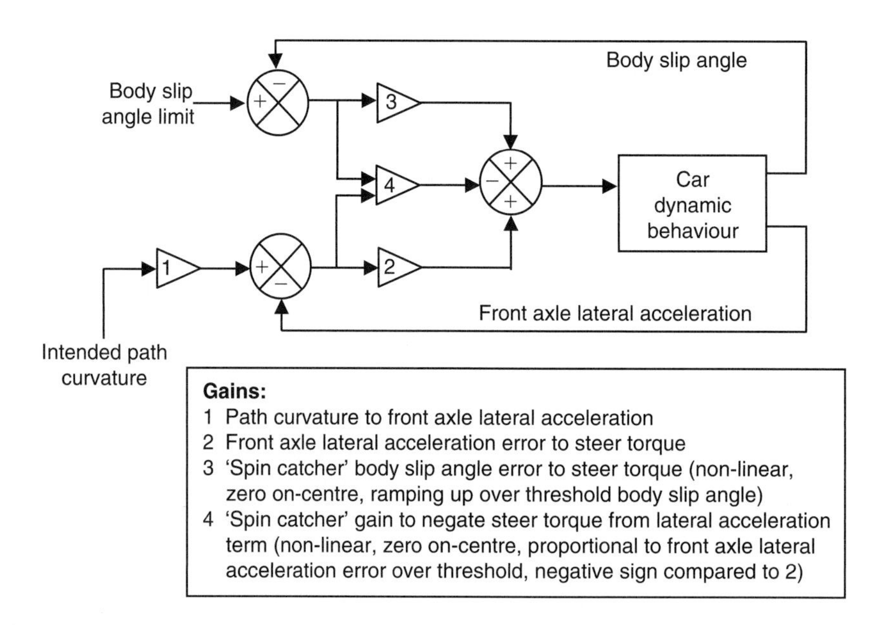

Fig. 15.1-48 A typical two-loop driver model.

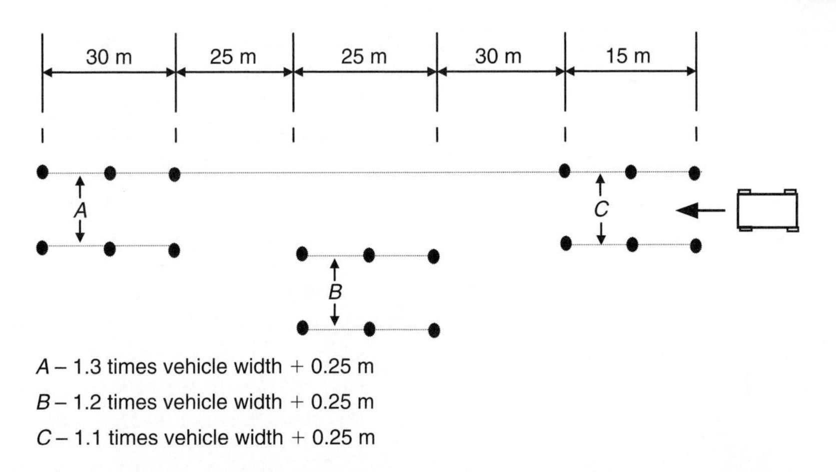

Fig. 15.1-49 Lane change test procedure. (This material has been reproduced from the Proceedings of the Institution of Mechanical Engineers, K2 Vol. 214 'The modelling and simulation of vehicle handling. Part 4: handling simulation', M.V. Blundell, page 74, by permission of the Council of the Institution of Mechanical Engineers.)

two axle parts each have 1 rotational degree of freedom relative to the body. Each of the four road wheel parts has 1 spin degree of freedom relative to the axles making a total of 12 degrees of freedom for the model.

When a simulation is run in MSC.ADAMS the program will also report the number of equations in the model. The software will formulate 15 equations for each part in the model and additional equations representing the constraints and forces in the model. On this basis the size of all the models is summarized in Table 15.1-6.

The size of the model and the number of equations is not the only issue when considering efficiency in vehicle modelling. Of perhaps more importance is the engineering significance of the model parameters. The roll stiffness model, for example, may be preferable to the lumped mass model. It is not only a simpler model but is also based on parameters such as roll stiffness that will have relevance to the practising vehicle dynamicist. The roll stiffness can be measured on an actual vehicle or estimated during vehicle design. This model does, however, incorporate rigid axles eliminating the independent suspension characteristics.

Measured outputs including lateral acceleration, roll angle and yaw rate can be compared with measurements taken from the vehicle during the same manoeuvre on the proving ground to assess the accuracy of the models. By way of example the yaw rate predicted by simulation with all four models is compared with measured track test data in Figs. 15.1-53–15.1-56.

Examination of the traces in Figs. 15.1-53–15.1-56 raises the question as to how an objective assessment of the accuracy of the simulations may be made. Accuracy is not a 'yes/no' quantity, but instead a varying absence of difference exists between predicted (calculated) behaviour and measured behaviour. Such a 'difference' is

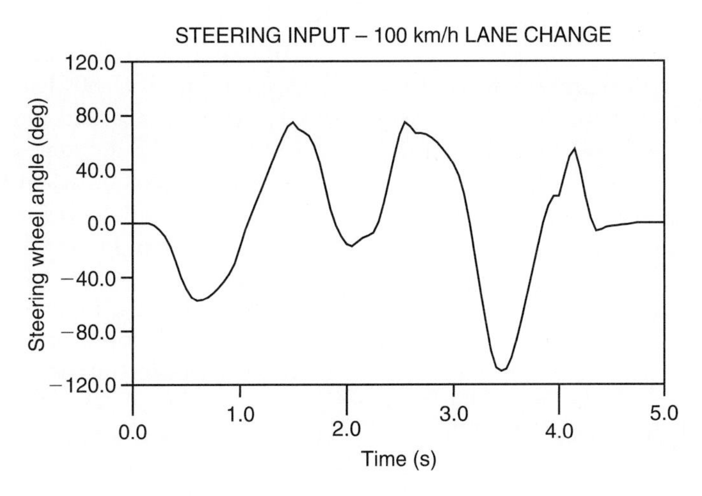

Fig. 15.1-50 Steering input for the lane change manoeuvre.

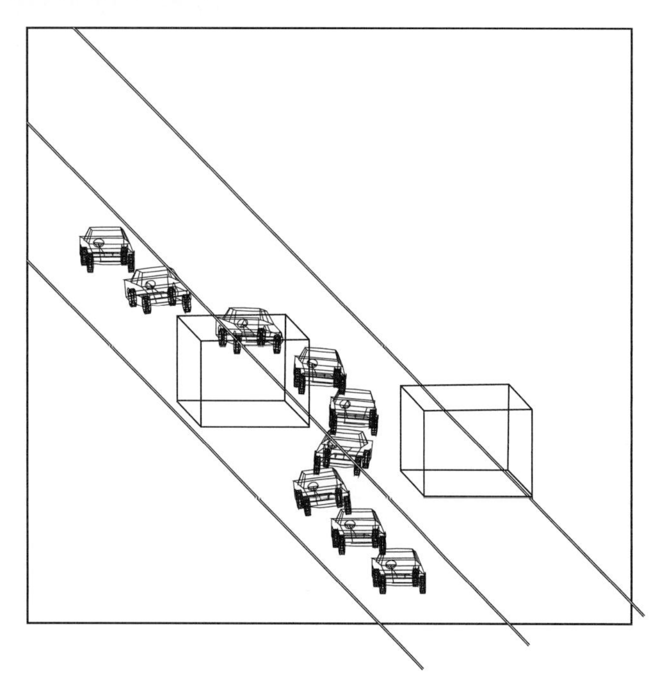

Fig. 15.1-51 Superimposed graphical animation of a double lane change manoeuvre.

commonly referred to as 'error'. This definition neatly sidesteps two other difficulties:

- Is the measured data what actually happens in the absence of measurement?
- Is the measured data what actually happens during service?

For example, the mass-loading effect of accelerometers may introduce inaccuracies at high frequencies and could mean that the system of interest behaves differently when being measured to when not. The accuracy of controlled measurements in discerning the behaviour of the system when in normal uncontrolled use is another matter entirely. Both topics are far from trivial.

In this case other questions arise such as:

- Does the model data accurately represent the vehicle conditions on the day of the test?
- Does the tyre test data obtained on a tyre test machine accurately represent the condition of the test surface and tyres used on the day of the test?
- How repeatable are the experimental test results used to make an assessment of model accuracy?
- Is there a model data input error common to all the models?

Comparing the performance of the equivalent roll stiffness model with that of the linkage model in Figs. 15.1-55 and 15.1-56 it is possible to look, for example, at the error measured between the experimental and simulated results for the peaks in the response or to sum the overall error from start to finish. On that basis it may seem desirable to somehow 'score' the models giving, say, the linkage model 8/10 and the roll stiffness model 7/10. In light of the above questions the validity of such an objective measure is debatable and it is probably more appropriate to simply state:

> *For this vehicle, this manoeuvre, the model data, and the available benchmark test data the equivalent roll stiffness model provides reliable predictions when compared with the linkage model for considerably less investment in model elaboration.*

Clearly it is also possible to use an understanding of the physics of the problem to aid the interpretation of model performance. An important aspect of the predictive models is whether the simplified suspension models correctly distribute load to each tyre and model the tyre position and orientation in a way that will allow a good tyre model to determine forces in the tyre contact patch that impart motion to the vehicle and produce the desired response. Taking this a step further we can see that if we use the equivalent roll stiffness and linkage models as the basis for further comparison it is possible in Figs. 15.1-57 and 15.1-58 to compare the vertical force in, for example, the front right and left tyres. The plots indicate the performance of the simple equivalent roll stiffness model in distributing the load during the manoeuvre. The weight transfer across the vehicle is also evident as is the fact that tyre contact with the ground is maintained throughout. It should also be noticed that in determining the load transfer to each wheel the equivalent roll stiffness model does not include the degrees of freedom that would allow the body to heave or pitch relative to the suspension systems.

In Figs. 15.1-59 and 15.1-60 a similar comparison between the two models is made, this time considering, for example, the slip and camber angles predicted in the front right tyre.

Although the prediction of slip angle agrees well it can be seen in Fig. 15.1-60 that the equivalent roll stiffness model with a maximum value of about 1.5 degrees underestimates the amount of camber angle produced during the simulation when compared with the linkage model where the camber angle approaches 5 degrees. Clearly the wheels in the effective roll stiffness model do not have a camber degree of freedom relative to the rigid axle parts and the camber angle produced here is purely due to tyre deflection.

It is perhaps fortuitous in this case that for a passenger car of the type used here the lateral tyre force produced due to slip angle is considerably more significant than that arising due to camber between the tyre and road surface. Further investigations can be carried out to establish the

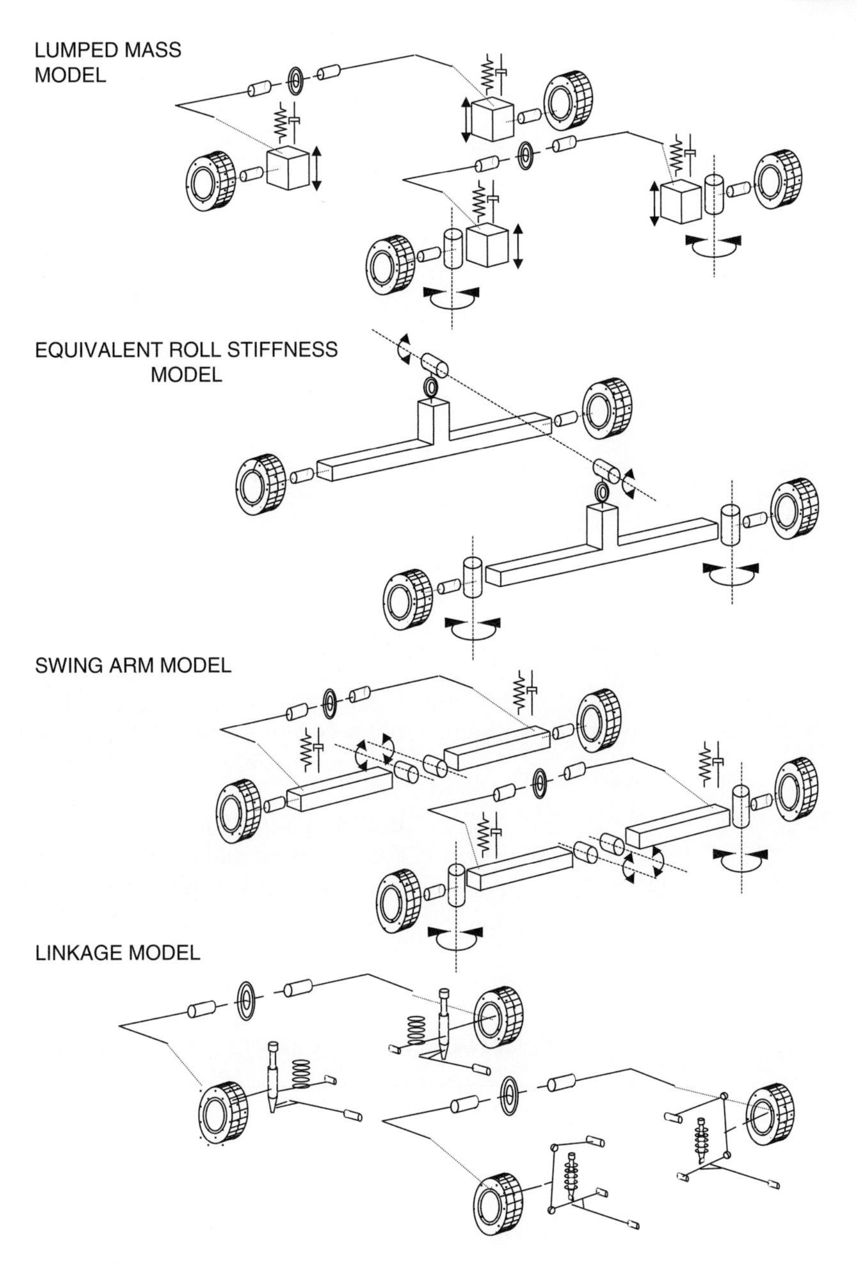

Fig. 15.1-52 Modelling of suspension systems

significance of a poor camber angle prediction input to the tyre model. In Fig. 15.1-61 the linkage model has been run using an interpolation tyre model where it has been possible to deactivate the generation of lateral force arising from camber angle. In this plot it can be seen that the prediction of yaw rate, for example, is not sensitive for this vehicle and this manoeuvre to the modelling of camber thrust.

To conclude this case study it is possible to consider an alternative modelling and simulation environment for the prediction of the full vehicle dynamics. As discussed earlier the incorporation of microprocessor control systems

Table 15.1-6 Vehicle model sizes

Model	Degrees of freedom	Number of equations
Linkage	78	961
Lumped mass	14	429
Swing arm	14	429
Roll stiffness	12	265

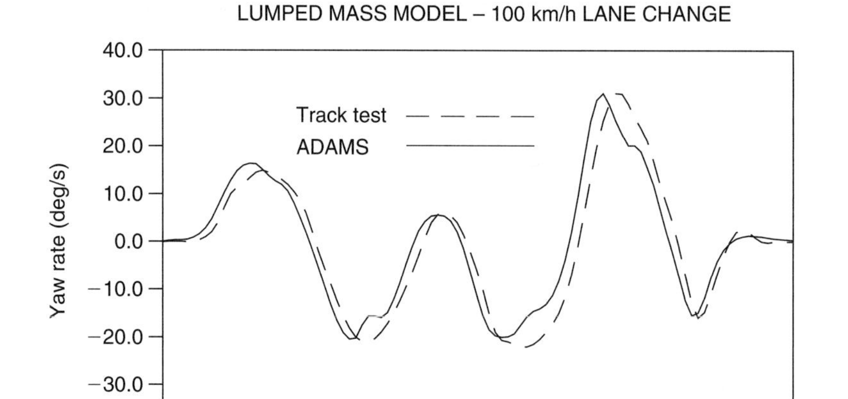

Fig. 15.1-53 Yaw rate comparison – lumped mass model and test. (This material has been reproduced from the Proceedings of the Institution of Mechanical Engineers, K2 Vol. 214 'The modelling and simulation of vehicle handling. Part 4: handling simulation', M.V. Blundell, page 80, by permission of the Council of the Institution of Mechanical Engineers.)

in a vehicle may involve the use of a simulation method that involves:

(i) the use of multibody systems software where the user must invest in the modelling of the control systems

(ii) the use of software such as MATLAB/Simulink where the user must invest in the implementation of a vehicle model or

(iii) a co-simulation involving parallel operation of the multibody systems and control simulation software

In this example the author[2] has chosen the second of the above options and a vehicle model (Fig. 15.1-62) is developed from first principles and implemented in Simulink. The model developed here is based on the same data used for this case study with 3 degrees of freedom: the

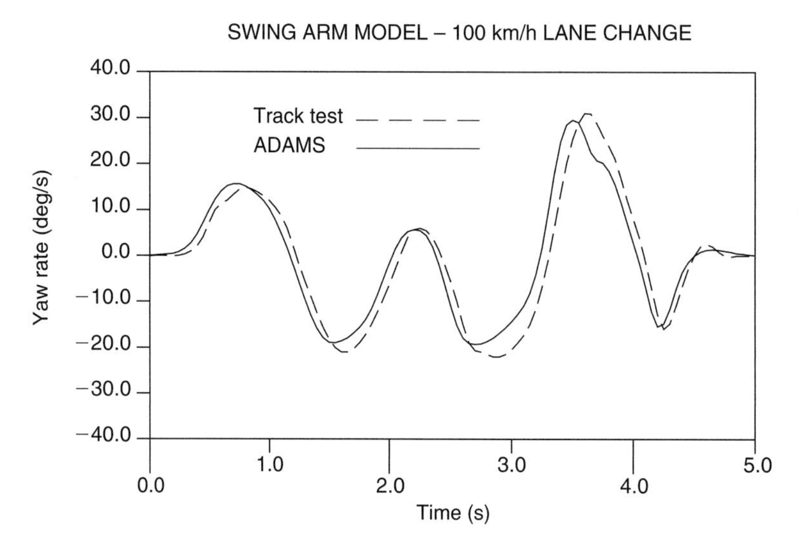

Fig. 15.1-54 Yaw rate comparison – swing arm model and test. (This material has been reproduced from the Proceedings of the Institution of Mechanical Engineers, K2 Vol. 214 'The modelling and simulation of vehicle handling. Part 4: handling simulation', M.V. Blundell, page 80, by permission of the Council of the Institution of Mechanical Engineers.)

[2] In their 2003 work, Wenzel and co-workers describe preliminary work undertaken in a collaborative research project with Jaguar Cars Ltd, Coventry, UK and funded by the Control Theory and Applications Centre, Coventry University, Coventry, UK. It forms the PhD programme for Thomas A. Wenzel.

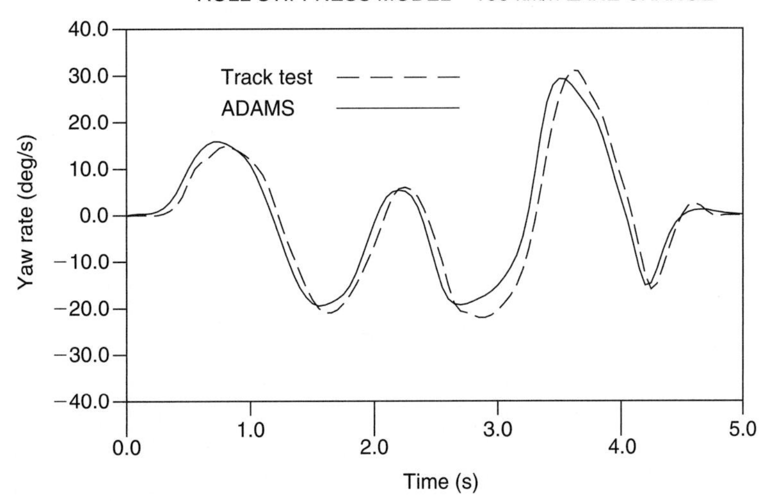

Fig. 15.1-55 Yaw rate comparison – roll stiffness model and test. (This material has been reproduced from the Proceedings of the Institution of Mechanical Engineers, K2 Vol. 214 'The modelling and simulation of vehicle handling. Part 4: handling simulation', M.V. Blundell, page 81, by permission of the Council of the Institution of Mechanical Engineers.)

longitudinal direction x, the lateral direction y and the yaw around the vertical axis z.

The vehicle parameters used in the following model include:

v_x = longitudinal velocity (m/s)
v_y = lateral velocity (m/s)
v_{cog} = centre of gravity velocity (m/s)
a_x = longitudinal acceleration (m/s^2)
a_y = lateral acceleration (m/s^2)
Γ = torque around z-axis (Nm)
δ = steer angle (rad)
β = side slip angle (rad)
α_{ij} = wheel slip angles (rad)
$\dot{\psi}$ = yaw rate (rad/s)
F_{zij} = vertical forces on each wheel (N)
ij = position: i = front(f)/rear(r), j = left(l)/right(r)

Note that steer angle δ and the velocity of the vehicle's centre of gravity v_{cog} are specified as model inputs.

The relationship between the dynamic vehicle parameters can be formulated as differential equations. Most of these can be found in the standard literature. Using formulas by Wong and Will and Z.ak the following

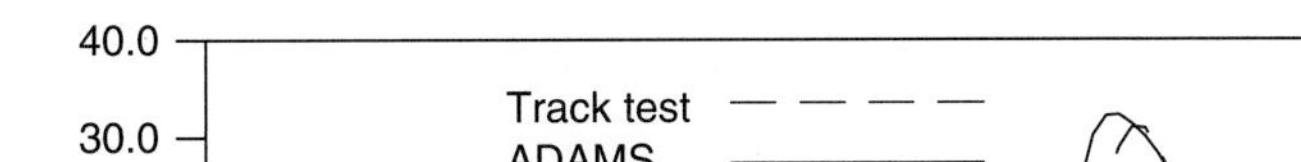

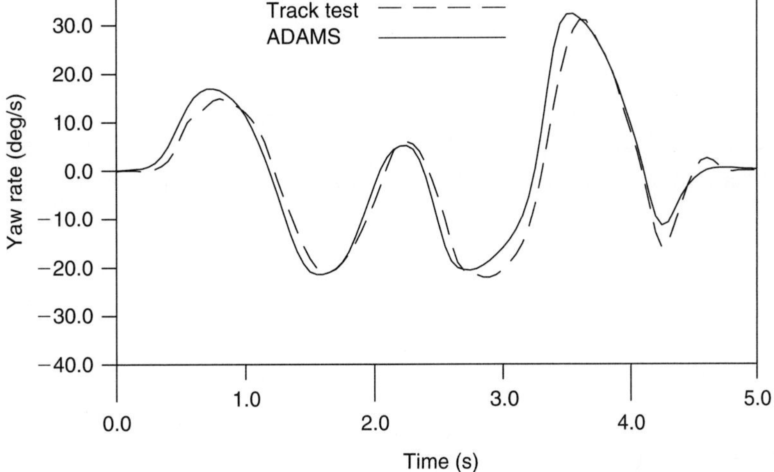

Fig. 15.1-56 Yaw rate comparison – linkage model and test. (This material has been reproduced from the Proceedings of the Institution of Mechanical Engineers, K2 Vol. 214 'The modelling and simulation of vehicle handling. Part 4: handling simulation', M.V. Blundell, page 81, by permission of the Council of the Institution of Mechanical Engineers.)

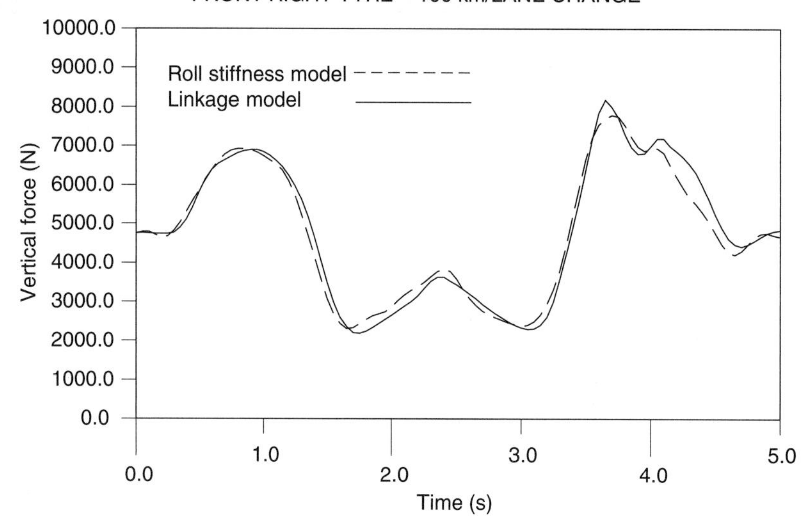

Fig. 15.1-57 Vertical tyre force comparison – linkage and roll stiffness models.

differential equations for acceleration, torque and yaw rate can be derived:

$$\dot{v}_x = \frac{1}{m}(F_{xfl}\cos\delta - F_{yfl}\sin\delta + F_{xfr}\cos\delta - F_{yfr}\sin\delta + F_{xrl} + F_{xrr}) + v_y\dot{\psi} \quad (15.1.28)$$

$$\dot{v}_y = \frac{1}{m}(F_{yfl}\cos\delta + F_{xfl}\sin\delta + F_{yfr}\cos\delta + F_{xfr}\sin\delta + F_{yrl} + F_{yrr}) - v_x\dot{\psi} \quad (15.1.29)$$

$$\Gamma = \frac{t_f}{2}F'_{xfl} - \frac{t_f}{2}F'_{xfr} + \frac{t_r}{2}F_{xrl} - \frac{t_r}{2}F_{xrr} + bF'_{yfl} + bF'_{yfr} - cF_{yrl} - cF_{yrr} + M_{zfl} + M_{zfr} + M_{zrl} + M_{zrr} \quad (15.1.30)$$

$$\ddot{\psi} = \frac{\Gamma}{J_z} \quad (15.1.31)$$

where the additional parameters are defined as:

F_{xij} = longitudinal forces on tyre ij (N)
F_{yij} = lateral forces on tyre ij (N)
F'_{xij} = longitudinal forces on tyre ij in the vehicle's co-ordinate system (N)
F'_{yij} = lateral forces on tyre ij in the vehicle's co-ordinate system (N)
M_{zij} = self-aligning moment on tyre ij (N m)
m = mass of vehicle (kg)
J_z = moment of inertia around vertical axis (N m^2)
t_f, t_r = front and rear track width (m)
b, c = position of centre of gravity between wheels (m)

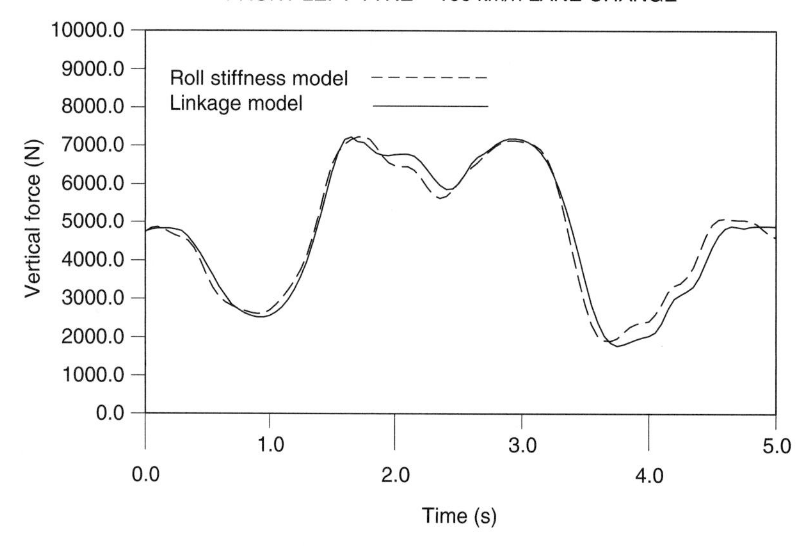

Fig. 15.1-58 Vertical tyre force comparison – linkage and roll stiffness models.

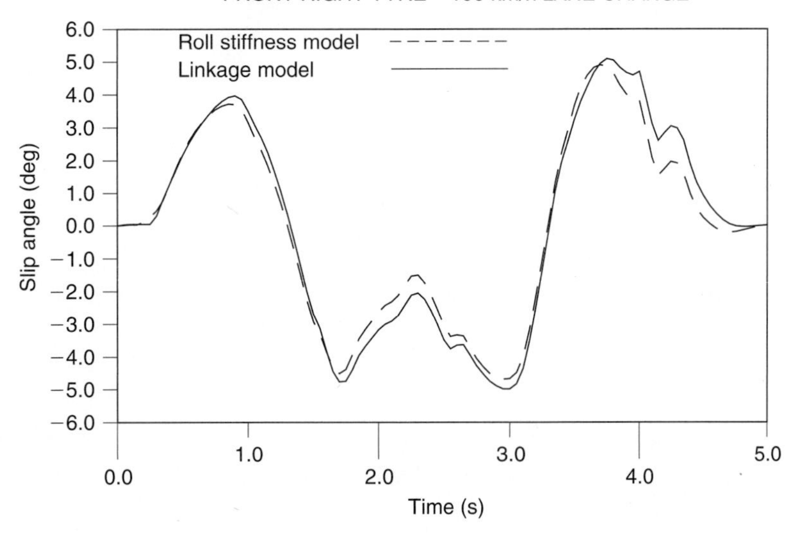

Fig. 15.1-59 Slip angle comparison – linkage and roll stiffness models.

Other important states are the wheel slip angles α_{ij} and the body slip angle β, defined as follows:

$$\alpha_{fl/r} = \delta - \arctan\left(\frac{v_y + b\dot{\psi}}{v_x \pm \frac{1}{2}t_f\dot{\psi}}\right) \quad (15.1.32)$$

$$\alpha_{rl/r} = \arctan\left(\frac{-v_y + c\,\dot{\psi}}{v_x \pm \frac{1}{2}t_r\,\dot{\psi}}\right) \quad (15.1.33)$$

$$\beta = \arctan\left(\frac{v_y}{v_x}\right) \quad (15.1.34)$$

In this model roll and pitch of the vehicle are neglected but weight transfer is included to determine the vertical load at each wheel as defined by Milliken and Milliken:

$$F_{zfl/r} = \left(\frac{1}{2}mg \pm m\frac{a_y h}{t}\right)\frac{c}{l} - ma_x\frac{h}{l} \quad (15.1.35)$$

$$F_{zrl/r} = \left(\frac{1}{2}mg \pm m\frac{a_y h}{t}\right)\frac{b}{l} + ma_x\frac{h}{l} \quad (15.1.36)$$

The additional parameters are the height h of the vehicle's centre of gravity, the wheelbase l and the gravitational acceleration g.

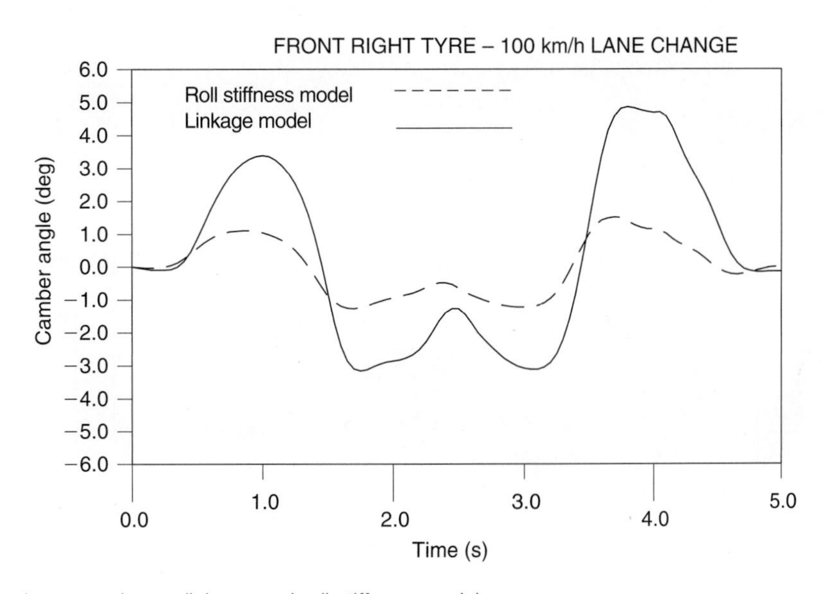

Fig. 15.1-60 Camber angle comparison – linkage and roll stiffness models.

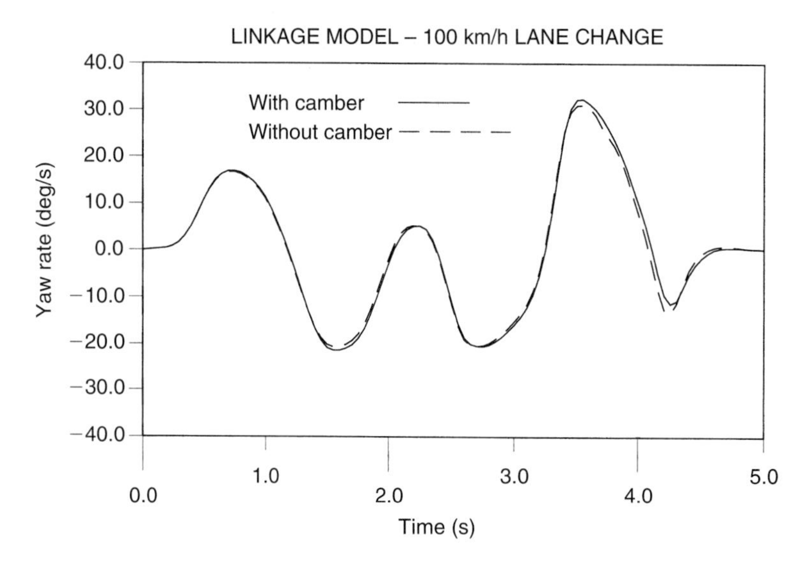

Fig. 15.1-61 Yaw rate comparison – Interpolation tyre model. (This material has been reproduced from the Proceedings of the Institution of Mechanical Engineers, K2 Vol. 214 'The modelling and simulation of vehicle handling. Part 4: handling simulation', M.V. Blundell, page 83, by permission of the Council of the Institution of Mechanical Engineers.)

In equations (15.1.37) and (15.1.38) it has to be considered that $a_x \neq \dot{v}_x$ and $a_y \neq \dot{v}_y$. The yaw motion of the vehicle has to be taken in account giving:

$$a_x = \dot{v}_x - v_y\dot{\psi} \tag{15.1.37}$$

$$a_y = \dot{v}_y - v_x\dot{\psi} \tag{15.1.38}$$

In this work the author has simulated a range of vehicle manoeuvres using both the 'Magic Formula' and Fiala tyre models. The example shown here is for the lane change manoeuvre used in this case study with a reduced steer input applied at the wheels as shown in the bottom of Fig. 15.1-63.

Also shown in Fig. 15.1-63 are the results from the Simulink model and a simulation run with the MSC. ADAMS linkage model. For this manoeuvre and vehicle data set the Simulink and MSC.ADAMS models can be seen to produce similar results.

In completing this case study there are some conclusions that can be drawn. For vehicle handling simulations it has been shown here that simple models such as the equivalent roll stiffness model can provide good levels of accuracy. It is known, however, that roll centres will 'migrate' as the vehicle rolls, particularly as the vehicle approaches limit conditions.

Using a multibody systems approach to develop a simple model may also throw up some surprises for the unsuspecting analyst. The equivalent roll stiffness model, for example, does not include heave and pitch degrees of freedom relative to the front and rear axles. During the simulation, however, the degrees of freedom exist for the body to heave and pitch relative to the ground inertial

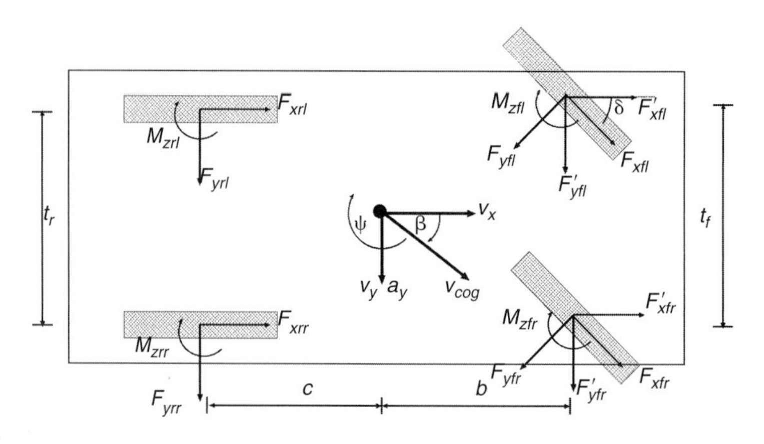

Fig. 15.1-62 Three-degree-of-freedom vehicle model.

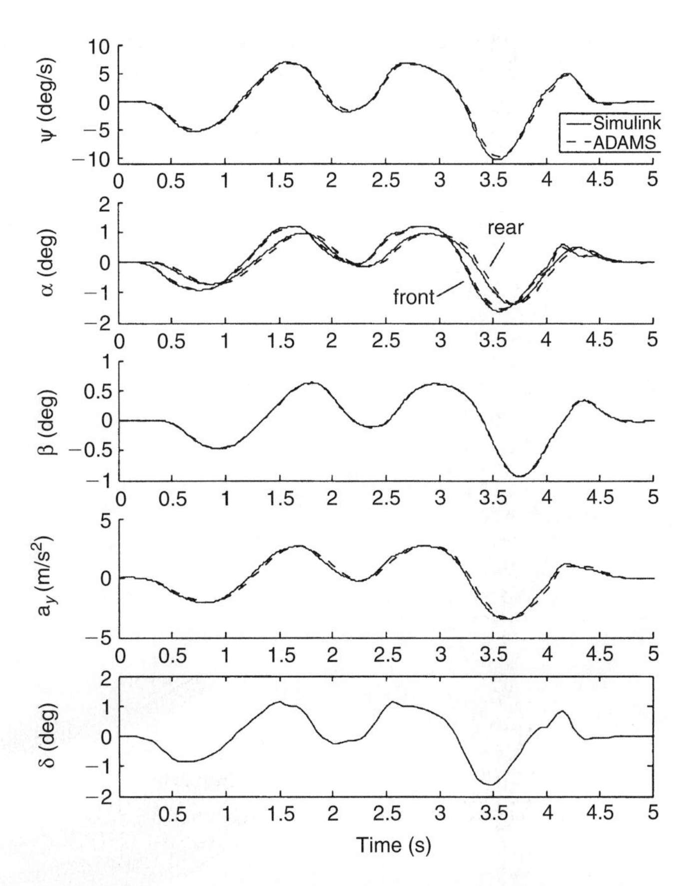

Fig. 15.1-63 Comparison of Simulink and MSC.ADAMS predictions of vehicle response.

frame. These degrees of freedom must still be solved and in this case are damped only by the inclusion of the tyre model. In the 3 degrees of freedom model these motions are ignored and solution is only performed on the degrees of freedom that have been modelled. While the main theme in this book is to demonstrate the use of multibody systems analysis the Matlab/Simulink model is useful here in providing the basis for additional modelling and simulation of the modern control systems involved in enhancing the stability and dynamics of the vehicle. The effort invested in this modelling approach also provides educational benefits reinforcing fundamental vehicle dynamics theory.

15.1.15 Summary

Many different possibilities exist for modelling the behaviour of the vehicle driver. That none has reached prominence suggests that none is correct for every occasion. In general, the road car vehicle dynamics task is about delivering faithful behaviour during accident evasion manoeuvres – where most drivers rarely venture. Positioning the vehicle in the linear region is relatively trivial and need not exercise most organizations unduly, but delivering a good response, maintaining yaw damping and keeping the demands on the driver low are of prime importance in the non-linear accident evasion regime. For this reason, controllers that take time to 'learn' the behaviour of the vehicle are inappropriate – road drivers do not get second attempts. For road vehicles, the closed loop controller based on front axle lateral acceleration gives good results and helps the analyst understand whether or not the vehicle is actually 'better' in the sense of giving an average driver the ability to complete a manoeuvre.

In motorsport applications, however, drivers are skilled and practised and so controllers with some feed-forward capability (to reflect 'learned' responses), plus closed loop control of body slip angle are appropriate to reflect the high skill level of the driver. Whether or not advanced gain scheduling models, such as the MRAS or Self-Tuning Regulator, are in use depends very much on whether or not data exists to support the verification of such a model. The authors preference is that 'it is better to be simple and wrong than complicated and wrong' – in other words, all other things being equal, the simplest model is the most useful since its shortcomings are more easily understood and judgements based on the results may be tempered accordingly. With elaborate schemes, particularly self-tuning ones, there is a strong desire to believe the complexity is in and of itself a guarantee of success.

In truth if a relatively simple and robust model cannot be made to give useful results it is more likely to show a lack of clarity in forming the question than a justification for further complexity.

Section Sixteen

Structural design

Chapter 16.1

16.1

Terminology and overview of vehicle structure types

Jason Brown, A.J. Robertson and Stan Serpento

16.1.1 Basic requirements of stiffness and strength

The purpose of the structure is to maintain the shape of the vehicle and to support the various loads applied to it. The structure usually accounts for a large proportion of the development and manufacturing cost in a new vehicle programme, and many different structural concepts are available to the designer. It is essential that the best one is chosen to ensure acceptable structural performance within other design constraints such as cost, volume and method of production, product application, etc.

Assessments of the performance of a vehicle structure are related to its *strength* and *stiffness*. A design aim is to achieve sufficient levels of these with as little mass as possible. Other criteria, such as crash performance, are not discussed here.

16.1.1.1 Strength

The strength requirement implies that no part of the structure will lose its function when it is subjected to road loads. Loss of function may be caused by instantaneous overloads due to extreme load cases, or by material fatigue. Instantaneous failure may be caused by (a) overstressing of components beyond the elastic limit, or (b) by buckling of items in compression or shear, or (c) by failure of joints. The life to initiation of fatigue cracks is highly dependent on design details, and can only be assessed when a detailed knowledge of the component is available. For this reason assessment of fatigue strength is usually deferred until after the conceptual design stage.

The strength may be alternatively defined as the maximum force which the structure can withstand (see Fig. 16.1-1). Different load cases cause different local component loads, but the structure must have sufficient strength for all load cases.

16.1.1.2 Stiffness

The stiffness K of the structure relates the deflection Δ produced when load P is applied, i.e. $P = K\Delta$. It applies only to structures in the elastic range and is the slope of the load vs deflection graph (see Fig. 16.1-1).

The stiffness of a vehicle structure has important influences on its handling and vibrational behaviour. It is important to ensure that deflections due to extreme loads are not so large as to impair the function of the vehicle, for example so that the doors will not close, or suspension geometry is altered. Low stiffness can lead to unacceptable vibrations, such as 'scuttle shake'.

Again, different load cases require different stiffness definitions, and some of these are often used as

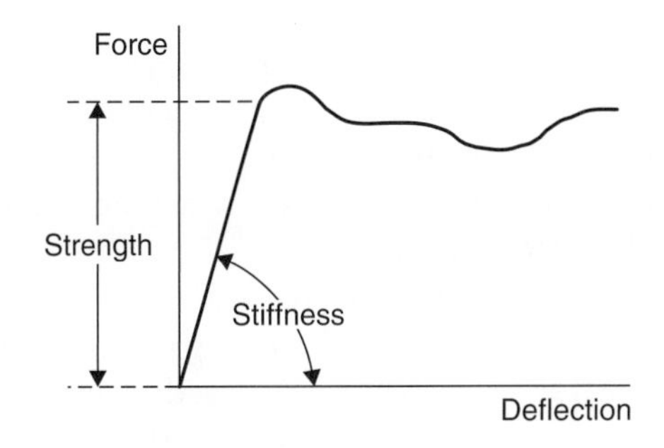

Fig. 16.1-1 The concepts of stiffness and strength.

Motor Vehicle Structures; ISBN: 9780750651349
Copyright © 2001 Jason Brown, A.J. Robertson and Stan Serpento. All rights of reproduction, in any form, reserved.

'benchmarks' of vehicle structural performance. The two most commonly used in this way are:

(a) *Bending stiffness* K_B, which relates the symmetrical vertical deflection of a point near the centre of the wheelbase to multiples of the total static loads on the vehicle. A simplified version of this is to relate the deflection to a single, symmetrically applied load near the centre of the wheelbase.

(b) *Torsion stiffness* K_T, relates the torsional deflection θ of the structure to an applied pure torque T about the longitudinal axis of the vehicle. The vehicle is subjected to the 'pure torsion load case' described in Section 16.1.4 (where the torque is applied as equal and opposite couples acting on suspension mounting points at the front and rear), and the twist θ is measured between the front and rear suspension mountings. Twist at intermediate points along the wheelbase is sometimes also measured in order to highlight regions of the structure needing stiffening.

These two cases apply completely different local loads to individual components within the vehicle. It is usually found that the torsion case is the most difficult to design for, so that the torsion stiffness is often used as a 'benchmark' to indicate the effectiveness of the vehicle structure.

16.1.1.3 Vibrational behaviour

The global vibrational characteristics of a vehicle are related to both its stiffness and mass distribution. The frequencies of the global bending and torsional vibration modes are commonly used as benchmarks for vehicle structural performance. These are not discussed in this book. However, bending and torsion stiffness K_B and K_T influence the vibrational behaviour of the structure, particularly its first natural frequency.

16.1.1.4 Selection of vehicle type and concept

In order to achieve a satisfactory structure, the following must be selected:

(a) The most appropriate structural type for the intended application.

(b) The correct layout of structural elements to ensure satisfactory load paths, without discontinuities, through the vehicle structure.

(c) Appropriate sizing of panels and sections, and good detail design of joints.

An assumption made in this book is that if satisfactory load paths (i.e. if equilibrium of edge forces between simple structural surfaces) are achieved, then the vehicle is likely to have the foundation for sufficient structural (and especially torsion) stiffness. Estimates of interface loads between major body components calculated by the simplified methods described are assumed to be sufficiently accurate for conceptual design, although structural members comprising load paths must still be sized appropriately for satisfactory results. Early estimates of stiffness can be obtained using the finite element method, but the results should be treated with caution because of simplifications in the idealization of the structure at this stage.

16.1.2 History and overview of vehicle structure types

Many different types of structure have been used in passenger cars over the years. This brief overview is not intended to be a detailed history of these, but to set a context. It covers only a selection of historical and modern structures to show the engineering factors which led to the adoption of the integral structure for mass produced vehicles, and other types for specialist vehicles.

16.1.2.1 History: the underfloor chassis frame

In the 1920s, when mass production had become well established, the standard car configuration was the separate 'body-on-chassis' construction. This had certain advantages, including manufacturing flexibility, allowing different body styles to be incorporated easily, and allowing the 'chassis' to be treated as a separate unit, incorporating all of the mechanical components. The shape of the chassis frame was ideally suited for mounting the semi-elliptic spring on the beam axle suspension system, which was universal at that time. Also, this arrangement was favoured because the industry at that time was divided into separate 'chassis' and 'body' manufacturers. Tradition played an additional part in the choice of this construction method.

The underfloor chassis frame, which was regarded as the structure of the car, consisted of a more or less flat 'ladder frame' (Fig. 16.1-2). This incorporated two open section (usually pressed C-section) sideframes running the full length of the vehicle, connected together by open section cross-members running laterally and riveted to the side frames at 90° joints. Such a frame belongs to a class of structures called *'grillages'*.

A grillage is a flat ('planar') structure subjected to loads normal to its plane (see Fig. 16.1-3). The *active* internal loads in an individual member of such a frame are (see inset):

(a) Bending about the in-plane lateral axis of the member.

(b) Torsion about the longitudinal axis of the member in the plane of the frame.

Fig. 16.1-2 Open section ladder frame chassis of the 1920s (courtesy of Vauxhall Archive Centre).

(c) Shear force in a direction normal to the plane of the frame.

Open section members, as used in 1920s and 1930s chassis frames, are particularly flexible locally in torsion. Further, the riveted T-joints were poor at transferring bending moments from the ends of members into torsion in the attached members and vice versa. Chassis frames from that era thus had very low torsion stiffness. Since, on rough roads, torsion is a very important loading, this situation was not very satisfactory. The depth of the 'structure' was limited to a shallow frame underneath the body, so that the bending stiffness was also relatively low.

Texts from the 1920s show that considerable design attention was paid only to the *bending* behaviour of the structure, mainly from the strength point of view.

The diagram in Fig. 16.1-4 (from Donkin's 1925 textbook on vehicle design) shows carefully drawn shear force and bending moment diagrams for the chassis frame, based on the static weight of the chassis, attached components, body, payload, etc. The bending moment diagram is compared with the distribution of bending strength in the chassis side members. Important to note, however, is the complete absence of any consideration of *torsion* behaviour of the structure. The importance of this was not fully understood by the engineering community until later.

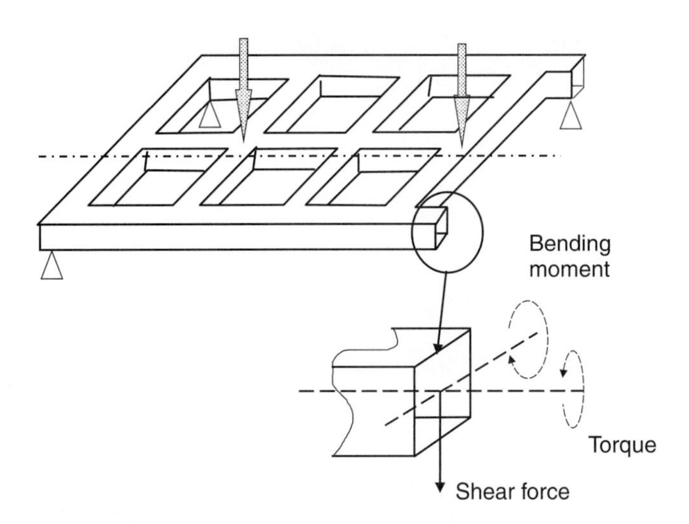

Fig. 16.1-3 Grillage frame.

Good torsional design is important to ensure satisfactory vehicle handling, to avoid undesirable vibrations, and to prevent problems of incompatibility between body and frame as described below. The torsion load case puts different local loads on the structural components from those experienced in the bending load case. Torsion stiffness is often used as one of the 'benchmarks' of the structural competence of a vehicle structure.

In view of the poor torsion performance of the early chassis frame, it is perhaps fortunate that car bodies in the 1920s (Fig. 16.1-5) were 'coachbuilt' by carpenters, out of timber, leading to body structures of very low stiffness. In the early part of the 1920s, the majority of passenger cars had open bodies which, as we will see later in the book, are intrinsically flexible. At that time, it was commonly *assumed* that the body carried none of the road loads (only self-weight of body, passengers and payload), and consequently it was not designed to be load bearing. This was particularly true for torsion loads.

Early experience with metal-clad bodies, particularly in 'sedan' form (i.e. with a roof), where torsion stiffness was built in fortuitously and inadvertently, led to problems of 'rattling' between the chassis and the body, and also 'squeaking' and cracking at various points within the body which were, unintentionally, carrying structural loads.

The root of these problems lay in the fact that the 'body-on-chassis' arrangement consists, in essence, of two structures (the body and the chassis) acting as torsion springs in *parallel*.

For springs in parallel, the load is shared between the springs in proportion to their relative stiffnesses. This is a classic case of a 'redundant' or 'statically indeterminate' structural system. In the simplified case where the body and chassis are connected only at their ends (as in Fig. 16.1-6):

$$T_{\mathrm{TOTAL}} = T_{\mathrm{BODY}} + T_{\mathrm{CHASSIS}}$$
$$K_{\mathrm{TOTAL}} = K_{\mathrm{BODY}} + K_{\mathrm{CHASSIS}}$$
$$T_{\mathrm{BODY}}/T_{\mathrm{CHASSIS}} = K_{\mathrm{BODY}}/K_{\mathrm{CHASSIS}}$$

where T = torque and K = torsional stiffness.

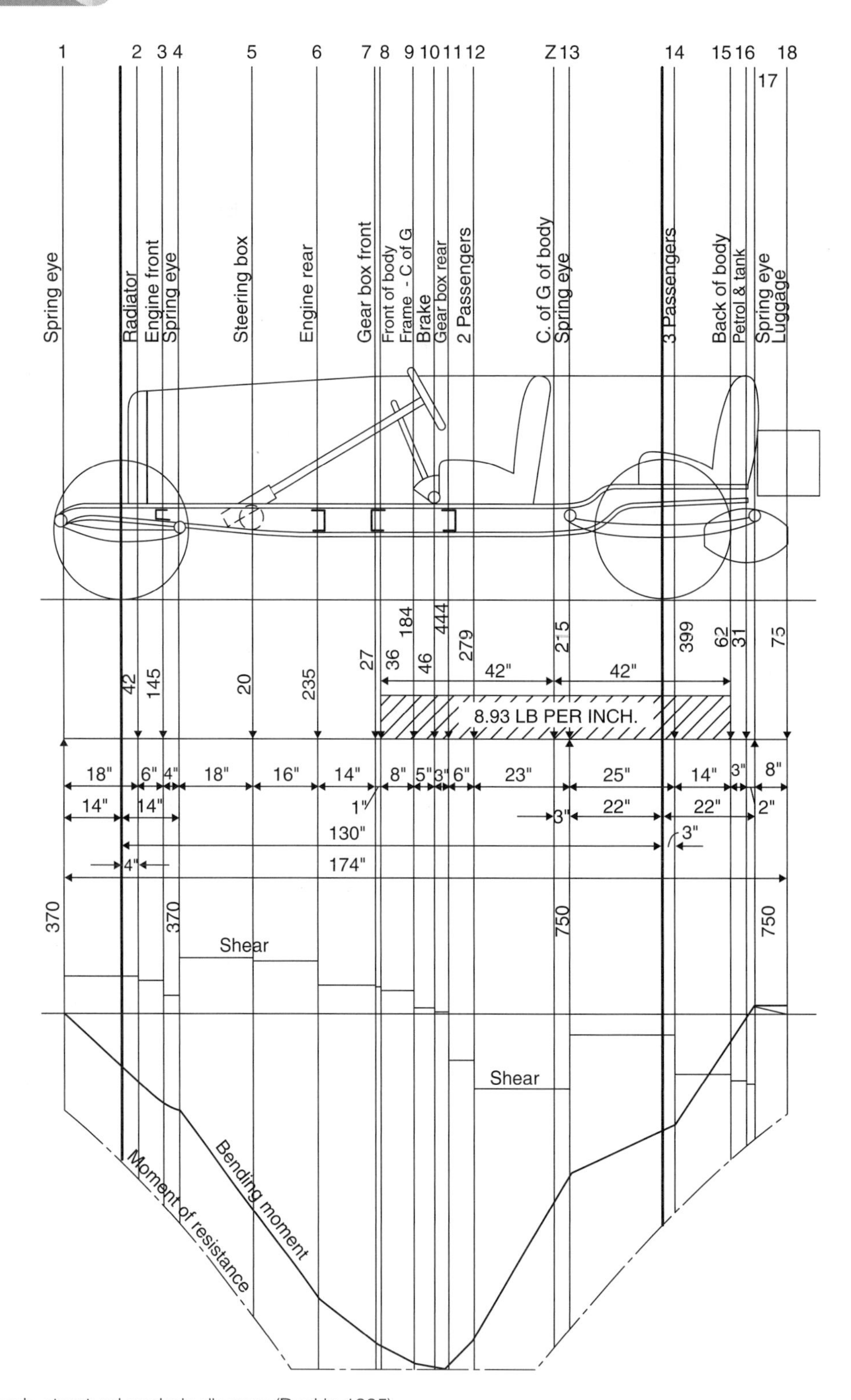

Fig. 16.1-4 Chassis structural analysis diagram (Donkin 1925).

Thus, in the case of a flexible body on a (relatively) stiff chassis frame, most of the torsion load would pass through the chassis. Conversely, if the body were stiff and the chassis flexible, then the body would carry a larger proportion of the torsion load.

As the 1920s progressed, this was implicitly recognized (based on practical experience) in the construction approach. Bodies were deliberately made flexible (particularly in torsion) by the use of flexible metal joints between the resilient timber body members, and by

Fig. 16.1-5 Car body manufacture in the 1920s (courtesy of Vauxhall Archive Centre).

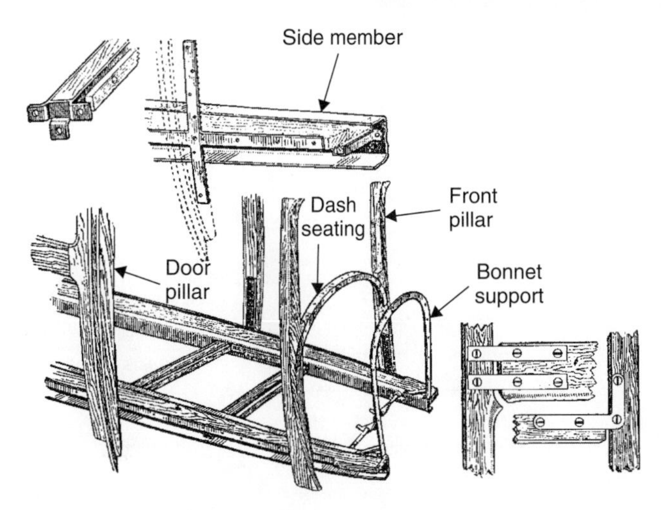

Fig. 16.1-7 Construction details of a timber framed 'fabric body'.

deliberate use of flexible materials for the outer skin of the body (Fig. 16.1-7).

It will be seen later in the book that closed shell structures such as closed car bodies are very effective in torsion, with the outer skin subjected locally to shear. In order to keep torsion stiffness (and hence loads) small, flexible materials such as fabric were used to form the outer skin of the body. In Europe, the 'Weymann fabric saloon' body was a well-known, and much copied, example of this.

An alternative approach was to use very thin aluminium cladding with deliberate structural discontinuities at key points to relieve the build-up of undesired stresses in the body.

The fabric-covered wood-framed car body was not amenable to large-scale mass production. As the 1920s gave way to the 1930s, the requirements of high volume production led to the widespread use of pressed steel car body technology. The bodies were formed out of steel sheets, stamped into shape, and welded or riveted together. This led to much greater stiffness in the body, particularly for torsion, because the steel panels were quite effective locally in shear. The overall configuration still remained as the 'body-on-separate- chassis'.

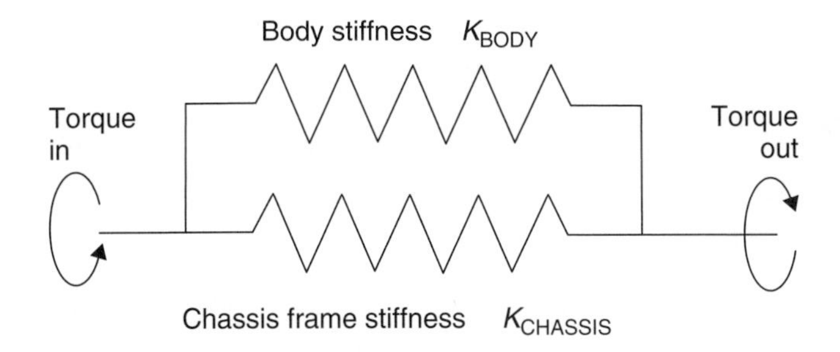

Fig. 16.1-6 Springs in parallel.

In the 1930s the subfloor chassis frame was still made of open section members, riveted together, and it was still regarded as the 'structure' of the vehicle.

From the 'springs in parallel' analogy, however, it can be seen that a much greater proportion of the load was now taken through the body, owing to its greater stiffness. This led to problems of 'fighting' between the body and the chassis frame (i.e. rattling, or damage to body mounts caused by undesired load transfer between the body and the chassis). Several approaches were tried to overcome this problem. These were used both individually and in combination with each other. They included:

(a) Flexible (elastomer) mountings were added between the chassis frame and the body. Laterally spaced pairs of these mountings acted as torsion springs about the longitudinal axis of the vehicle between the chassis and the body.

Consider a pair of body mounts positioned on either side of the body. If the linear stiffness of the individual elastomer body mounts is K_{LIN}, and they are separated laterally by body width B (see Fig. 16.1-8), then the torsional stiffness K_{MOUNT} of the pair of mounts about the vehicle longitudinal axis is:

$$K_{MOUNT} = K_{LIN}B^2/2$$

As used in 1930s vehicles this, in effect, made the 'load path' through the body more flexible, because the pairs of soft elastomer mounts and the body formed a chain of 'springs in *series*' (see Fig. 16.1-9). For such a system,

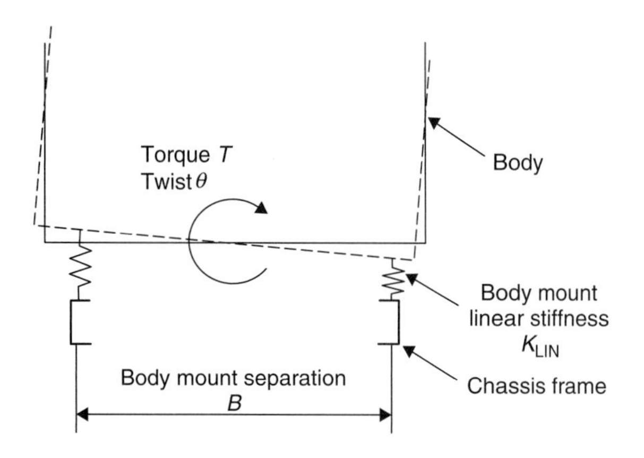

Fig. 16.1-8 Laterally positioned pair of body mountings.

the overall flexibility (the reciprocal of stiffness) is the sum of the individual flexibilities. As a result, the overall stiffness is lower than that of the individual elements in the series. Thus,

$$1/K_{\mathrm{TOTAL}} = K_{\mathrm{F\text{-}MOUNT}} + 1/K_{\mathrm{BODY}} + 1/K_{\mathrm{R\text{-}MOUNT}}$$

where K = torsion stiffness and $1/K$ = torsional flexibility.

Since the body-plus-mountings assembly was, structurally, still in parallel with the chassis, the effect of the reduced stiffness was to reduce the proportion of the torsion load carried by the body.

(b) The converse of the approach in (a) was to stiffen the chassis frame, thus encouraging it to carry more of the load. Open section, riveted chassis frame technology was still the norm in the 1930s, and so a method of increasing torsion stiffness, but still using open section members, was needed.

A common solution was the use of *cruciform bracing*. For this, a cross-shaped brace, made usually of open channel section members, was incorporated into the chassis frame as shown in Fig. 16.1-10. It was necessary for the ends of this to be well connected, in shear, to the chassis side members.

Fig. 16.1-11 shows how the cruciform brace works as a torsion structure. On the left, the 'input torque' is fed in as a couple consisting of two equal and opposite forces. This couple, or torque, is reacted by the couple composed of the equal and opposite forces on the right-hand side of the diagram.

The exploded view shows the local loads in the individual members. Member A has vertical loads downward at both ends. These are reacted by the upward force at point C. This in turn is reacted by an equal and opposite force at point C on member B, and the loads in member B are a mirror image of those in A.

Although the *overall* effect is a torsion carrying structure, *the individual members* (A and B) are subject only to bending and shear forces. Hence it was possible to use open sections. It is essential that there is a good, continuous, bending load path in both members A and B at point C (point of maximum bending moment).

By the mid-1930s, the need for reasonable torsion stiffness was well recognized. A paper of the time from which Fig. 16.1-10 is taken, gives overall torsion stiffness values between 1000 and 1750 N m/deg. for various cruciform-braced chassis frames. One of the best of the cruciform-braced underfloor open section chassis was that of the Lagonda V12 of the late 1930s. The frame consisted entirely of a substantial cross brace, with only small tie rods along the side to prevent 'wracking' distortion. Its torsion stiffness was measured to be a little over 2000 N m/deg.

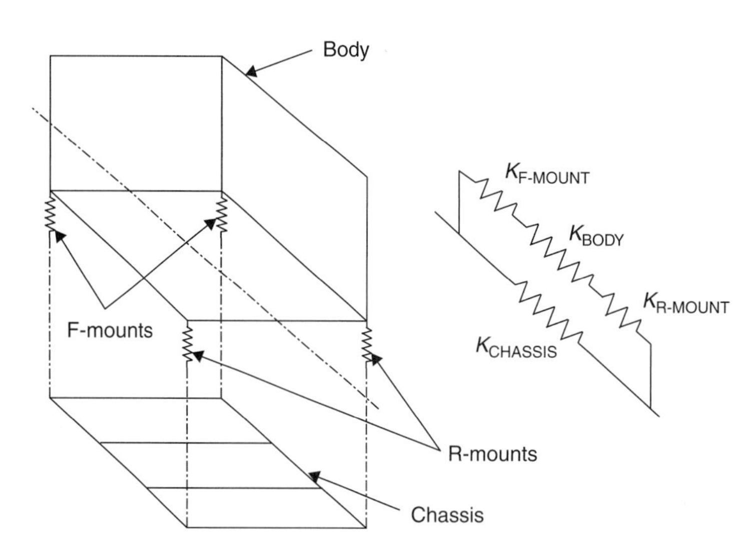

Fig. 16.1-9 Body and body mounts in series in torsion.

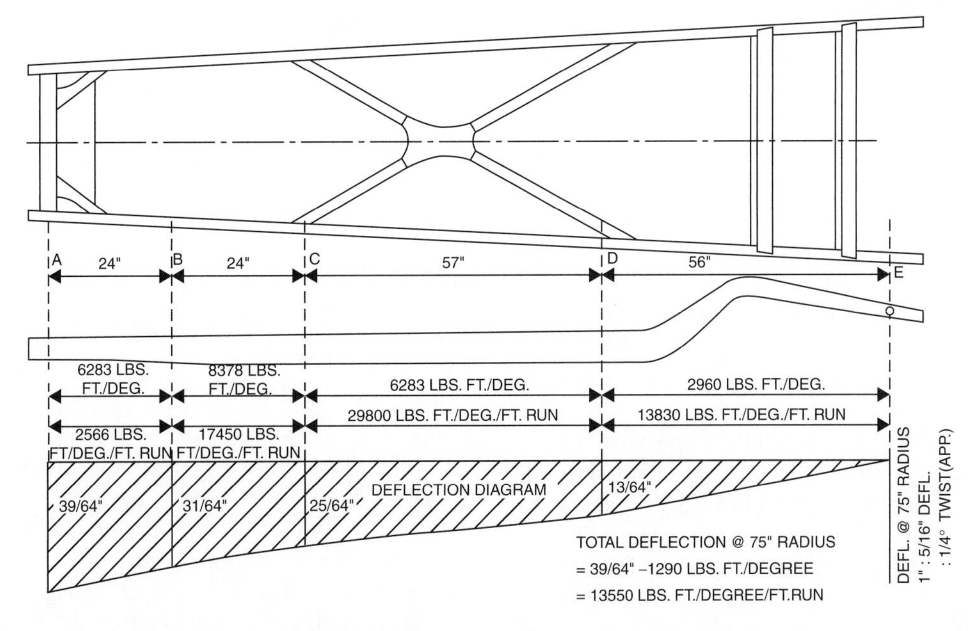

Fig. 16.1-10 Cruciform-braced chassis frame (Booth 1938 by permission Council of I.Mech.E.).

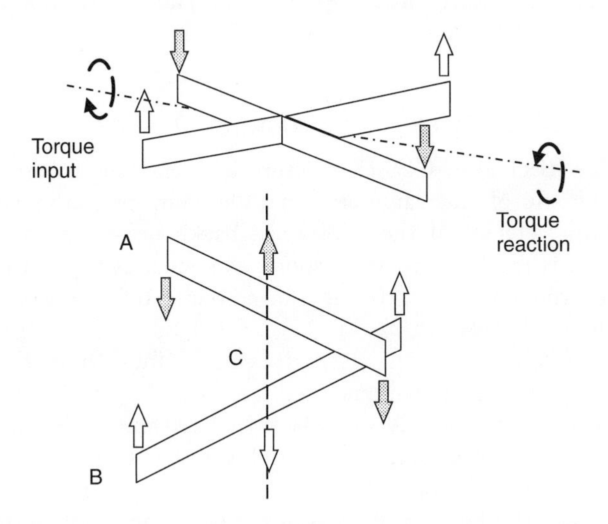

Fig. 16.1-11 Free body diagrams of cruciform brace members.

(d) Another way of improving chassis frame torsion stiffness was to incorporate *closed* (i.e. 'box section') cross-members. Closed section members are much stiffer in torsion than equivalent open section ones. Small closed section cross-members, as used in 1930s cars, whilst giving a considerable increase compared with previous chassis (which were extremely flexible in torsion), still gave overall results which would be considered low today.

(e) By the mid-1930s it was realized that the steel body was much stiffer than the chassis in both bending and torsion. Greater 'integration' of the body with the chassis frame was also used in some designs. A body attached to the chassis frame by a large number of screws, thus using some of the high torsion stiffness of the body, is shown in Fig. 16.1-12. This approach led eventually to the modern 'integral body' which is the major topic of the rest of this book.

(f) The ultimate version of the underfloor chassis frame using small section members is the 'twin tube' or 'multi tube' frame.

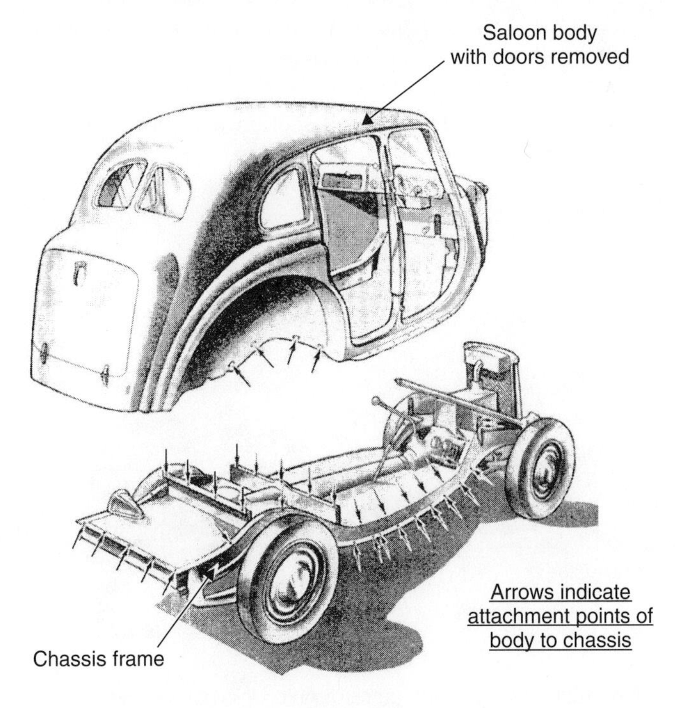

Fig. 16.1-12 Multiple attachments of body to frame.

This is still essentially a ladder type grillage frame, with side members connected by lateral cross-members. However, now both the side members and the cross-members consist of closed section tubes.

The advantage of this arrangement is that typically, for members of similar cross-sectional dimensions, a member with a closed section will be thousands of times stiffer in torsion than an equivalent open section member. Also adjacent members are usually welded together in this construction method, and such joints are much stiffer than riveted connections. The overall torsion stiffness of the chassis assembly improves accordingly.

This type of structure was used in specialist racing vehicles between the late 1930s and the 1950s. For example, the Auto Union racing car (1934–1937) shown in Fig. 16.1-13 used this system, as did many sports racing cars in the 1950s, such as Ferrari and Lister Jaguar. Examples of these may be seen in Costin and Phipps' book.

Separate ladder frames, with either open or closed section members, are still used widely on certain types of passenger car such as 'sport utility vehicles' (SUVs) and they are almost universal on commercial heavy goods vehicles.

16.1.2.2 Modern structure types

In more modern times, the closed tube (or closed box) torsion structure has been used to greater effect by using larger section, but thinner walled members. The torsion constant J for a thin-walled closed section member is proportional to the *square* of the area A_E *enclosed* by the walls of the section. Therefore

$J = 4A_E^2 t/S$ for a closed section with constant thickness walls

Fig. 16.1-13 Twin tube frame of Auto Union racing car (1934–1937) (courtesy of Deutsches Museum, München).

where t = wall thickness and S = distance around section perimeter.

Hence there is a great advantage in increasing the breadth and depth of the member. Additionally, a large depth will give good (i.e. stiff and strong) bending properties. The torsion *stiffness* K of the closed section backbone member is then:

$$K = GJ/L$$

where G = material shear modulus and L = length of member, so that:

$$T = K\theta$$

where T = applied torque and θ = torsional deflection (twist).

The development of better road holding, coming from a better understanding of suspension geometry, made greater body stiffness essential. This, and the push towards welded, pressed sheet steel body technology, led to the widespread use of the 'large section tube' concept in car structures in the post-World War II era. Some examples follow.

16.1.2.2.1 Backbone structure

The 'Backbone' chassis structure is a relatively modern example of the 'large section tube' concept (although Tatra vehicles of the 1930s had backbone structures). This is used on specialist sports cars such as the Lotus shown in Fig. 16.1-14. It still amounts to a 'separate chassis frame'.

The backbone chassis derives its stiffness from the large cross-sectional enclosed area of the 'backbone' member. A typical size might be around 200 mm × 150 mm. It will be seen later that, in tubular structures in torsion, the walls of the tube are in shear. Thus, in the case of the Lotus, the walls of the tube consist of shear panels. However, shear panels are not the only way of carrying in-plane shear loads. For example, a triangulated 'bay' of welded or brazed small tubes can also form a very effective and weight efficient shear carrying structure. It is possible to build an analogue of the 'backbone' chassis frame using triangulated small section tubes This approach is used in some specialist sports cars, such as the TVR shown in Fig. 16.1-15.

Such specialist vehicles often have bodies made of glass reinforced plastic. On many vehicles of this type, the combined torsion stiffness of the chassis and the attached body together is greater than the sum of the stiffnesses of the individual items. This reflects the fact that the connection between the two is not merely at the ends, as discussed earlier, but is made at many points, giving a combined structure which is highly statically indeterminate.

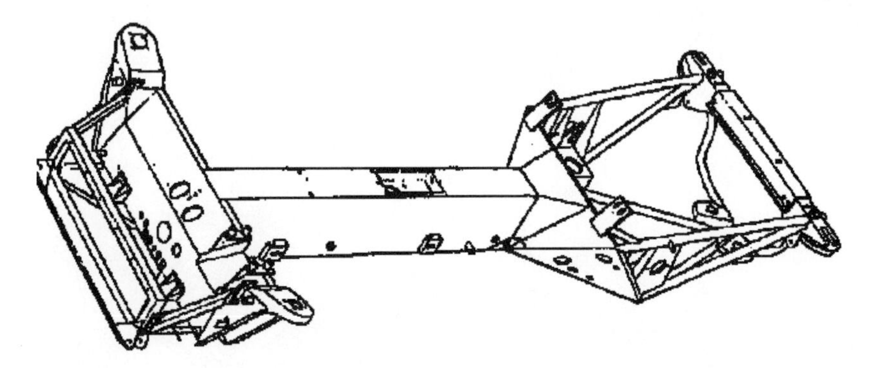

Fig. 16.1-14 Sheet steel backbone chassis (courtesy Lotus Cars Ltd).

16.1.2.2.2 Triangulated tube structure

The triangulated tube arrangement is not limited to backbone structures. Perhaps a more common approach using this principle, particularly for sports cars, is the 'bathtub' layout, in which the triangulated structure surrounds the outside of the body. A classic example is the Caterham shown in Fig. 16.1-16. This approach has the advantage that the coachwork can consist of thin sheet metal cladding, attached directly to the framework. If the vehicle is an open car, the large cockpit interrupts the 'closed box' needed for torsion stiffness. In such a case, torsion stiffness is sometimes restored by 'boxing in' either the transmission tunnel, or the dash/cowl area (or both) with triangulated bays. Stiffening of the edges of the passenger compartment top opening, particularly at the corners, can also bring some improvement to the torsional performance. The principles of the simple structural surfaces method (see later in Chapter 16.2) can still be applied to this type of structure if the triangulated bays (including edges) are treated as structural surfaces.

This method of construction is best suited to low volume production because of low tooling costs. It is not well suited to mass production due to complication and labour-intensive manufacture.

16.1.2.2.3 Incorporation of roll cage into structure

The ultimate way of using the 'tube' principle is to make the tube encompass the whole car body. A version of this is shown in the triangulated sports car structure shown in Fig. 16.1-17. The triangulated 'roll cage' now extends around the passenger compartment. The enclosed cross-sectional area of the body is thus very large, and hence the torsion constant is large also.

Roots and co-workers studied the torsion stiffness of a racing vehicle in which the roll cage was incorporated into the structure in a range of different ways. They

Fig. 16.1-15 Backbone chassis made of triangulated tubes (from author's collection, courtesy TVR Ltd).

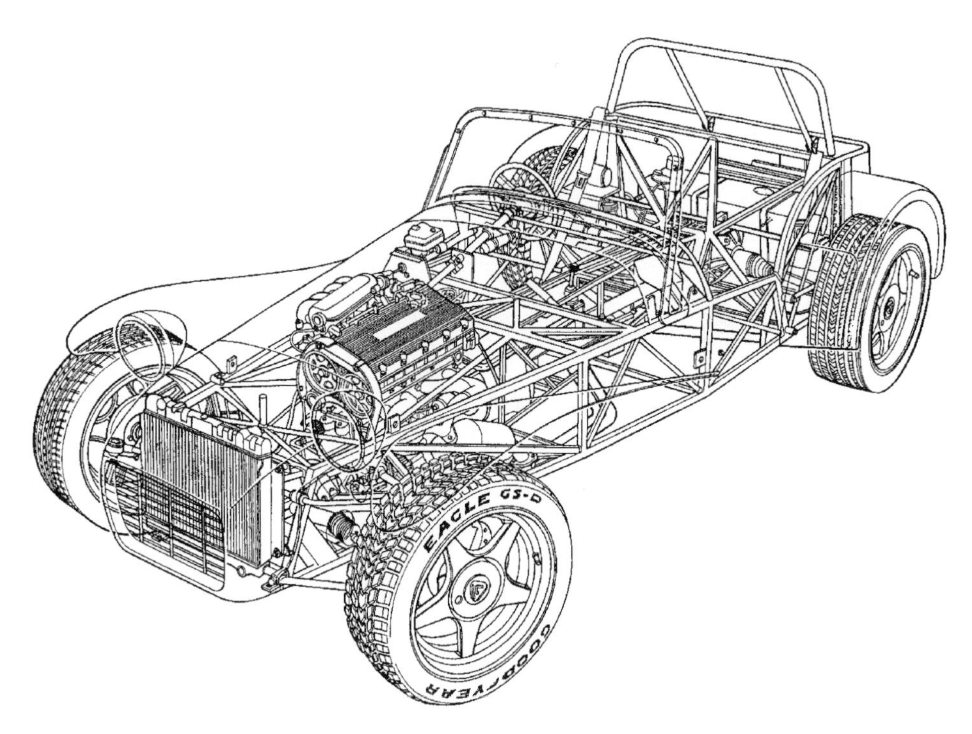

Fig. 16.1-16 Triangulated sports car structure (courtesy of Caterham Cars Ltd).

showed that the torsion stiffness can be increased by over 500 per cent as compared with the basic chassis frame. The contribution of the roll cage depended on (a) the degree of triangulation in the roll cage, and (b) on how well the roll cage was connected to the rest of the structure (i.e. on the continuity of load paths).

16.1.2.2.4 Pure monocoque

The logical conclusion of the 'closed box' approach is the 'monocoque' (French: 'single shell'). For this, the outer skin performs the dual role of the body surface and structure.

This is the automotive version of aircraft 'stressed skin' construction. It is very weight efficient.

The *pure* monocoque *car* structure is relatively rare. Its widespread use is restricted to racing cars as shown in Fig. 16.1-18. There are several reasons for this.

First, for the monocoque to work efficiently, it requires a totally *closed* tube. However, practical vehicles require openings for passenger entry, outward visibility, etc. This requires interruptions to the 'single shell' which

Fig. 16.1-17 Triangulated sports car structure with integrated roll cage (Roots *et al.* 1995 with original permission from TVR Ltd).

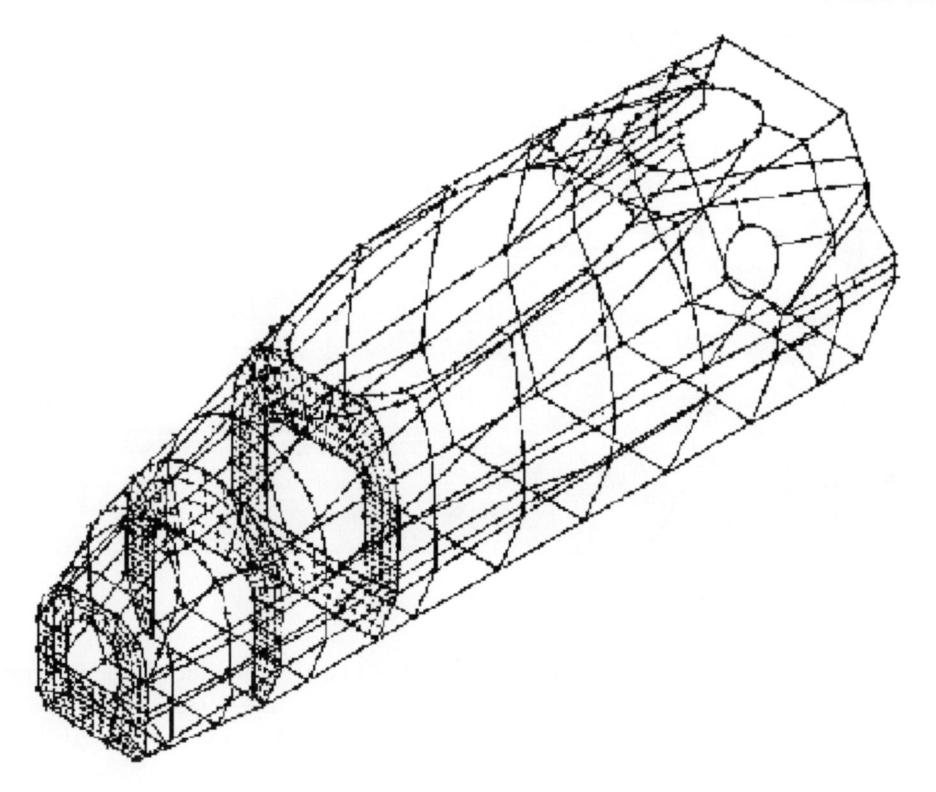

Fig. 16.1-18 Monocoque structure.

then reduce it to an open section, with consequent lowering of torsion stiffness. Also, the shell requires reinforcement: (a) to prevent buckling (stringers or sandwich skin construction often provide this), and (b) to carry out-of-plane loads, e.g. from the suspension (internal bulkheads are often used for this).

Typical Formula 1 racing car monocoques, usually made of carbon fibre composite sandwich material, can have torsion stiffness greater than 30 000 N m/deg. for the composite 'tub' alone. In such vehicles, the engine and gearbox also act as load bearing structures, in series with the monocoque.

16.1.2.2.5 Punt or platform structure

Other modern car chassis types include the 'punt structure'. This is usually of sheet metal construction, in which the floor members (rocker, cross-members, etc.) are of large closed section, with good joints between members. It is thus a grillage structure of members with high torsion and bending properties locally. In many cases (but not all), the upper body is treated as structurally insignificant. The punt structure is often used for low production volume vehicles, for which different body styles, or rapid model changes are required.

The Lotus Elise (aluminium, see Fig. 16.1-19) is an example of a punt structure.

This approach is often also used to create cabriolet or convertible versions of mass produced integral sedan car structures.

16.1.2.2.6 Perimeter space frame or 'birdcage' frame

Another modern structure is the perimeter or 'birdcage' frame. A typical example is the Audi A2 aluminum vehicle (Fig. 16.1-20).

In this type of structure, relatively small section tubular members are built into stiff jointed 'ring-beam' bays, welded together at joints or 'nodes'. We shall see later that ring beams are moderately effective at carrying local in-plane shear. For this, the edge members of each ring frame, and especially the corners, must be stiff locally in bending.

This choice of construction method is usually dictated by production requirements. In the case of the A2, the various beam sections are of extruded or cast aluminium (with some additional members of pressed sheet), and so they must be assembled into this structural concept using welded 'nodes' or joints.

The individual open-bay ring frame is not a very weight efficient shear structure. If the (very high) shear stiffness of the skin panels is incorporated into this type of body, it now becomes an 'integral' structure (see later), and a considerable increase in torsional rigidity is

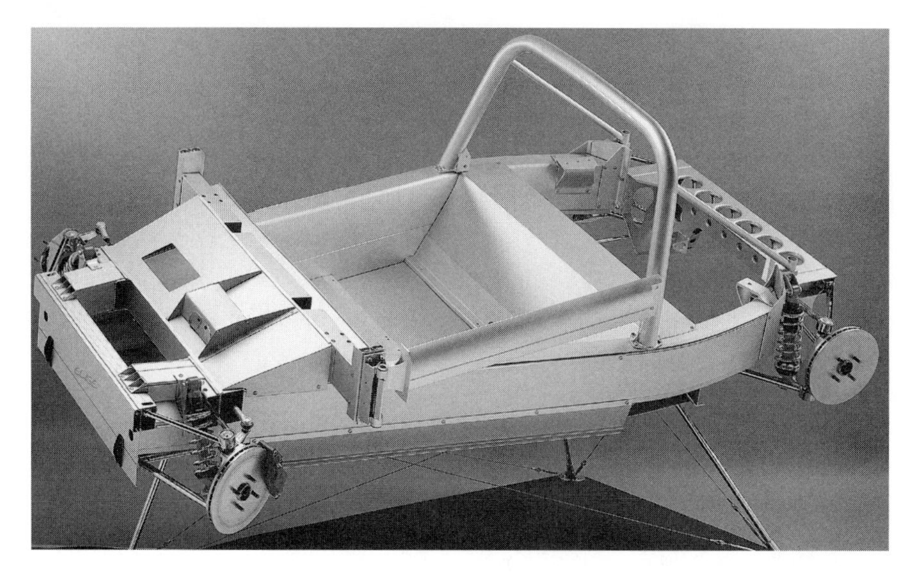

Fig. 16.1-19 Punt chassis (courtesy of Lotus Cars Ltd).

usually observed, depending on the stiffness of the attachment.

16.1.2.2.7 Integral or unitary body structure

The subject of this book, and the most widely used modern car structure type, is the 'integral' (or 'unitary'), spot welded, pressed steel sheet metal body. It is well suited to mass production methods. The body is self-supporting, so that the separate 'chassis' is omitted, with a saving in weight.

The first mass produced true integral car bodies were introduced in the 1930s. A notable example was the Citroën 11 CV, shown in Fig. 16.1-21, which was in production from 1934 to 1956.

An interesting study by Swallow in 1938 was made on another vehicle, the Hillman 10 HP, when the structure was changed from separate body-on-chassis (1938 model year) to integral construction (1939 model), the vehicles being otherwise largely identical. Amongst other things, the torsional behaviour was

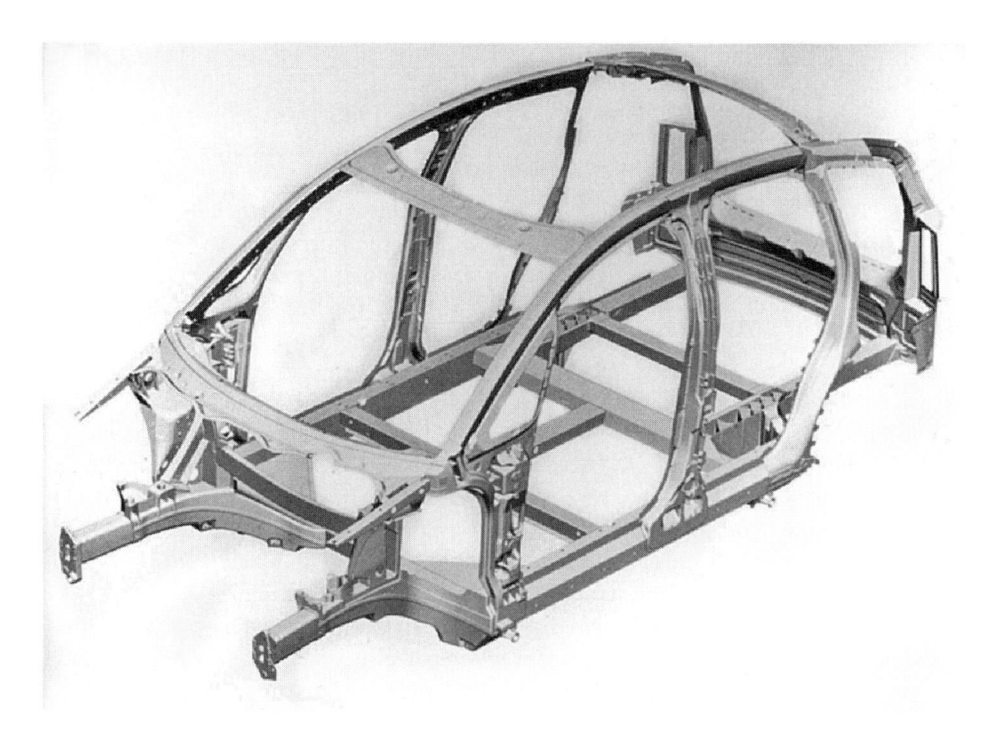

Fig. 16.1-20 Perimeter space frame (courtesy of Audi UK Ltd).

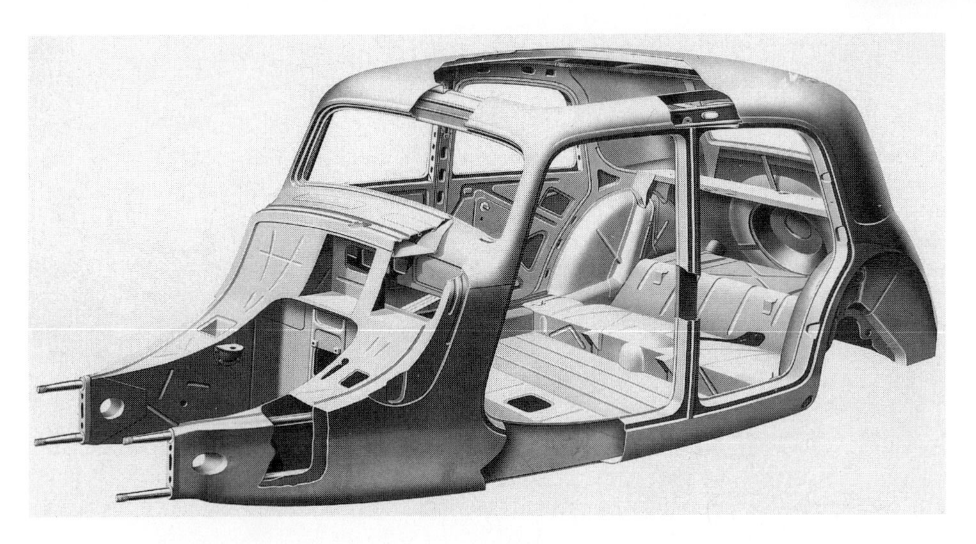

Fig. 16.1-21 Citroën 11 CV of 1934 (courtesy of Automobiles Citroën SA).

compared (Fig. 16.1-22). The torsion stiffness rose from 934 N m/deg. (689 lbft/deg.) for the chassis and body (together) for the earlier model to 3390 N m/deg. (2500 lbft/deg.) for the integral body.

An interesting 'hysteresis' effect is visible in the unloading curve for the separate frame vehicle, due to slippage between body and chassis at the mounting bolts. This is absent in the integral structure. Swallow also

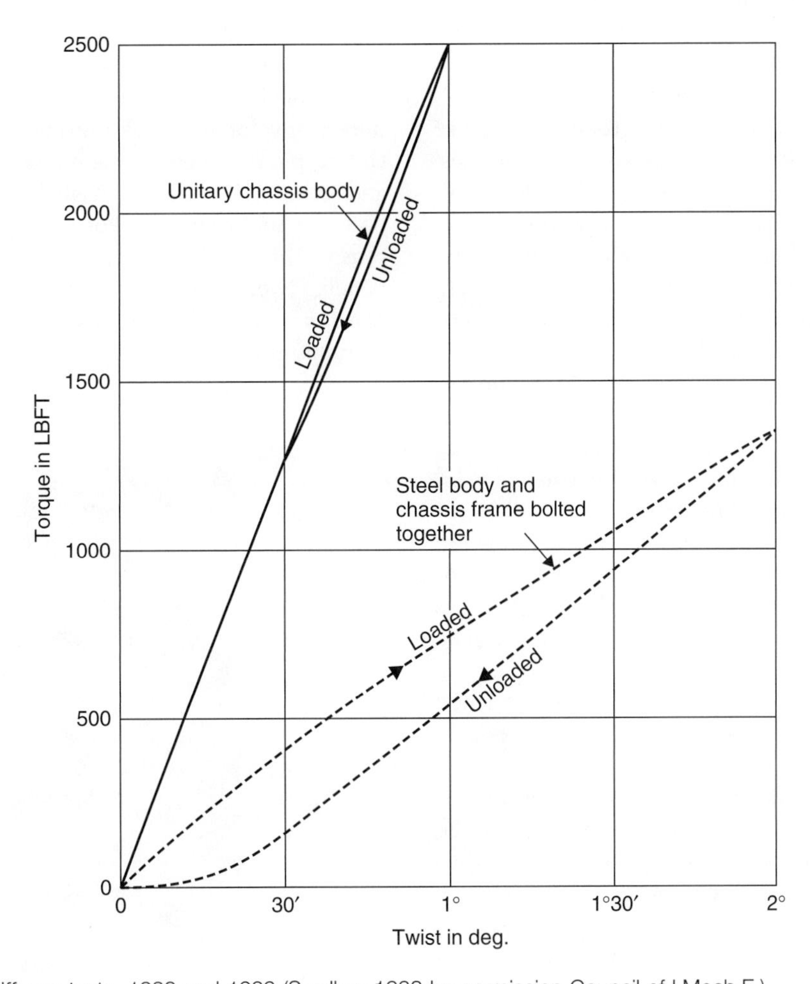

Fig. 16.1-22 Torsion stiffness tests, 1938 and 1939 (Swallow 1938 by permission Council of I.Mech.E.).

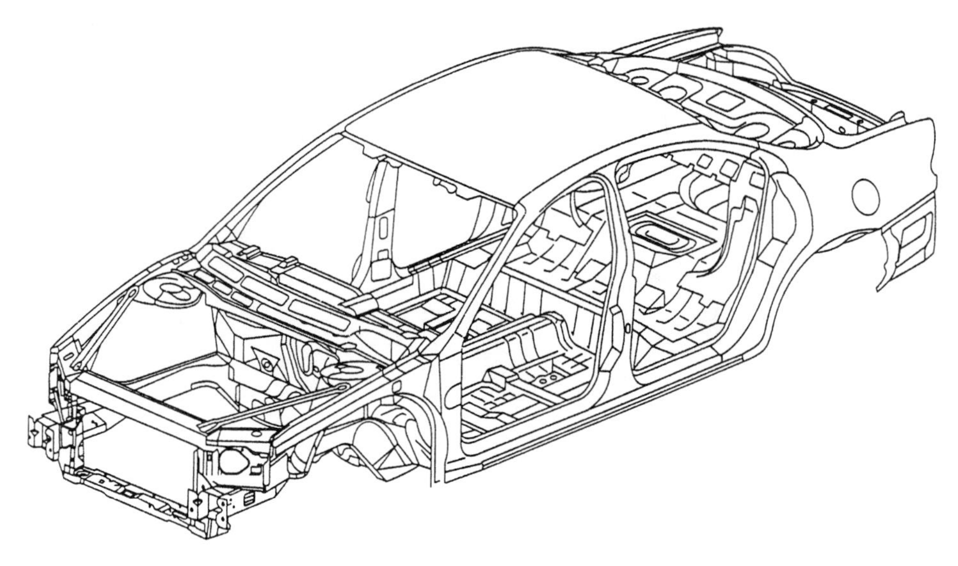

Fig. 16.1-23 Modern integral body-in-white (courtesy General Motors Corporation).

mentioned that spring rates for axle and engine suspensions, and for the suspension bushes had to be reduced due to increased 'harshness' (the passing of transient forces to the vehicle occupants) in the integral vehicle.

Fig. 16.1-23 shows a modern example of an integral 'body-in-white' (i.e. bare body shell). The integral body is really a mixture of the monocoque and the 'birdcage' types. The body forms a 'closed box' torsion structure (with consequent high stiffness). The walls, or 'surfaces' of the box, consist of the skin panels (such as the roof, floor, bulkheads, etc.) where possible. Elsewhere open bay ring frames (sideframe, windshield frames, etc.) form the surface of the box, wherever openings are required. Beam members are also used to carry out-of-plane loads, for example in the floor.

In the integral structure the panels and body components are stamped from sheet steel and fixed together mostly by spot welding, although clinching, laser welding or other methods are sometimes used for particular locations. The beam members are formed out of folded or pressed sheet steel shapes, welded together as shown in Fig. 16.1-24. These beams can be independent (e.g. B-pillar), or they can be formed as part of the larger panels (as in the case of the transmission tunnel), or they can be attached to panels by spot welding (e.g. floor cross-members, rockers). To avoid ugly 'sink marks', attached beam members are never spot welded to externally visible skin panels.

The ultralight steel auto steel body, ULSAB is a modernized version of this theme which may well show the way to near-future developments (Fig. 16.1-25). In this, hydroforming (the creation of complex cross-sections by forcing tubes into moulds by internal hydraulic pressure) is used widely as an alternative method of forming beam-like components. 'Laser welded blanks' (i.e. of tailored varying thickness) are also used widely, in addition to steel sandwich panels. Laser welding and adhesive bonding, both of which are stiffer than spot welding, are used extensively to join the panels together. The result is a structure which was recorded to be lighter and stiffer than the 'traditional' integral steel bodies it was compared to.

Although, at the time of writing, steel is used almost universally for high volume mass produced car bodies, the suppliers of competing materials, such as aluminium and composite plastics, have been developing integral body technologies also. For example, the aluminium intensive vehicle, AIV, is made of pressed sheet aluminium panels, 'weld-bonded' together. In their 1989 book, Nardini and Seeds have discussed the design issues for aluminium integral bodies.

Fig. 16.1-24 Integral body floor assembly, showing structural members.

Fig. 16.1-25 The ultralight steel auto body (ULSAB) (courtesy of the ULSAB consortium).

If properly constructed, the integral body is well capable of carrying torsion, bending and other loads. Because the structure comprises the outer surface of the body, it is much stiffer than most other vehicle structure types, for the reasons given above. Typical values of torsion stiffness for modern integral car bodies are approximately 8000–10 000 N m/deg. for 'everyday' sedans and higher (around 12 000–15 000 N m/deg. or more) for luxury vehicles.

Ideally the in-plane stiffness of the individual panels or surfaces must be used.

16.2

Chapter 16.2

Standard sedan (saloon) – baseline load paths

Jason Brown, A.J. Robertson and Stan Serpento

OBJECTIVES

- To demonstrate that a car structure can be represented by simple structural surfaces (SSSs).
- To introduce the load paths in a sedan structure for different load cases.

16.2.1 Introduction

The structures of different sedan passenger car structures vary according to size, vehicle layout ('package') and type and to the particular design and assembly methods of different manufacturers. Nevertheless, the nature of integral construction dictates that there will also be similarities.

In this chapter, a simplified car structure referred to as the 'standard sedan' is used as the basis for an introductory discussion of load paths in integral car structures. For further clarity, simplification of the payload and of the suspension input loads is also made.

The major load cases (especially bending and pure torsion) are described separately because they each make very different structural demands on the individual SSSs in the vehicle body. Actual road loads will be a combination of these cases, and the calculation results can be obtained as appropriate.

16.2.1.1 The standard sedan

The standard sedan, Fig. 16.2-1, consists of a 'closed box' passenger compartment, comprising floor, roof, sideframes, front and rear bulkheads and windscreen. For simplicity, all of these surfaces are assumed to be plane.

The suspension loads, at both front and rear, are carried on deep, stiff boom/panel cantilevers attached to the ends of the compartment. These are a simplified representation of the inner wing panels. The booms represent the lower rails (e.g. for engine mounting) and the upper flanges. To keep the model simple, the suspension tower loads are fed directly into the webs of the cantilevers.

Where it is necessary to carry out-of-plane loads, 'supplementary SSSs' (acting as beams) are provided, for example as floor cross-members and parcel shelves. To reduce complexity in the standard sedan, the parcel shelves are not considered to be part of the surface of the passenger compartment 'torsion box' for the torsion load case. Both parcel trays are, however, necessary for carrying loads in the bending load case, and for carrying loads into the 'torsion box' in the torsion case.

The 16 SSSs in the standard sedan, Fig. 16.2-1, are as follows:

1. Transverse floor beam (front) carrying the front passengers.
2. Transverse floor beam (rear) carrying the rear passengers.

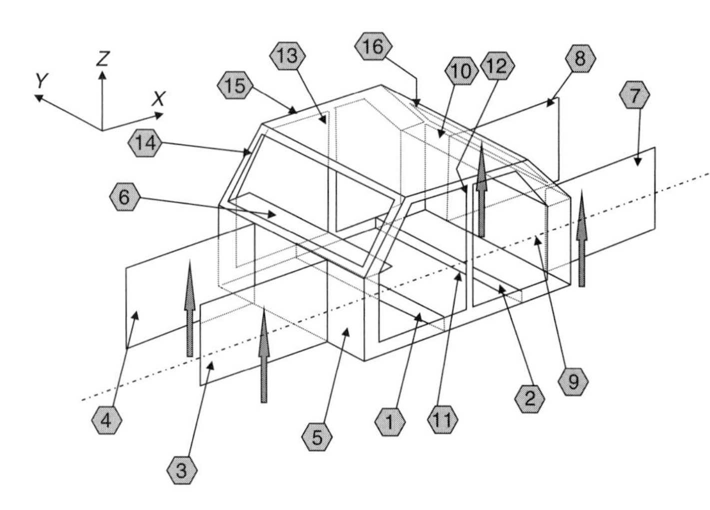

Fig. 16.2-1 Baseline model.

Motor Vehicle Structures; ISBN: 9780750651349
Copyright © 2001 Jason Brown, A.J. Robertson and Stan Serpento. All rights of reproduction, in any form, reserved.

3. and 4. Inner wing panels carrying the power-train and supported by the front suspension.

5. Dash panel–transverse panel between passengers and engine compartment.

6. Front parcel shelf.

7. and 8. Rear quarter panels carrying luggage loads and supported by the rear suspension.

9. Panel behind the rear seats.

10. Rear parcel shelf.

11. Floor panel

12. and 13. Left-hand and right-hand sideframes.

14. Windscreen frame.

15. Roof panel.

16. Backlight (rear window) frame.

These SSSs will be shown to be sufficient to carry the two fundamental load cases of bending and torsion. Some additional SSSs will be necessary for other load cases as described in section 16.2.4. Alternative SSSs may be necessary when modelling particular vehicles. The vehicle engineer must use his knowledge and make his/her own subjective assessment for the model that best represents a particular structure.

16.2.2 Bending load case for the standard sedan (saloon)

16.2.2.1 Significance of the bending load case

Fig. 16.2-2 shows the baseline loads that are considered for the bending case. The main loads of power-train F_{pt}, the front passengers/seats F_{pf}, the rear passengers/seats F_{pr}, and the luggage F_ℓ only are considered. The magnitude of the loads is the weight of the component factored by a dynamic load factor. It should be noted that all these loads are applied in the planes of SSSs. It is essential this condition is achieved in order to ensure sufficient strength and stiffness can be provided through the structure. The bending and shear loads on each component can be determined and from these satisfactory stress levels can be determined.

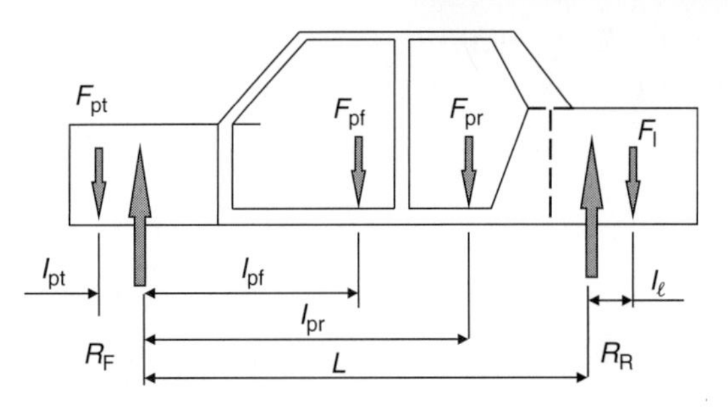

Fig. 16.2-3 Payload distribution.

16.2.2.2 Payload distribution

The passenger car structure when viewed in side elevation (Fig. 16.2-3) can be considered as a simply supported beam, the supports are at the front and rear axles. First, the masses of these components and their longitudinal and lateral positions in the vehicle must be known, then the front and rear suspension reaction forces can be obtained. Referring to Fig. 16.2-3, by taking moments about the rear suspension mounting the front suspension reaction is:

$$R_F = \frac{F_{pt}(L + l_{pt}) + F_{pf}(L - l_{pf}) + F_{pr}(L - l_{pr}) - F_l l_l}{L} \tag{16.2.1}$$

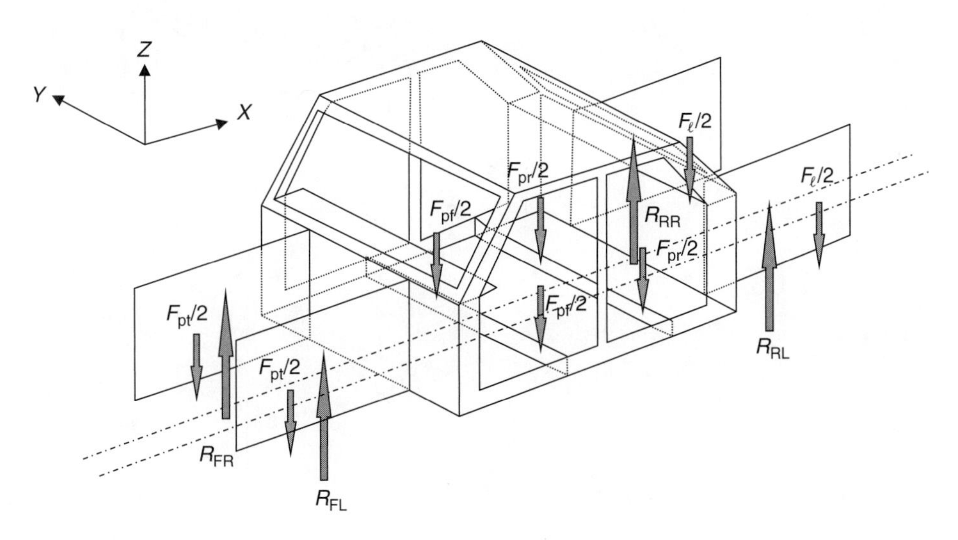

Fig. 16.2-2 Baseline model – bending loads.

where l_{pt}, l_{pf}, l_{pr} and l_l are defined in Fig. 16.2-3.

Similarly, taking moments about the front suspension mounting:

$$R_R = \frac{F_{pf}l_{pf} + F_{pr}l_{pr} + F_l(L + l_l) - F_{pt}l_{pt}}{L} \quad (16.2.2)$$

Check the answers by verifying $R_R + R_F = F_{pt} + F_{pf} + F_{pr} + F_l$.

When investigating a particular vehicle many more components can be considered making a more accurate model. Typical items that may be included are: front bumper, radiator, battery, instrument panel/steering column, exhaust, fuel tank, spare wheel, rear bumper and distributed loads due to the weight of the body structure. If these are to be included, the positions of all these components as well as their masses must be known. The suspension reactions can be calculated with a similar procedure.

16.2.2.3 Free body diagrams for the SSSs

Developing the model of Fig. 16.2-2 into the 'exploded' view shown in Fig. 16.2-4, it can be seen that we require edge loads and end loads to ensure all SSSs are in equilibrium. These edge/end loads are indicated by the forces P_1 to P_{13}.

When beginning the bending analysis it is essential to start at the central floor area. In this simple model it is assumed that the passenger loads are carried by the two transverse floor beams SSS (1) and (2). These floor beams are supported at each end by the side-frame forces P_1 and P_2. Note that there is an equal but opposite force acting on the sideframe.

Consider now the inner front wings SSS (3) and (4), the loads acting on these are the loads from the powertrain $P_t/2$ and from the front suspension R_{FL}. The applied loads $F_{pt}/2$ and R_{FL} are held in equilibrium by the end loads P_4 and P_5 and by the edge (shear) load P_3. The shear load P_3 reacts into the dash panel while the end load P_4 reacts into the front parcel shelf (6), and P_5 into the floor panel (11). These forces can be obtained by the equations of statics, i.e. resolving forces and taking moments.

When building the SSS model representing the vehicle it must always be remembered that the forces must act in the plane of an SSS. Failure to provide sufficient SSSs to satisfy this requirement soon reveals a weakness or unsatisfactory load path in the structure. Note, the horizontal SSS (6) is necessary in order to carry force P_4.

By working through the individual SSSs of this model shown in the next section it will be realized that there are sufficient forces to achieve equilibrium for each component and that all loads act in the planes of the SSSs.

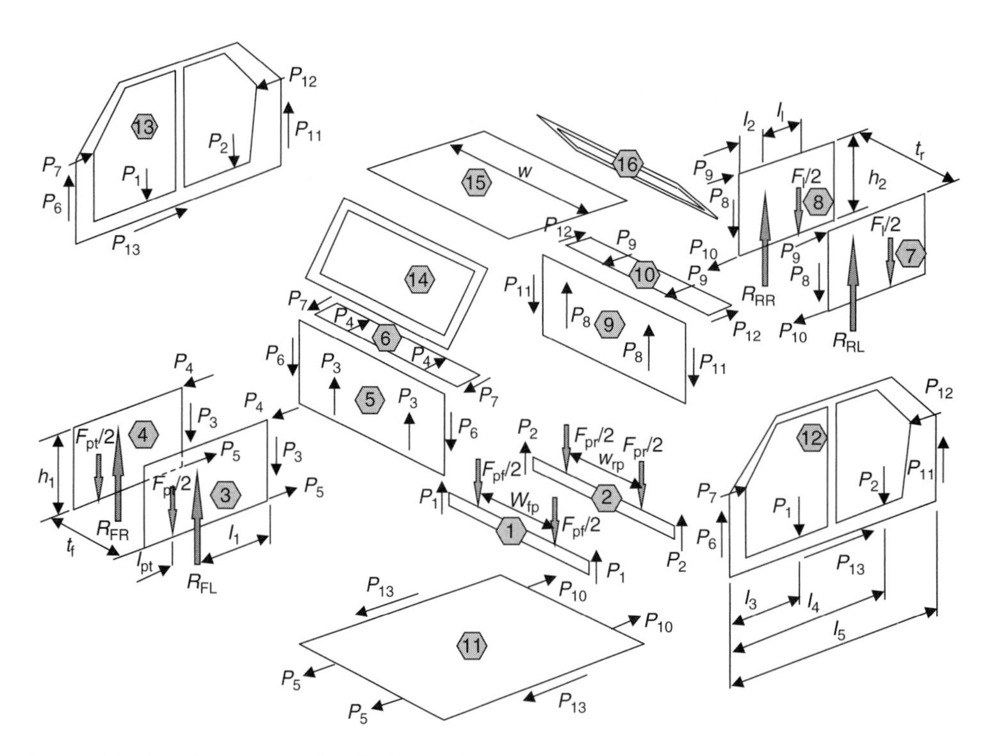

Fig. 16.2-4 Baseline model – bending case, end and edge loads.

16.2.2.4 Free body diagrams and equilibrium equations for each SSS

16.2.2.4.1 Transverse floor beam (front) (1)

Resolving forces vertically and by symmetry (loads are assumed to be applied symmetrically about the vehicle longitudinal centreline):

$$P_1 = F_{pf}/2 \tag{16.2.3}$$

16.2.2.4.2 Transverse floor beam (rear) (2)

Resolving forces vertically and by symmetry:

$$P_2 = F_{pr}/2 \tag{16.2.4}$$

16.2.2.4.3 Left and right front inner wing panel (3) and (4)

Resolving forces vertically for the left-hand panel:

$$P_3 = R_{FL} - F_{pt}/2 \tag{16.2.5}$$

A similar equation is obtained for the right-hand panel.

Taking moments about the rear lower corner:

$$P_4 = \{R_{FL}l_1 - F_{pt}(l_1 + l_{pt})/2\}/h_1 \tag{16.2.6}$$

Resolving forces horizontally:

$$P_5 = P_4 \tag{16.2.7}$$

16.2.2.4.4 Dash panel (5)

Equal and opposite reaction forces P_3 to those on the wing panels act on this SSS.

Resolving forces vertically and by symmetry:

$$P_6 = P_3 \tag{16.2.8}$$

16.2.2.4.5 Front parcel shelf (6)

Equal and opposite reaction forces P_4 to those on the inner wing panels act on this SSS.

Resolving forces horizontally and by symmetry:

$$P_7 = P_4 \tag{16.2.9}$$

16.2.2.4.6 Rear quarter panels (7) and (8)

Resolving vertically for the left-hand panel:

$$P_8 = R_{RL} - F_l/2 \tag{16.2.10}$$

Taking moments about the front lower corner:

$$P_9 = \{R_{RL}l_2 - F_l(l_1 + l_2)/2\}h \tag{16.2.11}$$

Resolving forces horizontally:

$$P_{10} = P_9 \tag{16.2.12}$$

Similar equations apply for the right-hand panel.

16.2.2.4.7 Panel behind the rear seats (9)

Resolving forces vertically and by symmetry:

$$P_{11} = P_8 \tag{16.2.13}$$

16.2.2.4.8 Rear parcel shelf (10)

Resolving forces horizontally and by symmetry:

$$P_{12} = P_9 \tag{16.2.14}$$

16.2.2.4.9 Floor panel (11)

Reaction forces P_5 from the inner front wing panels and forces P_{10} from the rear quarter panel are applied to this SSS. These will not necessarily be equal so additional forces P_{13} are required acting at the sides which react on the sideframes. It will be assumed these forces act in the direction shown in Fig. 16.2-4 although when numerically evaluated, these may be negative (i.e. in the opposite directions). Resolving forces horizontally:

$$2P_{13} = 2(P_{10} - P_5) \tag{16.2.15}$$

16.2.2.4.10 Left-hand and right-hand sideframes (12) and (13)

Both sideframes are loaded identically. Examining the forces acting on the sideframes shows that these have already been obtained from equations (16.2.3), (16.2.4), (16.2.8), (16.2.9), (16.2.13), (16.2.14) and (16.2.15). However, it is necessary to check that equilibrium conditions are satisfied by applying the equations of statics. Experience has shown that it is essential to make this equilibrium check as errors do occur with the use of the many equations.

Resolving forces vertically:

$$P_6 - P_1 - P_2 + P_{11} = 0 \tag{16.2.16}$$

Resolving forces horizontally:

$$P_7 + P_{13} - P_{12} = 0 \tag{16.2.17}$$

Moments may be taken about any point but in order to reduce the algebra it is better to take moments about a point where two forces act. For example, take moments about the lower corner of the windscreen pillar where P_6 and P_7 act. This simplifies the equation by eliminating two terms.

Moments about the lower corner of the windscreen:

$$P_1l_3 + P_2l_4 - P_{11}l_5 - P_{12}(h_2 - h_1) = 0 \tag{16.2.18}$$

In practice some rounding errors due to difficulties in defining the exact positions of each force may occur.

It should now be noted that windscreen frame (14), roof panel (15), and backlight (16) SSSs are not subject to any load for this bending case.

16.2.2.5 Shear force and bending moment diagrams in major components – design implications

Now that the forces on each SSS have been obtained the shear force and bending moment diagrams can be drawn.

Fig. 16.2-5(a) shows the loading, shear force and bending moment diagrams for the front transverse floor beam. The rear transverse floor beam although not shown is loaded in a similar manner. It should be realized that the beam is simply supported at its ends where it is attached to the sill member. Design of this joint must be suitable for carrying the vertical shear force P_1. The centre section has a constant bending moment and must be designed to provide suitable bending properties. Note that the positive bending moment means the beam is subject to a sagging moment.

In Fig. 16.2-5(b) the loading on the dash panel is shown, a similar condition applies to the panel behind the rear seats. In contrast to the floor beams this panel is subject to a negative bending moment or hogging moment. The panel behind the rear seats is also loaded in this manner. These panels are relatively deep so bending stresses and deflections will be small although stiffening at the top and bottom edges will be necessary to prevent buckling. The outer sections carrying the shear force will probably require swaging to prevent shear buckling.

The loading conditions on the front and rear parcel shelves are once again similar. The front shelf shown in Fig. 16.2-5(c) is loaded such that in plan view it is deflected towards the rear (+ve bending moment) while the rear shelf in Fig. 16.2-5(d) is deflected forward (–ve bending moment). Both these shelves must have good bending properties in the centre and adequate shear connections to the sideframe.

Considering the loads at the front and rear of the structure we have the conditions shown in Fig. 16.2-6. At (a) the loading on the front inner wing panel is shown and at (b) that on the rear quarter panel. With modern transverse engined cars the engine centre of gravity is forward of the front suspension so there is a hogging moment forward of the suspension but a sagging moment where the wing panel is attached to the dash panel etc. A similar situation occurs at the rear of the vehicle where the luggage is placed behind the rear suspension mountings causing a hogging moment. This may well change to a sagging moment at the attachment of the rear quarter panel to the panel behind the rear seats, depending on magnitudes of forces and dimensions. Both front inner wing panels and rear quarter panels must be designed to carry the indicated shear forces. The

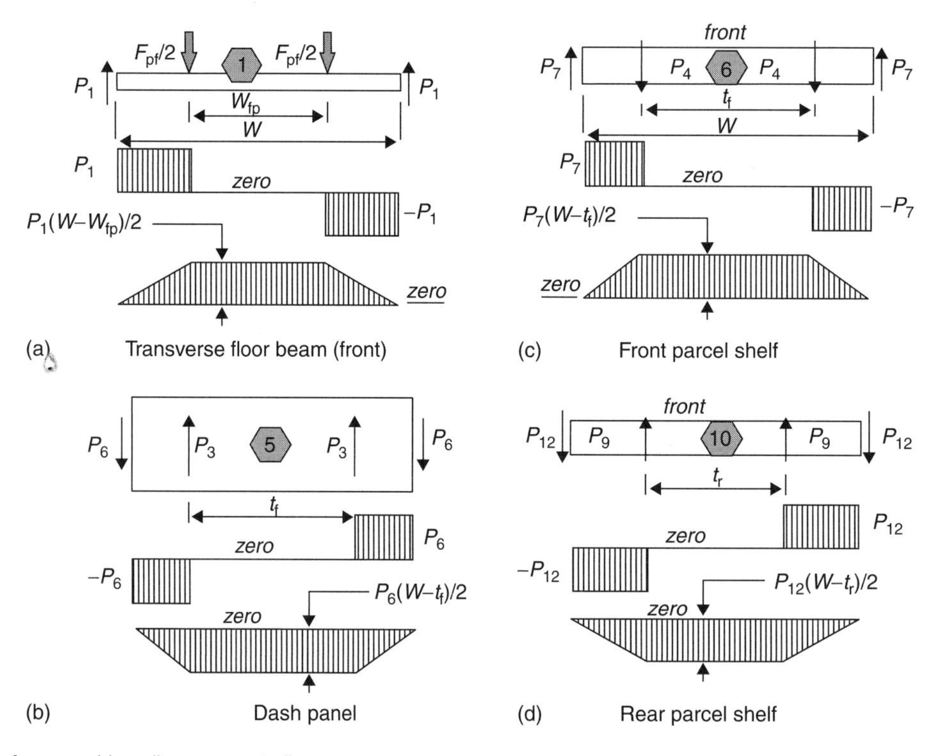

Fig. 16.2-5 Shear force and bending moment diagrams.

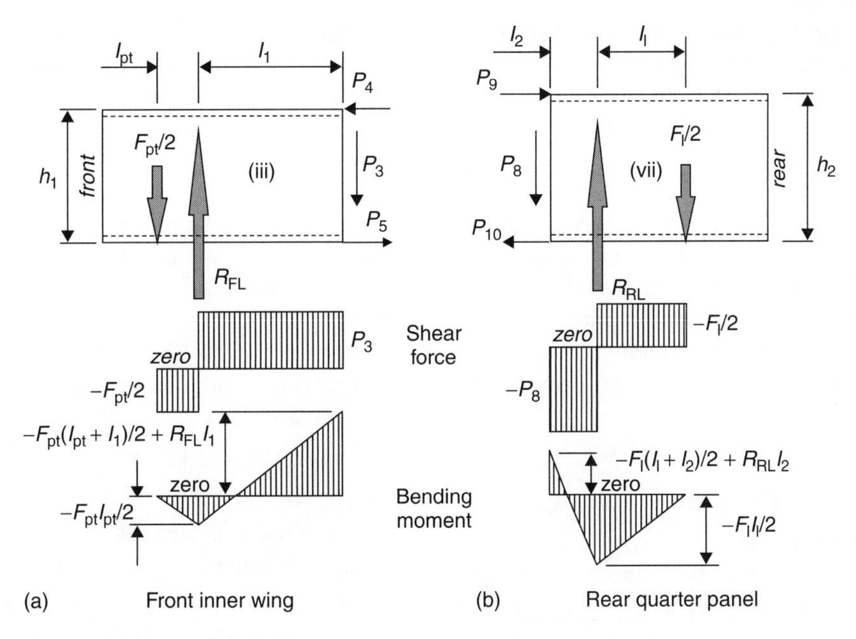

Fig. 16.2-6 Shear force and bending moment diagrams.

horizontal reaction forces P_4, P_5, P_9 and P_{10} for these SSSs must be distributed along the top and bottom edges by means of stiffeners as indicated by the dashed lines. Buckling of these deep members due to shear forces can be prevented by suitable stiffening swages.

The floor panel loading shown in Fig. 16.2-7(a) indicates that the outer sides are subject to shear forces. The shear force P_{13} is applied over a long length (l_5 shown in Fig. 16.2-4) hence the shear stresses are usually small. Local panel stiffening will be necessary for other reasons (e.g. to prevent panel vibrations and resist normal loads).

Although the diagram indicates bending moments these are not of significance as l_5 is large.

The end loads P_5 and P_{10} in practice cannot be applied as point loads to the front and rear edges of the floor. The stiffener necessary at the base of the front inner wing panel and the rear quarter panel will need to be extended along the floor panel as shown with dashed lines (Fig. 16.2-7(a)). The sideframe, Fig. 16.2-7(b), is the main structural member providing bending strength and stiffness. The internal load distribution through the elements of the sideframe is not considered here. From

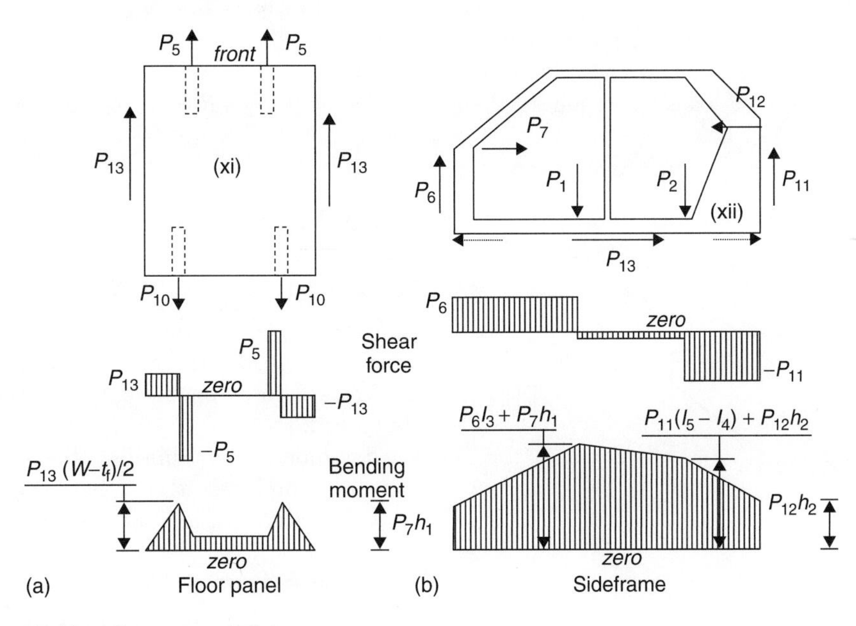

Fig. 16.2-7 Shear force and bending moment diagrams.

equation (16.2.15) the edge load P_{13} is the difference between P_{10} and P_5 which in turn can be shown equal to the difference between P_{12} and P_7 using equations (16.2.12)/(16.2.14) and (16.2.7)/(16.2.9). Therefore the front and rear ends of the sideframe may be considered to be subject to bending moments P_7h_1 and $P_{12}h_2$, respectively. The centre part of the sideframe is also subject to increased bending due to the shear forces P_6 and P_{11}. Therefore, the loading on the cantrail will be a combination of bending and compression and the loading on the sill a combination of bending and tension. Note also additional local bending will occur on the sill from the loads P_1 and P_2.

16.2.3 Torsion load case for the standard sedan

16.2.3.1 The pure torsion load case and its significance

On the road, the car is subjected to torsion when a wheel on one side strikes a bump or a pot-hole, causing different wheel reactions on each side of the axle. This is the vertical asymmetric load case which gives a combination of bending and torsion on the vehicle.

For calculation purposes, the torsion component of the asymmetric vertical case is considered *in isolation*, as the *pure torsion load case*. Equal and opposite loads R_{FT} are applied to the front left and right suspension towers, thus causing a couple T about the vehicle centreline. This is reacted by an equal and opposite couple at the rear suspension points so that the vehicle is in pure torsion (see Fig. 16.2-8). The SSS edge loads Q, resulting from this, are then calculated.

Clearly, the pure torsion load could not be experienced on the road, since there cannot be a negative wheel reaction. However, if road case loads are required, then the SSS edge loads Q from the pure torsion case could be combined with the edge loads P from the bending load case by suitable factoring and addition.

The significance of the pure torsion load case is that it applies edge forces on the individual SSSs that are completely different from those experienced in the bending case. The torsion stiffness (and hence also the torsional fundamental vibration frequency) of a vehicle body is often used as a benchmark of its structural competence.

The torsion case is found to be a stringent one. For torsion, the keys to a weight efficient integral sedan structure are:

1. a closed ('boxed') system of SSSs, in shear, in the passenger compartment, and
2. as in other load cases, continuity of the load paths at the dash, where the suspension loads are fed from the end structures into this 'torsion box'.

In this section, the baseline *standard sedan* with closed torsion box is discussed first. Later, the '*faux sedan*', with at least one missing SSS in the passenger compartment is considered. The missing surface(s) can have a disastrous effect on the torsion performance of the body. Remedies to the faux sedan's deficiencies are suggested.

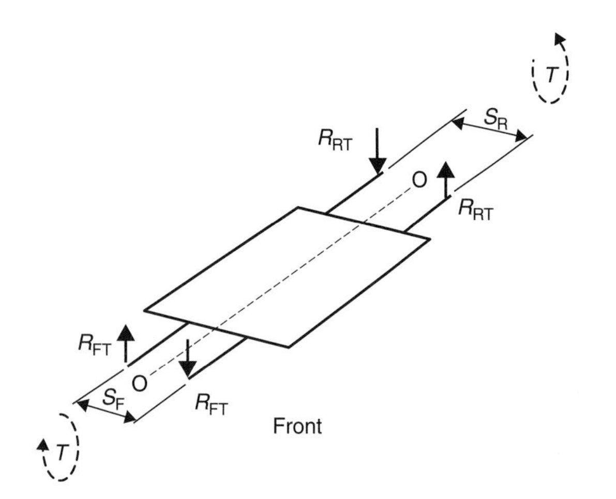

Fig. 16.2-8 Vehicle in pure torsion.

16.2.3.2 Overall equilibrium of vehicle in torsion

The torque T is applied about axis O–O as couple $R_{FT}S_F$ at the front suspension. This must be balanced by an equal and opposite couple $R_{RT}S_R$ due to the reaction forces R_{RT} at the rear suspension. See Fig. 16.2-8:

$$T = R_{FT}S_F = R_{RT}S_R$$

Hence

$$R_{FT} = T/S_F \quad R_{RT} = R_{FT}\,S_F/S_R = T/S_R$$

16.2.3.3 End structures

16.2.3.3.1 Front and rear inner fenders

On the right-hand fender as shown in Fig. 16.2-9, the suspension load acts upward. This is reacted by an equal downward force on the panel where it is joined to the bulkhead.

For moment equilibrium, the couple caused by the offset L_1 of forces R_{FT} is balanced by complementary shear forces P_{FT} at top and bottom of the panel:

$$R_{FT}L_1 = P_{FT}h_1 \quad \text{thus}$$
$$P_{FT} = R_{FT}L_1/h_1 = TL_1/(S_Fh_1)$$

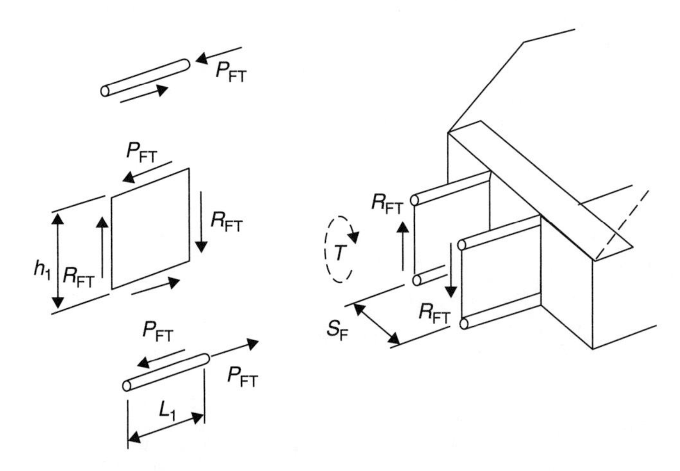

Fig. 16.2-9 Frontal structure.

Forces P_{FT} are reacted by equal forces, fed into the top and bottom flanges as shear flows (see later). These will in turn be reacted by axial forces P_{FT} in the flanges, as shown, where the flanges meet the passenger compartment.

The upper flange is of thin sheet material, so that the reaction P_{FT} will be concentrated at the junction between the web and the flange so that the latter can be treated as a 'boom'. The lower flange is usually the 'engine mounting rail' consisting of a substantial box member. This may also be treated as a boom here.

The left-hand fender will behave similarly, but with the forces in opposite directions.

In the standard sedan, the rear inner fenders will behave in an identical way to the front ones, with the forces in the appropriate directions (see Fig. 16.2-12). The vertical shear force on the rear fender is R_{RT} and the reaction forces in the top and bottom flanges P_{RT} are:

$$P_{RT} = R_{RT}L_2/h_2 = TL_2/(S_R h_2)$$

where L_2 = loaded length and h_2 = height of rear inner fender.

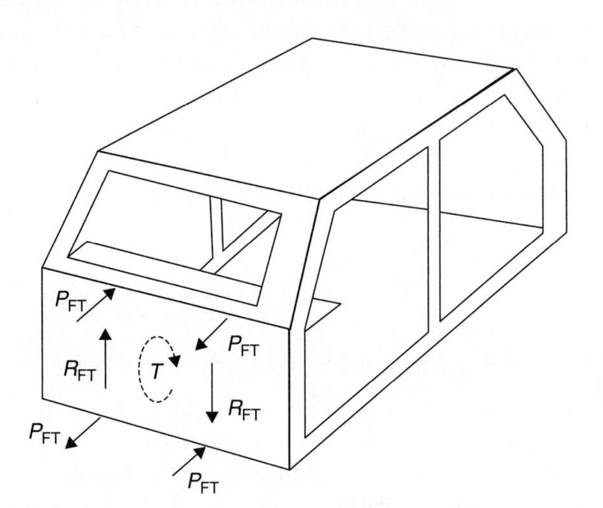

Fig. 16.2-10 Forces on dash panel.

16.2.3.3.2 Dash

Torque T is applied to the engine bulkhead by the reactions R_{FT} from the fender webs acting at a separation S_F giving a couple $P_{FT}S_F = T$ as shown in Fig. 16.2-10.

The reactions P_{FT} to the fender upper flange forces are carried by the parcel shelf which must be stiff in the appropriate direction. The reactions to the engine rail forces P_{FT} pass to the floor as in-plane forces in the directions shown in Fig. 16.2-12.

The rear fender forces are reacted in an identical way. Owing to the directions of the suspension forces in the torsion case, forces P_{RT} apply a couple to the parcel shelves of $P_{RT}S_R$. This is also the case for the floor.

16.2.3.3.3 Parcel shelf/upper dash

The parcel shelf acts as a beam, carrying the couple $P_{FT}S_F$ out to the sideframe at the mid A-pillars.

The end forces Q_{X1} form a couple to balance this couple, thus:

$$Q_{X1}B = P_{FT}S_F \quad \text{giving} \quad Q_{X1} = P_{FT}S_F/B \tag{16.2.19}$$

The shear forces in the parcel shelf (acting in a horizontal plane) and the bending moments (acting about a vertical axis) are as shown in Fig. 16.2-11. By similar reasoning the rear parcel shelf reaction forces Q_{X2} are:

$$Q_{X2} = P_{RT}S_R/B$$

16.2.3.4 Passenger compartment

A general view of the SSS edge forces in the passenger compartment is given in Fig. 16.2-12.

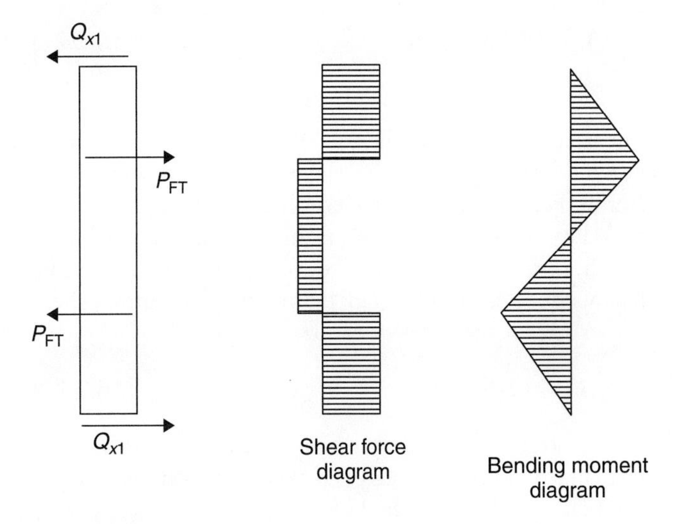

Fig. 16.2-11 Parcel shelf.

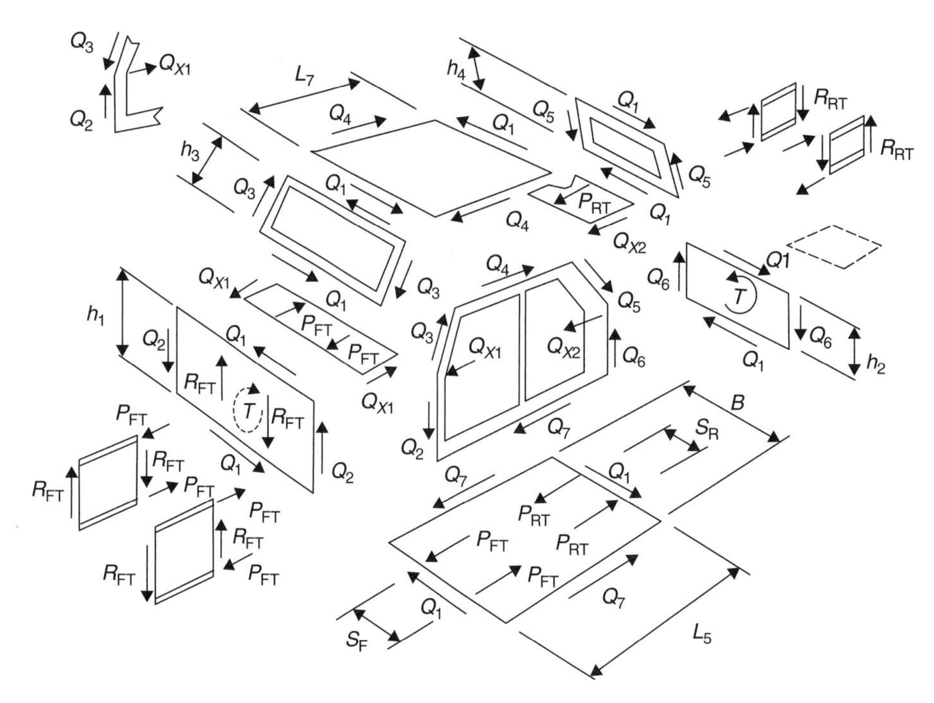

Fig. 16.2-12 Edge forces in the standard sedan in the torsion load case.

16.2.3.4.1 Engine bulkhead

The shear forces R_{FT} from the fender webs are reacted on the engine bulkhead. This creates a couple $T = R_{FT}S_F$.

This couple is reacted by shear forces Q_1 and Q_2 acting on the edges of the bulkhead, forming couples Q_1h_1 and Q_2B.

For lateral force equilibrium:

$$Q_{1\ \mathrm{TOP}} = Q_{1\ \mathrm{BOTTOM}} = Q_1$$

For lateral force equilibrium:

$$Q_{2\ \mathrm{LEFT}} = Q_{2\ \mathrm{RIGHT}} = Q_2$$

For moment equilibrium:

$$T = Q_1h_1 + Q_2B \tag{16.2.21}$$

16.2.3.4.2 Front windshield

Shear force Q_1 from the top of the engine bulkhead is reacted by an equal force on the bottom of the windshield frame. For horizontal force equilibrium, this will be balanced by an equal force Q_1 on the top edge of the frame.

These forces Q_1 form a couple Q_1h_3 which must be balanced by a couple Q_3B from complementary shear forces Q_3 on the sides of the frame. Thus

$$Q_1h_3 - Q_3B = 0 \tag{16.2.22}$$

This frame achieves its shear stiffness from the local in-plane bending stiffness (i.e. about an axis *normal* to the plane of the windshield frame). It is thus working as a 'ring beam', see section 16.2.3.6.

16.2.3.4.3 Roof

Shear force Q_1 is fed to the front of the roof from the top of the windshield frame. This will be balanced by edge force Q_1 at the rear. These forces form a couple Q_1L_7 which must be balanced by a complementary couple Q_4B from the edge forces Q_4 on the roof sides. Thus, for moment equilibrium:

$$Q_1L_7 - Q_4B = 0 \tag{16.2.23}$$

16.2.3.4.4 Backlight (rear window) frame

Force Q_1 from the rear of the roof is reacted by an equal and opposite force on the top of the rear window frame and this, in turn, is balanced by a force Q_1 on the bottom of the window frame. The couple Q_1h_4 from these forces is balanced by the complementary couple Q_5B from edge forces Q_5 on the sides of the backlight frame.

For moment equilibrium:

$$Q_1h_4 - Q_5B = 0 \tag{16.2.24}$$

16.2.3.4.5 Rear seat bulkhead

Shear force Q_1 is passed from the backlight frame to the top of the rear bulkhead. The top and bottom edge forces Q_1 on the rear bulkhead and forces Q_6 on its sides form couples which are balanced by the couple $T = R_{RT}S_R$ applied by the rear fender webs to this panel.

Thus for moment equilibrium:

$$T = Q_1 h_2 + Q_6 B \quad (16.2.25)$$

As before $Q_{1\ TOP}$ and $Q_{1\ BOTTOM}$ are equal and opposite, as are $Q_{2\ LEFT}$ and $Q_{2\ RIGHT}$. Because of the externally applied torque T, the edge forces Q_1, Q_6 do not form opposing couples (although *overall* the panel must be in moment equilibrium). As on all such surfaces with additional external forces or moments, force Q_1 is the *net* force on the top of this panel. The rear seat bulkhead is sometimes a continuous panel, or it may consist of a truss structure, formed by punching triangular holes leaving diagonal members forming a 'triangulated truss' SSS (Fig. 16.2-13).

16.2.3.4.6 Floor

This receives equal and opposite edge forces Q_1 from the front and rear bulkheads and Q_7 at the sides, from the sideframes. The forces Q_1 and Q_7 are oriented so as to form complementary couples $Q_1 L_5$ and $Q_7 B$, except that now additional forces P_{FT} and P_{RT} from the lower longitudinals of the front and rear fenders cause additional couples $P_{FT} S_F$ and $P_{RT} S_R$. All the pairs of forces Q_1, Q_7, P_{FT}, P_{RT} balance out for *force* equilibrium.

For *moment* equilibrium:

$$Q_1 L_5 - Q_7 B = P_{FT} S_F + P_{RT} S_R \quad (16.2.26)$$

Often, the engine rails run for some distance under the floor, and this gives a good connection path (in shear) for force P_{FT} between the engine rail and the floor.

16.2.3.4.7 Sideframes

The edges of the sideframe react edge forces Q_2 Q_3 Q_4 Q_5 Q_6 Q_7 from the surfaces attached to it, and Q_{x1} and Q_{x2} from the front and rear parcel shelves, as seen in Fig. 16.2-14. The sideframes on opposite sides of the vehicle experience identical edge loads, but in opposite directions. Force and moment equilibrium will be obeyed. Clearly, the sideframes are crucial in 'gathering' the edge forces from the other surfaces.

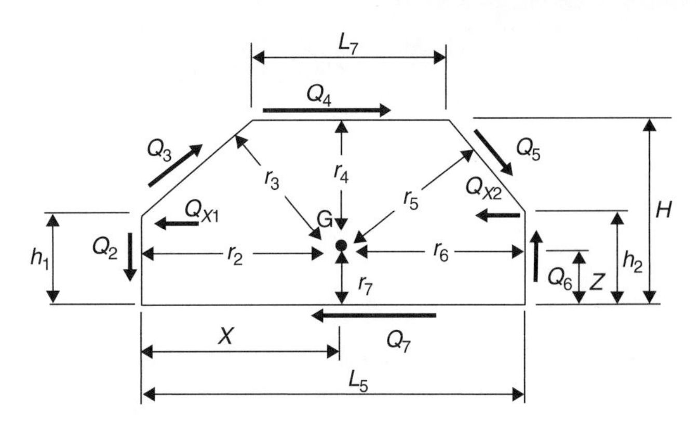

Fig. 16.2-14 Sideframe.

For moment equilibrium, take moments about an arbitrary point G distance X behind the lower A-pillar centreline and Z above the rocker centreline. If r_2 to r_7 are the moment arms of forces Q_2 to Q_7 about point G as in Fig. 16.2-14, then:

$$-r_2 Q_2 + r_3 Q_3 + r_4 Q_4 + r_5 Q_5 - r_6 Q_6 + r_7 Q_7 = Q_{X1}(h_1 - Z) + Q_{X2}(h_2 - Z)\ldots \quad (16.2.27)$$

The moments of most of the forces about G are shown in Table 16.2-1(a) (taking clockwise moments as positive).

The forces Q_3 and Q_5 act on the upper A- and C-pillars, which are at angles θ_A and θ_C from the vertical. These forces can be resolved into vertical and horizontal components. Noting that the line of action of Q_5 passes through the C-pillar waist joint (point E in Fig. 16.2-15) and that of Q_3 passes through the A-pillar waist joint (point B in Fig. 16.2-15), then the moments of these forces are given in Table 16.2-1(b).

Fig. 16.2-13 Rear seat aperture with diagonal braces (courtesy Vauxhall Heritage Archive).

Table 16.2-1 (a)

Edge force	Moment about G	Moment arm about G
Lower A-pillar	$-Q_2X$	$r_2 = X$
Header (roof rail)	$Q_4(H-Z)$	$r_4 = H-Z$
Lower C-pillar	$-Q_6(L_5-X)$	$r_6 = L_5-X$
Rocker	Q_7Z	$r_7 = Z$
Parcel shelf reactions	**Moment about G**	
Front	$-Q_{x1}(h_1-Z)$	h_1-Z
Rear	$-Q_{x2}(h_2-Z)$	h_2-Z

For sideframe equilibrium, all moments acting about G will sum to zero ($\Sigma M = 0$):

$$- Q_2X + Q_3\{(h_1 - Z)\sin(\theta_A) + X\cos(\theta_A)\} + Q_4(H-Z) + Q_5\{(h_2 - Z)\sin(\theta_C) + (L_5 - X)\cos(\theta_C)\} - Q_6(L_5 - X) + Q_7Z - Q_{X1}(h_1 - Z) - Q_{X2}(h_2 - Z) = 0$$

Careful inspection of this equation and of Fig. 16.2-14 reveals that the coefficients of the Qs must be their moment arms r_2, r_3, etc., about point G as given in equation (16.2.27). Moment arms r_2 to r_7 are listed in Tables 16.2-1(a) and (b).

16.2.3.5 Summary – baseline closed sedan

Front and rear inner fenders:

$$R_{FT} = T/S_F \qquad R_{RT} = T/S_R = R_{FT}S_F/S_R$$
$$P_{FT} = TL_1/S_Fh_1 \qquad P_{RT} = TL_2/S_Rh_2$$

Parcel shelves:

$$Q_{X1} = P_{FT}S_F/B = TL_1/h_1B \qquad (16.2.19)$$
$$Q_{X2} = P_{RT}S_R/B = TL_2/h_2B \qquad (16.2.20)$$

Passenger compartment (with some re-arrangements):

$$Q_1h_1 + Q_2B = T \qquad (16.2.21) \text{ front bulkhead}$$
$$-Q_1h_3 + Q_3B = 0 \qquad (16.2.22) \text{ windshield}$$
$$-Q_1L_7 + Q_4B = 0 \qquad (16.2.23) \text{ roof}$$
$$-Q_1h_4 + Q_5B = 0 \qquad (16.2.24) \text{ backlight}$$
$$Q_1h_2 + Q_6B = T \qquad (16.2.25) \text{ rear bulkhead}$$
$$Q_1L_5 - Q_7B = P_{FT}S_F + P_{RT}S_R \qquad (16.2.26) \text{ floor}$$
$$-r_2Q_2 + r_3Q_3 + r_4Q_4 + r_5Q_5 - r_6Q_6 + r_7Q_7 = Q_{X1}(h_1 - Z) + Q_{X2}(h_2 - Z) \qquad (16.2.27) \text{ sideframe}$$

The input moment is T and P_{FT}, P_{RT}, Q_{x1} and Q_{x2} are known in terms of T. The remaining seven unknown edge forces (Q_1 to Q_7) can be solved using the seven simultaneous equations (16.2.21)–(16.2.27) if T is known. This could be done by hand (the equations are linear and relatively 'sparse'). Alternatively, if the equations are rearranged slightly and put in matrix form, they become:

$$\begin{bmatrix} L_5 & 0 & 0 & 0 & 0 & 0 & -B \\ h_1 & B & 0 & 0 & 0 & 0 & 0 \\ -h_3 & 0 & B & 0 & 0 & 0 & 0 \\ -L_7 & 0 & 0 & B & 0 & 0 & 0 \\ -h_4 & 0 & 0 & 0 & B & 0 & 0 \\ h_2 & 0 & 0 & 0 & 0 & B & 0 \\ 0 & -r_2 & r_3 & r_4 & r_5 & -r_6 & r_7 \end{bmatrix} \begin{bmatrix} Q_1 \\ Q_2 \\ Q_3 \\ Q_4 \\ Q_5 \\ Q_6 \\ Q_7 \end{bmatrix} = \begin{bmatrix} P_{FT}S_F + P_{RT}S_R \\ T \\ 0 \\ 0 \\ 0 \\ T \\ Q_{X1}(h_1 - Z) + Q_{X2}(h_2 - Z) \end{bmatrix} \begin{matrix} (16.2.26) \text{ floor} \\ (16.2.21) \text{ front bulkhead} \\ (16.2.22) \text{ windshield} \\ (16.2.23) \text{ roof} \\ (16.2.24) \text{ backlight} \\ (16.2.25) \text{ rear bulkhead} \\ (16.2.27) \text{ sideframe} \end{matrix}$$

Equilibrium matrix   Edge forces   Input forces

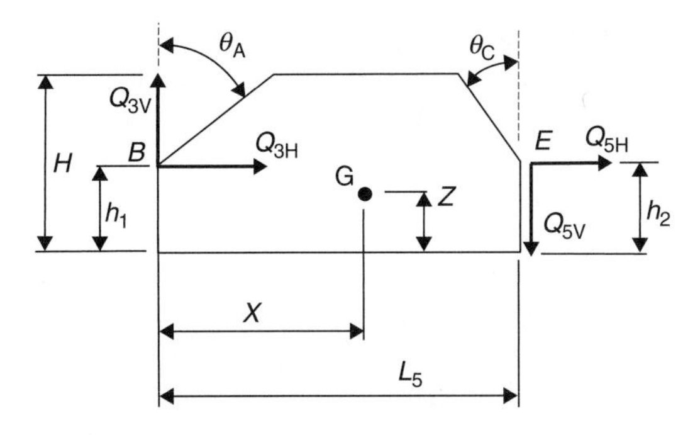

Fig. 16.2-15 Components of forces Q_3 and Q_5.

They can then be solved using a standard computer method (e.g. Gaussian reduction) available widely. Most 'spreadsheet' programs can do this. To ensure reliability of the solution, the matrix should be 'positive definite'. This requires all terms on the leading diagonal of the coefficient matrix to be positive and non-zero, and this is the reason for the rearrangement of the order in which the equations are listed.

Solution of the equations gives values for the edge forces Q_1 to Q_7.

The stresses in the shear panels and ring frames can then be estimated.

16.2.3.5.1 Solution check

It is easy to make errors in the setting up of the equilibrium equations. The solution for edge forces Q should be checked to ensure force balance in both horizontal and vertical directions on the sideframe. Referring to Figures 16.2-14 and 16.2-15:

$$\sum F_{\text{HORIZONTAL}} = 0 \ldots Q_{3\ \text{HORIZONTAL}} + Q_4 + Q_{5\ \text{HORIZONTAL}} - Q_7 - Q_{X1} - Q_{X2} = 0$$

$$\sum F_{\text{VERTICAL}} = 0 \ldots - Q_2 + Q_{3\ \text{VERTICAL}} - Q_{5\ \text{VERTICAL}} + Q_6 = 0$$

16.2.3.6 Some notes on the standard sedan in torsion

(a) A knowledge of the shear stress τ in the panels is required. The panel thickness t must be chosen so that shear stress is within permitted values. The average shear stress τ in the continuous shear panels can be calculated from the edge force Q by:

$$\tau_{AV} = \frac{Q}{\text{shear area}}$$

The force intensity on a panel edge is often expressed as shear force per unit length or 'shear flow' q. Shear flow is related to shear stress thus $q = t\tau$. As with shear stress systems, every shear flow has an equal 'complementary' shear flow q' at 90° to it.

For example in the roof (see Fig. 16.2-12):

Shear flow on front of roof $q_1 = Q_1/B$. Shear flow on side of roof $q_4 = Q_4/L_7$ but the latter would be the complementary shear flow to q_1. From equation (16.2.23) above:

$$Q_1 L_7 = Q_4 B$$

hence, substituting:

$$(q_1 B) L_7 = (q_4 L_7) B$$

thus

$$q_1 = q_4 \text{ (complementary shear flows)}$$

For a car of constant width B such as this one, the shear flows q_1 on panel edges across the car are all the same at $q_1 = Q_1/B$. For the panels subject to edge

Table 16.2-1 (b)

Force (Figure 16.2-15)	Component	Moment about G	Moment arm about G
Q_{3H}	$Q_3 \sin(\theta_A)$	$(h_1 - Z) Q_3 \sin(\theta_A)$	$r_{3H} = (h_1 - Z)$
Q_{3V}	$Q_3 \cos(\theta_A)$	$(X) Q_3 \cos(\theta_A)$	$r_{3V} = (X)$
Q_{5H}	$Q_5 \sin(\theta_C)$	$(h_2 - Z) Q_5 \sin(\theta_C)$	$r_{5H} = (h_2 - Z)$
Q_{5V}	$Q_5 \cos(\theta_C)$	$(L_5 - X) Q_5 \cos(\theta_C)$	$r_{5V} = (L_5 - X)$
Force	**Moment about G**		**Moment arm about G**
Q_3	$Q_3\{(h_1 - Z)\sin(\theta_A) + X\cos(\theta_A)\}$		$r_3 = (h_1 - Z)\sin(\theta_A) + X\cos(\theta_A)$
Q_5	$Q_5\{(h_2 - Z)\sin(\theta_C) + (L_5 - X)\cos(\theta_C)\}$		$r_5 = (h_2 - Z)\sin(\theta_C) + (L_5 - X)\cos(\theta_C)$

forces only (windshield, roof, backlight in this case) the complementary shear flows q_3 q_4 q_5 will be equal to this, and hence equal to each other. The shear flow around this part of the sideframe is thus constant. This reflects the Bredt–Batho theorem for shear flows in closed sections subject to torsion, because the passenger compartment can be thought of as a closed tube running across the car between the sideframes. For some of the other panels, such as bulkheads and floor, the edge forces calculated above are affected by additional external moments on the panel (e.g. $R_{TF}S_F$ on the front bulkhead). This masks the complementary shear flow effect in these cases. For example, the couple $T = R_{FT}S_F$ is introduced into the front bulkhead by forces R_{FT} (see Fig. 16.2-12) so that the shear force (and hence the local shear flow along the top and bottom edges) varies across the panel. Thus, force Q_1 is the *net* force on the top and bottom of this bulkhead.

The equality of complementary shear flows can also be used as a check on the solutions for the forces Q on the structural surfaces that have edge forces only (e.g. roof, windshield). Again, this check cannot be applied directly to structural surfaces (e.g. floor) which are subject to extra forces in addition to the edge forces.

(b) The 'shear panel' load path in the compartment depends on all SSSs in it carrying shear effectively.

The least effective SSSs in this respect are the ring frames, including the sideframes and the windshield frame. This type of structural surface achieves its in-plane shear stiffness by acting as a ring beam, with local bending of its edge beams about an axis normal to the plane of the frame. As discussed earlier, this leads to high local bending moments at the corners of the frame (see Fig. 16.2-16). The edge members are said to be in 'contraflexure' since the bending moment (and hence the curvature) changes sign part-way along the member.

The least effective of these frames is the slender upper A-pillar. Thickening the lower A-pillar will cause more of Q_{X1} etc., to be conveyed downward to the underbody (so-called 'semi-open' structure).

If the windshield glass is adhesive bonded to the frame, then the glass will act as a substantial load path (as a shear panel). This will relieve the corner bending moments on the frame (somewhat) and will make a significant increase in the shear stiffness of the window frame. Cars with bonded screens have shown large torsion stiffness increases (up to 60%). Care must be taken, however, to ensure that the load in the glass does not cause it to break in extreme load cases (e.g. vehicle corner bump case, or wheel jacking for puncture). The vehicle stiffness and member stresses (e.g. corner bending moment in window frame) should not degrade to unacceptable values in the event of a smashed windshield.

Windshields attached with (elastomer) gaskets give a much less significant contribution to the structure, due to the flexibility of the gasket.

(c) The sideframe consists of two (or sometimes three) rings, bordered by the A-, B- and C- (etc.) pillars. The share of the shear load Q_4 carried by each of these beams will be in proportion to their relative stiffness and hence will depend on their second moments of area, and lengths (see Fig. 16.2-17).

Calculating the share of the load in each pillar and hence the bending moments and the stresses in the ring frames accurately is complicated, particularly in the multiple ring case (sideframe), since they are statically indeterminate (three redundancies per 2D ring).

The sideframe is subject to overall shear, and since it is a (multi-bay) ring beam, its edge members experience 'contraflexure bending', giving high joint stresses as shown in Fig. 16.2-17.

(d) As noted above, the ring frames have high bending moments at their corners. Some pillars are shared by two adjacent ring frames (e.g. the upper A-pillar is shared by the windshield surround and sideframe. See

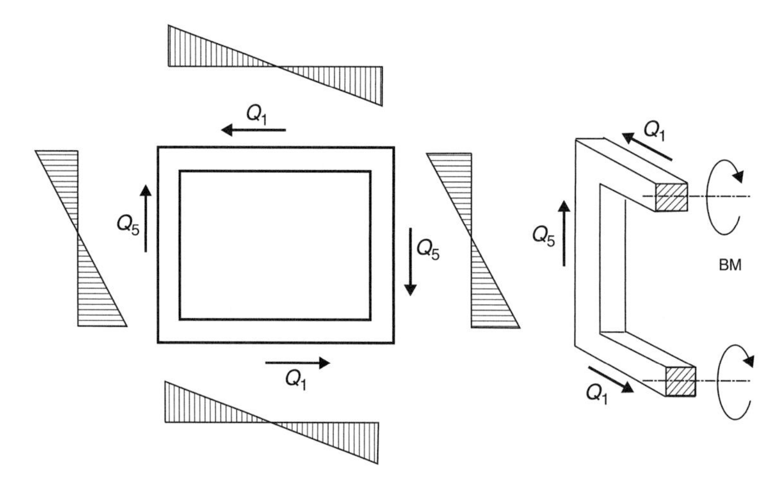

Fig. 16.2-16 Ring beam.

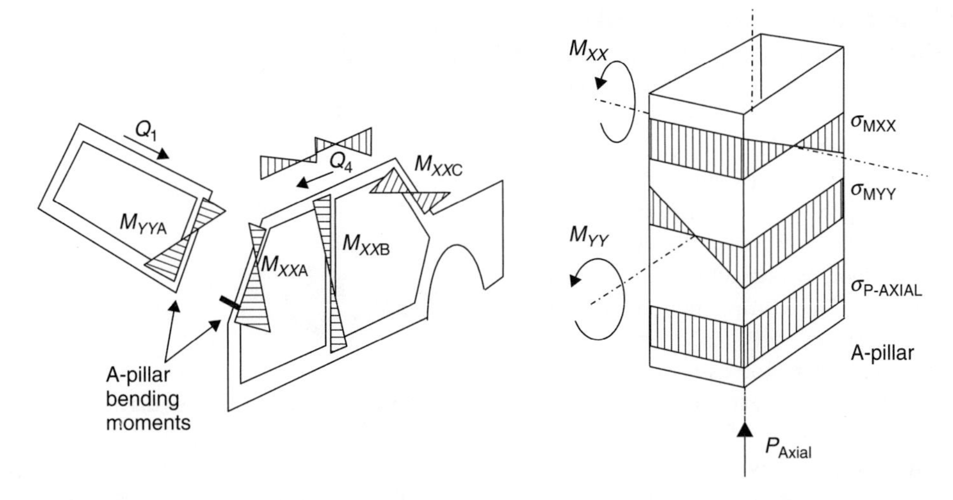

Fig. 16.2-17 Stresses in shared pillar.

Fig. 16.2-17). In this case the lateral load *in each direction* P_X and P_Y must be accounted for. They will lead to bending moments about two directions in the pillar. If the pillar is sloped in one of these frames (e.g. the A-pillar in the sideframe) then there will also be an axial component of force in the member. These moments and axial forces will cause a combined stress system in the pillar. Therefore

$$\sigma_{\text{max.}} = \frac{M_x y}{I_x} + \frac{M_y x}{I_y} + \frac{P}{A}$$

(e) A good load path from the inner fender upper booms to the sideframes, via the 'parcel shelf' is essential (unless the booms are offset to connect directly to the sideframe A-pillar). Thus the parcel shelf must be stiff in bending. It is a highly stressed component because of the high bending moments caused by forces P_{FT}.

(f) The connection of the parcel shelf to the A-pillar is another high stress area. This results from the high local bending moment in the A-pillar due to force Q_{x1}. Similarly, in the standard sedan, the middle of the C-pillar will be highly stressed due to bending moments caused by the 'hat shelf' end reactions Q_{x2}.

16.2.3.7 Structural problems in the torsion case

16.2.3.7.1 The 'faux sedan'

A 'closed box' of SSSs is required in the passenger compartment to maintain an effective load path of 'shear panels' to carry torsion.

If any one of the structural surfaces in the compartment is missing then the structure becomes an 'open box' and the shear panel load path for torsion breaks down.

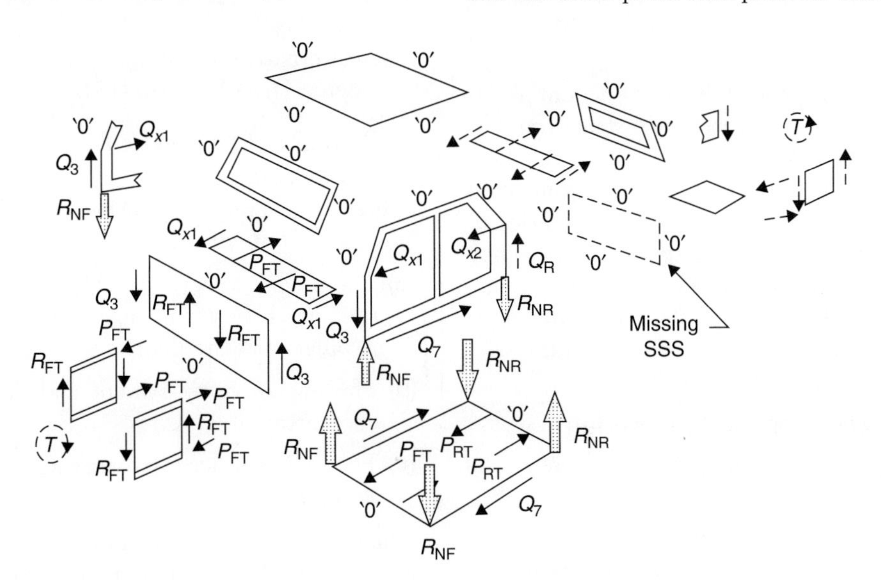

Fig. 16.2-18 Faux sedan in torsion.

For example, sometimes the rear seat bulkhead is omitted to allow a 'split rear seat' feature (see Fig. 16.2-18). In this case there is no ability to react edge forces Q_1 and Q_6 around the rear seat back, so that the shear forces Q_1 at the bottom of the rear window and on the rear of the floor must be zero. Hence (for force equilibrium) the edge forces Q_1 at the other ends of these surfaces become zero. This will 'propagate' all the way round the compartment so that all forces Q_1 (roof, windshield, front bulkhead, floor) must also be zero. The complementary forces on all of these structural surfaces will also become zero.

The only compartment edge forces remaining will be those which can be balanced by 'external' torques. Thus in the front bulkhead the torque T is balanced by the couple Q_2B from forces Q_2 at the A-pillar. The couple Q_7B from edge forces Q_7 on the floor are balanced by the couples $P_{FT}S_F$ and $P_{RT}S_R$.

It is assumed here that a load path is available to transmit the reaction torque T (from the rear suspension) to the sideframe as edge forces Q_R on the C-pillars via the *outer* rear fender and/or via floor members. (The rear seat bulkhead is no longer available for this.)

The moments on the sideframe due to forces Q_{x1} Q_{x2} Q_2 Q_R and Q_7 (see Fig. 16.2-18) are all in the same direction. These moments can only be balanced by moments from forces R_{NF} and R_{NR}. These cause reactions R_{NF}, R_{NR} *normal to the plane of the floor.* This causes the floor to twist out-of-plane. The SSS assumptions are not satisfied in this case. Similar twisting of other surfaces (e.g. front bulkhead/parcel shelf) will occur.

In practice, such a 'faux sedan' structure is much more flexible (and less weight efficient) in torsion than a 'closed box' sedan. The sideframes tend to act as 'levers' to twist the cowl/dash (engine bulkhead/parcel shelf) assembly.

High local stresses and large strains are experienced by the remaining loaded members (in the floor etc.) leading to: (a) poor fatigue life and (b) damage to the paint system with resulting early corrosion.

The same 'faux sedan' problem will be encountered if any one (or more) of the compartment structural surfaces are missing or of reduced structural integrity (e.g. roof, front bulkhead or floor missing, or poorly connected to adjacent components). Similarly loss of integrity of sideframe 'ring beam' members: e.g. poor quality joints, 'panoramic' A-pillars, pillarless sedans (to some extent), corroded rockers, etc. – especially if the weakening is at the high bending moment corners of the ring frames.

16.2.3.7.2 Remedies for the faux sedan

16.2.3.7.2.1 Replacement of missing shear panel with ring frame

The ideal remedy for the faux sedan is to modify it so as to restore the 'closed box' type structure with its weight efficient shear panel load paths.

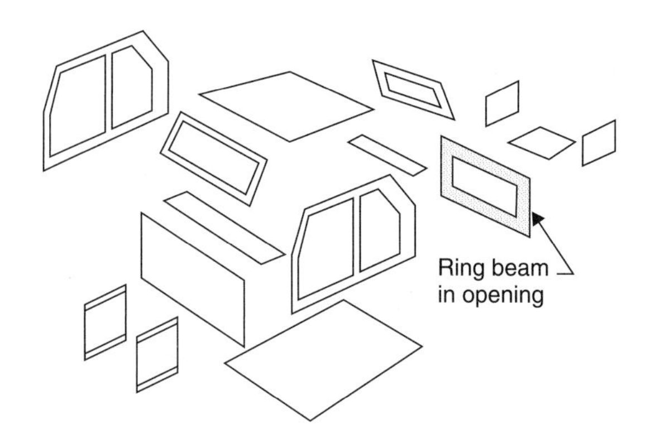

Fig. 16.2-19 Ring frame to remedy faux sedan problem.

This is possible, in some cases, by using a ring frame in place of the missing panel (Fig. 16.2-19). This restores shear integrity to the surface in question, whilst maintaining a substantial opening. For such a ring frame, care must be taken to ensure an effective path for local bending moments all round the frame *especially* at the corners. For example, in the rear seat bulkhead case, parcel shelf, side-wall beams and floor cross-member (which could be under the floor) must all have high stiffness for bending about axes normal to the plane of the frame, *and* they *must all* be well connected for bending at the corners (e.g. gusseted joints). C-pillars and parcel shelves are often not stiff in the required direction, and so require special design attention.

If a larger opening is required (e.g. hatchback, station wagon, opening rear screen with split seat) then a ring frame, running the full height of the sideframe (through the C-pillars, across the roof and across the floor), is a possibility. This works better if the ring frame is as planar as possible, and if the corner joints are good in bending. Even so, the result will not be as weight efficient as a true shear panel.

An alternative to the ring beam is a triangulated bay in the opening (see Fig. 16.2-13).

16.2.3.7.2.2 Provision of 'closed torque box' in part of structure

There are several areas in the integral car body that are 'box like'. Some of these are:

(a) The cowl/footwell assembly (the region enclosed by the parcel shelf, the engine bulkhead, the lower A-pillars and the floor).

(b) The engine compartment and/or the rear luggage compartment.

(c) The region under the rear seat, where there is often a step in the floor.

If any of these can be converted into a 'closed box' structure by the addition of shear panels or ring frames, *and* can be well connected to the sideframes, then this

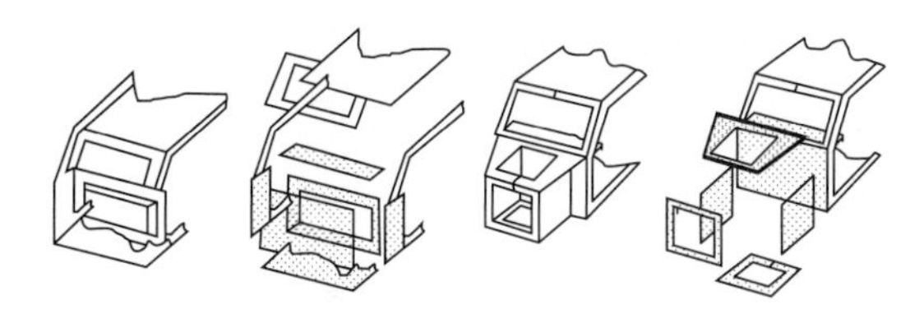

Fig. 16.2-20 Torsion stiffening by 'boxing in' localized regions of body.

will provide some torsion stiffness. The sideframes, acting as stiff levers, will make this torsion stiffness available to the whole body. The added members making up the box may already be present, providing other functions (e.g. instrument panel, floor cross-member, etc.). Examples are shown in Fig. 16.2-20.

This approach is likely to be less weight efficient, and more prone to high stress/strain problems (fatigue, corrosion) than the true 'closed' integral structure described in section 16.2.3.4.

16.2.3.7.2.3 Provision of true grillage structure in floor

A grillage is a flat structure which is stiff (for out-of-plane loads) in both bending and twist. Individual members in a true grillage experience local bending moments and torques (both about axes within the plane of the grillage) and shear forces normal to the plane. They must therefore be stiff for these loads (see Fig. 16.2-21).

Hence, they are likely to be closed section (box) members of considerable depth. There are a few versions of grillage structures with torsion stiffness for which individual members require only bending stiffness (e.g. cruciform grillages).

A degree of torsion stiffness could be restored to the faux sedan by building a torsionally stiff grillage into the floor. This is not very weight efficient. The result is called a 'semi-open' structure since structurally it behaves like an open ('convertible') car structure of the 'punt' type, the upper structure contributing little to the torsional performance. See the discussion of convertible and punt structures in Chapters 16.1.

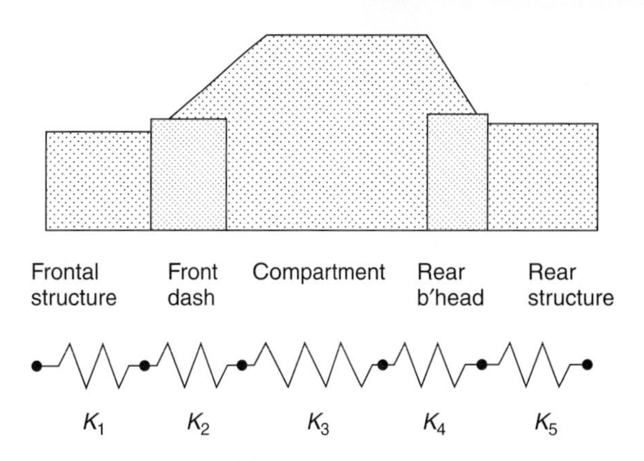

Fig. 16.2-22 Structures in series.

16.2.3.7.3 Effect of one poor subassembly on overall torsion stiffness

For the torsion case, the vehicle behaves as a set of five subassemblies in series in torsion as shown in Fig. 16.2-22: (1) frontal structure, (2) dash, (3) passenger compartment, (4) rear bulkhead/hat shelf, (5) rear structure.

For structures in series, the overall torsion stiffness K is given by:

$$1/K = 1/K_1 + 1/K_2 + 1/K_3 + 1/K_4 + 1/K_5$$

where K_1 to K_5 are the stiffnesses of the assemblies listed above.

In such a series, the overall flexibility $(1/K)$ is dominated by any member that has very low stiffness. For example, consider a vehicle where K_1 to K_5 all have the value 50 000 Nm/deg. Thus in this case:

$1/K = 5/50\,000$, hence overall torsion stiffness: $K = 10\,000$ Nm/deg.

Now suppose that the dash has reduced torsional stiffness, $K_2 = 5000$ Nm/deg. This might be because of a faulty load path (for example, insufficiently stiff parcel shelf). The causes of poor dash performance in torsion are very similar to those in the vehicle bending case (see discussion in section 16.2.2.5). The overall torsion stiffness of the vehicle is now given by:

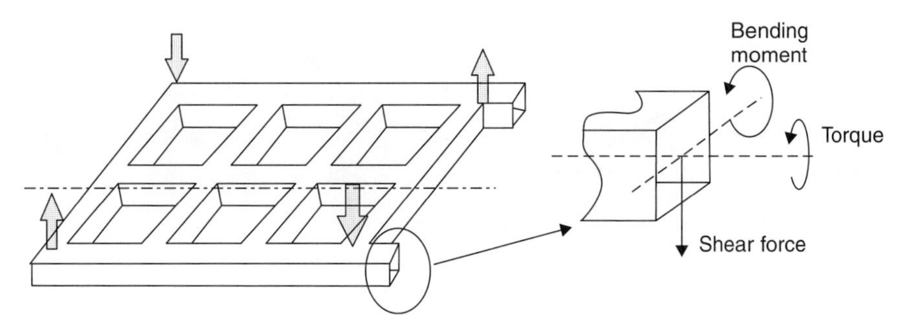

Fig. 16.2-21 Grillage structure.

$$1/K = 1/K_2 + (1/K_1 + 1/K_3 + 1/K_4 + 1/K_5)$$
$$= 1/5000 + 4/50\,000$$

Thus: $K = 3571$ Nm/deg.

The overall torsion stiffness has been reduced to about 36% of the original value. It is thus of paramount importance that all of the subassemblies have correct load-path design and that the connections between them are structurally sound.

16.2.4 Lateral loading case

When a vehicle travels on a curved path lateral forces are generated due to centrifugal acceleration. Inertia forces act at the centres of mass of the components which tend to throw them away from the centre of turn. These are balanced by lateral forces generated at the tyre to ground contact points which are transferred to the structure of the vehicle through the suspension. This condition is illustrated in Fig. 16.2-23 where the vehicle is considered moving forward and turning to the right.

The inertia forces on the power-train, the front and rear passengers and the luggage are shown acting towards the left of the vehicle. The balancing side forces are shown as R_{YF} and R_{YR} acting at the front and rear axles, respectively. As the lateral forces at the centres of mass act above the floor line these produce a rolling moment that is balanced by vertical loads R_{ZYF} and R_{ZYR} at the front and rear suspension mountings. From Fig. 16.2-23, it should be noted that these forces act downward on the right-hand side and upward on the left-hand side. These forces are sometimes known as the weight transfer due to cornering. It is important to realize that these are forces which act in *addition* to the vertical forces shown in Fig. 16.2-2. In this section only these forces are considered and they should then be added to the force system analysed in section 16.2.2.

16.2.4.1 Roll moment and distribution at front and rear suspensions

Taking moments about the vehicle centreline (see Fig. 16.2-24) at the plane of the floor to obtain: roll moment

$$M_R = F_{ypt}h_{pt} + F_{ypf}h_{pf} + F_{ypr}h_{pr} + F_{yl}h_l$$

This is the moment due to the forces acting through the centres of mass. This is balanced by the vertical reactions at front and rear suspension mounting points:

$$M_R = R_{ZYF}t_f + R_{ZYR}t_r$$

There are now unknowns, R_{ZYF} and R_{ZYR}, and only one equation. A ratio must be assumed between these two unknowns. The roll stiffness of the front suspension of most cars is greater than the roll stiffness of the rear suspension and the body roll angle at front and rear is assumed equal (the body is very stiff compared to the suspension), therefore it will be assumed that the front suspension provides nM of the roll moment, where n is in the range 0.5 – 0.7.

Therefore at the front suspension mounting:

$$R_{ZYF} = nM_R/t_f$$

and at the rear suspension mounting:

$$R_{ZYR} = (1-n)M_R/t_r$$

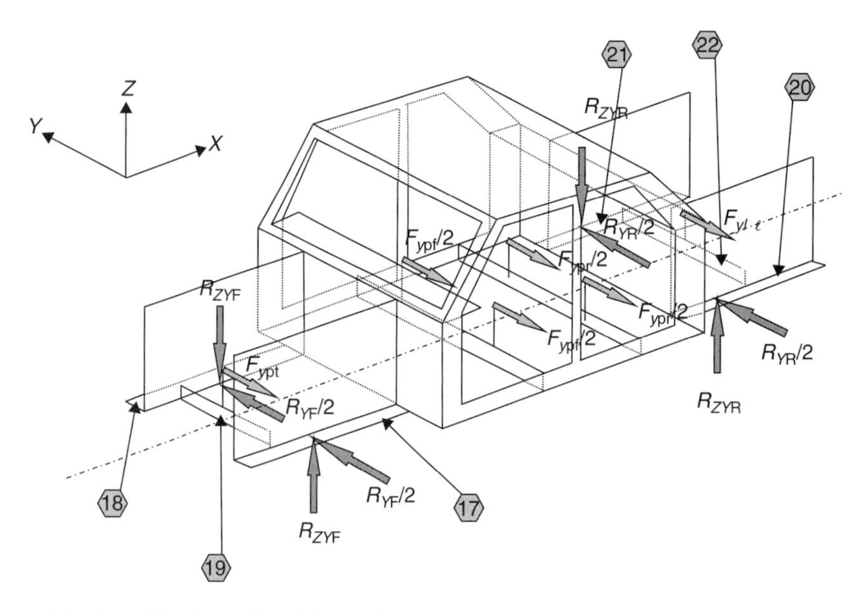

Fig. 16.2-23 Baseline model – lateral loading with additional SSSs.

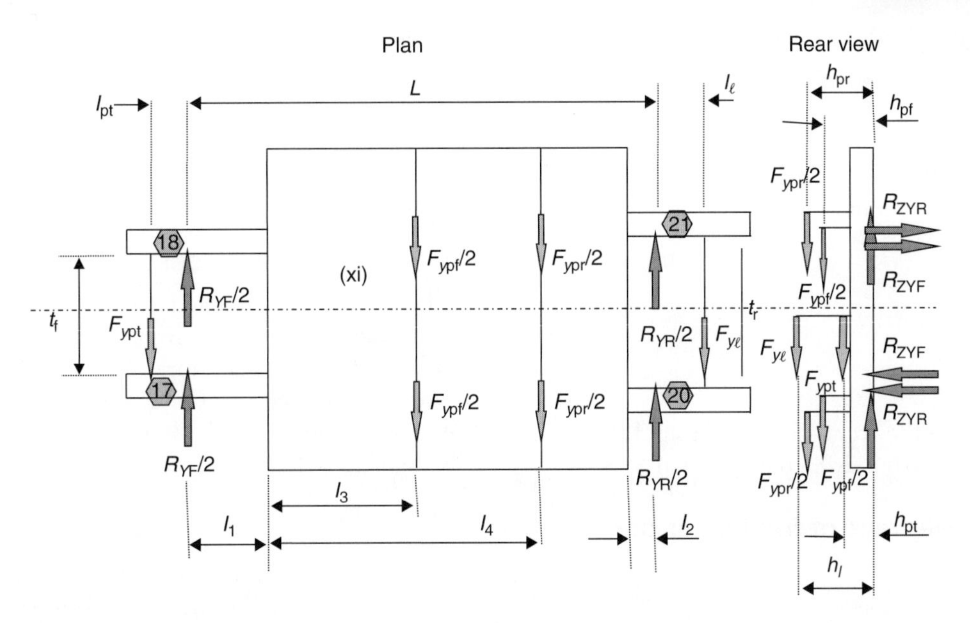

Fig. 16.2-24 Baseline model – lateral loading.

Returning to Fig. 16.2-24 and the plan view, take moments about the front suspension:

$$R_{YR} = \{F_{ypf}(l_1 + l_3) + F_{ypr}(l_1 + l_4) + F_{yl}(L + l_l) - F_{ypt}l_{pt}\}/L$$

Resolving lateral forces:

$$R_{YF} = \{F_{ypt} + F_{ypf} + F_{ypr} + F_{yl}\} - R_{YR}$$

16.2.4.2 Additional SSSs for lateral load case

As the lateral forces act through the centres of mass of the power-train and the luggage and are in front of and behind the centre floor additional SSSs (17) to (22) are required to those shown in Fig. 16.2-1 and are shown in Fig. 16.2-23. SSS (19) will transfer the power-train force as vertical forces in the planes of the front inner wing panels and it will be assumed the lateral force is shared equally between SSS (17) and SSS (18) (see Fig. 16.2-24). Similar conditions are assumed to act at the luggage floor beam. As the beams (19) and (22) cannot take moments about the vehicle z-axis, the beams (17), (18), (20) and (21) act as simple cantilevers protruding forward and rearward from the central floor.

Consider the cross-beams all shown in Fig. 16.2-25.

16.2.4.2.1 Engine beam (19)

By taking moments about one end

$$P'_{14} = F_{ypt}h_{pt}/t_f \tag{16.2.28}$$

Resolve forces laterally

$$P'_{15} = F_{ypt}/2 \tag{16.2.29}$$

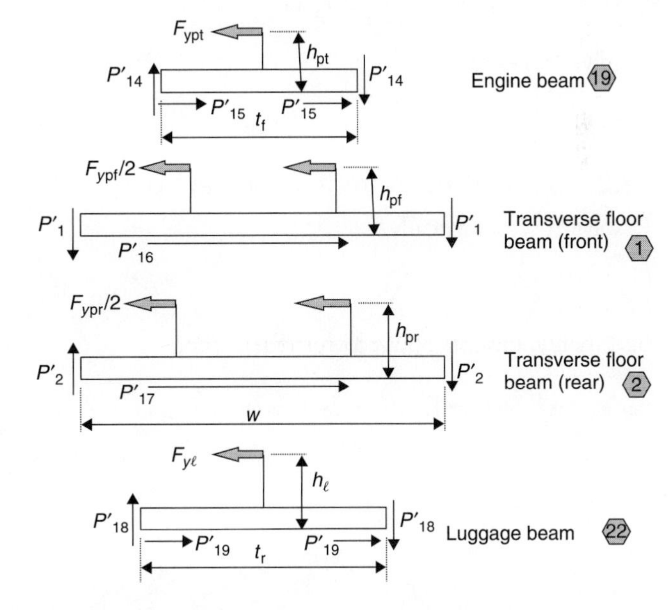

Fig. 16.2-25 Baseline model cross-beams – lateral loading.

16.2.4.2.2 Transverse floor beam (front) (1)

By taking moments about one end

$$P'_1 = F_{ypf}h_{pf}/w \tag{16.2.30}$$

Resolving forces laterally

$$P'_{16} = F_{ypf} \tag{16.2.31}$$

16.2.4.2.3 Transverse floor beam (rear) (2)

By taking moments about one end

$$P'_2 = F_{ypr}h_{pr}/w \tag{16.2.32}$$

Resolving forces laterally

$$P'_{17} = F_{ypr} \quad (16.2.33)$$

16.2.4.2.4 Luggage beam (22)

By taking moments about one end

$$P'_{18} = F_{yl}h_l/t_r \quad (16.2.34)$$

Resolving forces laterally

$$P'_{19} = F_{yl}/2 \quad (16.2.35)$$

Now moving to the front structure around the engine compartment (Fig. 16.2-26), consider each SSS.

16.2.4.2.5 Lower rails of front inner panels (17 and 18)

By taking moments about the rear end

$$M_1 = R_{YF}l_1/2 - P'_{15}(l_1 + l_{pt}) \quad (16.2.36)$$

Resolve forces laterally

$$P'_{20} = R_{YF}/2 - P'_{15} \quad (16.2.37)$$

16.2.4.2.6 Left-hand front inner wing panel (3)

Resolve forces vertically

$$P'_3 = R_{ZYF} - P'_{14} \quad (16.2.38)$$

By taking moments about lower rear corner

$$P'_4 = \{R_{ZYF}l_1 - P'_{14}(l_1 + l_{pt})\}/h_1 \quad (16.2.39)$$

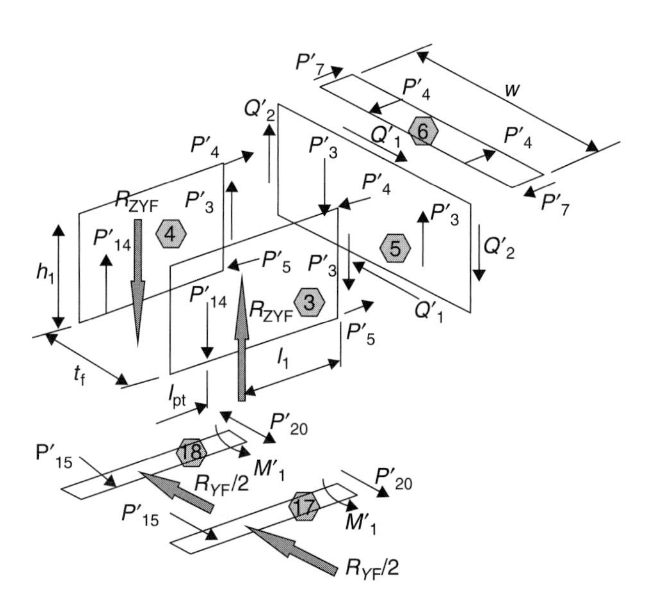

Fig. 16.2-26 Baseline model (front structure) – lateral loading.

Resolve forces horizontally

$$P'_5 = P'_4 \quad (16.2.40)$$

The right-hand front inner wing panel has similar load values but all are in the opposite sense giving similar equations.

16.2.4.2.7 Front parcel shelf (6)

This member is attached to the dash panel and the windscreen frame at the front edge but with no attachment at the rear. Therefore, no lateral loads can be applied. The forces from the inner wing panels P'_4 are reacted by the forces P_7' acting on the sideframe.
By taking moments

$$P'_7w - P'_4t_f = 0 \quad (16.2.41)$$

The rear structure is shown in Fig. 16.2-27 and has similar loading conditions to the front structure.

16.2.4.2.8 Lower rails of rear quarter panels (20 and 21)

Resolve forces laterally

$$P'_{21} = R_{YR}/2 - P'_{19} \quad (16.2.42)$$

By taking moments about the front end

$$M'_2 = R_{YR}l_2/2 - P'_{19}(l_2 + l_l) \quad (16.2.43)$$

16.2.4.2.9 Left-hand rear quarter panel (7)

Resolve forces vertically

$$P'_8 = R_{ZYR} - P'_{18} \quad (16.2.44)$$

By taking moments about lower front corner

$$P'_9 = \{R_{ZYR}l_2 - P'_{18}(l_2 + l_1)\}/h_2 \quad (16.2.45)$$

Resolve forces horizontally

$$P'_{10} = P'_9 \quad (16.2.46)$$

The right-hand rear quarter panel (8) has opposite forces acting on it and these will produce identical equations.

16.2.4.2.10 Rear parcel shelf (10)

Similar to the front parcel tray, no lateral forces act on this member.
By taking moments

$$P'_{12}w - P'_9t_r = 0 \quad (16.2.47)$$

By working through the equations (16.2.28) – (16.2.47) in the sequence shown, all the unknowns (i.e. $P'_1 \ldots P'_5$, $P'_7 \ldots P'_{10}$, P'_{12}, $P'_{14} \ldots P'_{21}$, M'_1 and M'_2) can be evaluated.

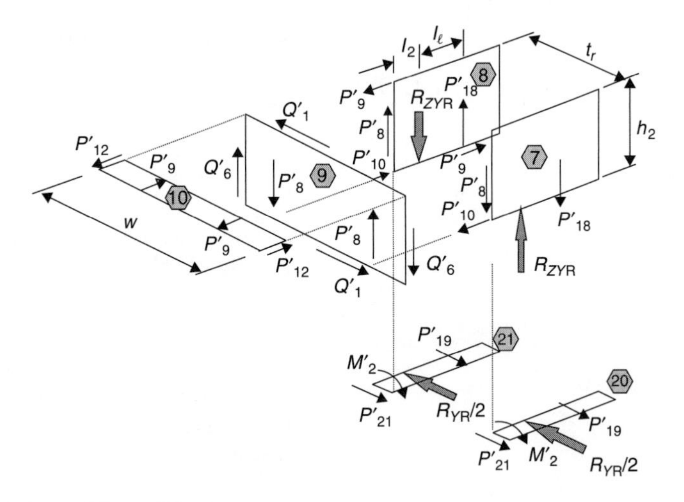

Fig. 16.2-27 Baseline model (rear structure) – lateral loading.

To evaluate the forces in the passenger compartment shown in Fig. 16.2-28 consider the seven SSSs forming the 'torsion box'. The right-hand sideframe is not shown as its loading condition is exactly opposite to the left-hand sideframe. It should be noted that the moments applied into the 'torsion box' via the dash panel (5) and the panel behind the rear seats (9) are not in the opposite sense as in the torsion case described in section 16.2.3. These two moments in this case are unequal and in the same sense, because they both contribute to balancing the moments due to the moments applied through the floor cross-beams. The effect of these unequal moments is to apply shear to all the SSSs forming the 'torsion box'.

16.2.4.2.11 Dash panel (5)

The loads P'_3 from the inner wing panels produce a moment that is balanced by edge loads at the sides, top and bottom. These loads Q'_1 and Q'_2 are assumed to act as shown.

Moment equation

$$Q'_1 h_1 + Q'_2 w - P'_3 t_f = 0 \qquad (16.2.48)$$

16.2.4.2.12 Windscreen frame (14)

This member must react the edge load Q'_1 from the dash panel and is held in equilibrium by the complementary shear forces Q'_3.

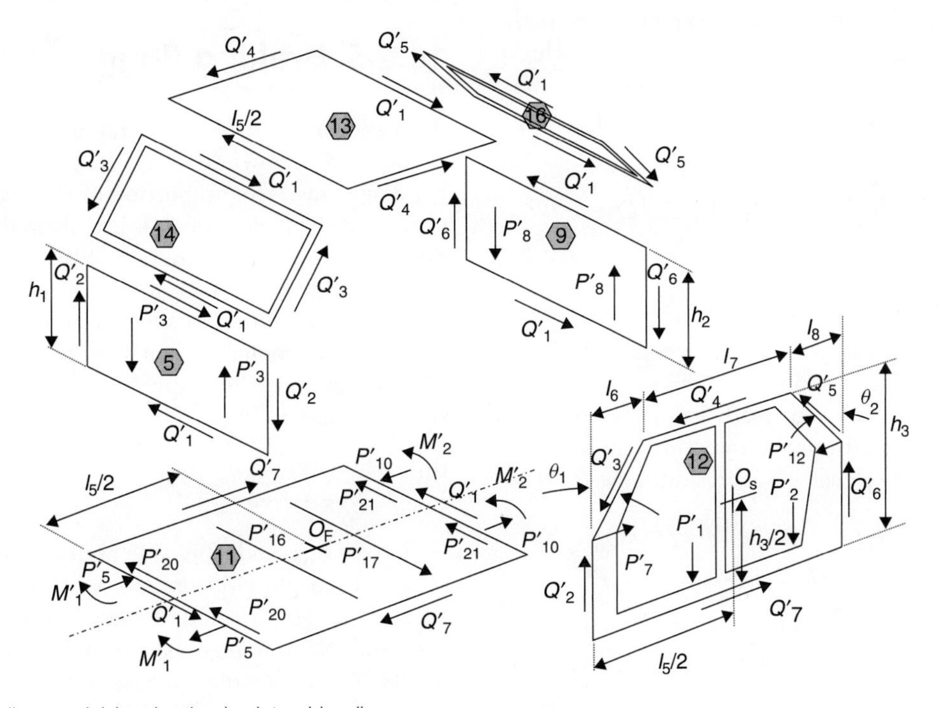

Fig. 16.2-28 Baseline model (torsion box) – lateral loading.

Moment equation

$$Q_1'(l_6^2 + (h_3 - h_1)^2)^{0.5} - Q_3'w = 0 \quad (16.2.49)$$

16.2.4.2.13 Roof (8)

This again is in complementary shear.

Moment equation

$$Q_1'l_7 - Q_4'w = 0 \quad (16.2.50)$$

16.2.4.2.14 Backlight frame (16)

Once again this is in complementary shear.

Moment equation

$$Q_1'(l_8^2 + (h_3 - h_2)^2)^{0.5} - Q_5'w = 0 \quad (16.2.51)$$

16.2.4.2.15 Panel behind the rear seats (9)

Loads P'_8 applied from the rear quarter panels apply a moment to this panel. Edge load Q'_1 is applied to the top edge from the backlight frame, therefore an equal and opposite force Q'_1 acts at the lower edge. All these forces apply moments in the same sense, therefore to maintain equilibrium Q'_6 at each side must act in the directions shown.

Moment equation

$$Q_6'w - Q_1'h_2 - P_8't_r = 0 \quad (16.2.52)$$

16.2.4.2.16 Floor panel (11)

There are a large number of loads acting on this SSS. Shear loads, tension/compression loads and moments act on the front and rear edges plus shear loads from the front and rear transverse floor beams. In order to achieve equilibrium, loads Q'_7 are required at the sides.

Moments about the centre of the floor (O_F)

$$(2P'_{20} - Q_1')l_5/2 + P_5't_f + 2M_1' - (2P'_{21} + Q_1')l_5/2 - 2M_2' - P'_{10}t_r + P_{16}(l_3 - l_5/2) + P_{17}(l_4 - l_5/2) + Q_7'w = 0 \quad (16.2.53)$$

16.2.4.2.17 Left-hand sideframe (12)

The forces acting on this SSS are shown (Fig. 16.2-28). These are the equal and opposite reactions to the loads on the previous six SSSs plus loads from the front and rear parcel shelves and from the transverse floor beams.

Moments about the centre of the sideframe (O_S)

$$Q_2'l_5/2 + P_1'(l_3 - l_5/2) + P_2'(l_4 - l_5/2) - Q_6l_5/2) + P_7'(h_1 - h_3/2) - P'_{12}(h_2 - h_3/2) - Q_4h_3/2 - Q_7h_3/2 - Q_3'(h_3/2)\sin\theta_1 - Q_3'(l_5/2 - l_6)\cos\theta_1 - Q_5'(h_3/2)\sin\theta_2 - Q_5'(l_7 + l_6 - l_5/2)\cos\theta_2 = 0 \quad (16.2.54)$$

Equations (16.2.48)–(16.2.54) are seven simultaneous equations with seven unknowns $Q'_1 \ldots Q'_7$, so their solutions can be obtained by standard mathematical methods. Equation (16.2.54) can be simplified by taking moments about the lower corner of the windscreen pillar. This will eliminate terms in Q'_2, Q'_3 and P'_7 but the numerical solutions will be unchanged. These equations provide the values of all the forces on the SSSs of the 'torsion box'. When evaluating these forces accidental errors may occur, therefore it is *imperative* that the sideframe equilibrium is verified. This is done by resolving forces vertically and horizontally using the following equations:

Resolve forces vertically

$$Q_2' - Q_3'\cos\theta_1 + Q_5'\cos\theta_2 + Q_6' - P_1' - P_2' = 0 \quad (16.2.55)$$

Resolve forces horizontally

$$P_7' - P'_{12} + Q_7' - Q_3'\sin\theta_1 - Q_4' - Q_5'\sin\theta_2 = 0 \quad (16.2.56)$$

Provided these equations are satisfied, the correct force conditions through the structure have been determined.

It should be observed that lateral loads cause additional bending on the front inner wing panels, rear quarter panels, parcel shelves and shear on the dash panel, windscreen frame, roof, backlight frame, and seat panel. The main floor and the sideframes have both additional shear and bending. These additional loads must be added by superposition to those obtained in section 16.2.2.

16.2.5 Braking (longitudinal) loads

A passenger car subject to braking conditions has additional loads shown in Fig. 16.2-29 over the normal bending loads. The proportion of the total braking force applied at the front wheels is usually in the order of 50 to 80%. This is due to the design of the braking system and because additional vertical load R_{ZXF} is applied to the front wheels. Modern braking systems may have variable brake proportioning to suit the decelerating condition. For this analysis the assumed proportion of braking on the front axle is:

$$n = \frac{R_{XF}}{R_{XF} + R_{XR}} \quad (16.2.57)$$

Forces R_{XF} and R_{XR} are applied at the front and rear tyre to ground contacts which are h_f below the base of the structure. The load transfer onto the front axle and off the rear axle is obtained by taking moments about the rear tyre/ground contact point (see Fig. 16.2-30):

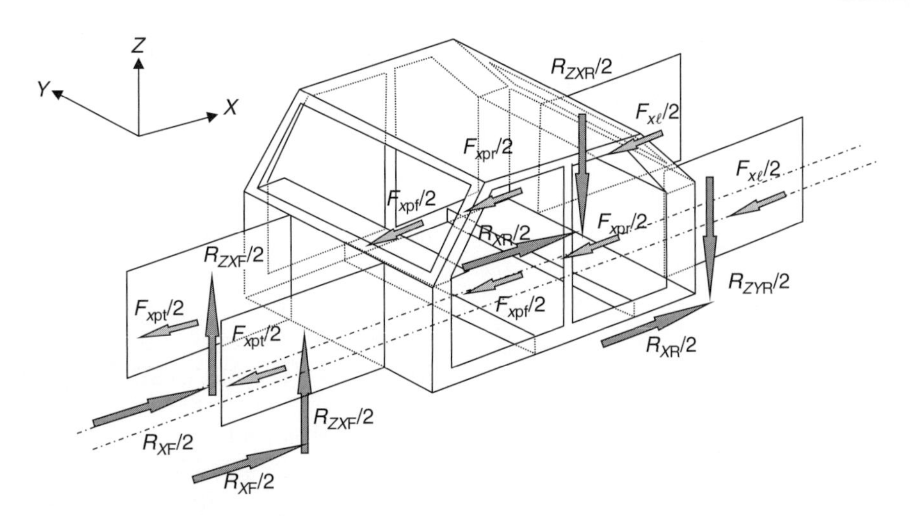

Fig. 16.2-29 Baseline model – braking loads.

$$\begin{aligned} R_{ZXF} &= R_{ZXR} \\ &= \{F_{xpt}(h_{pt} + h_f) + F_{xpf}(h_{pf} + h_f) \\ &\quad + F_{xpr}(h_{pr} + h_f) + F_{xl}(h_l + h_f)\}/L \end{aligned} \tag{16.2.58}$$

The loading on the front structure (Fig. 16.2-31) is now known so applying the equations of statics to each of the SSSs.

16.2.5.1 Inner wing panels (3) and (4)

The inertia forces from the power-train $F_{xpt}/2$ are assumed to act into each of the inner wing panels.

Resolve forces vertically $R_{ZXF}/2 - P''_3 0$ (16.2.59)

Resolve forces horizontally

$$R_{XF}/2 - F_{xpt}/2 - P''_4 + P''_5 = 0 \tag{16.2.60}$$

Moments about the lower rear corner

$$R_{ZXF}l_1/2 - F_{xpt}h_{pt}/2 - R_{XF}h_f/2 - P''_4 h_1 = 0 \tag{16.2.61}$$

Hence P''_3, P''_4 and P''_5 are obtained.

16.2.5.2 Dash panel (5)

Resolving forces vertically and by symmetry

$$2P''_6 - 2P''_3 = 0 \tag{16.2.62}$$

16.2.5.3 Front parcel shelf (6)

Resolving forces horizontally

$$2P''_7 - 2P''_4 = 0 \tag{16.2.63}$$

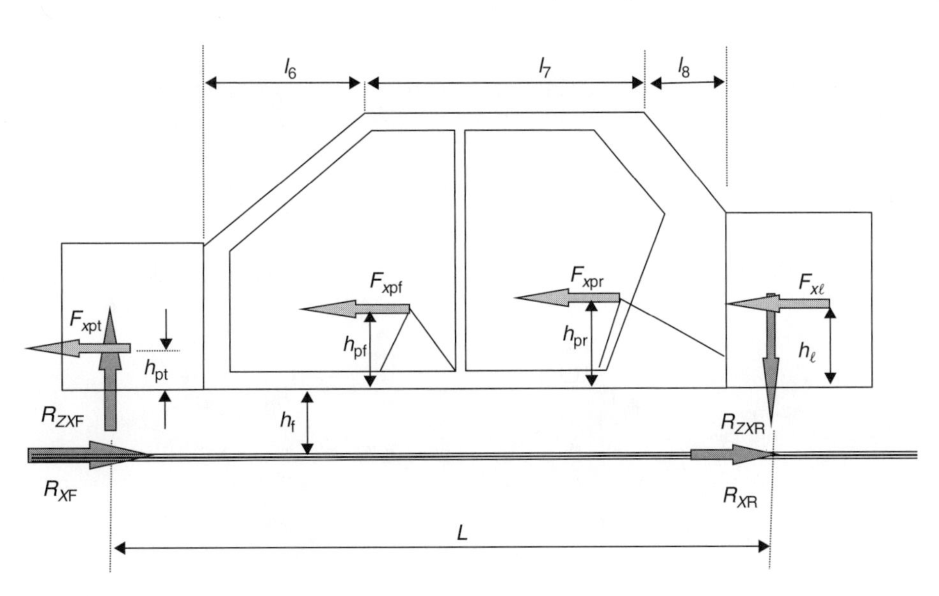

Fig. 16.2-30 Baseline model (side view) – braking loads.

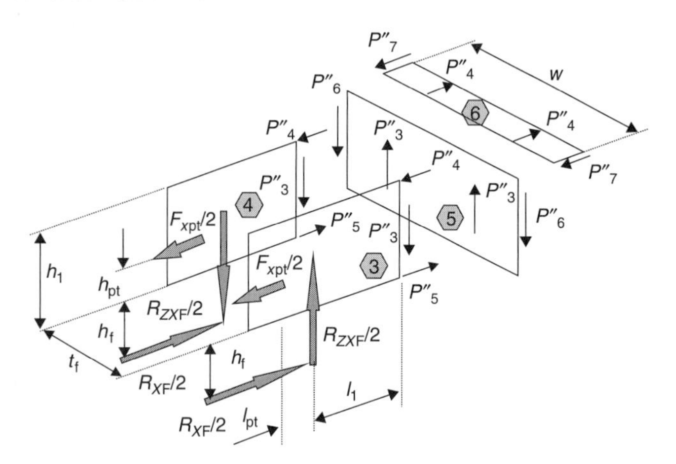

Fig. 16.2-31 Baseline model (front structure) – braking loads.

The rear structure shown in Fig. 16.2-32 is loaded such that the suspension and luggage loads all act on the rear quarter panel.

16.2.5.4 Rear quarter panels (7) and (8)

Resolving forces vertically

$$P''_8 - R_{ZXR}/2 = 0 \tag{16.2.64}$$

Resolving forces horizontally

$$P''_{10} + R_{XR}/2 - P''_9 - F_{xl}/2 = 0 \tag{16.2.65}$$

Moments about front lower corner

$$R_{ZXR}l_2/2 - R_{XR}hf/2 - F_{xl}h_l/2 - P''_9 h_2 = 0 \tag{16.2.66}$$

16.2.5.5 Panel behind rear seat (9)

This member provides reaction for the shear force P''_8 and an additional force P''_{26}. P''_{26} is caused by the inertia load of the rear passengers

$$P''_{26}(l_5 - l_4) - F_{xpr}h_{pr}/2 = 0 \tag{16.2.67}$$

Resolving forces vertically and by symmetry

$$P''_{11} + P''_{26} - P''_8 = 0 \tag{16.2.68}$$

16.2.5.6 Rear parcel shelf (10)

Resolving forces horizontally

$$2P''_{12} - 2P''_9 = 0 \tag{16.2.69}$$

16.2.5.7 Transverse floor beam (rear) (2)

As the inertia load from the rear seat passengers acts at height h_{pr} the vertical load on this member is:

$$P''_{25} = F_{xpr}h_{pr}/2(l_5 - l_4) \tag{16.2.70}$$

16.2.5.8 Transverse floor beam (front) (1)

Again the inertia load from the front passengers acts at a height h_{pf} resulting in a moment about the floor plane. In this situation an extra floor beam is required and so two front floor beams as shown with dashed lines in Fig. 16.2-33 are used to replace the original one:

$$P''_{22} = P''_{23} = F_{xpf}h_{pf}/2l_{10} \tag{16.2.71}$$

16.2.5.9 Floor panel (11)

The inertia forces from the front and rear passengers are transferred as shear to the floor:

$$P''_{24} = F_{xpf}/2 \tag{16.2.72}$$

$$P''_{27} = F_{xpr}/2 \tag{16.2.73}$$

Resolving forces horizontally and by symmetry

$$P''_{13} = P''_5 + P''_{10} + P''_{24} + P''_{27} \tag{16.2.74}$$

Finally, once again check the equilibrium of the side-frame by the three equations of statics.

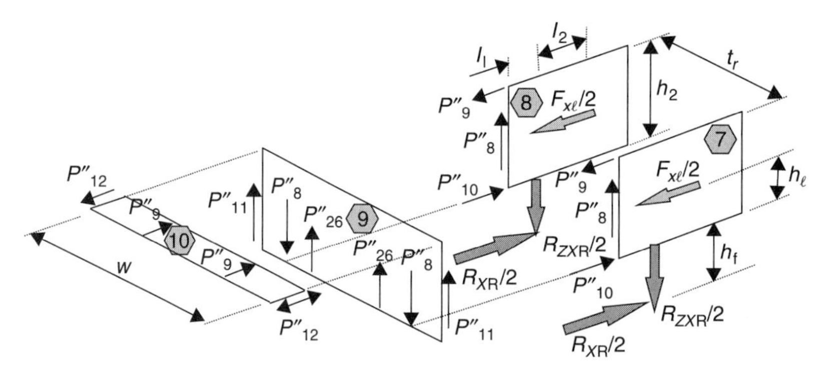

Fig. 16.2-32 Baseline model (rear structure) – braking loads.

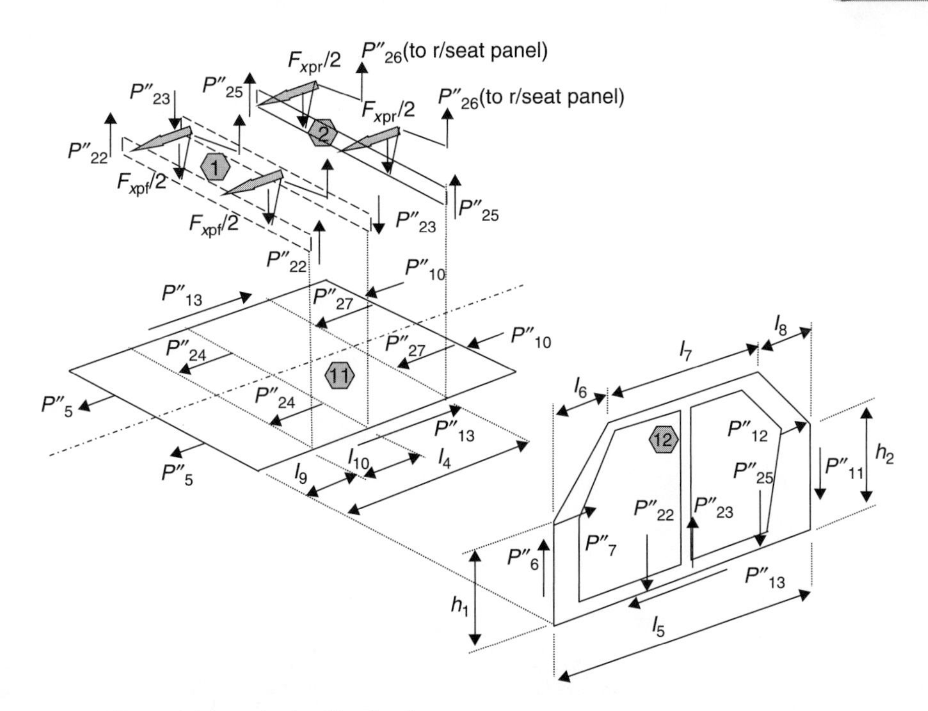

Fig. 16.2-33 Baseline model (floor, sideframe) – braking loads.

Resolving forces horizontally

$$P''_7 + P''_{12} - P''_{13} = 0 \tag{16.2.75}$$

Resolving forces vertically

$$P''_6 - P''_{11} - P''_{22} + P''_{23} - P''_{25} = 0 \tag{16.2.76}$$

Moments about lower front corner

$$P''_7 h_1 + P''_{22} l_9 - P''_{23}(l_9 + l_{10}) + P''_{25} l_4 + P''_{11} l_5 + P''_{12} h_2 = 0 \tag{16.2.77}$$

16.2.6 Summary and discussion

The main load conditions that act on a typical passenger car structure have been considered. The important cases that are the most straightforward to model are the bending and torsion cases. Other cases such as the cornering and the braking have also been modelled. Further cases such as the acceleration or tractive force case and the one wheel bump/pot-hole can also be modelled although not detailed here. The acceleration case is similar to the braking case except that the tractive force is only applied to either the front or the back wheels, not both, unless the vehicle is a four wheel drive vehicle. For the four wheel drive vehicle the proportion of tractive force on front and rear wheels, like in the braking case, will not necessarily be equal.

Although only a 'standard sedan' has been considered the analysis contained in this chapter has revealed that when considering the various load cases the SSSs that are required are quite numerous. In all of the cases considered care must be taken to ensure that there are suitable SSSs to carry the applied loads from one component to another. It should be noted that in the torsion case it is easy to find that insufficient SSSs have been specified as in the 'faux' sedan. Likewise the need for additional SSSs to carry lateral loads should be noted.

In modelling a structure when it is found that it is not reasonable to represent a component with an SSS and that the loads are not satisfactorily transferred through the structure then a weakness in the structure is revealed. This is one of the main advantages of the SSS method – it is useful in revealing if the structure has adequate load paths.

The load conditions on the main structural components have also been determined by the methods described in this chapter. The bending moments, shear forces, distributed edge loads and joint loads can be evaluated using the equations developed. In cases such as the A-pillar (windscreen side), the loading is sufficiently detailed that stress calculations can follow directly from this analysis.

The detailed model of any particular vehicle will require the judgement of the structural engineer. In an earlier work it was shown that the modelling of a particular vehicle has rather different SSSs and this will indicate the care that is necessary in modelling a structure. Reversing that procedure the SSS model can be used in developing the design of a structure. Having obtained suitable load paths through SSSs the designer can then detail subassemblies and components that have the required structural properties.

Section Seventeen

Vehicle safety

Chapter 17.1

17.1

Vehicle safety

T.K. Garrett, K. Newton and W. Steels

This chapter covers safety in the context of design for the avoidance of accidents (active safety), and for the protection of the occupants and pedestrians from serious injury if they are involved in accidents (passive safety). Active safety, which is obtained by optimising braking, ride, road holding, steering and handling in general, will be covered in Sections 17.1.12 *et seq*. Passive safety can be sub-divided into two categories: safety for the occupants and safety for pedestrians struck by the car.

As regards pedestrian safety, the design of the front end is all important. Ideally, the front would be a vertical plane surface with reduced stiffness at shoulder and hip level, so that it will do nothing worse than bruise pedestrians if the car strikes them. Clearly, however, this is impracticable, although many of the so-called people mover types of car go a long way towards the achievement of this ideal. Currently, bumper height is fixed by law, so the best that can be done to a conventional saloon car is to bring the front skirt panel forward, increase its stiffness and cover both the skirt and bumper with shock absorbent material.

The stiffness of all shock absorbent material in crash safety-related applications is critical: if too soft, the part of the human body involved will crash right through it on to the underlying structure; if too hard, its shock absorption capability will be reduced. Bumper stiffness involves a difficult compromise: if too stiff it breaks bones and, if too soft, it fails to perform its function of energy absorption.

Above the bumper, the panelling again should be a flat and almost vertical surface, to support the thigh and hip of the victim. The radius of transition from the front panel to the front edge of the bonnet should be large enough to avoid breakage of hip and thigh bones. Also, its height should be such as to ensure that the person struck is rolled over on to the bonnet. The victim will then tend to slide towards, and hit, the windscreen, the angle of which should be steep enough to prevent him from being deflected up over the roof of the saloon. This is because serious injury could occur if the victim were to strike the hard surface of the road behind, where he might be run over by a following vehicle. As he hits the laminated glass windscreen, the plastics inter-layer will cushion the shock although, nevertheless, there is risk of injury from slivers of glass.

Passive safety has been a preoccupation of many vehicle manufacturers for several decades. As regards actual legislation, however, the USA has so far been far ahead of Europe, even covering in the early 1950s details such as the avoidance of injury by protruding hardware both inside and outside the vehicle. Indeed, in Europe, the only safety legislation actually in force between 1947 and 1998 was initially that limiting the deflection of the steering wheel to 127 mm rearwards and then, more than ten years later, upwards too, as the car hit a rigid barrier at 30 mph (48.28 km/h). Legislation coming into effect from 1996 onwards, in Europe and the USA, is outlined in Table 17.1-1.

Standards have been applied also for the testing of various components. Among these are the steering column and wheel, dash fascias, seats, seat belts and their anchorages.

17.1.1 Crash testing

Not only have costly cars to be tested to destruction but also the equipment with which they are tested is costly too. Indeed, for test facilities alone, the investment required is about £50 million. Consequently, many manufacturers all over the world sub-contract some or all of their testing. The Motor Industry Research Association

The Motor Vehicle, 13th edn; ISBN: 9780750644495
Copyright © 2000 Elsevier Ltd; All rights of reproduction, in any form, reserved.

Table 17.1-1

Year	Europe	USA
1996	Directive 74/297/EEC	FMVSS 208 Occupant crash protection: frontal, lateral, dynamic roll-over. FMVSS 301 Fuel system integrity: frontal, lateral, rear impacts FMVSS 214 Side impact protection
Jan. 1997	Steering wheel intrusion	FMVSS 208 Temporary alternative for air bag depowering
Oct. 1998	Directive 96/79/EC (ECE 94) Front impact protection directive Directive 96/27/EC (ECE 95) Side impact protection	
Oct. 2000	Review and extension by EC of front and side impact directives Directive on pedestrian protection	
Sept. 2001	FMVSS 208 Requirements for smart air bags to be in place	
Oct. 2003	Front and side impact directives extended to cars approved before 1997	
Oct. 2005	Pedestrian protection directive extended to cars approved before Oct. 2000	

(MIRA) at Nuneaton has the advantage of being a totally independent organisation, and its resources for safety testing are claimed to be the best in the world. Furthermore, because their specialists in the various aspects of vehicle testing are constantly liaising with experts employed by its client manufacturers internationally, they are particularly advantageously placed as recipients of a vast pool of knowledge and experience, and also tend to be unfettered by the preconceived notions that one often finds in some individual manufacturing organisations.

MIRA was the first to develop a system on a CD-ROM for displaying the numerical and photographic data side by side on the monitor of a table top computer. This has the advantage that the numerical data, usually presented in graphical form, can be instantly correlated with what actually happened inside the car. Not only is this the simplest and most reliable way to understand the situation, but also it can save hours of manual correlation. From the enormous amount of data available in this laboratory, programs can be set up for crash testing by computer-aided simulation. This can save the time and enormous cost of repeated crash testing, leaving only the final proof testing to be done on the mechanical test rigs.

Another first scored by MIRA was the use of X-ray imaging of crash tests, at 1000 frames/second (fps) in 1.2×0.8 m format. As can be seen from Fig. 17.1-1, this type of imaging is especially useful for viewing regions where conventional cameras cannot access. It can show details of, for example, foot well intrusion and precisely how the legs of occupants might be injured in a crash. Although 1000 fps is the speed normally demanded for conventional video and film, MIRA has also some specialised cameras that can be run at 10 000 fps.

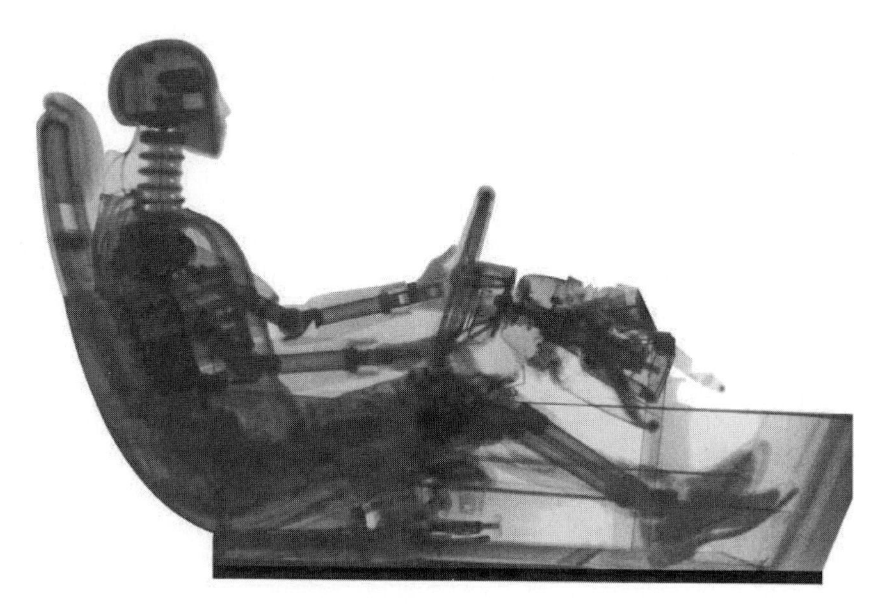

Fig. 17.1-1 MIRA has developed a method of X-ray imaging of tests at a rate of 1000 frames per second, which enables them to view regions which conventional cameras cannot acces, for example around the pedals.

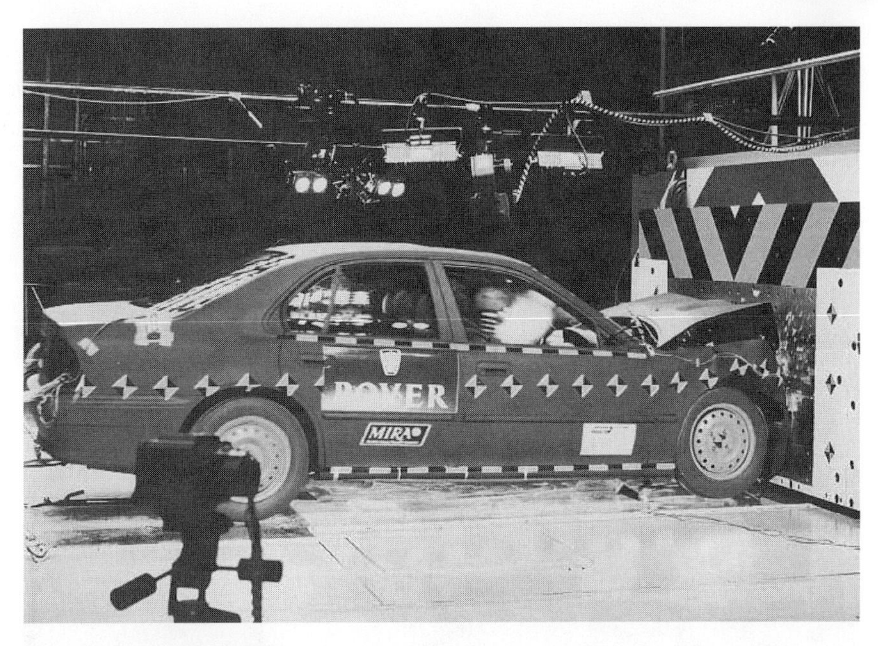

Fig. 17.1-2 For conventional frontal impact tests, the impacted face of the concrete block can be protected with steel plate to reduce the rate of erosion.

The previously mentioned European test introduced in 1974 entailed propelling the vehicle on to a rigid concrete barrier, Fig. 17.1-2, and measuring the steering wheel deflection. But, in a real crash on the road, even a brick wall at least deflects, or even disintegrates, when struck. For this reason and since vehicles rarely, if indeed ever, collide head-on into rigid concrete barriers, this test is useful only as a very rough guide to the survivability of its occupants in a severe accident.

To rectify this and other shortcomings, Directive 96/79/EEC was approved in 1996. This test is designed to assess the effects of the much more common type of crash, in which one vehicle hits another while overtaking a third. Consequently only 40% of the front end (between one side and a vertical line 10% from the centre) must impact on the block. Moreover, an aluminium honeycomb block, Fig. 17.1-3, is fixed to the face of the concrete barrier against which the car is propelled.

Fig. 17.1-3 Here a dummy front end of a car represented by an aluminium honeycomb structure is fixed to the concrete block to simulate the crushing of the other car involved in the front end impact.

Fig. 17.1-4 This aluminium honeycomb used in a side impact test is profiled to represent the vertically varying stiffnesses of the other car.

The function of the honeycomb is to simulate penetration into the front crush zone of a vehicle being struck. To represent the varying stiffnesses at different heights of the vehicle that is struck, the density of layers of the honeycomb structure is increased at the heights of the bumper and front edge of the bonnet lid, as can be seen from the set-up for side impact testing in Fig. 17.1-4.

This test calls for not only adult dummies to be belted into the front seats but also child dummies in the rear. Moreover, instead of assessing only the effects of the crash on the vehicle, those on the occupants must be measured too. Pass/fail criteria have been specified for different areas of each dummy.

Based on, but more severe than, the EEC 12 test in which a 40% offset is specified, is the NCAP test. This has been developed in the UK mainly for consumer information. For example, although the standard EEC 12 test calls for a vehicle approach speed of 57 km/h, the UK-NCAP (UK-New Car Assessment Programme) calls for 64 km/h and, since the kinetic energy of a moving mass is directly proportional to the square of the speed, this represents a significant increase in the severity of the crash.

Concerns have been expressed that although the NCAP test is basically sound, some shortcomings still remain to be obviated. For example, while a standard honeycomb is adequate for the majority of cars and light commercial vehicles, the heavier limousines tend to crash right through the honeycomb and impact on the concrete backing before rebounding, and thus are subjected to a two-stage rate of deceleration not normally experienced in practice. Another concern is that subjective assessments after the tests could lead to different engineers allocating different scores to the same car.

For the side impact test (Directive 96/27/EC), the side of the car to be tested has to be struck at a speed of 50 km/h by a mock-up representing the front end of another car. The mock-up is mounted on a sled and propelled along the track used for front end crash testing. However, this procedure is costly so, for development testing without incurring the costs associated with a full-scale test in accordance with the regulations, MIRA has developed what they term their HyGe test rig. A dummy front end can be mounted on a sled and propelled towards the side of the vehicle that is undergoing the test, Fig. 17.1-5.

This rig can be utilised also for a wide variety of other tests on either complete vehicles or sub-assemblies, or parts of vehicle structures. For the development of, for

Fig. 17.1-5 View from above of an impact test in which a sled, on which is mounted a dummy front end, is propelled towards the side of the target vehicle.

example, protection systems such as occupant-friendly trim and side air bags, sometimes termed *curtain air bags*, only the relevant part of the structure, which, of course, includes the doors, seats and dummies, need be mounted. HyGe simulation is used also for development work associated with the ECE 11 Latches and Hinges and ECE 17 Seat requirements. The relevant structures to be tested are mounted on the sled and propelled towards the rigid barrier to which can be fixed an appropriate target object. A maximum acceleration of 100*g* is obtainable, although 40*g* is more usual. The sled can be accelerated to speeds as high as 104 km/h, although 56 km/h is more commonly required (65 and 35 mph respectively).

The sled is powered by what, in effect, is a huge air gun, Fig. 17.1-6. Air is compressed up to 2760 kN/m^2 (4000 lb/in^2) into a holding cylinder. It is then suddenly released, to flow over a specially profiled firing pin, into the firing cylinder, where it forces the piston to thrust the sled towards its target. From the point at which the button is pressed to trigger the test, it is over all over within a tenth of a second!

The profiled firing pin controls the flow of air to the piston, so that its motion truly represents the impact pulse experienced in practice. Among the factors affecting the shape of the pulse required are the inertia of the whole vehicle and the reactions of not only its tyres but, in some instances, also the inertia and reactions of occupants. MIRA has a stock of 70 different pins, and makes another when ever a new pulse is required. Speeds of up to 56 mph (104.6 km/h) are attainable, although 35 mph (56.33 km/h) is more commonly demanded.

17.1.2 Protection of occupants

Manufacturers have always been more concerned with protecting the occupants of their cars than simply

Fig. 17.1-6 In the MIRA impact simulation laboratory, what is in effect a giant air gun projects the sled at speeds of up to 56 miles per hour towards its target.

passing mandatory tests. Therefore the majority undertake tests covering a wide variety of crash circumstances, Fig. 17.1-7. Speeds above 64 km/h are considered to be too high for the avoidance of fatal, or even serious, injury to be guaranteed. However, impacts at lower speeds also produce serious hazards. This is because the crush characteristics of the vehicle structure may be significantly different from those at high speeds. Furthermore, the effects of slack seat belts can be different at low speeds, and there is more time available for the occupant to accelerate before being restrained by them.

The key to meeting the legal requirements without prejudice to safety at all reasonable speeds therefore is precise control over both the structural collapse and restraint systems. Clearly, this will lead ultimately to the fitting of seat belt pre-tensioners and, possibly, even load-limiters. Furthermore, smart air bags will be needed (Section 17.1.10), the deployment of which is adjusted automatically to cater for type of crash, vehicle speed and weights of occupants.

Additional tests done by MIRA for various manufacturers include impacts between moving vehicles and poles, impacts of occupants on interior components, drop tests, air bag deployment, fire tests (ECE 34 Plastic tanks), Euro, US Federal and Australian seat belt anchorage tests, and the ECE 33 rear impact test. They also have an open air test site where tests can be done with fully fuelled vehicles impacting, for example, motorway central reservation barriers, other safety fences and items such as bridge supports, parapets, lighting columns and roadside emergency installations. Vehicle-to-vehicle crash tests can be done at closing speeds up to 160 km/h, although 113 km/h is the most commonly called for maximum speed. On this site, the only artificial elements of the tests are the dummy occupants.

17.1.3 Testing for occupant safety

For testing in general, a wide range of human dummies, termed Anthropomorphic Test Devices (ATDs) up to 14 stone (90 kg) and 6 ft 2½ in (1.89 m), is available, Fig. 17.1-8. General Motors, working in collaboration with medical institutions examining the tolerance of humans in crash situations, pioneered the development of sophisticated crash test dummies. In the USA, the

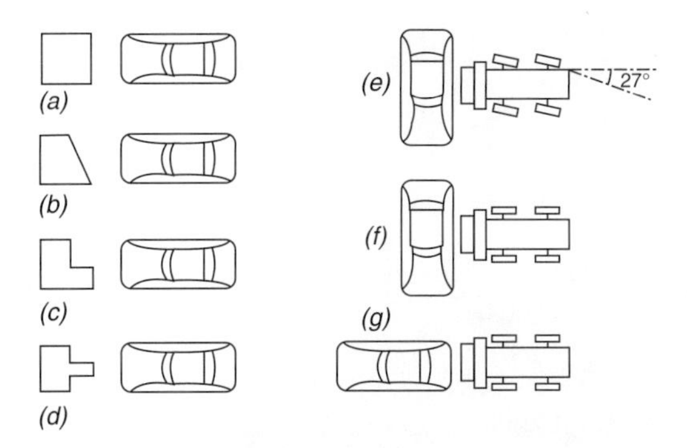

Fig. 17.1-7 Some of the tests done by manufacturers to ensure that the occupants of their vehicles will be, so far as is practicable, safe in the event of an accident. At (*a*) is the simple basic zero offset frontal impact, at (*b*) is a 30° offset, at (*c*) a 40% offset and at (*d*) a pole impact test. A side impact test for representing an impact between two vehicles moving along lines at right angles to each other is shown at (*e*) while, at (*f*), the vehicle that is struck is stationary. Finally, a rear end impact test is shown at (*g*).

earliest crash testing had been done with cadavers! The modern sophisticated dummies are particularly useful because, provided they are properly calibrated, they react consistently to crash conditions.

Anthropomorphic dummies are now available representing babies, children of different ages, and large and small adult males and females. The term anthropomorphic implies that, when subjected to impacts, their reactions and movements are representative of those of live humans. Consequently they have a steel skeleton in which the critical parts, such as the spine, necks and ribs, Fig. 17.1-1, are sprung and damped. The legs and arms articulate about their joints, like those of their human counterparts. To simulating the flesh, a foamed plastics material covers the whole of this basic structure. Bonded to it is a plastic covering the properties of which closely resemble those of human skin.

The dummies are even clothed in cotton garments so that their frictional performance on the seats and in the interiors of the vehicles closely represents what happens with real people. Their clothing is colour coded according to the position of the dummy in the vehicle. Size 11 leather shoes are fitted to protect their feet from damage! Dummies are not allowed to sit for more than about two hours, lest their foam covered buttocks compress, and thus reduce the height of the dummy in the seat. When out of use, they are stored in a closely controlled air-conditioned, humidity-regulated atmosphere.

European and US standard dummies for side impact tests differ. Therefore, MIRA has to carry in stock not only the European but also the American standard dummies. Clearly, since not only does each dummy cost about £30 000 and its instrumentation a further £30 000 but also exports of vehicles and components in both directions are involved, there is a need for harmonisation of the standards for both continents.

In the USA, two types of dummy are used for full frontal impact tests. These are Hybrid II and Hybrid III,

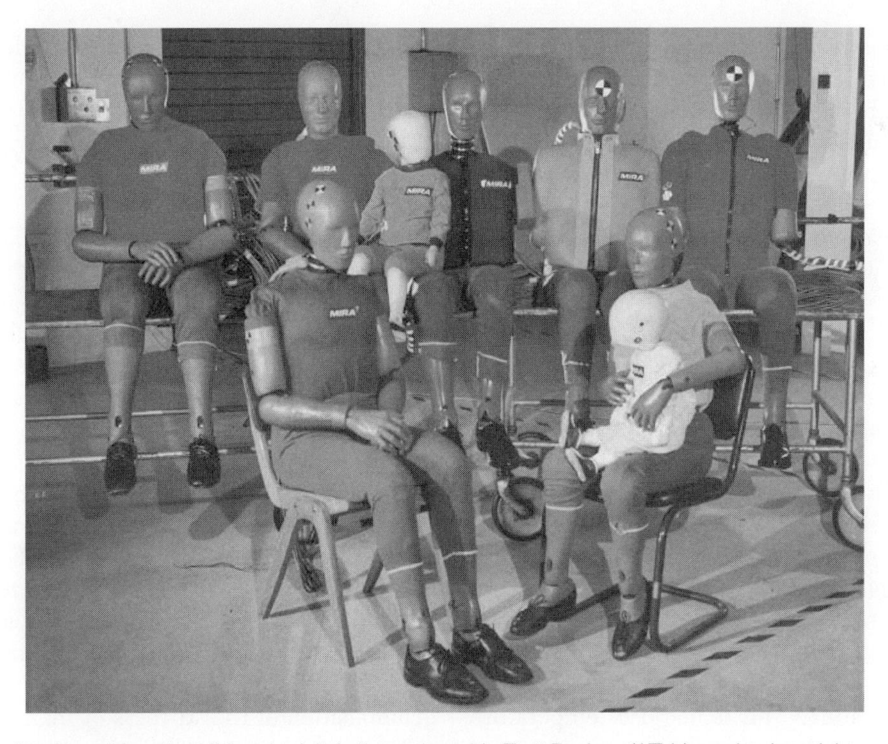

Fig. 17.1-8 A range of male and female child and adult Anthropomorphic Test Devices (ATVs) ranging in weight up to 14 stone (90 kg) is used by MIRA.

Fig. 17.1-9 Experts assessing the performance of a BMW bonnet when struck by a fully instrumented dummy head.

the latter being more sophisticated in that its lower leg can be instrumented. Both have a one-piece rib cage.

In general, both Euro and US frontal impact crash test dummies contain up to 33 transducers for measuring accelerations, displacements and forces. These include three acceleration sensors in the head, and three in the pelvis region. Also some of the dummies are equipped with as many as twenty acceleration and bending load sensors in their upper and lower legs.

Dummies used for side impact testing, SIDs, generally have either no arms or only upper arms. Euro SIDs have a spine and three damped steel ribs, all four components being equipped with acceleration sensors. In America, they have their own standard SIDs for crash tests but, for research into side impact, the SAE has developed what it terms its BioSID, which has a spine and five ribs. Among the sensors and transducers employed on all these SIDs are strain gauges and displacement sensors on each rib, to indicate deceleration, and three load cells in the abdomen.

After each test, the dummies are recalibrated at a controlled temperature. This is because temperature affects damping. Then, after every five to ten tests, they are stripped down and minutely examined for wear or damage that would affect their performance, and then recalibrated.

17.1.4 Protection of pedestrians from serious injury

Pedestrians struck by the front ends of cars are particularly vulnerable in three regions: the upper leg, the lower leg and the head. Heads of adults weigh, on average, 4.5 kg but, for the tests for satisfying the legal requirements, dummy heads weighing 4.8 kg are employed. For children, a dummy head weighing 2.5 kg is employed although, of course, in practice the actual weight depends on the age of the child. Similarly, both the height of initial impact on the leg and the precise nature of the wrap-round after impact depends on the height of the person struck. Wrap-round determines how far along the bonnet lid the head impacts: tests are usually done in zones 1000 mm, 1500 mm and 2000 mm from the front edge, Fig. 17.1-9. This, in turn, defines the area over which the combination of the stiffness of the bonnet lid and clearance beneath it must be adequate to prevent the head from deflecting the lid far enough to strike the engine or

Fig. 17.1-10 The Head Impact Test System (HITS) rig used by MIRA for assessing the occupant-friendliness of interior components and trim.

any other hard component. On the other hand, the lid must not be too unyielding, or it will itself cause injury.

To assess the effectiveness of lids in decelerating the heads gently enough to avoid brain damage, the dummy heads are fully instrumented with triaxial accelerometers. These indicate the rates of deceleration experienced by the dummy heads when they are propelled at predetermined speeds, by an air-actuated gun, at the lids. The same dummies, similarly propelled, are also employed to determine the forces to which heads of the occupants of the car are subjected when they strike windscreens or any other parts of the vehicle structure, Fig. 17.1-10.

Propulsion by air-actuated gun is also used to determine the deceleration and bending forces to which the upper (femural) and lower part of the leg are subjected in an accident. In this case, a dummy of only the part of the leg to be tested is propelled by the gun against the bumper. It comprises an instrumented steel core representing the relevant bone structure, covered with a special foam plastic to represent the flesh. The steel cores are instrumented for assessing the shear, bending and acceleration forces, while secured to their ends are masses for simulating the reactions of the parts of the body to which they would, in real life, be connected.

17.1.5 Active safety

Whereas passive safety entails introducing measures for protecting those involved from injury when an accident occurs, active safety entails, in effect, rendering the vehicle inherently safe before it occurs. If the ride is good and the road holding is such that the wheels never lose contact with the ground, the driver will more easily be able to maintain control in difficult circumstances, including in an emergency. Too little roll can give the driver a false sense of security when cornering, while too much could lead to instability.

As regards handling, there should be no sudden changes in either steering or braking characteristics. For example, some slight oversteer may be desirable, but it should not occur suddenly, nor should there be a change from under to oversteer, or *vice versa*, while the car is cornering. Tyres too play a major part. In extreme circumstances they will inevitably break away into a slide, but this should occur progressively: sudden breakaway while cornering can cause an inexperienced driver to lose control. Furthermore, the tread pattern should be such as to eject water rapidly from between the contact patch and road. This not only increases the speed of onset of aquaplaning, but also helps to maintain a high coefficient of friction between tyre and road. Incidentally, a useful rule of thumb for estimating aquaplaning speeds for tyres without treads is:

$$\text{Aquaplaning speed} = 9 \times \sqrt{\text{tyre pressure}}$$

It is based on experimental data and the fact that recommended tyre pressures are a function of, among other things, the load on the tyre and the area of its contact patch with the road.

Aquaplaning occurs when the film of water on the road is driven by the forward rolling motion of the wheels into the wedge-shaped-gap between the tyre and the

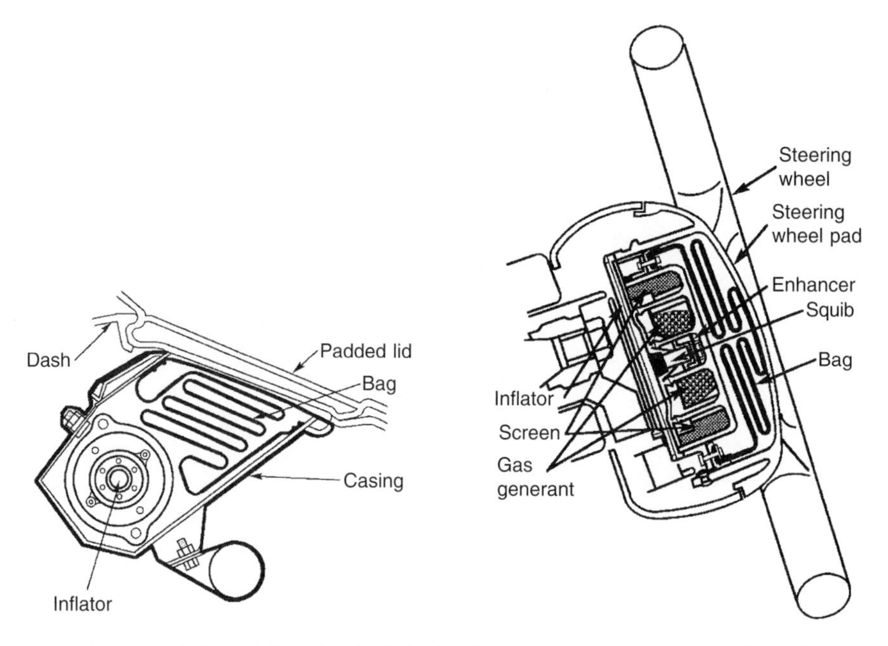

Fig. 17.1-11 Two Toyota gas bag installations: *left*, in the dash for the front seat passenger and, *right*, in the steering wheel hub, for the driver. In both instances, an electrically fired squib generates the heat to fire the pellets which generate the gas. As the bags inflate, they push away the padded trim panels beneath which they are housed.

leading edge of its contact patch with the road. At the critical aquaplaning speed, the pressure in this wedge of water has risen to the point at which it is high enough to support the vehicle. Therefore, the tyres then ride up on to the film of water, which, of course, has a coefficient of friction even lower than that of ice, so the car is floating and will respond to neither steering nor braking forces.

An important aspect of design for active safety is the minimisation of driver stress and fatigue. Another is provision for warning the driver of danger as early as possible before the situation becomes critical. To this end, good all-round visibility and efficient lighting at night are, for instance, two of the measures that can be taken. Others include the installation of devices such as electronic detection systems for warning the driver that he is becoming drowsy: some of these depend on the monitoring of eyelid movements and others of pulse and steering wheel movements. Thirdly, the design should be such that, should the car become involved in an accident, its occupants will be, so far as practicable, protected from injury due to collapse of the structure.

17.1.6 Structural safety and air bags

Since it is neither practicable nor desirable to build vehicles as strong as tanks, their basic structures must be designed to collapse in a controlled manner in an accident. A prime consideration is to prevent the steering wheel from being thrust back and crushing or penetrating the driver's chest or neck or, perhaps, even breaking his jaw. Among the measures originally adopted were the inclusion of telescopic or concertina type collapsible elements in the steering column. In some early instances, the lower end of the steering column tube was coarsely perforated, so that it would collapse when subjected to heavy axial loading.

Another of these measures was the incorporation of two universal joints, one at the lower end of a shortened steering column shaft and the other on the steering box, the section between them being set at an angle relative to the axis of the steering column. In the event of a front end impact, the section between the two universal joints would displace laterally instead of pushing the upper part of the column back towards the driver.

Subsequently, two further changes were made. One was to increase the area of the hub of the wheel, to reduce the intensity of loading locally on the chest. The other was to reduce the stiffness of the rim of the wheel, so that, if the driver was thrown forward on to it, it yielded rather than severely damaging his rib cage.

Later, gas-inflated bags were installed in the steering wheel hub, Fig. 17.1-11. These are supplementary safety devices, as they are effective only in conjunction with correctly adjusted seat belts. They can be inflated by air but, to obtain rapid deployment, inflation using chemicals producing nitrogen or other gases are more commonly used. Correctly tensioning the belt is important, otherwise it will fail to guide the driver in a manner such that his face comes down on to the air bag instead of slithering over it and striking hard objects beyond. In the

USA, failure of drivers to fasten seat belts has been the cause of serious injuries, which has led, unjustifiably, to doubts being expressed regarding the effectiveness of air bags.

For the protection of front seat passengers, air bags are installed behind a panel in the dash fascia, and side air bags may be embodied in the seat squabs. An advantage of the latter site is that it moves with the seat when its position is adjusted, so the bag can be smaller than if it were stowed, for example, in the door. Moreover, in the door, it could be more vulnerable to impact damage. Mercedes has developed what they term *window bags*, 2 m long, for the protection of the heads of all the passengers, which otherwise could be injured either by hitting the side window or by intrusion. These are stowed in the sides of the roof, and deploy in 25 ms. Bags suspended from the cant rail and extending the full length for protecting the passengers in both the front to the rear seats are sometimes called *curtain bags*.

In general, because the occupants' heads start further away from the bags than do their shoulders, side bags at or near shoulder height, for protection against side impacts, should open earlier than those for either window bags or those for frontal impacts. To meet this requirement, Toyota have developed a system in which pellets of a chemical that generates mostly argon gas are used for inflation. The sensors are mounted low in the centre pillars and the air bags are stowed in the front seat squabs.

Since the primary impact may be over within 10 ms, all the bags have to deploy within 20–30 ms. To obtain rapid deployment, most manufacturers employ pellets of sodium azide which, when heated, produce large quantities of nitrogen to inflate the bags. Sodium azide is a salt of hydroazic acid (N_3H_3). Initially, air bag deployment was mostly triggered by deceleration force acting on some very simple form of mechanism, such as a ball in a tube, mounted adjacent to, or within, the steering wheel hub. Subsequently, electrically fired gas generators have been triggered by computers in response to its receipt of appropriate deceleration signals. The deceleration sensors are usually mounted on a front transverse member of the vehicle structure. An advantage of this system is that the whole sub-assembly, including the gas generator, can be housed compactly within the steering wheel hub assembly, and the deceleration sensor can be placed in any position where it will be most effective, Figs. 17.1-11 and 17.1-12.

Perforations in all bags allow the gas to leak out at a rate that increases with internal pressure, thus modifying their spring rates so that the occupants' heads do not rebound violently. This at least reduces, and hopefully even completely obviates, the possibility of spinal whiplash damage. Moreover, the deflation and collapse of the bag, within a few ms after inflation, leaves the steering wheel relatively clear of obstruction so that the driver will have a better chance of regaining control after the impact. In the event of a multiple collision, the air bags are, of course, effective in only the first impact.

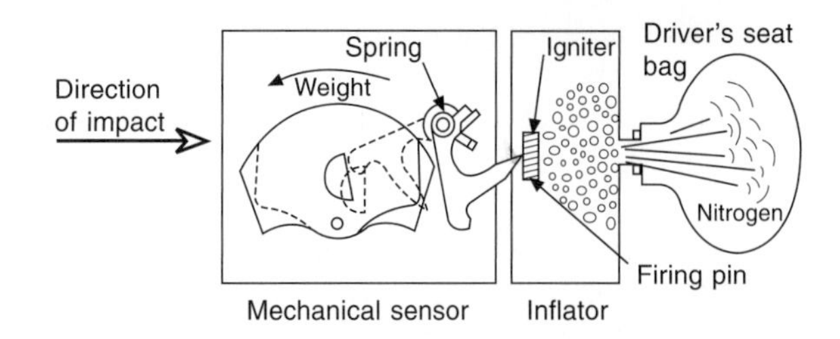

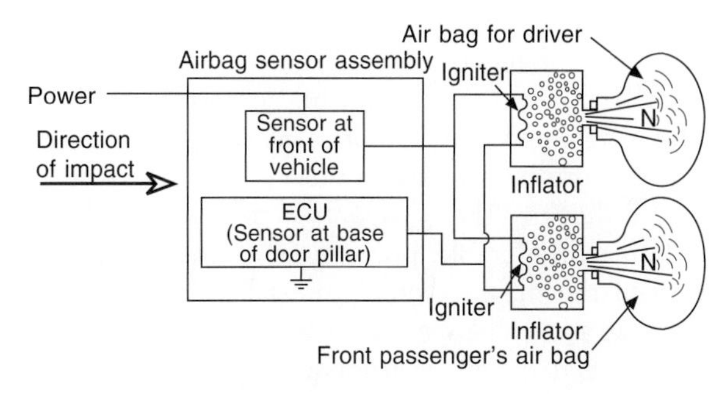

Fig. 17.1-12 *Top*, mechanically actuated bag firing mechanism: *bottom*, electrically actuated alternative. The latter has the advantages of greater compactness of the parts that may have to be accommodated in the steering wheel hub and the sensors and electronic control unit can be sited in the most appropriate positions.

17.1.7 Passenger compartment integrity

The compartment that houses the driver and passengers should remain intact after an accident. Four measures are necessary: one is to incorporate crush zones at the each end of the car; the second is to stiffen the door and its immediate surroundings so that, in the event of a side impact, it will not be penetrated or deflected violently inwards and strike the occupants; third, the door trim must be soft or side air bags must be installed so that, if the occupants are flung against it by the lateral acceleration, they will not be seriously injured; and fourth, the door frame and not only its joins but also those between the pillars and cant rail must be strong and stiff enough to react elastically to absorb the shock loading.

Basically, the occupants must be housed in what amounts to a strong cage, which will protect them also if the car rolls over. This generally entails the use of substantial fillets, and perhaps the fitting of reinforcement plates, at the joints between the pillars and the cant rails and sills. With the current need to reduce overall weight, the use of thin gauge high strength ductile steel, instead of the traditional thicker gauge high ductility material for structural members and some body panels can help to improve both crushability and integrity of structures.

It is important to design so that the loads due to an impact (whether front, rear or side) are, so far as practicable, spread uniformly throughout the whole structure and that the proportions of all the principal members of the cage containing the occupants are adequate to react those loads elastically. Diagonal and transverse members may have to be incorporated under the floor and, possibly, in the roof to transfer some of the loads from one side to the other especially, although not solely, for catering for side or offset frontal impacts.

If the shock to the occupants is to be reduced significantly, a considerable proportion of the total kinetic energy of the moving vehicle must be absorbed by the crush zone as it collapses. At the front, the space between the grille and engine is inadequate for absorbing that energy, except in very minor collisions. Consequently, in the more severe accidents the engine will be pushed back, and it is important to prevent it from thrusting the dash and toe board back until they strike the occupants and possibly trap them in their seats. Consequently, the engine is generally mounted in a manner such that it will be deflected downwards and slide under the toe board. In particular, if the engine is on a sub-frame, the attachment of the longitudinal members of that frame to the toe board and front floor can be designed to shear, to enable the whole installation to slide back under the floor. Even so, the dash and toe board structure must still be stiff enough to prevent significant engine intrusion into the saloon. At the rear, there is more space for a crush zone, but the fuel tank must not be ruptured, which is the reason for the modern trend towards installing fuel tanks much further forward than hitherto.

Ideally, the structure should collapse progressively at a constant rate, as if it were a sprung buffer, Fig. 17.1-13. One design method that has been successful is to bow the longitudinal members so that they either spread outwards or collapse progressively inwards when heavily loaded in compression. Another is to incorporate vertical swaged grooves in the side walls of straight members so that they collapse in a controlled fashion. Ideally, the swages would be distributed alternately, along each side, over the length of the longitudinal members of the frame or sub-frame. However, the zig-zag, or concertina type of collapse thus aimed at is extremely difficult to achieve in practice. Once the first kink has formed, usually at the foremost swage, the member is already bowed and therefore is more likely to continue to do so than to concertina. One manufacturer has notched the corners of the rectangular section longitudinal members to initiate progressive collapse. Each notch extends from the corner only a very short distance down one face and a long distance across the other face. However, one should be wary

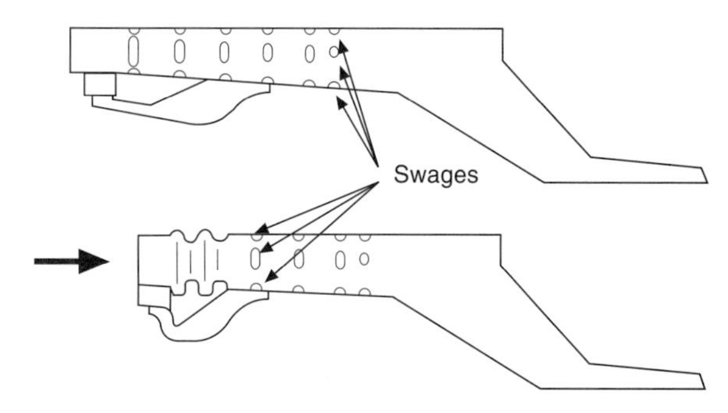

Fig. 17.1-13 Diagrammatic representation of front longitudinal frame member carrying the suspension and engine. The lengths of the swages, in each set of four (in the top, bottom and two sides of the frame), become progressively smaller, from the foremost to the rearmost, so that the frame will offer progressively increasing resistance to collapse in a frontal impact. The lower diagram shows it only partially collapsed.

of introducing notches in such structural members subject to fatigue loading, since cracks are liable to be generated by and spread from the stress concentrations thus induced.

It is preferable to encourage simple bowing by siting all the swages along either the outer or the inner face rather than the top and bottom of each member, to cause both to bow respectively either inwards or outwards. If both bow outwards, the restriction imposed by the body panelling attached to them will help considerably in providing a progressive reaction to the crushing force, If they bow inwards, they are similarly restricted, but perhaps by the presence of the engine between them. Inward bowing, however, tends to absorb more energy per unit of length of collapse. This might or might not be what is desired, hence crash testing is essential for proving designs.

An aspect that should not be overlooked is that swaging the sides of the longitudinal members will reduce their stiffness for reacting to side loads. This need not be serious if the ends of the vertical swages terminate short of the junction with the top and bottom plates, each of which will then become, in effect, a separate U-section member. The ends of the arms of each U terminate where the swages begin, Fig. 17.1-14. Incidentally, box section longitudinal members can be welded fabrications. Alternatively, they could be square section tubes, the swages being produced by hydroforming, using internal hydraulic pressure to expand the tube into a mould.

17.1.8 The problem of the small car

In an impact with a large car, a small car is inevitably at a disadvantage because the inertia of the former is greater than that of the small car. Moreover, the provision of

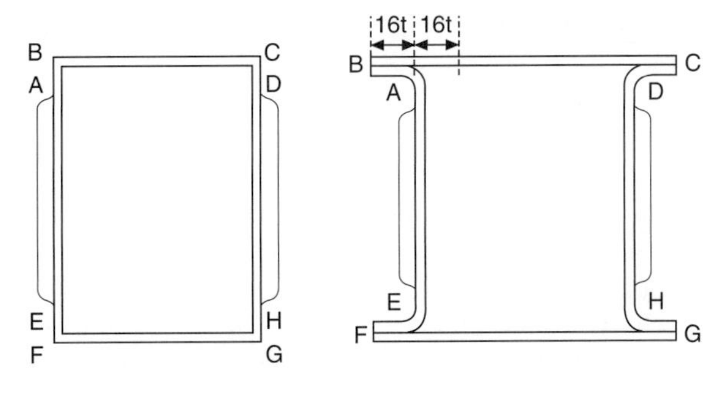

Fig. 17.1-14 Sections through two box section frame side members, one tubular and the other fabricated. Although the swages in their sides weaken them so far as taking side loads is concerned, these loads can be taken mainly in the sections ABCD and EFGH. A useful rule of thumb is that a length equal to 16 multiplied by the thickness of the metal represents the maximum length that is stable on each side of each angle under compression, the measurement being taken from the inner face in each corner or, for the fabricated section, the centres of the bends.

a crush zone of adequate length at both the front and back of the small vehicle is, of course, much more difficult. For this reason, the principle of designing for the engine so that, when thrust backwards, it slides down beneath the toe board and floor is the only practicable course. Furthermore, maximum use should be made of transverse members to distribute the loads appropriately between all the longitudinal members, including the body panelling, in a manner such that they are all equally stressed, as in Fig. 17.1-15.

An interesting feature in this illustration is the pair of gusset struts, one each side, between the front transverse member and each longitudinal side member. If an impact occurs as indicated by either of the two thick arrows, the corner affected by the impact will be pushed back. The gusset strut will stabilise the front end of the side member so that, assuming it is designed to collapse concertina fashion, it will not bow. Moreover, the transverse member will tend to pivot about the opposite corner, which will be stiffened by the gusset strut. It therefore will offer more resistance to the pivoting movement, and therefore a larger share of the impact loading will be transferred to that side than if there were no gusset member there. At the rear, the design is such that the spare wheel will help to take some of the loading from a rear end impact and transfer it to the main structure.

At the rear, the main requirement again is to utilise transverse members to the best advantage. Also important is a robust C-pillar and a good supporting structure for the rear axle. Double skinning the rear quarter panels can enormously strengthen that part of the structure, although this does raise problems as regards repair to minor damage. In general, the overall strength and integrity of the occupant cage may need to be higher than that of a car with long crush zones front and rear.

17.1.9 Side impacts

As regards side impacts, there is not enough space within doors to serve as a crush zone, so the emphasis is on the use of transverse members between the sills and cant rails to share the loading between the structural elements on both sides of the vehicle. Within the doors themselves, horizontal beams the ends of which are securely fixed to the front and rear frame members of each door are widely used. However, it is difficult to make them stiff enough to help much unless the frame and especially

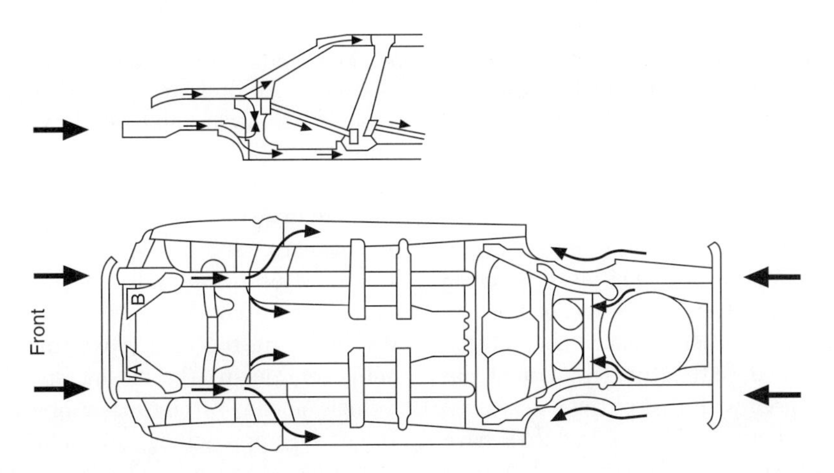

Fig. 17.1-15 *Below*: plan view of a Toyota frame designed to spread the loads imposed by front and rear end impacts uniformly throughout the structure. The combination of the front transverse member and the diagonal members, A and B, one on each side, triangulate the front end of the frame to constrain it to collapse concertina fashion, as shown in Fig. 17.1-13. *Scrap view above*: elevation of a different frame, showing how the loads are distributed as viewed in a vertical plane. The triangulation struts shown in this example are fitted in the door frames.

its waist and bottom rails are very stiff, so that vertical or diagonal beams can be fixed to them to support the centre of the horizontal ones. The longer the door, the more intractable is the problem. Of particular importance is that the B-pillar be strong enough to prevent it and, with it, both doors from being pushed inwards in a side impact situation.

In general, if the central portion of the outer panel of the door is thrust inwards, it will tend to pull not only its front and rear edges, but also the waist and bottom rails towards each other. Consequently, all these members must be adequately stiff. Another measure that has been adopted, for example by Volvo, is to fill the space between the outer and inner panels of the door with a plastics honeycomb. If the hexagonal elements of the honeycomb are fairly thick, the filling as a whole will offer significant resistance to penetration. Moreover, it also transfers some of the loading radially outwards to the door frame members and thus further reduces the tendency towards penetration of the door. It would appear, however, that structural stiffening alone will not be sufficient to satisfy future legislation, so the installation of side air bags to supplement the door stiffening measures will probably be inescapable. Arm rests which could be forced against the vulnerable areas of the lower ribs of the occupants, should not be installed.

17.1.10 Smart air bags

Some early work with air bags revealed shortcomings, but these have been overcome. First, the occupants of the car must be accommodated in fully supportive seats, with their seat belts fastened. Second, the bags in front of the driver and passengers must not deploy in any situation other than a serious frontal impact. Third, because the impact in a crash is usually over in about a tenth of a second, the deployment of the air bags must be accurately timed. Deployed too soon, they might strike the occupants' faces and cause the driver to lose control earlier than he might otherwise have done and, if too late, they may be ineffective.

Research to overcome these problems has demonstrated that first the precise shape of the impact acceleration pulse must be determined. This is a function of the crush characteristics of the front end of the car. Then the characteristic of the performance characteristics of the bags is ascertained, so that the deployment and collapse can be synchronised with that of the pulse. Gas-inflated bags deploy in about 20–30 ms, but they have perforations in them so that they subsequently deflate to enable the driver to maintain control after the impact. In any case, if they did not deflate, the heads of the occupants might bounce back from them, possibly causing neck injury.

An outcome of this research is the development of computerised controls for regulating not only the deployment, but also the tensioning of the seat belts. These are the smart air bags referred to in Table 17.1-1. Signals transmitted to the computer include seat belt tension, rapidity of brake application and the deceleration detected by a sensor mounted on a front transverse member of the structure of the vehicle.

If the belts are too loose, the occupants are accelerated forwards before being suddenly restrained by them, which can cause injury. Incidentally, the acceleration sensor for side air bag control is generally mounted at the base of the door pillar. Testing is now carried out initially using computer programs, which are followed by full-scale crash tests both to prove the validity of the computer modelling and to enable any fine tuning necessary to be done.

Smart air bags are still under development, so further sophistication can be expected. A recent advance has been the provision of sensors and a control system that will inhibit deployment of bags in front of empty seats. This will reduce costs for the owner, since only those for the occupied seats will need to be reinstated. A further refinement that has been proposed is automatic assessment of the size and weight of each occupant and his or her belt restraint status, and setting the deployment characteristics accordingly. Yet another factor that can be brought into the equation is the direction and severity of the crash.

Compartmented air bags have been produced that could be selectively inflated, according to the severity and direction of the impact, and perhaps the weight of the occupant of the seat, or whether a child seat has been fitted. Following instantaneous assessment of the weight of the occupant relative to vehicle speed or the severity of the impact and seat belt status, such a system might be able to inhibit deployment if it is unnecessary.

TRW Automotive has developed what they call a heated gas inflator (HGI), in the form of a vessel containing a weak mixture of hydrogen and air at a pressure of 175–310 bar, as a substitute for explosive pellet type inflators. This device would be difficult to accommodate in a steering wheel hub, but it would be suitable for passenger and side air bags. Two or more such devices might be used for multiple rates of deployment or for compartmented air bags.

Further in the future, we might see radar-based systems for gauging the closing speed of the car with the vehicle ahead, or any other object with which the car might be approaching, and setting the air bag control system appropriately. This could entail also the fitting of an acceleration sensor in the crush zone.

Current provisions for adjustment of the driver's seat and steering wheel could, if he had short legs and a long body, place him too close to the steering wheel for safety in the event of air bag deployment. It therefore could become desirable to mount the pedals on a base plate that could be moved horizontally to adjust its position relative to the seat. This, together with the usual seat adjustment facility, would give the driver the means of positioning himself optimally in the horizontal sense relative to his steering wheel and other controls. Vertical adjustment of either the steering wheel or seat might also be desirable, however.

17.1.11 Seat belts

Ideally everyone would have a safety harness of the two shoulder strap (four point) type, worn by aircraft pilots and rally drivers. However, this is commonly regarded as too restrictive to be acceptable by the motoring public. It is also more costly than the adjustable combined lap and single shoulder strap (three point) type harness. The latter type is satisfactory if the lap belt fits snugly round the pelvis, the upper anchorage for the shoulder strap is low enough to prevent strangulation or damage to the neck of the person it is supposed to protect, the whole harness is a reasonably close fit around the body, and the occupant can instantly release himself from it in the event of, for example, a fire hazard.

To obtain a snug fit without discomfort to the wearer, and so that the strap automatically winds back on to its reel when not in use, the harness reel is, of course, spring loaded. A further refinement is an acceleration sensor, either mechanical or electronic which, in the event of an accident, triggers a ratchet to lock the belt and thus hold the occupant firmly and snugly in his seat. If this trigger mechanism is too sensitive, the belt may lock unnecessarily and cause discomfort to the occupant: on the other hand, if not sensitive enough, it may fail the occupant in a real emergency.

The shape of the seat bucket must be such as to prevent him from sliding down through the lap belt, sometimes termed *submarining*, and the combined effects of both the lap and shoulder straps should restrain him from being shot either forwards or upwards out of the seat. Another requirement is that when the belt is retracted, the buckles must be in a position such that, when the occupant is seated, he can easily reach them. The upper buckle is usually mounted on the B-pillar, and has to be pulled down and snapped into the socket beside the seat pan Attachment to the B-pillar may present a problem, especially to cater for the seat's being slid forward for the benefit of driver with short legs. To overcome this problem, the belt is in some instances carried on the otherwise free end of a short arm pivoted to the B-pillar. Alternatively, either a manually or electrically actuated adjustment device may be installed.

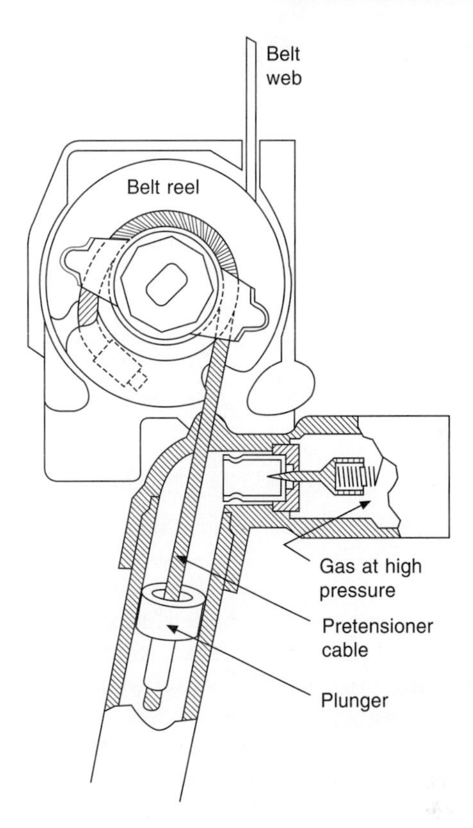

Fig. 17.1-16 Toyota automatic seat belt tensioner. In the event of an impact, gas is discharged from the horizontal cylinder on the right, into the chamber above the plunger, which it forces downwards. Dragging the pre-tensioner cable with it, the plunger thus pre-tensions the seat belt for the duration of the impact, after which the tension is progressively released as the gas escapes through the clearances round the cable and plunger.

As previously indicated, if belts are not fitted snugly, violent acceleration forces can propel the occupant forward at high speed until the slack in the belt is suddenly taken up and he strikes it with such force as to injure him. To avoid this situation without having to rely on the occupant's making the appropriate adjustment, Toyota offer a system incorporating an automatic device for increasing the pre-tension in an emergency, Fig. 17.1-16. When an electronic sensor detects an acceleration rapid enough to throw the occupants violently forward out of their seats, the electronic control causes gas at high pressure to be released into a cylinder which is part of the seat belt tensioning mechanism. A piston in this cylinder pulls a cable wound round the belt pulley, which it rotates to pull the harness tight. Subsequently, the gas escapes through the clearances around the piston and cable rapidly enough to release the tension so that the occupants can, for example in the event of a fire, immediately unlatch their harnesses and escape. Without seat belts, or even if they are inadequately tensioned, the occupants can be thrown violently in virtually any direction.

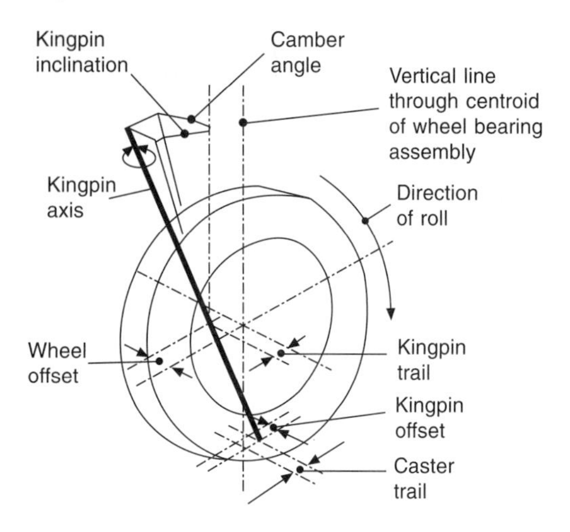

Fig. 17.1-17 Principal features of front steering geometry.

Small children are best belted into rearward-facing safety seats of appropriate sizes. Such seats can be anchored either to the adult seats or to the dash or backs of the front seats. This implies ensuring, at the design stage, that suitable anchorages are provided and that seat backs are strong enough to take the weights of both a front seat occupant and a child plus its safety seat. Where air bags are installed, care must be taken to ensure that they will not strike the children if they deploy.

17.1.12 Improvement of active safety

Active safety embraces the ergonomic design of the vehicle for ease of control by the driver without his becoming fatigued, as well as the more obvious features such as harmonisation of the steering, braking, tyres suspension and handling characteristics, to reduce the likelihood of his losing control. There are five main requirements:

1. That while the motorist is at the wheel, he can readily verify that driving conditions are safe
2. In every situation, all control responses should be proportional to the driver's input
3. All responses of the vehicle must be instant as well as accurately reflect the input
4. The vehicle must be dynamically stable
5. Drivers must be able to recognise when limits of stability are being approached.

Modern measures for improving dynamic stability include anti-lock brake systems (ABS), traction control systems (TRC), and vehicle stability control (VSC), sometimes called vehicle dynamics control (VDC). None of these, however, extends the critical limit at which the tyres lose their grip on the road. Under normal conditions, all three systems are dormant, automatically coming into operation only in emergency situations, when they are needed for the avoidance of an accident. Provided the vehicle accelerates, brakes and turns consistently with the driver's input, he should be able to avoid accidents in all normal driving conditions.

Different drivers behave in different ways in an emergency. Some will slam on the brakes, others will steer out of trouble, some will do both, while others will, if appropriate, accelerate to avoid the problem. These variations should be taken into consideration by the designer but, of course, it is impossible to cater for drivers who freeze and do nothing: only passive safety can help these.

17.1.13 Tyres, suspension and steering

The grip of the tyres on the road determines how rapidly the car will accelerate, where it will go and at what point it will stop. Very rough roads may cause the tyres to bounce clear of the surface and therefore to lose their grip. Smooth roads when wet will tend to have a low coefficient of friction, so the car may slide on a corner or the effectiveness of the brakes may be significantly reduced. Water on roads of any sort will lead to aquaplaning at some critical speed. The more efficient the tread in squeezing water from the contact patch between the tyre and the road, the higher will be the critical speed as regards aquaplaning.

As regards safety, the function of the suspension system is to keep all four tyres on the road, to maintain a flat and stable ride, to keep the attitude of the wheels relative to the road in the optimum position under all dynamic conditions, and to limit vehicle posture changes when cornering, braking and accelerating. In other words, it must reduce to a minimum changes in the position of the centre of gravity of the vehicle due to pitching and rolling.

Steering performance is affected by suspension layout, tyre characteristics and the centre of gravity of the vehicle, all of which affect the inherent tendency to over- or understeer. A summary of geometrical characteristics that influence steering is illustrated in Fig. 17.1-17. The steering should be firm and stable in the straight-ahead condition, and the feel, or feel-back, of the steering control is an important requirement as regards safety.

During departures from this central position, with both manual and power assisted systems, the feel-back should increase linearly, to give the driver a positive indication of the angle through which he has turned the wheel. Some designers favour an increase in the rate over the last degree or so to full lock, so that the driver has a positive indication that he is approaching the limit of wheel movement. Consistency of feel-back improves with increases in the stiffnesses of the links and mountings of steering mechanisms.

17.1.14 Electronic control systems in general

Electronic control can be exercised either by a central electronic unit (ECU), or individual electronic control units can be incorporated, as sub-systems to each of the controls, such as steering, brakes, etc., where they can be used to transmit information to, and receive it from, the other sub-systems. The latter generally offers the advantages of compactness and because, provided the central computer can be eliminated, the wiring harness can be simpler and installation easier.

Because electric motors are amenable to electronic control, they are now being considered for use as actuators. However, it might be more practicable to substitute electric motor driven for engine-driven hydraulic pumps, the former being potentially both lighter and more compact.

17.1.15 Electric power assisted steering

Electronically controlled electrohydraulic power-assisted steering systems have been developed by, for example, AB Automotive Electronics, Delphi, Echlin Automotive Systems and TRW Lucas Steering Systems. In general, the direct input signals are vehicle speed and the torque applied by the driver to the steering wheel. Power assistance is provided by a 12 V permanent magnet brushless electric motor which, in turn, drives a hydraulic pump to actuate the assistance mechanism. This type of motor is relatively quiet, powerful and compact. Moreover, by virtue of its low inertia, its responsiveness to changes in demand is good.

On the basis of signals indicating the temperature of the motor and current flowing through it, the ECU regulates the speed of the hydraulic pump to that appropriate for exercising control safely and efficiently. As the steering wheel is rotated further from the straight-ahead position, the ECU applies a progressively increasing current, and thus correspondingly increases the hydraulic assistance.

A closed centre hydraulic control valve and engine-driven pump take a constant supply of energy at a fixed level from the engine, so the advantage of the electronically controlled pump is that the power demanded is no more than is required to cater for the instantaneous operating conditions. At idle, for instance, the current can be as little as 0.5 A and, in most operating conditions, it will be around 1–2 A. Only under extreme conditions will the demand become higher. Consequently, energy requirement for power assistance is reduced to a minimum.

The implication, of course, is that, with the electronically controlled pump, the fuel consumption will be lower. For the manufacture one advantage is that a single power assistance sub-assembly can be common to the whole range of vehicles manufactured, from sub-compact to minivans. Another is that, for development work on the test track, the unit can be tuned with a laptop computer to try out various steering characteristics.

Perhaps the most significant benefit that can be obtained is that, by virtue of electronic control, it becomes possible to do things that would be impossible with mechanical or hydraulic control. For instance, the compromises inherent in conventional mechanical steering geometry can be obviated, because the electronic system could control each wheel independently.

Even the mechanical linkage between the steering wheel and gear could be eliminated and replaced by a drive-by-wire system. Aircraft are operated in this way, so worries about failure would appear to be unfounded. Various ways of getting around this problem, such as dual or triple control circuits, are available. With such systems, the computer monitors all the circuits and, if one is found to be malfunctioning, the computer switches it off and relies on those that are functioning normally. At the same time, a warning signal indicates to the driver that his steering system urgently needs attention.

17.1.16 Brakes

Except to cater for their deterioration in service, there is no point in installing brakes the torque capacity of which significantly exceeds the maximum adhesion limit of the tyres. For safety, the vehicle should slow and stop in a manner consistent with the input by the driver to the pedal. In emergency braking, the attitude of the vehicle should, as previously indicated, remain so far as practicable constant. In all circumstances while the vehicle is slowing or stopping, control should be easily maintained, and the performance of the brakes should not vary with the length of time they are applied. This is especially important in emergency situations and during descents of long, steep inclines because, under these conditions, the friction elements tend to become very hot and brake fade could therefore occur.

The attainment of all these aims is greatly facilitated if the braking effort is divided between the front and rear wheels in proportion to both the front–rear weight distribution and the limiting adhesion of the tyres, which may vary with their vertical and lateral deflections. There should be no lag between pedal and brake application, and the feel-back from the pedal should accurately reflect the degree of braking applied at the road surface. Finally, the performance of the brake linings, or pads, should be consistent.

17.1.17 Automatic braking and traction control

If the wheels lock, the coefficient of friction between the tyres and the road becomes lower than when they are rolling, and the vehicle is liable to become unstable and skid. To prevent the wheels from locking when the brakes are applied, the sensor in a simple ABS system signals to a computer the speed of rotation of the wheels. In the more primitive systems, as soon as the speed of any wheel is reduced to the point at which it is about to lock, the computer signals the brake control to reduce the hydraulic pressure to all four brakes. In modern advanced systems, however, the pressure is reduced for only the brake of the wheel that is about to lock. With either form of ABS, therefore, brake control in an emergency is greatly simplified: all the driver has to do is to push as hard as he can on his brake pedal.

Limited slip differentials help to prevent the total loss of traction that occurs with simple differential gears when a driven wheel on one side spins freely, for example on ice or in very soft ground. It also ensures that some torque is delivered to the inner wheel of a vehicle that is cornering tightly, and thus it increases the overall tractive potential.

Traction control systems prevent the wheels from spinning if the torque transmitted to any wheel rises above that which can be transmitted by the tyre. If one or more of the wheels spin, the consequent loss of coefficient of friction between their tyres and the road tends to cause the vehicle to become unstable and go out of control. The sensor for detecting the onset of wheel spin is usually common to both the ABS and TCS systems but, of course, for the latter function, it sends a signal of impending wheel spin, instead of wheel lock, to the electronic control. On receipt of such a signal, the computer orders application of the relevant brake until the tendency to spin is nullified, and thus maintains the vehicle in a stable condition.

With four-wheel drive (4WD), the torque output from the engine is distributed to four instead of two wheels, so the tractive force delivered through each of the tyres is halved. Consequently, the tendency to wheel spin is correspondingly reduced. To prevent torque wind-up to the drive-line, most modern 4WD systems have a differential or limited slip device between the gearbox output and the drive shafts to the front and rear wheels.

17.1.18 Recently introduced advanced systems

Four-wheel steering, although costly, has some advantages as regards stability and can increase ease of parking. With the application of computers and electronics to vehicle control systems, it is now possible to have active four-wheel steering. In other words, the four-wheel steer system can be made to adjust the angle of the rear wheels to compensate for any force input from one side, such as a sudden gust of wind or some other tendency to cause over- or understeer (Figs. 17.1-18 and 17.1-19). It also helps the driver to keep closely to his intended course.

17.1.19 Suspension control

Suspension performance can be improved too by an advanced safety measure. This is an electronic control system by means of which the characteristics of the dampers are adjusted automatically in relation to the speed of the vehicle and roughness of the road. One such system is the Toyota electronically modulated suspension (TEMS), which also includes a two-way switch on the dash for enabling the driver to select damping for either normal or sporting operation. For very many years manually adjustable dampers have, of course, been available, but very few drivers have the skill needed to make the appropriate adjustments manually.

For the future, a further development could be active suspension, in which electronically controlled hydraulic

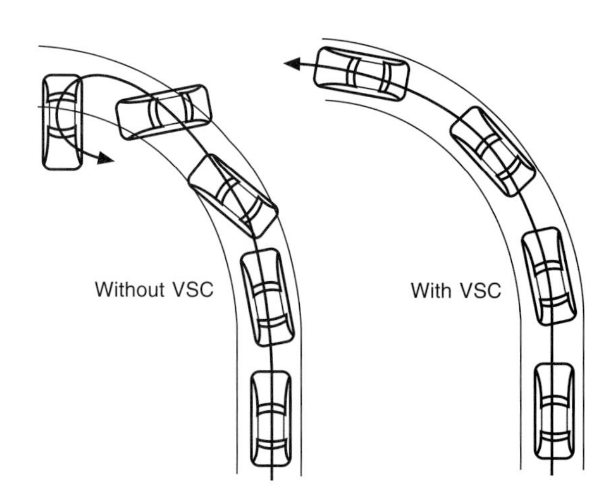

Fig. 17.1-18 *Left,* diagram showing the path typical of an oversteering car driven beyond the limit of adhesion of the wheels on the road, compared with, *right,* one driven in the same manner but equipped with vehicle stability control.

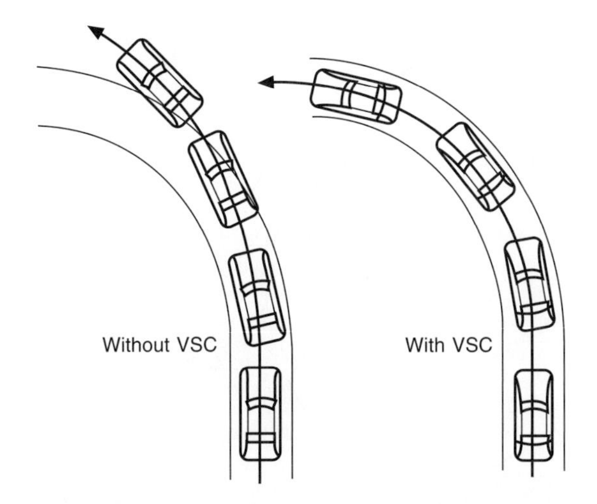

Fig. 17.1-19 *Left,* diagram showing the path typical of an understeering car driven beyond the limit of adhesion of the wheels on the road compared with, *right,* one driven in the same manner but equipped with vehicle stability control.

jacks keep the body at all times at a constant height and attitude relative to the road. Some of these systems dispense with suspension springs. However, it seems more likely that those used in conjunction with them will ultimately preferred since, if the static weight of the car is supported by springs, malfunction of the hydraulic jacks and their control system would not be so catastrophic. In general, active suspension has the advantage of utilising to the full the tyre performance potential. However, it is costly, it consumes energy and its durability and reliability remain open to question. Consequently, for general application, its future would appear to be in doubt.

17.1.20 Ergonomic considerations and safety

Since ergonomic measures are taken before the accident, we shall categorise them as active. The driver's seating position is of prime importance, in that he has to use his eyes to obtain at least 90% of the information he needs for driving safely. As regards the position of the seat itself, this must be such that his view of the scene outside the car is obstructed as little as possible by components such as rear view mirrors or windscreen pillars; similarly his view of the instruments, switches and other controls must be clear; at the same time, he must be able to reach all his controls easily, with a minimum of effort, and without being distracted from what is happening on the road ahead. Visibility of indicator and warning lamps under brilliant sunlight is a consideration sometimes overlooked. All controls should be easy to operate, and instrument and other indicators easy to read. In other words, the aim is at enabling the driver to remain relaxed and comfortable throughout his journey, and thus minimising fatigue.

Also important is his view of the four corners of his car. Radar devices to indicate the proximity of obstructions have been suggested. However, even if they could be offered at acceptable costs, their effectiveness in traffic moving at even relatively modest speeds would be open to question. Drivers' rear view mirrors in many instances fail to cover a range of vision wide enough to include cars overtaking from all possible angles. On the nearside, the view through the mirror should include the nearside wheel, for ease parking, as well as the road behind. In general, mirrors should not be so wide that they are in danger of being struck by the mirrors of passing cars or other items such as gate posts. Although some of the points raised here are relatively unlikely themselves to cause accidents involving death or injury, they are relevant as regards driver fatigue, which can have serious consequences.

Driver fatigue is affected by, among other things, the climate: in cold countries, an electric seat warming system may be desirable, but when it is very hot, ventilation of the seat cushion and squab, as well as the saloon, may be more appropriate. Air conditioning is even better, and can remove the need for seat conditioning in any climate. In hot countries, air conditioning is generally regarded as essential, as also, of course, is interior heating and ventilation in cold conditions.

17.1.21 Seating

For seat comfort, the more uniform is the pressure distribution over the cushion the better. This calls for measures to offset the natural tendency for the pressure to be highest under the hip bones and lowest beneath the thighs and coccyx, or tail bone. Even so, too high a pressure under the thighs may restrict the blood circulation to the legs and thus lead to severe discomfort after a short period at the wheel. On the other hand, too little support in this region can cause strain and tiredness of the legs.

As regards the squab, the most important requirement for comfort is support for the lumbar region of the occupant regardless of his or her size and, preferably, this support should be adjustable. Appropriate selection of the shape and position of the lumbar support can also reduce some of the pressure on the cushion. Provision for adjustment of the angle of the squab is, of course, also highly desirable. Another requirement, and one that is not widely appreciated, is cushioned, yet firm, support for the lower end of the spine, just above the coccyx. It is in this region that spinal damage is most likely to be sustained in the long term as a result of wear and tear. Because of variations in the sizes and physical proportions of drivers, biaxial adjustment (vertical and longitudinal) of the positions of both the seat and steering wheel can contribute significantly to comfort.

For the prevention of whiplash injury of the spine in a rear end impact, the conventional headrest is not fully effective. According to Volvo, the spine may be affected throughout its length, so the shape and restraint offered by the whole of the seat back is relevant. To this end Volvo have been concentrating on:

1. Controlled resilience of the squab cushion and installation of a new recliner system to reduce the severity of a rear end impact pulse on the spine.
2. Reducing to a minimum movement of all parts of the spine relative to each other, by providing good support from the base right up the spine to the head, so that, throughout the impact, the curvature of the spine changes as little as possible.
3. Reducing to a minimum the forward rebound of the occupant from the seat into the seat belt.

The outcome of design on these three principles is what Volvo calls the WHIPS seat. In a rear end impact with a conventional seat, the occupant is first accelerated forward. This causes him to sink into the resilient trim on the squab and then rebound forward against the seat belt. In the meantime, his head is supported by the head rest. With Volvo's new system, the forces between the body of the occupant and the seat squab activate the WHIPS system. The new recliner allows the squab to move backwards, and therefore reduces the forces between it and the body without increasing the distance between the head and its restraint. Additionally, the seat squab bends backwards to reduce further the g-force on the body. The overall result is a reduction also in the energy available to induce rebound.

A safety seat introduced by Saab has what they term the Pro-tech self-aligning head restraint. Their aim is at providing uniform support throughout the whole length of the spine. In a rear end impact, this seat back absorbs energy by allowing the lower part of the spine and back to sink into it in a controlled manner, an action likened by Saab to the catching of a ball in a padded, gloved hand. The rearward movement of the occupant's body is reacted by a pressure plate in the squab. This plate is connected to the lower end of a vertical pivoted lever, the upper end of which pushes the head rest forward until the moments about the pivot balance, to counteract the tendency to whiplash, Fig. 17.1-20. Saab claim that, with this system, the head rest can be set in a lower position than would otherwise be safe.

For the rear seats of some of their models, Saab provide vertically adjustable head rests which can be lowered by the driver when it is not carrying passengers in them. In this way he is assured of the best possible range of rearward vision. If left in the lowered position, however, they are very uncomfortable so, as soon as passengers enter the seats, they are obliged to elevate them to the position in which they will provide adequate protection from spinal whiplash.

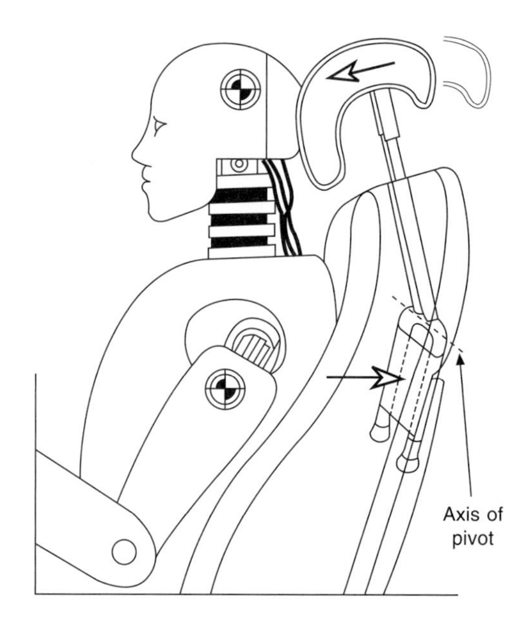

Fig. 17.1-20 The Saab seat designed to prevent spinal damage due to whiplash. At the lower end of the pivoted lever is a pad against which, during a rear end impact, the shoulders push to swing the head rest at the upper end of the lever forwards about the pivot to support the head. Stops, not shown here, limit the motion of the lever.

17.1.22 The pedal controls

Inappropriate pedal arrangement can have a significant effect on driver fatigue as well as directly cause accidents. To enable the driver to differentiate easily between the brake, clutch and accelerator controls, the pad on the accelerator pedal should be positioned directly beneath the ball of the driver's right foot, when resting in its natural position. It should be slightly convex so that the driver can easily pivot his foot about his heel to depress the pedal, with just enough friction between the sole of his foot and the pedal to enable him to maintain a steady throttle opening when needed. The range of angular movement must be large enough to avoid jerky operation of the throttle and, over virtually all its range, the feel-back should be linearly progressive and proportional to throttle opening. In some instances, when the throttle is just cracking open, the movement of the pedal relative to that of the throttle may be increased to avoid jerky take-off from rest.

For rapidity and ease use in an emergency, the brake pedal must be directly beneath the driver's left foot.

It should be significantly larger than the throttle pedal, so that the two are readily distinguishable from one another, and in order that the pressure per unit area of the sole needed for application of the brakes is not too high. Obviously, since the loads applied to the brake pedal are much greater than to the throttle control, it must be both stronger and stiffer. Brake pedal travel should be proportional to the degree of braking applied, and the length of travel must be large enough to provide the necessary feel-back, but without calling for excessive movement of the upper leg.

The clutch and brake controls should be well separated so that the driver can neither confuse one for the other nor accidentally push both down simultaneously. However, for comfort while cruising, there should be space adjacent to the break pedal for the driver to rest his foot clear of the pedals. Within reasonable bounds, the feel-back can be such as to give the driver a positive signal that the clutch is fully disengaged. For example, with a mechanically actuated clutch, a toggle linkage may be appropriate so that the resistance to pedal depression suddenly almost disappears when the clutch is fully disengaged.

Section Eighteen

Materials

Chapter 18.1

18.1

Design and material utilization

Geoffrey Davies

OBJECTIVE

To briefly review the historical development of the automotive body structure before considering how materials have helped to realize engineering and associated objectives. Specific examples are selected to illustrate the changes that have taken place in the selection of steel grades, the emergence of aluminium and the increasing trend to hybrid material combinations.

CONTENT

A brief outline is given on the evolution of various design concepts and materials utilization – reference is made to specific examples, where relevant changes have been made – the criteria typically used in optimizing design using FEM are defined – the utilization of coated and high strength steel (HSS) grade are introduced – alternative body architecture is considered – use of aluminium and other lightweight materials are referenced together with the use of hybrid structures.

18.1.1 Introduction

While highlighting significant developments in the design architecture of the automotive body structure, the emphasis of this chapter is more concerned with the selection and use of materials, and how engineers have utilized relevant properties to satisfy their selection criteria over a timescale which now covers at least 100 years. Early materials selection was fairly limited and chiefly dictated by cost as the demands of mass production grew. Later availability became an issue, as two world wars had a draining effect on resources. But perhaps the period of greatest interest has spanned the last 30 years. During this time engineers have had to respond to a series of different outside influences among them legislation on safety, emissions control and recycling. These have resulted in some conflicting results, for while weight reduction and use of lower density materials have extended the choice the increased variety and particularly the use of some plastics have caused additional headaches for the dismantling industry.

After a consideration of milestones in autobody design through the last 100 years, the rationale of the modern designer is introduced with an example (from BMW) of how this is influenced by material choice. Examples are then given of alternative approaches to design using lightweight alternatives to steel, the traditional choice. Reference is made here to broader initiatives such as the major aluminium programmes (ECV, ASVT) and steel (ULSAB), which have demonstrated the feasibility of the newer technology. The spin-off from these background 'enabling' projects has been adopted in many parallel model programmes in recent years and this is self-evident in some of the instances given in this chapter. The same applies to many international initiatives (e.g. BRITE Light Weight Vehicle Programme BE – 5652[1]) plus supplier/user projects which have provided increased confidence in longer range concerns, and although the immediate pay-off from these programmes is not always immediately obvious it is critically important that funding for these general 'feeder' activities continues.

18.1.2 Historical perspective and evolving materials technology

The progess made in the development of engineering structures over the last century has been dealt with expertly elsewhere[2] and with regard to recent model

Materials for Automobile Bodies; ISBN: 9780750656924
Copyright © 2003 Elsevier Ltd; All rights of reproduction, in any form, reserved.

programmes the significant use and benefits of FEM techniques (basic introduction summarized below) in shortening delivery times is emphatic.

However, as these become more complex, the more need there is for input detail such as material properties. As well as physical properties the need also exists for empirical data regarding material behaviour in diverse engineering situations and it is important that past designs and associated materials performance is analysed and 'rules' extracted for future design purposes and use in numerical form when required. In general terms the same developments have been evident on a worldwide scale, although size is a feature of American built vehicles, and the needs of mass production technology, reaching global proportions, have perhaps influenced the Japanese design philosophy (robot access, automation). Therefore although written from a UK perspective, with its foreign 'transplant' influences over the years the following content probably mirrors the worldwide trends and requirements for body materials in the future.

18.1.2.1 Body zones and terminology

First, it is necessary to clarify the terminology used to differentiate the various areas comprising the body. The body-in-white splits down into the main structure 'body-less-doors' and the 'bolt-on' or skin assemblies. These in turn break down into the inner panel, usually deep drawn to provide bulk shape and rigidity, plus the shallow skin panels which provide the outer contour of the body shape and require more aesthetic properties such as smooth blemish-free surface and scuff or dent resistance. The key elements of the main structure are the floor and main cage containing 'A', 'B/C' and 'D' posts or corner pillars and roof/cantrail surround, plus closed sections such as cross members, and front and rear longitudinal sections which provide essential impact resistance. The requirements of each zone are summarized in Table 18.1-1 together with recommendations for appropriate steels and possible alternatives:

Table 18.1-1 Requirements of different panels comprising the BIW structure

		Materials choice		
		Steels		Possible* alternatives
Zone/ assembly	Requirements	Type	YS MPa	(material/ form)
Main structure:				
• Front/rear longit mbrs	Impact resistance	HSS	300	DP600, AP
• 'A' post inner/outer	Rigidity, strength	HSS	300	AP, HT
• Cantrail	Rigidity, strength	HSS	260	AP, HT
• Main/rear floor	Moderate strength	HSS	180	AS
• Bodyside	Moderate strength, formability	HSS	180	AS, TWB
• Spare wheel well	Deep drawability	FS	140	SWS, SPA
• Wheelhouse, valance	Formability	FS	140	AS
'Bolt-on' assemblies				
Outer panels				
Door skins	'A' Class surface Dent resistance	FS	140	BH180, AS, SPA PLA-RRIM,
Bonnet	'A' Class surface Dent resistance	FS	140	BH180, AS, SPA PLA-SMC
Boot	'A' Class surface Dent resistance	FS	140	BH180, AS, SPA PLA-SMC
Roof	'A' Class surface Dent resistance	FS	140	BH180, HS
Inner panels				
Doors	Drawability	FS	140	TWB
Intrusion beams, rails	High impact strength	UHS	1200	AT, DP600+

* Code:
HSS = high strength steel; UHS = ultra HS steel; FS = forming steel; AS = aluminium sheet; AP = aluminium profile; SPA = superplastic Al; HT = hydroformed tube; HS = hydroformed sheet; BH180 = Bake hardened steel; Dpyyy = Dual phase steel; TWB = Tailor welded blank; PLA-xxx = Plastic-xxx type; RRIM = reinforced reaction injection moulded; SMC = sheet moulding compound; SWS = sandwich steel.

18.1.2.2 Distinction between body-on-chassis and unitary architecture

Prior to the 1930s the body-on-chassis was the most popular vehicle configuration, the upper passenger containing compartment being mounted on a stout chassis which also carried the power-train unit plus other essential suspension, braking and steering gear. The body and chassis arrangement provided some versatility of model change and facility flexibility within the limited confines of earlier factories. Bodywork such as that used for the Morris Oxford in the early 1920s

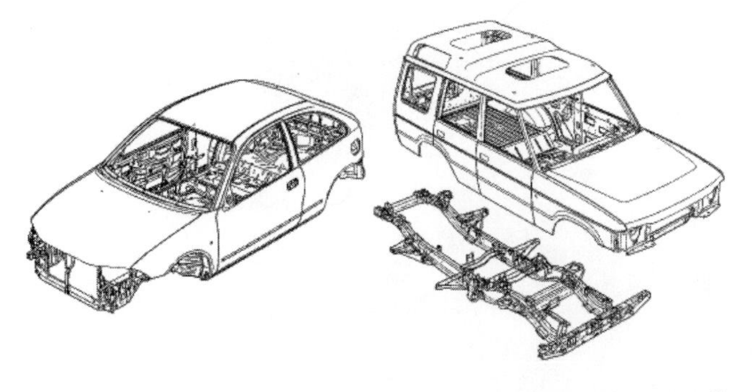

Fig. 18.1-1 Illustration of unitary and chassis body architecture.[3]

featured a wood, fabric and metal construction, the main change being to an all steel assembly in 1929 as the influence of the American Budd Company became obvious within Pressed Steel who supplied the body. The first significant aluminium body, the Pierce Arrow, also made its appearance in the early 1920s, but all steel construction had found favour in the USA because it was more suited to mass production, chiefly due to the ease of pressed panel production allied to the advantages of joining by spot welding. From an engineering point of view it also significantly increased torsional stiffness. A step change in design came with the integration of the chassis and body, claimed to have been introduced by Citroen in 1934 for its 11 CV model.[2] The difference in construction of the integrated or unitary construction compared with the chassis mounted body is illustrated by reference to the two modern day vehicles shown in Fig. 18.1-1.

18.1.2.3 Early materials and subsequent changes

Wood used in conjunction with fabric has been referred to already and was the construction of the bodywork of many cars in the 1920s before its replacement by steel. For outer panels this was of fairly thick gauge between 0.9 and 1.00 mm and much of it destined for the UK Midlands car plants was produced in the South Wales steelworks in ingot cast rimming or stabilized grades (Chapter 18.2). The rimming steels could be supplied in the 'annealed last' condition for deeper drawn internal parts but for surface critical panels a final skin pass was essential to optimize the paint finish. For complex and deeper drawn shapes the more expensive stabilized or aluminium killed material was used which conferred enhanced formability. Gradually a change took place – due to weight and cost reduction studies the average thickness of external panels reducing progressively to 0.8 mm in the 1950s/1960s and to the current level of 0.7 mm in use today for the production of the body of unitary construction shown in Fig. 18.1-1. Internal parts for structural members range from 0.7 to 2.0 mm, the scope for downgauging over the years being limited by stiffness constraints. Therefore although the thickness of strength related parts such as longitudinal members can be reduced by utilizing high strength grades on the basis of added impact resistance, as rigidity is a major design criterion, and the elastic modulus of steels is constant throughout the strength range, opportunities for substituting lighter gauges are limited. This situation can,

Land Rover Discovery Tdi (3 door)

Fig. 18.1-2 Selection of Land Rover vehicles and body types.

Land Rover Defender 90 hard top Tdi

Land Rover Defender 110 county station wagon Tdi

Fig. 18.1-2 *(contd).*

however, be improved by use of adhesives or peripheral laser welded joints and examples of the use of these techniques are given later in this chapter.

Although not introduced until 1948 the Land Rover provides a good example of a modern vehicle with a chassis of two standard lengths serving a myriad of agricultural and military purposes. Although answering the rugged off-road requirements of the 4 × 4 vehicle virtually any type of body shape could be tailormade and constructed without the need for a dedicated higher volume facility. When steel was difficult to obtain in sheet or coil form in 1948 the underbody frames were produced by welding together strips of steel cast off remnants and aluminium was used for many body panels.

Together with the BMW 328 Roadster (1936–1940) and the Dyna Panhard (1954), Rover and Land Rover were among the first users of aluminium in Europe, the ubiquitous Defender models using the 3xxx series alloys for flatter panels with the Al–Mg 5xxx series being used in other applications, a wealth of experience being gained in pressing, assembly and paint pre-treatment and finishing (Fig. 18.1-2). Although the chassis was cumbersome it was – and still is – ideal for mounting the extensive range of Land Rover Defender body variants.

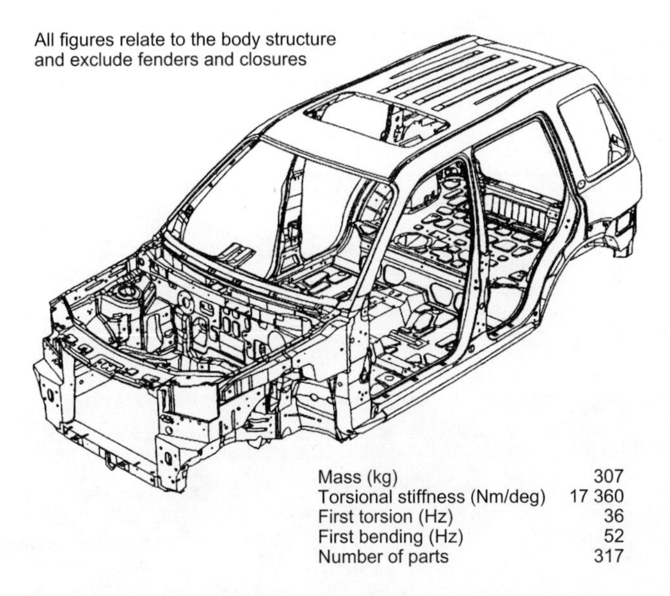

Fig. 18.1-3 Land Rover Freelander with monocoque body and plastic front wings. (Reprinted with permission from SAE paper 1999-01-3181 Copyright 1999 Society of Automotive Engineers Inc.)

Until this day the hot rolled grades of steel are used (typically HR 4) but it is easy to see why efforts are being made to downscale these relatively massive ladder frames with consideration being given to using newer material in thinner gauges, e.g. HSS up to 300 N/mm^2 (TRIP steels up to 590 N/mm^2 are now being used for 80 chassis parts on the Mitsubishi Paquera). Design modifications must be made to accommodate the thinner gauges and consideration has already been given to alternative material forms such as hydroformed sections (described later), as referenced by the ULSAB process, which could be used to bolster stiffness and crashworthiness. Although better suited to more conventional car body design, the incorporation of tailored blanks again offers an alternative approach giving the engineer strengthening exactly where required and a further opportunity for parts consolidation/reduced weight. This enduring type of rugged and versatile design has persisted as it answers the diverse needs of military purchasers but it is not surprising that as fleet average economy targets are considered more critically the monocoque is now becoming more stringent for the more volume-oriented 4 × 4 vehicles – as featured by the Land Rover Freelander (Fig. 18.1-3). Durability is satisfied by the use of hot-dip or iron–zinc alloy coating as steel substrates replace the use of expensive aluminium for outer panels and the model features another material innovation in the selection of polymer front wings.

Before leaving body-on-chassis design it should be mentioned that other types of chassis include the steel backbone type used by Lotus and the designs featuring triangular sectional arrays as shown in Fig. 18.1-4. These were steel square or tubular sections, and later Lotus adopted another chassis configuration termed the 'punt', also shown.

The Lotus Elise featured the punt, which has also been termed a spaceframe concept and as this is more of a transitionary structure this will be described in greater detail later on together with similar aluminium internally structured bodies.

It has been debated as to what exactly constitutes a chassis-less design as various forms can incorporate some features of the original underframe, e.g. subframes and longitudinal/sidemember sections. Engineers such as Garrett[4] claim that the ideal form of chassis-less construction emerged in the 1940s with the launching of the Austin A30 as shown in Fig. 18.1-5.

They would argue that using their aircraft design principles they were able to incorporate all the essential load bearing requirements into a relatively lightweight body without even building in partial box sections that were featured in 'integral' or 'unitary' designs, with elements of the chassis incorporated in the underbody. However, even with such box sections and subframes the easily spot welded and finished bodies provided a significant advance in bodyweight reduction while meeting most engineering and manufacturing criteria.

The unitary design (monocoque is referred to in the industry but some say this should be reserved for competition type bodies of tube configuration) is by far the most popular type of body and using the powerful FEM analytical programs that exist today (see below) the design can be optimized at the design stage to maximize the use of properties and thereby reduce the number of prototypes, rework and development time. The more numerical data that can be gathered at this stage related to materials behaviour the more efficient modelling will be, and this applies to other simulation processes besides those predicting dynamic and static behaviour such as

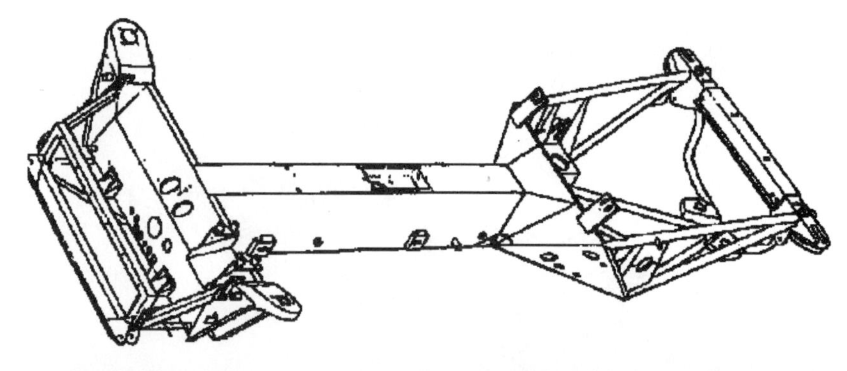

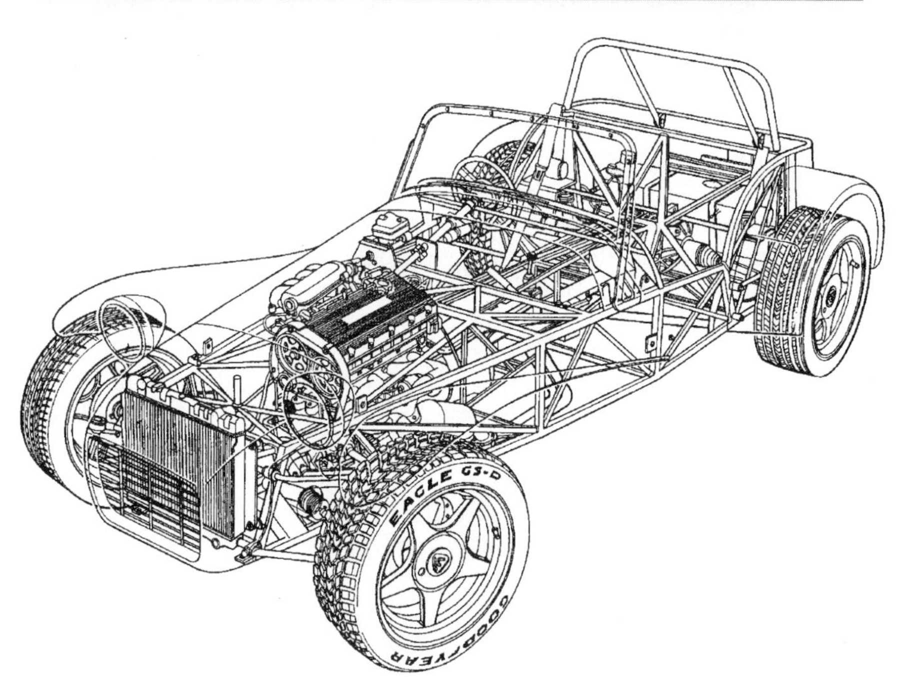

Fig. 18.1-4 Selection of alternative chassis designs.[2]

Fig. 18.1-4 *(contd).*

impact and torsional stiffness (demonstrated later). Forming is the obvious example but the complexity of accurately predicting thin shell behaviour during pressing brings in other variables as well as mechanical properties including friction, lubrication and topography.

18.1.3 Finite element analysis

For those readers requiring some basic understanding of FEA which is now a standard feature of computer-aided design (CAD) procedures used by body designers the following extract is presented from the publication 'Lightweight Electric/Hybrid Design' by Hodkinson and Fenton.[5]

This computerized structural analysis technique has become the key link between structural design and computer-aided drafting. However, because the small size of the elements usually prevents an overall view, and the automation of the analysis tend to mask the significance of the major structural scantlings, there is a temptation to by-pass the initial stages in structural design and perform the structural analysis on a structure which has been conceived purely as an envelope for the electromechanical systems, storage medium, passengers and cargo, rather than an optimized load-bearing structure. However, as well as fine-mesh analysis which gives an accurate stress and deflection prediction, course-mesh analysis can give a degree of structural feel useful in the later stages of conceptual design, as well as being a vital tool at the immediate pre-production stage.

One of the longest standing and largest FEA software houses is PAFEC who have recommended a logical approach to the analysis of structures, Fig. 18.1-6. This is

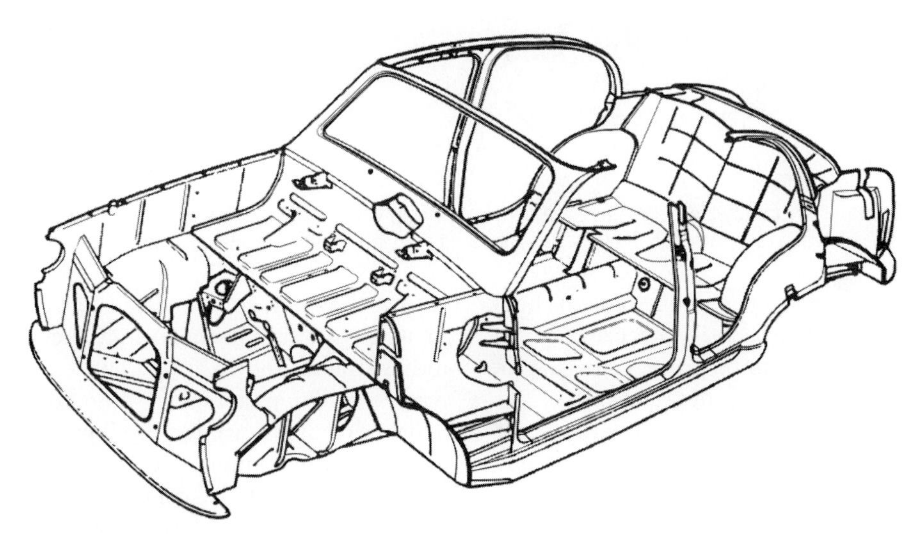

Fig. 18.1-5 Base structure of the Austin A30.

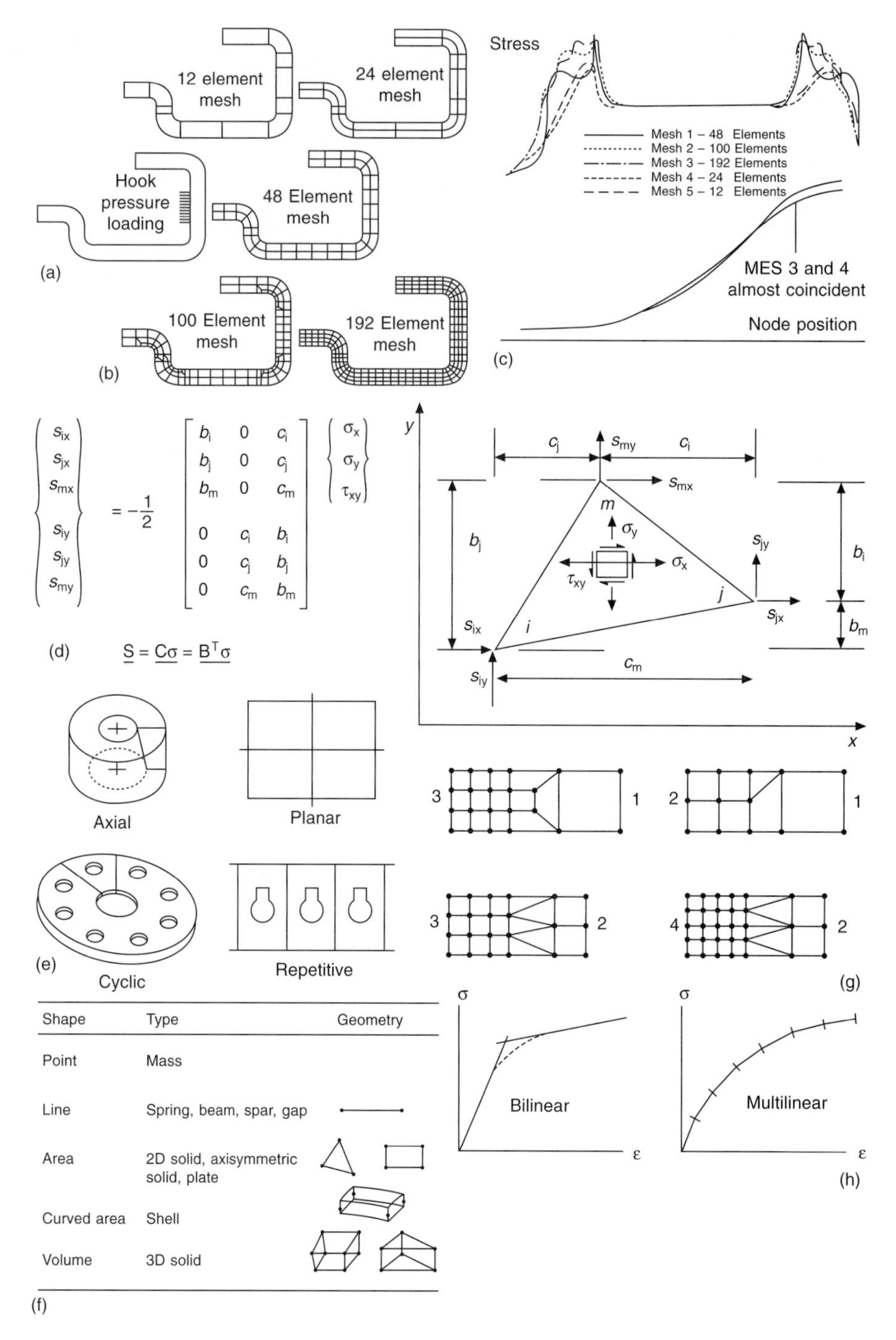

Fig. 18.1-6 Development of FEA: (a) towing hook as structural example; (b) various mesh densities; (c) FEA vs elasticity theory; (d) node equations in matrix form; (e) types of symmetry; (f) element shapes; (g) varying mesh densities; (h) stress–strain curve representation.

seen in the example of a constant-sectioned towing hook shown at (a). As the loading acts in the plane of the section the elements chosen can be plane. Choosing the optimum mesh density (size and distribution) of elements is a skill which is gradually learned with experience. Five meshes are chosen at (b) to show how different levels of accuracy can be obtained.

The next step is to calculate several values at various key points – using basic bending theory as a check. In this example nearly all the meshes give good displacement match with simple theory but the stress line-up is another story as shown at (c). The lesson is: where stresses vary rapidly in a region, more densely concentrated smaller elements are required; over-refinement could of course, strain computer resources.

Each element is connected to its neighbour at a number of discrete points, or nodes, rather than continuously joined along the boundaries. The method involves setting up relationships for nodal forces and displacements involving a finite number of simultaneous linear equations. Simplest plane elements are rectangles and triangles, and the relationships must ensure continuity of strain across the nodal boundaries. The view at (d) shows a force system for the nodes of a triangular element along with the dimensions for the nodes in the one plane. The figure shows how a matrix can be used to represent the coefficients of the terms of the simultaneous equations.

Another matrix can be made up to represent the stiffness of all the elements $[K]$ for use in the general equation of the so-called 'displacement method' of structural analysis:

$$[R] = [K] \cdot [r]$$

where $[R]$ and $[r]$ are matrices of external nodal forces and nodal displacements; the solution of this equation for the deflection of the overall structure involves the inversion of the stiffness matrix to obtain $[K]^{-1}$. Computer manipulation is ideal for this sort of calculation.

As well as for loads and displacements, FEA techniques, of course, cover temperature fields and many other variables and the structure, or medium, is divided up into elements connected at their nodes between which the element characteristics are described by equations. The discretization of the structure into elements is made such that the distribution of the field variable is adequately approximated by the chosen element breakdown. Equations for each element are assembled in matrix form to describe the behaviour of the whole system. Computer programs are available for both the generation of the meshes and the solution of the matrix equations, such that use of the method is now much simpler than it was during its formative years.

Economies can be made in the discretization by taking advantage of any symmetry in the structure to restrict the analysis to only one-half or even one-quarter – depending on degree. As well as planar symmetry, that due to axial, cyclic and repetitive configuration, seen at (e), should be considered. The latter can occur in a bus body, for example, where the structure is composed of identical bays corresponding to the side windows and corresponding ring frame.

Element shapes are tabulated in (f) – straight-sided plane elements being preferred for the economy of analysis in thin-wall structures. Element behaviour can be described in terms of 'membrane' (only in-plane loads represented), in bending only or as a combination entitled 'plate/shell'. The stage of element selection is the time for exploiting an understanding of basic structural principles; parts of the structure should be examined to see whether they would typically behave as a truss frame, beam or in plate bending, for example. Avoid the temptation to over-model a particular example, however, because number and size of elements are inversely related, as accuracy increases with increased number of elements.

Different sized elements should be used in a model – with high mesh densities in regions where a rapid change in the field variable is expected. Different ways of varying mesh density are shown at (g), in the case of square elements. All nodes must be interconnected and therefore the fifth option shown would be incorrect because of the discontinuities.

As element distortion increases under load, so the likelihood of errors increases, depending on the change in magnitude of the field variable in a particular region. Elements should thus be as regular as possible – with triangular ones tending to equilateral and rectangular ones tending to square. Some FEA packages will perform distortion checks by measuring the skewness of the elements when distorted under load. In structural loading beyond the elastic limit of the constituent material an idealized stress/strain curve must be supplied to the FEA program – usually involving a multilinear representation, (h).

When the structural displacements become so large that the stiffness matrix is no longer representational then a 'large-displacement' analysis is required. Programs can include the option of defining 'follower' nodal loads whereby these are automatically reorientated during the analysis to maintain their relative position. The program can also recalculate the stiffness matrices of the elements after adjusting the nodal coordinates with the calculated displacements. Instability and dynamic behaviour can also be simulated with the more complex programs.

The principal steps in the FEA process are: (i) idealization of the structure (discretization); (ii) evaluation of stiffness matrices for element groups; (iii) assembly of these matrices into a supermatrix; (iv) application of constraints and loads; (v) solving equations for nodal displacements; and (vi) finding member loading. For vehicle body design, programs are available which automate these steps, the input of the design engineer being, in programming, the analysis with respect to a new model introduction. The first stage is usually the obtaining of static and dynamic stiffness of the shell, followed by crash performance based on the first estimate of body member configurations. From then on it is normally a question of structural refinement and optimization based on load inputs generated in earlier model durability cycle testing. These will be conducted on relatively course mesh FEA models and allow section properties of pillars and rails to be optimized and panel thicknesses to be established.

In the next stage, projected torsional and bending stiffnesses are input as well as the dynamic frequencies in these modes. More sophisticated programs will generate new section and panel properties to meet these criteria. The inertias of mechanical running units, seating and trim can also be programmed in and the resulting model examined under special load cases such as pot-hole road obstacles. As structural data is refined and updated, a fine-mesh FEA simulation is prepared which takes in such detail as joint design and spot-weld configuration. With this model a so-called sensitivity analysis can be carried out to gauge the effect of each panel and rail on the overall behaviour of the structural shell.

Joint stiffness is a key factor in vehicle body analysis and modelling them normally involves modifying the local properties of the main beam elements of a structural shell. Because joints are line connections between panels, spot-welded together, they are difficult to represent by local FEA models. Combined FEA and EMA (experimental modal analysis) techniques have thus been proposed to 'update' shell models relating to joint configurations. Vibrating mode shapes in theory and practice can thus be compared. Measurement plots on physical models excited by vibrators are made to correspond with the node points of the FEA model and automatic techniques in the computer program can be used to update the key parameters for obtaining a convergency of mode shape and natural frequency.

An example car body FEA at Ford was described at one of the recent Boditek conferences, Fig. 18.1-7, outlining the steps in production of the FEA model at (a). An extension of the PDGS computer package used in body engineering by the company – called Finite-Element Analysis System(FAST) – can use the geometry of the design concept existing on the computer system for fixing of nodal points and definition of elements. It can check the occurrence of such errors as duplicated nodes or missing elements and even when element corners are numbered in the wrong order. The program also checks for misshapen elements and generally and substantially compresses the time to create the FEA model.

The researchers considered that upwards of 20 000 nodes are required to predict the overall behaviour of the body-in-white. After the first FEA was carried out, the deflections and stresses derived were fed back to PDGS-FAST for post-processing.

This allowed the mode of deformation to be viewed from any angle – with adjustable magnification of the deflections – and the facility to switch rapidly between stressed and unstressed states. This was useful in studying how best to reinforce part of a structure which

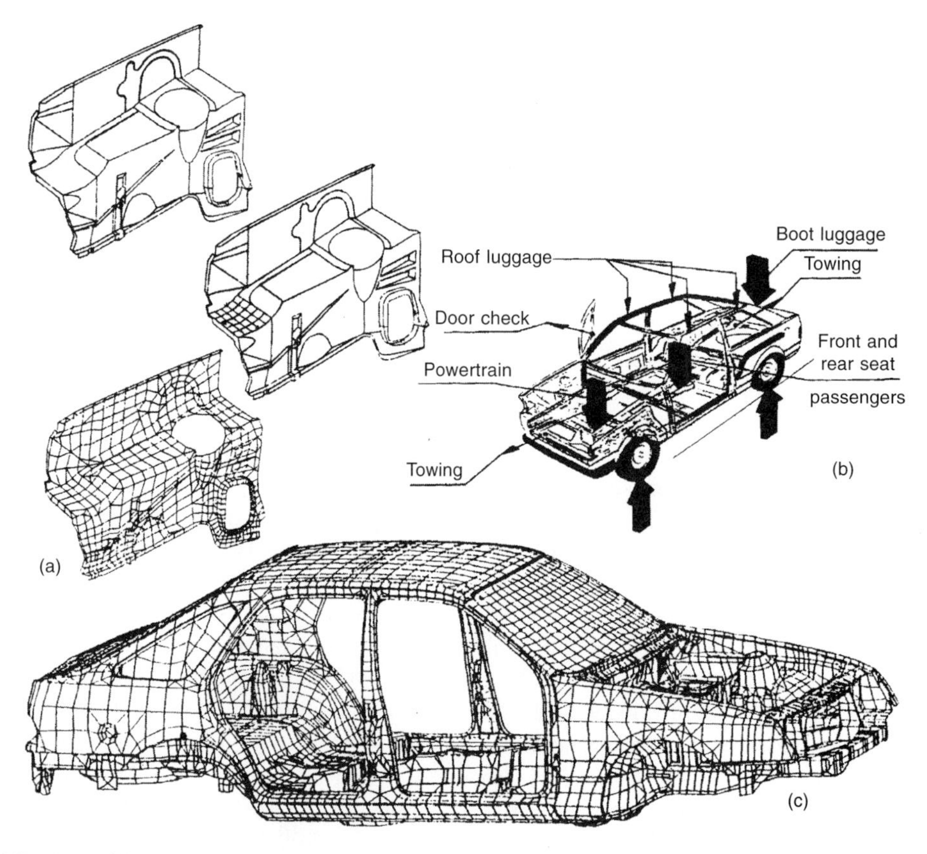

Fig. 18.1-7 FEA of Ford car: (a) steps in producing FEA model; (b) load inputs; (c) global model for body-in-white (BIW).

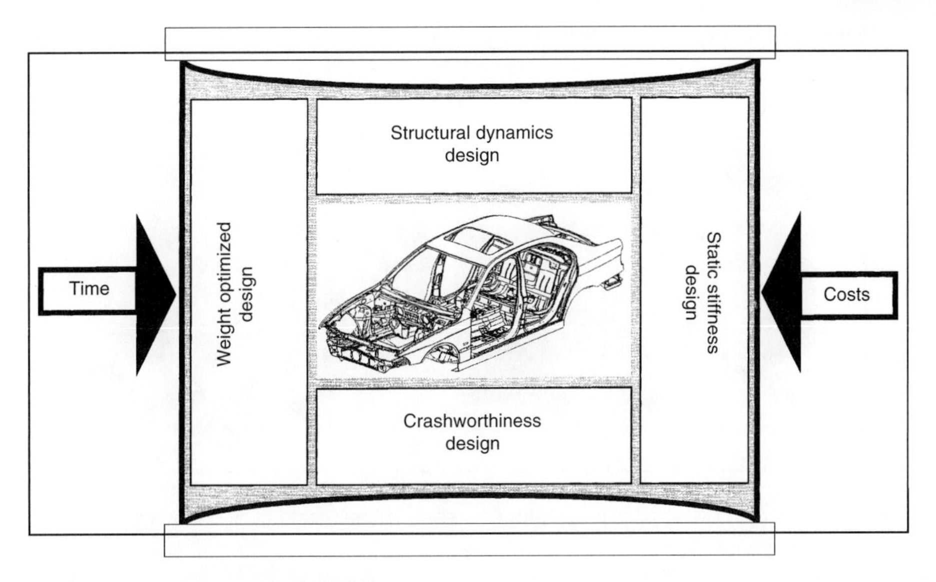

Fig. 18.1-8 Critical areas for body design (after Ludke[6]).

deforms in a complex fashion. Average stress values for each element can also be displayed numerically or by graduated shades of colour. Load inputs were as shown at (b) and the FE model for the BIW at (c).

18.1.4 One manufacturer's approach to current design

It is now timely to consider the more contemporary approach to design and reference is made to the approach that BMW have adopted to utilizing materials to optimize structural performance while at the same time satisfying prevailing safety, performance and environmental requirements. Their progress is illustrated by extracts taken from information presented recently by Bruno Ludke, BMW Body Design specialist, at recent international automotive conferences (Fig. 18.1-8).

18.1.4.1 Product requirements

In terms of lightweight bodyshell functional design Ludke[6] has identified four areas for critical consideration:

- Structural dynamics
- Static stiffness
- Crashworthiness
- Weight optimization

18.1.4.2 Structural dynamics

Improvements in performance including significant weight savings in the steel body achieved over recent model generations are described in the following sections and these are attributed to the effective application of FEM analysis and the interrelationship with material properties.

Structural dynamics is described[6] as the achievement of the desired level of comfort in terms of noise, vibration and harshness (NVH) for which the yardstick is taken as behaviour at idling speed – normally between 600 and 700 rpm. To ensure 'vibration-free' operation, the frequencies for the first bending and torsional natural modes of the complete vehicle must lie within a limited frequency range. The upper limit of this range is represented by the third engine order of the 6-cylinder engine and the lower limit by the second engine order of the 4-cylinder engines thus constituting an 'idle frequency window'. To attain the target frequencies of 26/29 Hz for the vehicle as a whole, the corresponding natural modes of the bodyshell must be twice as high and no local modes must occur below these frequencies, e.g. at the front or rear of the vehicle. The improvements achieved with the outstandingly popular 3, 5 and 7 series BMW models are illustrated schematically in Fig. 18.1-9 and the success in this area is attributed primarily to the application of FEM analysis and experimental modal analysis technique applied at the early stages of body shell development.

18.1.4.3 Design for static stiffness

Static design entails the optimization of torsional stiffness and strength under quasistatic loading conditions and good static stiffness values are fundamental requirements for target dynamic characteristics previously described. The variation in torsional stiffness with vehicle kerb weight (K_w) has been developed as shown in

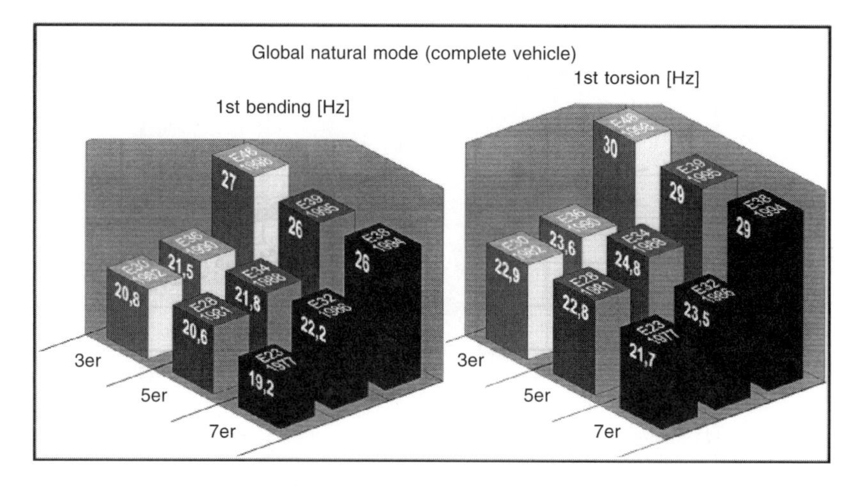

Fig. 18.1-9 Progressive improvement in dynamic stiffness with successive generations of BMW vehicles.

Fig. 18.1-10 for BMW models, the target C_t value being $15 \times K_w$. To avoid excessive loading of the windscreen and stone chipping damage resulting from excessive surface stresses, the inherent stiffness without glass must reach 66 per cent of the final stiffness. Specific design improvements were made in the latest models to key joints and structural members to increase torsional stiffness from 20 000 N m/deg to 28 500 N m/deg. Again the progression through successive BMW models is shown in Fig. 18.1-10 with a doubling of previous values.

18.1.4.4 Crashworthiness

All vehicle manufacturers are placing continued emphasis on occupant passive safety and here FEM simulation is of special importance, avoiding the need for expensive vehicle compliance tests during development. In the case of more recent models referred to above the stiffness and dynamic improvements form an excellent basis for crash optimization, and as requirements are aligned to 40 mph impacts the absorbed energy per structural unit (vehicle side) has risen by 80 per cent in comparison with predecessors. The shift in design requirements over the last 25 years is illustrated in Figs. 18.1-11 and 18.1-12 together with the configurations used for modelling 40 mph offset crash and side impact simulations.

18.1.4.5 Weight efficiency

Although a basic design requirement previously, the drive for lower weight vehicles, in the knowledge that 10 per cent reduction in vehicle mass leads to fuel savings of up to 6–7 per cent has intensified over the last 20 years. In the 1970s and 1980s the initial momentum swung towards aluminium as the industry attempted to confirm the fuel economy figures using the most radical materials solution available at the time. The ECV and ASV programmes together with the aluminium structured A8, A2, NSX and Z8 programmes described elsewhere in this chapter, have helped confirm more efficient

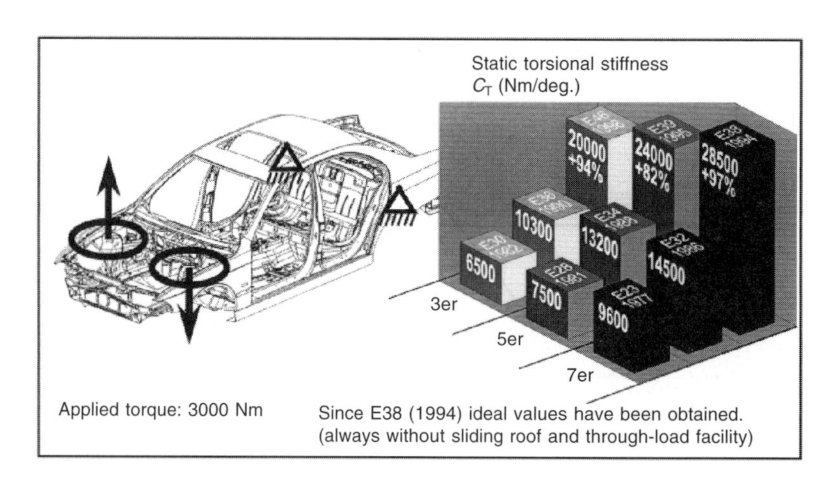

Fig. 18.1-10 Progressive improvement in torsional stiffness shown for successive BMW model generations.[6]

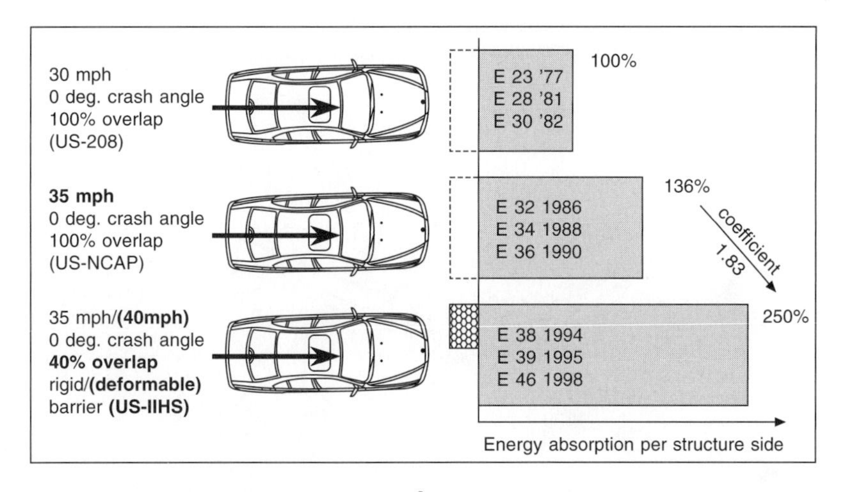

Fig. 18.1-11 Increasingly stringent objectives for crashworthiness.[6]

consumption figures but have also underlined the substantial changes in the supply and process chain manufacturing facilities required together with peripheral costs such as higher repair and subsequently insurance costs. It is these factors which may account for the slow emergence of aluminium as a significant body material despite more positive forecasts and the fact that most major organizations have now gained the technological and design experience (with low volume derivatives) to enter full-scale production.

The feeling also existed in the early 1990s that a lot more potential for weight reduction still existed with

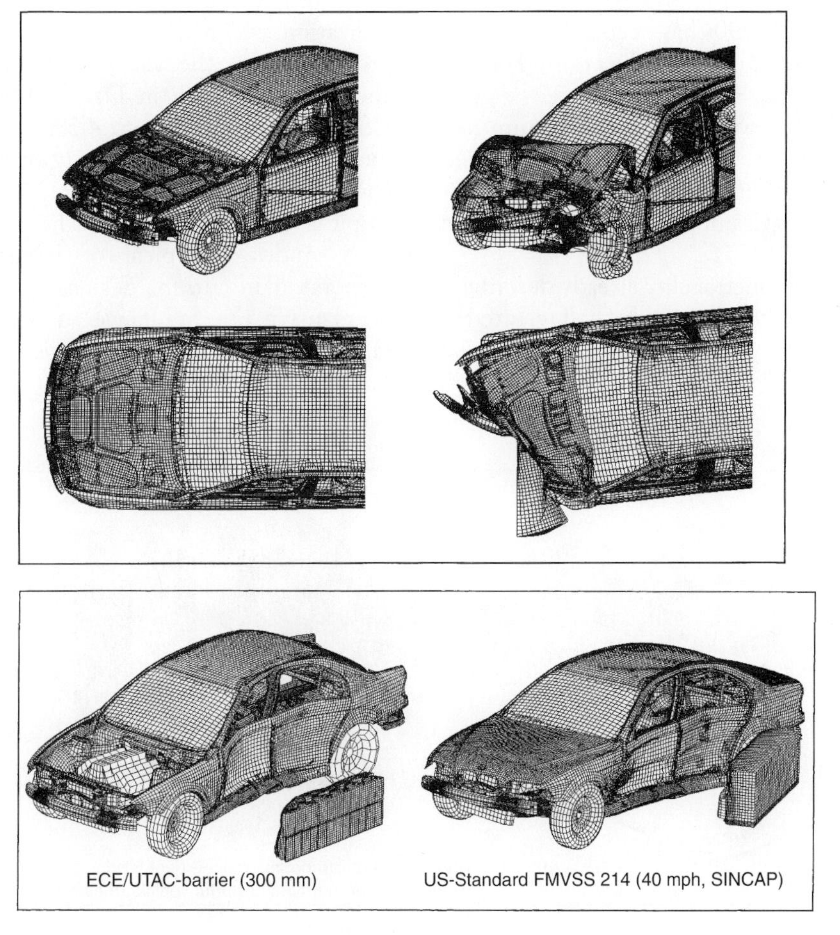

Fig. 18.1-12 Offset barrier (top) and side impact simulations (below).[6]

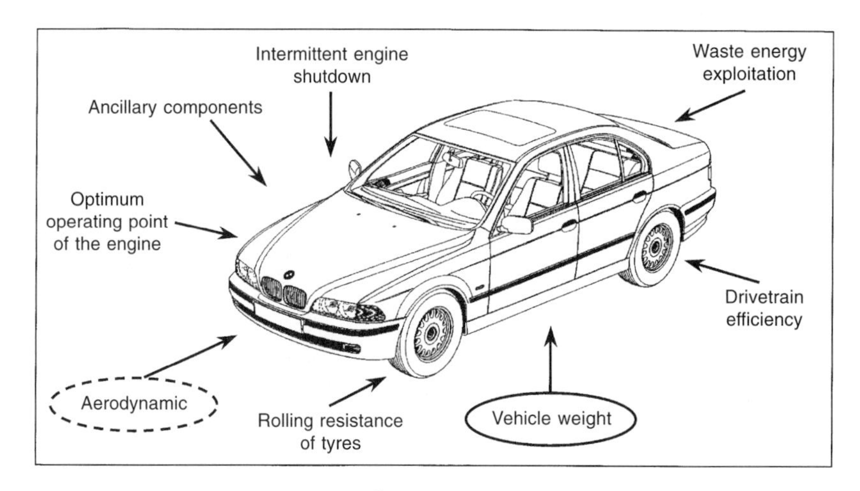

Fig. 18.1-13 Factors contributing to improved fuel economy.[6]

steel albeit in slightly different guises which would yield useful weight savings, and perhaps if mixed with lighter materials such as aluminium or plastic skins would allow most future objectives to be achieved. The more flexible facilities and experience gained over recent years could be adapted to accept newer high strength materials and different configurations such as TWBs and hydroformed tube sections. It was with this knowledge that the steel industries response in the 1990s has been a design study undertaken on behalf of 32 steel producers by Porsche Engineering Services.

The significance of bodyweight on fuel consumption, acceleration and emissions control has been outlined already, but to put these in perspective with other relevant factors some of these parameters are illustrated in Fig. 18.1-13.

Despite the improved functionality already described, the optimized unitary design of the body shell resulted in it making up a significantly reduced proportion of the kerb weight as shown in Fig. 18.1-14.

A factor '*L*' has been used by BMW to summarize the weight reduction improvement effected by design, which relates to structural performance and vehicle size and is shown in Fig. 18.1-15 together with the progressive achievement over the years and although relatively empirical does provide a measure of design optimization.

More specific materials related data was further presented by Ludke[7] who referred to the changes in HSS utilization which had accompanied the functional improvements in the various BMW models referred to above.[6,7] As shown in Fig. 18.1-16 the proportion of HSS was increased from 4.5 per cent for the 5-Series model to 50 per cent. This utilized the range of bake-hardening steels H180B to H300B together with isotropic and IF HSS grades. The increased utilization of HSS grades is typical of strengths now being incorporated in current designs by European body engineers using the full range of rephosphorized, IF HS, HSLA and bake hardening grades which are included in Euronorm 10292. This covers hot-dip galvanized grades,

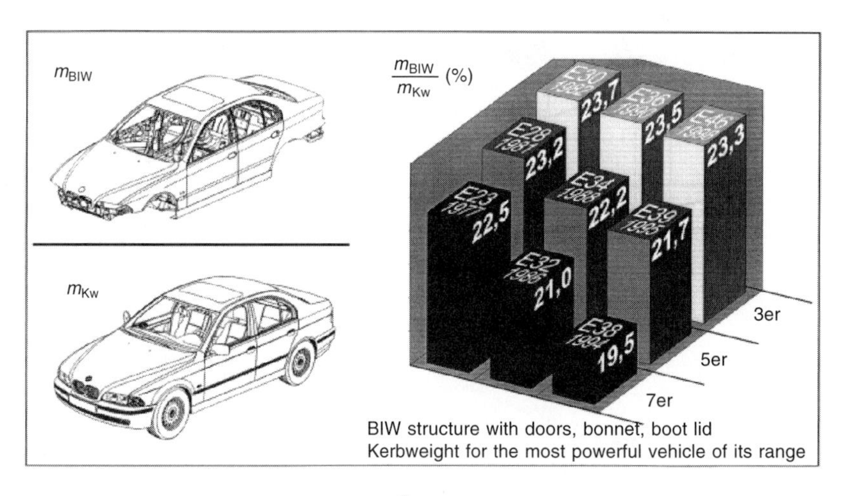

Fig. 18.1-14 Relationship of body-in-white weight to kerbweight.[6]

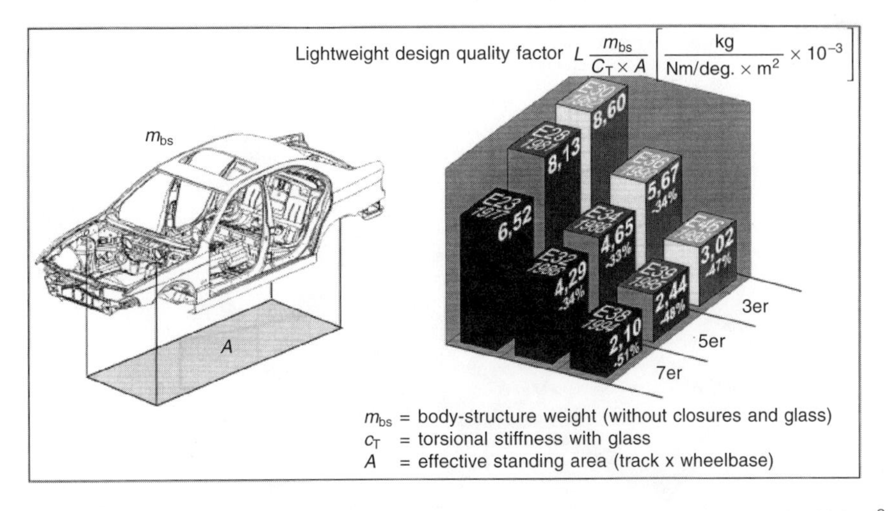

Fig. 18.1-15 Design efficiency as defined by functional optimization and size, over three generations of vehicles.[6]

again reflecting the improvements and change in durability required by today's structures, but an uncoated parallel standard is in preparation.

The principal parameters (referred to in Table 18.2-2) used in the analysis of the structural and panel components are shown in Fig. 18.1-17.

Similar analyses were carried out on bolt-on assemblies but totally different criteria apply as will become evident in the following section.

Evaluation of individual requirements of each part are made at an early simulation stage (Fig. 18.1-18). One method of enhancing stiffness is to use linear laser welding or apply a structural adhesive to inner flange or seam surfaces and this is a key development for the future. It is generally recognized that a compatible pretreatment such as those used in aircraft construction is required to ensure that any degradation of the bond does not occur in service and it is equally important to maintain coating integrity through forming and assembly to ensure a firm foundation for the adhesive system. As described previously the influence of material properties on impact and collapse characteristics is becoming more evident with the development of dual-phase and TRIP steels which, due to unique work hardening and ductility combinations (increased area under the stress/strain curve), offer increased energy absorption. Again enhancement of these properties should be possible in combination with adhesive application leading to a HSS utilization of 80–90 per cent (Fig. 18.1-19).

18.1.5 Panel dent resistance and stiffness testing

Optimized designs of outer body panels must also meet several other performance criteria including stiffness, oil

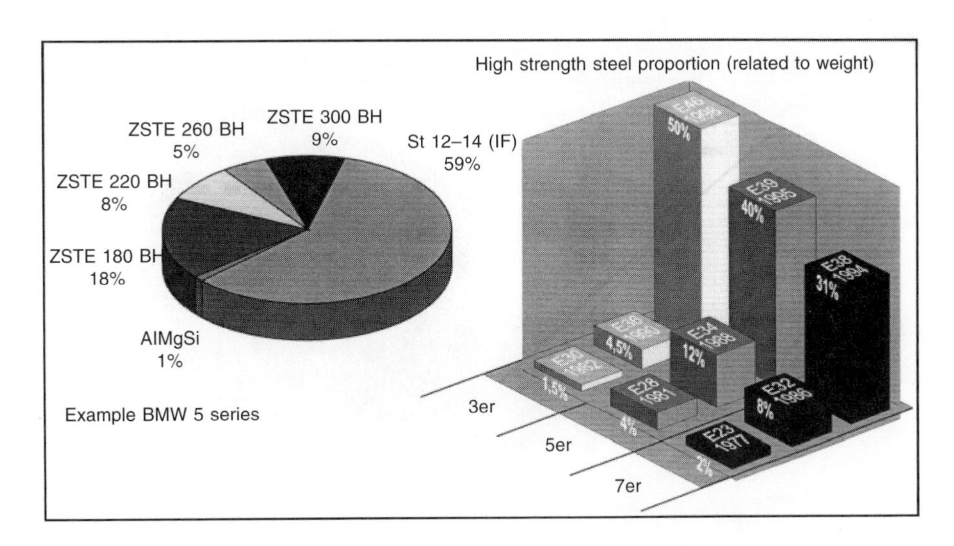

Fig. 18.1-16 Increasing proportion of high strength steel used in BMW body structures.[7]

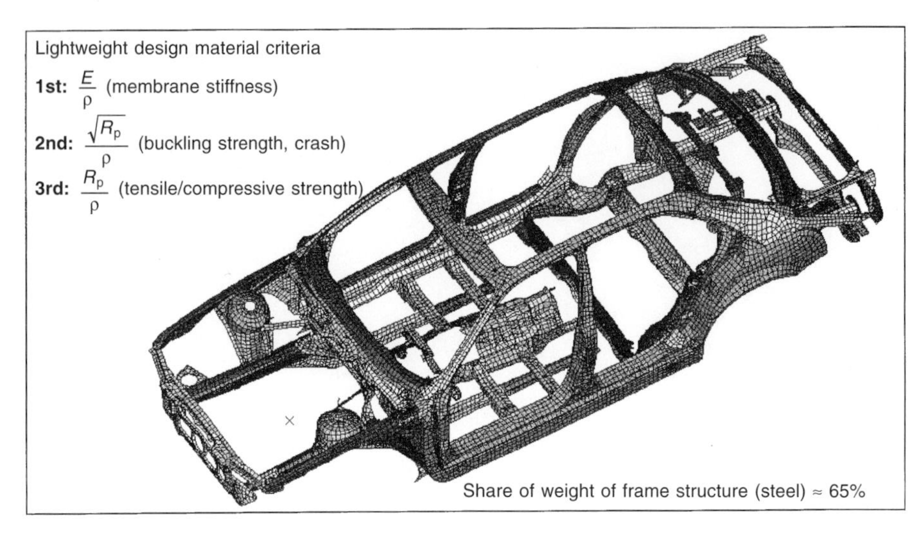

Fig. 18.1-17 Criteria used in the analysis of structural components by BMW.[6]

canning or critical buckling load and dent resistance. Stiffness is a fundamental concern for the perceived quality of a body panel. Along with oil canning (the 'popping' of a panel when pressed) it determines how the panel 'feels' to a customer. Dent resistance is important to avoid panel damage in-plant and minimize dents and dings on external parts in-service. Poor panel quality in used cars will generally depress resale values and possibly influence the decision to purchase a particular brand.

From a practical point of view, dents can be caused in a number of ways and on the full range of external body panels. Considering doors as an example, denting can occur from stone impacts (dynamic denting) or, to the frustration of the vehicle owner, from the careless opening of an adjacently parked vehicle door. Denting can occur where the door surface is smooth and may not have sufficient curvature to resist 'door slamming' (quasi-static denting) or along prominent feature lines where 'creasing' can occur.

Panel dent resistance and stiffness has been the subject of considerable research. Despite this there is no industry-wide, generally adopted method of testing. For quasi-static dent testing, a wide range of purpose built dent resistance/stiffness test equipment configurations have been employed within the automotive industry. In addition, the configuration of a tensile testing machine for compression testing and similar modified equipment has allowed suitable data to be obtained. Whichever system is used, the principle of force application resulting in deflection and ultimately plastic deformation of the panel remains the same. Variables can include method of load application (hydraulic or stepper motor), speed of load application, indenture shape and size and panel assembly conditions. Some reported methods of testing are based on repeated application and removal of

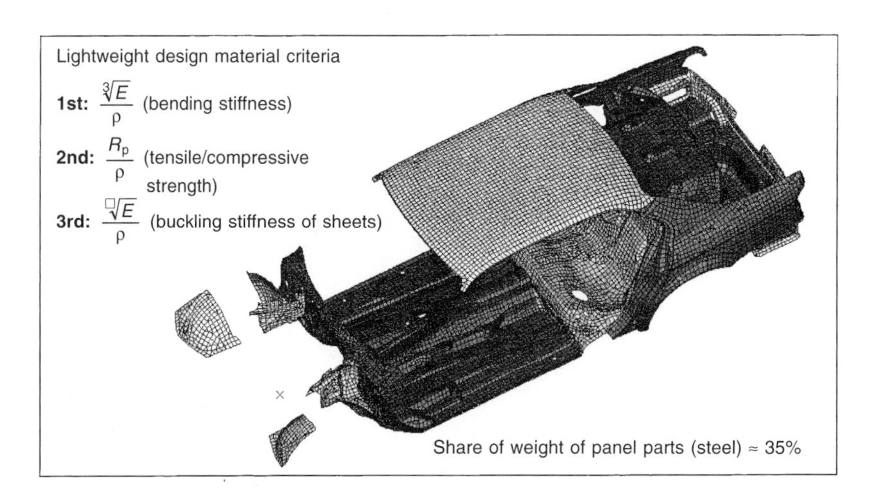

Fig. 18.1-18 Criteria used in the analysis of panel parts by BMW.

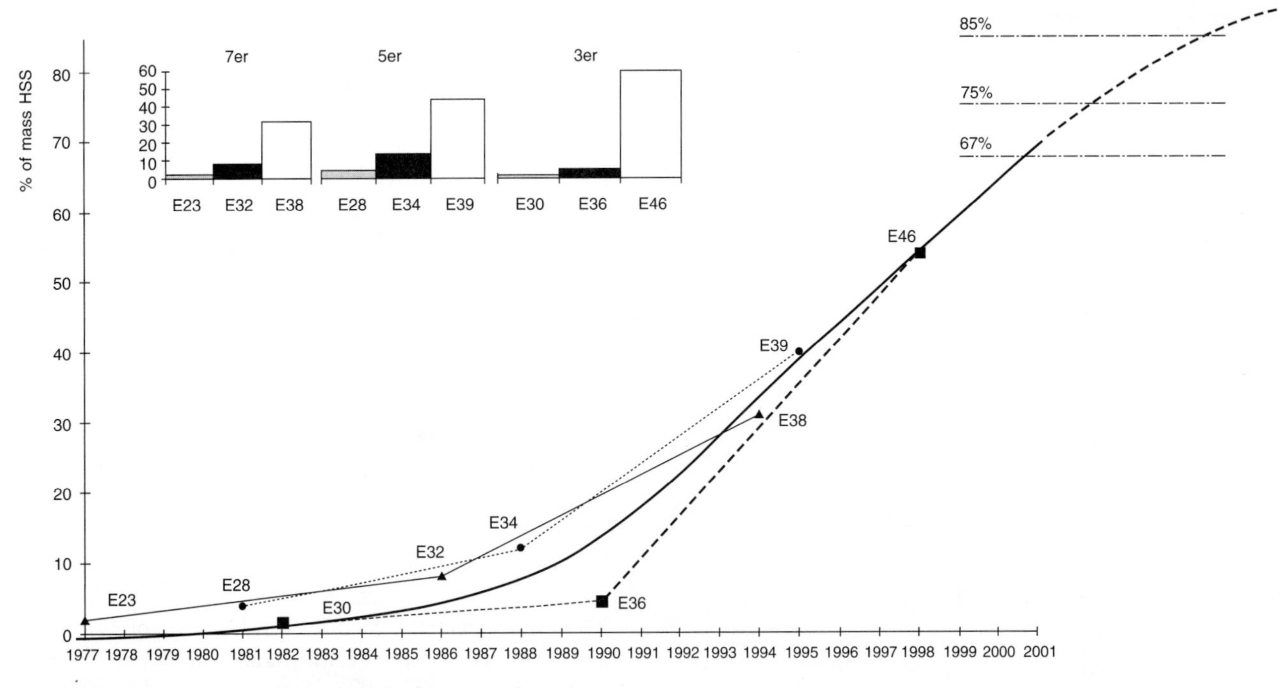

Fig. 18.1-19 Possible scenario of future HSS usage.[7]

force at increasing levels. Others involve the continued application of a steadily increasing force until denting occurs. In some cases, stiffness is assessed using the same basic test equipment but with a much larger radiussed loading head to prevent localized deformation. Force and displacement measurements are generally incorporated into a data acquisition system.

An illustration of a typical output from such a test would be as represented in Fig. 18.1-20. Initial stiffness is given by the slope of the curve in the first region, until the buckling load is reached. After 'oil canning', the panel continues to deflect elasticity, before the onset of plastic deformation in the material. When the load is reversed, the permanent deformation of the panel is indicated since the lower portion of the curve does not return to zero.

Experimental testing of dynamic dent resistance has previously concentrated on drop weight rigs, using various indentor masses and drop heights to achieve a range of denting energies. Even higher energies can be achieved through the use of a compressed air operated ball bearing gun. The key issue to be considered is that test conditions (denting force, impact speed, etc.) must be genuinely representative of those conditions existing in field, i.e. if the energy input to cause a perceptible dent in-service is 10 J, then the dent testing procedure should reflect this.

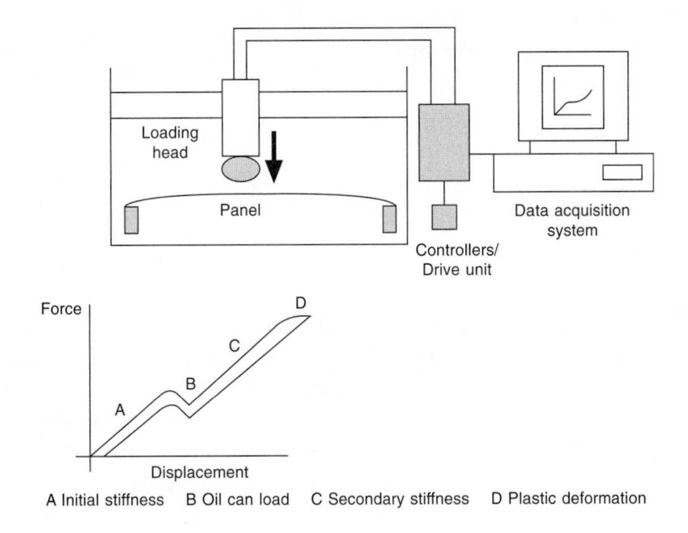

Fig. 18.1-20 Test rig for measurement of panel dent resistance and stiffness.

Test results generated from the above techniques will typically be compared against performance standards set by an individual manufacturer. Standards widely known include those published by the American Iron and Steel Institute which defines a minimum dent resistance of 9.7 J and a stiffness that should exceed 45 N/mm.

Based on testing using the practical techniques outlined, empirical formulae predicting the force and energy required to initiate a dent have been presented in recent years. Typically:

$$W = (K.YS^2.t^4)/S$$

where W is the denting energy, K is a constant, YS is the material yield strength, t is the panel thickness and S is the panel stiffness. Panel stiffness depends upon the elastic modulus, the panel thickness, shape and geometry and boundary conditions. The ability of plastic panels to meet light denting is a definite advantage (as long as 40 years ago Henry Ford could be seen striking a Ford development vehicle with plastic panels, to demonstrate the ability of plastic/composites to resist denting. Nonetheless, HSS offer more dent resistance at the same thickness as mild steel, or the opportunity for weight saving and equivalent dent resistance at reduced panel thickness.

Given the many iterations of automotive body panel design that can take place, it is usually late in the product development process that the first production representative parts are available for dent and stiffness testing. With press tooling already produced, it is generally only initial material properties that can be changed or local reinforcements added to improve the stiffness/dent resistance. It is not surprising, therefore, that currently much attention is being focussed on the use of analytical tools such as finite element analysis, for body panel performance predictions. Thus, given certain part geometry and dimensions, predictions of stiffness and dent resistance can be made. Based on material gauge and grade and in the case of metallic panels, strain levels in the material, optimization of the design can take place. Should the accuracy of such techniques be proven, the use of dent and stiffness testing equipment may in future be limited to selected verification of such performance predictions and quality control issues.

18.1.6 Fatigue

The behaviour of sheet materials under conditions of constantly fluctuating stress or strain is of critical importance to the life body structure, whether of high or low frequency. High cycle fatigue is more descriptive of conditions existing, say, in close proximity to the engine compartment, while low cycle conditions represent those induced by humps and bumps encountered in road running. Both are assessed very carefully in the initial engineering selection procedure, the high cycle behaviour being determined using Wohler S–N curves, often used as the input to CAD design programmes. Steels typically give a clearly defined fatigue limit below which components can be designed in relative safety but aluminium gives a steady stress reduction with time. A cautionary note in using cold work strengthening – low cycle fatigue can induce a progressive cyclic softening, which can counteract the strengthening developed by strain ageing as well as cold deformation.

The behaviour of a particular design is very difficult to predict due to the nature of materials characteristics combined with the complexity of design features comprising all body shapes, and which can result in stress concentrations. Therefore, despite extensive measurements and predictive programmes the only true way to determine the sensitivity of a structure to cyclic behaviour is rig testing. This can take the form of simple push–pull load application or extend to four poster simulated movements gathered under arduous track testing. Push–pull tests even of the simple tensile test type must be carried out carefully to avoid buckling effects which may limit the range of thicknesses on which these tests may be used. If investment can be made in the hydraulic facilities necessary for the full rig simulation, these are the only realistic means of detecting weaknesses prone to cyclic failure apart of course from the accelerated track tests over rough terrain.

Weaknesses can be identified by the application of stress lacquer techniques or similar, and modification carried out by localized strengthening. The effect of material properties is debatable as again body features are claimed to negate these, especially when considering spot welded joints. Many studies have shown that with HSS the notch effects associated with the weld geometry overpower any effect due to material strength.

A more lengthy description of the fatigue process and body design follows and is borrowed from *Lightweight Electric/Hybrid Vehicle Design*[5] and presents a concise summary of factors applicable to fatigue resistance, relevant to most body structures.

18.1.6.1 Designing against fatigue

Dynamic factors should also be built in for structural loading, to allow for travelling over rough roads. Combinations of inertia loads due to acceleration, braking, cornering and kerbing should also be considered. Considerable banks of road load data have been built up by testing organizations and written reports have been recorded by MIRA and others. As well as the normal loads which apply to two wheels riding a vertical

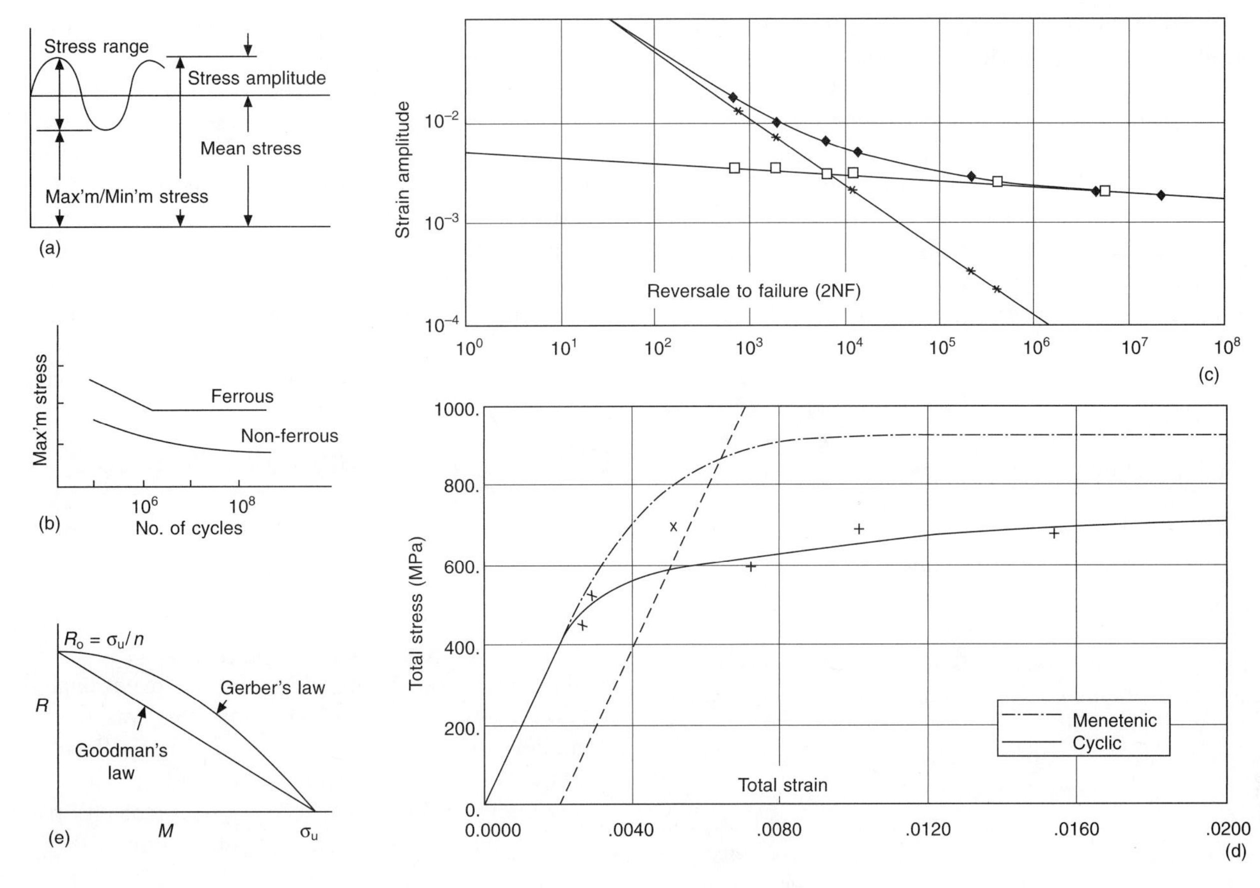

Fig. 18.1-21 Fatigue life evaluation: (a) terminology for cyclic stress; (b) *S–N* diagram; (c) strain/life curves; (d) dynamic stress/strain curves; (e) fatigue limit diagrams.

obstacle, the case of the single wheel bump, which causes twist of the structure, must be considered. The torque applied to the structure is assumed to be 1.5 × the static wheel load × half the track of the axle. Depending on the height of the bump, the individual static wheel load may itself vary up to the value of the total axle load.

As well as shock or impact loading, repetitive cyclic loading has to be considered in relation to the effective life of a structure. Fatigue failures, in contrast to those due to steady load, can of course occur at stresses much lower than the elastic limit of the structural materials, Fig. 18.1-21. Failure normally commences at a discontinuity or surface imperfection such as a crack which propagates under cyclic loading until it spreads across the section and leads to rupture. Even with ductile materials failure occurs without generally revealing plastic deformation. The view at (a) shows the terminology for describing stress level and the loading may be either complete cyclic reversal or fluctuation around a mean constant value. Fatigue life is defined as the number of cycles of stress the structure suffers up until failure. The plot of number of cycles is referred to as an *S–N* diagram, (b), and is available for different materials based on laboratory controlled endurance testing. Often they define an endurance range of limiting stress on a 10 million life cycle basis. A log–log scale is used to show the exponential relationship $S = C. Nx$ which usually exists, for C and x as constants, depending on the material and type of test, respectively. The graph shows a change in slope to zero at a given stress for ferrous materials – describing an absolute limit for an indefinitely large number of cycles. No such limit exists for non-ferrous metals and typically, for aluminium alloy, a 'fatigue limit' of 5 × 10^8 is defined. It has also become practice to obtain strain/life (c) and dynamic stress/strain (d) for materials under sinusoidal stroking in test machines. Total strain is derived from a combination of plastic and elastic strains and in design it is usual to use a stress/strain product from these curves rather than a handbook modulus figure. Stress concentration factors must also be used in design.

When designing with load histories collected from instrumented past vehicle designs of comparable specification, signal analysis using rainflow counting techniques is employed to identify number of occurrences in each load range. In service testing of axle beam loads it has been shown that cyclic loading has also occasional peaks, due to combined braking and kerbing, equivalent to four times the static wheel load. Predicted life based on specimen test data could be twice that obtained from service load data. Calculation of the damage contribution of the individual events counted in the rainflow analysis can be compared with conventional cyclic fatigue data to obtain the necessary factoring. In cases where complete load reversal does not take place and the load alternates between two stress values, a different (lower) limiting stress is valid. The largest stress amplitude which alternates about a given mean stress, which can be withstood 'infinitely', is called the fatigue limit. The greatest endurable stress amplitude can be determined from a fatigue limit diagram, (e), for any minimum or mean stress. Stress range R is the algebraic difference between the maximum and minimum values of the stress. Mean stress M is defined such that limiting stresses are $M+/-R/2$.

Fatigue limit in reverse bending is generally about 25% lower than in reversed tension and compression, due, it is said, to the stress gradient – and in reverse torsion it is about 0.55 times the tensile fatigue limit. Frequency of stress reversal also influences fatigue limit – becoming higher with increased frequency. An empirical formula due to Gerber can be used in the case of steels to estimate the maximum stress during each cycle at the fatigue limit as $R/2 + (\sigma_u 2 - nR\sigma_u)^{1/2}$ where σ_u is the ultimate tensile stress and n is a material constant = 1.5 for mild and 2.0 for high tensile steel. This formula can be used to show the maximum cyclic stress σ for mild steel increasing from one-third ultimate stress under reversed loading to 0.61 for repeated loading. A rearrangement and simplification of the formula by Goodman results in the linear relation $R = (\sigma_u/n)\ [1 - M/\sigma_u]$ where $M = \sigma - R/2$. The view in (e) also shows the relative curves in either a Goodman or Gerber diagram frequently used in fatigue analysis. If values of R and σ_u are found by fatigue tests then the fatigue limits under other conditions can be found from these diagrams.

Where a structural element is loaded for a series of cycles $n1$, $n2$... at different stress levels, with corresponding fatigue life at each level $N1$, $N2$... cycles, failure can be expected at $\sum n/N = 1$ according to Miner's law. Experiments have shown this factor to vary from 0.6 to 1.5 with higher values obtained for sequences of increasing loads.

18.1.7 Alternative body architecture

Before examples of more adventurous modern designs are presented, certain vehicles are now highlighted which illustrate further interesting steps in body and materials development. Having commenced with essentially steel bodies of unitary design, and it must be remembered that these still constitute the vast majority of volume cars produced, 'conventionally built' aluminium structures are considered, before moving to the spaceframe concept and finally the inevitable hybrid configurations. In this context hybrid means mixed material content and introduces the 'user friendly' advantages of polymers (low impact resistance and styling freedom) combined with the lightweight advantage of aluminium plus the practicality and safety connotations associated with steel. The latter has always been a strong argument of the anti-CAFÉ lobby in the USA who contend that the benefits in fuel economy accompanying hybrid lighter-weight bodies are achieved at the expense of vehicle safety and claim to have accident statistics to prove this.

18.1.7.1 The unitary aluminium body

The development of the all aluminium body is now more associated with the A8 and A2 spaceframe type of vehicle (described later) which constitutes a different type of concept, and the need for a fundamentally different type of design may become more obvious if the production of an aluminium body is considered with conventional production technology. Although some use was made of aluminium prior to 1900 for the Durkopp developed sports car[8] and later reference was made to the Pierce Arrow body (1909) which incorporated rear end panel, roof, firewall and doors in cast aluminium, the Dyna Panhard was probably the first aluminium bodied car to be mass produced in Europe in any numbers. The Honda NSX sports car represents the most recent conventional body built within the context of modern manufacturing and proves that although some equipment modifications were necessary assembly was possible in moderate numbers.

18.1.7.1.1 The Honda NSX

Following a consideration of specific strength, specific rigidity and equivalent rigidity compared with sheet steel and SMC[9] the decision was made to manufacture the BIW in aluminium to reduce the weight by about 140 kg. The rigidity of a car such as the NSX is critical to maintain steering stability, and to help improve this the sills were produced as extrusions with variable side wall thickness. The comparison is shown in Fig. 18.1-22.

To satisfy different requirements for strength, formability, weldability and coating, detailed preparatory background studies showed that different alloys should be used for different panel applications and these are indicated in Fig. 18.1-23.

It was found that wrinkling and shape control were the main problems on forming, attributed to lower modulus which resulted in more springback (compared to steel) and also the lower 'r' value. Twice the overcrowning allowance was required than for steel in the forming of door outer panels. Together with proportionally lower

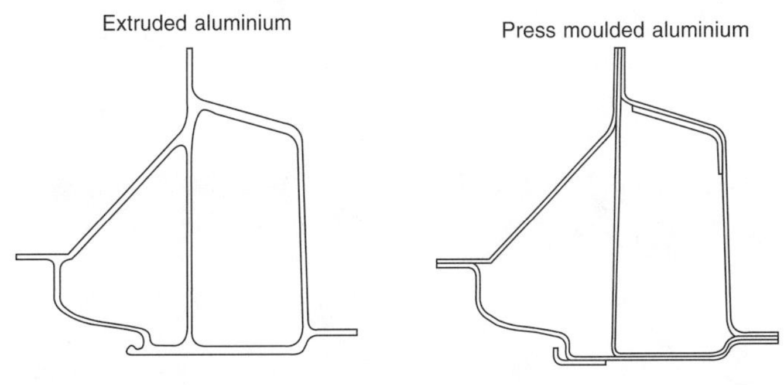

Fig. 18.1-22 Sill sections produced from pressed parts compared with extruded sections.[9]

forming limit curves, it was found that new disciplines in the form of die adjustments, crowning and lubrication were essential if the required shapes were to be mass produced.

Regarding welding, instantaneous welding currents of 20 000–50 000 amps were now necessary compared with 7000–12 000 amps for steel, and higher weld force values of 400–800 kgf compared with 200–300 kgf.

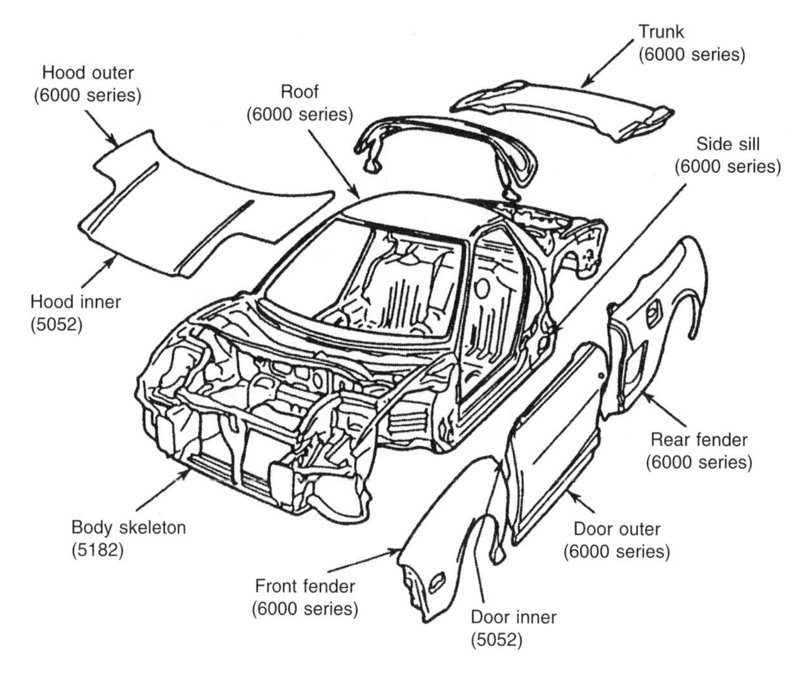

Fig. 18.1-23 Aluminium alloys used for NSX body panels.[9]

A special hand welding gun was devised having a built-in small transformer to reduce current loss. Spot welds were augmented with short MIG welding runs. Prior to painting with a four-coat system, a change to a chromate–chromium pretreatment was found more suitable than the usual zinc phosphate formulation. Dacromet was found effective in protecting small steel parts, e.g. bolts, from bi-metallic corrosion.

18.1.7.2 The pressed spaceframe (or base unit) concept – steel

The use of aluminium skin panels was extended to the Rover car range, the P4 (1954–1964 doors, bonnet and boot lid), and P6 (1964–1976 bonnet and boot lid; Fig. 18.1-24) although the SD1 body (Rover 3.5 1976–1984) reverted to steel. It was interesting to note that the P6 (Rover 2000) bonnet was changed to the 2117 Al–Cu grade to improve formability and obviate any signs of 'stretcher-strain' markings. However, the most significant design feature associated with the P6 was the appearance of the steel base unit in the 1964 P6 which featured a central frame to which pressed outer assemblies were rigidly and consistently bolted using drilled and tapped forged bosses.

This 'base unit' was then clad with steel fenders (wings) and doors while the bonnet and trunk lid (boot lid) were in aluminium. The advantage of this type of design is that in theory the cladding and external shape can be changed relatively frequently without changing the substructure, and repair simplified. This concept allowed clad panels in aluminium as an option and the same idea was adopted on the American Pontiac Fiero, where a steel substructure was clad in polymer skin panels again using adjustable box type attachment points

Fig. 18.1-24 Rover P6 spaceframe.

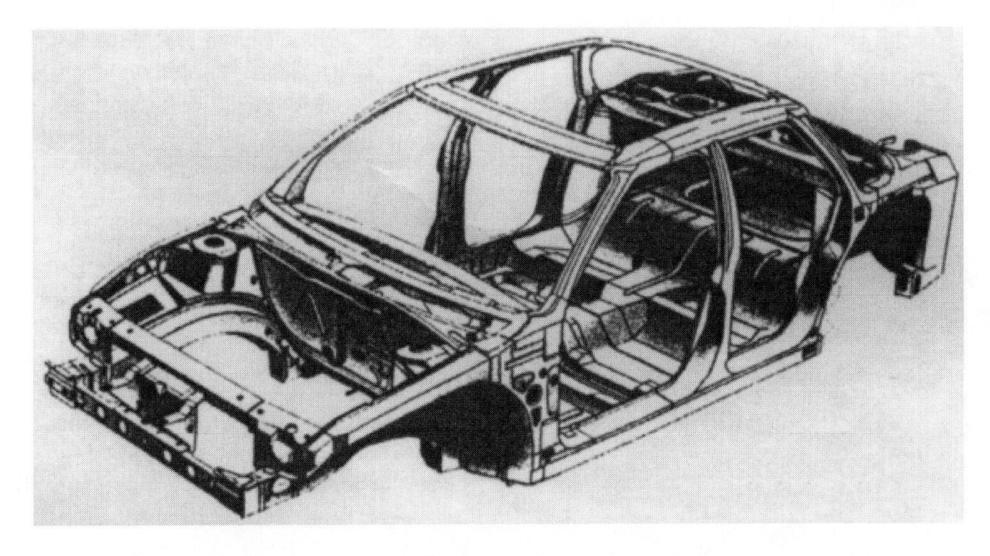

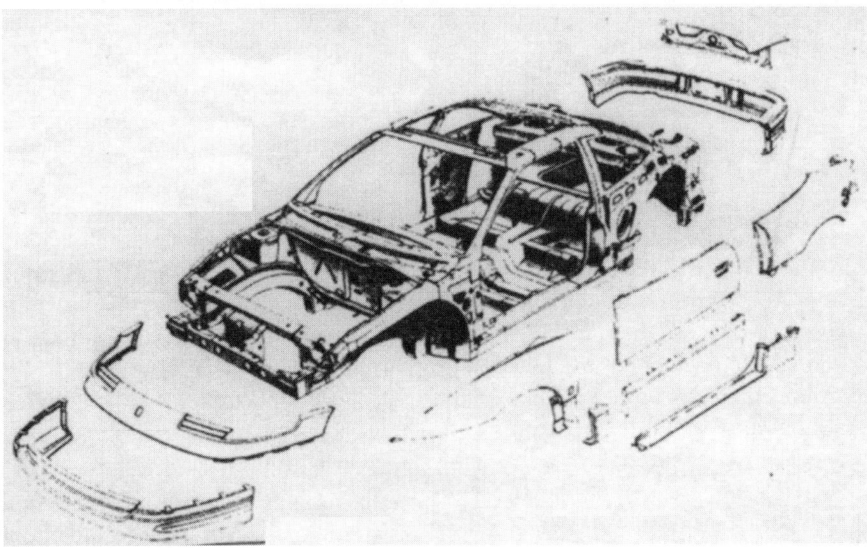

Fig. 18.1-25 Saturn spaceframe showing polymer panels on a four-door sedan.[10]

which could accommodate any differences in expansion between the two materials. The GM Saturn, shown in Fig. 18.1-25, has used the same type of pressed steel spaceframe (lower parts galvanized) for structural integrity and strength while clad in thermoplastic skin panels (doors, fenders, quarter panels and fascias) to enhance corrosion resistance and reduce damage from low speed impacts. The roof, bonnet and boot lid are retained in steel and the skin assemblies are painted in complete sets on support bucks in simulated on-car positions. The design technology was carried forward and developed with the Renault Espace which featured a steel substructure (in reality unibody structures with non-structural plastic cladding panels), but the distinction here was that the steel substructure was fully hot-dip galvanized prior to cladding with a polymer exterior, the penetration of zinc into crevices adding to the torsional stiffness of the main frame. This is theoretically the most effective type of corrosion protection allowing the encapsulation and full coverage of spot welds and cut edges, although thickness control can be variable and result in a weight penalty.

18.1.7.3 Examples of pressed aluminium spaceframes and associated designs

Again referring to examples of design innovation within the Rover vehicle range, the experimental ECV3 vehicle had demonstrated that the base unit concept could be extended further to provide an even lighter structure using aluminium pressed parts. The torsional stiffness in that case was improved by the use of adhesive in a weldbonding mode employing a specially developed pretreatment and prelubrication technology. The manufacturing feasibility of this approach was proven by

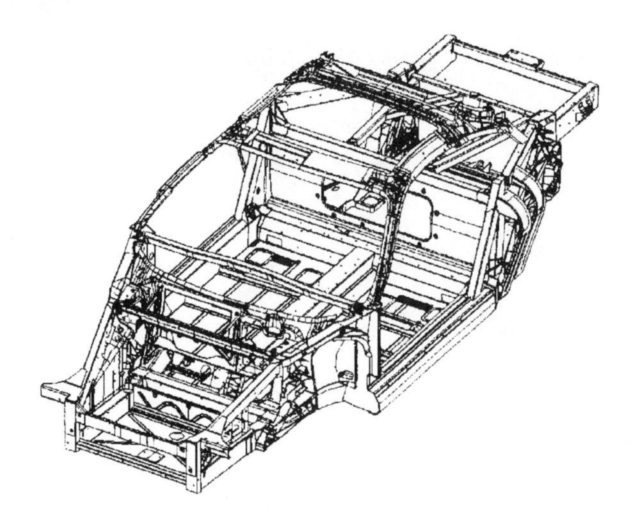

Fig. 18.1-26 Ferrari 360 Modena design.

the production of a small fleet of Rover Metros. The adhesively bonded aluminium spaceframe was clad in plastic, the horizontal panels being in a high modulus material to improve flexure and sagging effects, with the vertical panels in RRIM polyurethane to improve low speed impact and denting. Similar technology has now been transferred to production vehicles via ASV designs applied to the Jaguar 220, the XJ series and Lotus Elise.

Introduced in September 1995, the Lotus Elise featured a further type of structure termed 'the punt'. This followed joint design technology developed by Lotus Engineering and Hydro Aluminium Automotive Structures of Denmark and as shown in Fig. 18.1-26 features aluminium extrusions joined by a combination of adhesive bonding and mechanical fasteners. At 68 kg the spaceframe achieved a 50 per cent weight reduction compared with an equivalent steel construction and with bonded structures it was found that thinner sections could be used, and compared with spot welding or mechanical fastening no local stresses are produced. Excellent torsional rigidity at low mass results in good driving force and agility[11] and the aluminium structure absorbs additional energy in high speed impacts contributing to maximum occupant protection for the passenger cell. The complete vehicle is noteworthy for the use of extrusions for suspension uprights, door structures, pedal assemblies and dashboard fascia. Repair is considered with a replacement composite crash structure at the front and a mechanically fastened subframe at the rear onto which the rear suspension is mounted. In the case of major frame damage the complete spaceframe can also be replaced. The choice of alloys for extrusions and sheet is influenced by ease of recycling.

The Ferrari 360 Modena extends the aluminium spaceframe concept even further and comprises cast, extrusions and sheet.[12] A co-operative venture with Alcoa, the extruded and die cast components are made in Soest, Germany, a Ferrari supplier fabricates the sheet components and Ferrari supplies the sand castings including the integral parts of the spaceframe such as the front and rear shock towers. The spaceframe structure increases overall body stiffness (42 per cent in bending, 44 per cent in torsion) and safety while lowering the weight by 28 per cent and part count by 35 per cent compared to the steel predecessor. The F360 is claimed to be competitive in cost with a comparable steel body. This model is 10 per cent larger than the one it replaced. Materials used are summarized in Table 18.1-2.

The spaceframe comprises 42 per cent extruded components and 33 per cent cast components, the remaining 25 per cent being formed sheet parts and stampings. All critical loads are transferred to the spaceframe through six castings. Sand casting was selected on the basis of low part volume and minimum weight requirements and these parts also provide significant part consolidation.

Most joining operations were carried out by MIG welding and self-piercing rivets with special emphasis on the achievement of extremely accurate build tolerances. Consistent conditions are maintained by using a machining centre for the location of reference locators. The final spaceframe is shown in Fig. 18.1-27.

18.1.7.4 The ASF aluminium spaceframe utilizing castings and profiles Audi A8 and A2

A significant evolutionary step in the application of aluminium in autobody construction is the Audi A2, the first

Table 18.1-2 Ferrari 360 Modena materials

	Sand castings		Extrusions		Sheet components	
Alloy temper	**B356-T6**	**CZ29-T6**	**6260-T6**	**6063-T6**	**6022-T4**	**6022-T6**
0.2% proof stress (MPa)	170	125	200	160	130	275
UTS (MPa)	240	185	225	205	235	310
Elongation %	7	11	10	8	23	10

Fig. 18.1-27 Final 360 Modena spaceframe.

volume production vehicle to have body structure manufactured completely from aluminium. While the earlier 1994 A8(DT) version was clearly a major step forward in aluminium application, the spaceframe technology employed was designed for medium volumes (A8 annual was typically 15 000 per annum). With the A2, aimed annual volumes are 60–70 000 units per annum, and the technology employed is more suitable for higher production rates. Other noticeable differences to the original Audi A8 is the increased use of complex castings and the widespread application of rolled aluminium profiles, Fig. 18.1-28. Interesting components included the 'B' pillar complex casting which replaces typically eight steel pressings in traditional vehicles and a world first application of laser welding aluminium (around 30 m in total).

Following a number of years of research, Audi unveiled the aluminium intensive vehicle concept Audi 100 in 100 per cent aluminium at the Hanover Fair in 1985. This development progressed to the ASF or Audi Space Frame design in 1987 and finally the Audi A8 production model. The frame structure was formed from straight and curved box extruded sections joined into complex die cast components at highly stressed cornered connection points. The load bearing parts are integrated as a structure mainly through the MIG welding process, with stressed skin panels attached mainly by the punch riveting process. This was one of the first applications of such a process in the automotive industry and one of the main reasons for this was the 30 per cent higher strength of joints made using punched rivets compared to spot welding. Resistance spot welding was used for joints, which were not accessible for punch riveting. The final assembly of the body structure illustrates the three major differences between the ASF concept and traditional steel monocoque construction:

- Fabricated spaceframe with extrusions and castings
- Manufacture of hang-on parts (closures) including extrusions for stiffeners
- Combining the separate front and rear body sections to form the final body shell

At the time of release by Audi, the ASF was claimed to exceed the rigidity and safety levels of modern steel bodies while achieving a weight reduction of the order of 40 per cent.

In many ways it is likely that in future years the Audi A8 will be regarded as one of the key technical developments in autobody materials technology. However, six years later in 2000, Audi unveiled the next stage of their aluminium body development, perhaps the more important A2, Fig. 18.1-29.

While the original A8 was largely a hand-built car (a strategy that is acceptable for a production volume of 15 000 cars per year), the A2 was always intended to sell four times this number. This demanded a manufacturing concept integrating faster automated systems and techniques. The resulting A2 body structure is a highly innovative design taking elements of the A8's earlier concept but refining and adding technologies to them. In addition, the number of components has been reduced from 334 in the A8 to 225 in the A2. An excellent example of this part integration is the 'B' post component which in the A8 consisted of eight individual parts (extrusions, sheet, castings) integrated into one component, while the A2 'B' post consists of a single casting, Fig. 18.1-30.

The whole structure consists of 22 wt per cent aluminium cast elements, 18 wt per cent aluminium extrusions and 60 wt per cent aluminium sheet. The joining technologies used in the A8 have been refined for the A2. Spot welding and clinching were abandoned. The use of laser welding is of particular note, especially the floor pan

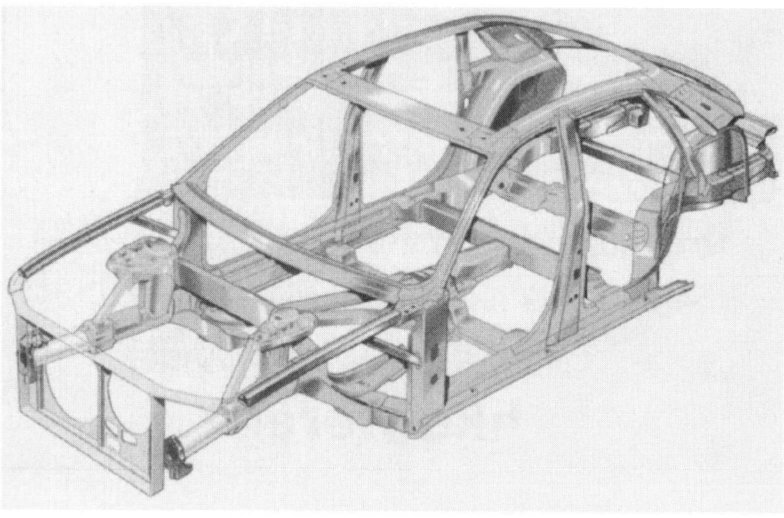

(a) 1994 (D2) version

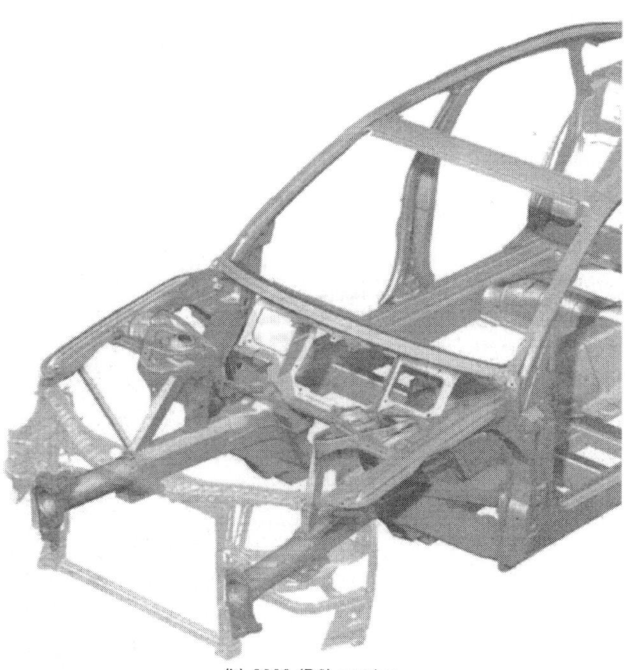

(b) 2002 (D3) version

Fig. 18.1-28 Audi A8 body structure.

laser welded to the spaceframe structure of extruded sections and pressure die casting. In total 30 m of laser welding is defined and the need for only one-sided access provides designers with extra styling freedom at the early concept stage of development. The most difficult aspect of laser welding is the tight tolerances for panel matching that are required (typically ±0.2 mm). Compared with the A8, the self-pierce riveting process has been used increasingly to join sheet metal and extrusions.

In the latest version of the A8, announced in late 2002, much of the design and manufacturing technology has been carried over from the A2. However, it does represent a step change from the previous model. It is still essentially of ASF construction but the number of parts has fallen from 334 (including hang-on parts) to 267 through larger format pressings such as the side frame plus extruded sections such as the 3 metre long hydroformed roof frame, and multi-functional large castings used for the 'B' post (and radiator tank). The 'B' post previously comprised eight parts (4254 g) but is now a single component with the weight now reduced by 600 g. Compared with a conventional steel body weight has been reduced by 40 per cent. One hundred and fifty-six robots ensure an automation level of 80 per cent with a claimed 50 per cent saving in the production cycle and other manufacturing advances include a hybrid laser

Fig. 18.1-29 Audi A2 body structure.

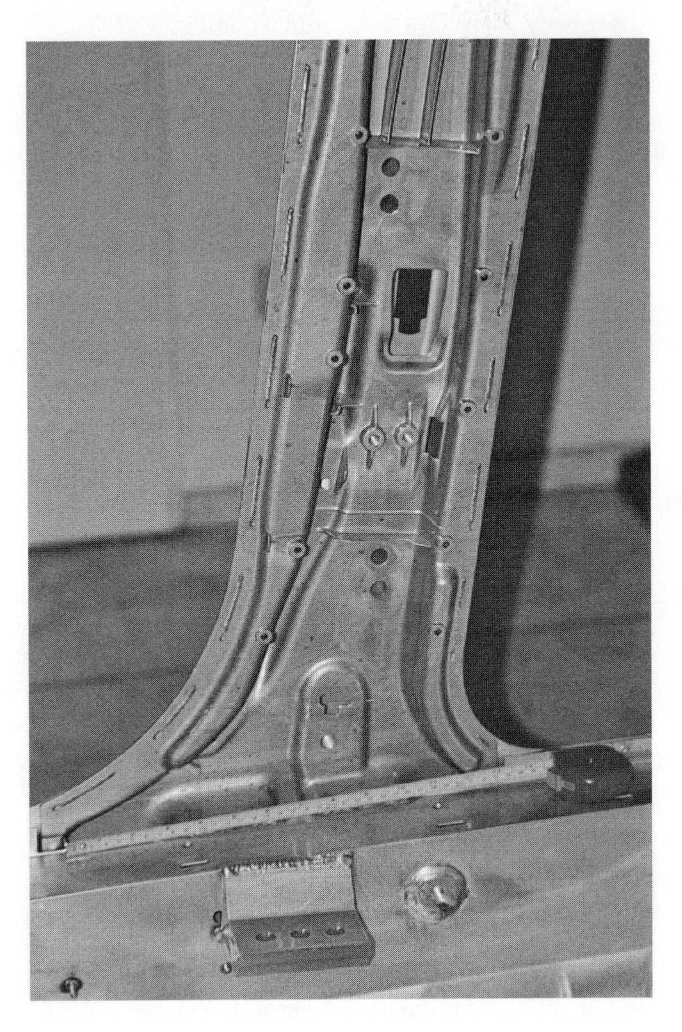

Fig. 18.1-30 Audi A2 B post casting.

Fig. 18.1-31 Aston Martin Vanquish.

MIG welding process achieving synergistic effects by combining both joining processes. A hybrid welding seam length of 4.5 metres is achieved per body. As well as the hybrid welding seams there are also 2400 punch rivets, 64 metres of MIG welding seams and 20 metres of laser welding on each A8 body. Concerning the peripheral joint on bolt-on panels, rollers secured to a robot arm bend the outer panel over the inner and create a strong joint with the application of a hem-bonding adhesive. The doors, bonnet and tailgate are hemmed in this way and the wheel arch to side frame is similarly processed. Induction curing is used to prevent movement between inner and outer panels at the body-in-white stage.

The new A8 body is shown alongside the original version in Fig. 18.1-28 and a lot of development time and effort have obviously been expended in optimizing an extremely efficient production process. The emphasis on laser and MIG welding plus mechanical fastening appears to have advantages over the ASVT technology where adhesive application to structural joints must prove more of a challenge in terms of application and resulting consistency and durability.

18.1.7.5 Examples of hybrid material designs

18.1.7.5.1 Aston Martin Vanquish

The new Aston Martin Vanquish,[13] Fig. 18.1-31, contains a mixture of innovative materials technologies in a low volume model (350 units/year). A number of the skin panels are pressed or superplastically formed from aluminium sheet.

The body structure is mounted on an aluminium bonded and riveted lower structure (similar to the Lotus Elise), but incorporates a mixture of carbon fibre and aluminium extrusions in the floor/tunnel construction as shown in Fig. 18.1-32.

In addition to the tunnel the windscreen pillars are also carbon fibre bonded to the central structure to create a high strength safety cell. A steel, aluminium and carbon fibre subframe carries the engine, transmission and front suspension and is bolted directly to the front bulkhead. As can be seen above the doors are fabricated from aluminium incorporating extruded aluminium side impact beams.

18.1.8 Integration of materials into designs

18.1.8.1 General

The above appraisal of designs shows how the choice of material is paramount to achieving today's objectives and how this choice is widening. Having minimized the weight of a specific design and assuming the best materials have been specified, the next consideration is to optimize material with regard to each link in the chain of processing operations necessary to produce a functional part, and each of these elements can strongly influence the selection. For instance, complex parts require maximum formability which requires a compromise with strength, realistically placing a maximum at around 300 N/mm^2 proof stress, although for simple sections such as door reinforcement beams levels of 1200 N/mm^2 may be specified. The constraints imposed by local steelmakers may obviate certain grades where for instance a bake-hardening or isotropic steel is required, and a restricted choice of coating types

Fig. 18.1-32 Internal structure of Vanquish showing carbon fibre composite and extrusion construction.

may be available. However, despite these minor restrictions, apart from obvious exceptions most manufacturers are maintaining a conservative steel grade policy, requiring only minimal changes in processes, and as has been seen above the use of predominantly aluminium structures is only evident by one or two of the more adventurous companies who can absorb the extra supply and manufacturing costs. The majority would still prefer the more cautious approach employing the advantages of aluminium for closure or 'bolt-on' parts and using the accompanying weight savings to satisfy legislative weight-band requirements or added sports car performance. Many manufacturers are, however, gaining valuable manufacturing experience by building low volume sports models in aluminium, e.g. NSX or BMW Z8 or specific parts Peugeot 607. Once the different disciplines demanded by this less robust material are fully understood and a way is found of absorbing the extra cost it may then find a wider usage. Plastics as referred to later in this section require much development in an engineering context and only very expensive derivatives fulfil impact and other functional requirements. Until the market price falls then use will be limited to exterior cladding and trim items. Thus, for the main body structure the increasing use of HSS will continue to develop and the trend for a typically progressive European car manufacturer such as BMW is shown in Fig. 18.1-14 – a weight saving of 10–15 per cent being achieved for selective parts via thickness reduction.

Magnesium is now starting to find favour as the quest for lower density materials intensifies, but has been used occasionally in the past, e.g. Austin Maestro gear box covers (Fig. 18.1-33). The latest interest is for vehicle cross-beams and similar aluminium products as described in the following sections.

18.1.8.2 Other materials used in body design

So far only primary materials have been considered but use now is being made of secondary forms of steel, aluminium and plastics. Examples are sandwich steel and similar aluminium products, described below, hydroformed steel and aluminium, and tailor welded blanks (TWBs). High performance and competition cars are also making extensive use of honeycomb materials which when consolidated with composite skin layers provide ultralight high strength impact and structural sections. Because of their exceptional strength to weight ratio these may be the future first choice of body material for electric and alternatively fuelled cars.

18.1.8.2.1 Tube hydroforming

As evident in the ULSAB programme, hydroformed tube has significant potential in the consideration of parts consolidation especially for the more rugged applications such as the 4 × 4 sector which also allows a little more freedom of construction. The background and other weight-saving technologies demonstrated by the ULSAB initiative are discussed in an earlier work but from a design viewpoint it is instructive to consider the recent study which evaluated the possible advantages in incorporating hydroformed structural elements within the Land Rover Freelander.

Described in a recent IBEC conference,[16] it is worth highlighting to show how potentially good ideas can be evaluated under realistic conditions, while defraying costs and resources of two major organizations. In this instance the design data for the recently developed Freelander was immediately to hand and could be relatively easily modified to allow an immediate comparison of new and conventional structures. The opportunity was

Fig. 18.1-33 Side view of Vanquish.

also presented to allow full vehicle testing of a new concept rather than the body-only exercise with the ULSAB sedan, relying on FE modelling to predict performance.

The Land Rover Freelander was chosen as the focus for this programme principally due to the maturity of the development programme for the vehicle and the design package which allowed application to either smaller or larger products. Although a Land Rover (hitherto body-on-chassis design), the body is of monocoque or unitary construction, and the incorporation of a rigid sectional product seemed a natural choice for a rugged off-road performer.

The final configuration of hydroformed components incorporated in the design is shown in Fig. 18.1-34 and followed an extremely detailed study. It is worth mentioning that the normal procedure is to work to a controlled pre-development plan whereby the features of a new design are compared with the original, a cost-effective manufacturing route defined and rigorous testing of new components undertaken. The whole process is regulated with frequent timing reviews and concurrence obtained before proceeding through successive 'gateways' or decision points. These pre-concept stages

Fig. 18.1-34 Freelander design in steel and proposed alternative hydroformed parts.[16]

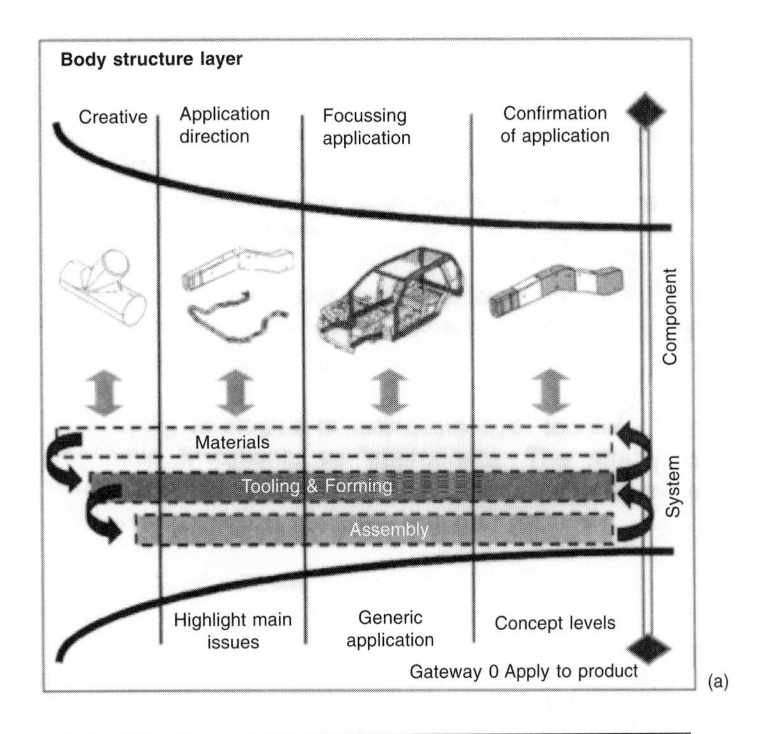

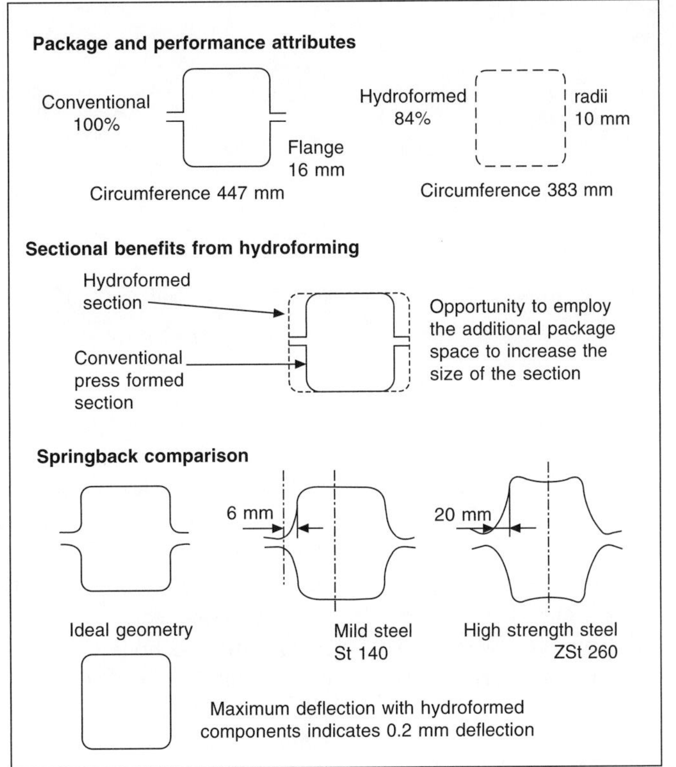

Fig. 18.1-35 Pre-development phases of ULSAB 40 and attributes of hydroformed sections.[16] (a) Pre-concept phases, (b) shape comparison with conventional sections. (Reprinted with permission from SAE paper 1999-01-3181 Copyright 1999 Society of Automotive Engineers Inc.)

constituting the 'creative' phase, gateways and process steps are illustrated in Fig. 18.1-35(a).

During this evaluation the advantages of the hydroformed sections will have been assessed, first, to confirm weight and space saving potential allowed by shape characteristics, Fig. 18.1-35(b) and second, as shown in Fig. 18.1-36 in comparison with other possible methods that could produce similar savings.

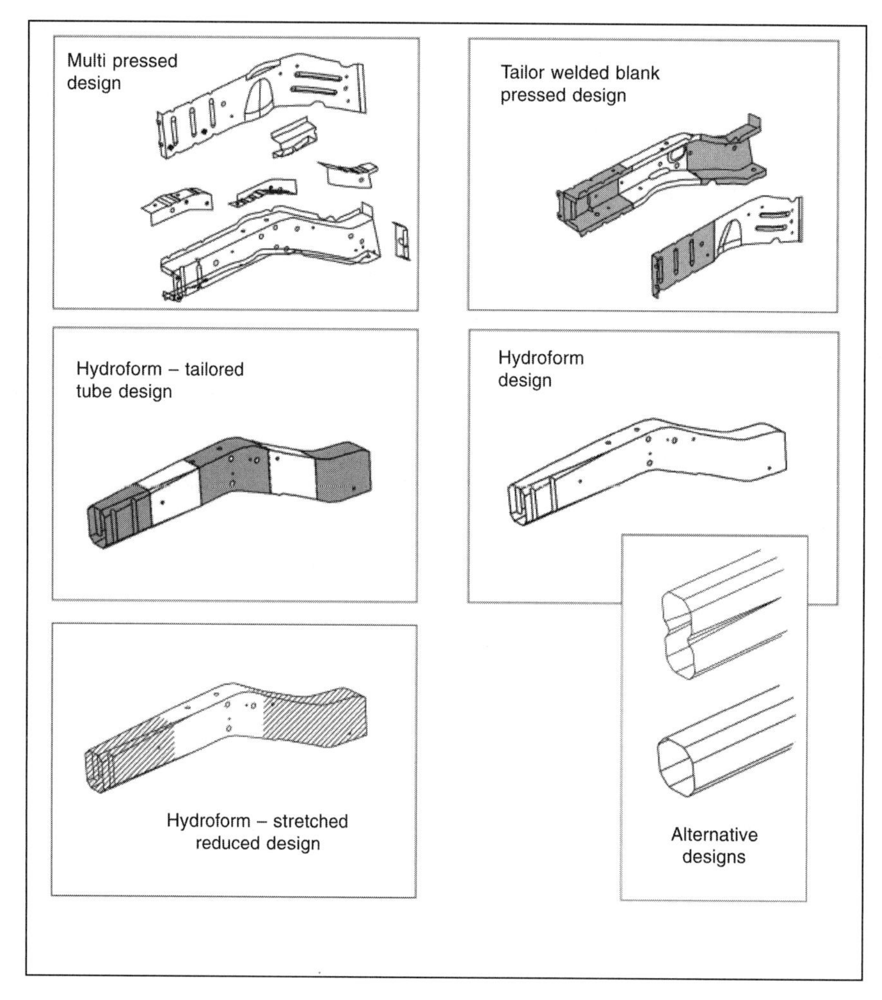

Fig. 18.1-36 Alternative forms of longitudinal section. (Reprinted with permission from SAE paper 1999-01-3181 Copyright 1999 Society of Automotive Engineers Inc.)

It was essential for comparisons involving joints and flange replacement that the welds were accurately modelled. For normal press steel box sections the assumption is made that flangeless sections are used and no allowance is made for distinguishing between alternative joining methods. However, it was critical for this new type of hydroformed joint that the joining method was represented more accurately and solid elements were used to represent adhesives and rigid bars positioned at the centre of flanges to simulate spot welds, Fig. 18.1-37.[10]

The various design iterations can then proceed to determine sections and joints which would probably benefit most from alternative hydroformed sections.

Comparison of hydroformed with conventional parts and the stages in the manufacture from tube are shown in Fig. 18.1-38.

The 'Application' phase comprised manufacture of the prototype parts illustrated in Fig. 18.1-38, using representative methods by a number of key tube hydroform suppliers, and finally a build and test programme to validate the advantages of the modified structure. The findings are summarized in Fig. 18.1-39.

The results shown in Fig. 18.1-39, and from crash and durability testing, demonstrated that the revised structure was equivalent in performance to the Freelander while torsional stiffness was markedly improved. However, starting a completely new model programme without the constraints of an existing body it is highly likely that far more significant weight savings and parts consolidation could have been achieved. Manufacturing feasibility was also demonstrated so opportunities can now be determined in the forward model programme.

18.1.8.2.2 Tailor welded blanks

The concept of producing composite blanks with tailored combinations of different thicknesses, strength grades and coated/uncoated steel provides the body engineer with the option of localized property variations wherever he wants them. Thus for longitudinal impact sections,

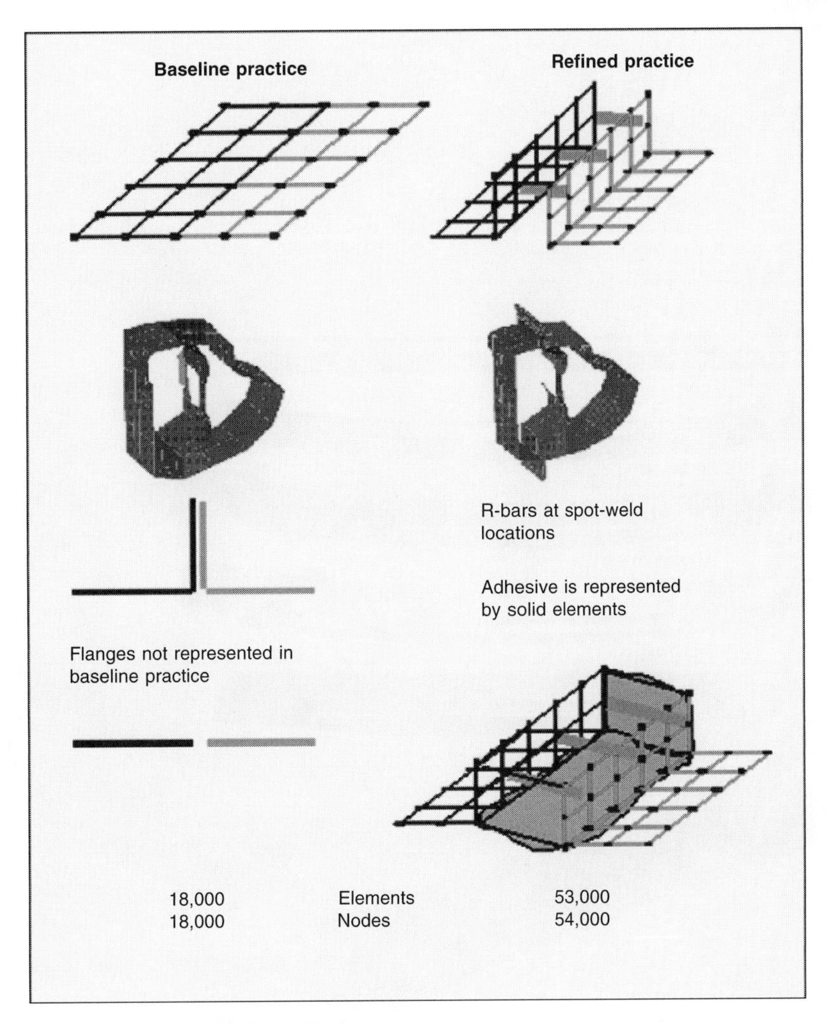

Fig. 18.1-37 FE representation of joints.[16] (Reprinted with permission from SAE paper 1999-01-3181 Copyright 1999 Society of Automotive Engineers Inc.)

controlled collapse may be induced as an alternative to 'bird beak' design, body side blanks may be blended to give formability in central areas and higher strength at pillar locations and door inner panels may be split to provide strengthening of the frontal area thereby dispensing with the need for reinforcement as illustrated in Fig. 18.1-40.

Thus increased scope exists for engineering solutions and parts consolidation which may offset the premium charged for the composite blank. This technology can be applied to steel or aluminium blanks. Composite steel blanks can be produced by mash or butt resistance welding but the finish containing a roughened fused weld zone is normally only suited to underbody parts. More often the blanks are now laser welded giving a narrow joint with a minimum of distortion and are widely used for most European models, typically for cross and longitudinal members, door inners and body-sides. Questions that must be addressed on order placement concern quality control procedures to ensure consistent weld quality, and liability in the case of failure of a structural part.

18.1.8.2.3 Sandwich materials

A material with extensive weight saving potential is sandwich steel. This consists of two thin sheet outers encapsulating a thicker polypropylene central layer. At present there is not an extensive supplier base for these materials, since the commercial and engineering viability of the materials is not proven. Some of the versions that are on the market cannot resist the elevated temperatures during the body structure painting process. As a result, this material type is only viable for components that are assembled into the body after the painting process. In addition, this material is not weldable and must be assembled into the BIW by a cold joining process of either adhesive bonding or mechanical fastening. The ULSAB programme identified and subsequently defined two components in

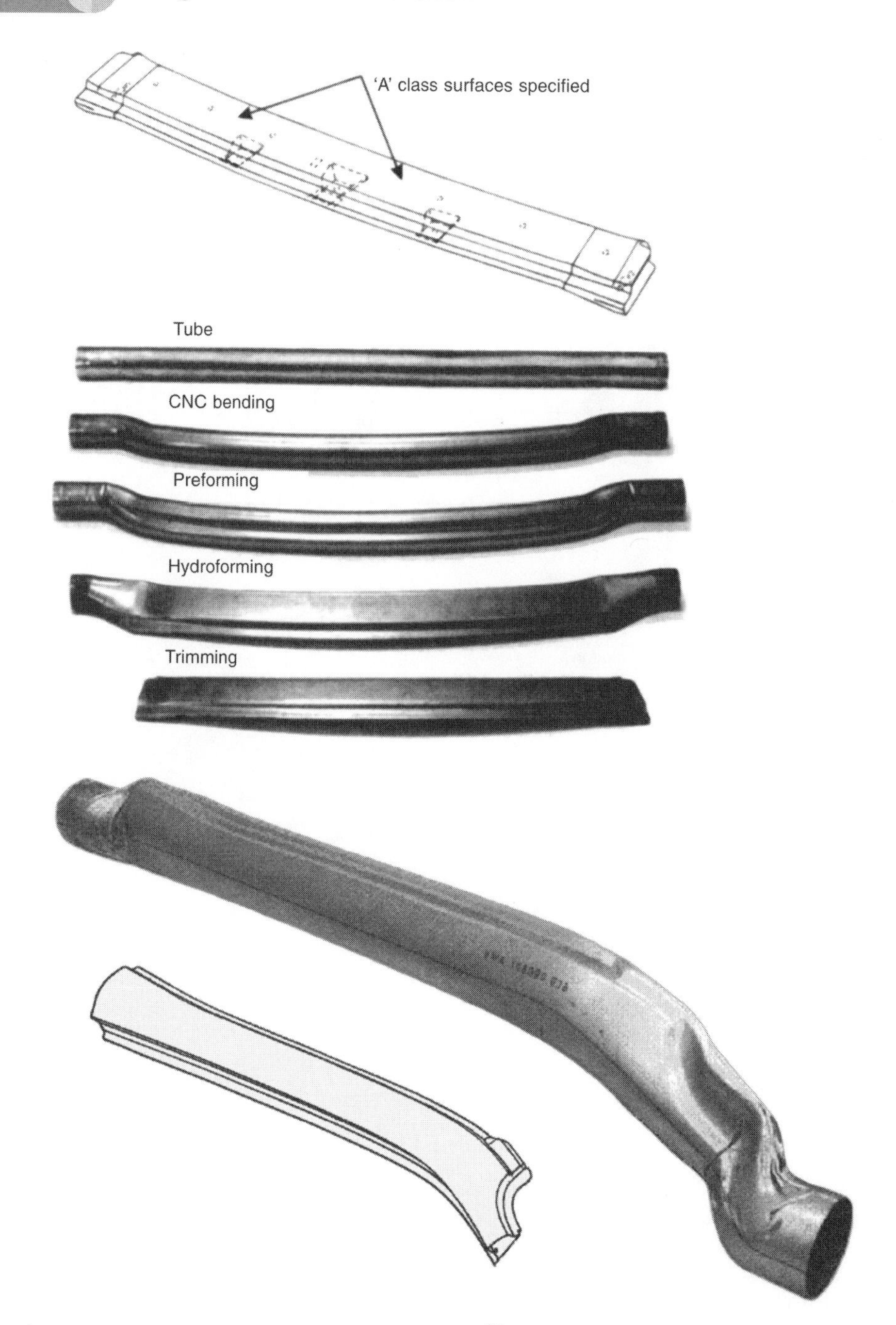

Fig. 18.1-38 Stages in component manufacture by tube hydroforming.[16]

a sandwich steel material: a dash panel insert and spare wheel well. The steel skin used for the spare wheel well has a yield strength of 240 MPa and a thickness of only 0.14 mm. The core thickness was 0.65 mm, i.e. a total sheet thickness approaching 0.9 mm. The dash panel steel was a forming grade material (yield strength 140 MPa) with a thickness of 0.12 mm and a core of 0.65 mm.

Even greater weight savings may be achievable through the use of an aluminium sheet version of the sandwich material. In this case typical thicknesses to achieve a similar level of bending stiffness to a steel panel would be 0.2 mm thick aluminium sheets surrounding a 0.8 mm thick thermoplastic core. Compared to steel this material offers weight saving opportunities even greater than aluminium, i.e. up to 68 per cent.

These sandwich materials could be argued to be good examples of the new type of hybrid materials technology that will be applied in the future, making use of the positive advantages of each material type, i.e. using the lightweight nature of the thermoplastic core and the stiffness, corrosion resistance and surface appearance of

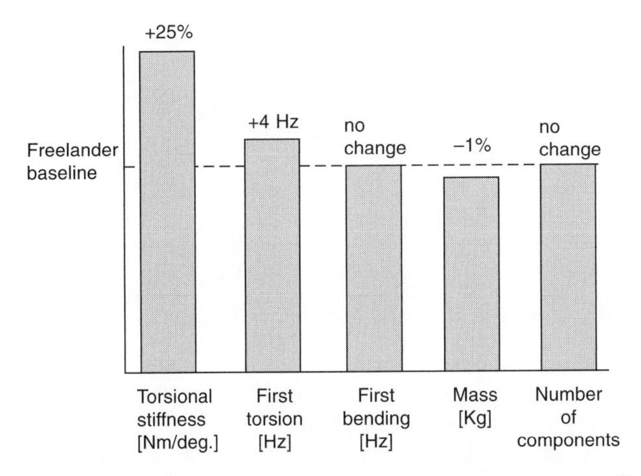

Fig. 18.1-39 Results from the application and proving phase.[16] (Reprinted with permission from SAE paper 1999-01 -3181 Copyright 1999 Society of Automotive Engineers Inc.)

the metallic outer layers. However, application of this hybrid or composite material brings its own inherent difficulties with regard to recycling.

Customers now demand levels of in-car refinement that were unheard of a decade ago and one technique used within automotive design is to apply significant quantities of bitumen-based damping materials to critical regions of the body structure and closures. The main drawbacks associated with this approach are the additional mass and cost.

Laminated materials consist of two layers of conventional sheet material (usually steel) sandwiching a very thin layer of viscoelastic resin. The combination of these materials results in good sound damping performance (x). This material has been used previously in non-autobody applications, e.g. engine camshaft covers and oil sumps. Attention is now being focussed on the application of these materials to panels such as the main floor and dash panels, which are typically covered in bitumen damping pads. Removal of these pads potentially offers weight and/or cost reduction opportunities along with NVH improvements.

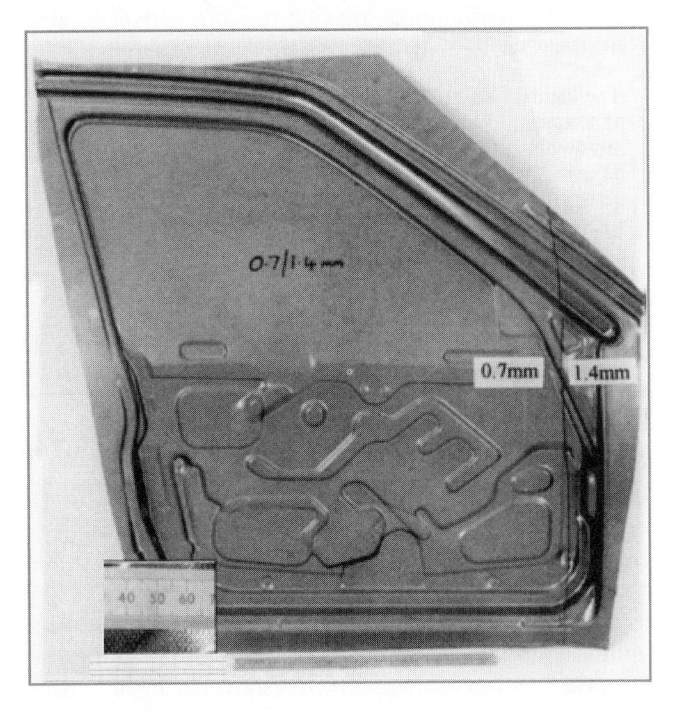

Fig. 18.1-40 Tailor welded door inner blank showing thicker frontal area allowing deletion of separate reinforcement panel.[17]

Japanese motor manufacturers have pioneered the application of this material in body structures; examples of volume production use include the firewall panel on the Lexus LS400 and the Honda Legend. Clearly, there are many considerations about the use of this material through the process chain, including formability performance, welding and joining and recyclability at the end of vehicle life.

18.1.9 Engineering requirements for plastic and composite components

The different types of plastic and respective manufacturing processes are referred to in Chapter 18.2 but reference should now be made to their engineering capabilities. Only very expensive derivatives such as carbon fibre composities fulfil impact and other essential structural needs. Until the market price falls the widespread use of polymers will be limited to exterior cladding and trim items.

Performance requirements for automotive body parts, and specifically plastics, are quite demanding. For example, vehicles must perform acceptably below –30 °C, and under solar heating conditions, exterior components can reach temperatures in excess of 90 °C. Panels must also be resistant to a wide range of chemicals and expected vehicle life can nowadays be in excess of 10 years or 100 000 miles during which material performance must be acceptable. These properties include:

- *Mechanical performance* – mechanical properties of relevance include tensile and shear strengths and modulus. Engineering thermoplastics typically have moduli of around 3 GPa and this relatively low value is related to the weak interchain bonds that hold the longer polymer chains together. In a thermoset where the chains are interlinked by strong chemical bonds, a higher modulus is exhibited (typically 4–5 GPa). Further increases can be achieved in composite materials through the addition of fibres, though the resultant modulus will still likely be less than that of metallic materials. As a result plastic and composite panels will usually be of greater thickness than metallic panels if a specific level of panel stiffness is required.

- *Impact* – impact performance (both low energy impacts or dent resistance and high energy in terms of crash performance) is a major consideration. To maintain polypropylene impact performance at low temperatures it is necessary to add additional components to the polymer blend to avoid brittle failure. For composites the fibre/matrix failure is the major energy absorption mechanism. In terms of dent resistance, polymeric panels deform in a different way to steel panels and many polymers can exhibit superior dent resistance performance by virtue of their low modulus. Material properties and their effect on dent resistance therefore become a prime consideration in panel design. The Land Rover Freelander 4 × 4 vehicle incorporates two new material applications for body panels, which as well as offering other benefits, provide improved dent resistance performance. On a vehicle designed for off-road use, the enhanced dent resistance of the plastic front fenders and zinc-coated HSS should provide significant customer benefits.
- *Temperature performance* – high temperature performance in service is critical. Since polymers tend to have a greater rate of thermal expansion than steel, it is possible to have visual quality problems in terms of buckling, warping or uneven panel gaps. This expansion must be allowed for at the design stage – by appropriate design of the fixing method. Composites such as SMC have expansion rates more similar to steel and therefore this issue is of less concern.
- *Durability in-service* – both UV resistance and solvent resistance are key performance measures for exterior panels in particular. Unlike metallic panels, polymers can be susceptible to UV degradation and the addition of stabilizers is necessary to the base polymer. Solvent resistance is also critical, e.g. petrol, and again it is necessary to use a protective coating to ensure thermoplastic materials do not suffer a loss in strength or stiffness due to absorption of solvents.

18.1.10 Cost analysis

Many of the technologies described herein are aimed at achieving a reduction in component weight. Indeed, the selection of material type is based in part on careful consideration of the improved fuel economy derived from the use of lightweight materials versus the increased costs that are often incurred. This will be apparent from the table of selection criteria used by a prominent motor manufacturer in Chapter 18.2. Many different cost models can be applied to the evaluation of material types in these applications,[8] but general trends can be identified, Fig. 18.1-41.

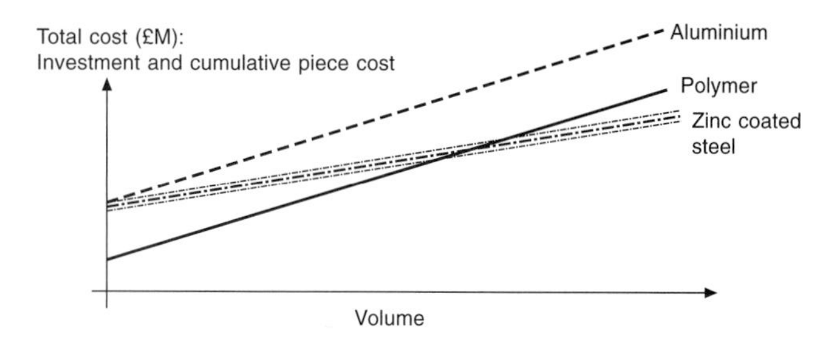

Fig. 18.1-41 General cost basis for automotive skin materials.

However, since material selection is very often based on cost analysis at the design and engineering phase of the chain, an overall appreciation of the relative cost balance of the various materials is included here.

For a particular panel, this may result in an increased cost for plastic compared to zinc coated steel, when manufacturing in excess of a certain annual volume. This is because although tooling costs for plastics are lower than for zinc-coated steel, raw material costs are higher. Thus, as total vehicle volume increases, the cost benefit derived from polymeric panels decreases until a certain break-even volume when steel becomes the most economical solution. This break-even volume is the subject of on-going debate, though is likely to be less than 200 000 cars. It is pertinent to note that most medium and high volume models involve the production of over 250 000 cars/year, explaining why use of plastics/aluminium has been mainly limited to low volume vehicles. Nonetheless, with improvements in technology, the cost advantage for polymeric panels may potentially shift to higher volumes, making the alternatives to zinc coated steel more attractive to the automotive industry.

The material costs must be considered only an approximate guide. Each material manufacturer will produce the common material grades at different cost levels depending on the exact specification of their production equipment. In addition geographical differences can exist, for example throughout the Eurozone (EZ) coatings have been considered to offer a cost advantage over galvanneal coatings in Germany while the reverse has traditionally been true in the UK. This may go some way to explain the pattern of material policy within European carmakers.

For a true comparison of the economics of body materials the input detail may also extend to include different design and manufacturing strategies. A more comprehensive cost analysis has been demonstrated by Dieffenbach.[15] First, he compares five different systems that could be employed to design and manufacture a mid-range sedan: steel and aluminium unibodies, steel and aluminium spaceframes, together with a composite structure, and a cost breakdown is shown in Table 18.1-3.

Again at low volumes costs reflect investment levels while at higher volumes material costs have a bigger

Table 18.1-3 Body-in-white cost analysis presented by Dieffenbach[14]

	Steel unibody	Aluminium unibody	Steel spaceframe	Aluminium spaceframe	Composite monocoque
Body-in-white cost analysis: key design inputs for selected case study alternatives					
Geometry					
Overall vehicle					
mass (kg)	315	188	302	188	235
Mass as % of steel unibody	100%	60%	96%	60%	75%
Spot joints (#)	3250	3400		1000	n/a
Seam joints (cm)	n/a	n/a	4000		6000
Piece count					
Total piece count (#)	204	224	137	137	41
Count as % of steel unibody	100%	110%	67%	67%	20%
Number of stampings	187	207	40	40	n/a
Number of castings	n/a	n/a	30	30	n/a
Number of roll/ hydroformings	n/a	n/a	50	n/a	n/a
Number of extrusions	n/a	n/a	n/a	50	n/a
Number of mouldings	n/a	n/a	n/a	n/a	7
Number of foam cores	n/a	n/a	n/a	n/a	34
Panels (inners/outers)	17	17	17	17	17
Materials					
Material prices ($/kg)	$0.77–0.92	$3.00–3.50	$0.77–2.20	$2.00–3.00	$3.13
Material density (g/cm^3)	7.85	2.70	7.85	2.70	1.59
	Stamping	**Casting**	**Hydro-forming**	**Extrusion**	**Moulding**
Body-in-white cost analysis: key fabrication input for selected case study alternatives					
Range of cycle times (s)	8–12	50–60	30–40	3–10	600–1200
Range of labourers/fab'n line	4–6	2	2	2	2
Range of machine costs ($M)	$1.3–7.5	$0.8–1.5	$1.0–2.0	$1.0–2.0	$0.5–1.0
Range of tool set costs ($M)	$0.2–6.0	$0.1–0.2	$0.1–0.5	$3k–7k	$0.1–$1.2

influence. Thus the trends shown in Table 18.1-3 are mirrored in these studies with steel characterized by high investment cost, lower material and faster production rate. Conversely moulded plastic has a lower investment cost, higher material cost and slower production rate. The composite monocoque has the lowest cost up to about 30 000 vehicles per year; from 30 000 to 60 000 vehicles per year the steel spaceframe shows the lowest cost. For higher volumes the steel unibody shows the lowest cost. The aluminium spaceframes or unibody do not show a cost advantage although the aluminium spaceframe competes fairly well (a 15 per cent cost penalty) and compares with the steel unibody at high volumes. For outer panel assembly sets compression moulded SMC has the lowest cost for volumes up to about 100 000 sets per year above which steel has the lowest costs.

The challenges for the future for each category include lower tooling costs and scrap production (down to 25 per cent) for steel unibodies, lower raw material costs, e.g. by continuous casting, for aluminium unibodies, full exploitation of 40 per cent mass reduction potential apparently available from the steel spaceframe, while the aluminium spaceframe would benefit by the adoption of SMC (or similar) cladding (24 per cent cheaper than aluminium) which would make it cost competitive up to 80 000 vehicles per year. The composite monocoque is characterized by relatively expensive materials and clearly the challenge here is to reduce raw material costs, especially for carbon fibre composites.

A second approach proposed by Dieffenbach[15] is to use a stainless steel spaceframe clad with self-coloured composite panels where economies are proposed by deletion of various levels of the painting operation. This idea highlights another method of utilizing materials development to reduce costs. Costs presented for steel vs stainless steel are shown in Table 18.1-4.

Therefore comparing costs can be an extremely complex process requiring an intimate knowledge of the expected design and production scenario before accurate forecasts can even be attempted.

It is important to appreciate that the application of new material technologies as a means of vehicle weight reduction will usually often be decided by the vehicle programme development manager who may be willing to pay a cost penalty to reduce weight. This penalty may be influenced by the need for the vehicle to remain in a certain weight class or to move the vehicle into a lower weight class. For example, in the USA higher profit luxury vehicles have a negative rating on the company's CAFÉ rating. Production of a large number of heavy vehicles in this class may incur a cost penalty and the programme manager may decide that the cost penalty of introducing a new materials technology will be compensated for by the ultimate weight positioning of the final vehicle.

In conclusion the main evolutionary phases of the automotive body structure have been reviewed, and the role of materials introduced with respect to properties and costs, and performance expected in service. We now move on to the production processes for each of these likely materials to understand more fully the strengths and weaknesses of each, enabling the exact specifications meeting design, process chain, and environmental requirements to be met at minimum cost, both direct and indirect.

Table 18.1-4 Relative costs of steel unibody vs stainless spaceframe.[15]

	Steel unibody	Stainless steel spaceframe
Structure	$748	$522
Panels	$191	$191
Assembly	$261	$115
Paint	$415	$314
Total	$1615	$1142

The stainless steel spaceframe is found to have a cost advantage of about $375 (23%) if paint is not included, and $475 (30%) if paint is included.

18.1.11 Learning points

1. Early chassis-based construction has now been replaced by body structures of unitary design. The spaceframe concept is increasingly popular allowing a mix of materials to be used, with ease of disassembly and repair.
2. Aluminium design using cast nodes, profiles and sheet has now been proven as a feasible design for volume production although material and vehicle insurance costs remain high.
3. FEM design techniques are now proving invaluable in reducing the timeframe of model development programmes. Parameters from a wide range of materials including HSS, aluminium and polymer variants can be used to help predict performance in dynamic situations, e.g. a crash. Lower strain rate programmes can help determine forming feasibility.
4. Contemporary design influences can introduce conflicting interests: ease of recyclability is not commensurate with increased use of plastics used to lighten body structure. There should be no threat to vehicle safety if larger, safer steel structures are gradually replaced by lighter alternative structures.
5. Specialized production techniques offering new forms of materials such as TWBs and hydroformed tube sections are allowing more freedom of design with opportunities for parts consolidation and weight reduction.
6. TWBs and use of lay-up techniques with composites such as carbon fibre now allow localized strengthening and stiffening of different body zones thereby shedding superfluous weight.
7. The combination of advanced composites and ultralightweight honeycomb structures could

provide the basis for future alternatively fuelled vehicles as demonstrated by current high performance vehicles.

8. Polymers offer the designer undoubted advantages extending the range of body shapes and exhibiting good low speed impact, scuff and dent resistance. However, the range of materials must be rationalized to allow simpler specification on drawings/electronic identification systems and ease the task of segregation for dismantlers.

References

1. BRITE EURAM II Low Weight Vehicle Project BE-5652 Contract No. BRE2-CT92-0264.
2. Brown, J.C., Robertson, A.J. and Serpento, S.T., *Motor Vehicle Structures*, Butterworth-Heinemann, 2002.
3. Davies, G.M., Walia, S. and Austin, M.D., 'The Application of Zinc Coated Steel in Future Automotive Body Structures', 5th International Conf. on Zinc Coated Steel Sheet, Birmingham, 1997.
4. Garrett, K., 'First Without Chassis?', *Car Design and Technology*, May 1992, pp. 56–61.
5. Hodkinson, R. and Fenton, J., *Lightweight Electric/Hybrid Vehicle Design*, SAE International, Butterworth-Heinemann, Oxford, 2001.
6. Ludke, B., 'Functional Design of a Lightweight Body-in-White. How to determine Body-in-White materials according to structural requirements'. VDI Berichte 1543 Symposium, 11 and 12 May 2000, Hamburg.
7. Ludke, B., 'Functional Design of a Lightweight Body-in-White for the new BMW generation', *Stahl und Eisen*, 119, 1999, No. 5, pp. 123–128.
8. Lewandowski, J., *Audi A8*, Delius Klasing Verlag, Bielefeld, 1995.
9. Muraoka, H., 'Development of an All-aluminium Body', *Journal of Materials Processing Technology*, 38, 1993, pp. 655–674.
10. Holt, D.J., 'Saturn: The Vehicle', *Automotive Engineering*, Nov. 1990, pp. 34–44.
11. Anon, 'Aluminium Spaceframe Makes the Lightest Lotus', *Materials World*, Dec. 1995, p. 584.
12. Novak, M. and Wenzel, H., 'Design Engineering and Production of the Alcoa Spaceframe for Ferrari's 360 Modena', SAE Paper 1999-01-3174.
13. Anon, 'The Aston Martin V12 Vanquish', *AutoTechnology*, Aug. 2001, Vol. No. 1, pp.28–29.
14. Dieffenbach, J.R., 'Challenging Today's Stamped Steel Unibody: Assessing Prospects for Steel, Aluminium and Polymer Composites', IBEC '97 Proceedings, Stuttgart, Germany, 30 Sept – 2 Oct 1997, pp. 113–118.
15. Dieffenbach, J.R., 'Not the Delorean Revisited: An Assessment of the Stainless Steel Body-in-White', S.A.E. Paper 1999-01-3239.
16. Walia, S. *et al.*, 'The Engineering of a Body Structure with Hydroformed Components', IBEC Paper 1999-01-3181, 1999.
17. Davies, G.M. and Waddell, W., 'Laser Welding Allows Optimum Door Design', *Metal Bulletin Monthly*, July 1995, pp. 40–41.

18.2

Chapter 18.2

Materials for consideration and use in automotive body structures

Geoffrey Davies

OBJECTIVE

To review the choice of materials suitable for body manufacture and provide an understanding of salient manufacturing processes, product parameters and associated terminology. This is essential if the limitations of specific materials are to be appreciated and advantages/disadvantages of competing materials are to be compared prior to the design selection stage.

CONTENT

The range of materials that can be realistically considered for body structures is reviewed – the critical need for consistency of properties and effect on productivity is emphasized, together with the associated advantages of continuous production methods – key stages of both steel and aluminium manufacture are described – the significance of the final skin pass is explained and methods used to vary the texture of work rolls and strip surface are described – types and strengthening mechanisms of high strength steel (HSS) grades in yield stress range 180–1200 MPa are graphically described and an introduction provided to the relevant polymer types and mode of manufacture.

18.2.1 Introduction

The main materials used in body construction are evident from Chapter 18.1, and, as indicated, early choices were governed by the increasing needs of mass production and later, during post-war years, availability, as suppliers struggled to resume production. It will also be obvious that nowadays the choice has broadened considerably as materials technology has responded to the needs of the automotive engineer and that a far more enlightened understanding of materials parameters is required if this enhanced range of properties is to be exploited to the maximum advantage. The situation has advanced significantly from the days when 'mild steel sheet' was the universal answer to most body parts applications. As apparent from the 'Introduction' any distinction between grades was then generally made on the basis of formability and in the case of the few aluminium specifications, differentiation was by temper (O = annealed, H = hard, etc.).

The metallurgy of both steel and aluminium alloys has now advanced significantly offering a wide choice of mechanical and physical properties together with other attributes. The metallics choice can now be widened and it is also necessary to consider plastics, where 20 different types can be used within the motor vehicle. The traditional requirements of the body engineer have always been strength, in both static and dynamic terms, and elastic modulus, which governs stiffness/rigidity and imparts stability of shape. To these can now be added drawability and work-hardening parameters which are important with respect to forming and stretching, respectively, the latter also having an important effect on energy absorption and impact resistance.

A number of surface parameters are now held to have considerable tribological (frictional) as well as cosmetic significance. Whereas the main requirement could be specified in terms of surface roughness, R_a, a range of 'deterministic' (pre-etched) rather than 'stochastic' (random, shot-blasted) finishes are now options. With the advent of CAD systems the design engineer can no longer rely on experience from previous models, or on a dedicated materials engineer for his materials choice. Instead he is often confronted with a series of predetermined choices, from which he must make his selection. For the correct choice at the engineering stage

Materials for Automobile Bodies; ISBN: 9780750656924
Copyright © 2003 Elsevier Ltd; All rights of reproduction, in any form, reserved.

therefore it is important that the automotive engineer fully understands the parameters presented to him and their relevance in terms of production and application, and controlling metallurgical characteristics. A summary of key parameters is presented in Table 18.2-1 showing some basic properties and relevance to the automotive engineer. The stage of the strip production process critical to the development of these parameters is also identified together with other influencing factors.

The importance of strength, ductility and surface finish will be self-evident from the preceding text but the '*r*' value provides a measure of the resistance of the material to thinning in the thickness plane during drawing via a favourable crystallographic texture. Likewise, during stretching a high work hardening '*n*' value spreads the strain over a more diffuse area thereby offsetting the tendency to form a local neck. A further essential consideration is cost and this is evident in Table 18.2-2 illustrating the type of data used by the design house of a key European vehicle manufacturer in its material selection process. The extended range of materials and interaction with engineering design parameters illustrates the wider vision of the contemporary designer together with his awareness of costs!

The key parameters and criteria applicable to the main structure and panels were defined by Ludke and were presented in Figs. 18.1-17 and 18.1-18. In addition to the parameters of engineering significance, it is now essential to consider ease of manufacturing in a much more detailed way as the stages representing the 'process chain' are each increasingly complex and require forethought as to the implications of the introduction of any new facet of materials change on productivity. Any responsible company now adopts a 'life cycle' approach to consideration of new materials and so to complete the selection process prior to the approval of a new material further criteria must be applied which rate acceptability according to emissions friendliness, ease of disposal and recyclability.

To illustrate this process a listing of realistic main contenders is presented below together with an impression of relevant ratings. This is not meant to be a definitive presentation, and due to differences in local references some minor anomalies will be noted regarding values presented for material properties compared with Table 18.2-2, but it does offer a concise summary of the wide spectrum of factors governing the selection of materials today, possible current ratings and methodology adopted by the larger organization.

This text is not only intended for the design engineer but anyone involved in the process or supply chain, whether in a technical or administrative capacity. It is in the author's experience that all those directly and indirectly involved with engineering and launch of automotive designs have benefited enormously by visiting suppliers and understanding the production route and terminology associated with the material they are using. The relevance of process features and tolerances applicable to certain aspects of the product can then be realistically appreciated. These are compiled from a volume car perspective and it is possible that the niche and specialist car sectors may reflect slightly different ratings. The content of this chapter may be biased to the main contender, steel, but prominent coverage is also given to aluminium and plastics, with reference to other materials as appropriate.

18.2.2 Material candidates and selection criteria

The range of body materials that can be considered for volume car body construction is shown in Table 18.2-3, and it will be apparent that the criteria used by a major manufacturer when considering a new design extend beyond the range of physical and mechanical properties on which selection was once based. Not all factors are

Table 18.2-1 Key design parameters and relevant processing details

Parameter	Relevance	Influencing factors and *key processing stages*
Strength	Design	Imparted by composition, deformation and grain size; *alloying during smelting and mechanical and thermal treatment.*
Ductility	Forming, collapse characteristics	Lean composition and optimum heat treatment; *careful analysis and extended annealing cycle.*
Drawability index '*r*' ('resistance to thinning')	Press forming	Crystallographic texture requiring *optimum rolling and annealing schedules.*
Work hardenability '*n*'	Stretch forming Energy absorption	Composition and grain size dependent; *casting and rolling.*
Surface finish	Lubricity during forming Painted appearance	Imparted by roll finish *at temper rolling stage.*

Table 18.2-2 Extended choice of materials and parameters used by a key car manufacturer. (Courtesy of B. Ludke, BMW Group).

		1	2	3	4	5	6	7	8	9	10	11	12	13	14	15	16	17	18	19
Material	UK equivalent	E-Modul	Dichte	$\frac{E}{\rho}$	$\frac{\left(\frac{E}{\rho}\right)}{Preis}$	$\sqrt{E}$	$\frac{\sqrt{E}}{\rho}$	$\frac{\left(\frac{\sqrt{E}}{\rho}\right)}{Preis}$	$\sqrt[3]{E}$	$\frac{\sqrt[3]{E}}{\rho}$	$\frac{\left(\frac{\sqrt[3]{E}}{\rho}\right)}{Preis}$	$R_{p0,2}$	$\frac{R_{p0.2}}{\rho}$	$\frac{\left(\frac{R_{p0.2}}{\rho}\right)}{Preis}$	$\sqrt{R_{p0,2}}$	$\frac{\sqrt{R_{p0.2}}}{\rho}$	$\frac{\left(\frac{\sqrt{R_{p0.2}}}{\rho}\right)}{Preis}$	$10^{-6}.K^{-1}$	A_5,A_{80}	Halbzeug Preis in DM/kg (1996)
FeP04 St 14	DCO 4 forming grade	210 000	7,85	26 752	22 293	458,3	58,4	48,6	59,4	7,6	6,3	185,0	23,6	19,6	13,6	1,7	1,4	11,0	>40	1,20
ZstE 300 P BH	HSS bake hardening 300 MPa YS grade	210 000	7,85	26 752	20 578	458,3	58,4	44,9	59,4	7,6	5,8	340,0	43,3	33,3	18,4	2,3	1,8	11,0	>28	1,30
S 420 MC ZstE 420 NbTi	HSS HSLA grade 420 MPa YS grade	210 000	7,85	26 752	19 816	458,3	58,4	43,2	59,4	7,6	5,6	480,0	61,1	45,3	21,9	2,8	2,1	11,0	>20	1,35
BTR 165 VHF – Stahl	Ultra high strength steel 1100 MPa YS	210 000	7,85	26 752	19 108	458,3	58,4	41,7	59,4	7,6	5,4	1100,0	140,1	100,1	33,2	4,2	3,0	11,0	10(A5)	1,40
AlMg5Mn 10%kV	Aluminium–magnesium wrought sheet for internal parts	70 000	2,70	25 926	4321	264,6	98,0	16,3	41,2	15,3	2,5	185,0	68,5	11,4	13,6	5,0	0,8	23,8	20(A5)	6,00
AlSi1.2Mg0.4 10%kV, 190°C, 0.5hr	Aluminium–silicon skin panel material paint bake hardened	70 000	2,70	25 926	3704	264,6	98,0	14,0	41,2	15,3	2,2	260,0	96,3	13,8	16,1	6,0	0,9	23,4	15(A5)	7,00
AZ 91T6 Magnesium alloy	Heat-treated magnesium alloy	45 000	1,75	25 714	4675	212,1	121,2	22,0	35,6	20,3	3,7	200,0	114,3	20,8	14,1	8,1	1,5	26,0	7(A5)	5,50
TiAl6V4 F89 Titanium alloy	Titanium alloy for automotive consideration	110 000	4,50	24 444	349	331,7	73,7	1,1	47,9	10,6	0,2	820,0	182,2	2,6	28,6	6,4	0,1	9,0	5(A5)	70,00
Kiefer – longitudinal	Pinewood grain longitudinal	12 000	0,50	24 000	6000	109,5	219,1	54,8	22,9	45,8	11,4	100,0	200,0	50,0	10,0	20,0	5,0	7,0		4,00
Kiefer – transverse	Pinewood grain transverse	12 000	0,50	24000	6000	109,5	219,1	54,8	22,9	45,8	11,4	3,0	6,0	1,5	1,7	3,5	0,9	40,0		4,00

Al_2O_3 (Keramic, massiv) 'spröde'	Fused alumina	370000	3,85	96104	481	608,3	158,0	0,8	71,8	18,6	0,1	500,0	129,9	0,6	22,4	5,8	0,0	8,6	0,0	200,00
GFK 55% force parallel to fibre	Glass reinforced plastic long. fibres	40000	1,95	20513	2051	200,0	102,6	10,3	34,2	17,5	1,8	950,0	487,2	48,7	30,8	15,8	1,6	6,0	2,0	10,00
GFK 55% force normal to fibre	Glass reinforced plastic transverse fibres	12000	1,95	6154	615	109,5	56,2	5,6	22,9	11,7	1,2	475,0	243,6	24,4	21,8	11,2	1,1	6,0	2,0	10,00
AFK 55%, TM – Typ parallel to fibres	Aramid fibre reinforced epoxy – long.	70000	1,35	51852	519	264,6	196,0	2,0	41,2	30,5	0,3	1500,0	1111,1	11,1	38,7	28,7	0,3	–3,0	2,0	100,00
	Aramid fibre reinforced epoxy – transverse	6000	1,35	4444	44	77,5	57,4	0,6	18,2	13,5	0,1	750,0	555,6	5,6	27,4	20,3	0,2	–3,0	2,0	100,00
CFK 55% force parallel to fibre	Carbon fibre reinforced epoxy – long.	110000	1,40	78571	1310	331,7	236,9	3,9	47,9	34,2	0,6	1100,0	785,7	13,1	33,2	23,7	0,4	0,0	1,0	60,00
CFK 55% force normal to fibre	Carbon fibre reinforced epoxy – transverse fibres	8000	1,40	5714	95	89,4	63,9	1,1	20,0	14,3	0,2	700,0	500,0	8,3	26,5	18,9	0,3	0,0	1,0	60,00
GF-PA-12 (54%) parallel to fibre	Glass reinforced polyester 54% – long.	35400	1,70	20824	1041	188,1	110,7	5,5	32,8	19,3	1,0	600,0	352,9	17,6	24,5	14,4	0,7	5,0	2,0	20,00
GF-PA-12 (54%) normal to fibre	Glass reinforced polyester 54% – transverse	4400	1,70	2588	129	66,3	39,0	2,0	16,4	9,6	0,5	65,0	38,2	1,9	8,1	4,7	0,2	5,0	2,0	20,00
Glas (massiv) 'spröde'	Fused glass	70000	2,50	28000	18667	264,6	105,8	70,6	41,2	16,5	11,0	1000,0	400,0	266,7	31,6	12,6	8,4	5,0	3,3	1,5

Hinweis: grau hinterlegte Materialien sind Basis für Anlage 2 bis 5.
|| = parallel zur Faser.
—|— = quer zur Faser.
Anlage 1: Materialeigenschaften.

Table 18.2-3 Main criteria and ratings for realistic selection of automotive body materials

Material	Design parameters					Ease of manufacturing* ('process chain')			Environmental** 'friendliness'		Cost
Criteria	YS MPa	UTS MPa	A_{80} min%	E.Mod GPa	D g/cc	Forming	Joining	Paint	CO_2 + emissions	Disposal (ELV)	Forming steel = 1
1. Forming grade steel EN 10130 DC04 + Z	140 min	270 min	40	210	7.87	8	9	9	7	9	1.0
2. HSS EN 10292 H300YD + Z	300 min	400 min	26	210	7.87	6	8	9	8	8.5	1.1
3. UHSS – martensitic	1050–1250	1350–1550	5	210	7.87	4	7	9	8	8.5	1.5
4. Aluminium 5xxx	110 min	240 min	23	69	2.69	6	5	8	9	9	4.0
5. Aluminium 6xxx	120 min	250 min	24	69	2.69	6	5	8	9	9	5.0
6. Magnesium sheet	160 min	240 min	7	45	1.75	4	4	7	9.5	6	4.0
7. Titanium sheet	880 min	924 min	5	110	4.50	6	5	7	9	6	60.0
8. GRP	950	400–1800	<2.0	40	1.95	8	7	8	8	5	8.0
9. Carbon fibre composite	1100	1200–2250	<2.0	120–250	1.60–1.90	8	7	8	9	5	50.0+

*Based on range 1 = difficult to process, 10 = few production problems.
**Ease with which prevailing legislation can be met: 10 = without difficulty, 1 = extensive development required.

shown and it is easy to subdivide any of the columns shown. As will be illustrated later, the legislative requirements concerning, for instance, emissions and end-of-life (ELV) disposal are now influencing the initial choice of material, and increasingly the process chain or successive stages of manufacture must be considered to ensure that minimum disruption is incurred which may have consequences in productivity and quality. Any allowances for new materials will have been thoroughly proven at the pre-development stage of production.

Steel is still the predominant material used for manufacture[1,2] and the generally high ratings levels shown under the 'Ease of manufacturing' column reflect the provision already made by the industry for compatible facilities. The lower ratings evident for processing of aluminium, for instance, do not indicate that newer materials are an inferior choice but are more indicative of the size of change and introduction of new practices necessary to accommodate them. However, changes are inevitable to ensure the different legislative requirements are met and Table 18.2-3 perhaps indicates the 'pain' necessary to implement these lighter, but sometimes problematic, alternative materials. Indicative figures are also included for polymeric materials frequently used in specialist car manufacture and carbon fibre composites used in competition vehicles.

18.2.2.1 Consistency: A prime requirement

Whatever the material with regard to the physical and mechanical properties, the one key requirement that is essential to maintain manufacturing productivity is consistency.

Once a piece of equipment has been set to operate within a given range of compositional, mechanical and dimensional properties, to ensure maximum output it must run continuously without disruption and the need for persistently reset process variables such as press and welding settings. Of course the tightest tolerances must be held with regard to tooling and machine efficiency but uniformity of material characteristics is essential if the full benefits of these facilities are to be realized. This is particularly true of state-of-the-art automated tri-axis progression presses and multistation robotic welding equipment where due to the momentum of the system and parts in process any delay due to defective processing can quickly lead to large quantities of scrap (or parts needing rework) being generated before the fault is detected. Rigid monitoring of the feedstock production process is therefore required with regard to both constituents and process, whereas for automotive assembly manufacture, statistical process control should be rigorously implemented, with all contributors to the supply chain demonstrating compliance with the requirements of BSEN ISO 9002, QS9000 and individual company quality approval procedures. Uniformity and the highest possible quality are the twin aims and continuous processing must be encouraged as batch manufacture of primary material, even with the best available controls, is by definition liable to local inhomogeneities in composition, temperature and other influences. Continuous processing, due to the scale of operation and improved operating efficiencies, must therefore be beneficial and although material might deviate slightly from the original specification, it can be utilized efficiently. With regard to steel manufacture, continuous casting and annealing processes have already been widely introduced (the advantages of which to the automotive industry are described later in this chapter), and casting to thin section dimensions is well on the way to reality. It is now interesting to note the coupling together of increasingly complex major operations as continuous processes.

Other more ambitious attempts have been made at amalgamation of pickling, rolling and annealing processes but are hampered by capacity imbalances. More widely, similar processing is being introduced in the aluminium industry where continuous annealing using rapid induction heating has been developed and continuous casting has been accepted practice for many years.

Uniformity of processing and product is therefore as significant as property levels and this will be the underlying theme throughout the following text. Before considering effect of materials in the context of design, manufacturing and service, however, it is essential that the manufacturing processes associated with each of the main materials is understood so that the implications of various grades, treatments and finishes can be appreciated in later sections.

As stated the individual materials are presented in order of the extent of utilization within the industry, commencing with steel.

18.2.2.2 Steel

Despite the quest for alternative materials, and the considerable amount of research that has been carried out to develop lighterweight materials in the last 30 years, most owners of the current generation of cars are driving an essentially steel structure, which required approximately half a tonne of flat rolled steel product to manufacture. As described almost 15 years ago,[2] this material has exceptional versatility in terms of formability, strength and cost, and the industry has responded quickly to recognize the changing engineering needs arising from legislative and environmental requirements. Put simply, advantages of steel as an autobody material include:

- Low cost
- Ease of forming

- Consistency of supply
- Corrosion resistance with zinc coatings
- Ease of joining
- Recyclable
- Good crash energy absorption

The main disadvantages of steel in autobody applications are:

- Heavier than alternative materials
- Corrosion if uncoated

However, both these factors have been addressed over the last 20 years through the development of a much wider range of sheet and strip products. Higher strength steels with a wide range of yield strength values – now extending to 1200 MPa – can now be supplied and, as will be seen later, designs can be suitably modified to either improve performance at existing thicknesses or downgauge with strength related parts. Although stiffness remains unaltered it is possible to offset decreased torsional rigidity, for example, by the application of structural adhesive in flange areas or elongated laser welded seams. The full range of steels used in automotive design, from the forming grades with a minimum yield of 140 N/mm^2 to ultra-high strength steels with values up to 1200 N/mm^2, is shown later in this chapter (Table 18.2-4) but an indication of types and properties is evident from Fig. 18.2-1.

At the lower end of the scale it is now possible to utilize very low carbon interstitial free 'IF' steels for some parts requiring exceptional deep drawing properties but it should be remembered that these are still very much the exception, and at yield/proof stress levels down to 120 N/mm^2 can depart from the accepted design minimum of 140 N/mm^2.

Likewise corrosion is much less of an issue than even 10 years ago. A range of zinc-coated steels, namely, electrogalvanized, hot-dip, alloyed and duplex, is now available, the preference of individual automotive companies being dictated by cost, historical preference and manufacturing policy. Generally, the same range of steel properties is available with these coated products as for normal forming and high strength grades but sometimes a slight reduction in ductility is associated with hot-dip galvanized sheet due to the effect of the heat treatment cycle.

18.2.2.2.1 Steel production and finishing processes

Processing improvements which have enabled the increased range of properties previously highlighted are summarized in the following sections and many are highlighted in the flow chart shown in Fig. 18.2-2, the sequence existing at a major Corus installation, but typical of most plants worldwide manufacturing automotive strip.

Regarding steel production, it is probably sufficient to know at this stage that most steel used for autobody

Table 18.2-4 High strength steel grades commonly available in Europe

Type	Range of yield stress MPa	Strengthening mechanism	Relevant standard
Low carbon mild steel sheet	140–180	Residual carbon, Mn, Si	EN 10130
Rephosphorized	180–300	Solid solution hardening	PrEN10xxxx EN 10292 (hot-dip zinc coated)
Isotropic	180–280	Si additions	PrEN10xxxx
Bake hardening	180–300	Strain age hardening	PrEN10xxxx EN 10292 (hot-dip zinc coated)
High strength low alloy	260–420	Grain refinement and precipitation hardening	PrEN10xxxx EN 10292 (hot-dip zinc coated)
Dual phase	450–600 (UTS)	Martensitic (hard) phase in ferritic ductile matrix	PrEn10xyz
TRIP steel	500–800	Transformation of retained austenite to martensite on deformation	PrEn10xyz
Complex and martensitic steels	800–1200	Bainitic/martensitic (hard) phases formed by controlled heat treatment	PrEn10xyz

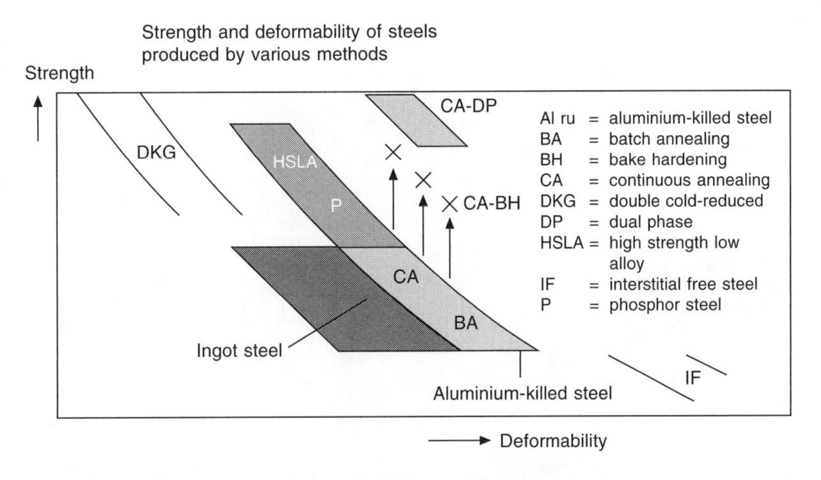

Fig. 18.2-1 Range of flat rolled steel products. (Courtesy of Corus.)

manufacture has been smelted from iron (produced in the blast furnace) and recycled scrap. Basic Oxygen Steelmaking (BOS) is the normal route whereby impurities are oxidized by the injection of oxygen through the bottom of the converter (included in Fig. 18.2-2), to produce refined material of composition typical of forming grades specified in Euronorm EN 10130. Normally these steels are aluminium killed (AK), an aluminium addition minimizing ageing effects by combining with nitrogen, and forming the characteristic 'pancake shaped', i.e. elongated, grains evident in the microstructure. Carbon levels are typically 0.03–0.05

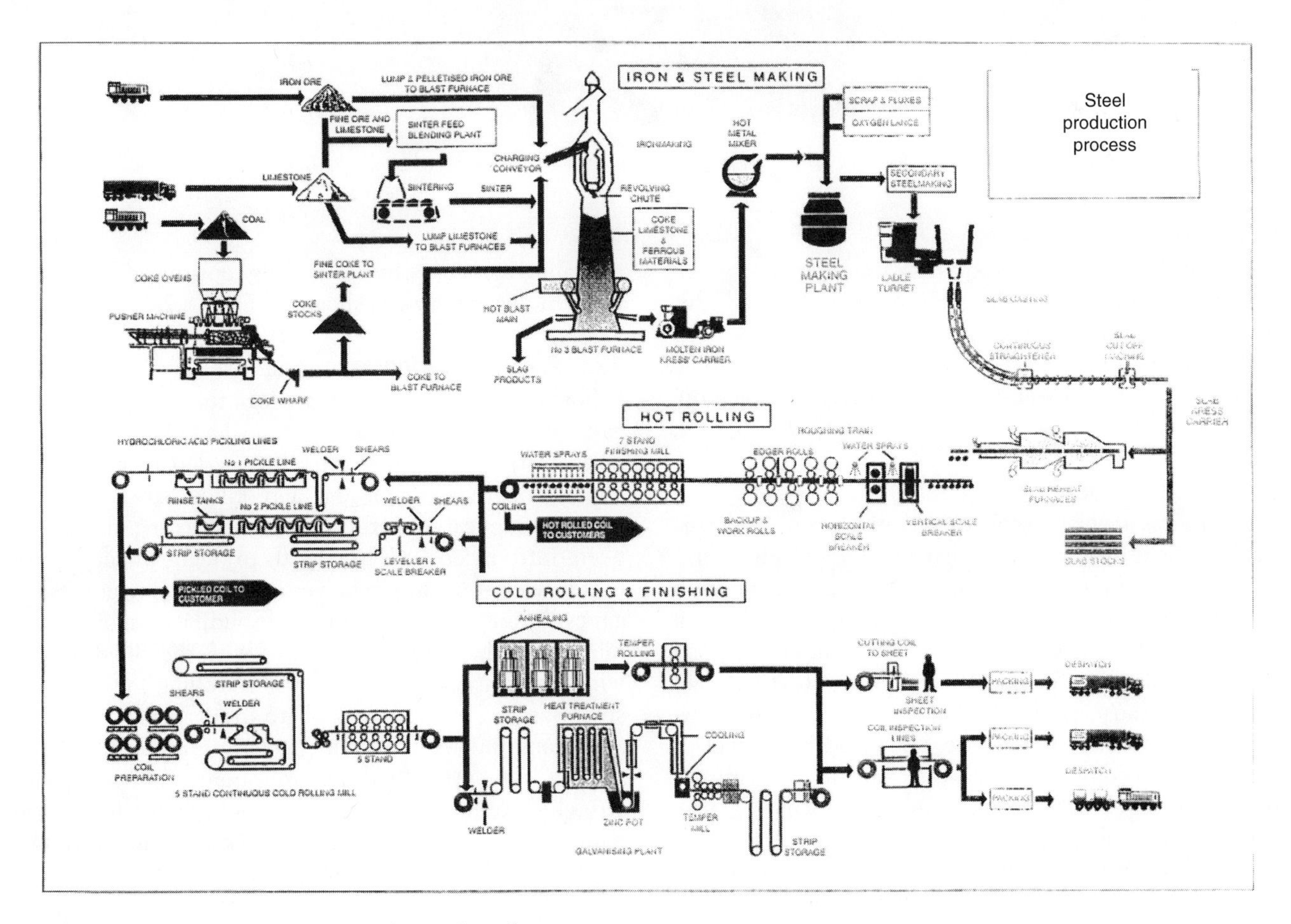

Fig. 18.2-2 (a) Typical sheet metal steel manufacturing process.

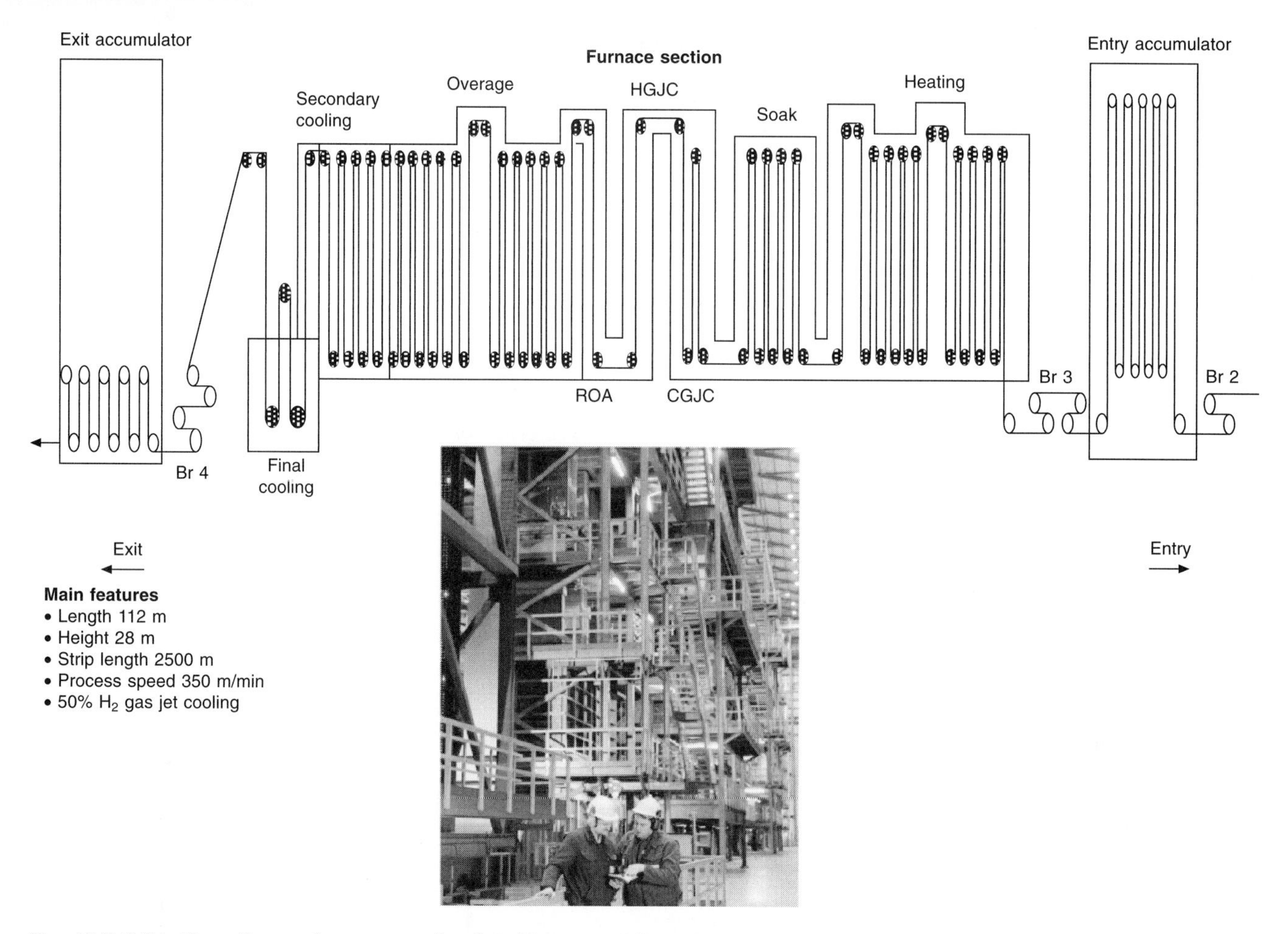

Fig. 18.2-2 (b) Alternative continuous annealing line. (Courtesy of Corus.)

per cent but for ultra-deep drawing coated or high strength grades where extra drawability is required, this is reduced to less than 0.0002 per cent. This is achieved by vacuum degassing of the molten steel prior to casting, resulting in the now well-known IF steels used for more complex shaped parts.

18.2.2.2.1.1 Vacuum degassing

This process involves the removal of gaseous and particulate inclusions, ensuring that very low levels of impurities are retained. Additions of titanium or niobium ensure interstitial elements such as carbon and nitrogen are reduced to extremely low levels and other compositional changes effected to optimize texture development, thus resulting in high '*r*' values – hence the term 'interstitial free' and the associated high level of formability accompanying IF variants. This is an important treatment for high strength steels such as grades H180–260YD included in EN 10292 which show higher '*r*' and '*n*' values than other grades classified with similar strength levels. Similarly IF substrates can boost the formability of hot-dip galvanized products where annealing cycles might not be fully optimized. A typical vacuum degassing rig is shown in Fig. 18.2-3.

18.2.2.2.1.2 Continuous casting

Following the steelmaking process the molten steel is cast into a shape suitable for re-rolling. The traditional ingot casting processing route which led to the differentiation between 'rimming' and 'killed' steels has now largely disappeared as the continuous casting of slab has been introduced. With regard to ingot manufacture better utilization was achieved with rimming grades, as gas evolution in the final stages of solidification offset the familiar 'V'-shaped pipe associated with the killed grades, which was later cropped off. On re-rolling the rimming steel was, however, susceptible to strain ageing in storage. As a consequence the yield strength increased and the surface was prone to the appearance of 'secondary stretcher-strain' markings, which could show through the painted finish. Now continuous slab production using the type of rig shown in Fig. 18.2-4 ensures that the maximum yield is now obtained and at least the same quality of material can be produced, but with a much higher

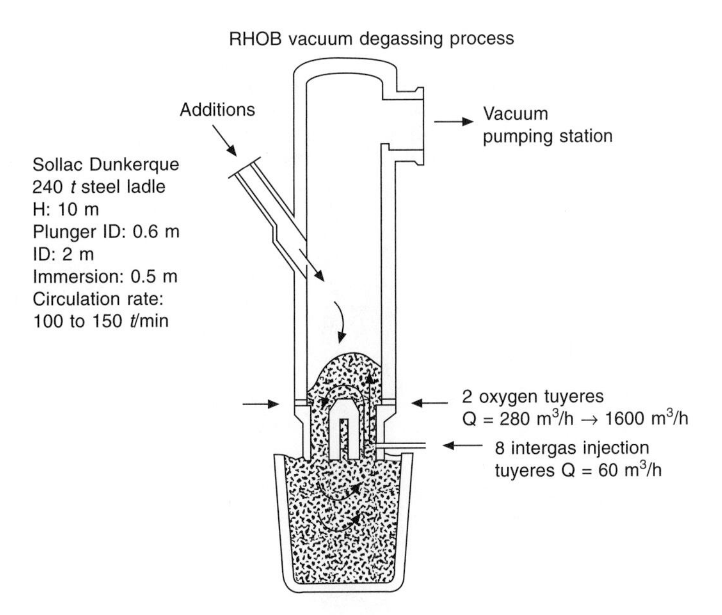

Fig. 18.2-3 Vacuum degassing in the steelmaking process.[3]

level of consistency regarding cleanliness and property variation.

As shown in Fig. 18.2-4, the ladle of steel is poured directly into a water-cooled copper mould, sticking being prevented by the reciprocating motion of the mould sides and a suitable dressing. The resulting slab thickness is typically around 250 mm but this has been reduced in some specialized 'mini mill' operations to 50 mm, thereby shortening the rolling process, and more recent advances aimed at direct strip manufacture make less than 2 mm material a realistic prospect.

18.2.2.2.1.3 Hot and cold rolling processes

After casting the slabs are progressive rolled down to the sheet thicknesses supplied in coil form to the automotive press shop (Fig. 18.2-5). Following reheating the slabs are hot rolled to produce a 'hot band' at 900–1200 °C, an intermediate form of now recrystallized material about 3 mm thick, but this has developed a thick layer of oxides or scale and although this is removed by 'pickling' in hydrochloric acid, the rough surface only renders it fit for selected underframe, chassis parts or bracketry. Process modifications have improved the quality of the hot rolled product and it is now possible to utilize this in thicknesses down to 1.6 mm, which gives a slight cost reduction to the part producer. Cold reduction is now essential to optimize dimensional accuracy, surface and properties, producing automotive material commonly 0.5–2.00 mm thick. This is normally carried out on a four- or five-stand sequence of four-high roll stations. The important influences on the final properties concerning formability are the grain size (controls '*n*' value) and crystallographic texture which is developed on subsequent annealing – which in turn controls the '*r*' value. The hot rolling coiling temperature, percentage final reduction and annealing temperature/rate critically affect these parameters, a high CR figure (of the order of 80 per cent) being commensurate with the optimum '*r*' value. This is controlled by the power of the mill equipment, and the roll diameter/configuration and final thickness. The thickness tolerance determines yield, and variation is corrected by automatic gauge control (AGC) systems working on an elaborate feedback system, which should achieve an accuracy of within 1 per cent.

Immediately after cold reduction the strip is then annealed to restore maximum ductility, and finally receives a skin pass (normally 0.8–1.2 per cent reduction) to impart the final surface texture and to remove any tendency for 'stretcher–strain' formation. The metallurgy and processing characteristics associated with this stage of steel manufacture are of extreme significance at this stage of manufacture of automotive grades of steel sheet and the mechanisms associated with skin passing are presented in detail after a consideration of annealing.

18.2.2.2.1.4 Continuous annealing

Although strand annealing has been used in the tinplate industry for annealing of thin strip,[4] and other similar research programmes were evident in the UK in the 1960s, continuous annealing process line (CAPL) technology (typical line illustrated in Fig. 18.2-2(b) has only been introduced for the processing of automotive sheet, as an alternative to batch annealing, over the last 15 years. Developed extensively in Japan, it is now used widely in Europe but the differences of the product compared with that of batch annealed coil need to be fully understood (Fig. 18.2-6).

As suggested previously, advantages should accrue in terms of consistency of composition and properties, but

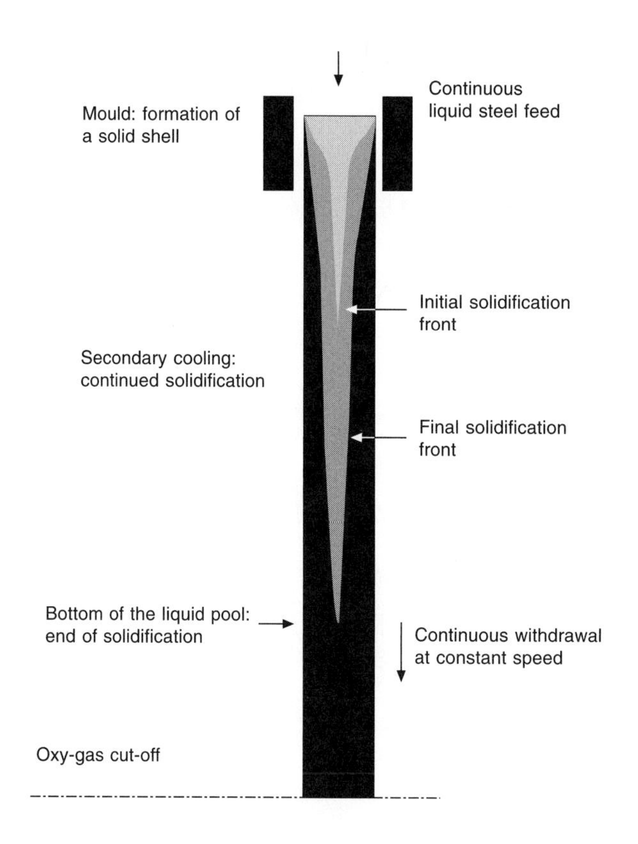

The principle of continuous casting

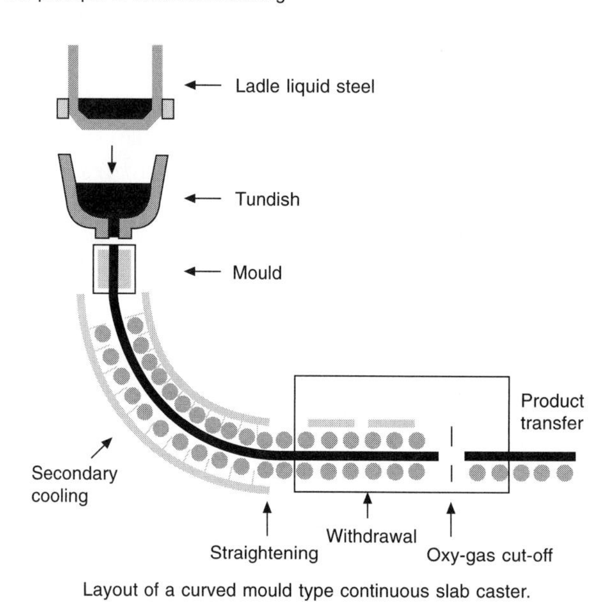

Layout of a curved mould type continuous slab caster.

Fig. 18.2-4 Continuous casting showing solidification and extraction processes.[3]

the length of the cycle being less than 10 minutes means that recrystallization and grain growth are not as complete as for the batch annealed (BA) product. This has been observed in the press shop where for lower grades of forming steels (DCO 2/3) the 'as received' properties have not matched the BA ductility levels. However, by correct allocation of material to the less demanding jobs and a reasonable degree of rework to tools, the CAPL products are now in regular use. The rapid annealing capability has also been used to tailor the properties of high strength steels by control of composition and temperature/time cycle with the result that bake-hardening

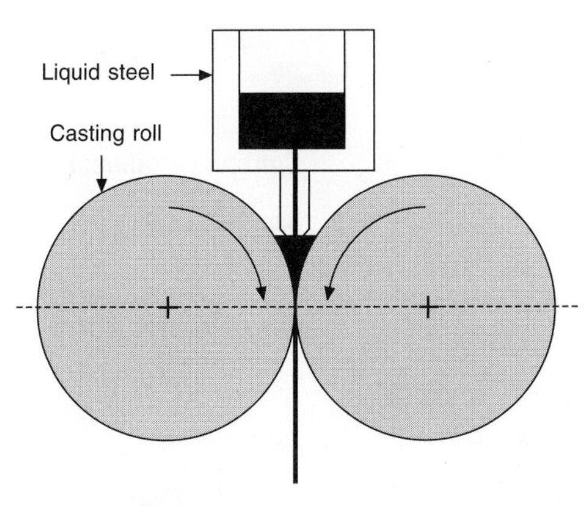

Fig. 18.2-5 Principle of direct process of casting thin strip from molten steel.

options and duplex/multiphase structured steels can be produced more readily.

Details of the differences between CAPL and BA processing are shown in Table 18.2-5.

Although requiring lower investment, the batch annealing processes commonly require 2–3 days for the more formable grades of steel as opposed to 10 minutes for the CAPL treatment, but even with the hydrogen atmosphere utilized by the Ebner process, which provides a higher heat transfer efficiency, the uniformity of properties is still a problem.

The lower processing time, while providing a vastly increased throughput via CAPL (Corus furnace section shown in Fig. 18.2-2(b)) limits the development of favourable crystallographic textures and grain size, and as a consequence '*r*' values are lower and drawability is impaired. Yield stress tends to be higher as it is related to grain size by the Petch equation[5] shown below,

$$\sigma_{LYS} = \sigma_i + k_y d^{-1/2}$$

where σ_i is the stress required to move free dislocations

$2d$ is the grain diameter

k_y is the effect of dislocation locking by impurity atoms

Relative properties for each type are shown in Table 18.2-6.

However, as stated, through tool and press adjustment and possibly the use of an enhanced lubricity mill oil, similar performance can generally be achieved for the same part.

18.2.2.2.1.5 Skin passing – effect on yield stress/yield point elongation

Unless IF technology is used during the steelmaking process, the presence of carbon and nitrogen interstitial atoms can lead to strain ageing on subsequent forming of the steel strip leading to the formation of stretcher-strain marks on the surface, accompanied by a well-defined yield point. This phenomenon is related to the pinning of residual dislocations by the interstitial atoms[6] and,

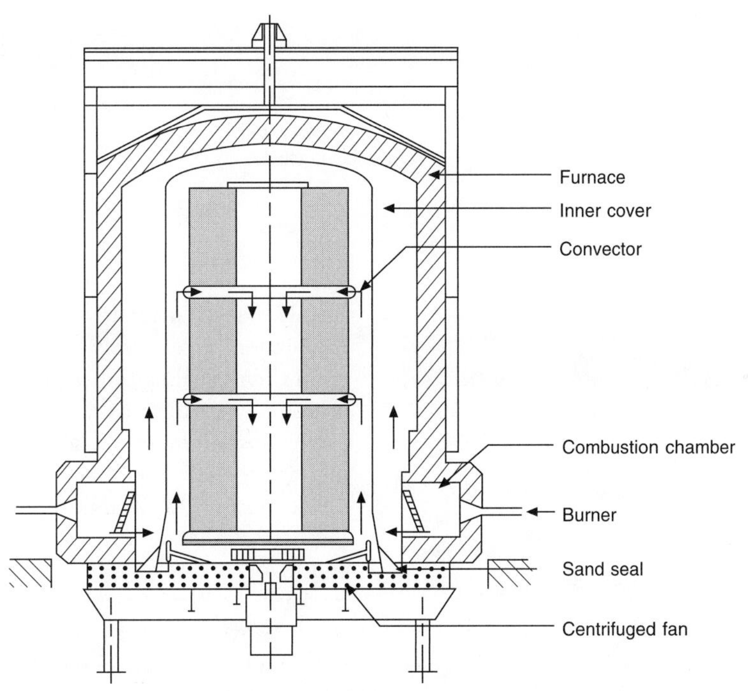

Fig. 18.2-6 Bell annealing furnace showing the vertical positioning of cold reduced coils.

Table 18.2-5 Comparison of batch and continuous annealing process characteristics[1]

Grade	Temp. (°C)	Batch annealing of coils	Continuous annealing
		Slow heating cycle = 30 ± 10 h Annealing temp. 710°C	Rapid heating cycle = 90 ± 30 s 30 to 60 s hold Annealing temp. 850°C
	500–550	AlN precipitation Recrystallization	
	550–600	Grain growth	Start of primary
	600–650	Texture reinforcement Grain growth	Recrystallization Start of AlN precipitation
	650–700	Texture reinforcement Solutioning and partial spheroidizing of Fe_3C Renitriding in the case of annealing in HNX* (N tied up by excess Al)	Grain growth impeded End of primary recrystallization
Aluminium-killed extra mild steel with nitrogen in solution	700–750	Start of secondary coarsening Coalescence of cementite Renitriding Loss of toughness (coarse grains)	End of AlN precipitation, grain growth impeded
	750–800		Grain growth impeded
	800–850	25 ± 5 h	Very slow grain growth, start of spheroidizing of cementite 10 ± 5 mn
	700–600	Formation of Fe_3C nuclei	Formation of Fe_3C nuclei
	600–200	Complete precipitation of dissolved carbon	Partial precipitation of dissolved carbon Residual C 4–15 ppm depending on the overageing cycle

Table 18.2-6 Relative properties for batch/continuous annealed steel[3]

Batch annealing and cooling of coils	Continuous annealing and cooling
Final properties	Final properties
Grain size 7–9 ASTM	Grain size 10–11 ASTM
YS: 160–190 MPa	YS: 230–270 MPa
UTS: 290–315 MPa	UTS: 330–400 MPa
El.: 40–44%	El.: 31–38%
r: 1.6–2.1	*r*: 1.0–1.4

simply stated, prior to skin passing the number of free dislocations is relatively few as they are locked on cooling from the annealing temperature. The 'locking atoms' are principally carbon, as any nitrogen has been combined with aluminium or an alternative addition, and these migrate to the interstices associated with the available dislocations. On straining a given amount, the stress required is that to move a few dislocations very quickly and involves a relatively high friction stress. Once these multiply on temper rolling the individual velocity and accompanying friction stress drops and the yield point falls to a lower level. The yield drop (discontinuous yield point) can therefore be explained in terms of differential strain rate effects.[6,7] On a macro scale the initial yielding of a steel in the annealed condition takes place on only a few fronts and results in the formation of coarse Lüders bands as shown in Fig. 18.2-7(b).

This effect leads to a 'flamboyant' surface marking on press formed panels and for this reason the sheet is subjected to 'skin passing' after annealing. As described by Butler[7] this produces numerous blocks of alternatively deformed and undeformed material which result in a slight roughening of the surface, but acceptable for painting (Fig. 18.2-8). When straining is resumed the velocity of each front and individual dislocation velocity is reduced, with an accompanying drop in friction stress. Therefore, the process, also known as temper rolling, results in an overall reduction in yield stress with the coarse markings now subdivided on a much finer scale virtually invisible to the naked eye. Yielding starts at a much lower stress and the accompanying stress/strain curve takes on a rounded smooth contour at lower strain levels, replacing discontinuous yielding as in Fig. 18.2-7(a).

18.2.2.2.2 Surface topography

As well as imparting a deformation of the order of 1 per cent to the strip to counter strain ageing effects, the process also dictates the final topography of the sheet. The type of finish embossed on the sheet surface by the work rolls that contact the sheet is becoming increasingly important as the lubrication characteristics during pressing and the finish developed during painting can both be optimized according to the final surface shape.

Traditionally the work rolls in the temper mill have been shot blasted and until the mid-1980s the optimum finish was defined in terms of R_a and peak density predominantly applied to a shot blasted texture. Other parameters were used in more detailed studies[8] such as Abbot curves but were difficult to apply on a routine basis especially in the workplace. Other surfaces from worldwide studies on sheet surfaces[9] examined using more sophisticated 3-D stylus plots of contour together with

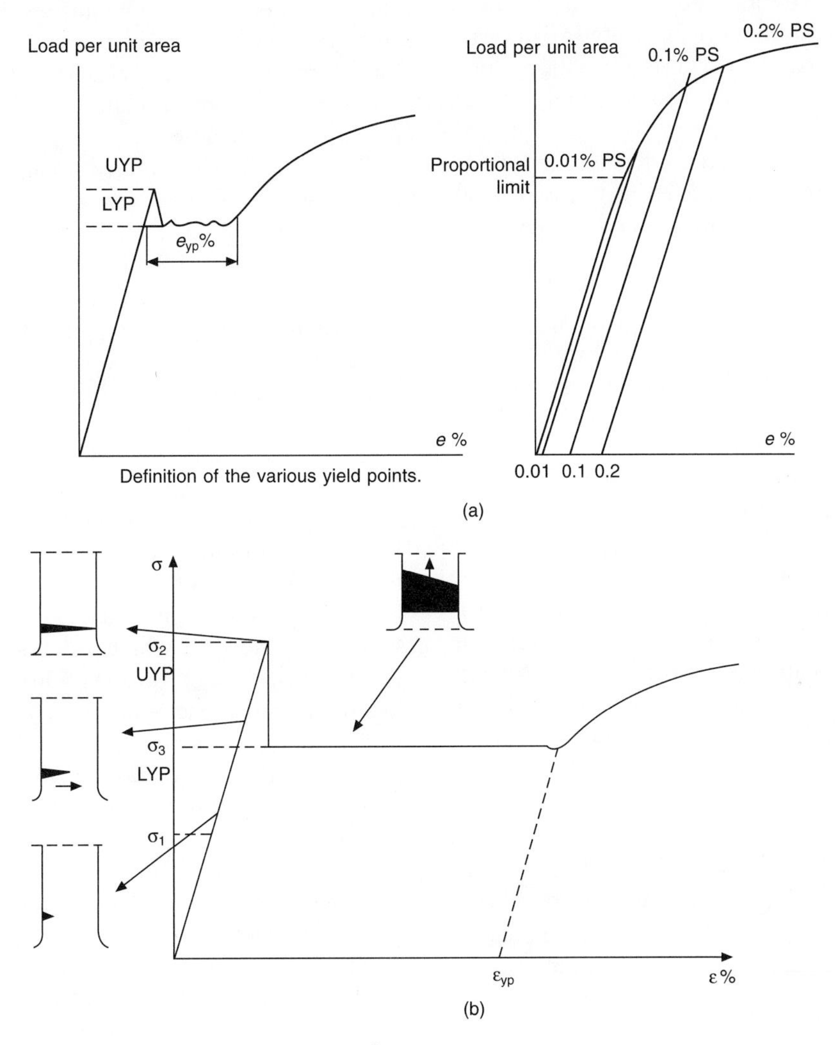

Fig. 18.2-7 (a) Effect of skin passing on yield point elongation.[3] (b) Lüders band formation showing development on relatively few fronts[3].

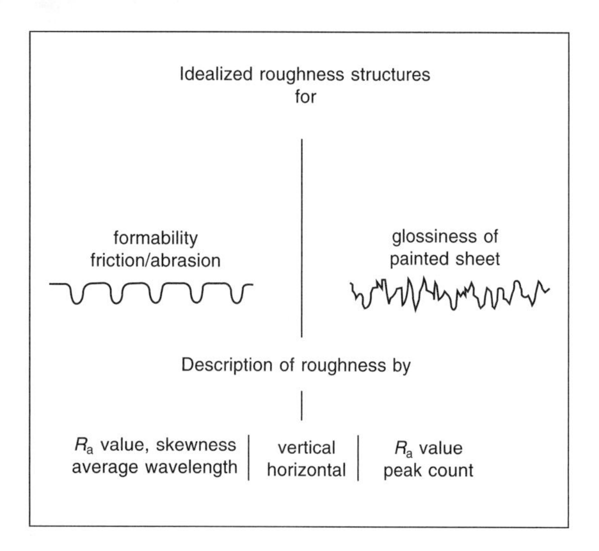

Fig. 18.2-8 Surface roughness profiles suited to forming and painting.

related parameters (skew, kurtosis) showed that the actual shape of the contours were important, plateau shapes resisting the collapse more associated with peaks (which can result in excessive debris formation), providing that channels were maintained to retain pressing lubricant.

For outer panels where maximum paint lustre was required a closer texture was needed in terms of peak spacing and the preferred finish as agreed by many automotive companies was as shown in Table 18.2-7.

In terms of consistency the shot blasted finish was not ideal as the actual surface contour varied with application mode and wear. Traditionally, a coarse texture from the tandem mill (main cold-reduction process) was required to avoid adjacent coil laps sticking on annealing and this was overlaid with the finer skin pass finish to provide the micro-texture producing the glossiness or lustre of the paint finish. The tandem mill texture was more open with peak spacings at 3–4 mm and relates to the coarse 'orange-peel' noted on many panel surfaces. With the advent of continuous annealing the tandem mill finish has less significance and the final finish is almost entirely due to the skin passing treatment. The control of texture using sand blasted temper rolls has always suffered due to lack of independent control over long and short wave contours and gradually development of alternative topographies has taken place over the last 30 years.

Table 18.2-7 Typical surface texture finish for automotive panels

Application	Outer panels (μm)	Inner panels (μm)
Range 2.5 mm cut-off	R_a 1.0–1.7	R_a 1.0–1.8

In 1971,[10] research had commenced on a more controlled system of surface preparation using electro-discharge texturing (EDT), the surface being eroded as in EDT machining by discharge of electrical energy through a di-electric, the work roll being one of the electrodes (Fig. 18.2-9). As well as a more repeatable process the peak count can be double that achieved with equivalent shot blasted finishes.

In 1982, more deterministic or specific roll surface patterns were developed beginning with Lasertex which, as the name suggested, was produced by subjecting the work roll surface to intermittent exposure employing a mechanical chopper to cut the laser beam (Fig. 18.2-10).

This created a regular array of circular channels providing lubrication while maintaining a central core to resist local deformation/wear. While this was excellent for oil retention for deep drawn parts, the periodicity of this array tended to show through higher gloss paint finishes and fluctuations in application only imparted a semi-deterministic pattern. A 'mirror finish' version of this was produced on a much finer scale, which was claimed to enhance paint finishes by higher reflectivity from the flat fraction of the surface but has not been introduced on a wide scale. Thus the search continued for either a more regulated version or an alternative method of application.

The EBT finish obtained using electron beam technology and employing better beam/workpiece synchronization results in a fully deterministic pattern as illustrated in Fig. 18.2-11.

The texture of the ERT rolls in the tandem and temper mill is generated by high energy and precisely positioned electron beam. The process takes place in a high vacuum chamber, a perfect atmosphere to avoid oxidation of the created crater surface.

The high-performance electron beam is focussed on the roll surface and melts the material. The locally created plasma blows the molten material aside, leaving behind a crater with a concentric rim (diameter from 50 to 250 μm, depth from 5 to 30 μm). The electron beam applies the homogeneous pattern to the total surface of the roll, which rotates at a constant speed (600 rpm). Due to the synchronization of electron beam and roll movement, the craters are generated on pre-defined positions on the roll surface. This technology leads to a complete and reproducible crater pattern. The total texturing process takes about 30 minutes. After the surface texturing process, the roll is returned to the mill and the texture is transferred onto the sheet surface at the final cold pass or temper reduction stage.

The process can be applied to both the last roll of the tandem rolling operation and the temper mill work rolls. Some suppliers can now supply a range (e.g. the 'Sibertex' range) of finishes which are combinations of stochastic

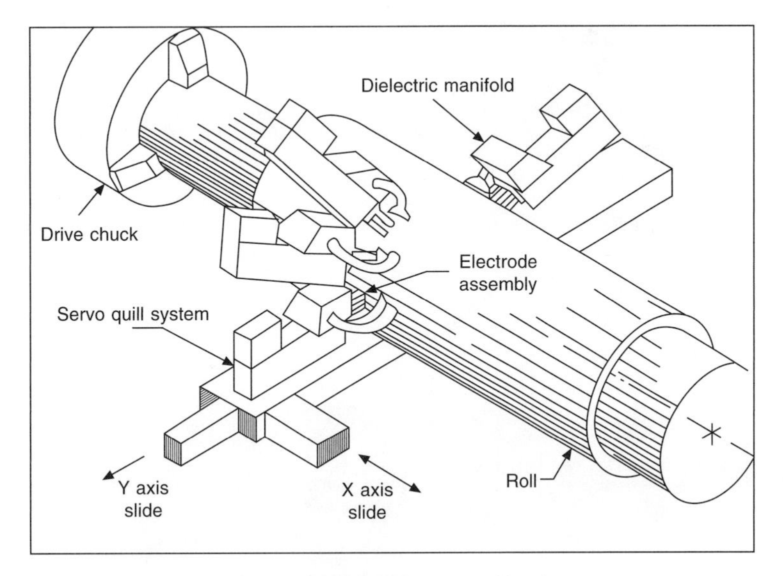

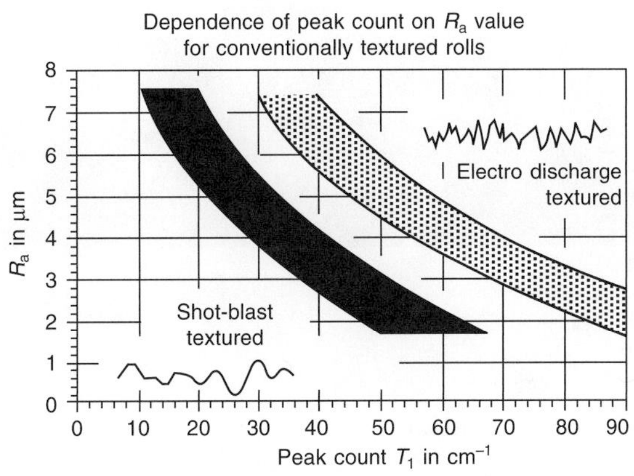

Fig. 18.2-9 Schematic diagram of an EDT system and comparison with shot blasting.

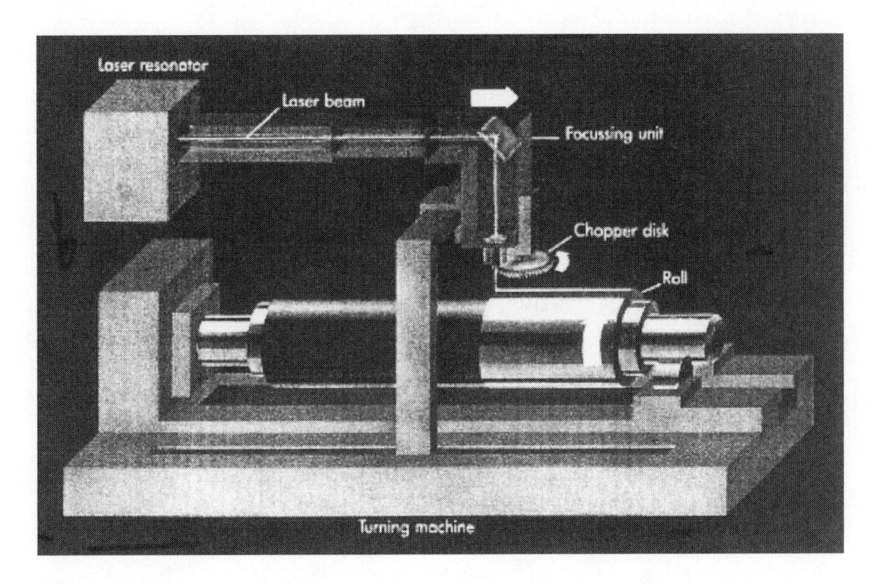

Fig. 18.2-10 Principle of the Lasertex method of roll texturing.

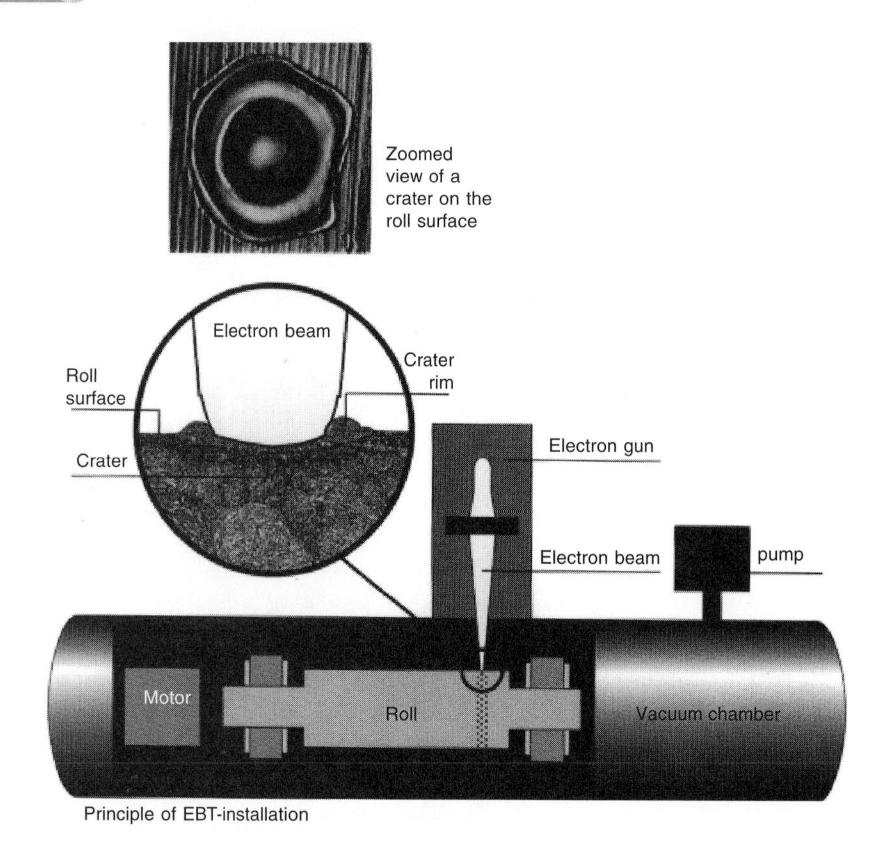

Fig. 18.2-11 Electron beam technology. (Courtesy Sidstahl.).

and deterministic topographies as shown schematically in Fig. 18.2-12, and a variety of finish combinations can be produced ranging from fully deterministic (regular) to stochastic (random) patterns as shown in Table 18.2-8.

Thus, in summary, a full range of surfaces is now possible which can be controlled to give optimum lubricant retention and paint reflectivity characteristics. Batch annealing relied on a coarse texture to prevent 'stickers' (binding together of adjacent laps) and this was then overlaid with a finer texture on skin passing which provided the microtexture required for optimum paint finish. The interdependence of the tandem mill coarse texture (or waviness) and temper mill texture has to some extent hampered the development of external paint finishes. Waviness is a further feature of batch annealed sheet affecting the uniformity of paint and can be related to an 'orange-peel' finish. Deterministic finishes such as those described above provide more control over the final texture than random shot blasted treatment and for continuously annealed strip (no coarse texture for sticker separation necessary) should provide a uniformly fine and mirror gloss paint finish – if that is what is required. Many argue that the majority of customers might favour a 'chunkier' coarse texture indicative of a substantial paint presence. Either way methods of independently controlling the fine and coarse finishes now exist but for maximum benefit this steelmaking capability should be universally available and

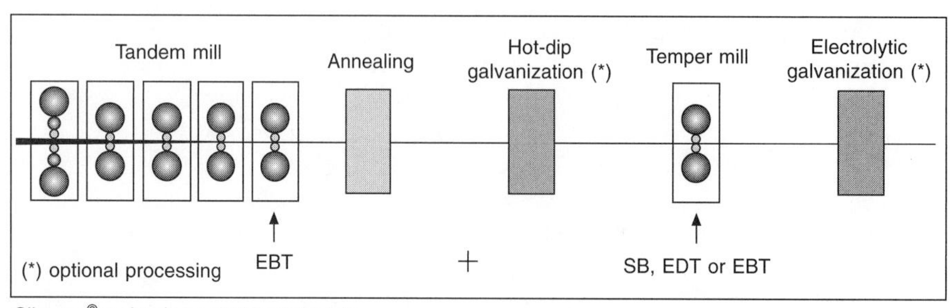

Fig. 18.2-12 Tandem and temper mill treatments offered by the Sibertex process. (Courtesy of Sidstahl.)

Table 18.2-8 Surface patterns available at specific manufacturing stages

Technology	Tandem roll	Temper roll
Shot blast (SB)	SB	SB
EDT	SB or EDT	EDT
Laser texturing	SB	Laser texturing
EBT-laser	SB	EBT
Simulation Sibetex©	EBT	SB, EDT or EBT

standards must be agreed at a national/international level to allow car manufacturers to specify the exact topography that best enhances their paint systems.

18.2.2.2.2.1 Effects in processing

As hinted at previously, the reason for modifying the surface is attributable to either benefits achieved in forming or influence over the final surface finish.

18.2.2.2.3 Higher strength steels

The application of higher strength steels in the automotive industry has been slow considering that basic grades have been available for at least 25 years, and this is due to a number of different preconceptions. Apart from concerns regarding the scope of application and the fact that high strength did not necessarily allow pro-rata reduction in thickness (due to the stiffness constraint, qualified in further detail in Chapter 18.1), formability was initially a key issue. Early grades, mainly rephosphorized and high strength low alloy (HSLA), were variable in properties and soon gained a reputation for tool wear and erratic performance, although as will be explained, process improvements have made these immeasurably better products. Together with exchange of experience worldwide, gathered and disseminated rapidly by organizations such as the International Deep Drawing Research Group, and by the adoption of design rules, harder wearing tool materials, surface treatments and compatible pressing facilities, these materials can be used with little extra effect. However, any attempt to run higher strength grades on existing or obsolete equipment will soon expose any weaknesses and result in high scrap rates/rework.

The following pages provide a basic understanding of the different types of higher strength steels by considering the simple metallurgy of mild steel and then introducing different strengthening mechanisms responsible for other grades which can range up to yield strength values of 1200 MPa. Most of these steels have related Euronorm or National Standards either published or in preparation.

The range of higher strength steels currently available is summarized in Table 18.2-4 which also highlights the strengthening mechanism and appropriate standard for that group of steels.

To further explain the strengthening modes associated with these steel grades a more graphical account is given in Table 18.2-9.

Thus it can be seen that a wide range of steels is currently available to the designer to help meet weight reduction targets and respond to challenges in safety engineering. The extent to which these special steels can be utilized and the changes necessary to accommodate them in manufacturing are considered in detail in Chapter 18.1. It should be stated that the increased corrosion risks posed by downgauging these grades is more than offset by the use of zinc-coated variants for most applications while also allowing vehicle durability targets to be maintained.

18.2.2.2.4 Stainless steel

As chromium is added to steels, the corrosion resistance increases due to the formation of a protective film of chromium oxide. The range and complexity of stainless steels is high and therefore detailed examination is outside the scope of this text. However, it can be stated that although stainless steels are not extensively used in current vehicles, they have nonetheless found applications in commercial vehicles, e.g. buses, and their potential for application has been advocated recently.[1] The main advantages of stainless steel as an autobody material are:

- Corrosion resistance
- Excellent formability
- Use of similar manufacturing infrastructure as mild steel

Disadvantages include:

- High cost
- Limited supply sources for automotive applications

Production of stainless steel is in many aspects similar to that described for mild steel.

The outstanding corrosion resistance of ferritic (13% Cr) and austenitic (18% Cr 8% Ni) steels and the potential they have for being used in the unpainted or partially painted condition has made them the subject of some very intense studies recently. When considered with especially attractive levels of forming parameters in sheet form, and reasonable on-costs, designers have been keen to re-examine the potential of this prestigious material which might give them a strong competitive advantage. It can be easily forgotten that these materials have high work-hardening rates and when considering press working with current facilities critical loads can be approached very quickly due to rapid work hardening (Fig. 18.2-13). This behaviour can also

Table 18.2-9 HSS strengthening mechanisms. (Illustrations courtesy of Thyssen Krupp Stahl.)

Rephosphorized steel is an example of a substitutional type where the larger atom is straining the lattice while smaller carbon, oxygen and nitrogen atoms occupy the interstices between the iron atoms

Solid-solution formation — Substitutional atom — Matrix atom — Interstitial atom

Bake-hardening effect — Interstitial atom — Edge dislocation

Interstitial-free 'IF' steels are vacuum degassed to remove the carbon and oxygen atoms which impede the movement of dislocations and therefore increase the ease of deformation (positive effect on forming, negative effect on dent resistance). IF high strength steels therefore combine the increased ductility associated with the ferritic matrix but gain enhanced strength from substitutional phosphorus, silicon and manganese additions.

Bake-hardening steels derive their increase in strength from a strain ageing process that takes place on paint baking at circa 180°C. Sufficient carbon is retained in solution during either batch or continuous annealing to allow migration to dislocations following cold deformation. These are effectively locked, requiring a higher subsequent stress to recommence deformation, thereby increasing dent resistance.

The strength of both rephosphorized and IF high strength steels can be enhanced by this mechanism but different modes of carbon retention are required to prevent premature diffusion of carbon at either room temperature in storage, or during the application of zinc by hot dipping. In continuous processing this can be achieved by incorporating an over ageing treatment and alloying whereby just enough carbon is retained in solution to allow the mechanism to occur at elevated temperatures. The degree of cold deformation will reduce the $\triangle BH$ response correspondingly (see the diagram below).

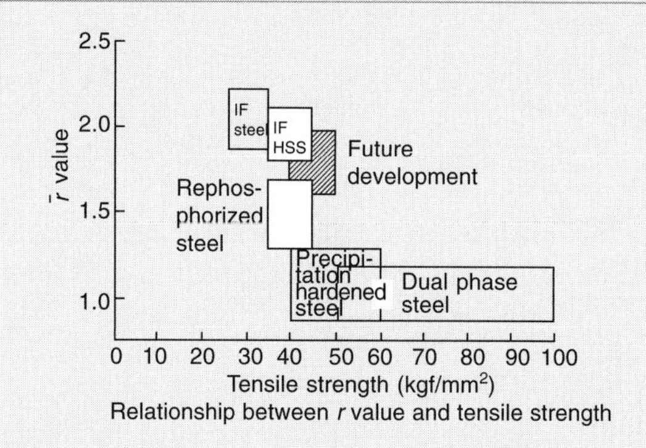

Relationship between *r* value and tensile strength

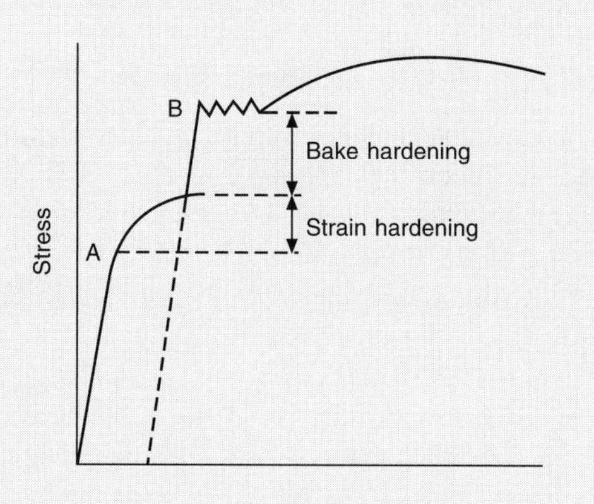

High strength low alloy steels gain their increased strength from the fine grain structure (smaller than ASTM No. 10) and fine dispersion of precipitates (e.g. niobium and titanium carbo-nitrides) both of which impede dislocation movement thereby increasing the flow stress.

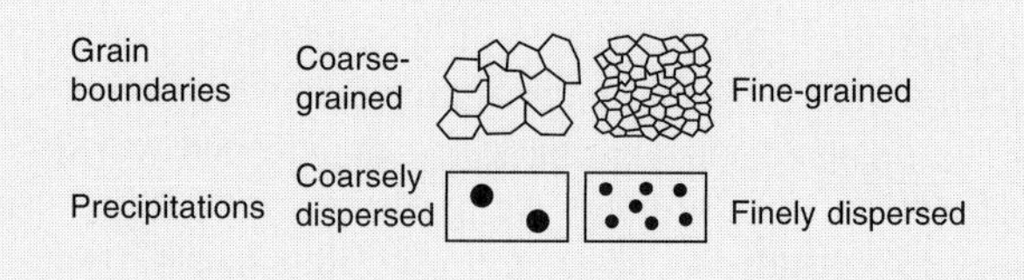

Multiphase steels derive their strength from thermo-mechanical processing, i.e. carefully balanced rolling, coiling and compositional control within the boundaries shown in the diagram opposite. Types of steel included in this category are dual phase, TRIP/TWIP, complex phase and martensitic phase as described below.

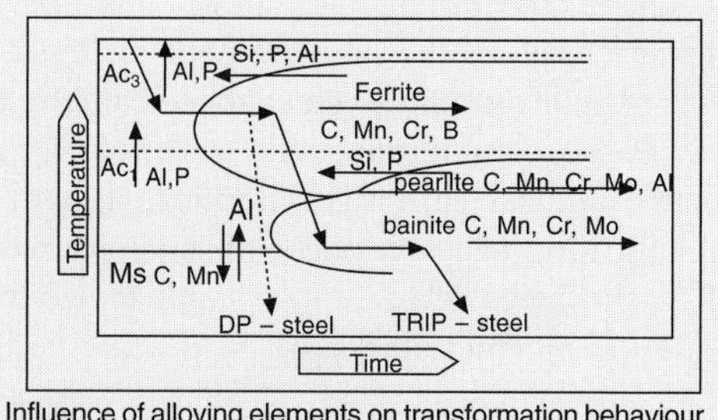

Influence of alloying elements on transformation behaviour

Table 18.2-9 *Cont'd*

Dual-phase steels normally contain a matrix of ductile ferrite plus a proportion of the hard martensite phase induced by alloying and heat treatment. The characteristic high work-hardening rate results from the generation and piling up of dislocations around the martensite fraction on straining. The combination of the high strength developed, associated with relatively high elongation values, enlarge the area under the stress/strain curve resulting in improved energy absorption compared with other steels of similar strength. These steels also exhibit bake hardenability but unlike normal BH steels the ΔBH increase is not limited by the amount of cold work received (see below).	DP R_m: 500–600
TRIP steels feature the transformation of metas-table austenite to martensite during deformation thereby imparting a similar (but increased) strengthening compared to dual phase. The mechanism is similar to D-P with dislocation pile-ups at the martensite/ferrite phase boundaries. (TWIP steels depend on the occurrence of mechanical twinning during deformation to achieve the necessary austenite phase change. These have a significantly different composition, e.g. 18% Mn, 3% Si and 3% Al and are under development for energy absorbing structural parts.)	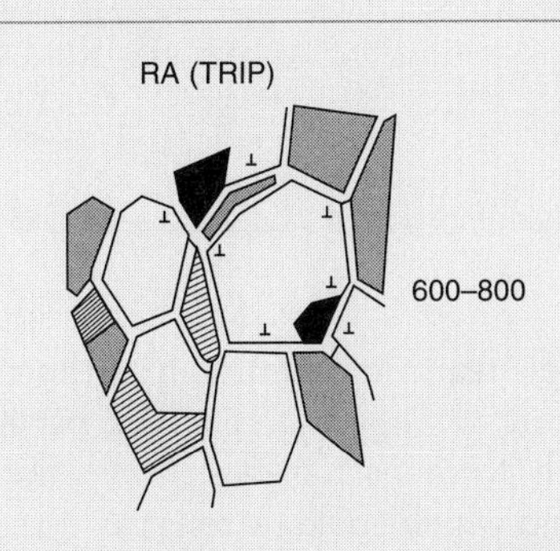
Complex steels are typically hot rolled, fine grain steels featuring ferrite, bainite and martensitic phases with a fine grained microstructure and uniformly dispersed superfine precipitates. Martensitic steels are hot rolled with extremely high strength levels imparted by the predominantly martensitic phase.	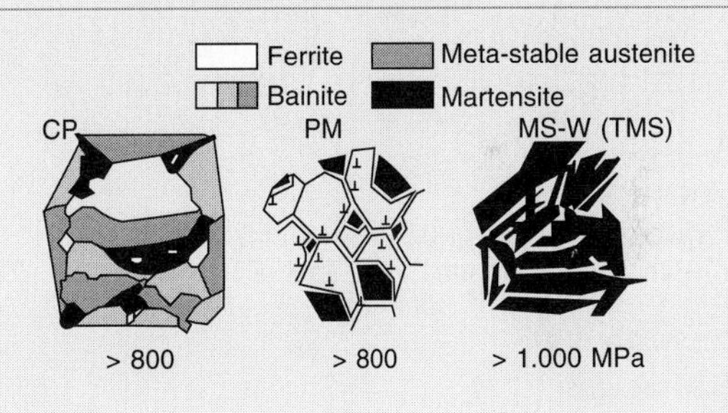

demand heavy-duty tooling materials and often interstage annealing to achieve deep drawn shapes, so despite the advantages of abbreviated paint processing overall costs for facilities and changed processes can be unfavourable. This assumes conventional pressed and hydroformed parts are used, but other recent studies[1] have made a strong case for an alternative body architecture utilizing a stainless steel spaceframe and various bolt-on assembly materials. Again reliance is placed on the need to paint only external surfaces.

18.2.3 Aluminium

In general terms, the attraction of aluminium is based on low density (2.69 g/cc) the relevance of which in automotive terms is discussed in Chapter 18.1. The historical rule thumb when considering structures or subassemblies equivalent to steel is that the weight can be approximately halved but the cost is doubled. Although the density is one-third of steel, the full down weighting potential cannot be realized as the modulus (69 GPa) is considerably lower than that of steel (210 GPa), and as stiffness is a primary influence for the design of most body parts some compensation must be made and thickness increased. Any comment on cost must be qualified by the fact that this can fluctuate with the rise and fall of the commodity markets and for planning purposes some means of stabilizing future costs by buying ahead or alternative strategy must be considered. The doubling in net terms compared with the steel cost also includes a factor for increased manufacturing costs such as those incurred by modification of welding equipment, faster electrode tip wear, cold joining and the need for additional changes to the paint process. Total ownership costs must also be considered and one disadvantage of the more

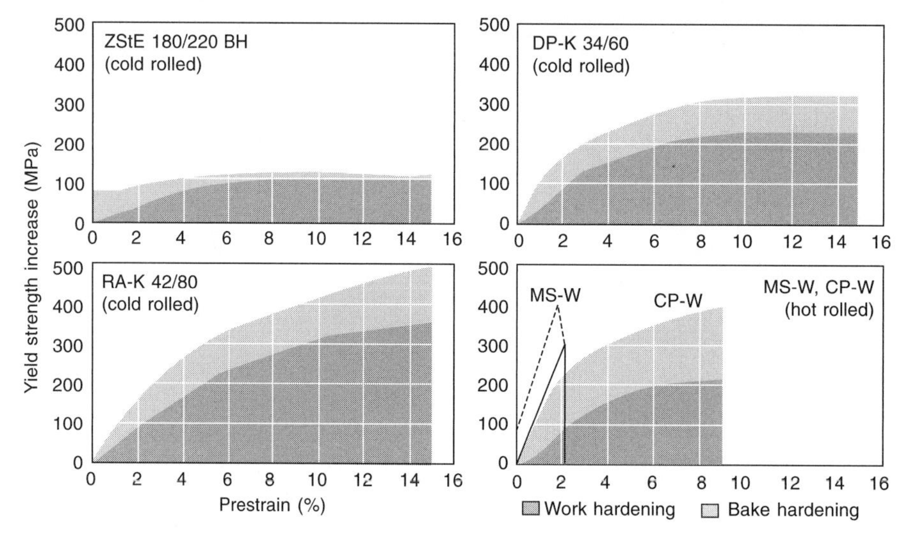

Fig. 18.2-13 Contribution of work hardening and bake hardening in HSS. (Diagram courtesy of TKS.)

recent models featuring aluminium is the expense likely to be incurred for repair of specialist parts such as cast nodes which form part of a complex integrated sub-structure, and which could be reflected in high insurance cover – and the necessity to locate a specialist repair shop. To summarize the major advantages and disadvantages of aluminium as an autobody material:

Advantages:
- Low density
- Corrosion resistance
- Strong supply base
- Recyclability

Disadvantages:
- High and fluctuating cost
- Poorer formability than steel
- Less readily welded than steel

18.2.3.1 Production process

Aluminium is the most prolific metal comprising the earth's crust (8 per cent as opposed to 5 per cent for iron) but has only in the last 100 years been smelted industrially, the Hall–Heroult process being used to extract the metal from alumina dissolved in molten cryolite (a fluoride of sodium and aluminium) by electrolysis using carbon anodes. A flow chart showing the production process is shown in Fig. 18.2-14.

Following casting the slabs are milled to remove the tenacious oxide film and annealed for up to 8 hours at a rolling temperature of 440–550 °C. They are then hot rolled to 10 mm then undergoing a continuous heat treatment before cold rolling finally on a four-high roll stand. The strip may then be straightened and cut to length.

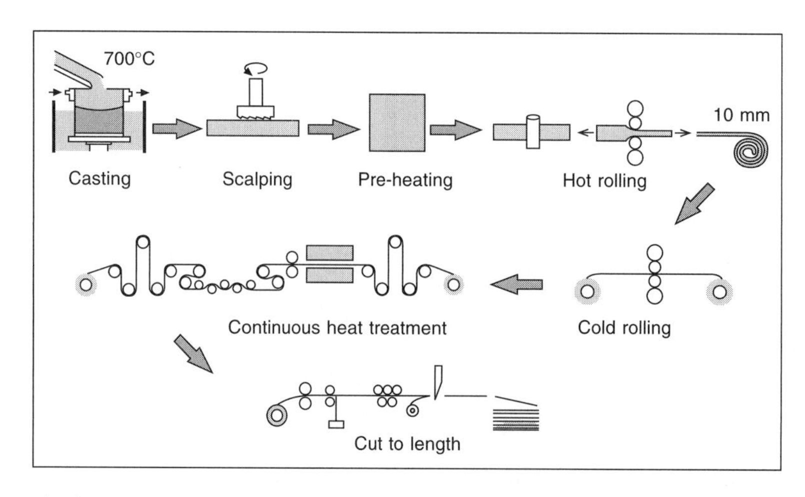

Fig. 18.2-14 Aluminium production process.

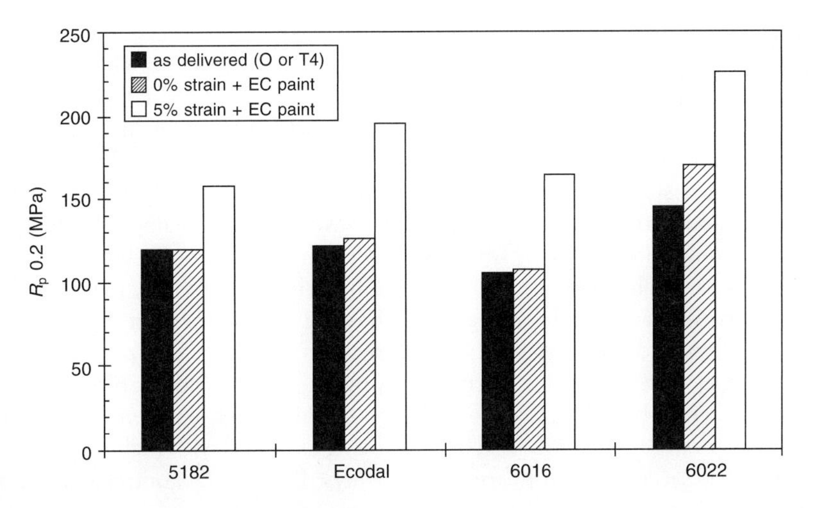

Fig. 18.2-15 Bake hardening response in 6000 series aluminium.

18.2.3.2 Alloys for use in body structures

The common alloys used for the manufacture of body panels are shown in Table 18.2-10 and are designated according to the internationally recognized four-digit system. The 5xxx series refers to aluminium–magnesium alloys while the 6xxx alloys refer to additions of magnesium plus silicon. The last two digits have no special significance beyond identifying different alloys; the second digit indicates alloy modifications and if zero indicates the original alloy.

The 5xxx 'wrought' series alloys have traditionally been used for panel production in the UK due to relatively low cost (3 × zinc-coated steel, compared with

Table 18.2-10 Automotive aluminium alloys in current use

Alloy AA DIN	AA6016 AlMg0.4Si1.2	AA6111 AlMg0.7Si0.9Cu0.7	AA6009 AlMg0.5Si0.8CuMn	AA5251 AlMg2Mn0.3	AA5754 AlMg3	AA5182 AlMg5Mn
Temper	T4	T4	T4	H22 (Grade 3)	0 / H111	0 / H111
UTS (MPa)	210	290	250	190	215	270
0.2 proof stress (MPa)	105	160	130	120	110	140
Elongation A80 (%)	26	25	24	18	23	24
r (mean value)	0.61	0.55	0.64		0.70	0.80
n 5% (mean value)	0.30	0.28	0.29		0.35	0.33
Advantages	Formability, no stretcher-strain marks, balanced properties	No stretcher-strain marks, improved bake-hardening response	No stretcher-strain marks, mechanical strength	Corrosion resistance, cost	Good formability	Very good formability
Disadvantages	Limited bake-hardening response at Rover paint temperature	Corrosion concerns, limited formability	Limited hemming and forming properties	Possible stretcher-strain marks (Lüders lines) after deep drawing		
Alloy type		BAKE HARDENING		NON-BAKE HARDENING		
Typical use	SKIN PANELS		INNER PANELS			

6xxx at 5 × zinc-coated steel) and formability. The main concern has been that they are prone to stretcher strain markings or Lüders bands which can appear as flamboyant 'Type A' coarse markings on the sheet surface coincident with yielding, or as finer more regular 'Type B' markings which appear during the plastic stage of deformation. Despite claimed rolling and heat treatment solutions these tend to reappear on forming and can show through the paint finish unless reworked by abrasive discing.

The 6xxx series alloys are characterized by higher yield strength than Al–Mg alloys and are heat treatable imparting a significant degree of bake hardening at temperatures approaching 200 °C (Fig. 18.2-15).

Despite increased cost the 6xxx series alloys (6016 in particular) are proving most versatile and are in use by the majority of car producers using aluminium in Europe, providing a combination of good stretching and drawing characteristics, dent resistance and consistent surface. With regard to the latter, as well as being stretcher-strain free the use of a 1.0 micrometer EDT textured finish (see previous reference to surface finishes) enables a similar quality of finish to be obtained as with outer panels in steel and this tends to be the universal specification even though advantages are being claimed for the EBT finish (see below). The other commonly used mill finish applied to internal panels and utility vehicles is less popular due to directionality effects on painting of vertical surfaces. For maximum economy, current designs often feature internal panels in 5xxx alloys with outers in 6xxx where critical quality is required.

Developments in alloys are summarized in Table 18.2-11 but include the emergence of an internal 6xxx quality, 6181A, and a 6022 alloy with a higher proof stress value than 6016 which may give further opportunities for downgauging providing forming and hemming performance can be sustained at realistic levels. It has been noted that the industry in the US has adopted the copper bearing 2036 alloy (not favoured in Europe for recycling reasons) for selected panels such as bonnets at gauges down to 0.8 mm compared with the more normal 1.2 mm in the UK and the potential for reducing thickness is now being explored with higher PS materials. Very high Al–Mg alloys (5.5% Mg content) are also being evaluated as elongation figures sometimes in excess of 30 per cent are achievable but high rolling load requirements make it very difficult to produce material with consistent properties.

As stated by Dieffenbach[1] the aluminium spaceframe represents the second leading body architecture and the history and current use of aluminium in design is presented in Chapter 18.1.

18.2.4 Magnesium

Magnesium is the lightest of all the engineering metals, having a density of only 1.74 g/m^3. It is 35 per cent

Table 18.2-11 Aluminium alloys under development

Alloy AA Rover/ Alusuisse DIN	AA6022 (AlMg0.6Si1.3)	AA6181A EcodalR-608 (AlMg0.8Si0.9)	AA5022 (AlMg4.5Cu)	AA5023 (AlMg5.5Cu)	Pe-600
Temper	T4	T4	O/H111	O/H111	O/H111
UTS (MPa)	270	230	275	285	270
0.2 proof stress (MPa)	150	125	135	130	140
Elongation A80 (%)	26	24	28	29	29
r (mean value)	0.60	0.65	0.70	0.70	0.72
n 5% (mean value)	0.26	0.28	0.34	0.36	0.34
Advantages	Improved bake-hardening response	Improved bake-hardening response	Improved formability	Improved formability	Improved formability
Disadvantages	Directional hemming properties	Limited hemming properties	Corrosion, susceptible to stretcher-strain	Corrosion, susceptible to stretcher-strain	
Alloy type	BAKE HARDENING		SLIGHTLY BAKE HARDENING		NON-BAKE HARDENING
Typical use	SKIN PANELS		INNER PANELS		

lighter than aluminium and over four times lighter than steel. It is produced through either the metallothermic reduction of magnesium oxide with silicon or the electrolysis of magnesium chloride melts from sea water. Each cubic metre of sea water contains approximately 1.3 kg of magnesium.

Common magnesium alloys are based on additions of magnesium, aluminium, manganese and zinc. Typical compositions and properties are shown in Table 18.2-12. The alloy designations are based on the following:

- The first two designatory letters indicate the principal alloying element (A for aluminium, E for rare earth element, H for thorium, K for zirconium, M for manganese, S for silicon, W for yttrium, Z for zinc).
- The two numbers indicate the percentages of these major alloying elements to the nearest percentage.
- A final letter indicates the number of the alloy with that particular principal alloying condition. Therefore, AZ91D is the fourth standardized 9% Al, 1% Zn alloy.

The higher elongation levels of the AM60 and AM50 alloys have meant that they may be preferred to AZ91.

High purity variants of these alloys with lower levels of heavy metal impurities (iron, copper and nickel) have vastly improved corrosion performance. The sand casting alloy AZ91C has now been largely replaced by its high purity variant AZ91E, which has a corrosion rate around 100 times better in salt-fog tests.

The major advantages of magnesium are:

- Very low density
- Ability to be thin cast
- Possible to integrate components in castings

Disadvantages include:

- Only viable as cast components (sheet and extruded magnesium not readily available)
- High cost at medium to high volumes

Table 18.2-12 Common automotive magnesium alloys

	AZ91	AM60	AM50
Composition			
% Al	9	6	5
% Zn	0.7		
% Mn	0.2	0.3	0.3
Typical RT properties			
UTS (MPa)	240	225	210
Yield strength (0.2% offset)	160	130	125
Fracture elongation	3	8	10

18.2.5 Polymers and composites

18.2.5.1 Introduction

Polymers used for autobody applications may be split into thermoplastics and thermosets. Thermoplastics are high molecular weight materials that soften or melt on the application of heat. Thermoset processing requires the non-reversible conversion of a low molecular weight base resin to a polymerized structure. The resultant material cannot be remelted or reformed. Composites consist of two or more distinct materials that when combined together produce properties that are not achievable by the individual components of that composite. In autobody applications, reinforced plastics are the major composite material. For example, the term fibreglass consists of a plastic resin reinforced with a fibrous glass component. The resin acts to define the shape of the part, hold the fibres in place and protects them from the damage. The major basic advantages of composites are their relatively high strength and low weight, excellent corrosion resistance, thermal properties and dimensional stability. The strength of a polymer composite will increase with the percentage of fibrous material and is affected by fibre orientation. Tailoring the fibre orientation and concentration can therefore allow for strength increase in the particular region of a component.

18.2.5.2 Thermoplastics

Thermoplastics can be divided into amorphous and crystalline varieties. In amorphous forms the molecules are orientated randomly. Typical amorphous thermoplastics include polyphenylene oxide (PPO), polycarbonate (PC) and acrylonitrile butadiene styrene (ABS). Advantages of amorphous thermoplastics include:

- Relatively dimensionally stable
- Lower mould shrinkage than crystalline thermoplastics
- Potential for application for structural foams

Disadvantages include:

- Poor wear abrasion and repeated impact
- Poor fatigue resistance
- Increased process times compared to crystalline thermoplastics

In a crystalline variety there will be regions of regularly orientated molecules depending on factors such as the processing techniques used, cooling rate, etc. Examples include nylon (PA), polypropylene (PP) and polyethylene (PE). Advantages of crystalline thermoplastics include:

- Good solvent, fatigue and wear resistance
- Higher design strain than amorphous grades
- High temperature properties improved by fibre reinforcement

Disadvantages include:

- Potentially high and variable shrinkage
- Difficult to adhesive bond
- Higher creep than amorphous thermoplastics

18.2.5.3 Thermosets

Thermosets are generally more brittle than thermoplastics so they are often used with fibre reinforcement of some type. Advantages of thermosets include:

- Lower sensitivity to temperature than thermoplastics
- Good dimensional stability
- Harder and more scratch resistant than thermoplastics

Disadvantages of thermosets include:

- Low toughness and strain at fracture
- Difficulties in recycling
- Difficult to obtain 'A' class finish

There is a wide range of different processing techniques that can be used to produce components from the above raw materials. A number of excellent texts exist to provide more detailed information on the basic processes.

The main problem with plastics concerns ELV (vehicle disposal). While the metallic content represents most of the 75 per cent recycled content, plastics are a main constituent of shredder fluff or ASR (auto shredder residue) which can only be disposed of by landfill, and until rationalization of the types of plastic used takes place, focussed on materials which are easily recycled, non-preferred types will be filtered from the initial approval process at the life-cycle analysis stage.

18.2.5.4 Polymer and composite processing

There are a number of ways of processing thermoplastic materials for automotive applications including extrusion, blow moulding, compression moulding, vacuum forming and injection moulding. However, some of these processes are more directly applicable for the production of autobody structures and closure parts than interior and exterior trim parts.

18.2.5.4.1 Injection moulding

This is one of the commonest routes for producing thermoplastic components and has been used in a number of autobody applications including the plastic fenders on the Land Rover Freelander and Renault Clio and the vertical panels on the BMW Z1. The process involves feeding polymer granules into a heated extruder barrel which heats the compound (Fig. 18.2-16). The resultant melt is injected into a chilled mould and pressure is maintained during cooling. The part is finally ejected. Advantages of the process include the relatively short production times and the ability of produce complex, precision parts. However, the pressures required during injection are high and necessitate the use of a precision tool which leads to high tooling costs and lead times.

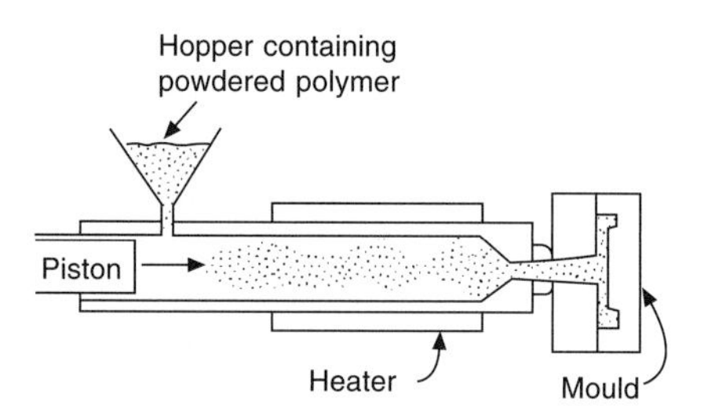

Fig. 18.2-16 Injection moulding process.

18.2.3.4.2 Glass mat thermoplastic compression moulding

Glass mat thermoplastic (GMT) is produced in sheet form that is cut into blanks, rather like traditional sheet metal. These blanks are pre-heated prior to loading into cooled tools within vertical presses. The tool is closed under high pressure and the material flows into the tool cavities. The main problem with GMT is the inability to achieve a truly 'A' class finish meaning its application is limited to internal applications, the most common being the GMT front end on many European production vehicles. The advantages of the process include the faster cycle times than SMC moulding, consistent quality and the potential to use modified metal stamping infrastructure.

Thermoset processing includes sheet moulding compound (SMC), resin transfer moulding (RTM) and reaction injection moulding (RIM).

SMC compression moulding processes are similar to those described for GMT stamping. The SMC sheet is taken unheated and placed in a heated tool at around 160 °C. This causes the resin to cross link and cure in the tool. The pressures are lower than those used for GMT and the resultant properties are higher with a modulus approaching twice that for GMT. Numerous SMC body panels have been used on production vehicles particularly in closure applications, including those from Ford, Lotus, Renault and Daimler Chrysler to reduce vehicle weight and investment cost. Bulk moulding compound is similar to SMC, but with bulk material replacing the sheet. The advantages of SMC include:

- Good surface finish possible
- Good accuracy of parts
- Viable for medium volumes

Limitations of the process include:

- Relatively high investment
- Not as significant weight saving as with thermoplastics
- Storage/shelf life of SMC

A more recent development in the field of SMC technology is the development of low-density SMC. By replacing the calcium carbonate and other SMC fillers with hollow glass microspheres, the density of SMC can be reduced from the traditionals SMCs' gravity of 1.9 to as low as 1.3, with a small reduction in stiffness. These low-density grades may be applicable to interior parts. In the longer term exterior grades may be possible but surface finish after repair is the major concern. When low-density materials are sanded and repaired the hollow glass sphere can be opened up and development of an effective surface sealer is required before this can be resolved.

Resin transfer moulding (RTM) is a low-pressure liquid moulding process. It has traditionally been used for parts with low to medium volume. The low pressures involved in the process allow the use of a low-cost tool, one of the major advantages of the process. Fibre reinforcement is placed in the tool cavity and the tool is closed. Clamping pressure is applied before the injection of the resin. Cure times tend to range from a few minutes to many minutes. Because of cycle time limitations the process is only practically viable for volumes of up to approximately 40 000 units.

18.2.5.5 Advanced composites

Additional detail on the technology with reference to the Fl McLaren production car is provided by Martin.[11] More than 95 percent of the McLaren Fl's body is constructed in high-performance advanced carbon (graphite) epoxy composite material. The material starts life in the tacky pliable condition in which the fibres, in this case carbon, are embedded in a partially cured resin. The weight of the cloth and the resin held within it in the prepregnated condition are controlled to very tight limits. Currently material weights used vary from a 150 gsm (g/m^2) twill weave 1k high strength carbon fibre to 660 gsm 12 k twill weave high strength carbon fibre. The 1 or 12 k reference defines the number of carbon filaments that make up one strand or tow of the woven cloth, i.e. 12 k means 12 000 filaments. In this state it is workable and thickness, stiffness and strength of the final structure can be controlled to very fine limits. Next, laminating takes place and after other specialized preparation processes, curing takes place in an autoclave programmed with two cure cycles. These are 125 °C 2 bar (250 °F at 30 psi) and 125 °C 5 bar (250 °F at 75 psi), and this phase takes 3 hours to complete. The advantage of this type of technology is that it allows the engineer to control the properties – including stiffness and strength in three dimensions – and develop these characteristics exactly where he wants them. Thus the maximum efficiency is obtained from each gram, i.e. the maximum structural and weight efficiency. The big advantage of this approach, apart from tailoring the properties in the required location, is that the strength to weight ratio is impressive and it is claimed that the same tensile strength as steel is obtained but at one-quarter of the weight. Unidirectional material in both the high strength and high modulus forms is used in specific areas where increased stiffness and reinforcement are required. This technique has allowed the impact resistance of carbon fibre composites evident in Formula 1 collisions to be transferred to production cars. As for the Formula 1 body described later, the lay-up technology can be used in conjunction with honeycomb panels to create a strong and stiff assembly.

SP Resin Infusion Technology (SPRINT)[12] provides an alternative approach which it is claimed is far less labour intensive than pre-impregnated reinforcement, liquid resin infusion and resin film infusion techniques, while providing a higher integrity product with high-quality external paint finish. The SPRINT material consists of two layers of dry fibre reinforcement either side of a precast, precatalysed resin film as shown in Fig. 18.2-17(a). The reinforcement can incorporate a wide variety of fibres such as glass, aramid and carbon, which can be in the form of random mats, woven fabrics or stitched fabrics. A number of resin systems such as epoxy, polyester and bismaelimide can also be used. Processing is straightforward with material laid up in a mould and vacuum bagged as for conventional prepreg. The nature of the product allows good consolidation and integrity, and flow of the material allows shape control, even into corners, without entrapment. Gel coats can be applied to give the desired finish. The comparative cost/weight effects are summarized in Fig. 18.2-17(b).

Specialist sports car manufacturers have already adopted this material and on the Ultima (shown in Fig. 18.2-18) – 45 kg has been saved compared to the GRP version. The SPRINT CBS material consists of a precombined series of materials with a surfacing ply on the outside, followed by a sandwich of carbon reinforcements either side of a thin syntactic core.

In conclusion, this appraisal of the process technology behind the manufacture of the main materials utilized in automotive body structures has allowed an understanding of the various aspects of process capability to be gained but also the development of parameters relevant to the production of body components.

18.2.6 Learning points

1. When selecting the materials for car body construction other factors such as environmental acceptability and ease of manufacture ('process chain effects') must now be considered alongside cost

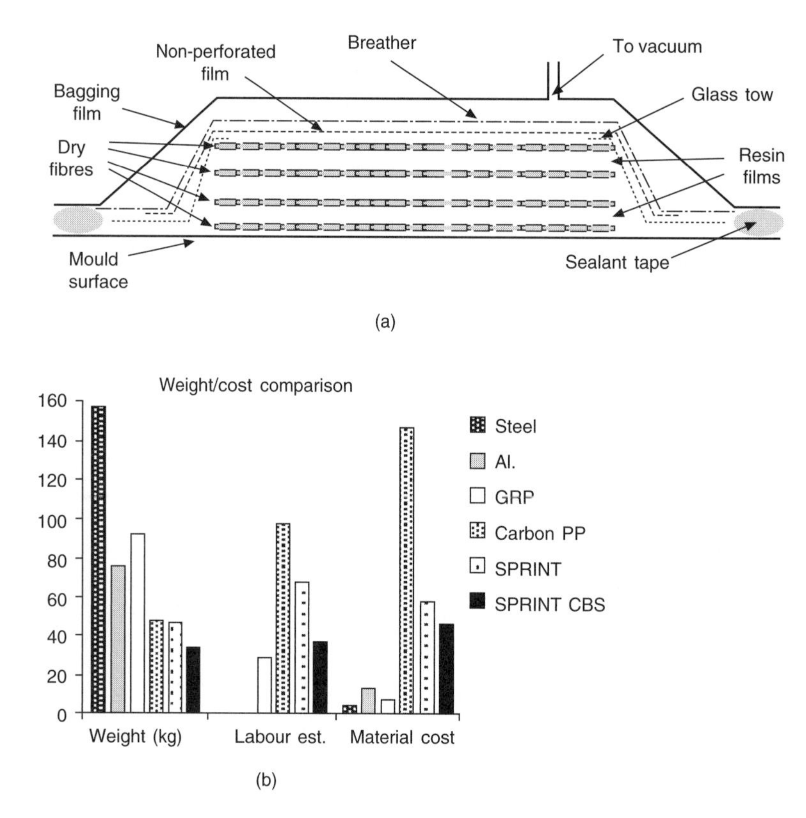

Fig. 18.2-17 (a) and (b) Nature of SPRINT material and comparative costs. (Courtesy of SP Systems.)

and the physical and mechanical properties traditionally used for engineering design.

2. Existing investment in automotive facilities and familiar design and manufacturing procedures associated with sheet steel will favour its continuing use as the predominant high volume material, and high strength and zinc-coated variants will allow medium-term lightweight and durability targets to be met.
3. Down weighting the body structure beyond 30 per cent will increasingly call for alternative materials but these will require more radical changes to both manufacturing and disposal (recycling) procedures.
4. From evidence to date aluminium is the most likely contender (demonstrated by the Audi A8/A2 spaceframe architecture) to replace steel, and recycling procedures are in place to absorb most scrap and ensure that up to 50 per cent of the original cost can be recovered. Improved pricing stability regarding the initial cost of aluminium strip is required.

Fig. 18.2-18 Ultima body and component structure.

5. Inevitably hybrid structures will increasingly find favour with mixed material application, e.g. steel/aluminium substructures with polymer skin panels, incorporation of front end panel and hardware parts, but selection and design should allow for ease of identification of component materials plus easy disassembly and recyclability.
6. More significant weight savings may be required to boost the performance of cars propelled by electricity or alternative fuel systems and ultralight-weight construction in sandwich or honeycomb forms utilizing aluminium and composite formats will be favoured. These will impose even greater constraints on process chain operations and recyclability than aluminium or composites.
7. As well as meeting the criteria used in the material selection summary in Table 18.2-3, product consistency is essential. Maximum productivity relies heavily on uniformity of properties, dimensions and finish.
8. Coincident development of continuous strip processes for casting, annealing and finishing with associated statistical process control at each stage of manufacture must help achieve increased product uniformity.
9. Process control during forming and painting should improve with the increasing number of deterministic surface finishes being promoted by European steel and aluminium producers (EDT, EBT, etc.). The uptake of this technology depends, however, on the market availability of such finishes and standards that allow precise definition of the required topography.
10. Polymer panels have the undoubted advantages of cost (especially at lower volumes) and shape versatility but recycling remains a major problem. More effort is required in rationalizing the number of materials with selection favouring the more easily reused/recycled material type.
11. Advanced composite manufacture has been regarded as labour intensive with high facility costs but 'one hit' preprepared resin/reinforcement materials appear to offer a more straightforward route allowing advantages for higher volume sports car production.

References

1. Dieffenbach, J.R., 'Not the Delorean Revisited: An Assessment of the Stainless Steel Body-in-White', SAE Paper 1999-01-3239, 1999.
2. Davies, G. and Easterlow, R., 'Automotive Design and Materials Selection', *Metals and Materials*, Jan. 1985, pp. 20–25.
3. Sollac, *Book of Steel*, Lavoisier Publishing, Paris, 1996.
4. Price, W.O.W. *et al.*, 'Conventional Strand Annealing', Special Report No. 79, The Iron and Steel Institute, London, 1963, pp. 71–81.
5. Petch, N.J., *JISI*, 174, 1953, pp. 25–28.
6. Kennett, S.J. and Owen, W.S., 'Some Metallurgical Aspects of the Annealing of Mild Steel Strip and Sheet', Special Report No. 79, The Iron and Steel Institute, London, 1963, pp. 1–9.
7. Butler, R.D. and Wilson, D.V., *JISI*, 201, 1963, pp.16–33.
8. Butler, R.D. and Pope, R., '*Sheet Metal Industries*', Sept. 1967, pp. 579–597.
9. Davies, G.M. and Moore, G.G., 'Trends in the Design and Manufacture of Automotive Structures', *Sheet Metal Industries*, Aug. 1982, pp. 623–628.
10. Pearce, R., *Sheet Metal Forming*, Kluwer Academic Publishers BV, Dordrecht.
11. Martin, P., 'McLaren F1's Composite Body', *Engineering Designer*, Nov./Dec. 1996, pp. 10–14.
12. Thomas, S.M., 'Automotive Infusion for Composites', *Materials World*, May 2001, pp. 19–21.
13. Gatenby, K.M. and Court, S.A., 'Aluminium–Magnesium Alloys for Automotive Applications – Design Considerations and Material Selection', IBEC '97 Proceedings, Stuttgart, Germany, 30 Sept. – 2 Oct. 1997, p. 139.

Section Nineteen

Aerodynamics

Chapter 19.1

19.1

Body design: Aerodynamics

Julian Happian-Smith

The aims of this chapter are to:

- review the role of the stylist and aerodynamist;
- review the basic aerodynamic concepts related to vehicles; and
- indicate the basic computations required for aerodynamic design.

19.1.1 Introduction

Throughout the history of the motor car there have been individual vehicles that have demonstrated strong aerodynamic influence upon their design. Until recently their flowing lines were primarily a statement of style and fashion with little regard for the economic benefits. It was only rising fuel prices, triggered by the fuel crisis of the early 1970s, that provided a serious drive towards fuel-efficient aerodynamic design. The three primary influences upon fuel efficiency are the mass of the vehicle, the efficiency of the engine and the aerodynamic drag. Only the aerodynamic design will be considered in this section but it is important to recognize the interactions between all three since it is their combined actions and interactions that influence the dynamic stability and hence the safety of the vehicle.

19.1.2 Aerodynamic forces

Aerodynamic research initially focused upon drag reduction, but it soon became apparent that the lift and side forces were also of great significance in terms of vehicle stability. An unfortunate side effect of some of the low drag shapes developed during the early 1980s was reduced stability, especially when driven in cross-wind conditions. Cross-wind effects are now routinely considered by designers but our understanding of the highly complex and often unsteady flows that are associated with the airflow over passenger cars remains sketchy. Experimental techniques and computational flow prediction methods still require substantial development if a sufficient understanding of the flow physics is to be achieved.

The aerodynamic forces and moments that act upon a vehicle are shown in coefficient form in Fig 19.1-1. The force and moment coefficients are defined respectively as

$$C_f = \frac{F}{\frac{1}{2}\rho v^2 A} \qquad C_m = \frac{M}{\frac{1}{2}\rho v^2 A l}$$

where F is force (lift, drag or side), M is a moment, ρ is air density, v is velocity, A is reference area and l is a reference length. Since the aerodynamic forces acting on a vehicle at any given speed are proportional to both the appropriate coefficient and to the reference area (usually frontal area) the product $C_f A$ is commonly used as the measure of aerodynamic performance, particularly for drag.

The forces may be considered to act along three, mutually perpendicular axes. Those forces are the drag, which is a measure of the aerodynamic force that resists the forward motion of the car, the lift which may act upwards or downwards; and the side force which only occurs in the event of a cross-wind or when the vehicle is in close proximity to another. The lift, drag and pitching moments are a measure of the tendency of those three

Introduction to Modern Vehicle Design; ISBN: 9780750650441
Copyright © 2001 Julian Happian-Smith. All rights of reproduction, in any form, reserved.

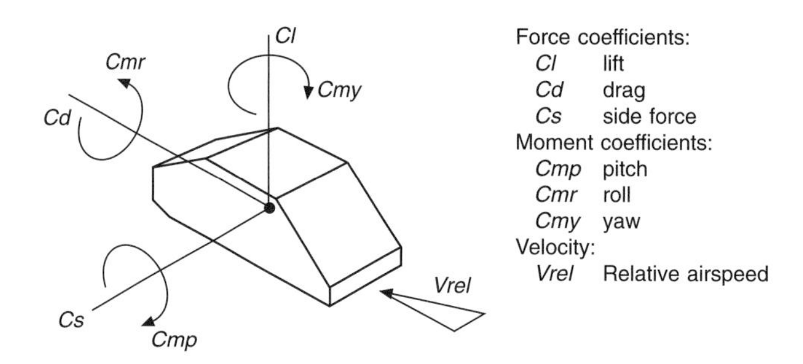

Fig. 19.1-1 Lift, drag, side force and moment axes.

forces to cause the car to rotate about some datum, usually the centre of gravity. The moment effect is most easily observed in cross-wind conditions when the effective aerodynamic side force acts forward of the centre of gravity, resulting in the vehicle tending to steer away from the wind. In extreme, gusting conditions the steering correction made by the driver can lead to a loss of control. Cross-wind effects will be considered further in Section 19.1.5.

19.1.3 Drag

The drag force is most easily understood if it is broken down into five constituent elements. The most significant of the five in relation to road vehicles is the *form drag* or *pressure drag* which is the component that is most closely identified with the external shape of the vehicle. As a vehicle moves forward the motion of the air around it gives rise to pressures that vary over the entire body surface as shown in Fig. 19.1-2a. If a small element of the surface area is considered then the force component acting along the axis of the car, the drag force, depends upon the magnitude of the pressure, the area of the element upon which it acts and the inclination of that surface element Fig. 19.1-2b. Thus it is possible for two different designs, each having a similar frontal area, to have very different values of form drag.

As air flows across the surface of the car frictional forces are generated giving rise to the second drag component which is usually referred to as *surface drag* or *skin friction drag*. If the viscosity of air is considered to be almost constant, the frictional forces at any point on the body surface depend upon the shear stresses generated in the boundary layer. The boundary layer is that layer of fluid close to the surface in which the air velocity changes from zero at the surface (relative to the vehicle) to its local maximum some distance from the surface. That maximum itself changes over the vehicle surface and it is directly related to the local pressure. Both the local velocity and the thickness and character of the boundary layer depend largely upon the size, shape and velocity of the vehicle.

A consequence of the constraints imposed by realistic passenger space and mechanical design requirements is the creation of a profile which in most situations is found to generate a force with a vertical component. That lift, whether positive (upwards) or negative, induces changes in the character of the flow which themselves create an *induced drag* force.

Practical requirements are also largely responsible for the creation of another drag source which is commonly referred to as *excrescence drag*. This is a consequence of all those components that disturb the otherwise smooth surface of the vehicle and which generate energy absorbing eddies and turbulence. Obvious contributors

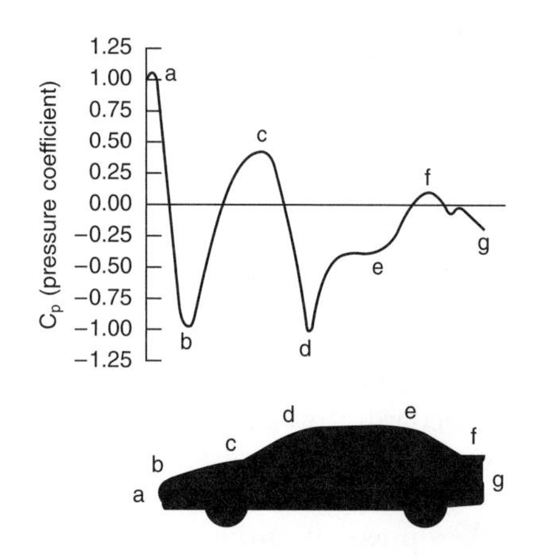

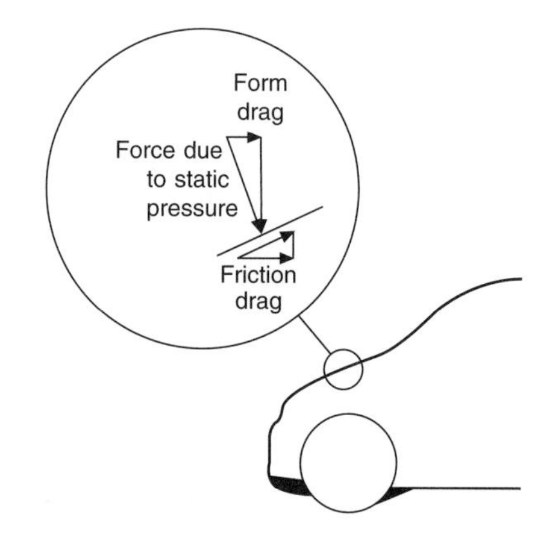

Fig. 19.1-2 (a) Typical static pressure coefficient distribution; (b) The force acting on a surface element.

include the wheels and wheel arches, wing mirrors, door handles, rain gutters and windscreen wiper blades but hidden features such as the exhaust system are also major drag sources.

Although some of these features individually create only small drag forces, their summative effect can be to increase the overall drag by as much as 50%. Interactions between the main flow and the flows about external devices such as door mirrors can further add to the drag. This source is usually called *interference drag.*

The last of the major influences upon vehicle drag is that arising from the cooling of the engine, the cooling of other mechanical components such as the brakes and from cabin ventilation flows. Together these *internal drag* sources may typically contribute in excess of 10% of the overall drag (e.g. Emmelmann, 1982).

19.1.4 Drag reduction

Under the heading of drag reduction the designer is concerned not only with the magnitude of the force itself but also with a number of important and directly related topics. Firstly there are the effects of wind noise. Aerodynamic noise is closely associated with drag creation mechanisms which often exhibit discrete frequencies and which tend to arise where the air flow separates from the vehicle surface. Flow separation is most likely to occur around sharp corners such as those at the rear face of each wing mirror and around the 'A' pillar of a typical passenger car. Because of the close relationship between drag and noise generation it is not surprising that drag reduction programmes have a direct and generally beneficial effect upon wind noise. Such mutual benefits are not true of the second related concern, that of dynamic stability. The rounded shapes that have come to characterize modern, low drag designs are particularly sensitive to cross-winds both in terms of the side forces that are generated and the yawing moments. Stability concerns also relate to the lift forces and the changes in those forces that may arise under typical atmospheric wind conditions.

The broad requirements for low drag design have been long understood. Recent trends in vehicle design reflect the gradual and detailed refinements that have become possible both as a result of increased technical understanding and of the improved manufacturing methods that have enabled more complex shapes to be produced at an acceptable cost. The centre-line pressure distribution arising from the airflow over a typical three-box (saloon) vehicle has been shown in Fig. 19.1-2a. A major drag source occurs at the very front of the car where the maximum pressure is recorded (Fig. 19.1-2a, point 'a') and this provides the largest single contribution to the form drag. This high-pressure, low-velocity flow rapidly accelerates around the front, upper corner (b) before slowing again with equal rapidity. The slowing air may not have sufficient momentum to carry it along the body surface against the combined resistance of the pressure gradient and the viscous frictional forces resulting in separation from the body surface and the creation of a zone of re-circulating flow which is itself associated with energy loss and hence drag. The lowering and rounding of the sharp, front corner together with the reduction or elimination of the flat, forward-facing surface at the very front of the car addresses both of these drag sources (Hucho, 1998). A second separation zone is observed at the base of the windscreen and here a practical solution to the problem is more difficult to achieve. The crucial influence upon this drag source is the screen rake. Research has clearly demonstrated the benefits of shallow screens but the raked angles desired for aerodynamic efficiency lead to problems not only of reduced cabin space and driver headroom but also to problems of internal, optical reflections from the screen and poor light transmission. Such problems can largely be overcome by the use of sophisticated optical coatings similar to those widely used on camera lenses but as yet there has been little use of such remedies by manufacturers. Fig. 19.1-3 demonstrates the benefits that may be achieved by changing the bonnet slope and the screen rake (based on the data of Carr (1968)).

There is further potential for flow separation at the screen/roof junction which similarly benefits from screen rake and increased corner radius to reduce the magnitude of the suction peak and the pressure gradients.

The airflow over the rear surfaces of the vehicle is more complex and the solutions required to minimize drag for practical shapes are less intuitive. In particular the essentially two-dimensional considerations that have been used to describe the air flow characteristics over the front of the vehicle are inadequate to describe the rear flows. Fig. 19.1-4 demonstrates two alternative flow structures that may occur at the rear of the vehicle. The first Fig. 19.1-4a occurs for 'squareback' shapes and is characterized by a large, low-pressure wake. Here the airflow is unable to follow the body surface around the sharp, rear corners. The drag that is associated with such flows depends upon the cross-sectional area at the tail, the pressure acting upon the body surface and, to a lesser extent, upon energy that is absorbed by the creation of eddies. Both the magnitude of the pressure and the energy and frequency associated with the eddy creation are governed largely by the speed of the vehicle and the height and width of the tail. A very different flow structure arises if the rear surface slopes more gently as is the case for hatchback, fastback and most notchback shapes (Fig. 19.1-4b). The centreline pressure distribution shown in Fig. 19.1-2a shows that the surface air pressure over the rear of the car is significantly lower than that of the surroundings. Along the sides of the car the

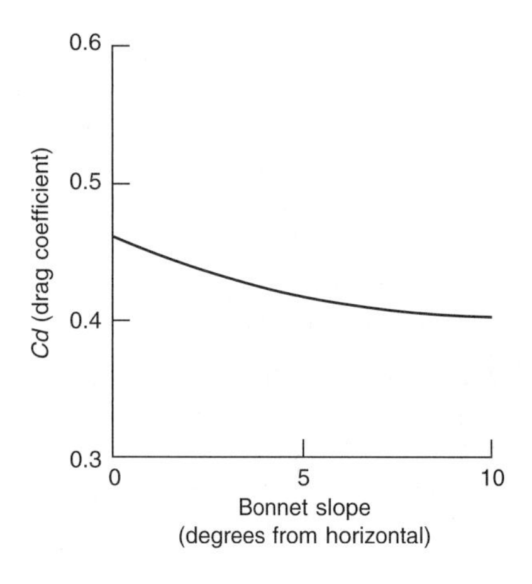

Fig. 19.1-3 Drag reduction by changes to front body shape.

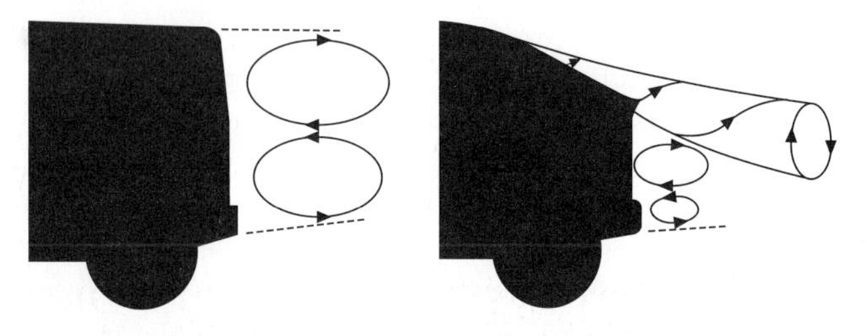

Fig. 19.1-4 (a) 'Squareback' large-scale flow separation. **(b)** 'Hatchback/fastback' vortex generation.

body curvature is much less and the pressures recorded here differ little from the ambient conditions. The low pressure over the upper surface draws the relatively higher pressure air along the sides of the car upwards and leads to the creation of intense, conical vortices at the 'C' pillars. These vortices increase the likelihood of the upper surface flow remaining attached to the surface even at backlight angles of over 30°. Air is thus drawn down over the rear of the car resulting in a reacting force that has components in both the lift and the drag directions. The backlight angle has been shown to be absolutely critical for vehicles of this type (Ahmed *et al.*, 1984). Fig. 19.1-5 demonstrates the change in the drag coefficient of a typical vehicle with changing backlight angle. As the angle increases from zero (typical squareback) towards 15° there is initially a slight drag reduction as the effective base area is reduced. Further increase in backlight angle reverses this trend as the drag inducing influence of the upper surface pressures and trailing vortex creation increase. As 30° is approached the drag is observed to increase particularly rapidly as these effects become stronger until at approximately 30° the drag dramatically drops to a much lower value. This sudden drop corresponds to the backlight angle at which the upper surface flow is no longer able to remain attached around the increasingly sharp top, rear corner and the flow reverts to a structure more akin to that of the initial squareback. In the light of the reasonably good aerodynamic performance of the squareback shape, it is not surprising that many recent, small hatchback designs have adopted the square profiles that maximize interior space with little aerodynamic penalty.

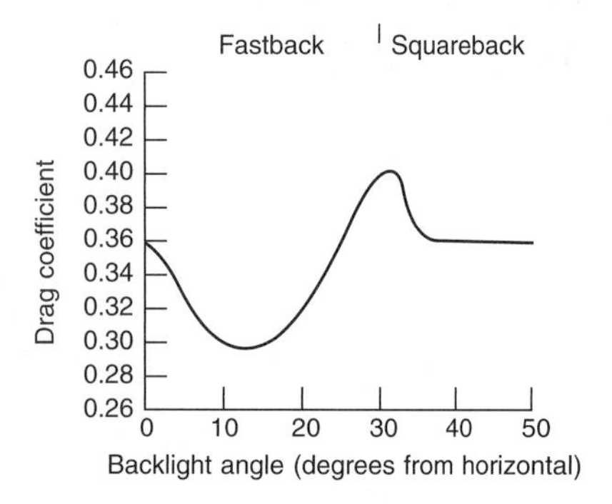

Fig. 19.1-5 The influence of backlight angle on drag coefficient.

The more traditional notchback or saloon form, not surprisingly, is influenced by all of the flow phenomena that have been discussed for the forms discussed above. As the overcar flow passes down the rear screen the conditions are similar to those of the hatchback and trailing; conical vortices may be created at or near the 'C' pillar. The inclination of the screen may be sufficient to cause the flow to separate from the rear window although in many cases the separation is followed by flow re-attachment along the boot lid. Research has shown that in this situation the critical angle is not that of the screen alone but the angle made between the rear corner of the roof and the tip of the boot (Nouzawa *et al.*, 1992). This suggests that the effect of the separation is to re-profile the rear surface to something approximating to a hatchback shape and consequently the variation in drag with this effective angle mimics that of a continuous, solid surfaced 'hatch'. It follows that to achieve the minimum drag condition that has been identified to

Fig. 19.1-6 High tail, low drag design.

correspond to a backlight angle of 15° (Fig. 19.1-5) it is necessary to raise the boot lid, and this has been a very clear trend in the design of medium and large saloon cars (Fig. 19.1-6). This has further benefits in terms of luggage space although rearward visibility is generally reduced. Rear end, boot-lid spoilers have a similar effect without the associated practical benefits. The base models produced by most manufacturers are usually designed to provide the best overall aerodynamic performance within the constraints imposed by other design considerations and the spoilers that feature on more upmarket models rarely provide further aerodynamic benefit.

Attention must also be paid to the sides of the car. One of the most effective drag reduction techniques is the adoption of boat-tailing which reduces the effective cross-sectional area at the rear of the car and hence reduces the volume enclosed within the wake (Fig. 19.1-7). In its most extreme configuration this results in the tail extending to a fine point, thus eliminating any wake flow, although the surface friction drag increases and the pressures over the extended surfaces may also contribute to the overall drag. Practical considerations prevent the adoption of such designs but it has long been known that the truncation of these tail forms results in little loss of aerodynamic efficiency (Hucho *et al.*, 1976).

Despite the efforts that have been made to smooth visible surfaces it is only recently that serious attempts have been made to smooth the underbody. The problems associated with underbody smoothing are considerable and numerous factors such as access for maintenance, clearance for suspension and wheel movement and the provision of air supplies for the cooling of the engine, brakes and exhaust must be given considerable weight in the design process. Just as the airflow at the extreme front and rear of the car were seen to be critical in relation to the overcar flow, so also it is necessary to give comparable consideration to the air flow as it passes under the nose of the vehicle and as it leaves at the rear. It comes as a surprise to many to learn that the sometimes large air dams that are fitted to most production vehicles can actually reduce the overall drag forces acting on the car despite the apparent bluntness that they create. The air dam performs two useful functions. The first is to reduce the lift force acting on the front axle by reducing the pressure beneath the front of the car. This is achieved by restricting the flow beneath the nose which accelerates with a corresponding drop in pressure. For passenger cars a neutral or very slight negative lift is desirable to maintain stability without an excessive increase in the steering forces required at high speed. For high-performance road cars, it may be preferred to create significant aerodynamic downforce to increase the adhesion of the tyres. The side effects of aerodynamic downforce generation such as increased drag and extreme steering sensitivity are generally undesirable in a family car. Lowering the stagnation point by the use of air dams has also been shown in many cases to reduce the overall drag despite the generation of an additional pressure drag component.

The shaping of the floorpan at the rear of the car also offers the potential for reduced drag (Fig. 19.1-8). As the flow diffuses (slows) along the length of the angled rear underbody the pressure rises, resulting in reduced form drag and also a reduced base area, although interactions between the overcar and undercar air flows can result in unexpected and sometimes detrimental effects. Such effects are hard to generalize and detailed experimental studies are currently required to determine the optimum geometry for individual vehicle designs, but typically it has been found that diffuser angles of approximately 15° seem to provide the greatest benefits (e.g. Howell, 1994).

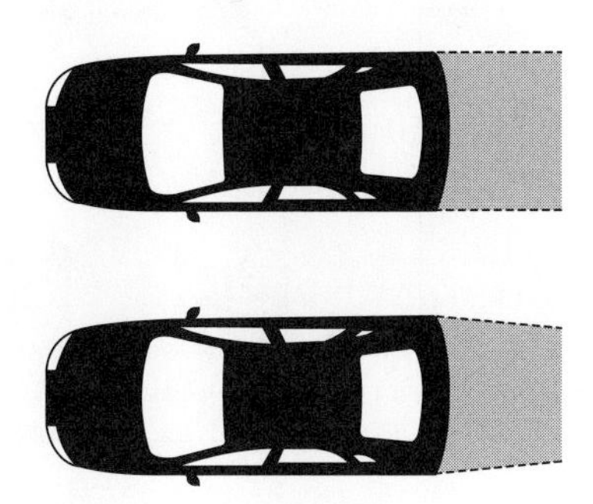

Fig. 19.1-7 Boat tailing: reduced wake.

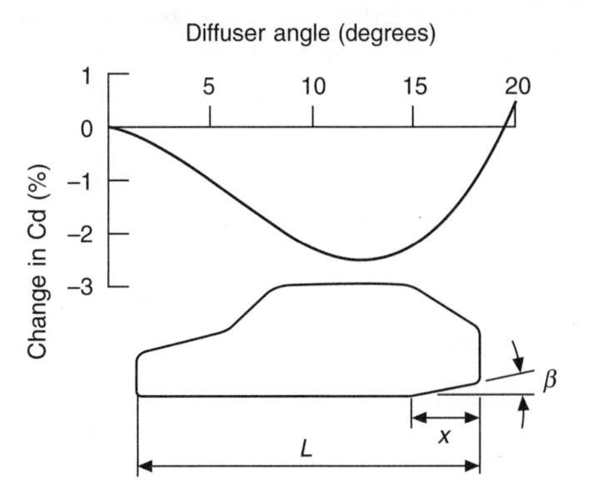

Fig. 19.1-8 Rear, underbody diffusion.

19.1.5 Stability and cross-winds

The aerodynamic stability of passenger cars has been broadly addressed as two independent concerns. The first relates to the 'feel' of a car as it travels in a straight line at high speed and in calm conditions and to lane change manoevrability. The second concerns the effects of steady cross-winds and transient gusts that are associated with atmospheric conditions and which may be exaggerated by local topographical influences such as embankments and bridges.

The sources of straight-line instability in calm conditions has proved to be one of the most difficult aerodynamic influences to identify. This is largely because of the complex interactions between the chassis dynamics and relatively small changes in the magnitude of lift forces and centre of pressure. Qualitative observations such as driver 'feel' and confidence have proved hard to quantify. New evidence suggests that stability and particularly lane change stability degrade with increases in the overall lift and with differences in lift between the front and rear axles (Howell, 1998).

The influence of cross-winds is more easily quantifiable. Steady-state cross-winds rarely present a safety hazard but their effect upon vehicle drag and wind noise is considerable. Most new vehicles will have been model-tested under yawed conditions in the wind tunnel at an early stage of their development but optimization for drag and wind noise is almost always based upon zero cross-wind assumptions. Some estimates suggest that the mean yaw angle experienced in the U.K. is approximately 5° and if that is correct then there is a strong case for optimizing the aerodynamic design for that condition.

The influence of transient cross-wind gusts such as those often experienced when passing bridge abutments, or when overtaking heavy vehicles in the presence of cross-winds is a phenomenon known to all drivers. To reduce the problems that are encountered by the driver under these conditions it is desirable to design the vehicle to minimize the side forces, yawing moments and yaw rates that occur as the vehicle is progressively and rapidly exposed to the cross wind. The low drag, rounded body shapes that have evolved in recent years can be particularly susceptible to cross-winds. Such designs are often associated with increased yaw sensitivity and commonly related changes of lift distribution under the influence of cross-winds can be particularly influential in terms of reduced vehicle stability. The influence of aerodynamics is likely to be further exaggerated by anticipated trends towards weight reduction in the search for improved fuel efficiency. Although methods for testing models under transient cross-wind conditions are under development, reliable data can, as yet, only be obtained by full-scale testing of production and pre-production vehicles. At this late stage in the vehicle development programme, the primary vehicle shape and tooling will have been defined so any remedial aerodynamic changes can only be achieved at very high cost or by the addition of secondary devices such as spoilers and mouldings; also an undesirable and costly solution. To evaluate the transient behaviour of a vehicle at a much earlier stage of its design it is necessary not only to develop model wind tunnel techniques to provide accurate and reliable data, but also most importantly to fully understand the flow mechanisms that give rise to the transient aerodynamic forces and moments. Initial results from recent developments in wind tunnel testing suggest that the side forces and yawing moments experienced in the true transient case exceed those that have been measured in steady state yaw tests (Docton, 1996).

19.1.6 Noise

Although some aerodynamic noise is created by ventilation flows through the cabin the most obtrusive noise is generally that created by the external flow around the vehicle. Considerable reductions have been made to cabin noise levels which may be attributed in part to improved air flows with reduced noise creation and also to improved sealing which has the effect both of reducing noise creation and insulating the occupants from the sound sources. Fig. 19.1-9 provides an approximate comparison between the different noise sources (engine, tyres and aerodynamics) that have been recorded in a small car moving at 150 km/h (based on the data of Piatek, 1986). The creation of aerodynamic noise is mostly associated with turbulence at or the body surface and moves to reduce drag have inevitably provided the additional benefit of noise reduction. Although there is a noise associated with the essentially random turbulence that occurs within a turbulent boundary layer, it is the sound associated with eddy creation at surface

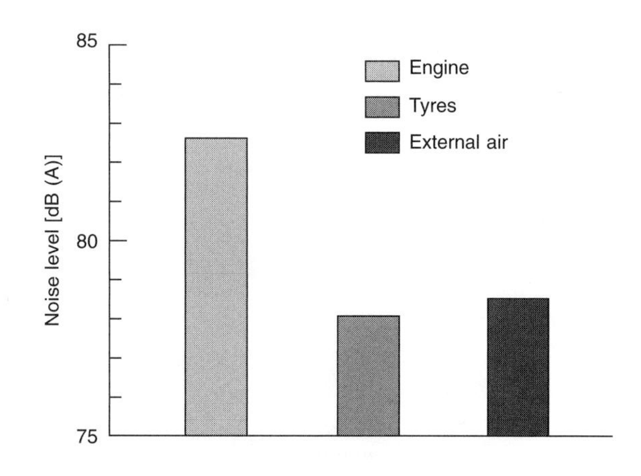

Fig. 19.1-9 Noise sources (Piatek, 1986).

discontinuities that has both the greatest magnitude and also the most clearly defined (and annoying) frequencies. Improvements in rain gutter design and the positioning of windscreen wipers reflect some of the moves that have been made to reduce noise creation and improved manufacturing techniques and quality control have also resulted in major noise reduction as a consequence of improved panel fit. Protrusions such as wing mirrors and small surface radii such as at the 'A' pillar remain areas of particular concern because of their proximity to the driver and because of the relatively poor sound insulation provided by windows. It has been demonstrated that it is the noise associated with vortex (eddy) creation that is the dominant aerodynamic noise source over almost the entire audible frequency range (Stapleford and Carr, 1971). One of the largest, single noise generators is the sun roof. Its large size results in low frequencies and large magnitudes and poorly designed units may even lead to discernible low frequency pressure pulsing in the cabin. Despite customer demand for low cabin noise there has been a parallel increase in the number of sun roofs that have been fitted to new cars. Open windows can create similar problems. Increased use of air conditioning is the best practical solution to this particular problem.

19.1.7 Underhood ventilation

The evidence from numerous researchers suggests that the engine cooling system is responsible for between 10% and 15% of the overall vehicle drag, so it is not surprising to note that considerable effort has been focused upon the optimization of these flows. Traditionally the cooling drag has been determined from wind tunnel drag measurements with and without the cooling intakes blanked-off. The results from those wind tunnel tests must be treated with caution since the closure of the intakes may alter the entire flow-field around a car. Underhood flow restrictions arising from the ever-increasing volume of ancillary equipment under the bonnet has further focused attention on cooling air flows, and this is now one of the primary applications for the developing use of computational flow simulation codes. Many of the sources of cooling drag are readily apparent such as the resistance created by the relatively dense radiator matrix and the drag associated with the tortuous flow through the engine bay. In general any smoothing of the flow path will reduce the drag, as will velocity reductions by diffusion upstream from the cooling system, although the implications of the latter upon the heat transfer must be considered. Less obvious but also significant is the interaction between the undercar flow and the cooling flow at its exit where high turbulence levels and flow separations may to occur. Careful design to control the cooling exit flow in terms of its speed and direction can reduce the drag associated with the merging flows but in general the aerodynamics are compromised to achieve the required cooling.

The potential for underhood drag reduction is greatest if the air flow can be controlled by the use of ducting to guide the air into and out from the radiator core. Approximate relationships between the slowing of the cooling airflow and the pressure loss coefficient, are widely described in the published texts (e.g. Barnard, 1996). The high blockage caused by the radiator core has the effect of dramatically reducing the air velocity through the radiator and thus much of the air that approaches the radiator spills around it. The relatively small mass flow that passes through the core can exhibit substantial non-uniformity which reduces the effectiveness of the cooling system. These problems can be much reduced if the flow is ducted into the radiator in such a way as to slow the flow in a controlled and efficient manner, and careful design of the degree of diffusion can greatly improve the efficiency of the cooling flow. Increasing the diffusion slows the air flowing through the radiator which reduces both the drag force and the heat transfer. Although the reduced heat transfer rate results in a requirement for a larger radiator core surface area, the drag reduction is proportionately greater than is the corresponding reduction in heat transfer. A low speed, large area core therefore creates less drag for a given heat transfer rate. Inevitably, compromises are necessary. The larger core adds weight and cost and the generally close proximity of the radiator to the intake leaves little scope for the use of long, idealized ducting. Too much diffusion will lead to flow separation within the intake which may result in severe flow non-uniformities across the face of the radiator. Gains are also available if the air is ducted away from the radiator in a similarly efficient manner, but in most cases the practical complexity of such a system and the requirement for a source of cooling air to the ancillaries has prevented such measures.

19.1.8 Cabin ventilation

Sealing between the body panels and particularly around the doors has achieved benefits in terms of noise reduction and aerodynamic drag, but the almost complete elimination of leakage flows has also led to changes in the design of passenger compartment ventilation. To achieve the required ventilation flow rates greater attention must be paid not only to the intake and exit locations but also to the velocity and path of the fresh air through the passenger compartment. The intake should be located in a zone of relatively high pressure and it should not be too close to the road surface where particulate and pollutant levels tend to be highest. The region immediately ahead of the windscreen adequately meets all of these requirements and is also conveniently located for air entry to the passenger compartment or air conditioning system. This location has been almost universally adopted. For the effective extraction of the ventilation air a zone of lower pressure should be sought. A location at the rear of the vehicle is usually selected and in many cases the air is directed through the parcel shelf and boot to exit through a controlled bleed in the boot seal. Increasing the pressure difference between the intake and exit provides the potential for high ventilation air flow rates but only at the expense of a flow rate that is sensitive to the velocity of the vehicle. This is particularly noticeable when the ventilation flow is heated and the temperature of the air changes with speed. A recent trend has been to use relatively low-pressure differences coupled with a greater degree of fan assistance to provide a more controllable and consistent internal flow whether for simple ventilation systems or for increasingly popular air conditioning systems.

19.1.9 Wind tunnel testing

Very few new cars are now developed without a significant programme of wind tunnel testing. There are almost as many different wind tunnel configurations as there are wind tunnels and comparative tests have consistently shown that the forces and moments obtained from different facilities can differ quite considerably. However, most manufacturers use only one or two different wind tunnels and the most important requirement is for repeatability and correct comparative measurements when aerodynamic changes are made. During the early stages in the design and development process most testing is performed using small scale models where 1/4 scale is the most popular. The use of small models allows numerous design features to be tested in a cost effective manner with adequate accuracy.

For truly accurate simulation of the full-scale flow it is necessary to achieve geometric and dynamic similarity. The latter requires the relative magnitudes of the inertia and viscous forces associated with the moving fluid to be modelled correctly and the ratio of those forces is given by a dimensionless parameter known as Reynolds number (Re):

$$\mathrm{Re} = \frac{\rho u d}{\mu}$$

where ρ is the fluid (air) density, u is the relative wind speed, d is a characteristic dimension and μ is the dynamic viscosity of the fluid. For testing in air this expression tells us that the required wind speed is inversely proportional to the scale of the model but in practice the velocities required to achieve accuracy (using the correct Reynolds number) for small scale models are not practical, and Reynolds number similarity is rarely achieved. Fortunately, the Reynolds numbers achieved even for these small models are sufficiently high to create representative, largely turbulent vehicle surface boundary layers, and the failure to achieve Reynolds number matching rarely results in major errors in the character of the flow. The highest wind speeds at which models can be tested in any particular wind tunnel are more likely to be limited by the ground speed than by the air speed. The forward motion of a vehicle results not only in relative motion between the vehicle and the surrounding air but also between the vehicle and the ground. In the wind tunnel it is therefore necessary to move the ground plane at the same speed as the bulk air flow, and this is usually achieved by the use of a moving belt beneath the model. At high speeds problems such as belt tracking and heating may limit the maximum running speed, although moving ground plane technology has improved rapidly in recent years with the developments driven largely by the motor racing industry for whom 'ground effect' is particularly important. A considerable volume of literature is available relating to the influence of fixed and moving ground planes upon the accuracy of automotive wind tunnel measurements (for example Howell, 1994, Bearman *et al.*, 1988).

The use of larger models has benefits in terms of Reynolds number modelling and also facilitates the modelling of detailed features with greater accuracy, but their use also requires larger wind tunnels with correspondingly higher operating and model construction costs.

The forces acting upon a wind tunnel model are usually measured directly using a force balance which may be a mechanical device or one of the increasingly common strain gauge types. The latter has clear benefits in terms of electronic data collection and their accuracy is now comparable to mechanical devices. Electronic systems are also essential if unsteady forces are to be investigated. Lift, drag and pitching moment measurements are routinely measured and most modern force balances also measure side force, yawing moment and rolling moment. These latter three components relate to the forces that are experienced in cross-wind conditions.

Although direct force measurements provide essential data, they generate only global information and provide little guidance as to the source of the measured changes or of the associated flow physics. That additional information requires detailed surface and wider flow-field measurements of pressure, velocity and flow direction if a more complete understanding is to be achieved. Such data are now becoming available even from transient flow studies (e.g. Ryan and Dominy, 1998), but the measurements that are necessary to obtain a detailed understanding of the flows remain surprisingly rare despite the availability of well-established measurement techniques.

19.1.10 Computational fluid dynamics

The greatest obstacle to the complete mapping of the flow-field by experimentation arises solely from time constraints. Recent developments in the numerical modelling of both external and internal flows now provide the engineer with a tool to provide a complete map of the flow field within a realistic timescale. Although the absolute accuracy of simulations is still questionable there is no doubt that, as a pointer to regions of interest in a particular flow, they have revolutionized experimental studies. The complexity of the flow around and through a complete vehicle is immensely intricate and despite the claims of some it is unlikely that within the next decade numerical simulations will achieve sufficient accuracy to replace wind tunnel testing as the primary tool for aerodynamic development.

The relationships between the pressure, viscous and momentum forces in a fluid flow are governed by the Navier–Stokes equations. For real flows, these equations can only be solved analytically for simple cases for which many of the terms can be neglected. For complex, three-dimensional flows such as those associated with road vehicles it is necessary to achieve an approximate solution using numerical methods. Although different approaches may be adopted for the simulation, there are aspects of the modelling that are common to all. Initially the entire flow field is divided into a very large number of cells. The boundaries of the flow field must be sufficiently far from the vehicle itself to prevent unrealistic constraints from being imposed upon the flow. From a pre-defined starting condition (e.g. a uniform flow velocity may be imposed far upstream from the model), the values of each of the relevant variables are determined for each cell. Using an iterative procedure those values are repeatedly re-calculated and updated until the governing equations are satisfied to an acceptable degree of accuracy. As a rule the accuracy of a simulation will be improved by reducing the volume of each cell although there are particular rules and constraints that must be followed near surfaces (e.g. Abbott and Basco (1989)).

Unlike the aerospace industry, where aerodynamics is arguably the single-most important technology, automotive manufacturers rarely have sufficient resources to develop CFD codes for their own specific applications and, in almost all cases, commercially available codes are used. A danger of this approach is that users who are not fully conversant with the subtleties of the numerical simulation can overlook minor and sometimes major shortcomings in their predictions.

References and further reading

Abbott, M.B., and Basco, D.R. (1989). *Computational Fluid Dynamics: an Introduction for Engineers*. Longman. ISBN 0-582-01365-8.

Ahmed, S.R., Ramm, G and Faltin, G. (1984) Some salient features of the time-averaged ground vehicle wake, SAE International Congress and Exposition, Detroit. Paper no. 840300.

Barnard, R.H., (1996). *Road Vehicle Aerodynamic Design*. Longman, ISBN 0-582-24522-2.

Bearman, P.W., DeBeer, D., Hamidy, E. and Harvey, J.K. (1988). The effect of moving floor on wind-tunnel simulation of road vehicles, SAE International Congress and Exposition, Detroit. Paper no. 880245.

Carr, G.W., (1968). The aerodynamics of basic shapes for road vehicles, Part 2, Saloon car bodies. MIRA report no. 1968/9.

Docton, M.K.R. (1996). The simulation of transient cross winds on passenger vehicles, Ph.D. Thesis, University of Durham.

Emmelmann, H-J. (1982). Aerodynamic development and conflicting goals of subcompacts – outlined on the Opel Corsa. *International Symposium on Vehicle Aerodynamics*, Wolfsburg.

Howell, J. (1994). The influence of ground simulation on the aerodynamics of simple car shapes with an underfloor diffuser. Proc. RAeS Conference on Vehicle Aerodynamics, Loughborough.

Howell, J. (1998). The Influence of Aerodynamic Lift in Lane Change Manoevrability. Second M.I.R.A. Conference on Vehicle Aerodynamics, Coventry.

Hucho, W.H. (ed.) (1998). *Aerodynamics of Road Vehicles: from Fluid Mechanics to Vehicle Engineering*, 4th edition, S.A.E., ISBN 0-7680-0029-7.

Hucho, W.H., Janssen, L.J. and Emmelman, H.J. (1976). The optimization of body details – a method for reducing the aerodynamic drag of road vehicles. SAE International Congress and Exposition, Detroit. Paper no. 760185.

Nouzawa, T., Hiasa, K., Nakamura, T., Kawamoto, K. and Sato, H. (1992). Unsteady-wake analysis of the aerodynamic drag on a hatchback model with critical afterbody geometry. SAE SP-908, paper 920202.

Piatek, R. (1986). Operation, safety and comfort: in '*Aerodynamics of Road Vehicles*', Butterworth-Heinemann (ed. Hucho, W.H., 1986).

Ryan, A., and Dominy, R.G. (1998). The aerodynamic forces induced on a passenger vehicle in response to a transient cross-wind gust at a relative incidence of 30°. SAE International Congress and Exposition, Detroit. Paper no. 980392.

Stapleford, W.R., and Carr, G.W. (1971). Aerodynamic noise in road vehicles, part 1: the relationship between aerodynamic noise and the nature of airflow, MIRA report no. 1971/2.

Recommended reading

Abbott, M.B., and Basco, D.R. (1989). *Computational Fluid Dynamics: an Introduction for Engineers*. Longman. ISBN 0-582-01365-8.

Barnard, R.H., (1996). *Road Vehicle Aerodynamic Design*. Longman, ISBN 0-582-24522-2.

Hucho, W.H., (ed.) (1998). *Aerodynamics of Road Vehicles: from Fluid Mechanics to Vehicle Engineering*, 4th edition, S.A.E., ISBN 0-7680-0029-7.

Section Twenty

Refinement

Chapter 20.1

20.1

Vehicle refinement: Purpose and targets

Matthew Harrison

20.1.1 Introduction and definitions

The book opens with one man's thesis:

> "Styling and value sell cars – Quality keeps them sold". Lee Iacocca, *Iacocca: An Autobiography**

To explain, it takes years and millions of US dollars to produce a new car. If the styling is attractive and the marketing is effective, and if the value for money is good then that car will probably sell moderately well for the first few months after launch. However, if the quality is bad then word will get round and sales will quickly drop off. It is vital that good sales are maintained for a significant period if the development costs for the car are going to be recouped.

Lee Iacocca was writing from Ford's and Chrysler's perspective. Both of these already had their own strong brand image and so branding was not included in his thesis, but it should be when comparing cars from different manufacturers.

It is useful to introduce some definitions at this point. The term *Vehicle Refinement*** covers:

- noise, vibration and harshness*** (NVH – a well-known umbrella term in the automotive industry);
- ride quality;
- driveability.

A refined vehicle has certain attributes,**** they being:

- high ride quality;
- good driveability;
- low wind noise;
- low road noise;
- low engine noise;
- idle refinement (low noise and vibration);
- cruising refinement (low noise and vibration, good ride quality);
- low transmission noise;
- low levels of shake and vibration;
- low levels of squeaks, rattles and tizzes;
- low level exterior noise of good quality;
- noise which is welcome as a 'feature'.

The term 'NVH' is usually taken to cover:

- noise suppression;
- noise design (altering the character of noise but not necessarily its level);
- vibration suppression;
- suppression of squeaks, rattles and 'tizzes'.

20.1.2 Scope of this section

The scope of this section is illustrated in Fig. 20.1-1.

It has been designed to cover the core science, engineering and technology required by the NVH engineer with added material on ride quality and driveability. Wherever possible, issues that affect the customer's buying decision have been emphasised. Refinement is a customer-facing subject as any refinement engineering

*Lee Iacocca, former President of the Ford Motor Company, former President of Chrysler in *Iacocca*: An *Autobiography*, Lee Iacocca with William Novak, Bantam 1984.

**Refine (vb) to make or become free from coarse characteristics; make or become elegant or polished (Collins English Dictionary). Refinement is both a process (the act of refining) and a description of the eventual state (fineness, polish, etc.).

***Harsh (adj) rough or grating to the senses (Collins English Dictionary).

****Attribute (n) a property, quality or feature belonging to or representative of a person or thing (Collins English Dictionary).

Vehicle Refinement; ISBN: 9780750661294
Copyright © 2004 Matthew Harrison. All rights of reproduction, in any form, reserved.

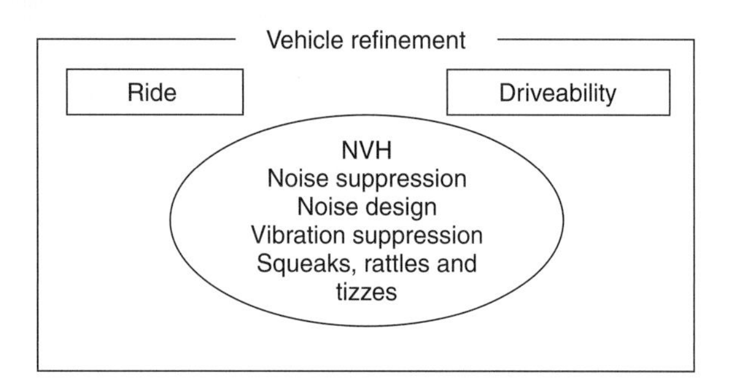

Fig. 20.1-1 The scope of the term 'Vehicle Refinement'.

undertaken affects the driving experience directly. Therefore, the reader is encouraged to think of refinement as being inextricably linked to the business of selling passenger cars (and increasingly this is being extended to vans, light trucks, trucks, buses, coaches and other road vehicles).

Some further definitions are offered for the sake of clarity. In this section the terms:

Noise shall be used to describe audible sound, with particular attention paid to frequencies in the range of 30–4000 Hz.

Vibration shall be used to describe tactile vibration, with particular attention paid to frequencies in the range of 30–200 Hz.

Other terms that will be encountered include:

Primary ride taken to be the rigid body motion of the passenger compartment relative to the road. Typical frequency range – 0–6 Hz.

Secondary ride taken to be the relatively large amplitude motion of sub-elements of the vehicle such as individual wheels, axles or elements of the powertrain. Typical frequency range – 6–30 Hz.

Structure-radiated noise being airborne noise radiated by a structural surface that is vibrating. Also known as 'structure borne noise'.

20.1.3 The purpose of vehicle refinement

Refinement helps manufacturers sell their vehicles. Brandl et al. (2000) published the results of a formal investigation of customer attitudes to vehicle refinement. Customers were asked to complete a questionnaire relating to their own vehicle. They were asked questions relating to their attitudes on vehicle prestige (brand by another name), performance, convenience, family friendliness, noise quality and cost. Although there was scatter in the results obtained, there was evidence of clustering with certain classes of customer showing predictable tastes with vehicle-refinement issues helping to define that taste.

Throughout the 1990s Rover Group in the UK (subsequently they traded under the MG badge) defined 'Refinement' as *the invisible feature* of their vehicles suggesting a strong commitment to it.

Refinement (or NVH) has always been a consideration for vehicle design and development. However, over the last 20–30 years it has assumed a greater importance (witness the advertisements by manufacturers that stress how quiet their cars are). Reasons for this include:

- *Legislation* Since the adoption by Member States of Council Directive 70/157/EEC (1970) limiting the permissible levels of noise emitted by accelerating vehicles, the sale of new non-conforming vehicles is prohibited in the European Community. Similar legislation has been adopted in many non-EU countries, and certainly in all the main automotive territorial markets.
- *Marketing to new customers* Refinement is a feature that may be used to distinguish a vehicle from its otherwise similar competitors, thus attracting customers not necessarily loyal to that particular brand.
- *Customer expectation* Customers have come to expect continuous improvement in the new vehicles that they buy. They expect their new purchase to be better equipped, more comfortable, and perform better than the vehicle they just traded in (which may only be a few years old). The new vehicle may be better in all respects than the old one on paper, but if it lacks refinement then it *will not feel better* and the customer will not be fully satisfied. They may choose to take their loyalty to another vehicle manufacturer next time.
- *Marketing to existing customers* 'Trade your old model in for this year's model: it is loaded with features, more comfortable and more *refined*.' The modern car industry needs turnover to survive. People need to be encouraged to trade in regularly. The increase in vehicle leasing schemes for retail customers encourages this turnover.

The importance of vehicle refinement can be tested using the following group exercise where team members put a monetary price on refinement. This is an absolute test on the value that they place on refinement.

20.1.4 How refinement can be achieved in the automotive industry

The traditional vehicle manufacturer is organised according to functional divisions being typically: Design,

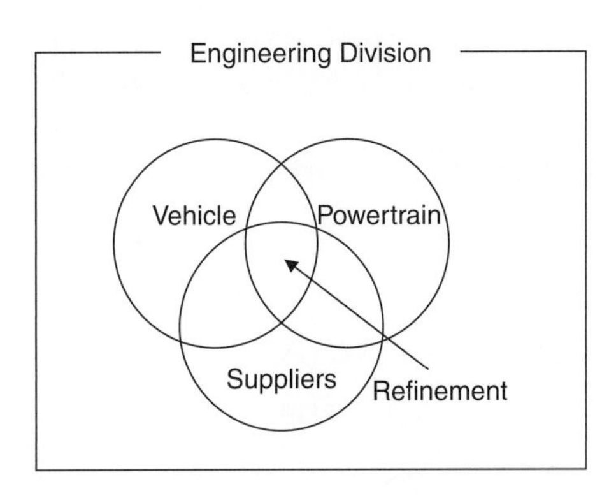

Fig. 20.1-2 Refinement within the traditional vehicle manufacturer.

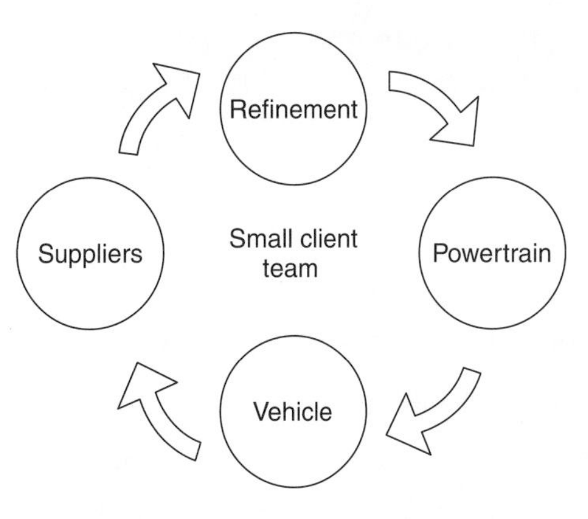

Fig. 20.1-3 Refinement in the 'Extended Enterprise'.

Engineering, Manufacturing, Marketing and Sales. Each division might be divided further into groups. For simplicity, an Engineering Division shall be divided into three groups: Powertrain, Vehicle and the Suppliers of components to the engineering effort as shown in Fig. 20.1-2.

The refinement sub-group straddles the interface between the three main groups, having influence on each one in turn but not enjoying any decision-making authority. For example, consider the engine mounts: they are attached to both the powertrain and the vehicle and they are manufactured by a third-party supplier. The refinement sub-group has an interest in their performance in order to improve NVH but no over-riding authority to broker compromises between the three main groups. Management of such an interface is never going to be easy, and inefficiencies, misunderstandings, mistakes or arguments that result may delay the development programme.

An alternative organisational structure is that of the 'Extended Enterprise' (Ashley, 1997) where suppliers assume greater responsibility for the design and development of their particular contribution to the whole vehicle. Such organisations are more fluid as illustrated in Fig. 20.1-3.

Design authority is pushed out from the small client team to the various engineering functions (including a refinement function). With this fluid structure, design information is visible to all interested parties and compromises are brokered 'out in the open'. Responsibility for delivering refinement targets, defined by the client team in a Product Design Specification (PDS), is shared by all. However, in many cases the bulk of the responsibility falls to the Supplier, with the refinement function checking for compliance with the PDS.

The wider adoption of the Extended Enterprise structure within the global automotive industry has led to new opportunities for refinement engineers, particularly within component supply organisations used to manufacture but now required to design and engineer as well.

The Extended Enterprise also leads to consolidation in the industry, with the big organisations getting bigger by acquisition (Hibbert, 1999).

20.1.5 The history of vehicle refinement: one representative 20-year example

In 1979, Vauxhall offered the Royale for sale in the UK – a 2.8-litre, six-cylinder, executive class (small) car for the on-the-road cost of £8354 (1979 prices). Motor magazine described it as being 'in general a refined car' (Vauxhall Royale – Star road test, Motor Magazine, 13 January 1979).

By 1989, that car had been replaced by the Vauxhall Senator 2.5i – a 2.5-litre, six-cylinder, executive class (small) car for the on-the-road cost of £16 529 (1989 prices). Autocar & Motor magazine attributed refinement as one of the car's strengths (Vauxhall Senator 2.5i – Test extra, Autocar & Motor, 21 September 1988).

By 1999, that car had been replaced by the Vauxhall Omega 2.5 V6 CD – a 2.5-litre, V6, executive class (small) car for the on-the-road cost of £21 145 (1999 prices). Autocar & Motor magazine described it as having 'one of the most refined, well mannered six-cylinder engines...' (Vauxhall Omega 2.5I V6, Autocar & Motor, 1 June 1994).

By 1999, Vauxhall had also launched a smaller vehicle with what seems on paper to be the same degree of sophistication – the 2.5-litre V6 Vectra GSi. Autocar & Motor magazine (New cars – Vauxhall Vectra GSi, Autocar & Motor – 14 July 1999) were not so complimentary about its refinement as they had been with the Omega.

The specifications and relative costs of these four cars are compared in Table 20.1-1 (non-SI units are retained for historic authenticity).

Table 20.1-1 Some Vauxhall executive cars (small) 1979–1999

Vehicle	Peak power (bhp @ rpm)	Peak torque (lbfti @ rpm)	Weight (kg)	Cost new	Estimate of real cost in 1999
1979 Royale (Ref. a)	140 @ 5200	161 @ 3400	1402	£8354	£28832*
1989 Senator 2.5i (Ref. b)	140 @ 5200	151 @ 4200	1465	£16528	£28333*
1999 Omega V6 2.5 CD (Ref. c)	168 @ 6000	167 @ 3200	1480	£21145	£21145
1999 Vectra V6 Gsi (Ref. d)	193 @ 6250	193 @ 3750	1431	£21700	£21700

Source: Autocar & Motor magazine

*An estimate of what the real cost of that particular vehicle would be in 1999 relative to the cost of other goods included in the Retail Price Index. It has been calculated by scaling the cost when new by data obtained for the percentage change in Retail Price Index in the period 1979–1999, shown in Fig. 20.1-4.

References
(a) Vauxhall Royale – Star road test
Motor Magazine, 13 January 1979
(b) Vauxhall Senator 2.5i – Test extra
Autocar & Motor, 21 September 1988
(c) Vauxhall Omega 2.5l V6
Autocar & Motor, 1 June 1994
(d) New cars – Vauxhall Vectra GSi
Autocar & Motor – 14 July 1999

This comparison of scaled historic costs with the costs in 1999 for new vehicles (99MY denotes the 1999 model year) is shown in more detail in Fig. 20.1-5.

The above comparisons tell us something about vehicle refinement over the twenty-year period 1979–1999, although any conclusions drawn apply strictly only to the development of one series of vehicles produced by one manufacturer on sale in one territory. It can be seen that:

1. The term 'Refinement' has been used by motor journalists since the 1970s (and indeed probably quite sometime before that). Therefore, the term has been placed in the mind of the consumer as being a relevant factor in the decision-making process of buying a car for a generation.
2. The real cost of buying an executive class (small) Vauxhall car that has been commended for its refinement dropped significantly after 1979 – Table 20.1-1 suggests that this drop might be as large as 25% by 1999. The drop seems to have occurred in the years between 1989 and 1999 (the scaled cost of the 1979 model year (79MY) Royale matches almost exactly the 1989 retail price for the Senator, but the scaled cost of the Senator is much greater than the 99MY Omega). During that 10-year period, the performance of the vehicle increased significantly in terms of specific engine torque and power (Table 20.1-1). In addition, vehicle emissions dropped because of tougher legislation. The 99MY cars are more refined

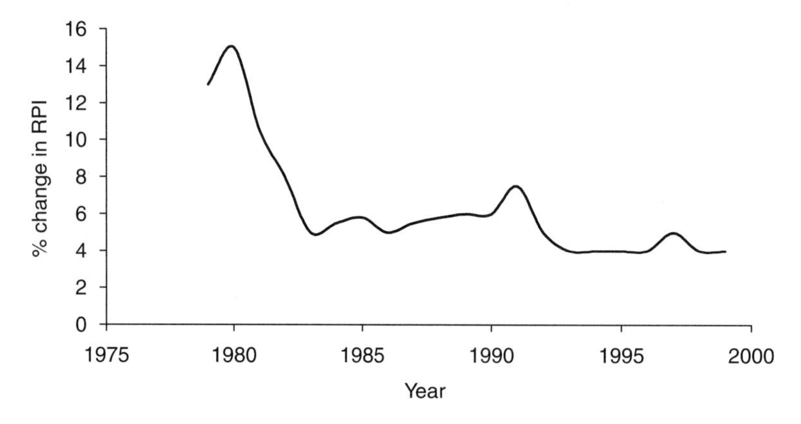

Fig. 20.1-4 Change in Retail Price Index in the UK.
Source: www.bizednet.bris.ac.uk – University of Bristol.

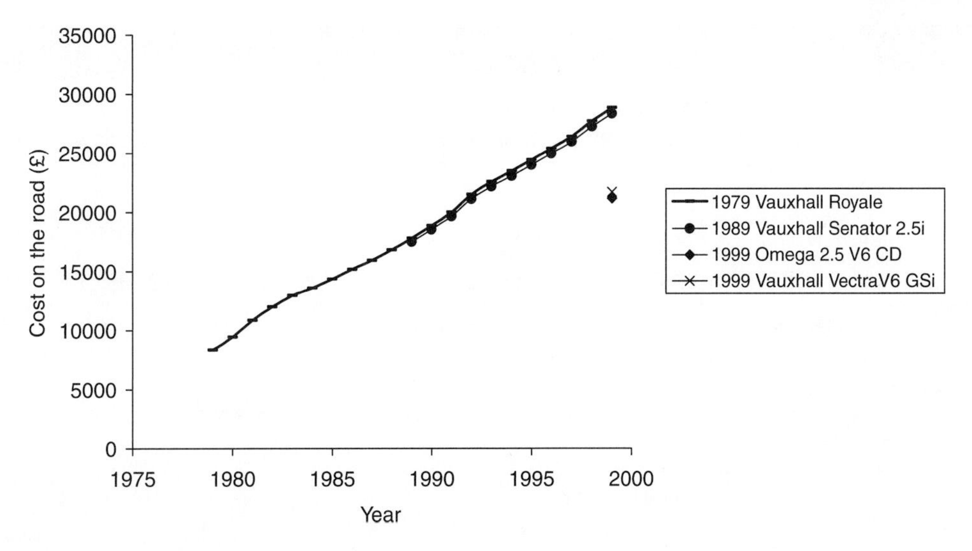

Fig. 20.1-5 Comparing the cost of 99MY vehicles with the historic costs of equivalent 79MY and 89MY, scaled according to the UK Retail Price Index.
Data sources: References a–d in Table 20.1-1 and www.bizednet.bris.ac.uk – University of Bristol.

having to meet noise drive-by levels of 74 dBA compared with 82 dBA in 1979 (8 dBA is a significant drop in noise, equivalent to almost a halving in loudness). Also 99MY vehicles are generally better equipped than their predecessors (air conditioning, satellite navigation, CD players, etc.).

3. The price of many new cars in the UK dropped by 10–15% in mid-2000 due to action by the UK Government and pressure from consumers. This makes the results discussed in (2) even more startling.

4. The consumer looking for a six-cylinder Vauxhall car in 1999 could choose between the Omega and the Vectra for virtually the same price. The Vectra seems to offer better performance, whilst the Omega is reported to offer more space and better refinement.

The situation described above is counter-intuitive – 99MY vehicles that have better performance, less emissions, better refinement and more equipment than 89MY vehicles but cost less in real terms. It is possible to speculate how the industry found itself in that position. It might have been due to:

- The effects of increased competition in a saturated market.
- Action by the UK Government (only applies to 2000MY prices).
- The effects of vehicle leasing schemes and manufacturer's vehicle finance schemes.
- The adoption of measures to improve organisational efficiency such as the 'Extended Enterprise' and technology employed in design and manufacturing activities.
- Macro political and economic factors affecting the automotive industry but not generated by that industry (globalisation, problems in Pacific Rim countries, etc.).

A similar trend can be shown with Ford, one of Vauxhall's traditional competitors. Data for six-cylinder executive (small) cars manufactured by Ford (broadly competing vehicles to the Senator/Omega/Vectra) are shown in Table 20.1-2 and these show similar trends to the data for the Vauxhall cars shown in Table 20.1-1.

20.1.6 Refinement targets

The setting of refinement targets is important for the successful operation of the so-called Extended Enterprise that is described in Section 22.1.1.4. Without these, individual system suppliers would determine their own interpretation of an appropriate level of refinement for their component and the final vehicle would most likely be truly refined only in some aspects and not in others. In addition, it should be noted that type approval testing (at present) is undertaken as 'whole vehicle' type approval (see Section 22.1.1.1) and therefore by definition it is only undertaken once the production intent vehicle is fully developed. If one component or sub-system causes the vehicle to fail its type approval test due to excessive noise then the cost implications are obviously serious.

The standard management tool for setting refinement (and other) targets is the PDS document. This is written by the brand holder and adherence to it becomes a conditions of contract for any supplier. A typical PDS will contain the following refinement targets:

- whole vehicle exterior noise targets;
- single component exterior noise targets;
- whole vehicle interior noise targets;
- ride quality targets (including tactile vibration targets).

Table 20.1-2 Some six-cylinder Ford executive cars (small) 1979–1999

Vehicle	Peak power (bhp @ rpm)	Peak torque (lbft @ rpm)	Weight (kg)	Cost new	Estimate of real cost in 1999*
1979 (no comparable model)					
1989 Granada 2.4l V6 Ghia (Ref. e)	130 @ 5800	142 @ 3000	1340	£16 045	£25 996
1999 Mondeo 2.5 litre V6 Ghia X (Ref. f)	168 @ 6250	162 @ 4250	1408	£21 680	£21 680
1999 Mondeo 2.5 litre V6 ST24 (Ref. g)	168 @ 6250	162 @ 4250	1410	£19 680	£19 680

*Calculated in the same manner as Table 20.1-1.
References
(e) Test Update: Ford Granada 2.4i Ghia
Autocar & Motor, 11 February 1987
(f) Autocar Twin Test: Ford Mondeo 24v Ghia X vs Peugeot 406 3.0 SVE
Autocar & Motor, 6 November 1996
(g) Autocar Twin Test: Ford Mondeo ST24 vs Vauxhall
Vectra Gsi
Autocar & Motor, 22 April 1998

The exterior noise targets will mostly be objective relating to the passing of the type approval noise test, although occasionally subjective criteria might be included in addition (exhaust noise quality in particular for sports cars). The interior noise, ride quality and tactile vibration targets will be a mix of objective and subjective criteria.

20.1.6.1 Whole vehicle exterior noise targets

Whole vehicle noise targets are set in terms of drive pass noise levels for the type approval test required in the territory in which the vehicle will be offered for sale. Separate noise targets are commonly set for each sub-system on the vehicle. A typical set is given in Table 20.1-3 for vehicles offered for sale in the EU.

It is not possible to determine a single set of such targets for a given territory as every vehicle will have its own unique noise signature. This is made clear in Tables 20.1-2 and 20.1-3 (see Section 22.1.1.6) where it is clear that the EU passenger car fleet exhibits type approval noise levels that vary by as much as 10 dB. Notwithstanding this, Table 20.1-3 is offered as a starting point. If a particular vehicle is diesel powered, perhaps the engine noise target should be increased by 1–2 dB. If the vehicle is a sports-utility then perhaps the tyre noise and transmission noise targets should be raised by 1–2 dB apiece.

20.1.6.2 Single component exterior noise targets

In order to minimise the risk that the whole vehicle exterior noise targets are not met at the final type approval

Table 20.1-3 Suggested target noise levels for achieving type approval under 9297/EEC. See Section 22.1.1.6 for further details

	Passenger car	Light truck	Heavy truck
	Target levels at 7.5 m, acceleration test (dBA)		
Engine	69	72	77
Exhaust	69	70	70
Intake	63	63	65
Tyres	68	69	75
Transmission	60	63	66
Other	60	72	65
Combined level	74.2	77.3	80.1

testing, the PDS will routinely include separate airborne noise targets for certain components or vehicle subsystems. The most common targets relate to engine-radiated noise, intake noise and exhaust noise.

20.1.6.2.1 Engine-radiated noise targets

These are normally set in terms of sound power level. European engine suppliers are now required by law to measure and declare the sound power emissions of their engines. This makes a comparison between competitor engines easy.

Engine-radiated noise targets are more important for diesel engines than for gasoline engines as the various type approval tests undertaken around the world demand only quite low engine speeds where gasoline engines are generally quiet and rarely cause failure of the test.

Again, it is not possible to adopt a single set of noise targets applicable to all potential engines, but an indicative set is offered in Table 20.1-4. This relates to a particular diesel engine used in light European trucks and sports-utility vehicles.

An alternative target for a 4.0-litre four-cylinder DI diesel is offered by Pettitt (1988) at 107 dB re 10^{-12} W. An alternative (and rather different) noise source ranking is offered for a six-cylinder diesel engine by Beidl et al. (1999).

20.1.6.2.2 Intake orifice-radiated noise targets

These are commonly set as maximum sound pressure levels to be recorded at a distance of 100 mm from the intake orifice at an angle of incidence of 90°. Common targets are:

- An overall A-weighted sound pressure level of 90 dBA at 1000 rev min^{-1} wide-open throttle (full load) rising at a rate of 5 dBA per 1000 rev min^{-1} to a maximum of 115 dBA at 6000 rev min^{-1}. The adoption of this target is likely to result in an intake level of 63 dBA during an EC type approval test (as required by the targets shown in Table 20.1-3) without relying on any attenuation offered by the vehicle bodyshell.
- A firing order sound pressure level of 105 dB (lin) at low drive-away engine speeds (full load), a level of 100 dB (lin) at moderate engine speeds and a level of 105 dB (lin) at high engine speeds.
- Sound pressure levels for other higher orders of 95 dB (lin) at low drive-away engine speeds (full load), a level of 90 dB (lin) at moderate speeds and a level of 95 dB (lin) at high engine speeds.

Table 20.1-4 Typical sound power targets in dB re 10^{-12} W for a 4.0-litre I4 NA (naturally aspirated) IDI (in-direct injection) diesel used in light trucks and sports-utility vehicles

Component	Sound power level (dB re 10^{-12} W)
Sump	102
Block	104
Head	93
Exhaust	102
Intake	97
Fuel injection pump	96
Total	108

20.1.6.2.3 Exhaust tailpipe-radiated noise targets

These are commonly set as maximum sound pressure levels to be recorded at a distance of 500 mm from the exhaust tailpipe at an angle of incidence of 45°. Common targets are:

- An overall A-weighted sound pressure level of 82 dBA at 1000 rev min^{-1} wide-open throttle (full load) rising at a rate of 5 dBA per 1000 rev min^{-1} to a maximum of 107 dBA at 6000 rev min^{-1}. The adoption of this target is likely to result in an exhaust level of 69 dBA during a type approval test (as required by the targets shown in Table 20.1-3).
- A firing order sound pressure level of 120 dB (lin) at low drive-away engine speeds (full load), a level of 100 dB (lin) at moderate speeds and a level of 115 dB (lin) at high engine speeds.
- Sound pressure levels for other higher orders of 105 dB (lin) at low drive-away engine speeds (full load), a level of 95 dB (lin) at moderate speeds and a level of 105 dB (lin) at high engine speeds.

20.1.6.3 Whole vehicle targets for interior noise

Interior noise levels are routinely measured at the driver's ear position (and elsewhere in the vehicle interior) in accordance with BS 6086 1981 (ISO 5128, 1980) (see Section 21.1.1.3 for details). Either of two basic schemes for objective interior noise targets can be adopted.

20.1.6.3.1 Interior noise targets: perceptible improvement in sound pressure level

In this scheme, sound pressure levels in the vehicle interior are measured in accordance with BS 6086 1981 (ISO 5128, 1980) in the competitor vehicle. Then a target is set for the vehicle under development in terms of relative improvement. Bies and Hansen (1996) summarise:

Change in apparent loudness	Change in sound pressure level (dB)
Just perceptible	−3
Clearly noticeable	−5
Half as loud	−10
Much quieter	−20

Recordings of interior noise can be analysed to obtain the following metrics:

- Overall sound pressure level. This would normally be A-weighted for sound pressure levels below 55 dB (re 20 micro-pascals) and C-weighted for sound pressure levels in the range of 55–85 dB (Bies and Hansen, 1996). Typical targets would be devised in terms of maximum sound pressure level at certain road speeds under cruise conditions and at certain engine speeds under full load acceleration.
- Sound pressure level at the engine firing frequency order.
- Sound pressure level at higher engine orders.

Perceptible improvement targets can be set for each of these metrics. Such targets are most likely to be 3 dB lower than the metrics measured in the competitor vehicle in order to achieve tangible improvement at minimum cost.

20.1.6.3.2 Interior noise targets: brand value

In this scheme, a definitive set of interior noise targets are adopted, irrespective of the relative performance of competitor vehicles, in order to make a particular brand statement. The targets might apply to either aural comfort, or more usually speech intelligibility. Such targets are obviously brand specific but some general rules do apply:

- American Standard ANSI S3.1–1977 shows that effective speech communication can take place at normal voice levels between two persons spaced 1 m apart (in a free acoustic field) providing the background noise level is less than 65 dBA (or the so-called speech interference level (SIL) being the arithmetic average sound pressure level in the 500-, 1000- and 2000-Hz octave bands is less than 58 dB (lin)). These two targets are the upper limits for just acceptable speech communication. Bies and Hansen (1996) describe this as being 95% sentence intelligibility or 60% word-out-of-context recognition.

  In accordance with this, and assuming a highly sound absorbing vehicle interior, a rational target for interior noise level with a vehicle under steady-state cruise conditions would be an interior sound pressure level of 65 dBA at the driver's ear. When intelligibility of the female voice is of particular importance, this target might be lowered to 60 dBA. When front seat to back seat communication is of particular importance, the target might be lowered to 55 dBA. With the general adoption of hands-free mobile phones in vehicles, a target in the range of 60–65 dBA is prudent. A survey of a 2003 MY executive class sedan revealed interior cruise noise levels in the range of 60–65 dBA at 80 km hr^{-1} on B-class roads. Speech communication in the front seats in this vehicle should be adequate but front to back seat communication would be compromised. By way of historical comparison, Rust and co-workers suggested an interior noise level of 70 dBA at 80 km hr^{-1} for a 1980s direct injection (DI) diesel.
- Priede (Priede, 1974; Priede and Anderton, 1984) showed that typically, full-load engine noise increases by 5dB per 1000 rev min^{-1} increase in engine speed. Challen and Crocker (1982) suggest that the maximum effect of engine load on engine noise levels is of the order of 1–2 dB. Priede (1974) suggests that intake and exhaust noise levels increase by 5dB per 1000 rev min^{-1} and it is well known that intake noise is strongly affected by load (±15 dB) whereas exhaust noise levels are not so strongly affected (±10 dB). Underwood (1973) suggested that tyre noise increases by 9–13 dB per doubling of vehicle speed.

In accordance with this guidance, any objective interior noise target for any metric should vary with engine speed, road speed and engine load as appropriate. A consistent set of interior noise targets for a mid-priced family car powered by a four-cylinder gasoline engine might be:

- 65 dBA at the driver's inner ear at 80 km hr^{-1} cruise and 70 dBA at 120 km hr^{-1}.
- Under full load acceleration in second or third gear, a sound pressure level at the driver's inner ear of 55 dBA at 1000 rev min^{-1} rising linearly to 80 dBA at 6000 rev min^{-1}. Under overrun (zero load) the targets should be 10 dB less.
- The level of any engine order should be at least 3 dB lower than the overall noise level.

For the case of an executive or luxury car, these targets should all be reduced by at least 5 dB.

Alternative brand-value interior noise targets can be set in terms of:

- (Zwicker) loudness in accordance with ISO 532 – 1975. This method is suitable for calculating the loudness level (in 'phons') of combinations of octave bands for random-like noise in which there are prominent tonal components. The loudness in phons is given by the set of equal loudness contours

Table 20.1-5 Common subjective rating scheme

1	2	3	4	5	6	7	8	9	10
Not acceptable			Objectionable	Requires improvement	Medium	Light	Very light	Trace	No Trace

(BS 3383 – 1988, ISO 226 – 1987). Each contour is labelled with X phons, X being the sound level at 1000 Hz for that particular contour.

- Articulation index (AI) in the range of 0.5–0.6. The calculation method is given in ANSI S3.5 – 1969 and the long-term rms. speech level in 1/3 octave bands is compared against the long-term rms. masking noise level, and the 1/3 octave results so generated are aggregated using separate weighting functions for each 1/3 octave band.
- Speech transmission index (STI) and the more rapid RASTI (rapid speech transmission index) (BS 6840: part 16 – 1989) are commonly used alternatives to articulation index. Rust and co-workers suggest a RASTI of 0.8 at 50 km hr^{-1} reducing linearly to 0.4 at 125 km hr^{-1}.

20.1.6.3.3 Interior noise: subjective targets

There are several different strategies for the subjective assessment of interior noise. These are discussed, along with target levels in Section 21.1.1.4. An engineering method for subjective appraisal involves a panel of people driving and riding in the vehicle(s) along a pre-determined test route on public roads and rating the following noise (and vibration) attributes:

- wind noise;
- road noise;
- engine noise;
- idle refinement;
- cruising refinement;
- transmission noise;
- general shakes and vibrations;
- squeaks, rattles and tizzes;
- ride quality;
- driveability;
- noise that is a 'feature' (sporty exhaust notes, etc.).

The ratings are made to a common scale from 1 to 10 as per Table 20.1-5

- A rating of less than 4 is unacceptable for any attribute.
- A rating of 5 or 6 is borderline.
- A rating of 7 or more on any attribute is acceptable.

Most new passenger cars are launched with a subjective rating of 7 or 8 on most attributes.

20.1.6.4 Targets for ride quality (including tactile vibration)

Ride quality is taken here to be the subjective response to a low-frequency vibration phenomenon. There are several different strategies for the assessment of in-vehicle vibration levels. A summary is measured vibration levels are rated according to objective criteria, and the most commonly used criteria are offered in:

- ISO 2631 Part 1 (1985);
- BS 6841 (1987);
- NASA discomfort level index (1984) (Leatherwood and Barker, 1984).

Experience of using all three has led to the conclusion that a family class vehicle for either the European or the US Federal markets will be ready for sale when the appropriately frequency-weighted seat rail vibration levels measured at 80 km hr^{-1} on a straight road with 5–10-year-old tarmac and a few spot repairs (in other words, a typical B-class inter-urban road) are:

- Close to the four-hour reduced comfort boundary in the vertical direction as defined in ISO 2631 Part 1 (1985).
- Have an rms. level less than 0.63 $m\,s^{-2}$ (classed as better than 'a little uncomfortable' according to BS 6841 (1987)).
- Have a NASA discomfort rating below 4.0 (Bosworth et al., 1995).

References

Ashley S November 1997 Keys to Chrysler's comeback, Mechanical Engineering.

Beidl CV, Rust A, and Rasser M 1999 Key steps and methods in the design and development of low noise engines SAE Paper No. 1999-01-1745.

Bies DA, and Hansen CH 1996 Engineering noise control – theory and practice, Second edition. E&FN Spon, London.

Bosworth R, Trinick J, Smith T, and Horswill S 1995 Rover's system approach to achieving first class ride comfort for the new Rover 400, IMEchE Paper No. C498/25/111/95.

Brandl FK, Biermayer W, Thomann S, Pfluger M, and von Hofe R 2000 Objective description of the required interior sound character for exclusive passenger cars IMechE Paper No. C577/013/2000.

Challen BJ, and Crocker MD 1982 A review of recent progress in diesel engine noise reduction SAE paper No. 820517.

Hibbert L November 1999 Last ones standing are the winners. Professional Engineering 3.

Leatherwood JD, and Barker LM 1984 A user-orientated and computerized model for estimating vehicle ride quality. NASA Technical Paper 2299.

Pettitt RA 1988 Noise reduction of a four litre direct injection diesel engine IMEchE Paper No. C22/88.

Priede T 1974 Effect of operating parameters on sources of vehicle noise, Symposium on noise in transportation. University of Southampton.

Priede T, and Anderton D 1984 Likely advances in mechanics, cooling, vibration and noise of automotive engines. Proceedings of the Institution of Mechanical Engineers Vol. 198D(No. 7):95–106.

Rust A, Schiffbaenker H, and Brandl FK 1985 Complete NVH optimisation of a passenger vehicle with DI diesel engine to meet subjective market demands and future legislative requirements SAE Paper No. 890125.

Underwood MCP 1973 A preliminary investigation into lorry tyre noise. Transport Research Laboratory Paper LR 601, Crowthorne, UK.

Section Twenty-One

Interior noise

Chapter 21.1

21.1

Interior noise: Assessment and control

Matthew Harrison

21.1.1 Subjective and objective methods of assessment

21.1.1.1 Background

Vehicle interior noise is a combination of:

- engine noise;
- road noise;
- intake noise;
- exhaust noise;
- aerodynamic noise;
- noise from components and ancillaries;
- brake noise;
- squeaks, rattles and 'tizzes'.

Apart from squeaks, rattles and tizzes that occur inside the passenger compartment, noise or vibration usually originates from outside, interacting with the vehicle structure in some way (and possibly hence with other noise sources) and then producing radiated sound inside the compartment. This process is illustrated in Fig. 21.1-1.

The interaction with the structure can be either as:

- An airborne noise path – airborne noise from outside the passenger compartment leaking in to cause airborne noise inside.
- A structure-borne noise path – vibration from outside causing the surfaces of the passenger compartment to vibrate and radiate noise.

Direct airborne noise paths are found where there is a lack of sealing between the interior and the exterior environments (around door seals, grommets in the bulkhead, etc.). Indirect airborne noise paths are found when airborne noise outside impinges on the surfaces of the passenger compartment, causing them to vibrate and radiate noise inside.

The interaction between the noise source and the structure has a filtering effect on the final interior noise level (see Fig. 21.1-1). For instance, in the case of indirect

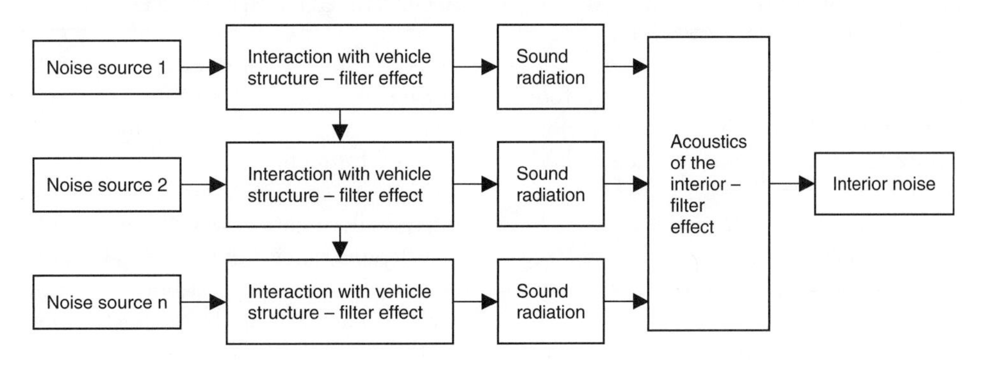

Fig. 21.1-1 Schematic showing the origin of interior noise.

Vehicle Refinement; ISBN: 9780750661294
Copyright © 2004 Matthew Harrison. All rights of reproduction, in any form, reserved.

airborne paths, the transmission of sound to the passenger compartment will be greatest at low frequencies due to the action of the mass law for transmission loss (TL) (see Section 21.1.10.5). In the case of structure-borne paths, the use of resilient mounts will isolate the passenger compartment at higher frequencies and so transmission will also be greatest at low frequencies.

From the above observations, one would expect a strong low-frequency component to interior noise levels. This is found in practice (see for instance Steel et al. [2000]) in spite of the filtering effect due to the fact that steel panels are rather poor radiators of sound at low frequencies (see Section 21.1.3.7).

One of the reasons for this is due to acoustic modes within the passenger compartment enhancing low-frequency noise levels. The so-called acoustic modes are set up at the natural frequencies of the space, frequencies given by (for a rectangular space):

$$f_{a,b,c} = \frac{c}{2\pi}\sqrt{\left(\frac{a\pi}{x}\right)^2 + \left(\frac{b\pi}{y}\right)^2 + \left(\frac{c\pi}{z}\right)^2} \qquad (21.1.1)$$

where

c = speed of sound in air (m s^{-1})
a, b, c = integer indices 1, 2, 3, ...
x, y, z = acoustic dimensions of the space (m)

For the modern generation of small European car, the longest acoustic dimension of the passenger compartment is that between the footwell and the rear screen. This dimension is typically a little more than the wheelbase of the vehicle, perhaps 2.5 m. The lowest frequency acoustic mode of the passenger compartment (the 1, 0, 0 mode, $a = 1, b = 0, c = 0$) therefore has pressure maxima (anti-nodes) at the footwells and at the rear screen and pressure minima at the midpoint of the wheelbase. Assuming $x = 2.5$ m this would occur at 69 Hz. Note that the driver's head position is usually just aft of the midpoint of the wheelbase and so the driver seldom enjoys the benefits of being precisely at the nodal position.

The 1, 0, 0 mode is commonly expected in the frequency range of 65–75 Hz. Our estimate of 69 Hz equates to the firing frequency of a four-cylinder, four-stroke engine operating at 2070 rev min^{-1}. European four-cylinder cars commonly exhibit low-frequency noise peaks (known as interior booms or sometimes as body booms) at engine speeds in the 2000–2500 rev min^{-1} range as a result of exciting the 1, 0, 0 mode.

The next longitudinal mode (2, 0, 0) would be expected at around 138 Hz. It often occurs at a lower frequency though in road cars.

The first transverse mode (0, 1, 0) can be expected at 123 Hz if the transverse acoustic dimension is assumed equal to the typical track of a small European car (1.4 m). The (0, 2, 0) mode would be expected at 246 Hz.

The first vertical mode (0, 0, 1) can be expected at 143 Hz if the vertical dimension is assumed to be the typical body height of a small European car (1.2 m). The (0, 0, 2) mode would be expected at 286 Hz.

The firing frequency of a four-cylinder, four-stroke engine lies in the range of 33–200 Hz for the corresponding speed range of 1000–6000 rev min^{-1}. In this range, using the rather over-simplified analysis used above, one might reasonably expect the following body booms:

- (1, 0, 0) at around 70 Hz/2100 rev min^{-1}
- (0, 1, 0) at around 120 Hz/3600 rev min^{-1}
- (2, 0, 0) at around 140 Hz/4200 rev min^{-1}
- (0, 0, 1) at around 140 Hz/4200 rev min^{-1}
- (1, 1, 0) at around 140 Hz/4200 rev min^{-1}
- (2, 1, 0) at around 185 Hz/5550 rev min^{-1}

The boom around 3600 rev min^{-1} is often the most annoying. This is because the lower speed boom is usually only transient as the vehicle will accelerate quickly through it. Also the high speed booms are seldom a problem as most drivers will not run their engines for long periods at 4000+ rev min^{-1} and will select a higher gear when they first hear such booms. However, drivers on high-speed roads may well find themselves having to cruise around 3600 rev min^{-1} in top gear with an annoying boom ever present in the passenger compartment. As the occupants tend to sit near to the sides of the passenger compartment, all find themselves at anti-nodes of the (0, 1, 0) boom and all suffer pressure maxima.

The passengers in the rear seats of most cars suffer from pressure maxima at most modes – longitudinal, lateral, vertical (and other non-orthogonal modes). This is not the case for the occupants of the front seats. For this reason, rear seat noise quality is often a greater issue for concern than the levels and quality of noise at the driving position.

21.1.1.2 On the balance between airborne and structure-borne noise

The vibration isolating effects of resilient components (engine mounts, subframe mounts, under-carpet treatments) tends to limit the significance of structure-borne noise to frequencies below around 500 Hz. At higher frequencies, noise received via airborne noise paths generally dominates the interior noise levels.

At frequencies below 500 Hz, airborne noise may remain a significant contributor to overall interior noise levels, particularly if the sealing of the passenger compartment (door seals, window seals, grommets in the bulkhead) is not perfect.

21.1.1.3 On the measurement of interior noise

A procedure for measuring vehicle interior noise is specified in BS 6086 – 1981 (ISO 5128 – 1980).

BS 6086 calls for:

- The measurement of sound pressure level, fast weighted, 'A'-weighted – 1/3 octave analysis if possible.
- The measurement of noise levels at more than one location, and at least at the driver's ear position and one point at the rear of the vehicle.
- Microphones to be horizontal and pointing with their direction of maximum sensitivity in the direction that an occupant would normally be looking.
- Microphones to be no closer than 0.15 m to walls or upholstery.
- Microphones should be mounted in such a way as to be unaffected by vibrations of the vehicle.
- Tests shall be carried out with the vehicle being stationary (at idle and at full engine speed, held for 5 seconds), at various steady speeds in the range of 60–120 km hr^{-1}, and with full-load acceleration from 45% of the maximum power speed to 90% of the maximum power speed with the transmission in the highest position without exceeding 120 km hr^{-1}. Many practitioners make full load accelerations from near idle to near maximum engine speed in both second and third gears as their standard interior noise acceleration tests.

Most organisations have their own test regimes, and in many cases these vary from BS 6086. For this reason it is vitally important to note the microphone positions and orientations used in any test and the exact test conditions used when reporting test results.

The results of transient tests are commonly presented in the form of order diagrams or waterfall plots. The results of steady-state/speed tests are commonly presented in the form of narrow band or third octave band spectra.

21.1.1.4 On the subjective assessment of interior noise

BS 6086 presents a method for the objective assessment of interior noise. These assessments are commonly used to guide the process of vehicle refinement. Notwithstanding this, it is clear that the subjective impression of noise – both noise level (as assessed under BS 6086) and noise quality – is a factor in the process of selling cars.

The subjective assessment of vehicle interior noise can be used for different purposes (Van der Auweraer et al., 1997):

- to judge sound quality;
- to diagnose the causes of poor sound quality.

The judgement of sound quality is best performed by a panel of potential customers. A diagnosis is best made by a panel of refinement specialists. Different methodologies can be used for assessment or diagnosis and those designed for assessment will be discussed first.

Subjective assessment of interior noise quality can be performed in two ways:

1. The driving method where panel members answer questions relating to their experience of driving a vehicle (and perhaps some competitor vehicles).
2. The listening room method where panel members assess the quality of recorded sounds in a dedicated listening room or over headphones.

The driving method is the most complex and costly choice as, in order to be statistically significant, it involves many people undertaking extended test drives (safely) along predetermined test routes. If it is undertaken, one of two strategies can be followed. The first is to get drivers to rank sounds that they hear in order of preference. So, if only one vehicle is being assessed, the panel member could rank the following sound quality descriptors according to which they best described their experience:

Quiet
Luxurious
Powerful
Sporty
Pleasant
Unpleasant
Noisy
Commonplace
Unusual

If competitor vehicles are being tested a relative ranking can be produced, for example: car A is quieter than car B; car B is more sporty than car A, etc. In an alternative method of ranking, panel members answer paired questions like: *which best describes the noise level – quiet or noisy?*

Instead of ranking tests, panel members can be asked to rate (usually on a scale of 1–10) each descriptor. A score of 0 on the 'quiet' descriptor would indicate a noisy vehicle and a score of 10 would indicate a very quiet vehicle. This magnitude estimation often proves difficult for untrained members of the public. An alternative is to set a scale for paired questions such as:

Is the vehicle

Quiet? Noisy?

Extremely, very, somewhat, neither, somewhat, very, extremely.

and a semantic differential system of questioning is obtained (Otto et al., 1999).

The listening room method is different. Here an artificial head (and torso) is placed in a passenger seat of the vehicle and binaural recordings of the interior noise are made. Thereafter, ranking, paired ranking, magnitude estimation or semantic differential questioning can be undertaken for a statistically significant number of panel members in a room (see Johnson (1995) for example). Each panel member can be tested separately, or juries often persons can be assembled. Otto et al. (1999) describe the process well.

Apart from safety and cost, one significant advantage that the listening room method has over the driving method is that the recorded sounds can be manipulated digitally and the altered sounds used to establish customer preferences without the cost of constructing prototype vehicles. Russell et al. (1988, 1992) describe an early system with more sophisticated versions being demonstrated by Maunder (1996) and Naylor and Willats (2000). Such advanced digital techniques can be used by panels of refinement specialists to diagnose the causes of poor sound quality. Otherwise, magnitude estimations for the driving method are commonly used when the panel has the opportunity to drive a candidate vehicle. Most involve a panel of people driving and riding in the car(s) along a predetermined test route on public roads and rating the following noise (and vibration) attributes:

- wind noise
- road noise
- engine noise
- idle refinement
- cruising refinement
- transmission noise
- general shakes and vibrations
- squeaks, rattles and tizzes
- ride quality
- driveability
- noise that is a 'feature' (sporty exhaust notes, etc.).

The ratings are made to a common scale from 1 to 10 as shown in Table 21.1-1.

- A rating of less than 4 is unacceptable for any attribute.
- A rating of 5 or 6 is borderline.
- A rating of 7 or more on any attribute is acceptable.

Most new cars are launched with a subjective rating of 7 or 8 on most attributes. Russell et al. (1992) supply supplementary attributes for rating noise from particular sources such as diesel engines (called dimensions in the reference):

- overall level;
- low-frequency content and 'boom';
- impulsiveness;
- harmonic content;
- strong tones;
- irregularity;
- high-frequency content.

21.1.2 Noise path analysis

21.1.2.1 Background

Interior noise levels can be controlled under certain circumstances by adding sound-absorbing material to the passenger compartment. The advantage in controlling sound by absorption in the vehicle interior is that it will take effect on noise in a certain frequency range regardless of origin or noise path providing that the receiver is at some distance from the source. The disadvantage is that its effect is usually rather small unless the vehicle had little sound-absorbing trim in the first place.

As an alternative, a noise path analysis may be used to determine the contributions to interior noise levels made by noise using different paths between the source(s) and the vehicle interior. Depending on the dominant noise path(s) identified, the following noise control options are available:

Structure-borne noise from the engine

- Improve the vibration isolation provided by engine mounts.
- Reduce the vibration produced by the engine.
- Add damping treatments to resonant portions of the firewall and floor.

Airborne noise from the engine

- Improve the TL of the firewall and/or floor by adding a barrier layer (Wentzel and Saha, 1995), commonly a mat of EVA, PVC or natural rubber (surface density 1–7 kg m^{-2}) glued to a decoupling layer of chip foam or something similar (volumetric density in the range of 20–60 kg m^{-3}) or a fibrous matting (density in the range of 60–80 kg m^{-3}).

Table 21.1-1 Common subjective rating scheme

1	2	3	4	5	6	7	8	9	10
Not acceptable			Objectionable	Requires improvement	Medium	Light	Very light	Trace	No trace

- Plug all gaps in the firewall caused by ill-fitting grommets, etc.
- Add sound-absorption treatment to the underside of the hood (bonnet).

Structure-borne noise from the road

- Change tyres.
- Change suspension bushes.
- Change subframe bushes (if an isolated subframe is used).
- Add damping treatments to resonant portions of the firewall and floor.

Airborne noise from the tyres

- Change tyres.
- Improve the TL of the firewall and/or floor by adding a barrier layer along with a decoupling layer.
- Improve the door seals if necessary.

Structure-borne noise from the exhaust

- Improve the vibration isolation afforded by the mounts by using more compliant mounts, fixed to high impedance points on the chassis and nodal points on the exhaust system.
- Improve the TL of the trunk and/or rear floor by adding a barrier layer along with a decoupling layer.
- Add damping treatments to resonant portions of the trunk and floor.
- Add a flexible coupling between the exit of the catalyst and the remainder of the system.

Airborne noise from the exhaust

- Improve the TL of the trunk and/or rear floor by adding a barrier layer along with a decoupling layer.
- Improve the door seals if necessary.

Structure-borne noise from the intake

- Mount the body-side elements of the intake system (filter box and snorkel usually) on resilient mounts.

Airborne noise from the intake

- Improve the TL of the firewall and/or floor by adding a barrier layer along with a decoupling layer.
- Improve the door seals if necessary.
- Add sound-absorption treatment to the underside of the hood.

Aerodynamic noise

- Re-contour wing mirrors, aerials, door-handles, etc.
- Improve the door seals if necessary.

Engine component noise

- Reduce at source by adopting a quieter component.
- Improve the TL of the firewall and/or floor by adding a barrier layer along with a decoupling layer.
- Add sound-absorption treatment to the underside of the hood.

21.1.2.2 Coherence methods for noise path analysis

In the first instance, readers will need to understand the significance of the terms coherence and frequency response function. To do this, they are directed to:

- a review of some background materials on systems in Appendix 21.1A;
- an explanation of the convolution integral in Appendix 21.1B;
- explanations of the covariance function, correlation and coherence given in Appendix 21.1C;
- the derivation of the frequency response function given in Appendix 21.1D;
- Sinha (1991), Fahy and Walker (1998) and texts similar to Weltner et al. (1986) for further reading.

This section studies one noise source identification method (or noise path analysis method) based on measured coherence – that proposed by Halvorsen and Bendat (1975).

The analysis starts with their linear single input, single output (two pole – Appendix 21.1A) problem as illustrated in Fig. 21.1-2.

If data were acquired at point A in Fig. 21.1-2, then

$$x(t) = u(t) + n(t) \qquad (21.1.2)$$

would be recorded, where u is the wanted input and n is some unwanted but unavoidable noise.

Equally, if data were acquired at point B in Fig. 21.1-2, then

$$y(t) = v(t) + m(t) \qquad (21.1.3)$$

would be recorded, where v is the wanted input and m is some unwanted but unavoidable noise.

Now one can write down the following relationships (where $G(f)$ is a one-sided spectrum):

$$G_{xx}(f) = G_{uu}(f) + G_{nn}(f) \qquad (21.1.4)$$

$$G_{yy}(f) = G_{vv}(f) + G_{mm}(f) \qquad (21.1.5)$$

$$G_{xy} = G_{uv}(f) \qquad (21.1.6)$$

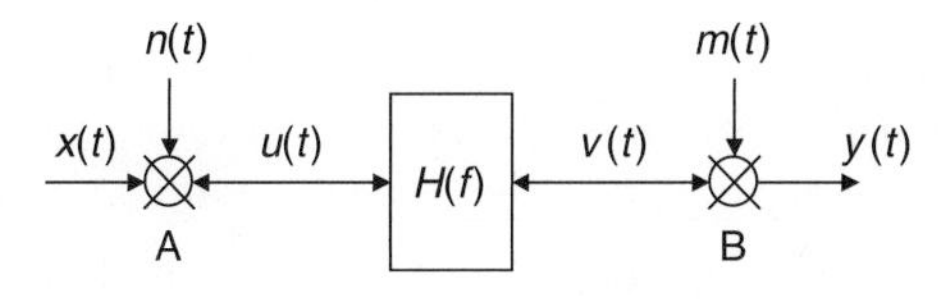

Fig. 21.1-2 Halvorsen and Bendat's single input/single output problem (1975).

because $n(t)$ and $m(t)$ are assumed to be uncorrelated with $u(t)$ and $v(t)$

$$G_{vv}(f) = |H(f)|^2 G_{uu}(f) \quad \text{from Appendix 21.1D} \tag{21.1.7}$$

$$G_{uv}(f) = H(f)G_{uu}(f) \quad \text{from Appendix 21.1D} \tag{21.1.8}$$

$$\gamma_{xy}^2(f) = \frac{|G_{xy}(f)|^2}{G_{xx}(f)G_{yy}(f)} \quad \text{from Appendix 21.1C} \tag{21.1.9}$$

The coherence function given by equation (21.1.9) is the measured coherence not the true coherence.

If the measured coherence is multiplied by the measured output power spectrum, the coherent output power spectrum is obtained:

$$\gamma_{xy}^2(f)G_{yy}(f) = \frac{|G_{xy}(f)|^2}{G_{xx}(f)} \tag{21.1.10}$$

Substituting equation (21.1.6) into equation (21.1.10)

$$\gamma_{xy}^2(f)G_{yy}(f) = \frac{|G_{uv}(f)|^2}{G_{xx}(f)} \tag{21.1.11}$$

Substituting equation (21.1.4) into equation (21.1.11)

$$\gamma_{xy}^2(f)G_{yy}(f) = \frac{|G_{uv}(f)|^2}{G_{uu}(f) + G_{nn}(f)} \tag{21.1.12}$$

Substituting equation (21.1.8) into equation (21.1.12)

$$\gamma_{xy}^2(f)G_{yy}(f) = \frac{|H(f)G_{uu}(f)|^2}{G_{uu}(f) + G_{nn}(f)} = \frac{|H(f)|^2 G_{uu}^2(f)}{G_{uu}(f) + G_{nn}(f)} \tag{21.1.13}$$

Substituting equation (21.1.7) into equation (21.1.13)

$$\gamma_{xy}^2(f)G_{yy}(f) = \frac{G_{vv}(f) \cdot G_{uu}(f)}{G_{uu}(f) + G_{nn}(f)} = \frac{G_{vv}(f)}{1 + \left(\frac{G_{nn}(f)}{G_{uu}(f)}\right)} \tag{21.1.14}$$

So, the measured coherent output power spectrum will yield a good measure of the true system output spectrum $G_{vv}(f)$ providing the input signal-to-noise ratio is high. This method is not commonly used for automotive noise path analysis for the following reasons:

- $n(t)$ and $u(t)$ are often correlated when they result from the same source – like the engine. For example, the vibration experienced at the engine mount under study will contain contributions (the noise $n(t)$) provided by the same engine via other nearby engine mounts.
- $m(t)$ and $v(t)$ are often correlated. For example, the sound in the cabin due to transmission of vibration power through one engine mount is partially correlated with the sound due to power transmitted through the other mounts.
- The noise paths are often non-linear, particularly when transmission of vibration power via rubber components is concerned.
- Delays between $u(t)$ and $v(t)$ result in low estimates of coherence due to a lack of properly synchronous sampling. The effects of delays can be minimised by using sample data lengths that are much longer than the longest delay.

See Piersol (1978) and Verhulst and Verheij (1979) for further solutions to these problems.

21.1.2.3 Standard methods for noise path analysis

There are standard measurement methods for noise path analysis in automotive vehicles that overcome (to some degree at least) the limitations discussed for the coherent output power method.

Both the LMS (1998) and the I-DEAS (MSX 1998) measurement systems commonly used in the automotive industry offer methods for noise path analysis. These are broadly similar, allowing the user to choose between:

- the complex stiffness method (I-DEAS call this the 'force vector method'); and
- the matrix inversion method (I-DEAS call this the 'full matrix method').

Both methods are based on the same principle: that the received sound pressure level (or vibration acceleration) during operational conditions is the superposition of partial results, each describing the contribution of individual transfer paths. Therefore:

$$r(f) = \sum_{i=1}^{n} \frac{R(f)}{S_i(f)} \cdot S_i(f) \tag{21.1.15}$$

where

$r(f)$ = received power spectral density
$R(f)/S_i(f)$ = frequency response function between the received power spectral density and the input power spectral density applied to transfer path i
$S_i(f)$ = input power spectral density of operational force or operational volume velocity applied to transfer path i.

The complex stiffness method (force-vector method) is suitable for occasions where the source of input power is connected to the receiver via sprung or compliant mounts and there is a reasonable differential movement across each mount. For example, the complex stiffness method is commonly applied to the engine mounting problem thus:

1. Identify all probable vibration paths between the source (the engine) and the receiver (a microphone positioned in the passenger compartment).
2. Organise a phase reference signal for use in subsequent measurements. This might be an electrical signal taken from the tachometer or the conditioned output from an accelerometer attached to the engine block.
3. On the test track, or the rolling-road dynamometer, measure acceleration levels in the three axes (x, y, z) on the engine side of the engine mounts. Analyse these levels either as power spectral densities or as 'order-levels'. Order levels may be determined either by synchronous sampling relative to the reference phase signal or they may be extracted from the time-averaged power spectral density (in which case the magnitude of the order level is taken as the average of three spectral lines, whereas the phase information is taken from the middle spectral line).
4. Repeat (3) for the body side of each engine mount. Ideally steps (3) and (4) should be undertaken simultaneously.
5. Remove the engine from the vehicle and measure the input accelerances in the three axes (x, y, z) on the body side of the mounts due to force excitation applied in each of the three axes in turn. The force excitation is applied with either a shaker or using a hammer fitted with a force transducer at its tip. The accelerance between the acceleration (m s^{-2}) at point i due to a force applied at point j is given by

$$\text{accelerance} = \frac{a_{ij}}{F_j} \qquad (21.1.16)$$

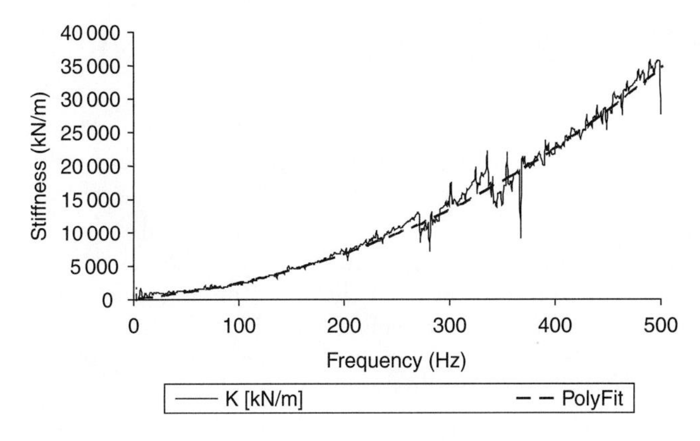

Fig. 21.1-3 Measured engine mount stiffness (Verstraeten, 2003).

6. With the engine still removed from the vehicle, measure the noise transfer functions (NTFs)

$$\text{NTF} = \frac{P}{F_i} \quad P \text{ is the sound pressure (Pa)} \qquad (21.1.17)$$

using the force excitation from (5) and a microphone in the passenger compartment.
7. Measure the dynamic stiffness of each engine mount along with its damping. These may be combined to give a complex stiffness (Cremer and Heckl, 1988)

$$\overline{K} = K(1 + i\eta) \qquad (21.1.18)$$

where K is the dynamic stiffness (N m^{-1}) and η is the loss factor (dimensionless). A typical example of the results from a measurement of engine mount stiffness is shown in Fig. 21.1-3.
8. Double integrate the engine side accelerations from (3) and the body side accelerations from (4) and estimate the differential displacements of the mounts. The force applied to the body at each mount is therefore

$$F_i(f) = \overline{K}(f)\cdot(x_s(f) - x_r(f)) \qquad (21.1.19)$$

where x_s is the displacement on the engine side of the mount (the source side) and x_r is the displacement on the body side (the receiver side).
9. Use the forces obtained from (8) along with the NTF from (6) to estimate the partial sound pressure in the passenger compartment due solely to vibration power transmitted across each engine mount.
10. Sum all of the partial sound pressure contributions from all of the identified noise paths on a polar plot (or Nyquist diagram) and compare the result with the measured total sound pressure for validation of the method.

The use of a polar plot for displaying the results is important as it shows both magnitude and phase information. Therefore, interference effects between the contributions from different paths may be investigated.

For noise paths (transfer paths) comprising only rigid (or fairly rigid) connections, the complex stiffness method is not suitable as there will be negligibly small relative displacement across the connections.

In such cases, the force imposed by the source at each input to the system is determined from the inverse of the full accelerance matrix multiplied by the accelerations measured on the receiver side of the first connection under operational conditions. This is the matrix inversion method (full matrix method), that is:

$$[F] = [A]^{-1}[\ddot{x}] \tag{21.1.20}$$

where

$$[A] = \begin{bmatrix} \frac{\ddot{x}_{11}}{F_1} & \frac{\ddot{x}_{12}}{F_2} & \cdots & \frac{\ddot{x}_{1n}}{F_n} \\ \frac{\ddot{x}_{21}}{F_1} & \frac{\ddot{x}_{22}}{F_2} & \cdots & \cdots \\ \vdots & \vdots & \vdots & \vdots \\ \frac{\ddot{x}_{m1}}{F_1} & \cdots & \cdots & \frac{\ddot{x}_{mn}}{F_n} \end{bmatrix}$$

is the accelerance matrix

$$[\ddot{x}] = \begin{bmatrix} \ddot{x}_1 \\ \ddot{x}_2 \\ \vdots \\ \ddot{x}_m \end{bmatrix} \quad \text{and} \quad [F] = \begin{bmatrix} F_1 \\ F_2 \\ \vdots \\ F_n \end{bmatrix}$$

For a unique solution, the number of measured responses m must be at least equal to the number of identified forces n. If $m > n$ then the equation is over-determined and a least square estimate is found.

The matrix inversion method may be applied to the engine mounting problem thus:

1. Identify all probable vibration paths between the source (the engine) and the receiver (a microphone positioned in the passenger compartment).
2. Organise a phase reference signal for use in subsequent measurements. This might be an electrical signal taken from the tachometer or the amplified output from an accelerometer attached to the engine block.
3. On the test track, or the rolling dynamometer, measure acceleration levels in the three axes (x, y, z) on the body side of the engine mounts. Analyse these levels either as power spectral densities or as 'order-levels'.
4. Remove the engine from the vehicle and measure the input accelerances in the three axes (x, y, z) on the body side of the mounts due to force excitation applied in each of the three axes in turn. The force excitation is applied with either a shaker or using a hammer fitted with a force transducer at its tip.
5. With the engine still removed from the vehicle, measure the NTF using the force excitation from (4) and a microphone in the passenger compartment.
6. Estimate the forces applied at each engine mount from the inverse of the full accelerance matrix obtained from (4) multiplied by the body side accelerations measured in (3).
7. Use the forces obtained from (6) along with the NTF from (5) to estimate the partial sound pressure in the passenger compartment due solely to vibration power transmitted across each engine mount.
8. Sum all of the partial sound pressure contributions from all of the identified noise paths on a polar plot (or Nyquist diagram) and compare the result with the measured total sound pressure for validation of the method.

21.1.2.4 Non-invasive methods for noise path analysis

The practical difficulty with both the complex stiffness and the matrix inversion method is that the source must at some point be disconnected from the receiver. There are at least two further methods in the literature that do not require this disconnection.

The first, the TopExpress MPSD system (Harper et al., 1993) simplifies the equations of motion by assuming that the various transmission paths can only interact via their attachments to the source and/or the structure as illustrated in Fig. 21.1-4. The simplified matrix equation takes account of forces exerted on the paths by the source and forces exerted on the source by the paths. The matrix equation may be solved for the forces at the support ends of the structure, and the measured NTF may then be used to calculate radiated sound.

Another non-invasive technique originates from TNO (Netherlands) and ISVR (Southampton) – the equivalent forces method (Janssens et al., 1999).

Here, transfer functions

$$A_{ij} = \frac{a_i}{F_j} \tag{21.1.21}$$

are measured at convenient positions on the machine along with NTF

$$H_{kj} = \frac{P_k}{F_j} = \frac{a_j}{Q_R} \tag{21.1.22}$$

where P_k is the acoustic pressure, F_j is the force and Q_R is the volume acceleration and a_j is the acceleration.

The inventive part of the method is an analytical step of determining the amplitude and phase of a set of equivalent forces $\{F_j\}_{EQ}$ which when applied at the positions j would reconstruct the vibration field due to the machine $\{a_i\}_{mach}$ as closely as possible

$$\{F_j\}_{EQ} = [A_{ij}]^{-1}\{a_i\}_{mach} \tag{21.1.23}$$

$$\{P_k\}_{path} = [H_{kj}]\{F_j\}_{EQ} \tag{21.1.24}$$

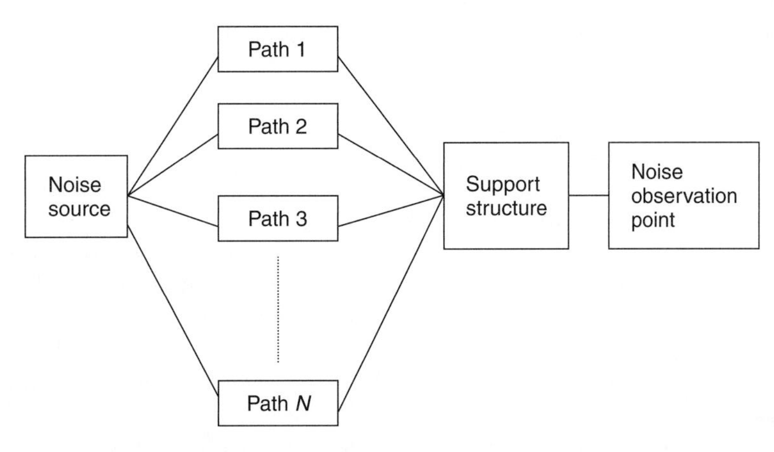

Fig. 21.1-4 Model of the TopExpress MPSD system: after Harper et al. (1993).

Because $\{F_j\}_{EQ}$ may be placed anywhere convenient on the system, they have no particular significance and are of little interest in themselves. For the standard (complex stiffness or matrix inversion) noise path analysis methods, the NTFs are measured at the connection between the source and the receiver and therefore $\{F_j\}$ has physical significance.

21.1.3 Measuring the sound power of IC engines and other vehicle noise sources

21.1.3.1 Near and far acoustic fields

Before moving on to discussing methods by which sound power may be measured, certain restrictions concerning the acoustic near field of the sound source must be clarified.

The sound field produced by a sound source radiating in a free field may be divided into three regions (Bies and Hansen, 1996):

1. the hydrodynamic near field;

2. the geometric (or Fresnel) near field;

3. the far field.

The hydrodynamic near field is that region immediately adjacent to the vibrating surface of the source. The thickness of the hydrodynamic near field is much less than one wavelength. In this region, the fluid motion is not directly associated with sound propagation. Tangential fluid motion may occur. The local particle velocities will be out of phase with the acoustic pressures. As the propagation of sound to the far field is associated with the in-phase components of pressure and particle velocity, it follows that measurements of the acoustic pressure amplitude in the hydrodynamic near field give no indication of the sound power radiated to the far field by the source.

The sound field adjacent to the hydrodynamic near field is known as the geometric near field. In this field the particle velocities are in phase with the pressure levels and so sound propagates through the geometric near field. However, the sound pressure levels do not decrease monotonically at the rate of 6 dB per doubling of distance from the source. In fact, variations in the surface velocity of the source produce interference patterns in the geometric near field which result in a distribution of sound pressure maxima and minima.

This effect is more noticeable in tonal noise than in broadband noise. Radiated sound power can be estimated from a sufficient number of pressure measurements made in the geometric near field. However, interpretation of the measurements for directivity information should be made only with caution.

The region of the sound field extending beyond the geometric near field is known as the far field. In open free-field space, sound pressure levels decrease monotonically at a rate of 6 dB per doubling of distance in the far field. Directivity in the far field is well defined. According to Bies (1976) and reported in Bies and Hansen (1996) the far field is characterised by the satisfaction of three criteria:

$$r \gg \lambda/2\pi \tag{21.1.25}$$

$$r \gg l \tag{21.1.26}$$

$$r \gg \frac{\pi l^2}{2\lambda} \tag{21.1.27}$$

where

r = distance between source and measurement position (m)

l = maximum source dimension (m)

λ = wavelength (m)

In this case the criteria ≫ is taken to mean a factor greater than 3.

It should be noted that the use of these criteria will ensure that a position is in the far field, but occasionally the far field will also occur nearer to the source. For instance, a large pulsating sphere only has a far field.

The adoption of these criteria can be difficult in anechoic chambers. For instance, a large source in a small anechoic chamber may only in theory produce far field conditions near the walls of the chamber. The absorption characteristics of the walls may disturb the field adjacent to the walls, and so the usable far field region may be quite small.

When the sound source is radiating in a reverberant or semi-reverberant space, the monotonic decrease in sound pressure level may be reduced in the far field due to the influence of the reverberant field.

21.1.3.2 Different methods for measuring sound power

In circumstances where the control of noise from existing machinery is to be considered, it is often advantageous to measure the sound power output of the source.

Sound power can be calculated or estimated from either:

1. pressure measurements in the far field;
2. pressure measurements in the geometric near field;
3. intensity measurements;
4. surface vibration velocity measurements.

The most precise laboratory methods involve far field pressure measurements made in anechoic or reverberant chambers. The next most precise method uses far field pressure measurements made in quiet semi-reverberant spaces in the field. Where background noise levels cause difficulties, a less precise method using near field pressure measurements may be used.

As an alternative to pressure measurements, realistic measures of sound power may be made in the field or laboratory with careful use of intensity techniques. In cases where there are several noise sources, or the use of pressure or intensity techniques is not practical, estimates of sound power can be made from measurements of surface vibration velocity.

21.1.3.3 Measurement of sound power in a free field using sound pressure techniques

The sound power (W, watts) of a source may be obtained from the integration of the intensity $I\,(\mathrm{W\,m^{-2}})$ over a notional spherical surface of area S ($\mathrm{m^2}$) surrounding it, that is:

$$W = IS = 4\pi r^2 I \quad (21.1.28)$$

The integration of the acoustic intensity over the spherical surface is achieved by determining the time averaged, squared acoustic pressures at a number of measurement points arranged to uniformly sample the integration surface.

In an environment in which there are no reflecting surfaces, the maximum intensity is given by Bies and Hansen (1996) as:

$$I_{\max} = \frac{1}{T}\int_0^T p\cdot u\,dt \quad (21.1.29)$$

where p is the fluctuating (acoustic) pressure (Pa) and u is the particle velocity ($\mathrm{m\,s^{-1}}$) and

$$I_{\max} \simeq \frac{\bar{p}^2_{\mathrm{rms}}}{\rho_0 c} \quad (21.1.30)$$

ρ_0 is the undisturbed density of fluid (air) ($\mathrm{kg\,m^{-3}}$)
c is the speed of sound in the fluid (air) ($\mathrm{m\,s^{-1}}$)

It can be shown that for free-field conditions at room temperature

$$L_p \simeq L_I \quad (21.1.31)$$

where the reference sound pressure level is 20×10^{-6} Pa and the reference intensity is $10^{-12}\,\mathrm{W\,m^{-2}}$.

Therefore, the spatial average intensity is:

$$\bar{L}_p \simeq \bar{L}_I = 10\log_{10}\left[\frac{1}{N}\sum_1^N 10^{L_{pi}/10}\right]\mathrm{dB} \quad (21.1.32)$$

where L_{pi} is the sound pressure level (dB) recorded within one portion of the equiparti-tioned surface of a sphere that completely encloses the sound source in the far field.

So, for a free-field

$$L_w = \bar{L}_p + 10\log_{10}(4\pi r^2)(\mathrm{dB})$$
$$L_w = \bar{L}_p + 20\log_{10}r + 10\log_{10}(4\pi)(\mathrm{dB})$$
$$L_w \simeq \bar{L}_p + 20\log_{10}r + 11(\mathrm{dB}) \quad (21.1.33)$$

For the case of a sound source in a free field but placed next to a flat surface such as a hard floor

$$L_w = \bar{L}_p + 10\log_{10}(2\pi r^2)(\mathrm{dB})$$
$$L_w \simeq \bar{L}_p + 20\log_{10}r + 8(\mathrm{dB}) \quad (21.1.34)$$

For the case of a sound source placed near the junction of two flat planes

$$L_w \simeq \bar{L}_p + 20\log_{10}r + 5(\text{dB}) \qquad (21.1.35)$$

For the case of a sound source placed near the junction of three flat planes

$$L_w \simeq \bar{L}_p + 20\log_{10}r + 2(\text{dB}) \qquad (21.1.36)$$

If the measurement surface is in the far field then the directivity index corresponding to location i is:

$$DI = L_{pi} - \bar{L}_p(\text{dB}) \qquad (21.1.37)$$

The requirements for a space in which such measurements of sound power may be made are as follows:

1. Any reverberant sound is negligible.
2. A semi-anechoic space may be used if convenient providing that a suitable correction is made.

The procedure for the measurement of sound power using pressure measurements in a free field is as follows:

1. Install the sound source in a sufficiently large anechoic or semi-anechoic space so that free-field conditions are met at a suitable distance from the absorbent surfaces of the space.
2. Calculate the minimum propagation distance from the source required to reliably expect free-field conditions.
3. Construct a notional spherical surface in the free-field with the acoustic (or geometric) centre of the source located at the centre of the sphere (a piece of string can be used for this!).
4. Divide the surface of the sphere into a reasonable number of equal area sections (perhaps 20).
5. Make time-averaged measurements of L_{pi}.
6. If any area (S_i) is of a different size to the usual (S_n) then apply this correction

$$L_{pi,\,corr} = L_{pi} + 10\log_{10}\left[\frac{S_i}{S_n}\right](\text{dB}) \qquad (21.1.38)$$

7. Calculate L_w as described earlier.

21.1.3.4 Measurement of sound power in a diffuse acoustic field

In a diffuse acoustic field, the net sound intensity at a point is zero. Therefore, the intensity-based methods used to determine sound power in a free field cannot be used. Instead, sound power is determined by using a number of pressure measurements to estimate the spatial energy average in the room combined with knowledge of the absorption characteristics of the space.

To provide a diffuse field a test room should fulfil the following characteristics:

- The room should be of adequate volume and suitable shape.
- The boundaries over the frequency range of interest can be considered acoustically hard.

The volume of the room should be large enough so that the number of normal modes of vibration in the octave of third octave frequency band of interest is enough to provide a satisfactory state of sound diffusion. It is common to expect at least 20 modes in the lowest frequency band where Morse and Bolt (1944) suggest that:

$$N = \frac{4\pi f^3 V}{3c^3} + \frac{\pi f^2 S}{4c^2} + \frac{fL}{8c} \qquad (21.1.39)$$

where

c = speed of sound (m s^{-1})
V = the room volume (m^3)
S = total room surface area (m^2)
L = total perimeter of the room (m) (in a rectangular room this is the sum of lengths of all the edges).

In order to estimate the average number of modes in a narrow frequency band, the derivatives of this equation can be used to yield a modal density, that is:

$$\frac{dN}{df} = \frac{4\pi f^2 V}{c^3} + \frac{\pi f S}{2c^2} + \frac{L}{8c} \qquad (21.1.40)$$

Modal density increases with the square of frequency and so a large number of modes are expected at high frequencies with only a small number of modes at low frequencies. Therefore, one can expect spatial variations in sound pressure levels at low frequencies, but at high frequencies the fluctuations become small and the field becomes more diffuse.

The shape of the test room should be such that the ratio of any two dimensions is not close to an integer. A common ratio is 2:3:5. The sound power of a source in a reverberant room is determined from the spatially averaged sound pressure level. A minimum number of six measurements would be expected for inclusion in the average. These measurements should be at locations:

- more than a quarter wavelength from any reflecting surface;
- more than one half wavelength apart;
- away from the source's direct field.

The microphone may be moved manually, or traversed linearly across the room or mounted on a continuously rotating circular boom. The number of measurement locations can be reduced using a rotating diffuser in the room particularly if the source emits narrow band or tonal noise. An accuracy of measurement of ±0.5dB is reasonable.

The simplest method of determining sound power in a reverberant room is to use the substitution method whereby the spatially averaged sound pressure level $\overline{L}_p$ resulting from the source of unknown power output level L_w is compared to the spatially averaged sound pressure level L'_p resulting from a source of known power output L'_w.

Therefore,

$$L_w = L'_w + (\overline{L}_p - L'_p) \qquad (21.1.41)$$

The absolute method may be used as an alternative to this, whereby the sound power is obtained from the spatially averaged sound pressure level $\overline{L}_p$ and the determination of the absorption characteristic of the room (Ver and Holmer, 1971):

$$L_w = \overline{L}_p + 10\log_{10} V - 10\log_{10} T_{60} + 10\log_{10}\left(1 + \frac{S\lambda}{8V}\right) - 13.9\ \text{dB, re}10^{-12}\text{W} \qquad (21.1.42)$$

where

V = Room volume (m^3)
T_{60} = Time taken for the sound pressure level to drop by 60 dB (s)
S = Surface area of the room (m^2)
λ = Wavelength corresponding to the centre frequency in the analysis band (m)

21.1.3.5 Measurement of sound power in a semi-reverberant far field

Most rooms containing noise sources are neither anechoic nor truly reverberant. When determining the sound power of a source in such a space no assumptions are made about the nature of the field (unlike before, where the field was assumed to be either all direct or all diffuse). The only requirements for the room are:

1. The sound source should be in its normal position.
2. The room should be large enough to allow measurements to be made in the far field.
3. The microphone should be kept at least one quarter wavelength away from any reflecting surface of the room.

If the source is located on a hard floor, and can be moved, and is at least a half wavelength from any other reflective surfaces, a version of the substitution method described earlier may be used to determine sound power. First the spatially averaged sound pressure level $\overline{L}_{p1}$ is obtained over a hemispherical surface around the source. Then the source is moved away and replaced with a reference source of power output L_{W2}, and the exact procedure is repeated to obtain $\overline{L}_{p2}$. Therefore, the sound power output L_{W1} of the source is:

$$L_{W1} = L_{W2} + (\overline{L}_{p1} - \overline{L}_{p2})\ (\text{dB}) \qquad (21.1.43)$$

The movement of the source is vital as it allows both sets of pressure measurements to be made at the exactly same locations and therefore with the same mix of direct and reverberant fields.

If the source cannot be moved, then the substitution method is excluded and hence a knowledge of the spatial distribution of direct and reverberant fields must be determined before sound power estimates are made.

The sound pressure level $\overline{L}_{p1}$ averaged over a portion of a sphere around a sound source of acoustic power output L_{W1} is:

$$\overline{L}_{p1} = L_{W1} + 10\log_{10}\left[\frac{D}{4\pi r^2} + \frac{4}{R}\right]\ (\text{dB}) \qquad (21.1.44)$$

where the directivity constant is:

$D = 1$ for a sphere
$D = 2$ for a hemisphere
$D = 4$ for a quarter of a sphere etc.
and R is the room constant (Kutruff, 1979); and

$$R = \frac{S\overline{\alpha}}{1 - \overline{\alpha}} \qquad (21.1.45)$$

where

S = total room area (m^2)
$\overline{\alpha}$ = average Sabine absorption in the room

The room constant may be taken to be the total absorption of the room measured in units of area (m^2).

If $\overline{L}_{p1}$ is obtained over a hemisphere around a reference source of known sound power output L_{W1}, then the contribution to sound pressure level over the test surface made by the direct field is:

$$\overline{L}_{pd} = L_{W1} + 10\log_{10}\left[\frac{2}{4\pi r^2}\right](\text{dB})$$

$$\overline{L}_{pd} = L_{W1} + 10\log_{10}\left[\frac{1}{r^2}\right] + 10\log_{10}\left[\frac{2}{4\pi}\right]\ \text{dB}$$

$$\overline{L}_{pd} = L_{W1} - 20\log_{10} r - 8\ \text{dB} \qquad (21.1.46)$$

or

$$\bar{L}_{pd} = L_{W1} + 10\log_{10}\left[\frac{1}{S_H}\right] \text{dB} \quad (21.1.47)$$

where S_H is the surface area of a hemisphere (m^2).

Now for a hemisphere

$$\bar{L}_{p1} = L_{W1} + 10\log_{10}\left[\frac{1}{S_H} + \frac{4}{R}\right] \quad (21.1.48)$$

$$10\log_{10}\left[\frac{1}{S_H} + \frac{4}{R}\right] = \bar{L}_{p1} - L_{W1} \quad (21.1.49)$$

$$\frac{1}{S_H} + \frac{4}{R} = 10^{[L_{p1} - L_{W1}]/10} \quad (21.1.50)$$

Now, it is known from equation (21.1.47) that:

$$\bar{L}_{pd} = L_{W1} + 10\log_{10}\left[\frac{1}{S_H}\right] \quad (21.1.47)$$

therefore,

$$L_{W1} = \bar{L}_{pd} - 10\log_{10}\left[\frac{1}{S_H}\right] \quad (21.1.51)$$

so,

$$\frac{1}{S_H} + \frac{4}{R} = 10^{[L_{p1} - (L_{pd} + \log_{10}(1/S_H))]/10}$$

$$\frac{1}{S_H} + \frac{4}{R} = \frac{10^{[L_{p1} - L_{pd}]/10}}{S_H}$$

$$\frac{4}{R} = \frac{1}{S_H}\left[10^{[L_{p1} - L_{pd}]/10} - 1\right] \quad (21.1.52)$$

Now that the room constant is known, $\bar{L}_p$ can be determined over a hemisphere around the sound source, and the sound power determined directly according to:

$$L_w = \bar{L}_p - 10\log_{10}\left[\frac{1}{2\pi r^2} + \frac{4}{R}\right] \quad (21.1.53)$$

A third alternative way of determining sound power levels involves obtaining the spatially averaged sound pressure levels $\bar{L}_{p1}$ and $\bar{L}_{p2}$ over two different hemispherical surfaces of areas S_1 and S_2 both of which have the acoustic source at their centres, that is:

$$\bar{L}_{p1} = L_w + 10\log_{10}\left[\frac{1}{S_1} + \frac{4}{R}\right] \quad (21.1.54)$$

$$\bar{L}_{p2} = L_w + 10\log_{10}\left[\frac{1}{S_2} + \frac{4}{R}\right] \quad (21.1.55)$$

L_w is to be determined. $\bar{L}_{p1}$ and $\bar{L}_{p2}$ are known, measured quantities. The term $4/R$ is of no inherent interest here, so it will be removed by algebraic manipulation, that is:

$$\frac{4}{R} = 10^{[L_{p1} - L_w]/10} - \frac{1}{S_1} \quad (21.1.56)$$

$$\frac{4}{R} = 10^{[L_{p2} - L_w]/10} - \frac{1}{S_2} \quad (21.1.57)$$

Therefore,

$$10^{[L_{p1} - L_w]/10} - \frac{1}{S_1} = 10^{[L_{p2} - L_w]/10} - \frac{1}{S_2} \quad (21.1.58)$$

$$10^{[L_{p1} - L_w]/10} - 10^{[L_{p2} - L_w]/10} = \frac{1}{S_1} - \frac{1}{S_2} \quad (21.1.59)$$

Multiply both sides by $10^{L_w/10}$,

$$10^{L_{p1}/10} - 10^{L_{p2}/10} = 10^{L_w/10}\left[\frac{1}{S_1} - \frac{1}{S_2}\right] \quad (21.1.60)$$

Take logarithms on both sides,

$$\log 10\left[10^{L_{p1}/10} - 10^{L_{p2}/10}\right] = \frac{L_w}{10} + \log_{10}\left[\frac{1}{S_1} - \frac{1}{S_2}\right] \quad (21.1.61)$$

Multiply both sides by 10

$$10\log_{10}\left(10^{L_{p1}/10} - 10^{L_{p2}/10}\right) = L_w + 10\log_{10}\left[\frac{1}{S_1} - \frac{1}{S_2}\right] \quad (21.1.62)$$

Now,

$$10\log_{10}\left(10^{L_{p1}/10} - 10^{L_{p2}/10}\right)$$

$$= 10\log_{10}\left[10^{L_{p2}/10}\left(10^{(L_{p1} - L_{p2})/10} - 1\right)\right]$$

$$\bar{L}_{p2} + 10\log_{10}\left[10^{(L_{p1} - L_{p2})/10} - 1\right]$$

$$= L_w + 10\log_{10}\left[\frac{1}{S_1} - \frac{1}{S_2}\right] \quad (21.1.63)$$

and finally

$$L_w = \bar{L}_{p2} - 10\log_{10}\left[\frac{1}{S_1} - \frac{1}{S_2}\right]$$

$$+ 10\log_{10}\left(10^{(L_{p1} - L_{p2})/10} - 1\right) \quad (21.1.64)$$

This equation can be used directly to determine the sound power from a source when

$$\bar{L}_{p1,2} = 10\log_{10}\left[\frac{1}{N}\sum_{i=1}^{N} 10^{(L_{pi}/10)}\right] \text{dB} \quad (21.1.65)$$

21.1.3.6 Measurement of sound power in the near field

All the earlier methods of determining sound power have required that

(i) the room is large enough for pressure measurements to be made in the far field;

(ii) background noise is negligible (i.e. the sound source produces an increase of at least 3 dB over background noise levels at all microphone positions).

Near field techniques can be used in cases that do not satisfy the criteria above. Near field techniques consist of making sound pressure level measurements in the near field of the source (typically 1 m away or less if reduced accuracy is acceptable) across a notional test surface (which is often parallel piped).

An average sound pressure level is obtained from N measurement locations, thus

$$\overline{L}_p = 10 \log_{10}\left[\frac{1}{N}\sum_{i=1}^{N} 10^{(L_{pi}/10)}\right] \quad (21.1.66)$$

and the following equation is used to determine an approximate value for sound power level (Pobol [1976], Jonasson and Elson [1981] and reported in Bies and Hansen [1996]):

$$L_w = \overline{L}_p + 10 \log_{10} S - \Delta_1 - \Delta_2 \quad (21.1.67)$$

where

S = area of test surface (m^2)

Δ_1 = correction factor to account for the absorption characteristics of the room

Δ_2 = correction factor to account for possible tangential wave propagation

Δ_1 can be obtained from

$$\Delta_1 = 10 \log_{10}\left[1 + \frac{4S_1}{A\overline{\alpha}}\right] \quad (21.1.68)$$

where

S_1 = test measurement surface area (m^2)

A = total area of room surfaces (m^2)

$\overline{\alpha}$ = mean acoustic Sabine absorption coefficient

Alternatively, L_{p1} and L_{p2} can be obtained for two concentric surfaces around the machine and

$$\Delta_1 = L_{p1} - L_{p2} - 10 \log_{10}\left[10^{(L_{p1}-L_{p2})/10} - 1\right] + 10 \log_{10}[1 - S_1/S_2] \quad (21.1.69)$$

Pobol (1976), reported in Bies and Hansen (1996), suggests the following values for Δ_1 in terms of the volume of the room over the area of the test surface shown in Table 21.1-2.

Table 21.1-2 Values for Δ_1 suggested by Pobol (1976)

	V/S (m)			
Usual room	20–50	50–90	90–3000	>3000
Highly reflective room	50–100	100–200	200–600	>600
Δ_1 dB	3	2	1	0

Johansson and Eston (1981), reported in Bies and Hansen (1996), suggest values for Δ_2 in terms of the ratio between the test surface areas S and to the area of the smallest parallelepiped surface Sm which just encloses the source. These are shown in Table 21.1-3.

Care should be taken to avoid errors due to the directional sensitivity of the microphone at higher frequencies. Note should be made as to whether a microphone has been calibrated for an expected angle or incidence (free-field calibration) which is a function of angle of incidence or whether a single random incidence calibration has been achieved. In either case, it is prudent to set an upper limit on frequency below which the effects of angle of incidence on microphone response are small.

21.1.3.7 Determination of sound power using surface vibration velocity measurements

The sound power being radiated by a vibrating structure can be estimated from a mean square vibration velocity averaged over the surface.

The radiated sound power is equal to:

$$W = \rho c S \sigma \langle v^2 \rangle \quad (21.1.70)$$

where

ρ = density of the air (1.2 kg m^{-3} is typical)

c = speed of sound (343 ms^{-1} is typical)

S = surface area (m^2)

σ = radiation ratio (efficiency = 100%, ratio = unity)

$\langle v \rangle$ = surface-averaged vibration velocity (m s^{-1})

Table 21.1-3 Values for Δ_2 suggested by Johansson and Eston (1981)

S/Sm	Δ_2 dB
1–1.1	3
1.1–1.4	2
1.4–2.5	1
>2.5	0

The sound power level (re 10^{-12} W) is thus:

$$L_W = 10\log_{10}\left[\frac{\rho c S\sigma\langle v^2\rangle}{10^{-12}}\right] \quad (21.1.71)$$

$$L_W = 10\log_{10}\left[\langle v^2\rangle\right] + 10\log_{10}S + 10\log_{10}\sigma + 10\log_{10}\left[\frac{\rho c}{10^{-12}}\right]$$

$$L_W \simeq 10\log_{10}\left[\langle v^2\rangle\right] + 10\log_{10}S + 10\log_{10}\sigma + 146\ \text{dB re}10^{-12}\text{W} \quad (21.1.72)$$

The determination of sound power from surface vibration measurements is an approximate method due to uncertainties in:

(i) the surface-averaged vibration velocity;

(ii) the radiation efficiency.

Ideally, the surface-averaged vibration velocity should be obtained using an accelero-meter and an integrating circuit, making a large number of narrow band measurements at points away from the edges of the structure, and averaging each band result as follows:

$$\overline{L}_{v_i} = 10\log_{10}\left[\frac{1}{N}\sum_{i=1}^{N}10^{L_{v_i}/10}\right] \quad (21.1.73)$$

However, care should be taken when using integrating circuits that the signal-to-noise ratio for the measurement chain is not too severely reduced. If this is the case, or if a reasonable quality integrating circuit is not available, then the vibration velocity can be estimated from the vibration acceleration as follows:

$$|v|e^{i\omega t} = \int_0^\infty |a|e^{i\omega t}dt \quad (21.1.74)$$

$$|v| = \frac{|a|}{i\omega} \quad (21.1.75)$$

$$|v^2| \approx \frac{a^2}{(2\pi f)^2} \quad (21.1.76)$$

f is the band centre frequency (Hz). If this approximate integration technique is used on third octave bands, then significant errors can occur at higher frequencies where the bandwidths are greater.

The radiation efficiency of a vibrating structure is notoriously difficult to determine. The key parameter is the critical frequency, which for uniform flat plates is:

$$f_c = c^2/1.8\,hc_L(\text{Hz}) \quad \text{(Fahy, 1985)} \quad (21.1.77)$$

where

h = plate thickness (m)
c_L = phase speed of longitudinal waves in plate (ms^{-1})
c = speed of sound in fluid (air) (ms^{-1})

The radiation efficiency is generally greater than 100% at frequencies around the critical frequency. At frequencies greatly above the critical frequency, the radiation efficiency is usually 100% whereas at frequencies greatly below the critical frequency the radiation efficiency can be as low as a fraction of 1%.

Radiation efficiencies can be predicted using boundary element techniques (SYSNOISE), or approximations may be obtained from generalised curves such as those presented by Vér and Holmer (1971), reported in Fahy (1985), reproduced in Fig. 21.1-5.

21.1.3.8 Determination of sound power using an intensity meter

It has already been stated that sound power is directly related to sound intensity by the bounding area, thus:

$$W = I\cdot S$$

The fact that for free-field conditions at room temperature

$$L_p \simeq L_i$$

was also employed when determining the sound power of a source using sound pressure measurements. This relationship breaks down for non-free-field conditions, but the relationship between sound power and sound intensity still holds. Therefore, if one can measure sound intensity directly then sound power may be determined even in non-free-field conditions.

Sound intensity is the long-term average integral of acoustic pressure and acoustic particle velocity

$$I = \lim_{T\to\infty}\frac{1}{T}\int_0^T p\cdot u\,dt \quad (21.1.29)$$

An intensity meter has been developed by Fahy and others (Fahy, 1989) which measures intensity directly by using two phase-matched microphones at a known spacing apart to infer acoustic particle velocity from the pressure gradient. It should be appreciated that sound intensity is a vector product having both magnitude and direction.

Sound power is determined from the intensity averaged over a notional surface placed around the source. The same restrictions on accuracy regarding the number

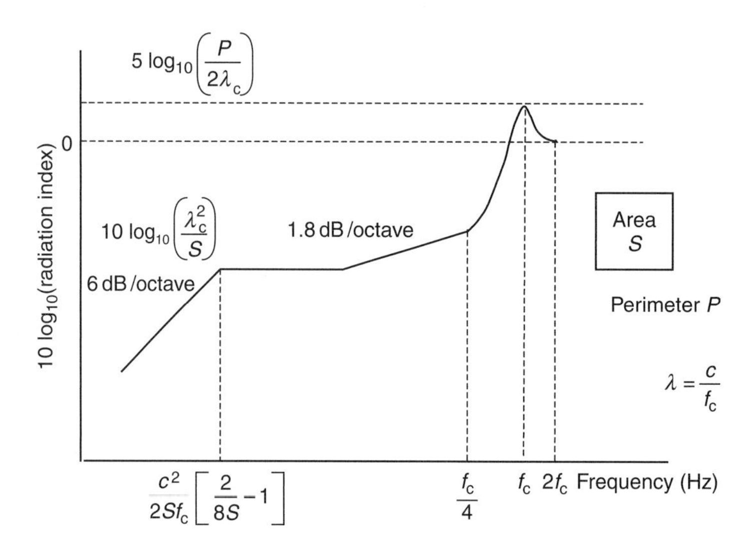

Fig. 21.1-5 Theoretical modal average radiation efficiency of a baffled rectangular panel: after Vér and Holmer (1971).

of measurement positions apply equally to intensity techniques as they do to sound pressure techniques.

The determination of sound power using an intensity meter has the following advantages over sound pressure techniques to the same aim:

1. Sound pressure techniques must assume free-field conditions where the sound is only travelling away from the source. The sound intensity meter determines the direction of propagation and therefore sound power may be determined even in the geometric near field.
2. Some areas of vibrating surfaces may act as radiators of sound while other areas may act as absorbers of sound. A sound intensity meter scanning a surface will detect such phenomena.
3. The output of an intensity meter gives magnitude and direction which can usefully be used for source location. The output from pressure measurements over a surface is less informative.

Although the intensity meter is a very flexible tool, there are conditions under which it will not perform well. These are mostly where pressures are large and the pressure gradient between the two microphones is small resulting in a poor estimate of particle velocity. Such a condition will occur near to highly reflective surfaces where the incident power is approximately equal to the reflected power. Therefore, the use of an intensity meter near to highly reflective surfaces should be avoided.

The other, unexpected, difficulty with sound intensity measurements is that due to the high detail of the output, a large number of measurements must be made at different locations to ensure that a realistic spatial average of sound intensity has been achieved. This problem gets worse the nearer to the source one gets. As long as background noise levels are not a problem, there are occasions where a more meaningful estimate of sound power can be obtained from a few quick far field sound pressure measurements rather than a detailed and laborious survey in the nearer field with an intensity meter.

21.1.3.9 Standard methods for measuring sound power under different circumstances

There are UK national (BS) and internationally (ISO) recognised standard methods for measuring sound power. They broadly follow the methods already discussed so far, and therefore will not be discussed further. The list of standards includes, but is not necessarily limited to (source www.bsi.org.uk).

BS 4196-0:1981 (ISO 3740:1980)
Sound power levels of noise sources. Guide for the use of basic standards and for the preparation of noise test codes.

BS 4196-1:1991 (EN 23741:1991 ISO 3741:1988)
Sound power levels of noise sources. Precision methods for determination of sound power levels for broad-band sources in reverberation rooms.

BS 4196-2:1991 (EN 23742:1991 ISO 3742:1988)
Sound power levels of noise sources. Precision methods for determination of sound power levels for discrete-frequency and narrow-band sources in reverberation rooms.

BS 4196-5:1981 (ISO 3745:1977)
Sound power levels of noise sources. Precision methods for determination of sound power levels for sources in anechoic and semi-anechoic rooms.

BS 4196-7:1988 (ISO 3747:1987)

Sound power levels of noise sources. Survey method for determination of sound power levels of noise sources using a reference sound source.

BS 4196-8:1991 (ISO 6926:1990)

Sound power levels of noise sources. Specification for the performance and calibration of reference sound sources.

BS EN ISO 3743-1:1995

Acoustics. Determination of sound power levels of noise sources. Engineering methods for small, movable sources in reverberant fields. Comparison for hard-walled test rooms.

BS EN ISO 3743-2:1997

Acoustics. Determination of sound power levels of noise sources. Engineering methods for small, movable sources in reverberant fields. Methods for special reverberation test rooms.

BS EN ISO 3744:1995

Acoustics. Determination of sound power levels of noise sources using sound pressure. Engineering method in an essentially free field over a reflecting plane.

BS EN ISO 3746:1996

Acoustics. Determination of sound power levels of noise sources using sound pressure. Survey method using an enveloping measurement surface over a reflecting plane.

BS EN ISO 9614-1:1995

Acoustics. Determination of sound power levels of noise sources using sound intensity. Measurement at discrete points.

BS EN ISO 9614-2:1997

Acoustics. Determination of sound power levels of noise sources using sound intensity. Measurement by scanning.

21.1.4 Engine noise

21.1.4.1 Introduction to engine noise

In this section, the term engine noise will be taken as the noise produced by a combination of the gas loads in the cylinders and the mechanical motions in the base engine. Intake and exhaust noise shall be considered as separate problems as will be the noise caused by engine ancillaries (alternators, fans, pumps, motors, etc.).

Engine noise is the sum of two elements:

1. combustion noise
2. mechanical noise.

The relative mix of the two will vary between engines but as a general rule:

- mechanical noise dominates the engine noise produced by spark ignition (gasoline) engines;
- combustion noise is a more significant contributor to the engine noise produced by compression ignition (diesel) engines.

Engine noise is dependent on engine speed and may also depend on engine load for some types of engine (the normally aspirated direct injection diesel and the gasoline engine in particular).

21.1.4.2 Combustion noise

Combustion noise results from gas forces in the cylinders applied to the structure of the engine, causing vibration to occur which is then radiated as noise. It is produced therefore by an indirect noise-generating mechanism.

The gas forces in each cylinder vary during the working cycle of the engine (two or four stroke). They are highest during the combustion period where the cylinder pressure is rising quickly.

The vibration response of the engine is greatest when the forcing caused by the rate of pressure rise is greatest. This is intuitively obvious: if the rate of pressure rise is zero, then the forces due to cylinder pressure will be in equilibrium with the restraining forces in the engine structure, and hence with no net force there will be no net acceleration of the structure. However, with a rapidly changing cylinder pressure, the response of the structure lags behind the causal force, equilibrium is never reached, and a net force results producing vibration. The more rapid the rate of change of pressure, the greater the net force and hence the greater the vibration and the noise. A more rapid rise in pressure also increases the high-frequency content of the force, and hence of both the vibration and the noise.

The tendency to produce combustion noise of different engine types can be reasonably ranked according to their typical rates of cylinder pressure rise during combustion.

Starting with the noisiest for combustion noise:

- NA, DI diesel (4+ bar/degree crank)
- NA, indirect injection (IDI) diesel (3–4 bar/degree crank)
- Turbocharged DI diesel (2–3 bar/degree crank)
- Gasoline engine (<2–3 bar/degree crank)

The spectrum of the cylinder pressure is a more useful/reliable indicator of combustion noise. Typical spectra for the NA-DI diesel engine at full load are shown in Fig. 21.1-6 (data taken from Nelson (1987), originally published in Russell (1979)).

The effect of increasing speed can be seen in Fig. 21.1-6 as:

- a shift in the spectrum towards the higher frequencies;
- an increase in spectral levels in each third octave band as a result of the shift to the higher frequencies.

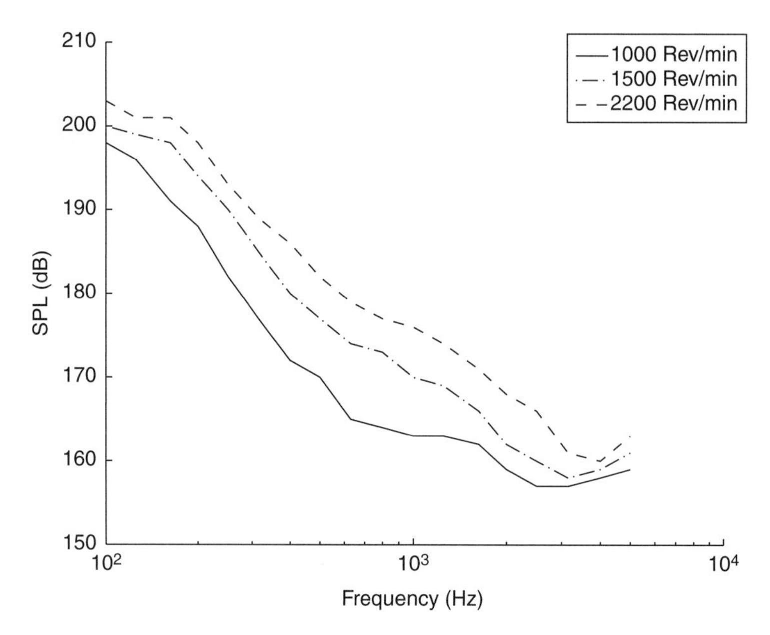

Fig. 21.1-6 Effect of engine speed on cylinder pressure spectra. NA-DI diesel engine at full load: data obtained from Nelson (1987) and Russell (1979).

It should be appreciated that the slope of the cylinder pressure spectrum provides an indication of the speed dependence of combustion noise. Typical slopes per decade (tenfold change in frequency) are given as (Lilly, 1984):

- NA-DI diesel 25–30 dB/decade
- NA-IDl diesel 40–50 dB/decade
- Turbo DI diesel 40–50 dB/decade
- Gasoline 50–60 dB/decade

The higher the slope, the greater the speed dependence. It can be seen that the noisiest engines for combustion noise have the lowest speed dependency. This explains why heavy trucks remain noisy even when used at low speeds, and why this effect is not noticed in gasoline-powered vehicles.

The portion of each spectrum in Fig. 21.1-6 in the 1–4-kHz range is responsible for the diesel knock commonly associated with diesel engines. The rate of pressure rise is greatest in diesel engines with the greatest ignition delay. This ignition delay is extended when the injection timing is advanced. With advanced injection, the fuel has more time to pre-mix with air before combustion occurs, yielding a larger pre-mixed charge which will burn quickly producing a rapid pressure rise.

So, advancing injection timing increases combustion noise. This effect is commonly used to separate combustion noise from mechanical noise. The injection timing can be slowly advanced until the change in exterior noise spectrum matches the change in cylinder pressure spectrum. At that point the exterior noise is dominated by combustion noise. This technique may not be practical with certain fuel injection equipment (it also may yield very high peak cylinder pressures) and so variants have been developed that require smaller swings in timing (Atkins and Challen (1979), reported in Nelson (1987)).

Of course the opposite applies and a retarding of injection produces a reduction in combustion noise. This noise control technique may be used only to a small extent due to the impact on engine performance and emissions (retarding the injection increases bsfc and the formation of smoke/soot/particulates). A compromise between smoke and noise emissions may be found by retarding the injection timing but increasing the injection rate (Glikin, 1985).

The modelling of the wave dynamics in the fuel injection system that strongly influences the rate of fuel injection is reported in Russell and Lee (1994) and the effect that the rate of delivery has on combustion is reported in Russell (1997). The recent advent of common rail diesel fuel injection equipment with solenoid controlled injectors allows more than one pulse of diesel fuel per cylinder per cycle and therefore fuel flow rate modulation as a means of noise control (as well as emissions control) can be explored readily.

21.1.4.3 Mechanical noise

The crank mechanism (pistons, conrods, crankshaft, bearings) experiences externally applied forces due to gas forces and internally generated forces due to its own inertia. The reaction of the engine structure to the sum of these forces produces mechanical noise by an indirect noise-generating mechanism.

Around TDC there is a rapid reversal in side force produced by the slider-crank mechanism. This produces piston slap as the piston impacts on the cylinder-liner. Piston slap is normally the dominant source of mechanical noise in the diesel engine (Lalor et al., 1980). There is side force throughout the cycle, along with other force reversals but the one at TDC yields the highest rate of change of side force. Piston slap noise increases with engine speed. It also increases with turbocharging. It is mostly controlled by reducing clearance between the piston and the cylinder-liner.

In gasoline engines, piston/liner clearances are relatively small, and mechanical noise tends to be dominated by impacts in the crankshaft bearings made through the oil film (Lalor et al., 1980). At low engine speeds these are magnified by increasing engine load. At high engine speeds, the inertia effects of the crank mechanism dominate so there is little load dependency.

Other sources of mechanical noise include:

- timing drive;
- valve train;
- fuel injection equipment.

21.1.4.4 The effects of engine speed and load on noise

The total noise emission (combustion and mechanical) for the DI diesel changes only slightly over the normal operating speed range. For the NA engine the slope is around 30 dB per decade and for the turbocharged engine around 20 dB per decade (Priede, 1975). There is modest load dependency for the NA-DI diesel engine (4–5 dB) and little for the turbo DI diesel unless excessive mechanical noise occurs due to the boost pressure.

Small high-speed NA-IDI diesel engines with smooth combustion show little load dependence and a greater speed dependence (around 40 dB per decade) than the DI diesel engines.

Gasoline engines have two sets of noise characteristics. At low speeds (up to say 2500 rev $\min^{-1}$) they have modest load dependence (around 5 dB increase in noise level due to increasing load) and slight speed dependence (20 dB per decade). At higher speeds there is little load dependence and greater speed dependence (50 dB per decade) due to the effects of inertial forces on the mechanical noise. This explains the sudden onset of roaring engine noise commonly experienced as gasoline engines are revved hard.

21.1.4.5 Measuring engine noise

The most universal parameter for quantifying the noise emission from any source (including the engine) is sound power. However, it is difficult to measure (see Section 21.1.3).

An alternative scheme is to measure sound pressure level at specified locations around the engine and use this for rating engine noise. The most commonly used standard method of this kind is detailed in SAE J1074. The important information in J1074 is as follows:

- The engine is tested either outdoors in a flat, open space or in an acoustically treated test cell that replicates the outdoor environment (commonly a semi-anechoic cell with large sound-absorbing wedges on the walls and ceiling and a flat concrete floor).
- The engine is either tested in its bare state (with just enough equipment to run – pumps and manifolds are fitted but the intake/exhaust noise is ducted away) or in its fully equipped state (everything fitted including ancillaries and sometimes full intake and exhaust systems).
- The engine is tested at the maximum power point, at the maximum torque point, at the point of maximum speed but minimum load and also at idle.
- Sound pressure levels (slow response, both 'A'- and 'C'-weightings) are measured at three positions for each engine operating condition. These are at 1.0 m from the longitudinal centres of the vertical planes forming the smallest rectangular box which completely encloses the bare engine. The measuring points are on both sides and in front of the engine at the height of the exhaust manifold and at least 1 m off the ground.
- The noise levels at the three specified locations are reported. Octave band results are also reported for the location with the highest 'A'-weighted level.
- A survey is made of 'A'-weighted sound pressure level at the same height and distance from the box as the specified locations. If the survey reveals readings more than 3 dB above the highest reading at the specified locations, then the survey readings are also reported.
- The reported results should be the averaged results of two or more test results within 2 dB of each other.

If a semi-anechoic cell is used it must be large enough to undertake the measurements and each microphone should be at least one-quarter wavelength from the walls and from the ceiling to avoid the near-acoustic fields of these absorbing surfaces (see Section 21.1.3.1).

21.1.4.6 Engine noise source ranking

Various noise source ranking techniques are discussed in Section 22.1.2. All are in common use for engine noise source ranking.

At higher frequencies (<300 Hz say) the shielding technique generally gives reliable results and is easy to use. However, it is far less reliable at low frequencies and commonly results in the under estimation of the contribution made by engine components at these frequencies (Crocker et al., 1980).

Therefore, at low frequencies, sound intensity (Crocker et al., 1980) or noise from vibration techniques might be preferred (Dixon and Phillips, 1998). The noise source rankings for contemporary powertrains are as follows, with the noisiest item at the top of the list (March and Croker, 1998).

	Diesel engines	Gasoline engines
	Oil pan	Transmission
	Other sources	Other sources
	Fuel system	Intake system
Similar effect	Cylinder block	Ancillary drive
	Ancillary drive	Oil pan
	Transmission	Exhaust manifold
	Intake system	Cam cover
	Exhaust manifold	Front cover
	Front cover	Fuel system
	Cam cover	Cylinder block

See also Section 20.1.6.2.1 for an alternative noise source ranking.

21.1.4.7 Engine noise control

The options for controlling engine noise are the usual ones available to the noise control engineer, namely:

- stiffen structures to push resonant frequencies above the highest forcing frequency;
- isolate components from sources of excitation;
- encapsulate noise sources with massive panels;
- add damping where resonances occur.

Engine-specific noise control measures include the following:

- Oil pan – the use of an isolating gasket between the oil pan and the crankcase. The adoption of structural aluminium oil pans to replace the traditional pressed steel components has made oil pan noise more significant in spite of improvements to the crankcase to reduce its noise radiation.
- Rocker cover – the use of rubber isolating gaskets (Querengasser et al., 1995).
- Fuel injection equipment – the adoption of common rail systems and unit injector systems which are more compact and quieter have brought about significant improvements (March and Croker, 1998).
- Cylinder block – engine blocks with separate crankshaft bearing endcaps at the bulkheads between cylinders exhibit the lowest frequency for the first bending mode of the engine. For larger engines this may be as low as 200 Hz. The use of a ladderframe or bedplate bottom end to join the bearing caps together with a locally stiff structure can push the first bending mode above 300 Hz (Querengasser et al., 1995), reduce the axial excursion of the endcaps at resonance (commonly around 1000 Hz) and generally reduce the low-frequency modal density (March and Croker, 1998). Careful design of the crankcase and the block to reduce the effect of panel modes is also beneficial for frequencies around 800+ Hz (Russell, 1972).
- Intake system – the avoidance of large planar surfaces on intake components can reduce noise emissions along with general stiffening of the structures.
- Noise shields – well-damped, isolated engine covers can reduce noise radiated by the engine structure (Russell, 1972).
- Engine bay enclosures – engines may be effectively enclosed within their engine bay in the vehicle, thus encapsulating the noise sources. Problems with ventilation and cooling are common (Thien et al., 1984).

21.1.5 Road noise

21.1.5.1 Introduction to road noise

The term road noise might be replaced by the more complete description 'road and tyre noise' as it is taken here to include:

- Interior noise resulting from the contact between the tyres and the road, being transmitted to the interior by both airborne and structure-borne paths. This is often labelled as road noise and is the subject here.
- Exterior noise resulting from the contact between the tyres and the road. This is often labelled as tyre noise and was the subject of an earlier section (Section 22.1.4).

21.1.5.2 Interior road noise

Vehicle interior road noise is mainly a low-frequency noise problem (<1000 Hz). Contributions are made by

- structure-borne noise paths through the vehicle suspension (<500 Hz);
- direct airborne noise paths from the tyre through the vehicle structure (>500 Hz); often confused with wind noise.

The structure-borne components tend to dominate overall noise levels except on the smoothest of road surfaces. The structure-borne element becomes clear when a vehicle is operated on a rolling road with the body jacked on air jacks and the suspension links disconnected (the so-called disconnect test). Interior noise levels below 500 Hz drop significantly during such tests whilst the levels of airborne tyre noise above 500 Hz remain fairly constant.

Interior road noise levels might peak at around 60–65 dBA at 250 Hz and fall off with increasing frequency. Levels at frequencies greater than 1000 Hz are typically 15 dBA lower than those at frequencies less than 300 Hz.

Interior road noise is typically measured with the vehicle operating at fixed speeds over different road surfaces. A standard measurement scheme is often employed such as BS 6086 (see Section 21.1.1.3). It is common to make measurements at the inner ear positions for all seats. Note that the 'A'-weighting scale is commonly employed. This has the benefit of avoiding the overload of the measurement chain when the vehicle passes over bumps in the road (the so-called bump thump). However, when measuring bump thump deliberately, the 'A'-weighting scale might be replaced by the 'C'-weighting scale, or no weighting may be applied at all in order to preserve signal-to-noise ratio at low frequencies.

Low frequency peaks in interior road noise spectra are commonly attributable to:

- Acoustic modes of the cockpit/interior space.
- Modes of vibration of the tyre structure, typically <200 Hz. These are often called breaker modes as they are modified by the breaker: a composite layer directly below the tread constructed from steel cords embedded in rubber. The breaker provides structural stiffness to the tyre.
- Acoustic modes of the tyre cavity (typically around 250 Hz).

21.1.5.3 Analysing structure-borne road noise

A disconnect test can separate structure-borne and airborne road noise in the laboratory. However, on the test track a noise path analysis (see Section 21.1.2) technique is required. The complex stiffness method is usually used. The noise path analysis for road noise is far more complicated than for powertrain noise as it is difficult to find a single suitable reference signal to relate the phase information to. As a result, measured force data are grouped into several sets of coherent signals (Vandenbroeck and Hendricx, 1994; Storer et al., 1998) using a technique known as principal component analysis.

Because of the complexities of noise path analysis for road noise, disconnect tests are commonly used. Shakers are attached to the wheel hubs and the noise levels (or vibration velocity amplitude at certain structural elements) produced inside the vehicle are measured. The effect of disconnecting elements of the suspension and hence severing potential noise paths can then be ascertained readily.

There is common use of measured NTF – the ratio of noise (Pa) to force input (N). With a shaker (and force gauge) attached at every suspension point in turn, the NTF to a microphone in the vehicle interior can be found using a two-channel FFT analyser. A typical NTF target for road noise is a maximum of 0.01 Pa/N.

21.1.5.4 Controlling interior road noise

Some design guidance has emerged for controlling interior road noise:

- Choose quieter tyres (!).
- Achieve an NTF of 0.01 Pa/N at all suspension attachment points. In practice this generally requires a body stiffness of more than 10 kN mm^{-1} (longitudinal and lateral bracing may be required at strut tops).
- Use suspension bushes with a stiffness that is between one-fifth and one-tenth of the body stiffness.
- Use a compliantly mounted subframe (Hardy, 1997) where the suspension is mounted to a modified subframe, which is resiliently mounted to the body (most resilience in the longitudinal direction). Too much lateral compliance affects vehicle handling. The beneficial effects of the compliant subframe are shown in Fig. 21.1-7, the data taken from Hardy (1997).
- Use of palliative treatments to reduce interior noise levels (see Sections 21.1.2.1, 21.1.9, and 21.1.10) such as isolated barrier materials.
- Reduce the noise caused by shock absorbers (Cheng and Akin, 1995).

21.1.6 A note on aerodynamic (wind) noise

Aerodynamic (wind) noise is a significant source of interior noise for many vehicles travelling at higher speeds. It is easily confused with road noise, being generally broadband in nature but with a strong low-frequency bias.

Aerodynamic noise is caused through a variety of mechanisms:

- Aerodynamic excitation of the so-called 'greenhouse surfaces' on the car (the glass-work and the roof panel) causing structure-radiated noise in the interior.

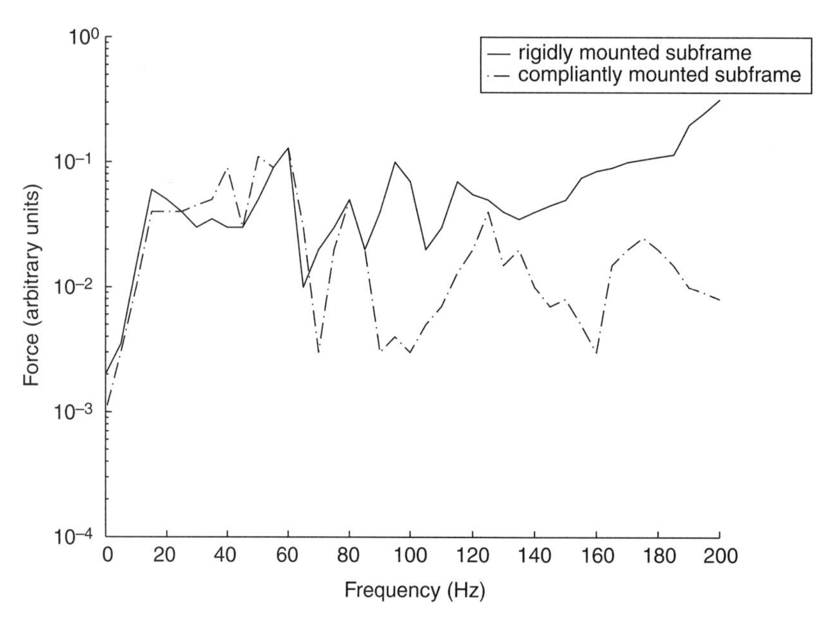

Fig. 21.1-7 The beneficial effects of the compliant subframe: data taken from Hardy (1997).

- Airflow over the underside of the vehicle causing transmission of airborne sound to the interior, particularly in the wheel-arch areas.
- Noise transmission through door and glass seals due to aspiration (leakage) or due to aerodynamic excitation of doors and glass caused by disturbed airflow over the seals.
- Vortex shedding over protrusions in the bodywork (such as aerials, roof bars, etc.) causing tonal airborne noise.
- Cavity flows through partially open windows and sunroofs causing intense low-frequency (below 25 Hz commonly) noise and buffeting.

Aerodynamic noise is best investigated at full scale in a windtunnel although some promising results have been obtained at model scale (Kim, 2003). The use of a windtunnel allows the car to be stationary, thus removing noise contributions from the powertrain and from the tyres. Frequently, aerodynamic noise development is undertaken on a subjective basis. Windows and doors are taped up, lead sheeting is added to wheel arches and to the underfloor, bodywork protrusions are removed and, at every stage, binaural recordings are made in the car using a dummy head system. The resulting recordings can be played to juries, or Zwicker loudness or Articulation Index are frequently used as a measure of improvement (see Section 20.1.6.3.2).

Publications by Anderson (2002) and by Coney et al. (1999) detail objective methods of assessment. These include the use of an intensity probe inside the vehicle to check for aspiration noise. An inexpensive and traditional alternative is to use a length of plastic tubing, to place one end in your ear and to pass the other end around the glasswork and door seals. The intensity probe can also be used outside the vehicle to map the contours of aerodynamic noise from bodywork protrusions and from the greenhouse surfaces. A form of noise path analysis is also possible by making measurements of structural vibration on the greenhouse surfaces and then combining these with NTFs measured between positions inside the interior and the various surfaces (see Section 21.1.2.3 for a discussion of the NTF). The character of the aerodynamic excitation can be determined using a combination of static pressure tappings and flush mounted microphones embedded into thin rubber mats that are simply laid on the greenhouse surfaces.

21.1.7 A note on brake noise

Brake noise has been an issue of concern to vehicle manufacturers for decades. There are several distinct categories of brake noise (Betella et al., 2002):

- Brake squeal (occurring at higher speeds and having a tonal character with components well above 1000 Hz).
- Brake moan (occurring at moderate speeds and characterised by frequency components around 100 Hz).
- Brake creep-groan (occurring at speeds less than walking pace and characterised by frequency components around 100 Hz).
- Brake judder (occurring at speeds less than walking pace and characterised by frequency components around 10 Hz).

The source for all four classes of brake noise is the pairing of friction surfaces at the pad and rotor disk. Brake squeal noise is radiated by the brake components themselves (the disk, the pads and the caliper assembly) as is clearly seen using laser vibrometry techniques. However, because the resonant frequencies of the brake components are typically so high (seldom less than 1000–2000 Hz) it is the vehicle suspension that acts as the resonant system in the case of brake moan, creep-groan and judder.

There are few options for the control of brake noise. Careful choice of friction pairs is the most common (it is the difference between static and dynamic coefficients of friction that causes creep-groan, judder and moan [Bettella et al., 2002]) whereas the use of layers of metal plates riveted together on the back of the brake pads can act as efficient dampers for brake squeal.

21.1.8 A note on squeak, rattle and tizz noises

The control of squeak, rattle and tizz noises is extremely important to the passenger car industry. For some car manufacturers these faint but annoying noises are the number one cause of dealer returns (vehicles that have been returned to the manufacturer because the retail customer has not been satisfied by the remedial works undertaken by the dealer and has effectively given up on the vehicle). The main reason for their resulting in high levels of dealer returns is that they are easy to detect but their source is usually hard to find. To further complicate the matter, once the source is found a cure is often elusive.

Tizz noises are caused by high-frequency tactile vibration (such as that occurs on the gear lever knob) and are relatively easy to find and cure. By comparison, squeaks (which can be individual tick noises or sequences of ticks that sound like a squeak) are caused by relative motion between material pairs and these pairs can be buried deep within the fabric of the car interior (such as somewhere inside the dashboard assembly). The noise source results from a stick-slip process caused by a difference in the values of static and dynamic coefficient of friction.

Squeaks are known to be influenced by (Juneja et al., 1999):

- material choice;
- surface finish;
- frequency of excitation;
- amplitude of excitation;
- interference levels between the two materials;
- normal loads;
- temperature;
- humidity.

The worst combination of these seems to be high amplitude and high-frequency excitation applied to a pair of materials with high degree of interference at low humidities. Typically material pairs are tested on a rig, where one material is kept still and the other is shaken, and the rig is routinely situated in a climatic chamber. Squeak levels are frequently measured in terms of loudness (see Section 20.1.6.3.2). This type of pragmatic testing is now commonplace with the suppliers of vehicle interior trim systems, and much of the contractual risk caused by dealer returns is placed on these organisations.

Rattle noises can be caused by loose fitting (and hard) interior trim items and also in the vehicle transmission line. Due to the rather chaotic nature of the rattle phenomena, statistical correlations between subjective response to rattle sounds and objective measures (Croker et al., 1990) such as sound pressure level, loudness or speech interference level tend to be inconclusive (see Sections 20.1.6.3.1 and 20.1.6.3.2 for details on these).

21.1.9 Control of sound through absorption within porous materials

21.1.9.1 Practical approach

Interior noise levels (particularly those related to body booms and those at high frequencies) may be controlled by the addition of sound-absorbing material to the passenger compartment.

For a diffuse space

$$T_{60} = \frac{0.161\,V}{S\bar{a}} \tag{21.1.78}$$

where

T_{60} = reverberation time at a particular frequency
V = volume in the space (m^3)
S = surface area in the space (m^2)
$\bar{a}$ = average Sabine absorptivity.

The reverberation time in the vehicle interior is not constant throughout the space and therefore the acoustic field is not diffuse, but accepting this lack of realism and to make the calculation simple, for the purpose of illustration only, the average T_{60} @ 1000 Hz for a small European wagon style car can be estimated, thus:

Estimated volume = 3.2 m^3
Estimated surface area of the interior space (without passengers)= 21.7 m^2

From Bies and Hansen (1996), maximum absorption coefficients as measured in an impedance tube are:

- 0.89 @ 1000 Hz for an unoccupied, well-upholstered seat;

- 0.61 @ 1000 Hz for an unoccupied, leather-upholstered seat;
- 0.03 @ 1000 Hz for glass.

From Haines (1987), maximum absorption coefficients (method of measurement unknown) for a molded glass fibre headliner are:

- 0.60 @ 1000 Hz for 13-mm thick liner.

From Saha and Baker (1987), maximum absorption coefficients (measured in a reverberation room) are

- 0.18 @ 1000 Hz for a 6-mm thick carpet; and
- 0.42 @ 1000 Hz for a 12-mm thick carpet.

For cloth seats and leather seats see Tables 21.1-4 and 21.1-5.

The surface average absorption coefficient is in this case for cloth seats is (without passengers) $\bar{a} = 0.41$. Therefore, the average T_{60} @ 1000 Hz would be 58 ms for upholstered seats (without passengers). The surface average absorption coefficient in this case for leather seats is (without passengers), $\bar{a} = 0.35$. Therefore, the average T_{60} @ 1000 Hz would be 68 ms for leather seats (without passengers).

The two reverberation times estimated above broadly agree with the typical, measured reverberation times shown (Quian and Vanbuskirk, 1995), these being in the range of 70–90 ms.

From Section 21.1.3.5 it is known that the contribution to interior sound pressure levels from the reverberant field in a semi-diffuse environment is given by:

$$\Delta Lp_{rev} = 10 \log_{10}\left(\frac{4}{R}\right) \qquad (21.1.79)$$

where the room constant R is:

$$R = \frac{S\bar{a}}{1-\bar{a}} \qquad (21.1.80)$$

Using the data from Tables 21.1-4 and 21.1-5:

$$R_{cloth} = 14.98$$

$$R_{leather} = 11.53$$

So the likely effect of fitting cloth seats rather than leather seats would be (without passengers)

$10 \log_{10}(R_{cloth}/R_{leather}) = 1.1$ dB reduction in the sound due to the reverberant field.

This seems a rather modest reduction in reverberant sound level (remember, the level due to direct sound is not considered in the above analysis), although Quian and Vanbuskirk (1995) suggest that making the change in seat covering *resulted in measurable and noticeable differences in interior sound levels during some modes of actual vehicle operation... Speech intelligibility was improved.*

One may decide whether the driver is likely to be influenced most by direct of reverberant sound using the relationship

$$\frac{I_r}{I_d} = \frac{16\pi r^2}{S\bar{a}} \qquad (21.1.81)$$

where I_r and I_d are the reverberant and direct sound intensities, respectively. It can be seen that for the earlier case with cloth seats, the reverberant and direct intensities are equal (according to this simple model, assuming a diffuse space) at a distance of 0.4 m. So, when the driver's ears are less than 0.4 m from a noise source (like the side windows and the roof) the direct field will dominate and noise levels will be little affected by changes in the total absorption of the space. However, at longer distances from sources (the windscreen, the rear screen, the footwell, the rear floor) an increase of absorption will reduce the interior noise levels to at least some degree.

Table 21.1-4 Sound absorption in a typical European wagon style passenger car with cloth seats (estimates for illustration only)

Surface	Surface area (m^2)	A	Sa
Headliner	3.5	0.6	2.1
Carpet	1.47	0.42	0.618
Footwell	1.05	0.42	0.441
Trunk	1.12	0.42	0.4704
Front screen	1.05	0.03	0.0315
Rear screen	1.05	0.03	0.0315
Seat squabs	1.96	0.89	1.7444
Seat backs	2.8	0.89	2.492
Side glass	3.0	0.03	0.09
Side trim	3.0	0.18*	0.54
Dash	0.84	0.18*	0.1512
Rear trim	0.84	0.18*	0.1512
Total	21.68	—	8.8612

*Estimated.

Table 21.1-5 Sound absorption in a typical European wagon style passenger car with leather seats (estimates for illustration only)

Surface	Surface area (m^2)	A	Sa
Headliner	3.5	0.6	2.1
Carpet	1.47	0.42	0.618
Footwell	1.05	0.42	0.441
Trunk	1.12	0.42	0.4704
Front screen	1.05	0.03	0.0315
Rear screen	1.05	0.03	0.0315
Seat squabs	1.96	0.61	1.1956
Seat backs	2.8	0.61	1.708
Side glass	3.0	0.03	0.09
Side trim	3.0	0.18*	0.54
Dash	0.84	0.18*	0.1512
Rear trim	0.84	0.18*	0.1512
Total	21.68	—	7.5284

*Estimated.

Sound-absorbing materials make a contribution to controlling the levels of higher-frequency noise in vehicle interiors. Their usefulness in controlling lower-frequency sound (say below 500 Hz) is limited as the thicknesses of absorbing material required would be bulky and add too much weight to the vehicle.

Note: With porous materials, the absorption coefficient α tends to increase rapidly at low frequencies and approach unity at moderate frequencies as shown in Fig. 21.1-8.

Thickness of material increases α up to a point. As a general rule, the material thickness should be at least one quater wavelength of the lowest frequency of interest. Placing an air gap between the rear of the material and its backing changes the apparent thickness. The low-frequency absorption increases with the increasing air layer thickness but the high-frequency absorption deteriorates. The maximum absorption occurs around the one quarter wavelength resonant frequency of the air layer.

The effectiveness of sound-absorbing materials can be assessed in two ways:

1. Calculating the effect by combining the known absorption performance of individual absorbing items.
2. Measuring the effectiveness of the whole sound-absorption package fitted to the vehicle.

The effectiveness of the whole vehicle sound package may be measured by:

1. Determining insertion loss, where the effect that the addition of each sound-absorbing component has on

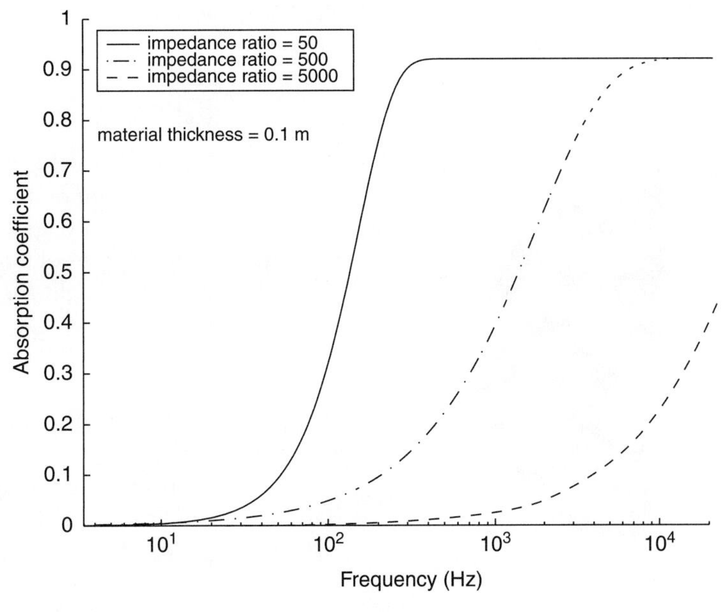

Fig. 21.1-8 Typical absorption characteristics of a porous material: adapted from data provided by Bies and Hansen (1996). Impedance ratio is the specific acoustic impedance of the material divided by the characteristic impedance of the air.

interior noise levels is measured during certain vehicle operating conditions.

2. Determining the reverberation time. Reverberation time (frequency variable) tends to be in the range $\ll$100 ms for typical sedan cars. There is significant spatial variation in reverberation times so it is commonly measured at many locations and an average is then sought.

The seats and the headliner contribute most to the absorption of sound (as shown in Fig. 21.1-9). This is due to their appropriate flow resistance (neither too high so as to appear reflective, nor too low so as to appear acoustically transparent) and their large surface areas.

Individual sound-absorbing components may be characterised by their characteristic specific acoustic impedance

$$z_c = \frac{p}{u'} = \sqrt{\kappa \rho'} \tag{21.1.82}$$

measured in either:

1. an impedance tube (Bies and Hansen, 1996; Chung and Blaser, 1980); or
2. a free field (Allard and Sieben, 1985).

A commonly quoted index is that of absorption coefficient. This varies with both frequency and angle of incidence of the impinging sound (as does the characteristic specific impedance of the material). It may be measured in many ways:

1. Random incidence α: Measured in the reverberation chamber room (see Section 21.1.3.4 for an introduction to such spaces).
2. Specific angle of incidence α: Measured in anechoic chamber (Ingard and Bolt, 1951).
3. Normal incidence α: Measured in the impedance tube (Bies and Hansen, 1996; Chung and Blaser, 1980)

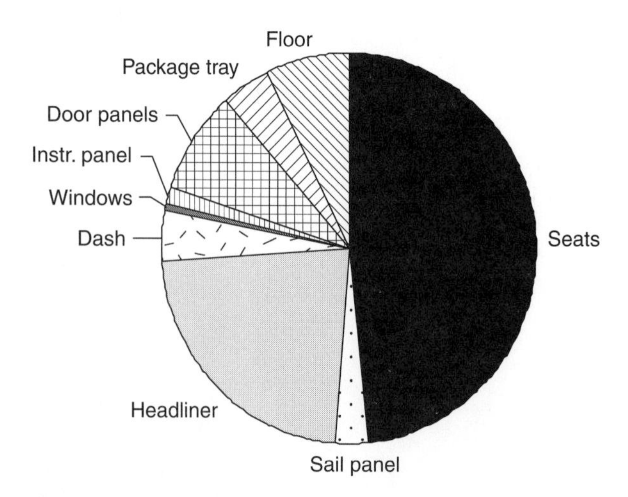

Fig. 21.1-9 The absorption of sound by vehicle interior trim: data taken from Qian and Vanbuskirk (1995).

The random impedance α seems to agree better with observed effects in road vehicles than the normal impedance α (Qian and Vanbuskirk, 1995).

21.1.9.2 Physical processes for sound absorption within porous materials

The physical mechanisms by which the sound is absorbed are widely acknowledged to be rather complex (Qian and Vanbuskirk, 1995). There is more than one mechanism and these are generally classified as follows (Fahy and Walker, 1998):

1. viscous losses due to oscillating flow within the internal spaces of the material;
2. heat conduction within the material;
3. vibration of the material (particularly in closed cell plastic foams).

Porous materials have rather complex geometric structures that defy mathematical description using deterministic models. Rather, bulk properties of the material are used to characterise its sound absorption (Fahy and Walker, 1998). Neglecting the vibration of the material itself, these are:

- flow resistivity (r);
- porosity (h);
- structure factor (s).

One may gain useful insight into the behaviour of sound within porous materials by inspecting a modified form of the linear plane wave acoustic equation:

$$\frac{\partial^2 p}{\partial t^2} = c^2 \frac{\partial^2 p}{\partial x^2} \tag{21.1.83}$$

This equation relates the spatial variation of acoustic pressure p with the temporal distribution via the speed of sound c (m s^{-1}). It only applies to a free field. In the restricted internal spaces of a porous material, additional relationships will be sought for the spatial distribution of acoustic pressure.

21.1.9.3 Flow resistivity

Consider the steady flow of air through a porous material. The air is flowing with a volume velocity per unit cross section area of u' (m s^{-1}) (Fahy and Walker, 1998)

$$u' = \frac{m^3/S}{m^2/l^2} = \frac{m^3}{s} \cdot \frac{l^2}{m^2} = \frac{m}{S} \tag{21.1.84}$$

The flow resistivity is defined as:

$$\frac{\partial p}{\partial x} = -ru' \qquad (21.1.85)$$

Bies and Hansen (1996) propose a method of measuring the flow resistivity (MKS rayls per metre) in accordance with ISO 9053-1991, by measuring the pressure drop across a sample of material in a tube through which there is a steady flow of air.

$$r = \frac{\rho \Delta P A}{\dot{m} l} \qquad (21.1.86)$$

ρ = density of gas (kg m^{-3})
ΔP = differential pressure (N m^{-2})
A = gross sectional area of material (m^2)
$\dot{m}$ = mass flow rate of air (kg s^{-1})
l = specimen thickness (m)

The limitation of using flow resistivities measured in this way is that one must assume that the DC flow resistivity is the same as the oscillatory acoustic flow resistivity. For low speed internal flows this assumption generally holds (Fahy and Walker, 1998).

An alternative method of measuring flow resistivity does not suffer from such limitations as it uses acoustic field in an impedance tube to produce oscillatory flow in the material (Ingard and Dear, 1985). In this technique, the base of the material sample is placed at a point in the tube that is an odd number of quarter wavelengths from the closed end of the tube. Two pressures are recorded, $p_1$and p_3 as shown in Fig. 21.1-10. Under these conditions,

$$\frac{r}{\rho_0 c_0} \approx \left|\frac{p_1}{p_3}\right| \qquad (21.1.87)$$

at low frequencies (Ingard and Dear, 1985).

The frequency of the single tone (best below 100 Hz) driving the loudspeaker is adjusted until this is the case. The reason for the $\frac{n\lambda}{4}$ $(n = 1, 3, 5, \ldots)$ separation is explained in Appendix 21.1E (see also Kinsler et al. (1982)).

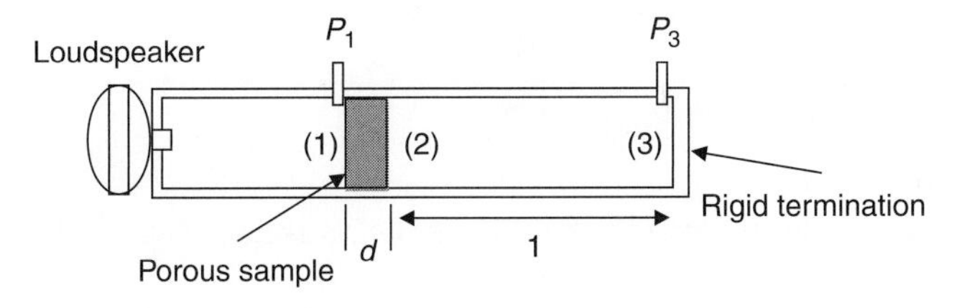

Fig. 21.1-10 Impedance tube method for measuring flow resistivity: after Ingard and Dear (1985).

For fibrous materials Bies proposes (reported in Bies and Hansen [1996]):

$$\frac{rl}{\rho c} = 27.3\left(\frac{\rho_m}{\rho_f}\right)^{1.53}\left(\frac{\mu}{d\rho c}\right)\left(\frac{l}{d}\right) \qquad (21.1.88)$$

ρ = density of gas (kg m^{-3})
ρ_m = porous material bulk density (kg m^{-3})
ρ_f = fibre material density (kg m^{-3})
μ = gas viscosity (N s m^{-1})
d = fibre diameter (m)
c = speed of sound in gas (m s^{-1})
l = specimen thickness (m)

Delaney and Bazely (reported in Bies and Hansen (1996)) offer a relationship between flow resistivity in the range of 10^3–5×10^4 MKS rayls per metre and the characteristic acoustic impedance of a porous material Z_M

$$Z_M = \rho c\left[1 + 0.0571X^{-0.754} - i0.087X^{-0.732}\right] \qquad (21.1.89)$$

$$X = \frac{\rho f}{r} \qquad (21.1.90)$$

where

ρ = gas density (kg m^{-3})
f = frequency (Hz)
r = flow resistivity (MKS rayls per metre)

21.1.9.4 Porosity

Porosity is defined as (Fahy and Walker, 1998):

$$h = \frac{\text{volume of voids in the material}}{\text{total volume of the material including voids}} \qquad (21.1.91)$$

For acoustic-absorbing materials $h \rightarrow 0.90$ or 0.95.

The linearised equation of mass conservation in free air is given as (Kinsler et al., 1982):

$$\frac{\partial u}{\partial x} = -\frac{1}{\kappa_0}\frac{\partial p}{\partial t} \qquad (21.1.92)$$

$$\kappa_0 = \text{bulk modulus} = \rho_0 c_0^2 \qquad (21.1.93)$$

$$\kappa_0 = \rho_0\left(\frac{\partial p}{\partial \rho}\right)_{\rho_0} \qquad (21.1.94)$$

$$p = \kappa s, \qquad s = \frac{\rho - \rho_0}{\rho_0} \qquad (21.1.95)$$

The effect of the porosity is to introduce a modified bulk modulus

$$\kappa = \frac{\kappa_0}{h} \qquad (21.1.96)$$

so that

$$\frac{\partial u'}{\partial x} = -\frac{1}{\kappa}\frac{\partial p}{\partial t} \qquad (21.1.97)$$

The derivation of the linearised mass conservation equation is given in Appendix 21.1F.

21.1.9.5 The structure factor

The structure factor (s) expresses the influence of the geometric form of the structure on the effective density of the fluid (Fahy and Walker, 1998).

There are four mechanisms for this:

1. Side pockets within the material (outside the flow stream) reduce the effective bulk modulus of the fluid. This produces a smaller acoustic pressure for a given strain gradient in the pore (see above equation) which leaves the impression that the effective density of the fluid is higher in the material than in free space.
2. Non-uniform pores causing sudden expansions that leave the impression of added mass.
3. Non-axial pore orientation. The in viscid momentum equation normal to the orientation of the pore is:

$$\frac{\partial p}{\partial h} = -\rho_0\frac{\partial u_n}{\partial t} \qquad (21.1.98)$$

or for orientation angle θ to the surface of the material:

$$\frac{\partial p}{\partial x}\cos\theta = -\rho_0\frac{\partial u'}{\partial t}\frac{1}{h\cos\theta}$$

$$\frac{\partial p}{\partial x} = \frac{-\rho_0}{h\cos^2\theta}\frac{\partial u'}{\partial t}$$

$$\frac{\partial p}{\partial x} = -\frac{s\rho_0}{h}\frac{\partial u'}{\partial t} \qquad (21.1.99)$$

$$s\propto\alpha\frac{1}{\cos^2\theta} \qquad (21.1.100)$$

The effective density is increased by factor (s). For a random orientation of pores $s = 3$. Generally s is in the range of 1.2–2.0 (Fahy and Walker, 1998).

4. The velocity profile across each pore acts to reduce volume acceleration produced by a given gradient and hence contributes to s.

21.1.9.6 The modified 1-D linear plane wave equation

The modified equation of motion becomes:

$$\frac{\partial p}{\partial x} = -\left(\frac{s\rho_0}{h}\right)\frac{\partial u'}{\partial t} - ru' \qquad (21.1.101)$$

Reminder: u' is the volume flow rate/unit cross section area *not* the flow velocity in the pores.

For simple harmonic motion of frequency ω

$$\frac{\partial p}{\partial x} = -\left[\frac{s\rho_0}{h} + \frac{r}{i\omega}\right]\frac{\partial u'}{\partial t} \qquad (21.1.102)$$

This is a modified Euler equation with the complex terms in the brackets being the effective density (Fahy and Walker, 1998). The derivation of the 1-D Euler equation is given in Appendix 21.1G.

Now, if this equation is combined with the linearised mass conservation equation (21.1.92)

$$\frac{\partial u'}{\partial x} = -\frac{1}{\kappa_0}\frac{\partial p}{\partial t} \qquad (21.1.92)$$

a modified wave equation is obtained in this manner:

- Differentiate both sides of the modified Euler equation with respect to x

$$\frac{\partial^2 p}{\partial x^2} = -\left[\frac{s\rho_0}{h} + \frac{r}{i\omega}\right]\frac{\partial}{\partial x}\left(\frac{\partial u'}{\partial t}\right)$$

- Now take the time derivative of the linearised mass conservation equation

$$-\frac{1}{\kappa_0}\frac{\partial^2 p}{\partial t^2} = \frac{\partial}{\partial t}\left(\frac{\partial u'}{\partial x}\right) = \frac{\partial}{\partial x}\left(\frac{\partial u'}{\partial t}\right)$$

- So, the combination gives:

$$\frac{\partial^2 p}{\partial x^2} = -\left[\frac{s\rho_0}{h} + \frac{r}{i\omega}\right] \times -\left[\frac{1}{\kappa_0}\frac{\partial^2 p}{\partial t^2}\right]$$

$$\kappa_0\frac{\partial^2 p}{\partial x^2} = \left[\frac{s\rho_0}{h} + \frac{r}{i\omega}\right]\frac{\partial^2 p}{\partial t^2}$$

$$\rho_0 c_0^2\frac{\partial^2 p}{\partial x^2} = \left[\frac{s\rho_0}{h} + \frac{r}{i\omega}\right]\frac{\partial^2 p}{\partial t^2}$$

$$\frac{c_0^2}{s}\frac{\partial^2 p}{\partial x^2} = \left[\frac{1}{h} + \frac{r}{i\rho_0 s\omega}\right]\frac{\partial^2 p}{\partial t^2}$$

- Finally the modified wave equation is

$$c_1^2\frac{\partial^2 p}{\partial x^2} = \frac{1}{h}\frac{\partial^2 p}{\partial t^2} + \left(\frac{r}{\rho_0 s}\right)\frac{\partial p}{\partial t} \qquad c_1^2 = \frac{c_0^2}{s} \qquad (21.1.103)$$

The effect of parameters s, r, h is to lower the speed of sound within the material (compared with the speed of sound in free space) and to attenuate the acoustic wave as it propagates. Therefore,

$$p(x,t) = Ae^{i\omega t}e^{-\gamma x} \qquad (21.1.104)$$

$\gamma = \alpha + i\beta$ is the propogation constant.

$$p(x,t) = Ae^{i(\omega t-\beta x)}e^{-\alpha x} \qquad (21.1.105)$$

α = attenuation constant

$$z_c = \frac{p}{u'} = \sqrt{\kappa\rho'} \qquad (21.1.106)$$

where z_c is the characteristic specific acoustic impedance of the gas in porous material; and

$$\gamma = i\omega\sqrt{\frac{\rho'}{\kappa}} \qquad (21.1.107)$$

$$\kappa = \frac{\rho_0 c_0^2}{h} \qquad (21.1.108)$$

$$\rho' = \frac{s\rho_0}{h} + \frac{r}{i\omega} \qquad (21.1.109)$$

21.1.10 Control of sound by minimising transmission through panels

21.1.10.1 Introduction

The encapsulation of a noise source using panels with high TL is a valuable tool for the refinement engineer. When used appropriately it can produce significant reductions in interior noise level (more than 10 dB). It is used during the shielding technique for noise source ranking (see Section 22.1.2).

The principles of encapsulation are also used when designing noise barrier panels to fit under carpet in order to isolate the passenger compartment from the noise in the engine bay. Although these are a powerful way of controlling interior noise levels, a note of caution is given that they are a heavy solution. One published benchmarking exercise (Wentzel and VanBuskirk, 1999) identified nearly 39 kg of such noise barrier materials in a sedan and nearly 48 kg in a mini-van.

Encapsulation techniques can, along with noise control at source, be used as part of a general noise control strategy. A noise control problem can be split into three components – a noise source, noise propagation and the reception of noise as illustrated in Fig. 21.1-11.

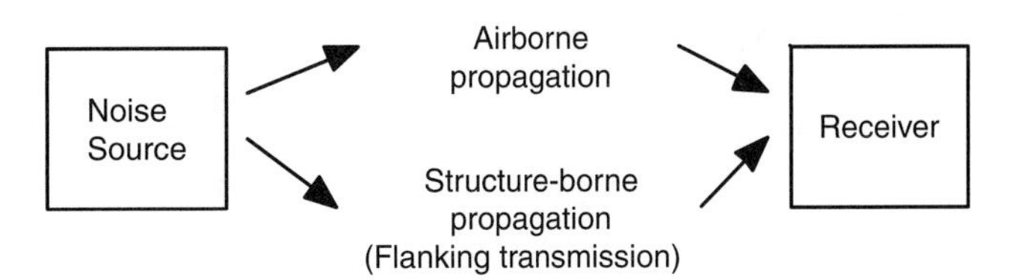

Fig. 21.1-11 A typical noise control problem.

Encapsulation allows for the interruption of the airborne noise path. It should be noted that encapsulation will only remain effective whilst any structure-borne paths (or other flanking transmission) remains insignificant. Therefore, in many cases the structure-borne paths must also be interrupted or controlled for encapsulation to be fully effective.

21.1.10.2 The measurement of the acoustic performance of enclosures

The effectiveness of an acoustic enclosure may be assessed according to a number of different parameters. The first is termed noise reduction (NR) and is simply the arithmetic difference between the sound pressure level at a point (or the average over a number of points) within an enclosure and the sound pressure level at a prescribed point outside the enclosure as shown in Fig. 21.1-12:

$$\mathrm{NR} = \mathrm{SPL}_1 - \mathrm{SPL}_2 \text{ (dB)} \qquad (21.1.110)$$

This method is experimentally convenient, and an adequate means of comparing the acoustic performance of two geometrically similar enclosures. However, it is of limited use as an absolute indicator of acoustic performance as the value for NR obtained is valid only for the precise microphone locations chosen.

A more generally applicable method of assessment uses TL as a parameter. TL in decibels is obtained from the ratio of incident and transmitted acoustic intensities across the boundary of the enclosure (Fig. 21.1-13):

$$\mathrm{TL} = 10\log_{10}\left[\frac{I_i}{I_t}\right]\left(\mathrm{dB}\right) \qquad (21.1.111)$$

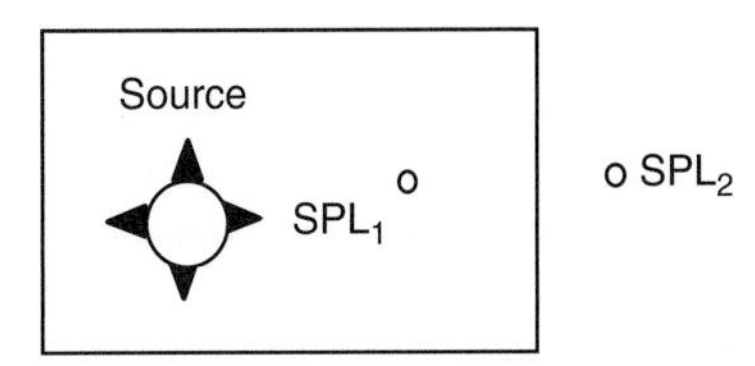

Fig. 21.1-12 Measuring noise reduction.

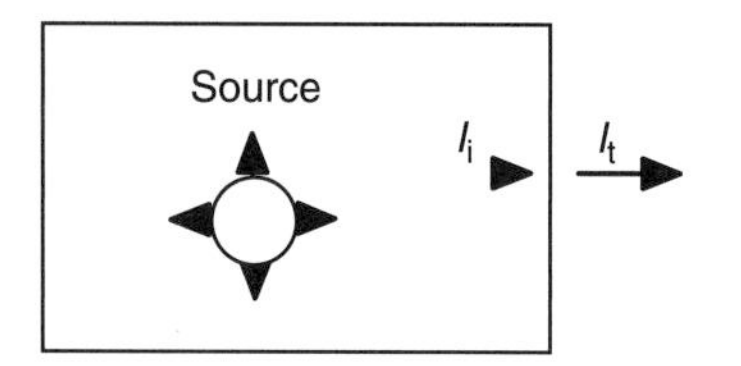

Fig. 21.1-13 Measuring TL.

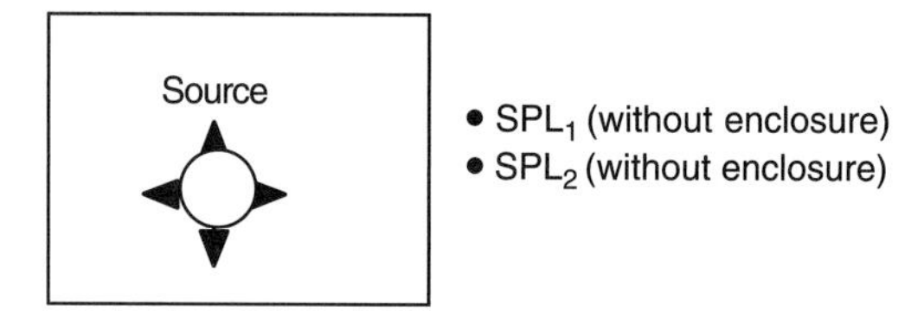

Fig. 21.1-14 Measuring insertion loss.

The TL across sections of enclosures may be obtained from measurements made in the field using an intensity meter, or the TL of single panels can be obtained from measurements made in a pair of special interconnected laboratory rooms known as transmission suite.

A third and intuitive method of assessment uses insertion loss as a parameter. Insertion loss is obtained by comparing sound pressure levels at a point in space with and without the enclosure in place (see Fig. 21.1-14). Therefore,

$$\mathrm{IL} = \mathrm{SPL}_1 - \mathrm{SPL}_2 \ \mathrm{dB} \tag{21.1.112}$$

Only TL is a general and reliable performance indicator for comparing the acoustic transmission through individual panels with the performance of an entire enclosure. The main reason for this is that both the alternative NR and insertion loss methods can produce results with negative values in the cases where the sound pressure levels outside the enclosure actually increase at certain frequencies once the enclosure is installed. This seemingly unexpected situation will occur when resonant acoustic modes within the enclosure result in standing wave ratios that are greater than the TL of the enclosure boundary. This phenomenon can catch the novice refinement engineer out.

21.1.10.3 Interpreting the acoustic performance data supplied by enclosure and panel manufacturers

Manufacturers of acoustic enclosures and high TL panels will often publish details of the acoustic performance of individual panels in terms of:

- TL;
- absorption coefficient;
- sound transmission class.

It is important to ascertain the conditions under which these data were obtained. They will generally be representative yet conservative values for results obtained from a number of laboratory tests made according to relevant material or international standards.

TL data may be obtained from tests performed in a transmission suite in accordance with ISO 140/1: 1990, ISO 140/3: 1990. The sound TL or sound reduction index (SRI) for a panel is described in terms of the ratio of sound power incident on the panel to the sound power transmitted through the panel. Then,

$$\mathrm{TL} = 10 \log_{10} \frac{\Pi_i}{\Pi_t} \ \mathrm{dB} \tag{21.1.113}$$

Sound TL is a quantity that depends only on the frequency of the sound and the properties of the panel. Consider a panel, with fixed sound TL mounted in a transmission suite (Fig. 21.1-15).

A given spatially averaged sound pressure level in the source room (L_i) gives rise to a spatially averaged velocity on the surface of the panel (u) and hence a spatially averaged sound pressure level in the receiving room (L_t) according to the relationship

$$L_t \propto \sigma S u^2 \tag{21.1.114}$$

where

S = panel area (m)
σ = acoustic radiation efficiency

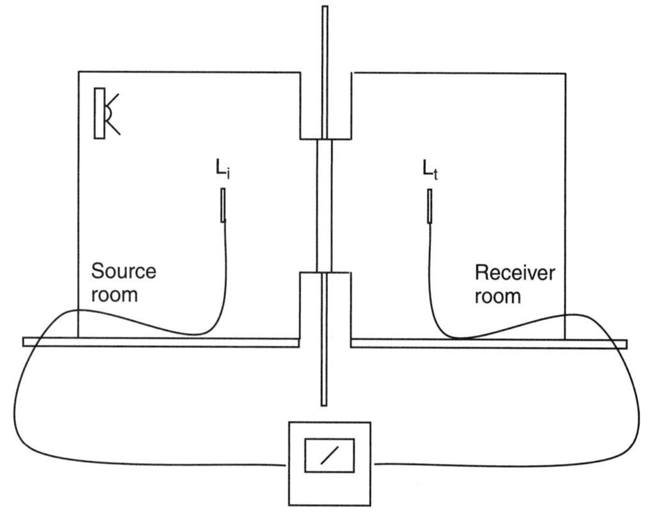

Fig. 21.1-15 The transmission suite.

Therefore, it can be seen that L_t is dependent on the area of the panel.

Also, L_t is found to be dependent on the absorption (A) of the receiving room for a given intensity (I) in the source room. A receiving room with little absorption will have a higher sound pressure level than one with a large absorption.

The sound TL for the panel and the NR in the suite can be thus related.

Power incident on the panel

$$\Pi_i = I_i S \tag{21.1.115}$$

Power absorbed by the receiving room

$$\Pi_t = I_t A \tag{21.1.116}$$

$$TL = 10 \log_{10} \frac{\Pi_i}{\Pi_t}$$

$$TL = 10 \log_{10} \left(\frac{I_i S}{I_t A} \right)$$

$$TL = L_i - L_t + 10 \log_{10} \left(\frac{S}{A} \right) \tag{21.1.117}$$

The difference in levels between the two rooms is known as the noise reduction or NR

$$NR = 10 \log_{10} \left(\frac{I_i}{I_t} \right) = L_i - L_t \tag{21.1.118}$$

$$TL = NR + 10 \log_{10} \left(\frac{S}{A} \right) \tag{21.1.119}$$

According to the relations developed by Sabine, the absorption of the receiving room is related to its reverberation time (T) – the time in seconds required for the level of the sound to drop by 60 dB. Therefore,

$$A = \frac{0.161\, V}{T} \tag{21.1.120}$$

where V is the volume of the receiving room (m^3). The reverberation time for each 1/3 octave band can readily be measured in the receiving room using a real-time analyser.

The transmission suite is constructed in such a way as to:

- restrict the transmission of sound to paths passing directly through the test panel;
- provide a source field which impinges with random angles of incidence on the test panel.

For such tests, a panel could be mounted and sealed in an aperture within the brick dividing wall between two rectangular reverberant rooms, both of which may be constructed from 215-mm brick with reinforced concrete floors and ceilings. In one particular case (by way of example only), the adjoining brick wall is 4.8-m wide and 3.1-m high with 550-mm nominal thickness, and forms the whole of the common area between the two rooms. One of the rooms is larger and is termed the receiving room. It has a depth of 20.2 m and a volume of 200 m^3. The adjoining room has a depth of 4 m and a volume of 60 m^3 and is known as the source room. The receiving room is isolated from the surrounding structure and adjoining room by the use of resilient mountings and seals.

Broadband white noise is produced in the source room using a minimum of two loudspeakers. The average sound pressure levels in each room are determined using distributed arrays of microphones, connected to a suitable third octave analyser. The difference in the averaged sound pressure levels forms the NR (shown as the upper curve in the left hand plot of Fig. 21.1-16). From this, and the knowledge of the 1/3 octave reverberation times (also shown in Fig. 21.1-16), the measured SRI or TL can be found (shown as the lower curve in the left-hand plot of Fig. 21.1-16).

The absorption coefficient of the test panel may be measured according to standard procedures in an impedance tube or in a reverberation chamber (Section 21.1.9.1). The impedance tube method is restricted to small samples of material and sound of normal incidence and therefore when testing the acoustic absorption of enclosure panels, a test based on measurements is made in a reverberation chamber.

It should be noted that both TL and absorption data are obtained from tests performed in reverberation chambers with diffuse fields where the sound impinges on the material samples with an essentially random angle of incidence. These ideal conditions may not be found in the case of a practical enclosure. In addition to this, the laboratory tests may have been performed on a single panel in isolation. The TL, and, to a lesser extent, absorption characteristics of a panel depend on the panel dimensions and on its mounting arrangement. Therefore, the published laboratory specifications may not be fully realistic in the case of an enclosure made up of a number of panels mounted in a variety of ways.

When examining the acoustic performance of enclosure panels it is important to consider absorption characteristics as fully as the TL qualities, if the detrimental effects of acoustic resonances within the enclosure are to be avoided. In addition to TL data, manufacturers of acoustic enclosures might also quote sound transmission class (STC) values for panels. The STC parameter was developed based on studies made with noise sources typical of multifamily dwellings, and provides a convenient and fairly successful single index specification of the transmission characteristics of a panel. To determine the STC of a panel, the TL is measured in the sixteen

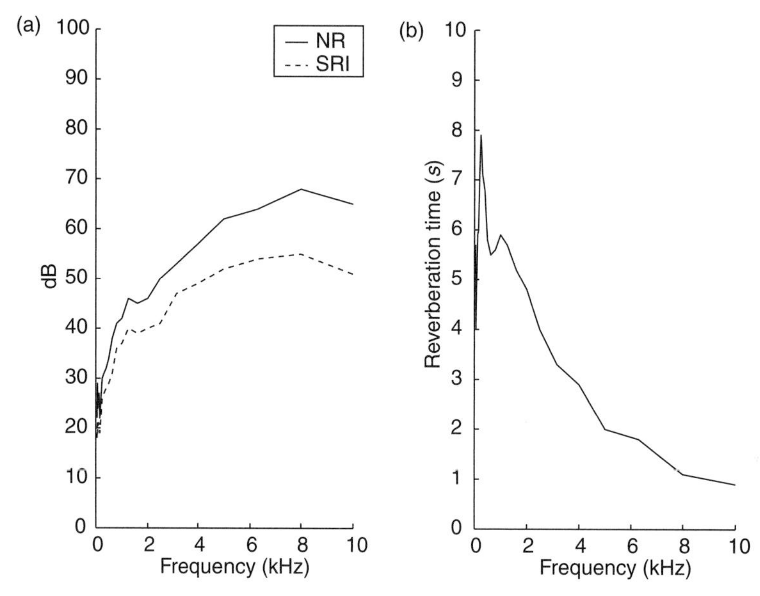

Fig. 21.1-16 Measured NR and SRI (TL) curves along with reverberation times.

contiguous 1/3 octave bands between 125 Hz and 4000 Hz inclusive. These measured values are then compared to a family of reference curves as shown in Fig. 21.1-17.

To determine the STC of a panel, the reference contour is chosen so that the maximum deficiency (deviation of the data below the contour) at any one frequency does not exceed 8 dB and the total deficiency at all frequencies does not exceed 32 dB. The STC of the panel is then the value of TL corresponding to the value of the chosen reference curve in the 500 Hz band. STC values are published widely in the literature, for example in Kinsler et al. (1982).

21.1.10.4 The significance of acoustic seals and controlling flanking transmission

The need for tight acoustic seals between adjoining panels cannot be overstated. Even the smallest air gap will limit the maximum TL that can be achieved, particularly at higher frequencies. A simple yet effective way of detecting sound leaking around a seal between panels is to use a length of narrow bore (5 mm) plastic tube, place one end in one ear and pass the other end

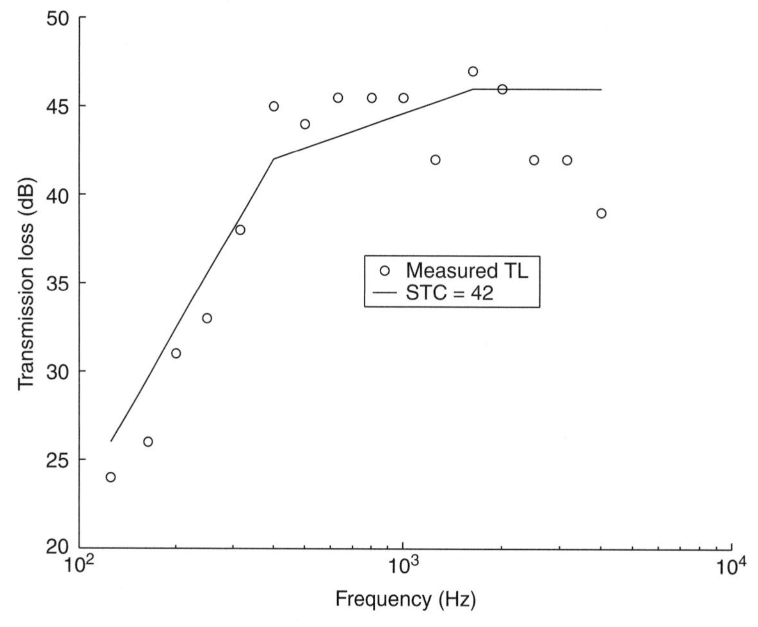

Fig. 21.1-17 STC curves: data after Kinsler et al. (1982).

over the edges of the panel. Any acoustic leaks become immediately apparent. The fitting of quality acoustic seals is one of the reasons for the relatively high cost of acoustic encapsulation. One of the major challenges in enclosure design is the provision of adequate ventilation and cooling without causing leakage of sound.

Flanking transmission must be controlled if an enclosure is to be fully effective. A common flanking path is due to vibration from the noise source being transmitted into the floor of the enclosure and either producing re-radiated sound outside the enclosure or forcing the walls of the enclosure into vibration and this seriously limits their TL characteristics. An obvious example is the interior noise due to an IC engine where structure-borne noise is in effect a flanking path.

Care should be taken that pipes and service ducts crossing the boundary of the enclosure are well isolated from the structure of the enclosure. Particular care should be taken if forced ventilation is being provided as in the case of an enclosure around an IC engine.

21.1.10.5 The transmission of sound through panels

Vibration may propagate through a structure in the form of compressional, shear or torsional waves. In structures constructed with thick members all three types of propagation may be significant. However, in thin panels pure compressional propagation is not likely, and in the audio frequency range such panels are usually excited by sound to form bending waves which are a combination of compression and shear motions. Bending waves result in a deflection of the panel in a direction that is normal to its surface. The spatial distribution of this deformation is a function of the bending wave speed.

The speed of propagation of sound in air c (m s^{-1}) is a function of the composition of the fluid and of temperature

$$c = \sqrt{\gamma RT} \qquad (21.1.121)$$

where

$\gamma = c_p/c_v$ ratio of specific heats
R = gas constant
T = temperature (K)

Interestingly, the speed of propagation of bending waves is dependent not only on the mechanical characteristics of the material (Poison's Ratio, Young's Modulus), but also on the shape and particularly the thickness of the panel. Thus for large elements, the bending stiffness B is:

$$B = EI \qquad (21.1.122)$$

where E is the Young's modulus (N m^{-2}) and l is the moment of inertia (m^4 or beware, commonly mm^4).

For plates of thickness h the moment of inertia per unit width l' is used

$$I' = \frac{h^3}{12} \qquad (21.1.123)$$

and the bending stiffness becomes:

$$B' = \frac{I'E}{1-\mu^2} \qquad (21.1.124)$$

The velocity at which one must travel to remain always at the same phase of one infinite sinusoidal wave is the phase velocity of bending waves

$$c_B = \sqrt[4]{\frac{B}{m'}}\sqrt[2]{\omega}(\mathrm{m\,s^{-1}}) \qquad (21.1.125)$$

where m' is the mass per unit length (ρS) and ω is the radial frequency (rad s^{-1}):

$$\omega = 2\pi f \quad f \text{ is the frequency (Hz)} \qquad (21.1.126)$$

For plates of thickness h, this becomes (providing that the wavelength $\lambda \gg 6h$)

$$c_B \approx \sqrt{1.8c_{L_I}hf} = c_{L_I}\sqrt{\frac{1.8h}{\lambda_{L_I}}}\;(\mathrm{m\,s^{-1}}) \qquad (21.1.127)$$

$$c_{L_i} = \sqrt{\frac{E}{\rho_m(1-\mu^2)}} \qquad (21.1.128)$$

where ρ_m is the material density and μ is Poisson's ratio.

Note that the phase velocity is frequency-dependent and therefore the wavefront distorts (dispersion) as higher-frequency components propagate with higher phase velocity than lower-frequency components.

One may demonstrate that energy propagates at the group velocity C_B (see p. 106 of Cremer and Heckl (1988) for details)

$$C_B = 2c_B(\mathrm{m\,s^{-1}}) \qquad (21.1.129)$$

It should be noted that practical panels are unlikely to be isotropic in construction and therefore bending wave speed could vary with direction making the panel orthotropic to some extent. However, the isotropic assumption is convenient and therefore will be pursued further here, particularly as the assumption tends to hold true as frequency increases and the flexural wavelength tends towards the characteristic dimension (usually thickness) of the panel.

In the case of isotropic panels there exists a frequency, named the critical frequency, at which the flexural wavelength in the panel matches the acoustic wavelength in the air. Orthotropic panels will have more than one critical frequency. The critical frequency or frequencies in either case are given by (Bies and Hansen, 1996):

$$f_c = \frac{c^2}{2\pi}\sqrt{\frac{m}{B'}}\ \text{Hz} \qquad (21.1.130)$$

where c is the speed of sound in air (m s^{-1}) and m is the surface density (kg m^{-2}).

At frequencies around the critical frequency an effect known as coincidence is noted. At coincidence frequencies the panel is strongly coupled to the fluid so that sound impinging on the panel from any angle of incidence will produce a strong flexural response in the panel (Fig. 21.1-18). Applying reciprocity indicates that the converse is true with a panel being a strong radiator of sound of any angle of emission at coincidence frequencies. It therefore follows that the coincidence effect greatly reduces the TL of a panel at frequencies near the critical frequencies. The response of the panel at the critical frequencies is a resonant phenomena and such a response is strongly dependent on the damping in the system.

At frequencies above the critical frequency an angle of incidence may be found so that the trace of the sound wave matches that of a flexural wave and good coupling results. At frequencies below the critical frequency, the wavelength of sound is longer than that of any flexural waves and poor coupling results due to local cancellation effects. Panels are therefore poor radiators of sound at low frequencies except at discontinuities or boundaries in their surfaces (such as at edges or ribs) where the cancellation effect is not present.

At these regions of localised coupling, the panel may be driven by sound of normal incidence. Curves showing the typical TL behaviour of isotropic and orthotropic panels are shown in Fig. 21.1-19.

At very low frequencies, the TL of an isotropic panel is controlled by the stiffness of the panel. At the frequency of the first panel resonance the TL dips and is in part controlled by the damping of the system. At moderate frequencies, TL is controlled by the mass or surface density of the panel and increases at a rate of 6 dB per octave (the so-called mass law relationship) up to the coincidence dip around the critical frequency. At frequencies above the critical frequency the TL is said to be damping-controlled and rises at a rate of 9 dB per octave.

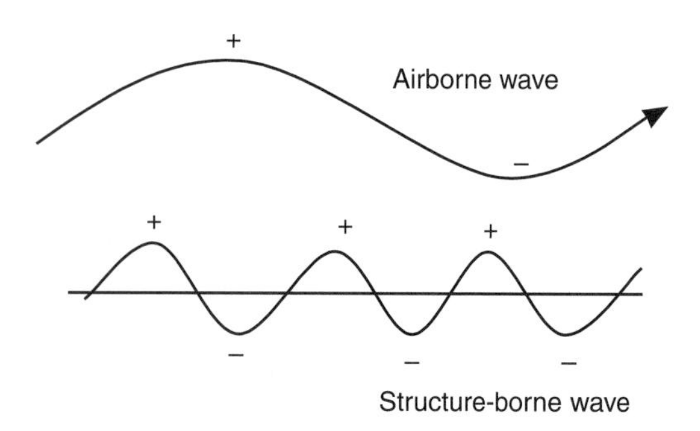

Fig. 21.1-18 Strong coupling between the airborne and structure-borne waves occurs around the coincidence frequency.

The corresponding typical TL curve for orthotropic panels is characterised by a wide coincidence region caused by the presence of more than one critical frequency. For this reason orthotropic panels should be avoided when noise control is important. However, the TL characteristics of a heavily damped orthotropic panel will tend towards that of an isotropic panel.

It should be noted that the range of frequencies for which the mass law operates is controlled by the panel stiffness. Stiffening a panel tends to move the first panel resonance f_0 up in frequency, and the critical frequency f_c down. For a panel of uniform thickness h (Bies and Hansen, 1996)

$$f_0 = 0.453 c_L h(a^{-2} + b^{-2})\ \text{Hz} \qquad (21.1.131)$$

$$f_c = \frac{c^2}{1.81 c_L h}\ \text{Hz} \qquad (21.1.132)$$

The mass law is dependent on the angle of incidence of the impinging sound. Sharp (reported in Bies and Hansen (1996)) suggests that for an infinite panel

$$TL_\theta = 10\log_{10}\left[1 + \left(\frac{\pi f m}{\rho c}\cos\theta\right)^2\right]\ \text{dB} \qquad (21.1.133)$$

where ρ is the density of air (kg m^{-3}).

Pierce (reported in Fahy (1985)) shows that the random incidence (diffuse field) TL (TL_D) is obtained from a weighted average of all angles of incidence:

$$TL_D = TL_N - 10\log_{10}(0.23 TL_N)\ \text{dB} \qquad (21.1.134)$$

where TL_N is the normal incidence TL (dB).

The field TL is not necessarily equal to the TL_D. Sharp (reported in Bies and Hansen (1996)) suggests the following relationship for the field TL:

$$TL = TL_N - 5\ \text{dB} \qquad (21.1.135)$$

The difference between TL and TL_D is probably due to the finite size of panels tested for the case of TL_D.

Sharp (reported in Bies and Hansen (1996)) suggests the following mass law:

$$TL = 20\log_{10}[(\pi f m/\rho c)] - 5\ \text{dB} \qquad (21.1.136)$$

(a)

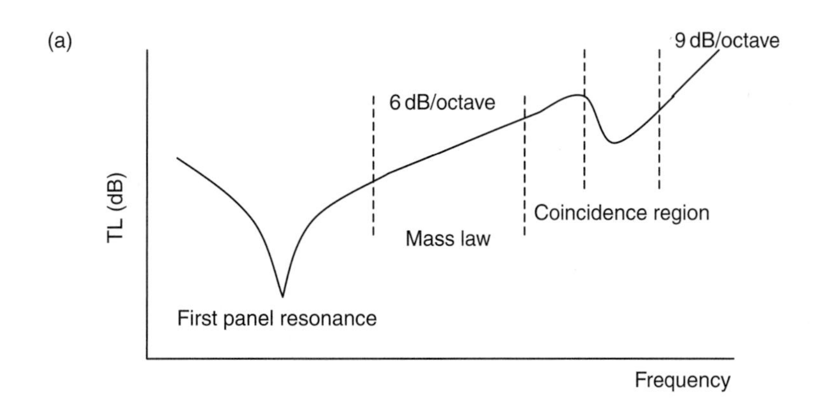

(b)

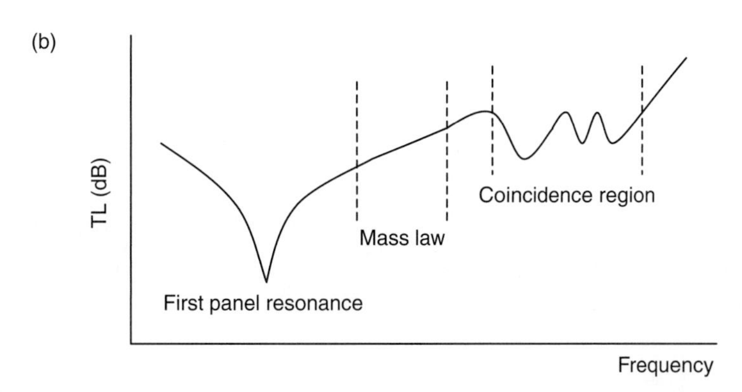

Fig. 21.1-19 Typical panel TL (after Bies and Hansen, 1996). (a) Isotropic panel and (b) orthotropic panel.

where f is the third octave centre frequency and for the case when $\frac{fm}{\rho c} > 1$.

Sharp (reported in Bies and Hansen (1996)) also suggests the following relationship for the field incidence TL of an isotropic panel above the critical frequency:

$$TL = 20\log_{10}\left(\frac{\pi f m}{\rho c}\right) + 10\log_{10}\left[\frac{2\eta f}{\pi f_c}\right]\text{dB} \qquad (21.1.137)$$

where η is the panel loss factor and f is the frequency (Hz) which this time is independent of bandwidth.

Higher levels of TL may be achieved if a double-skinned enclosure is used. This construction is usually more cost-effective than constructing a very massive single-walled enclosure. For best results the two skins should be mechanically and acoustically isolated from each other. Mechanical isolation can be achieved by mounting the two skins on separate beams or by using neoprene rubber between the skins and the common studs. Acoustic isolation can be achieved by filling the air gap with an absorptive material. The material should be at least $15/f$ metres thick with an impedance of 3–5 ρ_c (too high an impedance might result in a mechanical path through the material).

The TL characteristics of a single panel has been shown to be influenced by two frequency bands, the first centred on the lowest-order panel resonance and the second on the critical frequency. In the double leaf case, the influence of the lowest-order resonance of a single panel is replaced with the lowest-order acoustic resonance within the cavity at f_2 where (Sharp reported in Bies and Hansen (1996))

$$f_2 = \frac{c}{2L} \qquad (21.1.138)$$

L = longest cavity dimension (m)

The critical frequencies f_{c1} and f_{c2} are still important and added consideration must be given to the mass-air-mass resonance at f_0 of the two panels on the compliance of the cavity and a limiting frequency f_1 related to the width of the air gap (d) between panels. Sharp (reported in Bies and Hansen [1996]) gives the following relationships for panels which are totally mechanically and acoustically isolated from each other:

$$f_0 = \frac{1}{2\pi}\left(\frac{1.8\rho c^2(m_1 + m_2)}{d m_1 m_2}\right)^{1/2}\text{Hz} \qquad (21.1.139)$$

$$f_1 = \frac{c}{2\pi d}\text{Hz} \qquad (21.1.140)$$

Below $f_{c\,1,2}/2$

$$TL_M = 20\log_{10}\left[\frac{\pi f M}{\rho c}\right] - 5\text{ dB} \quad (21.1.141)$$

Above $f_{c\,1,2}$

$$TL_M = 20\log_{10}\left[\frac{\pi f M}{\rho c}\right] + 10\log_{10}\left[\frac{2\eta f}{\pi f_c}\right]\text{ dB} \quad (21.1.142)$$

where

$$M = m_1 + m_2 \quad (21.1.143)$$

Finally

$$TL = TL_M \quad f < f_0 \quad (21.1.144)$$

$$TL = TL_1 + TL_2 + 20\log_{10} fd - 29 \quad f_0 < f < f_1 \quad (21.1.145)$$

where TL_1 and TL_2 are found by using m_1 or m_2, respectively in the mass law:

$$TL = TL_1 + TL_2 + 6 \quad f < f_1 \quad (21.1.146)$$

These equations relate to the ideal case where the two leaves are mechanically isolated from one another and absorptive material is introduced between the leaves to eliminate the effect of acoustic resonances in the cavity. Sharp presents some algorithms to predict the effect that the mounting of the leaves (line, line-point or point-point) has on the ideal TL, which will not be presented here (Sharp reported in Bies and Hansen (1996)).

Fahy presents an alternative model for the TL of double-leaved panels in Fahy (1985). Here he presents a 1-D model that assumes mechanical isolation between the leaves, but takes account of the acoustic resonances in the air gap. It should be noted that Fahy's method produces predictions of normal incidence TL while Sharp's method relates to field TL. Fahy's equations are:

$$\omega_0 = \left[\frac{\rho_0 c^2}{d}\left(\frac{m_1 + m_3}{m_1 m_2}\right)\right]^{1/2} \quad (21.1.147)$$

(note the omission of the factor 1.8 which appears in Sharp's equation);

$$TL_{M,0} = 20\log_{10} M + 20\log_{10} f - 20\log_{10}\left(\frac{\rho_0 c}{\pi}\right)\text{ dB} \quad (21.1.148)$$

where $M = m_1 + m_2$ and $TL_{1,0}$ and $TL_{2,0}$ are obtained by substitution of m_1 and m_2 respectively.

Below ω_0

$$TL_0 = TL_{M,0}\text{dB} \quad (21.1.149)$$

At ω_0

$$TL_0 = TL(0, M, \omega') + 20\log_{10}\eta\text{ dB} \quad (21.1.150)$$

for the case of $m_1 = m_2$, and ω' is the natural frequency of the panels which Bies and Hansen (1996) suggest to be:

$$\omega'_{i,n} = \pi^2\sqrt{\frac{B'}{m}}\left[\frac{i^2}{a^2} + \frac{n^2}{b^2}\right]\ (\text{rad s}^{-1}) \quad i, n = 1, 2, 3, \ldots \quad (21.1.151)$$

where a, b are panel width and length, B' is the bending stiffness per unit width and the natural frequency occurs when $i = n = 1$.

Between ω_0 and $kd = \pi/4$ where

$$k = \frac{\omega}{c} \quad (21.1.152)$$

$$TL_0 = TL_{1,0} + TL_{2,0} + 20\log_{10}(2kd)\text{ dB} \quad (21.1.153)$$

At frequencies corresponding to acoustic anti-resonances of the air gap the TL is maximum at:

$$kd = (2n-1)\frac{\pi}{2} \quad n = 1, 2, 3, \ldots fc_{1,2} \quad (21.1.154)$$

$$TL \simeq TL_{1,0} + TL_{2,0} + 6\text{ dB} \quad (21.1.155)$$

At frequencies corresponding to resonances of the air gap, the TL dips down to the combined mass law for the two leaves:

$$TL_0 = 20\log_{10} M + 20\log_{10} f - 20\log_{10}\left(\frac{\rho_0 c}{\pi}\right)\text{ dB} \quad (21.1.156)$$

when

$$kd = n\pi$$

These equations can be used to form a generalised TL curve as shown in Fig. 21.1-20.

21.1.10.6 Sound inside and outside large enclosures

Enclosures are deemed to be large if they are not designed to be close fitting around a noise source. An

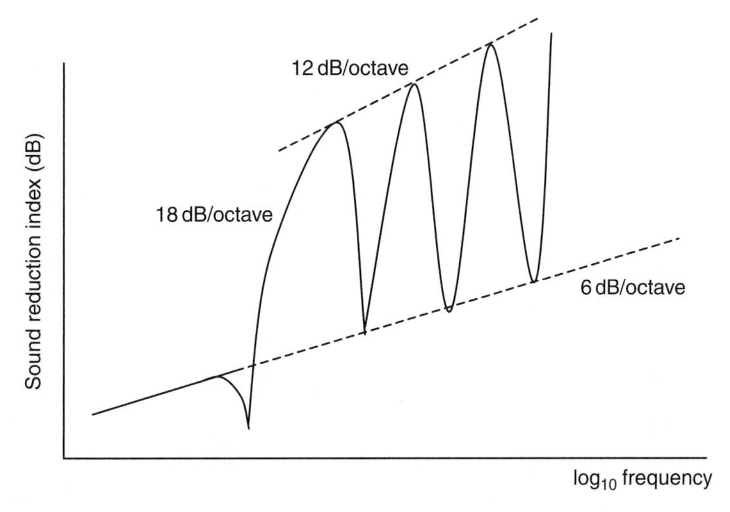

Fig. 21.1-20 Generalised TL curves: after Fahy (1985).

enclosure might be considered large if a complete wavelength of sound at a low frequency separated the nearest wall of the enclosure from the noise source.

As already discussed, the installation of an enclosure around a noise source will produce a reverberant field within the enclosure. The distribution of sound pressure level within the enclosure can be ascertained, given its absorption characteristics and a noise source of known power output.

The sound field outside the enclosure may be assumed to comprise a contribution from the direct field of the source reduced in amplitude by the normal TL (TL_N) of the boundary of the enclosure and a contribution from the reverberant field within the enclosure reduced in amplitude by the field TL of the enclosure.

Now

$$TL_n = -10 \log_{10}\tau_n \text{ dB} \quad (21.1.157)$$

$$TL = -10 \log_{10}\tau \text{ dB} \quad (21.1.158)$$

where τ_n and τ are transmission coefficients. Also

$$TL_N - TL = 5 \text{ dB} \quad (21.1.159)$$

therefore

$$\tau_n = 0.3\tau \quad (21.1.160)$$

Now, the total sound power radiated by the enclosure is given by (Bies and Hansen, 1996):

$$S_E \frac{\langle p_1^2 \rangle}{\rho c} = W_{\tau_n} + W(1-\bar{a}_i)\frac{S_E}{S_i \bar{a}_i}\tau \quad (21.1.161)$$

which is the fraction of sound power transmitted by the direct field added to the fraction of sound power transmitted by the reverberant field. In this equation

$\bar{a}_i$= average internal Sabine absorptivity

S_i = internal surface area including the surface area of the noise source

S_E = external surface area of the enclosure

This equation can be re-written as:

$$S_E \frac{\langle p_1^2 \rangle}{\rho c} = WT_E \quad (21.1.162)$$

where

$$\frac{\langle p_1^2 \rangle}{\rho c} = \text{sound intensity (Bies and Hansen, 1996)} \quad (21.1.163)$$

$$T_E = \tau[0.3 + S_E(1 - \bar{a}_i/S_i\bar{a}_i)] \quad (21.1.164)$$

The sound pressure directly outside the enclosure can therefore be obtained from (Bies and Hansen, 1996)

$$L_{p_1} = L_w - TL - 10 \log_{10}S_E + C \quad (21.1.165)$$

where

$$C = 10 \log_{10}\left[0.3 + S_E(1-\bar{a}_i)/(S_i\bar{a}_i)\right] \text{ dB} \quad (21.1.166)$$

It should be noted that this equation gives very approximate results only.

The sound pressure level at a point in free space some distance r away from the enclosure can be determined from (Bies and Hansen, 1996):

$$L_{WE} \simeq L_{p1} + 10\log_{10} S_E \text{ dB} \quad (21.1.167)$$

$$L_{p_2} = L_{WE} + 10\log_{10}\left[\frac{D_\theta}{4\pi r^2}\right] \text{ dB} \quad (21.1.168)$$

When the enclosure is positioned on a hard floor:

$$D_\theta = 2 \quad (21.1.169)$$

The sound pressure level at a point in a reverberant space can equally be found using (see Section 21.1.3.5)

$$L_{p_2} = L_{WE} = 10\log_{10}\left[\frac{D_\theta}{4\pi r^2} + \frac{4(1-\bar{a})}{S\bar{a}}\right] \text{ dB} \quad (21.1.170)$$

By performing this calculation twice, once with the enclosure in place and once without, it can be shown that

$$NR = TL - C \text{ dB} \quad (21.1.171)$$

A similar calculation may be performed to estimate the sound pressure field within an enclosure sited in a reverberant field. The power flow into the enclosure is equal to

$$W_i = S_E \frac{\langle p_1^2 \rangle}{4\rho c}\tau \quad (21.1.172)$$

so that

$$L_{Wi} = L_{p1} + 10\log_{10} S_E - TL - 6 \text{ dB} \quad (21.1.173)$$

so

$$L_{pi} = L_{Wi} + 10\log_{10}\left[\frac{1}{S_E} + \frac{4(1-\bar{a}_i)}{S_i\bar{a}_i}\right] \text{ dB} \quad (21.1.174)$$

It can be shown that

$$NR = TL - C \text{ dB} \quad (21.1.171)$$

The problem is not so simple when the direct field is dominant at one or more walls of the enclosure. In this case, the enclosure should be treated as a barrier.

21.1.10.7 Sound inside and outside close fitting enclosures

Jackson produced two useful papers on the performance of close fitting enclosures (Jackson, 1962; 1966). He developed a 1-D model of such an enclosure by treating the noise source and the enclosure as a pair of concentric pulsating boxes. The potentially complex three-dimensional problem was reduced to that of a pair of flat, infinite parallel plates separated by a distance l and immersed in air as shown in Fig. 21.1-21.

Jackson made the following assumptions:

1. That generally large radiating areas of machinery are involved in cases using close fitting enclosures which encourage the propagation of acoustic waves normal to their surface.
2. The enclosure does not touch any part of the body it encloses.
3. The presence of the enclosure does not affect the magnitude of vibration of the enclosed surfaces.
4. Direct transmission of vibration through the support of the sound source does not occur.

Jackson developed an equation describing the attenuation produced by such an enclosure, which is:

$$\left|\frac{Y_1}{Y_0}\right| = A = \left[1 - \frac{2\sin\theta(X\cos\theta - R\sin\theta)}{\rho c} + \frac{\sin^2\theta(X^2+R^2)}{\rho^2 c^2}\right]^{1/2} \quad (21.1.175)$$

where

$$X = \omega M - \frac{S}{\omega} \quad (21.1.176)$$

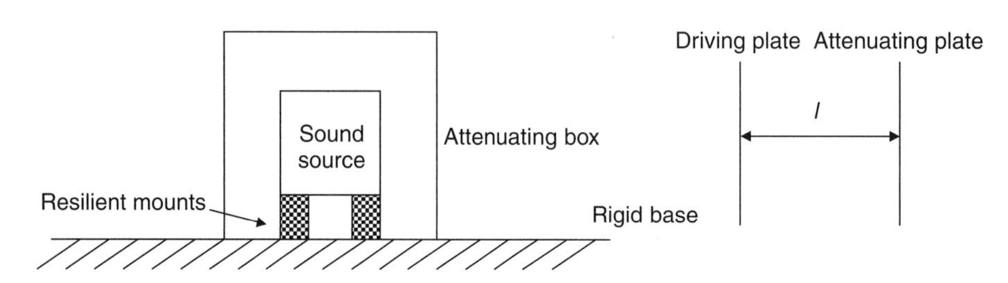

Fig. 21.1-21 Jackson's models (Jackson, 1962).

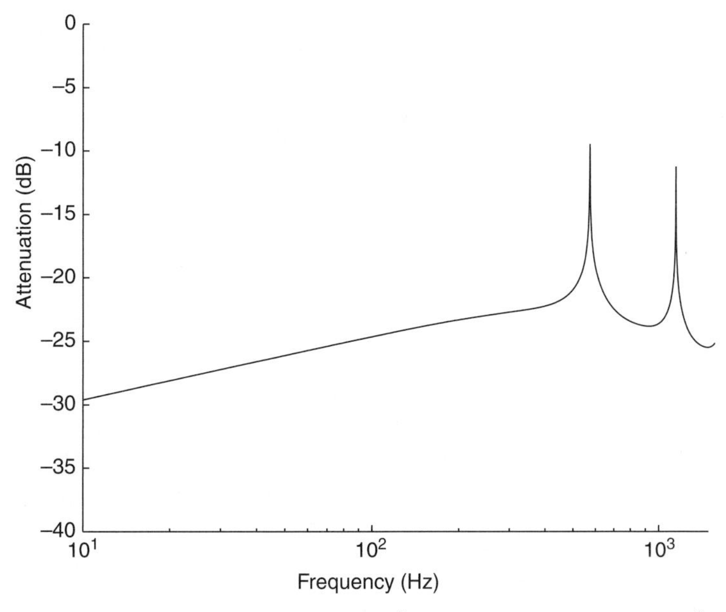

Fig. 21.1-22 Results for a steel box with a surface density of 16 kg m^{-2} and $l = 0.3$ m and s $= 1 \times 10^8$ N m^{-1} using Jackson's method (Jackson, 1966).

M= mass per unit area (kg m^{-2})
R = mechanical resistance (damping)

$$\theta = \frac{\omega L}{c}$$

L = distance between plates (m)
ρ = density of air (kg m^{-3})
ω = angular frequency (rad s^{-1})
S = uniform elastic restraint per unit area (stiffness)

Theoretical results are shown in Fig. 21.1-22.

Jackson concluded that, providing the vibration source is sealed from free space, good low-frequency performance is possible if the wall stiffness is made high. Also, in a sealed system in which little mechanical damping is associated with the mass, if no stiffness is present then a magnification of sound will occur at low frequencies due to the presence of the hood, an effect not predicted by the mass law.

Appendix 21.1A: Some background information on systems

In the simplest form a system may have only one input and one output. Such systems are often called single input, single output systems or two-pole systems (Sinha, 1991). In the more general case, systems may have several inputs and several outputs. These are called multivariable or multipole systems (as illustrated in Fig. A22.1-1).

A zero state system is one where all the initial conditions are zero.

Linear and non-linear systems

A system is said to be linear if, and only if, it satisfies the superposition theorem. Note that superposition consists of two basic but quite distinct concepts:

1. The property of additivity – the response of a zero state system to the sum of two inputs is equal to the sum of the responses to each of the inputs acting alone.
2. The property of homogeneity – requires that the effect of multiplying the input by a constant would be to multiply the output by the same constant.

A system is said to be non-linear if it does not satisfy the properties of additivity and homogeneity.

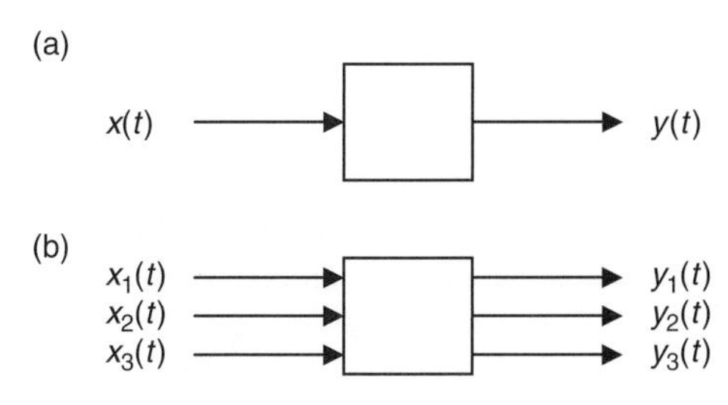

Fig. A21.1-1 (a) Two-pole system; (b) multipole system.

Differential equations

The input–output relations for networks that store energy (dynamic systems) are given by differential equations.

Consider a linear, continuous-time system where:

- The output is represented by $y(t)$;
- The input is represented by $x(t)$.

The two variables will be related by a differential equation of the form:

$$a_n\frac{d^ny}{dt^n}+a_{n-1}\frac{d^{n-1}y}{dt^{n-1}}+\cdots+a_1\frac{dy}{dt}+a_0y = b_m\frac{d^mx}{dt^m}+\cdots+b_0x \quad \text{(A21.1.1)}$$

This equation is called a 'linear differential equation of order n' and for most practical cases $n \geqslant m$.

It is convenient to replace $\frac{d}{dt}$ by the operator 'p' resulting in the equation

$$(a_np^n+a_{n-1}p^{n-1}+\cdots+a_1p+a_0)y(t) = (b_mp^m+\cdots+b_1p+b_0)x(t) \quad \text{(A21.1.2)}$$

which may be written compactly as

$$D(p)y(t) = N(p)x(t) \quad \text{(A21.1.3)}$$

where $D(p)$ and $N(p)$ are polynomials in the operator 'p'

$$D(p) = a_np^n+a_{n-1}p^{n-1}+\cdots+a_1p+a_0 \quad \text{(A21.1.4)}$$

$$N(p) = b_mp^m+b_{m-1}p^{m-1}+\cdots+b_1p+b_0 \quad \text{(A21.1.5)}$$

The operator 'p' does not satisfy the commutative property

$$py(t)\neq y(t)p \quad \text{(A21.1.6)}$$

The system operator $L(p)$ or transfer function is the ratio of the two polynomials $D(p)$ and $N(p)$

$$L(p) = \frac{N(p)}{D(p)} \quad \text{(A21.1.7)}$$

Any dynamic system described by a differential equation of order n can be solved uniquely only if at least n initial or boundary conditions are known.

As an example, consider that the unique solution to the differential equation characterising the input–output relationship of an electrical circuit can be obtained only if the initial values of the voltages across each capacitor and the current through each inductance are known.

The same rules apply to the well-known, second-order differential equation characterising the motion of a mass on a spring with a viscous damper

$$m\ddot{x}+c\dot{x}+kx = F(t) \quad \text{(A21.1.8)}$$

If two initial conditions are known, namely $x(0)$ and $\ddot{x}(0)$, the value of x at some instant t later may be found.

Appendix 21.1B: The convolution integral

Some aperiodic signals have unique properties and are known as singularity functions because they are either discontinuous or have discontinuous derivatives (Sinha, 1991). The simplest of these is the unit step function (as illustrated in Fig. B21.1-1), given the symbol $\gamma(t)$

The unit impulse function or delta function $\delta(t)$ is defined as the function which after integration yields the unity step function, so (Sinha, 1991)

$$\gamma(t) = \int_{-\infty}^{t}\delta(\tau)d\tau \quad \text{(B21.1.1)}$$

Alternatively,

$$\delta(t) = \frac{\gamma(t)}{dt} \quad \text{(B21.1.2)}$$

The impulse function must satisfy:

$$\delta(t) = 0, \text{ for } t \text{ not equal to zero} \quad \text{(B21.1.3)}$$

and

$$\int_{-\infty}^{\infty}\delta(t)dt = 1 \quad \text{(B21.1.4)}$$

Therefore, the area under the impulse function is unity and it occurs over an infinitesimal interval around $t = 0$. So, as the period dt tends towards zero, the height of the impulse function approaches infinity.

Also,

$$\frac{d\delta(t)}{dt} = \infty \quad \text{at} \quad t = 0 \text{ and is zero elsewhere.}$$

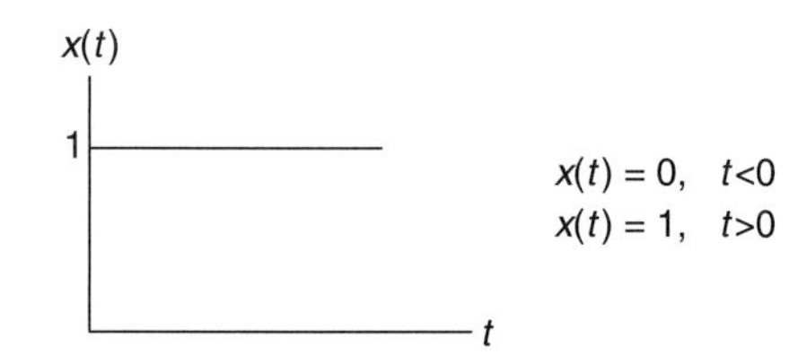

Fig. B21.1-1 The unit step function.

Consider a unit impulse that occurs at time $t = \tau$ as illustrated in Fig. B21.1-2

Remember that the definition of the unit impulse requires it to occur at time $t = 0$. Therefore, shift the time axis in the Fig. B21.1-2 by the amount required for $t = \tau = 0$. Therefore, the unit impulse occurring at time $t = \tau$ is assigned with the symbol $\delta(t - \tau)$.

Define the impulse–response function $h(t - \tau)$ of a system as the response $y(t)$ of the system at time t to a unit impulse $\delta(t - \tau)$ of duration $\rightarrow 0$ input sometime earlier at time $t = \tau$ and remember that the definition of the impulse function dictates that input time to be time $t = 0$.

Now, remember that the area under the unit impulse function is unity so it follows that the response $y(t)$ to a non-unitary impulse (i.e. a practical pulse, one with a finite duration Δ somewhat larger than zero) at time $t = \tau$ is given approximately by the product of the area of the non-unitary pulse and the impulse–response function:

$$y(t) \approx x(\tau)\Delta \cdot h(t - \tau) \qquad \text{(B21.1.5)}$$

In the limit as $\Delta \rightarrow 0$, and applying the superposition theorem for linear systems where the signal represented by a continuum of impulses is given by the sum of the individual responses to earlier impulses the convolution integral is obtained

$$y(t) = \int_{-\infty}^{\infty} x(\tau)h(t - \tau)\mathrm{d}\tau \qquad \text{(B21.1.6)}$$

An important application of the impulse function is the possibility of representing some arbitrary, continuous time signal of time $x(t)$ as a continuum of impulses as illustrated in Fig. B21.1-3.

One approximation to the smooth function above can be obtained by representing it as a sequence of rectangular pulses where the height of each pulse is made equal to the value of $x(t)$ at the centre of each pulse. The width of the pulse is Δ.

It follows that the approximation improves as the pulse width Δ tends to zero, i.e. as the pulse tends towards the unit impulse and at this point one can write:

$$x(t) = \int_{-\infty}^{\infty} x(\tau)\delta(t - \tau)\mathrm{d}\tau \qquad \text{(B21.1.7)}$$

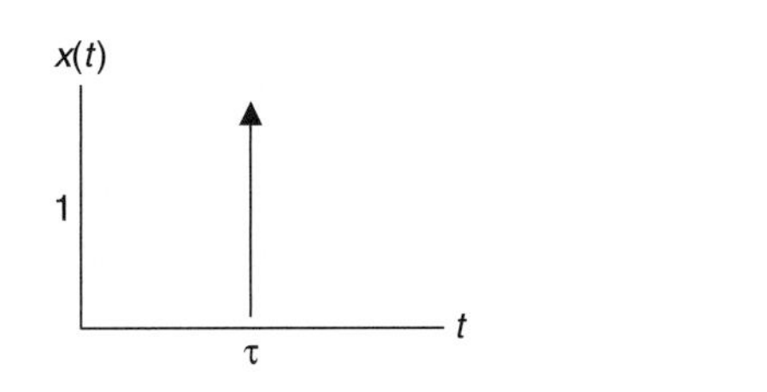

Fig. B21.1-2 The unit impulse at time $t = \tau$.

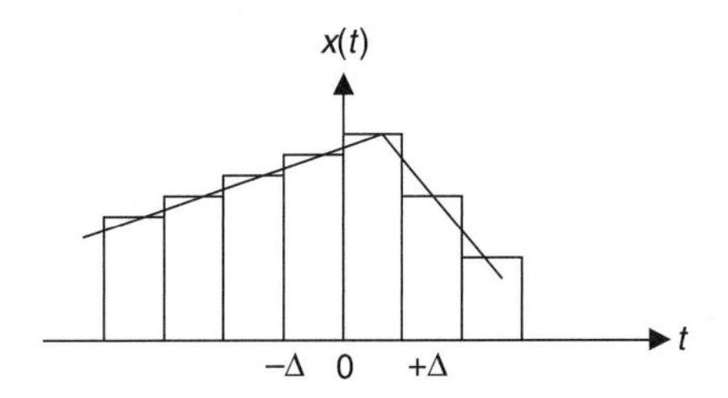

Fig. B21.1-3 $x(t)$ represented as a continuum of impulses: after (Sinha, 1991).

This equation is the result of an interesting property of the unit impulse function known as the sifting property, whereby a time-varying signal is described as the sum of a train of impulses, each one with a strength that is equal to the value of the signal at the time of the impulse.

Another important application of the impulse–response function is that it is directly related to the transfer function of a linear, time-invariant, continuous time system. There are two possible formal definitions of the transfer function (Sinha, 1991).

Definition 1 The transfer function of a linear, time-invariant, continuous time system is the Laplace transform of its impulse response.

Definition 2 The transfer function of a linear, time-invariant, continuous time system is the ratio of the Laplace transforms of the output and input under zero initial conditions.

Appendix 21.1C: The covariance function, correlation and coherence

Consider some probability attributes of a random variable X (Fahy and Walker, 1998). The distribution function $F(x)$ of a random variable X is given by

$$F(x) = \int_{-\infty}^{x} p(u)\mathrm{d}u \qquad \text{(C21.1.1)}$$

where p is the probability density function having the following attributes for a continuous distribution:

$$p(x) \geq 0$$

$$\int_{-\infty}^{\infty} p(x)\mathrm{d}x = 1$$

$$P[a\langle x\rangle b] = \int_{a}^{b} p(x)\mathrm{d}x$$

so

$$p(x) = \frac{\mathrm{d}F(x)}{\mathrm{d}x} \qquad \text{(C21.1.2)}$$

$F(x)$ is the probability of X taking a value up to and including x.

The expected value of X is defined as:

$$E[X] = \int_{-\infty}^{\infty} x \cdot p(x)dx \quad \text{(C21.1.3)}$$

which is also known as the mean value μ_x or the first moment of X.

If Y is a function of X, i.e. $Y = g(X)$

$$E[Y] = E[g(X)] = \int_{-\infty}^{\infty} g(x)p(x)dx \quad \text{(C21.1.4)}$$

Where W is a function of two variables, i.e. $W = g(X, Y)$

$$E[W] = \int_{-\infty}^{\infty}\int g(x,y)p(x,y)dxdy \quad \text{(C21.1.5)}$$

The second moment is given by:

$$E\left[X^2\right] = \int_{-\infty}^{\infty} x^2 p(x)dx \quad \text{(C21.1.6)}$$

This is a measure of the spread relative to the origin.

The spread relative to the mean is called the variance and is given by:

$$V(x) = E\left[(x-\mu_x)^2\right] = \int_{-\infty}^{\infty} (x-\mu_x)^2 p(x)dx \quad \text{(C21.1.7)}$$

The standard deviation is given by:

$$\sigma_x = \sqrt{V(x)} \quad \text{(C21.1.8)}$$

A random variable has a Gaussian distribution as illustrated in Fig. C21.1-1 if (Weltner et al. (1986) for example)

$$p(x) = \frac{1}{\sigma\sqrt{2\pi}} e^{-\frac{1}{2}\left(\frac{x-\mu}{\sigma}\right)^2} \quad \text{(C21.1.9)}$$

The second joint moment of two randomly distributed variables is:

$$E\left[(x-\mu_x)(y-\mu_y)\right] = \int_{-\infty}^{\infty}\int (x-\mu_x)(y-\mu_y)p(x,y)dxdy \quad \text{(C21.1.10)}$$

This is called the covariance function relating x and y.

Some useful definitions are (Fahy and Walker, 1998):

x and y are *uncorrelated* if $E(X,Y) = E(W) = E(X) \cdot E(Y)$

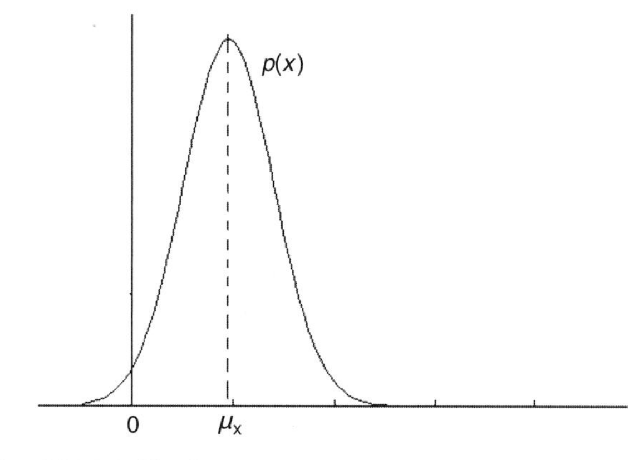

Fig. C21.1-1 The Gaussian distribution.

x and y are orthogonal if $E(W) = 0$ (i.e. X and Y do not coexist)

x and y are independent if $p(x,y) = p(x)p(y)$

The degree of correlation between two statistical data sets might be established using the three categories above or using the correlation coefficient. Therefore,

$$r = \frac{\sum xy - \frac{\sum x \sum y}{n}}{\sqrt{\left(\sum x^2 - \frac{(\sum x)^2}{n}\right)\left(\sum y^2 - \frac{(\sum y)^2}{n}\right)}}$$

$$-1 < r > 1 \quad \text{(C21.1.11)}$$

(see http://max.econ.hku.hk/stat/hyperstat/A56626.html for example)

x, y are the measured values. All sums are formed from $i = 1$ to $i = n$, where n is the number of measurements.

However, beware, there are many potential pitfalls when using correlation coefficients. A high correlation does not imply causation. Reasons for this include:

- x and y may seem well correlated (a value near -1 or $+1$) but this may be due to the effect both of them being related to the same third variable.
- x and y may seem to be poorly correlated but there might be a causal relationship between them – it might be that the relationship is not linear or is being confounded by the effect of another variable, or that the data range of x is rather small.

(see for example http://www.math.virginia.edu/~der/useml70/Chapter05/sld040.htm)

An alternative to the use of the correlation coefficient is the use of the autocovariance function with a random process (Fahy and Walker, 1998), that is:

$$R_{xx}(t_1,t_2) = E[(x(t_1) - \mu_x(t_1))(x(t_2)\mu_x(t_2))] \quad \text{(C21.1.12)}$$

This is a measure of the degree of association of the signal at time t_1 and the same signal at time t_2. Perhaps one could see it as a measure of how predictable future signal levels are based on a historic knowledge of that signal.

If the mean values are not subtracted the autocorrelation function is obtained

$$E[x(t_1)x(t_2)] \quad \text{(C21.1.13)}$$

With a stationary random process μ_x remains constant with time in the period $(t_1 - t_2)$

$$R_{xx}(t_2 - t_1) = E[(x(t_1) - \mu_x)(x(t_2) - \mu_x)] \quad \text{(C21.1.14)}$$

Commonly:

$t_2 = t_1 + \tau$ Where τ is the time lag

$t_1 = t$

so

$$R_{xx}(\tau) = E[(x(t) - \mu_x)(x(t+\tau) - \mu_x)] \quad \text{(C21.1.15)}$$

when

$$\tau = 0, \quad R_{xx} = V(x)$$

When $\tau \to \infty$, $R_{xx} \to 0$ as the two random samples tend to be less associated.

A typical autocorrelation function looks like that shown in Fig. C21.1-2.

When two random variables are involved the cross covariance function is obtained.

$$R_{xy}(\tau) = E\left[(x(t) - \mu_x)(y(t+\tau) - \mu_y)\right] \quad \text{(C21.1.16)}$$

Note that as $\tau \to \infty$, the mean values $\to 0$ for random signals. So

$$R_{xx}(\tau) = E[x(t)x(t+\tau)] \quad \text{(C21.1.17)}$$

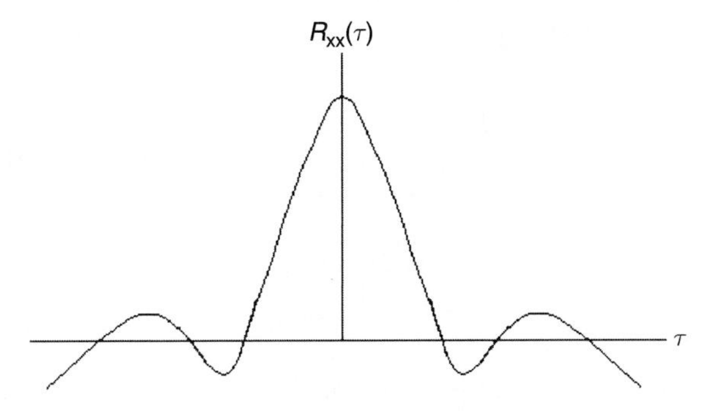

Fig. C21.1.2 A typical autocorrelation function.

$$R_{xy}(\tau) = E[x(t)y(t+\tau)] \quad \text{(C21.1.18)}$$

When $T \to 0$, $R_{xx}(T)$ is the variance of x (i.e. x^2 when the mean is zero).

The average power in the signal over period $T \to \infty$ is (Sinha, 1991)

$$P_{\lim T\to\infty} = \frac{1}{T}\int_{-T/2}^{T/2} x^2(t)\mathrm{d}t \quad \text{(C21.1.19)}$$

Now (Fahy and Walker, 1998) Parseval's theorem states that

$$\frac{1}{T}\int_{-T/2}^{T/2} x^2(t)\mathrm{d}t = \frac{1}{T}\int_{-\infty}^{\infty} x_T^2(t)\mathrm{d}t = \frac{1}{T}\int_{-\infty}^{\infty} |X_T(f)|^2\mathrm{d}f \quad \text{(C21.1.20)}$$

where x_T is the truncated data set for $x(t)$ between times $-T/2$ and $T/2$. Now, as $T \to \infty$

$$P_{\lim T\to\infty} = \frac{1}{T}\int_{-T/2}^{T/2} x^2(t)\mathrm{d}t = \int_{-\infty}^{\infty} \left|\frac{X_T(f)}{T}\right|^2 \mathrm{d}f \quad \text{(C21.1.21)}$$

So,

$$E\left[P_{\lim T\to\infty}\right] = E\left[\frac{1}{T}\int_{-T/2}^{T/2} x^2(t)\mathrm{d}t\right] = E\left[\int_{-\infty}^{\infty}\left|\frac{X_T(f)}{T}\right|^2\mathrm{d}f\right] \quad \text{(C21.1.22)}$$

Now the power spectral density function S_{xx} is given by (Fahy and Walker, 1998)

$$S_{xx}(f)_{\lim T\to\infty} = \frac{E[X_T(f)]^2}{T} \quad \text{(C21.1.23)}$$

So, from

$$R_{xx}(\tau) = E[x(t)x(t+\tau)] \quad \text{(C21.1.24)}$$

$$R_{xy}(\tau) = E[x(t)y(t+\tau)] \quad \text{(C21.1.25)}$$

one may write

$$S_{xx}(f) = \int_{-\infty}^{\infty} R_{xx}(\tau)e^{-i(2\pi f\tau)}\mathrm{d}\tau \quad \text{(C21.1.26)}$$

$$S_{xy}(f) = \int_{-\infty}^{\infty} R_{xy}(\tau)e^{-i(2\pi f\tau)}\mathrm{d}\tau \quad \text{(C21.1.27)}$$

$S_{xy}(f)$ is the cross spectral density function.

The auto and cross correlation functions may be readily obtained from the power and cross spectral densities, which are quantities commonly measured.

The normalised cross spectral density is called the coherence between signals $x(t)$ and $y(t)$. Therefore,

$$\gamma_{xy}^2(f) = \frac{|S_{xy}(f)|^2}{S_{xx}(f)S_{yy}(f)} \quad \text{(C21.1.28)}$$

When one-sided power and cross spectral densities are used,

$$\gamma_{xy}^2(f) = \frac{|G_{xy}(f)|^2}{G_{xx}(f)G_{yy}(f)} \quad \text{(C21.1.29)}$$

If $x(t)$ and $y(t)$ are linearly related and the only signals in a two-port system, then the coherence will be unity. The coherence will be zero if those two signals are completely unrelated by linear relationships.

Appendix 21.1D: The frequency response function

Take a linear, time-invariant system with input $x(t)$ and output $y(t)$.

The response due to an input starting at time t_0 is given by (Fahy and Walker, 1998) as:

$$y(t) = \int_{t_0}^{t} h(t-t_1)x(t_1)\mathrm{d}t_1 \quad \text{(D21.1.1)}$$

where $h(t)$ is the impulse response function of the system. This equation is the causal version of the convolution integral.

If $y(t)$ is stationary then

$$y(t) = \int_{-\infty}^{t} h(t-t_1)x(t_1)\mathrm{d}t_1 \quad \text{(D21.1.2)}$$

or

$$y(t) = \int_0^{\infty} h(\tau)x(t-\tau)\mathrm{d}\tau \quad \text{(D21.1.3)}$$

Now, for random signals as $T \to \infty$, the autocorrelation function R_{yy} is:

$$R_{yy}(\tau) = E\left[y(t)y(t+\tau)\right] \quad \text{(D21.1.4)}$$

one can substitute for $y(t)$ and for $y(t+T)$ to give

$$R_{yy}(\tau) = E\left[\left(\int_0^{\infty} h(\tau_1)x(t-\tau_1)\mathrm{d}\tau_1\right) \times \left(\int_0^{\infty} h(\tau_2)x(t+\tau-\tau_2)\mathrm{d}\tau_2\right)\right] \quad \text{(D21.1.5)}$$

Re-arranging:

$$R_{yy}(\tau) = E\left[\int_0^{\infty}\int_0^{\infty} h(\tau_1)h(\tau_2) \times x(t-\tau_1)x(t+\tau-\tau_2)\mathrm{d}\tau_1\mathrm{d}\tau_2\right] \quad \text{(D21.1.6)}$$

which is after substitution of

$$R_{xx}(\tau) = E[x(t)x(t+\tau)] \quad \text{(D21.1.7)}$$

$$R_{yy}(\tau) = \int_0^{\infty}\int_0^{\infty} h(\tau_1)h(\tau_2)R_{xx}(\tau+\tau_1-\tau_2)\mathrm{d}\tau_1\mathrm{d}\tau_2 \quad \text{(D21.1.8)}$$

The Fourier transform of this equation is:

$$S_{yy}(f) = |H(f)|^2 S_{xx}(f) \quad \text{(D21.1.9)}$$

where $H(f)$ is the system frequency response function, related to the system impulse-response function thus:

$$H(f) = \int_0^{\infty} h(\tau)e^{-i(2\pi f\tau)}\mathrm{d}\tau \quad \text{(D21.1.10)}$$

$S_{yy}(f)$ is the power spectral density function.

The above can be repeated, this time using the cross correlation function R_{xy} for random signals as $T \to \infty$,

$$R_{yy}(\tau) = E[x(t)y(t+\tau)] \quad \text{(D21.1.11)}$$

$$R_{xy}(\tau) = E\left[x(t)\int_0^{\infty} h(\tau_1)x(t+\tau-\tau_1)\mathrm{d}\tau_1\right] \quad \text{(D21.1.12)}$$

$$R_{xy}(\tau) = \int_0^{\infty} h(\tau_1)R_{xx}(\tau-\tau_1)\mathrm{d}\tau_1 \quad \text{(D21.1.13)}$$

$$S_{xy}(f) = H(f)S_{xx} \quad \text{(D21.1.14)}$$

$S_{xy}(f)$ is the cross spectral density function.

Appendix 21.1E: Plane waves in a tube with a termination impedance

If the frequency of sound in a tube of length L is sufficiently low so that only plane waves propagate, then the solution to the 1-D linear acoustic wave equation has the form (Kinsler et al., 1982):

$$p(x) = Ae^{i[wt+k(L-x)]} + Be^{i[wt-k(L-x)]} \quad \text{(E21.1.1)}$$

At $x = L$ the impedance of the acoustic wave equals the mechanical impedance of the termination Z_{mL}.

Now, taking the linearised inviscid Euler equation (derived in Appendix 21.1G)

$$-\frac{\partial p}{\partial x} = \rho_0 \frac{\partial u}{\partial t} \qquad \text{(E21.1.2)}$$

From this:

$$u(x,t) = -\frac{1}{\rho_0}\int \frac{\partial p}{\partial x} \mathrm{d}t \qquad \text{(E21.1.3)}$$

So, a relationship is found between pressure gradient and particle velocity.

For harmonic waves the integral with respect to time is given by Morse and Ingard (1968), $\frac{1}{i\omega}$.

So,

$$u(x,t) = \frac{1}{i\omega\rho_0}\frac{\partial p}{\partial x} \qquad \text{(E21.1.4)}$$

$$\frac{\partial p}{\partial x} = ik(L-x)Ae^{i[\omega t + k(L-x)]} - ik(L-x)Be^{i[\omega t - k(L-x)]} \qquad \text{(E21.1.5)}$$

As $k = \omega c$

$$u(x,t) = \frac{1}{\rho_0 c_0} Ae^{i[\omega t + k(L-x)]} - Be^{i[\omega t - k(L-x)]} \qquad \text{(E21.1.6)}$$

Now

$$Z_{\mathrm{ML}} = \frac{\mathrm{Force}}{u} \qquad \text{(E21.1.7)}$$

$$\mathrm{Force} = P(L,t).\, S$$

$$Z_{\mathrm{ML}} = \rho_0 c_0 S \frac{A+B}{A-B} \qquad \text{(E21.1.8)}$$

The input mechanical impedance is given by:

$$Z_{\mathrm{MO}} = \rho_0 c_0 S \frac{Ae^{ikL} + Be^{-ikL}}{Ae^{ikL} - Be^{-ikL}} \qquad \text{(E21.1.9)}$$

The two equations may be combined to eliminate both A and B from equation (E21.1.8)

$$x = \frac{A+B}{A-B} \qquad \frac{Z_{\mathrm{ML}}}{\rho_0 c_0 S} = x$$

$$xA - xB = A + B$$

$$xA - A = xB + B$$

$$B = \frac{(x-1)}{x+1} A \qquad \text{(E21.1.10)}$$

Substituting equation (E21.1.10) into equation (E21.1.9)

$$\frac{Z_{\mathrm{MO}}}{\rho_0 c_0 S} = Y \frac{Ae^{ikL} + \left(\frac{x-1}{x+1}\right)Ae^{-ikL}}{Ae^{ikL} - \left(\frac{x-1}{x+1}\right)Ae^{-ikL}}$$

Dividing both numerator and denominator by A

$$Y = \frac{e^{ikL} + \frac{x-1}{x+1}e^{-ikL}}{e^{ikL} - \left(\frac{x-1}{x+1}\right)e^{ikL}} \qquad \text{(E21.1.11)}$$

Multiplying both numerator and denominator of equation (E21.1.11) by (x+ 1)

$$Y = \frac{e^{ikL}x + e^{ikL} + e^{-ikL}x - e^{-ikL}}{e^{ikL}x + e^{ikL} - e^{-ikL}x + e^{-ikL}} \qquad \text{(E21.1 12)}$$

Now, there are these two standard relationships (see Weltner et al. [1986] for example)

$$\frac{e^{ix} - e^{-ix}}{2i} = \sin x \qquad \text{(E21.1.13)}$$

$$\frac{e^{ix} + e^{-ix}}{2} = \cos x \qquad \text{(E21.1.13a)}$$

Substituting equations (E21.1.13) and (E21.1.13a) into equation (E21.1.12)

$$Y = \frac{x\left(e^{ikL} + e^{-ikL}\right) + \left(e^{ikL} - e^{-ikL}\right)}{x\left(e^{ikL} - e^{-ikL}\right) + \left(e^{ikL} + e^{-ikL}\right)}$$

$$Y = \frac{2x \cos kL + i2 \sin kL}{2 \cos kL + i2x \sin kL} \qquad \text{(E21.1.14)}$$

Divide both the numerator and the denominator of equation (E21.1.14) by 2 cos kL to get

$$Y = \frac{x + i \tan kL}{1 + ix \tan kL}$$

$$\frac{Z_{\mathrm{MO}}}{\rho_0 c_0 S} = \frac{\frac{Z_{\mathrm{ML}}}{\rho_0 c_0 S} + i \tan kL}{1 + i\frac{Z_{\mathrm{ML}}}{\rho_0 c_0 S} \tan kL} \text{ [Kinsler et al., 1982]} \qquad \text{(E21.1.15)}$$

Now $\frac{Z_{\mathrm{ML}}}{\rho_0 c_0 S}$ is a complex term

$$\frac{Z_{\mathrm{ML}}}{\rho_0 c_0 S} = r + ix \qquad \text{(E21.1.16)}$$

r = acoustic (flow) resistance
x = auoustic (flow) resistance

Substituting equation (E21.1.16) into equation (E21.1.15) yields

$$\frac{Z_{MO}}{\rho_0 c_0 S} = \frac{r + ix + i\tan kL}{1 + i[(r+ix)\tan kL]} \quad \text{(E21.1.17)}$$

Now separate the real and the imaginary parts. Take the denomintor of equation (E21.1.17) first. Then

$$D = 1 + ir\tan kL - x\tan kL$$
$$D = (1 - x\tan kL) + ir\tan kL \quad \text{(E21.1.18)}$$

multiplying equation (E21.1.18) by its complex conjugate yields

$$D\times D^* = [(1 - x\tan kL) + ir\tan kL] \times [(1 - x\tan kL - ir\tan kL)]$$
$$D\times D^* = (1 - x\tan kL)^2 + r^2\tan^2 kL$$
$$D\times D^* = 1 - 2x\tan kL + x^2\tan^2 kL + r^2\tan kL$$
$$D\times D^* = (x^2 + r^2)\tan^2 kL - 2x\tan kL + 1 \quad \text{(E21.1.19)}$$

Multiplying the numerator of equation (E21.1.17) by the complex conjugate of equation (E21.1.18) gives:

$$N\times D^* = [r + i(x + \tan kL)] \times [(1 - x\tan kL) - ir\tan kL]$$
$$N\times D^* = (r - rx\tan kL) - (ir^2\tan kL) + i(x - x^2\tan kL + \tan kL - x\tan^2 kL) + (xr\tan kL + r\tan^2 kL)$$
$$N\times D^* = r + r\tan^2 kL + i[x - x^2\tan kL - x\tan^2 kL + \tan kL - r^2\tan kL]$$
$$N\times D^* = r(\tan^2 kL + 1) + i[x - (x^2 + r^2)\tan kL - x\tan^2 kL + \tan kL]$$
$$N\times D^* = r(\tan^2 kL + 1) + i[x - (x^2 + r^2 - 1)\tan kL - x\tan^2 kL] \quad \text{(E21.1.20)}$$

So equation (E21.1.17) becomes (using equations (E21.1.20) and (E21.1.19) for numerator and denominator respectively):

$$\frac{Z_{MO}}{\rho_0 c_0 S} = \frac{r(\tan^2 kL + 1)}{(x^2 + r^2)\tan^2 kL - 2x\tan kL + 1} + \frac{i[x - (x^2 + r^2 - 1)\tan kL - x\tan^2 kL]}{(x^2 + r^2)\tan^2 kL - 2x\tan kL + 1} \quad \text{(E21.1.21)}$$

Consider what happens to equation (E21.1.15) when there is a truly rigid termination to tube, i.e. $|Z_{ML}| \to \infty$

$$\frac{Z_{MO}}{\rho_0 c_0 S} = \frac{Z_R + i\tan kL}{1 + iZ_R\tan kL} \quad \left|\frac{Z_{ML}}{\rho_0 c_0 S}\right| = Z_R$$
$$\frac{Z_{MO}}{\rho_0 c_0 S} = \frac{(Z_R + i\tan kL)(1 - iZ_R\tan kL)}{1 + iZ_R^2\tan^2 kL}$$
$$\frac{Z_{MO}}{\rho_0 c_0 S} = Z_R - iZ_R^2\tan kL + i\tan kL + Z_R\tan^2 kL$$
$$\frac{Z_{MO}}{\rho_0 c_0 S} = \frac{Z_R(1 - \tan^2 kL) + i(1 - Z_R^2)\tan kL}{1 + Z_R^2\tan^2 kL}$$
$$\frac{Z_{MO}}{\rho_0 c_0 S} = \frac{Z_R(1 - \tan^2 kL)}{1 + Z_R^2\tan^2 kL} + \frac{i(1 - Z_R^2)\tan kL}{1 + Z_R^2\tan^2 kL} \quad \text{(E21.1.22)}$$

When $Z_R \to \infty$ the reactance $\to 0$ when

$$-i\frac{1}{\tan kL} \to 0 \quad \text{i.e.} \quad \cot kL = 0$$

i.e.

$$\text{for} \quad k_n L = (2n-1) - \tfrac{\pi}{2} \quad n = 1, 2, 3, 4, \ldots$$
$$\frac{2\pi f}{c}L = \frac{(2n-1)\pi}{2} \quad f = \frac{2n-1}{4}\frac{c}{L} \quad \text{(E21.1.23)}$$

Under these conditions, flow reactance is zero and the input mechanical impedance is a function of flow resistivity only.

Appendix 21.1F: The derivation of the linearised mass conservation equation (after Fahy and Walker, 1998)

The net rate of mass inflow into a 1-D control volume of cross-sectional area S (m^2) length dx (m) is:

$$\rho_{TOT}uS - \left[\rho_{TOT}u + \frac{\partial(\rho_{TOT}u)}{\partial x}dx\right]S = -\frac{\partial(\rho_{TOT}u)}{\partial x}dxS \quad \text{(F21.1.1)}$$

This net inflow of mass must be balanced by the increase in mass in the control volume (the principle of conservation of mass) which is given by:

Rate of increase in mass within control volume = $\frac{\partial \rho_{TOT}}{\partial t}dxS$

So

$$\frac{\partial \rho_{TOT}}{\partial t} dx\, S = -\frac{\partial(\rho_{TOT} u)}{\partial x} dx\, S \quad \text{(F21.1.2)}$$

$$\frac{\partial \rho_{TOT}}{\partial t} + \frac{\partial(\rho_{TOT} u)}{\partial x} = 0 \quad \text{(F21.1.3)}$$

This equation of mass conservation may be linearised thus:

$$\begin{aligned} \rho_{TOT} u &= (\rho_0 + \rho) u \\ \rho_{TOT} u &= \rho_0 u + \rho u \end{aligned} \quad \text{(F21.1.4)}$$

The ρu term is the product of two small quantities and one might choose to neglect it and so

$$\rho_{TOT} u \approx \rho_0 u \quad \text{(F21.1.5)}$$

As ρ_0 is not a function of x:

$$\frac{\partial(\rho_{TOT} u)}{\partial x} \approx \rho_0 \frac{\partial u}{\partial x} \quad \text{(F21.1.6)}$$

As ρ_0 is not a function of t either, so

$$\frac{\partial \rho_{TOT}}{\partial t} = \frac{\partial \rho}{\partial t} \quad \text{(F21.1.7)}$$

and the linearised mass conservation equation becomes

$$\frac{\partial \rho}{\partial t} + \rho_0 \frac{\partial u}{\partial x} = 0 \quad \text{(F21.1.8)}$$

Re-arranging

$$\frac{\partial u}{\partial x} = -\frac{1}{\rho_0} \frac{\partial \rho}{\partial t} \quad \text{(F21.1.9)}$$

Now $-\frac{1}{\rho_0} = -\frac{1}{\kappa_0}\left(\frac{\partial p}{\partial \rho}\right)$ from

$$\kappa_0 = \rho_0 \left(\frac{\partial p}{\partial \rho}\right)_{\rho_0} \frac{\partial u}{\partial x} = -\frac{1}{\kappa_0}\left(\frac{\partial p}{\partial \rho}\right)\left(\frac{\partial \rho}{\partial t}\right) \quad \text{(F21.1.10)}$$

So finally the linerised mass conservation equation is:

$$\frac{\partial u}{\partial x} = -\frac{1}{\kappa_0}\left(\frac{\partial p}{\partial t}\right) \quad \text{(F21.1.11)}$$

Appendix 21.1G: The derivation of the non-linear (and linearised) inviscid Euler equation

Take two plane surfaces – one at x and the other at $x + dx$, each one with unit area. There is an acoustic wave causing:

$$p = p_0 + p' \quad \text{(G21.1.1)}$$

$$\rho = \rho_0 + \rho' \quad \text{(G21.1.2)}$$

$$T = T_0 + T' \quad \text{(G21.1.3)}$$

The mass of air between the two plates is given by $(\rho_0 + \rho')dx$ (Morse and Ingard, 1968). The net force acting on this mass is $p(x) - p(x + dx)$. Using Newton's second law, this must be equal to the mass times the acceleration of the fluid.

$$p(x) - p(x + dx) = -\frac{\partial p}{\partial x} dx = \frac{du}{dt}(\rho_0 + \rho')dx \quad \text{(G21.1.4)}$$

Now the total differential $\frac{du}{dx}$ may be expressed in its partial differential form where $u = f(x, t)$ and the total change in u is given by the sum of the partial changes:

$$du = \frac{\partial u}{\partial x} dx + \frac{\partial u}{\partial t} dt \text{(Weltner et al. (1986) for example)}$$

So, dividing both sides by dt in the limit $dt \rightarrow 0$

$$\frac{du}{dt} = \frac{\partial u}{\partial x} \cdot \frac{dx}{dt} + \frac{\partial u}{\partial t} \cdot \frac{dt}{dt}$$

$$\frac{du}{dt} = u \frac{\partial u}{\partial x} + \frac{\partial u}{\partial t}$$

So, from equation (G21.1.4)

$$\begin{aligned} -\frac{\partial p}{\partial x} dx &= \left[u \frac{\partial u}{\partial x} + \frac{\partial u}{\partial t}\right](\rho_0 + \rho')dx \\ -\frac{\partial p}{\partial x} &= (\rho_0 + \rho')\left[\frac{\partial u}{\partial t} + \frac{\partial u}{\partial x}\right] \end{aligned} \quad \text{(G21.1.5)}$$

This is the non-linear inviscid Euler equation (Kinsler et al., 1982).

Now if $u \frac{\partial u}{\partial x} \ll \frac{\partial u}{\partial t}$ and the condensation $(s = \frac{\rho'}{\rho_0}) \ll 1$, the non-linear inviscid Euler equation reduces to its linearised form, being

$$-\frac{\partial p}{\partial x} = \rho_0 \frac{\partial u}{\partial t} \quad \text{(G21.1.6)}$$

From this

$$u(x, t) = -\frac{1}{\rho_0} \int \frac{\partial p}{\partial x} dt \quad \text{(G21.1.7)}$$

So, a relationship between pressure gradient and particle velocity is found.

References

BS 6086: 1981 (ISO 5128 – 1980), Method of measurement of noise inside motor vehicles, British Standards Institution, 1981.

J1074, Engine sound level measurement procedure, Society of Automotive Engineers, 1987

ISO 140/1, Acoustics: measurement of sound insulation in buildings and of building elements.

Part 1: Requirements for laboratories, International Standards Organisation, 1990

ISO 140/3, Acoustics: measurement of sound insulation in buildings and of building elements.

Part 3: Laboratory measurements of airborne sound insulation of building elements (amendment 1 to ISO 140/3 – 1978), International Standards Organisation, 1990

Allard, J.F., Sieben, B., Measurements of acoustic impedance in a free field with two microphones and a spectrum analyser, *Journal of Acoustical Society of America*, 77(4), 1617–1618, 1985

Anderson, C.G., A review of phenomena and associated experimental methods in automotive aero-acoustics, IMechE Paper No. C605/017/2002, 2002

Atkins, K.A., Challen, B.J., A practical approach to truck noise reduction, Institution of Mechanical Engineers, C131/79, 1979

Betella, M., Harrison, M.F., Sharp, R.S., Investigation of automotive creep groan noise a distributed-source excitation technique, *Journal of Sound and Vibration*, 255(3), 531–547, 2002

Bies, D.A., Uses of anechoic and reverberant rooms, *Noise Control Engineering Journal* 7, 154–163, 1976

Bies, D.A., Hansen, C.H., Engineering noise control – Theory and practice – Second edition, E&FN Spon, 1996

Cheng, J.G., Akin, T., Air and structure borne noise reduction of automotive dampers, SAE Paper No. 951256, Proceedings of the 1995 noise and vibration conference – Vol. 1, Society of Automotive Engineers, 1995

Chung, J.Y., Blaser, D.A., Transfer Function Method of Measuring In-duct Acoustic properties. II. Experiment, *Journal of Acoustical Society of America*, 68(3), 914–921, 1980

Coney, W.B., Her, J.Y., Tomaszewicz, K., Zhang, K.Y., Moore, J.A., Experimental evaluation of wind noise sources, SAE Paper No. 1999-01-1812, 1999

Cremer, L., Heckl, M., Structure borne sound – Structural vibrations and sound radiation at audio frequencies – Second edition, Springer Verlag, 1988

Crocker, M.J., Zockel, M., McGary, M., Reinhart, T., Noise source identification under steady and accelerating conditions on a turbocharged diesel engine, Vehicle Noise Regulation and Reduction - SP-456, Society of Automotive Engineers, Paper No. 800275, 1980

Croker, M.D., Greer, R.J., Hilbert, D., Granstrom, J., The development of transmission rattle indices, IMEchE Paper No. C420/024, Published in Quiet Revolutions – powertrain and vehicle noise refinement, international conference 9–11 October 1990, pp. 129–135, 1990

Dixon, J., Phillips, A.V., Power unit low frequency airborne noise, European Conference on Vehicle Noise and Vibration, Institution of Mechanical Engineers, C521/032/98, 1998

Fahy, F., Sound and structural vibration – radiation, transmission and response, Academic Press, 1985

Fahy, F.J., Sound intensity, Elsevier Applied Science, London, 1989

Fahy, F., Walker, J. (eds), Fundamentals of Noise and Vibration, E&FN Spon, 1998

Glikin, P.E., Fuel injection in diesel engines, Proceedings of the Institution of Mechanical Engineers, Vol. 199, D3, 161–174, 1985

Haines, J.C., Sound absorbing properties of molded fibreglass panels for use in vehicle noise control, SAE Paper No. 870987, Society of Automotive Engineers, 1987 Noise & Vibration Conference, 1987

Halvorsen, W.G., Bendat, J.S., Noise Source Identification using Coherent Output Power Spectra, *Journal of Sound and Vibration*, pp. 15–24, August 1975

Hardy, M., Whole vehicle refinement, Paper No. S477/001/97, IMechE Seminar – Automotive modeling and NVH – techniques and solutions, Institution of Mechanical Engineers, 1997

Harper, M.F.L., Dorling, C.M., Allwright, D.J., A non-intrusive method of signature analysis in systems with multiple noise paths. *International Journal of Modeling and Simulation*, 13(4), 146–151, 1993

Ingard, K.U., Bolt, R.H., A free field method of measuring the absorption coefficient of acoustic materials, *Journal of Acoustical Society of America*, 23(5), 509–516, 1951

Ingard, K.U., Dear, T.A., Measurement of acoustic flow resistance, *Journal of Sound and Vibration*, 103(4), 567–572, 1985

Jackson, R.S., The performance of acoustic hoods at low frequencies, Acustica, 12, 139–152, 1962

Jackson, R.S., Some aspects of the performance of acoustic hoods, *Journal of Sound and Vibration*, 3(1), 82–94, 1966

Janssens, M.H.A., Verheij, J.W., Thompson, D.J., The Use of an Equivalent Forces Method for the Experimental Quantification of Structural Sound Transmission in Ships, *Journal of Sound and Vibration*, 226(2), 305–328, 1999

Johnson, C.M., Vehicle exhaust sound specification development, SAE Paper No. 951259, Society of Automotive Engineers, 1995 Noise & Vibration Conference, 1995

Jonasson, H., Elson, L., Determination of sound power levels of external sources, Report SP-RAPP, National testing institute, Acoustics laboratory, Borus, Sweden, 1981

Juneja, V., Rediers, B., Kavarana, F., Kimball, J., Squeak studies on material pairs, SAE Paper No. 1999-01-1727, 1999

Kim, J., Vehicle cavity wind noise experiments on a scale model in a windtunnel, MSc Thesis, Cranfield University, 2003

Kinsler, L.E., Frey, A.R., Coppens, A.B., Sanders, J.V., Fundamentals of acoustics – Third edition, John Wiley & Sons Inc., 1982

Kuttruff, H., Room Acoustics – Second edition, E&FN Spon, 1979

Lalor, N., Grover, E.C., Priede, T., Engine noise due to mechanical impacts at pistons and bearings, Society of Automotive Engineers, Paper No. 800402, 1980

Lilly, L.R.C., Diesel engine reference book, Butterworths, London, 1984 (Second edition now available – Challen, Baranescu 1999)

LMS International – Application Note Transfer path analysis – the qualification and quantification of vibro-acoustic transfer paths, LMS International, 1998

March, J.P., Croker, M.D., Present and future perspectives of powertrain refinement, European Conference on

Vehicle Noise and Vibration, Institution of Mechanical Engineers, C521/023/98, 1998

Maunder, R.M.S., Practical applications of sound quality analysis techniques, Paper No. C498/13/067/95, Institution of Mechanical Engineers, 1996

Morse, P.M., Bolt, R.H., Sound waves in rooms, *Reviews of Modern Physics*, 16, 65–150, 1944

Morse, P.M., Ingard, K.U., Theoretical acoustics, McGraw-Hill Book Co, New York, 1968

MSX International, I-DEAS noise path analysis, MTS Systems Corporation, 1998

Naylor, S., Willats, R., The development of a sports tailpipe noise with predictions of its effect on interior vehicle sound quality, IMEchE Paper No. C577/002/2000, 2000

Nelson, P.M., (ed.) Transportation noise reference book, Butterworths, London, 1987

Otto, N., Amman, S., Eaton, C., Lake, S., Guidelines for jury evaluations of automotive sounds, SAE Paper No. 1999-01-1822, 1999

Piersol, A.G., use of coherence and phase data between two receivers in evaluation of noise environments, *Journal of Sound and Vibration*, 56(2), 215–228, 1978

Pobol, O., Method of measuring noise characteristics of textile machines, measurement techniques, USSR, 19, 1736–1739, 1976

Priede, T., The effect of operating parameters on sources of vehicle noise, *Journal of Sound and Vibration*, 43(2), 239–252, 1975

Qian, Y., Vanbuskirk, J.A., Sound absorption composites and their use in automotive interior sound control, Proceedings of the 1995 Noise and Vibration Conference, Traverse City Michigan, SAE P – 291 (Vol. 1), 1995

Querengasser, J., Meyer, J., Wolschendorf, J., Nehl, J., NVH optimisation of an in-line 4-cylinder powertrain, Proceedings of the 1995 noise and vibration conference – Vol. 1, (p-291) Society of Automotive Engineers, Paper No. 951294, 1995

Russell, M.F., Reduction of noise emissions from diesel engine surfaces. Society of Automotive Engineers, Paper No. 720135 1972

Russell, M.F., Automotive diesel engine noise analysis, diagnosis and control, Lucas Engineering Review – 7(4), 1979

Russell, M.F., Worley, S.A., Young, C.D., An analyser to estimate subjective reaction to diesel engine noise, IMEchE Paper No. C30/88, 1988

Russell, M.F., Sekowski, M., Nikokiroulis, N., Subjective assessment of diesel vehicle noise Paper No. C389/044, Institution of Mechanical Engineers, 1992

Russell, M.F., Lee, H.K., Modeling diesel injection rate to control combustion noise, Institution of Mechanical Engineers, C-487/027, 1994

Russell, M.F., The dependence of diesel combustion on injection rate, Institution of Mechanical Engineers, S-490/005/97, 1997

Saha, P., Baker, R.N., Sound absorption study for automotive carpet materials, SAE Paper No. 870988, Society of Automotive Engineers, 1987 Noise & Vibration Conference, 1987

Sinha, N.K., Linear Systems, John Wiley & Sons 1991

Steel, J.A., Fraser, G., Sendall, P., A study of exhaust noise in a motor vehicle using statistical energy analysis, Proceedings of the Institution of Mechanical Engineers, Vol. 214, Part D, 75–83, 2000

Storer, D., Gatti, S., Pisino, E., Characterizing the transfer of road-surface excited vibrations through vehicle suspension systems, Paper No. C521/015/98, Institution of Mechanical Engineers, 1998

Thien, G.E., Brandl, F.K., Kirchweger, K., Winklhofer, E., Cars with closed engine compartment – effect upon exterior noise and passenger comfort, Vehicle noise and vibration – IMechE conference publications, Institution of Mechanical Engineers, C133/84, 1984

Van der Auweraer, H., Wyckaert, K., Hendricx, W., From sound quality to the engineering of solutions for NVH problems: case studies, Acustica: acta acustica, 83, 796–804, 1997

Vandenbroeck, D., Hendricx, W.S.F., Interior road noise optimization in a multiple input environment, Proceedings of the International Conference on vehicle NVH and refinement, Birmingham, 3–5 May 1994, Paper No. C487/002/94, Institution of Mechanical Engineers, 1994

Vér, I.L., Holmer, C.I., In: Noise and Vibration Control (L.L. Beranek (ed.)), McGraw-Hill, New York, 1971

Verhulst, K., Verheij, J.W., Coherence Measurements in multi-delay systems, *Journal of Sound and Vibration*, 62(3), 460–463, 1979

Verstraeten, S., Dynamic bush properties and their effects on compliant subframe performance, MSc Thesis, Cranfield University, 2003

Weltner, K., Grosjean, J., Shuster, P., Weber, W.J., Mathematics for Engineers and Scientists, Stanley Thornes (Publishers) Limited, 1986

Wentzel, R.E., Saha, P., Empirically predicting the sound transmission loss of double-wall sound barrier assemblies, SAE Paper No. 951268, Society of Automotive Engineers, 1995 Noise & Vibration Conference, 1995

Wentzel, R.E., VanBuskirk, J., A dissipative approach to vehicle sound abatement, SAE Paper No. 1999-01-1668, 1999

Section Twenty-Two

Exterior noise

Chapter 22.1

22.1

Exterior noise: Assessment and control

Matthew Harrison

22.1.1 Pass-by noise homologation

22.1.1.1 Background to homologation

Most countries restrict the types of vehicle that can operate legally on their roads by some form of national legislation. Taking the United Kingdom (UK) as an example, the relevant legislation is the Road Vehicles (Construction and Use) Regulations 1986, sections of which have been amended many times since first coming into operation on 11 August 1986. The 1986 Regulations specify minimum requirements for the construction (and hence design), maintenance and use of road vehicles (including heavy goods vehicles, passenger service vehicles and track-laying vehicles) for the following:

- dimensions and manoeuvrability;
- brakes;
- wheels, springs, tyres and tracks;
- steering;
- vision;
- instruments and equipments;
- fuel;
- control of emissions (including noise);
- laden weight;
- the use of trailers;
- the avoidance of danger when in use.

Compliance with the requirements of the 1986 Regulations is ensured by a mandatory vehicle type approval scheme (Statutory Instrument No. 981, 1984; The Road Traffic Act, 1988) that covers:

- Mass-produced vehicles for sale or use in the UK.
- Low-volume-produced vehicles for sale or use in the UK.
- Single vehicles (produced in the UK or imported) that do not enjoy type approval by virtue of their unique construction or subsequent modification, or were imported from a country without suitable type approval.

In addition to national legislation, many countries place restrictions on type approval according to international agreements. In the UK (as in other European Community (EC)) countries, restrictions are placed in accordance with

- EC Directives.
- Economic Commission for Europe (UN–ECE) Regulations.

There are a significant number of UN–ECE Regulations, mostly concerning individual components or systems. The principal EC Directive for type approval is 70/156/EEC and this has been amended many times since first publication (notably 92/53/EEC). In the UK, the mechanism by which the Secretary of State regulates EC type approval is currently provided by the Motor Vehicles (EC Type Approval) Regulations 1998.

Homologation is the process by which approval for sale or use of a vehicle in a particular country is obtained. This varies from one country to other and may take the form of:

- Type approval (conducted by an independent body).
- Single vehicle approval (conducted by an independent body).
- Self certification by the manufacturer (who certifies that the vehicle complies with all the legislative requirements).

Automotive EC Directives require third-party approval which has three components to it:

1. Testing (to particular technical EC Directives).
2. Certification.

Vehicle Refinement; ISBN: 9780750661294
Copyright © 2004 Matthew Harrison. All rights of reproduction, in any form, reserved.

3. Product Conformity Assessment (assessing the ability to produce series products in conformity with the specification, performance and labelling/marking requirements of the type approval).

In the United Kingdom, the Vehicle Certification Agency (VCA – an executive agency of the Department of the Environment, Transport and the Regions – DETR) is the only Competent Authority (as defined in 70/156/EEC) for the type approval of vehicles under EC Directives. The VCA is therefore the only body in the UK to issue EC type approval certificates. However, an EC type approval certificate issued by a Competent Authority in another Member State is equally valid in all Member States including the UK. The VCA is also the only test authority in the UK for EC type approval. In other Member States, such as Germany, there are several test authorities but only one Government Authority may issue type approval certificates.

EC Component Type Approval may be obtained for generic components (to be fitted to any vehicle) after evaluation against particular technical EC Directives. EC Separate Technical Unit Type Approval may be obtained for components to be fitted to only one type of vehicle after evaluation against particular technical EC Directives.

EC approval of most new mass-produced road vehicles is based on whole vehicle approval. In such cases, a representative example of the production intent vehicle is tested/evaluated by the Competent Authority (they must at least attend all tests if they do not actually conduct them). The items tested include:

- seat belts
- lights
- tyres
- brakes
- emissions (including noise)
- seats and head restraints
- reward visibility
- vehicle impact test.

Compliance with around 45 technical standards is checked. VCA practises worst case selection in order to reduce the amount of testing needed across a range of product types. (*Source:* www.vca.gov.uk.)

22.1.1.2 EC noise homologation

Current EC automotive Directives are what are known as old approach whereby a third party tests for compliance with the requirements of particular technical Directives. In new approach Directives (that Member States have so far rejected for automotive applications) more obligation is placed on the manufacturer to ensure that the product complies with the appropriate requirements.

The original technical Directive relating to type approval and noise was 70/157/EEC. This defined limiting noise levels outside the vehicle during a particular form of acceleration test (the test being improved since then and now detailed in ISO 362:1998). An acceleration test was chosen to provide a vehicle operating condition with worst-case noise emissions. The original test was as follows:

- The test site, a section of level road (more than 3 m wide) in an open area of minimum radius 50 m, is constructed in accordance with ISO 10844. ISO 10844 defines the texture and porosity and hence the acoustic effect of the road surface. The road widens at one point to make a 20 m by 20 m test area of special road surface as shown in Fig. 22.1-1.
- A line is drawn down the centre line of the road. The driver follows that line throughout the test.
- One microphone position is marked on each side of the road, 7.5 m from the centre point of the test area and at a height of 1.2 m. A type 1 sound level meter is placed at one of the microphone positions.
- The test vehicle approaches the test area at a fixed speed and in a predetermined gear. For common EC passenger cars of category M1 with 5-speed manually operated transmissions the approach speed is 50 km hr^{-1} and the test is performed in both second and third gears.
- When the front of the vehicle enters the 20-m long test area, the driver depresses the throttle briskly and fully and holds it in that position until the rear of the vehicle leaves the test area.
- The maximum, fast-weighted, A-weighted, noise level recorded whilst the vehicle is within the test area is noted. The results are considered valid when four consecutive results are within 2 dB from each other.
- At least four valid levels are recorded on both sides of the vehicle.

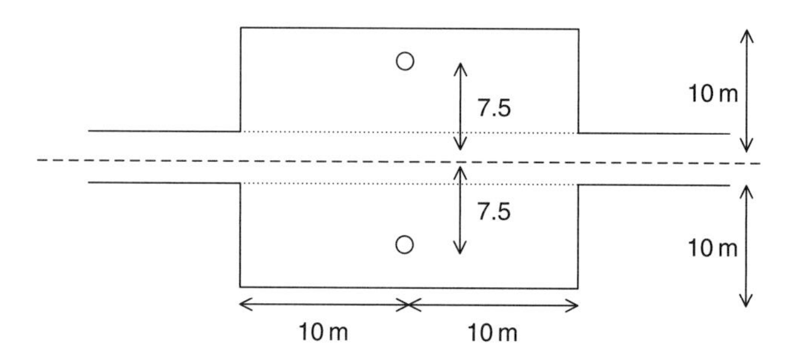

Fig. 22.1-1 The test area defined in ISO 362:1998.

- The results from each side of the vehicle are averaged separately. The intermediate result is the highest of the two averages.
- The final reported result for the 5-speed M1 car is the arithmetic average of the two intermediate results obtained from the tests for both gears.
- The reported level may be reduced by 1 dBA to account for the natural variability of the test.

The homologation noise levels for type approval have reduced significantly since their introduction in 70/157/EEC. The level for M1 passenger cars has been reduced by 8 dBA from 82 dBA to the current level of 74 dBA (92/97/EEC). The reduction has been in stages as shown in Table 22.1-1.

The current homologation levels correspond to the test detailed in 92/97/EEC. ISO 362:1998 is broadly similar but not identical in all details. In particular, 92/97/EEC requires additional measurements to be made 0.5 m (45° angle of incidence, 0.2 m above the ground) from the exhaust tailpipe of a stationary vehicle in order to facilitate subsequent checks of vehicles in use. The engine is run at three-quarters speed and the maximum sound pressure level (fast, A-weighted) is recorded. The results from three of these tests must be reported as part of the vehicle certification process. 92/97/EEC also includes a method for assessing compressed air noise.

22.1.1.3 Track and atmospheric effects

Directive 92/97/EEC introduced a requirement for a standard road surface to be adopted at all test tracks used for drive-pass noise homologation. 92/97/EEC is not completely prescriptive over the details of construction for such a standard track and so many Member States have adopted ISO 10844 as a specification. According to Sandberg (1991), the specification in ISO 10844 was aimed initially at achieving three goals:

1. To make the results achieved at any given track repeatable and reproducible at any other track.
2. To make the track surface highly reflective so that all noise sources on the vehicle (even those hidden from view under the vehicle) make a contribution to the overall drive-pass level and thus this would encourage manufacturers to treat the whole vehicle.
3. Minimise the noise radiated from the tyre/road contact as this is a fairly unavoidable source of noise and if it were to dominate, manufacturers would not be incentivised to treat the noise from other sources on the vehicle.

Before the widespread adoption of ISO 10844 surfaces, Dunne and Yarnold (1993) reported that they knew of 4 dBA variations in the drive-pass noise levels recorded for the same car at different test tracks around Europe.

Table 22.1-1 EC homologation noise levels for type approval

EC Directive	**70/157/EEC**	**77/212/EEC**	**81/334/EEC**	**84/424/EEC**	**92/97/EEC**
Enforced for new vehicles	*1976*	*1983*	*1984***	*1990*	*1996*
Category of vehicle	Maximum permissible noise level (dBA) at 7.5 m (acceleration test)				
Cars (<9 seats)	82	80	80	77	74*
Minibus <2 tonnes	84	81	81	78	76
Minibus 2–3.5 tonnes	84	81	81	79	77
Bus ≤150 kW	89	82	82	80	78
Bus >150 kW	91	85	85	83	80
Light truck <2 tonnes	84	81	81	79	76
Light truck 2–3.5 tonnes	84	81	81	80	77
Truck >3.5 tonnes					
≤75 kW	89	86	86	83	77
75–150 kW	89	86	86	85	78
>150 kW	91	88	88	86	80

* 75 dBA if fitted with a DI diesel engine or if very powerful 'supercar'.
** Imposed more stringent requirements for exhaust and intake systems.

Walker (1994) reports on an experimental investigation to test the effectiveness of these three goals in practice. Noise measurements were made on an ISO 10844 test track and on another track surfaced with high-drainage asphalt. An omni-directional noise source was positioned on the centre line of the track at heights between 100 mm and 400 mm above the surface. A microphone was positioned 7.5 m away from this at a height of 1.2 m from the surface. The comparison showed that:

- The ISO 10844 surface did indeed behave as a close approximation to a perfectly reflective (low absorption) surface at most frequencies of interest.
- The sound absorption characteristics of the ISO 10844 track are more uniform across the frequency spectrum than those of the high-drainage asphalt surface. This suggests consistent surface characteristics over the 7.5 m acoustic path length and should result in results being more repeatable and reproducible.
- Because of the relative smoothness of the ISO 10844 surface, low-frequency tyre noise was reduced compared with the high-drainage surface, but high-frequency tyre noise was increased.

Walker concluded that with all these taken into account, the ISO 10844 would produce drive-pass noise levels some 2 dB greater than those expected for a given car tested on the high-drainage asphalt surface.

In drafting Directive 92/97/EEC, Dunne and Yarnold (1993) report that the European Commission recognised that other factors such as meteorological conditions may influence the drive-pass test result.

22.1.1.4 Future developments in noise homologation in the EC

Although noise homologation limits have been reduced significantly since 1970, comprehensive noise testing alongside roads in Germany (Steven, 1995) has shown that noise levels produced by vehicles in normal use have reduced only slightly over the first twelve years of that period.

The main reason for this seems to be that the homologation process has forced vehicle manufacturers to reduce engine and intake/exhaust noise which are strongly engine-load-dependent and thus dominate the low-speed, high-load acceleration test. However, the process has had less impact on tyre noise and on the noise caused by vehicles at speeds much in excess of 50 km hr^{-1}. The higher speed noise is most important in the environmental impact of vehicles in normal use outside the urban setting.

It is intuitively obvious that once engine, intake and exhaust noise are reduced further, tyre noise will dominate even the low-speed acceleration test. Once that becomes the case (and it already is with many passenger cars fitted with wide tyres), the homologation level cannot be reduced further without reducing tyre noise first.

The EC have responded to this problem by proposing noise homologation levels for tyres (C30/8, 28/1/98). These have not yet been adopted. A new higher speed variant of the acceleration test is proposed, whereby a vehicle coasts through the 20-m test area at a constant approach speed but with the engine turned off. The approach speeds would vary in the range of 60–90 km hr^{-1} according to the type of tyre. The noise limits would vary according to the type of tyre and tyre width.

22.1.1.5 Noise homologation in the US and otherwise outside the EC

Most countries outside the EC have their own systems of vehicle type approval that include restrictions on noise levels. Most tests and limiting levels are based on those of the EC or of the United States.

There are two US noise homologation tests – SAE J986 AUG94 which has the vehicle entering a test area with predetermined vehicle speed and SAE J1030 FEB87 which has the vehicle leaving the test area with predetermined engine speed. Both tests feature vehicle deceleration as well as acceleration.

The acceleration part of SAE J986 AUG94 is broadly similar to that in the EC test except that the microphone is positioned at 15 m from the vehicle pathline (rather than 7.5 m) and the test area is much longer (53 m) with the aim of allowing the vehicle to reach its rated engine speed during the test. In the EC test, the vehicle will seldom reach its rated engine speed.

Generally, a vehicle that achieves noise homologation in the EC will achieve US Federal homologation with comparative ease. Japan and Switzerland have traditionally had homologation restrictions that are more onerous than those of the EC or the US.

22.1.1.6 The consequences of meeting homologation noise limits

The current homologation noise limits according to 92/97/EEC are very demanding. A great deal of engineering effort was devoted in the early 1990s to reduce noise levels from accelerating passenger cars by the 3 dBA required to meet a level of 74 dBA. Vehicle manufacturers are achieving type approval for their current vehicles, but it is difficult for those with higher powered engines (Sports Utilities, GTi models and the

Table 22.1-2 The ten lowest homologation levels in the UK

Manufacturer	Model	Description	Trans.	Capacity	Fuel	dBA
Daihatsu	Terios	1.3L Efi 4WD	A4	1298	Petrol	65
Daihatsu	YRV	1.3L Efi turbo	A4	1298	Petrol	66
Renault	Vel Satis	2.0 Turbo	A5	1998	Petrol	66.7
Mitsubishi	Carisma	1.6 Mirage	A4	1597	Petrol	67
Smart	Cabrio Hatchback	Cabrio & Passion	A6	599	Petrol	67
Nissan	Primera	2.0 4/5 door	M6	1998	Petrol	67
Mitsubishi	Galant – sedan	2.5 Elegance	A4	2498	Petrol	67
Nissan	Primera	2.0 Estate	A6	1998	Petrol	67
Nissan	Primera	2.0 4/5 door	A6	1998	Petrol	67
Mitsubishi	Galant – wagon	2.5 Elegance	A4	2498	Petrol	67

Source: www.vca.gov.uk. Accessed in November 2003.

like). Tables 22.1-2 and 22.1-3 give an indication of this. Note that most of the ten lowest homologation levels shown are for automatic cars and most of the highest homologation levels are recorded for sports-utility vehicles with large engines and manual transmissions.

It should be appreciated that the total wayside noise level during the acceleration test is made up of contributions from many noise sources (engine, intake, exhaust, tyres, etc.). Noise levels from each of these must be controlled in order to achieve type approval. Some suggested noise targets for each noise source are given in Table 22.1-4.

It should be noted that due advantage is taken in Table 22.1-4 of the 1 dBA reduction in reported values due to potential inaccuracy of the noise homologation test.

22.1.2 Noise source ranking

When a vehicle manufacturer is having difficulty homologating a vehicle for noise, it is important to know the

Table 22.1-3 The ten highest homologation levels in the UK

Manufacturer	Model	Description	Trans.	Capacity	Fuel	dBA
Chrysler Jeep	Grand Cherokee	4.7L Overland	A5	4701	Petrol	76
Vauxhall	Frontera	3.2l V6 24v	A4	3165	Petrol	76
Mitsubishi	Colt	1.3 Attivo 2	M5	1300	Petrol	76
Volkswagen	Touareg	2.5	M6	2461	Diesel	76
Mitsubishi	Shogun	3.2 Equippe LWB	M5	3200	Diesel	76
Land Rover	Discovery	2.5 TD5	A4	2495	Diesel	76
BMW	X5	X5 3.0d	M6	2993	Diesel	76
Land Rover	Defender Wagon	2.5 TD5 90	M5	2495	Diesel	76
Land Rover	Discovery	2.5 TD5	M5	2495	Diesel	76
BMW	X5	X5 3.0l	M6	2979	Petrol	76

Source: www.vca.gov.uk. Accessed in November 2003.

Table 22.1-4 Suggested target noise levels for achieving type approval under 9297/EEC

	Target levels at 7.5 m, acceleration test (dBA)		
	Passenger car	Light truck	Heavy truck
Engine	69	72	77
Exhaust	69	70	70
Intake	63	63	65
Tyres	68	69	75
Transmission	60	63	66
Other	60	72	65
Combined level	74.2	77.3	80.1

relative contributions made to the pass-by level by the various sources (intake, exhaust, tyres, etc.) so that the most significant sources can be controlled first. This is achieved using noise source ranking.

First, a list of potential noise sources should be compiled, and then the next task should be to rank all of the sources in order of significance. The term 'significance' can have several different meanings according to the case in hand. Significance could mean:

- overall linear sound power level or sound pressure level at a particular point in space;
- overall 'A'-weighted sound power level or sound pressure level at a particular point in space;
- sound pressure or power level in a certain frequency band;
- long-distance propagation;
- subjective rating.

The noise source ranking procedure may be undertaken in a variety of ways. The more popular methods include:

- Isolation – where each component of the machine is run in isolation, where possible, and its noise contribution is measured directly.
- Shielding – where the machine is run several times, and each time a different noise source is encapsulated in a sound-retaining structure. The reduction in noise level achieved yields that particular noise source's contribution to the total noise level. A practical example of this technique is given by Balcombe and Crowther (1993).
- Close microphone techniques – where the machine is run in an anechoic chamber and a microphone is passed in the near field over the surfaces of the machine to locate regions of high pressure level (an approximate indicator only usually used to quickly back-up a hunch as to where the noise is coming from).
- Sound intensity mapping – where the machine is run continuously, and an intensity probe is passed over an notional surface set around the machine. Regions of high intensity give an indication of the position and level of noise sources.
- Spatial Transformation of Sound Field (STSF) method. This is a relatively new technique that uses an array of microphones in a plane (or for stationary sound fields, a single scanning microphone) and the principle of near-field acoustical holography along with the Helmholtz Integral Equation to calculate the three dimensional sound field. Taylor and Bridgewater (1998) give an example of the technique in use. Kim and Lee (1990) give a more detailed description of the technical background.
- Noise from vibration – where the machine is run and measurements of vibration velocity are made on the surfaces of potential noise sources. Radiation efficiencies are calculated or assumed, and the noise contribution from each potential noise source is estimated.
- Modelling – where the noise contribution from each potential source is calculated using empirical or mathematical models.

22.1.2.1 Noise source ranking using shielding techniques

The most commonly used noise source ranking procedure in the automotive industry is still (arguably) the rather laborious shielding or encapsulation method. This usually consists of first eliminating the various noise sources, using acoustic treatments, and then removing the treatments systematically and measuring the noise contribution from each source as it is uncovered. The shielding technique is used for both vehicle and engine noise. The vehicle noise case shall be discussed further here.

The interior noise contribution from each source is tested using microphones placed at each seat in the vehicle, and the exterior noise contribution is assessed using the relevant standard drive-pass test. A typical noise-source ranking programme might take the following form:

- Perform baseline noise tests on the programme vehicle. Interior noise levels are recorded at each seat position with the vehicle accelerating uniformly in second or third gear, first with wide-open throttle and then part throttle. Exterior noise for the baseline vehicle is assessed using the relevant drive-pass standard.

- Fit a large and effective additional silencer(s) to the exhaust tailpipe (often called an infinite silencer). A series of absorptive silencers may be sufficient. The infinite silencer should be of low backpressure design so as not to restrict the performance of the vehicle. Care must be taken to have the outlet from the additional silencers as near to the original exhaust tailpipe position as possible.
- Lag the standard exhaust system silencers with a layer of glass fibre, and then cover in sheet lead. This is to reduce any contribution from exhaust shell noise (noise radiating from the structure of the exhaust system).
- Fit a large and effective silencer to the intake system. This silencer will often be strapped to the hood (bonnet) of the car, and can be constructed from perspex to allow for driver's visibility (large plastic drinking-water bottles are commonly adopted for use). The silencer system must be of low pressure loss design in order not to restrict the performance of the vehicle. It is vital that the outlet from the silencer is as near to the original snorkel position as possible.
- Encapsulate the engine bay with a thick plywood under tray, ensuring that the edges are well sealed. The engine encapsulation must be performed carefully if engine breathing and cooling problems are to be avoided.
- Encapsulate the drive train (if appropriate) using plywood, or lag with lead.
- Fit treadless tyres (or tyres known to be relatively quiet) to reduce tyre noise.
- Perform identical interior and exterior noise tests on the fully treated vehicle.
- Fit conventional tyres and re-test for tyre noise contribution. Once tested, re-fit the treadless tyres.
- Remove the transmission lagging and re-test for drive-line noise contribution. For speed, it is customary not to re-fit any acoustic treatments after testing.
- Remove the exhaust system lagging and test for exhaust shell noise contribution.
- Remove the engine bay encapsulation and re-test for engine noise contribution.
- Remove the infinite intake silencer and re-test for intake noise contribution.
- Remove the infinite exhaust silencer and re-test for exhaust noise contribution.

The apparent sound pressure level due to each noise source in isolation from the others is derived from the increase in overall sound pressure level as each additional source is uncovered. Therefore,

$$L_{Si} = 10 \log_{10}\left[10^{\frac{L_u}{10}} - 10^{\frac{L_S}{10}}\right] \text{dB} \tag{22.1.1}$$

where

L_{Si} is the apparent sound pressure level due to source i in isolation.
L_u is the total sound level recorded once source i is uncovered.
L_S is the sound level recorded before source i is uncovered.

This noise source ranking procedure is relatively simple, but involves a large number of tests. The results should be viewed only as being approximate, due to the potential for significant test-to-test variations.

It is important to reduce the number of variables during the tests, so it is wise to use the same test routine for all tests and try to perform the tests on or near the same day. Once the relative noise contribution from each noise source is determined, the degree of additional silencing required is assessed. Only when there is a potential for significant decrease in overall noise levels is the costly process of component development carried out. Some sample vehicle noise source ranking results are shown in Fig. 22.1-2.

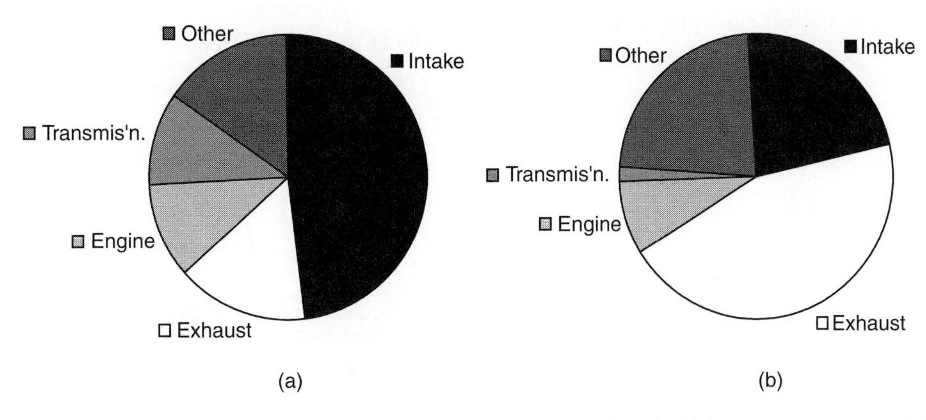

Fig. 22.1-2 Noise source ranking results adapted from Balcombe and Crowther (1993). (a) Results at the start of the pass-by test. (b) Results at the end of the pass-by test.

22.1.3 Air intake systems and exhaust systems: Performance and noise effects

22.1.3.1 Introduction

Intake and exhaust noise must feature prominently in this chapter as they both can be significant sources of exterior vehicle noise. Arguably, for the first sixty years of the motor car the exhaust tailpipe was the most significant cause of vehicle noise (even brief exposure to an unsilenced light aircraft provides a clear reminder of that). With the rise in popularity of front-wheel drive passenger cars, permitting the adoption of large-volume exhaust silencers (driven of course by the arrival of noise control legislation), the levels of exhaust tailpipe noise reduced considerably from around 1960. This uncovered hitherto unheard intake noise.

For some twenty years (from around 1980) the intake system was commonly the dominant noise source in family sedans operating at full load. It has only been much more recently that improvements in plastic moulding technologies (and the adoption of port fuel injection removing the need for a metallic intake manifold) have freed intake designers to add several silencing elements to even low-cost intake systems. As a result, the levels of intake noise now match those of the exhaust system and so both engine and tyre noise now commonly dominate the noise emissions from most family sedans.

This chapter is more heavily biased towards intake system design rather than exhaust system design. There are two reasons for this:

1. The careful design of the intake system for a naturally aspirated (NA) engine can raise volumetric efficiency by more than 10%, whereas the corresponding improvement caused by detailed exhaust system design is seldom more than 3–4%. Therefore, the intake designer must consider both engine performance and noise control, whereas (within back pressure limits) the exhaust designer is free to concentrate more on noise control.
2. The skills and techniques used for exhaust system design are only slightly modified versions of those acquired for intake system design.

This rather extended section will commence with an overview of intake system requirements and then move on to a detailed assessment of intake design for engine performance followed by intake design for noise control. Only a brief discussion of exhaust system design is offered as this concentrates on the main differences in approach for exhaust system design as compared with intake system design.

22.1.3.2 Intake noise – objectives

Some intake noise is desirable and can help create an impression of speed or power. Other noise is undesirable and can contribute to excessive interior noise levels or failure of the legislative drive-pass test.

Unwanted intake noise may be classified into two categories:

1. Exterior noise which contributes to the overall noise level during the type approval drive-pass test (see Section 22.1.1).
2. Interior noise which reduces the comfort of the driver or passengers, or interferes with speech communication (see Chapter 21.1).

22.1.3.3 The issues involved in intake system design

It is important to understand the constraints on design which include:

- wave action tuning of the intake system to improve engine performance;
- intake acoustic design;
- intake system static pressure loss;
- intake system mounting;
- packaging space (under-hood);
- turbo or super-charging;
- cost;
- controlling vibration in the system;
- intake snorkel positioning;
- intake water inhalation;
- manufacturing process;
- intake filter durability/servicing;

22.1.3.4 Intake systems

Intake systems fulfil a number of roles in an overall vehicle design which are:

- channel air to the engine;
- filter particulates;
- enhance engine performance through wave action tuning;
- reduce noise.

22.1.3.5 The intake system designer

The intake system designer is in a unique position within a vehicle development programme, falling within the interest areas of several design groups. His or her design

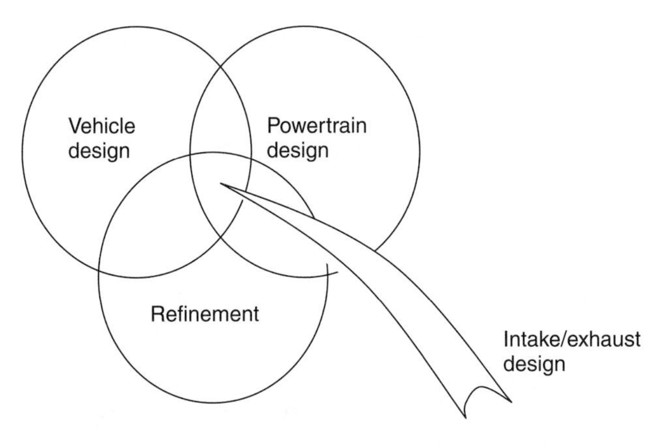

Fig. 22.1-3 The role of the intake system designer.

decisions will directly affect the work of other designers as illustrated in Fig. 22.1-3.

Changes to the system geometry or layout which are beneficial for noise control may well not be advantageous for power or pollutant emissions.

22.1.3.6 The development cycle for intake systems

The initial design and further development of a vehicle intake system takes place in step with the design and development cycle for the whole car.

The key activities are:

- *Concept stage*: Claim packaging volumes.
- *Engine prototype stages*: Develop engine breathing and wave action tuning through manifold design. Minimise system pressure loss. Develop air filtration. Benchmark orifice noise. Develop silencers and resonators.
- *Vehicle prototype stages*: Refine packaging. Test for exterior noise levels.
- *Pre-production vehicles*: Refine interior and exterior noise levels.
- *Pre-production assessment*: Address any problems of noise quality.

22.1.3.7 Principal intake system components

Filter box	housing for a paper filter element. Has a dirty side and a clean side.
Snorkel	dirty-side pipe feeding fresh air to the filter box.
Zip tube	clean-side tube feeding air into the plenum.
Plenum	collecting volume used to join the manifold runners together.
Runners	join each intake port to the intake plenum. These are also commonly known as primary lengths.

There are many issues that need to be considered when designing air intake systems. The following will be discussed here:

- location of the snorkel orifice;
- snorkel and filter box dimensions;
- intake system design for improved engine performance;
- sources of intake (and exhaust) noise;
- flow duct acoustics;
- acoustic performance of a simple intake system;
- conical snorkels;
- protrusion of the snorkel into the filter box;
- Helmholtz resonators;
- sidebranch (quarter wave) resonators;
- effect of flow on sidebranch resonators;
- intake system mounting.

22.1.3.8 Location of the snorkel orifice

Issues to consider are:

- Location is often chosen for packaging reasons only.
- A long snorkel will give good low-frequency attenuation but will also result in many dips in the attenuation at the resonant frequencies of the snorkel.
- Acoustic flow duct silencing predictions usually relate to an unflanged snorkel. Fitting the snorkel into a hole in the bodywork will act like a flange and alter the acoustic performance of the system.
- Positioning the snorkel orifice near to reflective surfaces may increase interior or exterior noise levels.
- The intake air needs to be relatively cool to preserve intake charge density.
- Possible water ingestion must be considered.
- A snorkel orifice position that is good for interior noise may not be appropriate for the control of drive-pass noise.
- The snorkel orifice position should not excite acoustic cavity modes, whether in the engine bay, the inner wing of the bodyshell or passenger cell.

22.1.3.9 Snorkel and filter box dimensions

Consider a snorkel position at the top of the front grille, above the radiator. The filter box is to be mounted on the inner wheel arch. This effectively fixes the snorkel length to, for example, 375 mm.

The snorkel cross-sectional area is often dictated by:

- the requirement for minimum mass flow rate required to preserve engine performance;
- packaging constraints.

The dimensions of the filter box are often dictated by:

- The dimensions of a preferred filter element;
- packaging constraints;
- the need for easy access for exchanging the filter element.

22.1.3.10 Intake and exhaust system design for improved engine performance

Common engine performance indicators include:

- The power curve being the power (usually brake power as measured at the output shaft) in kW plotted against engine speed for a particular engine load (frequently full load/wide-open throttle). Typical gasoline engines (also known as petrol engines or spark-ignition (SI) engines) of the I-4 (four cylinders, in line with each other down the engine), 1.6–2.0-litre class yield a maximum brake power of perhaps 80–90 kW at an engine speed of perhaps 5000–6000 rev min^{-1}.
- The torque curve being the torque (as measured at the output shaft) in N m plotted against engine speed for a particular engine load. Typical gasoline engines of the I-4, 1.6–2.0-litre class yield a maximum brake torque of perhaps 150–180 N m at an engine speed in the range of 2500–4000 rev min^{-1}.
- The brake-specific fuel consumption (bsfc) being the mass of fuel consumed in grams per kWh of power produced at a particular engine speed and load. The expected bsfc of a gasoline engine of the I-4, 1.6–2.0-litre class operating at full engine load is 250–300 g/kWh.

Indicated mean effective pressure (*imep*) is given by Bosch (1986) as:

$$imep(\text{bar}) = \frac{1200 P_I}{V_H N} \text{ for the four-stroke engine} \quad (22.1.2)$$

$$imep(\text{bar}) = \frac{600 P_I}{V_H N} \text{ for the two-stroke engine} \quad (22.1.3)$$

where

P_I = indicated power (kW)
V_H = displacement volume of engine (litres) = $A_p s \times 1000$
A_p = piston area (m^2)
s = stroke (m)
N = revolutions of the engine per minute

Typical values for NA gasoline engines are around the 10 bar mark. *imep* is given in Heywood (1988) as:

$$imep\ (\text{kPa}) = \frac{P_I \times n_r 10^3}{V_H \times n} = \frac{W_{c,I} \times 10^3}{V_d} \quad (22.1.4)$$

where

P_I = indicated power (kW)
n_r = number of revolutions per power stroke
V_H = displacement volume of engine (litres)
n = revolutions of the engine per second

The *imep* is the constant pressure which acting on the piston area throughout the stroke would produce the indicated work per cycle $W_{c,I}$.

The indicated work per cycle is given by the area of the pressure–volume (PV) diagram of the engine. Only if the area of compression and expansion stroke parts of the PV diagram are considered, the gross indicated work per cycle is found. If all parts of the PV diagram are used, then the net indicated work per cycle is found.

A summary of the engine requirements for excellent performance is offered here:

> *Excellent engine performance requires a combination of good airflow and good combustion characteristics.*
>
> (Advice given by Tim Drake, the author's colleague whilst at Lotus Engineering.)

This is shown quite clearly in the governing equations (Heywood, 1988):

$$P_I = \frac{\eta_f m_a N Q_{hv} (F/A)}{n_r} \quad (22.1.5)$$

$$T_I = \frac{P_I}{2\pi N} \quad (22.1.6)$$

where

T_I = indicated engine torque (N m)
P_I = indicated engine power (W)
m_a = mass of air inducted per cycle (kg)
F/A = fuel–air ratio $\dot{m}_f/\dot{m}_a$
$\dot{m}_f$ = mass flow rate of fuel (kg s^{-1})
$\dot{m}_a$ = mass flow rate of air (kg s^{-1})
N = number of revolutions of the crankshaft per second
n_r = number of revolutions of the crankshaft per power stroke (one for two-stroke, two for four-stroke)
Q_{hv} = heating value of fuel, typically 42–44 MJ kg^{-1} ($M = 1 \times 10^6$)
η_f = fuel conversion efficiency

$$\eta_f = \frac{W_c}{m_f Q_{hv}} = \frac{\frac{P n_r}{N}}{\left(\frac{\dot{m}_f n_r}{N}\right) Q_{hv}}$$

$$= \frac{P_I}{\dot{m}_f Q_{hv}} \text{ (Heywood, 1988)} \quad (22.1.7)$$

where

m_f = mass of fuel inducted per cycle (kg)

Now introducing the volumetric efficiency η_v for the four-stroke engine only (where there is a distinct induction stroke) (Heywood, 1988)

$$\eta_v = \frac{2\dot{m}_a}{\rho_{a,i} V_d N} \quad (22.1.8)$$

where

v_d = volume displaced by piston (m^3)
$\rho_{a,i}$ = inlet air density (kg m^{-3})

The following equations are obtained for four-stroke engines (Heywood, 1988):

$$P_I = \frac{\eta_f \eta_v N V_d Q_{hv} \rho_{a,i} (F/A)}{2} \quad (22.1.9)$$

$$T_I = \frac{\eta_f \eta_v V_d Q_{hv} \rho_{a,i} (F/A)}{4\pi} \quad (22.1.10)$$

$$imep = \eta_f \eta_v Q_{hv} \rho_{a,i} (F/A) \quad (22.1.11)$$

The mechanical efficiency η_m allows the conversion of indicated power to brake power P_b (the useful power as measured by a dynamometer)

$$\eta_m = \frac{P_b}{P_{ig}} = 1 - \frac{P_f}{P_{ig}} \quad (22.1.12)$$

where P_{ig} is the gross indicated power (found from the gross indicated work per cycle – the area under the PV diagram for compression and expansion strokes only) being the sum of brake power and friction power P_f. Therefore,

$$P_{ig} = P_b + P_f \quad (22.1.13)$$

Friction power includes the power used to overcome friction in the engine (the rubbing friction) and in pumping the gas in and out of the engine (the pumping loss). Therefore, the mechanical efficiency is a function of throttle setting and engine speed.

Brake mean effective pressure (*bmep*) is a useful measure of the torque output of an engine. It will be maximum at the engine speed corresponding to peak torque (Heywood, 1988) and will be 10–15% less at peak power.

Typical maximum *bmep* are (Heywood, 1988):

• NA gasoline engine	8–11 bar
• Turbo gasoline engine	12–17 bar
• NA diesels	7–9 bar
• Turbo-diesels	10–12 bar

The relationship between *bmep* and brake torque is (Heywood, 1988):

$$bmep\ (\text{kPa}) = \frac{2\pi n_r T(\text{Nm})}{V_d} \quad (22.1.14)$$

where V_d is in dm^3.

Note: If V_d were in m^3 then the resulting *bmep* would be in pascals.

The *bmep* is commonly used to assess the torque output or *load* on an engine. If *bmep* is measured to be less than the known maximum *bmep* expected from an engine, then the engine must be operating at part load. *bmep* tends to zero at idle (η_m tends to zero), with the useful work produced during the cycle being used up in overcoming pumping losses (Heywood, 1988).

Note that:

$$imep = bmep + fmep + pmep + amep \quad (22.1.15)$$

where

imep = indicated mep (bar)
bmep = brake mep (bar)
fmep = rubbing friction mep (bar)
pmep = pumping (or gas exchange) mep (bar)
amep = accessory (alternator, pumps etc.) mep (bar)

So, to use Tim Drake's criteria for excellent performance, the above equations show that the designer needs:

- to maximise η_v and $\rho_{a,i}$ in order to give good airflow;
- to maximise η_f and Q_{hv} along with (F/A) whilst minimising *bsfc*;

$$bsfc = \frac{\dot{m}_f}{P}\ \text{g J}^{-1} \text{ or g/kWh (Heywood, 1988)} \quad (22.1.16)$$

22.1.3.10.1 Some basic background terms

Each cylinder of an engine with poppet valves (conventional four-stroke engines) will have the following components:

- Intake system, usually comprising a dirty-side duct (or snorkel), supplying air to a filter (housed in a filter box), which in turn supplies air to a manifold.
- Intake manifold, distributing air (possibly air–fuel mixture) to each cylinder. Different configurations of intake manifold exist, most are variants of end-feed and centre-feed designs (Figs. 22.1-4 and 22.1-5 respectively).
- The intake manifold has two sections – a plenum being a volume commonly equal to the swept volume of the engine, and primary lengths (or runners) being the ducts that connect the plenum to the cylinder head.

Fig. 22.1-4 End-feed intake manifold.

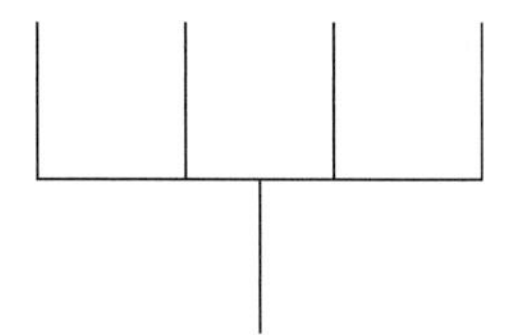

Fig. 22.1-5 Centre-feed intake manifold.

- Intake ports, being the lengths of duct concealed within the cylinder head connecting the manifold with the rear of the intake valves.
- Intake valves, the timing and lift of which are commonly controlled mechanically by a rotating camshaft.
- Exhaust valves, the timing and lift of which are commonly controlled mechanically by a rotating camshaft.
- Exhaust ports.
- Exhaust manifold, combining the exhaust flow from different cylinders into a common (or several common) exhaust tailpipe. The combination of flows may be sudden (four pipes combining as one in the four-into-one manifold) or gradual (two sets of two pipes combine, then the two larger pipes subsequently combine, forming the four-into-two-into-one manifold).

22.1.3.10.2 Valve lift and timing

Commonly, poppet valves have a maximum lift in the order of 10 mm. This is commonly timed to occur (being known as the maximum opening point – MOP) at around 100–110 after top dead centre (ATDC) for the intake valves and 110–120 after bottom dead centre (ABDC) for the exhaust valves.

The period during which a valve is open (often called the cam duration and measured in degrees of revolution of the crankshaft) is commonly measured between the top of ramp (TOR) of the valve rise and the TOR of the valve fall. The TOR marks a position on the cam profile where there is a transition between a gradual rise and a steeper rise. These ramps are used to smooth the path of the cam follower which rides on the cam and provides the mechanical link between the cam and the valve. The TOR is often used as a basis for valve timings. The four valve timings are:

1. IVO (inlet valve open);
2. IVC (inlet valve close);
3. EVO (exhaust valve open);
4. EVC (exhaust valve close).

Cam durations are commonly in the range of 220–250° of crankshaft rotation. The intake duration may be different from that of the exhaust. With longer durations, it is possible that EVO will occur before IVC (so both intake and exhaust valves will be open) and a period of valve overlap occurs. This is shown in Fig. 22.1-6.

Valve overlap is often quoted as an area (mm deg) and is generally less than 5 mm deg. Peak engine power increases with increasing valve overlap. However, high overlap causes combustion instability (exhaust gases

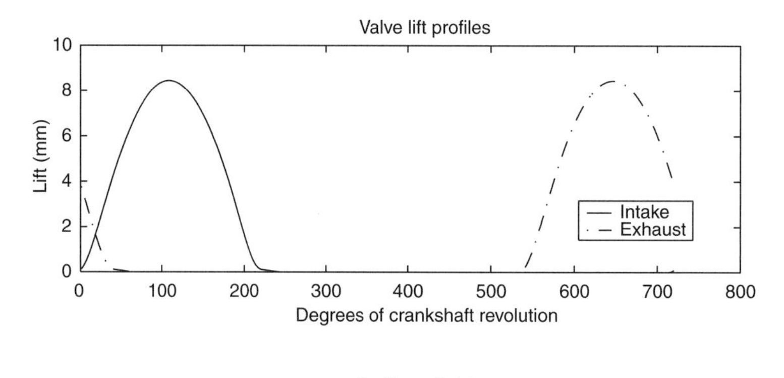

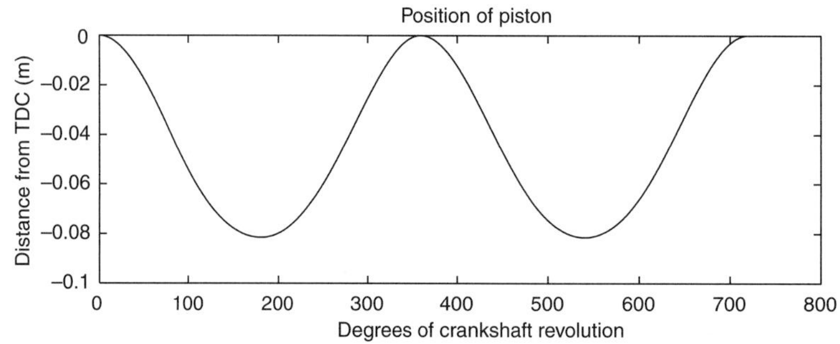

Fig. 22.1-6 Valve lift profiles and the instantaneous position of the piston.

drawn back into the cylinder) particularly at low engine speeds and this is important for stable idle. Combustion instability manifests itself as a cylinder-to-cylinder variation in IMEP. Low valve overlap is beneficial for low-speed torque. Twenty degree overlap is a typical compromise for producting gasoline engines.

22.1.3.10.3 On the flow through the intake valve

Knowledge of the gas dynamics at the intake valve is useful to the refinement engineer. A non-conservative form of the conservation of momentum equation for a fluid in three dimensions is (Hirsch, 1988)

$$\rho \frac{d\bar{u}}{dt} = -\nabla p \bar{\bar{I}} + \bar{\nabla} \cdot \bar{\bar{\tau}} + \rho \bar{f_e} \quad (22.1.17)$$

where

$$\text{grad}\, p = \nabla p = \frac{\partial p}{\partial x}\bar{i} + \frac{\partial p}{\partial y}\bar{j} + \frac{\partial p}{\partial z}\bar{k}$$

$\bar{i}, \bar{j}, \bar{k}$ are unit vectors
$\bar{\bar{I}}$ is the unit tensor
$\bar{\bar{\tau}}$ is the viscous shear stress tensor
$\bar{f_e}$ is the external force vector
p is pressure
ρ is density
$\bar{u}$ is the velocity vector

The total inertial term on the left-hand side of equation (22.1.17) can be re-written as the sum of linear, kinetic and rotational forces:

$$\rho \frac{d\bar{u}}{dt} = \rho \left[\frac{\partial \bar{u}}{\partial t} + \bar{\nabla}\left(\frac{u^2}{2}\right) - \left(\bar{u} \times \bar{\xi}\right) \right] \quad (22.1.18)$$

where $\bar{\xi}$ is known as the vorticity vector.

Simplifying the analysis to consider only one dimension x, and neglecting viscosity effects, external forces and vorticity, equation (22.1.17) reduces to the familiar non-linear inviscid Euler equation (see Appendix 21.1G for a derivation):

$$\frac{\partial u}{\partial t} + \frac{\partial}{\partial x}\left(\frac{u^2}{2}\right) = -\frac{1}{\rho}\frac{\partial p}{\partial x} \quad (22.1.19)$$

From the second law of thermodynamics (for instance Zemansky and Dittman (1997)):

$$T ds = dh - v dp \quad (22.1.20)$$

or

$$T \frac{ds}{dx} = \frac{dh}{dx} - v \frac{dp}{dx} \quad (22.1.21)$$

where T is the temperature, s is specific entropy, h is specific enthalpy and v is specific volume. This can be re-written as

$$T \frac{ds}{dx} = \frac{dh}{dx} - \frac{1}{\rho}\frac{dp}{dx} \quad (22.1.22)$$

Substituting equation (22.1.22) into equation (22.1.19) gives

$$\frac{\partial u}{\partial t} + \frac{\partial}{\partial x}\left(\frac{u^2}{2}\right) = T ds - dh \quad (22.1.23)$$

Now the total (or stagnation) enthalpy H is given as

$$H = h + \frac{u^2}{2} \quad (22.1.24)$$

and substituting the differential of equation (22.1.24) into equation (22.1.23)

$$\frac{\partial u}{\partial t} = T ds - dH \quad (22.1.25)$$

Now for the assumption of homentropic flow, equation (22.1.25) reduces to:

$$\frac{\partial u}{\partial t} + dH = 0 \quad (22.1.26)$$

Now along a streamline the stagnation enthalpy is constant, so

$$\frac{\partial u}{\partial t} + H = H_0 = \text{constant} \quad (22.1.27)$$

A scalar potential function ϕ can be declared so that

$$u = \bar{\nabla}\phi \quad (22.1.28)$$

$$u = \frac{\partial \phi}{\partial x} \text{ in a 1-D model} \quad (22.1.29)$$

and thereby create what is known as a potential flow model given by:

$$\frac{\partial \phi}{\partial t} + H = H_0 = \text{constant} \quad (22.1.30)$$

For an ideal gas, where a is the speed of sound (m s^{-1}) and γ is the ratio of specific heats c_p/c_v

$$h = \frac{a^2}{\gamma - 1} \quad (22.1.31)$$

and so using equation (22.1.24), equation (22.1.30) can be written for flow along a streamline (or a Fanno line) as:

$$\frac{\partial \phi}{\partial t} + \frac{a^2}{\gamma - 1} + \frac{u^2}{2} = \frac{a_0^2}{\gamma - 1} \quad (22.1.32)$$

Now as

$$\frac{\partial \phi}{\partial t} = \frac{\partial^2 u}{\partial x \partial t} \qquad (22.1.33)$$

most workers omit that term so that the 1-D, non-conservative, inviscid, irrotational, homentropic momentum equation, along a streamline (!) becomes:

$$a_0^2 = a^2 + \frac{\gamma - 1}{2} u^2 \qquad (22.1.34)$$

Benson (1982) famously calls this an energy equation (although this is probably a misnomer).

From the derivation of equation (22.1.34) a simple and very well-known intake valve flow model can be constructed that assumes a flow from a large reservoir (representing the manifold) into one of the cylinders via a single orifice of negligible length. This was first derived by Tsu in 1947. Stagnation conditions (subscript 0,1 meaning zero flow velocity in zone 1) are assumed in the cylinder.

For inflow to the cylinder via the intake valve, the intake manifold is assumed to constitute a sufficiently large volume for constant pressure conditions to occur, and the conditions in the manifold are given subscript '2'. For outflow (reverse flow through the intake valve) the subscripts are reversed.

From a principle of continuity of mass, assuming quasi steady flow it can be written that:

$$\dot{m} = \rho_2 u_2 A_m \qquad (22.1.35)$$

where

$\dot{m}$ = mass flow rate (kg s^{-1})
A_m = open flow area of the valve (given in Appendix 22.1A at the end of this chapter, after Heywood [1988])

The following isentropic relationships apply for an ideal gas:

$$\rho_2 = \rho_1 \left(\frac{p_2}{p_1}\right)^{1/\gamma} \qquad (22.1.36)$$

$$T_2 = T_1 \left(\frac{p_2}{p_1}\right)^{\gamma - 1/\gamma} \qquad (22.1.37)$$

Now also for an ideal gas, where R is the specific gas constant

$$p = \rho R T \qquad (22.1.38)$$

and

$$a = \sqrt{\gamma R T} \qquad (22.1.39)$$

so

$$\rho = \frac{\gamma P}{a^2} \qquad (22.1.40)$$

Substituting equation (22.1.40) into equation (22.1.36) and thus assuming isentropic expansion of the gas as it enters the cylinder:

$$\rho_2 = \frac{\gamma p_{01}}{a_{01}^2} \left(\frac{p_2}{p_{01}}\right)^{1/\gamma} \qquad (22.1.41)$$

Re-writing equation (22.1.34) with the appropriate subscripts

$$a_{01}^2 = a_2^2 + \frac{\gamma - 1}{2} u_2^2 \qquad (22.1.42)$$

Re-arranging equation (22.1.42)

$$u_2^2 = (a_{01}^2 - a_2^2) \frac{2}{\gamma - 1} \qquad (22.1.43)$$

Substituting equation (22.1.39) into equation (22.1.37) gives, with the appropriate subscripts,

$$T_2 = \frac{a_{01}^2}{\gamma R} \left(\frac{p_2}{p_{01}}\right)^{\gamma - 1/\gamma} \qquad (22.1.44)$$

Use of equation (22.1.39) once more to replace T_2 yields:

$$a_2^2 = a_{01}^2 \left(\frac{p_2}{p_{01}}\right)^{\gamma - 1/\gamma} \qquad (22.1.45)$$

Substituting equation (22.1.45) into equation (22.1.43):

$$u_2^2 = \left(a_{01}^2 - a_{01}^2 \left(\frac{p_2}{p_{01}}\right)^{\gamma - 1/\gamma}\right) \frac{2}{\gamma - 1}$$

$$u_2^2 = \frac{2 a_{01}^2}{\gamma - 1} \left(1 - \left(\frac{p_2}{p_{01}}\right)^{\gamma - 1/\gamma}\right) \qquad (22.1.46)$$

Now substituting both equation (22.1.46) and equation (22.1.41) into (22.1.35) gives:

$$\dot{m} = \frac{\gamma p_{01}}{a_{01}^2} \left(\frac{p_2}{p_{01}}\right)^{1/\gamma} \left[\frac{2 a_{01}^2}{\gamma - 1}\right]^{1/2} \left[1 - \left(\frac{p_2}{p_{01}}\right)^{\gamma - 1/\gamma}\right]^{1/2} A_m$$

$$\dot{m} = \frac{p_{01} A_m}{a_{01}} \left[\left(\frac{2\gamma^2}{\gamma - 1}\right) \left(\frac{p_2}{p_{01}}\right)^{2/\gamma} \left[1 - \left(\frac{p_2}{p_{01}}\right)^{\gamma - 1/\gamma}\right]\right]^{1/2} \qquad (22.1.47)$$

Equation (22.1.47) presents the mass flow rate as a function of the open valve area and the pressure ratio across the valve for subsonic flow through the orifice (valve). However, equation (22.1.47) will tend to predict much higher flow rates than are encountered with practical engines, due to the large number of simplifying assumptions that went into its derivation. One way to correct this effect is to introduce a discharge coefficient c_d where

$$c_d = \frac{A_e}{A_r} \tag{22.1.48}$$

and A_e is the effective area and A_r is some suitable reference area. The effective area (Annand and Roe, 1974) is the outlet area of an imaginary frictionless nozzle which would pass the required flow when drawing from a large constant pressure reservoir and discharging into another reservoir. The reference area can be the cross-sectional area of any suitable part of the real flow path such as the curtain area under the open valve.

Measured discharge coefficients can be used to calculate the effective area for a given reference area, and hence equation (22.1.47) becomes:

$$\dot{m} = \frac{p_{01}A_e}{a_{01}}\left[\left(\frac{2\gamma^2}{\gamma-1}\right)\left(\frac{p_2}{p_{01}}\right)^{2/\gamma}\left[1-\left(\frac{p_2}{p_{01}}\right)^{\gamma-1/\gamma}\right]\right]^{1/2} \tag{22.1.49}$$

The most well-known discharge coefficients for inflow through the intake valve are given in Annand and Roe (1974) for a reference area equal to the curtain area under the open valve

$$A_r = \pi D L_v \tag{22.1.50}$$

where D is the valve head diameter and L_v is the valve lift. These discharge coefficients are reproduced in Fig. 22.1-7.

At low valve lifts, the flow past the valve remains attached to both the valve head and the valve seat. At intermediate valve lifts, the flow is separated on one side but not on the other, producing a sudden drop in discharge coefficient that subsequently recovers with further valve lift. At high valve lifts, the flow is detached from both sides, and the so-called free-jet is formed.

The discharge coefficients for outflow through the inlet valve (reverse flow) are generally higher (around 0.7 up to $L_v/D = 0.2$, then falling to 0.5 at $L_v/D = 0.4$).

Flow loss coefficients (rather than discharge coefficients) are commonly used in commercial engine simulations (AVL, 2000). These are defined as the ratio between the actual mass flow and the loss-free isentropic mass flow for the same stagnation pressure and the same pressure ratio (AVL, 2000). The difference between a discharge coefficient and a flow loss coefficient is important. The discharge coefficient applies to flow between stagnant reservoirs passing through a frictionless nozzle. The flow loss coefficient applies to steady or pulsating flow through the cylinder head.

Flow loss coefficients (such as those shown in Figs. 22.1-8 and 22-1-9) are often measured using steady flow on a bench (Blair and Drouin, 1996). Sometimes, they are measured using pulsed flow (for instance, Fukutani and Watanabe (1982)) in order to improve the realism of the model represented by equation (22.1.49).

An example for an intake port is given in Fig. 22.1-8 and an example for an exhaust port in Fig. 22.1-9.

For the flow loss coefficients shown in Figs. 22.1-8 and 22.1-9

$$A_e = \text{coefficient} \times \frac{d_{vi}^2 \times \pi}{4} \tag{22.1.51}$$

where d_{vi} is the inner valve seat diameter (reference diameter corresponding to D in Figs. 22.1-8 and 22.1-9).

Equation (22.1.49) reduces to

$$\dot{m} = \frac{A_e \gamma p_{01}}{a_{01}}\left(\frac{2}{\gamma+1}\right)^{\gamma+1/2(\gamma-1)} \tag{22.1.52}$$

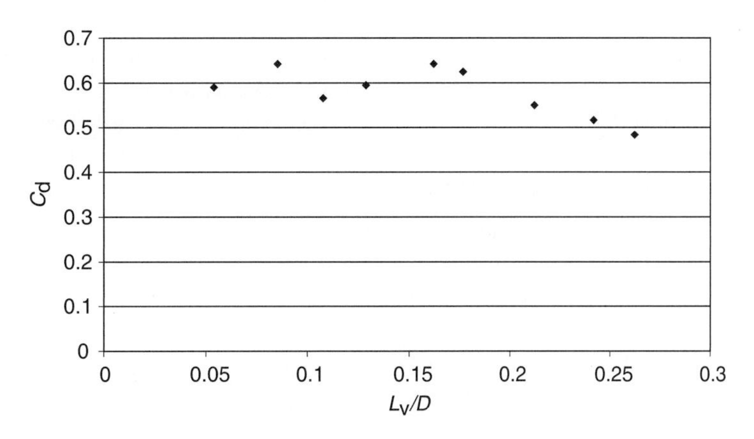

Fig. 22.1-7 Discharge coefficients with respect to reference area given by equation (22.1.50) for the case of inflow through the intake valve (Annand and Roe, 1974).

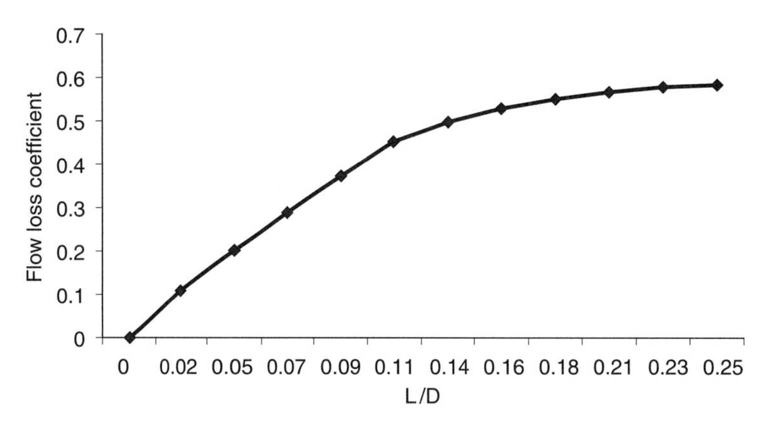

Fig. 22.1-8 Intake port flow loss coefficient (AVL, 2000).

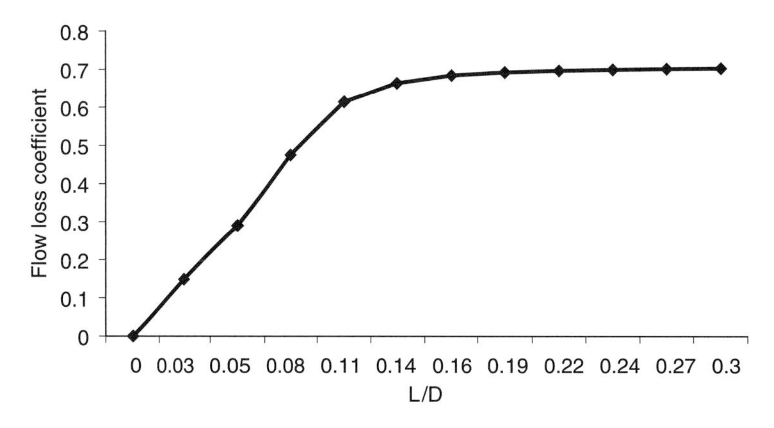

Fig. 22.1-9 Exhaust port flow loss coefficient (AVL, 2000).

for sonic flow through the orifice (valve) which is also known as choked flow as this represents the maximum rate of mass transfer possible through a single orifice of negligible length. Note that this is no longer a function of pressure ratio across the valve, but is solely a function of the conditions in the manifold and of the effective open area of the valve. The sonic condition is reached when

$$\frac{p_{01}}{p_2} \leq \left(\frac{2}{\gamma+1}\right)^{\gamma/\gamma-1} \qquad (22.1.53)$$

which is a pressure ratio of around 0.53.

22.1.3.10.4 A note on intake manifold design

The volumetric efficiency of a four-stroke engine can be improved through dynamic tuning whereby it is contrived to maximise the pressure in the intake port at around IVC (Ohata and Ishida, 1982). This of course maximises the pressure ratio in equation (22.1.49) and hence the mass flow rate, and hence the volumetric efficiency (see equation (22.1.8)).

Achieving the dynamic tuning effect is strongly dependent on intake manifold design, and in particular the length of the primary runners (Winterbone and Pearson, 1999). Generally, longer runner lengths are used to maximise low-speed torque, but often with a penalty of reducing high-speed power.

There is a connection between ignition timing and runner length. Fifteen degree changes in advance may be required at a given speed when changing runner length (Harrison, 2003).

In addition to optimising spark advance, one really needs to optimise the valve timing as well. IVC is important but good intake manifold design is more important. A good cam design cannot make up for a poor manifold.

The balance in flows between runners is important. With one restrictive intake runner, that cylinder will tend to run rich at high engine speeds. As a result, the other cylinders will run lean. Intake port velocity is typically 70–90 m s^{-1}.

22.1.3.10.5 The motion of the piston during the intake stroke

Earlier discussions show that volumetric efficiency is strongly influenced by the pressure ratio across the inlet valves at IVC. Consider what is happening in the cylinder around IVC.

Consider one cylinder of an engine with the following specification:

- 1.6 litre I4 four-stroke;
- stroke 81.5 mm;
- compression ratio 10.0;
- conrod length = 129 mm;
- piston pin offset = 1.0 mm.

From this data, it is easy to calculate the instantaneous volume in that cylinder as it changes with the rotation of the crankshaft. The results of such a calculation are shown in Fig. 22.1-10 where the volume has been normalised by the maximum volume (swept volume + clearance volume) and expressed as a percentage.

Inspecting Fig. 22.1-10 then at TDC, the volume is 10% of the maximum volume. This of course corresponds to the clearance volume in an engine with a compression ratio of 10. At the intake stroke BDC (540°) the volume is at its maximum (100%). The swept volume is the difference between the volume at BDC and the volume at TDC.

Inspection of the data used to generate Fig. 22.1-10 reveals that for this typical case, the cylinder volume is in the range of 99–100% of the maximum for 28° of crankshaft rotation in the range of 526–554°.

The volume lies in the range of 98–100% for the 40° of crankshaft rotation between 520° and 560°. So, for the 30–40° around BDC the volume in the cylinder is changing by less than 1–2%.

Typically, MOP for the intake valve is 100–110° ATDC (see Section 22.1.3.10.2). Take an intake cam with MOP at 102° ATDC (462°) and 230° duration. That places IVC at 37° ABDC (577°). For the case shown in Fig. 22.1-10 the cylinder volume at 577° is 93.6% of the maximum value at BDC. By the time intake MOP is reached, the cylinder volume is 70% of its maximum value at the BDC position.

A very short duration camshaft (MOP 102° duration 184°) will position IVC at a point when the cylinder volume is >99° of that at BDC. An MOP 102° duration 196° intake camshaft will position IVC at a point when the cylinder volume is >98% of that at BDC. With more conventional camshafts (duration >220°) the point when the cylinder volume is >98% of its maximum will occur 20–50° before IVC.

It is the pressure ratio across the valve during this part of the intake lift profile 20–50° before IVC, that most strongly influences volumetric efficiency. Winterbone and Pearson (1999) suggest that a significant period is some 50° before IVC.

Now the piston velocity around IVC will be calculated. This is easily done by dividing the distance that the piston travels in 1° crank by the time period for 1° crank. Fig. 22.1-11 is the result for 4000 rev min^{-1} using the same input data as for Fig. 22.1-10.

At BDC (540°) the piston velocity is zero; therefore there is a 90° phase lag relative to displacement. The acceleration will be at its maximum value at BDC (180° phase lag relative to displacement). At 577° the piston velocity in the case shown in Fig. 22.1-11 is 7.6 ms^{-1} compared with a maximum of 17.9 ms^{-1}.

The maximum piston velocity in this case occurs at 646° (at this point the piston is slightly more than halfway back to TDC – 106/180 through the compression stroke) and also at 434° (the piston is 74/180 through the intake stroke).

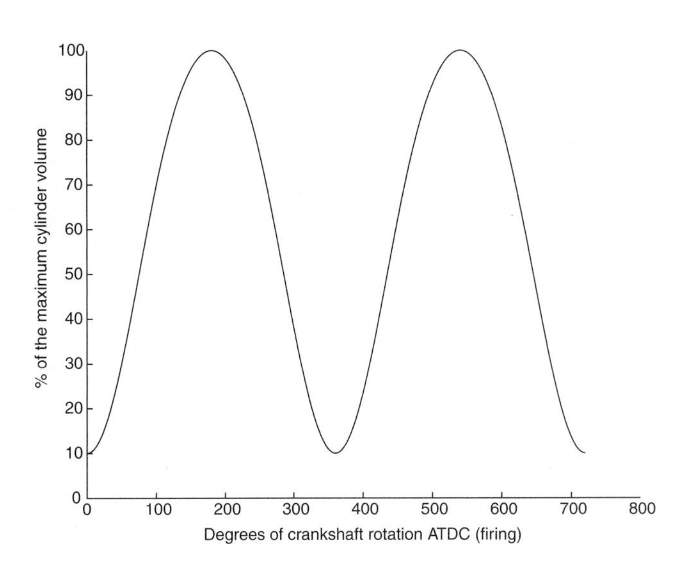

Fig. 22.1-10 Time history of the cylinder volume.

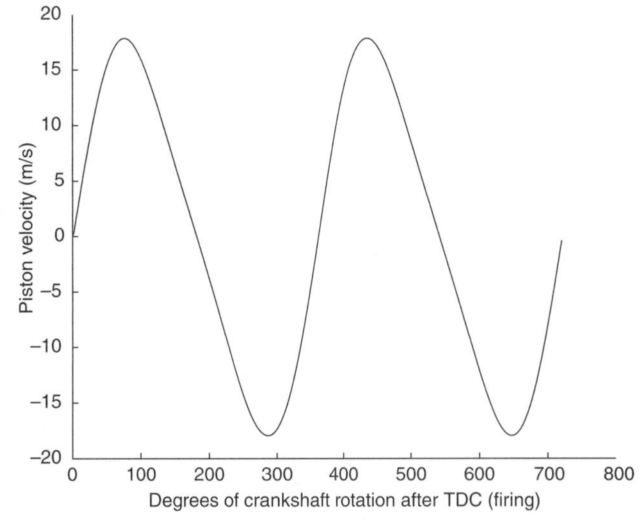

Fig. 22.1-11 Piston velocity, 4000 rev min^{-1}.

The mean piston speed for this case is:

$$2 \times \frac{\text{rev min}^{-1}}{60} \times \text{stroke} = 10.9\ \text{ms}^{-1}$$

In Section 22.1.3.12.12 it is shown that a typical gasoline engine producing 50 kW consumes of the order of 0.068 m^3 of air per second. If the engine here has 8 intake ports of diameter 30 mm then the average flow velocity through the ports is 12.0 ms^{-1} averaged over 720°. However, there is flow through the port for only say 230° of the 720° so the average velocity in the port is more likely to be 37.6 ms^{-1}. Peak flow velocities are of the order of 70-90 ms^{-1} for the case of a single intake valve per cylinder.

Note that there are several algorithms to describe the motion of the piston:

- Bosch (1986);
- Heywood (1988).

The results are subtly different for each one.

22.1.3.10.6 The pressure waves in the intake system

How can the volumetric efficiency be maximised by increasing the pressure in the intake port at around IVC? Fig. 22.1-12 shows the intake port pressure trace for a single cylinder racing engine fitted with a simple intake pipe which is a conical pipe some 150 mm long (the intake primary). The engine is at 3/4 of its rated speed, full load (wide-open throttle) and it is running 'on tune' (Dunkley, 1999).

In Fig. 22.1-12:

- 0° corresponds to ignition TDC.
- IVO occurs a little before 360°
- IVC occurs around 600° (>40° ABDC).

There are three distinct features on Fig. 22.1-12:

1. A period of pressure depression after IVO and lasting some 120° until the intake MOP.
2. A clear pressure peak around IVC. In this case the wave action in the port has a pressure amplitude of 0.6 bar (60 kPa, 190 dB re 20 μPa) at around IVC.
3. After IVC the pressure decays with (in this case) three successive oscillations before the intake valve opens once more.

The origin of the pressure peak at IVC can be explained thus:

- Just after IVO (350°) the intake valve opens, the piston accelerates downwards and air is drawn into the cylinder. If the intake primary length is sufficiently long (as it is in this case) this causes a pressure depression in the port just behind the valve.
- The pressure depression travels away from the valve at the speed of sound. It is reflected at the open end of the intake primary (with a phase shift – 180° for an unflanged open pipe) and the reflected pressure pulse travels back towards the valve as a positive pressure (pressure peak). The 180° phase shift is a result of the open pipe end being a pressure release surface. The net dynamic pressure at the open end must be equal to the prevailing static pressure just beyond the pipe end. The only way to produce this

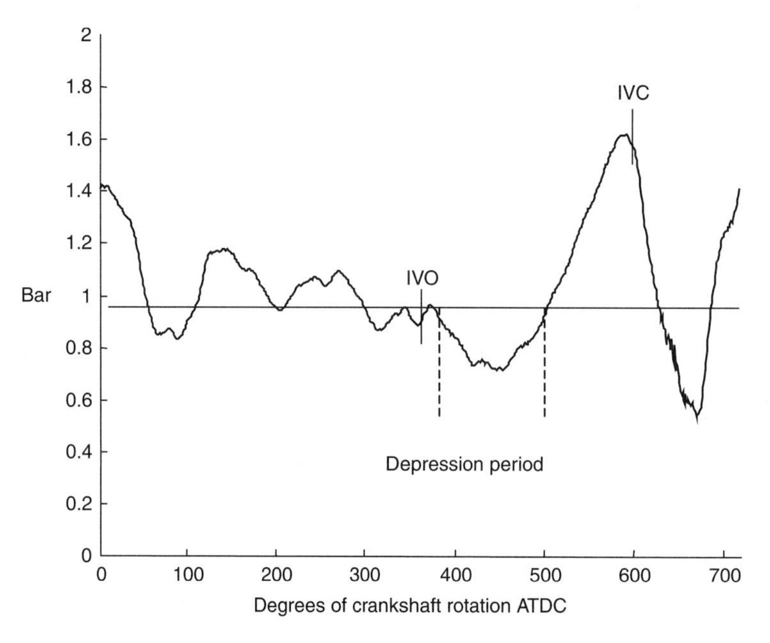

Fig. 22.1-12 Intake port pressure trace, 3/4 rated speed, full load (Dunkley, 1999).

net cancellation of the dynamic pressure is to have two waves travelling in opposite directions, in antiphase, thus cancelling each other out.

- By choosing a suitable primary length (and the dimensions of the rest of the intake system if fitted) one can phase the pressure peak to arrive back at the valve at around IVC. This is known to be happening in the case of Fig. 22.1-12.
- In the case shown in Fig. 22.1-12, the arrival of a positive pulse at the valve might account for 20 kPa of dynamic pressure at IVC but one can note a level of 60 kPa. The additional 40 kPa is the result of the so-called inertia effects or ram. During the pressure depression after IVO the flow through the valve builds up a certain momentum. As the valve closes towards IVC this momentum is transferred to static pressure as the air is brought to a halt. Providing that the acceleration of the air after IVO was efficient (a function of timing the IVO) then the ram effect produces a net gain in volumetric efficiency (usually only at high speeds) (Harrison and Dunkley, 2004).

Note that in the exhaust a similar phenomenon occurs except that a positive pressure peak is generated at the exhaust valve which is subsequently reflected back as a pressure depression that aids scavenging.

Consider Fig. 22.1-13 where the same engine is operating at a much lower speed, full load and 'off tune' (Dunkley, 1999). Comparing this pressure trace with that of the 'on tune' operating condition in Fig. 22.1-12:

- When, 'off tune' (in this case), the pressure depression period is much shorter than when 'on tune'. In this case this is due to flow reversal in the intake valve caused by the particular timing of IVO.
- There is a pressure depression (in this case) at IVC which will reduce volumetric efficiency. The magnitude of this depression is 1.5 bar, which is much less than the peak magnitude seen in Fig. 22.1-12.

So, in summary, inspection of Figs. 22.1-12 and 22.1-13 has identified two physical mechanisms by which the intake port pressure may be altered around IVC:

1. *Wave effects* whereby a travelling pressure pulse is timed to be positive at the valve near IVC at the 'on tune' engine speed. Away from this speed, a pressure depression might occur at IVC thus reducing volumetric efficiency.
2. *Inertia effects* where by flow momentum is transformed to pressure around IVC (usually most significant at higher engine speeds).

In practical production engines, there will be more than one cylinder. A different intake port pressure trace results from the wave effects due to the filling of adjacent

---

Group demonstration – acoustic resonance in a pipe
A loudspeaker, a signal generator, an amplifier, a long piece of pipe and a sound level meter are used in this demonstration. Acoustic resonances in the pipe appear as peaks in the noise level at the end of the pipe. These resonances are similar to those found in a straight-pipe intake system when the valve is closed that produce the oscillations seen in Fig. 22.1-12.

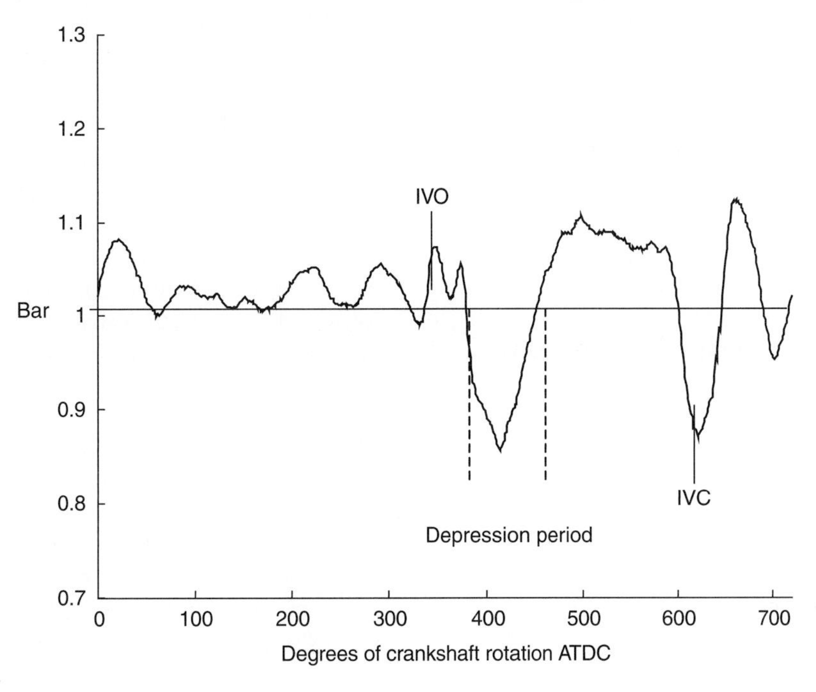

Fig. 22.1-13 Intake port pressure trace, 35% rated speed, full load (Dunkley, 1999).

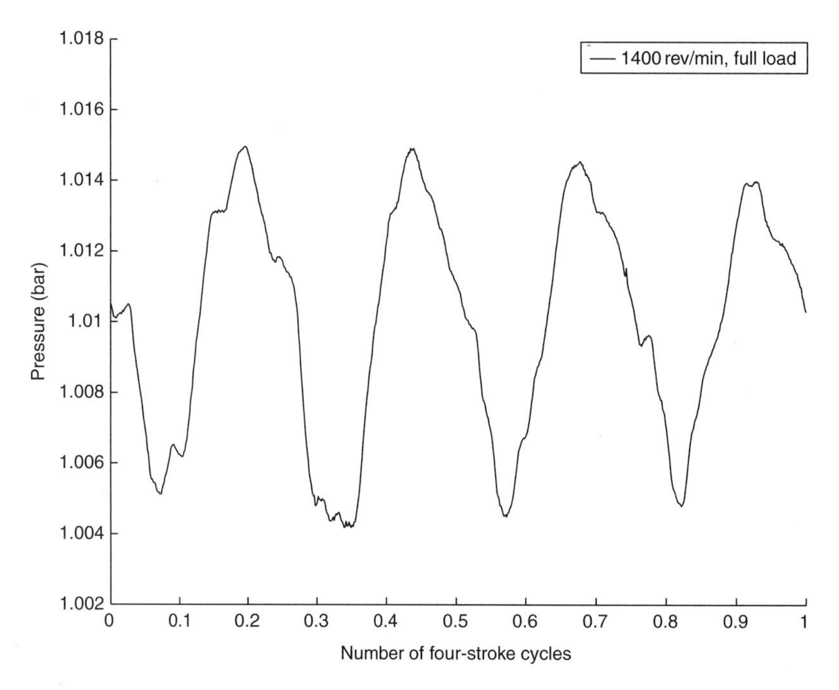

Fig. 22.1-14 Intake pressure trace for a four-cylinder engine.

cylinders. Fig. 22.1-14 shows such a trace for a four cylinder engine.

Ohata and Ishida (1982) measured the intake pressure trace in PORT # 2 of a four-cylinder engine with only one cylinder operating at any given time. When they summed the four pressure traces they found that the sum matched the pressure trace recorded in PORT # 2 when all four-cylinders were activated.

The Ohata and Ishida experiment supports the notion that the wave action in the intake system is linear as the additivity part of the theory of superposition seems to hold (see Chapter 21.1, Appendix 21.1A).

22.1.3.10.7 On turbo and supercharging the engine to improve performance

Torque is a useful description of engine performance (see Section 22.1.3.10). Low-speed torque strongly affects driveability. High-speed power affects maximum speed and high-speed torque affects elasticity in the gear selection.

Torque and power are simply related by engine speed N (rev/s)

$$T = \frac{P}{2\pi N}$$

The governing torque equation is:

$$T_I = \frac{\eta_f \eta_v V_d Q_{hv} \rho_{a,i}(F/A)}{4\pi}$$

where

T_I = indicated engine torque (Nm)
F/A = fuel air ratio $\dot{m}_f/\dot{m}_a$
Q_{hv} = heating value of fuel, typically 42–44 MJ kg^{-1} ($M = 1 \times 10^6$)
η_V = volumetric efficiency
η_f = fuel conversion efficiency
V_d = swept volume (m^3)

Assuming those variables connected with fuelling and combustion (η_f, Q_{HV}, F/A) are to be kept constant, inspection of the torque equation reveals the following mechanisms for increasing torque:

- increase η_v;
- increase V_d;
- increase $\rho_{a,i}$.

Increasing engine capacity has always been an option for increasing torque (witness the massive aero engines fitted to early twentieth-century racing cars). More recently, design skills have been applied with modern multi-point fuel injection technologies to raise η_v by careful tuning of the intake and exhaust manifolds and appropriate valve lift timing (and multi-valve engines).

Raising torque by increasing the intake air density is the subject here. It requires additional work on the intake gas beyond the pumping work found in the NA engine. This work may be supplied by either:

- *Supercharging* where a mechanical drive from the engine powers a positive displacement device (a sweep or a pump basically) or
- *Turbocharging* where a turbine runs off the engine's exhaust gas and is used to power a compressor (a high-speed impeller).

bmep is a useful measure of torque output. Typical maximum *bmep* are:

• NA gasoline	8–11 bar
• Turbo gasoline	12–17 bar
• NA diesel	7–9 bar
• Turbo diesel	10–12 bar

Based on these data, it appears that the turbocharger is being used to raise inlet air densities by a factor of ≈1.5 in the gasoline engine and by ≈1.3–1.4 in the diesel engine.

22.1.3.10.8 The rationale for turbocharging

Turbochargers are now routinely fitted to diesel engines. The reasons for this are:

- Specific output (power/swept volume) of the NA diesel is poor, leading to large, heavy expensive engines (due to the need to run lean to avoid excessive smoke (Heywood, 1988)).
- Power output of the diesel is smoke-limited. This restriction is relaxed if more air mass is added to a given cylinder size by turbocharging.
- There are no knock problems to overcome even when turbocharging diesel engines (unlike gasoline engines).
- Diesel engines are more costly, allowing the additional costs of a turbocharger to be absorbed (Watson and Janota, 1982).

The best-known reference work on turbocharging (Watson and Janota, 1982) is mostly devoted to diesel engines. The rationale for turbocharging the gasoline engine is, to be honest, rather less compelling than for the diesel engine.

If the designer needs/wants to boost the torque output from a restricted swept volume then they might choose to turbocharge. However, there are two limits to this. The first is that the increased air density at inlet, when combined with the correct ratio of fuel will produce higher cylinder pressures and temperatures thus increasing the tendency to knock (Heywood, 1988) – this can be controlled by reducing the compression ratio and accepting the loss of thermal efficiency caused. Even with the loss of efficiency, the work achieved per cycle is greater in the turbocharged engine than achieved by the NA engine because more fuel can be burnt whilst preserving the desired air–fuel ratio. The second limitation to turbocharging the gasoline engine is speed. The turbocharger is ideally matched at one engine operating point (speed/load combination) and achieving adequate performance at other operating points is difficult.

The turbocharged diesel engine may be more fuel efficient than its NA counterpart (unlikely to be the case in gasoline engines). This is because the turbocharged engine produces more torque at a given engine speed (Watson and Janota, 1982). As friction losses in the engine are speed-dependent, these form a smaller proportion of the turbocharged IMEP than that of the NA IMEP. Thus the overall efficiency of the turbocharged engine is greater.

Watson and Janota have an interesting perspective on turbocharging (Watson and Janota, 1982).

> *Turbocharging is a specific method of supercharging. An attempt is made to use the energy of the hot exhaust gas of the engine to drive the supercharging compressor. The user is not getting something for nothing, but is merely using energy that would normally go to waste; however it is clear that it is no longer necessary to debit the power requirement of the compressor from the indicated power of the engine.*

22.1.3.10.9 On turbocharger noise

The addition of a turbocharger to an engine is known to reduce the levels of intake noise. The silencing mechanism is believed to be simple with the compressor housing behaving as a small, reactive silencer element. The presence of the spinning rotor will have little effect except to add some flow-induced noise. Reductions in narrow-band intake noise of the order of 3–8 dB should be expected depending on the frequency content of the intake noise.

22.1.3.11 Sources of intake (and exhaust) noise

The noise due to the operation of the intake and exhaust systems can be classified as follows:

Primary noise sources being the unsteady mass flow through the valves, which causes pressure fluctuations in the manifold and these propagate to the intake orifice (or exhaust tailpipe) and are radiated as noise. It should be noted that the mechanism that causes primary (or engine breathing) noise is the same mechanism that is usefully harnessed to improve the volumetric efficiency of the NA engine by wave action tuning (see Section 22.1.3.10.6).

Harrison and Stanev (2004) propose the following hypothesis to explain the fluctuating pressure time history found in the intake manifold:

> *The early stages of the intake process are governed by the instantaneous values of the piston velocity and the open area under the valve. Thereafter, resonant wave action dominates the process. The depth of the early depression caused by the moving piston governs the intensity of the wave action that follows. A pressure*

ratio across the valve that is favourable to inflow is maintained and maximised when the open period of the valve is such to allow at least, but no more than, one complete oscillation of the pressure at its resonant frequency to occur whilst the valve is open.

Harrison and Dunkley (2004) also identified the role for intake flow momentum in the breathing performance of higher-speed engines.

Secondary noise sources being noise created by the motion of the flow through the intake and exhaust systems. This self-induced noise is commonly called flow noise.

Shell noise being the structure-radiated sound from the intake or exhaust tailpipes as excited by either primary or secondary noise sources.

The first two classifications are the subject of this section and are illustrated in Fig. 22.1-15.

When analysing the sound recorded at the intake orifice (or exhaust tailpipe), it is notoriously difficult to distinguish between sound that is due to primary noise sources and that due to secondary noise sources. There is a common misconception that whilst primary (engine breathing) noise is tonal and dominated by low-frequency components of the fundamental cycle frequency (which is true), flow noise is chaotic, broadband and mostly of high frequency (which is not true). In reality, flow noise in the intake or exhaust system is likely to be tonal, with both low- and high-frequency components.

The fact that engine breathing noise and flow noise can be confused and misdiagnosed is hardly surprising as they both result in similar effects: a level of sound power in the duct that propagates away from some source towards the open end of the pipe and along the way that sound power flux can be filtered, attenuated or amplified.

Both engine-breathing noise and flow noi... of aerodynamic noise. Before concentrating on th... ferences between the two, it is useful to briefly consider how they are similar. Howe (1975) provides the most convenient framework for this. Howe offered a contribution to the theory of sound generation by flow turbulence and vorticity by using the specific stagnation enthalpy as the fluctuating acoustic quantity rather than vorticity (Powell, 1964) or fluctuating momentum (Lighthill, 1952; 1954) as had been used before. This is particularly convenient for the case of internal combustion (IC) engine intake and exhaust noise as the engine-breathing process is well modelled by the component of enthalpy that is associated with inviscid irrotational flow (see Section 22.1.3.10.3). By contrast, the flow noise can be considered as having its origin in the component of enthalpy associated with:

- rotational flow;
- irreversibilities in the flow;
- viscous forces;
- local heat input or loss.

Intuitively the formation of vortices in the flow are the most likely manifestation of this component of enthalpy. These might be formed:

- at sudden expansions of the flow such as at the entrance to reactive silencer chambers;
- at points of flow separation such as at tight radius bends;
- at free jets such as at the exhaust tailpipe.

Although in one way Howe's model unifies both engine-breathing noise and flow noise as being related to enthalpy, it also leads to a useful point of distinction:

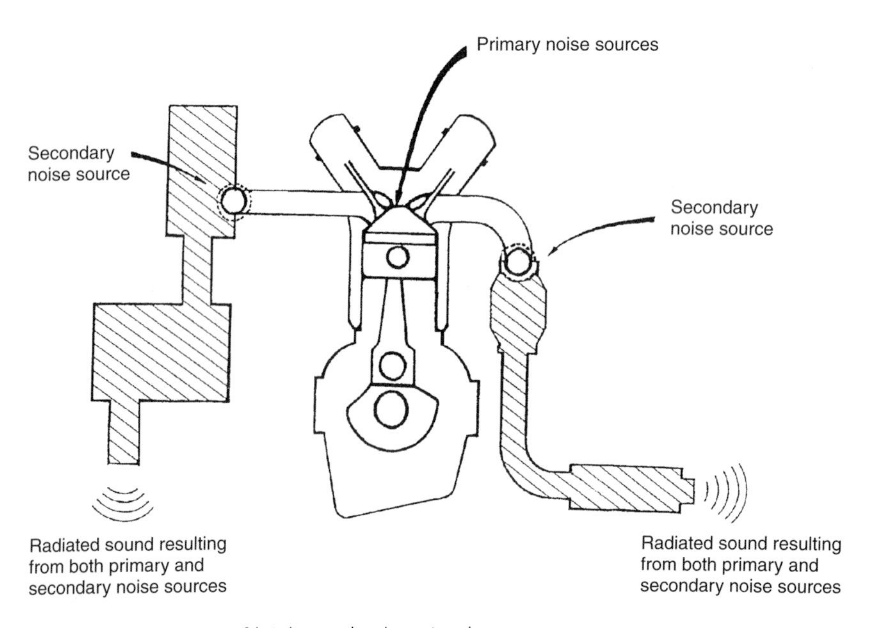

Fig. 22.1-15 Primary and secondary sources of intake and exhaust noise.

engine-breathing noise is associated most with volume velocity (or mass flux) sources, whereas flow noise is associated mostly with stresses at a boundary or aerodynamic forces or fluctuating pressures.

Davies (1996) presents two acoustic source/filter models for use in flow duct acoustics: one for excitation by fluctuating mass or volume velocity (Fig. 22.1-16a) and one for excitation by fluctuating pressure or aerodynamic force (Fig. 22.1-16b).

Davies shows that for a volume source the acoustic power W_m of the source is given by:

$$W_m = 0.5\mathrm{Re}\{p_1^* V_s\} = 0.5\left|V_s^2\right|\mathrm{Re}\{Z_1/(1+Z_1/Z_e)\}/S_s \qquad (22.1.54)$$

and for a fluctuating force

$$W_D = 0.5\mathrm{Re}\{f_1^* u_s\} = 0.5\left|f_s^2\right|\mathrm{Re}\left\{\frac{1}{Z_e S_s}\right\} \qquad (22.1.55)$$

where S_s is the associated surface area of the source.

These two equations imply that the sound power of the sources is a function of the termination impedance Z caused by the reflection coefficient at the open end and the transfer element T (being the acoustics of the flow duct network).

Harrison and Stanev (2004) used a volume velocity source located at the intake valve and linear models for one-dimensional acoustic propagation in flow ducts to successfully calculate the engine-breathing noise component in an IC engine inlet flow. The model can be used to identify the effects of:

- engine speed;
- valve timing, lift and open period;
- intake system acoustic resonances

in the prediction of engine-breathing noise.

In a similar way, Davies and Holland (1999) used a fluctuating pressure source positioned at the outlet of a reactive silencer chamber to predict a common component of flow noise (see Fig. 22.1-17).

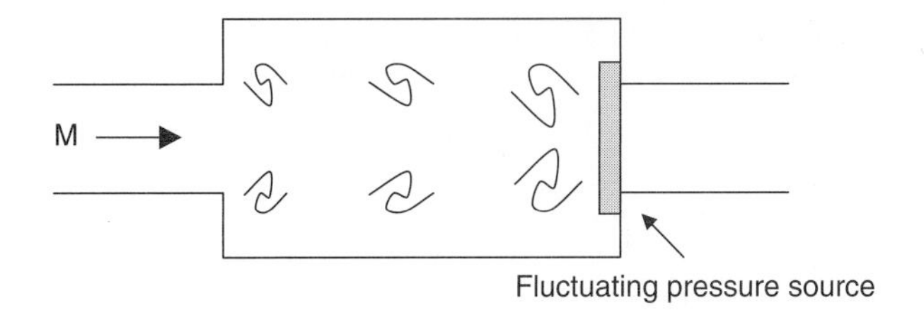

Fig. 22.1-17 Prediction of a common flow noise source (after (Davies and Holland, 1999)). M denotes a flow with a finite Mach number.

In a later publication, Davies and Holland (2001) describe the physical process at work in the pressure source. When the flow first enters the silencer chamber it leaves the downstream facing edge and separates, forming a thin shear layer or vortex sheet. Such sheets are very unstable and quickly develop waves that roll up to form a train of vortices with well-ordered spacings. When the vortices impact on the downstream face of the silencer, a fluctuation in pressure occurs. Acoustic energy propagating upstream from this source can affect the formation of vortices and a feedback mechanism is created. The feedback is strongly influenced by resonances downstream of the vortex generating expansion. Davies (1981) identified the influence of exhaust tailpipe and chamber resonances in the spectrum of flow noise generated in a simple reactive silencer element (see Fig. 22.1-18).

In addition, Davies (1981) highlighted the possibility that simple reactive silencer elements could act as amplifiers of sound rather than attenuators due to the feedback processes that generate flow noise. The use of a length of perforated pipe to bridge the gap between inlet and outlet of a simple silencer is effective at eliminating this amplification by suppressing the formation of vortices at the inlet. Providing the porosity of the perforate pipe is greater than 15% it will have a negligible effect on the attenuation of engine-breathing noise afforded by the silencer.

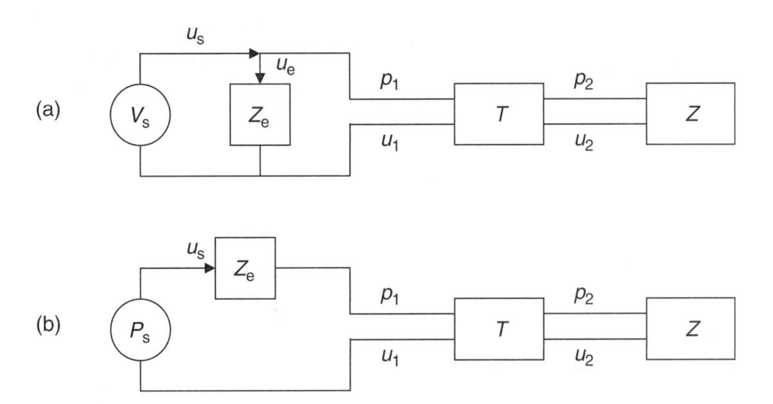

Fig. 22.1-16 Acoustic circuits for flow ducts. (a) Excitation by a fluctuating mass or volume velocity, (b) Excitation by fluctuating pressure or aerodynamic force (after Davies (1996)).

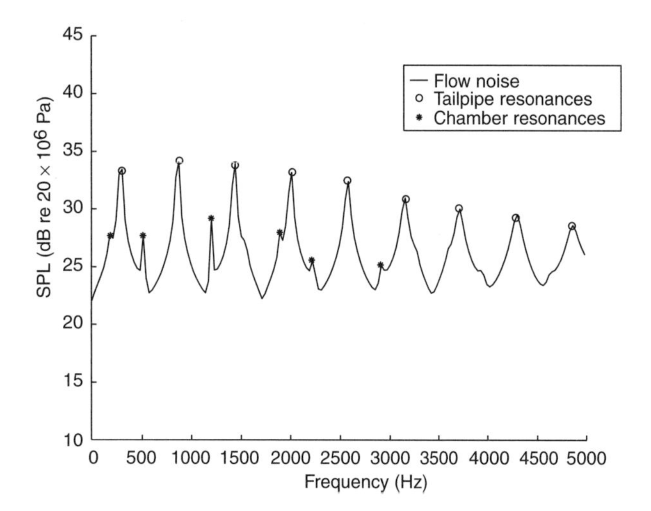

Fig. 22.1-18 Predicted levels of flow noise from a simple silencer element using the method of Davies and Holland (1999). The exhaust tailpipe and chamber resonances have been marked as in Davies (1981).

The above discussion paints a picture of volume velocity sources of engine-breathing noise located at the intake/exhaust valves superimposed with fluctuating pressure sources of flow noise distributed down the intake (and exhaust) system. A rational way to separate the contributions made to intake orifice (or exhaust tailpipe) noise by the two classes of sound source would be to measure the sound energy flux at several locations down the length of the intake (or exhaust) system. Near to the valves, the engine-breathing noise sources should dominate and the level due to this source can be predicted elsewhere in the system. Any local differences between predicted levels of engine-breathing noise energy flux and measured energy flux must be due to sources of flow noise.

Morfey (1971) showed that for a non-uniform flow, the acoustic intensity I which is the wave energy per unit area is given by

$$I = (1 + M^2)\langle pu\rangle + M\left[\left(\frac{\langle p^2\rangle}{\rho_0 c_0}\right) + \rho_0 c_0 \langle u^2\rangle\right] \tag{22.1.56}$$

where $\langle\,\rangle$ denotes a time average and M is the Mach number. The first term corresponds to the sound intensity associated with the wave motion itself and the second with that due to the convection of acoustic energy density by the mean flow.

With plane wave propagation this becomes (Davies, 1988)

$$I = \frac{1}{\rho_0 c_0}\left[(1+M)^2\left|p^+\right|^2 - (1-M)^2\left|p^-\right|^2\right] \tag{22.1.57}$$

Holland et al. (2002) demonstrate the use of an experimental method of measuring sound intensity flux down the exhaust systems of running IC engines. This seems the most promising way of separating engine-breathing noise from flow noise. Rather, simpler experimental methods have been used elsewhere (Selemet et al., 1999; Sievewright, 2000) although the distinction between primary and secondary noise sources is not possible with these (Kunz and Garcia, 1995).

The narrow band noise spectra shown in Figs. 22.1-19 and 22.1-20 give an indication of the typical level and spectral content of both intake and exhaust noise. The tonal quality of the noise, arising from the cyclic operation of the engine is obvious.

22.1.3.12 Flow duct acoustics

More space in this book is devoted to the control of intake noise than to the control of exhaust noise. Once learnt for the intake system, the design methods and supporting theory of acoustics can be readily transferred to the design and development of the exhaust system. The additional complicating factors for the exhaust system are:

- higher flow velocities;
- higher amplitude sound;
- steep temperature gradients;
- increased levels of flow-generated noise.

22.1.3.12.1 Basic design concepts

Basic intake and exhaust systems are made up of expansions, contractions and pipe protrusions. In the absence of temperature gradients or flow, these elements behave in a predictable manner as shown in Fig. 22.1-21.

Inspection of Fig. 22.1-21 leads to the following basic rules for the design of flow duct silencers (intake or exhaust):

- The sudden expansion of the gas at an area discontinuity strongly reflects acoustic waves back towards their source (the engine) and results in the attenuation of that part of the acoustic wave that finally radiates from the end of the system (snorkel or exhaust tailpipe noise).
- Lengths of duct (pipes, expansion chambers and the like) that are open at both ends have acoustic resonances that reduce the attenuation achieved at certain predictable resonant frequencies.
- Lengths of pipe that are open at one end and closed at the other act as resonators that increase the attenuation achieved at certain predictable resonant frequencies.

The silencing elements of expansions, contractions and sidebranches can be used to construct rather complex silencing units as shown in Fig. 22.1-22.

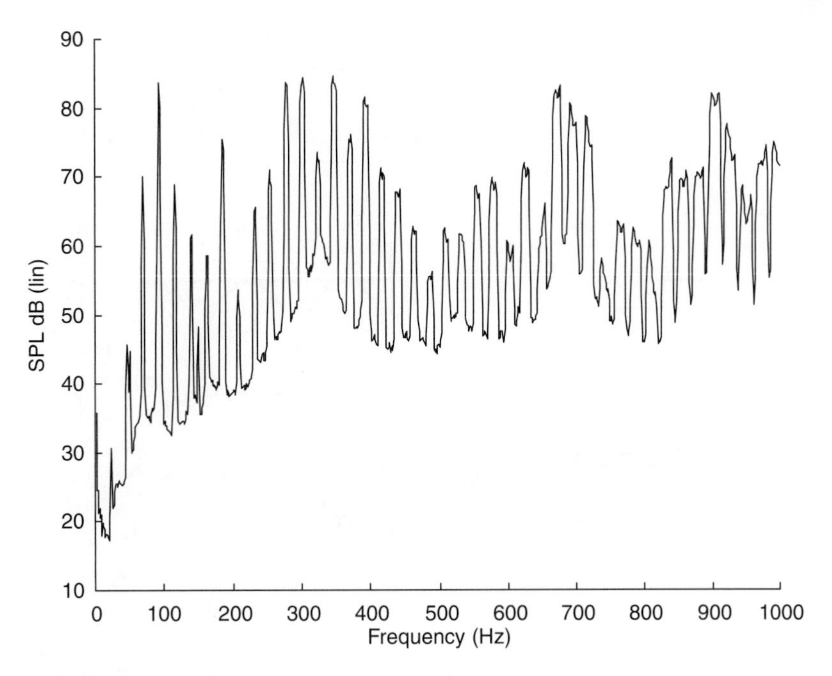

Fig. 22.1-19 Sound pressure levels recorded 100 mm from the intake orifice of a 1.6-litre four-cylinder gasoline engine running full load at 2800 rev min^{-1}.

22.1.3.12.2 Important design and development parameters

It is important to get the basic silencer expansion ratios and exhaust tailpipe (or snorkel) lengths right at the start of the design. It is not good practice to select a silencer by volume. Rather, a designer should determine the pipe diameter required to channel the flow efficiently and then, selecting an appropriate expansion ratio for the silencer, determine the cross-sectional area of the silencer. The length of the silencer will depend on its position in the system (packaging considerations, length of exhaust tailpipe required, etc.) and on the avoidance of critical resonant frequencies. These considerations taken together determine the location and

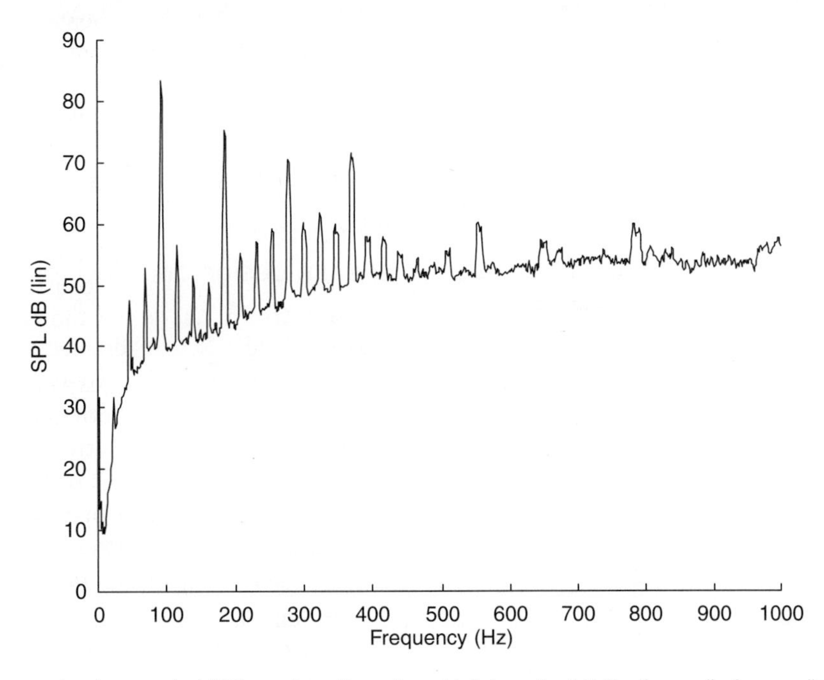

Fig. 22.1-20 Sound pressure levels recorded 500 mm from the exhaust tailpipe of a 1.6-litre four-cylinder gasoline engine running full load at 2800 rev min^{-1}.

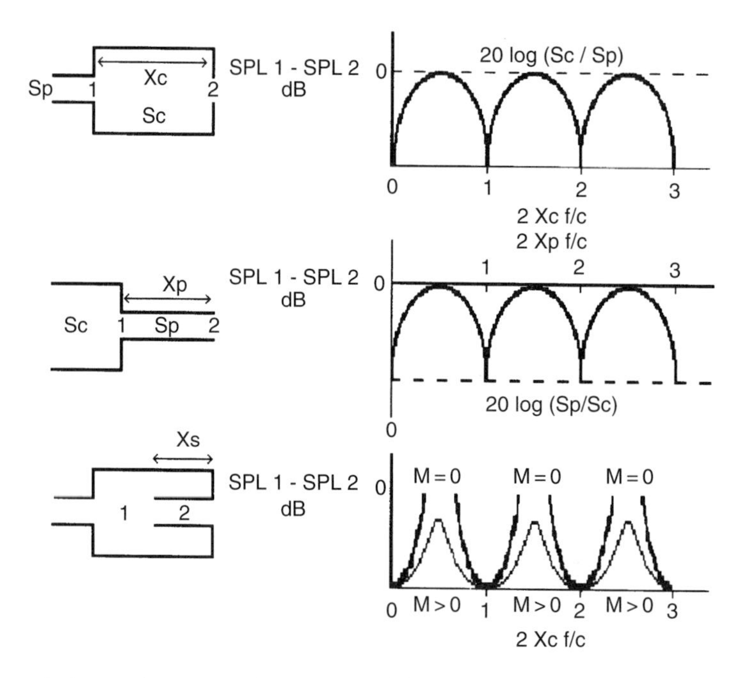

Fig. 22.1-21 The performance of silencer elements in the absence of flow or temperature gradients.

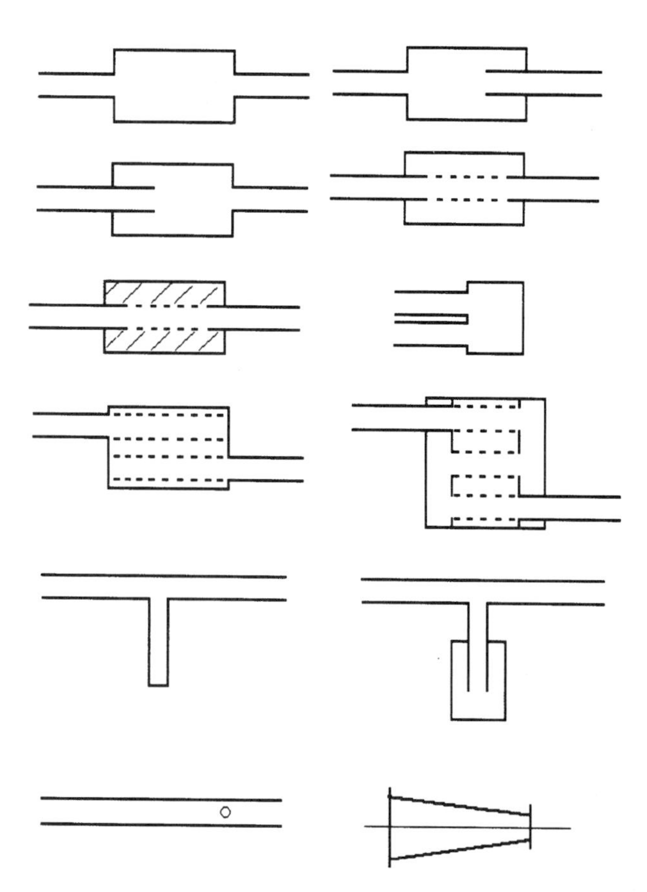

Fig. 22.1-22 Simple and complex silencing units used for intake and exhaust systems.

volume of the silencer in a rational way as illustrated in Fig. 22.1-23.

22.1.3.12.3 Design and development strategies

There is a spectrum of choice for design and development strategies as illustrated in Fig. 22.1-24. Empirical methods are at one extreme of the spectrum, relying on know-how and on the extensive testing of prototype silencers. The experienced silencer designer will develop their own 'favourite' silencers. However, acoustic theory allows them to tune such silencers to a particular engine or car without too much prototype testing. The theoretical tuning of the complex silencers shown in Fig. 22.1-22 defies simple hand calculations, and a computer model is required.

22.1.3.12.4 Traditional intake and exhaust system design and development techniques

Cut-and-try methods have been very successful in the past. They allow a designer, time and opportunity to balance the need for low noise emission against the desire for a particular quality of noise. However, there are a number of disadvantages to this approach namely:

- these are often time consuming;
- they may be costly due to the need for a large number of prototypes for testing;
- it is often difficult to be sure that an optimum solution has been obtained;
- the quality of the final product is highly dependent on the talents and experience of the development engineer.

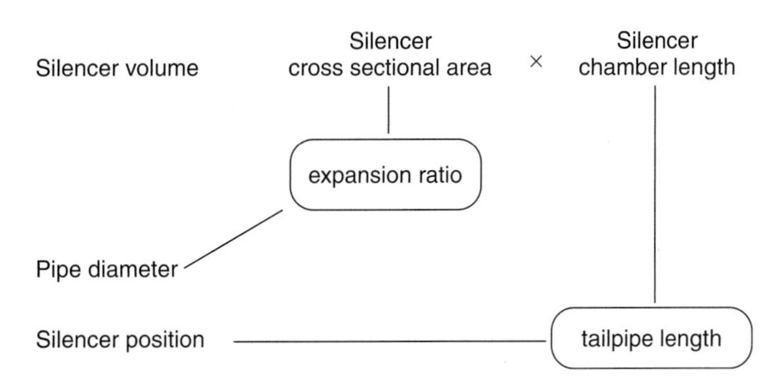

Fig. 22.1-23 Important parameter in flow duct acoustics.

Some rules of thumb for acoustic design of silencers:

Maximum attenuation	$20 \log_{10}$ (expansion ratio) dB
Intake filter box volume	3–5 times the engine swept volume
Exhaust silencer volume	5–10 times the engine swept volume
Intake snorkel length	300–400 mm (usually bell mounted)
Exhaust tailpipe length	less than 500 mm.
<120 Hz intake noise control	Helmholtz resonator onto the snorkel
120–250 Hz intake noise control	sidebranch resonator onto the snorkel
250–500 Hz intake noise control	tuning holes, conical snorkel
>500 Hz intake noise control	not usually a problem
> 500 Hz exhaust noise control	pack silencers with porous materials such as basalt or wire wool

22.1.3.12.5 Predicting the acoustic behaviour of complex flow duct systems

Adopting a suitable acoustic prediction method during the development programme can help reduce programme time and cost by minimising the number of prototypes that need to be manufactured and tested.

Acoustic prediction models fall into two groups:

1. Time domain predictions – where a prediction of the gas dynamic behaviour in the intake or exhaust manifold is extended through the entire intake system.
2. Frequency domain predictions – where the acoustic characteristics of the intake system are predicted in terms of their variation with frequency.

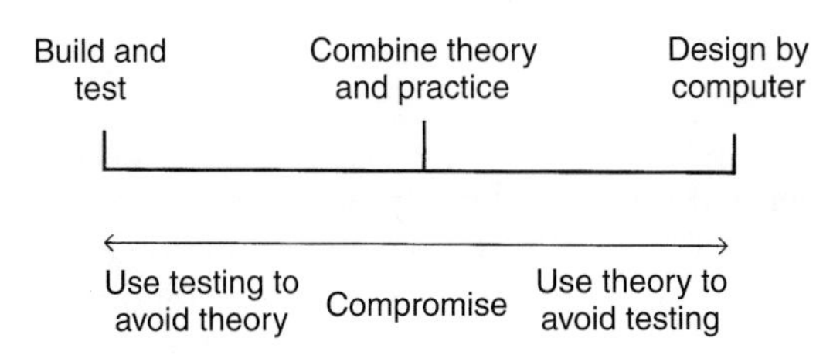

Fig. 22.1-24 Spectrum of choice for intake design strategy.

Without predicting radiated noise the acoustic performance of a system is assessed relative to a baseline, such as a straight pipe, or another reference system.

Frequency domain computer programs are popular due to their fast run-times and their ability to adequately model complex system geometries. Most frequency domain methods make the following assumptions:

- linear acoustic theory remains realistic;
- one-dimensional models remain realistic;
- all wave propagation is planar;
- all disturbances are isentropic.

22.1.3.12.6 Models for acoustic propagation in flow ducts

Acoustic motion in flow ducts is very complex. The overall motion of the fluid being excited by an acoustic source can however be described as the sum of a variety of modes of propagation.

These modes can be described as patterns of motion such as axial oscillation, sloshing and spiralling. All these modes are generally present in a flow duct acoustic field, but it is known that each mode will only propagate efficiently at frequencies below its own specific cut-off frequency.

This frequency is generally expressed in terms of Helmholtz number ka, where in this section

$$k = \text{wavenumber} = \frac{\omega}{c} \qquad (22.1.58)$$

a = duct radius and ω is the radial frequency in rad s^{-1} and c is the speed of sound for the fluid in m s^{-1}.

The most efficient mode of propagation is in the form of plane waves. Here, the pressure and velocity of the disturbance are constant across a given plane along the duct. It is assumed that below the first cut-off frequency given by:

$$Ka = 1.84$$

only plane waves propagate any great distance, while other modes decay rapidly away from their source and are referred to as evanescent waves.

For vehicle systems the first cut-off frequency is in the 2000+ Hz range, which for a four-stroke engine would represent the 24th engine order at 5000 rev min^{-1}. It is therefore a reasonable and convenient simplification to consider only plane wave propagation for vchicle intake and exhaust systems.

22.1.3.12.7 Limits of linear acoustic theory

Acoustic or linear theory remains appropriate as long as the waves travel along the uniform sections of duct between discontinuities without significant change in their shape. Typically, this is the case for pressure amplitudes in the region of 0.01–0.001 bar, with the limit falling with increasing frequency. Different investigations into the limits of the linear assumption in practical flow ducts are reported in Davies and Holland (2004) and Payri et al. (2000).

The plane wave restriction is useful as:

- it allows for a simpler means for including the effects of flow into the analysis;
- it allows measurements to be made at duct walls.

22.1.3.12.8 Acoustic plane waves in ducts

Standing waves occur as a result of the interference between waves travelling out of the source and waves reflected by each discontinuity in area or more generally each discontinuity in acoustic impedance. The concept of one forward-travelling wave and one backward-travelling wave is illustrated in Fig. 22.1-25. In the following,

$$p(x,t) = p^+(x,t) + p^-(x,t) \tag{22.1.59}$$

the magnitudes of p^+ and p^- remain effectively invariant between discontinuities, while the relative phase will vary in an organised manner.

However, $p(x, t)$ will vary along the duct due to the presence of standing waves. These so-called standing waves do not actually stand at all but are the result of the interference between waves travelling in opposite directions. Traversing a microphone down the pipe will show the standing wave pattern which may be viewed as the mode shape of a particular resonance.

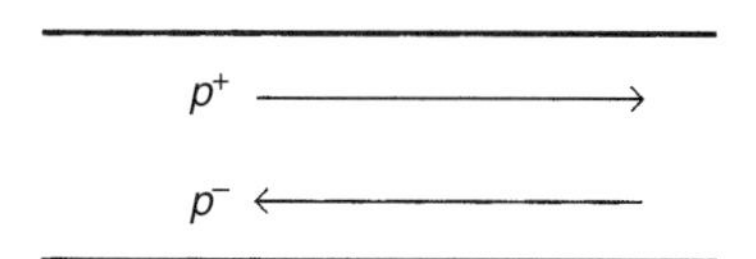

Fig. 22.1-25 Sound field in a duct.

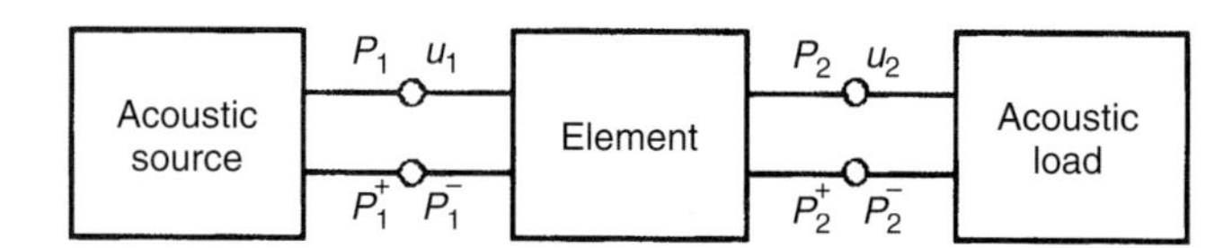

Fig. 22.1-26 The interaction between an acoustic source and an acoustic load (Davies and Harrison, 1997).

22.1.3.12.9 Acoustic sources and acoustic loads

The acoustic performance of a flow duct element depends on the acoustic source and the acoustic load acting on it as illustrated in Fig. 22.1-26 (Davies and Harrison, 1997).

Consider an exhaust system. The acoustic performance of the first silencer will depend on the acoustic load imposed on it. That load is the combined acoustic impedance of the remainder of the exhaust system downstream of the first silencer. Therefore, the acoustic performance of the first silencer depends on where in the system it is placed (hence a silencer that performs in a certain way on the flow bench, may behave differently on the vehicle). This is illustrated in Fig. 22.1-27 (Z is the acoustic impedance, the ratio of acoustic pressure to acoustic volume velocity).

22.1.3.12.10 The acoustics of an unflanged pipe with flow

The final acoustic termination in an intake system is at the snorkel orifice. The final acoustic termination for the exhaust system is the exhaust tailpipe. The inflow of air or outflow of exhaust gas produces a final acoustic load on the system that is well understood and takes the form of a reflection coefficient r:

$$r = \frac{p^-}{p^+} \tag{22.1.60}$$

$$\frac{Z}{\rho_0 c} = \frac{1+r}{1-r} \tag{22.1.62}$$

and

$$r = \mathrm{R}e^{i\theta} = -\mathrm{R}e^{-i2kl} \tag{22.1.62}$$

where R is the modulus, θ is the phase, k the wavenumber and l is an end correction (Davies et al., 1980; Davies, 1987). Ideally, on the assumption that the pressure waves remain plane, $r \rightarrow -1$, $R \rightarrow 1$ and $\theta \rightarrow \pi$. In reality, the plane wave assumption remains realistic only when the dimension a (pipe radius) remains a small

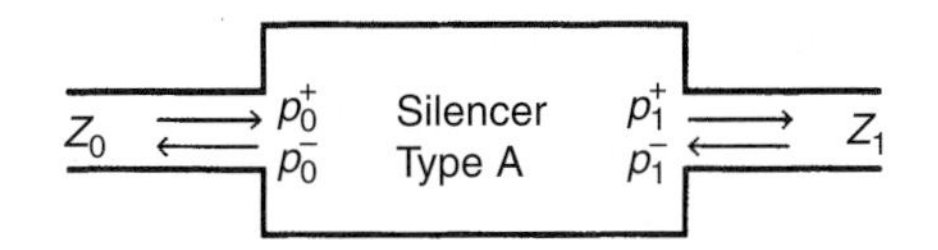

p_2 is different to p_1

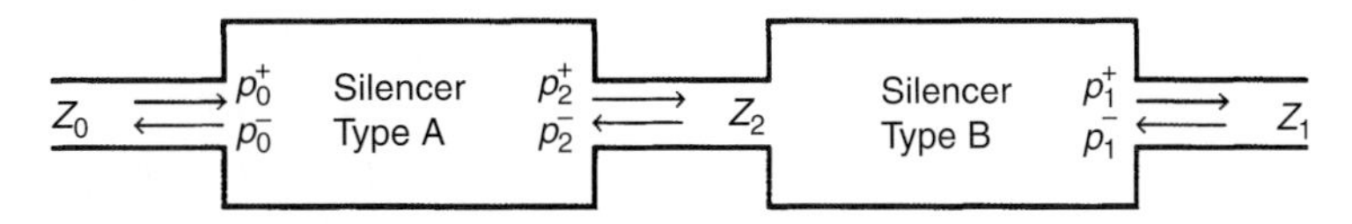

Fig. 22.1-27 The interacting sound fields in a twin-silencer exhaust system.

fraction of a wavelength; that is, for the limiting case when the Helmholtz number ka approaches zero (the plane wave cut-off occurs at $ka = 1.84$).

For zero outflow, the modulus of the reflection coefficient is unity at low Helmholtz number and reduces with increasing Helmholtz number (the higher-frequency sound transmitted is reflected less than the low-frequency sound – i.e. the high-frequency sound is radiated more than the low-frequency sound).

The effect of mean outflow is to increase R throughout the frequency range. The effect of mean inflow is to decrease R throughout the frequency range.

22.1.3.12.11 The general effects of temperature and flow speed

In addition to its effect on the reflection coefficient at an unflanged pipe termination, the effect of a steady mean flow on isentropic acoustic plane wave propagation in a duct of constant area is to modify the appropriate wavenumber k. With flow, the distance between the nodes of a standing wave is reduced by a factor $(1 - M^2)$ where M is the Mach number of the flow. In other words, the effective duct lengths are reduced.

The presence of a mean flow alters the values of the wavenumber by a factor $1/1 \pm M$ (Davies, 1988). The Mach number in the duct varies with the mass flow rate of gas (or the engine speed and load) and the gas density (or the gas composition and temperature).

The wavenumbers are also affected by changes in temperature through variations in the speed of sound (see also the effect of sound-absorbing materials – Section 21.1.9) as:

$$\beta = \text{complex wavenumber} = \left(\frac{\omega}{c}\right) - i\alpha \qquad (22.1.63)$$

where α is a visco-thermal attenuation coefficient which for plane wave propagation in a circular pipe of radius a is given by:

$$\alpha = \left(\frac{1}{ac}\right)\left(\frac{v\omega}{2}\right)^{1/2}\left[1 + (\gamma - 1)\left(\frac{1}{P_r}\right)^{1/2}\right] \qquad (22.1.64)$$

where v is the local kinematic viscocity and P_r is the Prandtl number for the gas in the pipe $P_r = \frac{\mu c_p}{k}$ where μ is the shear viscosity and k is the thermal conductivity (Davies, 1988).

The speed of sound varies with the relation

$$c = \sqrt{\gamma RT} \qquad (22.1.65)$$

where

R = specific gas constant,
T = absolute temperature (K)

As an added complication the values of γ and R also vary with gas composition and temperature, but for air in an intake, values of $\gamma = 1.4$ and $R = 290$ are reasonable, whilst for exhaust gas $\gamma = 1.35$ and $R = 285$ might be chosen.

For intake systems, it is reasonable to assume a sound speed in the order of 343 m s^{-1}. The final effect of flow on the acoustic performance of silencers is complex, and defies simple calculations.

22.1.3.12.12 Calculating mach numbers

Regrettably, the car fitted with an IC engine is a rather inefficient means of turning chemical energy into tractive effort.

In fact, as will be demonstrated here, under urban driving conditions, only 20% (or less) of the chemical energy in gasoline is made available for tractive effort at the wheels. For the case of the engine in isolation, Heywood calls this the thermal conversion efficiency (Heywood, 1988)

$$\eta_{tc} = \frac{W_c}{\eta_c m_f Q_{HV}} \qquad (22.1.66)$$

where

W_c = work output from the cycle
η_c = combustion efficiency
m_f = mass of fuel
Q_{HV} = heating value of fuel

To obtain the thermal conversion efficiency for a complete vehicle, the chain of efficiencies has to be analysed, first for the engine at full load (Bosch, 1986) and then subsequently for part load operation, and finally for the engine installed in the vehicle.

The chain of efficiencies for the engine itself, operating at full load throttle is given by:

$$\eta_{\text{engine}} = \eta_i \times \eta_d \times \eta_c \times \eta_m \quad (22.1.67)$$

where

η_i = ideal fuel–air cycle efficiency, which Taylor (1985) shows to be 46% for stoichiometric octane/air mixture and a compression ratio of 10.

η_d = diagram factor – the ratio of work output from the real cycle to work output from an ideal cycle. Taylor (1985) suggests 78% for stoichiometric octane/air mixture and a compression ratio of 10.

η_m = mechanical efficiency. Lumley (1999) suggests 85% for low engine speeds.

η_c = combustion efficiency. Heywood suggests 98% for stoichometric mixtures (Heywood, 1988).

Thus, the product of the engine efficiency chain for full load operation:

Engine efficiency = 0.46 × 0.78 × 0.85 × 0.98 = 30%

This figure is a little above the energy balance of 25–28% given for the gasoline engine by Heywood (1988). It is commonly held that 1/3 of the chemical energy ends up as useful work, 1/3 goes to heat up the cooling water and 1/3 goes down the exhaust tailpipe.

It is useful to try out some numbers now (Harrison, 2003). The ratio of energy put in to the system compared with energy converted to useful work is 100/30 = 3.333

Now, for gasoline

$$Q_{LHV} = 43.1\ \text{MJ/kg}^{-1}$$

Thus, the fuel required per second for 1 kW output from the cycle is = $(1000 \times 3.333)/43.1 \times 10^6$ = 0.000077 kg/s^{-1} = 278 g/kWh

In 1988, Heywood expected the values of bsfc to be around 270 g/kWh for gasoline engines and as low as 200 g/ kWh for diesel engines (Heywood, 1988). By the mid-1990s, port fuel injection resulted in engines operating in the 250–300 g/kWh range with bsfc at around 250 g/kWh (full load) at the medium engine speed ranges.

For a 50 kW output, and a bsfc of 278 g/kWh, fuel flows would be 0.278 × 50 = 14 kg/hr^{-1}. With the density of typical gasoline being 733 kg m^{-3} this equates to 19 litres of fuel burned per hour at 50 kW (4.2 gallons). At stoichiometric air-fuel mixtures, this requires 14 × 14.7 = 206 kg/hr^{-1} of air.

With the density of air at standard conditions being 1.19 kgm^{-3} this is equal to 173 m^3 of air an hour or 0.048 m^3 per second. With throttle diameters being commonly around 60 mm, the local flow speed through the wide-open throttle body would be around 17 ms^{-1} (0.05 Mach).

The method adopted here allows the intake designer to calculate intake Mach numbers using practical engine performance data (brake power and bsfc) with the need for only a limited set of additional data, all of which is freely available in the literature.

22.1.3.12.13 Describing the acoustic performance of flow duct systems

Referring to Fig. 22.1-28 and to Section 22.1.3.12.11, the acoustic field in the intake pipe is the sum of the positive and negative wave components p_1^+ and p_1^- respectively and is described by the equation

$$P_{\text{inlet}} = p_1^+ e^{i(\omega t - \beta_1^+ x)} + p_1^- e^{i(\omega t + \beta_1^- x)} \quad (22.1.68)$$

The acoustic field in the outlet pipe is:

$$P_{\text{outlet}} = p_2^+ e^{i(\omega t - \beta_2^+ x)} + p_2^- e^{i(\omega t - \beta_2^- x)} \quad (22.1.69)$$

The acoustic performance of the duct section can be characterised in terms of ratios of the complex wave components:

$$\text{Attenuation (dB)} = 20 \log_{10}\left(\frac{p_1^+}{p_2^+}\right) \quad (22.1.70)$$

$$\text{Reflection at plane (1)} = \frac{p_1^-}{p_1^+} \quad (22.1.71)$$

$$\frac{Z}{\rho_0 c} = \frac{1 + r}{1 - r} \quad (22.1.72)$$

Note that a dip in system attenuation will result in a peak in radiated noise. Therefore, much of the designer's work is in filling in the dips in attenuation at critical frequencies by the alteration of pipe lengths and the addition of resonators.

There are various methods available for bench testing the acoustic performance of flow duct systems:

- Pressure ratio measurements in duct.
- Radiated sound measurements.

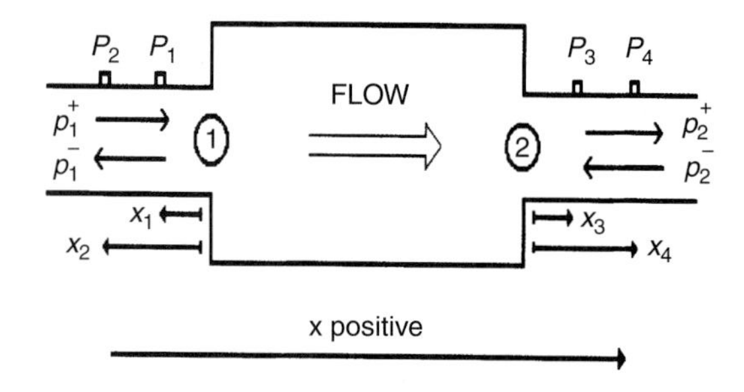

Fig. 22.1-28 Inlet and outlet acoustic fields in a silencer element.

- Wave decomposition techniques (Davies et al., 1999) – Measure sound pressures at four known positions and solve the four simultaneous equations that result:

$$P_1 = p^+_{up} e^{i(\omega t - \beta^+_{up} x_1)} + p^-_{up} e^{i(\omega t + \beta^-_{up} x_1)} \quad (22.1.73)$$

$$P_2 = p^+_{up} e^{i(\omega t - \beta^+_{up} x_2)} + p^-_{up} e^{i(\omega t + \beta^-_{up} x_2)} \quad (22.1.74)$$

$$P_3 = p^+_{down} e^{i(\omega t - \beta^+_{down} x_3)} + p^-_{down} e^{i(\omega t + \beta^-_{down} x_3)} \quad (22.1.75)$$

$$P_4 = p^+_{down} e^{i(\omega t - \beta^+_{down} x_4)} + p^-_{down} e^{i(\omega t + \beta^-_{down} x_4)} \quad (22.1.76)$$

22.1.3.13 Intake noise control: a case study

Frequently, the fundamental dimensions of an intake system (such as minimum pipe diameters and the choice of air filter element) are set in advance of noise control engineering by the engine development team. In this case study, those considerations lead to an expansion ratio of 18.33:1.00 between the snorkel and the filter box. The peak attenuation in this filter box can be estimated using:

$$\begin{aligned} \text{attenuation dB} &= 20 \log_{10} (\text{expansion ratio}) \\ &= 20 \log_{10} (18.33) = 25.27 \text{ dB} \end{aligned} \quad (22.1.77)$$

The peak attenuation will occur at frequencies away from resonances in the snorkel and in the filter box. These resonant frequencies can be estimated using the equation below:

$$\frac{2xf}{c} \approx n \quad n = 1, 2, 3, 4, \ldots \quad (22.1.78)$$

where

x = length
f = frequency (Hz)
c = speed of sound (m s^{-1})

The attenuation (predicted using a software known as APINEX) afforded by this simple intake system is shown in Fig. 22.1-29.

Inspection of Fig. 22.1-29 reveals:

- peak attenuation = 25 dB at 650 Hz;
- two dips in the attenuation relating to the snorkel length when n = 1 and 2 (filter box depth is too short for any chamber resonances <1000 Hz).

The attenuation characteristics of the expansion portion of the filter box are shown in Fig. 22.1-30. The attenuation characteristics of the contraction section of the filter box are shown in Fig. 22.1-31.

The effect of the intake snorkel is to have negative attenuation near its resonant frequencies. Contrasting the last two figures illustrates the point that it is the expansion within the filter box which provides the basic attenuation, while the snorkel and the filter box lengths tune the dips in the attenuation characteristics. Figs. 22.1-30 and 22.1-31 (and others in this section) were

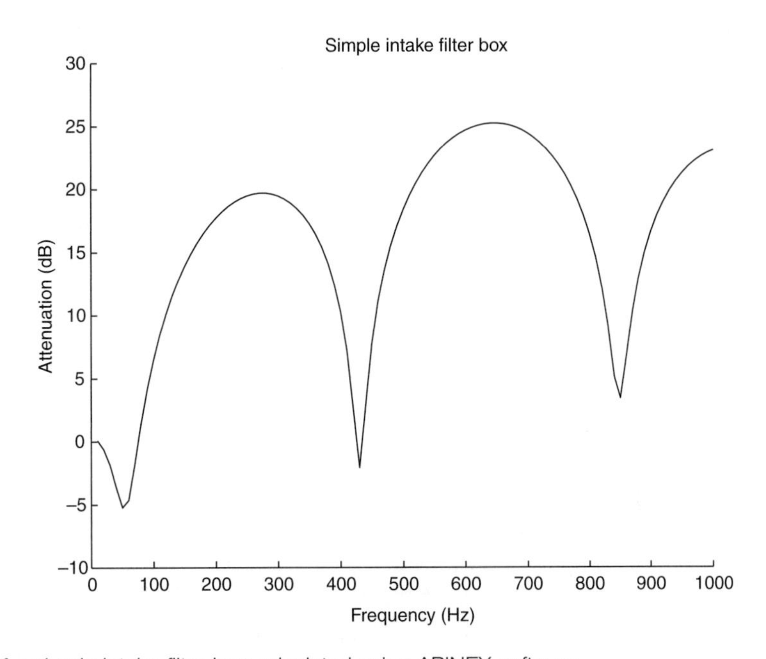

Fig. 22.1-29 Attenuation of a simple intake filter box calculated using APINEX software.

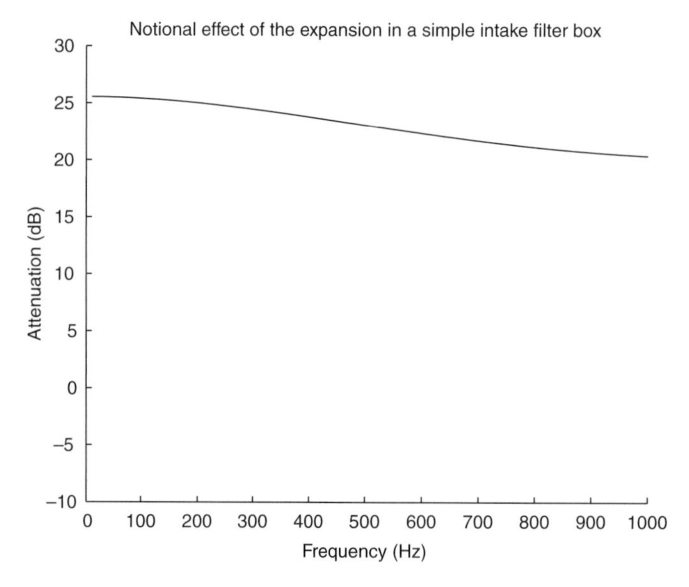

Fig. 22.1-30 Attenuation of the expansion section of a simple intake filter box calculated using APINEX software.

constructed using sophisticated flow duct acoustic modelling tools (Davies, 1988; Harrison et al., 2004). These results compare favourably with common understanding of how reactive silencers work (see Fig. 22.1-21).

22.1.3.13.1 Conical snorkels

A conical snorkel can be used to reduce the depth of the troughs in the attenuation due to the tuning effect of the snorkel length as shown in Fig. 22.1-32. The typical conical snorkel is tapered such that the snorkel orifice is reduced in area. Care must be taken not to seriously limit engine performance by throttling the intake flow as a result.

Note: It is generally assumed that all intake snorkels, tapered/conical or not have a slight flare at the intake orifice in order to improve the initial inlet flow.

22.1.3.13.2 Protrusions of the snorkel into the filter box

Protrusions of the snorkel into the base of the filter box act like sidebranch resonators. The effect of a 120-mm protrusion to the snorkel along the bottom of the filter box is shown in Fig. 22.1-33.

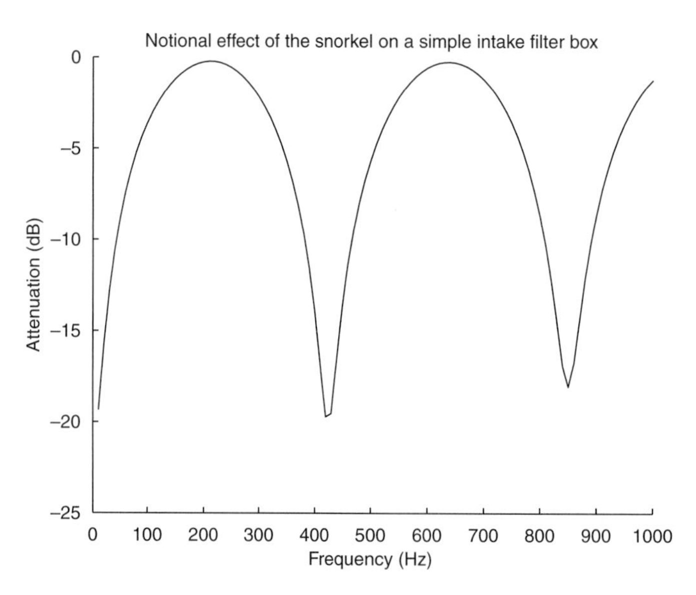

Fig. 22.1-31 Attenuation of the contraction section of a simple intake filter box calculated using APINEX software.

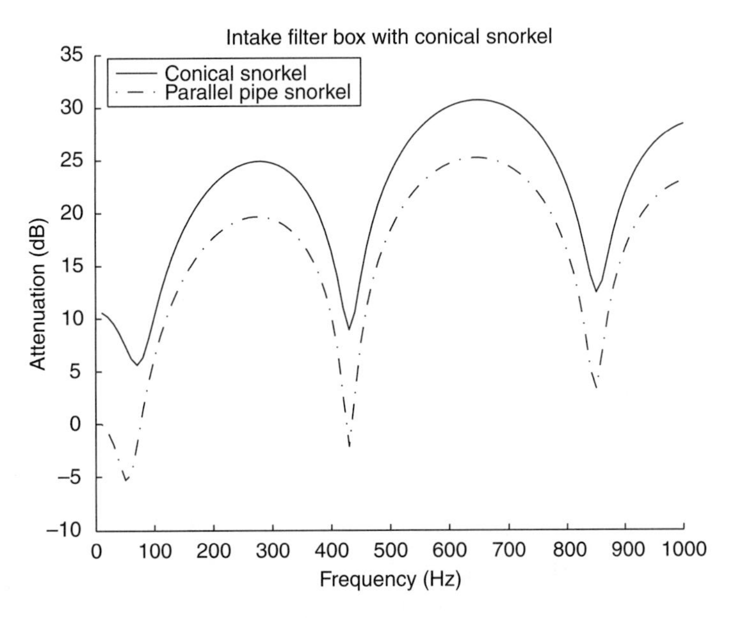

Fig. 22.1-32 Effect of adopting a conical snorkel. Calculated using APINEX software.

The frequencies at which the system attenuation is improved through the action of the sidebranches may be calculated for the zero flow case from:

$$\frac{4xf}{c} = n \quad n = 1, 3, 5, 7, 9, \ldots \qquad (22.1.79)$$

where x is the sidebranch length and f is the resonant frequency.

In practical cases, the attenuation shown in Fig. 22.1-33 would be limited to around 40 dB by the effects of flow noise.

22.1.3.13.3 Helmholtz resonators

These feature a 1–3-litre volume placed as a sidebranch to the snorkel. Connection to the snorkel is by a tuned length of pipe. These devices provide useful low-frequency attenuation with little extra static pressure loss. The effect of adding of a Helmholtz resonator, tuned to around 90 Hz to the snorkel (as illustrated in Fig. 22.1-34) is shown in Fig. 22.1-35.

The Helmholtz resonator is a reliable (and often the only practical) way of increasing intake system attenuation at frequencies below 120 Hz.

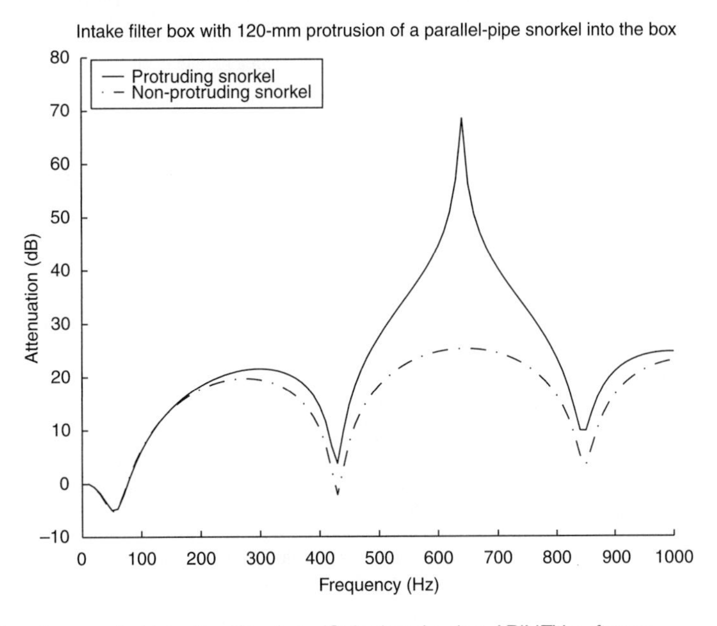

Fig. 22.1-33 Effect of protruding the snorkel into the filter box. Calculated using APINEX software.

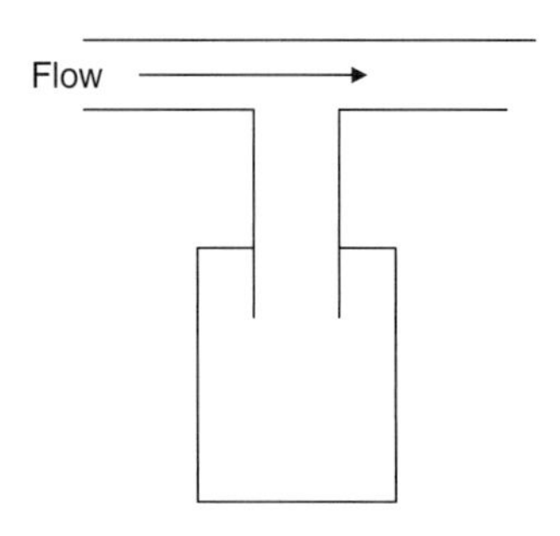

Fig. 22.1-34 The Helmholtz resonator.

22.1.3.13.4 Sidebranch resonators

A simple 300-mm sidebranch resonator (illustrated in Fig. 22.1-36) is fitted instead of the Helmholtz resonator, with the effect shown in Fig. 22.1-37. Fig. 22.1-38 shows a photograph of a typical production example of this. The frequencies at which the system attenuation is improved through the action of the sidebranches may be calculated for the zero flow case from:

$$\frac{4xf}{c} = n \quad n = 1, 3, 5, 7, 9, \ldots \qquad (22.1.80)$$

where x is the sidebranch length and f is the resonant frequency. In practical cases, the attenuation shown in Fig. 22.1-37 would be limited to around 40 dB by the effects of flow noise.

22.1.3.13.5 The effects of the manifold and the zip tube

The effect of adding a simple manifold and a zip tube to the filter box is shown in Fig. 22.1-39. The primary runner lengths of the manifold's branches act as sidebranch resonators when they are terminated by closed

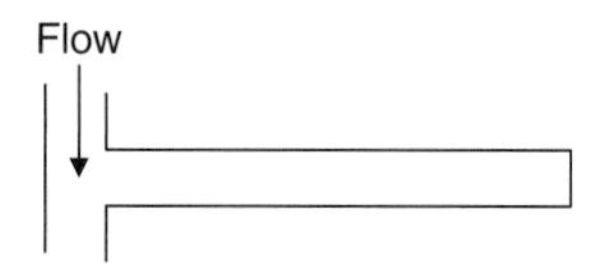

Fig. 22.1-36 The sidebranch resonator.

intake valves. Therefore, in the absence of cylinder-to-cylinder overlap in intake valve opening, at any given time only one valve will be open and the remainder of the branches act as sidebranch resonators to the branch with the open valve. This results in a strong peak in the attenuation spectrum that is related to the primary runner length.

Away from this strong peak, there are additional troughs in the attenuation caused by resonances of the zip tube. The frequencies at which these troughs occur may be controlled by varying the length of the zip tube as shown in Fig. 22.1-40.

Fig. 22.1-41 shows a line illustration of an intake system where the design has been optimised using linear acoustic theory.

22.1.3.13.6 Intake system mounting

The mounting of the intake system must give effective attenuation of:

- transmitted engine vibration;
- shell noise.

Care should be taken to avoid noise 'breaking out' through the walls of flexible hoses. If the filter box is mounted on the vehicle body, the mounting system must be flexible enough to allow for engine movement. The filter box should be fixed to a low-mobility piece of bodywork. Filter box mounts should be sufficiently

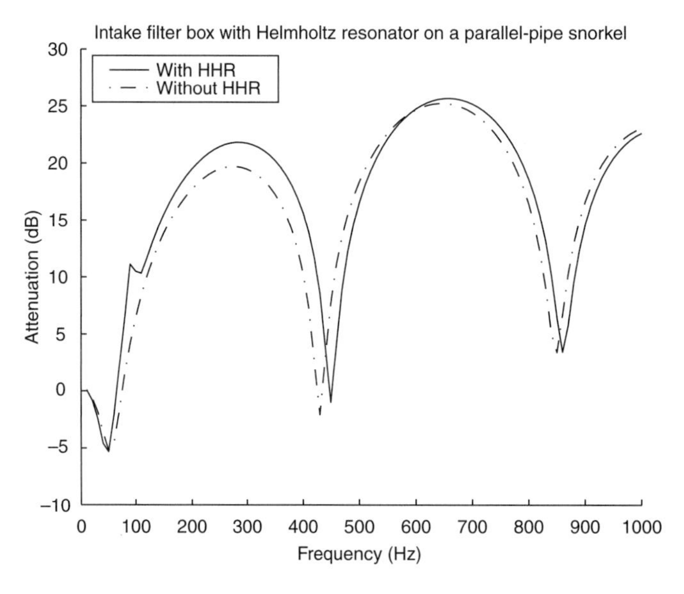

Fig. 22.1-35 Effect of an added Helmholtz resonator calculated using APINEX software.

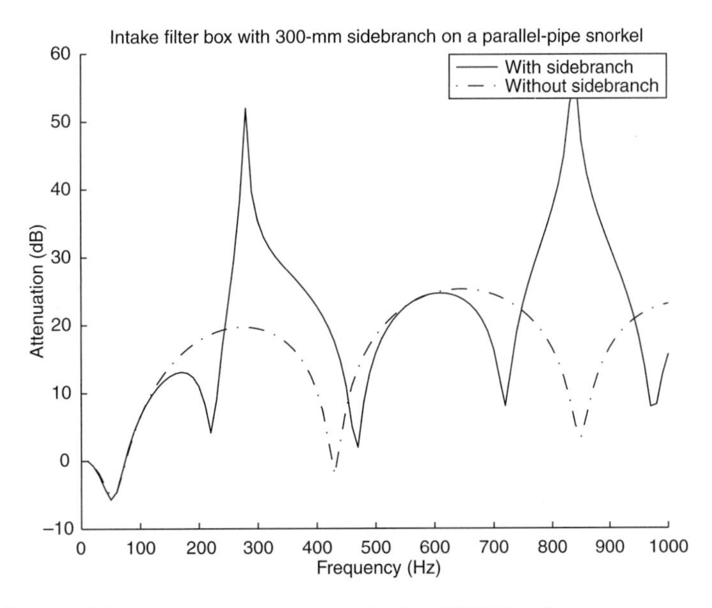

Fig. 22.1-37 The effect of adding a sidebranch resonator calculated using APINEX software.

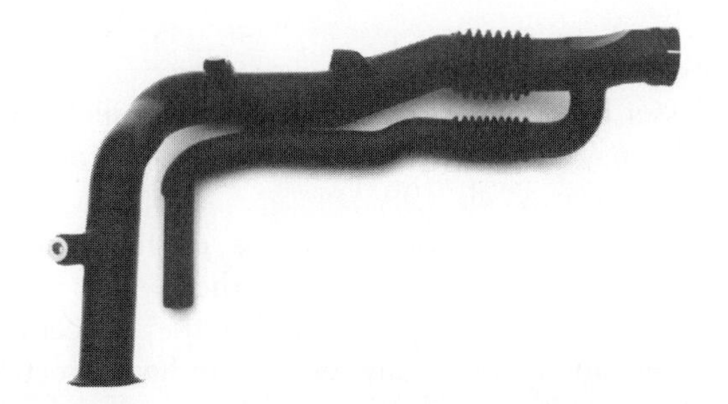

Fig. 22.1-38 Photograph of a production snorkel with sidebranch mounted.

compliant to give adequate vibration isolation. In terms of vehicle refinement, interior noise concerns should generally influence the mounting of the filter box.

22.1.3.14 Exhaust noise control

The earlier discussion of intake system design applies equally to the exhaust system with the following notable exceptions:

- The influence of the flow is much greater in the exhaust system due to the higher flow speeds.

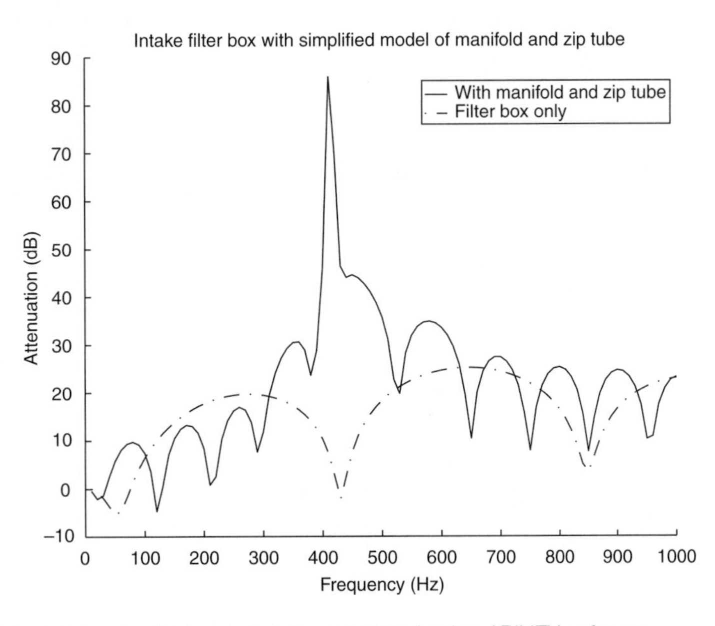

Fig. 22.1-39 The influence of the intake manifold and zip tube calculated using APINEX software.

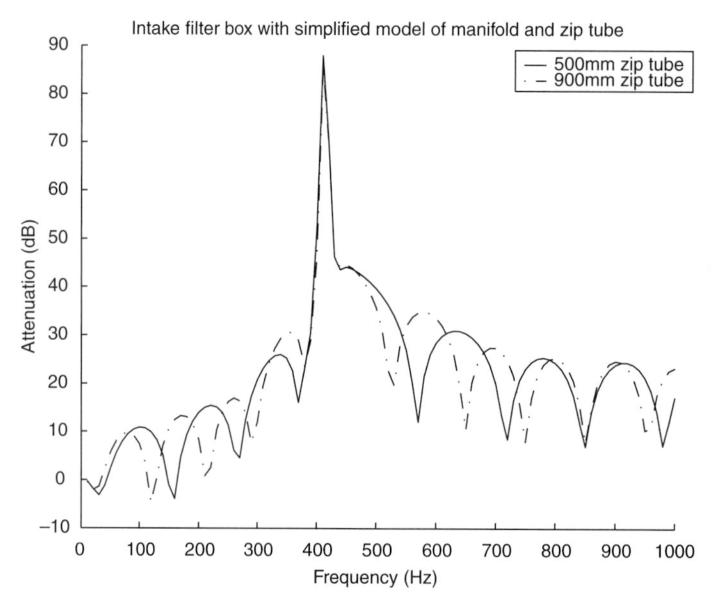

Fig. 22.1-40 Influence of zip tube length calculated using APINEX software.

- Temperature gradients are very steep, particularly in the exhaust manifold (300°C drop in gas temperature between the exhaust port and the outlet of the catalytic converter is typical).
- Durability/corrosion resistance is often a limiting factor in the design of silencers.
- Backpressure has a stronger influence on engine performance than the wave action in the exhaust manifold.
- Perforate tubes are commonly used (porosity less than 15%) to provide gradual expansion of the gas into the silencer, achieving noise reduction with a lesser backpressure penalty. Several parallel flow paths may be provided, each linked via perforated tube (Davies et al., 1997).
- Many silencer chambers can be provided within a single silencer shell by dividing the silencer using baffle plates. The flow path through the silencer may be convoluted with many reversals of flow direction. In this way, the flow path length is commonly much longer than the physical length of the exhaust system. This is sharp contrast to common practice

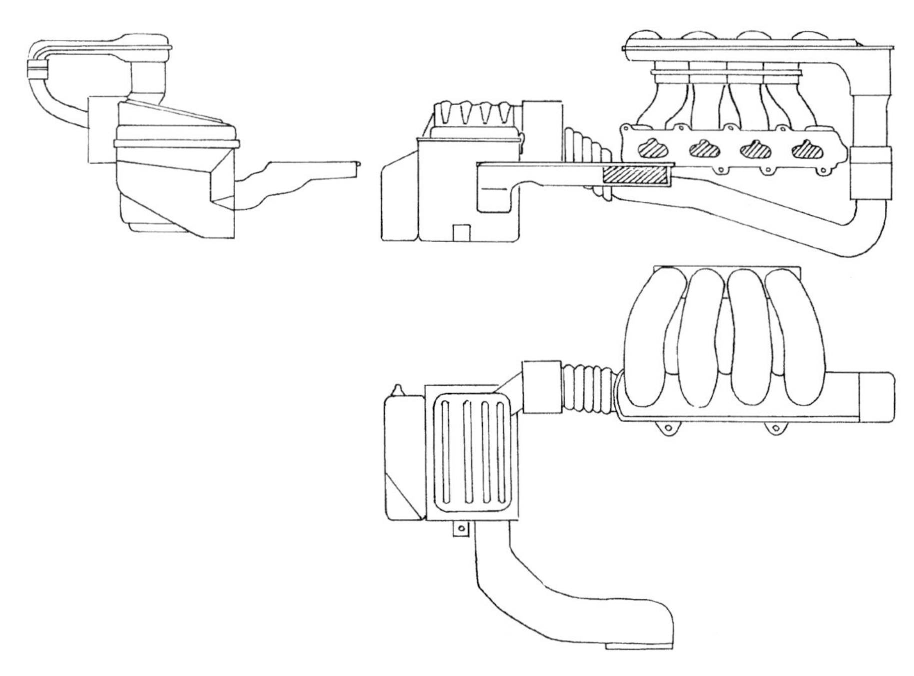

Fig. 22.1-41 An intake system where the design has been optimised using linear acoustic theory.

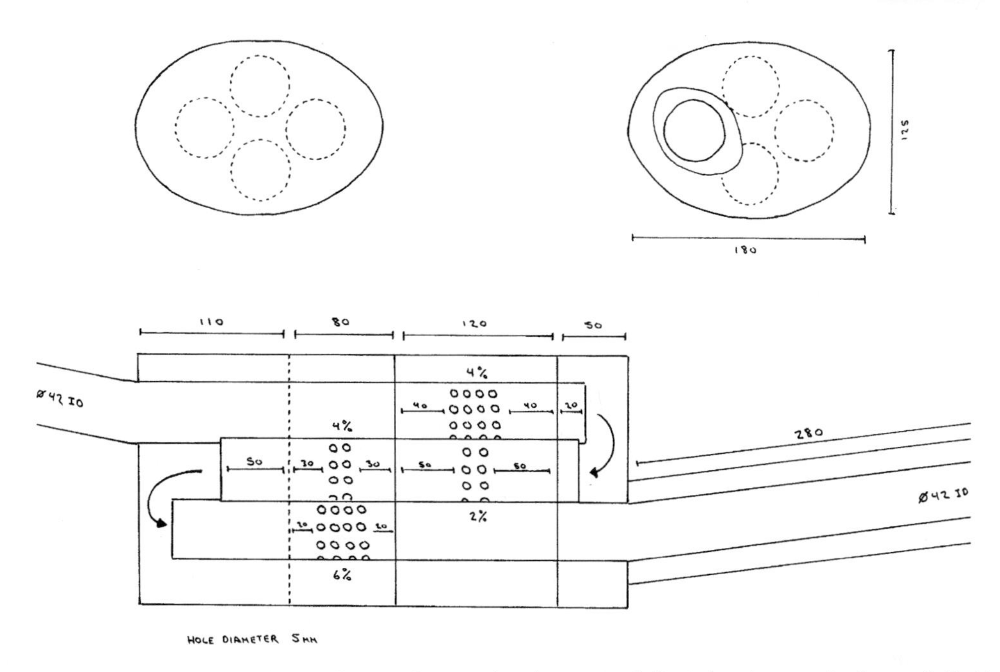

Fig. 22.1-42 The internals of a compact exhaust silencer element showing many distinct chambers and a flow path that is longer than the physical length of the element. Dimensions are in millimetres.

with intake systems. One example is given in the line illustration shown in Fig. 22.1-42.

- Noise radiated from the outer surfaces of the silencers may be significant (shell noise). This is commonly controlled using two-layer skins of thin-sheet steel for the silencer construction.
- Noise generated by the flow itself may be significant. This flow noise is generated within the silencers themselves, and it is common for a silencer to actually amplify sound as a result of flow acoustic coupling (Davies and Harrison, 1997).
- There is less use of sidebranch and Helmholtz resonators compared with common practice with intake systems.
- Silencers are often packed with sound-absorbing materials such as basalt, wire wool or glass fibre strands. This improves attenuation at high frequencies, but often at the expense of durability due to the retention of water within the materials.
- The mounting of the exhaust system below the vehicle is important to avoid the creation of paths for structure-borne noise to enter the vehicle.

22.1.3.14.1 On the flow through the exhaust valve

The same principles apply to flow through the exhaust valve as for the case of the inlet valve. However, the model that results in equation (22.1.49) is too simplistic to realistically model the high temperature, high speed flow through the exhaust valve. An alternative model that allows for sonic flow through the valve and/or through the port is given in both Benson (1982) and Winterbone and Pearson (1999). That model requires an iterative procedure to account for the entropy loss across the valve.

It is clear that, for practical engines, the performance may be optimised (or 'tuned') by varying:

- Valve timing and lift to influence intake and exhaust flow rates and promote effective scavenging.
- Inlet air density (by super or turbo charging) to maximise the air charge (and hence fuel charge).
- Port design to influence in-cylinder gas motion to improve combustion and reduce flow losses (often these are mutually exclusive).
- Varying the exhaust design to reduce backpressure.
- Varying the intake geometry (and usually to a lesser extent the exhaust manifold geometry) to maximise the pressure ratio across the valve at IVC.

22.1.3.14.2 Calculating exhaust temperatures

Just as for intake system design, flow duct acoustic modelling is useful for exhaust system design. However, in order to determine the important changes in the speed of sound down the length of an exhaust system, it is first necessary to calculate the gas temperature gradient. Common flow duct acoustic models require the user to declare exhaust manifold, and exhaust tailpipe gas (or ambient) temperatures and the temperature gradient is calculated accordingly. The exhaust manifold temperature may not be known early in a development cycle but it is possible to make realistic estimates of in-cylinder gas pressure and temperature at certain points in the cycle using suitable simplifying assumptions. These points are:

- firing TDC (top dead centre);
- peak pressure/temperature (assumed crank angle in the range of 10–30° ATDC);
- EVO (exhaust valve open);
- IVO (intake valve open);
- IVC (intake valve close);
- Ignition point.

At the heart of the approach, the temperature rise across the engine ∂T_E is estimated using an enthalpy balance (Weaving, 1990):

$$\partial h = Q_{LHV} m_f (1 - \eta_{th} - Q_c) \quad (22.1.81)$$

where

∂h = enthalpy rise across the engine (J)
Q_{LHV} = lower heating value of the fuel (assume 43.1×10^6 J/kg for gasoline)
m_f = mass of fuel trapped in the cylinder at IVC
η_{th} = thermal efficiency
Q_c = fraction of heat input lost to the cooling (in the range of 0.2–0.35)

$$\partial T_E = \frac{\partial h}{C_{Pe} m_f (1 + AFR)} \quad (22.1.82)$$

where

C_{Pe} = specific heat capacity of the exhaust gas (J/kgK)
AFR = air fuel ratio

It is clear from equations (22.1.81) and (22.1.82) that m_f cancels in the calculation of the temperature rise. However, as it is of general interest to the engine designer, it can be found from the bsfc, the brake power and the engine speed.

Thus

$$\text{kg/s fuel flow} = (bsfc/1000) \times \frac{1}{3600} \times P_b \quad (22.1.83)$$

where

$bsfc$ = brake specific fuel consumption (g/kWh)
P_b = brake power (per cylinder – kW)

$$m_f = \frac{\text{kg/s} \times 2}{\text{rpm}/60} \quad (22.1.84)$$

The thermal efficiency can be calculated from

$$\eta_{th} = 100/Q_{LHV} \times (bsfc \times 10^{-3}/(1000 \times 3600)) \quad (22.1.85)$$

A value for the specific heat capacity of exhaust gas at elevated temperature is needed. This can be found from a polynomial fit of published data for CO_2 (Rogers and Mayhew, 1980) assuming that CO_2 is the main constituent of exhaust gas:

$$C_{Pe(CO_2)} = -3.6459 \times 10^{-11} T^4 + 3.0779 \times 10^{-7} T^3 - 9.7959 \times 10^{-4} T^2 + 1.4606T + 485.6034 \quad (22.1.86)$$

The elevated temperature T(K) can be estimated from the following empirical relationships for peak-burnt gas temperature, using excess air ratio λ and ambient temperature T_{amb} (Benson and Whitehouse, 1979). If $\lambda > 1$ (lean mixture), then

$$T = T_{amb} + \frac{2500}{\lambda} \quad (22.1.87)$$

If $\lambda < 1$ (rich mixture)

$$T = T_{amb} + \frac{2500}{\lambda} - 700 \left(\frac{1}{\lambda} - 1\right) \quad (22.1.88)$$

So, from a knowledge of bsfc (typically in the range of 250–280 g/kWh for a gasoline engine operating at full load) and an estimate of the heat loss coefficient to the coolant, the temperature rise across the engine can be calculated. Therefore, temperature at EVC

$$T_{exh} - \partial T_e + T_{amb}$$

where T_{amb} is the ambient temperature (K).

Now, the ideal thermal efficiency for the gasoline engine is given by:

$$\eta_{ideal} = 1 - \frac{1}{r^{\gamma-1}} \quad (22.1.89)$$

where r is the compression ratio and γ is the ratio of specific heats. With knowledge of the compression ratio, and the estimate of the actual thermal efficiency from equation (22.1.85), equation (22.1.89) can be used in an iterative procedure to find the value of γ for which $\eta_{ideal} = \eta_{th} + Q_c$. That value of γ can be used to estimate the peak cylinder temperature assuming isentropic expansion during the expansion stroke: At 10–30° ATDC

$$T_{peak} = T_{exh} r^{\gamma-1} \quad (22.1.90)$$

The pressure of the reactants at TDC can be estimated using the following polytropic relationship for the compression stroke

At TDC

$$P_{TDC} = P_{amb} r^k \quad (22.1.91)$$

where the polytropic index k is in the range of 1–1.2.

The reactant temperature at TDC can be obtained assuming isentropic compression of air ($\gamma = 1.4$)

At TDC

$$T_{TDC} = T_{amb}\left(\frac{P_{TDC}}{P_{amb}}\right)^{\gamma-1/\gamma} \qquad (22.1.92)$$

The peak pressure can be obtained assuming an ideal gas, and that the combustion chamber volume at the crank-angle corresponding to peak pressure is the same as at TDC, and that the gas constant remains invariant with changing temperature:

At 10–30° ATDC

$$P_{peak} = P_{TDC}\left(\frac{T_{peak}}{T_{TDC}}\right) \qquad (22.1.93)$$

The exhaust pressure can be estimated by assuming isentropic expansion and using the value for γ for which $\eta_{ideal} = \eta_{th} + Q_c$

At EVO

$$P_{exh} \approx P_{peak}\left(\frac{1}{r \times \frac{evo}{180}}\right)^{\gamma} \qquad (22.1.94)$$

In summary, the pressure estimates are:

- at TDC, P_{TDC}
- at 10–30° ATDC, P_{peak}
- at EVO, P_{exh}
- at IVO,P_{amb}
- at IVC, P_{amb}
- at ignition, P_{TDC}.

And the temperature estimates are:

- at TDC, T_{TDC}
- at 10–30° ATDC, T_{peak}
- at IVO, T_{exh}
- at EVC, T_{exh}
- at IVC, T_{amb}
- at ignition, T_{TDC}.

In order to validate this approximate approach, the case of a 2.0-litre gasoline engine with a compression ratio of 10.5:1 is considered. The results of a full engine simulation (AVL Boost), using the integral of the first law of thermodynamics, a heat release model and Woschini's heat transfer relationships for the cylinder (Woschini, 1967) give a bsfc of 271 g/kWh, and the cylinder gas pressure and temperature curves shown in Fig. 22.1-43. The results from the simplified approach are seen to compare favourably.

22.1.4 Tyre noise

This section will deal with exterior noise resulting from the contact between the tyres and the road. This is often labelled tyre noise.

In a separate section (Section 21.1.5) the following is considered. Interior noise resulting from the contact between the tyres and the road, being transmitted to the interior by both airborne and structure-borne paths. This is often labelled road noise.

It is important to make a distinction between tyre and road noise for the reason that the motivation for controlling tyre noise is usually a desire to pass the drive-pass noise test for type approval (see Section 22.1.1), whereas the motivation for controlling road noise is usually to maximise passenger comfort and preserve the quality of speech communication (see Section 21.1.1).

22.1.4.1 Sources of airborne tyre noise

Airborne tyre noise has dominated the wayside noise levels caused by vehicles travelling at higher speeds for years, and more recently has begun to affect the low-speed acceleration tests used for type approval. As a result, a proposed EC Directive aims to reduce the problem by setting noise limits for different tyre types [C30/8, 28/1/98].

There is some debate over the sources of airborne tyre noise. The two noise-generating mechanisms given most attention are:

1. Noise generated when air is pumped in and out of tyre tread and road cavities during the contact process – the so-called air-pumping noise.
2. Noise generated by vibrations in the tyre caused by the contact process.

The most plausible explanation for the doubt over noise-generating mechanisms is that both may prove significant depending on:

- tyre construction and tread pattern;
- road surface;
- speed of the tyre.

The air-pumping mechanism has been shown to be significant for tyres with deep cross-grooves (known as cross-bar or cross-lug tyres) (Wilken et al., 1976). The effect of a single cross-groove cut into a treadless tyre was studied. Filling the groove with foam helped identify that the air-pumping mechanism is reinforced by acoustic resonance of the groove near its quarter wavelength frequency. Opening the closed end of the groove to circumferential grooves helped control this resonance.

The common observation that many treadless tyres are as noisy as tyres with treads suggests that tyre vibration also cause noise in addition to air pumping. With most modern tyre tread patterns that are not block like, the tyre vibration is commonly the dominant noise-generating mechanism.

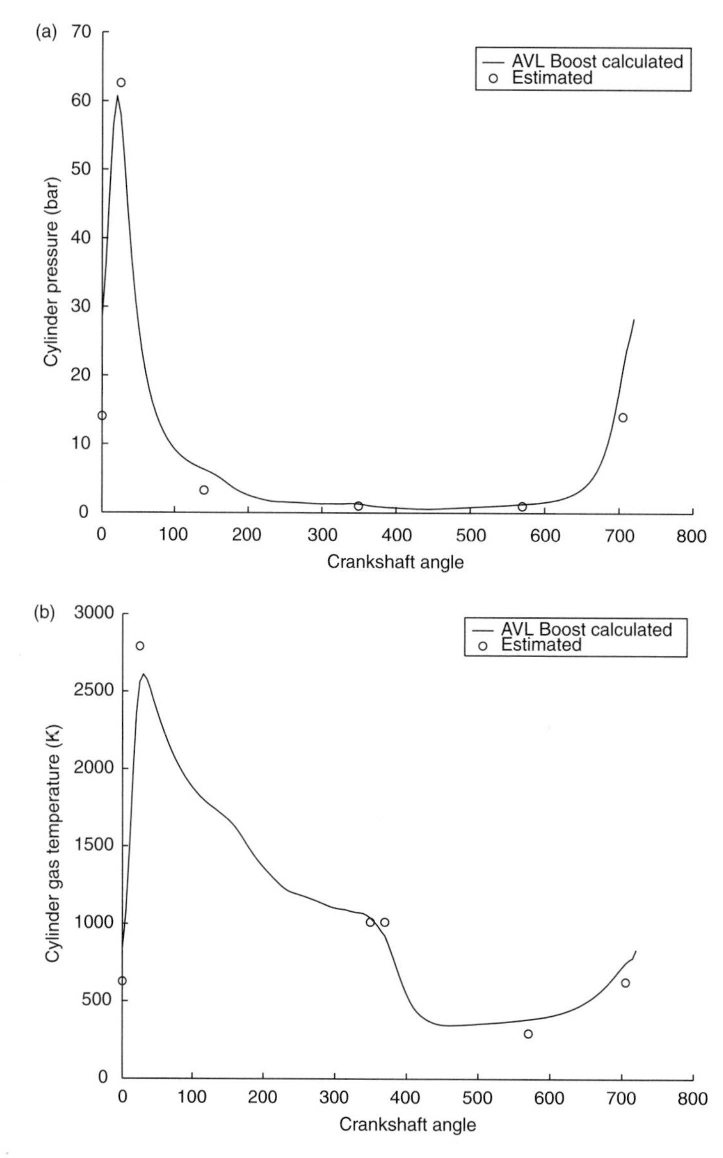

Fig. 22.1-43 Estimates of in-cylinder gas pressure and temperature made using simplified assumptions detailed in Section 22.1.3.14.2. Comparison is made with the results obtained from a full engine simulation of a 2.0-litre gasoline engine at 5000 rev min^{-1} using AVL ***Boost***

Comparisons made between noise measurements near to and far from a rotating tyre suggest that (reported in Nilsson (1976)):

- most of the noise originates near to the contact patch;
- the sound intensity is greatest at the entrance and exit surfaces of the contact patch;
- the exit of the contact patch is important for tonal components of tyre noise;
- the tyre sidewall is not a significant radiator of sound.

As a result of the work described above, subsequent investigations have concentrated on measuring vibration and noise levels within the tread of the tyre (Jennewein and Bergmann, 1985).

The tonal tyre noise originates from regularities in the tyre construction. The random tyre noise originates first by radial excitation due to roughness in the road but also from random tangential movements of the tread pattern (Nilsson, 1976).

Tonal tyre noise is more speed-dependent than random tyre noise. Random tyre noise is strongly affected by the characteristics of the road surface. A simple empirical relationship between noise levels at 7.5 m and the tyre noise caused by coasting vehicles is presented in (Nilsson, 1976). Therefore,

$$L_A = C + 10\log_{10}(V^n)\text{dB} \quad (22.1.95)$$

where

L_A = sound pressure level at 7.5 m dBA due solely to tyre noise V = vehicle coasting speed (km h^{-1})

Rib tyre on wet road	$C = 47,\ n = 1.7$
Smooth tyre on wet road	$C = 23,\ n = 2.7$
Smooth tyre on dry road	$C = 10,\ n = 3.4$
Regular tyre on dry road	$C = 18,\ n = 3.0$

Measurements of vibration acceleration made on the tread show it to be greatest during the contact process (Jennewein and Bergmann, 1985). Removal of the influence of accelerations due solely to the flattening of the tyre contour yields the following:

- Radial acceleration of the tread bottom (particularly in the run-in section of the contact patch) is most important at frequencies below 1000 Hz.
- Tangential vibration of the tread blocks is most important at frequencies above 1000 Hz (particularly in the run-out section of the contact patch).

Noise measurements made with tiny microphones placed in the tread grooves (Jennewein and Bergmann, 1985) show:

- As a tread block strikes the ground, a groove that did have both ends open has one end sealed now. This forms a one-quarter wavelength resonator with a resonant frequency commonly in the 1250 Hz third octave band. The resonance is excited by tread vibrations.
- As the tread block lifts at the trailing edge it forms a new resonator with the volume of air trapped in the groove behind. This second resonant frequency is typically in the range of 1500–2500 Hz.
- Different block arrangements create different types of resonator.
- There is a high amplification of sound (20 dB or more) within the resonators formed in the contact patch due to the leading and trailing edges of the contact patch acting as acoustic horns.

The various acoustic resonances are clearly seen in the noise spectrum measured at the contact patch and remain evident in the wayside noise.

22.1.4.2 The influence of the road surface on airborne tyre noise

It is commonly known that the characteristics of the road surface affect the wayside noise levels. As a result, a reference road surface is provided for use in type approval tests (ISO 10844: 1994) as described in Section 22.1.1.3.

The three road characteristics influencing tyre noise are:

1. surface roughness;
2. the ability to shed surface water;
3. sound absorption.

There are some published data for the surface roughness of different roads (Cebon, 1999). These are in the form of displacement spectra, with the highest wavenumber components (leading to the highest-frequency noise) having the lowest amplitude. The simplest descriptor of the spectral density is

$$S_u(\kappa) = S_0|\kappa|^{-n}\ \mathrm{m^3/cycle} \qquad (22.1.96)$$

where κ is the wavenumber (cycles/m) and $n = 2.5$ and S_0 varies according to the type of road as shown below.

Motorway	$S_0 = 3\text{–}50 \times 10^{-8}\ (\mathrm{m^{0.5}\ cycle^{1.5}})$
Principal road	$S_0 = 3\text{–}800 \times 10^{-8}\ (\mathrm{m^{0.5} cycle^{1.5}})$
Minor road	$S_0 = 50\text{–}3000 \times 10^{-8}\ (\mathrm{m^{0.5} cycle^{1.5}})$

The frequency associated with each wavenumber is the product of wavenumber and forward velocity ($\mathrm{ms^{-1}}$).

Two common ways to texture a road surface to improve safety in wet conditions are:

- to roll stone chippings into hot asphalt (HRA);
- to produce closely spaced ridges that run transverse across a concrete road using a wire brush.

The second method of texturing a road surface will produce wayside noise levels 5 dB greater than the first method (Wright, 1999).

One alternative road surface, thought to be 3.5 dBA quieter than HRA (DoT, Calculation of road traffic noise, 1988) is the use of porous asphalt (PA). This is a thick (100 mm) road with a porous layer of larger stones sitting on an impermeable base course of asphalt. It is free flowing for surface water. Unfortunately, it is costly, gives poor wear and tends to freeze. Alternatives to PA are now available (Wright, 1999) which are thinner and more hard wearing. They are generally 2–5 dBA quieter than HRA and cheaper than PA.

It is important to remember road safety when considering tyre noise. Although noise levels increase when a vehicle is accelerating rather than coasting (Donavan, 1993) and when tyre load increases, fortunately there does not seem to be any correlation between wet friction and noise (Sandberg et al.,1999).

22.1.4.3 Measuring airborne tyre noise

Many test schedules have been developed for measuring airborne tyre noise. Some important ones include:

- Wayside noise from a coasting vehicle (C30/8, 28/1/98, J57 SAE 1994).
- Sound intensity measured alongside a rolling tyre mounted on a vehicle in use on the road (Donavan, 1993; Bolton et al., 1995).

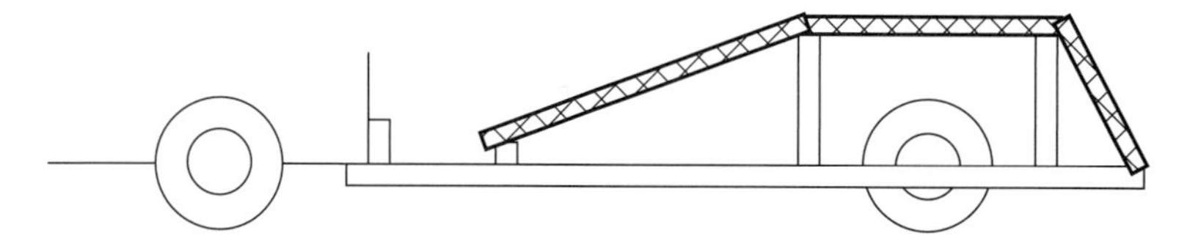

Fig. 22.1-44 Measuring tyre noise using an Austrian trailer: after Sandberg (1998).

- Sound pressure measured alongside a rolling tyre mounted on a special trailer. Various trailers are described in Sandberg (1998). One example, taken from Sandberg (1998) is sketched in Fig. 22.1-44.
- Measuring tyre noise on a rolling road rig in the laboratory (Pope and Reynolds, 1976; Sandberg and Ejsmont, 1993).

22.1.4.4 Controlling airborne tyre noise by design

Some tyre design guidance has emerged for reducing airborne tyre noise:

- Reduce the tread modulus (Muthukrishnan, 1990).
- Use a softer rubber on the tread to reduce the impact of the block at the leading edge of the contact patch (Jennewein and Bergmann, 1985).
- Avoiding tension in the tread blocks to decrease tangential vibration at the trailing edge of the contact patch (Jennewein and Bergmann, 1985).
- Avoid transverse grooves (Jennewein and Bergmann, 1985) to reduce the effect of acoustic cavity resonances. Short grooves with one end open are preferred.
- Provide frequency the modulation in the tyre by selectively arranging tread elements of various sizes – so-called pitch sequencing (Williams, 1995).

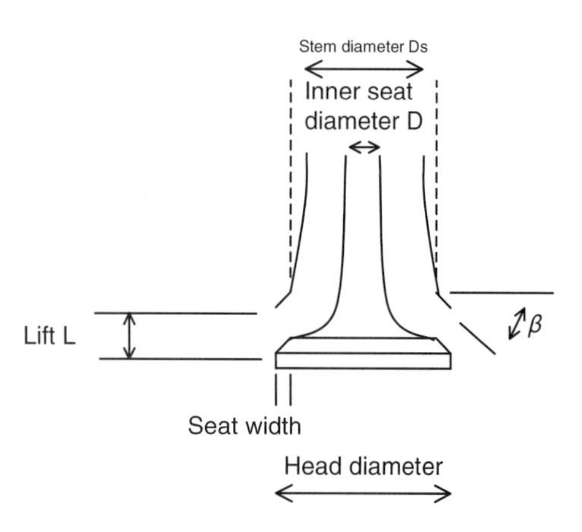

Fig. 22.1-A1 Valve geometry.

Appendix 22.1A : Valve and port geometry

At low valve lifts, the open area of the valve is given by (Heywood, 1988)

$$A_m = \pi L_v \cos\beta \left(D_v - 2w + \frac{L_v}{2}\sin 2\beta\right) \qquad \text{(A22.1.1)}$$

$$\text{for} \quad \frac{w}{\sin\beta\cos\beta} > L_v > 0$$

At intermediate valve lift,

$$A_m = \pi D_m [(L_v - w\tan\beta)^2 + w^2]^{1/2} \qquad \text{(A22.1.2)}$$

$$\text{for} \left[\left(\frac{D_p^2 - D_s^2}{4D_m}\right) - w^2\right]^{1/2} + w\tan\beta \geq L_v > \frac{w}{\sin\beta\cos\beta}$$

where D_p is the port diameter.

When the valve lift is sufficiently large, the minimum flow area actually becomes the port area minus the valve stem area

$$A_m = \frac{\pi}{4}(D_p^2 - D_s^2) \qquad \text{(A22.1.3)}$$

$$\text{for } L_v > \left[\left(\frac{D_p^2 - D_s^2}{4D_m}\right) - w^2\right] + w\tan\beta$$

References

70 /156/EEC, Council Directive of 6 February 1970 on the approximation of the laws of the Member States relating to the type-approval of motor-vehicles and their trailers, *official Journal of the European Communities*, No. L 42/1, 23.2., 1970

70 /157/EEC, Council Directive of 6 February 1970 on the approximation of the laws of the Member States relating to the permissible sound level and the exhaust system of motor-vehicles, *Official Journal of the European Communities*, No. L 42/16, 23.2., 1970

Statutory Instrument No. 981, 1984, The Motor Vehicles (Type Approval) (Great Britain) Regulations, HMSO, London, 1984

Statutory Instrument No. 1078, 1986, The Road Vehicles (Construction and Use) Regulations 1986, HMSO, London, 1986

SAE J1030 FEB87, Maximum sound level for passenger cars and light trucks, 1996 SAE Handbook, Society of Automotive Engineers, 1996

The Road Traffic Act, HMSO, London, 1988

Department of Transport Welsh Office, Calculation of road traffic noise, HMSO, London, 1988

92 /97/EEC, Council Directive of 10 November 1992 amending Directive 70/157/EEC on the approximation of the laws of the Member States relating to the permissible sound level and the exhaust system of motor-vehicles, *Official Journal of the European Communities*, No. L 371/35, 19.2., 1992

ISO 10844: 1994, Acoustics – Test surface for road vehicle noise measurements, International Organization for Standardization, 1994

J57, Sound level of highway truck tires, Society of Automotive Engineers, June 1994

SAE J986 AUG94, Sound level for passenger cars and light trucks, 1996 SAE Handbook, Society of Automotive Engineers, 1996

Statutory Instrument No. 2051, 1998, The Motor Vehicle (EC Type Approval) Regulations 1998, HMSO, London, 1998

C30/8, 28/1/98, Proposal for a European Parliament and Council Directive amending Council Directive 92/23/EEC relating to tyres for motor vehicles and their trailers and to their fitting, *Official Journal of the European Communities*, 1998

Bs Iso 362: 1998, Acoustics – Measurement of noise emitted by accelerating road vehicles – Engineering method, British Standards Institution, 1998

Annand, W.J.D., Roe, G.E., Gas flow in the internal combustion engine – power, performance, emission control and silencing, GT Foulis, 1974

GmbH, A.V.L., Boost user's manual – version 3, April 2000

Balcombe, D.R., Crowther, P., Practical development problems in achieving 74 dBA for cars, Proc. IOA, Vol. 15: Part 1, pp. 49–58, 1993

Benson, R.S., Whitehouse, N.D., Internal combustion engines–Vols. 1 and 2, Pergamon Press, 1979

Benson, R.S., The thermodynamics and gas dynamics of internal combustion engines – Vols. 1, Clarendon Press, Oxford, 1982

Blair, G.P., Drouin, F.M.M., Relationship between discharge coefficients and accuracy of engine simulation, Society of Automotive Engineers, Paper No. 962527, 1996

Bolton, J.S., Hall, H.R., Schumacher, R.F., Stott, J., Correlation of tire intensity levels and passby sound pressure levels, SAE Paper No. 951355, Proceedings of the 1995 noise and vibration conference–Vol. 2, Society of Automotive Engineers, 1995

Bosch, Automotive handbook – Second edition, Robert Bosch GmbH, 1986

Cebon, D., Handbook of vehicle–road interaction, Swets & Zeitlinger, 1999

Davies, P.O.A.L., Flow-acoustic coupling in ducts, *Journal of Sound and Vibration*, 77(2), 191–209, 1981

Davies, P.O.A.L., Plane wave reflection at flow intakes, *Journal of Sound and Vibration*, 115(3), 560–564, 1987

Davies, P.O.A.L., Practical flow duct acoustics, *Journal of Sound and Vibration*, 124(1), 91–115, 1988

Davies, P.O.A.L., Aeroacoutics and time varying systems, *Journal of Sound and Vibration*, 190(3), 345–362, 1996

Davies, P.O.A.L., Harrison, M.F., Predictive acoustic modelling applied to the control of intake/exhaust noise of internal combustion engines, *Journal of Sound and Vibration*, 202, 249–274, 1997

Davies, P.O.A.L., Holland, K.R., I.C. engine intake and exhaust noise assessment, *Journal of Sound and Vibration*, 223(3), 425–444, 1999

Davies, P.O.A.L., Holland, K.R., The observed aeroacoustic behaviour of some flow-excited expansion chambers, *Journal of Sound and Vibration*, 239(4), 695–708, 2001

Davies, P.O.A.L., Holland, K.R., The measurement and prediction of sound waves of arbitrary amplitude in practical flow ducts, *Journal of Sound and Vibration*, in press 2004

Davies, P.O.A.L., Harrison, M.F., Collins, H.J., Acoustic modelling of multiple path silencers with experimental validations, *Journal of Sound and Vibration*, 200(2) 195–225, 1997

Davies, P.O.A.L. Bento Coelho, J.L., Battacharya, M., Reflection coefficients for an unflanged pipe with flow, *Journal of Sound and Vibration*, 72(4), 543–546, 1980

Donavan, P.R., Tire-pavement interaction noise measurement under vehicle operating conditions of cruise and acceleration, SAE Paper No. 931276, Included in *Tires and Handling–PT-59* (Ellis Johnson (ed.)), Society of Automotive Engineers, 1993

Dunkley, A., The acoustics of IC engine manifolds, Cranfield University – MSc Thesis, 1999

Dunne, I.C., J.M., Yarnold, I.C., Vehicle noise legislation – an overview, Proc. IOA, Vol. 15: Part 1, pp. 1–8, 1993

Fukutani, I., Watanabe, E., An analysis of the volumetric efficiency characteristics of 4-stroke cycle engines using the mean inlet Mach number Mim, SAE Paper No. 790484, 1979

Fukutani, I., Watanabe, E., Air flow through poppet inlet valves - analysis of static and dynamic flow coefficients, Society of Automotive Engineers, Paper No. 820154, 1982

Harrison, M.F., Notes to accompany the course on Piston Engines – Thermofluids, MSc Automotive Product Engineering, Cranfield University, 2003

Harrison, M.F., Stanev, P.T., A linear acoustic model for intake wave dynamics in IC engines, *Journal of Sound and Vibration*, 269(1), 361–387, 2004

Harrison, M.F., Dunkley, A., The acoustics of racing engine intake systems, *Journal of Sound and Vibration*, 271, 959–984, 2004

Harrison, M.F., De Soto Beodo, I., Rubio Unzueta, P.L., A linear acoustic model for multi-cylinder I.C. engine intake manifolds, including the effects of the intake throttle, *Journal of Sound and Vibration*, in press, 2004

Heywood, J.B., Internal combustion engine fundamentals, McGraw Hill Book Company, 1988

Hirsch, C., Numerical computation of internal and external flows – Vol. 1, John Wiley & Sons, 1988

Holland, K.R., Davies, P.O.A.L., van der Walt, D.C., Sound power flux measurements in strongly excited flow ducts, *Journal of the Acoustical Society of America*, 112(6), 2863–2871, 2002

Howe, M.S., Contributions to the theory of aerodynamic noise, with applications to excess jet engine noise and the theory of the flute, *Journal of Fluid Mechanics*, 71, 625–673, 1975

Jennewein, M., Bergmann, M., Investigations concerning tyre/road noise sources and possibilities of noise reduction, Proceedings of the Institution of Mechanical Engineers, Vol. 199, D3, 199–205, 1985

Kim, G.T., Lee, B.H., 3-D sound source reconstruction and field prediction using the Helmholtz Integral Equation, *Journal of Sound and Vibration*, 136(2), 245–261, 1990

Kunz, F., Garcia, P., Simulation and measurement of hot exhaust gas flow noise with a cold air flow bench, SAE Paper No. 950546, 1995

Lighthill, M.J., On sound generated aerodynamically: I, general theory, Proceedings of the Royal Society of London, A211, 564–587, 1952

Lighthill, M.J., On sound generated aerodynamically: II, turbulence as a source of sound, Proceedings of the Royal Society of London, 222, 1–32, 1954

Lumley, J.L., Engines – an introduction, Cambridge University Press, 1999

Morfey, C.L., Acoustic energy in non-uniform flows, *Journal of Sound and Vibration*, 14(2), 159–170, 1971

Muthukrishnan, M., Effects of material properties on tire noise, SAE Paper No. 900762, Included in *Tires and Handling - PT-59* (Ellis Johnson (ed.)), Society of Automotive Engineers, 1990

Nilsson, N.A., On generating mechanisms for external tire noise, SAE Paper No. 762026, Proceedings of the SAE highway tire noise symposium, 10–12 November 1976, Society of Automotive Engineers, 1976

Ohata, A., Ishida, Y., Dynamic inlet pressure and volumetric efficiency of four cycle four cylinder engine, Society of Automotive Engineers, Paper No. 820407, 1982

Payri, F., Torregrosa, A.J., Payri, R., Evaluation through pressure and mass velocity distributions of the linear acoustical description of I.C. engine exhaust systems, Applied Acoustics, 60, 489–504, 2000

Pope, J., Reynolds, W.C., Tire noise generation: the roles of tire and road, SAE Paper No. 762023, Proceedings of the SAE highway tire noise symposium, 10–12 November 1976, Society of Automotive Engineers, 1976

Powell, A., Theory of vortex sound, *Journal of the Acoustical Society of America*, 36(1), 177–195, 1964

Rogers, G.F.C., Mayhew, Y.R., Thermodynamic and transport properties of fluids – Third edition, Basil Blackwell, 1980

Sandberg, U., Standardisation of a test track surface for use during vehicle noise testing, SAE Paper No. 911048, 1991

Sandberg, U., Ejsmont, J.A., The art of measuring noise from vehicle tyres, SAE Paper No. 931275, Included in *Tires and Handling – PT-59* (Ellis Johnson (ed.)), Society of Automotive Engineers, 1993

Sandberg, U., Noise trailers of the world–tools for tire/road noise measurements with the close-proximity method, Proceedings of the 1998 National Conference on noise control engineering, 5–8 April 1998

Sandberg, U., Ejsmont, J.A., Mioduszewski, P., Taryma, S., Noise–the challenge, Tire Technology International, 99, 98–101, 1999

Selemet, A., Kurniawan, D., Knotts, D., Novak, J.M., Study of whistles with a generic sidebranch, SAE Paper No. 1999-01-1814, 1999

Sievewright, G., Air flow noise of plastic air intake manifolds, SAE Paper No. 2000-01-0028, 2000

Steven, H., Effects of noise limits for powered vehicles on their emissions in real operation, SAE Paper No. 951257, 1995

Taylor, C.F., The internal combustion engine in theory and practice–1, Second edition, revised, MIT Press, 1985

Taylor, N.C., Bridgewater, A.P., Airborne road noise source definition, IMechE, C521/011, 1998 Tsung-chi Tsu Theory of the inlet and exhaust processes of internal combustion engines, NACA TN 1446, 1947

Walker, A.W., The effect of the ISO surface on noise passby levels, Reproduced in: *Noise and the automobile*, selected papers from Autotech 93, Birmingham, 16–19 November 1993, Published by Mechanical Engineering Publications Ltd, London, 1994

Watson, N., Janota, M.S., Turbocharging the internal combustion engine, Macmillan Press Ltd, 1982

Weaving, J. (ed.) Internal combustion engineering–science and technology, Elsevier Applied Science, 1990

Wilken, I.D., Oswald, L.J., Hickling, R., Research on individual noise source mechanisms of truck tyres: aeroacoustic sources, SAE Paper No. 762022, Proceedings of the SAE highway tire noise symposium, 10–12 November 1976, Society of Automotive Engineers, 1976

Williams, T.A., Tire tread pattern noise reduction through the application of pitch sequencing, SAE Paper No. 951352, Proceedings of the 1995 noise and vibration conference – Vol. 2, Society of Automotive Engineers, 1995

Winterbone, D.E., Pearson, R.J., Design techniques for engine manifolds – wave action methods for IC engines, Professional Engineering Publishing, 1999

Woschini, G., A universally applicable equation for the instantaneous heat transfer coefficient in the internal combustion engine, Society of Automotive Engineers, Paper No. 670931, 1967

Wright, M., Thin and quiet? An update on new quiet road surface products, Acoustics Bulletin, September/ October 1999, Institute of Acoustics, 1999

Zemansky, M.W., Dittman, R.H., Heat and thermodynamics (an intermediate textbook) – Seventh edition, McGraw Hill International Editions, 1997

Section Twenty-Three

Instrumentation and telematics

Chapter 23.1

23.1

Automotive instrumentation and telematics

William Ribbens

This chapter describes electronic instrumentation and the relatively new field of telematics. By the term *instrumentation,* we mean the equipment and devices that measure engine and other vehicle variables and parameters and display their status to the driver. By the term *telematics,* we refer to communication of all forms within the vehicle as well as communication to and from the vehicle. Communication within the vehicle takes the form of digital data links between various electronic subsystems. Communication to and from the vehicle spans all communication from voice and digital data via cell or satellite phone systems to digital data sent from land or satellite. Internet connections to an on-board PC (or the like) are included in those categories listed above. This chapter begins with a discussion of electronic instrumentation and concludes with telematics.

From about the late 1920s until the late 1950s, the standard automotive instrumentation included the speedometer, oil pressure gauge, coolant temperature gauge, battery charging rate gauge, and fuel quantity gauge. Strictly speaking, only the latter two are electrical instruments. In fact, this electrical instrumentation was generally regarded as a minor part of the automotive electrical system. By the late 1950s, however, the gauges for oil pressure, coolant temperature, and battery charging rate were replaced by warning lights that were turned on only if specified limits were exceeded. This was done primarily to reduce vehicle cost and because of the presumption that many people did not necessarily regularly monitor these instruments.

Automotive instrumentation was not really electronic until the 1970s. At that time, the availability of relatively low-cost solid-state electronics brought about a major change in automotive instrumentation; the use of low-cost electronics has increased with each new model year. This chapter presents a general overview of typical automotive electronic instrumentation.

In addition to providing measurements for display, modern automotive instrumentation performs limited diagnosis of problems with various subsystems. Whenever a problem is detected, a warning indicator alerts the driver of a problem and indicates the appropriate subsystem. For example, whenever self-diagnosis of the engine control system detects a problem, such as a loss of signal from a sensor, a lamp illuminates the "Check Engine" message on the instrument panel. Such warning messages alert the driver to seek repairs from authorized technicians who have the expertise and special equipment to perform necessary maintenance.

23.1.1 Modern automotive instrumentation

The evolution of instrumentation in automobiles has been influenced by electronic technological advances in much the same way as the engine control system, which has already been discussed. Of particular importance has been the advent of the microprocessor (MPU), solid-state display devices, and solid-state sensors. In order to put these developments into perspective, recall the general block diagram for instrumentation which is repeated here as Fig. 23.1-1.

In electronic instrumentation, a sensor is required to convert any nonelectrical signal to an equivalent voltage or current. Electronic signal processing is then performed on the sensor output to produce an electrical signal that is capable of driving the display device. The display device

Understanding Automotive Electronics; ISBN: 9780750675994
Copyright © 2003 Elsevier Ltd; All rights of reproduction, in any form, reserved.

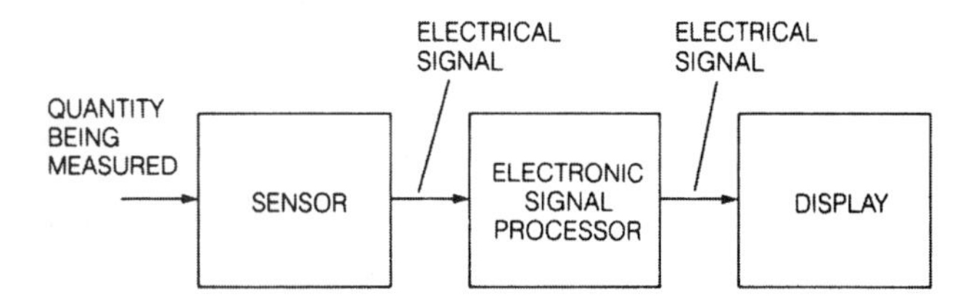

Fig. 23.1-1 General instrumentation block diagram.

is read by the vehicle driver. If a quantity to be measured is already in electrical form (e.g., the battery charging current) this signal can be used directly and no sensor is required.

In some modern automotive instrumentation, a microcomputer (or related digital subsystem) performs all of the signal processing operations for several measurements. The primary motivation for computer-based instrumentation is the great flexibility offered in the design of the instrument panel. A block diagram for such an instrumentation system is shown in Fig. 23.1-2.

All measurements from the various sensors and switches are processed in a special-purpose digital computer. The processed signals are routed to the appropriate display or warning message. It is common practice in modern automotive instrumentation to integrate the display or warning in a single module that may include both solid-state alphanumeric display, lamps for illuminating specific messages, and traditional electromechanical indicators. For convenience, this display will be termed the *instrument panel* (IP).

The inputs to the instrumentation computer include sensors (or switches) for measuring (or sensing) various vehicle variables as well as diagnostic inputs from the other critical electronic subsystems. The vehicle status sensors may include any of the following:

1. Fuel quantity
2. Fuel pump pressure
3. Fuel flow rate
4. Vehicle speed
5. Oil pressure
6. Oil quantity
7. Coolant temperature
8. Outside ambient temperature
9. Windshield washer fluid quantity
10. Brake fluid quantity

In addition to these variables, the input may include switches for detecting open doors and trunk, as well as IP selection switches for multifunction displays that permit the driver to select from various display modes or measurement units. For example, the driver may be able to select vehicle speed in miles per hour (mph) or kilometers per hour (kph).

An important function of modern instrumentation systems is to receive diagnostic information from certain subsystems and to display appropriate warning messages to the driver. The powertrain control system, for example, continuously performs self-diagnosis operations. If a problem has been detected, a fault code is set indicating the nature and location of the fault. This code is transmitted to the instrumentation system via a powertrain digital data line (PDDL in Fig. 23.1-2). This code is interpreted in the instrumentation computer and a "Check Engine" warning message is displayed. Similar diagnostic data are sent to the instrumentation system from each of the subsystems for which driver warning messages are deemed necessary (e.g., ABS, airbag, cruise control).

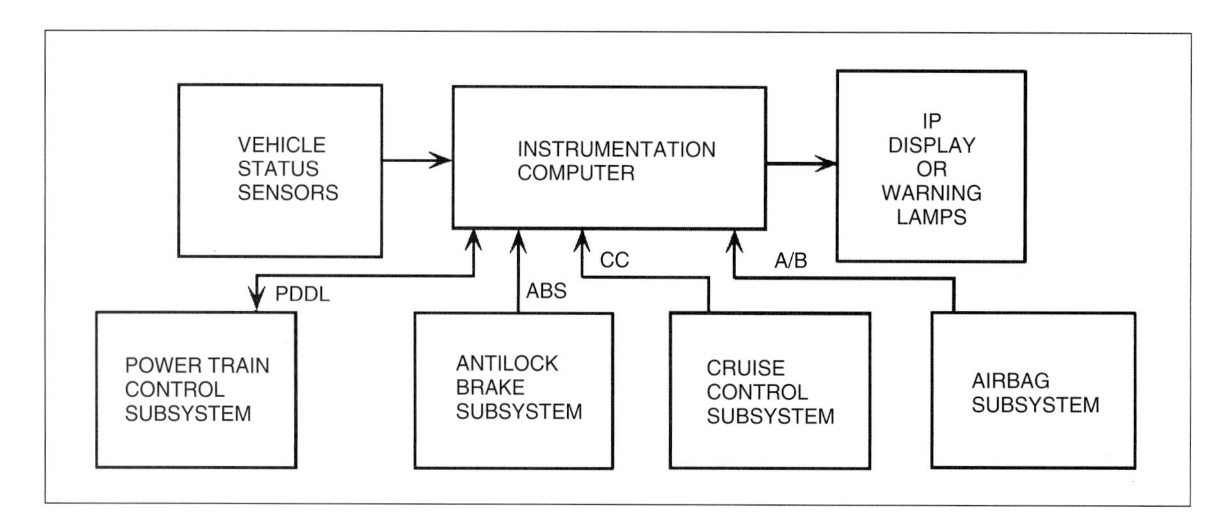

Fig. 23.1-2 Computer-based instrumentation system.

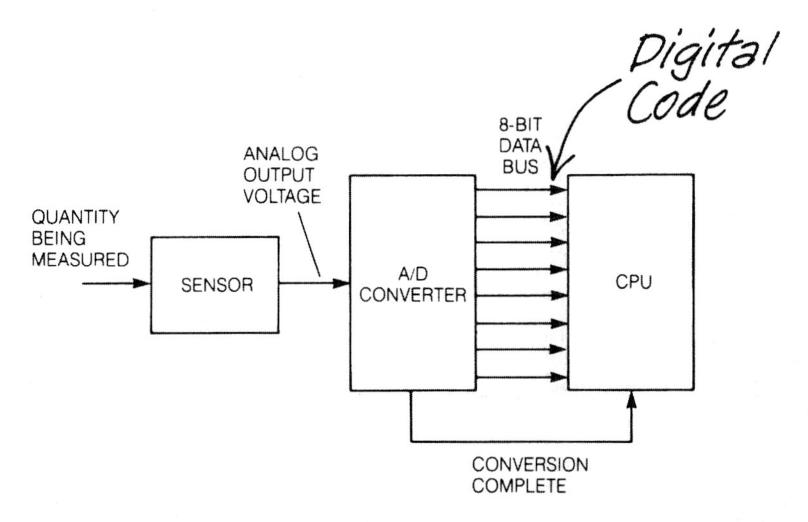

Fig. 23.1-3 Analog-to-digital conversion.

23.1.2 Input and output signal conversion

It should be emphasized that any single input can be either digital switched or analog, depending on the technology used for the sensor. A typical instrumentation computer is an integrated subsystem that is designed to accept all of these input formats. A typical system is designed with a separate input from each sensor or switch. An example of an analog input is the fuel quantity sensor, which is normally a potentiometer attached to a float, as described in detail later in this chapter. The measurement of vehicle speed given in Chapter 13.1 is an example of a measurement that is already in digital format.

The analog inputs must all be converted to digital format using an analog to digital (A/D) converter as illustrated in Fig. 23.1-3. The digital inputs are, of course, already in the desired format. The conversion process requires an amount of time that depends primarily on the A/D converter. After the conversion is complete, the digital output generated by the A/D converter is the closest possible approximation to the equivalent analog voltage, using an M-bit binary number (where M is chosen by the designer and is normally between 8 and 32). The A/D converter then signals the computer by changing the logic state on a separate lead (labeled "conversion complete" in Fig. 23.1-3) that is connected to the computer. The output voltage of each analog sensor for which the computer performs signal processing must be converted in this way. Once the

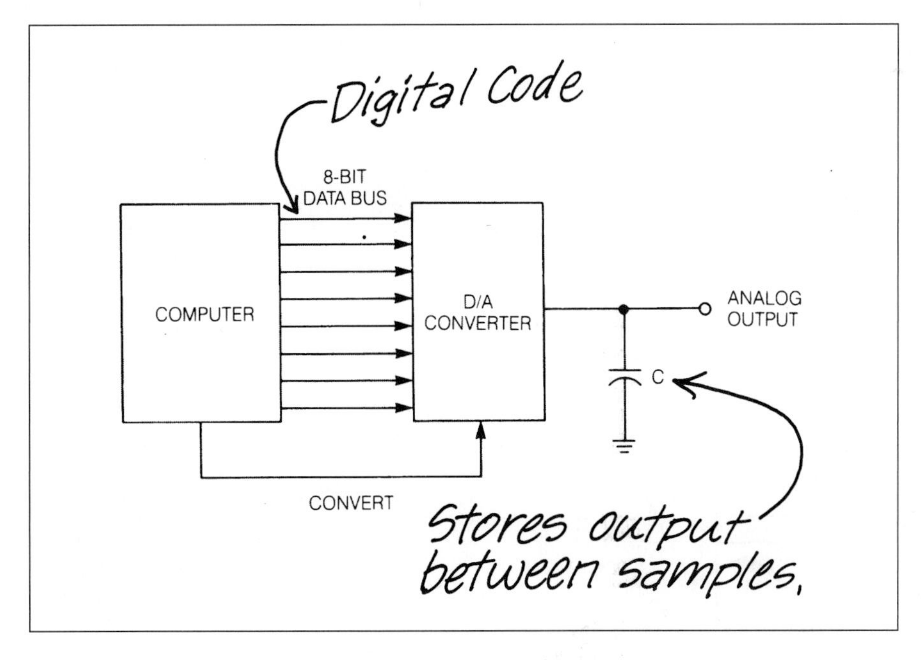

Fig. 23.1-4 Digital-to-analog conversion.

conversion is complete, the digital output is transferred into a register in the computer. If the output is to drive a digital display, this output can be used directly. However, if an analog display is used, the binary number must be converted to the appropriate analog signal by using a digital-to-analog (D/A) converter.

Fig. 23.1-4 illustrates a typical D/A converter used to transform digital computer output to an analog signal. The eight digital output leads ($M = 8$ in this example) transfer the results of the signal processing to a D/A converter. When the transfer is complete, the computer signals the D/A converter to start converting. The D/A output generates a voltage that is proportional to the binary number in the computer output. A low-pass filter (which could be as simple as a capacitor) is often connected across the D/A output to smooth the analog output between samples. The sampling of the sensor output, A/D conversion, digital signal processing, and D/A conversion all take place during the time slot allotted for the measurement of the variable in a sampling time sequence, to be discussed shortly.

23.1.2.1 Multiplexing

Of course, the computer can only deal with the measurement of a single quantity at any one time. Therefore, the computer input must be connected to only one sensor at a time, and the computer output must be connected only to the corresponding display. The computer performs any necessary signal processing on a particular sensor signal and then generates an output signal to the appropriate display device.

In Fig. 23.1-5 the various sensor outputs and display inputs are connected to a pair of multiposition rotary switches—one for the input and one for the output of the computer. The switches are functionally connected such that they rotate together. Whenever the input switch connects the computer input to the appropriate sensor for measuring some quantity, the output switch connects the computer output to the corresponding display or warning device. Thus, with the switches in a specific position, the automotive instrumentation system corresponds to the block diagram shown in Fig. 23.1-1. At that instant of time, the entire system is devoted to measurement of the quantity corresponding to the given switch position.

Typically, the computer controls the input and output switching operation. However, instead of a mechanical switch as shown in Fig. 23.1-5, the actual switching is done by means of a solid-state electronic switching device called a *multiplexer* (MUX) that selects one of several inputs for each output. Multiplexing can be done either with analog or digital signals. Fig. 23.1-6 illustrates a digital MUX configuration. Here it is assumed that there are four inputs to the MUX (corresponding to data from four sensors). It is further presumed that the data are available in 8-bit digital format. Each of the MUXs selects a single bit from each of the four inputs.

There must be eight such MUX circuits, each supplying one data bit. The output lines from each MUX are connected to a corresponding data bus line in the digital computer. Similarly, the output switching (which is often called demultiplexing, or DEMUX) is performed with a MUX connected in reverse, as shown in Fig. 23.1-7.

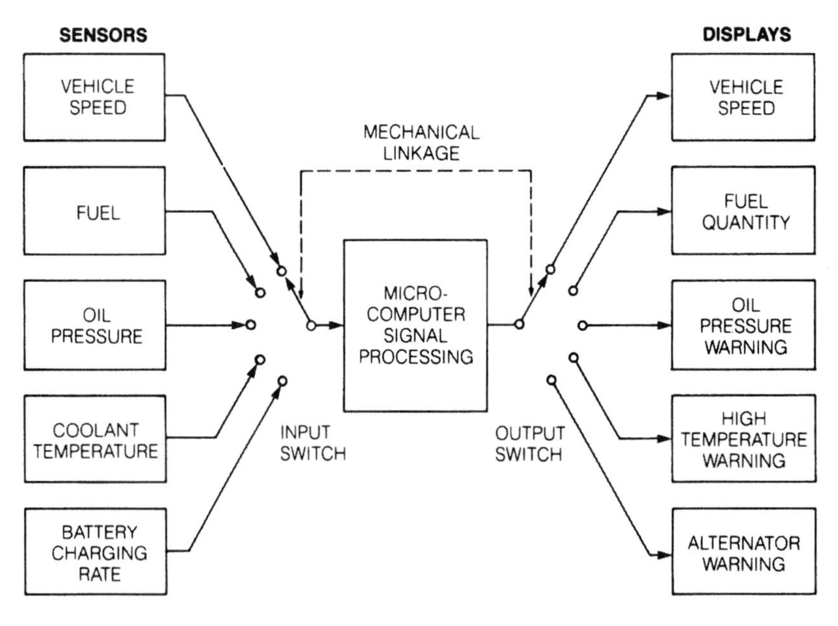

Fig. 23.1-5 Input/output switching scheme for sampling.

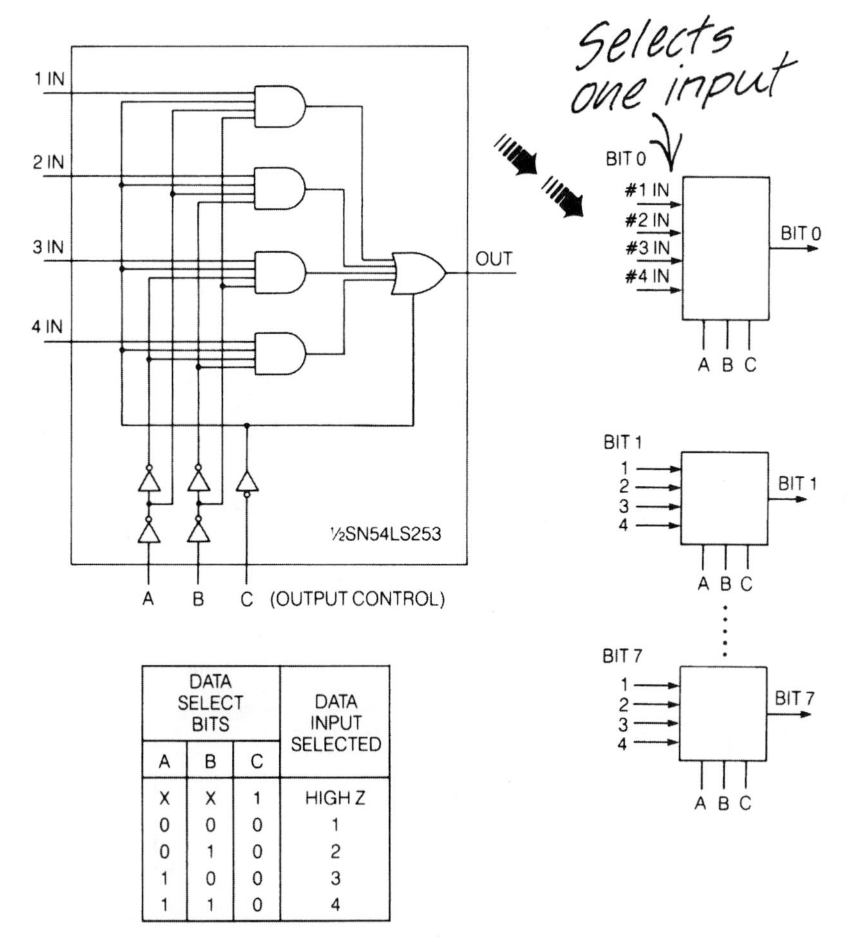

DATA SELECT BITS			DATA INPUT SELECTED
A	B	C	
X	X	1	HIGH Z
0	0	0	1
0	1	0	2
1	0	0	3
1	1	0	4

Fig. 23.1-6 Data multiplexer.

The MUX and DEMUX selection is controlled by the computer. Note that in Figs. 23.1-6 and 23.1-7, each bit of the digital code is multiplexed and demultiplexed.

23.1.3 Sampling

The measurement of any quantity takes place only when the input and output switches (MUX and DEMUX) functionally connect the corresponding sensor and display to the computer, respectively. There are several variables to be measured and displayed, but only one variable can be accommodated at any given instant. Once a quantity has been measured, the system must wait until the other variables have been measured before that particular variable is measured again. This process of measuring a quantity intermittently is called *sampling*, and the time between successive samples of the same quantity is called the *sample period*.

One possible scheme for measuring several variables by this process is to sample each quantity sequentially, giving each measurement a fixed time slot, t, out of the total sample period, T, as illustrated in Fig. 23.1-8. This method is satisfactory as long as the sample period is small compared with the time in which any quantity changes appreciably. Certain quantities, such as coolant temperature and fuel quantity, change very slowly with time. For such variables, a sample period of a few seconds or longer is often adequate.

On the other hand, variables such as vehicle speed, battery charge, and fuel consumption rate change relatively quickly and require a much shorter sample period, perhaps every second or every few tenths of a second. To accommodate the various rates of change of the automotive variables being measured, the sample period varies from one quantity to another. The most rapidly changing quantities are sampled with a very short sample period, whereas those that change slowly are sampled with a long sample period.

In addition to sample period, the time slot allotted for each quantity must be long enough to complete the measurement and any A/D or D/A conversion required. The computer program is designed with all of these factors in mind so that adequate time slots and sample periods are allowed for each variable. The computer then simply follows the program schedule.

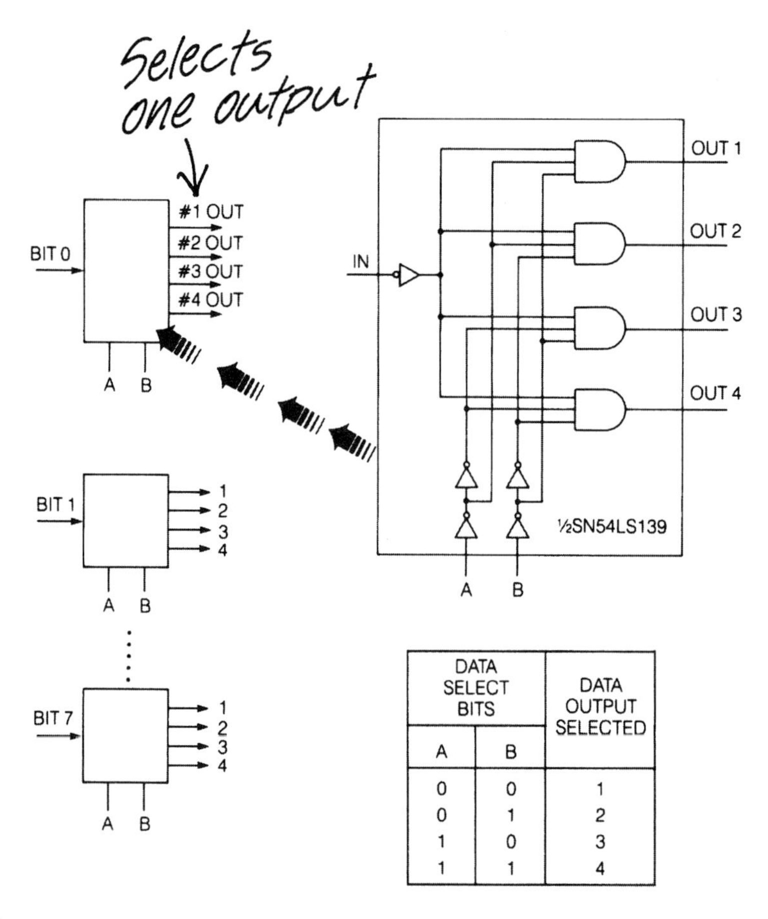

DATA SELECT BITS		DATA OUTPUT SELECTED
A	B	
0	0	1
0	1	2
1	0	3
1	1	4

Fig. 23.1-7 Data demultiplexer.

23.1.3.1 Advantages of computer-based instrumentation

One of the big advantages of computer-based instrumentation is its great flexibility. To change from the instrumentation for one vehicle or one model to another requires only a change of computer program. This change can often be implemented by replacing one read-only memory (ROM) with another. Remember that the program is permanently stored in a ROM that is typically packaged in a single integrated circuit package.

Another benefit of microcomputer-based electronic automotive instrumentation is its improved performance compared with conventional instrumentation. For example, the conventional fuel gauge system has errors that are associated with variations in the mechanical

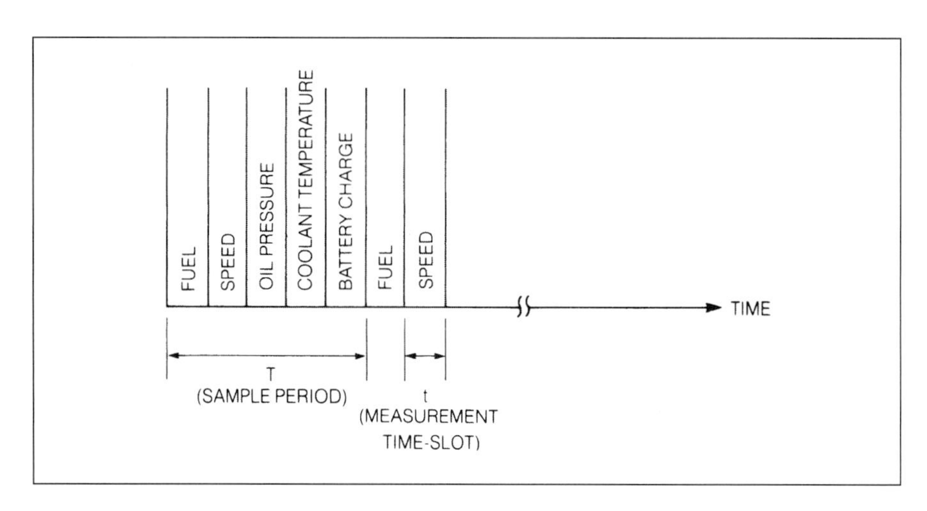

Fig. 23.1-8 Sequential sampling.

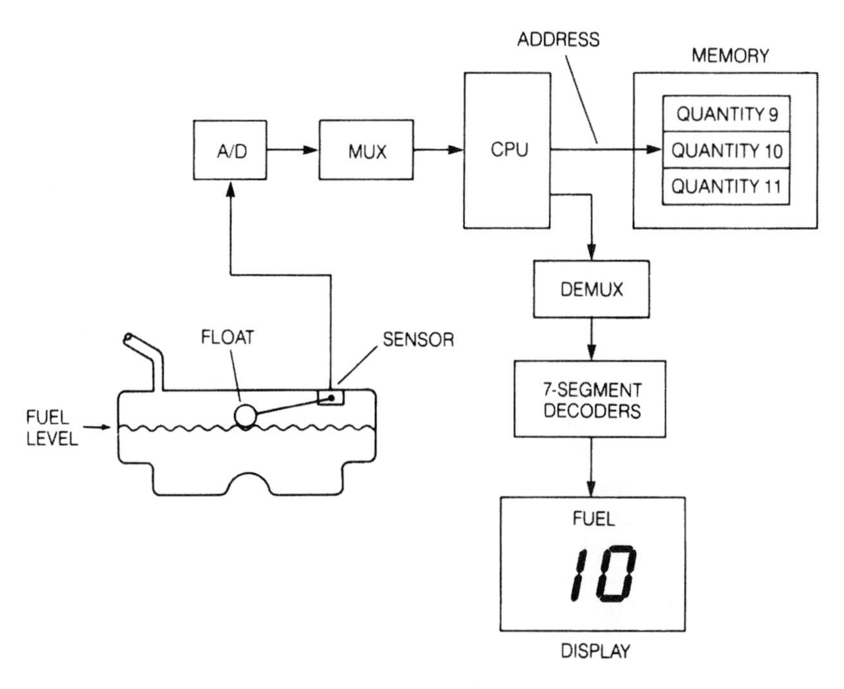

Fig. 23.1-9 Fuel quantity measurement.

and geometrical characteristics of the tank, the sender unit, the instrument voltage regulator, and the indicator (galvanometer). The electronic instrumentation system eliminates the error that results from imperfect voltage regulation. Generally speaking, the electronic fuel quantity measurement maintains calibration over essentially the entire range of automotive electrical system conditions. Moreover, it significantly improves the display accuracy by replacing the electromechanical galvanometer display with an all-electronic digital display.

23.1.4 Fuel quantity measurement

During a measurement of fuel quantity, the MUX switch functionally connects the computer input to the fuel quantity sensor, as shown in Fig. 23.1-9. This sensor output is converted to digital format and then sent to the computer for signal processing. (*Note:* In some automotive systems the analog sensor output is sent to the instrumentation subsystem, where the A/D conversion takes place.)

Several fuel quantity sensor configurations are available. Fig. 23.1-10 illustrates the type of sensor to be described, which is a potentiometer connected via mechanical linkage to a float. Normally, the sensor is mounted so that the float remains laterally near the center of the tank for all fuel levels. A constant current passes through the sensor potentiometer, since it is connected directly across the regulated voltage source. The potentiometer is used as a voltage divider so that the voltage at the wiper arm is related to the float position, which is determined by fuel level.

The sensor output voltage is not directly proportional to fuel quantity in gallons because of the complex shape of the fuel tank. The computer memory contains the functional relationship between sensor voltage (in binary number equivalent) and fuel quantity for the particular fuel tank used on the vehicle.

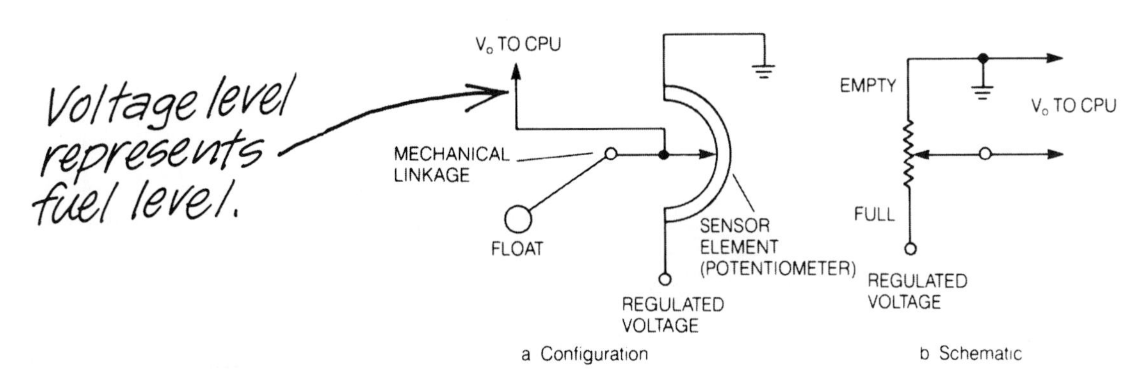

Fig. 23.1-10 Fuel quantity sensor.

The computer reads the binary number from the A/D converter that corresponds to sensor voltage and uses it to address a particular memory location. Another binary number corresponding to the actual fuel quantity in gallons for that sensor voltage is stored in that memory location. The computer then uses the number from memory to generate the appropriate display signal—either analog or digital, depending on display type—and sends that signal via DEMUX to the display.

Computer-based signal processing can also compensate for fuel slosh. As the car moves over the road, the fuel sloshes about and the float moves up and down around the average position that corresponds to the correct level for a stationary vehicle. The computer compensates for slosh by computing a running average. It does this by storing several samples over a few seconds and computing the arithmetic average of the sensor output. The oldest samples are continually discarded as new samples are obtained. The averaged output becomes the signal that drives the display. It should be noted that this is actually a form of digital filtering.

23.1.5 Coolant temperature measurement

Another important automotive parameter that is measured by the instrumentation is the coolant temperature. The measurement of this quantity is different from that of fuel quantity because usually it is not important for the driver to know the actual temperature at all times. For safe operation of the engine, the driver only needs to know that the coolant temperature is less than a critical value. A block diagram of the measuring system is shown in Fig. 23.1-11.

The coolant temperature sensor used in most cars is a solid-state sensor called a *thermistor*; the resistance of this sensor decreases with increasing temperature. Fig. 23.1-12 shows the circuit connection and a sketch of a typical sensor output voltage versus temperature curve.

The sensor output voltage is sampled during the appropriate time slot and is converted to a binary number equivalent by the A/D converter. The computer compares this binary number to the one stored in memory that corresponds to the high temperature limit. If the coolant temperature exceeds the limit, an output signal is generated that activates the warning indicator. If the limit is not exceeded, the output signal is not generated and the warning message is not activated. A proportional display of actual temperature can be used if the memory contains a cross-reference table between sensor output voltage and the corresponding temperature, similar to that described for the fuel quantity table.

23.1.6 Oil pressure measurement

Engine oil pressure measurement is similar to coolant temperature measurement in that it uses a warning

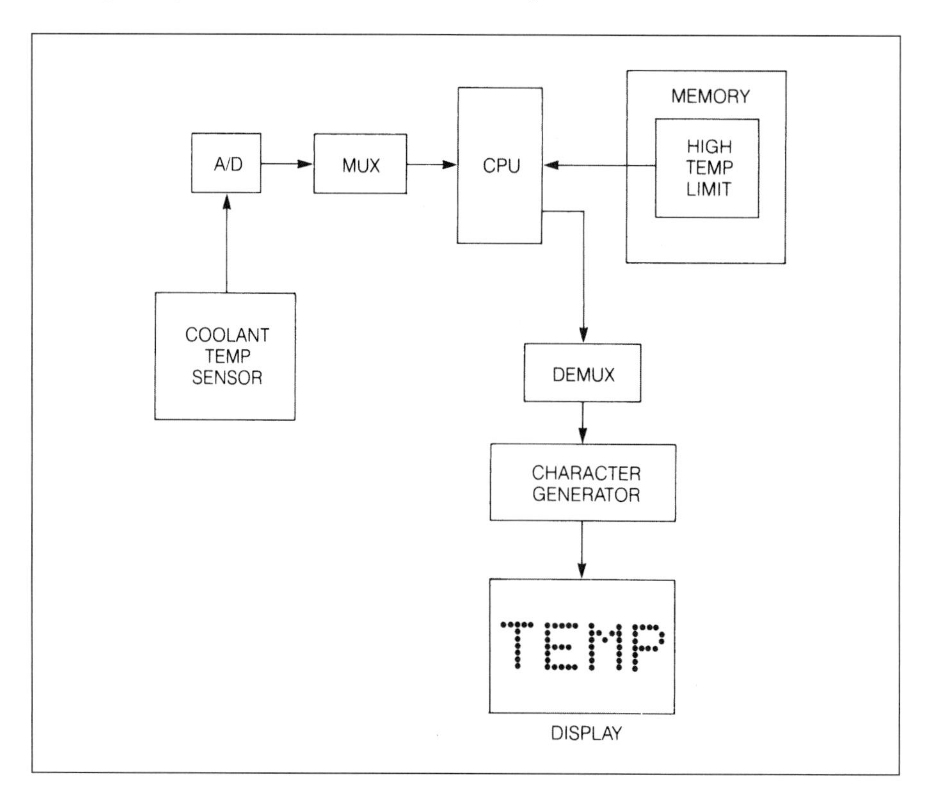

Fig. 23.1-11 Coolant temperature measurement.

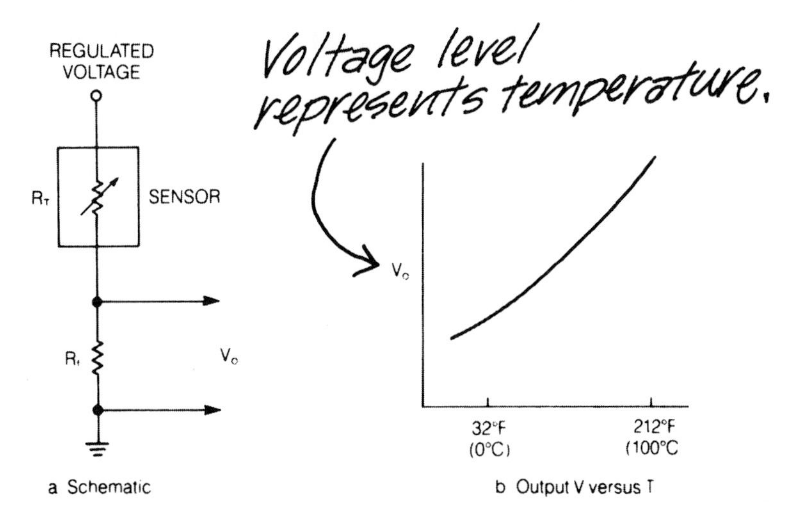

Fig. 23.1-12 Coolant temperature sensor.

message display rather than an indicated numerical value. Whenever the oil pressure is outside allowable limits, a warning message is displayed to the driver. In the case of oil pressure, it is important for the driver to know whenever the oil pressure falls below a lower limit. It is also possible for the oil pressure to go above an allowable upper limit; however, many manufacturers do not include a high oil pressure warning in the instrumentation.

The simplest oil pressure warning system involves a spring-loaded switch connected to a diaphragm. The switch assembly is mounted in one of the oil passageways such that the diaphragm is exposed directly to the oil pressure. The force developed on the diaphragm by the oil pressure is sufficient to overcome the spring and to hold the switch open as long as the oil pressure exceeds the lower limit. Whenever the oil pressure falls below this limit the spring force is sufficient to close the switch. Switch closure is used to switch on the low oil pressure warning message lamp.

One of the deficiencies of this simple switch-based oil pressure warning system is that it has a single fixed low oil pressure limit. In fact, the threshold oil pressure for safe operation varies with engine load. Whereas a relatively low oil pressure can protect bearing surfaces at low loads (e.g., at idle), a proportionately higher oil pressure threshold is required with increasing load (i.e., increasing horsepower and RPM).

An oil pressure instrument that operates with a load- or speed-dependent threshold requires an oil pressure sensor rather than a switch. Such an oil pressure warning system is illustrated in Fig. 23.1-13. This system uses a variable-resistance oil pressure sensor such as seen in Fig. 23.1-14. A voltage is developed across a fixed resistance connected in series with the sensor that is proportional to oil pressure. It should be noted that this assumed pressure sensor is hypothetical and used only for illustrative purposes.

During the measurement time slot, the oil pressure sensor voltage is sampled through the MUX switch and converted to a binary number in the A/D converter. The computer reads this binary number and compares it with the binary number in memory for the allowed oil pressure limits. The oil pressure limit is determined from load or crankshaft speed measurements that are already available in the engine control system. These measurement data can be sent to the instrument subsystem via

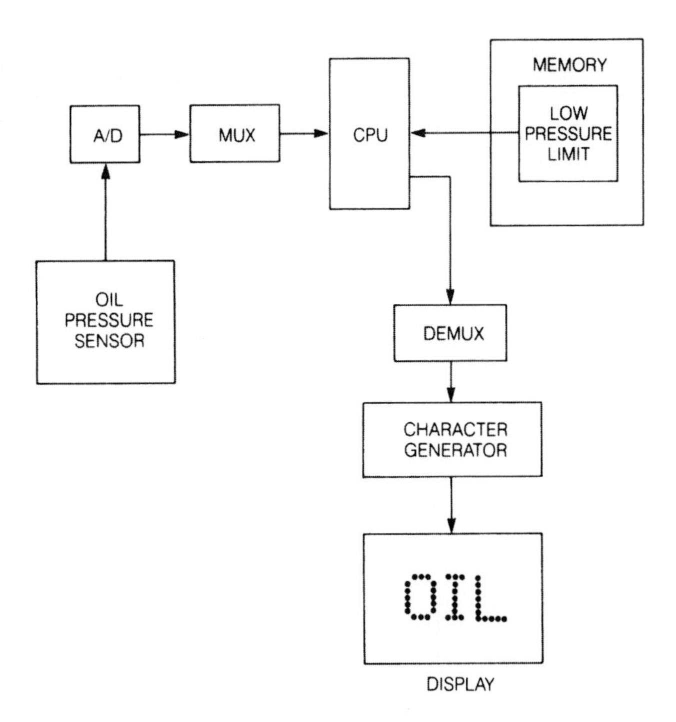

Fig. 23.1-13 Oil pressure measurement.

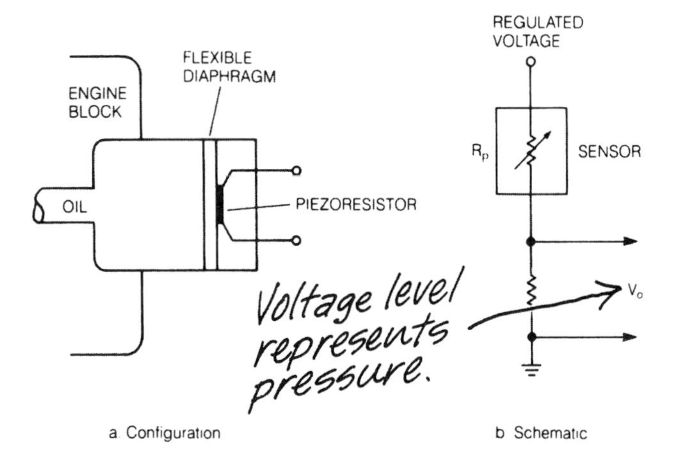

Fig. 23.1-14 Oil pressure sensor.

the PDDL (see Fig. 23.1-1). These measurements serve as the address for a ROM lookup table to find the oil pressure limit. If the oil pressure is below the allowed lower limit or above the allowed upper limit, an output signal is generated that activates the oil pressure warning light through the DEMUX.

It is also possible to use a proportional display of actual oil pressure if a cross-reference table, similar to the fuel quantity table, is used. A digital display can be driven directly from the computer. An analog display, such as the electric gauge, requires a D/A converter.

23.1.7 Vehicle speed measurement

An example of a digital speed sensor has already been described in Chapter 13.1 for a cruise control system. A sensor of this type is assumed to be used for car speed measurements. The output of this sensor is a binary number, *P*, that is proportional to car speed S. This binary number is contained in the output of a binary counter (see Chapter 13.1). A block diagram of the instrumentation for vehicle speed measurement that uses this digital speed sensor is shown in Fig. 23.1-15.

The computer reads the number *P* in the binary counter, then resets the counter to zero to prepare it for the next count. After performing computations and filtering, the computer generates a signal for the display to indicate the vehicle speed. A digital display can be directly driven by the computer. Either mph or kph may be selected. If an analog display is used, a D/A converter must drive the display. Both mph and kph usually are calibrated on an analog scale.

23.1.8 Display devices

One of the most important components of any measuring instrument is the display device. In automotive instrumentation, the display device must present the results of the measurement to the driver in a form that is easy to read and understand. The speedometer, ammeter,

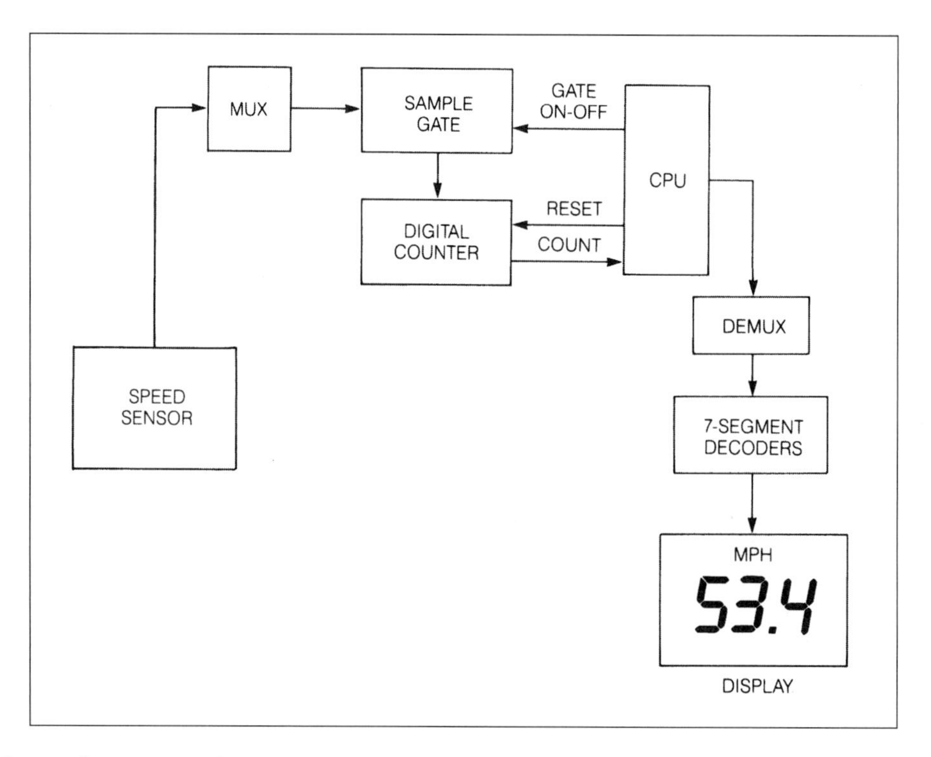

Fig. 23.1-15 Vehicle speed measurement.

and fuel quantity gauge were originally electromechanical devices. Then automotive manufacturers began using warning lamps instead of gauges to cut cost. A warning lamp can be considered as a type of electro-optical display.

Recent developments in solid-state technology in the field called optoelectronics have led to sophisticated electro-optical display devices that are capable of indicating alphanumeric data. This means that both numeric and alphabetic information can be used to display the results of measurements of automotive variables or parameters. This capability allows messages in English or other languages to be given to the driver. The input for these devices is an electronic digital signal, which makes these devices compatible with computer-based instrumentation, whereas electromechanical displays require a D/A converter.

Automobile manufacturers have considered many different types of electronic displays for automotive instrumentation, but only four have been really practical: light-emitting diode (LED), liquid crystal display (LCD), vacuum-fluorescent display (VFD), and the cathode ray tube (CRT). It now appears that the VFD will be the predominant type of instrumentation for at least the near future. Each of these types is discussed briefly to explain their uses in automotive applications.

23.1.9 LED

The LED is a semiconductor diode that is constructed in a manner and of a material so that light is emitted when an electrical current is passed through it. The semiconductor material most often used for an LED that emits red light is gallium arsenide phosphide. Light is emitted at the diode's PN junction when the positive carriers combine with the negative carriers at the junction. The diode is constructed so that the light generated at the junction can escape from the diode and be seen.

An LED display is normally made of small dots or rectangular segments arranged so that numbers and letters can be formed when selected dots or segments are turned on. The configuration for these segments is described in greater detail later in this chapter in the section on VFD. A single LED is not well suited for automotive display use because of its low brightness. Although it can be seen easily in darkness, it is difficult to impossible to see in bright sunlight. It also requires more electrical power than an LCD display; however, its power requirements are not great enough to be a problem for automotive use.

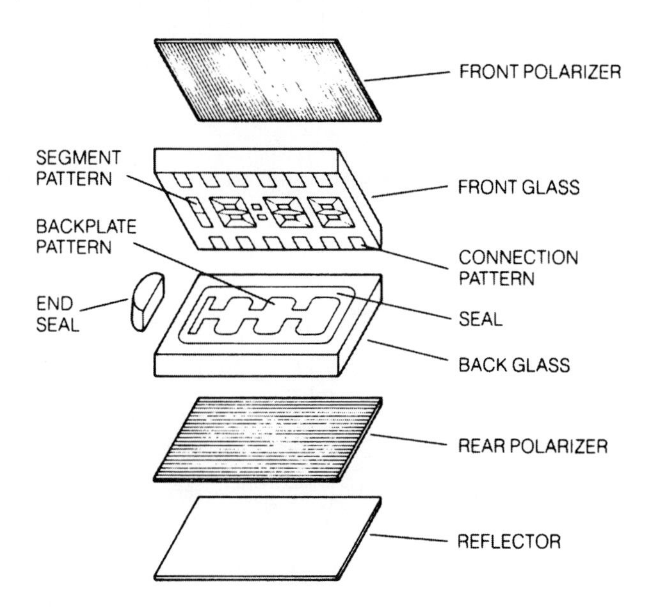

Fig. 23.1-16 Typical LCD construction.

23.1.10 LCD

The LCD display is commonly used in electronic digital watch displays because of its extremely low electrical power and relatively low voltage requirements. The heart of an LCD is a special liquid that is called a *twisted nematic liquid crystal.* This liquid has the capability of rotating the polarization of linearly polarized light. Linearly polarized light has all of the vibrations of the optical waves in the same direction. Light from the sun and from most artificial light sources is not polarized, and the waves vibrate randomly in many directions.

Nonpolarized light can be polarized by passing the light through a polarizing material. To illustrate, think of a picket fence with narrow gaps between the pickets. If a rope is passed between two of the pickets and its end is whipped up and down, the ripples in the rope will pass through the fence. The ripples represent light waves and the picket fence represents a polarizing material. If you whip the rope in any direction other than vertically, the ripples will not pass through.

Now visualize another picket fenced turned 90° so that the pickets are horizontal. Place this fence behind the vertical picket fence. This arrangement is called a cross-polarizer. If the rope is now whipped in any direction, no ripples will pass through both fences. Similarly, if a cross-polarizer is used for light, no light will pass through this structure.

The configuration of an LCD can be understood from the schematic drawings of Fig. 23.1-16. The liquid crystal is sandwiched between a pair of glass plates that have transparent, electrically conductive coatings. The transparent conductor is deposited on the front glass plate in the form of the character, or segment of a character, that is to be displayed. Next, a layer of dielectric (insulating) material is coated on the glass plate to produce the desired alignment of the liquid crystal molecules. The polarization of the molecules is vertical at the front, and they gradually rotate through the liquid crystal structure until the molecules at the back are horizontally polarized. Thus, the molecules of the liquid crystal rotate 90° from the front plate to the back plate so that their polarization matches that of the front and back polarizers with no voltage applied.

The operation of the LCD in the absence of applied voltage can be understood with reference to Fig. 23.1-17a. Ambient light enters through the front polarizer so that the light entering the front plate is vertically polarized. As it passes through the liquid crystal, the light polarization is changed by the orientation of the molecules. When the light reaches the back of the crystal, its polarization has been rotated 90° so that it is horizontally polarized. The light is reflected from the reflector at the rear. It passes back through the liquid crystal structure, the polarization again being rotated, and passes out of the front polarizer. Thus, a viewer sees reflected ambient light.

The effect of an applied voltage to the transmission of light through this device can be understood from Fig. 23.1-17b. A voltage applied to any of the segments of the display causes the liquid crystal molecules under those segments only to be aligned in a straight line rather than twisted. In this case, the light that enters the liquid crystal in the vicinity of the segments passes through the crystal structure without the polarization being rotated. Since the light has been vertically polarized by the front vertical polarizing plate, the light is blocked by the horizontal polarizer so it cannot reach the reflector. Thus, light that enters the cell in the vicinity of energized segments is not returned to the front face. These segments will appear dark to the viewer, the surrounding area will be light, and the segments will be visible in the presence of ambient light.

The LCD is an excellent display device because of its low power requirement and relatively low cost. However, a big disadvantage of the LCD for automotive application is the need for an external light source for viewing in the dark. Its characteristic is just the opposite of the LED; that is, the LCD is readable in the daytime, but not at night. For night driving, the display must be illuminated by small lamps inside the display. Another disadvantage is that the display does not work well at the low temperatures that are encountered during winter driving in some areas. These characteristics of the LCD have limited its use in automotive instrumentation.

23.1.10.1 Transmissive LCD

An LCD display can also function as an optical transmission device from a light source at the rear of the structure to the front face. A configuration such as this permits an LCD to display messages in low ambient light conditions (e.g., nighttime). In low ambient light conditions, a reflective LCD does not provide a visible display to the driver. The intensity of the back light is automatically adjusted to produce optimum illumination as a function of the signal from an ambient light-even sensor located inside the passenger compartment.

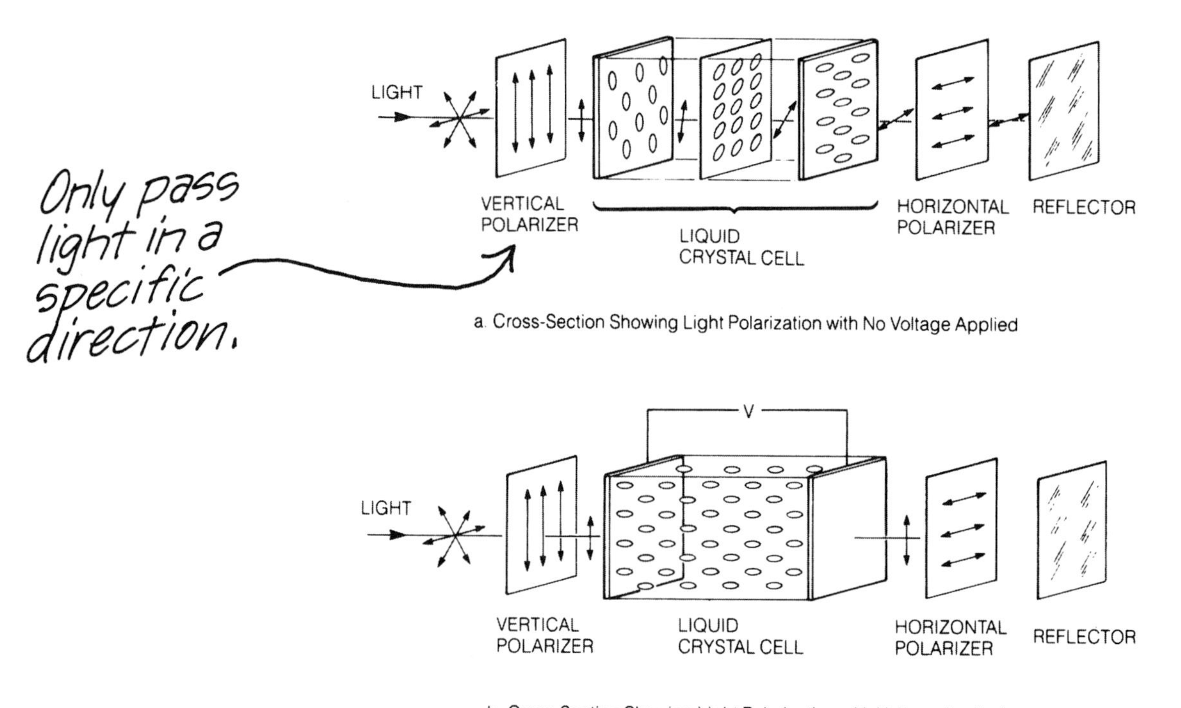

Fig. 23.1-17 Liquid crystal polarization.

Some display manufacturers produce an LCD that combines reflective and transmissive structures in a so-called transflexive LCD structure. The combination of these two basic LCD types into using a package permits optimal readability to be achieved for automotive displays over the entire range of ambient light conditions from bright sunny days to the darkest night conditions.

Another evolution of LCD technology has permitted automotive displays to be available in multiple colors. The LCD configuration described above is a black and white display. A suitable color filter placed in front of the mirror in a reflective LCD or in front of the back light in a transmissive LCD yields a color display, with the color being determined by the optical filter.

Still another evolution in LCD technology is the development of a very large array of programmable multicolor display. Such displays are capable of presenting complex programmable alphanumeric messages to the driver and can also present graphical data or pictorial displays (e.g., electronic maps). Since the array structure LCD is functionally similar to the CRT .

23.1.11 VFD

The VFD display has been widely used in automotive instrumentation, although the multicolor LCD is becoming the preferred choice for this application. This device generates light in much the same way as a television picture tube does; that is, a material called phosphor emits light when it is bombarded by energetic electrons. The display uses a filament coated with material that generates free electrons when the filament is heated. The electrons are accelerated toward the anode by a relatively high voltage. When these high-speed electrons strike the phosphor on the anode, the phosphor emits light. A common VFD has a phosphor that emits a blue-green light that provides good readability in the wide range of ambient light conditions that are present in an automobile. However, other colors (e.g., red or yellow) are available by using other phosphors.

The numeric characters are formed by shaping the anode segments in the form of a standard seven-segment character. The basic structure of a typical VFD is depicted in Fig. 23.1-18. The filament is a special type of resistance wire and is heated by passing an electrical current through it. The coating on the heated filament produces free electrons that are accelerated by the electric field produced by a voltage on the accelerating grid. This grid consists of a fine wire mesh that allows the electrons to pass through. The electrons pass through because they are attracted to the anode, which has a higher voltage than the grid. The high voltage is applied only to the anode of the segments needed to form the character to be displayed. The instrumentation computer selects the set of segments that are to emit light for any given message.

Since the ambient light in an automobile varies between sunlight and darkness, it is desirable to adjust the brightness of the display in accordance with the ambient light. The brightness is controlled by varying the voltage on the accelerating grid. The higher the voltage, the greater the energy of the electrons striking the phosphor and the brighter the light. Fig. 23.1-19 shows the brightness characteristics for a typical VFD device. A brightness of 200 foot-lamberts (fL) might be selected on a bright sunny day, whereas the brightness might be only 20 fL at night. The brightness can be set manually by the driver, or automatically. In the latter case, a photoresistor is used to vary the grid voltage in accordance with the

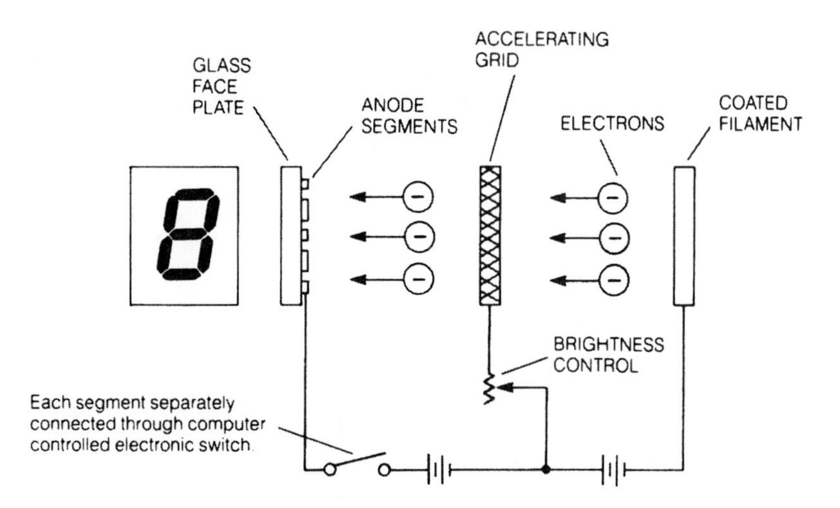

Fig. 23.1-18 Simplified vacuum-fluorescent display configuration.

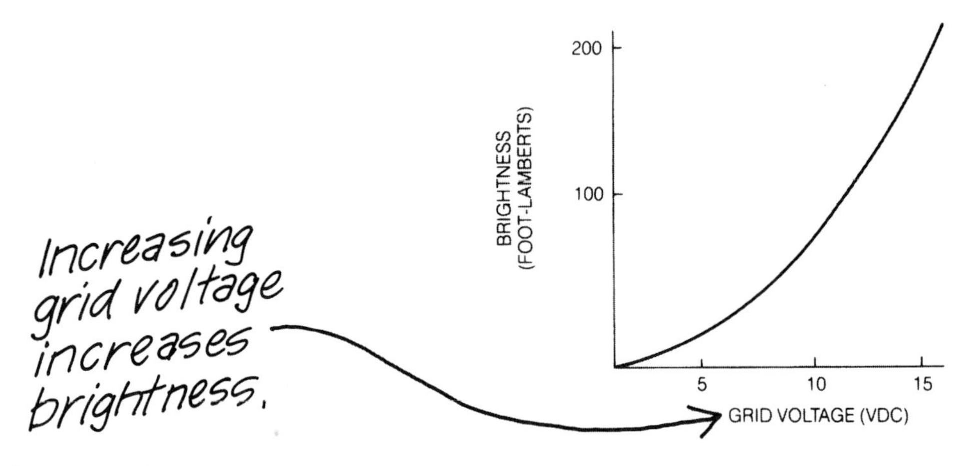

Fig. 23.1-19 Brightness control range for vacuum-fluorescent display.

amount of ambient light. A photoresistor is a device whose resistance varies in proportion to the amount of light striking it.

The VFD operates with relatively low power and operates over a wide temperature range. The most serious drawback for automotive application is its susceptibility to failure due to vibration and mechanical shock. However, this problem can be reduced by mounting the display on a shock-absorbing isolation mount.

23.1.12 CRT

The display devices that have been discussed to this point have one rather serious limitation. The characters that can be displayed are limited to those symbols that can be approximated by the segments that can be illuminated. Furthermore, illuminated warning messages such as "Check Engine" or "Oil Pressure" are *fixed* messages that are either displayed or not, depending on the engine

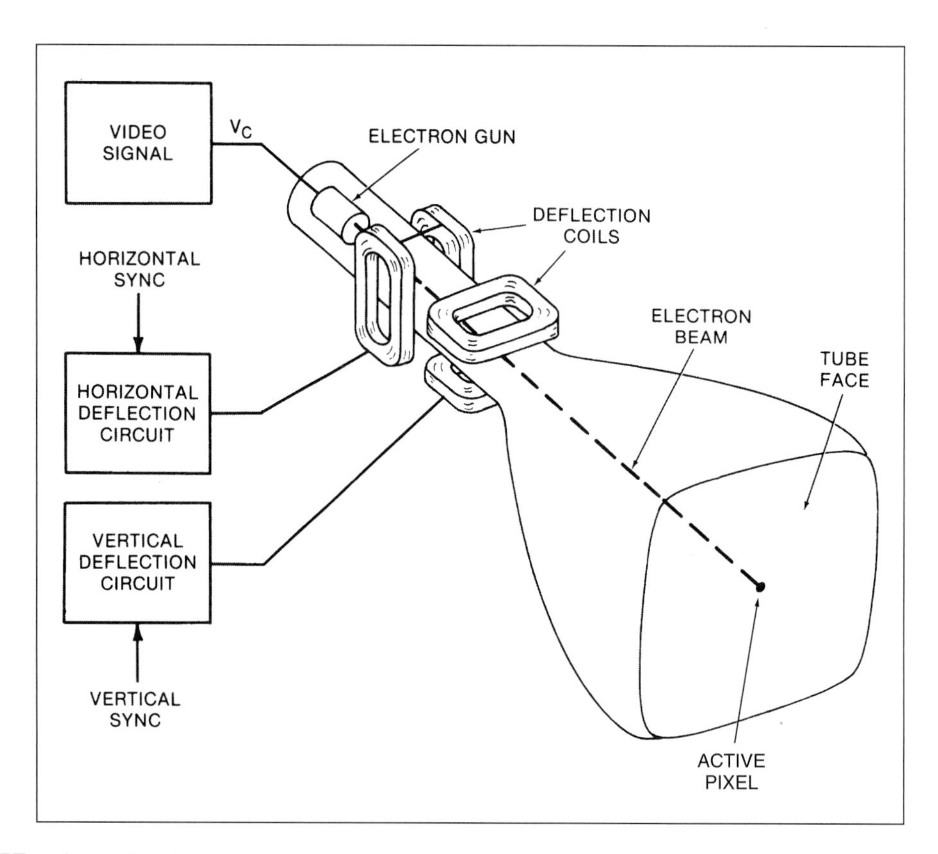

Fig. 23.1-20(a) CRT and associated circuitry.

conditions. The primary disadvantage of such ad hoc display devices is the limited flexibility of the displayed messages.

Arguably the display device with the greatest flexibility for presenting all types of data (including pictorial representations) is the CRT. The CRT is familiar to us all as the TV picture tube, although it has also been used for personal computer monitors. Recently a solid-state equivalent to the CRT has been developed that yields a flat panel display having all of the display capabilities of the CRT.

The CRT (or its solid-state equivalent) is being used increasingly for display purposes in the aerospace industry, where it is used to display aircraft attitude information (sometimes pictorially), aircraft engine or airframe parameters, navigational data, and warning messages. Clearly, the CRT-type display has great potential for automotive instrumentation display.

A technology that has the same flexibility of display as the CRT is the solid-state equivalent of a CRT. Such a display is sometimes called *a flat-panel display*. However, as it is functionally equivalent to the CRT and as the CRT is an existing, very mature technology, we first explain the CRT operation example. This solid-state equivalent of the CRT can be implemented in a number of technologies such as an LCD array.

Fig. 23.1-20 is a sketch of a typical black and white CRT. It is an evacuated glass tube that has a nominally flat surface that is coated with a phosphorescent material. This surface is the surface or face on which the displayed messages appear. At the rear is a somewhat complex structure called an *electron gun*. This device generates a stream of electrons that is accelerated toward the screen and brought to convergence at a spot on the screen. A system of coils in the form of electromagnets causes this convergence of electrons (or beam) and is referred to as the *magnetic focusing system*. The focused stream of electrons is called the *beam*.

The electron beam generates a spot of light at the point on the screen. The intensity of the light is proportional to the electron beam current. This current is controlled by the voltage (V_C), which is called the *video signal*, on an electrode that is located near the electron gun. A color CRT has three separate electron gun structures, with each focused on one of three dots at each picture location on the screen corresponding to red green blue (RGB).

A solid-state LCD display consists of an array of LCDs arranged in a matrix format as depicted in Fig. 23.1-20b. In this structure, only one LCD is active at any time. The color and intensity of the active LCD is controlled by circuitry connected within the microstructure. The active LCD is selected via horizontal and vertical detection circuitry. In the solid-state CRT equivalent, each pixel has its own address.

In the majority of applications (including TV), the CRT electron beam is scanned in a pattern known as a *raster by* means of specially located electromagnets (see Fig. 23.1-20). The magnetic fields created by the scanning coils deflect the beam horizontally and vertically. The amount of deflection is proportional to the current

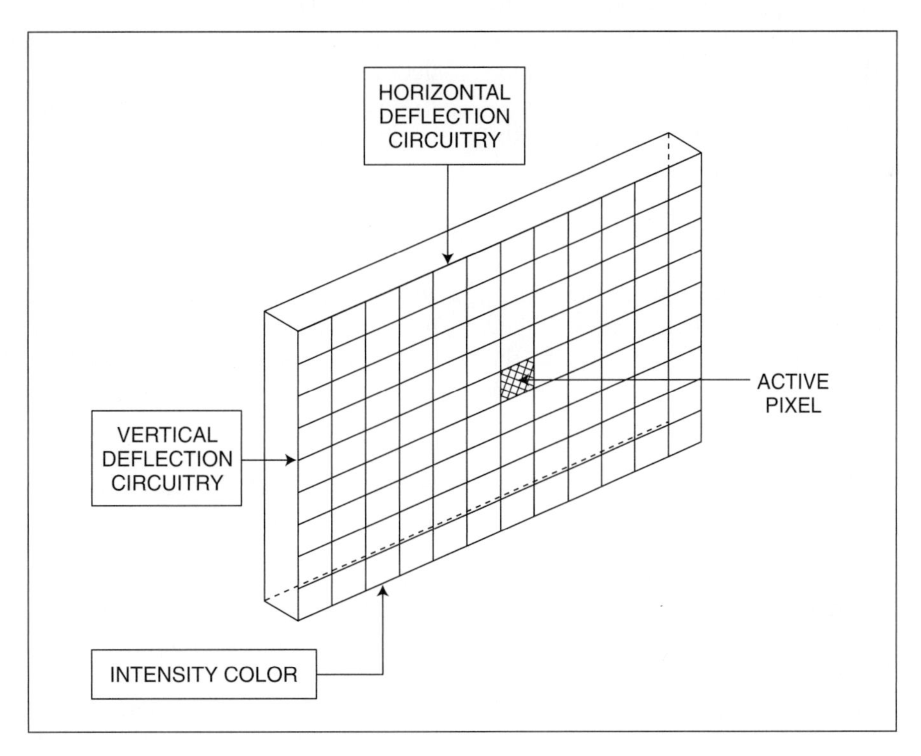

Fig. 23.1-20(b) Solid-state CRT.

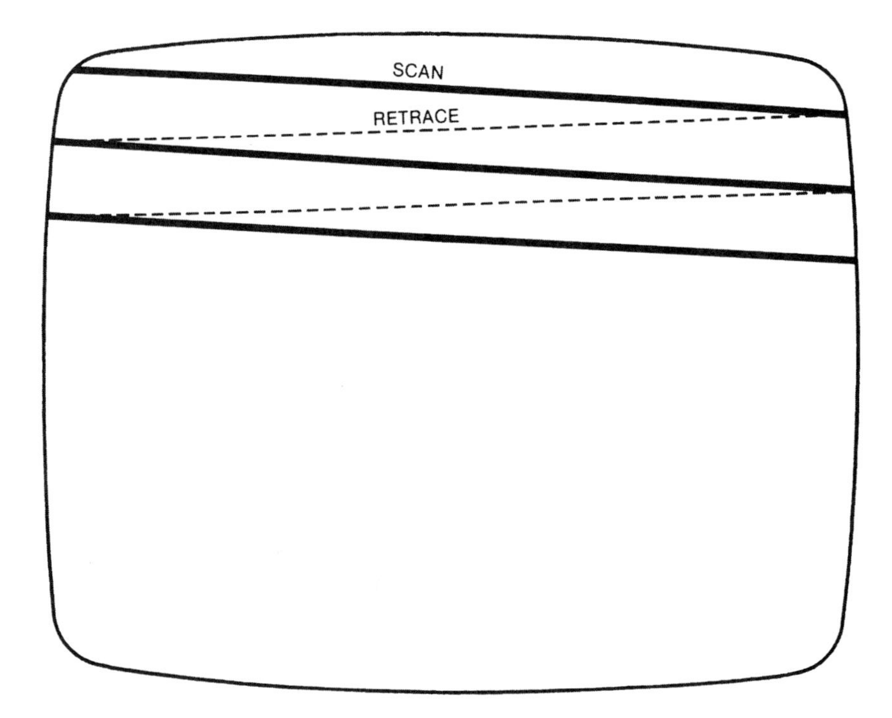

Fig. 23.1-21 Raster pattern.

flowing through the respective coils. The raster pattern traced by the beam is illustrated on the face of the CRT in Fig. 23.1-21. The raster begins at the upper left of the screen and sweeps rapidly to the right. During this scan, the intensity of each electron gun varies in proportion to the brightness (of each color) to be illuminated at each picture element (known as a pixel). The next line in the raster begins at the left of the screen slightly below the previous line. The standard raster for U.S. television consists of 525 lines completed 30 times per second.

The scanning motion is done in synchronism with the source of information being displayed. At the end of each horizontal scan line, a synchronizing pulse (called *horizontal sync*) causes the beam to deflect rapidly to the left and then to begin scanning at a constant rate to the right. A similar synchronizing pulse is generated at a time when the beam is at the bottom of the CRT. This pulse (called *vertical sync)* causes the beam to deflect rapidly to the top of the CRT face and then to begin scanning downward at a uniform speed.

23.1.12.1 Scan circuits

The raster scan for a CRT is accomplished by varying the current through the horizontal and vertical deflection coils (i.e., HDCs and VDCs). At the beginning of each horizontal line, the current through the HDC is such that the electron beam is at the left edge of the CRT face (as viewed from the front). Then this current increases with time such that the beam sweeps uniformly from left to right. At the time the beam is at the right edge, the current rapidly (ideally instantaneously) switches to the value corresponding to the left edge, and the scanning continues periodically. A graph of the beam horizontal position with respect to time resembles a sawtooth pattern and is called a sawtooth waveform. The frequency of this horizontal sweep signal (U.S. standard) is 15,750 Hz.

A similar sweep circuit causes the CRT beam to deflect vertical so that the entire screen is covered in $\frac{1}{30}$ sec. In actual fact, even number horizontal traces are scanned in $\frac{1}{60}$ sec and odd number lines in the next $\frac{1}{60}$ sec in a process called interlacing.

The horizontal and vertical signals can be generated using either analog or digital circuits, although modern CRT circuits are digital. One conceptually simple way to generate the horizontal and vertical sawtooth sweep waveforms uses a constant-frequency oscillator driving the trickle count input of a counter circuitry and D/A converter circuitry. Each cycle of the oscillator causes the counter to increment by one. In CRT systems, the counter output drives an A/D converter, creating an output signal having the required sawtooth waveform. The lowest count yields a current corresponding to the left edge of the CRT screen. The counter is automatically reset to this value once the electron beam position is at the right edge of the screen (controlled by horizontal synch pulses). Similar circuitry exists to drive vertical deflection.

Deflection of a solid-state (e.g., LCD) equivalent of the CRT is dependent on the wiring arrangement of the individual LCD elements. One scheme for achieving a solid-state raster scan display device is to construct an array of LCD elements as depicted in Fig. 23.1-20b. These elements are interconnected with two grids of wires, one running vertically and one running horizontally. Each vertical wire is connected to all of the LCD elements in a given column. Each horizontal wire similarly interconnects all of the LCD elements in a given row.

A given LCD element (forming one pixel of the display) is activated by an electrical signal applied to the vertical wire associated with its column and the horizontal wire associated with its row. Scanning along a given row is achieved by sending an appropriate signal to the wire associated with that row and then sequentially sending a similar signal to the wire corresponding to a column.

Circuitry for raster scanning this type of LCD array can be implemented with a constant frequency oscillator and counter (similar to the CRT circuitry). However, instead of using an analog/digital converter, the solid-state scanning circuitry uses decoding logic that activates one of its many leads corresponding to the digital count in the counter. This latter circuitry has a separate output for each grid wire in the horizontal scan or vertical scan, respectively.

A typical solid-state raster scan display might have 240 rows and 480 columns of LCD elements. The horizontal decoding circuit has 480 separate outputs, each connected to a column grid wire. The horizontal counter has a maximum count of 480. Similarly, the vertical decoder has 240 output leads each connected to a row grid wire.

The intensity and/or color of the active LCD is controlled by a signal connected to that LCD. The appropriate signal is known as the video signal.

In a typical CRT or solid-state equivalent display device, the video voltage and sync pulses are generated in a special circuit called the *CRT controller.* A simplified block diagram for a system incorporating a CRT type display with the associated controller is depicted in Fig. 23.1-22. The sensors and instrumentation computer, which are MPU based, shown at the left of this illustration have the same function as the corresponding components of the system in Fig. 23.1-2. The output of the instrumentation computer controls the CRT type display, working through the controller. The functional structure of the display is the same, regardless of whether the display is a CRT or the solid-state equivalent.

In the example architecture of Fig. 23.1-22, it is assumed that the instrumentation computer communicates with the CRT controller via data and address buses (DB and AB), and via a serial link along a line labeled UART (universal asynchronous receiver/transmitter). However, many other choices of data link are possible. The data that are sent over the DB are stored in a special memory called video RAM. This memory stores digital data that are to be displayed in alphanumeric or pictorial patterns on the CRT-type screen. The controller obtains data from the video RAM and converts them to the relevant video signal (V_C). At the same time, the controller generates the horizontal and vertical sync necessary to

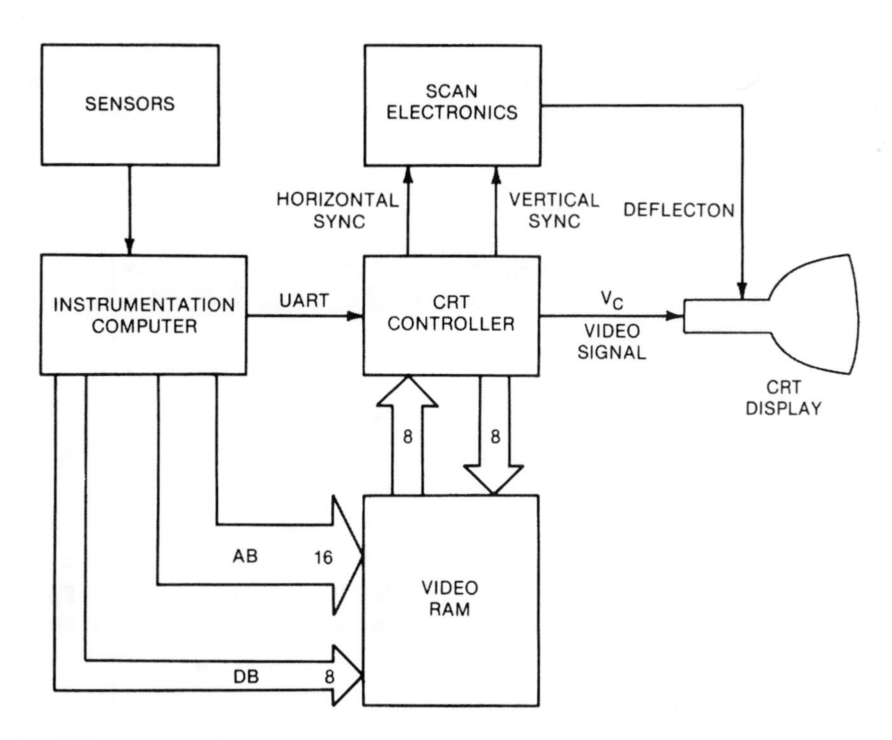

Fig. 23.1-22 Automotive CRT instrumentation system.

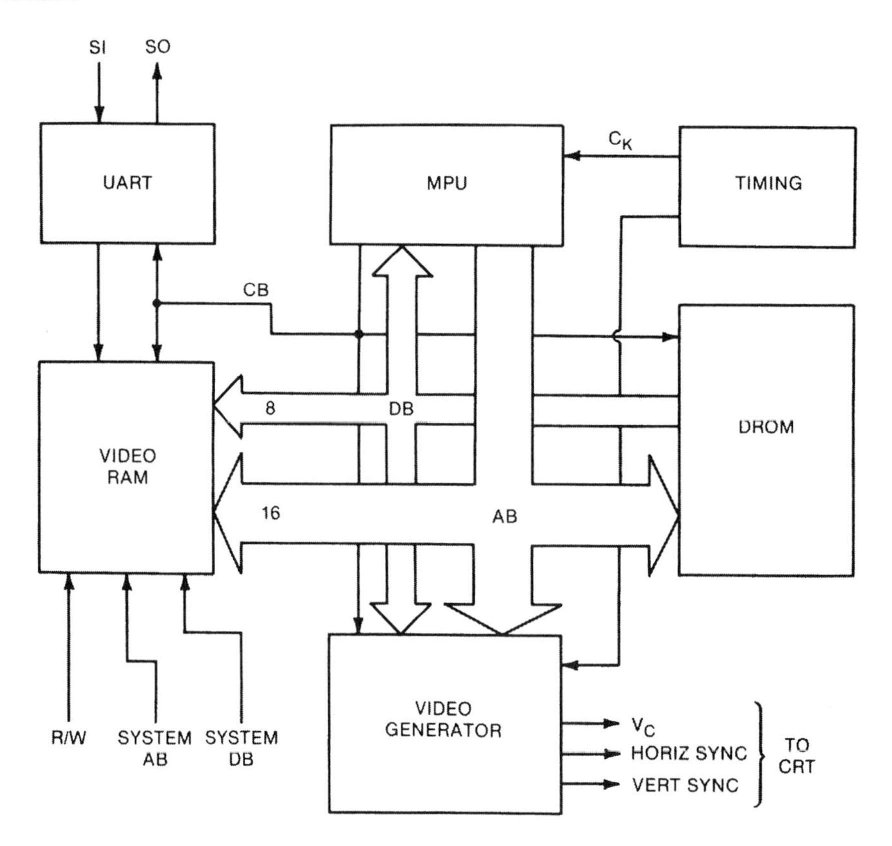

Fig. 23.1-23 CRT-type controller configuration.

operate the raster scan in synchronism with the video signal.

The video controller in the example system (Fig. 23.1-23) itself incorporates an MPU for controlling the CRT-type display. The data to be displayed are stored in the video RAM via the system buses under control of the instrumentation computer. The operation of the MPU is controlled by programs stored in a display ROM (DROM). This ROM might also store data that are required to generate particular characters. The various components of the display controller are internally connected by means of data and address buses similar to those used in the instrumentation computer.

The operation of the display controller is under control of the instrumentation computer. This computer transfers data that are to be displayed to the video RAM, and signals the CRT controller via the UART link. During the display time, the MPU operates under control of programs stored in the DROM. These programs cause the MPU to transfer data from the video RAM to the video generator in the correct sequence for display.

The details of the transfer of data to the video generator and the corresponding generation of video signals vary from system to system. In the hypothetical system seen in Fig. 23.1-22, the display is assumed to be an array of LCD elements arranged in 240 rows vertically by 480 columns horizontally. Here, the display generates the characters F and P (see Fig. 23.1-24). The dots are generated by switching on the active LCD element at the desired location. The LCD is switched by pulsing the video voltage at the time relative to the row and column of the active LCD at which a dot is to appear. The resolution of the display is one dot, which is often termed a picture element (i.e., a pixel).

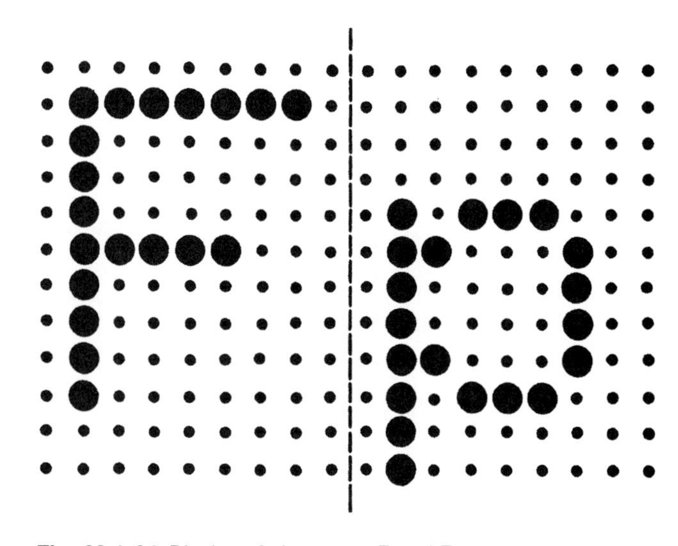

Fig. 23.1-24 Display of characters F and P.

A scheme for generating the suitable video signals for such a display is shown in a greatly simplified hypothetical example of a black and white solid-state display in the block diagram of Fig. 23.1-25. During the horizontal retrace time when the raster scan position is moving rapidly from right to left, the MPU (under program control) determines which data pattern is to be displayed during the next scanning line. The MPU maintains an internal record of the current active line on the display by counting vertical sync pulses. The actual bit pattern associated with the character being displayed along the active line on the CRT is loaded into the shift register. In our hypothetical example, these data come in eight separated 8-bit bytes from video RAM. Then during the scanning of the active line, the bit pattern is shifted out one bit at a time by a pulse signal, H_{ck}, at a frequency that is 240 times that of the horizontal sync frequency.

Each bit location in the shift register corresponds to a pixel location on the display. A "1" stored at any shift register location corresponds to a bright spot on the display. Thus, by placing a suitable pattern in the shift register for a particular line, it is possible to display complex alphanumeric or pictorial data on the display.

The enormous flexibility of the CRT type display offers the potential for a very sophisticated automotive instrumentation system. In addition to displaying the variables and parameters that have traditionally been available to the driver, the CRT type display can present engine data for diagnostic purposes vehicle comfort control system parameters, and entertainment system variables. The data required for such displays can, for example, be transmitted via a high-speed digital data (HSDD) link between the various on-board electronics systems. This CRT-type display has sufficient flexibility and detailed resolution that graphical data or electronic maps can be shown to the driver.

There are several reasons for using the serial HSDD link for transmitting data between the various systems rather than tying the internal data buses together. For example, it is desirable to protect any given system from a failure in another. A defect affecting the data bus of the comfort system could adversely affect the engine control system. In addition, each internal data bus tends to be busy handling internal traffic. Moreover, the transfer of data to the instrumentation computer can take place at relatively low data rates (for the diagnostic application outlined here).

Fig. 23.1-26 is a block diagram of an integrated vehicle instrumentation system in which all on-board electronic systems are coupled by an HSDD link. This system requires a keyboard (KB) or similar input device for operator control. The driver can, for example, select to display the entertainment system operation. This display mode permits the driver to select radio, tape, or CD, and

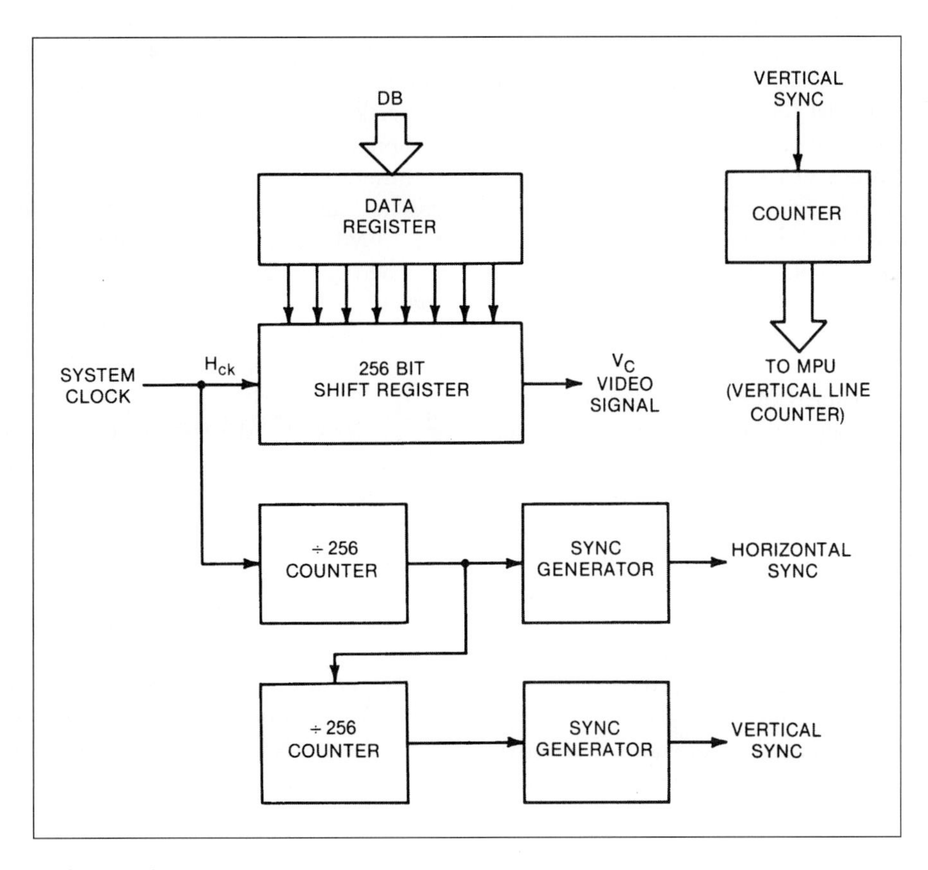

Fig. 23.1-25 Video signal generation.

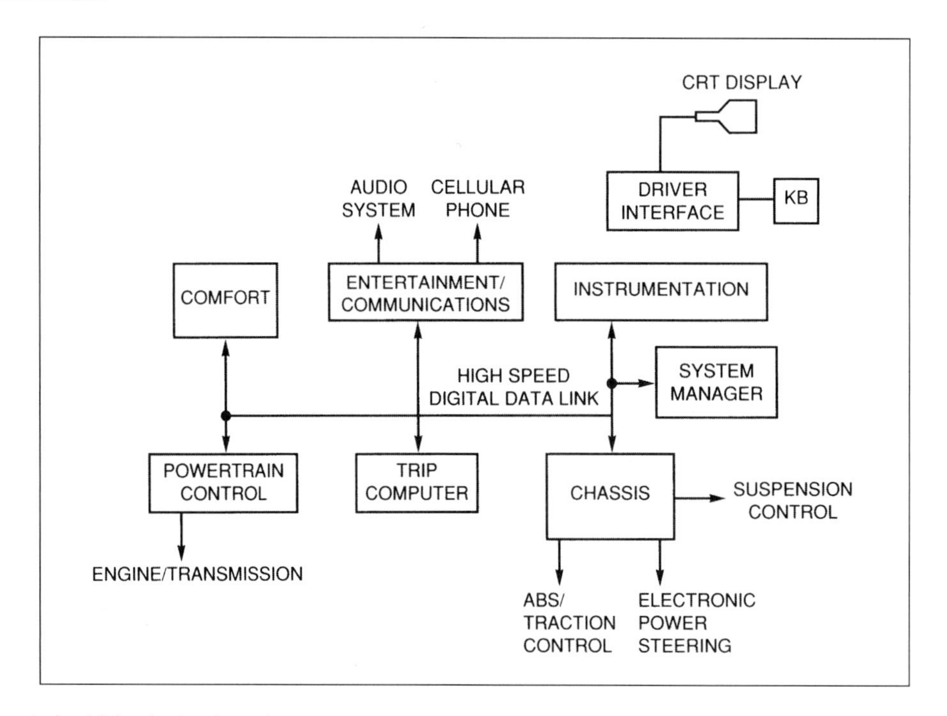

Fig. 23.1-26 Integrated vehicle electronic systems.

to tune the radio to the desired station and set the volume. In vehicle diagnostic mode, the CRT can be configured to display the parameters required by the mechanic for performing a diagnosis of any on-board electronic system.

In Fig. 23.1-26, several electronic systems are connected by the digital data link. Tying systems together this way has great potential performance benefits for the vehicle. Each automotive subsystem has its own primary variables, which are obtained through measurements via sensors. A primary variable in one subsystem might be a secondary variable in another system. It might not be cost effective to provide a sensor for a secondary variable to achieve the best possible performance in a stand-alone subsystem. However, if measurement data can be shared via the digital data link, then the secondary measurement is potentially available for use in optimizing performance. Furthermore, redundant sensors for measuring primary variables can be eliminated by an integrated electronics system for the vehicle. For example, wheel speed measurements are primary variables for ABS systems and are also useful in electronic transmission control.

This system manager is responsible for coordinating data transfer and regulating the use of the data bus so that no two systems are transmitting simultaneously.

Essentially, the digital data link provides a sophisticated communication system between various subsystems. Among the issues of importance for such a communication system are the protocol and message format. It is highly advantageous to have a standard protocol for all automobiles. The Society of Automotive Engineers (SAE) is working to develop a standard protocol for the HSDD. This link operates at a data rate of 1 megabit/sec and can be implemented with wire or optical fiber. Any of a number of bit-encoding schemes are useful for message formats, the details of which are unimportant for the present discussion.

Some form of network arbitration is required for determining priority of the use of the link whenever there is conflict between subsystems for its use. This feature is typically handled by the system manager.

The basic message structure is derived assuming that the majority of data on the link is regularly sent. This means that the content of each message is known (only the actual data varies).

23.1.12.2 CAN network

Automotive electronic subsystems have become numerous and interdependent, requiring subsystem intercommunication. This need for digital communication between all on-board digital systems has led to the creation of a standard automotive communication network known as Control Area Network (CAN). Originally developed for passenger car applications, CAN is a form of local area network that permits data to be shared.

In the CAN concept, each electronic subsystem incorporates communication hardware and software, permitting it to function as a communication module referred to as a gateway. CAN is based on the so-called broadcast communication mechanism in which communication is achieved by the sending gateway (i.e., subsystem) transmitting messages over the network (e.g., wire interconnect). Each message has a specific format (protocol) that includes a message identifier. The identifier defines the content of the message, its priority, and is unique within the network. In addition to the data and identifier, each message includes error-checking bits as well as beginning and end of file.

The CAN communication system has great flexibility, permitting new subsystems to be added to an existing system without modification, provided the new additions are all receivers. Each gateway (subsystem) can be upgraded with new hardware and software at any time with equipment that was not available at the time the car left the manufacturing plant or even when it left the dealer. Essentially, the CAN concept with its open standards frees the development of new telematics applications from the somewhat lengthy development cycle of a typical automobile model. Furthermore, it offers the potential for the aftermarket addition of new subsystems.

23.1.13 The glass cockpit

The development of a cost-effective solid-state equivalent of the CRT can have enormous application in automotive instrumentation. It can yield a completely reconfigurable display system similar to the multifunction display used in some modern transport aircraft. Such displays are termed a *glass cockpit* in aircraft parlance. It is also known as an electronic flight information system (EFIS).

A single CRT acting as a display for a digital instrumentation system has the capability of displaying any of several choices of data readout, including

1. Navigation data
2. Subsystem status parameters
3. Attitude (artificial horizon)
4. Air data (airspeed, altitude, etc.)

It can also be used for diagnosis of problems with various aircraft subsystems. In this case it can present a pictorial diagram of any aircraft subsystem (hydraulic system or electrical system) so that the flight crew does not have to resort to hunting through a manual for the aircraft to diagnose a problem with a subsystem.

One of the benefits of an automotive glass cockpit is its great flexibility. Any message in any format can be displayed. In fact, the format can be chosen by the driver via a set of switches or by a keypad arrangement. The driver selects a particular display format from a number of choices and the display will be reconfigured to his choice by software, that is, by the program stored in the instrumentation computer. A likely default choice would include a standard display having speed and fuel quantity and the capability of displaying warning messages to the driver.

Another benefit of the EFIS-type display is the capability of displaying diagnostic information to a service technician. The service technicians can select a display mode that presents fault codes from any vehicle subsystem whenever the car is taken for repairs or during routine maintenance operations.

Of particular importance is the capability of digital instrumentation to identify intermittent faults. The instrumentation computer can store fault codes with a time stamp that gives the time of occurrence to indicate to a service technician that a particular component or subsystem is experiencing intermittent failures. Such failures are extremely difficult to diagnose because they are often not present when the car is brought in for service. In this mode of operation the instrumentation computer along with its software reconfigurable display is serving a role somewhat analogous to a flight data recorder on an aircraft.

23.1.14 Trip information computer

One of the most popular electronic instruments for automobiles is the trip information system. This system has a number of interesting functions and can display many useful pieces of information, including the following:

1. Present fuel economy
2. Average fuel economy
3. Average speed
4. Present vehicle location (relative to total trip distance)
5. Total elapsed trip time
6. Fuel remaining
7. Miles to empty fuel tank
8. Estimated time of arrival
9. Time of day
10. Engine RPM
11. Engine temperature
12. Average fuel cost per mile

Additional functions can be performed, which no doubt will be part of future developments. However, we will discuss a representative system having features that are common to most available systems.

A block diagram of this system is shown in Fig. 23.1-27. Not shown in the block diagram are MUX, DEMUX, and A/D converter components, which are normally part of a computer-based instrument. This system can be implemented as a set of special functions of the main automotive instrumentation system, or it can be a stand-alone system employing its own computer.

The vehicle inputs to this system come from the three sensors that measure the following variables:

1. Quantity of fuel remaining in the tank

2. Instantaneous fuel flow rate

3. Vehicle speed

Other inputs that are obtained by the computer from other parts of the control system are

1. Odometer mileage

2. Time (from clock in the computer)

The driver enters inputs to the system through the keyboard. At the beginning of a trip, the driver initializes the system and enters the total trip distance and fuel cost. At any time during the trip, the driver can use the keyboard to ask for information to be displayed.

The system computes a particular trip parameter from the input data. For example, instantaneous fuel economy in miles per gallon (MPG) can be found by computing

$$\text{MPG} = S/F$$

where

S is the speed in miles per hour
F is the fuel consumption rate in gallons per hour

Of course, this computation varies markedly as operating conditions vary. At a steady cruising speed along a level highway with a constant wind, fuel economy is constant. If the driver then depresses the accelerator (e.g., to pass traffic), the fuel consumption rate temporarily increases faster than speed, and MPG is reduced for that time. Various averages can be computed such that instant fuel economy, short-term average fuel economy, or long-term average fuel economy can be displayed.

Another important trip parameter that this system can display is the miles to empty fuel tank, D. This can be found by calculating

$$D = \text{MPG} \cdot Q$$

where Q is the quantity of fuel remaining in gallons. Since D depends on MPG, it also changes as operating conditions change (e.g., during heavy acceleration). In such cases, the calculation of miles to empty based on the above simple equation is grossly incorrect. However, this calculation gives a correct estimate of the miles to empty for steady cruise along a highway in which operating conditions are constant.

Still another pair of parameters that can be calculated and displayed by this system are distance to destination, D_d, and estimated time of arrival, ETA. These can be found by computing

$$D_d = D_T - D_P$$

$$\text{ETA} = T_1 + (D_d/S)$$

where

D_T is the trip distance (entered by the driver)
D_P is the present position (in miles traveled since start)
S is the present vehicle speed
T_1 is the start time

The computer can calculate the present position D_P by subtracting the start mileage, D_1 (obtained from the odometer reading when the trip computer was initialized by the driver), from the present odometer mileage.

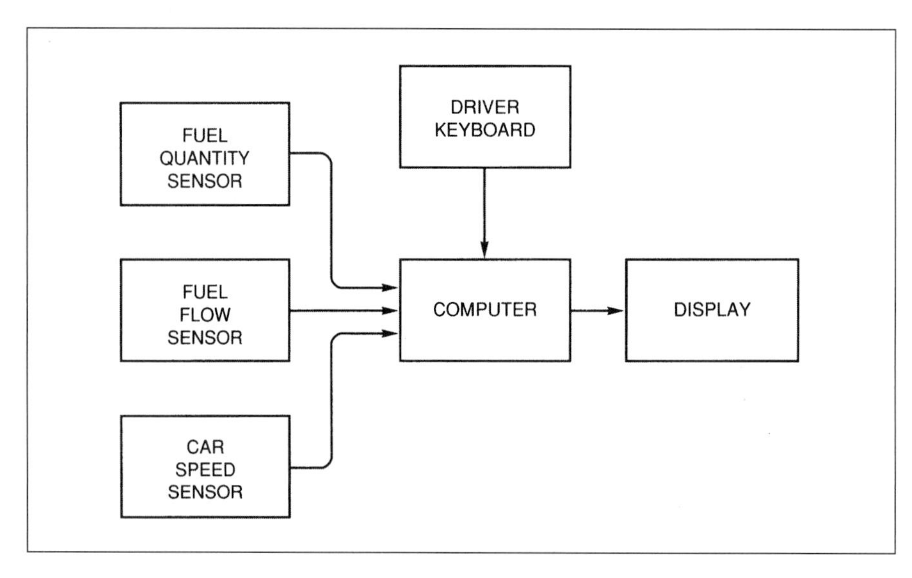

Fig. 23.1-27 Trip information system.

The average fuel cost per mile C can be found by calculating

$$C = (D_P/MPG) \cdot \text{fuel cost per gallon}$$

There are many other useful and interesting operations that can be performed by the variety of available systems. Actually, the number of such functions that can be performed is limited primarily by cost and by the availability of sensors to measure the required variables.

23.1.15 Telematics

Communications to and from an automobile has become routine as a result of both cell phone and satellite technology. In addition, the technology is evolving for area broadcast of road condition information on radio station subcarriers. Technology is also evolving that will permit Internet connections via cell phones, making the car in effect on Internet node. Automobile Internet connectivity opens a limitless range of services for the driver, from on-line navigation help to on-line diagnostic and/or road service for mechanical problems.

One of the major issues in telematics is how to present the information and services that are potentially available to the driver without distracting from the driving tasks. Of course, the various services can be made available to passengers without necessarily distracting the driver. For example, video monitors in rear seats can provide entertainment, game playing on any standard computer Internet terminal via on-board DVD, or wireless connection, be it cell phone or satellite links.

On the other hand, the use of any subsystem that provides information such as is described above is potentially distracting to the driver. The simple act of dialing a standard cell phone requires the use of at least one hand and at least a momentary look at the cell phone. Some state legislatures are passing laws prohibiting the driver's use of a standard cell phone while driving.

The driver's distraction through cell phone use is somewhat alleviated by voice-activated cell phone dialing in which the cell phone user verbally gives the phone number, speaking each digit separately. Included within the cell phone is a very sophisticated algorithm for recognizing speech. Speech recognition software identifies spoken words or numbers based on patterns in the waveform at the output of a microphone into which the user speaks. There are two major categories of speech recognition software: speaker dependent and speaker independent.

Speaker-dependent software recognizes the speech of a specific individual who must work with the system. The user is prompted to say a specific digit a number of times until the software can reliably identify the waveform patterns associated with that particular speaker. By this process, the system is "trained" to the individual user. It may not be capable of recognizing other users to whose speech it has been trained.

Speaker-independent voice recognition software can recognize spoken digits regardless of the user. It is generally more sophisticated than speaker-dependent speech recognition.

The cell phone connection can also be used to provide online navigation or other services by contacting a service with operators trained to provide this type of service. Alternatively, the cell phone can be used to provide an internet connection to an on-line navigation service that transmits data to the car for display on an electronic map.

The telematics technology is presently in its infancy and is certain to grow spectacularly in capability and flexibility, providing the motorist with virtually limitless services. Telematics functionality is probably limited more by imagination than by technology.

23.1.16 Automotive diagnostics

In certain automobile models, the instrumentation computer can perform the important function of diagnosis of the electronic engine control system. This diagnosis takes place at several different levels. One level is used during manufacturing to test the system, and another level is used by mechanics or interested car owners to diagnose engine control system problems. Some levels operate continuously and others are available only on request from an external device that is connected to the car data link for diagnostic purposes by a technician.

In the continuous monitor mode, the diagnosis takes place under computer control. The computer activates connections to the vehicle sensors and looks for an open- or short-circuited sensor. If such a condition is detected, a failure warning message is given to the driver on the alphanumeric display or by turning on a labeled warning light.

Index